T0177628

SEMICONDUCTOR PHYSICS

This text is the third volume of the series titled Electroscience.

1. S. Tiwari, *Quantum, statistical and information mechanics: A unified introduction*, Electroscience 1, Oxford University Press, ISBN 978-0-198-75985-0,
2. S. Tiwari, *Device physics: Fundamentals of electronics and optoelectronics*, Electroscience 2, Oxford University Press, ISBN 978-0-198-75984-3,
3. S. Tiwari, *Semiconductor physics: Principles, theory and nanoscale*, Electroscience 3, Oxford University Press, ISBN 978-0-198-75986-7 (2020) and
4. S. Tiwari, *Nanoscale device physics: Science and engineering fundamentals*, Electroscience 4, Oxford University Press, ISBN 978-0-198-75987-4 (2017).

These volumes comprise a sequence of undergraduate to graduate textbooks of the underlying physical foundations leading up to advanced devices of nanometer scale. Teaching slides are available on the book's companion website at www.oup.co.uk/companion/semiconductorphysics2020, and the solutions manual may be requested at global.oup.com/uk/academic/physics/admin/solutions.

Semiconductor Physics

Principles, Theory and Nanoscale

ELECTROSCIENCE SERIES, VOLUME 3

Sandip Tiwari

OXFORD
UNIVERSITY PRESS

Great Clarendon Street, Oxford, OX2 6DP,
United Kingdom

Oxford University Press is a department of the University of Oxford.
It furthers the University's objective of excellence in research, scholarship,
and education by publishing worldwide. Oxford is a registered trade mark of
Oxford University Press in the UK and in certain other countries

First Edition published in 2020

Impression: 1

Published in the United States of America by Oxford University Press
198 Madison Avenue, New York, NY 10016, United States of America

British Library Cataloguing in Publication Data
Data available

Library of Congress Control Number: 2019941496

ISBN 978–0–19–875986–7

Printed and bound by
CPI Group (UK) Ltd, Croydon, CR0 4YY

To

Pampam bhai (Anupam Mishra),

my childhood hero and beacon

(those visits to kabadi bazaar mattered singularly in this life);

and to the Mishra family

for showing the way to

love, grace, dignity, conviction and patience by example.

॥ फर्क ॥

मेरा और तुम्हारा
सारा फर्क,
इतने में है
कि तुम लिखते हो
मैं बोलता हूं,
और कितना फर्क हो जाता है इससे
तुम ढांकते हो, मैं खोलता हूँ ।

भवानी प्रसाद मिश्र

The difference

The entire difference
between you and me
is that you write
and I talk,
and how entirely different they are.
You cover it and I open it.

Bhawani Prasad Mishra

Acknowledgments

One's approach to life and, more specifically, to the choices one makes in the course of it, the questions one asks and the irritants one learns to ignore have a cumulative outcome. Life is defined by beauty, love, taste and the passion that propels one in pursuit of these lambent aspirations. In this, one's family and friends, early and advanced education, the company one keeps and the approaches one learns from the especially gifted people that one has the good fortune of befriending all make a difference. So do people who mentor and support through the vicissitudes of life. I was lucky to have had a good share of exemplary teachers in early years, and during my undergraduate education at the Indian Institute of Technology, Kanpur, as well as particularly enlightened colleagues during my years at the research laboratory at the International Business Machines Corporation (*IBM*). The latter was a community with a preponderance of exceptional people. Every encounter had something to learn from it. Science ruled. Nonsense was unacceptable. All these benefactors have influenced choices I have made, especially the nature of questions I became interested in and asked and the multifaceted approaches of different perspectives I used in answering them. That some parts of businesses can run with academic ideals for the greater good and that academe is so much beholden to business norms has been a surprising observation of mid-life. The loss of the research institutions, particularly the *IBM* and Bell laboratories, and the reduction in the diversity of thoughts and approaches that has followed, has changed the American story. These institutions were exemplary for most of their lives in their approach to discovery, research and development, understood their interconnectedness and had an intellectual commitment that overrode palace intrigues. The test of good management is in how an institution responds and negotiates through hard times. *IBM* maneuvered remarkably well for a fair period.

Foremost among the people I wish to acknowledge are the colleagues I had at Yorktown during Research's golden period. Among them, Paul Solomon, David Frank, Steve Laux, Massimo Fischetti, Wen Wang, Frank Stern, Rolf Landauer, Peter Price, Bob Dennard, Subu Iyer, Tom Jackson, Supratik Guha, Pat Mooney, Arvind Kumar, Chuck Black, Jeff Welser, Tom Theis, Doug Buchanan, Dan DiMaria, Jim Stathis, Emilio Mendez, Leo Easaki, Jim Misewich, Ravi Nair, Charles Bennett, Hans Rupprecht, Jerry Woodall, Peter Kirchner, Jeff Kash, Jimmy Tsang, Steve Wright, Dennis Rogers, Marshall Nathan, Frank Fang, Alan Fowler, Reuben Collins, Dick Rutz and many others shared technical and life-enriching wisdom.

This text being about semiconductors, I want to particularly acknowledge the long discussions with my contemporaries David Frank, Steve Laux and Max Fischetti, each with their own different interests but always ready to discuss the numerous interesting questions that kept unfolding as new technologies made new artificial creations in semiconductors with their own questions possible. Arvind Kumar later on joined this illustrious group. Paul Solomon straddled a sagacious understanding of devices and their materials' physics simultaneously. Among the senior colleagues, Frank Stern, with his Dirac-like devotion to choice of words and science, influenced all of us. His co-authored *Reviews of Modern Physics* book-sized article on two-dimensional systems continues to influence the community to this day. One had to be very careful in listening to Peter Price lest a gem of an insight—spoken in a few choice words—was missed. Soon after I left for academe, upon reading a letter of mine in the *Wall Street Journal* about the importance of bureaucracy in getting things done properly—I was responding to a published anti-government opinion that used developing nations as examples by pointing out that it was the select dedicated folks that partly saved the day when the politicians royally failed with the Indian partition and the accompanying migration—Peter wrote, "Sandip, the Op Ed page of *WSJ* is scarcely the place to look for rationality."

One notable aphorism from those days is as follows: "Beware the *PF*9s and *PF*10s." Before the internet, or the bitnet before that, there existed within *IBM* an internal network (*VNET*) for electronic mail, for accessing repositories of useful codes and technical documents that individuals had written, and for other network-wide computing tasks. *PF*9 and *PF*10 were programmable function keys for the "Receive" and the "Send" tasks in the mailing program. "*PF*9s and *PF*10s" was the euphemism for the doorkeepers, speed breakers and messengers—folks whose technical careers were short but who had the wherewithal to generate pointless work—from whom one needed to protect oneself to be a good scientist and engineer.

The conflict between a humane society and a living economy is a broader tension. It has always been. We fail at teaching our graduating students not to force fit "truth" to our biases. We succeed with too few. Part of the reason for this failure is that a business style—talking points, elevator pitch, slides to which one talks, spreadsheets, college bookstores as sweatshirt shops with little space for books, libraries as cafés with the loss of space and collection where one was exposed to printed thoughts that one was not actively searching for, and, underlying all this, finance over content—has broadly infected the educational institutions of this country. Science and technology are major social and economic forces. By not finding a balance between being the source of technological and financial success, which is initially limited to only a few who are educated and have power, and is economic, and in developing citizenship, which is a broader goal and is social, academia seeds a future downfall if it is not fixed. The standards need to be high, and, for this social commitment theme, legal acceptability or checking boxes of ethics as dictated by accreditation boards is not that standard. This is particularly important in science and engineering, where the student usually has a very different mental focus than the student of humanities.

The early thoughts on how to organize this writing occurred during a sabbatical leave at Harvard in 2006–2007. I even wrote some preliminary notes. But work could not begin in earnest until I managed to relinquish several responsibilities that work life brings. The first drafts started in Ithaca around 2010, but serious work had to wait for my next sabbatical leave in 2012–2013, which also gave a chance to try the material—of the fourth and this third volume—with different student audience. At the Indian Institute of Science, my hosts included Professors Navakant Bhat, Rudra Pratap and S. Shivashankar; at Stanford University, Professor Roger Howe; and, at Technische Universität München, Professor Paolo Lugli. The environment at these institutions was ideal for what I had in mind. And, in addition, it provided an opportunity to be in the company of several other faculty with a joyful outlook to science and life: Professor Ambarish Ghosh, Srinivasan Raghavan, Philip Wong, Yoshio Nishi, Walter Harrison, Christian Jirauschek, Wolfgang Porod and Peter and Johannes Russer—my immense gratitude to them for a stimulating year. The students who participated in these courses around the world have provided invaluable feedback that is reflected in the writing and rewriting.

Many colleagues have read and commented on parts or the whole, and this has helped with the exposition. To Tom Theis, Wolfgang Porod, Max Fischetti, Jerry Tersoff, Supratik Guha, Ed Yu, Federico Capasso, Siegfried Selberherr, Srikrishnaa Vadivel and Kunal Tiwari, my thanks for sharing their time and suggestions. Jack and Mary East—my dear friends—have given crucial support through their constant interest in the progress of this work. My wife Mari has kept a careful eye on the goings-on, helped keep my centrifugal propensities in check and given invaluable advice on the presentation. The very constructive exchange with Sonke Adlung, Harriet Konishi and Elizabeth Farrell at Oxford University Press has been immensely valuable in the creation of the final form.

The LaTeX class for Edward Tufte's style suggestions has been largely followed in these texts. The authors of such open source

When I first arrived at IBM Research in the beginning of the 80s, *APL*—A Programming Language—as an abstract and compact computational and graphical language tool was an eye opener. It had beauty, and it let you tackle data manipulation, matrix calculations such as for differential equations, searches, et cetera, all quite compactly, together with publishable graphics. But, as used to be said about *APL* programs' readability, "It was tough to write. It should be tough to understand." I found deciphering even my own codes that were just a few years old hard going. Now, when using Python, where, as with *MATLAB*, I can see through where these systems and their syntax built itself from, I do appreciate the readability, reusability and clarity that underlie the evolution. This goes together with the difficulties I have certainly faced with undoing the programming relearning that came from having been exposed to *FORTRAN* first. Those indiscriminate *GO TO* statements make a complete tangle of the ball of string that has an $\mathcal{O}(n^2)$ complexity.

resources, which here also includes Python for the calculations in the exercises and figures, perform an immense service to the society.

Over the years, I have been fortunate to have had generous and understanding mentors and supporters who have made the research and academic pursuits fulfilling. The research environment of those early years was a reflection of the focus on research with a perspective. The management ably fostered this.

In days past, on the board outside Professor Les Eastman's office, there used to be a press clipping circa the late 1970s, from the Universal Press Syndicate, with the following quote: "Our futures almost certainly depend less on what Ronald Reagan and Walter Mondale say and do than on what is going on inside the head of some young Cornell graduate waiting for a plane in Pittsburgh." The routes are now through Philadelphia, but the thought is still right. To students who have interacted with me through the classes goes the ultimate tip of the hat.

Science touches us all through its beauty. My colleagues, teachers, family, students and others who have, through occasional talks, conversations, writings, the way an argument was framed, a clever twist of reasoning, or even plowing through when a situation demands, have enlightened this recognition.

Sandip Tiwari
Ithaca, Orsay, Southport and Bhopal

Contents

Introduction to the series

These books are a labor of love and love of labor. They reflect a personal philosophy of education and affection for this small but vital subject area of electroscience, one that has given me satisfaction. This subdiscipline is also a domain where knowledge has evolved rapidly, leaving in its wake an unsatisfactory state in the coherence of content tying mathematical and physical descriptions to its practice.

Engineering-oriented science education, even though not really that different from science education itself, is difficult for two reasons. It aims to provide strong scientific foundations and also to make the student capable of practicing it for society's benefit. Adequate knowledge of design and technology to invent and optimize within constraints demands a fundamental understanding of the natural and physical world we live in. Only then can we create usefully with these evolving tools and technology. Three hundred years ago, calculus and Kepler's and Newton's laws may have been adequate. Two hundred years ago, this basic foundation was expanded to include a broader understanding of thermodynamics; the Lagrangian approach to classical mechanics; probability; and the early curiosity about the compositional origins of lightening. A hundred years ago, the foundational knowledge had expanded again to include a fair understanding of the periodic table, Hamiltonians, electricity, magnetism, and statistical mechanics, and, yet again, it was incomplete, as the development of new, non-classical approaches, such as Planck's introduction of quantum of action, and relativity, showed. Our understanding still remains very incomplete today even as evolution is gathering pace via science and engineering with non-carbon forms of intelligent machines quite imaginable. Reductive and constructive approaches, as before, pervade the pursuit of science and engineering. Understanding singularities, whether in black holes, in phase transitions with their information mechanics implications, or for solving near-infinite differential equations with near-infinite variables and constraints

A slight aside. Behind this education is an urge to understand our universe, the nature in which we exist, and perhaps through our acts bringing about small changes around us to make life and the world a bit better. When my eldest son came back home after first semester in college, he said, "I now get it. Biology is the emergence of chemistry, chemistry is the emergence of physics, and physics of mathematics." This is a modern take of Galileo Galeilei's statement in *Sidereus Nuncius*, "Nature is written in that great book which ever lies before our eyes. . . . . The book is written in the mathematical language, and the symbols are triangles, circles and other geometrical figures without whose helps it is humanly impossible to comprehend a single word of it, and without which one wanders in vain through a dark labyrinth." The symbols and this language, especially via Leibniz's calculus, have expanded tremendously. This Cartesian world view of physical reality, though powerful, is reductionist and incomplete in the Gödelian sense. Mathematics, music, paintings, and good writing are universal languages that reach out from the natural world to our senses. Objectivity is not and the networks they form, is central to science and engineering problems—connecting back the two ends of the string.

just the Cartesian physical objectivity with an in-built bias in the exclusivity of certain properties but must also expand to domains of other experiences for such a coming together to explain the world. When science uses objective measures of information, it is just that, a measure of that specific information content. The jump to knowledge—a phenomenological objective view—is much more. I would also add that engineering is very much a Martin Heidegger's *Being and Time*, where being-in-the-world is central, and Sartre's progression to existentialism with *Being and Nothingness* as humanism.

All this evolving knowledge and its usage would be deficient were we incapable of adequately using the tools available, which in their modern forms include software for the implementation of mathematics and their computational, observational and experiment-stimulating machinery and their operating software for designing and optimizing suitable answers to the questions posed. A physical understanding of the connections between different interactions, as well as the reasoning that leads one to identify the most primary of these interactions, is essential to utilizing them gainfully.

Another conundrum and this is particularly true for three volumes of this series is that much of this subject area is at the intersection of science and engineering with both important. A scientist is both an artist and a craftsman. The former in the sense of Edgar Degas who says "On voit comme on veut voir; c'est faux; et cette fausseté constitue l'art," Science in this art sense is a search for truth. Art is subjective. In the art of book writing, the choice of words and the exposition are our main tools for exploring the truth or maybe perhaps "what we want to see." The craft, on the other hand, has much objectivity to it. Objectivity can be tackled through the tools of mathematics. Both the art and the craft are important. The books are an attempt at finding that balance so that it appeals both to the foxes and the hedgehogs.

Translated, Degas is saying "People see what they want to see; it is false; and this falseness constitutes art."

Engineering education, with these continuing changes in fundamental understanding and its practice, raises difficult questions of content and delivery too under the constraint of a fixed time period for education. It has also raised serious humanist questions of affordability, even as engineering education claims to aim at frugality through less expensive scaled delivery mechanisms. Engineering, more than science, is beholden to societal needs. In growing fields, particularly the ones that that have the most immediate societal relevance, this rapidly brings content and finances into conflict. As the amount of engineering and science knowledge required rapidly increases, with the rapidly evolving technology, training becomes obsolete just as rapidly. In technical areas, whose educational needs expand suddenly because of their societal use and consequent professional needs, specialized course offerings proliferate rapidly. This puts pressure on the teaching of the foundational knowledge of the disciplines, and the time available for it. The inclusion of broad skill sets into the core curriculum is threatened by the need to teach an expanding number of specialized topics in an ever-shrinking amount of time. The pace of and need for change through new offerings or modifications to courses risk introducing disjointedness and decreasing rigor, because modifying and harmonizing a curriculum is a difficult and time-consuming task.

This series of books is an experiment in attempting to answer today's needs in my areas of interest while preserving thoroughness and rigor. It is an attempt at coherent systematic education with discipline, while maintaining reverence and a healthy disrespect for received wisdom.

The books aim to be conceptual not mechanical. This series is aimed ultimately at the electroscience of the nanoscale—the current interest of the semiconductors and devices stream—but which is also far more interdisciplinary than the norms suggest. Its objective is to have students understand electronic devices, in the modern sense of that term, which includes magnetic, optical, mechanical, and informational devices, as well as the implications of the use of such devices. It aims, in four semester-scale courses, to introduce the underlying science, starting with the fundamentals of quantum, statistical and informational mechanics and connecting these to an exposition of classical device physics, then dive deeper into the condensed matter physics of semiconductors, and finally address advanced themes regarding devices of nanometer scale: so, starting with the basics and ending with the integration of electronics, optics, magnetics and mechanics at the nanoscale.

A hallmark of the present times is introduction of new words when older ones lose their apparent luster or "branding." "Multidisciplinary" evolved to "interdisciplinary" with an expansion of indiscipline. "Transdisciplinary" must be trying to birth itself. Richard Feynman's statement, "In these days of specialization there are too few people who have such a deep understanding of two departments of our knowledge that they do not make a fools of themselves in one or the other" (from R. P. Feynman, "The meaning of it all: Thoughts of a citizen-scientist," Perseus ISBN 0-7382-0166-9 (1989), p. 9), is not inappropriate here.

The first book[1] of the series explores the quantum, statistical and information mechanics foundations for understanding semiconductors and the solid state. The second[2] discusses microscale electronic, optical and optoelectronic devices, for which mostly classical interpretation and understanding suffice. The third[3] builds advanced foundations utilizing quantum and causal approaches to explore electrons, phonons and photons and their interaction in the solid state, particularly in semiconductors, as relevant to devices and to the properties of matter used in devices. The fourth book[4] is a treatment of the nanoscale-specific physics of electronic, optical, magnetic and mechanical devices of engineering interest. The second and the third volumes are for subjects that can be taught in parallel but are necessary for the fourth. The value of this approach is that this sequence can be completed by the first year of graduate school or even the senior year of undergraduate studies, for a good student, while leaving room for much else that the student must learn. For those interested in electrosciences, this still includes electromagnetics, deeper understanding of lasers, analog, digital and high frequency circuits, and other directions. The fourth book was the first to come out because of the urgency

[1] S. Tiwari, "Quantum, statistical and information mechanics: A unified introduction," Electroscience 1, Oxford University Press, ISBN 978-0-198-75985-0 (forthcoming).

[2] S. Tiwari, "Device physics: Fundamentals of electronics and optoelectronics," Electroscience 2, Oxford University Press, ISBN 978-0-198-75984-3 (forthcoming).

[3] S. Tiwari, "Semiconductor physics: Principles, theory and nanoscale," Electroscience 3, Oxford University Press, ISBN 978-0-198-75986-7 (2020).

[4] S. Tiwari, "Nanoscale device physics: Science and engineering fundamentals," Electroscience 4, Oxford University Press, ISBN 978-0-198-75987-4 (2017).

I felt. The third book puts together the foundational learning for the modern insights of semiconductors.

I have always admired simplicity of exposition with a thorough discussion that even if simplified, is devoid of propaganda or the much too common modern practice of using templates where depth and nuances are lost and doubts and questions are not addressed. Also consistency is easily lost when modern tools, instead of a pencil and paper, are employed. The style of these books follows these beliefs. Notations, figures, the occasional use of color and other stylistic choices are consistent across the book series.

From early years, I have been a devotee of marginalia—much of the learning and independent thought have come from doodling in the margins and the back pages of notebooks. These books are organized so that the reader will feel encouraged to do so.

A list of very readable, in-depth sources, with my perspectives serving as a trigger for different contents within the book, is to be found at the end of each chapter, in the section titled "Concluding remarks and bibliographic notes." No attempt has been made to credit original discoverers or authors. These remarks and notes ascribe them, or they are to be found by following the references in these notes to their origins.

The exercises are formulated for use in self-study and in the class-room. A subject cannot really be learned by simply reading. Problems requiring application of the information learned and encouraging further thinking and learning are necessary. When we discover for ourselves, we learn best. The exercises here are meant to inform and to be instructive. They are also ranked for difficulty—those that need only a short time but test fundamental understanding are marked as **(S)**, for simple; those requiring considerable effort, bordering on being research problems, are rated **(A)**, for advanced; and those that are intermediate are rated **(M)**, for moderate.

Teaching slides are available on the companion website recorded in the front. The solutions manual may also be requested by providing information at the second link furnished in the front. Slides, when in the modern template-based style, can seriously hinder teaching when they become a tool for filtering key information and explanation while emphasizing summary points. The available presentation material is a tool to avoiding mistakes in writing out equations and to carefully and graphically explain the relationships that science and engineering unfold. They do not substitute for the book and the instructor needs to be diligent in making sure that important themes of teaching—probing, questioning, reasoning, explaining, exploring evidence—come out credibly. I am also happy

The emphasis on probing, questioning, reasoning, explaining and exploring cannot be emphasized enough. I use paradoxes, puzzles, gedanken experiments and real world analogies as common tools. A simple capacitor switched abruptly connected to an ideal source lets one explore dissipation and energy conversion in its broader sense. Displacement currents are real currents, a capacitor as an antenna can radiate, that this radiation proceeding to infinity has a real characteristic impedance, that an infinite L and C transmission line network ends up as a line with real impedance even if made of reactive elements, and that dissipation may arise in the material too, can all be followed through from the poor lowly capacitor. How did the energy appear throughout the capacitor lets one probe Maxwell's equations and electromagnetic propagation even when this current is asymptotically vanishing in a slow charging process. And farther on a tying in of all these connections between fundamental laws and physical behavior entwining Maxwell's electromagnetism, the quantum-mechanical origins of the materials' properties and the diverse meanings of entropy from statistical mechanics. When I tried to introduce a question drawing on the basic understanding of capacitors in a qualifying examination, a fellow member—a *PF9*/PF10 with administration responsibilities—spoke up that we teach our students to never connect ideal voltage sources to capacitors. Science is not religion. Looking for contradictions helps one find the invariants—the physical principles—that stand tall. It is through such probings and mental experiments that one learns and understands. There is nothing more satisfying in education than this peeling of onion from a simple question to deeper and deeper insights. This is what education is about. Curiosity should never be discouraged.

to hear and discuss the subtleties and the different viewpoints of principles, approaches and the deeper meanings of a derived result.

Lots of people can grasp things remarkably quickly. But grasping is not the same as understanding. Understanding is a much deeper network in the brain. I hope students will find in this sequence of books the ballast to propel their own interests through the understanding.

The books could have been shorter and crisper had there been more time. But, what time there was has given enormous pleasure—a time out for integrity in the presence of the incessant pressure of existence, particularly of life in modern academe. For this escape, my gratitude to this world. For making possible the following of my wishes to produce these songs as the shadow of a life in research, teaching and writing, I thank the Hitkarini Foundation.

सर्वजन हिताय । सर्वजन सुखाय ॥

Introduction

Semiconductors, as crystalline, polycrystalline or amorphous inorganic solids, as ordered or disordered organic solids or even in glassy and liquid forms, form a large set of materials useful in active and passive devices. The control of their properties arising in an interaction of particles—atoms, electrons, photons, their elementary one- and many-body excitations, transport and the exchange between different energy forms—has been a fruitful human endeavor since the birth of the transistor, where they found their first large-scale use. Integrated electronics, through its social and commercial informational ubiquity; optoelectronics, through lasers and photovoltaics; and thermoelectronics and magneto-electronics, with their use in energy transformation and signal detection, are but a few of these gainful uses. Nanoscale, within this milieu, opens up a variety of perturbative and significantly more substantial and sensitive effects. Some are very useful, and some can be quite a bother.

Dating back to the 1950s, there exist numerous good textbooks for the solid state. From these early years, J. M. Ziman's *Electrons and phonons* for the details and *Principles of the theory of solids* for a thoughtful broad discussion are particularly of note. Another one is Rudolf Peierls' *Quantum theory of solids*. Among solid-state texts, these remain particularly alive because much of their content is appropriate to electronics of semiconductors. They certainly treat several of the semiconductor-specific scattering and transport topics rather well. As optics—later rebranded as photonics—became important, the divergence in texts increased. Later solid-state texts, with their emphasis on metals, ferroelectricity, ferromagnetism, superconductivity, et cetera, inevitably gave short shrift to semiconductors. That there is a quite informal completeness, consistency and unity to the diversity in the foundations is something that, except for the early books, few capture. Nanoscale makes matters even more divergent. I have felt that a book that

Semiconductor Physics: Principles, Theory and Nanoscale. Sandip Tiwari.
 Published 2020 by Oxford University Press. DOI: 10.1093/oso/9780198759867.001.0001

brings together this unity and focuses on the foundations toward understanding why semiconductor matter behaves the way it does would be useful.

From an engineering perspective, and from that of science, information as the fountain from which much can be understood and explored, including the quantum-mechanical notions through the Bayesian interpretations, is a major change in our learning of recent times.

Not to belabor the point, but the Copenhagen interpretation—a duality—and the emphasis on an observation and therefore, secondarily, an observer both are causes of why folks see a hint of spookiness in quantum mechanics. Philosophically, I subscribe to the notion of deep truths—truths where a statement and its opposite are both true such as with wave and particle, or insistence on one's privacy yet wanting governments to give one strong security in this internet and information-centric age, or carbon as both a source of nature's suffering and a source of joy by and for humans—as well as observation as the action that unveils information. This is consistent with the earlier Bohrian notions but also does away with the trust that it expects, and perhaps the wonder it raises, when first introduced in an undergraduate classroom. Bohr had this inclination toward the complementarity of truth and clarity. He is known to have used the following story often. A young person was sent to another village to listen to a great rabbi. Upon his return, he reported, "The rabbi spoke three times. The first talk was brilliant, clear and simple. I understood everything. The second was even better, deep and subtle. I didn't understand much, but the rabbi understood it all. The third was just superb and unforgettable. I understood nothing and the rabbi himself didn't understand much either." Keeping an information-centric perspective helps do away with quite a bit of the metaphysics that developed over the decades around quantum mechanics.

The Fermi surface of a metal, which gets much attention in a solid-state text, is of enormous import, but it is more of an anachronistic appendage to semiconductor matters. The Fermi surface in a semiconductor, while important, is not as complicated. But there are many static and dynamic interactions, transformations and fluctuation effects that have enormous import and need emphasis. Included within this group are the topics of noise and dissipation as consequences of fluctuations, linear response and causality appearing in Kramers-Kronig-type relationships in multitudes of places beyond just the dielectric function, collective effects and interactions such as those of plasmons or polaritons, strain, semiconductor alloys, the nature of heterostructures and their periodic structures, of defects and multiparticle Auger interactions and of nonlinearities in energy coupling, such as those embodied in Onsager relationships, and even transport from classical to mesoscopic in off-equilibrium conditions.

Add to this collection of topics the consequences of nanoscale from surface to bulk, dimensionality change, collective behavior and, together, their effect on various interactions and transformations as additional subjects of modern importance. In teaching these, with the implicit understanding of nanoscale devices as the ultimate goal, one has to resort to a fair collection of diverse classic resources and combine them with one's own thoughts. This makes the task of getting across to the student the necessary physics for understanding devices difficult, with styles, nomenclature, incompleteness and substantive jumps abounding.

In keeping with the spirit of this textbook series, this volume is devoted to semiconductor-specific solid-state physics aimed at students of engineering, particularly electronics and materials science, but also with utility, because of the exposition, for those from physics and chemistry.

It is organized to certainly include the classical underpinnings ranging from bandstructure approaches to phonon behavior, scattering, approximations, et cetera, but it particularly stresses topics that are modern and aimed toward nanoscale. All are presented with principles and theory as areas of emphasis, expecting that review papers and other narrower but deeper treatises will become analyzable, understandable and critiqueable to those prepared from this approach.

The book reviews the essential basics and the tools of the trade first, including the quantum methods for ensembles and their approximations, before moving on to the approaches of bandstructure calculation as well as their limitations, which help us describe the behavior of electrons and phonons in a semiconductor. This serves to then develop the treatment of transport, including within it semi-classical, quantum and mesoscopic approaches under scattering and the limit of no scattering under equilibrium and off-equilibrium conditions. For semiconductors, particularly in newer applications, spin-orbit coupling manifests itself in several places, so care is taken to bring the insights from bandstructures to the interactions for semiconductors at different dimensionality.

This sets the stage for the atypical topics of emphasis of the text. The first of these is the discussion of electrons and phonon behavior at surfaces. This is then reformulated for interfaces. Here, heterostructures—what really happens physically at the boundaries—also appear as an important subject for analysis and discussion. Zinc blende, diamond and wurtzite, encompassing elemental and compound semiconductors, including the nitrides, are explored together in emphasizing the principles. The text also discusses the newer and perhaps presently unconventional semiconductors, such as monolayers, in the final chapter, where we return to the themes of the initial chapters in light of all the learning in-between. All this discussion has electrons and phonons as its center, where defect-catalyzed interactions and their variety of behavior under compositional changes are also important.

Photons, electron-photon interactions and radiative and non-radiative phonon-assisted processes are tackled to bring about the interactions in a broadband of energies, so including Auger processes.

This sets the stage for discussion of the next order of complexity in ensemble interactions. We start with a discussion of causality and response theory, and within it the different places where fluctuation-dissipation and Kramers-Kronig forms appear. Ensemble interactions, also in their coupled forms, such as excitons, polaritons and plasmons, follow next. This discussion of higher order interactions is expanded to the variety of manifestations of dissipative transport. Particularly important here is noise, which is central to the use of semiconductor devices at nanoscale. Another of these next order effects is strain, whose use is now pervasive in semiconductors. Spin again becomes quite central to this discussion through bandstructure, as it does for topological reasons.

The spin-topological connections and their device implications are tackled separately in S. Tiwari, "Nanoscale device physics: Science and engineering fundamentals," Electroscience 4, Oxford University Press, ISBN 978-0-198-75987-4 (2017).

The high permittivity of gate dielectrics often used with semiconductors appears with soft phonons and is an essential part of the tool set of semiconductor devices. We look at their behavior and the local and remote coupling effects arising in them.

Energy couplings and their transfer between various forms—heat to electric, and stress to electric—their off-equilibrium behavior and the role of Onsager relationships in these energy transformations is an essential set of topics in important areas of use of semiconductors, from thermoelectrics to piezoelectrics. These are discussed in sufficient detail for the reader to get good insight into the operating principles and how many of the effects undergo some change—sometimes small, sometimes large—at nanoscale.

We follow this broad swath of physics discussion by looking at periodic structures and the nature of the various excitations of interest in them. So, we discuss superlattices for electrons, phonons, plasmons and plasmon-polaritons, as well as the role of dimensionality within them.

The text intends to provide the reader with an in-depth discussion of semiconductors, aiming toward the nanoscale through this range of development of the subject. Readers who have had an introductory course in quantum and statistical approaches and have a general understanding of the operation of electronic and optical devices will benefit.

At the very least, readers must have internalized the meaning of equations at the end of the glossary and the principles of quantum and statistical mechanics and should be willing to pursue the appendices that sometimes serve as summary introductions of important ideas being employed in the main chapters.

The content here is quite comprehensive, as it tries to integrate a variety of ideas across the significant breadth of phenomena that one needs to understand in semiconductors. It is likely that, for some, it is more than can be tackled in a single semester, especially if the students have diverse educational backgrounds and disciplines. So, choices may need to be made. Mine have been to maintain balance between taste and the students' needs. And these have changed from year to year. This book, and this series, represent an attempt at a style where learning is also possible on one's own. The classroom is particularly useful in bringing about the connection of ideas, the emphasizing of principles, the creation of interesting segues where new thoughts can be explored, using the learning and the give and take that the classroom provides so well, and the stimulation of the students' spirit for adventure. The first two chapters of this text are an attempt at summarily introducing and reviewing major ideas, some part of the orthodoxy, but others,

Quantum and statistical mechanics treatment at the level of S. Tiwari, "Quantum, statistical and information mechanics: A unified introduction," Electroscience 1, Oxford University Press, ISBN 978-0-198-75985-0 (forthcoming), is expected. The reader will find the integrated treatment of information mechanics within this description quite useful because of the common themes that tie energy, entropy and information together, as well as the dominant usage of semiconductors in the processing of information.

An understanding of the operation of simple devices—p/n junctions, and unipolar and bipolar transistors—helps with understanding the relationship between the operational physics, such as that of high permittivity insulators, of noise, of strain or of heterostructures and the behavior of devices.

such as Fisher entropy and information, not. This content is the part of the book that one does have the freedom of referring back to if one starts with Chapter 3. Good books and teaching are like music, where the beauty and joy comes from the constant returning back to powerful ideas with variations, each with a little different way of looking at the subject—a different rhythm, a different harmony, a different timbre. Each class is then a different piece of music. The first two chapters facilitate this for the rest of the text so that the learning can be fulfilling to both the student and the teacher. This comment holds just as well for the appendices, where important notions are summarily emphasized. The rest of the book can then be managed in a reasonably demanding course where the students are expected to stay abreast with their reading and thinking.

My favorites for this returning back again and again in music is Verdi's *La forza del destino* as a simultaneous multipath arrival, but many of Schubert's and Chopin's piano—single instrument—pieces bring the path to the heart and the mind perhaps even more convincingly. The Giuseppe Verdi creation is itself a variation on a play by the Spanish master Angel Perz de Saavedra. Verdi endowed his beautiful home—Casa Verdi—in Milan as a retirement home for musicians who need such support late in their life. Music lives there in love and in peace. Verdi's is a life whose variations continue to live a hundred-plus years later. Variations are the most powerful, whether in music or in books and teaching, when they play out forever in time and space.

1
Hamiltonians and solution techniques

Nature is composed of objects—particles, solids or other assemblages in various representations—whose behavior—properties, evolution in time, consequences of stimulation and others—we attempt to explore, understand, design and predict in science and engineering. A major success of classical mechanics from the mid-17th century on was the ability to mathematically describe the evolution in time of the objective values of properties of interest. For example, if a system—a bounded object—of known spatial coordinates and velocity (or momentum), that is, one whose "state" was known, was stimulated under the action of a force, one could predict the future values in Euclidean space. Take this same mathematical construction—usually a set of differential equations—and one could build bigger and smaller objects and predict their evolution by changing the parameters of this differential construction.

Post-Copernicus—the importance of observation and prediction, and mathematical tools such as calculus—is the age of modern science. Prior to that, Euclidean geometry had been the dogma since about 300 *BC*. Likewise, dating from the same time period, Aristotle's views that there are four elements, that heavier objects fall faster or that Earth is the center of universe made up the dogma that was not to be questioned. The former was overthrown by Nikolai Lobatchevski, and the latter needed the Renaissance. This is almost two millennia of scientific darkness!

A space—the state space—could be described with the object at some location in it for each moment of time. The dynamic system's change of state could be described through the equations of motion using either the Lagrangian or the Hamiltonian function once an initial state had been described through the complete specification of the initial dynamic variables together with that of the action on the system. The complete quantitative description of a set of simultaneously measurable parameters—position and velocity (or momentum)—is an essential requirement of this classical determinism. At the quantum scale, simultaneous precise measurement of parameters such as position and momentum is not possible. *The state of the system is not characterized by a set of dynamic variables with specific values. Instead, the state is characterized by a statefunction.* The statefunction is composed of a set of chosen variables—the canonic variables—and the time dependence of this statefunction describes the dynamics of the system. This function of

Semiconductor Physics: Principles, Theory and Nanoscale. Sandip Tiwari.
 Published 2020 by Oxford University Press. DOI: 10.1093/oso/9780198759867.001.0001

time is our wavefunction of the quantum system. It has properties similar to those of waves but it also describes the state of the quantum object. The statistical/probabilistic nature of quantum is within this statefunction. Even for this quantum scale, one can write the Hamiltonian and the Lagrange functions by employing operators that correspond to an observable property—position, energy, momentum and others—and thus describe the evolution of the state.

What is extremely powerful in this approach of the Hamiltonian and Lagrangian functions is that these express the behavior of physical systems irrespective of whether they need to be treated classically or—in more depth—quantum-mechanically. They represent the principle of conservation of energy and the principle of least action as complementary articulations of nature's precept. Between the classical and quantum view, where the quantum view reduces to the classical in the limits, it is the quantum Heisenberg uncertainty, the quantum de Broglie wave-particle duality, the energetics of the interaction and the statistics of quantum to classical that make the enormous change we see in the real world happen. Atoms are stable—neutral—and an electric or magnetic field will have no effect were atoms just to be thought of as classical particles. Place them together in a solid and they form metals, semiconductors and insulators with a variety of seemingly magical properties that are quite different from those of the atom. Conduction and insulation both arise from the properties of the state. An electron in a propagating state leads to conduction. An electron in a bounded state leads to insulation. Largely unoccupied and largely occupied bands of states can both conduct, and conduction can be modulated! Tunneling can happen at microscopic scale. Control can be exercised at small energies—of the order of eVs of visible and infrared light, and therefore at useful bias voltages of a V in semiconductors—instead of Rydberg energies. And further modifications to properties become possible by making structures the size of an electron's wavelength. Quantum mechanics predicts this diverse complexity of solids, and specifically of semiconductors, and thereby makes it possible to use them judiciously.

For reasons of symmetry, explorations in quantum mechanics prefer the Hamiltonian methodology. If you can write the Hamiltonian, you have described the system and its evolution.

That writing the Hamiltonian suffices is written facetiously, of course. In theory, what one has to do is to use this Hamiltonian function in the classical approach, and the Hamiltonian operator in the quantum approach, to now solve the Hamiltonian equation describing the problem. In practice, only the simplest cases can be solved precisely. The rest need approximations.

A solid is a collection of particles—atoms and electrons as their simplest form in ordered or disordered arrangement—undergoing perturbations due to external stimulus because they exist at a temperature T connected to the rest of the universe as a reservoir with which it exchanges energy and particles. To understand semiconductors, therefore, one needs to be able to describe the interactive evolution of the system. An atom is the simplest, a molecule another level up and then there is the larger ensemble of the semiconductor solid involving an Avogadro-scale number of these particles. So. how does one describe an assembly of particles is an important question to start with.

When making predictions of what a collection of particles will do, classical analysis will resort to either the Lagrangian or the Hamiltonian formalism. These approaches employ a space—the phase space—whose spatial points each represent a specific arrangement of the individual particles. The space is built of the two canonic coordinates over the N dimension of the N particles. The evolution of this point follows from a mathematical operation on a single function leading to a description of the dynamic behavior. This function is the Lagrangian or the Hamiltonian. The Lagrangian employs a general position coordinate $\{q\} = q_1, \ldots, q_N$—not necessarily a set of Cartesian positions, whose choice is determined by convenience, and bundles it with a generalized velocity $\{\dot{q}\} = \dot{q}_1, \ldots, \dot{q}_N$—again, not necessarily a Cartesian velocity. The equation of motion of the system then follows from the Euler-Lagrange equations:

$$\frac{d}{dt}\frac{\partial \mathscr{L}}{\partial \dot{q}_i} - \frac{\partial \mathscr{L}}{\partial q_i} = 0 \;\; \forall \, i = 1, \ldots, N. \tag{1.1}$$

The Lagrange function $\mathscr{L} = T - V$ as a difference of the kinetic (T) and the potential energy (V) due to all sources—internal and external—captures the entire behavior of the particle set. The equations state the principle of least or stationary action. In the configuration space, the evolution between two fixed end points occurs when the action, which is the integral of the Lagrange function along the line connecting the two end points, is a minimum. This is akin to the more easily visualized picture that a function $f(x)$ is a minimum, a maximum or a saddle point if $df/dx = 0$. It is a stationary point: a minimum, a maximum or a saddle point in higher dimensions.

What we wrote here is the Lagrangian for classical mechanics. Lagrangians can be written for all the variety of physical phenomena. As shown in Table 1.1, what they have in common is a square gradient term, of energy, although there may not be an explicit energy connection. In the following chapter (Chapter 2), one sees the informational link to the observation of the physical phenomena—the data that contains the information and therefore the entropy and energy connection—that this represents.

This nomenclature of configuration and phase space arose in explorations of approaches for dynamic systems, where mechanical systems were the first ones of interest from the 18th century onwards. In phase space, every possible state of the system defined by the values that the parameters take defines a point and its evolution a trajectory. The position and the momentum are the variables of the phase space. One could also describe this evolution through position and velocity. Multiple particles will have a multidimensional space. The configuration of the system is writable in generalized coordinates. The vector space defined by these coordinates is the configuration space of the physical system.

Throughout this text, in order to limit the unwieldiness of an equation, we may employ the prime mark to indicate a derivative with respect to spatial coordinates, and a dot above for a derivative with respect to time. So, $q' \equiv dq/dz$ when using a prime, and $\dot{q} \equiv dq/dt$ when using a dot.

Lagrangians have been employed fruitfully in genetics, macroeconomics, machine learning and various other places where new formulations of energy that are not kinetic or potential but where one can see an intuitive energy interpretation. Lagrangian's beauty includes that it teaches us symmetries, conservation laws and other properties in addition to the equation of motion of the dynamical system.

System	Lagrangian $\mathscr{L}$	Comments
Classical mechanics	$(1/2)m(\partial q/\partial t)^2 - V$	Kinetic minus potential energy
Compressible fluid	$(1/2)\rho\left[(\partial q/\partial t)^2 - v^2\nabla^2 q\right]$	ρ: density, v: flow velocity
Diffusion	$-\nabla_{\mathbf{r}}^2\psi - O$	ψ: a concentration, O: other terms
Schrödinger's equation	$-(\hbar^2/2m)\nabla_{\mathbf{r}}^2\psi - O$	ψ: statefunction
Elastic wave equation	$(1/2)\rho(\partial^2 q/\partial t^2) - O$	ρ: density
Helmholtz equation	$\nabla_{\mathbf{r}}^2\psi$	ψ: a field
Lorentz transformation	$(\partial_i q_n)^2$	Integral invariance

Table 1.1: Lagrangians of some common physical situations. These are all energies, and their integral is the "action."

The Hamiltonian picture is more symmetrical, represents an identical description but, being symmetric, provides a more convenient method for analysis, particularly in quantum mechanics. The generalized position coordinate set $\{q\}$ is taken together with generalized momentum $\{p\} = p_1, \ldots, p_N$. Again, these are not necessarily the Cartesian positions or linear momenta. $p_i = \partial\mathscr{L}/\partial\dot{q}_i$. The symmetry of the Hamiltonian evolution appears in the form

$$\dot{p}_i = \frac{dp_i}{dt} = -\frac{\partial\mathscr{H}}{\partial q_i}, \text{ and } \dot{q}_i = \frac{dq_i}{dt} = \frac{\partial\mathscr{H}}{\partial p_i} \quad \forall\, i = 1, \ldots, N. \tag{1.2}$$

From this "canonical" form, one can also read that

$$\frac{d\mathscr{H}}{dt} = 0, \tag{1.3}$$

which is the law of conservation of energy.

The equivalence of the two approaches can be noted through the ability to determine the canonical conjugates p_i or $\dot{q}_i$ of the two approaches equivalently. The Hamiltonian may be found from the Lagrangian using $\mathscr{H} = \sum_i \dot{q}_i(\partial\mathscr{L}/\partial\dot{q}_i) - \mathscr{L}$. The Hamiltonian function $\mathscr{H} = \mathscr{H}(\{q\};\{p\})$ describes the total energy of the system. Its quantum-mechanical operator is $\hat{\mathscr{H}}$.

1.1 *Hamiltonian*

Being the total energy of the system, in classical mechanics, the Hamiltonian $\mathscr{H}$ is the sum of the kinetic energy T and the potential energy V of all the particles of the system; that is, $\mathscr{H} = T + V$. In quantum mechanics, $\hat{\mathscr{H}}$ is an operator, corresponding to the observation of energy of the system; operating on the wavefunction describing the system results in the total energy $E = T + V$. For every observable, one can write an operator that, upon operating on the wavefunction describing the system, leads to the observable.

More subtle and therefore more consequential is that one may define the Hamiltonian more generally and hence more powerfully than as just the sum written here.

Operators may be developed from an observable's functional form but require some subtlety so that symmetry and antisymmetry consequences are properly accounted for.

See S. Tiwari, "Quantum, statistical and information mechanics: A unified introduction," Electroscience 1, Oxford University Press, ISBN 978-0-198-75985-0 (forthcoming) for remarks on finding operators corresponding to an observable.

The Schrödinger equation,

$$-\frac{\hbar}{i}\frac{\partial}{\partial t}\psi = \hat{\mathscr{H}}\psi = E\psi, \tag{1.4}$$

incorporating both the time-dependent and the time-independent parts, describes the wavefunction of the system. A system is stationary if $\langle\psi\left|\hat{\mathscr{H}}\right|\psi\rangle$ is invariant in time. Its expected energy and the probability distribution $\langle\psi|\psi\rangle$ are constants of time. For any expectation value of any observable, say A, the expectation value ($\langle A\rangle$) follows:

A statistical—probabilistic—interpretation of stationarity—of various orders—will appear in our discussion of noise in Chapter 16. See Appendix A also if you want to look ahead.

$$-\frac{\hbar}{i}\frac{d}{dt}\langle\hat{A}\rangle = \langle[\hat{A},\hat{\mathscr{H}}]\rangle - \frac{\hbar}{i}\langle\frac{\partial}{\partial t}\hat{A}\rangle\Big|_{\mathscr{H}}. \tag{1.5}$$

For a particle of mass m in a potential V, the Hamiltonian is

$$\hat{\mathscr{H}} = -\frac{\hbar^2}{2m}\nabla^2 + \hat{V}, \tag{1.6}$$

where the first term for kinetic energy follows from the operator for momentum,

$$\hat{\mathbf{p}} = \frac{\hbar}{i}\nabla. \tag{1.7}$$

because kinetic energy $T = \mathbf{p}^2/2m$. Table 1.2 summarizes some of the Hamiltonian operators of interest in common and simple systems.

Since our interest in this text is in understanding the various ways that interactions occur in semiconductors, and the manifestations of these interactions in properties, the important underlying theme is a reasonable understanding of the semiconductor solid itself and therefore the predictive edifice for the collection of electrons and atoms therein. Reasonably accurate solutions of the

System	Hamiltonian $\hat{\mathscr{H}}$	Comments
$1D$ harmonic oscillator	$-\frac{\hbar^2}{2m}\frac{d^2}{dx^2} + \frac{1}{2}kx^2$	k: force constant
Rotation in a plane	$-\frac{\hbar^2}{2I}\frac{d^2}{d\phi^2}$	I: moment of inertia
Rotation on a sphere	$-\frac{\hbar^2}{2I}\Lambda^2$	Λ: Legendrian operator $\Lambda^2 = \frac{\partial^2}{\partial\theta^2} + \frac{\cos\theta}{\sin\theta} + \frac{\partial}{\partial\theta} + \frac{1}{\sin^2\theta}\frac{\partial^2}{\partial\phi^2}$
Hydrogenic atom	$-\frac{\hbar^2}{2\mu}\nabla^2 - \frac{1}{4\pi\epsilon_0}\frac{Ze^2}{r}$	Z: atomic number μ: reduced mass
Collection of charged particles	$-\sum_i \frac{\hbar^2}{2m_i}\nabla_i^2 + \sum_{i>j}\frac{1}{4\pi\epsilon_0}\frac{z_iz_je^2}{\lvert\mathbf{r}_i-\mathbf{r}_j\rvert}$	z_ie: charge of ith particle
Electric dipole in a field	$-\mathbf{p}\cdot\boldsymbol{\mathcal{E}}$	$\mathbf{p} = ze\langle\mathbf{r}\rangle$: electric dipole moment
Magnetic dipole in a field	$-\mathbf{m}\cdot\mathbf{H}$	$\mathbf{m}$: magnetic moment

Table 1.2: Hamiltonians of some often–encountered situations in semiconductor physics.

Hamiltonian for this collection and its interactions are necessary. And all this will require approximations, for obvious reasons of complexity therein.

We will build this understanding by starting with a discussion of the formulation and then, following some comments on the approach, proceed to its usage in model problems that are instructive. For example, we first look at systems with few electrons and few atoms such as a molecule, or even an atom with its collection of electrons, and then let the number N of particles of this ensemble expand to larger numbers. As we explore, what is important is to understand the reasoning behind the approximations that we make, so that we also know the limits of their validity.

What we are doing is building models based upon our understanding and interpretation of what is most important. Their test of success comprises predictions that come about to be true. Not all of the possibilities can be tested. There will be a range of variations of parameters where our predictions may be trusted with a good model. But it is still a model, an approximation and not the complete reality. So, these are all just different levels of sophistication of "toy models." This is how we should always look at our analytic or algorithmic interpretations.

1.2 *Preliminaries*

CALCULATION OF THE DYNAMICS of a single particle (an electron, for us) requires us to write the Hamiltonian with the potential energy of its interaction and solve the single particle Schrödinger equation

$$\hat{\mathscr{H}}|\psi\rangle = \left(-\frac{\hbar^2}{2m}\nabla^2 + \hat{V}\right)|\psi\rangle = E|\psi\rangle, \quad (1.8)$$

We will use the hat symbol, $\hat{}$, to identify an operator. Any observable—a physical measurable quantity—is associated with a self-adjoint linear operator. Operators yield the physical value, which is an eigenvalue of the set of possibilities for the system. The wavefunction of the system provides the probability amplitude of finding the system in that state. Pure states have unit norm, so they can be represented by unit-norm vectors. The operators are Hermitian, since operators must yield real eigenvalues, whose probability is in the wavefunction through the square of the amplitude of the orthonormal eigenfunctions. We will be a little loose in writing. Sometimes, an operator hat should be there but may be missing. Sometimes a wavefunction maybe written without the ket symbol $|\rangle$ denoting its vector nature, and sometimes it will. It should be clear from the context.

under the constraints of the boundary. In principle, this is straightforward. It may be as simple as the wave solution for $V=0$, or it may be computationally demanding when $V(\mathbf{r})$ takes on odd complexities. When we make this a few particle system, with $V(\mathbf{r}_1, \mathbf{r}_2, \ldots)$ (a potential that is a function of the position of the individual particles), it immediately becomes unwieldy, the dominant reason being the two-body nature of Coulomb interaction in a multi-component many-body problem. Only when particles are non-interacting, that is, when $V(\mathbf{r}_1, \mathbf{r}_2, \ldots) = V(\mathbf{r}_1) + V(\mathbf{r}_2) + \cdots$, does this problem reduce to a straightforwardly solvable form with

$$\hat{\mathscr{H}}(\mathbf{r}_1, \mathbf{r}_2, \ldots) = \mathcal{H}(\mathbf{r}_1) + \mathcal{H}(\mathbf{r}_2) + \cdots = \sum_i \left[-\frac{\hbar^2}{2m_0}\nabla_i^2 + \hat{V}(\mathbf{r}_i)\right], \quad (1.9)$$

a set of independent single particle equations. The wavefunction solution then is $|\psi\rangle = \prod_i |\psi_i\rangle$, where $|\psi_i\rangle$ is the eigenfunction solution of the partitioned Hamiltonian. The energy of the system is just the sum of eigenenergies of these non-interacting particles. *But, this works only if these particles—electrons here—can be treated as being non-interacting, a very rare situation.* If these particles were interacting, each would influence the other, and this picture is invalid, since interactions, even if infinitesimally small, will modify properties.

As an analogy, non-interacting classical gas molecules have very well-defined macroscopic properties, but, microscopically, the motion of each is affected by the collision interactions, even the elastic ones.

The semiconductor solid—the item of interest to us—is a collection of atoms where the atom itself is a collection of particles as a nucleus surrounded by electrons, with the entire solid being charge neutral. With M_i, Ze and $\mathbf{R}_i$ as the nuclei's mass, charge (Z being the atomic number) and position, respectively, and m_0, e and $\mathbf{r}_i$ as the corresponding electron parameters, this solid's Hamiltonian is

$$\hat{\mathscr{H}}_{xtal} = \sum_i \left[-\frac{\hbar^2}{2M_i} \nabla_i^2 \right] + \frac{1}{2} \sum_{i \neq j} \frac{1}{4\pi\epsilon_0} \frac{(Ze)^2}{\mathbf{R}_i - \mathbf{R}_j} + \sum_l \left[-\frac{\hbar^2}{2m_0} \nabla_l^2 \right] + \frac{1}{2} \sum_{i \neq j} \frac{1}{4\pi\epsilon_0} \frac{e^2}{\mathbf{r}_i - \mathbf{r}_j} - \sum_{i \neq j} \frac{1}{4\pi\epsilon_0} \frac{Ze^2}{\mathbf{r}_i - \mathbf{R}_j}. \tag{1.10}$$

This Hamiltonian is made up of energy terms representing the nuclei's kinetic energy (the first term), the potential energy from internuclear Coulomb interactions (the second term), electrons' kinetic energy (the third term), the potential energy from inter-electron Coulomb interactions (the fourth term) and the potential energy contribution of electron-nuclear Coulomb interactions (the fifth term). There are two summations over the kinetic energy, and three summations over the electrostatic interaction.

Since electrons are fermions, the total electronic wavefunction must be antisymmetric whenever the coordinates of two electrons are exchanged (an exchange interaction). The nucleus may be of different species, in which case, they are distinguishable. If they are of the same species, then nuclear spin will also matter. To manage a solution, we have to make judgments on what is important and how to judiciously incorporate it in a manageable calculation, and what is irrelevant, peripheral or a perturbation to be tackled secondarily.

Just a few of the electrons—the valence electrons of the outermost shells—of the semiconductor solid will be important to specific properties of interest to us. The inner ones stay confined with the nucleus, and we may treat them as staying rigidly along with it. We pull these electrons together with the nucleus into an ion—an ion core—and modify the nuclear charge. These inner electrons have now been incorporated into an ion. These ions are massive compared to the electron. This means that the ion motion, and its Coulomb interaction with each other, may also be treated as being small and so can be accounted for as a secondary perturbation if our principle interest is in the electron motion; thus, these terms are eliminated, although the ion motion's perturbation consequence will be included as a later thought. Our problem has now been reduced to solving the electrons' Hamiltonian:

Of course, even the notions of Ze, M_i and m_0 have much complexity buried in them. A remark on this "simple equation" is in order. Paul Dirac, in a 1929 paper in the Proceedings of the Royal Society, says as an introduction, "The general theory of quantum mechanics is now almost complete, the imperfections that still remain being in connection with the exact fitting of the theory with relativity ideas. ...The underlying physical laws necessary for the mathematical theory of a large part of physics and the whole of chemistry are thus completely known, and the difficulty is only that the exact application of these laws leads to equations much too complicated to be soluble. It therefore becomes desirable that approximate practical methods should be developed, which can lead to an explanation of the main features of complex atomic systems without too much computation" (P. A. M. Dirac, "Quantum mechanics of many-electron systems," Proceedings of the Royal Society of London, **123**, 714–733 (1929)). Papers and pencils have now been replaced by electronic computers and algorithms. But, nearly a hundred years after this foresighted publication, and nearly forty years after high temperature superconductivity, we don't have an acceptable explanation for the latter crystal phenomena. *There is much subtlety buried away at the low energy end and spread out at the high energy end of the universe.*

In a molecule, the nuclear spin consideration can be important. H and ^3He, for example are fermions, due to their 1/2 nuclear spin, while D, ^4He and H_2 are bosons. Many different properties arise from this difference. This nuclear spin aspect is the least of our worries in discussing a crystal, where the electron and the electron-dressed nucleus considerations will dominate.

In combining inner atomic electrons and the nucleus, we have modified the meaning of Z. It is now the net charge number of the core, not the atomic number.

$$\hat{\mathscr{H}} = \sum_i \left[-\frac{\hbar^2}{2m_i} \nabla_i^2 \right] + \frac{1}{2} \sum_{i \neq j} \frac{1}{4\pi\epsilon_0} \frac{e^2}{\mathbf{r}_i - \mathbf{r}_j} - \frac{1}{2} \sum_{i \neq j} \frac{1}{4\pi\epsilon_0} \frac{Z^* e^2}{\mathbf{r}_i - \mathbf{R}_j}. \quad (1.11)$$

The ion motion can, of course, be very important. The ion motion will cause electrons to scatter—exchange energy and momentum with them—and thus affect transport properties. But we may bring this into our description as a perturbation.

We have now reduced Equation 1.10—a relatively complete description of atom assembly—to Equation 1.11, still a very accurate description where the inter-ion Coulomb interaction and the electron-nucleus Coulombic interaction have been approximated into the third term. The Z^*—ionic charge number—has a new meaning. It is the dressed charge of the nuclei with their surrounding core electrons. And the slow motion of the ion vis-à-vis the electron will let us tackle this term as a perturbation.

An extension of this jellium and classical discussion is understanding the consequences of dopants; for example, in devices. The classical treatment of dopants in devices is as a continuum; that is, a jellium treatment. A uniform distribution of charge is assumed to arise in them, which in the quasineutral material is balanced by the electron charge cloud. If one makes the device small and have only a few of these dopants, then many of the assumptions underlying the description break down. There are not enough of them to appear as a continuum, and since they do have individual perturbation effects locally, the consequences show up in a small device.

The extent of the role of the core of the atom here is as a source of positive charge. If we could further approximate this charge, instead of being localized at $\mathbf{R}_j$, as being uniformly spread out—a continuum—then we have the *jellium model*. This jellium solid, if one also ignores all the quantum-mechanical constraints on the electron, is now just a classical electron particle gas in the solid.

Equation 1.11's second term—a many-body term—is one that requires much attention. Electrons interact with other electrons and have a Coulomb energy associated with that interaction. An electron is also a fermion. And an electron does not interact with itself. The first reflects an electromagnetic force effect. The second is a quantum-mechanical constraint. And the third reflects something much deeper with possibly many interpretations, although it is certainly tied to the first also. *An electron's response arises through its interaction with its surroundings.*

That an electron does not interact with itself can be viewed at many levels. Electric fields are polar vectors. Fields must terminate. There must be a surrounding, and that is how lowering (or even raising, as in single electron effects at nanoscale) of energy happens. An electric field at this level of interpretation arises as $\boldsymbol{\mathcal{E}} = \lim_{q\to 0}(-\boldsymbol{\nabla}_{\mathbf{r}} U_e/q)$. With no charge, there is no energy, and this equation now has a singularity. An electron needs the surrounding for it to be observable through its Coulomb interaction. Richard Feynman's discussion of the singularity and renormalization conundrum is especially powerful as enunciated in his Nobel lecture. It is a question he worried about, starting in his undergraduate years. The lecture can be found at https://www.nobelprize.org/nobel_prizes/physics/laureates/1965/feynman-lecture.html.

The three considerations are tackled by breaking this electron-electron term further. If we only consider classical electrostatic interaction energy, that is, the form $(1/2)(1/4\pi\epsilon_0) \int \rho(\mathbf{r})[\rho(\mathbf{r}')/|\mathbf{r} - \mathbf{r}'|]d\mathbf{r}\, d\mathbf{r}'$ in charge distribution, then we have used what is called a *Hartree energy term*, and a *Hartree equation* form will solve for it. The result has significant errors in it, since the second and third components have not been accounted for. So, we introduce corrections to the Hartree form.

Two electrons, when exchanged, being indistinguishable particles, and fermions, must have different wavefunctions, so the wavefunctions must be antisymmetric. We will see that a Slater determinant antisymmetrizes the many-body wavefunction. This takes care of this *exchange interaction* that has local and nonlocal contributions in it. The *Hartree-Fock equation* will apply this correction for us.

But the antisymmetrization process of the Slater determinant itself also employs one-electron wavefunctions. This requires us to have separability of Equation 1.11, that is, that an electron at any spatial coordinate in space is essentially independent of where the other electrons are. But, a repulsive electron-electron interaction prevents other electrons from approaching the electron at that spatial coordinate. *An electron in the jellium classical solid repels the other electrons electrostatically, thus exposing an equal and opposite positive background so that electric fields vanish far away.* The electrons correlate themselves in a way that screens the electric field. An electron here is surrounded by an equal and opposite charged hole in the electron density. So, there is an "exclusion" zone here. This is a *correlation interaction* representing the physical principle that an electron does not interact with itself and only with all else that surrounds it. The probability of an electron at this position depends very much on the location of the other electrons. Sometimes, this is also referred to as a *correlation hole,* which should not be confused with electron's quasiparticle "hole." This correlation hole is the Coulomb repulsion of other electrons from this electron's vicinity. It is a correlation charge hole. Figure 1.1 is a pictorial representation of such a correlation hole. In Sections 1.5 and 1.6, we will summarize the gradual increasing of accuracy in our search for the solution to this description of the solid. Since Equation 1.11 does not lend itself to easy decoupling, and this starting Equation 1.10 has 3× the total number of nuclei and electrons as its coupled degrees of freedom, we will first employ simple atom assemblies—molecules—to bring out the main physical features of the arguments that we will deploy.

This outline of the problem of energetics in the solid shows us the incremental path that we have to take to find satisfactory solutions. If a solution to a problem that is close enough to a new problem is known, we employ perturbation techniques to find the solution to the new problem. This following section summarizes a few of the common perturbation techniques that will be utilized throughout this text, including for this semiconductor crystal outlined so far.

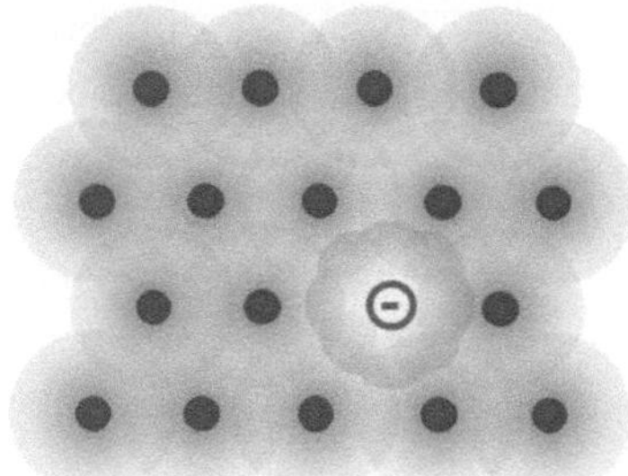

Figure 1.1: An electron in an ensemble of positive charge interacting with other electrons represented by variably filled areas, with atomic nuclei as the background. The lighter area surrounding the electron represents the exclusion zone of the Coulomb repulsive correlation. This is the correlation hole.

The Church thesis and its Turing machine form are examples of an algorithm for solving a problem by reducing it to a procedure that one steps through. Gamma functions—factorials for integer argument—can be solved for any n since $\Gamma(n+1) = n\Gamma(n)$, and $\Gamma(1) = 1$. This recursion approach is commonplace as a procedure for proof, although sometimes it is applied inappropriately heuristically. Gamma functions, as the integral analytic function, are

$$\Gamma(s) = \int_0^\infty \exp(-x) x^s \frac{dx}{x},$$

where $s = \sigma + it$ is complex, and have many quite amazing properties. They will appear for us during the use of Fermi integrals. Even more magical is the Riemann zeta function $\zeta(s)$, whose integral analytical form is

$$\zeta(s) = \frac{1}{\Gamma(s)} \int_0^\infty \frac{1}{\exp(x) - 1} x^s \frac{dx}{x}.$$

If $\sigma > 1$, this reduces to an infinite series, $\zeta(s) = \sum_{i=1}^{\infty} i^{-s}$. Riemann showed a relationship between the distribution of prime numbers and the non-trivial zeros of the zeta function. So, prime numbers are not randomly distributed, only pseudo-randomly. Other transcendent characteristics of this function include implications for Casimir forces, the cosmological constant, and the lack of Bose-Einstein condensation in two dimensions.

1.3 *Perturbation approaches*

IN OUTLINING SOME OF THE APPROACHES to solving Hamiltonians, our interest here is in dispensing with how one satisfactorily arrives at the solution to problems that we will encounter in this text. We have to set the problem up right; only then can we proceed to find the solution. Only under the most circumscribed of conditions in a many-body system may one find accurate direct solutions. Usually though, one has to transform a known related problem that has a known solution to the problem of interest as a perturbation. The Church and Turing forms provide important insight into information mechanics, particularly its deterministic form. The perturbation can be static, that is, time independent and steady state, or dynamic, where a time-dependent perturbation and a quantum system response unfold.

First, we take up the time-independent steady-state perturbation, and then we will take up the time-dependent perturbation. Later on, we will take on the adiabatic time-dependent perturbation, where a state evolves smoothly and continuously, maintaining its quantized identity. An example of time-independent perturbation is when two molecules come close enough, the energy changes. The properties of cohesion/adhesion and repulsion are also examples of the nature of this energetics, a quasi-steady state, where a perturbation causes changes in the energy landscape. Examples of a time-dependent perturbation are a photon exciting an electron, or an electron undergoing scattering during transport. An example of adiabatic perturbation is an electron in a confined quantized space as the size of the space is slowly changed electrically.

The prime number distribution ties it to the Lambert W function ($W(z)\exp[W(z)]=z$) and Ramanujan's series for $\zeta(3)$. Riemann's investigations in geometry inspired Einstein's relativity. A Riemann remark, paraphrased, that the geometry of physical space need not be a God-given Euclidean space but should be determined by experiment, not by hypothesis, stands as one of the most observant statements from a remarkable mathematician and a clergyman's son who, like Ramanujan, died too young. To balance these serious statements, here is a joke that punches at recursiveness. A Russian mathematician and a Russian engineer are visiting a European research institution. Smoking is still accepted. On the first day, when the engineer drops a lighted butt in the trash can sitting on the floor, the paper in it catches fire. So he uses the fire extinguisher to put it out. An identical sequence is repeated with the mathematician, who arrives later after a long night of work. The next day, upon cleaning, the janitor leaves the trash cans on their desks. Events repeat. The engineer uses the fire extinguisher again. The mathematician, however, places the trash can on the floor. *He had reduced it to a known problem.*

1.3.1 *Time-independent perturbation*

LET $\hat{\mathscr{H}}'$ BE A PERTURBATION; that is, let $\hat{\mathscr{H}} = \hat{\mathscr{H}}_0 + \hat{\mathscr{H}}'$ be the Hamiltonian of a quantum system whose statefunction is known in the perturbation's absence. The new statefunction is different. The new perturbation potential causes transitions between the states of the unperturbed Hamiltonian. Let $|u_i^0\rangle$ be the eigenfunctions for the Hamiltonian $\hat{\mathscr{H}}_0$ so that $\hat{\mathscr{H}}_0|u_i^0\rangle = E_i^0|u_i^0\rangle$. For the perturbation $\hat{\mathscr{H}}'$, the eigenenergy solutions exist if $\det|\mathscr{H} - E\mathscr{O}| = 0$. To determine the changes in energy and the eigenfunctions under perturbation, one uses the trick of separation of order where various combinations of terms leading to the same order of effect can be pooled together. Let

$$\hat{\mathscr{H}} = \hat{\mathscr{H}}_0 + \lambda\hat{\mathscr{H}}', \tag{1.12}$$

so that $\lambda = 0$ is the absence of perturbation, and $\lambda = 1$ is the complete turning on of perturbation. The eigenenergies and eigenfunctions of the perturbed system have changed and we will determine these as perturbational changes, by corrections of increasing order, using this λ. Let the eigenenergies and eigenfunctions for the Hamiltonian of Equation 1.12 be

The Hamiltonian equation, even if a simple equation to write summarily, can be quite difficult to solve. We will see this particularly when we dwell on bandstructure calculations where enormous-size matrices are encountered. Sometimes, it is prudent to solve a problem by writing its physical basis in its entirety, as in this equation. Sometimes, it is prudent to start from a known solution and find a solution with it perturbed. The latter is a linear view. The former is a nonlinear view. Each has its successes and failures. I have immense respect for the German organization of education: enough resources at every level, and all the education accessible to everybody. Early schools are equivalent. One doesn't choose a locality to live based on the local schools. Universities are essentially free. Technical professions are respected. A machinist is a precision worker—not a "blue collar"

$$E_n = |E_n^0\rangle + \lambda|E_n^1\rangle + \lambda^2|E_n^2\rangle + \cdots, \text{ and}$$

$$|u_n\rangle = |u_n^0\rangle + \lambda|u_n^1\rangle + \lambda^2|u_n^2\rangle + \cdots. \qquad (1.13)$$

The problem posed to us is

$$(\hat{\mathscr{H}}_0 + \lambda\hat{\mathscr{H}}')\left(|u_n^0\rangle + \lambda|u_n^1\rangle + \lambda^2|u_n^2\rangle + \cdots\right) = \left(E_n^0 + \lambda E_n^1 + \lambda^2 E_n^2 + \cdots\right)\left(|u_n^0\rangle + \lambda|u_n^1\rangle + \lambda^2|u_n^2\rangle + \cdots\right), \qquad (1.14)$$

which can be partitioned in the powers of order λ as

$$\lambda^0 : \hat{\mathscr{H}}_0|u_n^0\rangle = E_n^0|u_n^0\rangle,$$
$$\lambda^1 : \hat{\mathscr{H}}_0|u_n^1\rangle + \hat{\mathscr{H}}'|u_n^0\rangle = E_n^0|u_n^1\rangle + E_n^1|u_n^0\rangle,$$
$$\lambda^2 : \hat{\mathscr{H}}_0|u_n^2\rangle + \hat{\mathscr{H}}'|u_n^1\rangle = E_n^0|u_n^2\rangle + E_n^1|u_n^1\rangle + E_n^2|u_n^0\rangle, \qquad (1.15)$$

and so on. With $\lambda = 1$, we now have the posed problem but partitioned into different order corrections. The 0th order equation defines and describes the unperturbed system. The 1st order equation, with the lowest-order correction due to perturbation, has a term due to unperturbed Hamiltonian operating on the 1st order correction to the eigenfunction, and the perturbation Hamiltonian operating on the unperturbed eigenfunction. Both of these terms are of similar order correction. The 2nd order equation has three such terms pooling the same order of correction. The use of λ has let us achieve this deconvolving.

To obtain the first order correction, we use orthonormality by taking the inner product with the bra $\langle u_n^0|$ in Equation 1.15 of the λ^1 power:

$$\langle u_n^0|\hat{\mathscr{H}}_0|u_n^1\rangle + \langle u_n^0|\hat{\mathscr{H}}'|u_n^0\rangle = E_n^0\langle u_n^0|u_n^1\rangle + E_n^1|\langle u_n^0|u_n^0\rangle, \qquad (1.16)$$

where $\langle u_n^0|u_n^0\rangle = 1$, $\langle u_n^0|u_n^1\rangle$ is finite and non-zero, and $\langle u_n^0|\hat{\mathscr{H}}_0|u_n^1\rangle = E_n^0\langle u_n^0|u_n^1\rangle$, since $\hat{\mathscr{H}}_0$ is Hermitian. Therefore,

$$E_n^1 = \langle u_n^0|\hat{\mathscr{H}}'|u_n^0\rangle, \text{ and}$$
$$(\hat{\mathscr{H}}_0 - E_n^0)|u_n^1\rangle = -(\hat{\mathscr{H}}' - E_n^1)|u_n^0\rangle. \qquad (1.17)$$

The 1st order correction in the eigenenergy arose from the perturbation Hamiltonian and the unperturbed eigenstate. The 1st order correction to the eigenfunction needs a little reworking to write it in terms of the unperturbed orthonormal basis set of $|u_n^0\rangle$, which is a complete orthonormal basis set. We may write

$$|u_n^1\rangle = \sum_{i\neq n} c_i^1|u_i^0\rangle. \qquad (1.18)$$

It is useful to exclude the $i = n$ term in the expansion. As $(\hat{\mathscr{H}}_0 - E_n^0)|u_n^0\rangle = 0$, the existence of a term based on $|u_n^0\rangle$ in Equation 1.18 is

with all its subtle prejorativeness—just as the academic is, with the two following different paths of education, one for practice, and one for research and teaching. Both can raise a family with equal opportunity. This becomes possible by the system assuring proper investments across all the stages of education so that enough qualified folks appear at the end for the technical and scientific needs. University may be free, but the student must show that he/she belongs, through effort and examinations. It is not a surprise that there are more car manufacturers in Germany than in the *USA*. And science in Germany has revived exactingly well since the events and tragedy of the Second World War. This setting up the problem right and solving it all together works quite often. But there are situations where it is a luxury and the complexity is such that a local perturbation is desired to get a quick and good-enough answer.

In many situations, the different terms in any order of perturbation are comparable as a product of a large and a small entity. If the perturbation is very small, then one can see it as an operation of a small energy on the starting eigenfunction, and of a large energy operator operating on a small disturbance in the eigenfunction.

dispensable. So, we expand $|u_n^1\rangle$ in the 0th order orthonormal basis set and, to find the lth term of the correction, we take the inner product with the bra $\langle u_l^0|$:

$$\sum_{i\neq n}(E_i^0 - E_n^0)c_i^1\langle u_l^0|u_i^0\rangle = -\langle u_l^0|\mathscr{H}'|u_n^0\rangle + E_n^1\langle u_l^0|u_n^0\rangle$$

$$\therefore \quad (E_l^0 - E_n^0)c_l^1 = -\langle u_l^0|\mathscr{H}'|u_n^0\rangle \text{ for } l \neq n,$$

$$\therefore \quad c_i^1 = \frac{\langle u_i^0|\hat{\mathscr{H}}'|u_n^0\rangle}{(E_n^0 - E_i^0)} \text{ with } i=l, \text{ and}$$

$$|u_n^1\rangle = \sum_{i\neq n}\frac{\langle u_i^0|\mathscr{H}'|u_n^0\rangle}{E_n^0 - E_i^0}|u_i^0\rangle. \tag{1.19}$$

Both the eigenenergy correction (Equation 1.17) and the eigenfunction correction terms (Equation 1.19) are now known for the first order.

This formal approach is obviously extendable to higher orders recursively—more and more terms—and the writing of such an algorithm is quite straightforward. For the λ^2 set of terms,

$$\mathscr{H}_0|u_n^2\rangle + \hat{\mathscr{H}}'|u_n^1\rangle = E_n^0|u_n^2\rangle + E_n^1|u_n^1\rangle + E_n^2|u_n^0\rangle; \tag{1.20}$$

therefore,

$$\langle u_n^0|\mathscr{H}_0|u_n^2\rangle + \langle u_n^0|\mathscr{H}'|u_n^1\rangle = E_n^0\langle u_n^0|u_n^2\rangle + E_n^1\langle u_n^0|u_n^1\rangle + E_n^2\langle u_n^0|u_n^0\rangle. \tag{1.21}$$

Again, since $\mathscr{H}_0$ is Hermitian, $\langle u_n^0|\mathscr{H}_0|u_n^2\rangle = E_n^0\langle u_n^0|u_n^2\rangle$, we have

$$E_n^2 = \langle u_n^0|\mathscr{H}'|u_n^1\rangle - E_n^1\langle u_n^0|u_n^1\rangle, \tag{1.22}$$

and because $\langle u_n^0|u_n^1\rangle = \sum_{i\neq n} c_i^1\langle u_n^0|u_i^0\rangle = 0$, the 2nd order energy correction term is

$$E_n^2 = \langle u_n^0|\hat{\mathscr{H}}'|u_n^1\rangle = \sum_{i\neq n} c_i^1\langle u_n^0|\mathscr{H}'|u_i^0\rangle$$

$$= \sum_{i\neq n}\frac{\langle u_i^0|\hat{\mathscr{H}}'|u_n^0\rangle\langle u_n^0|\hat{\mathscr{H}}'|u_i^0\rangle}{(E_n^0 - E_i^0)} = \sum_{i\neq n}\frac{\left|\langle u_i^0|\mathscr{H}'|u_n^0\rangle\right|^2}{(E_n^0 - E_i^0)}. \tag{1.23}$$

The 1st order energy correction was the expectation of perturbation on the unperturbed state. The 2nd order energy correction is the perturbation on the 1st order corrected eigenfunction. One can proceed from this and now also build the 2nd order corrected eigenfunction by finding c_i^2.

A very simple consequence of these perturbational relationships is worth thinking through. In a semiconductor, one may represent the states as Bloch states that are propagating states spread out over

The last part of Equation 1.19 is a quantum-mechanical reflection on our classical intuition. Any classical state, under an energy perturbation, that is, the exercise of a force, changes. A mass moves. For any function $y = f(x)$ describing this classical picture, df is the marginal consequence, and df/dx the marginal rate. The lowest order correction for the change is the marginal efficiency of this perturbation. Force and acceleration—the rate change of velocity—are related, with a marginal efficiency determined by the inverse mass. This is an inertial mass. Mass is an emergent property, which superficially can be seen in the bundling of energy. Quantum-mechanically, how well two eigenstates will couple due to perturbation is again this derivative-like ratio of perturbational coupling energy and separation of the unperturbed states. And the statefunction reflects the statistical consequence of these couplings.

the entire semiconductor, and evanescent states that are localized, as at surfaces, defects, impurities, et cetera. The following chapters will expend considerable effort toward this analysis. An $E(n,\mathbf{k}) \equiv E_{n,\mathbf{k}}$ relationship, where n is a quantum number that identifies a band, and $\mathbf{k}$—another quantum number—which is the wavevector, describe the energies of the allowed eigenfunctions. Now suppose we apply a bias voltage V_{dc} (see Figure 1.2) to this semiconductor; that is, there is a net spatially invariant potential energy rise in the system. Do I now have to recalculate the $E(n,\mathbf{k})$? No. Why not? Because

$$E^1_{n,\mathbf{k}} = \langle u^0_{n,\mathbf{k}} | \hat{\mathscr{H}}' | u^0_{n,\mathbf{k}} \rangle = -eV_{dc} \langle u^0_{n,\mathbf{k}} | u^0_{n,\mathbf{k}} \rangle = -eV_{dc} \tag{1.24}$$

since $\hat{\mathscr{H}}_0$ and $-e\hat{V}_{dc}$ commute, that is, $[\hat{\mathscr{H}}_0, -e\hat{V}_{dc}] = 0$, so the same orthonormal basis set may be employed. What about the eigenfunction? Again,

$$|u^1_{n,\mathbf{k}}\rangle = \sum_{i \neq j} \frac{\langle u^0_{n,\mathbf{k}_i} | \hat{\mathscr{H}}' | u^0_{n,\mathbf{k}_j} \rangle}{(E^0_{n,\mathbf{k}_j} - E^0_{n,\mathbf{k}_i})} |u^0_{\mathbf{k}_i}\rangle = 0$$

$$\because \quad \langle u^0_{n,\mathbf{k}_i} | \hat{\mathscr{H}}' | u^0_{n,\mathbf{k}_j} \rangle = -eV_{dc} \langle u^0_{n,\mathbf{k}_i} | u^0_{n,\mathbf{k}_j} \rangle = 0 \;\; \forall\, i \neq j. \tag{1.25}$$

Here, we considered only the interactions within the band n, since the nearest states provide the strongest contribution, but the result is more general because of orthogonality. The implication is that the entire energy bandstructure may be shifted by this potential change and the eigenfunctions do not change. The device analysis is also in dynamic conditions where the time scale of electrodynamic response is much slower—adiabatic, which we discuss a little later—and hence one may again neglect any bandstructure consequence of the electrically applied stimuli's consequences. This underlies all the drawing of the band diagrams, where the conduction and valence bandedge lines represent the extrema of the bands.

There is one complication in the previous argument that we sidestepped but should address. The electron's state, absent magnetic interaction, with $s = 1/2$, and therefore the secondary spin quantum number of $\pm 1/2$, is degenerate in energy. The equation forms in Equations 1.17 and 1.23 actually blow up if one considers the interaction between these degenerate states. This, of course, is unphysical. The series does not converge. The perturbation expansion has this problem with states very close in energy. This can be effectively and efficiently tackled by treating the nearly degenerate states in the same way as we treated $|u_n\rangle$ in the perturbation expansion. This means that the 0th order state is being allowed to be an arbitrary linear combination of the degenerate states.

(a)
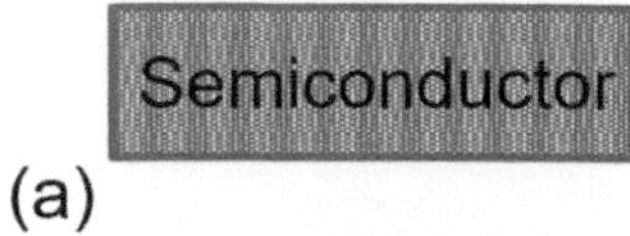

(b)
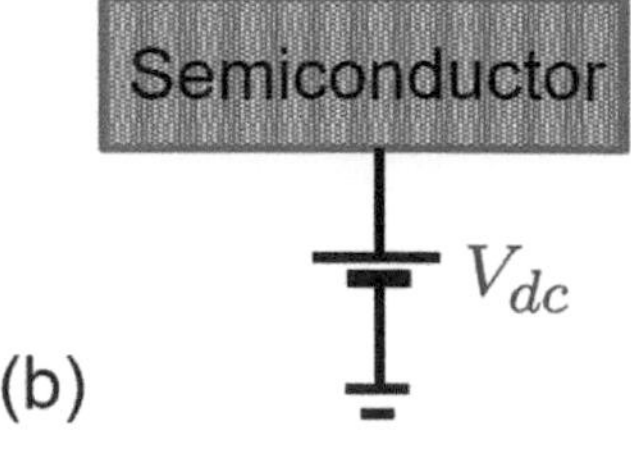

Figure 1.2: (a) A semiconductor whose bandstructure describing electron states has been determined to be $E(n,\mathbf{k}) \equiv E_{n,\mathbf{k}}$ with the semiconductor grounded (thermal equilibrium). (b) This semiconductor shown under the static bias voltage V_{dc}. Perturbation tells us that this just shifts the bandstructure—the eigenenergy solution—by $-eV_{dc}$.

Good thing too. Imagine having to calculate this bandstructure under all potentials and their distribution in real space in a device. There is more to this that we will see in our discussion in Chapter 4 of the parameter we call the effective mass, and its effectiveness.

Let there be N nearly degenerate states. We choose an orthonormal set of basis state in N; that is,

$$\langle \phi_j | \phi_i \rangle = \delta_{ji} \;\; \forall \;\; i,j \in N. \tag{1.26}$$

So, these $|\phi_i\rangle$'s together with their coefficients α_i let us expand the degenerate states into a non-degenerate basis. Now take the 0th and 1st order terms of the Schrödinger/Hamiltonian equation of the problem. Orthonormality lets us write

$$\sum_{i \in N} \langle \phi_j | \hat{\mathscr{H}}_0 + \hat{\mathscr{H}}' | \phi_i \rangle \alpha_i = E \alpha_j. \tag{1.27}$$

The number of degenerate states determines the number of solutions to this eigenvalue equation set, writable as

$$\begin{bmatrix} \mathscr{H}_{11} & \cdots & \mathscr{H}_{1N} \\ \vdots & \vdots & \vdots \\ \mathscr{H}_{N1} & \cdots & \mathscr{H}_{NN} \end{bmatrix} \begin{bmatrix} \alpha_1 \\ \vdots \\ \alpha_N \end{bmatrix} = E \begin{bmatrix} \alpha_1 \\ \vdots \\ \alpha_N \end{bmatrix}, \tag{1.28}$$

where $\mathscr{H}_{ji} = \langle \phi_j | \hat{\mathscr{H}}_0 + \hat{\mathscr{H}}' | \phi_i \rangle$ is the matrix element for the complete Hamiltonian. The solution follows from the condition $\det|\mathscr{H}_{ji} - E\mathscr{O}| = 0$. α_i are now known, and using this orthonormal set to replace the degenerate set in the Equation 1.14 tackles the problem in first order perturbation.

In S. Tiwari, "Quantum, statistical and information mechanics: A unified introduction," Electroscience 1, Oxford University Press, ISBN 978-0-198-75985-0 (forthcoming), we have tackled these various perturbation situations: 1st, 2nd, degenerate and more; for example, Rabi oscillations and the evolution of two-level systems when one applies a static perturbation at an instant in time. This discussion is very germane here, and a reading of it is strongly recommended for insight.

1.3.2 *Time-dependent perturbation*

In semiconductor problems, non-steady-state interactions—a shining of light, an electron scattering due to a Coulomb impurity, interface roughness, phonon excitation or even bandedge fluctuations arising from atomic motion due to thermal energy—are ubiquitous. Perturbation with time dependence allows us to view the quantum evolution under these conditions.

We employ a 2-level system (see Figure 1.3), but in a way that is extendable to the full basis set because the time dependence of the evolution still conforms to the same picture. The basis eigenfunction set consists of kets $|m\rangle$ and $|k\rangle$ with eigenenergies E_m and E_k, respectively. The statefunction solution for the unperturbed state is $|\psi\rangle = c_m|m\rangle + c_k|k\rangle$, with $\langle m|k\rangle = 0$, $\hat{\mathscr{H}}_0|m\rangle = E_m|m\rangle$ and $\hat{\mathscr{H}}_0\langle k\rangle = E_k|k\rangle$. With perturbation,

$$\hat{\mathscr{H}} = \hat{\mathscr{H}}_0 + \hat{\mathscr{H}}', \text{ and}$$
$$-\frac{\hbar}{i}\frac{\partial}{\partial t}|\psi\rangle = \hat{\mathscr{H}}|\psi\rangle, \tag{1.29}$$

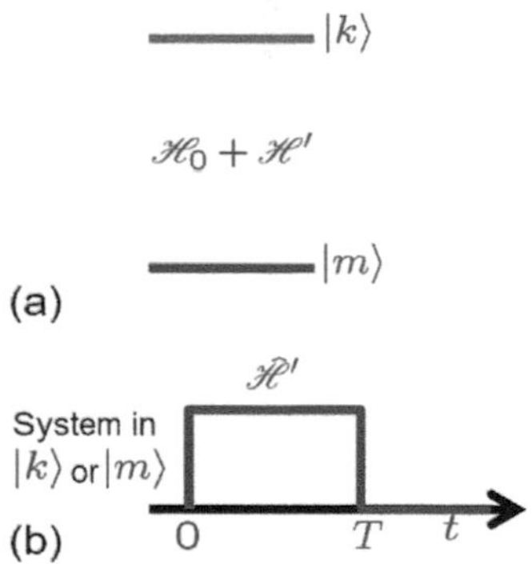

Figure 1.3: (a) A two-level quantum system. (b) This system initially in one or the other eigenstate is subjected to a perturbation for a time duration T. Consequently, it evolves.

describe the evolution of the statefunction. The state function evolves as

$$|\psi(t)\rangle = c_m(t)|m\rangle + c_k(t)|k\rangle, \tag{1.30}$$

which is prescribed by our restriction of the two-level orthonormal basis. If $\hat{\mathscr{H}}' = 0$—an unperturbed stationary state—$c_l(t) = c_l(0)\exp(-iE_l t/\hbar)$, where $l = k, m$ holds for the stationary state.

If a perturbation is applied, using the condition of orthonormality one can view the evolution for the coefficients through

$$-\frac{\hbar}{i}\frac{d}{dt}\begin{bmatrix} c_m(t) \\ c_k(t) \end{bmatrix} = \begin{bmatrix} \mathscr{H}_{mm} & \mathscr{H}'_{mk} \\ \mathscr{H}'_{km} & \mathscr{H}_{kk} \end{bmatrix}\begin{bmatrix} c_m(t) \\ c_k(t) \end{bmatrix}, \tag{1.31}$$

where $\mathscr{H}'_{mk} = \mathscr{H}'_{km}$ are real values of the energy observable. The statefunction $|\psi\rangle$ follows from

$$-\frac{\hbar}{i}\frac{\partial|\psi\rangle}{\partial t} = \hat{\mathscr{H}}|\psi\rangle = E|\psi\rangle, \tag{1.32}$$

that is,

$$\begin{bmatrix} c_m(t) \\ c_k(t) \end{bmatrix} = \exp\left(-i\frac{Et}{\hbar}\right)\begin{bmatrix} c_m(0) \\ c_k(0) \end{bmatrix}, \tag{1.33}$$

where

$$E\begin{bmatrix} c_m(t) \\ c_k(t) \end{bmatrix} = \begin{bmatrix} \mathscr{H}_{mm} & \mathscr{H}'_{mk} \\ \mathscr{H}'_{km} & \mathscr{H}_{kk} \end{bmatrix}\begin{bmatrix} c_m(0) \\ c_k(0) \end{bmatrix}. \tag{1.34}$$

Let $E = E_-, E_+$ be the perturbed eigenenergies; then, this two-level system, under this static perturbation turned on at $t = 0$, has the eigenfunction solutions

$$\begin{aligned} |\psi_-(t)\rangle &= \left[c_m^-(0)|m\rangle + c_k^-(0)|k\rangle\right]\exp\left(-i\frac{E_- t}{\hbar}\right) \\ &= (\cos\theta|m\rangle + \sin\theta|k\rangle)\exp\left(-i\frac{E_- t}{\hbar}\right), \text{ and} \\ |\psi_+(t)\rangle &= \left[c_m^+(0)|m\rangle + c_k^+(0)|k\rangle\right]\exp\left(-i\frac{E_+ t}{\hbar}\right) \\ &= (-\sin\theta|m\rangle + \cos\theta|k\rangle)\exp\left(-i\frac{E_+ t}{\hbar}\right). \end{aligned} \tag{1.35}$$

Here, the alternative set of equations are written with $c_m^-(0) = \cos\theta$, $c_k^-(0) = \sin\theta$, $c_m^+(0) = -\sin\theta$ and $c_k^+(0) = \cos\theta$. This maintains orthonormality and establishes a starting phase. If $\theta = 0$, then the system has been prepared in $|m\rangle$ before the turning on of the perturbation. The statefunction solution with perturbation is

$$\begin{aligned} |\psi(t)\rangle &= d_-|\psi_-(t)\rangle + d_+|\psi_+(t)\rangle \\ &= d_-|\psi_-(0)\rangle\exp\left(-i\frac{E_- t}{\hbar}\right) + d_+|\psi_+(0)\rangle\exp\left(-i\frac{E_+ t}{\hbar}\right). \end{aligned} \tag{1.36}$$

Choosing the amplitudes $d_- = \cos\theta$ and $d_+ = \sin\theta$ leads to

$$\begin{aligned}|\psi(t)\rangle &= \left[\cos^2\theta \exp\left(-i\frac{E_- t}{\hbar}\right) + \sin^2\theta \exp\left(+i\frac{E_+ t}{\hbar}\right)\right]|m\rangle \\ &\quad + \sin\theta\cos\theta\left[\exp\left(-i\frac{E_- t}{\hbar}\right) - \exp\left(+i\frac{E_+ t}{\hbar}\right)\right]|l\rangle \\ &= c_m(t)|m\rangle + c_l(t)|l\rangle. \end{aligned} \tag{1.37}$$

This is an oscillatory, not stationary, solution. If the system were initialized in $|m\rangle$, the probability of the system being found in $|k\rangle$ at time t in the presence of this steady-state perturbation would be

$$\begin{aligned}|c_k(t)|^2 &= \sin^2(2\theta)\sin^2\left[\frac{(E_+ - E_-)t}{2\hbar}\right] \\ &= \frac{4\left|\mathscr{H}'_{mk}\right|^2}{(\mathscr{H}_{kk} - \mathscr{H}_{mm})^2 + 4\left|\mathscr{H}'_{mk}\right|^2}\sin^2\left(\frac{\Omega t}{2}\right), \end{aligned} \tag{1.38}$$

with

$$\Omega = \left[\left(\frac{\mathscr{H}_{kk} - \mathscr{H}_{mm}}{\hbar}\right)^2 + \frac{4\left|\mathscr{H}'_{mk}\right|^2}{\hbar^2}\right]^{1/2} \tag{1.39}$$

as the oscillation frequency—the Rabi frequency (Figure 1.4). The two-level system would have stayed in the prepared state absent perturbation. With static perturbation, it now oscillates at the slower frequency of $\Omega = (E_+ - E_-)/2\hbar$ determined by the eigenenergies of the statefunction under perturbation. The perturbation energy determines the cycling depending on the magnitude of coupling.

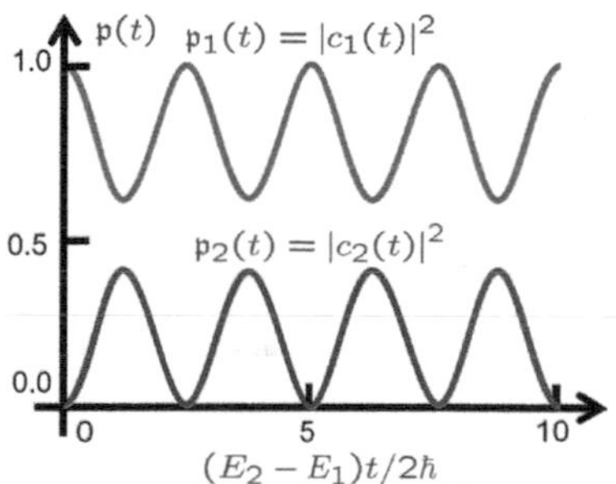

Figure 1.4: Rabi oscillation in a 2-level system under a perturbation.

What if the perturbation was for a short time duration T—a scattering event—as in Figure 1.3? We make this a harmonic perturbation. Our stimulus to this 2-level system is

$$\begin{aligned}\mathscr{H}'_{mk}(t) &= 0 \text{ for } t \leq 0, \\ \mathscr{H}'_{mk}(t) &= 2\mathscr{H}'_{mk}\sin(\omega t), \text{ that is,} \\ &= i\mathscr{H}'_{mk}\left[\exp(-i\omega t) - \exp(i\omega t)\right] \text{ for } t > 0, \end{aligned} \tag{1.40}$$

where we will make this duration finite while looking at the solution of the evolution. The time dependence follows from Equation 1.31.

Take the case of a system starting in the eigenfunction state $|m\rangle$. So, where $c_m(0) = 1$ and $c_k(0) = 0$, Equation 1.31 states (note $\mathscr{H}'_{km} = \mathscr{H}'_{mk}$ because the operator is Hermitian)

$$-\frac{\hbar}{i}\frac{d}{dt}c_k(t) = \mathscr{H}'_{km}c_m(0) + \mathscr{H}_{kk}c_k(0) = \mathscr{H}'_{mk}c_m(0) + \mathscr{H}_{kk}c_k(0), \tag{1.41}$$

leading to

$$-\frac{\hbar}{i}c_k(t) = \int_0^t \mathscr{H}'_{mk}(\tau)\exp\left(i\omega_{km}\tau\right)d\tau, \tag{1.42}$$

with the oscillation time dependence of the perturbation explicitly included through the time variable τ in the integral spanning $t = 0$ to $t = t$. Also, remember the phase terms in the basis set. The solution is

$$\begin{aligned} c_k(t) &= \frac{\mathscr{H}'_{mk}}{\hbar} \int_0^t \{\exp[i(\omega_{km} - \omega)\tau] - \exp[i(\omega_{km} + \omega)\tau]\} \, d\tau \\ &= \frac{\mathscr{H}'_{mk}}{\hbar} \left\{ \frac{\exp[i(\omega_{km} - \omega)t] - 1}{i(\omega_{km} - \omega)} \right. \\ &\quad \left. - \frac{\exp[i(\omega_{km} + \omega)t] - 1}{i(\omega_{km} + \omega)} \right\}. \end{aligned} \tag{1.43}$$

Here, $\omega_{km} = (E_k - E_m)/\hbar$. The largest contribution to amplitude evolution in time comes from where the denominator is the smallest. The perturbation frequency closest to the interlevel frequency has the largest effect. If $\omega_{km} \approx \omega$, the first term will dominate. Since we started with the lower level filled, the amplitude of the higher level is

$$\begin{aligned} c_k(t) &= \frac{2\mathscr{H}'_{mk}}{\hbar} \frac{\sin[(\omega_{km} - \omega)t/2]}{(\omega_{km} - \omega)} \exp\left[i(\omega_{km} - \omega)\frac{t}{2} \right] \\ \therefore \quad |c_k(t)|^2 &= \frac{4|\mathscr{H}'_{mk}|^2}{\hbar^2} \frac{\sin^2[(\omega_{km} - \omega)t/2]}{(\omega_{km} - \omega)^2}. \end{aligned} \tag{1.44}$$

This is the lowest-order correction for the probability of finding the system in $|k\rangle$ at time $t = T$, and let us say that we remove the perturbation at that point, leaving the system in that stationary state, is

$$|c_k(t)|^2 = \frac{4|\mathscr{H}'_{mk}|^2}{\hbar^2} \frac{\sin^2[(\omega_{km} - \omega)T/2]}{(\omega_{km} - \omega)^2} \quad \text{for } t \geq T. \tag{1.45}$$

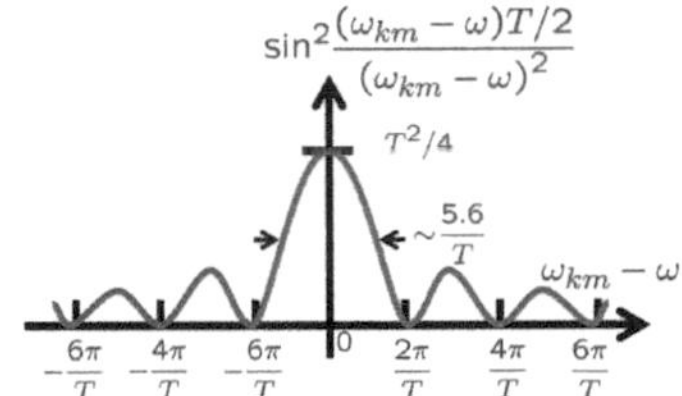

Figure 1.5: A plot of the term $\sin^2[(\omega_{km} - \omega)(T/2)]/(\omega_{km} - \omega)^2$, which is proportional to the probability of finding the 2-level system in the eigenfunction state $|k\rangle$ following an application of perturbation for a time duration T.

This response has a form that is the square of a sinc function. Figure 1.5 shows the normalized response. It has a peak when the frequencies/energies are precisely matched; that is, $\omega_{km} = \omega$. The peak of the normalized fraction is of magnitude $T^2/4$. The half width of the main peak is $\sim 5.6/T$. The transition probability per unit time $|c_{km}|^2/T$—a scattering rate—which we denote by S, is

$$S_{mk} = \frac{4|\mathscr{H}'_{mk}(0)|^2}{\hbar^2} \frac{\sin^2\left[(\omega_{km} - \omega)\frac{T}{2}\right]}{(\omega_{km} - \omega)^2 T}. \tag{1.46}$$

If T is large enough, the sinc function asymptotes to the Dirac delta function, that is, for large-enough T over which the perturbation appears,

$$S_{mk} = \frac{2\pi}{\hbar} |\mathscr{H}'_{mk}(0)|^2 \delta(E_k - E_m) \tag{1.47}$$

for $t \geq T$. The entire complexity of transitions under time-dependent finite time perturbation, under certain constraints, can be reduced

Not necessarily because of Matthew's principle, this relationship of Equation 1.47 is often referred to as Fermi's golden rule. It appeared in Fermi's quantum mechanics lectures. The compact and to-the-point lecture notes from his University of Chicago days—*Notes on quantum mechanics* from the University of Chicago Press is the 1954 version—are very worthwhile reading. Dirac had gotten there twenty years earlier. But the first order and second order perturbations to a time-normalized transition probability; that is, transition probability per unit time, which has a very simple form. If one knows the perturbation Hamilitonian, and two states between which this interaction's transition rate is to be ascertained, Equation 1.47 ascertains it. This is the Golden rule.

referred to as Golden rule 1 and Golden rule 2 in Fermi's notes struck a cord, and somewhere along the way his name got associated. We will call it just the Golden rule. It is of the first kind. We extracted it in the lowest order term of change. Enrico Fermi was an exceptional scientist, equally adept at experiment and theory. Fermi had a good sense of humor. In Rome, his Physics Institute was in the same compound as other senior government offices, where nobody worked on Sundays. Fermi was known to drive in wearing a hat and declaring himself a driver of dignitaries so that he could continue his experiments. His Nobel prize for discovering transuranic elements, which he named Ausonium ($Z=93$) and Hesperium ($Z=94$), is one of those for wrong reasons. The elements resulting from his slow neutron bombardment experiments were fission products, not heavier. But, like Bethe, who worked with him in Rome, there was plenty of other work for which the Nobel was deserved.

There are a number of interesting features embedded here. The peak is proportional to $T^2/4$, and the full width at half maximum is $\sim 5.6/T$. The area is proportional to T, and the width to time duration's inverse ($1/T$). The uncertainty principle is reflected in this spread. The central peak contains about 90 % of the area. When time is large enough, this function narrows further to the Dirac delta. It has a peak that corresponds to $T^2/4$, so, for matched conditions, only short times are needed. But, uncertainty relationships—embedded in our calculation of this relationship—must still hold and do. In addition, $\Delta E \Delta t \geq \hbar/2$ or $\Delta\omega\Delta T \approx 1$ still applies, as reflected in the central peak's areal argument. So, very short times, for example as in semiconductor-specific energy transition problems, will reduce the probability in state $|k\rangle$. But, as a first order term, this transition probability through the Dirac delta relationship will still be useful. If matching is poor, it may even vanish under certain conditions. If $\omega_{km} - \omega = 2\pi/T$, the transition vanishes, since a full cycle of interaction brings back the system to its original state. For all the problems that we are interested in, the time scales of interactions are large enough that this relationship written in Dirac delta form suffices. The equation can also be extended when there are spreads in frequencies of excitation, or spreads—as in bandstructure—of states of transition. For these,

See Appendix B where a number of computationally useful functions that one encounters in a variety of forms are summarized. The Dirac function can be written in a variety of ways and is often a very convenient manipulation tool, as is his bra and ket notation for vectors.

$$S = \frac{1}{T}\sum_k |c_k(t \geq T)|^2 = \frac{1}{T}\int |c_k(t \geq T)|^2 \mathscr{G}(k)\, dE_k, \qquad (1.48)$$

which reduces to

$$S = \frac{2\pi}{\hbar}\left|\mathscr{H}'_{if}(0)\right|^2 \mathscr{G}(E_f)\delta(E_f - E_i). \qquad (1.49)$$

From two levels to nearly continuous distribution simply follows as an extension through density of states. And if it is between two different distributions of density of states, then joint density of states will appear.

1.3.3 *Scattering by the perturbation*

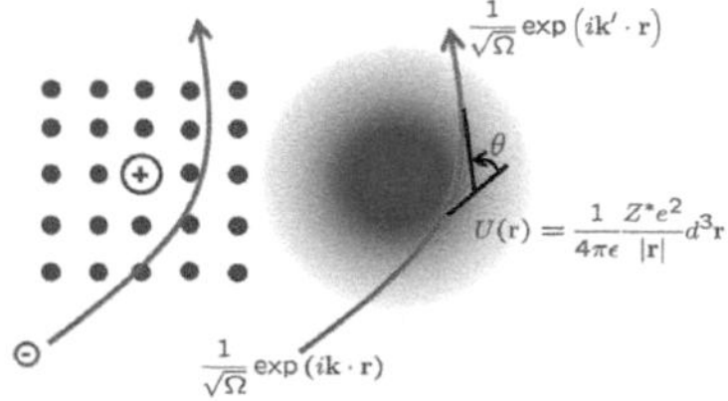

Figure 1.6: Coulomb scattering causing an electron wave to scatter—change momentum (magnitude and direction)—as it travels in a crystal. Coulomb energy is gained and then lost as the scattering takes place with a small net change.

The utility of this Golden rule can be illustrated through Coulomb scattering, as shown in Figure 1.6. Take the electron as a plane wave encountering the Coulomb attraction from a positive charge. This could be in free space, but, for us, this is particularly germane due to its correspondence to impurity, which has a charge in the semiconductor, causing a strong scattering for an electron. This electron is a nearly free electron traveling around, and the wavefunction of $\psi = (1/\sqrt{\Omega}) \exp(i\mathbf{k}\cdot\mathbf{r})$ for the spatial component is an adequate representation. Here, Ω represents the volume for normalization of the probability. Due to the charge, the perturbation potential is $U(\mathbf{r}) = (1/4\pi\epsilon)Z^*e^2/|\mathbf{r}|$. Z^*e here is the dressed static charge of the impurity. The electron's wavevector changes to $\mathbf{k}'$ as a result of this scattering. Under this perturbation, we have

We will generally use Ω for volume, and V for potential, except in rare situations, to avoid confusion. Ω_0 will be the volume of a unit cell.

$$\begin{aligned}\mathscr{H}'_{\mathbf{kk}'} = \langle \mathbf{k}'|\hat{\mathscr{H}}'|\mathbf{k}\rangle &= \frac{1}{\Omega}\int U(\mathbf{r}) \exp\left[i(\mathbf{k}-\mathbf{k}')\cdot\mathbf{r}\right] d^3\mathbf{r} \\ &= \frac{1}{\Omega} U_{\mathbf{kk}'},\end{aligned} \tag{1.50}$$

where this last simple notation tells us that it is the Fourier component—in reciprocal space—for the continuous electrostatic potential. Any scattering into a solid angle $d\theta$ depends on the density of states of the available states there. Classically, this is a continuous distribution, as all positions and momenta are possible. Quantum-mechanically, it will depend on the availability of states. Connecting quantum to classical with isotropicity, we may write $dE = v d(\hbar k)$, so the density of states for scattering is

Later on, in Chapter 10, we will see anisotropic consequences as a result of this state argument, since semiconductors in general are not isotropic, neither do they necessarily have a continuous state distribution.

$$d\mathscr{G} = \frac{\Omega}{(2\pi)^3} d\theta \frac{k^2\, dk}{v d(\hbar k)} = \frac{k^2 \Omega}{8\pi^3 \hbar v} d\theta. \tag{1.51}$$

The interaction happens during transit near the impurity, and the scattering or transition rate, writing it both classically through scattering cross-section and quantum-mechanically through the Golden rule, is

$$S_\theta = \frac{d\sigma}{\Omega} v = \frac{v}{\Omega}\frac{d\sigma}{d\theta} d\theta = \frac{2\pi}{\hbar}\left|\frac{1}{\Omega}U_{\mathbf{kk}'}\right|^2 \frac{k^2\Omega}{8\pi^3\hbar v} d\theta. \tag{1.52}$$

Therefore, with $d\sigma/d\theta$ from this relationship and using the momentum correspondence of $\hbar k = m^* v$, where m^* is an effective mass, we have

$$\frac{d\sigma}{d\theta} = \frac{m^{*2}}{4\pi^2\hbar^4}\frac{k^2}{v^2}|U_{\mathbf{kk}'}|^2. \tag{1.53}$$

The Fourier component of the Coulomb perturbation is

$$\begin{aligned}|U_{\mathbf{kk}'}|^2 &= \int \frac{1}{4\pi\epsilon}\frac{Z^*e^2}{|\mathbf{r}|} \exp\left[(\mathbf{k}-\mathbf{k}')\cdot\mathbf{r}\right] d^3\mathbf{r} \\ &= \frac{1}{4\pi\epsilon}\frac{Z^*e^2}{|\mathbf{k}-\mathbf{k}'|^2} = \frac{Z^*e^2}{4k^2\sin^2(\theta/2)}.\end{aligned} \tag{1.54}$$

This last equation now gives a direct correspondence between the classical and quantum-mechanical pictures of a nearly free electron scattering due to a charge in an isotropic condition. We could use the Golden rule to determine what this scattering rate will be in the angle $d\theta$. A classical scattering cross-section can be fitted to it. And, by correspondence, the quantum scattering will gracefully transform into a classical scattering relationship.

This discussion of states and their transformation due to interactions sets up a reasonable starting point for discussing the nature of the electrons and the atom systems and their analysis.

Many of the parameters—nearly constant—that one often utilizes in the physical world are the result of fast and slow at work. The friction coefficient—static and dynamic—arises in the electromagnetic and quantum-constrained interactions in the interface region between two objects. The couplings under static and dynamic conditions are different, but both are to a broadband of vibrational losses for the atoms of the objects. An electron undergoes fast scattering events, many of which are random, as it moves through the matter. This too is a broadband event in the frequency domain, with energy loss to the environment. Thermal equilibrium or steady state comes about because of the accumulation of the fast-and-slow events.

1.4 *Fast and slow, and the Born-Oppenheimer/adiabatic approximation*

SIMULTANEOUS PRESENCE OF FAST AND SLOW interactions is quite commonplace. A simple example of this is the variety of circumstances where one introduces a frictional damping. Brownian motion, conductance, et cetera, are illustrations of fast events—in these cases, random scattering encounters of a particle or an electron in a solid—in the presence of a slow external stimulus—the flow of the liquid and the particles, or of electrons—under an external cause of potential or kinetic energy change. For semiconductors, an illustration of this complexity is fast-moving electrons in the midst of the vibration of atoms around their equilibrium positions. The atomic motion—dressed nuclei, that is, nuclei and the core electrons vibrating around their equilibrium—is slow, since the mass is large. An electron transits a few atom distance ($\sim nm$) at a speed of $\sim 10^7$ cm/s in ~ 10 fs. Atomic vibration—a deformation, where the frequencies are of the order of a few THz—must lead to a change in the allowed states of the electron. So, if there is a scattering interaction between the electron and the perturbation due to this deformation, there will be the fast scattering event coupling to a slower deformation. The Golden rule lets us tackle the fast through the transition/scattering rate. The slow will follow for us from adiabatic approximation, which has its origins in the Born-Oppenheimer approximation in quantum mechanics' earliest application to the study of molecules. We will be particularly interested in this simultaneous presence of fast and slow processes because of its importance to transport and transitions. And we can then suitably put the two approaches together to understand fast and slow.

Adiabatic, a word of Greek origin, translates as "not to be passed through."

The *adiabatic process*, or *adiabatic approximation*, is an important analysis tool in quantum conditions. Both of these terms are also used in classical conditions but with subtle differences in meaning that need some elaboration. In classical mechanics, the adiabatic

In electronics, adiabatic circuits, by suppressing entropy production and by recovering energy, can consume vanishingly low energy. But they are slow, so this quantum-classical difference can cause plenty of tangle.

process is a process in which no heat is exchanged. The system can be viewed as one that is thermally isolated or in which change is taking place rapidly enough that transfer of energy as heat is absent. In the thermodynamic view, this lets one analyze conditions of rapid change—such as compression or expansion in a mechanical engine—and determine limits such as the Carnot efficiency. Absence of heat exchange means no entropy change. In the quantum view, adiabatic approximation implies a change that is sufficiently slow so that the eigenfunction evolves slowly from one to another, slowly enough that it is a tight coupling between one state to another state; that is, it remains reversible throughout the process. The state of the system remains the eigenstate of the instantaneous Hamiltonian. It is in this sense of reversibility that it corresponds to the classical use. However, in classical adiabatic conditions, the process needs to be rapid to eliminate entropy change; in quantum conditions, it needs to be quasistatic to allow state-to-state coupling. If the latter were not, a starting state would couple to a band of states—the final new state being a superposition in the new system—destroying reversibility.

In S. Tiwari, "Quantum, statistical and information mechanics: A unified introduction," Electroscience 1, Oxford University Press, ISBN 978-0-198-75985-0 (forthcoming), we looked at the reflection and transmission of an electron plane wave whose energy is larger than the barrier energy. An abrupt barrier—sudden change—causes reflection and transmission. A gradually changing barrier, where $\ell \gg 2\pi/k$, with ℓ as a length scale of change of the barrier's changing energy, and k as the wave vector of the incident wave, has the reflection suppressed. The incident wave of wavevector $\mathbf{k}$ will adiabatically adjust to a wavevector $\mathbf{k}'$ in the barrier region.

So, in quantum mechanics, an adiabatic process is a process where a system undergoing change—with modification to its energy levels—continues to remain in a single definite state. It maintains, for example, its quantum numbers. If a square well has a particle in the ith level, it remains there as the square well shape is changed adiabatically. The wavefunction adapts to the slowly changing parameters that define and mold the system.

The Born-Oppenheimer approximation is an example of resorting to adiabatic process in molecules and solids. When we reduced Equation 1.10 to Equation 1.11 to describe the atomic solid by decoupling the ionic motion—the slow process—from the electrons'—the fast process—Hamiltonian, we employed the adiabatic approximation. The deformation-induced scattering can now be added on as a perturbation to the solution. Another example is the simple system that we will take up next for analysis—a 2-electron and 2-atom system (Section 1.5)—where both an electronic and a nuclear part appear. Both have motional components. For any change from the initial state to a final state, one must consider both the electron and the nuclear part. The mass of the electron is significantly smaller than that of the nuclei, so nuclei can be viewed as moving so slowly that the electron distribution adjusts to them instantaneously responding to the changing potential. The consequence—a Born-Oppenheimer approximation—is that a fixed electronic wavefunction is calculable for any fixed nuclear locale. This is to say that, in the absence of

An N-particle system has $3N$ degrees of motional freedom; 3 will be translational, leaving $3N-3$ for assorted other possibilities. If we look at just what 2 atoms can do, of the 6 motional degrees of freedom, 3 are for translational movement of molecules in real space. This leaves 3. A complete and independent set of these is vibration along the axis, and 2 more for the rotational freedom in two orthogonal planes that intersect along the molecular axis. Water with 3 atoms has a larger collection of such modes. In a kitchen microwave, the 2.45 GHz frequency $\equiv 12\ cm$ of free space wavelength or a 10 μeV energy photon is absorbed by the water molecule in this motional freedom.

degeneracy, one may split the eigenfunction describing both the electrons and the nuclei as separable products; that is,

$$|\psi\rangle = |\phi(\{\mathbf{r}\}, \{\mathbf{R}\})\rangle |\chi(\{\mathbf{R}\})\rangle, \tag{1.55}$$

where $\{\mathbf{R}\} = (\mathbf{R}_1, \ldots, \mathbf{R}_N)$ are all the nuclear displacements, and $\{\mathbf{r}\} = \mathbf{r}_1, \ldots, \mathbf{r}_N$ all the electron coordinates, which may be generalized to include spins if appropriate. This separability establishes that the electronic motion instantaneously responds to the atomic and nuclear motion—that is, it is a function of these coordinates—and is separable from the nuclear response, which can be written as a function of its own coordinates alone. The electron charge cloud follows rapidly the slow response of the electron and nuclear particles of the system. The energy of all possible arrangements is calculable in principle within the constraints of the accuracy of the method for the Hamiltonian's solution—Hartree-Fock—there. We will discuss Hartree and Hartree-Fock approaches to solving the Hamiltonian shortly. The equilibrium position is relatively accurate through the distribution of separations, even if the derivative around this equilibrium is not.

The Born-Oppenheimer approximation is very useful but, nevertheless, imperfect. There is inevitably some mixing between states. An electron falling behind nuclear motion may be closer to another state—a mixing has happened—and therefore transitions that are not allowed may become possible, small shifts in energies may come about, or degeneracies may be removed.

As an example of separation of the nuclear part, the Born-Oppenheimer approximation's use and its analysis, take the example of electromagnetic absorption by a molecule, which can also be extended to the crystal and is particularly useful for the nanoscale. Figure 1.7 shows the underlying process of an interaction of a molecule absorbing a radiation photon. This is an illustration of the Franck-Condon shift where fast and slow again makes its appearance. The Franck-Condon principle states that since the electron mass is much smaller than the nuclear mass, electronic transitions can be treated with a stationary nuclear framework. This is pretty much the Born-Oppenheimer approximation. The Franck-Condon approach is used for molecules, but it is also pertinent to the electron in the crystal with atoms attached to each other.

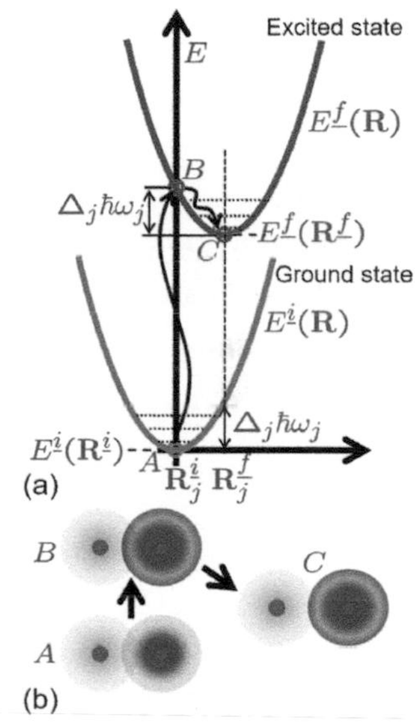

Figure 1.7: Franck-Condon shift, an example of fast-and-slow change, shown in configuration coordinate diagram. (a) A sequence of processes. An optical transition from A to B—a resonant absorption—is fast and accompanied by an ionic distortion since electrons move rapidly but the nucleus does not. This metastable excited state slowly relaxes to the excited state C with a Franck-Condon shift of $\Delta_j\hbar\omega_j$. (b) A physical interpretation at the atomic level. In the state C, the nuclei of the excited state are farther away. Eventually, another optical emission takes place, leaving the nuclei still far apart. And eventually they slowly come back to the starting state A. This picture of configuration coordinates is useful in understanding some of the deep levels in semiconductors— vacancies, atoms displaced, et cetera—where lattice distortion accompanies the electron capture or emission process.

A *configuration coordinate* diagram shows the energetic changes as multiple coordinates; for example, the geometric spacing of the center of motion, and the relative spacings of components, together with the momenta of a system, also undergo a change. Figure 1.7 shows the energetics as a function of nuclear position for a system undergoing transition by interaction with a photon: the initial and final state electronic energy as a function of the displacement $\mathbf{R}_j$ of the jth nuclei. The electron energies $E(\mathbf{R})$ as a function of the atomic/nuclear displacement within the harmonic approximation for two different nuclear positions of the jth nuclei are shown here for both the initial state and the final state. The vibrational part, due to the nuclear motion, occurs slowly. The electron state change occurs rapidly. The electronic and vibrational parts of the response can be separated. The vibration part, in the linear limit, is a simple harmonic oscillator, whose center will change as time proceeds, and this process will involve phonon emission in the solid. The different excitation states of the electron are the different dotted lines in the figure in the harmonic well.

These energies can be written as

$$E_{-}^{i}(\mathbf{R}) = E_{-}^{i}(\mathbf{R}_{-}^{i}) + \sum_{j=1}^{3N} \frac{1}{2} m\hbar\omega_j^2 (\mathbf{R}_j - \mathbf{R}_j^i)^2, \text{ and}$$

$$E_{-}^{f}(\mathbf{R}) = E_{-}^{f}(\mathbf{R}_{-}^{f}) + \sum_{j=1}^{3N} \frac{1}{2} m\hbar\omega_j^2 (\mathbf{R}_j - \mathbf{R}_j^f)^2 . \qquad (1.56)$$

The electronic energies differ in energy linearly with the displacement $\mathbf{R}_j$ if the coupling between electronic and nuclear motion is also linear under small displacement. The nuclear contribution oscillates around mean displacement position ($\mathbf{R}_j^i$ and $\mathbf{R}_j^f$). The change in mean position, representative of the electron-nuclei interaction, can be related through a parameter written here without proof as

$$\Delta_j = m\left(\frac{\omega_j}{2\hbar}\right)|\mathbf{R}_j^i - \mathbf{R}_j^f|^2 . \qquad (1.57)$$

Δ_j—the Huang-Rhys factor—is a dimensionless factor tying the strength of the coupling of the electronic states to the nuclear motional freedom.

The photon absorption induces an electronic transition from point A, with the system evolving, without nuclear positional change, to point B. Nuclear position now changes slowly to $\mathbf{R}_j^f$ (point C) via phonon emission, with the system still staying excited. The net difference between the final and initial state energies is

$$E_{-}^{f}(\mathbf{R}) - E_{-}^{i}(\mathbf{R}) = E_{-}^{f}(\mathbf{R}_{-}^{f}) - E_{-}^{i}(\mathbf{R}_{-}^{i}) + \sum_{j=1}^{3N} \Delta_j \hbar\omega_j + \sum_{j=1}^{3N} \sqrt{2m\hbar\omega_j^3 \Delta_j}|\mathbf{R}_j| . \qquad (1.58)$$

Here, the first summation is the Franck-Condon energy corresponding to the net relaxation of the molecule or crystal, and the last term is the result of electron and nuclei motion interaction, which is the electron-phonon coupling.

The absorption of the photon—a fast process—causes a change in energy without a change in nuclear configuration, which thereon relaxes through the transfer of energy to the vibrations. We could separate the terms, since the nuclear part of the wavefunction could be written as a product term solely in terms of nuclear position in the wavefunction of Equation 1.55. We could analytically resolve this because of the separability of the fast-and-slow process that adiabatic approximation could be applied to. This example also introduces us to the use of configuration diagrams when local positional changes have an effect on energy. We will encounter this a few times, particularly so in understanding defects, where the local crystal environment will have a consequence for electronic energy interactions, just as it did in this molecular example.

The heating of a water molecule in the microwave depends on an electronic transition that then shakes the water molecules, which is an effective way to heat food.

1.5 *A 2-electron and 2-atom system*

WE START OUR EXPLORATION OF SOLIDS with the simplest of cases of matter assembly: a 2-electron, 2-atom system—the hydrogen molecule—to understand the energetics and the different possibilities of the eigenstates as starting points. There is much here from which we can draw implications for the solid forms of interest to us. In this 2-electron, 2-atom example, the Hamiltonian operator, with the electron's kinetic and Coulomb potential energies included, is

$$\hat{\mathscr{H}} = -\frac{\hbar^2}{2m_0}\nabla_1^2 - \frac{\hbar^2}{2m_0}\nabla_2^2 - \frac{1}{4\pi\epsilon_0}\frac{e^2}{r_{1A}} - \frac{1}{4\pi\epsilon_0}\frac{e^2}{r_{2A}} - \frac{1}{4\pi\epsilon_0}\frac{e^2}{r_{1B}} - \frac{1}{4\pi\epsilon_0}\frac{e^2}{r_{2B}} + \frac{1}{4\pi\epsilon_0}\frac{e^2}{r_{12}} + \frac{1}{4\pi\epsilon_0}\frac{e^2}{R}, \quad (1.59)$$

where A and B denote the sites of the two atoms that are R apart and 1 and 2 represent the two electrons of mass m_0, so that r_{1A} means the separation of electron 1 of charge $-e$ from a residual core of charge $+e$ ($Z = 1$ in Equation 1.11) and r_{12} is the separation between the two electrons. We wish to find the lowest energies. It stands to reason that these will be reconstituted from the lowest energy states of the atoms from which the molecule is formed as the interaction evolves. Let $|u_{A\uparrow}\rangle$, $|u_{A\downarrow}\rangle$, $|u_{B\uparrow}\rangle$ and $|u_{B\downarrow}\rangle$ represent the 1s orbital wavefunction of the two atoms A and B of this hydrogen molecule, with spin up $\uparrow \equiv m_s = +1/2$ and spin down $\downarrow \equiv m_s = -1/2$ possibilities. These are the four possible spin orbitals.

In following through by the writing of Coulomb energy terms in this form, energy terms will appear as a Coulomb integral. An electron in an atomic orbital $|u_1\rangle$ has a charge density of $-e\langle u_1|u_1\rangle$ or a charge of $-e\langle u_1|u_1\rangle d^3r_1$ in volume d^3r_1. A second electron has a charge $-e\langle u_2|u_2\rangle d^3r_2$. For a separation r_{12}, the potential energy of the Coulomb interaction—the total electrostatic interaction between the two elemental charges—is the integral over the entire space of each of these volume elements. This is the Coulomb integral. It is a net increase in energy because charges are of the same sign.

Let the two electrons be represented by 1 and 2, respectively. So, $|u_{A\uparrow}(1)u_{B\uparrow}(2)\rangle$ speaks to electron 1 on atom A, and electron 2 on atom B, where both electrons have $m_s = +1/2$. We may exchange the electrons between the two atoms, in which case wavefunction $|u_{A\uparrow}(2)u_{B\uparrow}(1)\rangle$ is also a possibility. There are many more such possibilities—a total of ${}^4C_2 = 6$—that represent choosing any 2 out of 4 available spin orbitals. Two of these—$|u_{A\uparrow}u_{A\downarrow}\rangle$ and $|u_{B\uparrow}u_{B\downarrow}\rangle$—

This is the Pauli exclusion principle telling us that no two eigenfunction solutions for the fermions can be the same.

represent the 2 electrons on one atom (A or B). The other is bereft. So, we have a combination where one is now a negative ion, and the other a positive ion. We exclude these from our discussion. They are certainly possible under energetic circumstances, but we are interested in solutions where both centers are still neutral.

So, we need determinental functions that exclude the degenerate possibilities, for example, under exchange of the two electrons on the two centers ($|u_{A\uparrow}(1)u_{B\uparrow}(2)\rangle$ and $|u_{A\uparrow}(2)u_{B\uparrow}(1)\rangle$), and form a non-degenerate set for a stationary solution.

Let $\{|u_i^0\rangle\}$ be the set of orthonormal basis functions. We build a linear combination $|\psi\rangle = \sum_i c_i|u_i^0\rangle$. Let $\mathcal{O}_{jk} = \langle u_j^0|u_k^0\rangle$. This is an overlap matrix element. If $\mathcal{H}_{jk} = \langle u_k^0|\hat{\mathcal{H}}|u_j^0\rangle$, then stationarity requires

$$\sum_i c_i(\mathcal{H}_{jk} - E\mathcal{O}_{ji}) = 0, \tag{1.60}$$

by variational principle. For a unique stationary solution to exist for this set of equations, the secular equation

$$\det|\mathcal{H}_{jk} - E\mathcal{O}_{ji}| = 0 \tag{1.61}$$

must be satisfied. For N basis functions, there are N roots that are the eigenvalues. Each of these eigenvalues is associated with a combination of c_i's of Equation 1.60—a linear combination. For us, here these are to be built from $|u_{A\uparrow}u_{B\uparrow}\rangle$, $|u_{A\uparrow}u_{B\downarrow}\rangle$, $|u_{A\downarrow}u_{B\uparrow}\rangle$ and $|u_{A\downarrow}u_{B\downarrow}\rangle$. We have now discarded the identification of each electron, as it is implicit in this choice set, where M_s—the sum of the secondary spin number along the axis of quantization—changes from 1, to two with 0, and the last one with −1. This middle set of $M_s = 0$ leads to the linear combination through sum and difference, which are distinguishable, but which will also lead to a degenerate energy. Our four solutions, unnormalized, but explicitly including the spin and the electron and atom identity, are

$$\begin{aligned}
|u_{A\uparrow}u_{B\uparrow}\rangle &= [u_A(1)u_B(2) - u_A(2)u_B(1)]|\uparrow(1)\rangle|\uparrow(2)\rangle,\\
|u_{A\downarrow}u_{B\downarrow}\rangle &= [u_A(1)u_B(2) - u_A(2)u_B(1)]|\downarrow(1)\rangle|\downarrow(2)\rangle,\\
|u_{A\uparrow}u_{B\downarrow}\rangle + |u_{A\downarrow}u_{B\uparrow}\rangle &= [u_A(1)u_B(2) - u_A(2)u_B(1)]\\
&\quad\times[|\uparrow(1)\rangle|\downarrow(2)\rangle + |\downarrow(1)\rangle|\uparrow(2)\rangle], \text{ and}\\
|u_{A\uparrow}u_{B\downarrow}\rangle - |u_{A\downarrow}u_{B\uparrow}\rangle &= [u_A(1)u_B(2) + u_A(2)u_B(1)]\\
&\quad\times[|\uparrow(1)\rangle|\downarrow(2)\rangle - |\downarrow(1)\rangle|\uparrow(2)\rangle],
\end{aligned} \tag{1.62}$$

showing the separation of spatial and spin coordinates with their changing symmetries that make these combinations different from each other. The top three of these have antisymmetric spatial coordinates and spin coordinates are $M_s = 1, -1$ and 0.

A stationary state is one whose probability density ($\langle\psi|\psi\rangle$) is invariant in time. See the probability discussion of random processes in Appendix C.

See Appendix D for a discussion of variational principle and its usage. The reader will also find Appendix E and Appendix F as a short encapsulation of the important notions of thermodynamics and important distribution functions. Spin is discussed in Appendix G.

The simplest illustration of the removal of degeneracy in the presence of interaction is that of bonding and antibonding states. Let two systems share identical Hamiltonians. Independent, they have identical energy. Let $|l\rangle$ and $|r\rangle$ be the eigenfunctions that have this same energy E as their solution. Bring the two closer to cause each to perturb the other. Equation 1.61 gives the solution for the energy. Absent perturbation, so $\mathcal{O}_{ji}$s vanishing, the energy E is the solution for the two non-interacting systems in the $|l\rangle$ and $|r\rangle$ state. With perturbation, the energies change, and we have the solutions

$$\begin{aligned}
E^+ &= E + \Delta, \text{ with}\\
|+\rangle &= \frac{1}{\sqrt{2}}(|r\rangle - |l\rangle), \text{ and}\\
E^- &= E - \Delta, \text{ with}\\
|-\rangle &= \frac{1}{\sqrt{2}}(|r\rangle + |l\rangle),
\end{aligned}$$

where Δ is the energy change arising due to the perturbation. This perturbed system does not have $|l\rangle$ or $|r\rangle$ as its eigenfunction or E as its eigenenergy. The energies have changed to E^+ and E^-, and the new eigenfunctions are linear combinations of unperturbed eigenfunctions. $|l\rangle$ and $|r\rangle$ have hybridized. Degeneracy has been lifted. the higher energy state is the antibonding state. It is spatially antisymmetric. The lower energy state is the bonding state. It is spatially symmetric. This is a "molecular" description. States were localized. Such a "molecular" model is very useful in understanding many defects in semiconductors.

The eigenvalue of these three is identical, arising in the spatial dependences. Spatially antisymmetric functions form the triplet. The last wavefunction, also an eigenfunction and spatially symmetric, forms a singlet state.

With these functions known, one may calculate the energies following normalization, since the Hamiltonian of this problem only involves spatial dependence. This is shown in Figure 1.8 as a function of changing R. The singlet spatially symmetric solution has a lower energy, while the triplet spatially antisymmetric solution has higher energy. The former is our bonding state, and the latter the anti-bonding state. This approach is an illustration of the tight binding approach. We built a tight molecular construction using it. It is also known as linear combination of atomic orbitals (*LCAO*), since it built the hybrids, or the evolved eigenfunctions starting from the original atomic orbitals as the orthonormal set.

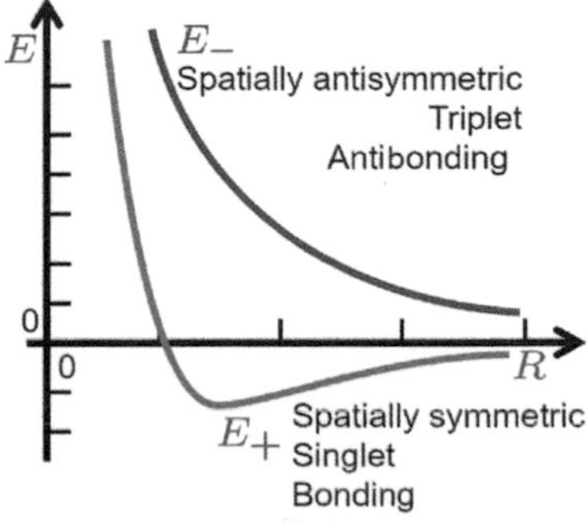

Figure 1.8: The energy, as a function of interatomic spacing, of a model hydrogen molecule for the four wavefunction solutions. The antisymmetric solution has a higher energy than the symmetric singlet solution.

Bringing two atoms together here has led to a lowering of energy in the bonding state and has resulted in a stable molecule with one bonded spatially symmetric solution. It has illustrated to us a methodology that will be a stepping stone to more complicated constructs. For us, a very instructive one is of N electrons together with the nuclei.

1.6 *N non-interacting electrons in the presence of nuclei*

IF ONE HAS N NON-INTERACTING ELECTRONS, and they are also independent of each other, that is, no Pauli-exclusion and fermionic constraints, the Hamiltonian may be written as $\hat{\mathscr{H}} = \sum_1^N \hat{\mathcal{H}}(\mathbf{z}_i)$, where the Hamiltonian of the ith electron—a sum of the kinetic $((-\hbar^2/2m_0)\nabla_i^2)$ and the potential energy (V) form—is a function of both the position ($\mathbf{r}_i$) and the spin ζ_i coordinates; that is, $\mathbf{z}_i = \mathbf{r}_i, \zeta_i$. These N Hamiltonians are identical, and their solutions degenerate, that is,

Spin often will become important because both Pauli exclusion and magnetic energetics relate to the spin's important role. See Appendix G for a discussion of the spin and spin matrices to represent spin coordinates.

$$\hat{\mathcal{H}}(\mathbf{z})|u_i(\mathbf{z}_i)\rangle = E_i|u_i(\mathbf{z}_i)\rangle \ \ \forall\, i = 1, \ldots, N, \tag{1.63}$$

so that the net energy of the N electrons is $E = \sum_{i=1}^N E_i$ and the wavefunction is also a product of the one-electron states, that is, $|\psi\rangle = \prod_{i=1}^N u_i(\mathbf{z}_i)$. The problem with this solution is that the wavefunction $|\psi\rangle$ is not antisymmetric, and it does not represent a collective ensemble of N non-interacting electrons that are not independent.

Another way of saying this is that we have found a solution for N one-electron one-nucleus systems that are all very far apart from each other so that they do not interact and are also independent of each other.

The wavefunction $|\psi\rangle$ for the ensemble Hamiltonian composed of these one-electron states must be an antisymmetric function for the fermion electron, since Pauli exclusion applies. This is the *Slater determinant*

$$|\psi\rangle = \frac{1}{\sqrt{N!}} \begin{vmatrix} |u_1(\mathbf{z}_1)\rangle & \cdots & |u_1(\mathbf{z}_N)\rangle \\ |u_2(\mathbf{z}_1)\rangle & \cdots & |u_2(\mathbf{z}_N)\rangle \\ \vdots & \vdots & \vdots \\ |u_N(\mathbf{z}_1)\rangle & \cdots & |u_N(\mathbf{z}_N)\rangle \end{vmatrix}, \tag{1.64}$$

A Slater determinant obeys antisymmetry by construction. Determinants change sign when rows or columns are interchanged. The Slater determinant also keeps the electrons indistinguishable.

with spin and which is orthonormal. The eigenenergy is $E = \sum_i^N E_i$. In this form, when two electrons are taken in identical wavefunction form, the determinant vanishes, as expected from Pauli exclusion, and the electrons are also indistinguishable. If the energy, and so also the Hamiltonian $\mathcal{H}(\mathbf{z})$, are independent of the spin, then one may separate the spin part from the spatial dependence; that is,

Indistinguishability here means that each electron is associated with each electron wavefunction.

$$|u_{i\sigma}(\mathbf{z})\rangle = |u_i(\mathbf{r})\rangle|v_\sigma(\zeta)\rangle. \tag{1.65}$$

Electrons of opposite spin may coexist in the same orbital state, so the ground state of this N non-interacting electron system is one where the entire gamut of lowest one-electron states is filled with two electrons, each of opposite spin per state.

This multiple electron question shows up as a problem of lowest energy in bonding in chemistry. Spin up and spin down are valid in a bonding state since the resulting eigenfunction is antisymmetric.

An important point of note here is that one could follow through this way and obtain the wavefunction for the system as a whole using the Slater determinant, because the electrons were non-interacting and because the total Hamiltonian could be written as a sum of each electron's Hamiltonian. And each electron's Hamiltonian was a function of its coordinate interacting only with nuclei. All other energetics, such as electron-electron interaction energies, for example, were, by fiat, zero. The independence let us write the total Hamiltonian as a sum of each electron's, and hence the wavefunction solutions found from Equation 1.63 served to build the system's wavefunction. The hydrogen molecule model in Section 1.5 is an example of this approach applied to a 2-electron, 2-nuclei problem.

1.7 *N interacting electrons in the presence of nuclei*

IN AN N-INTERACTING-ELECTRON SYSTEM, together with the atoms from which the electrons arose, the wavefunction of the system is a function of the coordinates of all electrons and of other particles, such as the cores, with which they meaningfully interact. Even with the simplification of these N electrons in a continuum approximation averaging the positively charged ion background, because of the interaction between all the electrons, the Hamiltonian is not separable as it was in previous case. We need to approximate the effect of interaction of an electron with all others by a potential that is only a function of the electron coordinate. Only then is it possible to treat the electrons one at a time separately. This is an example of a *self-consistent field approximation*—finding a mean field in which interaction may be tackled for specific particle with attention to only its coordinates—in numerous places and is useful for our N-interacting-electron system. Since the self-consistent approach requires the interaction potential and a knowledge of the states, with each dependent on the other, one must compute the solution self-consistently, iterating until one finds a solution that is satisfactorily accurate.

Here, this self-consistent field approximation implies that, for the specific electron for which one is computing, the potential in which it is present can be calculated by freezing all the other electrons and taking their averaged distribution by a centrosymmetric potential source. Solving the Schrödinger equation gives an updated state description. And this becomes a starting point for improving the accuracy of description of other electrons, again using the frozen representation for the rest. One cycle of this procedure updates all the electrons' orbital descriptions, and the procedure may be repeated. The procedure has limitations but is quite useful. As the number of particles increases, it becomes increasingly more accurate. It is therefore quite useful for the determination of many of the properties that are of interest to us.

1.7.1 *Hartree approximation*

WITH N ELECTRONS AND THEIR NUCLEI as an ensemble, the Hamiltonian may be written as

$$\hat{\mathscr{H}} = \sum_{i=1}^{N} \hat{\mathcal{H}}(\mathbf{z}_i) + \frac{1}{2}\sum_{i \neq j} \frac{1}{4\pi\epsilon_0} \frac{e^2}{|\mathbf{r}_i - \mathbf{r}_j|} + \hat{V}_{NN}. \tag{1.66}$$

The first term is as before (a sum of kinetic energy and the Coulomb interaction with the nuclei/ion), the second is the electron-electron Coulomb interaction while compensating for duplication and only including separate electrons, and the last is the Coulomb interaction between the nuclei $\hat{V}_{NN} = \sum_{i=1}^{N} \hat{V}(\mathbf{R}_i)$. It is the second term that mixes up the electron coordinates because of the electron-electron Coulomb interaction that makes the solution not decomposable to the form discussed in Section 1.6, where this interaction was excluded by making the independent electron approximation. So, this wavefunction solution is a function in the coordinates $(\mathbf{r}_1, \ldots, \mathbf{r}_N)$ has now become non-trivial.

Hartree's insight is that one may tackle each electron separately by viewing it as moving in the field of the nuclei (or ions, in the simplification for solids) and in the average field due to the other electrons. This self-consistent field approximation—the *Hartree approximation*—corresponds to finding the solution of

$$\left[\hat{\mathcal{H}}(\mathbf{z}_i) + \sum_{i \neq j} \int \frac{1}{4\pi\epsilon_0} \frac{e^2}{|\mathbf{r}_i - \mathbf{r}_j|} \langle u_j(\mathbf{z}_j)|u_i(\mathbf{z}_i)\rangle \langle u_j((\mathbf{z}_j)|u_j((\mathbf{z}_j)\rangle d\mathbf{z}_j \right] |u_i(\mathbf{z}_i)\rangle = E_i |u_i(\mathbf{z}_i)\rangle \;\; \forall\, i = 1, \ldots, N. \tag{1.67}$$

The problem has again been reduced to N equations, one for each electron, where each is in a field due to the other $N-1$ electrons.

On the left-hand side of Equation 1.67, we add and subtract an unphysical *self-interaction Coulomb term* for $i=j$, and one obtains the form

$$\left[\hat{\mathcal{H}}(\mathbf{z}) + \hat{V}_H(\mathbf{z}) - \hat{V}_{si}(\mathbf{z})\right] |u_i(\mathbf{z})\rangle = E_i|u_i(\mathbf{z})\rangle, \tag{1.68}$$

an equation that is now identical for all the electrons $i=1,2,\ldots,N$, where

$$\begin{aligned} V_H(\mathbf{z}) &= \int \frac{1}{4\pi\epsilon_0} \frac{e^2}{|\mathbf{r}-\mathbf{r}'|} n(\mathbf{z}')\, d\mathbf{z}', \\ n(\mathbf{z}) &= \sum_i o_i \langle |u_i(\mathbf{z})|u_i(\mathbf{z})\rangle, \text{ and} \\ V_{si}(\mathbf{z}) &= \int \frac{1}{4\pi\epsilon_0} \frac{e^2}{|\mathbf{r}-\mathbf{r}'|} \langle |u_i(\mathbf{z}')|u_i(\mathbf{z}')\rangle\, d\mathbf{z}'. \end{aligned} \tag{1.69}$$

This self-interaction term is introduced only for convenience of calculation. An electron, of course, does not interact with itself. But, through this subterfuge, one can write a Hartree potential as an integration over the ensemble.

Here, V_H is the *Hartree potential*—a Coulomb energy term arising in the electron interactions that includes the unphysical $i=j$ contribution and written in terms of electron density, which is a summation over the product of probability density of the ith state and its occupation factor (o_i, where o_i is 1 if an electron is present, and 0 if it is absent). The last term of V_{si} is a self-interaction term that is now being subtracted to compensate for what was artificially introduced in the Hartree potential.

This Hartree potential approach is quite a good starting point for describing an atom's electronic picture, or a many-electron picture of a solid, since the set of equations represented by Equation 1.67 are now solvable with the Hartree potential calculable through simple averaging in Equation 1.69. V_H represents a mean field effect, but one that includes the unphysical self-interaction term.

In the Hartree approximation, one starts with a trial wavefunction $|\psi\rangle$ composed of independent electrons; that is,

$$|\psi\rangle = \prod_{i=1}^{N} |u_i(\mathbf{z}_i)\rangle, \tag{1.70}$$

ignoring the antisymmetry. The desired solution is the one that has the minimum for $\langle\psi|\mathscr{H}|\psi\rangle$. This requires the variational expectation to vanish; that is,

$$\langle\delta\psi|\hat{\mathscr{H}}|\psi\rangle = 0. \tag{1.71}$$

See Appendix D, where the approach of using the variational principle to minimize energy by varying from a good guess of a starting trial solution is discussed.

So, the variational treatment of $|u_i\rangle$ leads to the solution of Equations 1.66 and 1.67, our equations of the Hartree formulation. We have reduced the problem to solving N single particle equations with unknown $n(\mathbf{z}') - \langle u_i(\mathbf{z}')|u_i(\mathbf{z}')\rangle$. An iterative approach that brings about self-consistency between the density's implication for potential (Equations 1.69) and the governing Hamiltonian (Equation 1.68) with the use of equilibrium statistics leading back to the density will tackle it.

But there are serious shortcomings. The most important is that these equations have no constraint that reflects Pauli exclusion (Equation 1.70 is not antisymmetric); that is, exchange.

1.7.2 *Hartree-Fock approximation*

THE HARTREE-FOCK APPROXIMATION ANTISYMMETRIZES the wavefunction. It starts with the Slater determinant, with its orthonormality for spin orbitals of independent electrons as the initial trial wavefunction. For the Hamiltonian of the Hartree equation (Equation 1.66) for this N-interacting-electron system, the energy solution is

$$E = \sum_i o_i \langle u_i | \hat{\mathcal{H}} | u_i \rangle + \frac{1}{2} \sum_{i,j} o_i o_j \left(\langle u_i u_j \left| \frac{1}{4\pi\epsilon_0} \frac{e^2}{|\mathbf{r}_i - \mathbf{r}_j|} \right| u_i u_j \rangle - \langle u_i u_j \left| \frac{1}{4\pi\epsilon_0} \frac{e^2}{|\mathbf{r}_i - \mathbf{r}_j|} \right| u_j u_i \rangle \right) + V_{NN}. \tag{1.72}$$

In the Hartree approximation, we started with the trial $|\psi\rangle$ as the product of independent electron wavefunctions. The Slater determinant gives us the coupled solution that adheres to all the quantum-mechanical constraints. This antisymmetrization is the Hartree-Fock approximation. Both Hartree and Hartree-Fock are, however, still approximate with Hartree-Fock, an improvement toward accuracy. Multiple electron assemblies such as atoms and molecules—systems with restricted numbers—show this.

Note the antisymmetrization in the second term. Here, we have used the generalized notation

$$\langle u_k u_l \left| \frac{1}{4\pi\epsilon_0} \frac{e^2}{|\mathbf{r} - \mathbf{r}'|} \right| u_i u_j \rangle = \int u_k^*(\mathbf{z}) u_l^*(\mathbf{z}') \frac{1}{4\pi\epsilon_0} \frac{e^2}{|\mathbf{r} - \mathbf{r}'|} u_i(\mathbf{z}) u_j(\mathbf{z}')\, d\mathbf{z}\, d\mathbf{z}' \tag{1.73}$$

for brevity.

Minimization of E for all $|u_i\rangle$ under the constraint of their orthonormality may be accomplished using Lagrangian multipliers; that is, we require

$$\delta E - \sum_{i,j} \lambda_{ij} \int \delta u_j^*(\mathbf{z}) u_i(\mathbf{z})\, d\mathbf{z} = 0 \ \forall\ \delta u_j^*. \tag{1.74}$$

A short summary of the method of Lagrangian multipliers can be found in Appendix D. This approach to finding solutions under constraints is an essential instrument from the tool set of mathematics that we employ throughout.

The one-particle equation set that this corresponds to is

$$\left[\hat{\mathcal{H}}(\mathbf{z}) + \sum_j o_j \langle u_j(\mathbf{z}') \left| \frac{1}{4\pi\epsilon_0} \frac{e^2}{|\mathbf{r} - \mathbf{r}'|} \right| u_j(\mathbf{z}') \rangle \right] |u_i(\mathbf{z})\rangle - \sum_j o_j \langle u_j(\mathbf{z}') \left| \frac{1}{4\pi\epsilon_0} \frac{e^2}{|\mathbf{r} - \mathbf{r}'|} \right| u_i(\mathbf{z}) \rangle |u_j(\mathbf{z})\rangle = \sum_j \lambda_{ij} |u_j(\mathbf{z})\rangle. \tag{1.75}$$

The first of the summation interaction terms on the left is just the Hartree potential V_H that we encountered before. The second summation term on the left is an *exchange term*. Since the Slater determinant undergoes only a phase factor change under unitary transformation, the equations remain the same structurally under diagonalization of this equation set. So, one may reform Equation 1.75 to a diagonal form where $\lambda_{ij} = E_i\delta_{ij}$. We now have the equation set

$$\begin{aligned}&\left[\hat{\mathcal{H}}(\mathbf{z}) + V_H(\mathbf{z})\right]|u_i(\mathbf{z})\rangle \\ &\quad + \int\left[-\frac{1}{4\pi\epsilon_0}\frac{e^2}{|\mathbf{r}-\mathbf{r}'|}\sum_j o_j\langle u_j(\mathbf{z}')|u_j(\mathbf{z})\rangle\right]|u_i(\mathbf{z}')\rangle\, d\mathbf{z}' \\ &= E_i|u_i(\mathbf{z})\rangle \;\; \forall\, i=1,\ldots,N. \end{aligned} \tag{1.76}$$

The last term on the left in this equation is a nonlocal exchange interaction. The equation form can be written more meaningfully—and simply—as

$$\left[\hat{\mathcal{H}}(\mathbf{z}) + V_H(\mathbf{z}) + V_x(\mathbf{z})\right]|u_i(\mathbf{z})\rangle = E_i|u_i(\mathbf{z})\rangle \;\; \forall\, i=1,\ldots,N, \tag{1.77}$$

where

$$V_x(\mathbf{z}) = \frac{1}{u_i(\mathbf{z})}\int\left[-\frac{1}{4\pi\epsilon_0}\frac{e^2}{|\mathbf{r}-\mathbf{r}'|}\sum_j o_j\langle u_j(\mathbf{z}')|u_j(\mathbf{z})\rangle\right]|u_i(\mathbf{z}')\rangle\, d\mathbf{z}' \tag{1.78}$$

In Equation 1.76, the $i = j$ contribution arising in the last term on the left is precisely V_{si}. But this Hartree-Fock approximation has reformed it into a correction term V_x arising in exchange for $i \neq j$. It is more accurate, even if less intuitive, and it has pulled in the nonlocal exchange's energetic consequence. This Hartree-Fock equation set can tackle the spin orbital as factorized by Equation 1.65. It accomplishes this by transforming the integration over $\mathbf{z}'$ to $\mathbf{r}'$, while V_H gets doubled for spin degeneracy and the exchange term is unchanged since their contribution to Equation 1.75 vanishes. In the Hartree approximation (Equation 1.68), we had to explicitly exclude the $i=j$ term in the summation. In the Hartree-Fock approximation, we do not have to exclude this, since the exchange term sums cancel with the $i=j$ term. When N is large, with electron contributions scaling as $1/N$, the distinction between Hartree and Hartree-Fock rapidly vanishes.

But, at small N, so few electron and few atom systems, such as nanostructures, these approaches and their judicious correction for $i=j$—within the self-consistent field approximation—will have noticeable consequences. Even though the Hartree-Fock approach should be expected to be more accurate by accounting for exchange, there will still be major shortcomings.

The next major shortcoming to consider is that we have not accounted for *correlation*. Correlation here is the notion that the mere presence of an electron causes a redistribution of the other electrons, due to electron-electron repulsion creating a "Coulomb hole." This issue of correlations and exchange-correlation holes is tackled in the next subsection.

We will encounter the word "correlation" often. Even Pauli exclusion is a form of correlation. Two electrons in a non-quantum mechanical view move independently of each other. But there is a low probability of them being in the same location, due to the Coulomb interaction. This is a charge correlation. The change in localization of electrons, such as in the hydrogen molecule, through a wavefunction that is a superposition of different configurations is a configurational correlation. Fermions avoiding each other for the same spin and clustering together when in opposite spin is a spin correlation. Any interaction between an electron and another electron that is not due to the Coulomb interaction is a quantum-mechanical consequence that is a correlation effect in this view. It appears in numerous forms, and we will see these throughout our discussion.

1.7.3 *Correlations*

DISCREPANCIES ARISE in a number of considerations neglected up to this point. In an atom or molecule, a principal one is from the relativistic effect in the core electrons with their large kinetic energy. Another one is due to *correlation energy*, which is important to atoms, molecules and atomic assemblies where many electrons will exist. Hartree-Fock ignores any local changes in the distribution of an electron, since it force fits a mean effect arising in the others.

Take the case of a molecule. When an electron is in the vicinity of another electron, Hartree-Fock accounts for it for the whole orbital as an average. This neglects any local electron-electron effect. This neglecting of electron correlations in its configuration form due to the mean field formalism makes energy calculation inaccurate at long separations and in the curvature at equilibrium. So, as seen in Figure 1.9, the poor representation of local distortion results in accurate representation of local equilibrium geometry but poor calculation of properties such as force constants, vibration frequencies, et cetera. Note that subsumed in the Hartree-Fock approximation also is the Born-Oppenheimer adiabatic approximation. The molecular potential energy is a function of relative nuclei locale. And any calculation where electrons follow any nuclear movement instantaneously will have increasing errors in the calculation of dynamic parameters. In a solid, this same correlation will cause similar inaccuracies when the number of interacting electrons is small and local electron-electron interaction important.

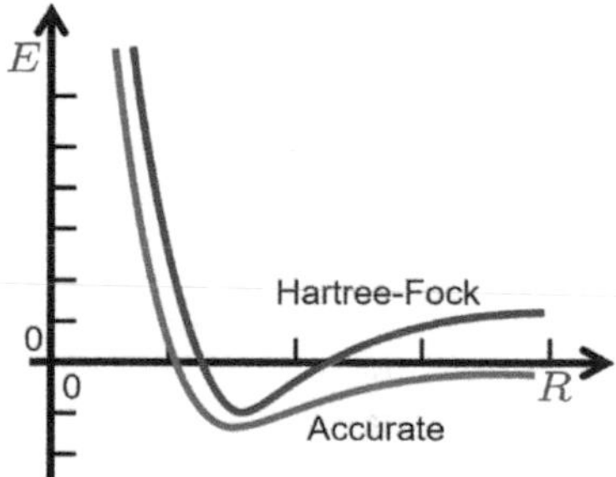

Figure 1.9: Molecular potential energy in Hartree-Frock approximation, where correlation is not accounted for. The minimum is geometrically close, but the derivatives are not, and the dissociation limit of $R \rightarrow \infty$ inaccurate.

We will modify the Hartree-Fock approach by accounting for this configuration interaction of correlation.

But, we also note that the Hartree-Fock equation is actually quite accurate and successful in a number of situations of interest to us. It is, for example, solvable and accurate for a free electron gas with a uniform compensating background. With one-electron wavefunction as a plane wave, the exchange interaction is calculable. The $1/r$ dependence then leads to a total energy that is proportional to

$n^{1/3}$, where $n = N/\Omega$ is the electron density for N electrons in volume Ω. The mean separation between the particles varies as $n^{-1/3}$ and the plane wave gives an equal probability everywhere, so any distortion effects naturally disappear with increasing particle population. We return to this problem shortly because of this construct's peculiarity.

The correlation effects have another important consequence: the existence of holes and mounds. A simple example is from spin correlation. Since two electrons of the same spin may not be found at the same point, if one looked at the probability of finding a second electron of identical spin, it will vanish at the point of no separation. The wavefunction of the second electron asymptotically vanishes in the vicinity of the first electron's locale. This is a *Fermi hole*. Likewise, for opposite spins, the second electron's probability is enhanced. Pauli exclusion, or exchange, has resulted in this spin correlation effect. In a configuration of atoms, the Hund rule's of maximum multiplicity is indirectly due to this spin correlation's consequence.

Hund's rule of multiplicity—an observation—is that when an atom or molecule has open electronic shells, then, in any given electronic configuration, the lowest energy form corresponds to the largest spin multiplicity. That is, the state with the largest total spin ($\sum_{m_s}$) is the one with the most stable configuration. So, with multiple orbitals of equal energy, electrons prefer to occupy them singly before pairing. Pairing reduces total spin. In silicon, which has $3s^2 3p^2$ as the outer orbital configuration, the $3p^2$ may exist as ${}^1D$, ${}^3P$ and ${}^1S$—multiplicity being $2S+1$ indicated in the superscript of $J = L + S$, with L the orbital quantum number—${}^3P$, with $M_L = 1$ and $M_S = 1$—is the favored ground state. This is Hund's rule 1; rule 2 is an observation on the consequences of electron repulsion, for which the silicon triplet state has no choice. Hund's rule 2 states that, in as much as it is possible to be consistent with rule 1, the configuration that maximizes orbital angular momentum is favored. Hund's rule 3 minimizes spin-orbit coupling and so appears for larger atoms with d and f electrons. The first two rules really are because of the Coulomb energy implication of the arrangements. Hund's rule can be seen classically in orbital motion. Two pendulums interacting with each other lock orbital and spinning motion.

Since the Hartree-Fock approximation does not include correlation effects, *configuration interaction* needs to be computed as a modification. The eigenstates of the system of N interacting electrons can be built by expansion of the Slater determinant constructed off an infinite set of orthonormal one-particle spin orbitals whose starting point is the Hartree-Fock ground state determinant. Let $|u_{Sn}\rangle$ represent the basis of the Slater determinant from the infinite orthonormal single particle spin orbitals. We have the wavefunction

$$|\psi\rangle = \sum_{n=1}^{N} c_n |u_{Sn}\rangle. \tag{1.79}$$

If we think of a free electron gas, with the nuclear charges smeared as a positive background—a jellium as a continuum—balanced by an opposite and uniform electron density, then the mean effect is a zero potential. The one-particle Hartree equation (Equation 1.67) is then just the free electron wave

$$u_k(\mathbf{r}) = \frac{1}{\sqrt{\Omega}} \exp(i\mathbf{k}\cdot\mathbf{r}), \quad \text{with } E_k = \frac{\hbar^2 \mathbf{k}^2}{2m_0}, \tag{1.80}$$

See Appendix H for a discussion of allowed **k** and the distribution of states in the reciprocal and real space.

where Ω is the free space volume, and the time dependence is implicit. A ground state has two electrons of opposite spin in these one-particle states. A Fermi level with energy E_F and wavevector $\mathbf{k}_F$ defines the highest state up to which these are filled at absolute zero temperature. Since the volume of **k**-space is $(4/3)\pi k_F^3$, with $2\pi/\Omega^{1/3}$ as the separation between the points of **k**-space that electrons with opposite spins may occupy,

$$N = 2\frac{(4/3)\pi k_F^3}{(2\pi)^3/\Omega} = \frac{k_F^3}{3\pi^2}\Omega \quad \therefore \quad n = \frac{N}{\Omega} = \frac{k_F^3}{3\pi^2}. \tag{1.81}$$

An electron has an effective space of a length scale $r_e = (3/4\pi n)^{-1/3}$, where n is the electron density, and its correspondence with the Fermi wavevector follows as

$$k_F = \frac{1}{\alpha r_e}, \text{ with } \alpha = \left(\frac{4}{9\pi}\right)^{1/3} \approx 0.521, \tag{1.82}$$

which is an indicator of the high level of filling and which is the reason why a jellium description becomes quite valid. The plane waves are eigenstates of the Hartree-Fock operator that had led to the exchange correction term of V_x (Equation 1.78) operating on the state $|u_i(\mathbf{z})\rangle$. This can be viewed as a one-body effective exchange potential operating on the one-particle eigenstate. For the plane waves, this energy function for a wavevector $\mathbf{k}$ is

$$V_{x,\mathbf{k}} = -\frac{4\pi e^2}{\Omega}\sum_{\mathbf{k}'}\frac{1}{4\pi\epsilon_0}o_{\mathbf{k}'}\frac{1}{|\mathbf{k}-\mathbf{k}'|}. \tag{1.83}$$

This term gives the exchange energy by summing over all the other states that are also occupied by electrons. Normalizing by the number of electrons N and avoiding double counting of the interaction by dividing by 2 gives the averaged exchange energy per electron of

$$E_x = -\frac{3}{4\pi}\frac{1}{4\pi\epsilon_0}\frac{e^2}{\alpha r_e} \approx -\frac{1}{4\pi\epsilon_0}\frac{0.458}{r_e} \tag{1.84}$$

for the free electron gas in a plane wave approximation for a metal.

The calculation that this all represents is that the electron doesn't interact with itself but with all others, and we have to self-consistently determine energy under this situation. The electron at $\mathbf{r}$ feels the field from other electrons, but, due to electron-electron repulsion, its presence in our calculation at $\mathbf{r}$ is also repelling these other electrons. So, it has a created a hole in the electron distribution around itself. This is a Coulomb hole due to exchange correlation—an exchange hole. Its presence is also changing the screening of the electron-electron interactions. Figure 1.1, in Section 1.2, is not an unreasonable representation. The exchange hole lowers the net energy. Charge neutrality also means that the electron and the Coulomb hole compensate each other locally. So, in this volume region, net charge still vanished, and the system is neutral. For free metal conducting systems, it has a fair and well-formed description.

V_x of Equation 1.78 is a one-body exchange potential on the one-body eigenstate $|u_i(\mathbf{z})\rangle$. One may view it as an electrostatic potential that arose due to the occupation density

$$o_{HF}(\mathbf{z},\mathbf{z}') = \sum_j o_j \frac{u_j^*(\mathbf{z}')u_j(\mathbf{z})u_i(\mathbf{z}')}{u_i(\mathbf{z})}. \tag{1.85}$$

Integrated over $\mathbf{z}'$, this must be unity, since it represents the existence of this occupied state for which the calculation is being performed. In any N-electron system, this electron at $\mathbf{r}$ is interacting with $N - 1$ other electrons. The Hartree potential V_H of Equation 1.69 contains N electrons and one exchange hole. Equation 1.85 says that if $o_j = 1 \ \forall\, j$, then the hole is localized on the electron—a delta function ($\delta(\mathbf{z} - \mathbf{z}')$). But, our previous paragraph argues that this cannot be the case. So, there exists a *broadening*. It is this broadening that is of the order of $\lambda_F = 2\pi/k_F = 2\pi\alpha r_e$ for the free electron gas. The exchange hole in "free" electron metals—alkali being the closest approximation—spreads out a bit beyond the nearest neighbor.

The spreading just beyond makes sense and should be general. Two electrons of same spin cannot be in the same position. The configuration interaction correction to Hartree-Fock used the Slater determinant, which takes this exclusion to heart.

This exchange-correlation hole can now be easily interpreted and understood. The electron density is the probability of finding the electron per unit volume. It is the number of occupied states, and if we normalized it to the states, it is the fraction. With $\mathbf{r}_i$ as the electron positions,

$$n(\mathbf{r}) = \langle\psi \sum_{i=1}^{N} \delta(\mathbf{r} - \mathbf{r}_i)|\psi\rangle, \tag{1.86}$$

where $|\psi\rangle$ is the N-particle wavefunction. Let $n(\mathbf{r},\mathbf{r}')$ be a pair correlation of the squared probability of finding two electrons, one at $\mathbf{r}$ and another at $\mathbf{r}'$, that is,

$$n(\mathbf{r},\mathbf{r}') = \langle\psi \sum_{i\neq j} \delta(\mathbf{r} - \mathbf{r}_i)\delta(\mathbf{r}' - \mathbf{r}_j)|\psi\rangle. \tag{1.87}$$

The term includes any contributions of correlations between electrons. The system Coulomb energy is

$$V_{Coul} = \langle\psi|\frac{1}{2}\sum_{i\neq j} \frac{1}{4\pi\epsilon_0}\frac{e^2}{|\mathbf{r}_i - \mathbf{r}_j|}|\psi\rangle = \frac{e^2}{2}\frac{1}{4\pi\epsilon_0}\int_\Omega \frac{n(\mathbf{r},\mathbf{r}')}{|\mathbf{r}_i - \mathbf{r}_j|}\,d\mathbf{r}\,d\mathbf{r}'. \tag{1.88}$$

The Hartree and Hartree-Fock approach didn't account for correlation. This means that

$$n(\mathbf{r},\mathbf{r}') = n(\mathbf{r})n(\mathbf{r}') \tag{1.89}$$

for the Hartree and Hartree-Fock treatments.

When configuration interaction—correlation—is included, one may rewrite the pair correlation function in a first order expansion as

$$n(\mathbf{r},\mathbf{r}') = n(\mathbf{r})n(\mathbf{r}')\left[1+\alpha(\mathbf{r},\mathbf{r}')\right], \tag{1.90}$$

where $\alpha(\mathbf{r},\mathbf{r}')$ is a correlation parameter that contains correlation's consequences. Since an electron at $\mathbf{r}$ interacts with $N-1$ other electrons in the $\mathbf{r}'$-space, $\int n(\mathbf{r},\mathbf{r}')d\mathbf{r}' = N-1$. Therefore, because of Equation 1.90,

$$\int \alpha(\mathbf{r},\mathbf{r}')n(\mathbf{r}')\,d\mathbf{r}' = -1. \tag{1.91}$$

This is the mathematical expression for stating that the electron at $\mathbf{r}$ has an exchange-correlation hole enveloping it. It arises both due to exchange and due to correlation, and it gives us an intuitive way of looking at exchange and correlation.

To distinguish exchange's and correlation's consequences in the creation of the hole, consider spin and we can look at what happens with aligned and anti-aligned spins. These are both conditions where Equation 1.91 is still valid. If a system had N electrons composed of $N_\uparrow$ (spin up) and $N_\downarrow$ (spin down) electrons, the electron at $\mathbf{r}$ with up spin will interact with $N_\uparrow - 1$ of up-spin electrons and $N_\downarrow$ of down-spin electrons. Therefore, the integral of Equation 1.91 split up is

$$\begin{aligned} &\int \alpha_{\uparrow\uparrow}(\mathbf{r},\mathbf{r}')n_\uparrow(\mathbf{r}')\,d\mathbf{r}' = -1, \quad \text{as before, and} \\ &\int \alpha_{\uparrow\downarrow}(\mathbf{r},\mathbf{r}')n_\uparrow(\mathbf{r}')\,d\mathbf{r}' = 0. \end{aligned} \tag{1.92}$$

An exchange hole exists (with this integral of -1) for the up-spin electron at $\mathbf{r}$, even with the correlation effect present. And, for interaction with electrons of opposite spin—no correlation—the local screening hole will have to be compensated for by the charge on the surface of the system so that the second part of Equation 1.92 is satisfied. At nanoscale, this effect will be of significance.

This discussion suffices for now to indicate that predictive description of large-N systems, such as solids, will require care. The energy state, the transitions under perturbations, will relate to the Hamiltonian description and its solution under the constraints of the circumstances. Bandstructure—energy states of the electrons—calculation will require related care. We will return to this calculation to summarize the different approaches—their salient points and applicability and limitations—in Chapter 4. Here, we continue with our discussion of approximation methods and now look at screening by the mobile charge.

1.8 Screening

HOW DO WE TACKLE SPATIAL VARIATION in electronic charge? It exists, since perturbations exist. The simplest type of perturbation may be a space charge, where electrons will locally rearrange themselves to minimize the energy of interaction by attempting to screen the perturbation. A positive space charge will attract, and a negative space charge will repel. The simple charge-induced perturbation may be static, but it could also be dynamic if it arises in an oscillatory phenomenon interacting with the screening electrons. If the electron states are filled up to some states up in energy—the Fermi energy of $\sim E_F$ with a Fermi wavevector of $\mathbf{k}_F$—then it is the electrons around the Fermi energy that are most likely to respond, since both filled and empty states are available around it. We have now created not only a dynamic condition—depending on the frequency scale—but also one where the consequences will be felt nonlocally, as the electrons at Fermi energy provide an oscillatory response. All these interactions will be mediated by permittivity, and which kind—static, intermediate or high frequency—will depend on the conditions of perturbation. A few remarks on the screening are therefore in order to understand how the particles moving around in the crystal respond.

The permittivity reflects the medium's ability to withstand the applied electric field as represented by displacement, or equivalently, polarization. In vacuum, this is quite clear. $\epsilon = \epsilon_0$, which does not depend on the frequency of the applied electromagnetic stimulus. In an atom, isolated in vacuum, when determining the orbitals, et cetera, without stimulus, it is again this $\epsilon = \epsilon_0$. Apply an electric field, and the atom responds by polarizing—slightly or significantly—and the response is concentrated in the outer orbitals, with electrons in the core orbitals shielded by the valence. Now, the permittivity needs some care and thought. In a semiconductor, the electrons or their anti-quasiparticle hole, sample the environment of the crystal. The binding energy, that is, the ionization energy of donors and acceptors (shallow hydrogenic) now must be the permittivity of the crystal. And it is the static permittivity, since this particle's binding exists in an unstimulated environment. Place the donor under very confined conditions in the semiconductor, and the permittivity must account for the change of the environment. In an unconfined crystal, the permittivity will change as a function of frequency, since the medium's response is changing. If the nearly free electron and an electromagnetic stimulus are interacting in the crystal environment, then this interaction will need to account for the frequency dependences and the time extent of the interaction, where phonons may also be important. We tackle this later. If an electromagnetic signal causes an electron transition from within the core, then, due to where it is from and the rapidity with which the change takes place up in the atom's higher orbitals—still localized—the permittivity is still free space permittivity. But, an excited electron localized at an atom relaxing into a delocalized state in the crystal environment will now need a more complicated permittivity analysis. So, use permittivity with care. The solution is generalizable. In electrical engineering texts, it appears as a question of how electrons screen a potential disturbance, that is, a field, such as when the jellium of

1.8.1 Debye-Hückel and Thomas-Fermi screening

THE STATIC SCREENING PROBLEM—largely an electronic many-body problem with electrons interacting with fields arising from other charges—goes back to Debye and Hückel, who explored it for the case of electrons interacting with other electrons. Figure 1.1 was an example showing an electron with the exclusion zone due to correlation around it arising from Coulomb repulsion. This picture can be seen—within the jellium approximation—as an illustration of Debye-Hückel screening.

The Poisson equation, with an electron located at $\mathbf{r}_0$, a charge distribution due to electrons of $-en(\mathbf{r}, \mathbf{r}_0)$, and a uniform positive neutralizing background of concentration en (the ionic jellium), is

$$\nabla^2 V(\mathbf{r}) = -\frac{1}{\epsilon}\left[-e^2\delta(\mathbf{r}-\mathbf{r}_0) - e^2 n(\mathbf{r}-\mathbf{r}_0) + e^2 n\right], \qquad (1.93)$$

where V is the electrostatic potential. To include the correlation effect, we write the pair correlation function $g(\mathbf{r}|\mathbf{r}_0) = n(\mathbf{r}|\mathbf{r}_0)/n_0$. This function gives the probability of finding an electron at $\mathbf{r}$, given that there is another electron at $\mathbf{r}_0$. This pair distribution function vanishes at $\mathbf{r}=\mathbf{r}_0$, that is, the probability of an electron vanishes, and it asymptotes to 1 at infinity. We now rewrite this equation as

$$\nabla^2 V(\mathbf{r}) = -\frac{e^2}{\epsilon}\left\{\delta(\mathbf{r}-\mathbf{r}_0) + n\left[g(\mathbf{r}|\mathbf{r}_0) - 1\right]\right\}. \tag{1.94}$$

We need to find the pair correlation function that solves this many-body simplified problem. This is possible at many different levels of accuracy. The equation as written holds true whether we need to include quantum constraints or not for physical charged particles. The quantum character of electrons introduces just additional non-triviality.

First, consider non-quantum classical conditions. The Boltzmann distribution applies. At very small $\mathbf{r}$ referenced to $\mathbf{r}_0$, this approximation will fail, but, at far enough distances where one may linearize the correlation function (a Poisson-Boltzmann function) of

$$g(\mathbf{r}) = g_{PB}(\mathbf{r}) = \exp\left[-\frac{V(\mathbf{r})}{k_B T}\right], \tag{1.95}$$

the solution will be quite accurate. Here, the position of perturbation at $\mathbf{r}_0$ is implicitly understood. With the linearization, the Poisson equation reduces to

$$\nabla^2 V(\mathbf{r}) = -\frac{e^2}{\epsilon}\delta(\mathbf{r}) + \frac{ne^2}{\epsilon k_B T}V(\mathbf{r}), \tag{1.96}$$

whose solution is

$$V(\mathbf{r}) = \frac{e^2}{4\pi\epsilon r}\exp\left(-\frac{r}{\lambda_{DH}}\right), \tag{1.97}$$

a form similar to that of the Yukawa potential encountered with massive bosons and is the static and spherically symmetric solution of the Klein-Gordon equation.

$$\lambda_{DH} = \left(\frac{\epsilon k_B T}{ne^2}\right)^{1/2} = \lambda_D \tag{1.98}$$

is the Debye-Hückel or just plain Debye screening length. This approximation is a linear screening approximation from that Boltzmann expression.

Now assume that the linearization is acceptable, but pair correlation as employed is not. We should still be able to use the thermal equilibrium condition, which brings about the equilibration of electrochemical potential. Electrostatic potential and chemical potential compensate each other. So, now, we have a screening length that is

$$\lambda_{scr} = \left[\frac{\epsilon\,(\partial E_F/\partial n)\,|_T}{e^2}\right]^{1/2}. \tag{1.99}$$

positive charge is uncovered. Edges of transition region are an example. So is the case where suddenly one has a sudden change in doping.

Appendix E and F have a short primer on thermodynamics and the statistical implications reflected in the distributions functions. A Boltzmann distribution will be a reasonable approximation where classical conditions are a good description. Electron description in non-degenerate semiconductor conditions is one example, even if electrons are quantum particles, and this quantum aspect is very necessary in describing the nature of their states in the semiconductor. Boltzmann distribution works pretty well here, since a deep potential also increases the probability of finding a particle there. The exponential arrives from the phase space.

Debye (Petrus Josephys Wilhemlmus Debije) is being given top billing. Electrical engineering's semiconductor device literature, with atoms as an afterthought except in reliability or processing discussions, largely ignores Hückel. Hückel, of course, finds a pride of place in chemistry. So does Debye, whom we first encounter through the Debye model for the low-frequency phonon contribution to specific heat, but which is only one of many significant contributions. He was Sommerfeld's student before Sommerfeld's Munich period. As with many of the European scientists who came of age in the pre-war years, there exists considerable tension and ambivalence in matters of life where science and society intersect. Debye became the head of the Kaiser-Wilhelm Institute in Berlin when Einstein left for the *USA* in the 1930s, and Debye himself moved to the *USA* just before the Second World War. He was the head of the Deutsche Physikalische Gesellschaft (the German physical society) from 1937 to 1939 and was among those who helped Lise Meitner escape, but one can also find letters that pay obeisance to powers that be, which in this case was Adolf Hitler, and an untenable situation brought on by his daughter's decision to stay back in Germany. Even Fermi was a member of the Fascist party.

If the conditions are degenerate, and a Fermi gas description is more appropriate, then we obtain the Thomas-Fermi screening length of

$$\lambda_{TF} = \left(\frac{2\epsilon E_F}{3ne^2} \right)^{1/2}, \tag{1.100}$$

for three-dimensional conditions, since, in degenerate conditions, $E_F \propto n^{2/3}$.

To paraphrase the great English mathematician, George Hardy, from his essay "A mathematician's apology," pure mathematics is the most beautiful mathematics, since it has no usefulness. And it is because of this very uselessness that the pursuit of pure mathematics cannot be misused to cause harm.

Figure 1.10 shows the magnitude of the screening length scale in *Si* as a function of the carrier concentration. At small carrier concentrations, the electrons are tens of *nm* or even more distant from the potential disturbance. At 10^{17} cm^{-3} carrier concentration, this screening length is of the order of 10 *nm*, and it is in the Debye-Hückel limit. As the semiconductor becomes degenerate (the effective density of states $\mathcal{N}_c \approx 2.8 \times 10^{19}$ cm^{-3} at room temperature), gradually, with degeneracy, the screening length scale bends over to the Thomas-Fermi limit. At the highest concentrations possible in *Si*, this screening length scale is a fair fraction of a *nm*, so spread over several atom spacings.

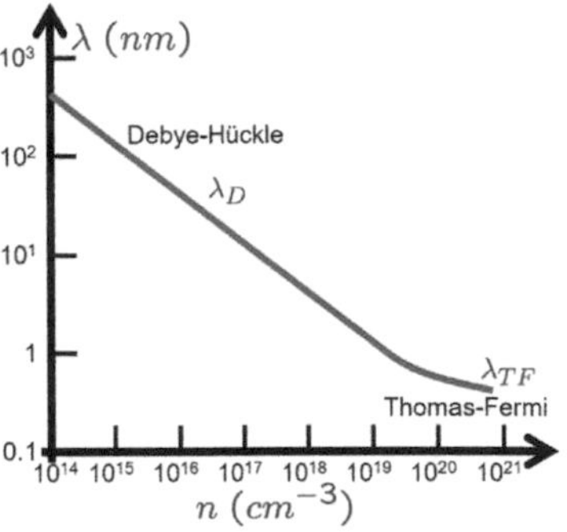

Figure 1.10: Electron screening length in *Si* at 300 K as a function of carrier concentration. The screening length scale can span over 100s of *nm* in the Debye-Hückel limit to sub-*nm* in the Thomas-Fermi limit.

1.8.2 *Static versus dynamic screening, and a note on permittivity*

INTERACTIONS CAN BE SLOW AND FAST. How screening will happen will depend very much on the pace of this interaction. A spatially fixed charge with electrons around it screening the perturbation is a static perturbation. The permittivity mediates it, and the electrons screen, present in these surroundings, through the static dielectric response. *The Debye and Thomas-Fermi screening are very applicable to such static circumstances, and it is the static dielectric constant that is applicable.* There are, however, circumstances where this will need modification. An electron in a confined condition, that is, with surrounding potential barriers that keep it in narrow atomic-scale regions, also feels the barrier and its behavior is not that of the electron in a crystalline surrounding of long-range periodicity. Its probability densities have changed, as have the state description and the energy and the wavevectors. The permittivity will change and needs to account for the change of the surroundings at such small dimensions, just as the eigenfuction description changes, leading to changes in energies and even the applicability of the mass assigned to the state of the electron.

Interactions can also be fast. Consider the absorption of phonons representing the quantization of crystal vibrations. A phonon energy of 50 *meV* is a 10 *THz* oscillating quantum. Since the time lengths of interactions at any energy change of ΔE also has time interaction of $\hbar/\Delta E$, that optical phonon interaction time is 100 *fs*. These are dynamic conditions, and the permittivity, which reflects the medium's permission to allow electrical changes to take place, will also have a frequency dependence. It is clear from this argument that screening's mediation in the interaction will be influenced by permittivity, which depends on the polarization response of the medium and the extent of the medium interacting, and all these changes occurring simultaneously will need to be reconciled. Polarization arises in many sources, electrons are moving around and respond to the fields, and an oscillating field causes the charge cloud to respond too by oscillating under forced conditions. These are plasmons affecting the permittivity locally. Under certain conditions, there will also exist long-range consequences spatially. Vibrating atoms—ion and charge with their different inertia—also respond, and this leads to an ionic response. So, the general nature of the permittivity is to drop with increasing frequency. There are, however, regions in between where resonances in the forcing function's frequency and the polarizing response may cause large-scale changes of increase or decrease either side of the resonance. In these situations, we will have to modify substantially the nature of screening—as in the Debye or Thomas-Fermi static interpretation—and incorporate the permittivity's behavior at high frequency, even as there is a background asymptotic behavior that is incorporated in the static response.

The reader might wish to ponder why this term is called permittivity, or, for that matter, the origin of the word "displacement," in these field relationships. Permittivity, tied to the nature of the polar character of electrical fields, is a negpermittivity. A higher permittivity, arising with higher polarization, reduces electric field.

There is one other aspect of permittivity and electrons that needs emphasis in a Hamiltonian and perturbation discussion related to the conditions in time and environment that the mathematical description must account for. It is related to the time scale of rapidity of the interaction. An example is the absorption of light in a semiconductor with an electron moving from the valence band—a bonding-based quasi-continuum—to the conduction band—an antibonding quasi-continuum. The electron occupies states of these quasi-continuums that are defined by the electron's environment of being in the midst of this bonded collective of atoms. This forces its $E(\mathbf{k})$ dispersion, but the electron that transitions has changed states defined by the crystal because of the interaction with the photon, by absorbing the energy and the vanishingly small momentum. The crystal here is only a phase space locale for the states. The photon-electron interaction happens with the electron's free space mass, while the states reflect the effective mass that reflects the $E(\mathbf{k})$ dispersion. This argument also holds for the photon processes where phonons—the quantized atomic vibrations—are also involved as in indirect bandgap materials.

A different—and yet analogous—circumstance is when an electron confined very locally in an impurity that is not shallow—a deep level—has an electron move from one energy state to another energy state, while still remaining confined. An example is a $Au^{2+} \mapsto Au^{3+} + e^{-}$ transition. The electron was in a local state, not spread out over the crystal. It was confined to the atom. The electron mass relevant in the description here is the free space mass, and the permittivity is free space permittivity. On the other hand, if the impurity is a shallow hydrogenic state, that is, a few *meV* separated from other conducting states, with properties very similar to that of the host crystal atoms that gave rise to those conducting states, then the electron really feels its presence in the crystal. It is not that localized, and an effective mass of the crystal, and a permittivity of the crystal, will be a more apt set of parameters. Now, one can imagine potential impurities and defects, where the behavior may very well be in between—neither entirely localized nor entirely delocalized—and one may have to use either an interpolation or a more rigorous description.

This dynamic behavior of permittivity and of electrons, in a large-gap material, is more easily deconvolved. If permittivity changes are arising due to plasmonic response—the response of the charge cloud—then the plasmons can be incorporated into the electron-phonon scattering through an effective treatment of the dynamic screening. In large-gap materials, there is just one of the bands contributing to the conducting carriers that one needs to worry about. This treatment will also have to change when one confines carriers to a plane or to a quantum wire, because carriers are not free to move in all the directions for screening. Graphene is a zero bandgap material. Now this dynamic screening for electron-phonon scattering will become considerably more complicated.

1.9 *Summary*

THIS CHAPTER WAS AN INTRODUCTION to several of the common principles, techniques and approximations that will be employed throughout the text, with an emphasis on their implications, context and physical meaning so that we may employ them with due care and restraint. In this approach, our quantum view, its emergence into the classical view under many of the natural world's sizes—in dimensions as well as the number of participating entities and their interactions—and others such as the statistical and informational views, all have an important role. This last theme—of information—will be deployed in Chapter 2, where we also bring in thoughts from statistical mechanics, which itself is spread throughout the text, toward understanding two of the most important pervasive themes in any study of natural phenomena: those of entropy and energy. The present chapter particularly stressed the quantum underpinnings toward understanding statics and dynamics at the quantum scale, with emphasis on many particles coming together in an ensemble, be it an atom, a molecule or an atomic assembly, and the methods for solving the Hamiltonians; that is, the energetics.

Understanding the energetics of multiparticle systems is essential to developing an understanding of the properties during interactions that define the internal properties and response of a semiconductor. Much skepticism, care and *understanding*—not just grasping the ideas—is important for prudent use, treatment and reaching a result that holds validity over a range of conditions. As an introductory chapter integrating partly the material that students will need to learn and the outlines of the underlying physical principles and techniques that will be employed throughout this text, we started with a discussion of Hamiltonians and Lagrangians as functional tools for unraveling the energetics. We sketched the broader nature of the Hamiltonian description of electrons and atoms in an assembly such as a metal, a semiconductor or a molecule, and then reduced it to the problem of understanding the electrons' interactions. Before embarking on the approximate solution techniques of this problem, we segued into the different perturbation approaches that are sprinkled throughout the text, since problems in general, and certainly the many-body problems, cannot usually be exactly solved. But the solutions can be approached via perturbation techniques. We illustrated the first order perturbation approach and applied it to both a time-dependent and a space-dependent perturbation. The former was useful in showing the Golden rule and the limits of its applicability. The latter was useful in illustrating scattering's quantum-mechanical origin and its classical fitting. Another important approximation technique was the use of the adiabatic or Born-Oppenheimer approximation, and, through it, the approach to situations where fast and slow phenomena interact. The adiabatic approximation is very important to calculating phonon-based; that is, atomic movement-based interactions with those of electrons. The former are slow, and the latter are fast. Energies may exist in multiple modalities—atomic bonding and vibrational and electronic kinetics, for example—and we outlined how a configuration coordinate diagram lets us see the slow and fast together in this energy transformation.

At this point, we returned back to solving the multiple electron problem to bring out the nature of many-body interactions. The

Understanding is an internalization—deeper than *knowing*, *being aware of* or other, similar terms. Approximation and model usage requires understanding and skepticism. In this, it is no different than in the populated world around us, where much marketing, manipulation and myth building abounds. Take history. The New Testament is so different from the Old Testament. In India, a different narration around Rama, a revered godly king, brings out violence and McCarthyism. Passage of time or distant lands is not necessary for creating mythologies. Take a statistic from the Second World War of the 20th century.

	Major event	Innocent deaths
Stalin	Russian famine	$6–7 \times 10^6$
Hitler	Death camps	2.8×10^6
Tôjô	East Indies	$2.4–4.0 \times 10^6$
Churchill	Bengal famine	$2.5–5.5 \times 10^6$
Truman	Atom bomb	$0.13–0.22 \times 10^6$

Stalin cannot be criticized in Russia, neither can Churchill in the West. The Soviet state's industrial and poor folks' transformation, as well as the war's transformational fight, is Stalin's contribution. Churchill's is his steadfastness in the war. Stalin's was the brutal killing of innocents in his home country. Churchill's was a tribal and racist view of freedom—Woodrow Wilson-like—as a white European prerogative. Even de Gaulle of France marched right back into Indochina, culminating in the Vietnam War. Narratives should always be looked at with caution. Stalin and Churchill stood up for their lands, and for that both should be lauded, but not worshipped. Indians still remember that, for the false promise of freedom, nearly 75,000 young men fell even in the First World War and have been forgotten, several during Churchill's Gallipoli folly. No site marks the forced fighters—the unknown soldiers—of the third world, even as a famous Western journalist declares himself and his kin the greatest generation. It turns out the

Hartree approximation is one where only Coulomb interaction is tackled and the electron as a point particle is a secondary thought in what is essentially a classical calculation. That the electron is a quantum particle—a fermion—and therefore requires a different wavefunction under exchange was brought in through the use of the Slater determinant and the evolution to a Hartree-Fock approximation. Exchange interaction is the result of Coulomb interactions between the electrons under the quantum constraints from spin. The Coulomb interaction becomes spin dependent under the constraint that the wavefunction of any pair of electrons must be antisymmetric with respect to any interchange of spatial coordinates and spins. When spins are parallel, the coordinate part must be antisymmetric. So, parallel spin pairing reduces the probability of two electrons of being spatially close, compared to the probability when possessing antiparallel spin. Parallel spin electrons, when more separated in space, have less repulsion and this lowers the energy of electrostatic interaction. Spin and orbit also interact, and this we will look at carefully in our discussion of valence bandstructure as well as defect-mediated point perturbations. In situations where the spin-orbit energetics is important, the velocity, as well as the structure of the wavefunction solution in a crystal assembly, which leads to the description of the motion of electrons on the atomic scale, affects the interaction and the electron g factor. We ignore nuclear spins, since nuclear magnetic moment is small ($\sim 2000\times$ smaller) than that of an electron, and its consequences are through perturbations in semiconductors where spin-dependent transport and other phenomena are important.

The final and very important, particularly so for nanoscale, interaction is that of correlation. An electron does not interact with itself. It only interacts with others. So, accounting for an electron in a Hamiltonian in the middle of other electrons is complex. If we take the electron away, it is a different problem. If we place the electron in, then it is also a different problem, since now the other electrons are responding to the presence of this electron. Mean field, as in the first case, is not representative completely, since the presence of the electron matters in the arrangements of others. Neither is the latter, since what the true energy picture needs is the arrangement of electrons where this electron takes into account exchange and correlation. Two spin-up electrons cannot be present simultaneously in identical space, but electrons with opposite spins can. The spin-up second electron has vanishing presence, while the spin-down electron's presence has been accentuated. The first has a hole, while the second formed a

man was also a sexual predator. This is the difference between "getting it" and "understanding it," where using the learning to solving a general—and not special—problem matters. Use of approximations and models requires tremendous care.

The g factor should be distinguished from the gyromagnetic ratio, which is the ratio of the magnetic moment to the angular momentum. g is dimensionless. The electron has charge and spin, but it is not quite appropriate to view it as an object with literal rotation about an axis. The g factor is the dimensionless number that modifies the gyromagnetic ratio as determined by the classical definition.

Spin-dependent transport is an important subject area for devices and is discussed in depth in S. Tiwari, "Nanoscale device physics: Science and engineering fundamentals," Electroscience 4, Oxford University Press, ISBN 978-0-198-75987-4 (2017).

mound. This configuration also needs to be accounted for, and we outlined how one may do this approximately.

Another important analytic theme related to how these electrons are behaving in the solid is how they self-consistently respond to the created conditions. Electrons screen because of the Coulomb interaction, but under all the rest of the constraints that we just discussed. If the perturbation is static, and the electron population small—non-degenerate—then the Debye-Hückel, or Debye, for short, length scale suffices in how the potential perturbation is screened. If it is large and degeneracy prevails, we observed Thomas-Fermi screening. These will all be mediated by static permittivity. And if it is a rapid perturbation, we must also bring dynamic permittivity: the high-frequency aspects of the electronic or ionic or other polarization responses. And, into this, one must also take into account the nature of the behavior of the electron. Is it feeling this polarization environment or not? So, both the permittivity and the mass must reflect the realities of the dynamic perturbation.

1.10 *Concluding remarks and bibliographic notes*

THIS CHAPTER WAS AN INTRODUCTION to several of the common techniques, principles and approximations that will be employed throughout the text, with an emphasis on their implications, context and physical meaning, so that we may employ them with due care and restraint. Solid state has a longer history and wider context than the subject of semiconductors, and the objective here was to introduce a few of the main techniques and the scope of the nature of the techniques that are particularly important for semiconductors.

Solid state has been the subject of numerous texts. A number of books have been standard bearers; historically, first and before all, are the conceptual and analytic discussions by Ziman. The first[1] is a very readable discussion at the senior undergraduate level, with an emphasis on scattering and transport as well as a semiconductor bent. The second[2], although it has much in common with the first book, has a more diverse treatment toward solid state, with magnetism, ferroelectricity and superconductivity as the ending points. This book is now in its second edition, having been revised in the early 1970s. Both of these books are worth reading so many decades after their writing. Another text, from the same time period—well, a little earlier—is the text by Peierls[3], which too has a treatment of phenomena from electrical and thermal conductivity, working from the behavior of electrons and phonons and ending in

[1] J. M. Ziman, "Electrons and phonons," Oxford (1960)

[2] J. M. Ziman, "Principles of the theory of solids," Cambridge, ISBN 0-521-29733-8 (1964)

[3] R. E. Peierls, "Quantum theory of solids," Oxford, ISBN 19-850781-X (1955)

broader solid-state topics as they were understood during that time period. A more advanced treatment from this period is the book by Pines[4]. This is a mathematically detailed text. One additional book very worthy of note, similar to Pines in its advanced treatment, is by Kittel[5]. Both Pines' and Kittel's first editions appeared in 1963. I reference these books since they have stood the test of time and are worth reading to get a perspective, from quite different ways of looking, by many of the luminaries of the early days of the marriage of quantum and the solid state.

[4] D. Pines, "Elementary excitations in solids," Perseus, ISBN 0-7382-0115-4 (1999)

[5] C. Kittel, "Quantum theory of solids," John Wiley, ISBN 0-471-62412-8 (1987)

A set of solid-state texts that have established themselves as standard texts in the undergraduate and early graduate curricula around the world, from this side of the Atlantic, are those of Kittel[6], whose first edition appeared in 1963, and Ashcroft and Mermin[7], whose first edition appeared in 1976. They are different in style from each other, but both have a very carefully and clearly written exposition.

[6] C. Kittel, "Introduction to solid state physics ," Wiley, ISBN 13 978-0471415268 (2004)

[7] N. Ashcroft and D. Mermin, "Solid state physics," Saunders, ISBN 13 978-0030839931 (2003)

Quantum mechanics is a subject with an even vaster collection of texts. Two that have become standards are one at the introductory level, by Griffiths[8], and one that is a little more advanced (intermediate), by Sakurai[9], both of which have gone through several incarnations. These texts are quite lucid in their exposition of the perturbation theory, the Golden rule, and the Golden rule's limitations. A mathematically sophisticated treatment is in the series of books by Landau and Lifshitz, which all physics students have since they encompass much of the formalism of physics through the 1960s. The volume devoted to quantum mechanics[10] is a translation by J. B. Sykes and J. S. Bell. Any book that Bell spent time translating has to stand head and shoulders above the rest. Elsewhere, Bell also likes the text by Gottfried[11], whose first edition is from 1966 and whose copy at *CERN* Bell found very well worn, and worth discussing in a work entitled *Speakables and unspeakables in quantum mechanics*, a subject that Bell had much to contribute to through his Bell inequalities that are so illuminating.

[8] D. J. Griffiths, "Introduction to quantum mechanics," Pearson, ISBN 0-13-191175-9 (2005)

[9] J. J. Sakurai, "Modern quantum mechanics," Addison-Wesley, ISBN 0-201-53929-2 (1967)

[10] L. D. Landau and E. M. Lifshitz, "Quantum mechanics," Butterworth-Heinemann, ISBN 13 978-0750635394 (2003)

[11] K. S. Gottfried, "Quantum mechanics," ISBN 0-387- 95576-3, Springer (2003)

A book from the early times with an excellent discussion of the finer points embedded within the formulation of quantum mechanics and its application to the description of solids is the book by Slater[12]. Another book mixing solid-state and quantum matters and which is a favorite of mine for its lucidity, a stronger bending toward semiconductors, and restrained and yet thorough discussions is the one by Harrison[13].

[12] J. C. Slater, "Quantum theory of atomic structure," 1, McGraw-Hill (1960)

[13] W. Harrison, "Sold state theory," Dover, ISBN 0-486-63948-7 (1979)

There are a few additional books that are quite representative of the physical intuition necessary in this transition from our observational classical thinking to the reality of the quantum-mechanical.

Dyakanov[14] discusses the different spin-based issues that appear in semiconductors. Spin has consequences through the Pauli principle and through exchange interactions. In semiconductors, the major manifestations include the spin-orbit interaction and the role it plays in optical transitions.

[14] M. I. Dyakanov, "Spin physics in semiconductors," Springer, ISBN 978-3-540-78819-5 (2008)

Hartree, Hartree-Fock and correlations have occupied considerable space in the discussion of this chapter, since these really represent the major collection of ways that we treat a multi-electron assembly, and, as the ensembles get smaller, from nanoscale to a molecule, the consequences as the quantum nature manifests itself in different pronounced ways. A good book for this discussion is the one by Delerue and Lannoo[15]. It discusses the general modeling techniques, their usage in quantum-confined systems and the variety of properties that result. Its early exposition is quite close to the several points we have emphasized, but it goes quite a bit beyond. Another good text for understanding the Hartree-correlation spectrum of subjects is the text by Kohanoff[16]. It develops the subject all the way through to density functional theory and Car-Parrinello techniques that we did not dwell on. Density functional theory will appear in a minor form in the discussion of bandstructures (Chapter 4).

[15] C. Delerue and M. Lannoo, "Nanostructures," Springer, ISBN 3-540-20694-9 (2004)

[16] J. J. Kohanoff, "Electronic structure calculations for solids and molecules," ISBN 13 978-0521815918, Cambridge (2006)

1.11 *Exercises*

1. The Maxwell's equations can be transformed into a simpler group under source-free free space conditions; that is, with $\mathbf{J} = 0$, $\rho = 0$, $\mathbf{D} = \epsilon_0 \boldsymbol{\mathcal{E}}$, $\mathbf{B} = \mu_0 \mathbf{H}$ and $1/c^2 = \mu_0 \epsilon_0$. For this simplified free space source-free form,

$$
\begin{aligned}
\text{General form} &\mapsto \text{Source free and free space form,} \\
\nabla \cdot \mathbf{D} = \rho &\mapsto \nabla \cdot \boldsymbol{\mathcal{E}} = 0, \\
\nabla \cdot \mathbf{B} = 0 &\mapsto \nabla \cdot \mathbf{B} = 0, \\
\nabla \times \boldsymbol{\mathcal{E}} = -\frac{\partial \mathbf{B}}{\partial t} &\mapsto \nabla \times \boldsymbol{\mathcal{E}} = -\frac{\partial \mathbf{B}}{\partial t}, \quad \text{and} \\
\nabla \times \mathbf{H} = \mathbf{J} + \frac{\partial \mathbf{D}}{\partial t} &\mapsto \nabla \times \mathbf{B} = \frac{1}{c^2} \frac{\partial \boldsymbol{\mathcal{E}}}{\partial t}.
\end{aligned}
$$

The Lagrangian function for the free space problem is

$$
\mathscr{L} = \frac{1}{2}\epsilon_0 \mathcal{E}^2 - \frac{1}{2}\mu_0 H^2 = \frac{1}{2}\epsilon_0 \mathcal{E}^2 - \frac{1}{2\mu_0} B^2.
$$

Show that, in the presence of external sources ρ and $\mathbf{J}$, and a generalized medium, the Lagrangian function has the form

$$\mathscr{L} = \frac{1}{2}\epsilon\mathcal{E}^2 - \frac{1}{2}\mu H^2 + \alpha\rho\phi + \beta\mathbf{J}\cdot\mathbf{A}.$$

See Appendix D for a broader discussion of the Lagrangians and variational methods.

Find α and β so that Euler-Lagrange equations reproduce the Maxwell's equations in the presence of sources. **[M]**

2. Consider a Hamiltonian operator $\hat{\mathscr{H}}$ that has discrete eigenvalues. It is also Hermitian, so

$$\langle\psi|\hat{\mathscr{H}}|\phi\rangle = \langle\phi|\hat{\mathscr{H}}|\psi\rangle^*$$

by definition. Show that

- the eigenvalues for this Hamiltonian are real, and that
- the eigenfunctions of $\hat{\mathscr{H}}$ that correspond to different eigenvalues must be orthogonal to each other. **[S]**

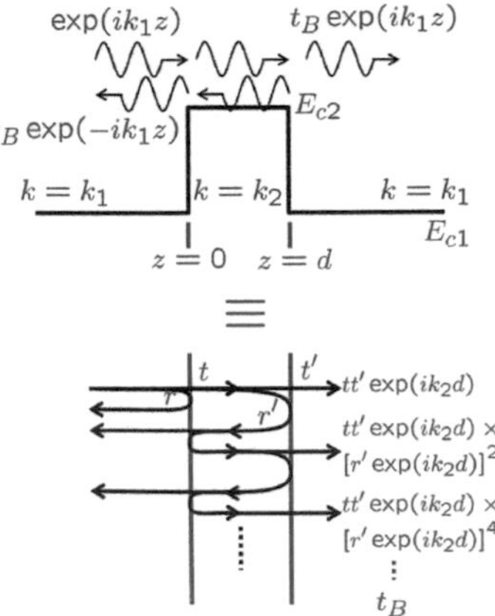

Figure 1.11: Multiple transmissions and reflections at a potential barrier, leading to net transmission and reflection.

3. This problem is to emphasize the power of the Golden rule, and a view of scattering that we mentioned but did not discuss in much detail—one that is particularly apropos of nanoscale devices with a finite and low number of scattering events, and also in mesoscopic transport. We will look at transmission and reflection at a barrier, the working example of which is shown in Figure 1.11, by two methods. We consider just a one-dimensional structure where waves transmit or reflect back, here due to an incident wave, $\exp(ik_iz)$. One can look at the net effect of the transmitted wave $t_B\exp(ik_iz)$ and the reflected wave $r_B\exp(ik_iz)$ as arising from multiple transmissions and reflections as the wave rattles back and forth between the two non-adiabatic discontinuities—the net effect being a convergent series, as shown in the figure. Show that the transmission coefficient $\mathscr{T}_B$ and the reflection coefficient $\mathscr{R}_B$ arising from the barrier can be written as

$$\mathscr{T}_B = |t_B|^2 = \frac{\mathscr{T}^2}{1+R^2-2R\cos(2k_2d)},$$

$$\mathscr{R}_B = |r_B|^2 = \frac{2\mathscr{R}-2\mathscr{R}\cos(2k_2d)}{1+\mathscr{R}^2-2\mathscr{R}\cos(2k_2d)},$$

which add to unity and where $\mathscr{T}$ and $\mathscr{R}$ are the transmission and reflection coefficients for individual step. Now use the Golden rule to calculate $\mathscr{R}_B$ and compare with this result, remarking on the conditions under which the two are in accord. **[S]**

The barrier is a perturbation!

4. For a system of particles of mass m in state ψ, the particle flux (number per unit time per unit perpendicular-to-motion area) is given by

$$\mathbf{S} = \frac{\hbar}{2im}\left(\psi^*\nabla\psi - \psi\nabla\psi^*\right).$$

- Show the validity of this expression, starting from the momentum operator, and
- show that, in the classical limit—a beam of free particles at velocity v—the one-dimensional expression gives

$$S = v \times \text{particle density}. \qquad \textbf{[S]}$$

This is how current (both quantum-mechanical tunneling and normal) can be calculated across a potential barrier. This problem shows that the correspondence principle holds here (quantum-mechanical expressions reducing to classical ones in the asymptotic limit).

5. Following Appendix E, show that, for bosons with $E_k = \hbar^2 k^2/2m$—the classical limit from quantum with $\hbar \rightarrow 1$—that the chemical potential μ is always negative. **[S]**

6. Again, following Appendix E, find the asymptotic behavior of the chemical potential as a function of temperature ($\mu(T)$) and show that the distribution function reduces to the Boltzmann distribution as $T \rightarrow \infty$. **[S]**

7. If the energy of all configurations is the same, show that the entropy as arrived at in Appendix E is equal to the logarithm of the number of configurations. **[S]**

8. Again, following Appendix E, show that a boson gas with n particles in state $|k\rangle$ will have an entropy of

$$S = -k_B \sum_k \left[n_k \ln n_k - (1 + n_k) \ln(1 + n_k) \right],$$

and that, for fermion gas, it will be

$$S = -k_B \sum_k \left[n_k \ln n_k - (1 - n_k) \ln(1 - n_k) \right]. \qquad \textbf{[S]}$$

9. The notions of Appendix E can also be applied to photons as bosons. Does a collection of photons—as a gas—have entropy? In thermodynamic equilibrium, as in blackbody radiation, what is the chemical potential of a photon? **[M]**

10. Let us make some order of magnitude energy estimates based on quasi-classical-quantum fitting to see the adiabatic approximation's use. We explore the situation of the applicability of electrons not undergoing transitions between stationary states. Take a molecular system. Such a system will have three different types of motion: electronic, nuclear vibration and rotation. If a is an interatomic distance, it is a length scale for electron movement, so $E_e \approx \hbar^2/2ma^2$ is an electronic energy scale. Estimate this for $a = 0.1 \quad nm$. The nuclear motion, vibrational, has an energy estimate of $E_{vib} = \hbar\omega_{\mathbf{q}}$. A mass M, moving a distance a at a frequency $\omega_{\mathbf{q}}$, has an energy of $\sim M\omega_{\mathbf{q}}^2 a^2$. Such a motion—of distance a—would remove the atom from the molecule. It is

bond-breaking energy, since chemical bonding is the binding of atoms as molecules through shared electrons. So, it is of the same order as E_e, that is, $E_e \approx M\omega_{\mathbf{q}}^2 a^2$. Hence, $E_{vib} = \hbar\omega_{\mathbf{q}} \approx \sqrt{m/M}E_e$. Rotation, being related to angular momentum, is quantized by the action $\hbar$. So, if I is the inertia, and L the angular momentum, then the rotational energy is $E_r = L^2/I \approx \hbar^2/Ma^2 = (m/M)E_e$. The energies are related as $E_r \approx \kappa^2 E_v \approx \kappa^4 E_e$. Take the example of an N_2 molecule and estimate these energies, that is, the binding energy of electron in the atom, the vibrational excitation energy and the rotational energy separation. **[S]**

11. This problem is an exploration of the Slater determinant in search of its hidden secrets. We have seen that an n-electron Schrödinger equation became separable in position $\mathbf{r}_i$ if the potential energies could also be separated in the spatial coordinates, so, with

$$V(\mathbf{r}_1, \ldots, \mathbf{r}_n) = \sum_{i=1}^{n} V(\mathbf{r}_i),$$

the eigenfunction of Equation 1.8 becomes

$$|\psi(\mathbf{r}_1, \ldots, \mathbf{r}_n)\rangle = \prod_{i=1}^{n} |u_i(\mathbf{r}_i)\rangle, \text{ and } E = \sum_{i=1}^{n} E_i,$$

where $|u_i\rangle$ satisfies the one-independent-electron Schrödinger equation

$$\left[-\frac{\hbar^2}{2m_0}\nabla_{\mathbf{r}_i} + V(\mathbf{r}_i)\right]|u_i\rangle = E_i|u_i\rangle.$$

The Slater determinant incorporates the antisymmetry dictated by Pauli exclusion. This Hamiltonian does not operate on the spin coordinate σ of the electron. So, when we write

$$|u_i(\mathbf{r}_i, \sigma_i)\rangle = |u_{i\sigma}(\boldsymbol{\tau}_i)\rangle = u_i(\mathbf{r}_i)\xi_i(\boldsymbol{\sigma}_i),$$

this Pauli-conditioned form also satisfies the Hamiltonian. This form, where we now have

$$|\psi_\sigma(\boldsymbol{\tau})\rangle = \prod_{i=1}^{N} |u_{i\sigma}(\boldsymbol{\tau}_i)\rangle,$$

is this solution, using the same argument of product in the independent electron approximation. These can be written as determinants of all possible configurations

$$|\Psi\rangle = \mathscr{A} \begin{vmatrix} |u_{1\sigma}(\boldsymbol{\tau}_1)\rangle & \cdots & |u_{1\sigma}(\boldsymbol{\tau}_N)\rangle \\ |u_{2\sigma}(\boldsymbol{\tau}_1)\rangle & \cdots & |u_{2\sigma}(\boldsymbol{\tau}_N)\rangle \\ \vdots & \vdots & \vdots \\ |u_{N\sigma}(\boldsymbol{\tau}_1)\rangle & \cdots & |u_{N\sigma}(\boldsymbol{\tau}_N)\rangle \end{vmatrix}.$$

Show that, with $|u_i\rangle$ orthogonal, a normalized $|\psi\rangle$ requires that $\mathcal{A} = 1/\sqrt{N!}$. **[M]**

12. To understand the implication of spin and Pauli exclusion, another exercise of interest is reworking our interpretation of the singlet and triplet construction for the hydrogen molecule. Let A and B be the atoms, and 1 and 2 the coordinates of the electrons. Using $|u_A\rangle$ and $|u_B\rangle$ as the basis orbital functions for trial functions,

$$|\psi_s\rangle = |u_A(1)\rangle|u_B(2)\rangle + |u_B(1)\rangle|u_A(2)\rangle, \text{ and}$$
$$|\psi_a\rangle = |u_A(1)\rangle|u_B(2)\rangle - |u_B(1)\rangle|u_A(2)\rangle,$$

where the former is symmetric, and the latter, asymmetric. Now consider new wavefunctions incorporating spin, $|u_\alpha\rangle$ and $|u_\beta\rangle$, and show that these may be written in the form

$$|\psi\rangle = [u_A(1)u_B(2) \pm u_B(1)u_A(2)]\,\xi(1,2),$$

with $\xi(1,2)$ as the spin functions of the two electrons. If $\hat{s}_z$ is an operator for the z component of the spin of the electron, then

$$\hat{s}_z|u_\alpha\rangle = \frac{1}{2}|u_\alpha\rangle, \text{ and } \hat{s}_z|u_\beta\rangle = -\frac{1}{2}|u_\beta\rangle;$$

lets us then construct the two-electron spin function $|\xi(1,2)\rangle$. Show that this $|\xi\rangle$ is the eigenfunction of the square of the total spin ($|S|^2$) with the eigenvalues of 0 or 2. These lead to the Heitler-London functions $|\psi_s\rangle$ and $|\psi_a\rangle$ corresponding to the singlet and the triplet states. **[M]**

13. Consider an assembly of atoms, subscripted as A and B, so of only two types. The internuclear repulsion is $\sum_{A<B} Z_A Z_B e^2/|\mathbf{R}_B - \mathbf{R}_A|$. When solving for the eigenenergy E of the electron, we include this repulsion energy even if it has internuclear spatial parameters. It sets the potential energy in the Schrödinger equation for nuclear motion. Take a center-of-mass form Schrödinger formulation for a diatomic molecule at rest. The reduced mass is $\mu = M_A M_B/(M_A + M_B)$, and the equation is

$$\left[-\frac{\hbar^2}{2\mu}\nabla_{\mathbf{r}}^2 + V(R)\right]|\psi\rangle = E|\psi\rangle,$$

where $R = |\mathbf{R}_B - \mathbf{R}_A|$, and V is the potential ("mechanical") from the interatomic forces. Let $V = (1/2)k_s(R - R_0)^2$, where R_0 is the equilibrium value of internuclear distance, and k_s is a force constant. Find the eigenfunctions $|\psi^{\underline{n}}\rangle$ and the eigenenergies E. **[M]**

14. The Schrödinger equation for H_2^+, the simplest molecule (a one-electron, two-nuclei problem) beyond the simplest atom (H) as a one-electron, one-nuclei problem), can be written as

$$\left[-\frac{\hbar^2}{2m_0}\nabla_{\mathbf{r}}^2 - \frac{e^2}{4\pi\epsilon_0}\left(-\frac{1}{r_A} - \frac{1}{r_B} + \frac{1}{R}\right)\right]|\psi(\mathbf{r})\rangle = E(R)|\psi(\mathbf{r})\rangle.$$

Here, r_A and r_B are the distances of the electron from the protons A and B that are separated by R. Using spherical coordinates,

$$\xi = \frac{r_A + r_B}{R}, \text{ with } 1 \leq \xi \leq \infty,$$

$$\eta = \frac{r_A - r_B}{R}, \text{ with } -1 \leq \eta \leq 1,$$

and ϕ is the azimuthal along about the axis, with

$$\psi(\xi,\eta,\phi) = X(\xi)Y(\eta)\Phi(\phi),$$

showing that the equation becomes separable. Find these differential equations for $X(\xi)$, $Y(\eta)$ and $\Phi(\phi)$.

15. An electron is in a one-dimensional system in interaction with two atoms. An approximation of the interaction between the atoms and the potential is

$$V(r-R) = -\beta\delta(r-R),$$

where R is the position of an atom. Using the adiabatic approximation, find

- the transcendental equation that relates k and r where the electronic binding energy is $\hbar^2k^2/2m_0$,

- the potential between the two atoms in and the limit of small r; that is, when

$$V(r \to 0) \approx V(r=0) - \alpha r.$$

Find $V(r=0)$, α, and

- show that the potential between the two atoms at large r is of the form

$$V(r \to \infty) = -\gamma \exp(-2k_\infty r).$$

Determine γ. **[M]**

16. Show that
   - in a system of two identical particles, each of which can be in one of n different quantum states, there are $(1/2)n(n+1)$ symmetric and $(1/2)n(n-1)$ antisymmetric states of the system and that

- if particles have a spin of I, then the ratio of symmetric to antisymmetric spin states is $(I+1):I$. **[S]**

17. Two non-interacting particles have a mass m and exist in a one-dimensional potential well of $V = 0$, a length of $2a$, and infinite barrier regions. Determine
    - the four lowest energy levels of the system and
    - determine the degeneracies of these energies if the two particles are
      - identical, with spin 1/2,
      - not identical, but with spin 1/2, and
      - identical, with spin 1. **[S]**

18. There exist two identical non-interacting particles in an isotropic harmonic potential. Take the three lowest energy levels, and show that
    - the degeneracies are 1, 12 and 39, if the particles are of spin 1/2, and that
    - the degeneracies are 6, 27 and 99 if the particles are of spin 1. **[S]**

19. Radiative transitions from an excited state to the ground state exists with a probability per unit time of γ in a system of atoms. Show that the power spectrum of the radiation is Lorentzian and that the angular frequency width at half amplitude ($\Delta\omega_{1/2}$) is γ. **[S]**

20. A short-range potential scatters particles moving along the z axis. If the wavefunction of the particle at large distances from this potential perturbation is of the form

$$\psi = \exp(ikz) + \frac{1}{r} f(\theta,\varphi) \exp(ikr),$$

then show that the differential scattering cross-section follows as

$$\frac{d\sigma}{d\Omega} = |f(\theta,\varphi)|^2.$$

[S]

2
Entropy, information and energy

JOHN WHEELER EMPLOYED THE PITHY phrase "it from bit" to emphasize the information-theoretic foundation of all things physical. This text employs these information-centric notions as the foundation for quantum and statistical description toward the semiconductor-specific solid state, and this book series emphasizes it as an important theme on equal footing with energy, entropy and others that find a central role in the science and engineering pedagogy. So, we are employing tools from the information-statistical-quantum mechanics arsenal to map the objective description. Entropy as a measure of incompleteness of knowledge ties to energy dissipation whenever this incompleteness of knowledge—entropy—changes. This is nonconservative forces at work. Energy, through the consequence of quantization and statistical distribution and as a constraint in state changes, can now be tied together with entropy in an information-centric approach.

The complement of this phrase "bit from it" also holds as an example of "deep truth."

Quantum-mechanically, a statefunction—in its various complexities, such as whether it includes or does not include the spin of an electron—conveys information embedded in it. Use the Hermitian operator of an observable, and one obtains an eigenvalue; we interpret that as a collapse of the statefunction of the system to an eigenfunction of the system. Unperturbed, the system now stays in this eigenfunction state. But, the spin eigenvalues cannot be found if the spin is not embedded in the statefunction. So, a statefunction in this view is a representation from which information can be extracted—not that different from a statistical probabilistic distribution containing information, but one that can only be accessed by making an observation.

Observation—a measurement—is the act of acquiring information and also of triggering the process that produces the information.

Cause and chance, or determinacy and random walk, as general guiding thoughts, are sprinkled throughout this text. These map to energy, and entropy, respectively. The former has clarity, while the latter not as much.

Semiconductor Physics: Principles, Theory and Nanoscale. Sandip Tiwari.
 Published 2020 by Oxford University Press. DOI: 10.1093/oso/9780198759867.001.0001

A phenomenon in the physical space has now been transformed into an informational space, which is a space containing data in its various forms. Information and observation are fundamentally linked through this action. Acquisition of information requires observation, and, reciprocally, the act of observation leads to information. More than the statefunction, it is the information that is at the heart of this physical-mathematical action to describe "reality." Information, or the lack of it, as with our statistical description of the distribution of the properties of an object, is interpretable through statistical mechanics.

Traditionally, this information foundation is not how one brings together the quantum-mechanical and statistical-mechanical approaches in the study of physical phenomena. But it is intimately linked, as many—including Wheeler, Brillouin, Shannon, Landauer, Bennett, Feynman, Solomonoff, Kolmogorov and even Maxwell, through his demon, and Boltzmann, through his debates with himself and others, like Mach, Ostwald and Helm, who doubted him—have posited in various forms over a century.

We emphasize here some of the deeper connections between probabilities used as the statistical tool when there are unknowns by discussing the classical notions from an information perspective. We show that the often-used behavioral relationships have an information-centric foundation by deploying the less used form of entropy and information: Fisher entropy and information.

Entropy is the variable introduced for circumstances involving unknowns. On equal footing—if not higher—is the energy variable, upon which we place constraints in sum and in exchange. Energy is much more clearly understood. Free energy is exchanged and remains a constant over the forms it can be exchanged. Kinetic—the motional form—and potential forms that arise as electromagnetic and gravitational in the quantum-constrained matter form—are allowed to interchange in the low energy conditions where non-relativistic conditions prevail. This chapter brings forth the intimate links between these two very powerful ideas through their intimate links to information.

2.1 *Entropy*

THE CONVENTIONAL INTRODUCTORY DISCUSSIONS OF THERMODYNAMICS, statistical mechanics, the states of matter, and information mechanics can be viewed quite generally from approaches based on probabilities—the mathematical approach to analyzing chance in the Bernoulli sense.

Gerolamo Cardano, a Renaissance-period credited with intellectual and gambler, is often credited for measuring and analyzing random events to understand "good luck and fair wager." His book entitled *Book on games of chance* was one of many, most of which were lost to history, since this was the Church- and Aristotle-dominated period. Two of his children were put to death by the church, while the youngest survived as a freelance torturer for the Inquisition. Cardono's formula calculates correct odds. Galileo and Kepler—Renaissance-period scientists—also contributed. In the mid-1600s, Antoine Gombaud, a French writer, posed a dice problem to Blaise Pascal, who, with Pierre de Fermat, laid the foundations for exploring randomness. The Bernoulli family—royalty in the mathematical world with thirteen celebrated mathematicians—had the biggest early impact. The family had moved to Basel in Switzerland because of the Spanish Fury—another of the never-ending religious shenanigans over the ages—from the Netherlands. Jacob (a.k.a. Jacques), Jean, Jean's son, Daniel, and Daniel's son, Nicolaus, are the most known. As with all royalty, with their power-centric desires, there was serious rivalry—in this case, between Jacob and Jean. Jacob, besides probability, is to be credited for the law of large numbers, the number e, which we call Euler's constant, and the logarithmic spiral—a beautiful spiral with increasing radius when followed clockwise—about which he said "Eadem mutato resurgo," translated as "although changed, I rise again the same." Daniel Bernoulli queried why people prefer low-risk bets rather than the more profitable ones and posited that it is the expected utility rather than the expected payoff—the worth of money depends on how much one has—that drives people. This is opposite to the traditional view. Nicolaus is known for the Saint Petersburg paradox. A player bets on the number of tosses required before a coin first turns up heads. The player starts with a fixed amount and then receives 2^n units if the head appears on the nth toss. The expectation value of gains is $\frac{1}{2}2 + \frac{1}{4}4 + \cdots = 1 + 1 + \cdots$;

Entropy, viewed from a probability perspective, is the amount of uncertainty of a distribution. Reworded, this uncertainty in the expected outcome of an observational experiment, when many outcomes are possible, is an "entropy." In this chapter, you will see the view that the entropy of a probability distribution may be interpreted as a measure of uncertainty and as a measure of information. The amount of information that one obtains upon an observation is the amount of uncertainty that existed about the outcome of the observation before the observation was made.

that is, infinite in the limit. A short cryptic answer to this paradox is in the concluding section. Probability's serious development happened in Russia and Eastern Europe in parallel with quantum mechanics' in the west. Probabilities and their abstractness entangled with reality attract analytic minds. Richard Feynman went through a period with Vegas interests. Nick the Greek, a successful gambler there, explained to Feynman how he would win against unfavorable odds: "I don't bet on the table. I bet with people around the table who have prejudices: superstitious ideas about luck numbers."

Entropy is usually introduced in physics studies first through thermodynamics ($dS = \delta Q_{rev}/T$, where δQ_{rev} is an infinitesimally small amount of heat exchanged across the system boundary reversibly), following Claussius. Heat capacities, latent heats, et cetera all follow through this, as does the famous $dU = \delta W + \delta Q$ in an infinitesimal process in a closed system. Pressure, volume, temperature, entropy, et cetera are the parameters of interest.

Boltzmann, who introduced the statistical approach, explored entropy through a microscopic viewpoint by treating the collection as a statistical ensemble representable through a distribution function for the large number of particles in it. All different possibilities for this particle collective—microstates—are postulated to be equally likely. The energies—or velocities—of the particles, continuous in the classical description, now enter the description as constraint, along with the number of particles that are being described. Boltzmann entropy is written as

$$S = k_B \ln \Omega, \tag{2.1}$$

where Ω is the number of accessible microstates. Note that $1/\Omega$ is the probability of a system being in a specific microstate where all the microstates are equally likely.

k_B is a constant chosen to fit with the temperature definition through the connection between entropy and reversible heat exchange. Boltzmann's grave in Vienna's central cemetery—the Zentralfriedhof is still active—has $S = k \ln W$ as its epitaph. Vienna has its priorities right. His grave along the main avenue with other stalwarts such as Brahms, Beethoven, Schubert, the Strausses and Schoenberg (Mozart is a maybe, with a guess for a grave that was moved), among others. It is a resounding place, with Mozart accompanying from Sankt Marxer Friedhof about four miles away in a region of diffuse boundaries where science and music find common ground in the high human pursuits. For those who love classical music, being in Vienna, Salzburg, München, Berlin, Paris, Bratislava, Budapest or Prague, almost any major city in Europe, is heaven. And these cities also appreciate their philosophers, scientists and engineers.

Later on, in the Planck view of radiation and of the quantum, these became discrete. One could now view the microstates—possible configurations each particle is in—of the distribution under the energy constraint and the distribution number. Through Boltzmann entropy, one can again relate macroscopic properties of classical particles such as inert gas atoms in thermal equilibrium. The Boltzmann H-factor, as well as his namesake equation, describe the propensity through an evolution of the distribution. The entropy of an assembly is related to the number of microstates of the assembly that befit the constraint of energy, volume and number of particles. This description brings together thermodynamics and statistical mechanics via Boltzmann and Gibbs.

Entropy increase in spontaneous changes is now interpretable as a tendency toward disorder rather than order, because there are more ways to disorder than order. Entropy here is a classical notional measure and the description of the state of randomness—lack of information—for a collection of particles. The Bose-Einstein and Fermi-Dirac distributions, among others, arise in quantum constraints. The von Neumann entropy,

$$S = -\text{Tr}(\rho \ln \rho), \tag{2.2}$$

where ρ is the density matrix of the quantum system, is an extension of the classical description to quantum mechanics.

For bits in a stream, electrical engineers reach out to Shannon entropy,

$$H = \sum_{i=1}^{N} \mathfrak{p}_i \log_2 \mathfrak{p}_i, \tag{2.3}$$

(where $\mathfrak{p}_i$ is the probability of the occurrence of the ith event), which can also be written in a continuous form. Shannon entropy is a measure of order—a negentropy—in the sense that it captures the separation of the arrangement of bits from randomness. H can be, but is not always, the equivalent of Boltzmann thermodynamic entropy.

There are additional informational entropies, particularly important among which are algorithmic entropy as well as an entropy related to entanglement in the quantum systems that are also important in modern science and engineering studies. Since absence of information exists in a multitude of ways, there are plenty of ways that one can partially identify and quantify measures related to this absence. These are all entropies. And there are many more of these, and probably more to come.

This also implies that higher temperatures are easier to generate than lower temperatures.

The axiomatic foundations to probability (see Appendix B of S. Tiwari, "Nanoscale device physics: Science and engineering fundamentals," Electroscience 4, Oxford University Press, ISBN 978-0-198-75987-4 (2017)) are due to Andrey Kolmogorov and were developed at around the same time as those of quantum mechanics. Probability is a measure of uncertainty. When new information becomes available, and one takes it into account, our belief of the possibilities changes. This is a subjective view, as seen through Bayes, and with more rigor, following Laplace, for hidden variables. In repeating experiments, one finds frequencies of occurrences, and one may find distributions, and measures of confidence, in them. This is the frequentist view. Both views have uncertainty in information embedded in them. But the subjective approach depends on incorporating new information into change expectations, and the frequentist approach depends on randomized experiments to probe the statistics and confidence in outcomes. Uniform priors and randomization are the starting points of these viewpoints. Schism between these two schools of thought is natural. Harold Jeffreys and Ronald Fisher spearheaded these two viewpoints—rather vehemently (statistics for blood!)—at Cambridge from the 1930s to the 1950s. While there will be discrepancies, given the lack of all information between the two doctrines, both are useful, and I view these differences as examples of deep truth and complementarity.

2.2 *Fisher entropy*

WE WILL EMPLOY THE INFORMATION-CENTRIC FISHER ENTROPY to illustrate it as a powerful tool for analyzing and quantifying physical phenomena that is on par with the more commonly employed Boltzmann entropy, which predates Fisher entropy, and Shannon entropy, which follows Fisher entropy, and their measures. Fisher entropy helps us emphasize a variety of important consequences that one traditionally derives in elementary statistical mechanics and thermodynamics probings.

An amusing set of restatements of the laws of thermodynamics from *The American Scientist* are as follows:
1st law: You can't win, you can only break even.
2nd law: You can only break even at absolute zero.
3rd law: You cannot reach absolute zero.

Let there be N observed values of the parameter y. Vectorially, $\mathbf{y} = y_1, \ldots, y_N$. Let ϑ be a parameter of unknown but definite value that parameterizes the data. The variance σ^2 or the standard deviation σ are examples of such a parameter for a normal distribution. With this parameter, the data then obeys

$$\mathbf{y} = \vartheta + \mathbf{x}, \tag{2.4}$$

The Fisher view is a frequentist view—to be distinguished from the Bayesian view—of statistical observation and interpretation.

where $\mathbf{x} = x_1, \ldots, x_N$ is the noise of the data. The observed data may be used to form an estimate of ϑ. This ϑ is a function that is optimal for all the data. As an illustration, if one is interested in a mean over the data, that is, $N^{-1}\sum_i^N y_i$, then using the optimal function $\boldsymbol{\vartheta}(\mathbf{y})$—an estimator—leads to a mean that is a superior estimate of the parameter ϑ—a function such as the mean—than any of the data of observables. Here, $\mathbf{x}$ again is the noise intrinsic to the parameter ϑ.

This system is closed, $\mathbf{y}$, ϑ and $\mathbf{x}$ characterize it and it is closed in the following sense.

In the quantum fluctuation problem, ϑ may be an ideal position, and $\mathbf{x}$ the quantum fluctuation. Quantum fluctuation is a form of noise—random chance in action.

If $q_n(\mathbf{x})$, where $n = 1, \ldots, N$ are the canonic variables of N different phenomena, then the Lagrangian $\mathscr{L}$ leads to one differential equation for each phenomenon, and $q_n(\mathbf{x})$ contributes to the total Lagrangian only through one term, which is a function of the two canonic variables $q_n(\mathbf{x})$ and $q_n'(\mathbf{x})$. This solution for phenomenon n is independent of the amplitudes $q_m(\mathbf{x})$ of some other phenomenon m. Because the total Lagrangian is additive, this same solution arises for the different amplitudes of different phenomena. This is isolation in the thermodynamic and statistical sense.

As another illustration, in a Gaussian distribution, σ is the parameter ϑ, $\mathbf{y}$ is the set of observations, and $\mathbf{x}$ is the canonical variable of the system.

The optimal estimator must be unbiased. So,

$$\langle \boldsymbol{\vartheta}(\mathbf{y}) - \vartheta \rangle = \int \left[\boldsymbol{\vartheta}(\mathbf{y}) - \vartheta\right] \mathfrak{p}(\mathbf{y}|\vartheta) d\mathbf{y} = 0. \tag{2.5}$$

Requiring that it be unbiased assures that it applies to all possible phenomena. No favorite has been chosen. There is no prejudice, so one ends with an unbiased estimation. Bias overfits and will lead to error, so being unbiased minimizes error.

Here, $\mathfrak{p}(\mathbf{y}|\vartheta)$ is the probability of observing $\mathbf{y}$, knowing the parameter value ϑ. It describes the fluctuations in $\mathbf{y}$, and the likelihood of any $\mathbf{y}$. Equation 2.5, differentiated w.r.t. ϑ, leads to

$$\int (\boldsymbol{\vartheta} - \vartheta) \frac{\partial \mathfrak{p}}{\partial \vartheta}\, d\mathbf{y} - \int \mathfrak{p}\, d\mathbf{y} = 0, \tag{2.6}$$

where $(\mathbf{y}|\vartheta)$—the variable upon which the probability depends—has not been explicitly written. Since $\partial\mathfrak{p}/\partial\vartheta = \mathfrak{p}\,\partial \ln \mathfrak{p}/\partial\vartheta$, and probability is normalized, we have

$$\int \mathfrak{p} \frac{\partial \ln \mathfrak{p}}{\partial \vartheta} (\boldsymbol{\vartheta} - \vartheta)\, d\mathbf{y} = 1$$

$$\therefore \quad \int \left[\frac{\partial \ln \mathfrak{p}}{\partial \vartheta}\sqrt{\mathfrak{p}}\right] (\boldsymbol{\vartheta} - \vartheta)\sqrt{\mathfrak{p}}\, d\mathbf{y} = 1. \tag{2.7}$$

We now apply Schwarz inequality, leading to

$$\left[\int \left(\frac{\partial \ln \mathfrak{p}}{\partial \vartheta}\right)^2 \mathfrak{p}\, d\mathbf{y}\right] \times \left[\int (\boldsymbol{\vartheta} - \vartheta)^2 \mathfrak{p}\, d\mathbf{y}\right] \geq 1. \tag{2.8}$$

Schwarz inequality states that if $\psi_1(x)$ and $\psi_2(x)$ are two functions that are integrable over the interval (a, b), then

$$|\langle\psi_1|\psi_2\rangle|^2 \leq \langle\psi_1|\psi_1\rangle\langle\psi_2|\psi_2\rangle.$$

The norm of the inner product in the vector space is at most as large as the product of the norms of the vectors. The equality happens when the vectors are linearly dependent on each other. We first encountered this in S. Tiwari, "Quantum, statistical and information mechanics: A unified introduction," Electroscience 1, Oxford University Press, ISBN 978-0-198-75985-0 (forthcoming) in the discussion of Heisenberg's uncertainty principle.

The factor on the left is the Fisher information $I \equiv I(\vartheta)$ (this can also be written in summational form for discrete data) writable in the form

$$I = \int \mathfrak{p} \left(\frac{\partial \ln \mathfrak{p}}{\partial \vartheta} \right)^2 d\mathbf{y} \tag{2.9}$$

where $\mathfrak{p} \equiv \mathfrak{p}(\mathbf{y}|\vartheta)$. If one writes probabilities in terms of amplitudes, so $\mathfrak{p}(y) = \mathfrak{q}^2(y)$, then Fisher information can also be written as

$$I = 4 \int \left(\frac{\partial \mathfrak{q}(y)}{\partial y} \right)^2 dy = 4 \int \mathfrak{q}'^2(y)\, dy, \tag{2.10}$$

a form that does away with the probability divisor and shows that I is a measure of the squared gradient content of amplitude. This is also the reason why Lagrangians appear as squared gradients. The Fisher information can be written in terms of expectations:

$$I = \langle (\partial \ln \mathfrak{p} / \partial \vartheta)^2 \rangle. \tag{2.11}$$

The second factor in the Equation 2.8 can be seen to be the mean square error

$$\int (\boldsymbol{\vartheta} - \vartheta)^2 \mathfrak{p}\, d\mathbf{y} = \langle (\boldsymbol{\vartheta} - \vartheta)^2 \rangle = \varepsilon^2. \tag{2.12}$$

The probability of likelihood $\mathfrak{p}(\mathbf{y}|\vartheta)$ includes all the fluctuations that arise internally, such as uncertainty, and those that arise externally, such as those due to particle exchange, heat exchange, et cetera. And we have now derived, because of Equation 2.8, that the best possible estimator $\boldsymbol{\vartheta}(\mathbf{y})$ has the mean square error $\varepsilon^2 = 1/I$. This estimator is the most efficient estimator so long as all estimators are unbiased, that is, $\langle \boldsymbol{\vartheta}(\mathbf{y}) \rangle = \vartheta$. In any estimation,

$$\varepsilon^2 I \geq 1 \tag{2.13}$$

is the Cramer-Rao inequality. The relationship expresses the reciprocity between mean square error and Fisher information in the observed data.

Fisher information is a measure of the ability to estimate a parameter that characterizes a statistical distribution such as of the observations made. It also measures the state of disorder of a system or a phenomenon. It is in this latter sense that it is a measure of entropy. Fisher information is thus a measure of indeterminacy.

Fisher information as an entropy or a measure of disorder can be understood by considering the implication of the derivative of the probability distribution function. High disorder implies lack of predictability. In such a condition, the probability distribution function is more uniform over a range of the values of the measurement y of the system. This probability distribution function has a small

In the form of the relationship of I as a functional of $\mathfrak{p}$, one can see parallels to the Boltzmann H-factor. For a collection of independent classical particles, Boltzmann's H-factor is

$$H(t) = \int_0^\infty f(E,t) \left[\ln \left(\frac{f(E,t)}{E^{1/2}} \right) - 1 \right] dE,$$

where $f(E,t)$ is the particle energy distribution function in time. In an isolated collection—inert gas molecules, for example, as an approximation—this H-factor is at a minimum when the particles obey a Maxwell Boltzmann distribution, and that of any other distribution with the same total kinetic energy will be higher. If collisions are allowed, any starting particle distribution will asymptotically approach the minimum H and a Maxwell-Boltzmann distribution. Shannon's H—the averaged information content with its negative sign—has its antecedents in the uncertainty that is represented in the discrete counterpart of Boltzmann's H-factor. Fisher information, like Shannon information, is a negentropy, that is, a measure of how far away one is from the randomness of the restricted informational space under consideration.

gradient, it is broader and smoother and its Fisher information is small. When a system has a preference for specific values of a measurement ($\mathbf{y}$), then around these values the derivative is large. This is a system of lower disorder, and the Fisher information is larger. This correspondence to order-disorder is why the Fisher information measure is also a Fisher entropy measure. It also follows a variety of properties, such as the monotonic increase in time of the Boltzmann entropy for a closed system.

Fisher information also provides a deeper understanding of natural phenomena. We illustrate this through multiple examples that are central to many of the discussions of semiconductors and their usage.

A measure of information is needed in the argument normally forwarded to place a bound on the speed of light as a bound on energy propagation. Central to the argument is how this Fisher information behaves temporally. Consider a distribution of particles that are highly localized in position $\mathbf{y}$ because of constraints placed, for example, through no particle-exchange boundaries. The parameter for mean position (ϑ) will now be closer to the observed $\mathbf{y}$ then when the boundaries are moved farther away in time. With the movement of the boundary, the error has increased ($\delta\varepsilon^2(t) \geq 0$), and therefore the Fisher information must follow

$$\delta I(t) \leq 0. \tag{2.14}$$

The change in Fisher information is negative in time. Increasing disorder occurs with lowering of the Fisher information. This is in the same form as the second law, which, written in the Boltzmann H-form for negentropy also states

$$\delta H(t) \leq 0, \text{ with } H = \int \mathfrak{p}(\mathbf{y})\, d\mathbf{y}. \tag{2.15}$$

This H too is another form of smoothness for the probability $\mathfrak{p}(\mathbf{y})$. For a Gaussian $\mathfrak{p}(\mathbf{y})$, this H equals $\ln(1/\sigma)$ plus a constant and decreases logarithmically as σ increases. Fisher information, on the other hand, decreases directly for such a Gaussian probability distribution. While Shannon entropy is a negentropy that deals with a stream of 1s and 0s, and their distributions, Fisher information tackles—because of this relationship to how the probability changes with the $\mathbf{y}$ even for a 1-and-0s stream—the relationships between them over a distribution. So, Fisher information is not just for Boltzmann-type conditions or the randomness of each bit of a bitstream—it has significant relevance for understanding information content in patterns and so for understanding natural and artificial neural networks.

The immutability of the speed of light comes from special relativity. Light is a noun that we associate with electromagnetic waves or photons, its massless particle. But the speed of light is not just the speed of light; it is also the speed of light, besides other things—for example, for gravitons, another massless particle, and gravitational waves. So, Maxwell's equations is not the only place with "speed of light"'s implication. Special relativity is about space and time unified into space-time, which, together with space and time, guides nature.

When there exist no physical constraints, I will decrease until it reaches its asymptotic limit of zero. This is a condition where $\mathfrak{p}(\mathbf{y}|\vartheta)$ is a constant. If there are constraints on the parameter $\mathbf{y}$, then the Fisher information reaches a finite minimum that arises in this constraint. So,

$$I = \int \frac{(\partial \mathfrak{p}(\mathbf{y})/\partial \mathbf{y})^2}{\mathfrak{p}(\mathbf{y})}\, d\mathbf{y} = \mathit{minimum}. \tag{2.16}$$

This is the equilibrium state. It is a stationary state where the probability distribution is not changing anymore. The time to stationary state may be finite, as this speaks to how the disorder appears under the interactions—the fluctuations—causing it. A similar form for the Boltzmann negentropy is

$$H = \int \mathfrak{p}(\mathbf{y}) \ln \mathfrak{p}(\mathbf{y})\, d\mathbf{y} = \mathit{minimum}, \tag{2.17}$$

which is a statement of the second law of thermodynamics.

Both H and I have been written to be subject to this minimization in equilibrium. In the Boltzmann H theorem, applicable to thermodynamic classical situations, the stationary state is reached through a path under the constraint of the negentropy equation as time evolves. Only those probabilities that conform to Equation 2.15 are allowed. This is the statement of the Boltzmann H theorem. In quantum-mechanical situations, with real potential that is a constant of time, the equilibrium probability is the same as the initial condition probability. This is a non-classical situation where the Boltzmann H theorem does not apply and the minimum Boltzmann negentropy does not necessarily apply. Fisher information minimization leads to an appropriate measure of disorder in these conditions. Fisher information from independent system parameters also adds, just as Boltzmann entropy and Shannon entropy do. This follows directly from the definitions, because of statistical independence.

Shannon entropy, written without losing generality as $H = -\int \mathfrak{p}(y) \ln \mathfrak{p}(y)\, dy$, is a measure of the probability distribution $\mathfrak{p}(y)$'s smoothness, but it is a global measure. It does give information, but Fisher entropy, as a functional based on a derivative, is a local measure. So, when minimization constraints are applied, while the Shannon measure leads to just exponential solutions and algebraic equations, the Fisher form leads to second order differential equations. Most of nature's equations—Maxwell, Schrödinger, Fokker-Planck and others that we employ in this text—are, more often than not, of the second order. We will now show that the Fisher form's local emphasis is of more natural importance.

The separation between two different probability distribution functions, say $\mathfrak{p}(y)$ and $\mathfrak{h}(y)$ (a hypothesis), is measured by the functional

$$KL(\mathfrak{p},\mathfrak{h}) = \int \mathfrak{p}(y) \ln \frac{\mathfrak{p}(y)}{\mathfrak{h}(y)}\, dy, \tag{2.18}$$

the Kullback-Leibler distance or entropy. This is a cross-entropy or a relative entropy (of $\mathfrak{p}$ w.r.t. $\mathfrak{h}$). If $\mathfrak{h}$ is a constant, KL reduces to H, and if $\mathfrak{p}(y) = \mathfrak{h}(y)$, it vanishes. When the distribution is multidimensional, the Kullback-Leibler measure is a mutual information.

2.3 *Principle of minimum negentropy or maximum entropy*

To ESTIMATE AN UNKNOWN PROBABILITY $\mathfrak{p}(y)$ in the presence of incompleteness of information, one often uses the principle of maximum entropy or, equivalently, minimum negentropy. The Kullback-Leibler distance (or entropy) is a measure of the separation between the unknown probability $\mathfrak{p}(y)$ and a hypothesis $\mathfrak{h}(y)$, which must be minimized under the constraints. So, under general constraint conditions,

$$\int \mathfrak{p}(y) \ln \frac{\mathfrak{p}(y)}{\mathfrak{h}(y)}\, dy + \sum_{i=1}^{N} \lambda_i \left[\int \mathfrak{p}(y) k_i(y)\, dy - K_i \right] = \mathit{minimum}. \tag{2.19}$$

Here, K_i represents data, and k_i a constraint kernel (velocity, position, kinetic energy, etc.) that are known through an observation and the objective representation. The solution of $\mathfrak{p}(y)$ here is *maximally probable.*

Take an example of the kinetic energy T as the one and only constraint. This forces

$$\int \mathfrak{p}(y) \ln \frac{\mathfrak{p}(y)}{\mathfrak{h}(y)}\, dy + \lambda \int \mathfrak{p}(y) T(y)\, dy = \mathit{minimum}. \tag{2.20}$$

The Lagrangian is

$$\mathscr{L} = \mathfrak{p} \ln \frac{\mathfrak{p}(y)}{\mathfrak{h}(y)} + \lambda \mathfrak{p}(y) T(y). \tag{2.21}$$

The Euler-Lagrange equation then implies that either $\partial \mathscr{L}/\partial \mathfrak{p} = 0$ or

$$1 + \ln \mathfrak{p} - \ln \mathfrak{h} + \lambda T = 0, \tag{2.22}$$

whose "solution" is

$$\tilde{\mathfrak{p}}(y) = \mathfrak{h}(y) \exp[-1 - \lambda T(y)]. \tag{2.23}$$

While a simple exponential solution can exist in specific circumstances, in general, the solution does not have to be this simple form. Quantum-mechanical problems, for example, a particle in a potential well, or the two-slit experiment, will have a complexity. Constructive and destructive interferences will show up in this solution form. Even in simpler classical problems, such as of the statistical distribution of kinetic energy or the velocity of particles, as in the Maxwell-Boltzmann problem, there will be interesting complexities related to the class of problem. ϑ, as the maximal estimation, will then depend on what the distribution is for: a single particle or a multiparticle unit. So, the simple exponential is the classic Boltzmann single particle distribution, but multiparticle constraints will give rise to different and more complex forms, as we see next, along with other forays that directly follow from the Fisher information narrative.

2.4 *Examples of Fisher information applied to particles*

WE NOW CONSIDER A FEW EXAMPLES of the application of the Fisher information for particles. To introduce constraint into Equation 2.16, we employ a linear Lagrangian constraint. For particles considered in conditions where potential doesn't vary with time, the energy constraint arises through kinetic energy. So, the Lagrangian form for the problem is the general equation

$$\int \frac{[\partial \mathfrak{p}(y)/\partial y]^2}{\mathfrak{p}(y)}\, dy + \lambda \int T(y)\mathfrak{p}(y)\, dy = minimum$$

$$\text{or} \quad \int \frac{[\partial \mathfrak{p}(y)/\partial y]^2}{\mathfrak{p}(y)}\, dy + \lambda \langle T(y) \rangle = minimum. \qquad (2.24)$$

If y is velocity, then $T(y) = (1/2)my^2$. If the average kinetic energy is known, this equation will solve for λ, using the Lagrangian approach. The normalization of the probability function can also be explicitly incorporated. If this average kinetic energy is not known, then λ is a factor (negative) that needs to be properly imposed. The former is a classical problem; the latter is a uncertainty-constrained quantum problem. When this minimization equation is multidimensional, then the solution follows from separation, using the marginalization of the kinetic energy in the independent dimensions. So, now let us look at the quantum-constrained conditions first, and the classical ones second.

2.4.1 *One-dimensional Schrödinger equation*

CONSIDER A PARTICLE of mass m in a time-invariant potential of $V(y)$, where y is the position. One measurement is made for the position y in order to estimate the particle's mean position ϑ. Let $E(y)$ be the total energy, and $V(y)$ be the potential energy, so that $T(y) = E(y) - V(y)$:

$$\langle T(y)\rangle = \int \mathfrak{p}(y)[E(y) - V(y)]\,dy = \int \mathfrak{p}(y)[E - V(y)]\,dy, \qquad (2.25)$$

where $E = \int E(y)\mathfrak{p}(y)\,dy$, together with the normalization constraint $\mathfrak{p}(y)dy = 1$. Equation 2.24 now forces

$$\int \frac{\partial \mathfrak{p}(y)/\partial y}{\mathfrak{p}(y)}\,dy + \lambda \int [E - V(y)]\mathfrak{p}(y)\,dy = \mathit{minimum}. \qquad (2.26)$$

The Euler-Lagrange equation for this variational problem is

$$\frac{d}{dy}\left(\frac{\partial \mathscr{L}}{\partial \mathfrak{p}'}\right) = \frac{\partial \mathscr{L}}{\partial \mathfrak{p}}, \text{ where } \mathfrak{p}' = \frac{\partial \mathfrak{p}(y)}{\partial y}, \text{ and}$$

$$\mathscr{L} = \frac{[\partial \mathfrak{p}(y)/\partial y]^2}{\mathfrak{p}} + \lambda[E - V(y)]\mathfrak{p}, \qquad (2.27)$$

leading to

$$\frac{\partial^2 \mathfrak{p}(y)}{\partial y^2} + \lambda \mathfrak{q}(y)[E - V(y)] = 0, \qquad (2.28)$$

where $\mathfrak{p}(y) = \mathfrak{q}^2(y)$. This equation is of the Schrödinger form. We do not know the probability density function, so the expectation for kinetic energy is also not known. The parameter λ—negative—therefore needs to be imposed externally. If we choose $\lambda = -2m/\hbar^2$, we have

$$-\frac{\hbar^2}{2m}\frac{\partial^2}{dy^2}\psi(y) - V(y)\psi(y) = E\psi(y), \qquad (2.29)$$

where we have replaced $\mathfrak{q}(y) = \psi(y)$, consistent with the $\mathfrak{p}(y) = \mathfrak{q}^2(y)$ requirement as probability amplitude and the meaning we assign to the wavefunction. This is the time-independent, one-dimensional form of the Schrödinger equation.

A generalization of this using marginals of the kinetic energy term is the three-dimensional form. $\mathfrak{q}(y)$ has taken the meaning of probability amplitude in a purely real form. The complex form also can be derived using two variables: y and an internal state variable that takes on one of two possibilities, that is, $\mathfrak{p}(y, i) = \mathfrak{p}_i(y)$, where $i = 1, 2$ denotes the joint probability of y and i with $\mathfrak{p}(y) = \mathfrak{p}_1(y) + \mathfrak{p}_2(y)$.

It is always insightful when one can arrive at relationships through multiple paths. Freeman Dyson described, in an article in *The American Journal of Physics* in 1990, his memory of a proof by Richard Feynman of two of Maxwell's equations ($\nabla \cdot \mathbf{H} = 0$, and $\nabla \times \boldsymbol{\mathcal{E}} = -\partial \mathbf{H}/\partial t$), starting from the use of the quantum commutation relationship. This can be achieved through either the Lagrangian or the Newtonian approach. It is an illustration of the properties of acceleration-independent force. What is interesting, and this tends to be true and gives insights well beyond when one encounters such situations, is that it also speaks to relativistic issues. These equations happen to be compatible with both Galilean and Lorentz invariance—allowing the existence of vector and scalar potentials—but the other two do not. Relativity has raised its head.

2.4.2 *Maxwell-Boltzmann distribution*

OUR SECOND ILLUSTRATION OF FISHER ENTROPY is the Maxwell-Boltzmann distribution—the distribution function in thermal equilibrium of classical non-interacting particles in the absence of any potential—so, an ensemble of inert particles of a Maxwell-Boltzmann gas. Let ϑ be the parameter for the unknown. We probe for the velocity distribution, so ϑ may appear as root mean square velocity in the form that we teach.

Let v_x, v_y and v_z be the three Cartesian components of the velocity of a randomly chosen particle's velocity. Equipartition of energy says that each degree of motional freedom is of identical magnitude in an equilibrium state. Each x-, y- and z-directed motion has the same energy that we choose to be $k_BT/2$. We have

T subscripted or with a parameter identifies the kinetic energy. T stand-alone is temperature. This should be clear from the context.

$$\langle T_z \rangle = \langle T_y \rangle = \langle T_x \rangle = \int \frac{1}{2} m v_x^2 \mathfrak{p}(v_x)\, dv_x = \frac{1}{2} k_B T. \tag{2.30}$$

$\mathfrak{p}()$ identifies the equilibrium state. This is a Lagrangian constraint of equality. λ needs to be solved for. Since the three independent motional directions are independent, their marginal constraints are separable, and one needs to solve only the one-dimensional version of the problem. The probability distribution is sought; its expectation is not known and is therefore minimized. We express this as

$$\int \frac{[\partial \mathfrak{p}(v_x)/\partial v_x]^2}{\mathfrak{p}(v_x)}\, dv_x + \lambda_1 \left[\int \frac{1}{2} m v_x^2 \mathfrak{p}(v_x)\, dv_x - \frac{1}{2} k_B T \right] + \lambda_2 \left[\int \mathfrak{p}(v_x) dv_x - 1 \right] = \textit{minimum}. \tag{2.31}$$

The Lagrangian is

$$\mathscr{L} = \frac{[\partial \mathfrak{p}(v_x)/\partial v_x]^2}{\mathfrak{p}(v_x)} + \lambda_1 \frac{1}{2} m v_x^2 \mathfrak{p}(v_x) + \lambda_2 \mathfrak{p}(v_x), \tag{2.32}$$

with the solution

$$2 \frac{d}{dv_x} \left\{ \frac{[\partial \mathfrak{p}(v_x)/\partial v_x]}{\mathfrak{p}(v_x)} \right\} + \left\{ \frac{[\partial \mathfrak{p}(v_x)/\partial v_x]}{\mathfrak{p}(v_x)} \right\}^2 - \frac{1}{2} \lambda_1 m v_x^2 - \lambda_2 = 0. \tag{2.33}$$

For simplicity of writing, $h(v_x) = [\partial \mathfrak{p}(v_x)/\partial v_x]/\mathfrak{p}(v_x)$ reduces Equation 2.33 to the form

$$2 \frac{\partial h(v_x)}{\partial v_x} + h^2(v_x) - \frac{1}{2} \lambda_1 m v_x^2 - \lambda_2 = 0, \tag{2.34}$$

with the solution

$$\mathfrak{p}(v_x) = A \exp \left[\int h(v_x)\, dv_x \right], \tag{2.35}$$

where A is a normalization constant.

Equation 2.34 is a Riccati equation, whose solutions are found through power series.

The lowest order solution is $h(v_x) = a + bv_x$, with a and b as constants. The constraints of Equation 2.31 force $a = 0$, so

$$\mathfrak{p}(v_x) = A\exp(Bv_x^2), \tag{2.36}$$

again with A and B as constants. These constants follow from the probability distribution being constant and the equipartition constraints as

$$\mathfrak{p}(v_x) = \left(\frac{m}{2\pi k_B T}\right)^{1/2} \exp\left[-\frac{(1/2)mv_x^2}{k_B T}\right], \tag{2.37}$$

whose generalization is

$$\begin{aligned}\mathfrak{p}(v_x, v_y, v_z) &= \mathfrak{p}(v_x)\mathfrak{p}(v_y)\mathfrak{p}(v_z)\\ &= \left(\frac{m}{2\pi k_B T}\right)^{3/2} \exp\left[-\frac{(1/2)m(v_x^2 + v_y^2 + v_z^2)}{k_B T}\right].\end{aligned} \tag{2.38}$$

To find $\mathfrak{p}(v)$, since $v = (v_x^2 + v_y^2 + v_z^2)^{1/2}$, we change the coordinates from Cartesian to polar by $(v_x, v_y, v_z) \mapsto (v, \theta, \phi)$ and integrate over the polar angle θ (π) and over the azimuthal angle ϕ (2π), which results in

$$\mathfrak{p}(v) = \sqrt{\frac{2}{\pi}}\left(\frac{m}{k_B T}\right)^{3/2} v^2 \exp\left[-\frac{(1/2)mv^2}{k_B T}\right]. \tag{2.39}$$

This is the lowest order solution—the Maxwell-Boltzmann distribution—for velocity, and it is also rewritable in a kinetic energy form through the quadratic dependence.

This Fisher information method is much more instructive, however, than the traditional (see Appendices E and F) approach. The classical Maxwell-Boltzmann solution is for ϑ of order 1, having been arrived at from the consideration of the first solution of the Riccati equation—a single particle being sampled—and the best estimate solution for that problem. The Riccati equation has many other solutions (infinite!). Take the trial solution of $h(v_x) = bv_x + c/v_x$, which satisfies Equation 2.34 for $c = 2$, and b as another normalizable constant. This solution has the form

$$\mathfrak{p}(v_x) = A\exp\left(\frac{b}{2}v_x^2 + 2\ln v_x\right), \tag{2.40}$$

which, following substitutions and normalization to extract constants A and b, gives

$$\mathfrak{p}(v_x) = \frac{1}{\sqrt{2\pi}}\left(\frac{3m}{k_B T}\right)^{3/2} v_x^2 \exp\left(-\frac{3mv_x^2}{2k_B T}\right), \tag{2.41}$$

and its generalized form

$$\mathfrak{p}(v) = \frac{54}{105}\left(\frac{27}{2\pi}\right)^{1/2}\left(\frac{m}{k_B T}\right)^{9/2} v^8 \exp\left(-\frac{3mv^2}{2k_B T}\right). \tag{2.42}$$

What gives? The units are one hint. Equation 2.42 is in per third power in velocity compared to the first. The other is the peak and shape of the distribution. For Equation 2.39, the peak occurs at $\sqrt{2k_BT/m}$; for Equation 2.42, it is at $\sqrt{8k_BT/3m}$, which is slightly higher, but not that much different. As Figure 2.1 shows, though, the spreads and skews of the distribution are considerably different.

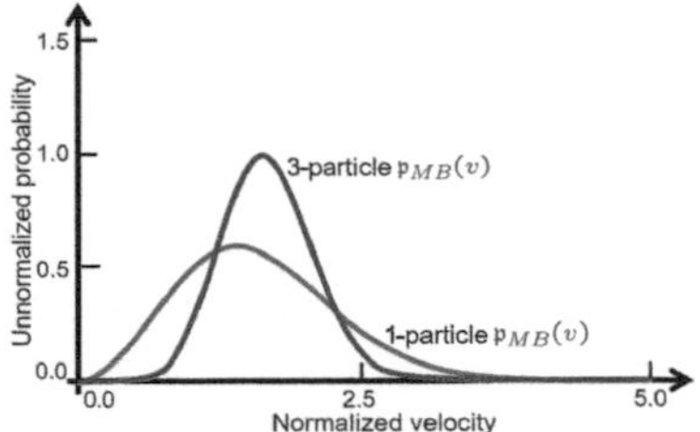

Figure 2.1: Two different solutions of the Riccati equation for the distribution function. The plot with the lower peak velocity is the classical Maxwell-Boltzmann solution for random selection of single particles in a sea of classical particles. The second solution is for a random selection of 3 particles, again from a sea of classical particles.

For velocity v as a root mean square velocity over n particles, the probability law is

$$\mathfrak{p}(v) = \frac{2}{\Gamma(3n/2)} \frac{n^{3n/2}}{\left(\sqrt{2}\sigma\right)^{3n}} v^{3n-1} \exp\left(-\frac{nv^2}{2\sigma^2}\right), \tag{2.43}$$

which follows from the argument that v^2 is the sum from $3n$ squared terms that are Cartesian independent terms. It arises in the χ^2 probability law and random variable transformation.

Equation 2.39 is the solution to $n = 1$ law, and Equation 2.42 is the solution to $n = 3$ law. The first considers a one-particle distribution (estimator $\vartheta(1)$), and the second considers a three-particle distribution (estimator $\vartheta(3)$). So, the solutions reflect the choice of distribution when one particle is selected in each observation, or three particles are simultaneously selected in each observation. The Fisher information for the first case is $\vartheta(1) = 3m/k_BT$ and, for the second case, is $\vartheta(3) = 27m/k_BT$, which is larger. I must increase with n. Information increases in a collection where additional patterns arise. It reflects a narrower breadth around the mean as n increases. Figure 2.1 reflects this; so does the information content, and the decrease in fluctuations around the mean.

One particular application of Fisher information that is dear to me is the use of the Dirac quantum-mechanical equation to show—the proof is longer and not really that related to the subject of this text—that the rate of change of the Shannon entropy H is

$$\frac{\partial H}{\partial t} \leq c\sqrt{I}. \tag{2.44}$$

One can assign a number of meanings to this expression. The most important of these is that the rate of gain of Shannon information in b/s acquirable is limited by the speed of light *and by Fisher information capacity*. The information in a power spectrum is $I = 4\int d\omega/S(\omega)$. And we have seen the relationship between information and error in Equation 2.12. Deeper down, this relationship is also stating that our current state of learning determines an upper limit to how rapidly we may learn.

2.4.3 *Diffusion*

RANDOM WALK, whose one example is diffusion, is another ubiquitous problem related to the estimation in the presence of incompleteness that arises in a distribution representing many possibilities. So, we tackle this as one final problem to illustrate Fisher information's usage to elicit the diffusion equation. We will tackle this in one dimension, and, similar to the case for the Maxwell-Boltzmann distribution, its three-dimensional form follows from coordinate independence.

Let $x(t)$ be the position at any time t, so $x(0)$ is the coordinate at $t = 0$. With t time elapsed,

$$x = x(0) + \Delta x, \quad \text{with } \Delta x = \int_0^t \Delta v(t')dt', \tag{2.45}$$

with $\Delta v(t')$ as a velocity with its fluctuation at time $t = t'$. We estimate $\mathfrak{p}(\eta, \zeta)$, where η is velocity Δv at $t = t'$, and ζ is Δv at another time, $t = t''$. Equilibrium in the velocity distribution exists axiomatically at t' and t''. This means that the averaged kinetic energies—averaging over the space of $(1/2)m(\Delta v)^2$—must be the same, and if ergodicity is also true, then this energy for this degree of freedom is $(1/2)k_BT$ for classical particles. So, our marginal constraints are

Ergodicity is discussed in Chapter 16. It implies that, for this ensemble, the time and spatial averages lead to the same measure for the function of interest.

$$\iint \frac{1}{2}m\eta^2 \mathfrak{p}(\eta,\zeta)\, d\eta\, d\zeta = \frac{1}{2}k_BT, \text{ and}$$
$$\iint \frac{1}{2}m\zeta^2 \mathfrak{p}(\eta,\zeta)\, d\eta\, d\zeta = \frac{1}{2}k_BT. \tag{2.46}$$

We need to now employ the Fisher marginal constraints arising in η and ζ. $\langle T_\eta\rangle = \langle(1/2)m\eta^2\rangle$, and $\langle T_\zeta\rangle = \langle(1/2)m\zeta^2\rangle$. So, the constraining is

$$\iint \frac{\mathfrak{p}_\eta^2+\mathfrak{p}_\zeta^2}{\mathfrak{p}^2}\, d\eta\, d\zeta + \lambda_1 \iint T_\eta \mathfrak{p}(\eta,\zeta)\, d\eta\, d\zeta$$
$$+ \lambda_2 \iint T_\zeta \mathfrak{p}(\eta,\zeta)\, d\eta\, d\zeta = minimum, \text{ with}$$
$$\mathfrak{p}_\eta = \frac{\partial \mathfrak{p}}{\partial \eta}, \text{ and } \mathfrak{p}_\zeta = \frac{\partial \mathfrak{p}}{\partial \zeta}. \tag{2.47}$$

The principle of minimum negentropy stated in Equation 2.19 implies here a separable solution so that

$$\mathfrak{p}(\eta,\zeta) = \mathfrak{p}_1(\eta)\mathfrak{p}_2(\zeta). \tag{2.48}$$

The Euler-Lagrange equation to this two-constraints problem is

$$\frac{\partial}{\partial\eta}\left(\frac{\partial\mathscr{L}}{\partial\mathfrak{p}_\eta}\right) + \frac{\partial}{\partial\zeta}\left(\frac{\partial\mathscr{L}}{\partial\mathfrak{p}_\zeta}\right) = \frac{\partial\mathscr{L}}{\partial\mathfrak{p}}, \tag{2.49}$$

where

$$\mathscr{L} = \frac{\mathfrak{p}_\eta^2+\mathfrak{p}_\zeta^2}{\mathfrak{p}^2} + \lambda_1 T_\eta(\eta)\mathfrak{p} + \lambda_2 T_\zeta(\zeta)\mathfrak{p}. \tag{2.50}$$

Equation 2.47 then reduces to

$$\frac{2}{\mathfrak{p}}\frac{\partial^2\mathfrak{p}}{\partial\eta^2} + \frac{2}{\mathfrak{p}}\frac{\partial^2\mathfrak{p}}{\partial\zeta^2} - \frac{1}{\mathfrak{p}^2}\left(\frac{\partial\mathfrak{p}}{\partial\eta}\right)^2 - \frac{1}{\mathfrak{p}^2}\left(\frac{\partial\mathfrak{p}}{\partial\zeta}\right)^2$$
$$- \lambda_1 T_\eta(\eta) - \lambda_2 T_\zeta(\zeta) = 0. \tag{2.51}$$

Using a probability product trial solution ($\mathfrak{p}(\eta,\zeta) = \mathfrak{p}_1(\eta)\mathfrak{p}_2(\zeta)$), two separate equations in probabilities follow:

$$\frac{2}{\mathfrak{p}_i}\frac{\partial^2\mathfrak{p}}{\partial\xi_i^2} - \left(\frac{\partial\mathfrak{p}}{\partial\xi_i}\right)^2 - \lambda_i T_{\xi_i} = 0, \text{ with} \tag{2.52}$$

$\xi_i = \eta$ for $i = 1$, and $\xi_i = \zeta$ for $i = 2$. The separated solutions, under the marginal constraints, appear as separate equations of

$$\int \frac{1}{p_i}\left(\frac{\partial p}{\partial \xi_i}\right)^2 d\xi_i + \lambda_i \int T_{\xi_i} p_i(\xi_i)\, d\xi_i = \textit{minimum} \text{ for } i = 1, 2. \quad (2.53)$$

The separation of probability (Equation 2.48), together with the minimization Equation 2.47, imply, again via separation, that both $\eta = \Delta v(t')$ and $\zeta = \Delta v(t'')$ are independent of each other. This also states that

$$\langle \Delta v(t)\Delta v(t')\rangle = \frac{k_B T}{m}\delta(t-t'), \text{ and } \langle \Delta v(t)\rangle = 0. \quad (2.54)$$

So, from our starting equations of this subsection, we arrive at

$$\langle \Delta x\rangle = 0, \text{ and } \langle \Delta x^2\rangle = \frac{k_B T}{m}t. \quad (2.55)$$

Δx is, for all times, the sum of the infinite independent variable $\Delta v(t')$, with time continuous. Δx is therefore Gaussian, from the central limit theorem, independent of the nature of Δv's probability law, and therefore also for Equation 2.31. The probability density for Δx therefore follows as

$$p_{\Delta x}(x) = \frac{1}{\sqrt{2\pi (k_B T/m)t}} \exp\left(-\frac{x^2}{2(k_B T/m)t}\right). \quad (2.56)$$

Since the position x depends on Δv over $(0, t)$ and is therefore a random number arising in Δx, and x_0 is independent of Δx, by convolution,

$$p(x,t) = \int p_0(y) \frac{1}{\sqrt{2\pi (k_B T/m)t}} \exp\left[-\frac{(x-y)^2}{2(k_B T/m)t}\right] dy. \quad (2.57)$$

This equation, following differentiation, gives

$$\frac{\partial p(x,t)}{\partial t} = a\frac{\partial^2 p(x,t)}{\partial x^2}, \text{ with } a = \frac{k_B T}{2m}. \quad (2.58)$$

This is the diffusion equation for the distribution of the particle in position and time coordinates. We have arrived at it through the independence of velocities at different times that arose in their constraints being marginals, that is, non-joint, and deploying the Fisher entropy arsenal.

2.5 *Summary*

ENERGY AND ENTROPY are two quite intertwined ideas. We need the concept of energy as the property—arising in potential and kinetic forms—that undergoes exchange as a system evolves. These energy forms exist in our description in a variety of ways. A chemical bond is a configuration that arises under quantum-mechanical constraint even as the electromagnetic rules apply.

A mechanical energy description—classical or semi-classical—is the manifestation of this bond as a storehouse of this energy and, being in the connected world, because of temperature and because it is stable, with the harmonic oscillator as the simplest descriptor, leads to oscillations with periodic potential and kinetic energy exchange. Different states of matter arise where these energies are being exchanged, or from which they are even lost in a non-conservative form, since the losses are to a broad band. Entropy, although classically through Boltzmann, arose as the description for what a collection of particles would do, ending up in thermodynamic equilibrium where a function—the Boltzmann entropy—is a maximum. In general, entropy is a somewhat dissatisfying state of lack of precise information of a system. If one doesn't have complete information for a system, one doesn't really know what one doesn't know, so it is quite out of bounds for us to completely summarize it in any objective way. This is the conundrum. But, information is physical, and therefore entropy and energy are related, whether one views it from the classical Claussius-to-Boltzmann and Gibbs view, from the quantum view or for black holes and their event horizons as they radiate, even if this last description is still a work in progress.

The lore is that von Neumann, during a discussion with Shannon, suggested that he refer to the informational content of a bitstream as an entropy. I would not be surprised if the suggestion was for negentropy, which is what it is, but the negative of the dissatisfying state of a lack of precise information is too obfuscating. The von Neumann entropy and, particularly with entangled quantum systems, this idea of information turn out to be not something that can be thought through in any simple way. von Neumann knew this, since he calls the observer and the observed a prime example of entanglement.

We focused in this chapter on one particular entropy: Fisher entropy, which does not receive much attention in the literature but is among the most meaningful ways to make the connection between entropy as the fount of information to the description of the physical world. Fisher entropy embodies beauty and taste. Being based on a derivative, it is a measure that emphasizes locality. Application of a minimization constraint to it therefore leads to second-order equations, and that is what many of nature's equations are. And this is unlike the other entropy measures, which, being global in nature, when applied across the distribution function via probability, lead to algebraic and exponential forms. That the statistical nature of events—cause and chance at play—and our deterministic analytic models formed as equations can both be seen together is quite a powerful truth at play.

Fisher entropy may be a good example of beauty and taste. But it is not a general guarantee. Statistics, being quite mathematical, is often used nefariously. Even Fisher himself was a fervent eugenicist.

In adopting this approach in this early chapter, this writing was a short digression from the norm of the way physical theories are taught, to illustrate the connections between the mathematical viewpoints and the physical viewpoints. Having multiple viewpoints of what we observe provides deeper insights into the connections present therein, and this chapter illustrated these connections.

As illustrations, we showed, through the Lagrangian constraint, Fisher entropy leading to the one-dimensional Schrödinger equation. We expanded this to show estimation over particle ensembles to derive the Maxwell-Boltzmann distribution in order 1 but also pointed out, through the derivation, that higher order solutions also occur when the distribution is over a multiple particle collection instead of a single independent particle collection. And we concluded these examples by illustrating the process of diffusion. Seemingly quite different and important equations describing the particle statics and dynamics can be arrived at through the use of Fisher entropy.

2.6 *Concluding remarks and bibliographic notes*

ENTROPY HAS OCCUPIED SCIENCE'S attention for many hundreds of years since the initial thermomechanical forays where energy also appeared. Entropy, largely as a measure of incompleteness of information, appears in many forms. This chapter employed Fisher entropy to illustrate the entropy-energy-information connections and used it to illustrate its applicability to arrive at equations that we employ commonly for modeling and predicting physical phenomena.

An excellent source for understanding entropy from the classical viewpoint, that is, employing thermodynamic and statistical meanings, is by Dugdale[1]. The book discusses the various laws of thermodynamics, together with the distribution functions, and the implications of low temperature, very meaningfully. Another historically important work—a classic text—is by Tolman[2]. Of course, the reader will find a wide variety of books in this subject area as our understanding continues to expand, and, of late, information, and the other twists to entropy, have become important.

Two scientists have very strongly influenced the interplay of entropy, energy and their use in understanding the physical phenomenon and have written their coherent thoughts in these past decades.

The first is E. T. Jaynes[3]. Jaynes' exposition of the principles embedded in the meanings of entropy, energy, information and their implications for any statistical inference are path setting and date back to the 1950s.

The second is B. Roy Frieden, whose book[4], while its aim is to develop tools for measurement theory, does an excellent analysis and derivation of a variety of classical and modern equations, using just the tools of the Fisher-based method of estimators. Frieden has also coedited a book with R. A. Gatenby[5] that explores the application of Fisher entropy beyond information and thermal physics to biology, finance and other domains.

[1] J. S. Dugdale, "Entropy and its physical meaning," Taylor and Francis, ISBN 0-7484-0568-2 (1996)

[2] R. S. Tolman, "The principles of statistical mechanics," Dover, ISBN 13 978-0486638966 (2010). This book's first edition is from Oxford in 1938

[3] E. T. Jaynes, "Information theory and statistical mechanics," Physical Review, **106**, 620–630 (1957), and "Information theory and statistical mechanics II," Physical Review, **108**, 171–190 (1957)

[4] B. R. Frieden, "Science from Fisher information," Cambridge, ISBN 13 978-0521009119 (2004)

[5] B. R. Frieden and R. A. Gatenby, "Exploratory data analysis using Fisher information," Springer ISBN 13 978-1-84628-506-6 (2007)

Finally, an end note to the Saint Petersburg paradox posed in the sidebar. The resolution is that "infinity" is a fair fee for entering the game. This is theoretically correct but really not that practical.

2.7 *Exercises*

1. The Shannon information measure H is additive for two mutually isolated systems. But H is not the only functional of a probability law obeying additivity. Show that the Fisher information measure I also obeys additivity. **[S]**

2. Take the amplitude function $\mathfrak{q}(x)$, where $\mathfrak{p}(x) = \mathfrak{q}^2(x)$. Show that this implies that

$$I = 4 \int \mathfrak{q}'^2(x)\, dx,$$

and that one can define a quadratic measure for the displacement between the amplitude function $\mathfrak{q}(x)$ and its displaced version $\mathfrak{q}^2(x + \Delta x)$ as

$$L^2 = \frac{1}{4} \Delta x^2 I.$$ **[S]**

3. The Boltzmann entropy obeys the second law of thermodynamics, that is, $dH(t)/dt \geq 0$. This is what the Boltzmann H theorem calls out. Show that the Fisher entropy I follows

$$\frac{dI(t)}{dt} \leq 0.$$ **[S]**

4. The relation $dH/dt \geq 0$ sets a lower bound for the closed system. We are interested in finding if there is an upper bound for the closed system that obeys the conservation of flow equation (continuity),

$$\frac{d\mathfrak{p}(\mathbf{r}|t)}{dt} + \nabla_{\mathbf{r}} \cdot \mathfrak{P}(\mathbf{r}, t) = 0,$$

where $\mathfrak{P}(\mathbf{r}, t)$ is a measure of the probability flow. Show that

$$\left(\frac{dH}{dt}\right)^2 \leq I \int \frac{\mathfrak{P} \cdot \mathfrak{P}}{\mathfrak{p}}\, d\mathbf{r}, \quad \text{where}$$
$$I = I(t) = \int \frac{\nabla_{\mathbf{r}}\mathfrak{p}(\mathbf{r}|t) \cdot \nabla_{\mathbf{r}}\mathfrak{p}(\mathbf{r}|t)}{\mathfrak{p}(\mathbf{r}|t)} d\mathbf{r}.$$

The entropy change in a small time interval is bounded by the square root of the Fisher information capacity for the position measurement. **[M]**

5. Hydrogen has an ionization energy of 13.6 eV. Estimate the temperature where a noticeable fraction of hydrogen is ionized because of thermal agitation. **[S]**

6. Lithium—a metal—has a Fermi energy of 4.7 eV away from the conduction bandedge. What is the fraction of electrons that are thermally excited at 300 K? **[S]**

3
Waves and particles in the crystal

We begin the exploration of semiconductor physics by taking first a broader look at the nature of the behavior of the duality of particles and waves in solids, and semiconductors specifically. This will give us a good starting understanding of electrons, atoms and electromagnetics or photons as their particles under the quantum constraints when brought together in the solid form. We are interested in understanding how we may see their behavior in the solid and how can one simplify, without losing the essentials, the description of their behavior with each other. We must therefore be able to sufficiently describe the solid, which we restrict to its periodic atomic arrangement, and then see within it the state description of the electrons and of the atoms so that we will be able to explore their interaction together with those of photons.

Our discussion therefore will start with a look at what it means to be a wave—as in an electromagnetic or quantum wave—or to be a particle, and the nature of this duality that must have many connecting themes so that classical to quantum correspondence holds. To understand the solid, we then look at the lattice as the periodic mathematical construct that fills space, with the unit cells as building blocks that can recreate identical environments surrounding them through translations. This lets one work with this smaller point construction space that suffices to describe the larger assembly. For reasons of symmetry, there are limitations to the periodic arrangements that completely fill a space. The allowed space-filling periodic arrangements constitute the Bravais lattices. If we associate specific atoms as a basis with this lattice construct, we have a crystalline solid. Three-dimensional crystals have 14 such Bravais possibilities. The cubic arrangement, for example, can be simple cubic, body-centered cubic (with a lattice point occupied by an atom in its center), or face-centered cubic (with lattice points at the center of faces). In nature, this cubic crystal appears in the three

Semiconductor Physics: Principles, Theory and Nanoscale. Sandip Tiwari.
 Published 2020 by Oxford University Press. DOI: 10.1093/oso/9780198759867.001.0001

different Bravais lattice forms for a number of solids. The face-centered form is a dense form of arrangement; so is the hexagonal-close-packed form. The difference between the two is the three-dimensional assembly by the arrangement of the layers across three layers. Semiconductors of most interest to us are based on these two packings and will appear, along with two-dimensional monolayer assemblies. The periodicity of these assemblies allows one to describe these semiconducting solids through unit cells, or through other reduced forms of less symmetry. The 14 Bravais lattices map to 7 crystal forms.

The electron wavefunction lends itself best for description through the wavevector, and for this it is convenient to view it through the reciprocal space of the crystal. So, we will translate our periodic real space lattice to a periodic reciprocal space lattice and understand its properties. Just as a unit cell in its various forms with different symmetries serves to describe the entire three-dimensional arrangement, one can also find equivalent zones for the reciprocal cell where the allowed wavevectors, one of the quantized parameters characterizing the state of the electron, can be pulled in, and herein the allowed energies of the electron described. This approach lets us discuss the 1st Brillouin zone, the simplest form for reciprocal space description, and the Jones zone, with all the valence electrons of the unit cell, which form the states of interest to us in the solid, since they are now allowed to spread out over the solid in real space and can be characterized.

Having thus set up the description of the real space and reciprocal space, we explore the properties of the electrons in periodic potentials through toy models that let us thus understand the various relationships between the real and the reciprocal space, understand the nature of the behavior of the electron as a wave spread out across the crystal, and clarify the distinctions between a free electron and an electron in the midst of periodic potentials and confined in a solid where it may be nearly free; that is, free-like by some measures but not by other equally important ones. This clarifies the notions related to the nearly free electron in the solid. It has now become possible for an electron that was confined to a very stable atom to move around in a crystal, but with a multitude of restrictions that these toy model approaches help clarify.

Atoms in this periodic arrangement also do not stay still. Thermal energy and atoms' bonding causes motion around equilibrium, and our periodic arrangement lets us analyze the modes of these oscillations characterized by a quasiparticle: the phonon. Again, we employ a toy model of a two-different-atom basis for the crystal. Motion can be longitudinal and transverse, and, since the particle is a boson, the occupation of these modes is via Bose-Einstein statistics. The use of a two-atom basis, and incorporating into this description the ionicity of the crystals, permit us to understand the nature of the oscillating dipole, and the difference between longitudinal and transverse forms. The dielectric function of the crystal arises in how the ionic and the electronic polarization—the two forms that will respond in the displacement—of the system respond to the fields. So, the atomic/ionic oscillations and the dielectric response and frequency can be connected to each other.

This is the frame for the objective representation of the crystalline semiconductor solid. Waves are seen best via the reciprocal space, the atom placement via the real space, and their interaction through both. For atoms in a crystal, interacting with each other leads to their contributing electron states that spread out in the crystal. Atoms bonded together in a crystal lets them vibrate, and this is a wave-like phenomena best seen through phonon states. The reciprocal space gives us a clear way of describing the allowed energy levels and other quantized properties of the electron and phonon states. The real space gives us a clear way of describing the bonding and the physical representation of the solid.

3.1 *Waves and particles: Classical and quantum views*

THE PROBABILITY OF FINDING A PARTICLE—an electron of mass m_0—represented by the wavefunction $\psi(\mathbf{r})$ in any volume d^3r is $\langle\psi(\mathbf{r})|\psi(\mathbf{r})\rangle\, d^3r = |\psi(\mathbf{r})|^2\, d^3r$, subject to the condition that it is somewhere; that is, $\int |\psi(\mathbf{r})|^2\, d^3r = 1$. The Schrödinger equation of this quantum wave description in free space, that is, for $\hat{\mathscr{H}} = \hat{T} = \hat{\mathbf{p}}^2/2m_0$ ($\hat{T}$ being the kinetic energy operator), is

$$-\frac{\hbar}{i}\frac{\partial}{\partial t}|\psi\rangle = -\frac{\hbar^2}{2m_0}\nabla_r^2|\psi\rangle = E|\psi\rangle, \tag{3.1}$$

whose solution is

$$|\psi\rangle = \mathcal{A}\exp[i(\mathbf{k}\cdot\mathbf{r} - \omega t)], \tag{3.2}$$

with the eigenenergy

$$E = \frac{\hbar^2(\mathbf{k}\cdot\mathbf{k})}{2m_0} = \frac{\hbar^2\mathbf{k}^2}{2m_0}. \tag{3.3}$$

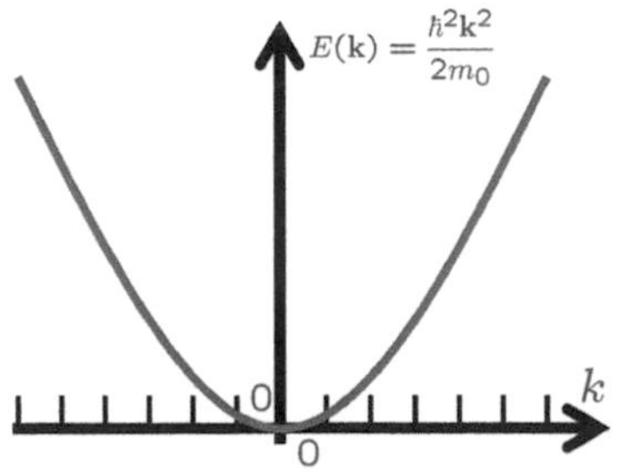

Figure 3.1: A free electron's energy dispersion in free space.

The kinetic energy operator $\hat{T}$ followed from the momentum operator $\hat{\mathbf{p}} = (\hbar/i)\nabla_{\mathbf{r}}$ and the energy–momentum relationship. Figure 3.1 shows this energy-wavevector ($E(\mathbf{k})$ or E-$\mathbf{k}$) behavior for a free electron. This is the free electron energy dispersion. It represents under non-relativistic conditions an inertial mass m_0 and a kinetic energy that varies as the square of the wavevector. This *free electron* also has a momentum that we can find from the wavefunction.

$$\hat{\mathbf{p}}|\psi\rangle = \hbar\mathbf{k}|\psi\rangle, \tag{3.4}$$

that is, $\hbar\mathbf{k}$. The wavevector specifies it.

While an electron has a mass, a photon doesn't. Since $E^2 = p^2c^2 + m_0^2c^4$, and the same wavefunction form still holds for the photon (no potential), the photon momentum is still $\hbar\mathbf{k}$, and therefore its energy $E = pc = \hbar c|\mathbf{k}|$. The photon is the quantum particle of an electromagnetic wave. The wavefunction ($\propto \exp[i(\mathbf{k}\cdot\mathbf{r} - \omega t)]$) that we have written just says that the particle can be found with equal probability anywhere, that is, the electromagnetic wave permeates the free space. The classical equation for the electromagnetic wave—consider the simplest case of free space—follows from Maxwell's equations as

$$\frac{1}{c^2}\frac{\partial^2\psi(\mathbf{r},t)}{\partial t^2} = \nabla_{\mathbf{r}}^2\psi(\mathbf{r},t). \tag{3.5}$$

This classical wave equation too describes how the wave is evolving in time, as did the Schrödinger equation (Equation 3.1). Seen through fields, the electromagnetic analog of the wavefunction $\mathcal{E} = \mathcal{E}_0\exp[i(\mathbf{k}\cdot\mathbf{r} - \omega t)]$) gives us the analogies of Table 3.1 between the classical electromagnetic wave and the quantum photon particle interpretation.

In the background to these equations is de Broglie's hypothesis that one can associate a wave-like quantum property of wavelength ($\lambda = h/p$) to a classical momentum property of a particle. The relationship $\omega(k) = \hbar k^2/2m_0$ is the *dispersion relation* of this wave. This particle ≡ wave is not localized. In the wave picture, its energy is distributed over the entire free space, and, in the quantum picture, it has a probability of being everywhere.

Our understanding of the particle is that of an object that can be described by some range of position and momentum (or velocity), with the range vanishing in the classical description. By this, we mean that, at a time $t = t_0$ in the range Δt, we will find it in some specific range $\Delta\mathbf{r}$ in position at $\mathbf{r} = \mathbf{r}_0$. The localization—in our quantum view—arises when the wave has a narrower spread of excitations $\Delta\mathbf{k}$ around the wavevector $\mathbf{k}$. This now is a wavepacket—the electromagnetic analogy will be the formation of a wavepacket by the superposition of plane waves—that we may write as

$$\psi(x,t) = \frac{1}{\sqrt{2\pi}}\int_{-\infty}^{\infty} u(k)\exp[i(kx - \omega t)]\,dk, \tag{3.6}$$

It is really the time-dependent equation that is the wave-equation formulation of Schrödinger. It teaches us the dynamic evolution of the wavefunction. The second part of Equation 3.1 is only a statement of Hamilton's energy formulation, that is, $\hat{\mathscr{H}}|\psi\rangle = E|\psi\rangle$.

Classical	Quantum
EM wave	Photon
$E \propto \mathcal{E}_0^2$	$E = n\hbar\omega$
$\nu\lambda = \omega/k = c$	$\lambda = h/p = 2\pi/k$

Table 3.1: A few notable associations in the classical and quantum views of electromagnetism. $\mathcal{E}_0$ is the amplitude of the electric field, ν is the frequency (cycles per second), ω is the radial frequency, and n is the number of photons. The amplitude of the electromagnetic (*EM*) wave is associated with the energy content of the wave. Quantum-mechanically, a photon has the energy $E = \hbar ck = \hbar\omega$. n of these photons will have n times that energy, which is spread out in the space that shows up as the field $\mathcal{E}_0$.

Take an electron moving at about $10^5\ m/s$ (a velocity magnitude that we associate with the motion of an electron in a crystal, ignoring the peripheral details of the crystal's presence buried in its effective mass there); then, the electron has a momentum of $p = m_0 v = 9.1\times10^{-31}\times10^5 = 9.1\times10^{-26}\ kg\cdot m/s$. The de Broglie wavelength $\lambda_{deB} = h/p = 6.626\times10^{-34}/9.1\times10^{-26} \approx 7\ nm$. Nanoscale and quantum are intimately entwined in this magnitude and its consequences.

In the classical description, the position and momentum (or velocity) canonic variables can be precisely determined simultaneously.

where we have, for convenience, written it for a wavepacket traveling in one dimension. Let this wavepacket be centered at k_0 in the wavevector, and ω_0 in frequency. We Taylor expand, including only the first term in the expansion to $\omega(k) \approx w_0 + \omega_0'(k - k_0)$, where $\omega_0' = d\omega/dk$ at $k = k_0$. Using a change of variables, $s = k - k_0$,

$$\begin{aligned}\psi(x,t) &\approx \frac{1}{\sqrt{2\pi}} \int_{-\infty}^{\infty} u(k_0 + s) \exp\left\{i\left[(k_0 + s)x - (\omega_0 + \omega_0' s)t\right]\right\} ds \\ &= \frac{1}{\sqrt{2\pi}} \exp\left[-i(\omega_0 t - k_0\omega_0' t)\right] \\ &\times \int_{-\infty}^{\infty} u(k_0 + s) \exp\left[i(k_0 + s)(x - \omega_0' t)\right] ds. \qquad (3.7)\end{aligned}$$

This is the evolution of the wavefunction $|\psi(x,t)\rangle$ in time from

$$\begin{aligned}\psi(x,0) &= \frac{1}{\sqrt{2\pi}} \int_{-\infty}^{\infty} u(k_0 + s) \exp[i(k_0 + s)x]\, ds \\ &= \frac{1}{\sqrt{2\pi}} \int_{-\infty}^{\infty} u(k) \exp(ikx)\, dk \qquad (3.8)\end{aligned}$$

at $t = 0$. What this arbitrary time and beginning time wavefunction says is that, with time, $x \mapsto x - \omega_0' t$, and one can write

$$\psi(x,t) \approx \exp\left[i(-\omega_0 + k_0\omega_0')t\right] \psi(x - \omega_0' t, 0). \qquad (3.9)$$

So, the probability $|\psi|^2$ function moves in space at a speed of ω_0'. This is the velocity of this group of waves centered around (k_0, ω_0)—the group velocity v_g— of

$$v_g = \frac{d\omega}{dk}, \qquad (3.10)$$

or, more generally,

$$\mathbf{v}_g = \nabla_{\mathbf{k}}\omega, \text{ or } \mathbf{v}_g = \frac{1}{\hbar}\nabla_{\mathbf{k}}E. \qquad (3.11)$$

This is the velocity with which the energy associated with this group of waves constituting the wavefunction $|\psi(x,t)\rangle$ travels. Note that, for the wave description, we found $\omega = \hbar k^2/2m_0$, so $v_g = d\omega/dk = \hbar k/m_0$ and $v_\phi = \omega/k = \hbar k/2m_0$. A particle representing a bundle of excitations with energy travels at v_g. The phase velocity is a velocity representing the speed of phase change of a stationary state.

This group of waves also has a phase velocity for the continuum distribution of waves constituting it. The phase velocity is $v_\phi = \omega/k$. We really don't need to resort to this wavepacket based argument to establish the relationship of group velocity. We will revisit this when discussing band diagrams as a linear response term where a state evolves to a different momentum and different energy, which directly leads to a group velocity. The next order term of this description lets us relate the changes to the effective mass—the mass of the electron occupying a state in the crystal—and the f-sum rule, which, through oscillator treatment, brings in how the different states couple to each other via the operation of the momentum. Oscillator strength is discussed in Appendix I and will appear often in the text.

So far, we have looked at a free particle and its wave characteristics. In semiconductors, we are interested in how these particles behave within that environment. The electron of free space conforms to wave properties. What does its behavior look like when it is in a crystal, that is, when it arises from the atoms that constitute a periodic array with all their interactions within? To handle this, we first build the edifice to represent the crystal and then look at how the electron picture, of the electrons that we care about in structures employing condensed matter, as a particle or a wave, exist within this arrangement—and, from this, how other interactions affect it when we subject the crystal to external forces in the presence of internal forces. So, how the crystal, with the interacting atoms localized at a lattice of sites, makes the carriers behave is of interest.

We begin by exploring how one can mathematically and physically treat the crystalline semiconductor solid with atoms and electrons in it, as a stepping stone to the exploration of interactions and other responses of interest.

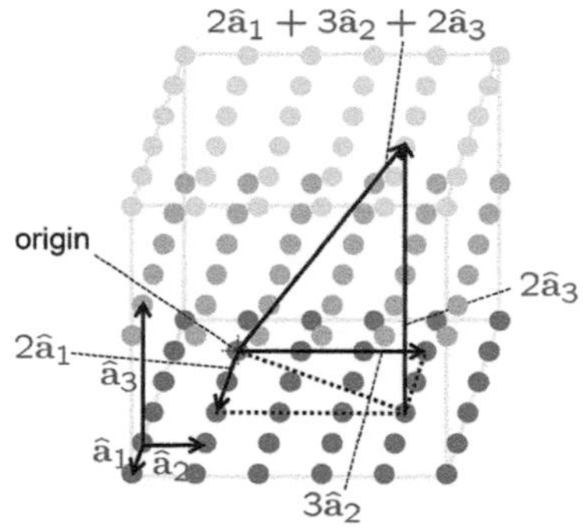

Figure 3.2: Lattice points in a three-dimensional lattice. $\hat{\mathbf{a}}_3$ is a unit vector out of the plane to the next neighbor lattice site, and $\hat{\mathbf{a}}_1$ and $\hat{\mathbf{a}}_2$ are in the plane. A lattice site at translated coordinates of $2\hat{\mathbf{a}}_1 + 3\hat{\mathbf{a}}_2 + 2\hat{\mathbf{a}}_3$ is shown from an origin located inside the lattice.

3.2 *Lattice and crystal*

THE LATTICE is a mathematical construct—an infinite array of points—in which the points have surroundings that are identical to those of all other similar points. This means that, no matter where one is at a small set or one of these lattice points, distances and angles to all other points from that point are the same as from any other similar point.

A Bravais lattice is generated by a primitive translation operator,

$$\mathbf{T}_m = m_1\hat{\mathbf{a}}_1 + m_2\hat{\mathbf{a}}_2 + m_3\hat{\mathbf{a}}_3, \tag{3.12}$$

where m_i are integers, positive, zero or negative, and $\hat{\mathbf{a}}_1$, $\hat{\mathbf{a}}_2$ and $\hat{\mathbf{a}}_3$ are three vectors, which are our units of translation. The infinite number of end points of the translational vector are the points of the lattice. For a hypothetical three-dimensional lattice, Figure 3.2 shows example vectors and translations that generate the infinite array.

These vectors as units of translation will sometimes have the magnitude subsumed as here and sometimes will be normalized. This should be clear from the context of usage.

Finding the smallest vector set that creates the infinite lattice out of translational operations can be tricky in many cases. For example, in Figure 3.3, a honeycomb structure, do $\hat{\mathbf{a}}_1$ and $\hat{\mathbf{a}}_2$ define the translation that results in the Bravais lattice? No. While the array of points has the same appearance from A and C, the view from point B is rotated π radians. To create the infinite assembly of the array of points, we will need to define two basis points, and then two vectors will suffice in creating the lattice. If the nearest neighbor distance is δ, then $\hat{\mathbf{a}}_1 = \delta(\sqrt{3}/2, 1/2)$, and $\hat{\mathbf{a}}_2 = \delta(\sqrt{3}/2, -1/2)$, and the basis of $\pm\delta(1/2\sqrt{3}, 0)$ will suffice.

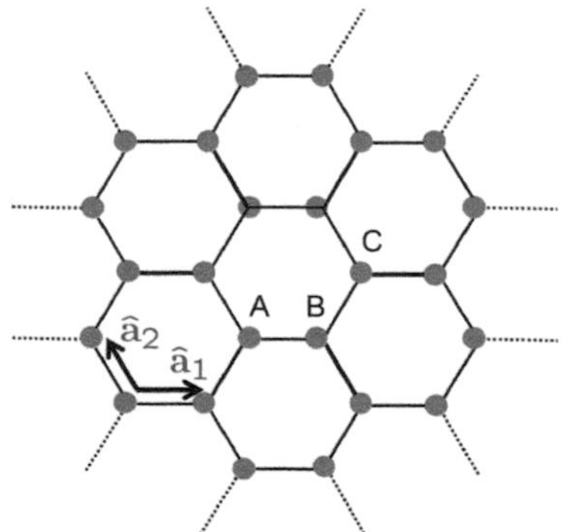

Figure 3.3: A honeycomb lattice. The crystalline form of graphene has this lattice structure.

A parallelepiped formed by $\hat{\mathbf{a}}_1$, $\hat{\mathbf{a}}_2$ and $\hat{\mathbf{a}}_3$, of the volume $\hat{\mathbf{a}}_1\cdot(\hat{\mathbf{a}}_2 \times \hat{\mathbf{a}}_3)$, when continuously translated parallel to itself by

$\mathbf{T}_m$, fills the infinite lattice. A *unit cell* is any cell that fills up the lattice through the translations. It may, however, be primitive or non-primitive. A *primitive unit cell* is the smallest part of the lattice that, if repeated, would reconstruct the entire crystal structure and consists of one and only one lattice point. Lattice points can be "shared." In a square lattice, a lattice point at the intersection of 4 unit cells is considered to be only 1/4th in each cell. In a simple cubic lattice, a lattice point is shared between 8 unit cells and contributes 1/8th to each cell for a total of 1. So, if the unit cell is composed of 8 vertex points, then it is a primitive unit cell. In Figure 3.2, one could have made multiple choices of the example vectors forming primitive cells. The figure also shows a non-primitive example that contains more than 1 lattice site but, when repeated using the translational operator, will still create the infinite lattice and fill the volume. For most examples of interest to us in crystals, the unit cell will contain more than one lattice point. These are *non-primitive unit cells.*

The lattice and the size of the atoms occupying the lattice site define the packing density, that is, the volume that is occupied by the atoms. Two forms of packing—the face-centered cubic (*FCC*) and the hexagonal-closed-packed (*HCP*)—have the highest density. The occupying atoms touch nearest neighbors in the densest form. If they are all the same atom, and therefore all the same size, they will be points on the surface of a sphere, with the center being another lattice point. Figure 3.4 shows an illustration of this high packing and its equivalent by showing a top view of vertical layering of equal-sized atoms. Figure 3.4(a), which is the *FCC* form, shows the face-centered cubic formation across multiple layers, which maintains the $\pi/3$ rotational symmetry that must exist in a single layer. Note that the third layer, the topmost layer, is displaced from the first, while still sitting above the interstitial of the 3-atom equilateral assembly of the second layer. Figure 3.4(b) shows the same for *HCP*, but now the third layer is precisely on top of the first layer.

The *FCC* and *HCP* are the densest forms, for the same reason: a sphere is the largest volumetric form with the smallest surface area. Within a layer, six nearest neighbors exist. The next layer has occupation above the interstitial region. This arrangement has still the highest density if the second layer consists of another atomic species.

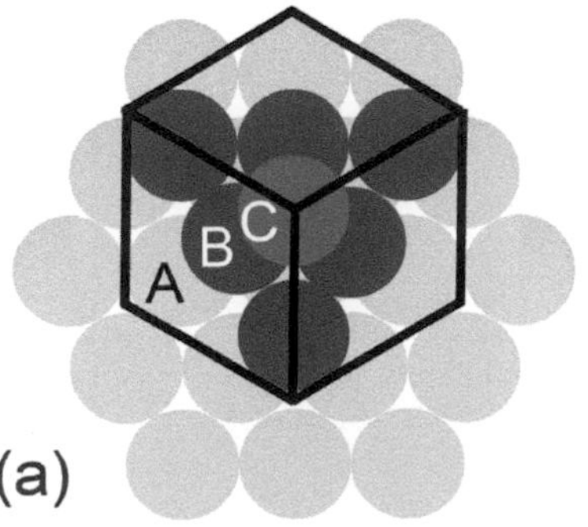

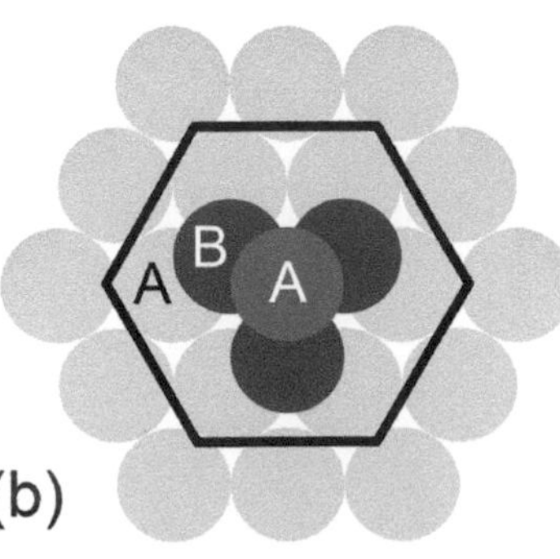

Figure 3.4: The highest packing possible in a single-atom-type assembly is one atom touching six other atoms on the diameter of the sphere. The figure shows a top view across three layers. Part (a) shows *FCC* packing, and (b) shows *HCP* packing.

Symmetry places restrictions on the unit cell lengths and inter-axial angles of a lattice. If one applies a rotational symmetry operation to an arbitrary lattice, the only rotations consistent with the requirement that the unit cell fill all space are C_n rotations with $n \subset \{1,2,3,4,6\}$. This results in 7 allowed crystal systems. Figure 3.5 shows the definitions of some of the notations, and Table 3.2 the 7 crystal systems and 14 Bravais lattices that this corresponds to, with the combinations of the six parameters $\hat{\mathbf{a}}, \hat{\mathbf{b}}, \hat{\mathbf{c}}, \alpha, \beta$ and γ that fill the space by translation.

In Table 3.2, the group notations are for reference purpose so that you can interpret the meaning of different symmetries that they identify when you see different crystallographic point groups

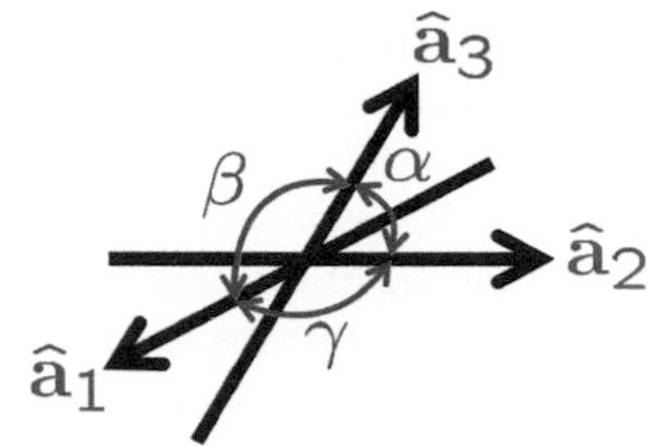

Figure 3.5: Definition of the unit cell vectors and angles of the coordinate system.

We are glossing over the very interesting complexities of quasicrystals here.

Crystal	Schoenflies/ International notation	Lattice properties	Bravais lattices
Triclinic	E or i/ 1 or $\bar{1}$	$a_1 \neq a_2 \neq a_3$ $\alpha \neq \beta \neq \gamma$	
Monoclinic	C_2 or σ/ 2 or $\bar{2}$	$a_1 \neq a_2 \neq a_3$ $\alpha = \beta = \pi/2 \neq \gamma$ 1st setting $\alpha = \gamma = \pi/2 \neq \beta$ 2nd setting	Simple, End-centered
Orthorhombic	Two C_2 or σ/ Two 2 or $\bar{2}$	$a_1 \neq a_2 \neq a_3$ $\alpha = \beta = \gamma = \pi/2$	Simple, Body-centered, End-centered, Face-centered
Tetragonal	C_4 or S_4/ 4 or $\bar{4}$	$a_1 = a_2 \neq a_3$ $\alpha = \beta = \gamma = \pi/2$	Simple, Body-centered
Cubic	Four 3-fold axes/ Four 3-fold axes/	$a_1 = a_2 = a_3$ $\alpha = \beta = \gamma = \pi/2$	Simple, Body-centered, Face-centered
Hexagonal	C_6 or S_3/ 6 or $\bar{6}$	$a_1 = a_2 \neq a_3$ $\alpha = \beta = \pi/2; \gamma = 2\pi/3$	
Trigonal/ Rhombohedral	C_3 or S_6/ 3 or $\bar{3}$	Same as hexagonal $a_1 = a_2 = a_3$ $\alpha = \beta = \gamma < 2\pi/3$ and $\neq \pi/2$	

Table 3.2: Crystal structure and Bravais lattices. The table identifies the 7 crystal systems together with the 14 Bravais lattices of three-dimensional crystals. Points that are behind a plane are shown in a lighter shade, and the table shows both the Shoenflies and international notations for identifying their symmetries.

with different materials of interest. For us, this is quite secondary. The Bravais lattices that these 7 crystal forms correspond to and that represent the variety seen in crystalline materials, quasicrystals excluded, are shown in the last column of the table. The various symmetries of bonding configuration possible define the various possibilities that the crystal structure may take.

The primitive or non-primitive identity of the unit cell now depends on choices we make. A primitive unit cell containing one lattice point is the spatial region that, when translated by all the Bravais lattice vectors, exactly fills the space. To illustrate this, consider the cubic lattice, which can appear in nature as simple cubic, face-centered cubic (*FCC*) or body-centered cubic (*BCC*). The *FCC* (and wurtzite, arising in the hexagonal) forms are of particular interest to us since they are the forms many semiconductors take. It is easy to see the simple cubic as a primitive form. The smallest cube with 8 lattice points as the corners has one total lattice point as the net number of lattice points per cell. It is therefore primitive. The primitive unit cells of the *FCC* and *BCC* Bravais lattice, which will have additional lattice point contribution from the face centers or the body centers and therefore are non-primitive, will be different. These are shown in Figure 3.6. Translated by $\hat{\mathbf{a}}_1$, $\hat{\mathbf{a}}_2$ and $\hat{\mathbf{a}}_3$ integrally, these primitive unit cells too fill the infinite lattice, as do the non-primitive through another set of translation vectors.

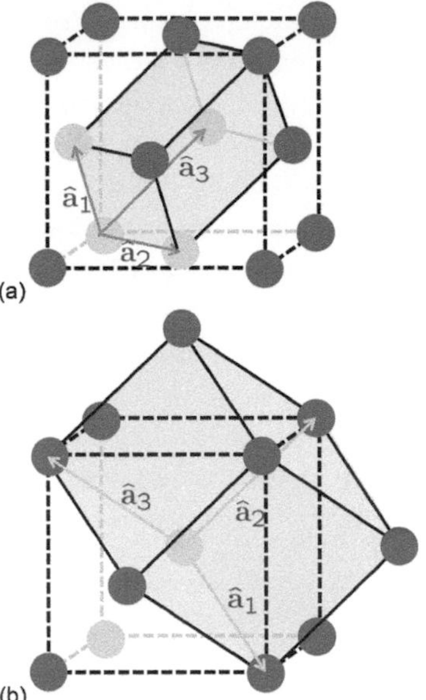

Figure 3.6: The primitive unit cell vectors for *FCC* (a) and *BCC* (b) lattices.

Most common semiconductors are diamond-like (a basis of two atoms of the same type), zinc-blende-like (a basis of two different atoms) or wurtzite-like (again, a two-atom basis). The former two have the translational symmetry of *FCC*, so of the cubic form. Wurtzite has that of the hexagonal form. Figure 3.7 shows these examples and other semiconductor forms.

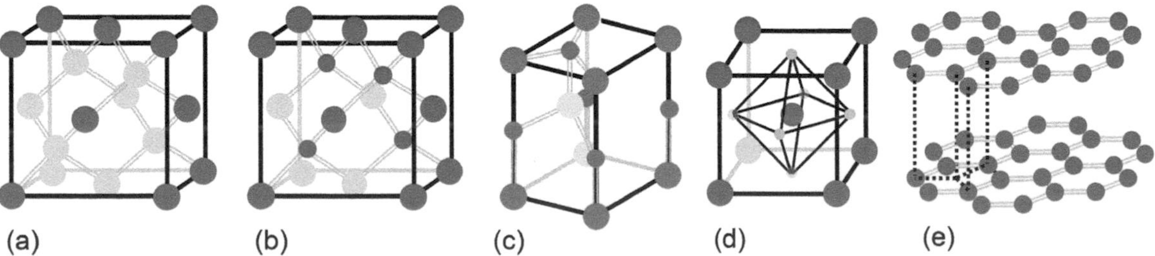

Figure 3.7: The non-primitive unit cells of some common semiconductors. *Si* and *Ge*, for example, occur in the diamond form shown in (a), *GaAs*, *InP*, *InAs*, et cetera are in the zinc blende form shown in (b), *GaN*, *AlN*, et cetera appear in the wurtzite form shown in (c), and a number of compounds of the ABO_3 form that are semiconducting, ferromagnetic, ferroelectric, et cetera appear in the perovskite form shown in (d). Two-dimensional semiconductors appear in various forms. The example here shows two layers of graphene in their natural stacking arrangement in (e).

We have now noted how the crystal description can be accomplished through primitive and non-primitive descriptions. These forms, unfortunately, don't quite easily show the symmetries that are so intimately related to the properties of the solid. A compact form of the unit cell that is primitive and displays the symmetry of the crystal is the Wigner-Seitz cell. This symmetry and compactness makes it very appealing for describing both the crystal and its analog in the reciprocal space, where it is referred as the first Brillouin zone, for the states that electron occupy. To construct it, start at any lattice point as the origin, draw vectors to neighboring lattice points and construct planes perpendicular to and passing through the midpoints; that is, the perpendicular plane that bisects the line connecting the neighboring point. The Wigner-Seitz cell is the cell with the smallest volume about the origin bounded by the planes. So, in the two-dimensional lattice example of Figure 3.8, the cell is the bounded area shown.

It is this bisection that connects this approach to the reciprocal space, the Fourier transform of the periodic real lattice space: a direction connected to a plane. This is therefore tied to the wavevector and hence the wave phenomena of the electron.

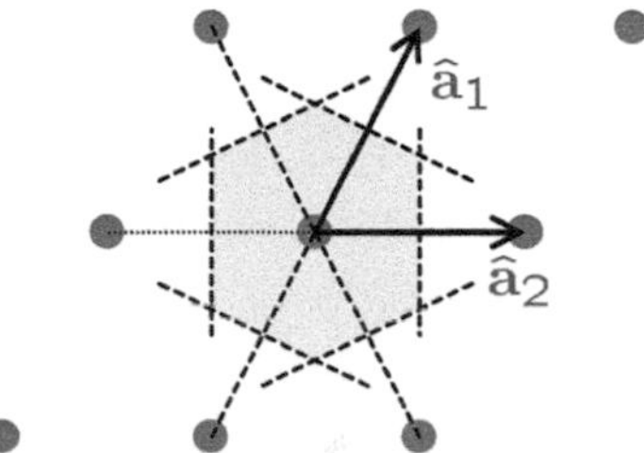

Figure 3.8: The Wigner-Seitz cell of a two-dimensional lattice whose primitive unit cell is a parallelogram.

Constructing a Wigner-Seitz cell is therefore a straightforward procedure; it is the visualization in three dimensions that becomes somewhat trickier. For the *BCC*, we show this bisecting in Figure 3.9(a)–(d), by truncating from the body center to the nearest neighbors. There are two lattice points per *BCC* (the body-center lattice point belongs entirely, and the eight corner lattice points each contribute 1/8th). The nearest neighbors to consider are those for the 8 corner points and the 6 points—above, below and 4 in the plane of the neighboring unit cells—in order to form the smallest repetitive unit. This results in a truncated octahedron that has 8 hexagonal faces in the 8 diagonal directions and 6 squares in the planes parallel to the cube's surfaces. This truncated octahedron fills the entire real space through translation. Figure 3.9(f) shows the Wigner-Seitz cell for an *FCC* lattice.

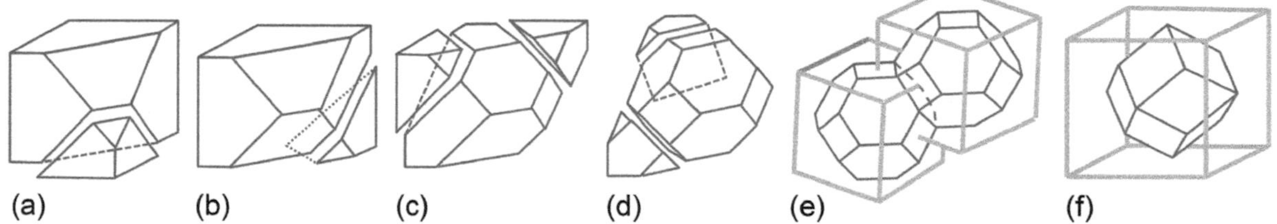

Figure 3.9: Construction of the Wigner-Seitz cell of a *BCC* lattice. Parts (a) through (d) show the bisection of the connection to the 8 corner lattice points, and the 6 body-center lattice points of the nearest 6 *BCC* cells. The resulting truncated octahedron can be repeated as shown in (e) to form the infinite lattice through translation. Here, a repetition in the $\langle 111 \rangle$ direction is shown. (f) shows the Wigner-Seitz cell of the *FCC* lattice, obtained again through the bisection process.

3.3 *Reciprocal lattice, Brillouin zones and real-reciprocal relationships*

SYMMETRY CONSTRAINS THE ELECTRON WAVEFUNCTIONS in the crystal to have the periodicity of the crystal structure. The periodic vibrations of atoms too will have to reflect the periodicity of the crystal structure.

For waves, the periodicity of position is also suitably seen through the wavevector, which reflects the spatial periodicity of the wave. So, in a one-dimensional periodic lattice, $\psi(x,t) \propto \exp[i(kx - \omega t)]$, at any instant of time, is a spatial wave with $\psi \propto \exp(ikx)$. The phase shifts by 2π every $2\pi/k$ in inverse spatial units. The wave properties of the allowed states can be more easily understood through the nature of k and dependences on it, since kinetic energies are related to momentum through the de Broglie relationship on wavelength, and the wavelength is $\lambda = 2\pi/k$. The wavevector therefore becomes tied to momentum.

We saw for the free electron the momentum as $\mathbf{p} = \hbar\mathbf{k}$ with $\mathbf{k}$ continuous. In a non-free environment—in any environment with spatially varying potential—this will have to change, but we will find that there is a meaningful momentum, which is not the actual momentum that one can usefully attach.

A reciprocal lattice is this mathematical construct in terms of the wavevector, which is an inverse transform of space lattice. With space, through the lattice construction, which is periodic, the reciprocal space too is periodic.

Let $\hat{\mathbf{b}}_1$, $\hat{\mathbf{b}}_2$ and $\hat{\mathbf{b}}_3$ be the basis vectors of the reciprocal lattice, so that any wavevector $\mathbf{k}$ can be written in terms of the basis vectors of the reciprocal space. These reciprocal lattice basis vectors in terms of the primitive lattice vectors, $\hat{\mathbf{a}}_1$, $\hat{\mathbf{a}}_2$ and $\hat{\mathbf{a}}_3$ of the 14 Bravais lattices, forming the volume $\hat{\mathbf{a}}_1 \cdot (\hat{\mathbf{a}}_2 \times \hat{\mathbf{a}}_3)$, follow the relationship

$$\begin{aligned}
\hat{\mathbf{b}}_1 &= 2\pi \frac{\hat{\mathbf{a}}_2 \times \hat{\mathbf{a}}_3}{\hat{\mathbf{a}}_1 \cdot (\hat{\mathbf{a}}_2 \times \hat{\mathbf{a}}_3)}, \\
\hat{\mathbf{b}}_2 &= 2\pi \frac{\hat{\mathbf{a}}_3 \times \hat{\mathbf{a}}_1}{\hat{\mathbf{a}}_1 \cdot (\hat{\mathbf{a}}_2 \times \hat{\mathbf{a}}_3)}, \text{ and} \\
\hat{\mathbf{b}}_3 &= 2\pi \frac{\hat{\mathbf{a}}_1 \times \hat{\mathbf{a}}_2}{\hat{\mathbf{a}}_1 \cdot (\hat{\mathbf{a}}_2 \times \hat{\mathbf{a}}_3)}, \text{ that is, cycles, and} \\
\hat{\mathbf{a}}_j \cdot \hat{\mathbf{b}}_l &= 2\pi\delta_{jl},
\end{aligned} \tag{3.13}$$

with j and l as the indexes for the real space and reciprocal space unit vectors. The reciprocal lattice vectors are perpendicular to the two differently indexed vectors of the real space.

This is what the bisecting plane does!

Similar to the translational operator $\mathbf{T}_m$ for the real space lattice (Equation 3.12), the translational operator for the reciprocal lattice is

$$\mathbf{K}_{m'} = m'_1\hat{\mathbf{b}}_1 + m'_2\hat{\mathbf{b}}_2 + m'_3\hat{\mathbf{b}}_3. \tag{3.14}$$

This results in

$$\mathbf{K}_{m'} \cdot \mathbf{T}_m = 2\pi(m_1' m_1 + m_2' m_2 + m_3' m_3). \tag{3.15}$$

The basis vectors of the reciprocal lattice define the primitive unit cell of the reciprocal cell. It is the Wigner-Seitz cell of the reciprocal lattice and has a volume in **k**-space of

$$\delta\Omega_k = \hat{\mathbf{b}}_1 \cdot (\hat{\mathbf{b}}_2 \times \hat{\mathbf{b}}_3). \tag{3.16}$$

While $\exp[i\mathbf{k} \cdot \mathbf{T}_m]$ is not unity in general, the specific reciprocal lattice points represented by $\mathbf{k} = \mathbf{K}_{m'}$ follow

$$\exp[i\mathbf{K}_{m'} \cdot \mathbf{T}_m] = 1. \tag{3.17}$$

So, the reciprocal space (**k**-space), where, in general, $\exp(i\mathbf{k} \cdot \mathbf{T}_m) \neq 1$, there are specific points, corresponding to the real space where $\exp[i(\mathbf{k} = \mathbf{K}_{m'}) \cdot \mathbf{T}_m] = 1$, with the basis vectors $\hat{\mathbf{b}}_1$, $\hat{\mathbf{b}}_2$ and $\hat{\mathbf{b}}_3$ integer translated. This is the origin of Equation 3.13. The physical meaning of this phase-related property is that the periodicity of the real space causes specific phase matching attributes to appear in the reciprocal space, which will have consequences. The uniqueness of the real space lattice points also maps to the uniqueness of the reciprocal space points. Wave interferences—transmissions and reflections—arise from perturbations and phase changes along the wave propagation, so the periodic boundaries of the lattice also cause interference in the wave motion—for electrons and phonons. The wavevector representation in the reciprocal space will show such effects more clearly.

Since **k** is a vector, and since $\mathbf{K}_{m'}$ followed from the phase relationship, it reflects a direction in the real space. A wave $\exp i(kx - \omega t)$ is traveling in the direction $\hat{\mathbf{x}}$, so **k** here points in the $\hat{\mathbf{k}}_x$ direction, which is normal to the y-z plane. A point in the **k**-space is a direction in the real space.

The construction of the reciprocal space and Equation 3.13 follows from the use of the $\exp(i\mathbf{K}_{m'} \cdot \mathbf{T}_m) = 1$ relationship. In the simplest of cases—a one-dimensional lattice with a as the spacing between the lattice points—the reciprocal lattice following this equation consists of the points that correspond to $\mathbf{K}_{m'} \cdot \mathbf{T}_m = 2n\pi$, where n is an integer, leading to $K_{m'} = 2m'\pi/a$. The Brillouin zones are the symmetric zones formed by bisection. The Brillouin zone construction follows the same approach of bisection that was employed for the Wigner-Seitz cell construction of the real space. The first one of these, the 1st Brillouin zone ($m' = 1$), where we will largely explore the electron state behavior, then stretches from $-\pi/a$ to π/a. The construction by bisection is illustrated in Figure 3.10(a) and (b) for a hexagonal and a square lattice, respectively. For drawing the first Brillouin zones, the vector connecting to the nearest neighbor collection of sites is bisected, and the region enclosed is this 1st Brillouin zone. When the next neighbor group is taken and the procedure repeated, with the region of the 1st Brillouin zone excluded, this gives the 2nd Brillouin zone. All

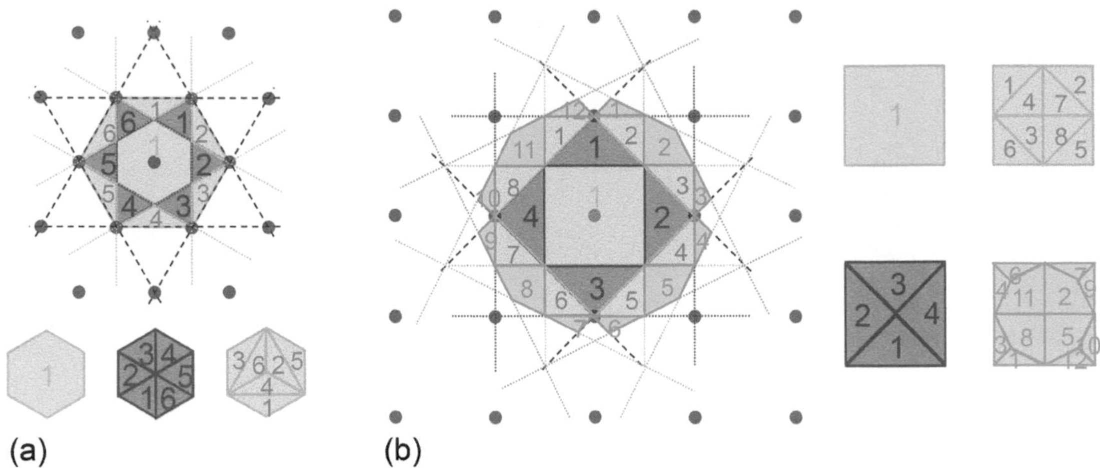

Figure 3.10: Illustration of Brillouin zone construction. Part (a) shows it for a hexagonal lattice, and (b) for a square lattice. In both, the construction follows by taking nearest neighbor sites, in ascending groups of nearest, next-nearest- and so on, neighbors and bisecting the vector to them. The 1st Brillouin zone is the region enclosed with the nearest neighbor group bisection. Then, the second group gives a region out of which the first Brillouin zone is excluded to form the second Brillouin zone. All Brillouin zones are of equal length (one-dimensional lattice), area (two-dimensional lattice) or volume (three-dimensional lattice), and this is seen from the first three Brillouin zones for the hexagonal lattice, and the first four Brillouin zones shown for the square lattice in the figure.

Brillouin zones have the same extent. Here, this is shown for the first 3 Brillouin zones for the hexagonal lattice, and the first 4 for the square lattice.

Since the basis vectors in real space are of the order of 0.6 *nm*, the reciprocal space basis vectors are of the order of $2\pi/0.6\ nm \approx 10^{10}\ m^{-1}$. The allowed **k**s are much closer than this since they are determined by the extent of the crystal, which is much larger than the interatomic spacing. We will presently show that the 1st Brillouin zone will usually suffice to describe the electron behavior in the crystal.

For the more complex real three-dimensional forms of semiconductors, we may construct the symmetric smallest unit cell for reciprocal cell—this 1st Brillouin zone—in the same way as for the real space so that much of what happens can be represented within it for convenience. This Wigner-Seitz cell in the reciprocal space is the first Brillouin zone. Since the relationship of Equation 3.17 is a relationship that establishes orthogonality and bisection, the consequence is that the Wigner-Seitz cell of the reciprocal space, the Brillouin zone and the Wigner-Seitz cell of the real space have correspondences. In the case of *BCC*, the real space Wigner-Seitz cell is a truncated octahedron. For the reciprocal space, for this cell, the first Brillouin zone is a rhombic dodecahedron. Similarly, in the case of *FCC*, the real space Wigner-Seitz cell is a rhombic dodecahedron, so, for the reciprocal space, for this symmetric cell, the first Brillouin zone is a truncated octahedron. Figure 3.11(a) shows the smallest such unit—the first Brillouin zone for *FCC*—which is the unit cell from which the non-primitive zinc blende or diamond crystal forms appear. Figure 3.11(b) shows it for the *HCP*—which is the unit cell from which the non-primitive wurtzite crystal form appears. If $\mathbf{k}$ is restricted to the 1st Brillouin zone, any vector $\mathbf{k}' = \mathbf{k} + \mathbf{K}_{m'}$ allows one to fold it back to the first 1st Brillouin zone. This 1st Brillouin zone too has symmetry, and one would employ a number of locales within this to describe the $(E, \mathbf{k})$ states available in it. In Figure 3.11(a), the center [000] is the Γ point. The locales $\langle 100 \rangle$—six of them—are the X points, and the locales $\langle 111 \rangle$ are the L points. The K point is the [110]. The direction from Γ to X is the Δ direction. The one from Γ to L is the Λ direction, and the direction from Γ to K is the Σ direction. Table 3.3 summarizes some the characteristics of these notable points that we will refer to often.

The *FCC*'s 1st Brillouin zone will be of interest for us for diamond and zinc blende crystals (*Si*, *GaAs*, etc.) and the *HCP*'s

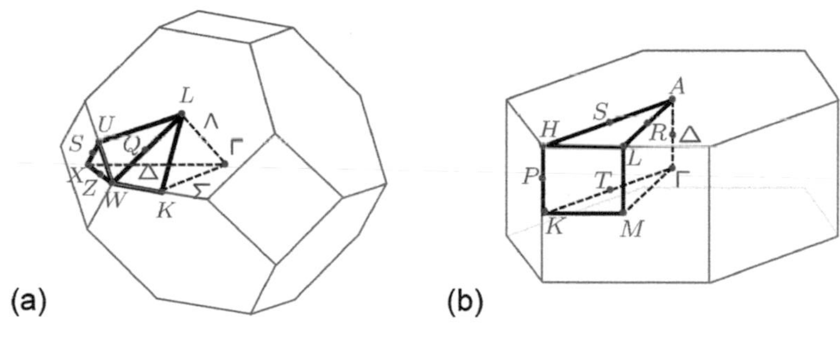

Figure 3.11: Part (a) shows the 1st Brillouin zone of the *FCC* lattice. The figure also shows the smallest irreducible region within this symmetric cell, identifying some of the significant reciprocal lattice points and directions. Λ, Δ and Σ are directions that tag Γ to L, Γ to X, and Γ to K. The 1st Brillouin zone of the *FCC* lattice is identical in shape to the Wigner-Seitz cell of the *BCC* lattice. Part (b) shows the 1st Brillouin zone of the hexagonal lattice, again together with the minimum irreducible region within the symmetric cell, and a few of the significant reciprocal lattice points.

Symmetry point	Wavevector $\mathbf{k}$	Degeneracy
Γ	0	1
L	$\pm(\pi/a)\langle 111 \rangle$, $\pm(\pi/a)\langle \bar{1}11 \rangle$, $\pm(\pi/a)\langle 1\bar{1}1 \rangle$ and $\pm(\pi/a)\langle 11\bar{1} \rangle$	8
X	$\pm(2\pi/a)\langle 100 \rangle$, $\pm(2\pi/a)\langle 010 \rangle$ and $\pm(2\pi/a)\langle 001 \rangle$	6
K	$\pm(3\pi/2a)\langle 110 \rangle$, $\pm(3\pi/2a)\langle \bar{1}10 \rangle$, $\pm(3\pi/2a)\langle 011 \rangle$, $\pm(3\pi/2a)\langle 0\bar{1}1 \rangle$, $\pm(3\pi/2a)\langle 101 \rangle$ and $\pm(3\pi/2a)\langle 10\bar{1} \rangle$	12
W	$\pm(\pi/a)\langle 210 \rangle$, $\pm(\pi/a)\langle 021 \rangle$, $\pm(\pi/a)\langle 102 \rangle$	6

Table 3.3: Some of the important symmetry points and their degeneracy in the 1st Brillouin zone of an *FCC* lattice.

1st Brillouin zone will be of interest for the wurtzite crystals (*GaN*, *AlN*, etc.).

This above points to the important relationship that exists between the reciprocal lattice vector $\mathbf{K} = \mathbf{K}_{m'}$ and the planes of corresponding direct lattice. A lattice plane is uniquely defined by three non-collinear sites. Each plane consists of infinite number of sites. Figure 3.12 shows an example simplified to two dimensions for ease of viewing using two non-collinear sites. Imagine the third dimension as being perpendicular to this plane. Each $\mathbf{K}$ of the reciprocal lattice is normal to some set of planes in the direct lattice, and the length of $\mathbf{K}$ is inversely proportional to the spacing between the planes of this set. For

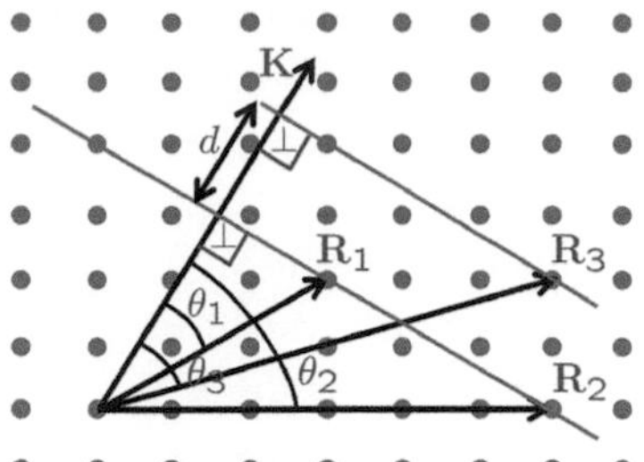

Figure 3.12: Relating reciprocal lattice vectors with real space planes and their orientation relationships. The third real space dimension is not shown.

$$\mathbf{R} = m_1\hat{\mathbf{a}}_1 + m_2\hat{\mathbf{a}}_2 + m_3\hat{\mathbf{a}}_3 \text{ where } m_1, m_2, m_3 \in \{0, 1, 2, \ldots\}, \quad (3.18)$$

$$\mathbf{K} = m'_1\hat{\mathbf{b}}_1 + m'_2\hat{\mathbf{b}}_2 + m'_3\hat{\mathbf{b}}_3 \text{ where } m'_1, m'_2, m'_3 \in \{0, 1, 2, \ldots\}, \quad (3.19)$$

and the $\hat{\mathbf{b}}$s, follow Equation 3.13. The unit vector $\hat{\mathbf{b}}_1$ is normal to the plane formed by the unit vectors $\hat{\mathbf{a}}_2$ and $\hat{\mathbf{a}}_3$, $\hat{\mathbf{b}}_2$ is normal to the plane of $\hat{\mathbf{a}}_3$ and $\hat{\mathbf{a}}_1$, and $\hat{\mathbf{b}}_3$ is normal to the plane of $\hat{\mathbf{a}}_1$ and $\hat{\mathbf{a}}_2$. The indices (m'_1, m'_2, m'_3) are the lowest common integer factors.

The product

$$\mathbf{K} \cdot \mathbf{R} = 2\pi(m_1m'_1 + m_2m'_2 + m_3m'_3) = 2\pi N, \quad (3.20)$$

where N is an integer. Therefore, $\exp(\mathbf{K} \cdot \mathbf{R}) = 1$, which is how we started our defining discussion of the reciprocal space. Figure 3.12 also shows the relationship of the projections, relating points in parallel real space planes. The planes through these lines will be normal to these two dimensions. For a point $\mathbf{R}_1$ or point $\mathbf{R}_2$ in the same plane,

$$|\mathbf{R}_1|\cos\theta_1 = |\mathbf{R}_2|\cos\theta_2 = \frac{2\pi N}{|\mathbf{K}|}. \quad (3.21)$$

For $\mathbf{R}_2$, what this figure implies is that

$$\mathbf{R}_2 = (m_1 - pm'_3)\hat{\mathbf{a}}_1 + (m_2 - pm'_3)\hat{\mathbf{a}}_2 + [m_3 + p(m'_1 + m'_2)]\hat{\mathbf{a}}_3, \quad (3.22)$$

where p is an integer. An infinite set of such points exists on the same plane for a constant N. The next adjacent plane orthogonal to $\mathbf{K}$ is defined by

$$|\mathbf{R}_3|\cos\theta_3 = \frac{2\pi(N+1)}{|\mathbf{K}|}. \quad (3.23)$$

The spacing between the planes (d) is

$$d = \frac{2\pi(N+1)}{|\mathbf{K}|} - \frac{2\pi N}{|\mathbf{K}|} = \frac{2\pi}{|\mathbf{K}|}. \quad (3.24)$$

The use of lattice vectors to designate planes in a direct lattice, that is, the use of the normal vector to the planes of the direct lattice plane, is equivalent to the use of $(m'_1 m'_2 m'_3)$, which we employed in the reciprocal space to identify reciprocal space lattice points. These are the Miller indices.

Claim: The reciprocal lattice coordinates correspond to a direction in real space that is normal to a plane.

Proof: For a given $\mathbf{K}$, choose m_is so that one plane of the set with

$$\mathbf{K} \cdot \mathbf{R} = 2\pi \left(m'_1 m_1 + m'_2 m_2 + m'_3 m_3 \right) = 2\pi N \tag{3.25}$$

intersects in the direction of $\hat{\mathbf{a}}_i$ at $m_i \hat{\mathbf{a}}_i$ for $i = 1, 2, 3$, that is,

$$\begin{aligned} \mathbf{K} \cdot m_1 \hat{\mathbf{a}}_1 &= 2\pi m'_1 m_1, \\ \mathbf{K} \cdot m_2 \hat{\mathbf{a}}_2 &= 2\pi m'_2 m_2, \text{ and} \\ \mathbf{K} \cdot m_3 \hat{\mathbf{a}}_3 &= 2\pi m'_3 m_3 = 2\pi N, \end{aligned} \tag{3.26}$$

with

$$m_1 = \frac{N}{m'_1}, \; m_2 = \frac{N}{m'_2}, \text{ and } m_3 = \frac{N}{m'_3}. \tag{3.27}$$

The intercepts of the planes are inversely proportional to the integral components of the reciprocal lattice vector.

Miller indices are thus particularly useful for identifying planes and directions. For example, in the case of a cubic lattice, Miller indices correspond to the three orthogonal simple cubic vectors. A direct lattice plane is a point in the reciprocal lattice. $(m'_1 m'_2 m'_3)$, a point in the reciprocal lattice space, indicates a plane in the real lattice space that intercepts at m_1 units in the $\hat{a}_1$ direction, m_2 units in the $\hat{a}_2$ direction, and m_3 units in the $\hat{a}_3$ direction. If the intercept is at infinity between a plane and a direct lattice vector, the Miller index, which is proportional to $1/m_i$, following Equation 3.27, is 0.

By convention, planes in the direct lattice are denoted by parentheses (round brackets). For example, (001) is a plane that intercepts the third direction at the unit vector and does not intercept in the other two directions. Directions in direct lattices and planes in reciprocal lattices are denoted by square brackets. For example, [111] is along the diagonal of the cube and is a point at [111] in the reciprocal lattice. Families of planes are denoted by curly brackets, so $\{100\} \equiv (100), (010), (001), (\bar{1}00), (0\bar{1}0)$ and $(00\bar{1})$, which are the six faces of the cube with unit intercepts along the orthogonal unit vectors. Families of directions are denoted by angular brackets, so $\langle 100 \rangle \equiv [100], [010], [001], [\bar{1}00], [0\bar{1}0]$ and $[00\bar{1}]$. These are directions in real space and points in reciprocal space. Figure 3.13 illustrates different planes in the real space for a cubic lattice. Normals to these planes will define a point in the reciprocal space.

The Fourier transform techniques for periodic signals—here periodic in space—give us a convenient analogy for tackling this problem. The reciprocal space is the space of the Fourier transform, where periodicity of real space has mapped to periodicity of reciprocal space ($\mathbf{R} \mapsto \mathbf{K}$), with $\exp\left[i(\mathbf{K}_{m'} \cdot \mathbf{T}_m\right) = 1$ prescribing the Fourier transformation constraint. There are multiple ways to interpret Fourier transformation: frequency-, wavevector-, time-space varying positive integration (single sideband), or integration from negative infinity to positive infinity (double sideband) with an apportioning of the round-trip path to the starting state (a 2π that is split as $\sqrt{2\pi}$ or not). I subscribe to maintaining symmetry.

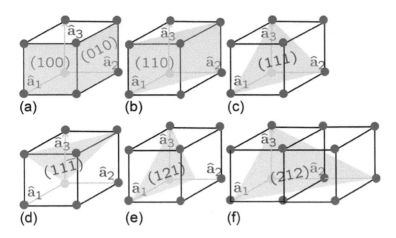

Figure 3.13: Examples of planes and directions in Miller index notations for a cubic lattice.

3.4 *Electrons in the periodic potential and the reduced zone*

IN THE RECIPROCAL SPACE of wavevector $\mathbf{k}$, $\mathbf{K}_{m'}$ represents lattice points that have periodicity defined by the basis vectors of the reciprocal space. In the real space, $\mathbf{T}_m$ defines lattice points that are periodic analogs in the real space. We have explored the free electron through its wavefunction description in the beginning of this chapter. We noted an energy dispersion of this one electron existing in free space following an E-$\mathbf{k}$ relationship that was sketched in Figure 3.1. It still holds even if we describe the positions in real space with a real space lattice description. Note that we have not introduced any potential perturbations. It is still free space, except where we have introduced a periodicity in description through the lattice. How does one represent this behavior of a free electron in the reciprocal space with this lattice-based gedankenerfahrung? Our solid will come about by the introduction of the atoms at the lattice sites.

The lattice is a mathematical construct that we will use as a stepping stone for building a crystal, where atoms will introduce physical interactions as a periodic potential.

Since $\exp[i\mathbf{K}_{m'} \cdot \mathbf{T}_m] = 1$, the single one-electron dispersion of Figure 3.1, when viewed in the reciprocal lattice, has infinite copies of itself, as seen in Figure 3.14, because $E(\mathbf{k}) = E(\mathbf{k} - K_{m'})$.

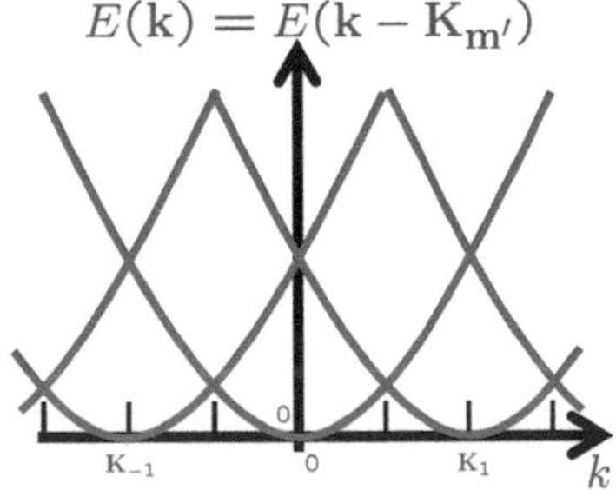

Figure 3.14: With the introduction of the lattice and its reciprocal lattice, the one-electron dispersion of real space is replaced by an infinite number of copies in the reciprocal space. Each of these is a valid solution, since $\exp[i\mathbf{K}_{m'} \cdot \mathbf{T}_m] = 1$.

The $\mathbf{k}$-space periodicity also implies that *all the information is contained in the primitive unit cell of the reciprocal lattice;* that is, in the first Brillouin zone. The periodicity also brings with it one additional important property. Let there be N_j primitive unit cells in the jth direction with the unit vector $\hat{\mathbf{a}}_j$ for $j = 1, 2, 3$. $N = N_1 N_2 N_3$ is the total number of primitive unit cells. Any propagating solution—not a standing wave—should reflect the translational symmetry of the lattice, so a solution $|\phi\rangle$ satisfies

$$\phi(\mathbf{r} + N_j \hat{\mathbf{a}}_j) = \phi(\mathbf{r}). \tag{3.28}$$

This boundary condition—the Born-von Karman boundary condition—then also forces the constraint

$$\exp(iN_j\mathbf{k}\cdot\hat{\mathbf{a}}_j) = 1 \text{ for } j = 1,2,3. \tag{3.29}$$

A trivial example of this is the plane wave solution $\exp[i(\mathbf{k}\cdot\mathbf{r}-\omega t)]$.

Since $\exp(i\mathbf{K}_{m'}\cdot\mathbf{T}_m) = 1$, and $\hat{\mathbf{a}}_j\cdot\hat{\mathbf{b}}_l = 2\pi\delta_{jl}$, the **k**s allowed are constrained to

$$\mathbf{k} = \sum_{j=1}^{3}\frac{m_j}{N_j}\hat{\mathbf{b}}_j. \tag{3.30}$$

For every unit change of the integer m_j, a new state is generated. The volume in the **k**-space associated with this state is

$$\frac{\hat{\mathbf{b}}_1}{N_1}\cdot\left(\frac{\hat{\mathbf{b}}_2}{N_2}\times\frac{\hat{\mathbf{b}}_3}{N_3}\right) = \frac{1}{N}\hat{\mathbf{b}}_1\cdot(\hat{\mathbf{b}}_2\times\hat{\mathbf{b}}_3). \tag{3.31}$$

Since $\hat{\mathbf{b}}_1\cdot(\hat{\mathbf{b}}_2\times\hat{\mathbf{b}}_3)$ is the volume of the first Brillouin zone, *the first Brillouin zone contains the same number of* **k** *states as the number of primitive unit cells in the crystal.*

The ability to represent all the states of the real space in the first Brillouin zone of the reciprocal space makes the description of crystalline solids enormously convenient.

The reader should see in this description—of real and reciprocal—the connection to Fourier transformation. We are performing periodic sampling, and the Fourier transformations' properties are manifesting. The generalized periodic function that will be the solution for periodic potentials—the Bloch function—is a Fourier function.

So, we can now extend this description to the periodic potential of the crystal environment.

First, consider a hypothetical empty crystal: a crystal with no potential perturbation from the basis at the lattice sites. The electron Hamiltonian accounts for all the kinetic and potential energy terms arising from forces and interactions with other particles. If we assume that we have a very dilute gas of electrons, so that they behave as independent particles—as if other electrons didn't exist—then the electron in this crystal is a free electron and the Hamiltonian operator is

$$\hat{\mathscr{H}} = -\frac{\hbar^2}{2m_0}\nabla^2 + \hat{V} = -\frac{\hbar^2}{2m_0}\nabla^2, \tag{3.32}$$

and

$$\hat{\mathscr{H}}\psi = E\psi \quad \therefore -\frac{\hbar^2}{2m_0}\nabla^2\psi = E\psi \tag{3.33}$$

is the model problem for this potential-free periodic crystal. Table 3.4 describes some of the differences in characteristics of the free electron in free space versus the free electron in perturbation-free periodic structure.

Figure 3.14 shows a pictorial description of the solutions for a free electron in free space, and a free electron in potential-free periodic space. The free electron in free space has a second power E-**k** relationship defining the energy and wavevectors allowed.

Table 3.4: A free electron in free space and in a periodic structure where no potential perturbations exist.

Free electron "Sommerfeld"-like	Free electron in a perturbation-free periodic structure
$\psi_k = \mathcal{A}\exp(i\mathbf{k}\cdot\mathbf{r})$	$\psi_k = \mathcal{A}\exp[i(\mathbf{k}-\mathbf{K}_{\mathbf{m}'})\cdot\mathbf{r}]$
$E(\mathbf{k}) = \hbar^2\mathbf{k}^2/2m_0$	$E(\mathbf{k}) = E(\mathbf{k}-\mathbf{K}_{\mathbf{m}'}) = \hbar^2\mathbf{k}^2/2m_0$
$\mathbf{p} = \hbar\mathbf{k}$	$\mathbf{p}(\mathbf{k}) = \mathbf{p}(\mathbf{k}-\mathbf{K}_{\mathbf{m}'}) = \hbar\mathbf{k}$

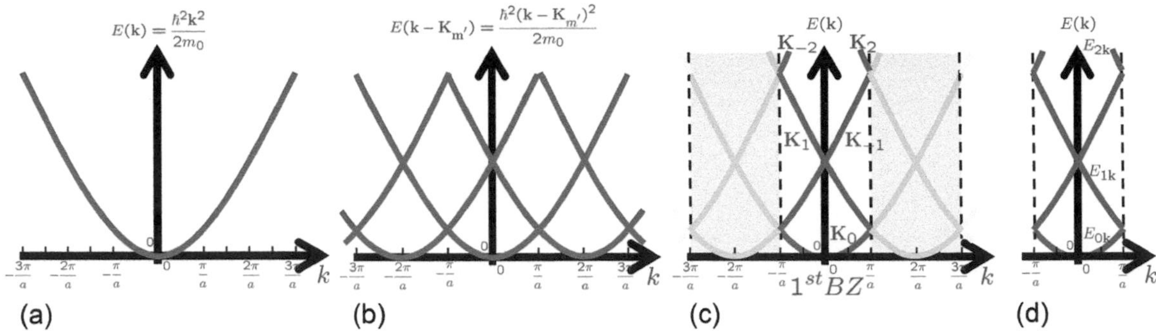

Figure 3.15: Part (a) shows the $E(\mathbf{k})$ dispersion of a free electron in free space. Parts (b) through (d) show the dispersion for a potential-free periodic crystal. The electron is still free, although it is in a periodic crystal with no potential. (b) shows $E(\mathbf{k})$ in the repeated, or periodic or extended, zone view and (c) shows the reduced zone scheme. Part (c) shows the identity of, via the $\mathbf{K}_{m'}$ notation, the Brillouin zone (BZ) from which the band arose. Since the first Brillouin zone has complete information, and the number of points in it are the number of primitive unit cells in the real space structure, Part (d) as a reduced zone picture suffices to describe the energy dispersion. There are now two quantum numbers to identify states: the wavevector and the band index.

In the hypothetical periodic crystal, there are repeated energy solutions allowed where, at each $\mathbf{k} - \mathbf{K}_{m'}$, there exist states with free electron energy whose band is centered at $\mathbf{k} = \mathbf{K}_{m'}$ for electrons with wavefunctions proportional to $\exp\{i[(\mathbf{k}-\mathbf{K}_{m'})\cdot\mathbf{r}-\omega t]\}$.

Figure 3.15(a) shows the energy-wavevector relationship of the free electron. There is nothing unusual here. It is the dispersion relationship of a wave given by a wavefunction proportional to $\exp[i(\mathbf{k}\cdot\mathbf{r}-\omega t)]$. This is the extended zone dispersion of the free electron in this one-dimensional zero potential space. Figure 3.15(b) shows the free electron in a potential-free periodic space. The allowed $\mathbf{k}$ states are separated by $2\pi/L$—a small number for large L—with the electron bounded over the span L of this solid cavity. $L \rightarrow \infty$ makes $\mathbf{k}$ asymptotic to continuity. The π/a markings are unique reciprocal space points derived from the $\exp(i\mathbf{K}_{m'}\cdot\mathbf{T}_m) = 1$ constraint and are useful for drawing the primitive unit cells of the reciprocal lattice, with the first Brillouin zone being the primitive version of the Wigner-Seitz type. Figure 3.15(b) is the repeated—or periodic—zone picture, whose energy dispersion curves we have established are also a solution because $E(\mathbf{k}) = E(\mathbf{k} - \mathbf{K}_{m'})$. Since we have established that the first Brillouin zone contains all the information and that the number of states in the first Brillouin zone reflect the number of primitive unit cells, it is convenient to view all the states within the first Brillouin zone (Figure 3.15(c)). The energy of the states for **k**s lying in the 1st Brillouin zone, but arising in the bands whose lowest energy reaches the minimum in the two Brillouin zones adjacent to the first Brillouin zone, are higher than the energies of the states arising from the band of the first Brillouin zone. Similarly, further up are bands with energies from the Brillouin zones of higher $\mathbf{K}_{m'}$. In Figure 3.15(c), these are identified by $\mathbf{K}_{m'}$ to identify the Brillouin zone that they arose from. All the energy states of the repeated zone are still here in the first zone. The nuance is that as we go higher in energy, we are now identifying those energies by the different bands. So, in Figure 3.15(d) we indicate this via an additional quantum number, n, which is a band index associated with the wavevector being referenced from a different Brillouin zone. *Two quantum numbers, n and $\mathbf{k}$, suffice to identify the states.* The states beyond the first Brillouin zone have been reduced to the 1st Brillouin zone, so this is the reduced zone picture, and we now have multiple bands that can be seen, indexed here as 0, 1 and 2 in the notation $E_{n\mathbf{k}}$ to identify their energy span of the states.

Take now the zinc blende crystal with its *FCC* Bravais lattice (in *GaAs*, this is a *Ga FCC* interpenetrating along the diagonals in an *As FCC*, and the unit cell has 2 basis atoms) and consider the free electron states in it, so still keep the problem potential-free. The periodicity is now different, and the calculation a little more exacting, but one will see the $(n, \mathbf{k})$ dependence of energy in it too. Figure 3.16 shows this. All these energy changes are still parabolic in **k** with the free electron mass. But since the different wavevector magnitudes are in different directions drawn along a line, the projections show up as visually different. If there were only one electron per primitive cell, only the lowest band—parabolic of course, but looking slightly different toward L, X or K—is filled. Higher bands are occupied as more electrons—still free, independent electrons—are contributed per unit cell.

Note that, so far, all these allowed E-$\mathbf{k}$ states are still free electron states in the zero-perturbation periodic structure. What happens if there is some periodic potential perturbation? The perturbation, even if very small, is a source of interference, so long as it is not adiabatic, which would suppress reflections. In a waveguide, the existence of boundary conditions—metal in microwaves, or an index change in fibers—places constraints on the wave modes that can exist within. Modes do not propagate in the directions toward

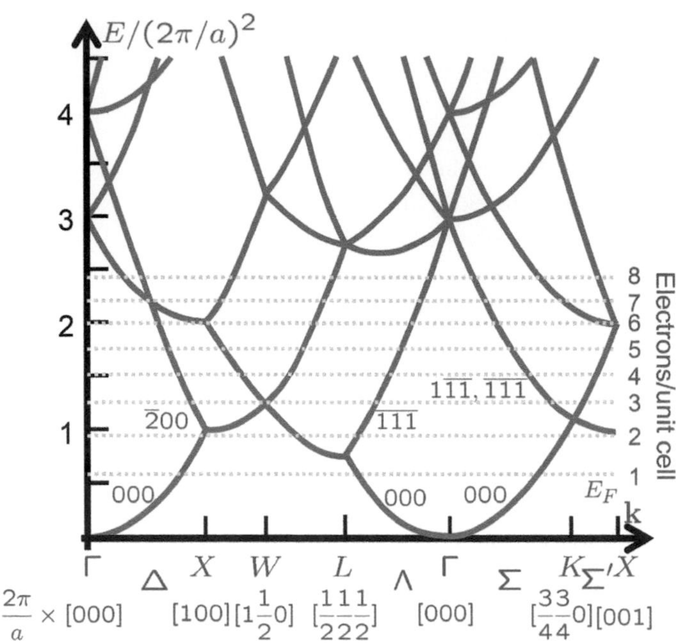

Figure 3.16: The free electron $E(n, \mathbf{k})$ bandstates diagram of a free potential zinc blende crystal, that is, with an *FCC* lattice with the free electron bands translated into the first Brillouin zone. The lower bands also identify the translation vectors for some of the lower bands. Δ, Λ, Σ and Σ′ identify points that belong on the line connecting Γ to *X*, Γ to *L*, Γ to *X*, and *X* to *K* and so are useful as indicators of the direction. Depending on the number of electrons arising per unit primitive cell, the Fermi energy (E_F) will have filled states up to the dotted lines at absolute zero.

these confining boundaries. The lowest frequency transverse electric mode for the microwave waveguide, with its highly conductive wall, arises because the longest wavelength solution is a half wave whose electric field in the plane of the metal at the boundary vanishes. This half wave represents a standing wave because the metal does not support an electric field. The energy cannot propagate beyond, and therefore conditions must come about that make the energy propagation disappear throughout in this direction. The standing wave here can be viewed as two counter-propagating waves—the half-wave sinusoidal electric field is a sum of two exponentials with imaginary arguments.

So, as soon as we introduce a periodic perturbation, it will cause interference. Diffraction of the wave of wavevector **k** occurs when $(\mathbf{k}+\mathbf{K}_{m'})^2 = k^2$. In a one-dimensional crystal, this is at $k = \pm K_{m'}/2 = \pm m'\pi/a$, where m' is an integer. So,

$$|\psi^+\rangle \propto \exp\left[i\left(\frac{\pi z}{a} - \omega t\right)\right] + \exp\left[i\left(-\frac{\pi z}{a} - \omega t\right)\right]$$
$$= 2\cos\left(\frac{\pi z}{a}\right)\exp(-i\omega t) \tag{3.34}$$

describes a standing wave arising from $m' = 1$ with an infinitesimally small perturbation. For a linear plane wave, in this gedanken-erfahrung crystal, we have a charge modulation with the electron charge varying as $\rho(+) \propto |\psi^+|^2 \propto \cos^2(\pi z/a)$. This specific condition of periodicity prohibits energy transmission in this direction at this specific **k**. An energy gap exists at this wavevector in an energy range for the electron as shown in Figure 3.17. The size of the energy gap is related to the size of the potential perturbation. The energy gap vanishes as the perturbation vanishes, and we are back to the description of the free electron in a free crystal with no potential perturbation of Table 3.4.

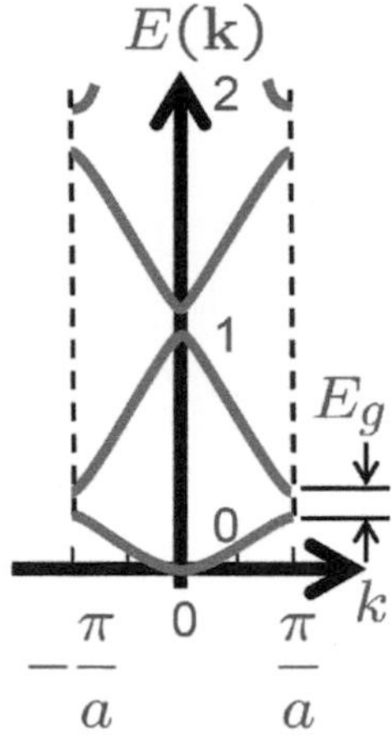

Figure 3.17: The appearance of a bandgap at the zone edge, due to periodic perturbation with a spatial periodicity of a.

A schematic view of this description is as shown in Figure 3.18, where there exists a perturbation arising from the crystal assembly. The electron states that we are interested in arise from the interactions of the valence states of the atom, while the bound states arise from core states of the atoms. Periodicity causes the charge accumulation and depletion, the accumulation occurs as $\cos^2(m'\pi z/a)$ with maxima at $z = 0, \pm a, \pm 2a, \ldots$, and depletion occurs as $\sin^2(m'\pi z/a)$ with minima at $z = 0, \pm a/2, \pm 3a/2, \ldots$. The charge density is highest in the region proximate to the nuclei, and lowest in between.

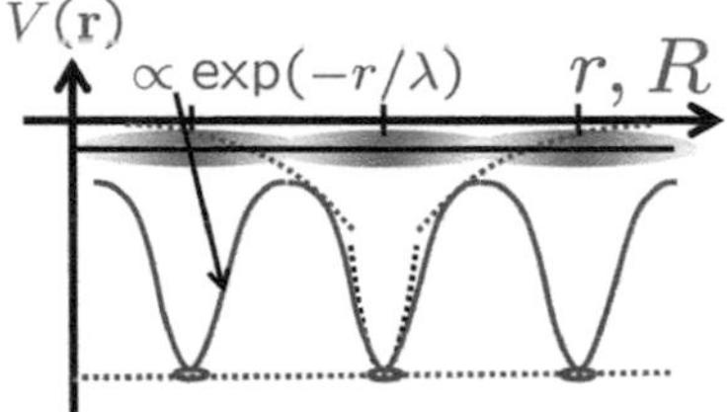

Figure 3.18: An illustrative view of charge density and potential in a semiconducting periodic chain of atoms. The dressed nuclei exercise a periodic Coulomb attraction that leads to a charge density distribution of the conducting electrons, with a maximum centered around the equilibrium position of the atoms. Farther away, this Coulomb potential decays exponentially.

This periodic perturbation gives rise to bands of allowed states of $(E, \mathbf{k})$ for electrons. To find them, we need to calculate much more rigorously by incorporating the periodic potential within the crystal. The different approaches in use draw on ease of use, computational efficiency and accuracy of the predictions of parameters of interest. We will explore this in Chapter 4.

3.4.1 *Bloch's theorem*

UNDERLYING ALL THESE APPROACHES is the use of Bloch's theorem, which provides a convenient means of incorporating periodicity in the quantum-mechanical solution. Bloch's theorem draws on the approach of Fourier analysis to bring in the periodicity in the wave solution. Instead of the plane wave that we started with, we will find that the solution is a function with spatially periodic modulation. The eigenvalues, that is, $(E, \mathbf{k})$, of this solution comprise the dispersion relationship we are interested in.

To find the eigenstate ψ of the one-independent-electron Hamiltonian with a spatial periodicity of $\mathbf{R}$ in a Bravais lattice, we start with the Schrödinger equation for the problem of

$$\hat{\mathscr{H}} = -\frac{\hbar^2}{2m_0}\nabla^2 + V(\mathbf{r}), \text{ where } V(\mathbf{r}+\mathbf{R}) = V(\mathbf{r}). \tag{3.35}$$

Bloch's theorem states that the solution is of the form

$$\psi_{n\mathbf{k}}(\mathbf{r}) = \exp(i\,\mathbf{k}\cdot\mathbf{r})u_{n\mathbf{k}}(\mathbf{r}), \text{ where } u_{n\mathbf{k}}(\mathbf{r}+\mathbf{R}) = u_{n\mathbf{k}}(\mathbf{r}) \tag{3.36}$$

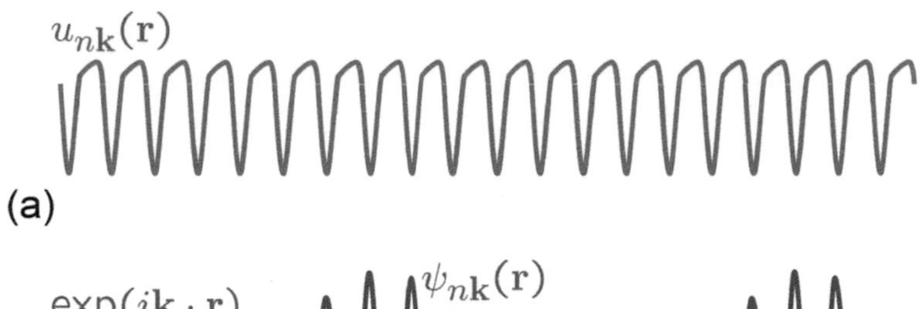

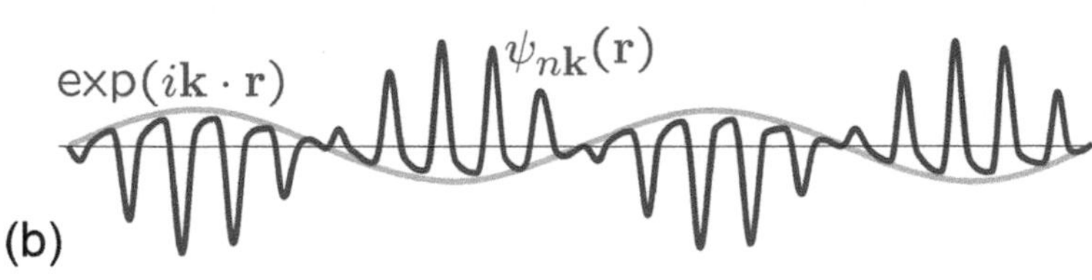

Figure 3.19: Part (a) shows a periodic function $u_{n\mathbf{k}}(\mathbf{r})$. (b) shows a plane wave $(\exp(i\mathbf{k}\cdot\mathbf{r}))$ modulating the periodic function resulting in the Bloch function $\psi(x)$. Part (b) shows only the real parts of the two functions.

for all $\mathbf{R}$ in the Bravais lattice. $u_{n\mathbf{k}}(\mathbf{r})$ is the modulation function of the plane wave. Figure 3.19 is a representative pictorial description that can be viewed together with Figure 3.18 for the charge density and the potential in the crystal. One could also see this as the periodic function $u_{n\mathbf{k}}(\mathbf{r})$ being modulated by a plane wave that has a much longer period.

The periodicity implies that

$$\psi_{n\mathbf{k}}(\mathbf{r}+\mathbf{R}) = \exp(i\,\mathbf{k}\cdot\mathbf{R})\psi_{n\mathbf{k}}(\mathbf{r}). \tag{3.37}$$

An alternate statement with the same implications is that the eigenstates of $\hat{\mathscr{H}}$ can be chosen so that associated with each ψ is a wavevector such that

$$\psi(\mathbf{r}+\mathbf{R}) = \exp(i\,\mathbf{k}\cdot\mathbf{R})\psi(\mathbf{r}) \tag{3.38}$$

for all $\mathbf{R}$ in the Bravais lattice. Periodicity in the lattice precipitates the periodicity in the eigenfunction solution. Figure 3.18 reflects this, since the charge density is proportional to $\langle\psi|\psi\rangle$. Absent periodicity, one would have a solution that decays away in the high potential region and builds up in the low potential regions.

We can demonstrate the truth of this Bloch theorem statement by invoking symmetry. Let $\hat{T}_{\mathbf{R}}$ be a translational operator that shifts the argument of any function $f(\mathbf{r})$ by $\mathbf{R}$, that is,

$$\hat{T}_{\mathbf{R}} f(\mathbf{r}) = f(\mathbf{r}+\mathbf{R}). \tag{3.39}$$

The Hamiltonian is periodic, that is,

$$\begin{aligned}\hat{T}_{\mathbf{R}}\hat{\mathscr{H}}\psi &= \hat{\mathscr{H}}(\mathbf{r}+\mathbf{R})\psi(\mathbf{r}+\mathbf{R}) \\ &= \hat{\mathscr{H}}(\mathbf{r})\psi(\mathbf{r}+\mathbf{R}) \\ &= \hat{\mathscr{H}}(\mathbf{r})\hat{T}_{\mathbf{R}}\psi(\mathbf{r}) \\ &= \hat{\mathscr{H}}\hat{T}_{\mathbf{R}}\psi.\end{aligned} \tag{3.40}$$

Bloch remarked somewhere that once he saw this as a Fourier problem, the solution was obvious. Felix Bloch was another of the illustrious émigré physicists escaping Hitler in the 1930s and bringing a rigorous and thoughtful style to these shores, in this case, Stanford. He was Heisenberg's first PhD student. A good undergraduate friend who was in one of Bloch's graduate classes at Stanford related the following incident. Bloch's style of teaching was to face the board and follow his thoughts, letting them develop analytically on the board. Board and chalk replaced paper and pen, with the students as spectators. At one point, a student spoke up and raised a question. No response. A few minutes later, the student spoke up again. Bloch turned around, looked at the student, kept looking for a while, and then turned around and continued with his thinking and writing. Another story that I heard is from a Nobel winner who spent time at Stanford as a postdoctoral fellow and as a young faculty colleague of Bloch. In this version, Bloch is said to have remarked to a student that if that was the student's question, then the student should not be in that class. Bloch had also worked with Pauli, and there is a bit of Pauliana in these stories.

This is a general result valid for any function ψ, so

$$\hat{T}_{\mathbf{R}}\hat{\mathscr{H}}\psi = \hat{\mathscr{H}}\hat{T}_{\mathbf{R}}\psi \tag{3.41}$$

in general, which says that $\hat{T}$ and $\hat{\mathscr{H}}$ commute, that is, $[\hat{T}, \mathscr{H}] = 0$. Now,

$$\hat{T}_{\mathbf{R}}\hat{T}_{\mathbf{R}'}\psi(\mathbf{r}) = \psi(\mathbf{r}+\mathbf{R}'+\mathbf{R}) = \psi(\mathbf{r}+\mathbf{R}+\mathbf{R}') = \hat{T}_{\mathbf{R}'}\hat{T}_{\mathbf{R}}\psi(\mathbf{r}), \tag{3.42}$$

that is,

$$\hat{T}_{\mathbf{R}}\hat{T}_{\mathbf{R}'} = \hat{T}_{\mathbf{R}'}\hat{T}_{\mathbf{R}} = \hat{T}_{\mathbf{R}+\mathbf{R}'}. \tag{3.43}$$

The translational operator $\hat{T}_{\mathbf{R}}$ commutes with the Hamiltonian $\hat{\mathscr{H}}$ for all Bravais lattice vectors $\mathbf{R}$. Therefore, eigenstates of $\hat{\mathscr{H}}$ are also simultaneously eigenstates of $\hat{T}_{\mathbf{R}}$, that is,

$$\begin{aligned} \hat{\mathscr{H}}\psi &= E\psi, \text{ and} \\ \hat{T}_{\mathbf{R}}\psi &= t(\mathbf{R})\psi. \end{aligned} \tag{3.44}$$

The translational operator, repeatedly applied, leads to eigenvalues $t()$ of

$$\hat{T}_{\mathbf{R}'}\hat{T}_{\mathbf{R}}\psi = t(\mathbf{R})\hat{T}_{\mathbf{R}'}\psi = t(\mathbf{R})t(\mathbf{R}')\psi \tag{3.45}$$

and

$$\hat{T}_{\mathbf{R}'}\hat{T}_{\mathbf{R}}\psi = \hat{T}_{\mathbf{R}+\mathbf{R}'}\psi = t(\mathbf{R}+\mathbf{R}')\psi, \tag{3.46}$$

implying, for the translational eigenvalues

$$t(\mathbf{R}+\mathbf{R}') = t(\mathbf{R})t(\mathbf{R}'). \tag{3.47}$$

The eigenvalue of a translation operator representing a set of repeated operations is the product of the individual operations of translation. Let $\hat{\mathbf{a}}_j$, with $j = 1, 2, 3$, be the three primitive vectors of the Bravais lattice; then

$$t(\hat{\mathbf{a}}_j) = \exp(i2\pi x_j) \tag{3.48}$$

satisfies the product condition. x_j here can be complex, and this follows from the boundary condition and the requirement that the eigenvalue be an observable, that is, a real quantity. Successive application of Equation 3.47 on a lattice vector $\mathbf{R} = m_1\hat{\mathbf{a}}_1 + m_2\hat{\mathbf{a}}_2 + m_3\hat{\mathbf{a}}_3$ leads to

$$\begin{aligned} t(\mathbf{R}) &= t(\hat{\mathbf{a}}_1)^{m_1} t(\hat{\mathbf{a}}_2)^{m_2} t(\hat{\mathbf{a}}_3)^{m_3} \\ &= \exp[i2\pi(m_1x_1 + m_2x_2 + m_3x_3)] \\ &= \exp(i\mathbf{k}\cdot\mathbf{R}), \text{ where} \\ \mathbf{k} &= x_1\hat{\mathbf{b}}_1 + x_2\hat{\mathbf{b}}_2 + x_3\hat{\mathbf{b}}_3. \end{aligned} \tag{3.49}$$

We have now shown that

$$\hat{T}_{\mathbf{R}}\psi = \psi(\mathbf{r}+\mathbf{R}) = t(\mathbf{R})\psi = \exp(i\mathbf{k}\cdot\mathbf{R})\psi(\mathbf{r}). \tag{3.50}$$

A function of the form $\exp(i\mathbf{k}\cdot\mathbf{R})\psi(\mathbf{r})$, a Bloch function, is also an eigenfunction of the translational operator. This Bloch function, which maintains the translational symmetry, is simultaneously an eigenfunction of the periodic Hamiltonian where the periodicity is embedded in the potential $V(\mathbf{r}+\mathbf{R}) = V(\mathbf{r})$.

Bloch's theorem states that the eigenstate ψ of the one-electron Hamiltonian $\hat{\mathscr{H}} = -(\hbar^2/2m_0)\nabla^2 + V(\mathbf{r})$, where $V(\mathbf{r}+\mathbf{R}) = V(\mathbf{r})$ for $\mathbf{R}$, in a Bravais lattice can be chosen to be of the form of a plane wave times a function with the periodicity of the Bravais lattice:

$$\psi_{n\mathbf{k}}(\mathbf{r}) = \exp(i\mathbf{k}\cdot\mathbf{r})u_{n\mathbf{k}}(\mathbf{r}), \tag{3.51}$$

where

$$u_{n\mathbf{k}}(\mathbf{r}+\mathbf{R}) = u_{n\mathbf{k}}(\mathbf{r}) \tag{3.52}$$

for all $\mathbf{R}$ in the Bravais lattice. $u_{n\mathbf{k}}(\mathbf{r})$ is an overlap function modulating the plane wave, and it too is periodic. The periodic boundary condition in $\mathbf{R}$ has now led to eigenvalues taking on discrete values that are being labeled via n. This is the band index that we had introduced earlier to identify the Brillouin zone centering and corresponds to the real space periodicity in $\mathbf{R}$. Equivalently, the eigenstates of $\hat{\mathscr{H}}$ can be chosen so that associated with each eigenfunction ψ is a wavevector such that

We will use the band index n where necessary, such as when states in two different bands have to be dealt with simultaneously; otherwise, we will leave it out to keep notational simplicity, keeping it as something to be understood from context.

$$\psi(\mathbf{r}+\mathbf{R}) = \exp(i\mathbf{k}\cdot\mathbf{R})\psi(\mathbf{r}) \tag{3.53}$$

for every $\mathbf{R}$ in the Bravais lattice.

Bloch's theorem's correspondence to Fourier expansion—an expansion in a plane wave basis—can be seen through the following argument. The modulating function of the plane wave in the Bloch function $\psi_{\mathbf{k}}(\mathbf{r}) = \exp(i\mathbf{k}\cdot\mathbf{r})$ can be Fourier expanded with the reciprocal space periodicity, so

$$u_{\mathbf{k}}(\mathbf{r}) = \sum_{\mathbf{K}_i} u_{\mathbf{k}}(\mathbf{K}_i)\exp(i\mathbf{K}_i\cdot\mathbf{r}), \tag{3.54}$$

which leads to

$$\begin{aligned}\psi_{\mathbf{k}}(\mathbf{r}) &= \exp(i\mathbf{k}\cdot\mathbf{r})\sum_{\mathbf{K}_i} u_{\mathbf{k}}(\mathbf{K}_i)\exp(i\mathbf{K}_i\cdot\mathbf{r}) \\ &= \sum_{\mathbf{K}_i} u_{\mathbf{k}}(\mathbf{K}_i)\exp[i(\mathbf{k}+\mathbf{K}_i)\cdot\mathbf{r}].\end{aligned} \tag{3.55}$$

The Bloch function is a Fourier series with terms resulting from the modulation function expanded in a basis set with the lattice periodicity. The wavevector $\mathbf{k}$ in Bloch's theorem has a similar role for motion in periodic potential as the free electron wavevector $\mathbf{k}$ has for the free electron wavefunction. The wavevector $\mathbf{k}$ in the periodic structure is associated with the crystal momentum—it reflects the properties resulting from the presence of the crystal. It was the lattice periodicity that resulted in the eigenfunction being a Bloch function.

3.5 *Free electron versus Bloch electron*

IN THE CRYSTAL, the potential $V(\mathbf{r})$ is a non-constant position-dependent potential. The crystal Hamiltonian $\hat{\mathscr{H}}$ and the electron momentum $\hat{\mathbf{p}} = (\hbar/i)\nabla$ do not commute, unlike the case for a free electron, where $V(\mathbf{r}) = 0$:

$V(\mathbf{r}) = 0$ is a special case of $V(\mathbf{r})$ being a constant, where the commutation relationship will hold.

$$\begin{aligned}\hat{\mathbf{p}}\psi_{n\mathbf{k}} &= \frac{\hbar}{i}\nabla\psi_{n\mathbf{k}} = \frac{\hbar}{i}\nabla\left[\exp(i\mathbf{k}\cdot\mathbf{r})u_{n\mathbf{k}}(\mathbf{r})\right] \\ &= \hbar\mathbf{k}\psi_{n\mathbf{k}} + \exp(i\mathbf{k}\cdot\mathbf{r})\frac{\hbar}{i}\nabla u_{n\mathbf{k}} \neq \hbar\mathbf{k}\psi_{n\mathbf{k}}.\end{aligned} \tag{3.56}$$

$\hbar\mathbf{k}$ is not an eigenvalue of the momentum operator for this Bloch function when $V(\mathbf{r})$ changes with $\mathbf{r}$. In Equation 3.56, the second term—a function of the modulation function—reflects the presence of the crystal's potential. *$\hbar\mathbf{k}$ is now the crystal momentum of the electron, and not its momentum* $\mathbf{p}$. This crystal momentum $\hbar\mathbf{k}$ reflects the crystal's periodicity and the potential—the effect of the different interactions between the electron and the periodic internal forces of the crystal.

So, the meaning is that while the rate of change of momentum $\mathbf{p}$ of the electron is the result of the total force on the electron—any external forces resulting from electrical or magnetic fields, as well as internal forces arising from the periodic atomic assembly that the electron exists in—the change in the crystal momentum $\hbar\mathbf{k}$ is determined only by the external forces. If an external electrical field $\boldsymbol{\mathcal{E}}$ and magnetic induction $\mathbf{B}$ were present,

$$\frac{\partial}{\partial t}(\hbar\mathbf{k}) = -e\left[\boldsymbol{\mathcal{E}}(\mathbf{r},t) + \mathbf{v}_{\mathbf{n}}(\mathbf{k}) \times \mathbf{B}(\mathbf{r},t)\right]. \tag{3.57}$$

No periodic crystal fields are involved, and, by deploying the energy-crystal momentum (E-$\mathbf{k}$) relationships, one can directly determine the response of electrons or collection of electrons to external stimulus without the complexity of the local periodic changes. Bloch's theorem gives us the means to fold the periodic effects within the (E-$\mathbf{k}$) relationships, and they are included within the prescription of Equation 3.57. In the first term of this equation, $\hbar\mathbf{k}$, the crystal momentum, is a quantum number characteristic of the translational symmetry of the periodic potential. The electron momentum $\mathbf{p}$, on the other hand, is a quantum number characteristic of the full translational symmetry of space. In our analysis of how electrons or their collection behave in the crystal under applied field, we will care about $\hbar\mathbf{k}$ or $\mathbf{k}$.

Since a Bloch state consists of a modulation function ($u_{n\mathbf{k}}(\mathbf{r})$, which is also periodic) modulating a plane wave $\exp(i\mathbf{k}\cdot\mathbf{r})$, it is also a wavepacket. So, the velocity of the state is

$$\mathbf{v} = \mathbf{v}_g = \frac{1}{\hbar}\nabla_{\mathbf{k}}E, \tag{3.58}$$

in agreement with the introductory Equation 3.11. The real space motion of the electron is being described through the reciprocal space operation on the $E(n,\mathbf{k})$ relationship that the Bloch states follow.

Now, let $\mathbf{F}$ be a force applied to the band electron. The work performed over time δt leads to an energy change of

$$\delta E = \mathbf{F}\cdot\mathbf{v}\delta t. \tag{3.59}$$

Also,

$$\delta E = \frac{dE}{d\mathbf{k}}\cdot\delta\mathbf{k} = \hbar\mathbf{v}\cdot\delta\mathbf{k}. \tag{3.60}$$

These two energy change equations, in the limit $\delta t \to 0$, give

$$\mathbf{F} = \frac{d\hbar\mathbf{k}}{dt}. \tag{3.61}$$

Equation 3.61 tells us that, in the periodic structure, an application of an external force leads to a momentum change of the electron. And this momentum is $\hbar\mathbf{k}$. It is the momentum of the electron in the crystal's periodic environment, and therefore it is called the crystal momentum.

It is also useful to assign a reciprocal-space-derived meaning of a mass—an inertial constant—that relates external applied forces. The rate of change of the velocity of the electron, under an external force, is

$$\frac{d\mathbf{v}}{dt} = \frac{1}{\hbar}\frac{d^2E}{d\mathbf{k}dt} = \frac{1}{\hbar^2}\frac{d^2E}{d\mathbf{k}^2}\frac{d\hbar\mathbf{k}}{dt},$$

$$\therefore \quad \frac{\hbar^2}{d^2E/d\mathbf{k}^2}\frac{d\mathbf{v}}{dt} = \frac{d\hbar\mathbf{k}}{dt} = \mathbf{F}. \tag{3.62}$$

This last equation gives a Newtonian meaning of mass to the electron in the crystal of

$$m^* = \frac{\hbar^2}{d^2E/d\mathbf{k}^2}, \quad \text{or} \quad \frac{1}{m^*} = \frac{1}{\hbar^2}\nabla_{\mathbf{k}}^2 E \tag{3.63}$$

for a generalized isotropic condition. The anisotropic form will just take the different directions into account in the second order gradienting of the energy. The effective mass now is a convenient way to describe the motion of band electrons in an external field.

Similar to how we discussed the wavefunction, energy and momentum of the free electron in the Sommerfeld picture and our gedanken free electron in a zero-perturbation periodic structure in Table 3.4, we can write the meaning and constraints of the Bloch electron in a perturbing periodic potential. This is shown in Table 3.5.

We will dwell on masses—effective of various forms—as well as where their usage is suitable, and where not, in a number of chapters in the text. Periodicity, Bloch propagational nature, and the specificity of the force at play will all matter for the suitability of the concept of effective mass.

	Free electron ("Sommerfeld"-like)	Bloch electron in non-zero perturbation periodic structure
Quantum number (excluding spin)	$\mathbf{k}$ $\hbar\mathbf{k}$: momentum	$\mathbf{k}, n$ $\hbar\mathbf{k}$: crystal momentum n: band index
Range of quantum number	$\mathbf{k}$ over $\mathbf{k}$-space consistent with periodic boundary conditions	For each $(\mathbf{k}, n)$ over wavevectors in a primitive cell of the reciprocal lattice consistent with periodic conditions conditions $n \in 0, 1, 2, \ldots$
Energy ($E(\mathbf{k})$)	$\hbar^2\mathbf{k}^2/2m_0$	$E_n(\mathbf{k})$ may not have an explicit simple form $E_n(\mathbf{k}+\mathbf{K}_{\mathbf{m}'}) = E_n(\mathbf{k})$
Velocity	$\mathbf{v} = \hbar\mathbf{k}/m_0 = (1/\hbar)\nabla_{\mathbf{k}}E$ for any $\mathbf{k}$	$\mathbf{v}_n(\mathbf{k}) = (1/\hbar)\nabla_{\mathbf{k}}E_n$ for any $(n, \mathbf{k})$
Wavefunction	$\psi_{\mathbf{k}} = (1/\sqrt{\Omega})\exp(i\mathbf{k}\cdot\mathbf{r})$	$\psi_{n\mathbf{k}}(\mathbf{r}) = \exp(i\mathbf{k}\cdot\mathbf{r})u_{n\mathbf{k}}(\mathbf{r})$ $u_{n\mathbf{k}}(\mathbf{r})$ has no simple explicit form $u_{n\mathbf{k}}(\mathbf{r}+\mathbf{R}) = u_{n\mathbf{k}}(\mathbf{r})$

Table 3.5: Meaning and constraints of a free electron versus that of a Bloch electron in a perturbing periodic potential.

3.6 *A toy model of periodic potential perturbation*

We now tackle the periodic perturbation problem more formally to gain physical insights. First, we will look at this in the free electron basis, by which we mean that the basis states for the evolution of states in the crystal comprise the orthonormal, infinite plane wave set. We have found, absent perturbation, that

$$\hat{\mathscr{H}} = \hat{T}\left(= \frac{\hat{p}^2}{2m_0}\right) + \hat{V}(= 0) \tag{3.64}$$

has the solution of free electron basis states

$$|\mathbf{k}\rangle = \frac{1}{\Omega^{1/2}} \exp(i\mathbf{k} \cdot \mathbf{r}). \tag{3.65}$$

With perturbation, we may write the eigenfunction solution of the new Hamiltonian as a linear combination of the plane wave states

$$|\psi(\mathbf{r})\rangle = \sum_{\mathbf{k}} c_{\mathbf{k}} |\mathbf{k}\rangle, \tag{3.66}$$

with the summation over allowed **k**s. A solution to $(\hat{\mathscr{H}} - E)|\psi(\mathbf{r})\rangle = 0$ requires the secular determinant to vanish, that is,

$$\det|\langle\mathbf{k}'|\hat{\mathscr{H}}|\mathbf{k}\rangle - E\langle\mathbf{k}'|\mathbf{k}\rangle| = \det|\mathscr{H}_{\mathbf{kk}'} - E\mathscr{O}_{\mathbf{kk}'}| = 0, \tag{3.67}$$

again with all the allowed waves. This is the consequence of the collection of algebraic equations whose number is the number of allowed waves and hence the different $|\mathbf{k}\rangle$s. If we choose the dominant ones, that is, the ones with the higher $c_{\mathbf{k}}$s, we get an approximate solution. In this calculation, to get the entire bandstructure, the Ω is the entire crystal's volume. $\hat{\mathscr{H}}_{\mathbf{kk}'} = \langle\mathbf{k}'|\hat{\mathscr{H}}|\mathbf{k}\rangle$ is the matrix element. $\langle\mathbf{k}'|\mathbf{k}\rangle$ is the overlap integral. We use the symbol $\mathscr{O}_{\mathbf{kk}'}$ for overlap.

In real calculations, one finds the plane wave method wanting for this reason. A large number of the terms have to be included, which means that this approach, although certainly not incorrect, is inefficient. In Chapter 4, we will see ways around this.

We fold this collection of states into the first Brillouin zone for convenience, and we also normalize the volume for convenience.

This determinant equation simplifies because the basis set is an orthogonal basis set. The plane waves here are the specific embodiment of the Bloch function—a constant modulation—so they conform to the properties arising from the translation operation $\hat{T}_{\mathbf{R}}$ or its reciprocal space equivalent, $\hat{T}_{\mathbf{K}}$. $|\mathbf{k}\rangle$ and $|\mathbf{k}'\rangle$ are orthonormal, and $\mathscr{O}_{\mathbf{kk}'} = \langle\mathbf{k}|\mathbf{k}'\rangle$ vanishes, except for $\mathbf{k}' = \mathbf{k} + \mathbf{K}_i$, which satisfies the translational property. So, in Equation 3.68, given the orthonormality of $|\mathbf{k}\rangle$ and $|\mathbf{k}'\rangle$, which is equivalent to saying that they belong to an irreducible representation of $\hat{T}$, except when $\mathbf{k}' = \mathbf{k} + \mathbf{K}_i$, one may write Equation 3.68 as

$$|\psi_{\mathbf{k}}(\mathbf{r})\rangle = \sum_{\mathbf{K}_i} c_{\mathbf{K}_i} |\mathbf{k} + \mathbf{K}_i\rangle. \tag{3.68}$$

We now have the secular determinant as

$$\det|\mathscr{H}_{\mathbf{KK}'} - E\mathscr{O}_{\mathbf{KK}'}| = 0, \tag{3.69}$$

where

$$\begin{aligned} \mathscr{H}_{\mathbf{KK}'} &= \langle\mathbf{k} + \mathbf{K}'|\hat{\mathscr{H}}|\mathbf{k} + \mathbf{K}\rangle, \text{ and} \\ \mathscr{O}_{\mathbf{KK}'} &= \langle\mathbf{k} + \mathbf{K}'|\mathbf{k} + \mathbf{K}\rangle. \end{aligned} \tag{3.70}$$

So, only diagonal terms exist, and the problem becomes solvable. In $\mathscr{O}_{\mathbf{KK}'}$, **k** and $-$**k** will cancel, so

$$\mathcal{O}_{\mathbf{KK}'} = \langle \mathbf{k}+\mathbf{K}'|\mathbf{k}+\mathbf{K}\rangle = \langle \mathbf{K}'|\mathbf{K}\rangle = \delta_{\mathbf{K},\mathbf{K}'}. \tag{3.71}$$

Throughout the text, in subscripting in terms such as in δ_{ij} or $\mathcal{V}_{xy}$, we will sometimes use a comma and sometimes not. The comma will be used when its absence has the potential for causing a misunderstanding.

For the other term,

$$\begin{aligned}\mathscr{H}_{\mathbf{KK}'} &= \langle \mathbf{k}+\mathbf{K}'|\hat{T}+\hat{V}|\mathbf{k}+\mathbf{K}\rangle \\ &= \langle \mathbf{k}+\mathbf{K}'|\hat{T}|\mathbf{k}+\mathbf{K}\rangle + \langle \mathbf{k}+\mathbf{K}'|\hat{V}|\mathbf{k}+\mathbf{K}\rangle. \end{aligned} \tag{3.72}$$

We may expand the **r**-dependent potential energy operator in a Fourier expansion with a $|\mathbf{K}''\rangle$ basis over the crystal; that is,

$$V(\mathbf{r}) = \sum_{\mathbf{K}''} \mathcal{V}_{\mathbf{K}''}|\mathbf{K}''\rangle. \tag{3.73}$$

The term $\mathcal{V}_{\mathbf{K}''}$ here is the Fourier coefficient—a structure factor—obtained from the crystal real space, that is,

$$\mathcal{V}_{\mathbf{K}''} = \sum_{\mathbf{R}} V_{\mathbf{K}''\mathbf{R}} \exp(-i\mathbf{K}''\cdot\mathbf{R}). \tag{3.74}$$

In the potential term, arising from substituting Equation 3.73 in the potential part of Equation 3.72, we have a plane wave $|\mathbf{k}+\mathbf{K}'+\mathbf{K}''\rangle$ arising from the two exponential factors corresponding to $|\mathbf{K}''\rangle$ and $|\mathbf{k}+\mathbf{K}'\rangle$, respectively. This has a normalization of $\Omega^{-1/2}$ of volume. This potential term will be, for us, a reference shift for the kinetic energy that we are really not interested in.

The kinetic energy term is the part of the total energy corresponding to the movement that we are particularly interested in:

$$\langle \mathbf{k}+\mathbf{K}'|\hat{T}|\mathbf{k}+\mathbf{K}\rangle = \langle \mathbf{k}+\mathbf{K}|\hat{T}|\mathbf{k}+\mathbf{K}'\rangle = E_{\mathbf{k}+\mathbf{K}'}|\mathbf{k}+\mathbf{K}'\rangle. \tag{3.75}$$

$E_{\mathbf{k}} \neq E_{\mathbf{k}+\mathbf{K}'}$—they are different eigenvalues. So,

$$\begin{aligned}\mathscr{H}_{\mathbf{KK}'} &= E_{\mathbf{k}+\mathbf{K}'}\langle \mathbf{k}+\mathbf{K}'|\mathbf{k}+\mathbf{K}\rangle \\ &\quad + \Omega^{-1/2}\sum_{\mathbf{K}''} \mathcal{V}_{\mathbf{K}''}\langle \mathbf{k}+\mathbf{K}'+\mathbf{K}''|\mathbf{k}+\mathbf{K}\rangle \\ &= E_{\mathbf{k}+\mathbf{K}'}\delta_{\mathbf{K},\mathbf{K}'} + \Omega^{-1/2}\sum_{\mathbf{K}''} \mathcal{V}_{\mathbf{K}''}\delta_{\mathbf{K},\mathbf{K}'+\mathbf{K}''}. \end{aligned} \tag{3.76}$$

Since Equations 3.71 and 3.76 specify the two contributions of the secular determinant equation of Equation 3.69, in principle we now have the algorithm for solving the problem.

We can show the efficacy through a toy example. Take a basis set consisting of only two plane waves. So, in Equation 3.68, consider only two terms: one corresponding to $|\mathbf{k}\rangle = |\mathbf{0}\rangle$, and another due to a vector of the reciprocal lattice, so $|\mathbf{k}\rangle = |\mathbf{K}\rangle$. The secular determinant then has four terms arising in **00**, **0K**, **K0** and **KK**. The overlap for the first and the last is unity; the other two vanish. The different terms are

$$\begin{aligned}
\mathcal{O}_{00} &= \mathcal{O}_{\mathbf{KK}} = 1, \\
\mathcal{O}_{0\mathbf{K}} &= \mathcal{O}_{\mathbf{K}0} = 0, \\
\mathcal{H}_{00} &= E_{\mathbf{k}} + \Omega^{-1/2}\mathcal{V}_0, \\
\mathcal{H}_{0\mathbf{K}} &= \Omega^{-1/2}\sum_{\mathbf{K}''}\mathcal{V}_{\mathbf{K}''}\delta_{0,\mathbf{K}'+\mathbf{K}''} \\
&= \Omega^{-1/2}\mathcal{V}_{-\mathbf{K}} = \Omega^{-1/2}\mathcal{V}^*_{\mathbf{K}}, \\
\mathcal{H}_{\mathbf{K}0} &= \Omega^{-1/2}\mathcal{V}_{\mathbf{K}}, \text{ and} \\
\mathcal{H}_{\mathbf{KK}} &= E_{\mathbf{k}+\mathbf{K}} + \Omega^{-1/2}\sum_{\mathbf{K}''}\mathcal{V}_{\mathbf{K}''}\delta_{\mathbf{K},\mathbf{K}+\mathbf{K}''} \\
&= E_{\mathbf{k}+\mathbf{K}} + \Omega^{-1/2}\mathcal{V}_0. \qquad (3.77)
\end{aligned}$$

The secular determinant is

$$\begin{vmatrix} E_{\mathbf{k}} + \Omega^{-1/2}\mathcal{V}_0 - E & \Omega^{-1/2}\mathcal{V}_0 \\ \Omega^{-1/2}\mathcal{V}_{\mathbf{K}} & E_{\mathbf{k}+\mathbf{K}} + \Omega^{-1/2}\mathcal{V}_0 - E \end{vmatrix} = 0. \qquad (3.78)$$

Note $\Omega^{-1/2}\mathcal{V}_0$ is a constant—a shift—and we have formed a quadratic equation set of

$$\begin{aligned}
&E_{\mathbf{k}}E_{\mathbf{k}+\mathbf{K}} - (E - \Omega^{-1/2}\mathcal{V}_0)(E_{\mathbf{k}} + E_{\mathbf{k}+\mathbf{K}}) \\
&\quad + (E - \Omega^{-1/2}\mathcal{V}_0)^2 - \Omega^{-1}|\mathcal{V}_{\mathbf{K}}|^2 = 0, \qquad (3.79)
\end{aligned}$$

whose solution is

$$\begin{aligned}
E_{\pm} = {} & \Omega^{-1/2}\mathcal{V}_0 + \frac{1}{2}E_{\mathbf{k}}E_{\mathbf{k}+\mathbf{K}} \mp \Omega^{-1/2}\mathcal{V}_0 \\
& \pm \frac{1}{2}\Big[(E_{\mathbf{k}} + E_{\mathbf{k}+\mathbf{K}})^2 - 4(E_{\mathbf{k}}E_{\mathbf{k}+\mathbf{K}} - 4\Omega^{-1}|\mathcal{V}_{\mathbf{K}}|^2)\Big]^{1/2}. \qquad (3.80)
\end{aligned}$$

Two free electron states of energies $E_{\mathbf{k}}$ and $E_{\mathbf{k}+\mathbf{K}}$ arising in the two-plane-wave basis, because of potential perturbation, have now evolved to energy states $E_{\pm}$ with a difference in energy given by

$$\begin{aligned}
E_+ - E_- &= \Big[(E_{\mathbf{k}} + E_{\mathbf{k}+\mathbf{K}})^2 - 4(E_{\mathbf{k}}E_{\mathbf{k}+\mathbf{K}} - 4\Omega^{-1}|\mathcal{V}_{\mathbf{K}}|^2)\Big]^{1/2} \\
&= \Big[(E_{\mathbf{k}} - E_{\mathbf{k}+\mathbf{K}})^2 + 4\Omega^{-1}|\mathcal{V}_{\mathbf{K}}|^2)\Big]^{1/2}. \qquad (3.81)
\end{aligned}$$

An $E_{\mathbf{k}} - E_{\mathbf{k}+\mathbf{K}} \gg 4\Omega^{-1}|\mathcal{V}_{\mathbf{K}}|^2$ results in a vanishingly small change in energy between the waves $|\mathbf{k}\rangle$ and $|\mathbf{k}+\mathbf{K}\rangle$. This is what would happen far away from the Brillouin zone edge. So, the energy picture of the free electron wave doesn't become too inaccurate when away from the Brillouin zone boundary. But, near the Brillouin zone edge $E_{\mathbf{k}} - E_{\mathbf{k}+\mathbf{K}} \ll 4\Omega^{-1}|\mathcal{V}_{\mathbf{K}}|^2$ is small, and the modification becomes significant. A bandgap appears. At the Brillouin zone edge itself, this gap $E_+ - E_- = 2\Omega^{-1/2}|\mathcal{V}_{\mathbf{K}}| = E_g$. Pictorially, this is shown in Figure 3.20 for the two-wave basis.

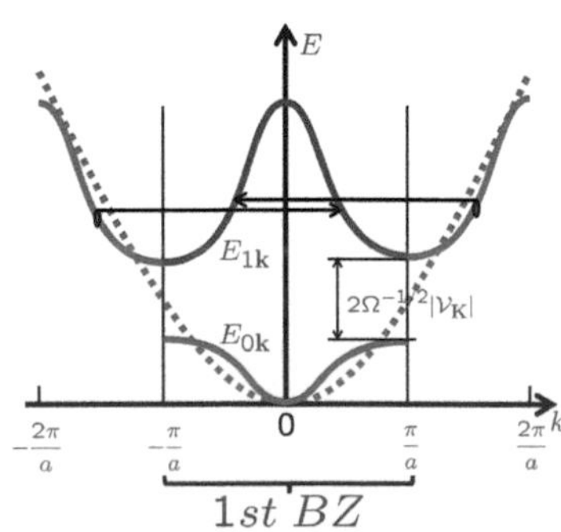

Figure 3.20: The appearance of bandgap at the Brillouin zone (BZ) edge in the two-zone model. The figure shows the first and second free electron bands in the extended zone, and the folding back by translation.

A few comments are in order. We have arrived at a picture with two bands—they map to the conduction band and the valence band, depending on the occupation of the states by electrons—with a bandgap intervening between these band of states. What we have described is a wave-basis solution to determining the bandgap valid for all **k**s, but under a two-wave-basis assumption. Equation 3.69 will determine it, and we have the procedure for determining $\mathscr{H}_{\mathbf{KK}'}$ and $\mathcal{O}_{\mathbf{KK}'}$ precisely. What we have done is—in this 2-wave basis—picked the dominant term in a two-zone structure and found its consequence. This is an example of finite basis state expansion. A real crystal structure will have to include much more than this simple procedural demonstration in 2-basis states. Many more will interact; the bands will distort much more. And we have to find an efficient basis. For example, atomic states are described by the orbital functions, and they may be a more suitable basis for states that are more localized.

A brief note on how this toy model maps to a more comprehensive discussion to come (Chapter 4) is in order. A crystal introduces perturbation. So, we perform a two-basis state expansion for this perturbed problem as

$$\psi(\mathbf{r}) = c_1 u_1^0(\mathbf{r}) + c_2 u_2^0(\mathbf{r}), \tag{3.82}$$

which is posited as a solution to the Hamiltonian problem of $\hat{\mathscr{H}}\psi(\mathbf{r}) = E\psi(\mathbf{r})$, where $\hat{\mathscr{H}} = \hat{\mathscr{H}}_0 + \hat{V}$, with the known solution as $\hat{\mathscr{H}}_0 u_{1,2}^0(\mathbf{r}) = E_{1,2}^0 u_{1,2}^0(\mathbf{r})$. The estimate of the energy then follows from

$$\begin{bmatrix} E_1^0 + V_{11} & V_{12} \\ V_{21} & E_2^0 + V_{22} \end{bmatrix} \begin{bmatrix} c_1 \\ c_2 \end{bmatrix} = E \begin{bmatrix} c_1 \\ c_2 \end{bmatrix}, \tag{3.83}$$

which requires

$$\begin{vmatrix} E_1^0 + V_{11} - E & V_{12} \\ V_{21} & E_2^0 + V_{22} - E \end{vmatrix} = 0 \tag{3.84}$$

for a solution to exist. This gives

$$E_{1,2} = \frac{E_2^0 + V_{22} + E_1^0 + V_{11}}{2} \pm \left[\left(\frac{E_2^0 + V_{22} - E_1^0 - V_{11}}{2} \right)^2 + |V_{12}|^2 \right]^{1/2}, \tag{3.85}$$

in conformity with the implications of Equation 3.81. If $E_2^0 + V_{22} - E_1^0 - V_{11} \gg |V_{12}|$, then expanding to the second order of $|V_{12}|$, the energies are

$$E_1 \approx E_1^0 + V_{11} + \frac{|V_{12}|^2}{E_1^0 + V_{11} - E_2^0 - V_{22}}, \text{ and}$$

$$E_2 \approx E_2^0 + V_{22} + \frac{|V_{12}|^2}{E_2^0 + V_{22} - E_1^0 - V_{11}}. \tag{3.86}$$

Beyond the E-$\mathbf{k}$ relationship that this represents, we now also have the coefficients for lower and higher energy wavefunctions as

$$\begin{bmatrix} c_1^- \\ c_2^- \end{bmatrix} = \begin{bmatrix} 1 \\ V_{12}/(E_1^0 + V_{11} - E_2^0 - V_{22}) \end{bmatrix}, \text{ and}$$
$$\begin{bmatrix} c_1^+ \\ c_2^+ \end{bmatrix} = \begin{bmatrix} 1 \\ V_{21}/(E_2^0 + V_{22} - E_1^0 - V_{11}) \end{bmatrix}. \quad (3.87)$$

We can reiterate to get a more accurate description of the band effect by doing a multi-basis set expansion. If the perturbation is small, so that $E_2^0 \approx E_1^0$, then this is very necessary to even achieve a semi-accurate calculated result. A 3-basis set solution will require

We now have to employ the degenerate perturbation approach discussed in Chapter 1.

$$\begin{vmatrix} E_1^0 + V_{11} - E & V_{12} & V_{13} \\ V_{21} & E_2^0 + V_{22} - E & V_{23} \\ V_{31} & V_{32} & E_3^0 + V_{33} - E \end{vmatrix} = 0. \quad (3.88)$$

Increasing the accuracy will require us to go to higher order terms. But, again, using perturbation methods for non-degenerate and degenerate calculation will let us determine the energies as also the coefficients of basis eigenfunctions for determining the wavefunction.

What we have now seen is how the bands appeared under potential perturbation, using a toy model approach that employed the simplest—a two-plane-wave basis set—to explore. It gives us some intuition as one builds toward real semiconductors.

3.7 *Bands and bandgap's nature from the toy model*

THIS PROCEDURE FROM Section 3.6, by introducing the generalization for perturbation, even if simplified, is quite instructive in telling us consequences of structural and parametric changes. For example, what happens when we reduce the length scale of periodicity? Consider again the one-dimensional model. So, we take a simple one-dimensional lattice, which is reduced from the three-dimensional form with the potential Fourier term of

We will return to one-dimension arrangements often. They are simpler and can even be analytically tractable, yet they are also instructive, because they are often generalizable.

$$\mathcal{V}_{\mathbf{k}} = \langle \mathbf{k} | \hat{V}(\mathbf{r}) = \Omega^{-1/2} \int_\Omega \exp(-i\mathbf{k} \cdot \mathbf{r}) V(\mathbf{r}) \, d^3r \quad (3.89)$$

with

$$E_g = 2\Omega^{-1} \left| \int_\Omega \exp(-i\mathbf{k} \cdot \mathbf{r}) V(\mathbf{r}) \, d^3r \right|, \quad (3.90)$$

which reduces to

$$E_g = 2L^{-1}\left|\int_0^L \exp\left(-i\frac{2\pi z}{a}\right) V(z)\,dz\right|$$

$$= 2L^{-1}\int_0^L \cos\left(\frac{2\pi z}{a}\right) V(z)dz \qquad (3.91)$$

in one dimension. Here, we have one-dimensional crystal of length L and a period a. The width of the Brillouin zone is $2\pi/a$—the 1st Brillouin zone stretches from $-\pi/a$ to $+\pi/a$—and the exponential reduced to a simple form because the cosine function is odd.

Equation 3.91 prescribes the bandgap dependence on lattice parameters. For $a \to \infty$, the energy states asymptotically approach the atomic state—all at similar energy since there is no interaction between them. Figure 3.21 shows this conceptually with a large as two states $|u_1\rangle$ and $|u_2\rangle$ from each of the atoms. As the atoms are brought together, the energies spread. Bands form, and a bandgap exists between the collection of the states that have spread. A gap continues to exist, as in Figure 3.21(a), which is semiconductor-like if the number of electrons per primitive cell is such that the bottom band is filled. But, it can also overlap, as in Figure 3.21(b). In this case, no bandgap exists for equilibrium spacing. So, changing spacing between atoms, such as by applying strain, will shift bands and change bandgaps. Also, depending on the natures of the states—their contributions to $\mathscr{H}$ and $\mathcal{O}$—the shifts and the spreads will be different with respect to the magnitude of the shifts and their signs.

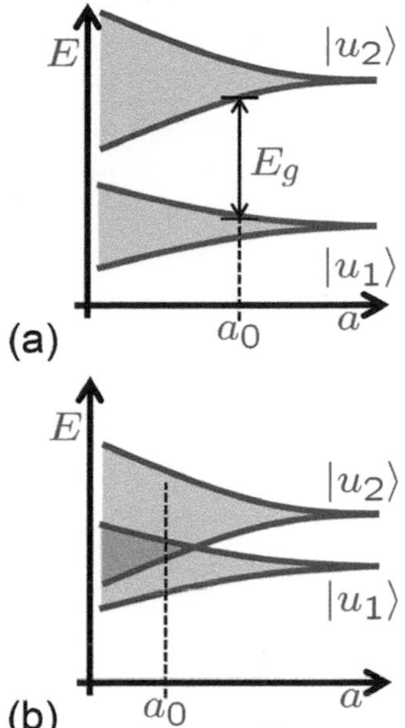

Figure 3.21: As a is made smaller, bands form and the bandgap between the bands is shown in (a) at equilibrium spacing of a. The bands can overlap too, as in (b) at the equilibrium spacing. They can even cross over and form a bandgap with even more state mixing.

The efficiency of the calculation will depend, in various ways, on how suitable the basis was as a component of the final solution, as well as the implementational efficiency match between the algorithm and the hardware.

We will return to a modification of this picture when we discuss tight binding approaches to bandstructure calculation and see the formation of these bands—the upper ones are antibonding states, and the bottom ones are bonding states—as well as state mixing and equilibrium unit cell size.

3.8 *Nearly free electron models*

WHEN THE PERIODIC PERTURBATION is weak, the bandgap and bandedge energy perturbation, such as those appearing in Figure 3.20, are small, and the behavior is nearly free-electron-like for both the conduction band and the valence band. Take the example of *GaAs*, with its two atoms in each cell. There is a total of eight valence electrons—the electrons in outermost orbit that constitute the non-core part. Two electrons of opposite spin can be degenerate in energy, so four valence bands result. The net width of the valence band, with all **k**s allowed, is the energy up

Until now, our discussion was about electron's states in the crystal. They are being occupied now. Bands can be partially or fully filled. Depending on how many electrons are available, bands might be partially or fully filled. For semiconductors, it is the bandgap-sized separation of filled and unfilled bands at usable energies that makes them interesting.

to which all electron states are filled; that is, the Fermi energy (E_F) at absolute zero temperature. The valence bandwidth follows this experimentally, and what this says is that this nearly free electron model is reasonable for the valence bands. If it is for the valence band, then it is also for the conduction band. Thus, we will see that $E_{\mathbf{k}} \approx \hbar^2\mathbf{k}^2/2m^*$ is an adequate description for most purposes. In particular since we are dealing mostly with electrons at the band minimum in the conduction band and at the band maximum in the valence bands, where perturbation—recall our Fourier argument—is smallest. It is a relationship that is still parabolic in the **k**—free-electron-like—but, the energy dispersion arising as a result of the periodic potential perturbation is reflected in a mass m^* that is not the free electron mass any more. *This is now a nearly free electron description of the electron in the semiconductor crystal.*

3.9 *Jones zone*

THE JONES ZONE IS ANOTHER VERY USEFUL WAY to look at the energy zones of a crystalline solid. *The Jones zone is the smallest reciprocal lattice volume where an energy gap exists at all points on its surface.* It may be the same as the Brillouin zone, but, in general, it is not and may be composed of partial Brillouin zone boundaries from several of the Brillouin zones. The volume of a Jones zone can be expressed in terms of the electronic states per unit cell. For face-centered cubic crystals, the Jones zone is composed of the first four Brillouin zones, and it contains eight electrons; that is, the entire valence bandstructure.

In the reduced zone picture, by writing $\breve{\mathbf{k}} = \mathbf{k} + \mathbf{K}$, where $\breve{\mathbf{k}}$ is the unrestricted wavevector and **k** is restricted to the first Brillouin zone, our nearly free electron picture says that

$$E_{\mathbf{k}} \approx \frac{\hbar^2\breve{\mathbf{k}}^2}{2m^*} \tag{3.92}$$

is the range of energy spread over four bands. Take a hypothetical three-dimensional crystal of unit cell of size a. $\mathbf{K} = [000]2\pi/a$ defines the first band, with $\breve{\mathbf{k}} = \mathbf{k}$. The second valence band arises in the next Brillouin zone; that is, with the smallest non-zero reciprocal lattice vectors. Along $\langle 111\rangle$, this is the point $\mathbf{K}_1 = [111]2\pi/a$ and its symmetry points. Along $\langle 100\rangle$, this is the point $\mathbf{K}_2 = [200]2\pi/a$ and its symmetry points.

Take the second band of this model example. At the zone boundary along $\langle 111\rangle$, $\breve{k} = K_1/2 = \sqrt{3}\pi/a$, so $k = -K_1/2 = -\sqrt{3}\pi/a$. As $\breve{k}$ increases, k decreases and vanishes when $\breve{k} = 2\sqrt{3}\pi/a$. At the zone boundary along $\langle 100\rangle$, $\breve{k} = K_2/2 = 2\pi/a$, so $k = -K_2/2 = -2\pi/a$. Again, as $\breve{k}$ increases, k decreases and vanishes when $\breve{k} = 4\pi/a$. These are parabolic bands, since $\mathbf{K}_1$ and $\mathbf{K}_2$ are aligned and subtract from $\breve{\mathbf{k}}$. In the Jones zone (Figure 3.22 shows this for *FCC* with eight electrons per unit cell), the bands reach the surface along the $\mathbf{K}_1$ and $\mathbf{K}_2$ directions.

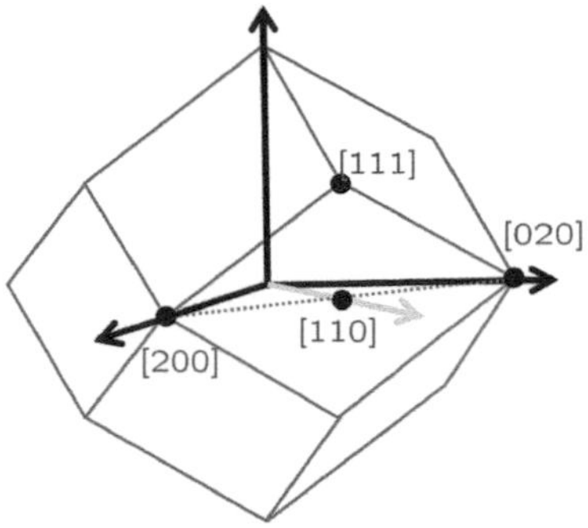

The third and fourth bands arise in combinations of $\mathbf{k}$, $\mathbf{K}_1$ and $\mathbf{K}_2$, where they do not have to be aligned. $\breve{\mathbf{k}}$ now stays near the zone boundary for all $\mathbf{k}$s. The small change in $\breve{\mathbf{k}}$ can be seen from the following argument. At the center of a face, $|\breve{\mathbf{k}}| = |\mathbf{K}_1 - \mathbf{K}_2/2| = 2\sqrt{2}\pi/a$ is the smallest $\breve{k}$ (also energy). Along the $\langle 100\rangle$ direction, $|\breve{\mathbf{k}}| = |\mathbf{K}_1 - \mathbf{k}| = 2\sqrt{3}\pi/a$ (at $\mathbf{k} = 0$, $\breve{k} = \mathbf{K}_1$). So, $|\breve{k}|$ has changed by $(\sqrt{3} - \sqrt{2})2\pi/a$, even as $\breve{k}$ has traversed the zone in this direction. These two bands are not free-electron-like. The Jones zone picture lets us see where the interfering effects are such that the nearly free electron pictures starts to fail. Effective masses can turn negative, for example, as the band curvature changes sign!

Figure 3.22: The Jones zone of an *FCC* crystal with eight electrons per cell. This Jones zone consists of the first four Brillouin zones. It is the smallest reciprocal space volume with energy gaps on the surface.

3.10 *Symmetries*

SYMMETRIES, TOO, PRESCRIBE ADDITIONAL PROPERTIES. Note, for example, that the Bloch wavefunction description leads to the following equation for the repeating positional part:

$$\left[\frac{\hat{\mathbf{p}}^2}{2m_0} + V(\mathbf{r})\right]\psi_{n\mathbf{k}} = E_{n\mathbf{k}}\psi_{n\mathbf{k}}$$

$$\therefore \quad \left[-\frac{1}{2m_0}\left(\hat{\mathbf{p}}^2 + 2\hbar\mathbf{k}\cdot\hat{\mathbf{p}} + \hbar^2\mathbf{k}^2\right) + V(\mathbf{r})\right]u_{n\mathbf{k}}(\mathbf{r}) = E_{n\mathbf{k}}u_{n\mathbf{k}}(\mathbf{r}). \quad (3.93)$$

Grouping of symmetry operations is an important mathematical theme that leads to very important implications for properties. Group theory is a major branch of mathematics. Groups of rotations, reflections and their combinations are point groups. Keeping one point fixed is reflected in these symmetries. The space group operation is the point group operation, together with translational symmetry operations.

Periodic translation symmetry is not the only symmetry that the lattice has. Rotations through specific angles is another one. Having a cubic lattice, for example, implies a fourfold rotation symmetry. The energy bands $E_{n\mathbf{k}}$ have the same symmetries as the crystal because of the reciprocal nature. Consider an operator $\hat{\theta}$ for rotation, so that $\mathbf{r}' = \hat{\theta}\mathbf{r}$. Scalar products are invariant under rotation, so $\hat{\mathbf{p}}'^2 = \hat{\theta}\hat{\mathbf{p}}\cdot\hat{\mathbf{p}} = \hat{\mathbf{p}}^2$. So, $\mathbf{k}' = \mathbf{k}$, and $\mathbf{k}'.\mathbf{p}' = \mathbf{k}\cdot\mathbf{p}$, which, in turn, means that, under rotation, Equation 3.93 transforms to

$$\left[-\frac{1}{2m_0}\left(\hat{\mathbf{p}}^2 + 2\hbar\mathbf{k}\cdot\hat{\mathbf{p}} + \hbar^2\mathbf{k}^2\right) + V(\mathbf{r})\right]u_{n\mathbf{k}'}(\hat{\theta}\mathbf{r}) = E_{n\mathbf{k}}u_{n\mathbf{k}}(\hat{\theta}\mathbf{r}). \quad (3.94)$$

$u_{n\mathbf{k}}(\hat{\theta}\mathbf{r})$ and $u_{n\mathbf{k}}(\mathbf{r})$ are the solutions to the same equations. So, $E_{n\mathbf{k}'} = E_{n\mathbf{k}}$, with $\mathbf{k}' = \hat{\theta}\mathbf{k}$. Since $\hat{\theta}E_{n\mathbf{k}} = E_{n\mathbf{k}}$, the reciprocal space has the same rotational symmetries as the real space.

The symmetry under inversion, $\mathbf{k} \mapsto -\mathbf{k}$ leading to $E_n(\mathbf{k}) \mapsto E_n(-\mathbf{k})$, holds true if the system has inversion symmetry, that is $V(-\mathbf{r}) = V(\mathbf{r})$. Even if inversion symmetry is absent, and the forces are time-reversal invariant, as they are for Coulomb forces, but not magnetic, then $E_{n\uparrow}(\mathbf{k}) = E_{n\downarrow}(-\mathbf{k})$. $\uparrow$ and $\downarrow$ here the spin configurations, and these are related by time-reversal symmetry. We discuss this a bit more during our discussion of spin-orbit coupling, which is important for valence bands, in Chapter 4.

The physical properties of a crystal let us probe symmetry properties. For example, centrosymmetric materials such as *Ge* and *Si* have no piezoelectricity. In contrast, *GaAs* does, since it is non-centrosymmetric because it consists of two different elements in an interpenetrating *FCC* lattice.

Neumann's principle states that symmetry elements of any physical property of a crystal must include the symmetry elements of the point group of the crystal. So, symmetry elements of a physical property include those of the point group. But the properties may be more extensive.

3.11 *Atomic motion: Phonons*

THE NUCLEI AND THE CORE ELECTRONS, hitherto neglected, are also in motion. In Chapter 1, in the discussion of the simplification of the crystal Hamiltonian (Equation 1.11), we moved the consequences to the last term—a perturbation term for the electron energy—under the assumption that this is a slow motion because of the large mass of the nuclei that the electron charge cloud can largely follow. The ions move around an equilibrium position $\mathbf{R}_i^0$ that we have ascribed to the lattice point. Equilibrium due to restorative forces under any disturbance means that we can view this motion around the equilibrium as a harmonic oscillator. The Hamiltonian is

$$\hat{\mathscr{H}}_{\mathbf{q}} = \sum_i \left[-\frac{\hbar^2}{2M_i} \nabla_{\mathbf{r}}^2 \right] + \sum_{i,j} D_{ij}(\mathbf{R}_i - \mathbf{R}_j)\mathbf{u}_i \cdot \mathbf{u}_j + \hat{\mathscr{H}}_{\mathbf{q}0}(\overline{\mathbf{R}}_i^0) + \hat{\mathscr{H}}_{\mathbf{q}}'. \quad (3.95)$$

Here, the first term is the kinetic energy of ions, the second term is the restorative energy—the increase in potential energy arising in any increase from equilibrium displacement ($D_{ij}(\mathbf{R}_i - \mathbf{R}_j)$) is a restoring force per unit displacement (displacements being $\mathbf{u}_i$ and $\mathbf{u}_j$) that we usually ascribe as a spring constant k_s—the third term is a constant shift related to equilibrium position separation, and the last term captures anharmonic effects. Assuming small disturbance—linearization—we ignore the last term. The second-to-last term is what brought the crystal together, lowering its energy from the separated ensemble of ions; it is a constant shift, so it is not important to our discussion. The displacement of ions is captured by the first two terms where motion of the ions is around an equilibrium position.

How many solutions—modes–of such motion exist? If there is one particle, it has 3 spatial degrees of freedom. In Cartesian coordinates, $\hat{\mathbf{x}}$, $\hat{\mathbf{y}}$ and $\hat{\mathbf{z}}$ describe three independent directions of motion. One could equivalently choose polar or spherical independent coordinates. But they are all still 3 independent spatial coordinates. Two independent atoms will have $2 \times 3 = 6$ degrees of freedom. If one couples these—a bonded two-atom molecule—it still has 6 degrees of freedom. All we have done is introduced an internal constraint of a bond between the two atoms of this two-particle system. Of these 6 motional degrees of freedom, 3 are of translation in the $\hat{\mathbf{x}}$, $\hat{\mathbf{y}}$ and $\hat{\mathbf{z}}$ direction. Two independent motional freedoms can also be ascribed to rotation: one along the axis of the bond, and one at an arbitrary orientation normal to this axis, is one choice. This leaves one additional degree of freedom not yet ascribed to this two-atom molecule. This is one of oscillation, with the atoms moving around the equilibrium position.

Mode is a loose term used to identify a solution such as that represented through the eigenfunction and eigenenergy of the state of a system described by a governing equation. The motion of a pendulum composed of a small weight at the end of a weightless rigid string ℓ units long hanging from a point in gravity g has a mode with the radial oscillation frequency $\omega_q = (1/2\pi)\sqrt{g/\ell}$ for small amplitude. This is not the only mode. The weight can also spin back and forth as it performs the small orbital motion. The electron states described in the energy dispersion with the various n and $\mathbf{k}$ are all electron modes.

Now, consider N atoms uncoupled. There are then $3N$ spatial motion degrees of freedom. If these are all coupled as in a crystal, there will be 3 degrees of freedom for the crystal translation, and 3 degrees of freedom for rotation around the 3 independent axes chosen, so there are $3N - 6$ degrees of freedom for oscillations. The crystal consisting of N atoms has $3N - 6$ modes of oscillations, that is, of phonons as the quasiparticles representing these oscillations. If N is large, as with the electrons in crystals, one may view these as nearly continuous. As with electrons ($E = \hbar\omega_{\mathbf{k}}$), there will be specific energies ($E = \hbar\omega_{\mathbf{q}}$) and wavevectors $\mathbf{q}$ allowed. For phonons, these will break in general to longitudinal (displacement parallel) and transverse (displacement normal) to the chosen directions in the unit cells. In an assembly of a single atom type, that is, a single atom basis per unit cell of the crystal, we will have one longitudinal polarization of motion and two transverse polarizations of motion. If the consecutive displacement has a small change in phase, the motion is acoustic, while if they are opposite in displacement, they are optical. The $3N - 6$ total modes are spread out as the allowed vibrational states across the acoustic and the optical branches of the phonons. We will now place more substance in this discussion as the number of basis atoms change, and in following chapters, when one encounters interfaces and surfaces, as also through what electron-phonon and electromagnetic-phonon interaction will be like for transport and light-matter interaction problems.

Consider a unit cell with more than one atom as the basis. We can write the motion in general as

$$\mathbf{u}(\omega_{\mathbf{q}}, \mathbf{q}) = \mathbf{u}_0 \exp\left[i(\mathbf{q} \cdot \mathbf{R} - \omega_{\mathbf{q}} t)\right]. \tag{3.96}$$

More specifically, the ith ion responds in motion as

$$\mathbf{R}_i - \mathbf{R}_i^0 = \mathbf{u}_i(\omega_{\mathbf{q}}, \mathbf{q}) = \mathbf{u}_{i0} \exp\left[i(\mathbf{q} \cdot (\mathbf{R}_i - \mathbf{R}_i^0) - \omega_{\mathbf{q}} t)\right]. \quad (3.97)$$

The crystal waves have dispersion, as sketched in Figure 3.23, which shows, for a general case, six distinct branches along which are the points of the variety of eigenmodes that satisfy Equation 3.95. As with energy diagrams, in a large assembly, these mode points are close enough together to be viewable as a quasi-continuous function until reduced to an assembly of only a few unit cells. The nature of longitudinal and transverse, and optical and acoustic, motion around equilibrium can be seen in Figure 3.24 for a small $\mathbf{q}$ for a two-atom basis chain.

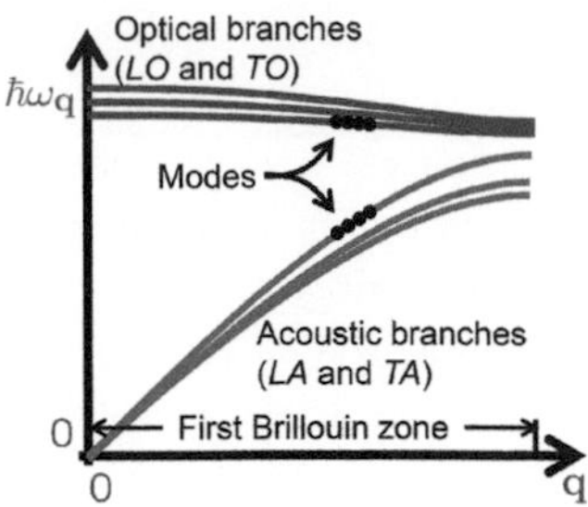

Figure 3.23: Crystal wave dispersion for a crystal with more than one atom basis set per primitive unit cell; *LA*, longitudinal acoustic; *LO*, longitudinal optical; *TA*, transverse acoustic; *TO*, transverse optical.

Each mode has an energy, because it is a simple harmonic oscillator, of

$$E(\omega_{\mathbf{q}}, \mathbf{q}) = \left[\langle n(\omega_{\mathbf{q}}, \mathbf{q})\rangle + \frac{1}{2}\right]\hbar\omega_{\mathbf{q}}, \quad (3.98)$$

with $\langle n(\omega_{\mathbf{q}}, \mathbf{q})\rangle$ as the average number of the vibration's quanta that have been excited. The quasiparticle that we have associated with this vibrational mode, with its distinct energy, is the phonon. Each phonon has an energy $\hbar\omega_{\mathbf{q}}$ for a wavevector $\mathbf{q}$.

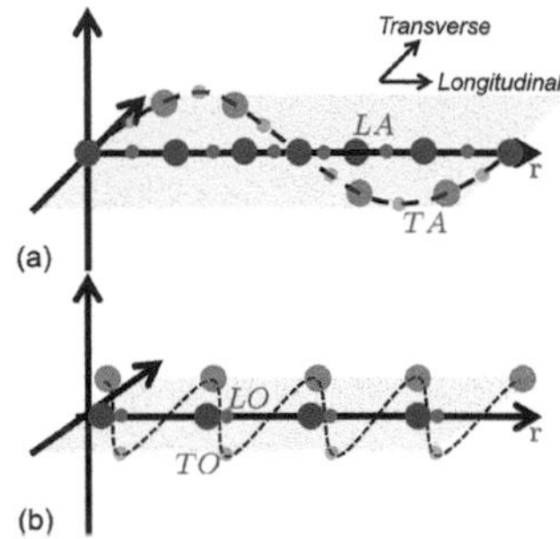

Figure 3.24: Part (a) shows the longitudinal and transverse acoustic modes of the two-atom one-dimensional crystal. Part (b) shows the higher frequency longitudinal and transverse optical modes of the crystal; *LA*, longitudinal acoustic; *LO*, longitudinal optical; *TA*, transverse acoustic; *TO*, transverse optical.

Phonons are bosons, and, given that $\omega_{\mathbf{q}}$ and $\mathbf{q}$ are related through the dispersion relationship,

$$n(\omega_{\mathbf{q}}) = \frac{1}{\exp(\hbar\omega_{\mathbf{q}}/k_B T) - 1}. \quad (3.99)$$

More than one phonon mode at any allowed $(\omega_{\mathbf{q}}, \mathbf{q})$ is permissible. Phonon wavevector $\mathbf{q}$ itself is periodic, analogous to the electron wavevector $\mathbf{k}$, with similar limitations and representation in the Brillouin zone arising in the periodicity. If there are N unit cells in a chain L long, then $\mathbf{q}$ must have a limit of the Brillouin zone boundary at the longest. The $\mathbf{q}$s in this hypothetical structure are $2\pi/L$ apart. Any perturbation in this crystal, for example, a change in mass or strength of coupling locally, will lead to localized modes in the vicinity. This localized mode will have an energy different from that of the crystal modes. Long wavelength acoustic modes (short $\mathbf{q}$) have linear dispersion, as seen in Figure 3.23: $\omega_{\mathbf{q}} = v_s q$, v_s being the velocity of sound. Optical modes tend to have an energy that is fairly wavevector independent.

That phonons are bosons is a point that requires some care and is delicate. Quantization is a result of the quantization of the field. For photons, field oscillation is in two directions, whether electric or magnetic, and the two directions of circular polarization are what we map this to as the basis and associate a spin of 1. This also reflects the unification of electricity and magnetism in the photon. Phonons, if they are longitudinal, just provide a scalar oscillating pressure; there is no angular momentum and no spin or orbital quantization. But transverse motion such as in an isotropic medium, which is doubly degenerate, may be interpreted via quantization along the direction of the propagation, where a quantum number of 1 will result. In either case, the phonons are still bosons.

We make a few short remarks on the density of states of phonon modes. Appendix H, which discusses density of states for electrons, is still pertinent for phonons. The longest wavelength phonon is determined by the extent of the dimensions, and the shortest by atomic spacing. The density of states in $\mathbf{q}$-space is similar to that in the $\mathbf{k}$ space, since it is determined by the periodicity of the lattice. So,

$$\mathscr{G}_{\mathbf{q}}(\mathbf{q}) = 4\pi \mathbf{q}^2 \frac{1}{(2\pi)^3} \quad (3.100)$$

per unit volume. The energy density of states for the phonons—written usually via the frequency in $E_{\mathbf{q}} = \hbar\omega_{\mathbf{q}}$—is

$$\mathscr{G}_{\mathbf{q}}(\omega_{\mathbf{q}}) = 4\pi \mathbf{q}^2 \frac{1}{(2\pi)^3} \frac{dq}{d\omega_{\mathbf{q}}}, \quad (3.101)$$

normalized in volume, where we have used the transformation resulting from $\mathscr{G}_{\mathbf{q}}(\omega_{\mathbf{q}})d\omega_{\mathbf{q}} = \mathscr{G}_{\mathbf{q}}(\mathbf{q})\, dq$, since the same phonon modes are involved.

Take the acoustic mode case. At high phonon wavevectors, there is also higher nonlinearity and a saturation of energy. So, the density of states in energy will reflect this. There is also both a minimum and a maximum phonon wavevector magnitude. For any reasonable sized volume, the lowest energy will be vanishing—a long wavevector following the dispersion curve—but there will be a maximum in the frequency allowed. There are a variety of fits provided for this conventionally. The Debye model, for example, is an attempt at modeling thermal properties by approximating the real dispersion curves. We will just generalize to specifically the difficulties that arise in real crystals, such as zinc blende or wurtzite, where different directions are different in characteristics and what we have are anisotropic oscillators. This is to say that the constant energy surface is not a sphere but will be warped. Since

$$d\omega_{\mathbf{q}} = |\boldsymbol{\nabla}_{\mathbf{q}}\omega_{\mathbf{q}}| d\mathbf{q}_{\perp} \;\; \therefore \;\; d\omega_{\mathbf{q}} = |\mathbf{v}_{gs}| d\mathbf{q}_{\perp}, \quad (3.102)$$

where $\mathbf{v}_{gs}$ is the group velocity of phonons. We may write in general

$$\mathscr{G}_{\mathbf{q}}(\omega_{\mathbf{q}}) = \frac{1}{(2\pi)^3} \int \frac{1}{|\mathbf{v}_{gs}|} \hat{\mathbf{n}} \cdot d\mathbf{S}, \quad (3.103)$$

an integral in area in direction of $\mathbf{q}_{\perp}$. This calculation must be performed on all dispersion curves and the results summed to get the net density of states.

3.11.1 *Toy model: Two-atom basis chain*

NOW LET US PLACE some more semiconductor-oriented specifics in this description to elucidate details of expected behavior. Consider a two-atom basis linear atom chain as a crystal, as shown in Figure 3.25. First, let it be covalent. In the Born-Oppenheimer approximation, we are assuming that the electrons can follow

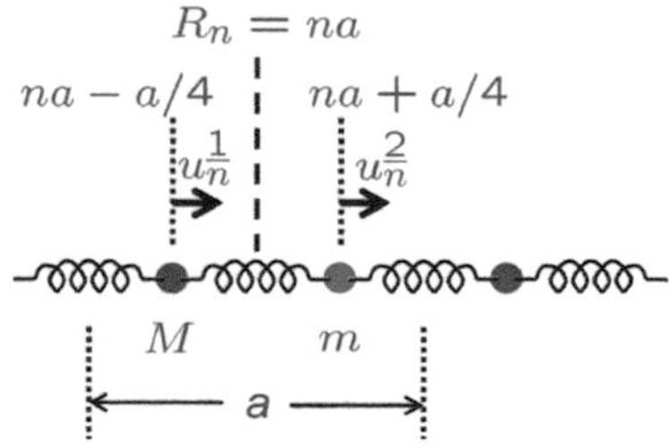

Figure 3.25: A two-atom-basis one-dimensional crystal. The atoms have masses M and m, respectively. $\mathbf{R}_n$ is the unit cell locale.

the nuclear motion. We will see the implications of perturbations arising in this oscillation-induced deformation, as well as its dipolar frequency dependence, during our discussion of optical scattering in Chapter 10, and electromagnetic interaction and dielectric response function for materials in Chapter 12.

We will presently modify the analysis from the covalency beginnings by introducing electron transfer, that is, ionicity, through a dipolar charge of $\pm e^*$ on the ions. This is the Born effective charge. During the discussion of scattering, frequency dependence of polarization and other consequences, it will enter in significant ways.

Because of the harmonicity, the two atoms on an average positioned at $na - a/4$ for one with mass M and $na + a/4$ for the one with mass m move around it with a displacement $u_n^{\underline{1}}$ and $u_n^{\underline{2}}$. The unit cell is not primitive; it has a real space basis vector $\mathbf{R} = a\hat{\mathbf{a}}$ so that $\mathbf{R}_n = na\hat{\mathbf{a}}$. The springs, of spring constant k_s, represent the forces arising from the change in potential away from equilibrium, which causes the return ($k_s = \partial^2 U/\partial z^2$ are related to the deformation D_{ij} in Equation 3.95). The underlined superscript in the displacement u identifies the separate basis of mass M and m. For these two types of atom in an infinite chain,

$$u_n^{\underline{1}} = u_{n0}^{\underline{1}} \exp\left\{ i \left[qa\left(n - \frac{1}{4}\right) - \omega_q t \right] \right\}, \text{ and}$$
$$u_n^{\underline{2}} = u_{n0}^{\underline{2}} \exp\left\{ i \left[qa\left(n + \frac{1}{4}\right) - \omega_q t \right] \right\}. \quad (3.104)$$

In the lowest order—no nonlinearity—approximation for stability, that is, harmonicity,

$$\begin{aligned} M\frac{d^2u_n^{\underline{1}}}{dt^2} &= k_s(u_n^{\underline{2}} - u_n^{\underline{1}}) - k_s(u_n^{\underline{1}} - u_{n-1}^{\underline{2}}), \text{ and} \\ m\frac{d^2u_n^{\underline{2}}}{dt^2} &= k_s(u_{n+1}^{\underline{1}} - u_n^{\underline{2}}) - k_s(u_n^{\underline{2}} - u_n^{\underline{1}}), \text{ so that} \\ M\frac{d^2u_n^{\underline{1}}}{dt^2} &= k_s(u_n^{\underline{2}} + u_{n-1}^{\underline{2}}) - 2k_s u_n^{\underline{1}}, \text{ and} \\ m\frac{d^2u_n^{\underline{2}}}{dt^2} &= k_s(u_{n+1}^{\underline{1}} + u_n^{\underline{1}}) - 2k_s u_n^{\underline{2}}, \end{aligned} \quad (3.105)$$

are the coupled linear differential equations describing the motion. Substituting for the displacements, the eigenenergy solution follow as

$$\begin{aligned} -M\omega_q^2 u_{n0}^{\underline{1}} &= k_s u_{n0}^{\underline{2}} \left[\exp\left(i\frac{qa}{2}\right) + \exp\left(-i\frac{qa}{2}\right) \right] - 2k_s u_{n0}^{\underline{1}} \\ &= 2k_s u_{n0}^{\underline{2}} \cos\left(i\frac{qa}{2}\right) - 2k_s u_{n0}^{\underline{1}}, \text{ and} \\ -m\omega_q^2 u_{n0}^{\underline{2}} &= k_s u_{n0}^{\underline{1}} \left[\exp\left(i\frac{qa}{2}\right) + \exp\left(-i\frac{qa}{2}\right) \right] - 2k_s u_{n0}^{\underline{2}} \\ &= 2k_s u_{n0}^{\underline{1}} \cos\left(i\frac{qa}{2}\right) - 2k_s u_{n0}^{\underline{2}} \end{aligned} \quad (3.106)$$

so that

$$\omega_\pm^2 = \frac{k_s}{mM} \left\{ (m+M) \pm \left[(m+M)^2 - 2mM\left(1 - \cos\frac{qa}{2}\right) \right]^{1/2} \right\} \quad (3.107)$$

are the radial frequencies. $\omega_-(q)$ is the acoustic branch, and $\omega_+(q)$ the optical branch.

We had a two-atom basis in an infinite chain with a single dimension of motion, so we ended up with two branches reflecting the two degrees of motional freedom, and an infinite number of modes spread out along each of these branches for an infinite chain. The motional degrees of freedom were longitudinal, so the solution is for longitudinal motion. With transverse motion also allowed, Figure 3.24 showed both a longitudinal and a transverse mode of the acoustic and optical branches. The branches are longitudinal optical (*LO*) for the positive sign, and longitudinal acoustic (*LA*) for the negative sign in Equation 3.107. Being a periodic lattice, many of the comments based on periodicity made for describing electrons in the crystal apply. The first Brillouin zone with wavevectors in $-\pi/a \leq q \leq \pi/a$ suffices to describe the dispersion. The maximum frequency for the *LO* is $[2k_s(m+M)/mM]^{1/2}$. It occurs at the zone center. The minimum frequency is at the zone edge at $(2k_s/m)^{1/2}$. The *LA* branch, which starts at a vanishing frequency, reaches its maximum at $(2k_s/M)^{1/2}$ at the zone edge. So, there is a phonon bandgap in this model. We have sidestepped transverse motion in this analysis.

A branch with zero gap is an acoustic branch. This energy going to zero at the shortest wavevector is a reflection of Goldstone's theorem's consequence for broken symmetry. Making a one-dimensional chain is a breaking of the continuous symmetry that existed in the atoms as points. Goldstone's theorem tells us that such a broken continuous symmetry will lead to short-range interactions that will have no gap. The acoustic branch extends propagation such as of sound waves all the way down to the static limit. By analogy, the optical branch is the higher frequency branch. For semiconductors, this is a frequency in the *THz* range, which is far into the infrared frequency range. Optical phonons will therefore lead to interactions and plenty of nonlinearity and absorption in that frequency part of the spectrum.

Now let us make this chain polar by introducing the effective Born charge of $\pm e^*$ for the dipole to reflect the polarity of the ionic fraction of the bonding. Let $\boldsymbol{\mathcal{E}}$ be the electric field. Our Equation 3.105 set changes with a force $\pm e^*\boldsymbol{\mathcal{E}}$. Let the positive sign be for the atom indexed $\underline{1}$ (mass M is now a cation), and negative for the atom indexed $\underline{2}$ (mass m is now an anion). Our modification to Equation 3.105 is

$$
\begin{aligned}
M\frac{d^2u_n^{\underline{1}}}{dt^2} &= k_s(u_n^{\underline{2}} + u_{n-1}^{\underline{2}}) - 2k_s u_n^{\underline{1}} + e^*\mathcal{E}, \\
\therefore \quad -M\omega_q^2 u_n^{\underline{1}} &= k_s\left[1+\exp(iqa)\right]u_{n-1}^{\underline{2}} - 2k_s u_n^{\underline{1}} + e^*\mathcal{E}, \text{ and} \\
m\frac{d^2u_n^{\underline{2}}}{dt^2} &= k_s(u_{n+1}^{\underline{1}} + u_n^{\underline{1}}) - 2k_s u_n^{\underline{2}} - e^*\mathcal{E}, \\
\therefore \quad -m\omega_q^2 u_n^{\underline{2}} &= k_s\left[1+\exp(-iqa)\right]u_{n+1}^{\underline{1}} - 2k_s u_n^{\underline{2}} - e^*\mathcal{E}.
\end{aligned} \tag{3.108}
$$

At long wavelengths ($q \to 0$), one doesn't need to index for each basis atom, since they are displaced identically. So, for $q \to 0$, without indexing the unit cell, we have

$$
\begin{aligned}
-M\omega_q^2 u^{\underline{1}} &= 2k_s(u^{\underline{2}} - u^{\underline{1}}) + e^*\mathcal{E}, \text{ and} \\
-m\omega_q^2 u^{\underline{2}} &= 2k_s(u^{\underline{1}} - u^{\underline{2}}) - e^*\mathcal{E}.
\end{aligned} \tag{3.109}
$$

These equations physically imply that the sum vanishes, that is, $-M\omega_q^2 u^{\underline{1}} - m\omega_q^2 u^{\underline{2}} = 0$, or that the displacements are inversely related to the masses with a sign reversal. This is displacement that is opposite in sign; that is, out of phase. This is the characteristic of the optical branch, as was seen in Figure 3.24(b). This condition for $q \rightarrow 0$ leads us to

$$\begin{aligned} -M\omega_q^2 u_-^1 &= 2k_s \left(-\frac{M}{m} u_-^1 - u_-^1 \right) + e^* \mathcal{E} \\ &= -2k_s \left(\frac{M}{m} + 1 \right) u_-^1 + e^* \mathcal{E}, \text{ and} \\ -m\omega_q^2 u_-^2 &= 2k_s \left(u_-^1 - u_-^2 \right) - e^* \mathcal{E} \\ &= -2k_s \left(\frac{m}{M} + 1 \right) u_-^2 - e^* \mathcal{E}. \end{aligned} \tag{3.110}$$

We define

$$\omega_0^2 = 2k_s \left(\frac{1}{M} + \frac{1}{m} \right), \tag{3.111}$$

and now the displacements can be related at the long wavelengths as

$$\begin{aligned} -(\omega^2 - \omega_0^2) u_-^1 &= \frac{e^* \mathcal{E}}{M}, \text{ and} \\ -(\omega^2 - \omega_0^2) u_-^2 &= -\frac{e^* \mathcal{E}}{m}. \end{aligned} \tag{3.112}$$

ω_0 is a resonant frequency when $e^* \rightarrow 0$. Absent the Coulomb dipolar consequence when a field is present, which is a covalent condition in a Born-Oppenheimer approximation condition, the infinite one-dimensional two-atom crystal resonates at a frequency determined by the geometric mass. When this crystal is ionic, there will be a shift in this frequency that we will follow up on in Chapter 12.

This response tells us one significant physical property. There exists an electric dipole polarization in this lattice. We generalize and use this solution for three dimensions so that one can consider longitudinal and transverse modes with similar underlying restorative constraints.

At low frequencies, that is, $q \rightarrow 0$, the response is with an effective charge e^* separated by $u_-^1 - u_-^2$ over the number of diatomic N pairs that exist per unit volume, that is, $Ne^*(u_-^1 - u_-^2)$. So, the polarization is

$$P = \frac{Ne^*(u_-^1 - u_-^2)}{\epsilon(\infty)} = \frac{1}{\epsilon(\infty)} \frac{Ne^{*2}}{\omega_0^2 - \omega^2} \left(\frac{1}{M} + \frac{1}{m} \right) \mathcal{E}, \tag{3.113}$$

Why the ∞ frequency for the permittivity? It is ∞ because there is the contribution arising in the core and the nucleus response during this displacement onto which the effective charge transfer has been planted to calculate the consequence in polarization. Far infrared frequency suffices as a proxy.

whose frequency dependence shows how the forced oscillator responds to an electric field.

Our analysis has been for longitudinal conditions, but it is generalizable to transverse motion. If a transverse field were present, we will have a similar response, with ω_0 being ω_{TO}, the transverse resonance frequency in the $e^* = 0$ limit. In analogy with the longitudinal discussion, a transverse acoustic mode is one where the displacement change between the two adjacent basis atoms is small, and the transverse optical mode is one where the displacement change for this adjacent atom will be large—out of phase—so that the corresponding polarization is large. The applied field and the phonon oscillations couple strongly around the resonance, and this coupled condition—a coupled energy excitation—is described by a polariton, which we will visit later in this text. For this transverse field coupling, which is absent for the longitudinal mode that is normal to it, we have

$$P(\omega_{TO}^2 - \omega^2) = \frac{Ne^{*2}}{\epsilon(\infty)} \left(\frac{1}{M} + \frac{1}{m} \right) \mathcal{E}, \tag{3.114}$$

with

$$\omega_{TO}^2 = 2k_s \left(\frac{1}{m} + \frac{1}{M} \right). \tag{3.115}$$

The important point to note here is that $e^{*2} = 0$, that is, Coulomb effects absent, corresponds to transverse optical frequency in the analysis underlying Equation 3.111. The difference between the *LO* and *TO* resonance frequency arises in the e^{*2} factor. The resonance frequency corresponds to the transverse optical resonance frequency, and now general electric field effects can be included.

We incorporate this behavior to determine the dielectric function that informs us of the electrical interaction response of an electromagnetic wave in a material. Starting from Maxwell's equations, absent sources, since $\nabla \times \boldsymbol{\mathcal{E}} = -\partial \mathbf{B}/\partial t$, we have

$$\nabla \times (\nabla \times \boldsymbol{\mathcal{E}}) = -\frac{\partial}{\partial t} \nabla \times \mathbf{B}, \tag{3.116}$$

Also, with linearity,

$$\nabla \times \mathbf{B} = \mu \nabla \times \mathbf{H} = \mu \frac{\partial \mathbf{D}}{\partial t}, \tag{3.117}$$

so

$$\begin{aligned} \nabla \times (\nabla \times \boldsymbol{\mathcal{E}}) &= -\frac{\partial}{\partial t} \mu \frac{\partial \mathbf{D}}{\partial t} \\ \therefore \quad \nabla(\nabla \cdot \boldsymbol{\mathcal{E}}) - \nabla^2 \boldsymbol{\mathcal{E}} &= -\mu \frac{\partial^2 \mathbf{D}}{\partial t^2} \\ \therefore \quad -\nabla^2 \boldsymbol{\mathcal{E}} &= -\mu \frac{\partial^2 \mathbf{D}}{\partial t^2} = -\mu \frac{\partial^2}{\partial t^2} (\epsilon_0 \boldsymbol{\mathcal{E}} + \mathbf{P}). \end{aligned} \tag{3.118}$$

This last equation is a time-independent wave equation—a Helmholtz equation—where the wave has the oscillatory dependence that exists in the form $\exp[i(qz - \omega t)]$, with $\hat{\mathbf{z}}$ as the direction of the chain. Substituting, we have the set of equations

The general form of a Helmholtz equation is $\nabla^2 \psi + \lambda \psi = 0$. A partial differential equation that can be separated into a space part and a time part, such as the wave equation, can be recast into this form. Our wave problem is the steady-state part (no oscillatory factor in the solution).

$$
\begin{aligned}
(\omega^2 \mu \epsilon_0 - q^2)\mathcal{E} + \mu \omega^2 P = 0, \text{ and} \\
\frac{Ne^{*2}}{\epsilon(\infty)/\epsilon_0}\left(\frac{1}{M} + \frac{1}{m}\right)\epsilon - (\omega_{TO}^2 - \omega^2)P = 0,
\end{aligned} \tag{3.119}
$$

where the second equation is a recasting of Equation 3.114—the forced oscillator equation. A solution exists iff

$$
\begin{vmatrix}
\omega^2 \mu \epsilon_0 - q^2 & \mu \omega^2 \\
\frac{Ne^{*2}}{\epsilon(\infty)/\epsilon_0}\left(\frac{1}{M} + \frac{1}{m}\right) & -(\omega_{TO}^2 - \omega^2)
\end{vmatrix} = 0 \tag{3.120}
$$

Consider the $q = 0$ mode. A trivial solution is $\omega = 0$, and the second solution is

$$
\omega^2 = \omega_{TO}^2 + \frac{Ne^{*2}}{\epsilon(\infty)}\left(\frac{1}{M} + \frac{1}{m}\right) = \omega_{LO}^2. \tag{3.121}
$$

This frequency is ω_{LO} since a propagating solution must exist for a longitudinal electromagnetic wave to propagate.

The polarization that we have included is the ionic polarization. In general, in the system, there will also be an electronic contribution from the free particles. Let that be P_e, and let P_i denote the computed ionic component. We have

$$
\begin{aligned}
\epsilon(\omega) &= \frac{D(\omega)}{\mathcal{E}(\omega)} = \epsilon_0 + \frac{P_e(\omega)}{\mathcal{E}(\omega)} + \frac{P_i(\omega)}{\mathcal{E}(\omega)} \\
&= \epsilon_0 + \frac{P_e(\omega)}{\mathcal{E}(\omega)} + \frac{Ne^{*2}}{\epsilon(\infty)}\left(\frac{1}{M} + \frac{1}{m}\right).
\end{aligned} \tag{3.122}
$$

The ionic response vanishes at $\omega \rightarrow \infty$, so

$$
\begin{aligned}
\epsilon(\infty) &= \epsilon_0 + \frac{P_e(\omega)}{\mathcal{E}(\omega)}. \\
\therefore \quad \epsilon(\omega) &= \epsilon(\infty) + \frac{P_i(\omega)}{\mathcal{E}(\omega)} \\
&= \epsilon(\infty) + \frac{1}{\omega_{TO}^2 - \omega^2}\frac{Ne^{*2}}{\epsilon(\infty)}\left(\frac{1}{M} + \frac{1}{m}\right).
\end{aligned} \tag{3.123}
$$

This lets us write the static dielectric constant and the frequency-dependent dielectric function as

$$
\begin{aligned}
\epsilon(0) &= \epsilon(\infty) + \frac{1}{\omega_{TO}^2}\frac{Ne^{*2}}{\epsilon(\infty)}\left(\frac{1}{M} + \frac{1}{m}\right), \text{ and} \\
\epsilon(\omega) &= \epsilon(\infty) + \frac{\epsilon(0) - \epsilon(\infty)}{1 - \omega^2/\omega_{TO}^2}.
\end{aligned} \tag{3.124}
$$

For LO electromagnetic propagation to exist, LO phonon frequency must be such that $\epsilon(\omega_{LO}) = 0$. So,

$$
\begin{aligned}
\epsilon(\omega_{LO}) &= \frac{\epsilon(0) - \epsilon(\infty)}{1 - \omega_{LO}^2/\omega_{TO}^2} = 0 \\
\therefore \quad \omega_{LO} &= \omega_{TO}\left[\frac{\epsilon(0)}{\epsilon(\infty)}\right]^{1/2}.
\end{aligned} \tag{3.125}
$$

With these relationships—they represent a manifestation of the Kramers-Kronig relationships that we will discuss later—between the dielectric function, and *LO* and *TO* phonon frequencies, it follows that

$$\begin{aligned}\epsilon(\omega) &= \epsilon(\infty) + \frac{(\omega_{LO}/\omega_{TO})^2\epsilon(\infty) - \epsilon(\infty)}{1-\omega^2/\omega_{TO}^2} \\ &= \epsilon(\infty)\left[1+\frac{(\omega_{LO}/\omega_{TO})^2-1}{1-\omega^2/\omega_{TO}^2}\right] \\ &= \epsilon(\infty)\frac{\omega_{LO}^2-\omega^2}{\omega_{TO}^2-\omega^2}. \end{aligned} \tag{3.126}$$

So,

$$\begin{aligned}\frac{\epsilon(\omega)}{\epsilon(\infty)} &= \frac{\omega_{LO}^2-\omega^2}{\omega_{TO}^2-\omega^2}, \text{ and} \\ \frac{\epsilon(0)}{\epsilon(\infty)} &= \frac{\omega_{LO}^2}{\omega_{TO}^2}, \end{aligned} \tag{3.127}$$

which is the Lyddane-Sachs-Teller relationship. We now have a direct connection between the dielectric function under static and high-frequency conditions. When $\omega = \omega_{LO}$, the dielectric constant vanishes.

Since

$$\omega_{TO}^2 + \frac{Ne^{*2}}{\epsilon(\infty)/\epsilon_0}\left(\frac{1}{M}+\frac{1}{m}\right) = \omega_{LO}^2, \tag{3.128}$$

with $e^* = 0$, that is, covalency, for zone center phonons, $\omega_{TO} = \omega_{LO}$. This is the case for semiconductors such as *Si*. For *GaAs* and other covalent semiconductors, there will be a frequency gap between ω_{TO} and ω_{LO} due to e^*. We will discuss this in Chapter 12.

3.12 *Summary*

THIS CHAPTER LAYS THE GROUNDWORK necessary to mathematically describe the crystalline solid that is to be our semiconductor and in which we are going to explore the variety of interactions and cause and chance behaviors that physics builds insights into. Atoms constitute this solid, and, at the energy scales of the condensed matter phenomena of interest, it suffices to look at the solid—crystalline, in our case—as a collection of atoms bonded to each other in a periodic arrangement where one may view the assembly and states within it that the electrons can exist in from a quantum-mechanical view.

We started with a classical-quantum discussion, where, by looking at an electromagnetic wave and the particle-quantum wave duality, we noted the connections between the interpretations of energy and the flow of energy. The electron wavefunction lends itself to a variety of corresponding interpretations to the electromagnetic wave description. The latter has the electron as an example for particle representation; the former has the photon as a particle example. We have to resort to the quantum view, since the description of what is happening in the atom, or in the collection of atoms together, needs that depth for an adequate description. Our first step toward describing the solid was to first observe that there are 14 Bravais lattices—mathematical arrangements of points—that fill space, and, by bonding, the solid appears as an arrangement of atoms, where these atoms can be seen as occupying the lattice sites as their equilibrium positions. There are 7 such crystal arrangements, and solids appear in all the 14 different lattice arrangements. Zinc blende and diamond, which are interpenetrating *FCC* crystals, wurtzite, which is hexagonal close packed, perovskite, which many semiconducting transition element oxides of the form ABO_3 have, and hexagonal sheets, are some of the forms that we encounter for semiconductors. The smallest volume, exhibiting the symmetries, that fills the space is the Wigner-Seitz cell. Since the solid is periodic, the electron wavefunction too is periodic.

We avoided any discussion of quasicrystals, which are very interesting in their own right, since none are known for being semiconducting. Quasicrystals too fill space completely and are made possible by a subtle twist in the rotational argument.

Waves are suitably represented through wavevectors. A periodic lattice lets us also draw a periodic reciprocal lattice. The waves are now describable conveniently in this reciprocal space. So, we looked at how we construct the reciprocal space lattice and its relationship to the real space lattice. Some of the conclusions from here were how we will represent planes, directions of planes, distances between planes and the mapping of these planes to the reciprocal lattice. So, while both real space and reciprocal space are equivalent views, for some items of interest, for example, those related to spatial arrangements, the real space is the convenient approach, while, for others, such as the properties—dispersion being an important one of interest—associated with these waves, the reciprocal space is the more convenient approach.

Having built this mathematical description, we embarked on understanding the description of electrons—as waves—and of atomic motion around their equilibrium—as vibrations to be quantified by phonons—through toy models. The simplest of this was to have a free potential crystal, that is, a crystal where there exists periodicity because there are atoms a apart but where these atoms cause no potential perturbation. This is just free electron states in a periodic arrangement. It lets us understand the representation of the allowed $E(\mathbf{k})$ states of the electron in a variety of ways as a function of **k**, whose most convenient form was the reduced zone. This reduced zone is the 1st Brillouin zone of the reciprocal lattice. It is the analog of the Wigner-Seitz cell in the reciprocal representation. A wavevector also tells us the direction of motion of the wave. The toy model showed us that one finds specific wavevectors—zone boundaries—as places of degeneracy where counter-propagating waves cancel; absence of propagation is absence of propagating states; that is, a bandgap. The reduced zone picture also showed the bringing in of an index to the band of states allowed, which, together with the wavevector, identifies two of the quantum numbers of an electron's state in the solid. The third is spin, and, except in unusual circumstances, the energies are degenerate for spin.

We followed the toy model with the introduction of more rigor through Bloch's theorem, which then let us write the Bloch function as a wavefunction for the electron states. We came to it through a combination of symmetry of translation arguments, and this leveraging of the spatial periodicity was discernible through Fourier expansion. The Bloch function in the form $\psi_{n\mathbf{k}}(\mathbf{r}) = u_{n\mathbf{k}}(\mathbf{r}) \exp(i\mathbf{k}\cdot\mathbf{r})$ represents the eigenfunction of an independent electron in a periodic arrangement stretched out over the entire space. Both $\psi_{n\mathbf{k}}(\mathbf{r})$ and $u_{n\mathbf{k}}(\mathbf{r})$ have the periodicity $\mathbf{R}$ of the lattice. This is a plane wave solution if there is no potential perturbation. This was our toy model case. And, in that case, $u_{n\mathbf{k}}(\mathbf{r})$—the modulation function—is unity. With the potential arising in a variety of causes—periodically, due to the presence of the atoms periodically—the form of $u_{n\mathbf{k}}(\mathbf{r})$ will be non-trivial. This is a modulation function expanded in the basis set chosen. An electron's probability density is higher closer to the atoms, due to the Coulomb attraction, even as the Bloch function is spread out across the entire crystal. $u_{n\mathbf{k}}(\mathbf{r})$ may therefore be very rapidly modulating near the atomic locales.

The meaning of the wavevector **k** of the Bloch function is analogous to that of the free electron, except that this **k** now represents the consequences of the presence of the electron in the crystal. The Hamiltonian of the free space is very different from that of the crystal. $\hbar\mathbf{k}$ is the crystal momentum, by which we mean the momentum of the electron in this state when it is present in the crystal, and $E(n, \mathbf{k})$ is the energy dispersion of the electrons in the crystal.

We extended our toy model picture by introducing perturbation of the original potential-free periodic structure, and this allowed us to show the nature of the connection between the bandgap that appears and the perturbation. Additional consequences of the periodicity, the perturbation and the close proximity of the atoms are the breaking of degeneracies and the spreading of the energies of the states. So, we could see now that the bandgap appears together with bands, and, since states are conserved, we can give meaning to the states, and the ones that we are interested in will be from the valence of the atoms. Valence bands and states are those that are largely filled, and conduction bands and states are those that are largely empty. An ideal semiconductor is one with this bandgap, where the bands below the bandgap—the valence band—are all filled up, and the ones above—the conduction band—are all empty in an ideal arrangement at absolute zero temperature. We also introduced the idea of the Jones zone for viewing the energies as an extension of the Brillouin zone view. Unlike the 1st Brillouin zone, whose construction is the creation of the smallest reciprocal space region that completely reproduces through translation the entire reciprocal space, the Jones zone, also in the reciprocal space, is the smallest region in which all the states from the atoms in a unit that went into creating the conduction and the valence bands also exist uniquely without any folding. So, a bandgap exists on the entirety of the Jones zone surface. For *FCC* in a two-atom basis for our semiconductor, the Jones zone then includes the first four Brillouin zones and will contain eight electrons.

We also employed periodicity to understand the vibrational motion of atoms around their equilibrium, exploring the number of such vibrational modes and their correspondence to the degrees of freedom of motion in the collective assembly, and their description through the quasiparticle phonon. Our solution to a toy model composed of a two-atom basis showed us the appearance of optical and acoustic branches, with longitudinal and transverse displacements. The optical branches are modes with high energy where the displacements change rapidly and are opposite in phase at the atomic length scale, while the acoustic branches consist of modes where the phase change is gradual. Acoustic modes exist to vanishingly low energy, while optical modes have high energy regardless of the wavevectors of the modes.

The dielectric function is a property that is important to any analysis of electromagnetic interaction with matter. If one introduces ionicity e^* in these vibrating masses, starting from Maxwell's equations one could also relate how the polarization will change as a function of frequency. And it is the polarization that is reflected in the dielectric response of a medium. We added an electronic term to account for the polarization from conducting charge to this response, and by noting that, at high frequencies beyond the vibrational frequencies, the ionic polarization response will vanish, we could relate transverse and longitudinal optical mode frequencies. The considerations also let us write the dielectric function as a function of frequency and relate the longitudinal and transverse optical responses to a transverse electromagnetic radiation.

3.13 *Concluding remarks and bibliographic notes*

THE FOUNDATIONS OF QUANTUM MECHANICS, and the underlying particle-wave, reality, local and nonlocal, and other conflicting notions that appear as one transplants our classical observations to the quantum, create a number of seemingly "paradoxical" conundrums that we have sidestepped in this chapter, where we used the wave particle description to extract a conceptual preliminary understanding to explore the behavior of semiconductors. Semiconductors are a subset of solid state, where conductivity and other properties can be strongly modulated because of the existence of a bandgap, that is technologically important.

Numerous texts explore this introductory understanding physically. The three volumes of Feynman's lectures, ably transformed by co-writers Leighton and Sands[1], is most certainly a genuinely intuitive introduction to physics, and its approach to viewing and understanding the world around us ranges from the classical approaches of many centuries ago to the modern notions of today. Its introduction to quantum ideas are among the smoothest, and it also introduces many other complex connections that physics builds for us. Also very suitable for an undergraduate-level exploration of quantum mechanics is the text by Griffiths[2]. Atkins[3] has created a quite beautifully illustrated handbook with short descriptions of concepts of quantum mechanics without much recourse to mathematics. For understanding wave phenomena in the classical electromagnetic sense, the book by Haus[4] is also a favorite. The text by Harrison[5] is another excellent source for learning the essentials of quantum mechanics that are useful in engineering.

McKelvey's[6] and Blakemore's[7] textbooks are two well-written texts with introductions to crystal structure, reciprocal space and lattice dynamics, that is, atomic vibrations. These subjects, as well as the free-potential crystal description, can also be seen in the introductory chapters of a well-written engineering-centric book by Wolfe, Holonyak and Stillman[8]. A more in-depth discussion of the crystalline structure and the group-theoretic foundations can be found in advanced texts devoted to materials science and physics. An example of this approach is a standard from times past by Madelung[9]. Another book that we will refer to quite often, along

[1] R. P. Feynman, R. B. Leighton, and M. Sands, "The Feynman lectures on physics," Addison-Wesley, ISBN 13 978-0201500646 (2005)

[2] D. J. Griffiths, "Introduction to quantum mechanics," Pearson, ISBN 13 978-0131118928 (2004)

[3] P. W. Atkins, "Quanta,", Oxford, ISBN 0-19-855572-5 (1991)

[4] H. A. Haus, "Waves in fields in optoelectronics," Prentice Hall, ISBN 0-13-946053-5 (1984)

[5] W. A. Harrison, "Applied quantum mechanics,", World Scientific, ISBN 9810243758 (2000)

[6] J. P. McKelvey, "Solid state and semiconductor physics," Krieger, ISBN 0-89874-396-0 (1982)

[7] J. S. Blakemore, "Solid state physics," Cambridge, ISBN 0-521-30932-8 (1989)

[8] C. M. Wolfe, N. Holoynak, and G. E. Stillman, "Physical properties of semiconductors," Prentice Hall, ISBN 0-13-669961-8 (1989)

[9] O. Madelung, "Introduction to solid-state theory," Springer, ISBN 3-540-08516-5 (1981)

with the Wolfe-Holonyak-Stillman one, is by Yu and Cardona[10], a book that is now in its fourth edition, with a rich content of semiconductor-specific information and advanced techniques. Fischetti and Vandenberghe[11] approach the semiconductor theme, again at an advanced level, with specific focus on developing the understanding of transport. Together with Yu and Cardona, this book is a very thorough and advanced treatment of semiconductor physics.

[10] P. P. Yu and M. Cardona, "Fundamentals of semiconductors," Springer ISBN 978-3-642-00709-5 (2010)

[11] M. V. Fischetti, W. G. Vandenberghe, "Advanced physics of electron transport in semiconductors and nanostructures," Springer, ISBN 978-3-319-01100-4 (2016)

A book devoted to phonons and their behavior in semiconductors is by Stroscio and Dutta[12]. Phonon interactions are sprinkled throughout this text, and the Stroscio-Dutta book is a good source to refer back to for the various coupling mechanisms between different energy mechanisms and particularly in confinement.

[12] M. A. Stroscio and M. Dutta, "Phonons in nanostructures," Cambridge, ISBN 0-521-79279-7 (2001)

3.14 *Exercises*

1. This is a problem to tie electromagnetics and quantum mechanics together through an exploration of nonlinearity. Beyond what field strengths of electric field $\mathcal{E}$ and magnetic flux B does nonlinearity appear for Maxwell's equations? At high-enough fields, quantum-mechanical vacuum (and virtual particles) must be perturbed.
   - Take the classical properties of particles (charge e, mass m and the speed of light c) and of free space (permittivity and permeability), and show that there is a unique combination with the right dimensions to make the field ($\mathcal{E}$ or B, which has the same units in Gaussian units). The lightest particle sets the minimum field. Evaluate it for the electron.
   - This last estimate is wrong because quantum mechanics matters. So, introduce $\hbar$ to the list of the relevant properties. Show now that there is no unique combination with the units of $\mathcal{E}$.
   - Now consider the virtual possibilities. Find the field that will accelerate a virtual particle of momentum $\sim mc$ to an energy $\sim mc^2$. This critical field, too, is determined by the lightest particle. Evaluate it for an electron.
   - Show that this last solution is equivalent in form to the one before. **[M]**
2. Show that
   - an infinite point lattice can only show $2-$, $3-$, $4-$, or 6-fold rotational symmetry,

- if $[uvw]$ is an axis of a zone, and (hkl) is a face in the zone, then $hu + kv + lw = 0$, and

- in a cubic system, the angle ϕ in between the normals to the face $(h_1k_1l_1)$ and $(h_2k_2l_2)$ is given by

$$\phi = \cos^{-1} \frac{h_1h_2 + k_1k_2 + l_1l_2}{\left[\left(h_1^2 + k_1^2 + l_1^2\right)\left(h_2^2 + k_2^2 + l_2^2\right)\right]}. \quad \text{[S]}$$

3. Show that, in a hexagonal-close-packed structure,
   - the ratio $c/a = 2\sqrt{6}/3 \approx 1.633$.

   Find also the angles
   - α between (0001) and $(10\bar{1}1)$,
   - β between (0001) and $(11\bar{2}1)$, and
   - γ between $(10\bar{1}1)$ and $(01\bar{1}0)$. [S]

4. What is the fractional space filled by an arrangement of spheres that is
   - primitive cubic,
   - body-centered and face-centered cubic, and
   - a diamond structure. [S]

5. What fraction of a tetrahedron, with four spheres placed at the corners, is filled by the spheres? What causes the difficulty in filling the space densely? [S]

6. Primitive vectors for the lattice construction are not unique. Show that, for new vectors

$$\hat{\mathbf{a}}_i' = \sum_j S_{ij}\hat{\mathbf{a}}_j$$

   to also be primitive, the sufficient condition is $\det|S_{ij}| = \pm 1$. [S]

There is a correspondence here with orthonormal basis set completeness. You can find the proof for that in many quantum mechanics-texts. Slater's text *Quantum Theory of Atomic Structure*, for example, discusses this.

7. One possibility for the primitive translation vectors of the hexagonal space lattice is

$$\begin{aligned}
\hat{\mathbf{a}}_1 &= (3^{1/2}a/2)\hat{\mathbf{x}} + (a/2)\hat{\mathbf{y}},\\
\hat{\mathbf{a}}_2 &= -(3^{1/2}a/2)\hat{\mathbf{x}} + (a/2)\hat{\mathbf{y}}, \text{ and}\\
\hat{\mathbf{a}}_3 &= c\hat{\mathbf{z}}.
\end{aligned}$$

   - Show that the volume of the primitive cell is $(3^{1/2}/2)a^2c$.
   - Show that the primitive translations of the reciprocal lattice are

$$\begin{aligned}
\hat{\mathbf{b}}_1 &= (2\pi/3^{1/2}a)\hat{\mathbf{x}} + (2\pi/a)\hat{\mathbf{y}},\\
\hat{\mathbf{b}}_2 &= -(2\pi/3^{1/2}a)\hat{\mathbf{x}} + (2\pi/a)\hat{\mathbf{y}}, \text{ and}\\
\hat{\mathbf{b}}_3 &= (2\pi/c)\hat{\mathbf{z}}
\end{aligned}$$

so that the lattice is its own reciprocal, but with a rotation of axes.

- Describe and sketch the first Brillouin zone of the hexagonal space lattice. **[S]**

8. Figure 3.26 shows a two-dimensional array of identical atoms. Using various sections of the drawing, point out the following:
   - the primitive translations of the lattice,
   - the basis for the periodic structure,
   - the Wigner-Seitz unit cell,
   - the primitive translations of the reciprocal lattice, and
   - the first Brillouin zone. **[S]**

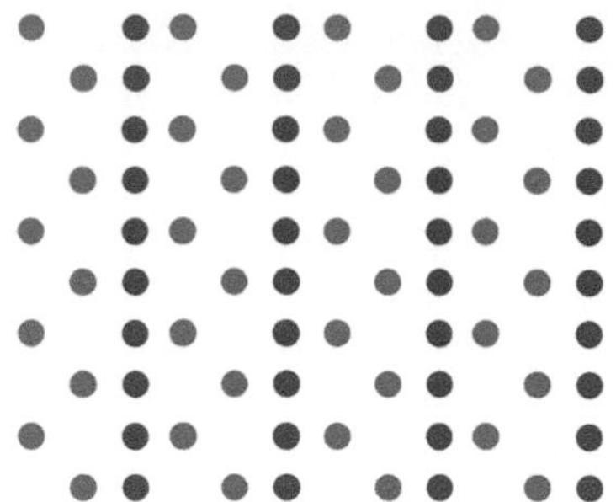

Figure 3.26: A two-dimensional lattice.

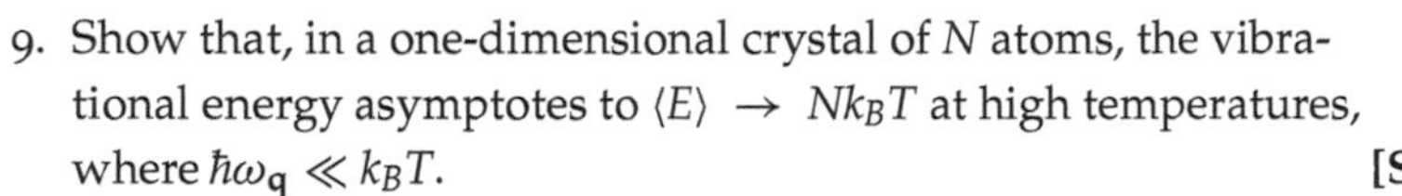

9. Show that, in a one-dimensional crystal of N atoms, the vibrational energy asymptotes to $\langle E\rangle \rightarrow Nk_BT$ at high temperatures, where $\hbar\omega_{\mathbf{q}} \ll k_BT$. **[S]**

10. For an ionic crystal, that is, one consisting of positively and negatively charged atoms that are spherically symmetric, bound together using Coulomb forces and kept separated by a repulsion that classical theory introduces *ad hoc*, the interaction energy between ions i and j, with the ions charged $+e$ and $-e$, is

$$U_{ij} = \pm\frac{e^2}{\rho_{ij}} + \frac{b}{\rho_{ij}^n}.$$

Here, $\rho_{ij} = \alpha_{ij}r$ is a normalized measure of the nearest neighbor separation of r. The energy U_i of the ith ion, by summing over ions $j \neq i$, in the field of all other ions is

$$U_i = -\frac{Ae^2}{r} + \frac{B}{r^n},$$

with $A = \sum_{j\neq i} \pm\alpha_{ij}^{-1}$ (the reference ion i is assumed negative), and $B = b\sum_{j\neq i}\alpha_{ij}^{-n}$. This factor A is called the Madelung constant. The total lattice energy $U(r)$ with $2N$ ions is

$$U(r) = NU_i = -N\left(\frac{Ae^2}{r} - \frac{B}{r^n}\right),$$

so long as N is large and we ignore surface effects.
Show that the lattice energy $U(r_0)$, where r_0 is an equilibrium separation between ions, is given by

$$U(r_0) = -\frac{NAe^2}{r_0}\left(1 - \frac{1}{n}\right).$$ **[M]**

11. Asymmetric molecules such as NH_3 and H_2O come about because of a polarizable central atom in rigid models such as that in Exercise 10. The energy due to the dipole moment $\mathbf{p}$ in a field $\boldsymbol{\mathcal{E}}$ is $(1/2)\mathbf{p}\cdot\boldsymbol{\mathcal{E}}$. Let α be the polarizability of the central atom. If $\boldsymbol{\mathcal{E}}$ induces this dipole moment, then $\mathbf{p} = \alpha\boldsymbol{\mathcal{E}}$.
   - Relate the bond angle β in stable equilibrium to the polarizability α for molecules of composition AB_2 and another one of composition AB_3.
   - If, for H_2O, $r_{OH} = 0.096\ nm$, and, for NH_3, $r_{NH} = 0.101\ nm$, then what is the polarizability for O and N? **[M]**

Polarizability is explored in depth in Chapter 12. Here, we can just view it as a distortion of charge cloud in the presence of an electric field that causes a dipole to appear, since, on average, the valence electron charge cloud is now slightly shifted in position compared to the core.

12. Draw the reciprocal lattice for a two-dimensional square lattice, and show Fermi surfaces in the first Brillouin zone when the atoms have 1-, 2-, 3- and 4-electron atoms. **[S]**

13. Show that a harmonic oscillator has a total energy of

$$E = \frac{1}{2}\hbar\omega \coth\frac{\hbar\omega}{2k_BT}.$$ **[M]**

14. Show that, under the constraints of Fermi-Dirac statistics, a free electron ensemble at $0\ K$ has a mean kinetic energy of $(3/5)E_F$. When the temperature is finite, this average is

$$\langle T\rangle = \frac{3}{5}E_F(0)\left[1 + \frac{5\pi^2}{12}\left(\frac{k_BT}{E_F(0)}\right)^2\right].$$

   Assuming this relation, find the ratio of the specific heat c_V for a Fermi-Dirac constrained condition versus classical conditions when $E_F = 7.0\ eV$. **[S]**

15. Find the pressure exerted by the gas of electrons obeying Fermi-Dirac statistics. What is it for copper? Now consider a non-degenerate semiconductor, say Si, with $10^{18}\ cm^{-3}$ concentration at $300\ K$. How different is it? **[S]**

16. A cubic lattice has N atoms per unit volume, and each atom has Z valence electrons. What is the Fermi wavevector radius in the free electron approximation? **[S]**

17. What specifically do we mean when we say that the electron's crystal momentum is $\hbar\mathbf{k}$? In what way is it not the electron's momentum? **[S]**

18. Should the group velocity $((1/\hbar)\nabla_{\mathbf{k}}E)$ of the electron in any band always go to zero at the Brillouin zone boundary? Give a short pro or con argument. **[S]**

19. Show that, in a cube of volume a^3 with periodic boundary conditions, the wavevector satisfies $k_x = 2\pi n_x/a$, $k_y = 2\pi n_y/a$, and $k_z = 2\pi n_z/a$, where n_x, n_y and n_z are integers. From this, show that
    - the density of states follows
    $$\frac{dn_x\, dn_y\, dn_z}{a^3} = \frac{dk_x\, dk_y\, dk_z}{(2\pi)^3}.$$
    - Also show that, for a spherical cavity,
    $$\frac{dk_x\, dk_y\, dk_z}{(2\pi)^3} = \frac{4\pi\nu^2}{c^3}\, d\nu,$$
    where ν is frequency and we have assumed this problem to be that for an electromagnetic wave and a photon. **[S]**

20. Why does the energy spacing between allowed eigenstates vary as $\propto L^{-2}$ where L is a linear dimension of the material? **[S]**

21. For conduction band minima or valence band maxima, does $E \propto \mathbf{k} \cdot \mathbf{k}$ have to be the lowest order dependence? Why? **[S]**

22. A one-dimensional periodic lattice has a period R. Show that the free electron wavefunctions are degenerate at the Brillouin zone boundaries. Now, if one introduces a small periodic potential V at each atom site inhabiting this lattice, show that the wavefunction solution at the boundary is proportional to $\sin(n\pi z/R)$ and $\cos(n\pi z/R)$, with n as an integer. **[S]**

23. Why is the appearance of an energy gap at the Brillouin zone boundary of a lattice in one dimension the equivalent of a Bragg reflection for the electron waves? **[S]**

24. A one-dimensional periodic lattice, of period R, has the following potential:
    $$V = V_0 \text{ for } -a \leq z \leq 0,$$
    $$V = 0 \text{ for } 0 \leq z \leq R - a, \text{ and}$$
    $$V(z+R) = V(z).$$
    What are the energies at the zone boundary when $V_0 = 0.1$, $R = 8$, and $a = 3$ in normalized units, at
    - the top of the first band, and
    - the bottom of the second band? **[S]**

25. A one-dimensional crystal has an energy dispersion given by
    $$E(k) = \frac{\hbar^2 k^2}{2m_0^*} - \alpha k^4.$$

Find α, the effective mass at the zone center and the zone edge (π/a), the maximum group velocity and the corresponding energy. **[S]**

26. A simple cubic crystal with a unit size a has the energy dispersion given by

$$E = -E_0[\cos(k_x a)\cos(k_y a) + \cos(k_z a)].$$

Evaluate the effective mass tensor at the zone center, the face center and the corners. **[S]**

27. Consider a linear lattice with a as the translational basis and the crystal formed from it with the periodic potential

$$V(z) = V_0 + V_1 \cos\left(\frac{2\pi z}{a}\right) + V_2 \cos\left(\frac{4\pi z}{a}\right) + \cdots.$$

- What are the conditions for the free electron approximation to work?
- Draw the three lowest energy bands in the first Brillouin zone.
- Calculate the first order energy gap at $k = \pi/a$ and $k = 0$. Where do these occur, and why? **[S]**

28. The conduction band minima appear at all sort of points, particularly L, near X, and other points in the Brillouin zone across various group IV and III-V semiconductors. But valence band maxima—light hole, heavy hole and split off—appear at the zone center. Can you think of a conceptual reason? **[S]**

29. Show the equivalence of the two formulations of Bloch's theorem,

$$\psi(\mathbf{k}, \mathbf{r} + \mathbf{R}_i) = \psi(\mathbf{k}, \mathbf{r}) \exp(i\mathbf{k} \cdot \mathbf{R}_i)$$

and

$$\psi(\mathbf{k}, \mathbf{r}) = u_{\mathbf{k}}(\mathbf{r}) \exp(i\mathbf{k} \cdot \mathbf{r}), \quad \text{where} \quad u_{\mathbf{k}}(\mathbf{r} + \mathbf{R}_i) = u_{\mathbf{k}}(\mathbf{r}).$$ **[S]**

30. Why does SiH_4 exist as a stable molecule while Si_2 does not? **[S]**

31. The periodic Dirac delta potential

$$V(z) = V_0 \sum_{n=-\infty}^{\infty} \delta(z - na)$$

is a toy model potential for a linear atomic chain. a is the periodicity of the lattice. Find the form of the Bloch wave. Note that the spike implies that the first derivative of the wavefunction will be discontinuous. Show also a graphic procedure for obtaining $E(\mathbf{k})$. **[S]**

32. A crystal has a static dielectric constant $\epsilon_r(0) = 5.9$, and a dielectric constant in the near infrared, approximated as $\epsilon_r(\infty) = 2.25$. The reflectivity vanishes at 30.6 μm. Calculate the longitudinal and the transverse phonon frequencies at $\mathbf{q} = 0$, expressing them in units of eV, K and s^{-1}. **[S]**

33. Take into account only first order perturbations in the density of a free electron gas due to an oscillatory applied potential, and show that the response is described by a dielectric constant of

$$\epsilon(\omega, \mathbf{q}) = 1 - \frac{e^2}{q^2\epsilon_0}\sum_{\mathbf{k}}\frac{f_0\left[E(\mathbf{k}+\mathbf{q})\right] - f_0\left[E(\mathbf{k})\right]}{E(\mathbf{k}+\mathbf{q}) - E(\mathbf{k}) - \hbar\omega}. \qquad \textbf{[M]}$$

34. If one assumes the result (or derives it), and then takes the static asymptotic limit $\omega \rightarrow 0$, show that a static perturbation produces a dielectric constant which will completely screen the external field of a long wavelength phonon; that is, of $\mathbf{q} \rightarrow 0$. This dielectric constant is

$$\epsilon(\mathbf{q}, 0)/\epsilon_0 \rightarrow 1 + \frac{\lambda^2}{\mathbf{q}^2}, \quad \text{where } \lambda^2 = e^2 N(E_F).$$

Here, $N(E_F)$ is the density of states on the Fermi surface. This is precisely the *Thomas Fermi* approximation. Using this approximation, show that a point charge of $+Ze$ in an electron gas exhibits a perturbing potential of

$$V(r) = \frac{1}{4\pi\epsilon_0}\frac{Ze^2}{r}\exp(-\lambda r).$$

Note that if we include into this the semiconductor environments' polarization, we arrive at

$$V(r) = \frac{1}{4\pi\epsilon}\frac{Ze^2}{r}\exp(-\lambda r). \qquad \textbf{[M]}$$

35. Give a physical reason why $\omega_{\mathbf{q}}$ of the acoustic branch asymptotes to 0 as $\mathbf{q} \rightarrow 0$? **[S]**

36. Find an expression for the "mass" of a phonon of average thermal energy at 300 K, and compare this value with that of an electron's mass, ignoring dispersion. **[S]**

Note that Planck-de Broglie relationships give us a way of connecting energy, momentum and wavevectors.

37. A one-dimensional crystal consists of identical masses M connected by two different spring constants k_{s1} and k_{s2}. Plot the dispersion curves for the acoustic and optical branches, and show that the characteristics frequencies are related as

$$\omega_q^2 = \frac{k_{s1} + k_{s2}}{M}\left\{1 \pm \left[1 - \frac{4k_{s1}k_{s2}\sin^2(qa/2)}{(k_{s1} + k_{s2})^2}\right]\right\}. \qquad \textbf{[S]}$$

38. We will explore the question of damping in phonon oscillations by removing the linear spring constant approximation and adding anharmonicity. Consider an infinite linear chain of atoms of mass m, a apart and connected via spring constants k_s. But there also exists a damping force of $F = -\Gamma du_n/dt$, where u_n is the displacement of the nth atom from its equilibrium position $\mathbf{R}_n$. So, the damping is proportional to the velocity. Assume that the damping $\Gamma^2 \ll k_s/m$, and discuss specifically for the zone center and zone edge modes the following:
    - How does this damping affect the energy-wavevector relationship; that is, $\omega(\mathbf{q})$?
    - What is the relaxation time of the modes? **[M]**

39. Another interesting exercise in exploring nonlinearity in harmonic oscillators is to add a weak anharmonic term to the problem and understand the consequences as derived from perturbation theory. Assume a particle of mass m is in a harmonic oscillator potential with frequency ω, but with a weak anharmonic term:
$$V(z) = \frac{1}{2}m\omega^2 z^2 + \epsilon\left(\frac{z}{l}\right)^2$$
with
$$l = \sqrt{\frac{\hbar}{m\omega}}.$$
Using perturbation theory, find the first order approximation for the wavefunction, and both the first and the second order approximations for the energy levels. **[S]**

40. Determine and plot in the $\langle 100 \rangle$ directions the three-dimensional vibration bands for a simple cubic Bravais lattice with a basis of one atom by considering interactions between nearest neighbors. **[M]**

41. The phonon dispersion in a two-atom basis system with one force constant may work for longitudinal modes but becomes quite inaccurate for transverse modes. Take the force constants k_{s1} and k_{s2} for the masses M_1 and M_2 of a chain and show that the phonon solutions are given by
$$\omega_q^4 - \frac{M_1 + M_2}{M_1 M_2}(k_{s1} + k_{s2})\omega_q^2 + \frac{2k_{s1}k_{s2}}{M_1 M_2}\left[1 - \cos\left(\frac{qa}{2}\right)\right] = 0$$
and
$$\frac{u_2}{u_1} = \frac{\left[k_{s1} + k_{s2}\exp(iqa/2)\right]/\sqrt{M_1 M_2}}{[(k_{s1} + k_{s2})/M_2] - \omega_q^2}.$$
[S]

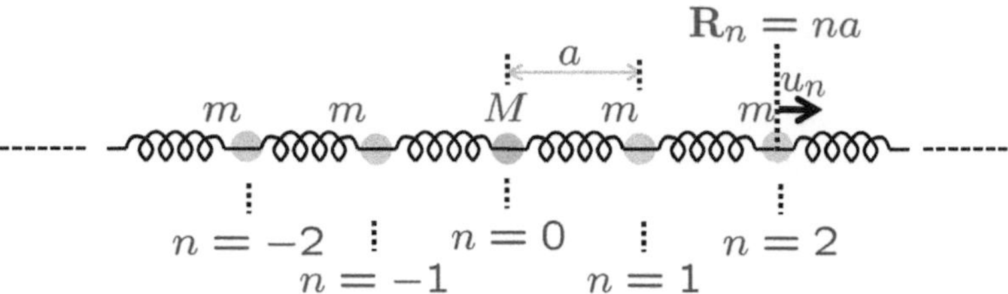

Figure 3.27: A linear bonded chain of m mass atoms with one atom of mass M as a defect, with other parameters constant.

42. Just as phonon dynamics are affected by the serious disruption caused by the surface and consequent translation symmetry breakdown, they are affected by other unintentional or intentional perturbations in the crystal. We will consider a single atom linear chain in which we replace one of the atoms by another of larger mass M while other parameters, such as the spring constant k_s, remain the same. Figure 3.27 shows this mass at $n = 0$. Assume that the displacement is described by the ansatz

$$u_n = u_0 \exp\left[-q(\omega)|n| - i\omega t\right]$$

for the displacement of the nth atom. Normally, we would have i in front of q here, with q real for oscillating mode. With this redefinition of q (i folded in and normalization), all those possibilities still exist (except that real and imaginary are reversed).

- Calculate the eigenfrequency of this chain.
- Is your solution valid for all masses? If not, it may be because the ansatz used here is not a fair choice.
- For the solution, how does the amplitude decay with distance?

[M]

"Ansatz", a German word, means "an approach" or "an attempt." An ansatz for us is the establishment of a starting point—of an equation, a theorem, a function or a value—that describes a problem or a solution, taking into account its boundary conditions. The ansatz is an assumption that will speak to its validity and constraints upon completion of the mathematical manipulation.

In writing the input to the exponential as $\left[-q(\omega)|n| - i\omega t\right]$, we have normalized the wavevector here to a, which is the translationally invariant distance.

4 Bandstructures

ELECTRONS AND PHONONS, one the carrier of charge and the other representing the vibrational excitation of the atoms of the crystal, are two particles or quasiparticles that we must be able to describe in the periodic crystal accurately. The wave-particle duality is reflected in the description of their states, whose most important characteristic is captured for us through the energy ($E = \hbar\omega = T + V$), a sum of the potential and kinetic forms for the electrons, which can traverse or stay bounded in the crystal, and the energy (call it $E_{\mathbf{q}} = \hbar\omega_{\mathbf{q}} = T_q + V_q$) of the vibrational modes. For the former, we have also ascribed the wavevector $\mathbf{k}$ and a crystal momentum $\hbar\mathbf{k}$, and, for the latter, we have ascribed a wavevector $\mathbf{q}$ and a momentum $\hbar\mathbf{q}$. We used toy models in Chapter 3. Semiconductors will be considerably more complex, given the nature of the atoms, their ionicity, the valence states, the residual core, the nature of interactions, the nature of bonding and various other factors. So, the description of real semiconductors requires more substance than our preliminary discussion of the essential concepts.

For the electrons, the wavefunction is a Bloch function. These Bloch functions are plane waves if the modulation function $u_{n\mathbf{k}}(\mathbf{r}) = 1$, as in the hypothetical zero-potential perturbation semiconductor crystal of Chapter 3. The two-wave solution was an example where this perturbation was a delta function in real space, with the reciprocal space periodicity of $\mathbf{K}$. In the realistic crystal, the perturbation will be a spread spatially, with the interaction strong when the position coordinate is close to the core, and less so farther away. If one were still using a plane wave basis, then many, many more terms will be needed in a Fourier expansion to capture the spread of variation in real space. So, the toy model techniques that we employed to understand the wave-particle description of a semiconductor bandstructure are quite wanting in describing the electron bandstructure of a real crystal. This difficulty can be

Semiconductor Physics: Principles, Theory and Nanoscale. Sandip Tiwari.
 DOI: 10.1093/oso/9780198759867.001.0001

attacked through a number of techniques employing choices in the basis functions from which to build the solution. Each has its merits and shortfalls.

Tight binding is the approach where one uses atomic orbitals. It is particularly useful when bonding is strong and where the final states are more atomic-orbital-like, such as the states in valence bands. But imagine the challenge of building a nearly free electron—the spread-out modulation function $u_{n\mathbf{k}}(\mathbf{r})$ of a conduction band electron that also modulates rapidly near the core—out of the orbital basis. Tight binding helps us look at hybrid orbitals (and the valence band), which will appear often as we attempt to understand point defects and point perturbations in general because of the intimate evolution from atomic orbitals to bonding hybrids. Interestingly, the plane wave method, which, as we just implied, is more suitable for the nearly free electron representation, can be modified through an orthogonalization to make it very suitable around the core region. This leads us to discuss the use of orthogonalized plane waves as a method. Here, one can build the wavefunction without having to include a large number of terms in the construction. This discussion then takes us on to a discussion of pseudopotential, where we will "concoct" a wavefunction in the core region—make a pseudopotential—that captures the kinetics accurately. This is a numerically efficient method for electron bandstructure calculation. Another modern technique is the use of a functional—the density functional—that is very adept at simplifying the solving of the crystal Hamiltonian and tackling the erring ways of the Hartree potential discussed in Chapter 1. Another method, equally useful, and also more physically meaningful and with numerous overtones, is the $\mathbf{k} \cdot \mathbf{p}$ method. We will use this to understand the meanings of the ideas of effective mass, of the strength of oscillators (the lowest order stable systems due to the restorative nature of the second power in canonic shifts from equilibrium), of anharmonicity at bandedges and of valence bandstructure, particularly due to the spin-orbit coupling.

The calculation of phonon bandstructure too, likewise, will have considerable changes, given the limitations of the one dimension and other simplifications that were adopted in their introduction. We will largely summarize these results, since many of the important and foundational thoughts about the longitudinal and transverse and the optical and acoustic modes have been dealt with in Chapter 3.

The introduction of these various techniques, their implications and their pitfalls is the scope of this chapter, where we end up with a realistic representation of the bandstructures of semiconductors.

By bandstructure here, we mean the structure of the bands of electron states as well as of the phonon states.

4.1 *Bonding and binding*

CHAPTERS 1 AND 3 have given us a preliminary understanding of how one may model when electrons are brought together. A 2-electron, 2-atom system—the hydrogen molecule—forms a spatially symmetric singlet state and a spatially antisymmetric triplet state. The simple $1s^1$ configuration of a hydrogen atom, where there are two possible s energy states with up and down spin, with the electron filling one, when two of the atoms are put together, now has $2 \times 2 = 4$ states, and the degeneracy has been lifted. One state, the singlet, has lowered in energy. Because the molecule has two electrons, and the singlet state accommodates two electrons with up and down spin, it forms a stable low energy bonding state. The triplet state, a higher energy state, is an antibonding state that the molecule may be excited to but is not an equilibrium state. Four degenerate states have evolved into a combination of 1 (singlet) and 3 (triplet). Electrons as fermions are in states that have different wavefunctions including their spin. When one assembles many atoms with their electrons together in a crystalline solid, the problem has evolved to one of a larger scale. States will evolve, as they did with hydrogen molecule. Periodicity will place constraints. Chapter 3 set up the preliminary framework and physical intuition of what may happen and how one may understand it.

The formation of molecules is a tight binding where the energies and wavefunctions of the atomic orbitals undergo change, forming the bonding and antibonding states. At its simplest, if one took an atom and brought a number of them together, the energies of the states—valence and core—will evolve, and degeneracies will be lifted. Figure 4.1 is an illustration of bringing a group of atoms together. When they are very far apart, in the left part, the energy description derived from the atomic description is a good description, and each atom looks the same, as shown on the left, with energies shown along the ordinate at the origin of the abscissa. As one brings them together, the degeneracy of these states will be lifted. The total number of states arising from these three energies is invariant. In a two-atom system, one may see something akin to the hydrogen molecule, or maybe not depending on the interactions, but some general features will hold. Consider the level at energy E_3 with an eigenfunction $|u_3\rangle$. This has the strongest interaction. The

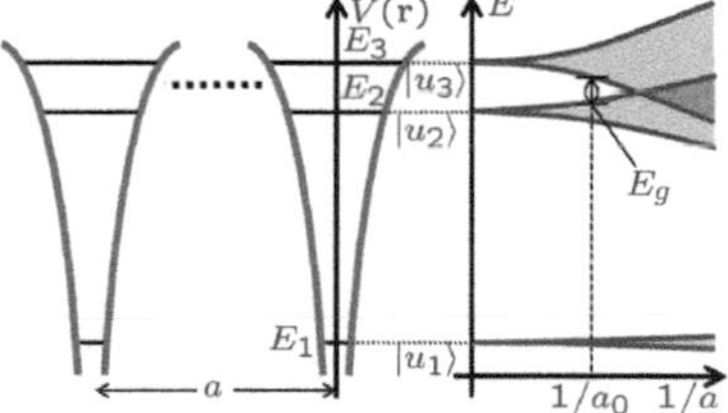

Figure 4.1: Atoms with three energy levels (for example, 1s, 2s and 2p) as they are brought together and evolve in weak coupling. Bands of states form, and degeneracy has been lifted.

degeneracy lifting will have the strongest energy spread because the interaction is the strongest. The antibonding states and the bonding states will spread, forming a band as many of these atoms interact. The next energy level down (E_2) too will have interaction, but it is slightly weaker and the spread a little smaller. The level at E_1, meant to represent a core level, will have very weak interaction and little spread. So, depending on the spreading of these bands, one may see an overlap with degeneracy still lifted, or one may see separate bands. As shown in the figure here, at a_0 separation the bands have a separation of E_g, the bandgap. We will see soon that it is possible for the bonding and antibonding states and their spread, as they arise from these different orbitals being brought together, to overlap and separate out again. Now there has been mixing of the states.

4.2 *Tight binding*

TIGHT BINDING GETS ITS NAME FROM the tight coupling between atoms. It was developed originally for molecules, where the covalent bonding is particularly tight, and the atomic orbitals of the atoms are a suitable starting point for the basis. The linear combination of atomic orbitals provides then a suitable approach to describing the bonding. In the crystal, this coupling it is not so tight anymore. If the crystal has a very strong covalent bond, the forming of these bonds from the atomic orbitals will give quite accurate solutions, at least for the valence band states, which are strongly localized at the atom. For the conduction band states, however, as they are more delocalized, it is generally necessary to employ other methods because the atomic orbitals are a poor starting basis. But there are strongly bonded materials, such as diamond and graphene, where the s and p orbitals of carbon—as orbitals or as sp^2 hybrids—do work accurately, with the out-of-plane p_z orbital feeling fairly weak perturbation. The tight binding method is also often referred to as the linear combination of atomic orbitals, and one can see in this discussion the origin of the term.

Tight binding is a good starting point for discussing crystal bandstructure calculations, for the following reasons. First, we understand bonding, antibonding and orbitals well. Second, it is quite important for understanding the valence band states with their heavy-hole and light-hole bands and split-off bands. In these latter bands, the magnetic interaction arising in the coupling between spin and orbital angular momentum leads to very significant effects. And, third, point perturbations—vacancies, interstitials and substitutional impurities—when they are very localized, are local interactions—molecular-like—and therefore using molecular energy calculations, that is, tight binding, becomes a reasonable way to approach the analysis.

Let $|m\rangle$ represent atomic orbitals: $|3s\rangle$ and $|3p\rangle$, for example, in the case of silicon. These are the eigenfunctions that the Bloch functions will asymptote to for vanishing interatomic interaction; that is, large atomic spacing. In this case,

$$|\psi(\mathbf{r})\rangle = \sum_m c_m |m\rangle \tag{4.1}$$

is the posited solution, and this solution follows from $\mathscr{H}_{mm'}$ and $\mathcal{O}_{mm'}$. If we were to form the basis in one orbital per site, with N lattice sites, this leads to an $N \times N$ secular determinant. This can be simplified. First, $|m\rangle$ may be written in terms of linear combinations of irreducible basis $|i\rangle$ of the space group. As an example, p orbitals may be written in terms of three orthogonal irreducible bases (p_x, p_y and p_z). So, $\mathscr{H}_{mm'}$ and $\mathcal{O}_{mm'}$ may now be written in terms of $\mathscr{H}_{ii'}$ and $\mathcal{O}_{ii'}$, where, due to orthogonality, many terms vanish unless $|i\rangle$ and $|i'\rangle$ are of the same irreducible representation. Bases forming irreducible representations of translation subgroup (T) suffice to construct representations of the entire space group (G). They are irreducible and complete. $\hat{T}_{\mathbf{k}}$ is one dimensional. The consequence of an operation by $\hat{T}_{\mathbf{m}}$ is a modification by $\exp(-i\mathbf{k}\cdot\mathbf{R}_m)$. The linear combinations of these, too, are bases of $\hat{T}_{\mathbf{k}}$. So, they are also bases of $|\psi_{\mathbf{k}}(\mathbf{r})\rangle$, that is, we now have a Bloch function

$$|\psi_{\mathbf{k}}(\mathbf{r})\rangle = \sum_m \exp(-i\mathbf{k}\cdot\mathbf{R}_m)|m\rangle. \tag{4.2}$$

We now form linear combinations $\sum_{\mathbf{k}} \psi_{\mathbf{k}}(\mathbf{r})$, where all different Bloch functions of different $\mathbf{k}$ are orthogonal. With diagonalization,

$$E_{\mathbf{k}} = \frac{\mathscr{H}_{\mathbf{kk}}}{\mathcal{O}_{\mathbf{kk}}} = \frac{\langle\psi_{\mathbf{k}}|\hat{\mathscr{H}}|\psi_{\mathbf{k}}\rangle}{\langle\psi_{\mathbf{k}}|\psi_{\mathbf{k}}\rangle}, \tag{4.3}$$

assuming one orbital per site. In silicon, where s, p_x, p_y and p_z will all contribute, each one will lead to its specific Bloch sum $\psi_{\mathbf{k}}(\mathbf{r})$. This will require further symmetrization.

4.2.1 *A one-dimensional tight binding toy model*

WE ILLUSTRATE TIGHT BINDING using a one-dimensional toy model. In Chapter 20, we will discuss a more complete usage of tight binding in calculating graphene's bandstructure. In the toy model, being a one-dimensional crystal, Equations 4.2 and 4.3

suffice. $\mathbf{R}_m = m\hat{\mathbf{a}}$, with $m = 1, 2, \ldots, N$. $\mathbf{k} = h\hat{\mathbf{b}}/N$, with $h = 1, 2, \ldots, N$, and $\hat{\mathbf{b}}$ the reciprocal space unit wavevector. Take N as even for convenience and without loss of generality. We may spread h symmetrically, because h is periodic in N. So,

$$h = -\frac{N}{2}, -\frac{N}{2}+1, \ldots, \frac{N}{2}-1, \frac{N}{2}, \tag{4.4}$$

and since $\hat{\mathbf{a}} \cdot \hat{\mathbf{b}} = 2\pi$,

$$\psi_{\mathbf{k}}(\mathbf{r}) = \sum_m \exp\left(-i\frac{2\pi hm}{N}\right)|m\rangle. \tag{4.5}$$

Consider only nearest neighbor interactions with

This approximation to overlap and nearest neighbor coupling is the Hückel approximation.

$$\begin{aligned} \langle m'|m\rangle &= \delta_{mm'}, \\ \langle m'|\hat{\mathscr{H}}|m\rangle &= E_0 \quad \text{for } m' = m, \\ &= \beta \quad \text{for } m' = m \pm 1, \quad \text{and, for all else,} \\ &= 0. \end{aligned} \tag{4.6}$$

E_0 is the eigenvalue of the Hamiltonian for the atomic orbital $|m\rangle$. By nature of atomic spacing being significant, we have also assumed that overlap vanishes, and that perturbation arises from nearest neighbor interactions only. $\langle\psi_{\mathbf{k}}|\psi_{\mathbf{k}}\rangle$ is just the sum of m 1s, so N. Therefore,

$$\begin{aligned} E_{\mathbf{k}} &= \frac{1}{N}\langle\psi_{\mathbf{k}}|\hat{\mathscr{H}}|\psi_{\mathbf{k}}\rangle \\ &= \frac{1}{N}\sum_{m'm}\exp\left(-i\frac{2\pi hm'}{N}\right)\exp\left(i\frac{2\pi hm}{N}\right)\langle m'|\hat{\mathscr{H}}|m\rangle \\ &= \frac{1}{N}NE_0 + \frac{1}{N}\sum_m \beta\left[\exp\left(-i\frac{2\pi h(m+1)}{N}\right)\exp\left(i\frac{2\pi hm}{N}\right)\right. \\ &\quad \left. + \exp\left(-i\frac{2\pi h(m-1)}{N}\right)\exp\left(i\frac{2\pi hm}{N}\right)\right] \\ &= E_0 + \frac{1}{N}2\beta\sum_m \cos\left(\frac{2\pi h}{N}\right) \\ &= E_0 + 2\beta\frac{1}{N}\cos\left(\frac{2\pi h}{N}\right), \\ &\quad \text{with } h = -\frac{N}{2}, -\frac{N}{2}+1, \cdots, \frac{N}{2}-1, \frac{N}{2}. \end{aligned} \tag{4.7}$$

Take the example of $N = 6$; h then takes on values of $-2, -1, 0, 1, 2$ and 3. This is a benzene-like ring structure, since no constraint was placed for end points. Two of these pairs, $(2, -2)$ and $(-1, 1)$, are degenerate, and $h = -3, 0, 3$ is non-degenerate. So, four terms need to be calculated, as summarized in Table 4.1 and plotted in Figure 4.2. With $N = 6$, this energy picture is a toy model for a benzene-like example employing carbon's $2p$ states, where β parameterizes the next neighbor coupling.

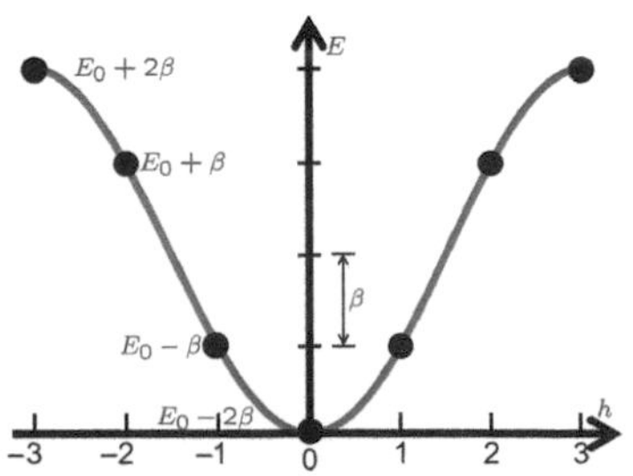

Figure 4.2: Energy levels, marked by solid circles, for a chain of $N = 6$ in the Hückel approximation for tight binding.

A more complete example of a tight binding calculation can be seen in Section 20.3, where graphene is employed as an example of a monolayer material.

4.2.2 *Reflections on tight binding*

THIS DISCUSSION OF TIGHT BINDING reflects the use of atomic orbitals as the starting point for calculating the bandstructure. In diamond and zinc blende lattices, this would imply the use of $|s\rangle$, $|p_x\rangle$, $|p_y\rangle$ and $|p_z\rangle$ as the starting points through the sp^3 hybridization. Since localized atomic orbitals comprise the basis function, one should expect an accurate description of the valence bandstructure easier to achieve since valence electrons are mostly localized in the bonds. The corollary to this is that an accurate description of the delocalized conduction bandstructure would be harder to achieve using tight binding.

As an aside, the aspect of computation as an arbiter is something that the reader will have noticed I abhor. There should be good arguments—not necessarily definitive but in support of plausibility—that appeal to reason. "Shut up and calculate" is often an edict when metaphysical discussions regarding the Copenhagen interpretation, the introduction of an observer, or even the Bayesian interpretation for quantum mechanics ensue. This phrase and the philosophy is amusingly dealt with by David Mermin in the Reference Frame column "Could Feynman have said this?" in the May 2004 edition of *Physics Today* (**57**, 10). He cites the phrase as an example of the Matthew effect, which I refer to in one of these marginalia notes in one of the other texts.

But since tight binding is intuitive to comprehend, it is a convenient tool for understanding the variety of general features one sees in semiconductor bandstructures, and one is not left to wonder or to accept it as "this is what the computation says." Tight binding is quite instructive in this regard.

Generally, for atoms, the *s* orbitals are closer to the nucleus than the *p* orbitals. *s* orbitals are not polar while *p* orbitals are polar. Semiconductor crystals are either covalent, as with single element semiconductors, such as those of group *IV*, or ionic, as with those that are of group *III-V* or *II-VI*. Ionicity implies transfer of charge—a polarization—with spatial movement of electron probability densities. Table 4.2 summarizes some of the ionicities of semiconductors of interest.

Ionicity is a term of pedagogical convenience that is bereft of a general definition, since it is employed for molecules and the different states of matter: gases, liquids, and solids with the electron charge cloud being considerably fuzzy. Ionicity to us is a parameterization of charge distribution such as used by Pauling and Phillips. We look upon it as a normalized scale of charge transfer.

Consider a hypothetical *Si*- or *Ge*-like crystal with inversion symmetry with the states evolving from the outermost filled orbitals of *s*, with its orbital quantization of $l=0$, and *p*, with $l=1$. This latter then has azimuthal quantization of $m_l = -1, 0, 1$. Tight binding's implication for the evolution of states through the crystal Hamiltonian consists of building states by mixing $|s\rangle$ and $|p\rangle$ orbitals. The former is symmetric, and the latter is oriented along the three principal Cartesian coordinates. In the conduction

h	Degeneracy	$\cos(2\pi h/t)$	$E_{\mathbf{k}}$
0	Singlet	$\cos(0)$	$E_0 + 2\beta$
1	Doublet	$\cos(2\pi/6)$	$E_0 + \beta$
2	Doublet	$\cos(4\pi/6)$	$E_0 - \beta$
3	Singlet	$\cos(6\pi/6)$	$E_0 - 2\beta$

Table 4.1: Energy levels in a linear atomic chain with $N=6$. The linear chain then ties back in a loop. β is the energy of next neighbor interaction, and E_0 is self-energy.

band, the more delocalized states, when they are near the center of the zone, will be $|s\rangle$-like. This follows from the symmetry argument. But, away from the center, they will arise from the more $|p\rangle$-like character. Both the ionicity, which represents the charge redistribution, and the orbital probability densities imply different features that one observes in semiconductor bandstructures.

	Ionicity
Si	0.00
Ge	0.00
SiC	0.18
GaAs	0.31
AlAs	0.42
InAs	0.46
GaSb	0.36
AlP	0.39
InSb	0.32
GaP	0.33
InP	0.42
ZnS	0.62
ZnSe	0.63
CdSe	0.70

Table 4.2: Ionicity of a select set of semiconductors.

The valence band—states closer to the nucleus—is very interesting, and shows a lot of complexity arising in the conservation principles and the localized core electron interactions close to the atom rather than the overlap of wavefunctions of electrons from separate atoms. We will discuss these at length. A number of important consequences can be mentioned here. The first point relates to the localization and evolution of the new states from the orbital wavefunctions. The second relates to the consequence of spin—and its magnetic manifestation—interacting with the nucleus charge while moving, as in the orbit of an atom or in the localized conditions of the crystal.

Regarding the first, the sp^3 hybrid is the way the s and p states evolve to tetrahedrally arranged bonding states with opposite spin electrons in each bond. These hybrids, $|h_1\rangle, \ldots, |h_4\rangle$, are of the form

$$|h_1\rangle = \frac{1}{2}\left[|s\rangle + |p_x\rangle + |p_y\rangle + |p_z\rangle\right], \quad \text{through}$$
$$|h_4\rangle = \frac{1}{2}\left[|s\rangle - |p_x\rangle - |p_y\rangle + |p_z\rangle\right], \tag{4.8}$$

where the negative signs circulate between the different $|p\rangle$s to form a non-degenerate set. These hybrids are oriented tetrahedrally, as shown in Figure 4.3(a). Since superposition of any p orbitals is also a p orbital, just oriented differently, we have

$$|h\rangle = \frac{1}{2}\left[|s\rangle + |p\rangle\right], \tag{4.9}$$

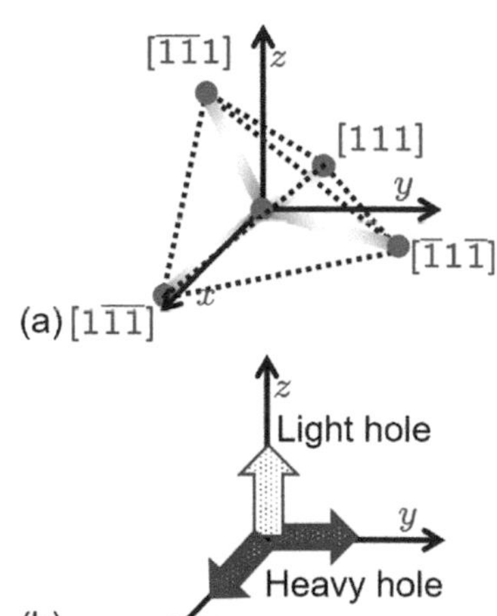

Figure 4.3: (a) A representation of the tetrahedral bonding in a cubic semiconductor crystal. (b) The polarization of the hole states. In (b), the heavy-hole polarization shows the consequence of two perpendicular p orbitals, and the light-hole state shows the consequence of the third p orbital, which has increased mixing with the s orbital.

oriented in the $\langle 111\rangle$ directions and orthogonal to each other. s orbitals are homopolar, while p orbitals are polarized; the hybrid leans toward the positive. Formally, the tight binding constructed using the atomic orbitals or through the hybrid are equivalent. So, a tetrahedral bonding will be based on eight hybrids and result in eight Bloch basis states. The diagonalization of the Hamiltonian will therefore lead to eight energy bands—bonding (valence) and antibonding (conduction), so, four for the valence band, and four for the conduction band. This construction of Bloch states could as well have been performed using s and p orbitals. The hybrid is a convenient view that fits with our chemical bonding picture and is closer to the physical arrangements of the electron probability distributions in the real crystal. It will computationally also be more efficient, due to this nearness. Quantum-mechanically, one may as well have stayed with the s and p orbitals. These will still lead to the eight Bloch basis states for a two-atom basis in the unit cell. Accurate calculation of the conduction bandstructure in tight binding will also require the incorporation of higher orbitals, such as 4s in the case of *Si*.

In the valence band states, this tight coupling in the hybrid (or atomic orbital) basis leads to the polarization of the p states, as shown in Figure 4.3(b). The in-plane p orbitals give rise to two heavy-hole polarizations, and the out-of-plane polarized p orbitals, with more mixing with the s, gives a lighter hole polarization.

Figure 4.4 is the pictorial view of this discussion, with reference to atoms with $|2s\rangle$ and $|2p\rangle$ orbitals. Part (a) shows our bonding-antibonding, hybrid orbital and band view in a simple way for the semiconductors where there is sp^3 hybridization and therefore tetrahedral bonding. The $|s\rangle$ and $|p\rangle$ orbitals of the atoms hybridize, forming the sp^3 hybrids $|h_1\rangle$ through $|h_4\rangle$ that comprise the bonding framework of the crystal. One could as well have seen the $|s\rangle$s′ and $|p\rangle$s′ degeneracy being broken when the atoms are brought together, forming this quantum assembly leading to the conduction and the valence bands. Parts (b) and (c) show the probability density representation of representative antibonding and bonding states of these bands, such as the bandedge state, for example. Part (d) shows this evolution in energy of the states—as represented by the figures in (a) through (c)—as one brings about the formation of the crystalline assembly. At large as, the orbitals are largely unperturbed, as one brings them together, bonding and antibonding

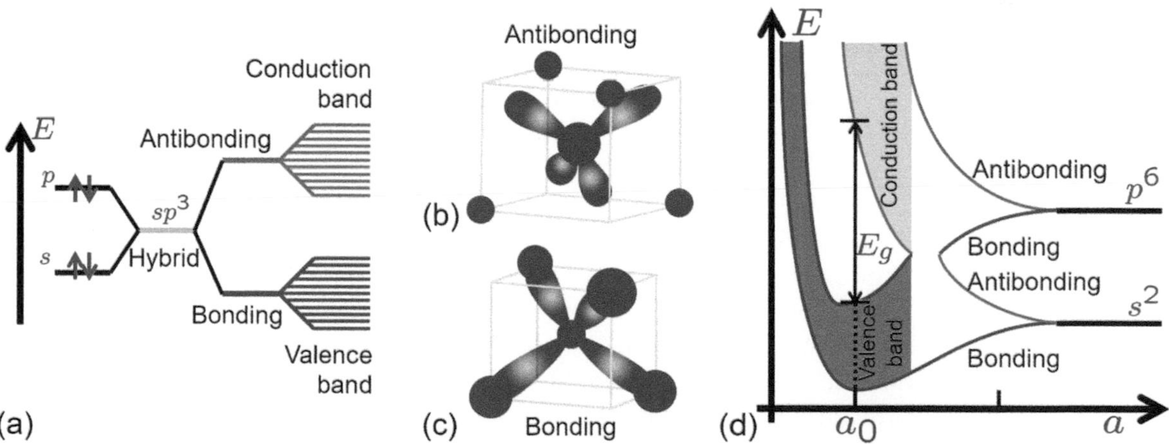

Figure 4.4: A representation of the formation of bonds and bands. (a) The creation of hybrids from the atomic orbitals, as well as bonding and antibonding states and their evolution into conduction and valence bands. (b) and (c) The representational probability distributions of antibonding and bonding states. (d) The forming of bonding and antibonding bands and their evolution to conduction and valence bands with a bandgap as atoms are brought together.

states form and spread out into bands, as we saw earlier in Figures 4.4 and 4.1. Even shorter interatomic spacing causes overlap of the bonding and antibonding states that arose from the $|s\rangle$ and $|p\rangle$ states. They merge, and, when there is even shorter spacing—at the equilibrium, where the system has the lowest energy—a bandgap appears. The valence band states are separated from the conduction band states. All the states as shown in this figure arose from the intermixing and evolution from the $|s\rangle$ and $|p\rangle$ states. In general, both the valence band and the conduction band will have symmetries that are *s*-like, *p*-like or a mix, but there will be a large bonding-like nature to the valence band states, that is, localization, and a large antibonding nature to the conduction band; that is, delocalization. Conduction will take place in both of these bands so long as filled and empty states exist, due to the Bloch nature of these states—the pictures in Figure 4.4(b) and (c) are connected spatially in real space, only the valence bands are more bound than the conduction band states. In Figure 4.4(b), the antibonding state is drawn to have spherical symmetry at the corners of the cube. These states will appear at and near Γ in reciprocal space with isotropic symmetry of transport. But they could have orientational features due to their *p*-like nature, and then, as in *Si* or *Ge*, the conduction band minimum will not have isotropy.

There is one additional notation that you will encounter in this text. When *s* orbitals in a crystal lead to the formation of a band due to the breaking of the translation symmetry, we will sometimes refer to it as the σ band. For *p* orbitals, it is the π band. We are showing deference to history here. This notation will appear, for example, in the discussion of monolayer crystals (Chapter 20) such as graphene, where the p_z orbital leads cleanly to the antibonding π band. σ also should not be confused with conductivity, or $\boldsymbol{\sigma}$ for spin matrix.

Figure 4.4(b) and (c) also point to what would happen if an atom was missing or replaced by another atom with different valence orbitals of importance, for example *Ti* with *d* orbitals, or of occupation, for example *S* with two more electrons in the *p* orbitals, compared to group *IV* atoms. There is now a local perturbation. In these situations, this local molecule-like representation and usage gives much more insight, since it emphasizes localization.

Chapter 7 will deploy this bonding-antibonding molecular representation to understanding localized states.

In cubic semiconductors, the *s* bonding state of the valence band is much lower in energy, since it is much more localized and homopolar. So, the edge states of the valence band arise in the p_x-, p_y- and p_z-based bonding states. These are Cartesian oriented. Figure 4.3(b) shows what a hole state polarization would be like along any $\langle 001\rangle$ direction. The valence band maximum by symmetry should be at the zone center. In any such orientation, since the composing *p* states are also orthogonal, two in-the-plane orbitals will contribute to a degenerate state, and one out-of plane orbital will form another state. So, there will be two bands for which the spin complexity is not incorporated. The two perpendicular orbitals from in-plane *p* orbitals are heavy-hole bands—heavy since the overlap is small due their original orthogonality. The third band will be a light-hole band with large curvature and increased overlap. When moving away from the Γ

point, the heavy-hole Bloch function is still very much p-like. But the light-hole Bloch function has considerable mixing with the s. This keeps the curvature high. Based on this argument, one will also see that as one deviates from the axial directions in the plane, the heavy-hole bands should do so with considerable anisotropy, but the light-hole bands should remain more isotropic. These differences are shown in Figure 4.5, which also shows a split-off band to be discussed shortly and in more detail later in the chapter after developing the mathematical formalism to look at spin-orbit interactions. The hole bands bring out the angular-momentum-dictated features because these states and the electrons in them are quite localized and under the strong influence of the nucleus and atom-driven quantum characteristics.

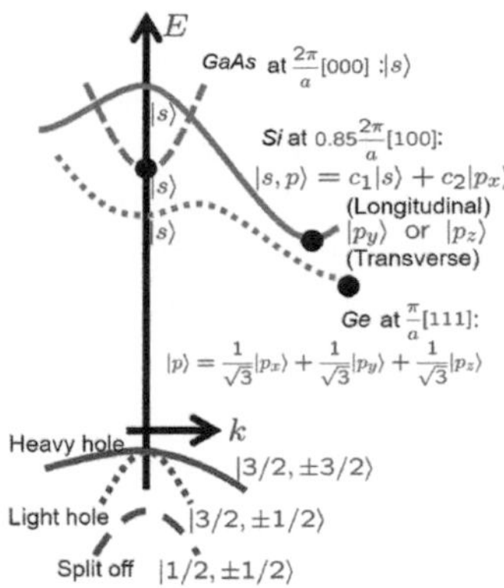

Figure 4.5: Tight binding and its bandstructure consequences. *Si*'s conduction band minimum is a mix of $|s\rangle$ and the different $|p\rangle$s *Ge*'s is a mix of the different $|p\rangle$s, and *GaAs*'s is made up of quite symmetric $|s\rangle$s. Valence band states—more strongly localized to the nucleus—have different angular momentum polarization and have the highest energy at the zone center. A split-off band also arises due to spin-orbit coupling energy. Conduction band states, since they are very delocalized, have vanishing spin-dependent energy. Most semiconductors, unless there are intentionally magnetic, will only show spin's effect in the conduction band through the state degeneracy.

The conduction band states are highly delocalized even if one has built them here through the atomic orbitals. Now the ionicity or covalency and the size of the atoms starts to matter substantially, even if angular momentum does not. Take *Si*, as sketched in Figure 4.5; it has a conduction band minimum along the Δ direction close to the X point. *Si* has $3s$ and $3p$ partially filled outer orbitals. It is also covalent. On the other hand, *Ge* has a conduction band minimum along the Λ direction at the L point. *Ge* has $4s$ and $4p$ orbitals partially filled, so farther out. Both are covalent, so charge transfer has little perturbational effect. When an atom is small, both the s and the p states are mixing and contributing, and are important, but it is p_x, p_y or p_z that is dominant, depending on the orientation. This is reflected in the minimum appearing along the Δ direction and near the X point, showing the Cartesian orientations of the minimum. The importance of the s is reflected in the plane perpendicular to this direction at this point. This orthogonal direction is affected strongly by the homopolar nature of the contributing orbital function. The transverse mass is smaller while the longitudinal mass is larger. *Ge* being a larger atom, the $4s$ orbital's influence is weaker, p_x, p_y and p_z contribute equally and the minimum is now at $\langle 111\rangle$, that is, the L point. The equienergy surface is still an ellipsoid arising in the s contribution transverse to this direction. *GaAs* has the conduction band minimum at the zone center. Although *GaAs* has a *Ge*-like atomic number average, it is ionic with a significant charge redistribution. And the consequences are both an increase in the bandgap, because of the larger Coulomb perturbation, and a shift in the locale of the conduction minimum.

So, for *Si*, along the direction $\langle 100\rangle$, the states are now a mix of $|s\rangle$ and $|p\rangle$. And, at the minimum, in the longitudinal direction, they are quite $|p\rangle$-like—or, more precisely, $|p_x\rangle$−, $|p_y\rangle$− and $|p_z\rangle$-like—along the six minima of the $\langle 100\rangle$ direction. Orthogonally, they are more $|s\rangle$-like. So, there appears a longitudinal mass and a transverse mass. For *Ge*, because the Hamiltonian energetics is now different, the atom is larger, the orbitals are farther away from the core, and the conduction minimum now arises in the $|p_x\rangle$, $|p_y\rangle$ and $|p_z\rangle$ mixing in equal strength. The minimum is at the L point, at the zone edge, so there are eight half ellipsoids or an equivalent of four full ellipsoids.

The p orbitals have orbital angular momentum quantization. Electrons also have spin quantization. And we incorporate the total angular momentum quantization through $\mathbf{J} = \mathbf{L} + \mathbf{S}$, where the total angular momentum $\mathbf{J}^2$ and its axial component $\mathbf{J}_z$ must be quantized, as is true for the orbital and the spin angular momentum. This conservation means that we now end up with states—valence states that are closer to the core—that have $j = 3/2$ and $j = 1/2$, with m_js of $3/2, 1/2, -1/2$ and $-3/2$. The orbital and the spin angular momenta also lead to an additional energy. The result is that there are a set of states of $|j, m_j\rangle = |3/2, \pm 3/2\rangle$ that form one band, a set of states $|3/2, \pm 1/2\rangle$ that form another band and, finally, a set of states $|1/2, \pm 1/2\rangle$ that form a third band. The first, with its specific polarization $m_j = \pm 3/2$ arising from the angular momentum quantization, is a heavy-hole band; the second, with $m_j = \pm 1/2$, is a light-hole band and the third, where spin-orbit interaction displaces the energy further, is the split-off band, with $j = 1/2$ and, therefore, $m_j = \pm 1/2$.

This valence picture holds true for the most common diamond (inversion symmetric) and zinc blende (inversion asymmetric) semiconductors, as well as for the inversion asymmetric wurtzite semiconductors. The conduction band picture, however, with the energy of the states higher and the electrons further away from the nucleus, is more strongly beholden to the stronger interaction between the outer states of the different atoms of the assembly. Conduction bands can differ substantially. The increased ionicity of *GaAs* makes it a direct bandgap semiconductor, while *AlAs*—*Al* being smaller than *Ga*—is an indirect bandgap semiconductor. We will return to the details of this discussion after having explored these various methods for bandstructure calculations.

4.3 *Orthogonalized plane waves method*

The problem with plane wave solution's numerical inappropriateness comes about from its strong inadequacy for short waves; that is, large reciprocal wavevectors. An electron strongly localized to lattice sites would then be quite inadequately described, since a large number of plane wave states are needed to describe this localization of the wavepacket. Herring proposed a clever way to circumvent this difficulty, which is a difficulty of electron oscillations—localization— in the core region. The electron wavefunction, with the orthogonalization, now corresponds more closely to that of the valence electrons, and it works well in most situations where the core valence states are not very extended out. d and f orbitals do stretch out, so the approach is more limited in applicability for transition metals. The basis set integrates the rapid local variations with slow, nonlocal variations by introducing plane waves that are orthogonalized to the core states. The method, therefore, is still a plane wave method, but it reproduces the oscillatory nature well.

Let t represent a core state, and j a specific lattice site, so that

$$|\sigma\rangle = |tj\rangle = |\psi_t(\mathbf{r} - \mathbf{r}_j)\rangle \tag{4.10}$$

represents a core state at locale j with energy $E_{tj} \equiv E_t$. Different eigenfunctions on the same atom, that is, at specific j but differing t, are orthogonal. The case with the same t but differing j, too, has a small overlap, and has vanishingly small overlap for core states, so it may be treated as orthogonal, that is,

$$\langle t'j'|tj\rangle = \delta_{t't}\delta_{j'j} \;\; \therefore \;\; \langle\sigma'|\sigma\rangle = \delta_{\sigma'\sigma}. \tag{4.11}$$

We now introduce the modified wavefunction

$$|\phi_{\mathbf{k}}\rangle \equiv |\mathbf{k}\rangle - \sum_{\sigma} |\sigma\rangle\langle\sigma|\mathbf{k}\rangle, \tag{4.12}$$

which has the property

$$\begin{aligned}\langle\sigma'|\phi_{\mathbf{k}}\rangle &= \langle\sigma'|\mathbf{k}\rangle - \sum_{\sigma}\langle\sigma'|\sigma\rangle\langle\sigma||\mathbf{k}\rangle \\ &= \langle\sigma'|\mathbf{k}\rangle - \sum_{\sigma}\delta_{\sigma'\sigma}\langle\sigma|\mathbf{k}\rangle \\ &= 0,\end{aligned} \tag{4.13}$$

that is, it is orthogonal. Using a tool of convenience, the projection operator

$$\hat{\mathscr{P}} = \sum_{\sigma} |\sigma\rangle\langle\sigma|, \tag{4.14}$$

we may write

$$|\phi_{\mathbf{k}}\rangle = |\mathbf{k}\rangle - \sum_{\sigma} |\sigma\rangle\langle\sigma|\mathbf{k}\rangle = (1 - \hat{\mathscr{P}})|\mathbf{k}\rangle, \tag{4.15}$$

an artful form, where the meaning of the projection in view of Equations 4.12 and 4.14 is clear. A generalized Fourier series is an expansion over any set of orthogonal functions—as it is in Bloch functions—and we use this projection to create a Fourier expansion of $|\mathbf{k}\rangle$ over orthogonal core states. Figure 4.6 shows a pictorial description of the procedure, showing the plane-wave-like nonlocal behavior and the core-like oscillatory local behavior.

We can now expand the crystal Bloch wavefunction in the plane wave basis, together with this orthogonalization, as

$$|\psi_{\mathbf{k}}\rangle = \sum_{\mathbf{K}_i} c_{\mathbf{K}_i} |\phi_{\mathbf{k}+\mathbf{K}_i}\rangle = \sum_{\mathbf{K}_i} c_{\mathbf{K}_i} (1 - \hat{\mathscr{P}})|\mathbf{k} + \mathbf{K}_i\rangle. \tag{4.16}$$

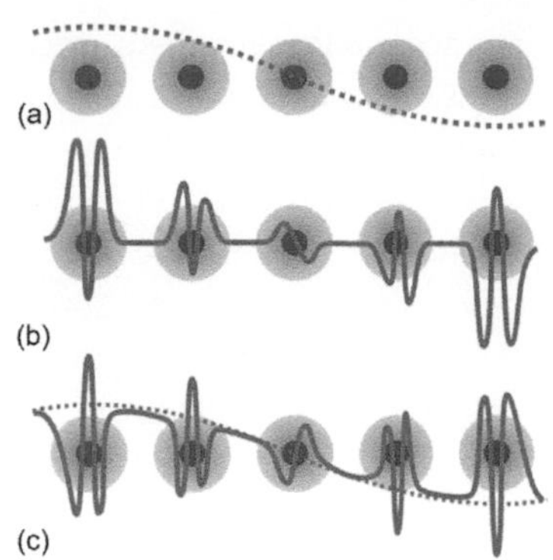

Figure 4.6: A pictorial depiction of the orthogonalized plane wave procedure. Part (a) shows a plane wave, (b) shows wavefunctions of the core, and (c) shows the orthogonalized plane wavefunction. The ion core is in the background to show the locale.

Since the projection and the orthogonalization removed the core, now only a few terms of $\mathbf{K}_i$ are needed for an effective calculation.

4.4 *Pseudopotential method*

THE ORTHOGONALIZED PLANE WAVE METHOD IS A STEPPING STONE to the pseudopotential approach, a more useful incarnation. The pseudopotential method attempts to make a judicious separation of the localized part of the wavefunction description from the nearly or completely delocalized valence and conducting parts. The Hamiltonian equation then can be reconstructed in a form that is easier to tackle, employing a plane wave basis for the bandstructure. This is to say that one wishes to replace the crystal wavefunction $|\psi_{\mathbf{k}}\rangle$ of the crystal Hamiltonian equation $\hat{\mathscr{H}}|\psi_{\mathbf{k}}\rangle = E_{\mathbf{k}}|\psi_{\mathbf{k}}\rangle$ with a form that is simpler than that of Equation 4.16. The wavefunction and the Hamiltonian will be "pseudo", and we will remove $1-\hat{\mathscr{P}}$ in the process. The energies will still be the solution to the problem of interest, even if the wavefunction is of the pseudo ilk.

Let $|\varphi_{\mathbf{k}}\rangle$ be this pseudo wavefunction, defined as

$$|\varphi_{\mathbf{k}}\rangle \equiv \sum_{\mathbf{K}_i} c_{\mathbf{K}_i} |\mathbf{k} + \mathbf{K}_i\rangle, \tag{4.17}$$

which is formulated to reproduce the crystal wavefunction as

$$\begin{aligned} |\psi_{\mathbf{k}}\rangle &= \sum_{\mathbf{K}_i} c_{\mathbf{K}_i} (1 - \hat{\mathscr{P}})|\mathbf{k} + \mathbf{K}_i\rangle \\ &= (1 - \hat{\mathscr{P}}) \sum_{\mathbf{K}_i} c_{\mathbf{K}_i} |\mathbf{k} + \mathbf{K}_i\rangle = (1 - \hat{\mathscr{P}})|\varphi_{\mathbf{k}}\rangle. \end{aligned} \tag{4.18}$$

So,

$$\begin{aligned} \hat{\mathscr{H}}|\psi_{\mathbf{k}}\rangle = \hat{\mathscr{H}}(1 - \hat{\mathscr{P}})|\varphi_{\mathbf{k}}\rangle &= E_{\mathbf{k}}(1 - \hat{\mathscr{P}})|\varphi_{\mathbf{k}}\rangle \\ \therefore\ (\hat{\mathscr{H}} - \hat{\mathscr{H}}\hat{\mathscr{P}} + E_{\mathbf{k}}\hat{\mathscr{P}})|\varphi_{\mathbf{k}}\rangle &= E_{\mathbf{k}}|\varphi_{\mathbf{k}}\rangle, \text{ or} \\ \hat{\mathfrak{H}}|\varphi_{\mathbf{k}}\rangle &= E_{\mathbf{k}}|\varphi_{\mathbf{k}}\rangle. \end{aligned} \tag{4.19}$$

The left-hand side has a pseudo-Hamiltonian $\hat{\mathfrak{H}}$ that works with a pseudopotential and results in the eigenenergy of the crystal Hamiltonian $E_{\mathbf{k}}$:

$$\begin{aligned}\hat{\mathfrak{H}} = \hat{\mathscr{H}} - \hat{\mathscr{H}}\hat{\mathscr{P}} + E_{\mathbf{k}}\hat{\mathscr{P}} &= \hat{\mathscr{H}} + (E_{\mathbf{k}} - \hat{\mathscr{H}})\hat{\mathscr{P}} \\ &= \hat{T} + \hat{V} + (E_{\mathbf{k}} - \hat{\mathscr{H}})\hat{\mathscr{P}} \\ &\equiv \hat{T} + \hat{W},\end{aligned} \tag{4.20}$$

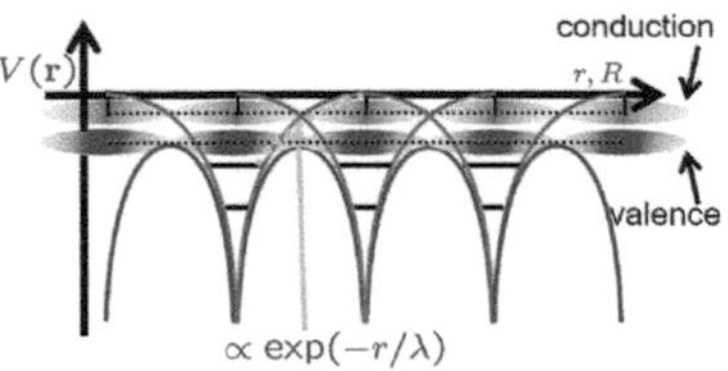

Figure 4.7: A schematic drawing of potentials and energies in the crystal, together with probability densities of states.

where $\hat{T}$ is the kinetic energy operator, $\hat{V}$ is the potential energy operator and

$$\hat{W} = \hat{V} + (E_{\mathbf{k}} - \hat{\mathscr{H}})\hat{\mathscr{P}} \tag{4.21}$$

is now a pseudopotential that goes together with the kinetic energy.

The reasoning of why this is significant is the following. V, the potential, is negative, attractive, spatially large and changing, as schematically drawn in Figure 4.7. An equation written in a form where the perturbation is small is quantitatively more tractable. The considerations applying to the modification arising from putting these all together in a crystal is captured in

$$(E_{\mathbf{k}} - \hat{\mathscr{H}})\hat{\mathscr{P}} = \sum_{\sigma}(E_{\mathbf{k}} - \hat{\mathscr{H}})|\sigma\rangle\langle\sigma|. \tag{4.22}$$

The magnitude of $E_{\mathbf{k}}$ is smaller than that of the core eignenergy E_{σ}; that is, the electron has a larger negative energy when bound to the atom, and it is freer in the crystal.

So, the term in Equation 4.22 is positive and it is large. The consequence is that the pseudopotential $\hat{W}$ is made small by largely canceling out $\hat{V}$. By reforming the problem to a smaller perturbation problem through the introduction of the difference, we make it easier to tackle. Figure 4.8 shows a schematic view of what this pseudopotential method achieves by modification of the Hamiltonian equation to the pseudo form. The equation

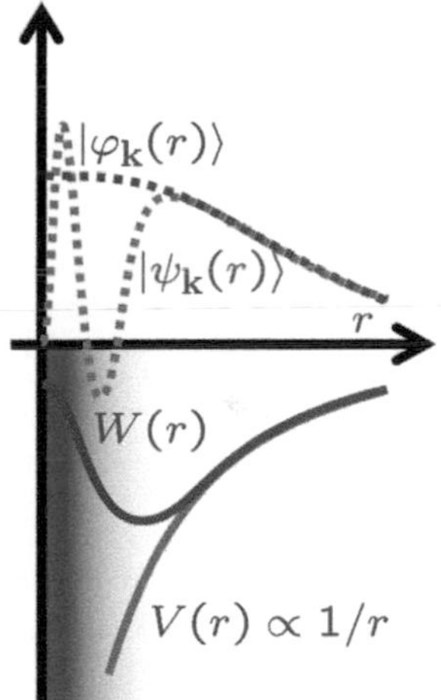

Figure 4.8: The reformulation of the crystal potential problem to that of pseudopotential. The original problem had large changes in the wavefunction $|\psi_{\mathbf{k}}(r)\rangle$ near the core under the strong $\propto 1/r$ Coulomb potential. With a soft core, that is, with a graded pseudopotential W, the reformulation of the problem happens with the wavefunction $|\varphi_{\mathbf{k}}(r)\rangle$ as its solution. If the wavefunction solution for the pseudopotential matches in magnitude and radial derivative outside the core, the derived results for valence electrons, bonds and crystal structure should be accurate.

$$\hat{\mathfrak{H}}|\varphi_{\mathbf{k}}\rangle = (\hat{T} + \hat{W})|\varphi_{\mathbf{k}}\rangle = E_{\mathbf{k}}|\varphi_{\mathbf{k}}\rangle \tag{4.23}$$

is a nearly free electron equation. $\hat{V}(\mathbf{r})$ is a local potential—electron exchange interactions are explicitly included in it. $\hat{W}$, however, is nonlocal and contains the operator $\mathscr{P}$. This is a complication in the pseudopotential. Another is that, of course, $E_{\mathbf{k}}$, the quantity desired through calculation, is a component of the pseudopotential. Because $|E_{\sigma}| \gg |E_{\mathbf{k}}|$, this is a not a serious limitation in the iterative process. An approximate guess of $E_{\mathbf{k}}$ suffices as a starting point. Indeed, taking it as the Fermi energy E_F gives an accurate estimation around it, and usually that is what we care about most.

4.5 *Density functional method*

DENSITY FUNCTIONAL THEORY is the most recent powerful—and pervasive—tool for determining many of the condensed matter characteristics—atomic and electronic structure, phonon spectra, et cetera—involving the trick of using a functional that simplifies the many-body calculation substantially. We will only summarize here the enormous simplification that density functional theory provides.

We are usually, at least as a starting point, interested in the ground state of a system for which we can write the Hamiltonian. Equation 1.10 was our full Hamiltonian of the crystal, and we have now spent a hundred-plus pages working out some of the enormous store of consequences from it, all still in a reasonable but still approximate way. Density functional theory leads to a good approximation that is efficient computationally because of how it can tackle Equation 1.10. In the Born-Oppenheimer approximation, we simplified the equation to Equation 1.11, which consists of terms representing electrons' kinetic energy, electrons' potential energy arising through the Coulomb interaction with screened ions, and the interelectron Coulomb repulsive potential energy. Chapter 1 expended considerable energy evaluating this last term by approximating it as a Hartree term, then added exchange to it in the Hartree-Fock approach, and finally brought in the correlation correction. We could tackle some of these for the atomic and molecular systems, but an inhomogeneous electron gas makes the problem a very difficult one. One of the Hohenberg-Kohn theorems' consequences is that the exchange and correlation effects depend only on the electron charge density of a system. This is an enormous simplification—yet rigorous—since the sum of the two potentials—interelectron and electron-screened ions—can now be expressed as a Hartree term, which is a Poisson equation problem, together with an additional term for exchange correlation that is a functional of electron density.

Hohenberg-Kohn theorems state that, for any system consisting of electrons moving under the influence of an external potential, the potential and the total energy can be expressed via a unique functional of the electron density. The ground state energy can then be obtained variationally, and the density that minimizes the total energy is the exact ground state density.

This functional, and the use of the electron-density-based functional, now reduce the problem to a single particle Schrödinger equation, also known as the Kohn-Sham equation. A number of excellent texts discuss the rigors of the density functional method, and we leave the reader interested in pursuing them to follow these sources mentioned at the end of the chapter.

There are many additional methods in use for calculation of bandstructure. These include the cellular, the augmented plane wave and the $\mathbf{k} \cdot \mathbf{p}$ method. The augmented plane wave approach uses plane waves outside core regions, and a superposition of atomic orbitals close to $\mathbf{R}_m$s, so near the core region. The $\mathbf{k}\cdot\mathbf{p}$ method uses superposition of bandedge wavefunctions. This is particularly useful for valence band states, so for light-hole and heavy-hole bands, and this is the one we stress the most, since it is very instructive for understanding the nature of states in different bands and their interactions.

4.6 k·p *method*

LIKE THE PSEUDOPOTENTIAL APPROACH, ANOTHER NUMERICALLY EFFICIENT and particularly useful method for understanding bandstructure-based parameters and what happens in the valence band, especially through its perturbation approach, is the $\mathbf{k}\cdot\mathbf{p}$ method. For bandstructure, its utility is in using the zone center energy gap and optical matrix elements, or, equivalently, oscillator strength, toward a good extrapolation over the Brillouin zone. And this is particularly powerful for the valence band.

We start by showing its usefulness in understanding parameters, group velocity in our example, before turning to bandstructure implications such as state interactions represented in effective mass or changes in energy of states.

The Bloch function is the solution to the crystal Hamiltonian. The $\mathbf{k}\cdot\mathbf{p}$ method focuses on $u_{n\mathbf{k}}(\mathbf{r})$, which is the periodic modulation function of the plane wave that, together with it, constitute the Bloch function. Let $\mathbf{k}'=\mathbf{k}+\Delta\mathbf{k}$ be the wavevector of a state in band n, separated from a neighboring state at $\mathbf{k}$. The Bloch function is a solution to the equation $\hat{H}\psi_{n\mathbf{k}'}=E_{n\mathbf{k}'}\psi_{n\mathbf{k}'}$. The $\mathbf{k}\cdot\mathbf{p}$ method views and employs the utility of looking at this same problem through the implications for the periodic modulation part of the Bloch function; that is, $u_{n\mathbf{k}'}$. The crystal Hamiltonian can be deconstructed as follows:

$$\hat{\mathscr{H}}\psi_{n\mathbf{k}'}(\mathbf{r})=E_n\psi_{n\mathbf{k}}(\mathbf{r}'),$$

$$\left[\frac{\hat{p}^2}{2m_0}+\hat{V}(\mathbf{r})\right]\exp(i\mathbf{k}'\cdot\mathbf{r})u_{n\mathbf{k}'}(\mathbf{r})=E_{n\mathbf{k}'}\psi_{n\mathbf{k}'}(\mathbf{r}),$$

$$\left[-\frac{\hbar^2}{2m_0}\nabla_{\mathbf{r}}^2+\hat{V}(\mathbf{r})\right]\exp(i\mathbf{k}'\cdot\mathbf{r})u_{n\mathbf{k}'}(\mathbf{r})=E_{n\mathbf{k}'}\exp(i\mathbf{k}'\cdot\mathbf{r})u_{n\mathbf{k}'}(\mathbf{r}),$$

$$\begin{aligned}&-\frac{\hbar^2}{2m_0}\nabla_{\mathbf{r}}\left[i\mathbf{k}'\exp(i\mathbf{k}'\cdot\mathbf{r})u_{n\mathbf{k}'}(\mathbf{r})+\exp(i\mathbf{k}'\cdot\mathbf{r})\nabla_{\mathbf{r}}u_{n\mathbf{k}'}(\mathbf{r})\right]\\&+\hat{V}(\mathbf{r})\exp(i\mathbf{k}'\cdot\mathbf{r})u_{n\mathbf{k}'}(\mathbf{r})\\&=E_{n\mathbf{k}'}\exp(i\mathbf{k}'\cdot\mathbf{r})u_{n\mathbf{k}'}(\mathbf{r}),\end{aligned}$$

$$\therefore\quad -\frac{\hbar^2}{2m_0}\left[(i\mathbf{k}')^2+2i\mathbf{k}'\cdot\nabla_{\mathbf{r}}+\nabla_{\mathbf{r}}^2+\hat{V}(\mathbf{r})\right]u_{n\mathbf{k}'}(\mathbf{r})=E_{n\mathbf{k}'}u_{n\mathbf{k}'}(\mathbf{r}),$$

$$\left[-\frac{\hbar^2}{2m_0}\nabla_{\mathbf{r}}^2 + \frac{\hbar^2\mathbf{k}'^2}{2m_0} + \frac{\hbar\mathbf{k}'}{m_0}\cdot\frac{\hbar}{i}\nabla_{\mathbf{r}} + \hat{V}(\mathbf{r})\right]u_{n\mathbf{k}'}(\mathbf{r}) = E_{n\mathbf{k}'}u_{n\mathbf{k}'}(\mathbf{r}),$$

$$\text{that is,}\quad \left[\frac{(\hat{p}+\hbar\mathbf{k}')^2}{2m_0} + \hat{V}(\mathbf{r})\right]u_{n\mathbf{k}'}(\mathbf{r}) = E_{n\mathbf{k}'}u_{n\mathbf{k}'}(\mathbf{r}),$$

$$\text{or}\quad \left[\frac{(\hat{p}+\hbar\mathbf{k})^2}{2m_0} + \frac{(\hat{p}+\hbar\mathbf{k})\cdot\hbar\Delta\mathbf{k}}{m_0} + \frac{\hbar^2\Delta\mathbf{k}^2}{2m_0} + \hat{V}(\mathbf{r})\right]u_{n\mathbf{k}'}(\mathbf{r}) = E_{n\mathbf{k}'}u_{n\mathbf{k}'}(\mathbf{r}). \tag{4.24}$$

We now have perturbation terms arising in $\Delta\mathbf{k}$. The second order term in $\Delta\mathbf{k}^2$ can be ignored in the limit of small deviations. All the eigenfunctions are orthonormal, since they are solutions to the unperturbed Hamiltonian $\hat{\mathscr{H}}$.

The energy of the state can also be written through Taylor series expansion as

$$E_{n\mathbf{k}'} = E_{n\mathbf{k}} + \sum_{i=1}^{3}\frac{\partial E_{n\mathbf{k}}}{\partial k_i}\Delta k_i + \frac{1}{2}\sum_{i,j=1}^{3}\frac{\partial^2 E_{n\mathbf{k}}}{\partial k_i \partial k_j}\Delta k_i \Delta k_j + \cdots, \tag{4.25}$$

where i, j are the chosen spatial coordinates. The first order term implies (we will see the implication of the second term in what follows)

$$\frac{\partial E_{n\mathbf{k}}}{\partial k_i} = \frac{\hbar}{m_0}\langle u_{n\mathbf{k}}|\frac{\hbar\Delta\mathbf{k}\cdot(\mathbf{p}_i+\hbar\mathbf{k})}{2m_0}|u_{n\mathbf{k}}\rangle = \frac{\hbar}{m_0}\langle\psi_{n\mathbf{k}}|\hat{p}_i|\psi_{n\mathbf{k}}\rangle. \tag{4.26}$$

Therefore, the group velocity of the electron in the $(n,\mathbf{k})$ state is

$$\begin{aligned} v_{gn\mathbf{k}} &= \frac{1}{\hbar}\frac{\partial E_{n\mathbf{k}}}{\partial k_i} \text{ in the } i\text{th direction, and} \\ &= \frac{1}{\hbar}\nabla_{\mathbf{k}}E_{n\mathbf{k}} \text{ in general.} \end{aligned} \tag{4.27}$$

As a state changes in wavevector or crystal momentum, it changes its energy, and the change in energy as the electron propagates is given by this group velocity.

Our argument here is employing non-degenerate energy levels. It also holds for degenerate energy levels but is mathematically unwieldy in that case. We are skipping it as an unnecessary distraction. We have now seen this same result of group velocity through multiple approaches. And recall also its connections from the Fisher information discussion.

The consequences of interaction between states can also be seen through another reformulation of the $\mathbf{k}\cdot\mathbf{p}$ formalism, and here we will see the significance of the second term. Using the Bloch function $\psi_{n\mathbf{k}}(\mathbf{r}) = \exp(i\mathbf{k}\cdot\mathbf{r})u_{n\mathbf{k}}(\mathbf{r})$, the Schrödinger equation may be reduced to another $\mathbf{k}\cdot\mathbf{p}$ form:

$$\begin{aligned} &\hat{\mathscr{H}}\psi_{n\mathbf{k}}(\mathbf{r}) = E_n\psi_{n\mathbf{k}}(\mathbf{r}) \\ \therefore\ &\left(-\frac{\hbar^2}{2m_0}\nabla^2 + \hat{V}\right)\exp(i\mathbf{k}\cdot\mathbf{r})u_{n\mathbf{k}}(\mathbf{r}) = E_{n\mathbf{k}}\exp(i\mathbf{k}\cdot\mathbf{r})u_{n\mathbf{k}}(\mathbf{r}) \\ \therefore\ &-\frac{\hbar^2}{2m_0}\nabla\cdot\left[\nabla\exp(i\mathbf{k}\cdot\mathbf{r})u_{n\mathbf{k}}(\mathbf{r})\right] \\ &+\hat{V}\exp(i\mathbf{k}\cdot\mathbf{r})u_{n\mathbf{k}}(\mathbf{r}) = E_{n\mathbf{k}}\exp(i\mathbf{k}\cdot\mathbf{r})u_{n\mathbf{k}}(\mathbf{r}) \end{aligned}$$

$$
\begin{aligned}
\therefore\ & -\frac{\hbar^2}{2m_0}\nabla\cdot\left[i\mathbf{k}\exp(i\mathbf{k}\cdot\mathbf{r})u_{n\mathbf{k}}(\mathbf{r}) + \exp(i\mathbf{k}\cdot\mathbf{r})\nabla u_{n\mathbf{k}}(\mathbf{r})\right] \\
& + \hat{V}\exp(i\mathbf{k}\cdot\mathbf{r})u_{n\mathbf{k}}(\mathbf{r}) = E_{n\mathbf{k}}\exp(i\mathbf{k}\cdot\mathbf{r})u_{n\mathbf{k}}(\mathbf{r}) \\
\therefore\ & -\frac{\hbar^2}{2m_0}\left[-\mathbf{k}^2u_{n\mathbf{k}}(\mathbf{r}) + 2i\mathbf{k}\cdot\nabla u_{n\mathbf{k}}(\mathbf{r}) + \nabla^2u_{n\mathbf{k}}(\mathbf{r})\right] \\
& + \hat{V}u_{n\mathbf{k}}(\mathbf{r}) = E_{n\mathbf{k}}u_{n\mathbf{k}}(\mathbf{r}) \\
\therefore\ & \left(-\frac{\hbar^2\nabla^2}{2m_0} + \frac{\hbar}{m_0}\mathbf{k}\cdot\frac{\hbar}{i}\nabla + \frac{\hbar^2\mathbf{k}^2}{2m_0} + V\right)u_{n\mathbf{k}}(\mathbf{r}) = E_{n\mathbf{k}}u_{n\mathbf{k}}(\mathbf{r}) \\
\therefore\ & \left(\frac{\hat{\mathbf{p}}^2}{2m_0} + \frac{\hbar}{m_0}\mathbf{k}\cdot\mathbf{p} + \frac{\hbar^2\mathbf{k}^2}{2m_0} + \hat{V}\right)u_{n\mathbf{k}}(\mathbf{r}) = E_{n\mathbf{k}}u_{n\mathbf{k}}(\mathbf{r}).
\end{aligned}
\tag{4.28}
$$

At the zone center $\mathbf{k} = 0$, Equation 4.28 is the simple form

$$
\left(\frac{\hat{\mathbf{p}}^2}{2m_0} + \hat{V}\right)u_{n0}(\mathbf{r}) = E_{n\mathbf{k}}u_{n0}(\mathbf{r}), \tag{4.29}
$$

which, for the Bloch solution, is a wave of zero wavevector. This equation, with $u_{n0}(\mathbf{r})$ periodic under the translational operator $\hat{T}_{\mathbf{R}}$, and its similar form for any other specific point $\mathbf{k}$, are easier to solve than the original form that we started with. E_{n0} and u_{n0} are now known, so the two middle terms of Equation 4.28 are perturbation terms. For small changes in $\mathbf{k}$ from the known symmetric solution points, band dispersion can be accurately determined. Since it works well at $\mathbf{k}=0$, it is particularly useful for hole bands.

The $\mathbf{k}\cdot\mathbf{p}$ method is also a convenient tool for showing how effective mass is an important parameter that allows us to exclude the crystal potential consequences in determining the response to external electrochemical forces on a system and for showing its underpinnings in the inter-state interactions through the $\mathbf{k}\cdot\mathbf{p}$ perturbation. Know that $\mathbf{k}$ here represents the crystal wavevector (a proxy for the crystal momentum operator) and the consequences of periodic potential, while $\mathbf{p}$ is the momentum operator. We expand the eigenfunction and the eigenenergy in the vicinity of the known locale. Equation 4.29 is of the form

$$
(\hat{\mathscr{H}}_0 + \hat{\mathscr{H}}_1 + \hat{\mathscr{H}}_2)u_{n\mathbf{k}}(\mathbf{r}) = E_{n\mathbf{k}}u_{n\mathbf{k}}(\mathbf{r}), \tag{4.30}
$$

where $\hat{\mathscr{H}}_1 = (\hbar/m_0)\mathbf{k}\cdot\mathbf{p}$, and $\hat{\mathscr{H}}_2 = \hbar^2\mathbf{k}^2/2m_0$. In the lowest order,

$$
u_{n\mathbf{k}}(\mathbf{r}) = u_{n0}(\mathbf{r}), \text{ and } E_{n\mathbf{k}} = E_{n0}. \tag{4.31}
$$

In the first order,

$$
u_{n\mathbf{k}} = u_{n0} + \frac{\hbar}{m_0}\sum_{l\neq n}\frac{\langle u_{l0}(\mathbf{r})|\mathbf{k}\cdot\mathbf{p}|u_{n0}(\mathbf{r})\rangle}{E_{n0}-E_{l0}}u_{l0}(\mathbf{r}), \text{ and}
$$

$$E_{n\mathbf{k}} = E_{n0} + \frac{\hbar}{m_0}\langle u_{n0}|\mathbf{k}\cdot\mathbf{p}|u_{n0}\rangle$$
$$= E_{n0} + \frac{\hbar}{m_0}\mathbf{k}\cdot\langle u_{n0}|\mathbf{p}|u_{n0}\rangle. \quad (4.32)$$

Inversion symmetry implies symmetrical Bloch functions. **p** is antisymmetric. So, this first order energy correction must be absent in materials such as *Ge* and *Si*. The energy correction will be present in compound semiconductors such as *GaAs* as well as in wurtzite structures such as *GaN*. Here, a term proportional to the **k** will appear. This has consequences for the valence band maximum and at the *X* point in the conduction band. The Bloch function correction will then be present irrespective of the presence or absence of this inversion symmetry. The momentum matrix element will connect $|s\rangle$- and $|p\rangle$-like states.

The second order correction is more important for energy than for the Bloch function and is

$$E_{n\mathbf{k}} = E_{n0} + \frac{\hbar^2\mathbf{k}^2}{2m_0} + \frac{\hbar^2}{m_0^2}\sum_{l\neq n}\frac{\langle u_{l0}|\mathbf{k}\cdot\mathbf{p}|u_{n0}\rangle^2}{E_{n0}-E_{l0}}$$
$$= E_{n0} + \frac{\hbar^2\mathbf{k}^2}{2m_0} + \frac{\hbar^2}{m_0^2}\sum_{l\neq n}\frac{|\mathbf{k}\cdot\langle u_{l0}|\mathbf{p}|u_{n0}\rangle|^2}{E_{n0}-E_{l0}}. \quad (4.33)$$

In writing an effective mass as

$$E_{n\mathbf{k}} = E_{n0} + \sum_{i,j}\frac{\hbar^2}{2m_{ij}^*}k_ik_j, \quad (4.34)$$

the effective mass m_{ij}^* is

$$\frac{1}{m_{ij}^*} = \frac{1}{m_0}\delta_{ij} + \frac{2}{m_0}\sum_{l\neq n}\frac{\langle u_{n0}(\mathbf{r})|p_i|u_{l0}(\mathbf{r})\rangle\langle u_{l0}(\mathbf{r})|p_j|u_{n0}(\mathbf{r})\rangle}{E_{n0}-E_{l0}}. \quad (4.35)$$

For energy, this states

$$E_{n\mathbf{k}} = E_{n0} + \frac{\hbar^2}{2}\sum_{i,j}k_im_{ij}^*k_j = E_{n0} + \frac{\hbar^2}{2}\mathbf{k}\cdot\frac{1}{\mathbb{M}^*}\cdot\mathbf{k}, \quad (4.36)$$

where $1/\mathbb{M}^*$ is a second order effective mass tensor of the form

$$\frac{1}{\mathbb{M}^*} = \begin{bmatrix} 1/m_1^* & 0 & 0 \\ 0 & 1/m_2^* & 0 \\ 0 & 0 & 1/m_3^* \end{bmatrix}. \quad (4.37)$$

This equation describes to us, conditional on the symmetries, the general form that $\mathbf{k}\cdot\mathbf{p}$ tells us, for the band minimum or maximum, the energy-crystal momentum form. Interactions between states within the same band and states in different bands cause the mass to change. $\langle u_{n0}|\mathbf{k}\cdot\mathbf{p}|u_{l0}\rangle$ is significant only if **p** causes a significant coupling matrix element. Energy separation between states within the band and in the other bands affect it through the denominator in Equation 4.35. The higher the energy difference is, the smaller is the consequence for effective mass. Effective mass decreases if the lth band has energies lower than the nth, because a positive term results. Bands with energy higher than E_{n0} will increase, and this can be to an extent where the effective mass becomes negative. Other bands, where states are closer in energy, will have a strong effect. Hole bands illustrate a multitude of these consequences for magnitude and sign.

To see the implications of the couplings arising in $\mathbf{k} \cdot \mathbf{p}$, take Si's conduction band first. There are six minima along $\langle 100 \rangle$ axes (the Δ direction) about 15 % in from the X point. The lowest conduction minimum in the Δ_1 band is separated from the next by 0.53 eV. Non-degenerate perturbation therefore suffices. Tackling close to conduction band minimum with the i, j indices, with i as longitudinal and j as transverse, Equation 4.37 maps to

$$\begin{aligned} \frac{1}{m_l^*} &= \frac{1}{m_0} + \frac{2}{m_0^2} \sum_{l \neq \Delta_1} \frac{\langle u_{\Delta_1 \mathbf{k}_0} | p_z | u_{l\mathbf{k}_0} \rangle^2}{E_{\Delta_1 \mathbf{k}_0} - E_{l\mathbf{k}_0}}, \quad \text{and} \\ \frac{1}{m_t^*} &= \frac{1}{m_0} + \frac{2}{m_0^2} \sum_{l \neq \Delta_1} \frac{\langle u_{\Delta_1 \mathbf{k}_0} | p_x | u_{l\mathbf{k}_0} \rangle^2}{E_{\Delta_1 \mathbf{k}_0} - E_{l\mathbf{k}_0}}. \end{aligned} \tag{4.38}$$

This reflects energy dispersion as

$$E(\mathbf{k}) = \frac{\hbar^2 (k_z - k_{min})^2}{2m_l^*} + \frac{\hbar^2 (k_x^2 + k_y^2)}{2m_t^*}. \tag{4.39}$$

As one proceeds higher in energy, non-parabolicity appears through the $\mathbf{k} \cdot \mathbf{p}$ interaction. Bands also show band warping, which we will discuss for valence bands. These are distortions as one changes directions. To see the band warping, one must include the second conduction band interactions. In the valence bands, the consequence of secondary band interactions is very significant.

The valence band consequences are richer since there are three bands—light hole, heavy hole and split off—which we will index as $n = 1, 2, 3$ and which need to be accounted for. We discuss spin-orbit—a magnetic energetic interaction—consequence reflected particularly in the split-off band in Section 4.7. For now, consider the three bands as degenerate at the Γ point. Degenerate perturbation theory needs to be applied with Bloch functions at $\mathbf{k}$ built out of linear combinations from those at the $\mathbf{k} = 0$ point. So,

$$u_{n\mathbf{k}}(\mathbf{r}) = \exp(i\mathbf{k} \cdot \mathbf{r}) \sum_{n=1}^{3} c_n^0 u_{n0}(\mathbf{r}), \tag{4.40}$$

and the $\hat{\mathscr{H}}_1$ perturbation leads to the secular equations

$$\sum_{j=1}^{3}[D_{nj}^{\alpha\beta}k_\alpha k_\beta - E_{\mathbf{k}}^2\delta_{nj}]c_j^0 = 0 \text{ for } n = 1,2,3, \tag{4.41}$$

where α and β are the coordinate directions (x, y and z), and

$$D_{nj}^{\alpha\beta} = \frac{\hbar^2}{m_0^2}\sum_{l\neq n}\frac{\langle u_{n0}(\mathbf{r})|p_\alpha|u_{l0}(\mathbf{r})\rangle\langle u_{l0}(\mathbf{r})|p_\beta|u_{n0}(\mathbf{r})\rangle}{E_{n0}-E_{l0}}. \tag{4.42}$$

Summation now needs to be over all the remote bands $l \neq n$. The cubic system's symmetry results can be parameterized via

$$L = \frac{\hbar^2}{2m_0} + \frac{\hbar^2}{m_0^2}\sum_l \frac{\langle u_{10}(\mathbf{r})|p_x|u_{l0}(\mathbf{r})\rangle\langle u_{l0}(\mathbf{r})|p_x|u_{10}(\mathbf{r})\rangle}{E_{n0}-E_{l0}},$$

$$M = \frac{\hbar^2}{2m_0} + \frac{\hbar^2}{m_0^2}\sum_l \frac{\langle u_{10}(\mathbf{r})|p_y|u_{l0}(\mathbf{r})\rangle\langle u_{l0}(\mathbf{r})|p_y|u_{10}(\mathbf{r})\rangle}{E_{n0}-E_{l0}}, \text{ and}$$

$$N = \frac{\hbar^2}{2m_0} + \frac{\hbar^2}{m_0^2}\sum_l \left[\frac{\langle u_{10}(\mathbf{r})|p_x|u_{l0}(\mathbf{r})\rangle\langle u_{l0}(\mathbf{r})|p_x|u_{20}(\mathbf{r})\rangle}{E_{n0}-E_{l0}} + \frac{\langle u_{10}(\mathbf{r})|p_y|u_{l0}(\mathbf{r})\rangle\langle u_{l0}(\mathbf{r})|p_x|u_{20}(\mathbf{r})\rangle}{E_{n0}-E_{l0}}\right] \tag{4.43}$$

to the Hamiltonian matrix with the $\mathbf{k}\cdot\mathbf{p}$ perturbation as

$$\begin{bmatrix} Lk_x^2 + M(k_y^2+k_z^2) & Nk_xk_y & Nk_xk_z \\ Nk_xk_y & Lk_y^2 + M(k_x^2+k_z^2) & Nk_yk_z \\ Nk_xk_z & Nk_yk_z & Lk_z^2 + M(k_x^2+k_y^2) \end{bmatrix}, \tag{4.44}$$

where valence bandedge energy has been chosen to be zero. This Hamiltonian is the Luttinger Hamiltonian without spin-orbit coupling. For diamond and zinc blende structures, the Bloch amplitudes u_{n0} here are of the p orbital type. The valence band wavefunctions at the Γ symmetry point ($|e_1(\mathbf{r})\rangle$, $|e_2(\mathbf{r})\rangle$ and $|e_3(\mathbf{r})\rangle$) are composed of p atomic orbitals with eigenvalue 1 for the orbital angular momentum $\mathbf{L}$.

As was seen in Figure 4.3, take any direction of $\langle 100\rangle$, and one sees the heavy-hole band arising in the parameterized M, and the light-hole band arising in the parameterization L. If one looked along the $\langle 111\rangle$ direction, the heavy-hole band will have an energy varying as $(L+2M-N)k^2/3$, and the light-hole band as $(L+2M+2N)k^2/3$. Along $\langle 110\rangle$, these three valence bands are non-degenerate, and the energies are $(L+M\pm N)k^2/2$ and Mk^2. This shows anisotropy and band warping and is a behavior that we had expected from the tight binding discussion.

So far, we have stressed here the importance of viewing these valence band states through the tight binding or the p orbital

Joaquin Mazdkak Luttinger is from the postwar period, when science in the United States came into its own on a par with engineering. This period was a happy gathering of European expatriate discipline and the free spirit of the America born. Rabi, although born in Poland, grew up in the *USA*. Fermi, Bethe, Onsager, Franck, Debye, even Pauli for a while, and many others comprised the expatriates, while Schwinger, Feynman, Gell-Mann and many others comprised the natives. In condensed matter, with its many-body intricacies, there was a particular happy coming together of subjects and spirit at the confluence of science and engineering. Bell Labs, with Bardeen, Anderson, Herring and others, *IBM*, with Landauer, Gutzwiller and Hubbard, and universities, with Kohn, Luttinger and very concentrated interaction over summers between industry and academia, were seminal in generating the marketing-free major successes of postwar science and engineering.

basis. The conduction band states, on the other hand, are more delocalized. In this same viewpoint, they will have more of an s and p mixed characteristic. We have now also introduced the angular momentum as an important contributor for determining the nature of these states. Spin angular momentum will also matter and this will soon be discussed. Even if we ignore the energy consequences of spin's interaction, one direct consequence is the doubling of all the states (a degeneracy, or a degeneracy that will be broken) in the conduction band and the valence band. The energy interaction will also be significant, and this will be part of the spin-orbit discussion.

Oscillator strength is a very powerful idea. Interaction between states is strongest, and any change due to perturbation of a system is strongest, when the coupling is most efficient. The perturbation should match well in real and imaginary coordinates for system consequence to be strongest, or, looked at in a complementary way, the real and imaginary responses reflect two faces of the input's consequence. This is the Kramers-Kronig relationship that will occupy us in Chapter 14. Response is also a manifestation of canonic matching. A stable system's lowest order response is that of a harmonic oscillator. This is the first term in the Taylor expansion—a square dependence—where the energetic change leads to a restorative force independent of the sign of perturbation. Any oscillator, at its resonance, shows a peak in the coupling and a change of phase. The response matches at the oscillation frequency. Oscillator strengths therefore show up, or one can place the system response in its terms, for physical systems. For semiconductors, some of the main analytic notions are discussed in Appendix I.

Before tackling spin's consequence, we should spend some effort on understanding the effective mass, its differences from the real mass and, through this, an understanding of the local-nonlocal spatial interactions' consequences. We therefore now introduce another twist of Fourier transformation and of Wannier functions that emphasize locality versus Bloch functions' spread over the crystal.

4.7 *Effective mass theorem and Wannier functions*

We have introduced effective mass in the $\mathbf{k} \cdot \mathbf{p}$ approach as a way of incorporating the interaction of states through the oscillator strengths and representing the net response in an $E(\mathbf{k})$ picture. This effective mass in this $E(\mathbf{k})$ picture encapsulates the effect of interaction between states, and these states that are interacting are the Bloch states that are spread out over the entire crystal. These are extended states and feel the periodicity of the crystal potential. The effective mass arose in the behavior of the electron in the crystal; that is, its periodicity and therefore the consequences that the Hamiltonian represented for the presence of the electrons in its crystalline surroundings. This capturing of the presence of the crystal in the allowed behavior of the electron of the $E(\mathbf{k})$ picture means that the *effective mass should also only be used under conditions where the electron feels those surroundings during its response.* A core electron, for example, does not feel the surroundings. And one can also immediately visualize conditions, such as confinement where the localization of the nearly free electron is small enough, say, well below the de Broglie wavelength scale, that the use of an effective mass of the bulk crystal would be quite wrong. A similar comment will apply for the time domain. When changes are very rapid, a change of state may happen entirely through local change of the modulation function.

A localized electron doesn't sample the semiconductor environment. Excitations too may be localized. So, clearly, effective mass would be inappropriate. A similarly important issue arises in what the permittivity should be. Recall our discussion of Chapter 1. For core electrons, it is the free space permittivity. A nearly free electron in the bulk sees the consequences of polarization of the crystal, and it will be the permittivity of the material subject to the frequency that will determine what the polarization is. There will also be questions related to this permittivity depending on which electrons are responding and what their environment is. Again, core electrons will behave quite differently from valence band electrons or conduction band electrons. So, care is needed in several of these circumstances.

The introduction of effective mass makes the analysis of a large set of conditions that we are interested in for semiconductors and their devices very convenient, since it helps reduce the problem of the description of the electron under the crystal's influence and the influence from external stimulus, and often other extended—that is, over several unit cells—crystal perturbations to a simpler equation where the crystal's consequence has been subsumed into the effective mass employed in a simpler equation. This is the effective mass equation. We discuss the effective mass theorem, and relate to it the Bloch function's Fourier complement, the Wannier function, to elucidate the behavior of the electron in the periodic potential of the crystal and the interaction of states.

The question we are faced with is that if we have an external perturbation ($\mathscr{H}'$), how will a Bloch electron's response evolve in a crystal? We approach the problem by exploring the crystal description using the Fourier transform of the Bloch function. This function, the Wannier function, is

$$|w_n(\mathbf{r}-\mathbf{R}_j)\rangle = \frac{1}{\sqrt{N}}\sum_{\mathbf{k}} \exp\left(-i\mathbf{k}\cdot\mathbf{R}_j\right)|\psi_{n\mathbf{k}}(\mathbf{r})\rangle. \tag{4.45}$$

Because of the symmetry of this mathematical transformation, the Bloch function, written in terms of the Wannier function, is

$$|\psi_{n\mathbf{k}}(\mathbf{r})\rangle = \frac{1}{\sqrt{N}}\sum_{j} \exp\left(i\mathbf{k}\cdot\mathbf{R}_j\right)|w_n(\mathbf{r}-\mathbf{R}_j)\rangle. \tag{4.46}$$

A Bloch function is spread out over the entire crystal, and we have identified it through the quantum number assignment of the band (n) and the wavevector ($\mathbf{k}$). This Fourier transformation means that the Wannier function as a Fourier amplitude is localized on the atom, and now the quantum assignments for it are the band (n) and the lattice coordinate ($\mathbf{R}_j$). Because it has now been written in terms of the position coordinate, it becomes very useful in dealing with processes that involve states that are localized.

With $\hat{\mathscr{H}}'$ as the perturbation, our problem is to solve

$$\left[\hat{\mathscr{H}}_0 + \hat{\mathscr{H}}'(\mathbf{r})\right]|\psi\rangle = E|\psi\rangle, \text{ with } \hat{\mathscr{H}}_0 = -\frac{\hbar^2}{2m_0}\nabla^2 + \hat{V}(\mathbf{r}). \tag{4.47}$$

Our approach will be to show that this problem description will become the equivalent of

$$\left[-\frac{\hbar^2}{2m^*}\nabla^2 + \hat{\mathscr{H}}'(\mathbf{r})\right]|\varphi(\mathbf{r})\rangle = E|\varphi(\mathbf{r})\rangle, \tag{4.48}$$

where $|\varphi(\mathbf{r})\rangle$ is a wavefunction. Note that $V(r)$ has disappeared, and, simultaneously, the mass has transformed: $m_0 \mapsto m^*$. The description of Equation 4.47, with an electron mass m_0 interacting in the potential of the crystal with added perturbations, will become equivalent to, that is will have the same eigenenergy solution as, the solution of Equation 4.48, where an effective mass m^* is employed together with the perturbation $\hat{\mathscr{H}}'$. $|\varphi(\mathbf{r})\rangle$ is a wavefunction that is the solution to this problem with the same eigenenergy E as that of the original Bloch description.

In this relationship of Equation 4.48, we now see the starting point of most, but not all, of the semiconductor quantum-mechanical response calculations when the electron senses the crystalline environment.

The Wannier function of Equation 4.46, as a Fourier transform, is also orthonormal, just as Bloch functions are. This orthonormality holds true for all the states, so also for different bands:

$$\begin{aligned}
\langle w_{n',j'}|w_{n,j}\rangle &= \int w^*_{n'}(\mathbf{r}-\mathbf{R}_{j'})w_n(\mathbf{r}-\mathbf{R}_j)\,d^3r \\
&= \frac{1}{N}\sum_{\mathbf{k},\mathbf{k}'}\int \exp\left[i(\mathbf{k}'\cdot\mathbf{R}_{j'}-\mathbf{k}\cdot\mathbf{R}_j)\right]\psi_{n'\mathbf{k}'}(\mathbf{r})\psi_{n\mathbf{k}}(\mathbf{r})\,d^3r \\
&= \frac{1}{N}\sum_{\mathbf{k}}\exp\left[i(\mathbf{k}'\cdot\mathbf{R}_{j'}-\mathbf{k}\cdot\mathbf{R}_j)\right]\delta_{n,n'} \\
&= \delta_{j,j'}\delta_{n,n'}.
\end{aligned} \tag{4.49}$$

We start with a bandedge Bloch function, as we did in the $\mathbf{k}\cdot\mathbf{p}$ approach. The Bloch function $\psi_{n\mathbf{k}} = u_{n0}\exp(i\mathbf{k}\cdot\mathbf{r})$, transformed, leads to the Wannier function in the form

$$|w_n(\mathbf{r}-\mathbf{R}_j)\rangle = \frac{1}{\sqrt{N}}u_{n0}\sum_{\mathbf{k}}\exp\left[i\mathbf{k}\cdot(\mathbf{r}-\mathbf{R}_j)\right]. \tag{4.50}$$

Since

$$\sum_{\mathbf{k}}\exp\left[i\mathbf{k}\cdot(\mathbf{r}-\mathbf{R}_j)\right] = L^3\delta(\mathbf{r}-\mathbf{R}_j), \tag{4.51}$$

written in terms of the Dirac δ, and

$$\frac{1}{L^3}\int \exp\left[i(\mathbf{k}-\mathbf{k}')\cdot\mathbf{r}\right]d^3\mathbf{r} = \delta_{\mathbf{k},\mathbf{k}'}, \tag{4.52}$$

written in terms of Kronecker δ, it follows that

$$|w_n(\mathbf{r}-\mathbf{R}_j)\rangle = \frac{1}{\sqrt{N}}u_{n0}L^3\delta(\mathbf{r}-\mathbf{R}_j). \tag{4.53}$$

The Bloch function is spread out over the crystal. The Wannier function is localized on the atom.

This localization of the Wannier function leads to its particularly suitability in understanding many defects such as deep donors and acceptors, et cetera, where the electron becomes localized, and the Bloch functions are not a very suitable starting set for the perturbation problem.

Since Wannier functions are an orthonormal set, we may expand the Bloch function in the Wannier basis, that is,

$$\therefore\ \ \psi(\mathbf{r}) = \sum_n\sum_j \varphi_n(\mathbf{R}_j)w_n(\mathbf{r}-\mathbf{R}_j). \tag{4.54}$$

$\varphi_n(\mathbf{R}_j)$ is to be viewed as the coefficient in the Wannier-basis expansion of the Bloch function. Now take a specific band. Absent perturbation, we have $\hat{\mathscr{H}}_0\psi_{\mathbf{k}}(\mathbf{r}) = E_0(\mathbf{k})\psi_{\mathbf{k}}(\mathbf{r})$. For the perturbed problem, we need to solve

$$[\hat{\mathscr{H}}_0 + \hat{\mathscr{H}}'(\mathbf{r})]\psi = E\psi. \tag{4.55}$$

We know that

$$\psi(\mathbf{r}) = \sum_{j'} \varphi(\mathbf{R}_{j'}) w(\mathbf{r} - \mathbf{R}_{j'}), \tag{4.56}$$

so we may simplify by multiplying with the conjugate of $w(\mathbf{r} - \mathbf{R}_{j'})$ and integrate over the real space. We have

$$\begin{aligned}
&\int \sum_{j'} w^*(\mathbf{r} - \mathbf{R}_j)\hat{\mathscr{H}}_0 \varphi(\mathbf{R}_{j'}) w(\mathbf{r} - \mathbf{R}_{j'})\, d^3\mathbf{r} \\
&+ \int \sum_{j'} w^*(\mathbf{r} - \mathbf{R}_j)\hat{\mathscr{H}}' \varphi(\mathbf{R}_{j'}) w(\mathbf{r} - \mathbf{R}_{j'})\, d^3\mathbf{r} \\
&= \int E(\mathbf{k}) \sum_{j'} w^*(\mathbf{r} - \mathbf{R}_j) \varphi(\mathbf{R}_{j'}) w(\mathbf{r} - \mathbf{R}_{j'})\, d^3\mathbf{r}, \\
\text{or } &\sum_{j'} \mathscr{H}_{0jj'} \varphi(\mathbf{R}_{j'}) + \sum_{j'} \mathscr{H}'_{jj'} \varphi(\mathbf{R}_{j'}) = E(\mathbf{k}) \varphi(\mathbf{R}_j),
\end{aligned} \tag{4.57}$$

since $\langle w_{n',j'} | w_{n,j} \rangle = \delta_{j,j'}\delta_{n,n'}$. Since the perturbation varies slowly spatially and Wannier functions are very localized, the overlap contribution to $\mathscr{H}'_{jj'}$ is significant only in the neighborhood of $\mathbf{R}_j$. The unperturbed Hamiltonian $\hat{\mathscr{H}}_0$ is translationally invariant in $\mathbf{R}$. To simplify, we change the coordinate origin by translating $\mathbf{r} - \mathbf{R}_j \mapsto \mathbf{r}$, so that

$$\mathscr{H}_{0jj'} = \int w^*(\mathbf{r}) \mathscr{H}_0 w(\mathbf{r} - \mathbf{R}_{j'} + \mathbf{R}_j)\, d^3\mathbf{r} = h_0(\mathbf{R}_j - \mathbf{R}_{j'}), \tag{4.58}$$

where h_0 is introduced to characterize a scalar energy contribution in spatially normalized units. This reduces Equation 4.57 to the form

$$\sum_{j'} h_0(\mathbf{R}_j - \mathbf{R}_{j'}) \varphi(\mathbf{R}_{j'}) + \mathscr{H}'(\mathbf{R}_j) \varphi(\mathbf{R}_j) = E(\mathbf{k}) \varphi(\mathbf{R}_j). \tag{4.59}$$

Translational invariance also lets $(\mathbf{R}_j - \mathbf{R}_{j'}) \mapsto \mathbf{R}_{j'}$, so that we can rewrite Equation 4.59 in the translated form as

$$\sum_{j'} h_0(\mathbf{R}_{j'}) \varphi(\mathbf{R}_j - \mathbf{R}_{j'}) + \mathscr{H}'(\mathbf{R}_j) \varphi(\mathbf{R}_j) = E(\mathbf{k}) \varphi(\mathbf{R}_j). \tag{4.60}$$

This is now our energy equation using the Wannier basis. Note that, with the change in the basis, the first term—a term that maps to the electron Hamiltonian of the crystal in Bloch basis—has become a localized energy contribution due to the deployment of the Wannier form.

We may now expand on this connection of the energy relationship in Wannier form to that from the Bloch form:

$$\begin{aligned} E_0(\mathbf{k}) &= \langle \psi^*_{\mathbf{k}}(\mathbf{r}) | \hat{\mathscr{H}}_0 | \psi_{\mathbf{k}}(\mathbf{r}) \rangle \\ &= \frac{1}{N} \sum_j \sum_{j'} \exp\left[-i\mathbf{k} \cdot (\mathbf{R}_j - \mathbf{R}_{j'})\right] \mathscr{H}_{0jj'} \\ &= \frac{1}{N} \sum_j \sum_{j'} \exp\left[-i\mathbf{k} \cdot (\mathbf{R}_j - \mathbf{R}_{j'})\right] h_0(\mathbf{R}_j - \mathbf{R}_{j'}). \end{aligned} \tag{4.61}$$

Since j and j' here are just different indices for the same lattice, they have the same equivalent contribution to the summation; that is,

$$\begin{aligned} E_0(\mathbf{k}) &= \frac{1}{N} \sum_j N \exp(-i\mathbf{k} \cdot \mathbf{R}_j) h_0(\mathbf{R}_j) \\ &= \sum_j \exp(-i\mathbf{k} \cdot \mathbf{R}_j) h_0(\mathbf{R}_j). \end{aligned} \tag{4.62}$$

This is the electron state's unperturbed energy relationship, with the wavevector describing the energy band. The energy band is a periodic function in the reciprocal space that we reduced to the first Brillouin zone for the sake of representational convenience. Since the spatial period $\mathbf{R}$ is related to the $\mathbf{R}_j$s in reciprocal space, this energy band can be expanded in the lattice vectors. Equation 4.62 is showing us this Fourier expansion in terms of the lattice coordinates $\mathbf{R}_j$. The equivalent of this for real space is

$$h_0(\mathbf{R}_j) = \frac{1}{N} \sum_{\mathbf{k}} E_0(\mathbf{k}) \exp(i\mathbf{k} \cdot \mathbf{R}_j). \tag{4.63}$$

We have made these correspondences exploiting periodicities and reciprocities. Equation 4.63 is assigning a meaning to h_0 as an energy contribution extracted because the lattice and its reciprocal are periodic, and the Wannier function, through its locality, allowed us to pull it out.

If we now Taylor expand the Wannier expansion coefficients $\varphi(\mathbf{r} - \mathbf{R}_j)$,

$$\varphi(\mathbf{r} - \mathbf{R}_j) = \varphi(\mathbf{r}) - \mathbf{R}_j \cdot \frac{d}{d\mathbf{r}} \varphi(\mathbf{r}) + \frac{1}{2} \mathbf{R}_j^2 \cdot \frac{d^2}{d\mathbf{r}^2} \varphi(\mathbf{r}) - \cdots, \tag{4.64}$$

which, written more generally in three-dimensional notation, is

$$\begin{aligned} \varphi(\mathbf{r} - \mathbf{R}_j) &= \varphi(\mathbf{r}) - \mathbf{R}_j \cdot \nabla\varphi(\mathbf{r}) + \frac{1}{2} (\mathbf{R}_j \cdot \nabla)[(\mathbf{R}_j \cdot \nabla)\varphi(\mathbf{r})] - \cdots \\ &= \exp(-\mathbf{R}_j \cdot \nabla)\varphi(\mathbf{r}). \end{aligned} \tag{4.65}$$

This is a relationship for translation of the Wannier coefficient term. Employing this, we can write

$$\sum_{j'} h_0(\mathbf{R}_{j'}) \varphi(\mathbf{r} - \mathbf{R}_{j'}) = \sum_{j'} h_0(\mathbf{R}_{j'}) \exp(-\mathbf{R}_{j'} \cdot \nabla)\varphi(\mathbf{r}). \tag{4.66}$$

Equation 4.63 also leads to

$$E_0(\mathbf{k})\varphi(\mathbf{r}) = \sum_{j'} h_0(\mathbf{R}_j) \exp(-i\mathbf{k}\cdot\mathbf{R}_{j'})\varphi(\mathbf{r}). \tag{4.67}$$

The implication of Equations 4.66 and 4.67 through the right-hand side is an equivalence of $\mathbf{k}$ and $-i\boldsymbol{\nabla}$, that is,

$$\sum_{j'} h_0(\mathbf{R}_{j'})\varphi(\mathbf{r}-\mathbf{R}_{j'}) = E_0(-i\boldsymbol{\nabla})\varphi(\mathbf{r}). \tag{4.68}$$

Summations over the energetic contributions over the crystal are related to the energy and a gradient operation on the Wannier coefficients.

We had started this analysis using Equation 4.59 to describe the energy. The derived equivalence can now be employed there. Take $\mathbf{R}_j$ as $\mathbf{r}$, so that

$$\begin{aligned} \sum_{j'} h_0(\mathbf{R}_{j'})\varphi(\mathbf{r}-\mathbf{R}_{j'}) + \mathscr{H}'(\mathbf{R}_j)\varphi(\mathbf{r}) &= E(\mathbf{k})\varphi(\mathbf{r}), \\ \text{or } E_0(-i\boldsymbol{\nabla})\varphi(\mathbf{r}) + \mathscr{H}'(\mathbf{r})\varphi(\mathbf{r}) &= E(\mathbf{k})\varphi(\mathbf{r}) \\ \therefore\ \left[E_0(-i\boldsymbol{\nabla}) + \mathscr{H}'(\mathbf{r})\right]\varphi(\mathbf{r}) &= E(\mathbf{k})\varphi(\mathbf{r}). \end{aligned} \tag{4.69}$$

This is the effective mass equation. $\varphi(\mathbf{r})$—the Wannier coefficient—can be viewed as an *envelope function*. The envelope function has removed the high-frequency positional oscillation that existed in the Bloch function, and it continues to capture the locality or spread of the electron wavefunction. This makes the envelope function extremely useful in situations—a large variety of them, particularly when confinement without atomic-scale localization is important—where one must resort to a bandstructure-based analysis.

The envelope function here is formally identical to that used for electromagnetic waves' propagation.

Equation 4.69 has connected a known bandstructure with its electron state description as a function of the wavevector and described to us how the presence of a perturbation changes the energy. If the perturbation is absent, $E_0(-i\boldsymbol{\nabla})$ as an operator operating on the envelope function tells us the unperturbed energy state description of the electrons. $E_0(\mathbf{k})$ is the electron energy as a function of the wavevector in the absence of perturbation. And if we wish to find it in the presence of perturbation, it follows as $E(\mathbf{k})$, using this Equation 4.69.

We arrived at this description by starting with the state function $\psi(\mathbf{r})$, Fourier transforming to Wannier function, which has $\varphi(\mathbf{R}_j)$ as the coefficients, that is, with $\mathbf{r} \mapsto \mathbf{R}_j$. Since $\varphi(\mathbf{r})$ is slowly varying in $\mathbf{r}$, the coefficients $\varphi(\mathbf{R}_j)$ are also slowly varying in spacing $\mathbf{R}_{j+1} - \mathbf{R}_j$. This slowly varying argument is directly related to the electron behavior being nonlocalized, and therefore Equation 4.69 must only be employed where the electron probability is spread out.

Equation 4.69 is the general form for the effective mass equation. It can be further simplified and used in conditions where the departure from equilibrium is not very pronounced. If the semiconductor is isotropic, that is if one can write $E_0(\mathbf{k}) = \hbar^2\mathbf{k}^2/2m^*$, then

$$\left[-\frac{\hbar^2}{2m^*}\nabla^2 + \mathscr{H}'(\mathbf{r})\right]\varphi(\mathbf{r}) = E(\mathbf{k})\varphi(\mathbf{r}), \tag{4.70}$$

an equation which in appearance is very Schrödinger-equation-like. But it is not the same. $V(\mathbf{r})$, the crystal potential, has disappeared. The wavefunction here is an envelope function that is spread out over many lattice sites. And the energy bandstructure $E_0(\mathbf{k})$ has been transformed into the first term form because of the use of this envelope function through the $\mathbf{k} \rightarrow -i\nabla$ equivalence under these constraints. The true electron mass m_0 of the Schrödinger equation has been replaced by an effective mass. For the general case of anisotropic mass, that is $1/\mathbb{M} = (1/\hbar^2)\nabla_{\mathbf{k}}E_0(\mathbf{k})$, this equation will take an equivalent form, with $1/m^* \mapsto 1/\mathbb{M}$. The equation in this form then also gives the kinetic energy of the electron through the first term as it exists in the crystal. So, it will include the crystal's effect. This then corresponds to the crystal momentum. Effective mass should be understood and interpreted drawing on this electron-in-a-crystal discussion, which in turn drew on $\mathbf{k}\cdot\mathbf{p}$ and the effective mass theorem discussion.

In literature, and in this text too from time to time, we may call this simplified form of the effective mass equation the independent electron Schrödinger equation in the crystal.

An important note regarding the prolific use of different masses is that these are instruments of convenience. A density of states mass is an inertial term with units of mass that allows one to write the density of states as it actually exists into a simpler form. The same is true for conductivity mass (or others). Electrons are in different locales in the energy-wavevector space, and the response to stimuli will be different depending on those locales. A conductivity mass then becomes a forced function that lets us write a simple equation of the response to a stimulus such as an applied field. The effective mass here is specifically the mass that captures the behavior of the $E(\mathbf{k})$ states in the crystal and applies to it. Density or conductivity mass can be written in terms of it but needs to account for the variety of degeneracies or anisotropies that will exist in the crystal.

4.8 *Valence bands*

THE VALENCE BAND STATES are states that evolved in the semiconductor crystal dominantly from the filled valence states. The filled valence states were from the outer orbital of the atoms that are interacting across the assembly. These and the core states are states where the electrons are closest to the nucleus of the atoms. This is also the reason why the tight binding method discussed in Section 4.2 has superior accuracy and calculational convergence for the valence band description. And this is also the reason why pseudopotentials are more suitable for the conduction band description, where plane wave and bound state descriptions can both be simultaneously incorporated. Another point worth noting here is that these separations between the valence and the conduction bands by a bandgap in-between corresponds to the bonding and antibonding that we have mentioned in the discussion of waves and particles in Chapter 3 for the molecule. The valence states are the bonded collection, and conduction states are the antibonded collection.

4.8.1 *Valence bands without spin-orbit coupling*

FIRST, CONSIDER AN INTRODUCTORY DESCRIPTION of the valence band without considering the interaction between the valence electron's spin and its orbit. The $|p\rangle$ states are a good basis to start from, following our tight binding discussion. We have the three states $|p_x\rangle$, $|p_y\rangle$ and $|p_z\rangle$ to build from. At the degenerate point, let us write the resulting state just as $|X\rangle$, $|Y\rangle$ and $|Z\rangle$. This is the traditional atomic physics notation to describe the p_x, p_y and p_z origins. These states have the same parity—they transform identically—as the $|p\rangle$ states, and it is odd. We need to use second order degenerate perturbation, and the energy is of the form

The conduction electron's is an antibonded state, nearly free, and the spin and orbit coupling is vanishingly absent. It can become important in magnetic semiconductors—semiconductors such as $Ga_{1-x}Mn_xAs$ or $Zn_{1-x}Cr_xTe$—but it cannot be ignored for the valence band, where the bonded states have considerable local interaction.

$$E_{n\mathbf{k}} = E_v + \frac{\hbar^2 k^2}{2m_0} + \mathcal{O}_{n\mathbf{k}}(\mathbf{k}\cdot\mathbf{p}), \quad n = 1,2,3, \tag{4.71}$$

where E_v is a reference. The last term is to denote the second order perturbation to be calculated employing the $\mathbf{k}\cdot\mathbf{p}$ methodology. Three bases lead to a 3×3 Hamiltonian with the elements

$$\mathcal{H}_{ij} = \left(E_v + \frac{\hbar^2 k^2}{2m_0}\right)\delta_{ij} + \frac{\hbar^2}{m_0^2}\sum_{\alpha>3}\frac{(\mathbf{k}\cdot\mathbf{p}_{i\alpha})(\mathbf{k}\cdot\mathbf{p}_{\alpha j})}{E_v - E_\alpha}, \tag{4.72}$$

where $i,j \equiv x,y,z$, and α represents the other orientations. Symmetry leads to

$$\begin{aligned}
\mathcal{H}_{11} &= \langle X|\mathcal{H}|X\rangle \\
&= E_v + \sum_{j=x,y,z}\left(\frac{\hbar^2}{2m_0} + \frac{\hbar^2}{2m_0}\sum_{\alpha>3}\frac{|\langle X|p_j|\alpha\rangle|^2}{E_v - E_\alpha}\right)k_j^2 \\
&= E_1 + Lk_x^2 + M(k_y^2 + k_z^2), \quad \text{with} \\
L &= \frac{\hbar^2}{2m_0} + \frac{\hbar^2}{2m_0}\sum_{\alpha>3}\frac{|\langle X|p_x|\alpha\rangle|^2}{E_v - E_\alpha}, \\
M &= \frac{\hbar^2}{2m_0} + \frac{\hbar^2}{2m_0}\sum_{\alpha>3}\frac{|\langle X|p_y|\alpha\rangle|^2}{E_v - E_\alpha} \quad \text{and} \\
\mathcal{H}_{12} &= \langle X|\mathcal{H}|Y\rangle = Nk_xk_y, \quad \text{with} \\
N &= \frac{\hbar^2}{2m_0}\sum_{\alpha>3}\frac{\langle X|p_x|\alpha\rangle\langle\alpha|p_x|Y\rangle + \langle X|p_y|\alpha\rangle\langle\alpha|p_y|Y\rangle}{E_v - E_\alpha},
\end{aligned} \tag{4.73}$$

where the symmetry of $|\langle X|p_y|\alpha\rangle|^2 = |\langle X|p_z|\alpha\rangle|^2$ has been used.

This form is the 3×3 Luttinger Hamiltonian, arrived at through the use of symmetry in the cubic crystal, and written in terms of the coefficients L, M and N as

$$\begin{bmatrix} E_v + Lk_x^2 + M(k_y^2 + k_z^2) & Nk_xk_y & Nk_xk_z \\ Nk_yk_x & E_v + Lk_y^2 + M(k_z^2 + k_x^2) & Nk_yk_z \\ Nk_zk_x & Nk_zk_y & E_v + Lk_z^2 + M(k_x^2 + k_y^2) \end{bmatrix}. \tag{4.74}$$

These coefficients that do need to be evaluated give the valence band dispersion in an analytic parabolic form. At the zone center, the three solution bands that follow from the diagonalization of this Hamiltonian are degenerate.

The expression in Equation 4.74 is identical to that of Equation 4.44, with the E_v bandedge energy as reference. We arrived at it through $\mathbf{k} \cdot \mathbf{p}$ and have ascribed meaning to the coefficients L, M and N.

Along the $\langle 100 \rangle$ orientations, the light-hole band follows the $E_{lh} = Lk_z^2$ form. The heavy-hole band is a degenerate two-band set in the form $E_{hh} = Mk_z^2$.

Along the $\langle 110 \rangle$ orientations, the bands are not degenerate and have the forms $E(k) = Mk^2$, and $E(k) = (L + M \pm N)k^2/2$.

Along the $\langle 111 \rangle$ orientations, the light-hole band energy is given by $E_{lh}(k) = (L + 2M + 2N)k^2/3$, and the heavy-hole bands are again degenerate, with an energy $E_{hh}(k) = (L + 2M - N)k^2/2$.

Note in the form of these solutions that—in the absence of a spin-orbit term—the light-hole band follows from the basis state that is parallel to the quantization axis of interest. The heavy-hole band, on the other hand, is composed of the other orthogonal basis. The heavy hole has a projection of its orbital angular momentum on the direction of $\mathbf{k}$, derived from its $|p\rangle$ starting basis, equal to ± 1. This gives it helicity. The light hole, on the other hand, has a vanishing projection.

4.8.2 *Valence bands with spin-orbit coupling*

ELECTRODYNAMICS TEACHES US THE IMPORTANCE OF FRAMES OF REFERENCE when viewing electric and magnetic fields. A charge that is at rest with respect to a collection of other charges that are also at rest sees an electric field and no magnetic field. If the charge was moving, in this charge's reference frame, the collection of other charges would now be observed as moving charges, that is, a current, and therefore it would feel a magnetic field. In general, in any movement with respect to any other charge or sets of charges, a current and hence a magnetic field will be experienced. What was electric in a stationary reference frame became magnetic in a moving frame.

Classically, a moving observer of velocity $\mathbf{v}$ in an external electric field $\boldsymbol{\mathcal{E}}$ experiences a magnetic field of $\mathbf{B} = (1/c)\boldsymbol{\mathcal{E}} \times \mathbf{v}$ under non-relativistic conditions. This is $\mathbf{B} = (1/m_0 c)\boldsymbol{\mathcal{E}} \times \mathbf{p}$, with the relativistic factors arising in terms of order $(v/c)^2$ being neglected. The moving electron in its rest frame feels the magnetic field arising in the Lorentz transformation of the static electric field. Orbital motion is quantized through orbital angular momentum. Orbital motion with its axial magnetic field arising in the angular orbital motion will couple to spin angular momentum—and therefore its field—through an $\mathbf{L} \cdot \mathbf{S}$-dependent form. For now, consider the consequence of this classical magnetic field. It will interact with the magnetic moment of the spin. An order of magnitude ignoring relativistic and acceleration consequences is the energy

$$-\boldsymbol{\mu} \cdot \mathbf{B} = -\frac{e}{m_0 c}\mathbf{S} \cdot \mathbf{B} = -\frac{e}{m_0^2 c^2}\mathbf{S} \cdot (\boldsymbol{\mathcal{E}} \times \mathbf{p}), \tag{4.75}$$

where $\boldsymbol{\mu}$ is the magnetic moment, and $\mathbf{S}$ is the spin angular momentum ($\mathbf{S} = (\hbar/2)\boldsymbol{\sigma}$). This energy—a spin-orbit energy—is only approximately right, since it doesn't account for the changing frame of reference resulting in a time transformation and hence a precession frequency of the electron spin in the changing magnetic field. This is a factor of 2—the Thomas factor. In the situation of the states of the atom or its assembly, the orbital electron or its perturbed collection in the crystal, we have the nucleus with an unscreened charge of Ze. An electron in an orbit around the nucleus—or even further away—is moving in an electric field arising from the nucleus and its surroundings. Effectively, this is a screened charge (Z^*e, or sometimes simply e^*), where the screening is by electrons of the core electrons and other localized electronic charge that is further out in the Coulomb field. Conduction band electrons see much less of this screened charge—they are nearly free and the shielding by valence electrons fairly complete—but the valence electrons feel this screened charge.

The electron has a magnetic moment arising from its spin and hence the magnetic field seen by the electron under motion in this very nearly stationary screened charge results in a spin-orbit interaction.

Electrons closer to the nucleus have a much higher velocity in their orbits. So, valence band electrons see pronounced spin-orbit interaction. The magnetic field $\mathbf{B}$ is perpendicular to the plane of the orbit in which $\mathbf{v}$ resides, so it is parallel to the orbital angular momentum $\mathbf{L}$. So, the energy corresponding to the magnetic moment interaction corresponds to $\pm\mu_B B$, where μ_B is the Bohr magneton. This implies an energy that is $\lambda\mathbf{L} \cdot \mathbf{S}$, where λ is a constant related to the electron-atom interaction and the field's

The Thomas factor has an interesting story. In the early days of quantum mechanics, each unexpected occurrence (first a speculation, then a discovery) turned into a Nobel Prize. But the spin 1/2 of an electron did not. It is credited to Uhlenbeck and Goudsmit, who noted it to explain the multiplet splitting of the excited state of hydrogen arising in a fourth degree of freedom: the spin. But, apparently, Kronig too had dwelt on this quantization. Goudsmit was invited to Copenhagen by Bohr and met Thomas, who could explain the factor of 2 on relativistic grounds. In a note to Goudsmit, Thomas writes, "I think you and Uhlenbeck have been very lucky to get your spinning electron published and talked about before Pauli heard of it. It appears that more than a year ago Kronig believed in the spinning electron and worked out something; the first person he showed it to was Pauli. Pauli ridiculed the whole thing so much that the first person became also the last and no one else heard anything of it which all goes to show that the infallibility of the Deity does not extend to his self-styled vicar on earth" (see the handwritten letter in S. S. Goudsmit, "The discovery of the electron spin," http://lorentz.leidenuniv.nl/history/spin/goudsmit.html, accessed 22 July 2019). To balance, Wolfgang Pauli, besides the eponymous exclusion principle, has his spin matrices. Heisenberg wrote that while he could think on only one problem at a time, Pauli had to have two. Max Born thought of Pauli as second only to Einstein. Pauli's list (and training) of postdoctoral fellows is legendary. Pauli's comment after listening to a talk, "It was not even wrong," is quite acerbic but to the point. Projects for which the path is clear do not count as research. Research entails questions whose resolution requires parting of the fog. Everybody has something to be sorry about because of the fog of all kinds. Einstein's was urging the building of the atomic bomb. And Einstein was a pacifist! Goudsmit's recounting of this spin is a very thoughtful and calm discussion of the meaning of doing science, and the satisfaction one derives from it. It doesn't have to be the most important problem. Minor contributions and having fun matters, and chance and luck are pervasive.

relationship to the angular momentum. The heavier the atom—the larger the Z^*e—the larger the spin-orbit interaction. In atoms, this is directly seen in the fine structure where magnetic field dependence is observed and in the electron g factor.

This argument is not unreasonable for non-relativistic conditions. For relativistic conditions, one must employ the Dirac equation—the linearized relativistic generalization of the Schrödinger equation—which will show up only sporadically as a theme in the low energy conditions of the semiconductor. At the atomic level, and hence for the electron states strongly affected by nucleus, this change will be consequential. The relativistic expression for kinetic energy is $\hat{T} = \hat{\mathbf{p}}^2c^2 + m_0^2c^4$. $\hat{\mathbf{p}}$ is the canonical momentum, so, with electrical and magnetic potentials to be included, one makes the change $\hat{T} \mapsto \hat{T} - e\phi$, where ϕ is the electric potential, and $\hat{\mathbf{p}} \mapsto \hat{\mathbf{p}} - (e/c)\hat{\mathbf{A}}$, so that

$$(\hat{T} - e\phi)^2 = (\hat{\mathbf{p}}c - e\hat{\mathbf{A}})^2 + m_0^2c^4. \tag{4.76}$$

The nuclei can be considered stationary, given the vast difference in the motion of mobile carriers versus the nuclei. So, we treat nuclei as being immobile. Spin-orbit interaction will occur even if the particle is not quantum-mechanical, so long as it spins. If the moving particle has a magnetic moment, this interaction arising in magnetic moment interacting with a magnetic field, which itself is a manifestation arising in a moving frame, will happen. The electron accelerates in the electric field of the nucleus—has orbital motion—and the reference frame is non-inertial, so there is another complexity of acceleration. This acceleration reduces the interaction energy by a half. This explains the fine structure splitting in atoms precisely and is also the origin of spin being 1/2, which is an unusual fractional quantum number.

With $\hat{T}$ and $\hat{\mathbf{p}}$ in the operator form as time and space derivatives of the wavefunction, a relativistic wave equation follows where external electric and magnetic fields are both present.

The force-free form of the wave equation is

$$\left(\hat{T}^2 - c^2\sum_{\mu}\hat{p}_{\mu}^2 - m_0^2c^4\right)|\psi\rangle = 0, \quad \text{where } \mu = x, y, z. \tag{4.77}$$

A rewritten form of this Equation 4.77 is

$$\left(\hat{T} - c\sum_{\mu}\alpha_{\mu}\hat{p}_{\mu} - \beta m_0c^2\right)\left(\hat{T} + c\sum_{\mu}\alpha_{\mu}\hat{p}_{\mu} - \beta m_0c^2\right)|\psi\rangle = 0, \tag{4.78}$$

subject to

$$\begin{aligned}\alpha_{\mu}\alpha_{\mu'} + \alpha_{\mu'}\alpha_{\mu} &= 2\delta_{\mu\mu'},\\ \alpha_{\mu}\beta + \beta\alpha_{\mu} &= 0, \text{ and}\\ \beta^2 &= 1.\end{aligned} \tag{4.79}$$

The lowest order term of Equation 4.78—so, first order in the time derivative, as with the Schrödinger equation—arises in the first part and is

$$\left(\hat{T} - c\sum_{\mu}\alpha_{\mu}\hat{p}_{\mu} - \beta m_0c^2\right)|\psi\rangle = 0, \tag{4.80}$$

the Dirac equation. This equation generalizes the Schrödinger equation to the relativistic covariant form

$$-\frac{\hbar}{i}\frac{\partial}{\partial t}|\psi\rangle = (c\boldsymbol{\alpha}\cdot\hat{\mathbf{p}} + \beta m_0c^2)|\psi\rangle = \hat{\mathscr{H}}|\psi\rangle. \tag{4.81}$$

Now, for an electron, the wavevector $|\psi\rangle$ is the four-component form for a spin-1/2 particle. The nonlinear form of the starting equation, written using the kinetic and momentum transformations for electromagnetic conditions, is

$$\left[\hat{T} - e\phi - c\boldsymbol{\alpha} \cdot (\hat{\mathbf{p}} - \frac{e}{c}\hat{\mathbf{A}}) - \beta m_0 c^2\right] \times \left[\hat{T} - e\phi + c\boldsymbol{\alpha} \cdot (\hat{\mathbf{p}} - \frac{e}{c}\hat{\mathbf{A}}) - \beta m_0 c^2\right] |\psi\rangle = 0. \quad (4.82)$$

If both kinetic (T) and potential (V) energies are small compared to m_0c^2, then Equation 4.82 reduces to

$$\left[\frac{1}{2m_0}(\hat{\mathbf{p}} - \frac{e}{c}\hat{\mathbf{A}})^2 + e\phi - \frac{e\hbar}{2m_0 c}\boldsymbol{\sigma} \cdot \mathbf{B} + \frac{e\hbar}{4m_0^2c^2}\boldsymbol{\sigma} \cdot \boldsymbol{\mathcal{E}} \times \hat{\mathbf{p}}\right] |\psi\rangle = \hat{\mathscr{W}}|\psi\rangle. \quad (4.83)$$

The total energy is $\mathscr{W} + m_0c^2$. In this equation, the first two terms are what will appear in the Schrödinger equation when external fields are present. The third term is the result of the energy change $-\hat{\boldsymbol{\mu}} \cdot \mathbf{B}$ of the magnetic dipole. This dipole has a moment operator of $\hat{\boldsymbol{\mu}} = (e\hbar/2m_0c)\boldsymbol{\sigma} = (e/m_0c)\mathbf{S}$ related to spin. The fourth term of the equation is a relativistic correction. The last term is the spin-orbit term of interest to us.

In the orbital motion of an electron in the electric field of the nucleus, $\boldsymbol{\mathcal{E}} = -(1/e)\hat{\mathbf{r}}\, dV/dr$. Using $\mathbf{S} = (\hbar/2)\boldsymbol{\sigma}$, we get

$$-\frac{e\hbar}{4m_0^2c^2}\boldsymbol{\sigma} \cdot \boldsymbol{\mathcal{E}} \times \hat{\mathbf{p}} = \frac{e}{2m_0^2c^2}\mathbf{S} \cdot \left(-\frac{1}{e}\frac{dV}{dr}\hat{\mathbf{r}}\frac{dV}{dr} \times \mathbf{p}\right) = \frac{1}{2m_0^2c^2}\frac{1}{r}\frac{dV}{dr}(\mathbf{L} \cdot \mathbf{S}), \quad (4.84)$$

with $\mathbf{L}$ as the orbital angular momentum. This is the spin-orbit interaction energy, and one can see that it is in the form $\lambda\mathbf{L} \cdot \mathbf{S}$.

So, our perturbational expansion of the Dirac equation for exploring the spin-orbit energy and effect is

$$\hat{\mathscr{H}} = \frac{\hat{p}^2}{2m_0} + V + \frac{\hat{p}^4}{8m_0^3c^2} + \frac{\hbar^2}{8m_0^2c^2}\nabla^2 V + \frac{\hbar}{4m_0^2c^2}\boldsymbol{\sigma} \cdot (\nabla V \times \hat{\mathbf{p}}). \quad (4.85)$$

The first two terms are the non-relativistic Hamiltonian terms. The third is a kinetic energy correction for relativistic conditions as an approximation to $\sqrt{p^2c^2 + m_0^2c^4} - m_0c^2$. The fourth is the correction term arising in the potential. And the final term is the spin-orbit interaction term, where $\boldsymbol{\sigma}$ is the spin matrix. The components of $\boldsymbol{\sigma}$ are the Pauli spin matrices

See Appendix G for a discussion of spin coordinates and spin matrices.

$$\sigma_x = \begin{bmatrix} 0 & 1 \\ 1 & 0 \end{bmatrix}, \quad \sigma_y = \begin{bmatrix} 0 & -i \\ i & 0 \end{bmatrix}, \quad \text{and} \quad \sigma_z = \begin{bmatrix} 1 & 0 \\ 0 & -1 \end{bmatrix}. \quad (4.86)$$

The symmetry of the spin matrices is reflected in the spin-orbit perturbation Hamiltonian.

These three correction terms are of about the same order of magnitude in an atom.

Equation 4.76 implies that spin-orbit coupling will break the spin degeneracy. For atoms, the degeneracy of opposite spin is lifted for the same spatial orbital wavefunction due to this. But this is not necessarily so for semiconductors and solids in general, due to crystal's symmetries.

A wavefunction $\psi(r,s)$ and its conjugate $\psi^*(r,s)$ differ through simultaneous conjugation of the wavevector and the spin coordinate because time-reversal symmetry preserves *Kramer degeneracy*. This forces $E_{k,\uparrow} = E_{-k,\downarrow}$, where k, $-k$, $\uparrow$ and $\downarrow$ correspond to the quantum numbers for the Bloch eigenstate. When a crystal has inversion symmetry, that is, symmetry is maintained under $\mathbf{r} \mapsto -\mathbf{r}$ (also called centrosymmetric; *Ge* and *Si* are examples of two interpenetrating *FCC* crystals that are centrosymmetric), then it follows that, for such crystals with inversion symmetry,

$$\begin{aligned} E_{k\uparrow} &= E_{-k\uparrow}, \\ E_{k\downarrow} &= E_{-k\downarrow}, \text{ and} \\ \therefore \quad E_{k\uparrow} &= E_{k\downarrow}. \end{aligned} \tag{4.87}$$

Spin degeneracy is maintained by crystals that have inversion symmetry, such as *FCC* and *HCP* crystals. But, even for these crystals, on the surface, this degeneracy will break, that is, $E_{k_\parallel\uparrow} \neq E_{k_\parallel\downarrow}$.

A few other properties also follow from this and symmetry. Take a one-dimensional arrangement:

$$\nabla_{\mathbf{r}} E(\mathbf{k}) = \lim_{\delta\mathbf{k}\to 0} \frac{E(\delta\mathbf{k}) - E(-\delta\mathbf{k})}{|2\delta\mathbf{k}|} = 0, \tag{4.88}$$

so $E(\mathbf{k})$ is either a maximum or a minimum at $\mathbf{k} = 0$. Furthermore, translation symmetry implies that

$$\begin{aligned} E\left(-\frac{\pi}{a} + \delta k\right) &= E\left(-\frac{\pi}{a} + \delta k + \frac{2\pi}{a}\right) \\ &= E\left(\frac{\pi}{a} + \delta k\right), \end{aligned} \tag{4.89}$$

and, in the presence of time-reversal symmetry (so, excluding the spin-orbit consequences),

$$E\left(-\frac{\pi}{a} + \delta k\right) = E\left(-\frac{\pi}{a} - \delta k\right), \tag{4.90}$$

which implies that

$$\begin{aligned} \nabla_{\mathbf{r}} E(\mathbf{k})|_{k=\pm\pi/a} &= \lim_{\delta\mathbf{k}\to 0} \frac{E\left(\pm(\pi/a) + \delta k\right) - E\left(\pm(\pi/a) - \delta k\right)}{2\delta k} \\ &= 0, \end{aligned} \tag{4.91}$$

Consider an atom with Z atomic number. The spin-orbit term $\Delta_{so} \approx (\hbar/m_0^2c^2)pV/r$. This is about $\Delta_{so} \approx \hbar^2 V/m_0^2c^2r^2$. With the potential $V \approx Ze^2/r$, $r \approx a_B/Z = \hbar^2 Z m_0 e^2$, so that $\Delta_{so} \approx Z^4 (e^2/\hbar c)^2 m_0 e^4/\hbar^2$. The electron probability density is proportional to $1/Z^2$, so the energy of spin-orbit interaction with changing Z is $\Delta_{s0}/Z^2 \approx (Ze^2/\hbar c)^2 (m_0 e^4/\hbar^2)$. The last term is $\sim 10\,eV$, and $e^2/\hbar c$ is the fine structure constant. So, in atoms, $10 \times Z^2/137^2$ is the order of magnitude of spin-orbit energy correction. For *Si*, $Z^2/137^2$ is about 0.01; for *Ga*, it is 0.05 and, for *As*, about 0.058.

Kramer's degeneracy theorem states that each energy eigenstate of a time-reversal symmetric system of 1/2 spin has at least one more eigenstate of the same energy. If $\hat{t}$ is an operator that transforms $t \mapsto -t$, and if it commutes with the Hamiltonian, so $[\mathscr{H}, \hat{t}] = 0$, then, for any eigenstate $|u_n\rangle$, $\hat{t}|u_n\rangle$ is also an eigenstate with the same energy. This eigenstate $\hat{t}|u_n\rangle$ cannot be identical to $|u_n\rangle$ for a 1/2 integer system, since time reversal reverses angular momentum.

Note, *Si* or *Ge* are diamond, and *GaAs* and many of the other compound semiconductors are zinc blende. These are two interpenetrating *FCC* lattices.

that is, that $E(\mathbf{k})$ is a maximum or a minimum at $k = \pm\pi/a$. $E(\mathbf{k})$, therefore is a maximum or a minimum at the zone center and the zone edge for inversion and time inversion symmetry.

For semiconductors, in general, the spin-orbit interaction will depend on the velocity, the presence or absence of inversion symmetry, and hence the nature of the Bloch function. Due to the large difference in the locales of the valence and the conduction electron states, a significant difference should be expected. Conduction band electrons will have spin-degenerate states, but valence band electrons, which are quite localized and under the stronger influence of the nucleus, will have spin-orbit consequences.

The basis orbital wavefunctions are spherical harmonics for $l = 1$; these states, in quantum number notation, are written as $|lm_l\rangle$, where $m_l = -1, 0, 1$, and the orbital vectors are representable as

$$\begin{aligned} |11\rangle &= -(x+iy)/\sqrt{2(x^2+y^2+z^2)}, \\ |10\rangle &= z/\sqrt{2(x^2+y^2+z^2)}, \quad \text{and} \\ |1\,{-1}\rangle &= (x-iy)/\sqrt{2(x^2+y^2+z^2)}. \end{aligned} \tag{4.92}$$

In the following, we ignore the normalizing factor, to keep expressions simpler. The spin-orbit interaction expression can be recast to the form

$$\hat{\mathscr{H}}_{so} = \lambda \mathbf{L} \cdot \mathbf{S}, \tag{4.93}$$

with λ as the spin-orbit coupling. The crystal Hamiltonian modified from the $\mathbf{k} \cdot \mathbf{p}$ version of Equation 4.28 is now

$$\hat{\mathscr{H}} = \frac{\hat{p}^2}{2m_0} + V(\mathbf{r}) + \frac{\hbar^2\mathbf{k}^2}{2m_0} + \frac{\hbar}{m_0}\mathbf{k} \cdot \mathbf{p} + \lambda \mathbf{L} \cdot \mathbf{S}, \tag{4.94}$$

where the last term is the magnetic energy perturbation. We have now six basis states: heavy hole with spin up, heavy hole with spin down, light hole with spin up, light hole with spin down, a split-off state with spin up and a split-off state with spin down.

We need to build eigenfunctions for $\mathbf{L} \cdot \mathbf{S}$. The spin states are $|s, m_s\rangle$, where $s = 1/2$ and $m_s = \pm 1/2$, often written as $|\uparrow\rangle$ and $|\downarrow\rangle$, and

$$\mathbf{L} \cdot \mathbf{S} = \frac{1}{2}(\mathbf{J}^2 - \mathbf{L}^2 - \mathbf{S}^2), \tag{4.95}$$

where $\mathbf{J} = \mathbf{L} + \mathbf{S}$ is the total angular momentum. This means that the eigenstates of the total angular momentum and J_z are also the eigenstates of $\mathbf{L} \cdot \mathbf{S}$ and therefore the spin-orbit Hamiltonian. The eigenvalue of $\mathbf{L} \cdot \mathbf{S}$, following Equation 4.95, is $(\hbar/2)[j(j+1) - l(l+1) - s(s+1)]$. The eigenfunctions of $\mathbf{J}$ and J_z follow from the following argument. j can take on the values $l+s$ and $l-s$; that is, 3/2 and 1/2 for $l=1$ and $s=1/2$. The quantization of J_z leads to m_j taking on $2j+1$ values (similar to m_l for L_z of $2l+1$, and m_s for S_z of $2s+1 = 2$). So, $j = 3/2$ leads to 4 eigenvalues ($m_j\hbar/2s$ for J_z, where $m_j = +3/2, +1/2, -1/2, -3/2$), and $j=1/2$ leads to 2 eigenvalues, with $m_j = 1/2, -1/2$. These eigenfunctions of $\mathbf{J}$ and J_z are linear combinations of orbital and spin angular momenta. For the $j = 3/2$ and $j = 1/2$ sets—writing the expanded vectors as $|jm_j\rangle \equiv |lm_lm_s\rangle$—we have

$$|jm_j\rangle = \begin{cases} |3/2,3/2\rangle = |11\uparrow\rangle, \\ |3/2,1/2\rangle = \frac{1}{\sqrt{3}}|11\downarrow\rangle + \frac{\sqrt{2}}{\sqrt{3}}|10\uparrow\rangle, \\ |3/2,-1/2\rangle = \frac{1}{\sqrt{3}}|1-1\uparrow\rangle + \frac{\sqrt{2}}{\sqrt{3}}|10\downarrow\rangle, \text{ and} \\ |3/2,-3/2\rangle = |1-1\downarrow\rangle \end{cases}$$

$$= \begin{cases} |1/2,1/2\rangle = \frac{1}{\sqrt{3}}|10\uparrow\rangle - \frac{\sqrt{2}}{\sqrt{3}}|11\downarrow\rangle, \text{ and} \\ |1/2,-1/2\rangle = \frac{1}{\sqrt{3}}|10\downarrow\rangle - \frac{\sqrt{2}}{\sqrt{3}}|1-1\uparrow\rangle. \end{cases} \tag{4.96}$$

Note that $j = 3/2$ states will be split from $j = 1/2$ states through the spin-orbit perturbation: a splitting of $\Delta_{so} = 3/2\lambda\hbar^2/2$. Valence band states now have an energy-momentum description tied to their localization and therefore through the magnetic energy in the orbital motion with spin. We now have six valence bands, with doubling caused by the up and down spins from the p orbital basis. Writing these states as $|X\uparrow\rangle$, $|X\downarrow\rangle$, $|Y\uparrow\rangle$, $|Y\downarrow\rangle$, $|Z\uparrow\rangle$ and $|Z\downarrow\rangle$, expanded, two of these states—with the others following the same logic—are

$$|X\uparrow\rangle = \frac{1}{\sqrt{2}}\left[-|\frac{3}{2}\frac{3}{2}\rangle + \frac{1}{\sqrt{3}}|\frac{3}{2}-\frac{1}{2}\rangle - \frac{\sqrt{2}}{\sqrt{3}}|\frac{1}{2}-\frac{1}{2}\rangle\right], \text{ and}$$

$$|X\downarrow\rangle = \frac{1}{\sqrt{2}}\left[-|\frac{3}{2}-\frac{3}{2}\rangle - \frac{1}{\sqrt{3}}|\frac{3}{2}\frac{1}{2}\rangle - \frac{\sqrt{2}}{\sqrt{3}}|\frac{1}{2}\frac{1}{2}\rangle\right], \tag{4.97}$$

which are all orthonormal. The non-vanishing matrix elements are

$$\langle X\uparrow|\hat{\mathscr{H}}_{so}|Y\uparrow\rangle = -i\frac{\Delta_{s0}}{3}, \quad \langle X\uparrow|\hat{\mathscr{H}}_{so}|Z\downarrow\rangle = \frac{\Delta_{s0}}{3},$$

$$\langle Y\uparrow|\hat{\mathscr{H}}_{so}|Z\uparrow\rangle = -i\frac{\Delta_{s0}}{3}, \quad \langle X\downarrow|\hat{\mathscr{H}}_{so}|Y\downarrow\rangle = i\frac{\Delta_{s0}}{3},$$

$$\langle X\downarrow|\hat{\mathscr{H}}_{so}|Z\uparrow\rangle = \frac{\Delta_{s0}}{3}, \text{ and} \langle Y\downarrow|\hat{\mathscr{H}}_{so}|Z\uparrow\rangle = -i\frac{\Delta_{s0}}{3}. \tag{4.98}$$

The spin-orbit splitting does not change significantly from those of atoms. In zinc blende structures with their two-atom basis, the splitting is weighted by the electron probability distribution, with anions, where the electron charge cloud has a higher probability density, dominating. In general, it is a parameter for the $\mathbf{k}\cdot\mathbf{p}$

calculation from Equation 4.94, where the spin-orbit term operates on the orbital degree of freedom. Matrix elements between opposite spins vanish. The six eigenstates of the angular momentum basis lead to the Luttinger Hamiltonian relationship with spin-orbit coupling that prescribes the energies:

$$\begin{bmatrix} -P-Q & S & -R & 0 & \frac{1}{\sqrt{2}}S & -\sqrt{2}R \\ S^\dagger & -P+Q & 0 & -R & \sqrt{2}Q & -\frac{\sqrt{3}}{\sqrt{2}}S \\ -R^\dagger & 0 & -P+Q & -S & -\frac{\sqrt{3}}{\sqrt{2}}S^\dagger & -\sqrt{2}Q \\ 0 & -R^\dagger & -S^\dagger & -P-Q & \sqrt{2}R^\dagger & \frac{1}{\sqrt{2}}S^\dagger \\ \frac{1}{\sqrt{2}}S^\dagger & \sqrt{2}Q^\dagger & -\frac{\sqrt{3}}{\sqrt{2}}S & \sqrt{2}R & -P-\Delta_{so} & 0 \\ -\sqrt{2}R^\dagger & -\frac{\sqrt{3}}{\sqrt{2}}S^\dagger & -\sqrt{Q}^\dagger & \frac{1}{\sqrt{2}}S & 0 & -P-\Delta_{so} \end{bmatrix} \begin{bmatrix} |HH\uparrow\rangle \\ |LH\uparrow\rangle \\ |LH\downarrow\rangle \\ |HH\downarrow\rangle \\ |SO\uparrow\rangle \\ |SO\downarrow\rangle \end{bmatrix} = E \begin{bmatrix} |HH\uparrow\rangle \\ |LH\uparrow\rangle \\ |LH\downarrow\rangle \\ |HH\downarrow\rangle \\ |SO\uparrow\rangle \\ |SO\downarrow\rangle \end{bmatrix}. \tag{4.99}$$

Note that, at the zone center, the six bases are eigenfunctions with the eigenenergies of 0, 0, 0, 0, $-\Delta_{so}$ and $-\Delta_{so}$, respectively. Corresponding to the states description of Equation 4.96, these are the different heavy-hole, light-hole and split-off states of up and down spins, as the state vector summarizes. Here, the modification to the earlier spin-less Luttinger calculation (Equations 4.44 and 4.74) consists of the parameters

$$\begin{aligned} P &= \frac{\hbar^2}{2m_0}\gamma_1(k_x^2+k_y^2+k_z^2), \\ Q &= \frac{\hbar^2}{2m_0}\gamma_2(k_x^2+k_y^2-2k_z^2), \\ S &= \frac{\hbar^2}{m_0}\sqrt{3}\gamma_3(k_x-ik_y)k_z^2, \text{ and} \\ R &= \frac{\hbar^2}{2m_0}\sqrt{3}\left[\gamma_2(k_x^2-k_y^2)-2i\gamma_3k_xk_y\right], \end{aligned} \tag{4.100}$$

with the γ parameters related to the earlier calculation as

$$-\frac{\hbar^2}{2m_0}\gamma_1 = \frac{1}{3}(L+2M), \quad -\frac{\hbar^2}{2m_0}\gamma_2 = \frac{1}{6}(L-M), \text{ and } \quad -\frac{\hbar^2}{2m_0}\gamma_3 = \frac{N}{6}. \tag{4.101}$$

Semiconductor devices on these cubic crystals are generally oriented with transport in the [110] direction. So, transport of heavy holes is particularly handicapped by this choice, but light holes compensate.

Figure 4.9 shows an example of this six-parameter Luttinger $\mathbf{k}\cdot\mathbf{p}$ calculation for *Si*. A split-off band is formed about 44 *meV* below the heavy-hole $j = 3/2$ and the light-hole $j = 1/2$ bands, with the heavy hole and the light hole degenerate at the zone center. At the zone center, the heavy-hole mass is $0.29m_0$ along $\langle 100\rangle$, and $0.61m_0$ along the $\langle 110\rangle$ direction; that is, it is highly anisotropic. The light-hole mass is $0.15m_0$ along $\langle 100\rangle$, and $0.20m_0$ along $\langle 110\rangle$. And as one goes higher in energy, these bands warp. The heavy-hole band is highly anisotropic, with the highest mass along the $\langle 110\rangle$ directions. The light-hole band is less anisotropic.

	Δ_{so} eV	γ_1	γ_2	γ_3
Si	0.044	4.22	0.39	1.44
Ge	0.29	13.4	4.24	5.69
GaAs	0.34	6.98	2.06	2.93
InAs	0.38	20.0	8.5	9.2
InSb	0.85	34.8	15.5	16.5

Table 4.3: Selected band parameters for some of the important semiconductors.

Some of the parameters, relevant to these valence band calculations are summarized in Table 4.3. Note the similarity of the

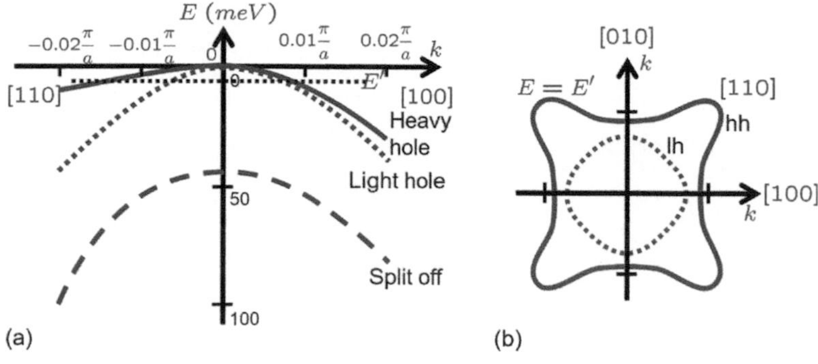

Figure 4.9: (a) Silicon's valence bandstructure near the zone center and at low energies. The heavy-hole band arises in $|3/2, \pm 3/2\rangle$ states, the light-hole band arises in $|3/2, \pm 1/2\rangle$ states and the split-off band in $|1/2, \pm 1/2\rangle$ states. The split-off band is about 40 *meV* separated from the other bands at the zone center. (b) The constant energy contour in plane on a (001) surface at the low energy $E = E'$ shown in (a). The heavy-hole band is highly anisotropic, and the light-hole band less so.

Δ_{s0} for *Ge* and *GaAs*, since the anion is about the same size. *Si* being a smaller atomic number atom, the split-off energy is quite small, while *Sb*, which is a row down in the periodic table from *As*, makes the split-off band energy much larger—larger than even its bandgap.

4.9 *Bandgaps*

GROUP *IV*, *IIIV* AND *IIVI* compounds are the natural combinations for achieving a bandgap where one of the bands will be filled at absolute zero. These will have a variety of distribution in the bandgaps—very small ($\sim 0.2\ eV$) as in *InSb* or very large ($\sim 5.5\ eV$) in diamond or even $\sim 6\ eV$ in *AlN*—depending on the perturbation interactions, which are guided by the ionicity and covalency resulting from the valence structure and the crystalline symmetries and parameters. Note here that we usually think of diamond as an insulator, even if it has a smaller bandgap than *AlN*. Both of these can be made conducting through donors; it is just that, at moderate temperatures, the thermal direct transitions will have a very low probability for transfer from the valence state to the conduction state.

It is useful to keep a perspective of this range of *eV*s and whether the semiconductor is direct or indirect, since this has a lowest order implication for the possible uses that the semiconductor can be put to.

Figure 4.10 shows the bandgap and the direct and indirect nature of the bandgap of common semiconductors—the elemental, binary and ternary forms—as a function of the lattice constant. These are bandgaps for the most common form when the crystal exists without any crystal distortion from its natural normal pressure and temperature conditions. Nitrides are generally in the wurtzite form, and the others in zinc blende or diamond. In nanowire form, as

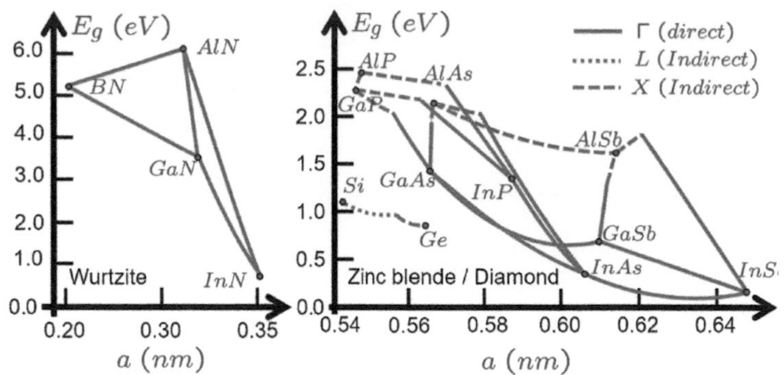

Figure 4.10: The bandgap as a function of the lattice constant for prominent semiconductors—the elemental, binary and compositionally mixed ternary forms—in their common bulk form. Nitrides—wurtzites—are shown on the left. Zinc blende and diamond crystal structure semiconductors are shown on the right. Note the direct and indirect conduction minima represent a direct and indirect bandgap, respectively.

grown, some nitrides can be zinc blende, and the others wurtzite, and normally zinc blende crystals can become wurtzite too. This has, of course, implications for bandstructure.

Nitrides have a large span in bandgap, and because they are composed of at least one small atom (N), many have the large bandgap and small lattice constant. The nitrides are shown on the left. Note also the direct and indirect conduction minimum dependence as a function of lattice constant for several of the examples. Direct bandgap semiconductors tend to be optically efficient. So, if the bandgap is in a desired range, it becomes possible to employ them gainfully for optoelectronic structures—light emission and light capture, as in photovoltaics, lasers, et cetera—and if they are indirect bandgap, they can still have utility for light capture if photoelectric conversion can be made efficient and the capturing happens in the desired blackbody radiation spectrum range of the sun. If the material is indirect, it is likely to have a higher lifetime, which makes bipolar electronic processes more conducive. The ability to vary the bandgap through compositional mixing means that one can employ the alloy composition changes as an additional crystal-potential-dependent force on mobile carriers in the structure. If one makes junctions between different semiconductor combinations abrupt, without stress, so at the same lattice constant, or with stress, while keeping the energy in this elastic deformation limited to keep the material stable, it gives one the means to preferentially control one carrier over the other. We will discuss this variety of semiconductor physics aspects in later chapters.

4.10 *Gapless semiconductors*

BASED ON OUR STARTING DEFINITION of what constitutes being a semiconductor—when ideal, at absolute zero, filled bands are separated from unfilled bands—a gapless semiconductor is an oxymoron. However, given that one can open up a gap due to the nature of the states of the bandstructure and the ability to use quantum confinement in de Broglie length scale thin layers, this naming is not without merit.

In the presence of very strong spin-orbit coupling, as in *HgTe*, which is composed of large atomic number elements, it is possible for the light-hole band to invert, that is, for the light-hole mass to become negative. This is a bulk band inversion. The light-hole band is now above the lowest conduction band. The heavy-hole band is still the valence band. And both these bands are degenerate at $\mathbf{k} = 0$. We now have a semiconductor that has no energy gap, or a negative bandgap! When one quantum confines *HgTe*, for example using *CdTe*-mixed cladding regions, it is possible to move the bands out higher in energy—conduction band up and valence band down on the electron energy plot—and a normal band insulation can be opened up. Small bandgaps, with adequate control of dark and leakage currents and noise, have several uses related to the detection of long wavelength electromagnetic waves.

Chapter 3 of S. Tiwari, "Nanoscale device physics: Science and engineering fundamentals," Electroscience 4, Oxford University Press, ISBN 978-0-198-75987-4 (2017), has a discussion of the topological uses of such materials. Traditionally, materials based on *HgTe* and *CdTe* compositional mixes have been used in infrared detectors, where their small bandgap is very useful for light detection. Infrared detectors, and their arrays, look farther out in the wavelength, so even in the dark, and are important tools of defense and astronomy.

4.11 *Example electron bandstructures*

WE CAN NOW RELATE the wave-particle discussion of Chapter 3, and the discussion of some of these methods of bandstructure calculations and their implications in how the conduction and valence states change, to the specific semiconductor bandstructures of interest throughout this text. These exemplars and the discussion related to them lets us draw implications for some of the important characteristics that will appear throughout this text and in uses of semiconductors.

Figure 4.5 let us comment on the general nature of the conduction and valence band states near the bandedge. Semiconductors can be direct bandgap and indirect bandgap. Direct means that the conduction band minimum and the valence band maximum are both at the zone center. Indirect semiconductors have the conduction band minimum away from the zone center. The conduction bandedge at the zone center arises from the $|s\rangle$-like states. The lowest indirect bandgap states in conduction band arise from a mix of the $|s\rangle$ and $|p\rangle$ states. These may be minimum at L as in *Ge*, or near X as in *Si*. The valence band states arise predominantly from the mixing of $|p\rangle$ states and $|s\rangle$ states. The heavy-hole states and the light-hole states have different contributions from the orbital and the spin angular momentum and these angular momenta feature will determine the transition rules

The reader will find a Python 3.5 code *pseudo.silconPyTh35.py*, for calculating silicon bandstructure, on the book's companion website mentioned on page *iv*. This code employs pseudopotentials.

when interactions occur with other particles such as photons. Also note here that the spin-orbit coupling caused a split-off band. So, the spin itself can have a major effect in the bandstructure.

Figure 4.11 shows, on a larger energy scale and along the major axes and points of the first Brillouin zone, the bandstructure of three sets of instructive semiconductors. Figure 4.11(a) and (b) are for *Ge* and *Si*, two elemental semiconductors that are

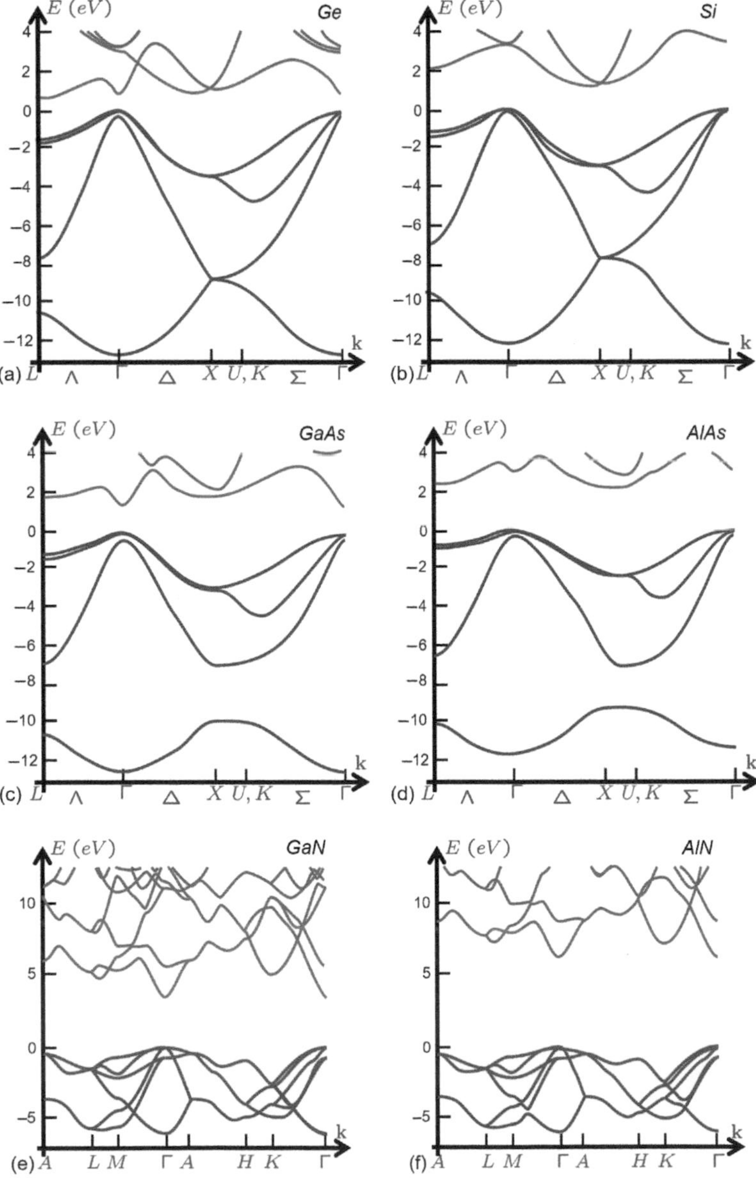

Figure 4.11: Bandstructures over comparative energy ranges for a number of illustrative semiconductors. Parts (a) and (b) show the bandstructure for *Ge* and *Si*, respectively. These are indirect. Parts (c) and (d) show bandstructure for *GaAs* and *AlAs*, respectively, with a change in cation and a simultaneous change from direct to indirect bandgap. Parts (e) and (f) show the bandstructure for a different anion (*GaN* and *AlN*) as also a different crystal structure (wurtzite instead of zinc blende). It is instructive to see the differences that have appeared in the zinc blende bandstructures here in the crystal compared to the free potential bandstructure of Figure 3.16.

centrosymmetric, covalent and indirect bandgap. Figure 4.11(c) and (d) are for *GaAs* and *AlAs*, where *GaAs* is direct bandgap while *AlAs*, with the change in the cation to a smaller atom and increased ionicity with the bond lengths very closely the same, is indirect bandgap. Figure 4.11(e) and (f) take these two same cations but change the anion to look at the bandstructure of *GaP* and *AlP*, both of which are indirect bandgap.

Split-off bands exist in all of them. *Si*, being the lightest atom, has the smallest splitting. One can also see the strong anisotropy of heavy- and light-hole bands. The heavy-hole band has the largest anisotropy, as seen through the difference between the Λ direction and the Δ and Σ directions. The light-hole anisotropy is particularly acute along the Σ direction. The symmetries in the reciprocal space are also quite interesting to see, as the perturbations cause the bands to change. *Ge* is in-between *Ga* and *As* with respect to atomic number. Many of the features of the different minima at L, Γ and X in *Ge* are present in *GaAs*, except for where the lowest conduction band minimum is in the Brillouin zone. *Si*, on the other hand, is quite different. The conduction band minimum in *Si* is slightly in from the X point in the Δ direction—the direction between the Γ and X points. At the zone center, *Si* has a fairly large bandgap, while *Ge*, with an L minimum, does show a valley at the center. All these bandstructures show that, at a zone edge, either a maximum or a minimum and group velocity exist, as expected from symmetry arguments.

Hot electron lasers have been heroically demonstrated in *Ge* because of the possibility of inversion through this central valley.

A number of compound semiconductors are direct bandgap, and, in the *eV* range, *InP* is an important one, as are ternary and quarternary compounds, where the bandgap of much smaller bandgap semiconductors such as *InAs* can be stretched out by growing them on other compound substrates. Note that larger bandgap compounds such as *GaP* and *AlP* are indirect. The size of anion and the ionicity matters. *P*, much smaller than *As*, makes the semiconductor indirect. With *AlAs* as indirect, *AlP*'s indirectness is not a surprise. But one sees that *GaAs* became indirect *GaP* when the smaller anion was employed. This conclusion is quite general with zinc blende's symmetry.

Take *GaN* and *AlN* in Figure 4.11(e) and (f). These are wurtzite crystals. The bandgap is large, and they are also direct bandgap. They also have the split-off band. Complexing them as ternaries with *InN* gives considerable freedom in changing a variety of properties, including those of heterostructures, and exploiting piezoelectricity. This set of properties together permits high breakdown fields, high carrier concentrations at interfaces and good mobility. These properties are conducive to high power operation at high frequency. The direct bandgap transitions likewise make it

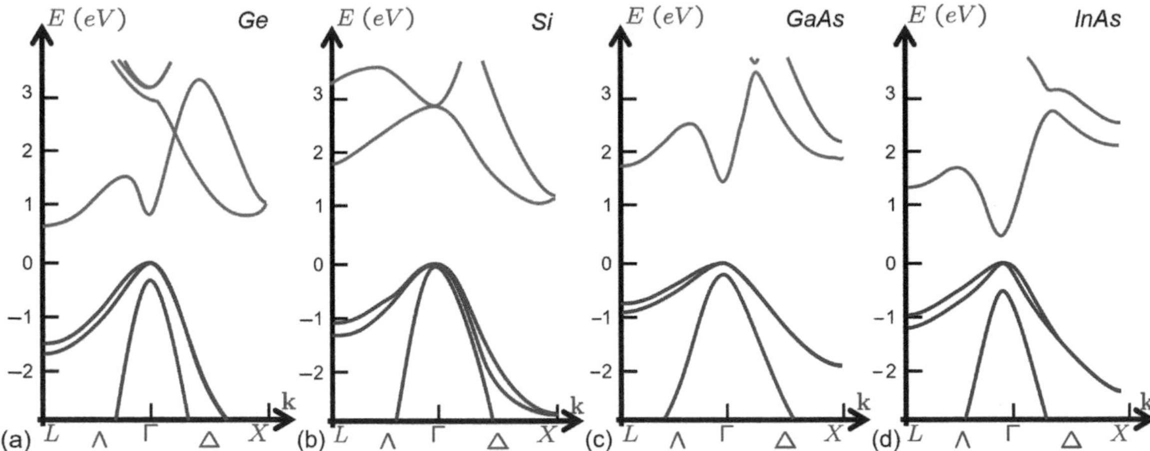

Figure 4.12: The conduction and valence bands of silicon and germanium on a comparative energy scale.

possible to achieve visible and ultraviolet emission. And because of the stronger bond, the higher energy light does not cause defect generation.

Since a few of the semiconductors will appear often and it is the states within a few *eV*s from the bandedge that are important in most uses of the structures, we particularly focus on these—*Ge*, *Si*, *GaAs* and *InAs*—through Figure 4.12 on identical energy scales and the L–Γ–X cuts, so along the Λ direction and the Δ direction where the minima appear.

The nature of different indirect bandgap between *Ge* and *Si* is now visible in the (a) and (b) parts, and the change to direct bandgap in *GaAs* and *InAs*, which have a larger and smaller bandgap, respectively, is now visible in these. *Ge*'s conduction minimum order is L-X-Γ, *Si*'s is X(a fraction, thereof)-L, while *GaAs*'s is Γ-L-X. In *AlAs* it is the X valley that will drop to the lowest, similar to that of *Si*.

Note also how dissimilar *InAs* is to *GaAs*, even if the ordering of valleys is similar. It has a small bandgap, and a low conduction mass at the bandedge, which is at the zone center, and the secondary valley at L is nearly an eV away. Carriers can be accelerated up in the same valley before the density of states suddenly jumps due to the states of L, and those even further up of the X valley.

GaAs is shown with its zone center direct bandgap. The next highest conduction valley in *GaAs* is L, followed by X. So, one may view *GaAs*, a direct bandgap material, as one whose higher conduction band states are more *Ge*-like before becoming *Si*-like. When one forms a crystal by introducing *AlAs* with *GaAs*, as the molefraction of *AlAs* is increased, the L state become lowest in energy before X takes over. Mixing of these states and their interaction with donors causes numerous effects—effects such as shallow donors such as silicon becoming deeper, and configurational effects due to the nature of local bonding. In direct transitions between states in the conduction band and the valence band, the polarization of the valence band states will matter in determining the transitions that are allowed without polarization change; that is, maintaining the conservation of angular momentum arising from electron spin and photon polarization.

Silicon has a bandgap of 1.1 eV at room temperature. The minimum in the conduction band is a fraction (~ 15 %) in from the X point. So, the constant energy surfaces are ellipsoids, and there can be six of these along the two directions of the major Cartesian axes. Germanium has a minimum at the L point in reciprocal space. So, the constant energy surface of germanium consists of half ellipsoids—eight of them. Direct bandgap semiconductors tend to have quite isotropic conduction constant energy surfaces at low conduction band energies.

4.12 *Density of states and van Hove singularities*

THESE BANDSTRUCTURES ALSO SHOW that while using and sometimes force-fitting isotropic mass, such as for the density of states at the minimum, may be acceptable, in general, the higher in energy one goes, the worse is this assumption.

Figure 4.13 shows the band dispersion and the density of states in silicon as a function of energy. Note that, starting as a square-root like dependence, which is a Taylor series manifestation of the vanishing of states at the bandedge, the density of states fluctuates considerably once it gets up into the range of a few eVs. This anisotropy effect is quite pronounced and visible when a dimension is quantized, such as in inversion regions. Another implication of this is that as one goes up in energy—as carriers get hot—there are plenty of states for scattering to take place to. Semiconductors start to look alike—*InAs* excepting up to about an eV—and therefore higher energy properties that are dominated by scattering should not be too dissimilar.

The density of states in multi-dimensions in the isotropic approximation, so useful at low energies, is discussed in Appendix H and is usually the form employed in introductory courses. See, for example, S. Tiwari, "Quantum, statistical and information mechanics: A unified introduction," Electroscience 1, Oxford University Press, ISBN 978-0-198-75985-0 (forthcoming).

Another feature of these density of states of import is the existence of divergences that are the van Hove singularities.

See Appendix H for a discussion of density of states.

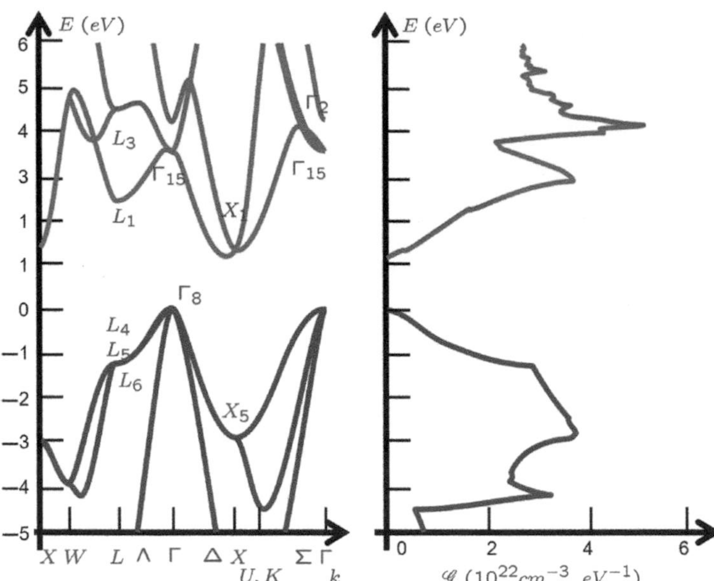

Figure 4.13: An expanded energy view of the conduction and valence bands of *Si*, together with density of states.

Take a one-dimensional structure and one specific band,

$$\mathscr{G}(E) = \int \frac{2}{2\pi}\delta(E - E_0)\, dk = \frac{2}{\pi}\int \frac{\delta(E - E_0)}{|dE(k)/dk|}\, dE$$

$$= \frac{2}{\pi}\frac{1}{|dE(k)/dk|}. \tag{4.102}$$

Since $E(k)$ is a periodic function of $\mathbf{k}$, dE/dk must vanish at the Brillouin zone edge for each band. The density of states diverges here. If the denominator $dE(k)/dk$ approaches the zone edge linearly in k, then the divergence approaches as

$$\mathscr{G}(E) \approx \frac{1}{k - k|_{edge}} \approx \frac{1}{E_{edge} - E}, \tag{4.103}$$

where we have used the second power dependence approach in k. For dE/dk to vanish in one dimension, the energy must be a maximum or minimum. The van Hove singularity in one dimension conditions is at energy bands' limits.

Van Hove singularities, however, can appear in a variety of places in two and three dimensions. Figure 4.13 shows plenty of instances of this. The general equation of density of states following Appendix H is

$$\mathscr{G}(E) = \int \frac{2}{(2\pi)^{\nu}}\delta(E - E_0)\, dk, \tag{4.104}$$

where ν is the dimensionality. This integral can be expressed over the energy surface at a specific E_k, with E representing the free ranging energy. Let

$$\delta(E - E_k) = \frac{\Theta(E - E_k) - \Theta(E - E_k - dE)}{dE}. \tag{4.105}$$

The normal to the energy surface is $\hat{\mathbf{n}} = \nabla_{\mathbf{k}} E / |\nabla_{\mathbf{k}} E|$. Since $dE_{\mathbf{k}+d\mathbf{k}} = E_{\mathbf{k}} + dE$, the separation between the $E_{\mathbf{k}}$ surface and $E_{\mathbf{k}+d\mathbf{k}}$ is $d\mathbf{k} \cdot \hat{\mathbf{n}}$. Taylor expanding $E_{\mathbf{k}+d\mathbf{k}}$,

$$\begin{aligned} E_{\mathbf{k}+d\mathbf{k}} &= E_{\mathbf{k}} + d\mathbf{k} \cdot \nabla_{\mathbf{k}} E, \\ \therefore \; dE &= d\mathbf{k} \cdot \nabla_{\mathbf{k}} E, \\ \therefore \; d\mathbf{k} \cdot \nabla_{\mathbf{k}} E &= \frac{dE}{|\nabla_{\mathbf{k}} E_{\mathbf{k}}|}. \end{aligned} \tag{4.106}$$

Let dS_E be an elemental area on the energy surface; then, Equation 4.104 can be rewritten as

$$\mathscr{G} = \frac{2}{(2\pi)^{\nu}} \int \frac{dS_E}{|\nabla_{\mathbf{k}} E|}, \tag{4.107}$$

with the integral being a $(\nu-1)$-dimensional integral over the energy surface.

Equation 4.107 says that a singularity exists, with the one-dimensional situation excluded, for all points in the reciprocal space where the group velocity $\mathbf{v}_g$ vanishes. It happens at maxima and minima, as with one-dimensional situations, but also at saddle points. Examples of saddle points are specific points where the energy may be maximum along one direction but a minimum along some other direction. For two-dimensional situations, the density of states varies as $\mathscr{G} \propto \ln|(E/E_s) - 1|$, where E_s is the specific energy point of divergence and, for three-dimensional situations, it varies as the square root, that is, $\mathscr{G} \propto \sqrt{E}$, whose trivial cases are the conduction band minimum and the valence band maximum.

Van Hove singularities, because these are regions of changes in slope marked by sharp corners, are a very convenient tool for connecting experimental measurements to bandstructure calculations.

4.13 *Example phonon bandstructures*

WE CAN NOW EXTEND the general discussion of the phonon bandstructure calculation of Chapter 3 to detail what happens in semiconductors. Let us pick the same semiconductors as the ones whose electron bandstructure was sketched in Figure 4.11: so, *Si* and *Ge*, to compare two indirect single element diamond structure semiconductors; *GaAs* and *AlAs*, to compare zinc blende structure semiconductors where now *GaAs* is direct and *AlAs* is indrect; and *GaN* and *AlN* from the wurtzite structure family. Figure 4.14 shows these.

The ordinate is in units of *meV* (100 *meV* ≡ 24.2 *THz*). *Si*'s is a stronger bond than *Ge*'s and this is reflected in the optical phonon energies. *Ge* has softer phonon modes. The *LO* modes are relatively flat with an energy of the order of 54 *meV* at *X* and which rises to >60 *meV* at the zone center. *Ge*'s optical phonons are nearly 40 % lower in energy. The longitudinal acoustic modes also have a larger energy than the transverse acoustic mode. There exists no phonon bandgap for the crystal.

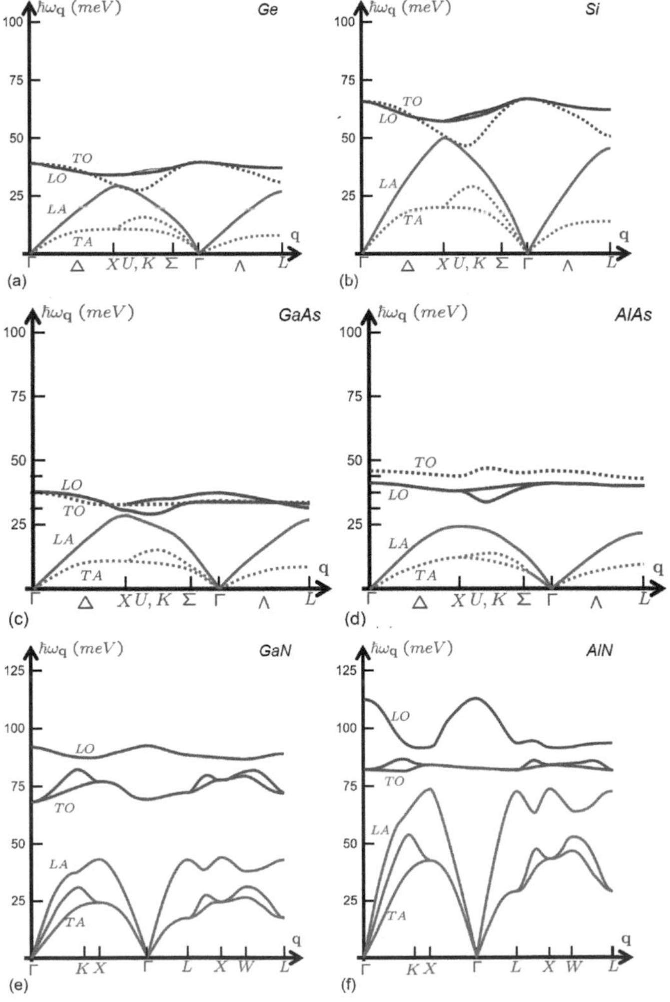

Figure 4.14: Phonon bandstructures over comparative energy ranges for a number of illustrative semiconductors. Parts (a) and (b) show the phonon bands for *Ge* and *Si*, respectively. These are covalent materials, with *Si* being one with a strong bond. Parts (c) and (d) show phonon bands for *GaAs* and *AlAs* with a change in cation. Parts (e) and (f) show the bandstructure for a different anion (*GaN* and *AlN*), as well as a different crystal structure (wurtzite instead of zinc blende).

GaAs and *AlAs* too have similar phonon behavior. *AlAs*'s are the stronger bonds, with *Al* a smaller atom. *GaAs*'s phonon dispersion looks quite a bit like that of *Ge*, which the two elements straddle in the periodic table.

The bonds of *GaN* and *AlN* are much stronger. The *LO* energy at the zone center is about 93 meV, while the *TO* energy is about 68 meV and they stay rather flat. *AlN*'s energies are a bit larger for the optical modes and considerably larger at their highest for the acoustic modes. Closer mass alignment makes the acoustic propagation much easier. What is particularly noteworthy is that *GaN* has a significant phonon gap between about 40 and 60 meV.

I find it interesting that I am not aware of any attempt at taking advantage of the *GaN* phonon gap. Suppressed thermal conductivity is of interest in a variety of circumstances, including in thermoelectric conversion.

4.14 *Summary*

THIS CHAPTER WAS A DISCUSSION of methods of bandstructure calculations, what these $E(n, \mathbf{k})$ states are like for common semiconductors, the physical behavior they represent and our parametric characterization, as well as the variety that becomes possible in the conduction and valence bands. We also ended with a summary of the phonon bandstructure. A number of important points are noteworthy from the discussion.

The tight binding approach is a technique that was developed for molecular bonding, where the basis functions are the atomic orbitals of constituting atoms. Many crystals, although not as tightly bound anymore, can also be described well using the tight binding approach. Graphene and nanotubes, which are crystals but also molecules, if one looks at them as finite-sized collections, of which fullerene would be the clearest example, are most apt places for its usage, and we will also employ this approach in Chapter 20. Here, we used tight binding to show the development of states under interaction in a one-dimensional chain under only nearest neighbor interaction. We obtained bonding states as singlets and doublets and, by extrapolation, multiples reflecting the degeneracies of the system. Crystals are translationally invariant structures with atoms bonded. Since the bases are atomic orbitals, tight binding has much more accuracy and numerical efficiency in tackling the valence band—bonding—states. So, these atomic orbitals let us see the origin of hybrid states such as the sp^3 hybridization that is common in diamond, zinc blende and wurtzite semiconductors. This hybrid construction will appear again in a discussion of defects when one of the atoms goes missing or gets misplaced, but, here, we used it to look at its connections to the valence band states and, in particular, to get from it a perspective of light and heavy holes. We

A missing or misplaced atom within the crystal starts to look like an antimolecule, or a defect in the region of order.

extended this discussion to also view the conduction bandstructure and how the symmetries of the $|s\rangle$ and $|p\rangle$ states show up in the reciprocal space for the bandedge states, with the direct bandgap Γ point being highly symmetric and $|s\rangle$-like, while points near X and L being a combination of $|s\rangle$ and $|p\rangle$. This discussion of orbital origin also let us introduce thoughts related to the interaction between spin and orbital angular momentum, which appears in a significant way in the valence band and leads to a third band near the valence band maximum.

We followed this with a twist on the plane wave discussion of Chapter 3, where we reworked the need to have short waves to adequately describe the wavefunction near the core, to an orthogonalized representation. Since this rapid change—long wavevector—is the difficulty of describing localization, the orthogonalization is close to the response to be expected from the core states. By using plane waves that are orthogonalized to the core states, one could reproduce the oscillatory nature, and yet keep the plane wave representation. In order to achieve this, we introduced a projection operator, and then only a few plane waves are needed to adequately build the Bloch function.

The projection operator was a convenient means to build the most common of bandstructure methods, that of pseudopotentials. Pseudoptentials cleverly separate the localized region of the wavefunction from the delocalized part. The projection operator served to show how, by using such an approach, the Hamiltonian can be mapped to a pseudo-Hamiltonian with a pseudopotential. This pseudopotential is much more slowly varying than the crystal potential in the core region, and the equation that needs to be solved has a smaller perturbation term. Pseudopotential techniques are now ubiquitous in solid state and for quite a bit of molecular chemistry. Another technique of similar importance is that of using density functionals, where a functional makes it possible to perform a Hartree calculation—appended with correlation—accurately.

We spent considerable effort on understanding and using the $\mathbf{k}\cdot\mathbf{p}$ method. It is very instructional and very conducive to gaining physical understanding. The $\mathbf{k}\cdot\mathbf{p}$ method works with $u_{n\mathbf{k}}(\mathbf{r})$, the periodic modulation function of the plane wave of the Bloch function. Starting from the crystal Hamiltonian and its solution—the Bloch function—we recast the description in a form where new states $|\mathbf{k}'\rangle$ can be written using perturbation on a known solution. We could demonstrate that the first order term on the changes in the energy with the wavevector, that is, group velocity, was precisely as expected. This Hamiltonian expansion operating on the periodic modulation function is in a form that now lets us reconstruct the energy and the modulation function through perturbation theory. Energies could now be written in a form with reference and a $\mathbf{k}$ dependence. So, we could assign "mass" that relates the energy to the crystal momentum through it. This is the effective mass. The free electron mass has now been modified through a perturbation correction arising in the $\mathbf{k} \cdot \mathbf{p}$ interaction. The mass, in general, is anisotropic, different in different directions, and an effective mass tensor is a convenient way to describe it. We gave meaning to the longitudinal and the transverse perturbation correction terms through this approach.

The interactions that one has to worry about are of all the states that can have a contribution through the $\mathbf{k} \cdot \mathbf{p}$ perturbation. In the conduction band of large gap materials, it will be the interaction from states within a band. In small bandgap materials, the interactions will also have to invoke conduction band and valence band state interaction. With valence bands being considerably more complex, generally with three bands—light, heavy and split off—interacting, the $\mathbf{k} \cdot \mathbf{p}$ method is particularly useful. We parameterized the $\mathbf{k} \cdot \mathbf{p}$ perturbation terms and showed how anisotropy appears, as well as how non-degeneracy plays out along the important directions, and showed the simpler form of the Luttinger Hamiltonian. Many of these aspects are a consequence of the angular momentum that the $|p\rangle$ states have. This discussion had still not included spin-orbit interaction, which we tackled later.

We moved from bandstructure calculation techniques to a discussion of the effective mass theorem and of Wannier functions, both of which are important to understanding and judiciously applying this notion of effective mass that takes into account the crystal potential. The Wannier function is the spatial Fourier transform of the Bloch function. While Bloch functions are spread out over the crystal, the Wannier function is localized and it characterizes the state in band n spatially while centered at the equilibrium position of the atoms. Wannier functions and Bloch functions, being Fourier transforms of each other, have all the associated properties of correspondence that exist in Fourier transforms. Wannier functions, for example, are also orthonormal. Using this Wannier function and expansion coefficients—an envelope function—we could derive the effective mass theorem. The effective mass theorem is a transform from the Schrödinger equation describing the energetics of the crystal but without the crystal potential in it. The effective mass and the "effective mass Hamiltonian" operating on the Wannier envelope function solves for eigenenergy. For the effective mass theorem to be valid, it must be applied to a perturbation phenomenon where the effective mass description—an electron wavefunction spread out and feeling the periodicity of the crystal potential—must be valid. So, as we will see later, if one had a shallow hydrogenic donor as a perturbing donor state donating an electron to the conduction band where it is spread out over the crystal through the Bloch state it is, it is valid. But, using it for a deep donor, where the interactions are very localized near the core, is not valid. It is also not valid if a transition from the valence band to the conduction band occurs by absorbing a photon since, during this transition, the electron really did not move spatially as such and only shifted energy at a specific **k**.

We returned back to the description of valence bands at this point to re-emphasize the richness of what happens as a result of their closer localization near the core. We described the origins and the mathematical formulation of the spin-orbit interaction, and, using this, reformulated the Luttinger Hamiltonian. The solution of this—again, with parameterization—let us now describe the nature of the behavior of the light-hole, the heavy-hole, the split-off bands, their anisotropy and their changes in different directions. This will become important and particularly interesting when we discuss strain in Chapter 17.

As an aside, and a consequence of the spin-orbit coupling, we had a short discussion of bandstructures where the energy placement of specific conduction and valence bands can be reversed. The $|p\rangle$-like valence band is higher than the $|s\rangle$-like conduction band. *HgTe* is an example of this. Of course, such semiconductors can be converted to normal semiconductors with bandgap when they are confined and, as a result, energies are shifted by the confinement momentum.

Finally, we ended this chapter with a few exemplary bandstructures: electronic direct and indirect bandgap; diamond, zinc blende and wurtzite crystals; and the phonon bandstructure of these same semiconductor examples. This information let us make a number of comments about the origins of many of the features we observe in these characteristics.

4.15 *Concluding remarks and bibliographic notes*

UNDERSTANDING THE ALLOWED STATES and the behavior of electrons and phonons in these states under the wide variety of perturbations, and how the electrons, phonons and photons interact with each other, provides the foundation for mathematically describing a semiconductor's behavior. This chapter was devoted to bandstructures. There is a wide collection of books, each with different qualifications, that are very appropriate as references for further reading and dwelling on details. In the order of the topics of this chapter, here are a few select examples.

The "Handbook on Semiconductors" series of books is an extensive and comprehensive source for an understanding and the state of semiconductors up until the early 1980s. Many new developments have taken place since then, particularly as nanoscale and new calculation techniques and practical technologies, and new materials and new uses, became of interest, but the underlying knowledge of the understanding, especially of the bulk semiconductors, has not changed significantly. The first volume in this series, which is edited by Paul[1], is one of my favorites, with a number of contributions by luminaries of that period. The energy band theory discussion, for example, is by Kane, and the discussion of pseudopotentials is by Cohen and Chelikowsky.

A book with a good discussion of bandstructure calculations using various techniques based on plane waves is by Marder[2]. Two additional books—both of recent years—are by Balkanski and Wallis[3] and by Cohen and Louie[4]. Cohen and Louie have a dedicated extended chapter on density functional theory—an approach we shortchanged in an attempt to maintain focus on $\mathbf{k} \cdot \mathbf{p}$'s instructiveness—and its use in calculations beyond bandstructures.

Yu and Cardona's text[5] has been referenced before. Chapter 2 of Yu and Cardona's text is a fairly complete description of how group theory, nearly free electron models, pseudopotentials, the $\mathbf{k} \cdot \mathbf{p}$ method and tight binding are to be employed in obtaining semiconductor bandstructures. This book is now in its fourth edition, which speaks to its completeness, with a lot of specific information useful in semiconductor calculations.

An advanced source for pursuing tight binding methods is by Kohanoff[6]. Kohanoff also discusses pseudopotentials, and molecular dynamics techniques that we did not discuss.

For understanding pseudopotential techniques, see the book by Cohen and Chelikowsky[7]. Be it known that many of the semiconductor bandstructures that one sees in texts originated in Professor Cohen's and Professor Chelikowsky's work.

A very thorough and modern text—also very advanced—discussing electronic structure techniques is the book by Martin[8]. This book also discusses the Kohn-Sham density functional approach. The text by Pisani[9] uses tight binding/linear combinations of atomic orbitals to approaching the density functional

[1] W. Paul (ed.), "Band theory and transport properties," **1**, North-Holland, ISBN 0-444-85346-4 (1982)

[2] M. P. Marder, "Condensed matter physics," Wiley, ISBN 978-0-470-61798-4 (2010)

[3] M. Balkanski and R. F. Wallis, "Semiconductor physics and applications," Oxford, ISBN 978-0-19-851740-5 (2007)

[4] M. L. Cohen and S. G. Louie, "Fundamentals of condensed matter physics," Cambridge, ISBN 978-0-521-51331-9 (2016)

[5] P. Y. Yu and M. Cardona, "Fundamentals of semiconductors," Springer, ISBN 978-3-642-00709-5 (2010)

[6] J. Kohanoff, "Electronic structure calculations for solids and molecules," Cambridge, ISBN 13-978-0521815918 (2006)

[7] M. L. Cohen and J. R. Chelikowsky, "Electronic structure and optical properties of semiconductors," Springer, ISBN 13-978-3-642-97082-5 (1988)

[8] R. M. Martin, "Electronic structure," ISBN 0-521-78285-6 (2004)

[9] C. Pisani, "Quantum-mechanical ab-initio calculation of the properties of crystalline materials," Springer, ISBN 13 978-3-540-61645-0 (1996)

calculation. The original papers related to the density functional approach[10,11] provide a comprehensive treatment and justification underlying the density functional theory. A comprehensive text for density functional theory is by Engel and Dreizler[12]. A more recent book that spans the applications to materials properties is by Feliciano Giustino[13].

Hamaguchi[14] is particularly good in his discussion of the Wannier function and the effective mass theorem.

An excellent source for a discussion of spin-orbit interactions and $\mathbf{k}\cdot\mathbf{p}$ methods, including a discussion of effective mass and oscillator strength, is the compact writing of Ridley[15]. A more advanced text, again with a quantum emphasis, is the work by Fischetti and Vandenberghe[16] that we have referenced before.

[10] P. Hohenberg and W. Kohn, "Inhomogeneous electron gas," Physical Review, **136**, B864 (1964)

[11] W. Kohn and L. J. Sham, "Self-consistent equations including exchange and correlation effects," Physical Review, **140**, A1133 (1965)

[12] E. Engel and R. M. Dreizler, "Density functional theory: An advanced course," Springer, ISBN 13 978-3642140891 (2011)

[13] Giustino, F., "Materials modeling using density functional theory," Oxford, ISBN 978-0-19-966243-2 (2014)

[14] C. Hamaguchi, "Basic semiconductor physics," Springer, ISBN 978-3-642-03302-5 (2010)

[15] B. K. Ridley, "Quantum processes in semiconductors," Oxford, ISBN 0-19-851170-1 (1988)

[16] M. V. Fischetti and W. G. Vandenberghe, "Advanced physics of electron transport in semiconductors and nanostructures," Springer, ISBN 978-3-319-01101-1 (2016)

4.16 Exercises

1. Show that the Schrödinger equation for the periodic crystal may be written in the form

$$\left[\frac{(\mathbf{p}+\hbar\mathbf{k})^2}{2m_0}+V(\mathbf{r})\right]u_{n\mathbf{k}}(\mathbf{r})=E_n(\mathbf{k})u_{n\mathbf{k}}(\mathbf{r}).$$

   Using this form, with the energy $E_n(\mathbf{k})$ near $\mathbf{k}=0$ written in terms of the momentum matrix elements, find the components of the reciprocal mass tensor terms using the momentum matrix elements to demonstrate that the
   - interaction between two bands leads to a lower band with a hole-like effective mass and a higher band that has an electron-like effective mass, and
   - that the two masses are equal. **[M]**

2. Take a body-centered cubic lattice with $|s\rangle$-like functions for the atom orbitals. Using tight binding,
   - demonstrate that, at $\mathbf{k}=0$, the energy surfaces are spherically symmetric, and
   - find the effective mass at $\mathbf{k}=0$. **[S]**

3. As a precursor to understanding surfaces, consider a linear chain with one end free. This end represents a surface. Using tight binding, show that now one can have allowed energies in the range between the normal bands. **[S]**

4. What do you think will happen if the end atom of Exercise 3 was an impurity, that is, different from the rest of the one-dimensional crystal? **[S]**

5. A one-dimensional crystal has the potential

$$V(z) = -3 - 2\cos 2z.$$

Assume that wavefunctions are the solutions of the simple harmonic oscillator problem, that is,

$$\psi(z) = \exp(-\alpha z^2),$$

with α an adjustable parameter that is used variationally to minimize energy. Find the eigenvalues for the lowest energy band. **[S]**

6. Prove that the Wannier functions and momentum eigenfunctions are Fourier transforms of each other. **[S]**

7. Show that the Wannier functions centered at different atomic sites are orthogonal. **[S]**

8. When a magnetic field **H** is applied to a metal, the **k** changes and follows an orbit in **k**-space. This orbit is at the intersection of a plane perpendicular to **H** and the energy surface on which the **k** lies. Show that an effective mass that describes this motion is

$$\frac{1}{m^*_{cycl}} = \frac{2\pi}{\hbar^2}\frac{dE}{dA_k},$$

where A_k is the area of the orbit in **k**-space. **[S]**

Magnetic field, by itself, cannot impart energy since it causes motional change—in velocity's direction—orthogonally.

9. Show that the conductivity tensor diagonal is unchanged under the application of a magnetic field. This implies that there is no transverse magnetoresistance in metals. **[S]**

10. The bottom of the conduction band of a semiconductor has a reciprocal mass tensor of the form

$$\frac{1}{\mathbb{M}^*} = \begin{bmatrix} 1/m^*_{xx} & 0 & 0 \\ 0 & 1/m^*_{yy} & 1/m^*_{yz} \\ 0 & 1/m^*_{zy} & 1/m^*_{zz} \end{bmatrix},$$

where $m^*_{yz} = m^*_{zy}$. What is the constant energy surface at the bottom of the band like? **[S]**

Bismuth, a metal, has this form of conduction band. It has one of the lowest thermal conductivities and one of the highest Hall coefficients among metals. When thin, it becomes semiconducting.

11. In a direct-gap semiconductor, as one moves away from the zone center, the energy dispersion becomes increasingly non-parabolic. This is often represented through a non-parabolicity parameter α in

$$E(\mathbf{k}) \approx \gamma(\mathbf{k})[1 + \alpha\gamma(\mathbf{k})], \text{ where } \gamma(\mathbf{k}) = \frac{\hbar^2\mathbf{k}^2}{2m^*}.$$

- The heavy- and the light-hole bands, warped, have an approximate dispersion near $\mathbf{k} = 0$ of

$$E(k) = -\frac{\hbar^2}{2m_0}\left\{Ak^2 \pm \left[B^2k^4 + C^2(k_x^2k_y^2 + k_y^2k_z^2 + k_z^2k_x^2)\right]^{1/2}\right\},$$

where the averaged A, B and C parameters and the resultant masses for *Si* and *Ge* are listed in Table 4.4. Calculate the effective masses for heavy- and light-hole bands along [100], [110] and [111] directions, based on the parameters and the nonparabolicity.

	Si	*Ge*
A	4.0	13.1
B	1.1	8.3
C	4.1	12.5
m^*_{lh}	$0.49m_0$	$0.28m_0$
m^*_{hh}	$0.16m_0$	$0.944m_0$

Table 4.4: Averaged A, B and C parameters and the resulting averaged effective masses for heavy and light holes for *Si* and *Ge*.

- Again, using this approximate $\mathbf{k}\cdot\mathbf{p}$ correction, derive an expression for the electron group velocity $\mathbf{v}_g = (1/\hbar)\nabla_{\mathbf{k}}E(\mathbf{k})$.

[M]

12. Germanium's valence band states are degenerate at $\mathbf{k}=0$. Near this $\mathbf{k} = 0$, the secular determinant is of the form

$$\begin{vmatrix} Ak_x^2 + B(k_y^2 + k_z^2) - E & Ck_xk_y & Ck_xk_z \\ Ck_yk_x & Ak_y^2 + B(k_z^2 + k_x^2) - E & Ck_yk_z \\ Ck_zk_x & Ck_zk_y & Ak_z^2 + B(k_x^2 + k_y^2) - E \end{vmatrix} = 0.$$

A, B and C relate to the reciprocal mass tensor.

- What is the form of the energy bands near $\mathbf{k} = 0$, along [100] and [111] directions?

- Show the surfaces of constant energy near $\mathbf{k} = 0$ are not spherical. **[S]**

13. A two-dimensional lattice of lattice constant a, absent any perturbation in the crystal, has degeneracy at $(-\pi/a, \pi/a)$ of the Brillouin zone. If there is periodic potential,

- show that the degeneracy is removed, and

- find the symmetry of the new states. **[S]**

5 Semiconductor surfaces

SURFACES AND INTERFACES break symmetry. Important changes in properties must therefore result. For us, so far, we have derived the energy-momentum relationships as ones that are of importance to us in describing electrons and phonons in the semiconductors. The bulk behavior must undergo change at the surface, since translational symmetry was central to our building of Bloch functions as well as correspondingly describing the oscillation, so of atoms around an equilibrium position.

In S. Tiwari, "Nanoscale device physics: Science and engineering fundamentals," Electroscience 4, Oxford University Press, ISBN 978-0-198-75987-4 (2017), we dwell on the thermodynamic link between the breaking of symmetry and the appearance of new phases with property changes. In bulk, the most important of these is thermodynamic phase transition. Peierls instability, also a transition, has major consequences for molecules and for solids. A Peierls transition can lead to metal-insulator transition in partially filled valence materials. Noether's first theorem, that differentiable symmetry of action leads to a conservation law, is another monumental observation that ties in to broken symmetry.

Devices require surfaces and interfaces for particle and energy exchange for the transformations they perform and for return of the signals to the world that is their environment. In addition to the symmetry change, there will also exist local changes spread out in a region at the interface that is also under the influence of different perturbations. Nature abhors discontinuous changes. With an order of $\lesssim 10\,eV$ of energy magnitudes in systems of interest to us, there will be a spread of a region of few atoms or more where the arrangements realign in the atomic positions as well as in how the interactions take place between them. Electron and phonon states will be affected, atoms will rearrange themselves to lower energy and there will be interfacial transmission and reflection effects. Each of these, and others, will be of importance to the properties of the material and the functioning of devices. This chapter discusses this electronic, phononic and atomic behavior at surfaces.

Any absolute discontinuous change will require infinite energy—more than there may exist in the universe.

We will develop first a physical understanding of the surface and interface. We will follow this with a semi-classical view using a one-dimensional toy model to explore how bulk electron states defer from surface states. This will let us see how atom orbital states evolve to bulk states in the translationally symmetric region to the nature of the states at the surface where the symmetry is broken. As with bulk states, we are interested in the energy, symmetry and distribution of these states. Such an approach also lets us note

Semiconductor Physics: Principles, Theory and Nanoscale. Sandip Tiwari.
 Published 2020 by Oxford University Press. DOI: 10.1093/oso/9780198759867.001.0001

useful comments regarding the common notions of workfunction, for example, that we use to characterize materials and when materials are put together with others. We also look at the surface from a free energy perspective to see how surface rearrangements take place, and see the evolution of the bonding states as a surface reconstructs. We will continue to use toy models to see the nature of vibrational modes—of phonons—that exist at the surface. Electron and phonon surface modes bring up very interesting questions of localization, that is, discreteness, versus delocalization, that is, propagation. With this background, we will have sufficient understanding to discuss semiconductor interfaces and the junctions that are formed by materials, many with crystalline continuity, in Chapter 6.

5.1 *Implications of surface and interface*

OUR TRADITIONAL TREATMENT of an infinite crystal in order to describe the propagating modes of electrons and phonons is a convenient starting point for understanding crystalline semiconductors. The approach gives us a means for describing the propagating three-, two- and one-dimensional modes. But the making of a small device—any device—breaks this symmetry. Devices have plenty of boundaries with changing symmetries. In quantum-confined nanoscale structure too, this symmetry is broken. What happens under these conditions? Intuitively, one would expect there to be both propagating and confining modes in the region where this symmetry is broken. For example, in a three-dimensional structure, with a surface, propagation of electrons must end as they encounter the surface traveling from within the bulk. Our traditional view is to say that there is a large energy barrier at the surface because an electron outside the crystal—say, in an infinite vacuum and at rest with no other electrons around—has an energy referenced as the vacuum energy level E_{vac}. Inside the crystal, the electron energy is significantly lowered. Indeed, the electron at the electrochemical potential—the Fermi energy—needs an energy called the workfunction to extract it as the difference in the two energies.

Recourse to workfunction requires caution. It is a property assigned to the bulk. Surface doesn't enter into this description, yet the electron needs to go through a surface to get to vacuum.

We will see that this picture is approximately correct but also misses several important details that we will deal with.

But this picture does say that, quantum-mechanically, we will find that the electron wavefunction decays rapidly perpendicular to the surface into the vacuum region. So, we might ask, what about along the surface? In this idealized picture of a surface cut out of the bulk, the translational symmetry still exists in the plane. Propagating modes, then, must continue to exist along the surface. We will see that, yes, this is true, but one will see a variety of unexpected behavior. Bulk propagating states may continue along the surface. Surface states may exist within the bulk's bandgap region—states that will be confined to the surface region but which follow our intuitive description to be available for propagation in the plane. If one has translational symmetry breaking or a potential symmetry breaking between crystalline semiconductors, such as at a heterostructure, one may see no surface states in the bandgap but see a discontinuity in the conduction bandedge and valence bandedge. If the materials were two dimensional or one dimensional, the dimensionality may also have its own consequences.

In general, because of the presence of propagating and confining states at the surface, numerous important property changes will result. Confined states will cause recombination and generation of carriers—coupling, and causing transitions that link valence and conduction band states. If one places a metal, one will see energy barriers—modeled through the barrier height—between the semiconductor and the metal that certainly will break the bulk-based vacuum level picture described above. Metal-semiconductor rectification or ohmic conduction will depend specifically on what happens at this interface and the region around it. States at surfaces will have their own unusual scattering characteristics. Atomic vibration modes—phonons—too will have their own interface and surface-constrained characteristics that will affect the scattering behavior arising in the perturbation of the electron characteristics by the phonons.

This behavior at surface is very important because devices, once made small, are a collection of interfaces and surfaces. Carrier transport, heat transport and other interactions will all arise from them.

5.2 *A semi-classical view*

FIGURE 5.1 SHOWS a semi-classical description of the surface and a few of the energies that we employ to describe. φ_w is the workfunction. This is the energy needed to take an electron from the hypothetical state at the electrochemical energy to vacuum energy. The electron affinity (χ) can refer to either the conduction bandedge—this is the usual way—or the valence bandedge. In general, when we employ the notation χ for electron affinity, we mean the energy from the lowest energy in the conduction band.

The states that exist in the crystalline assembly, including with its surfaces, evolve from states that existed in the contributive collection. One way to look at this is that states that arise on the surface emerged from conduction and valence band states of the bulk, due to the perturbation arising from surface. The states of the conduction and valence band, as we saw earlier, arose from the states that existed in the atoms due to the perturbation arising in the collective crystalline ensemble. The important point that this argument makes is that no additional states have been created. States of changed properties appeared as modifications when a perturbation was introduced due to the physical modification exercised. The number of electronic states in the crystal is just the number of states in all the atoms that are there in the crystal. This number of states are conserved.

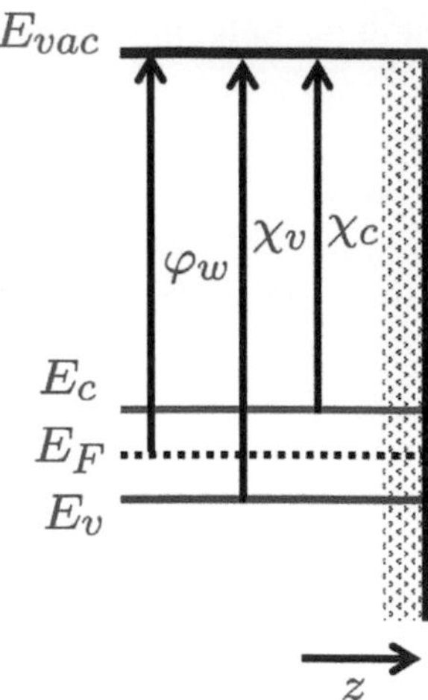

Figure 5.1: The semi-classical view of some of the reference energies in bulk and vacuum. E_{vac} is the vacuum energy, with φ_w being the workfunction—energy needed to take an electron from the electrochemical energy (Fermi energy of E_F, which is often also represented by the chemical potential μ in physics texts). Electronegativity is the energy to do the same from an electron at rest at the conduction bandedge (χ_c) or the valence bandedge (χ_v). Removing an electron implies taking it out from the surface, which is a complication for this semi-classical view and is not represented in this picture except as the textured surface region.

5.3 *A one-dimensional surface toy model*

WE WILL USE A ONE-DIMENSIONAL TOY MODEL to draw the implications that arise in broken symmetry and the perturbation introduced by the surface. The surface for a one-dimensional chain is a point. The surface of a three-dimensional chain with a broken symmetry in the z-direction is a plane whose orthonormal vector is $\hat{\mathbf{z}}$. Three-dimensional situations are a bit more complex; in-plane propagating states are potentially possible, while they are not for the point of the one-dimensional example. But, we will see that many of the important essentials will be captured by the one-dimensional model. We will be able to see from these, in analogy with bulk propagating states, surface states and surface bands. Some of these may be partly filled. This means the existence of the chemical potential (and electrochemical potential) in the surface band—or a discrete collection, if of limited numbers—inside the bandgap of the bulk and aligning with that of the bulk in thermal equilibrium. In turn, this leads to band bending and Fermi level pinning, which many semiconductors suffer from and sometimes gainfully utilize. We will be able to generalize this picture to interfaces, where charge realignment occurs as the two electrochemical potentials realign. When dissimilar materials form the interface, different interface states—different from that with vacuum—arise, leading to the interesting Schottky barriers and ohmic contacts that one gets with metals, and the conduction, valence and bandgap discontinuities that one sees in many of the heterostructures.

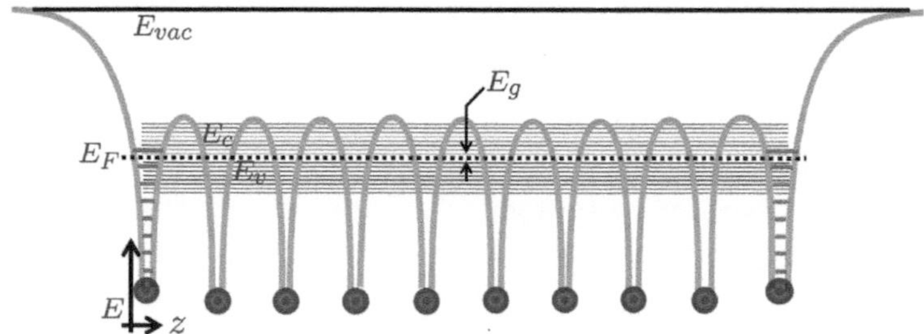

Figure 5.2: A collective chain of atoms leading to a formation of the propagating band states, together with a simplified representation of a surface where the translational symmetry breaks.

Consider the slightly more detailed view of the energy picture in Figure 5.2. This figure shows the crystal potential and energy of states along one of the coordinate axes—the z-axis—of a crystalline hypothetical semiconductor atomic assembly. In particular, it shows the periodic potential, whose $E(\mathbf{k})$ states represented in the conduction and valence bands we discussed in previous chapters. The bandedge of these bands are E_c for the conduction band, E_v for the valence band and a bandgap E_g in between. There is also our view of the vacuum energy E_{vac}, akin to the classical view. But, here we have also included a possible consequence of what happens as one approaches the surface. Energy equilibrium, with the change in symmetry conditions at the surface, must mean that the surface cannot be similar to the bulk. The potentials are different—there exists a quite different edge potential—and surface states too may come about because one has terminated the surface with a large potential change. This latter is a condition that must in turn cause confinement effects arising in continuity of probabilities and probability currents (momentum) for finite potential and energies. Intuitively, it says that states arising from valence band state interactions will rise in energy due to the confinement. But the situation will be more complicated. It is the conduction and valence states, or the more basic atomic tightly bound states, that will give rise to a spread of states within the band, as well as into the conduction band and the valence bands.

Bloch's theorem tells us that there are Fourier eigenfunctions of the form

$$\psi_{n\mathbf{k}}(\mathbf{r}) = u_{n\mathbf{k}}(\mathbf{r}) \exp(i\,\mathbf{k}\cdot\mathbf{r}) \tag{5.1}$$

that satisfy the symmetry within the crystal and from which one may construct the solution for the constrained problem in space.

Evanescent waves are unphysical in an infinite crystal. So, when this Bloch function is employed for bulk crystal analysis, one obtains real wavevector solutions. But, if we break symmetry, at the surface it is certainly possible to match two exponentials, one of which may be a decaying function—an evanescent wave—into the vacuum. This will still have a real energy solution. In general, it may be a localized state at the surface—localized away from vacuum as well as from the crystal—that is, confined to the surface, but it may also be a state at the surface that is localized away from vacuum and propagating in the crystal. Figure 5.2 attempts to represent this.

Let us consider a one-dimensional chain of N atoms spaced a apart with states of interest arising in, say, s orbitals. The Hamiltonian equation is

s orbitals in a crystal will lead to the formation of a σ band by breaking degeneracy, and p orbitals to π bands. We will see some of these implications in a slightly more complicated "toy" model that mimics graphene—a two-dimensional hexagon sheet.

$$\hat{\mathscr{H}}|\psi(z)\rangle = \left[-\frac{\hbar^2}{2m_0}\nabla^2 + \hat{V}(z)\right]|\psi(z)\rangle = E|\psi(z)\rangle,$$
$$\text{or} \quad \left\{\frac{\hbar^2}{2m_0}\nabla^2 + \left[E - \hat{V}(z)\right]\right\}|\psi(z)\rangle = 0, \tag{5.2}$$

where $V(z)$ is the periodic crystal potential. Our basis states are assumed to be the s states that arise from the solution to the individual atoms, which is

$$\left\{\frac{\hbar^2}{2m_0}\nabla^2 + [E_0 - U(z - z_n)]\right\}|\phi(z - z_n)\rangle = 0. \tag{5.3}$$

Here, $|\phi(z - z_n)\rangle$ is the eigenfunction for the atom whose expectation locale is z_n, and $U(z - z_n)$ is the potential for the non-interacting atom at $z = z_n$. The wavefunction solution of the ensemble is composed of the atomic solutions that form the orthonormal basis states, that is,

$$|\psi(z)\rangle = \sum_n c_n|\phi(z - z_n)\rangle, \tag{5.4}$$

where c_n are coefficients whose magnitude determines the probabilistic contribution. Basis states are known, perturbation is known and the Hamiltonian is known, so we should be able to write the algorithm to solve this. We will simplify in the spirit of the "toyness." The near end atoms of the chain are unique in their interactions, and the rest of chain has its own unique symmetry. So, the overlap and the perturbation contributions of these will be different. For overlap, we write

$$\langle\phi_m|\phi_n\rangle = \delta_{mn} + \beta\delta_{m,n\pm1}, \tag{5.5}$$

that is, the overlap is unity if the atoms are the same, and β if they are displaced by unity. This takes care of self and nearest neighbor normalization. Now, consider the consequence of the perturbation $V - U$, the difference in the potential assembly of atoms and the atoms. We write this as

Note here the correspondence between this toy model and the linear chain that was tied back as a tight binding toy example in Subsection 4.2.1. The ends had been removed there by "wrapping" it back.

$$\langle \phi_m | V - U | \phi_n \rangle = \begin{cases} -\alpha & \forall\, n = m \notin \{1, N\}, \\ -\alpha' & \forall\, n = m \in \{1, N\}, \\ -\gamma & \forall\, m = n \pm 1, \text{ and} \\ 0 & \text{otherwise.} \end{cases} \tag{5.6}$$

The two equations—Equation 5.2 for the assembly and Equation 5.3 for the individual atoms—pose constraints on solutions and how we may proceed, since they directly affect the expansion coefficients of the basis states. We will constrain ourselves to problems where $E - E_0$ and β are small. Essentially, this is to say that the atomic energies and the ensemble energies are not too different—the bands form but do not displace too much—and that nearest neighbor overlap is small enough. This is to some extent a toy model convenience in order to explore the implications. We parameterize the energies by setting

$$\begin{aligned} (E - E_0) + \alpha &= \epsilon, \text{ and} \\ \alpha - \alpha' &= \epsilon_0. \end{aligned} \tag{5.7}$$

What these state is that the perturbation in energy due to a self-contribution $E - E_0$ and the assembling contribution of α for the non-edge entities is a change ϵ and that the self-contributions from the end terms of the perturbations differ from that of the bulk by ϵ_0. So, the first is a parameterization for the changes due to assembling, and the second is to distinguish self-effects between edges, which we are specifically interested in, and the bulk. With this toy manipulation, we may determine the secular determinant, which must be zero for the solution to exist. This leads, in turn, to three equations representing the conditions at two edges, and the rest as a symmetric region of

$$\begin{aligned} (\epsilon - \epsilon_0)c_1 + \gamma c_2 &= 0, \\ \gamma c_{N-1} + (\epsilon - \epsilon_0)c_N &= 0, \text{ and} \\ \gamma c_{n-1} + \epsilon c_n + \gamma c_{n+1} &= 0 \;\; \forall\, n \in \{2, \ldots, N-1\}. \end{aligned} \tag{5.8}$$

We take the Fourier function form for the coefficients as

$$c_n = A \exp(ikna) + B \exp(-ikna). \tag{5.9}$$

In Equation 5.8, the last of the equations connecting these different coefficients in the "infinite"-limit form—an average that is the first term of the Fourier transform—gives

$$\epsilon = 2\gamma \cos(ka). \tag{5.10}$$

This, coupled with α—the crystal perturbation's averaged consequence in the 0th order term, which will lower it in energy—is the change in energy $E - E_0$.

The first two equations of Equation 5.8 describe the lowest order consequences of broken symmetry. From these matrix element relationships, we obtain the constraint

$$y = -\frac{\epsilon_0}{\gamma} = \frac{-\sin(Nka) \pm \sin(ka)}{\sin[(N-1)ka]}, \tag{5.11}$$

which describes the form of solution that connects y to ka and N—the number of contributing sites with single states, so the number of states.

Figure 5.3 shows the form of the solution. This equation form follows the same approach as that used to find the energies of a quantum well with a finite barrier—a form where one may not be able to write an implicit solution but can find the solution by parameterizing and using intersections representing equality. The region $-1 < y < 1$ represents an energy normalized to the energy perturbation contribution from the next neighbor. The next neighbor contribution averages to a constant over much of the chain but deviates at the edges.

From the $N = 10$ electronic states, $N = 10$ electronic states appear in the assembly. The s states, from which these arise, are isotropic and have a $\gamma = -\langle\phi_m|V - U|\phi_n\rangle$ for $m = n \pm 1$ that is negative. This means that it is the positive y that is of interest to us. For $0 < y < 1$, the solution lies in $0 < ka < \pi$, that is, propagative waves with positive energy. When $y > 1$, there continue to be several of these propagating solutions—$N - 2$ real solutions—but, as Figure 5.3 indicates, there are also solutions with ka imaginary. This is the region shown here as κa extending orthogonally at $ka = \pi$. So, for $y > 1$, there are still N solutions, but, at energy $y = 1^+$, one is non-propagating, and, a little beyond that, two of the modes are non-propagating.

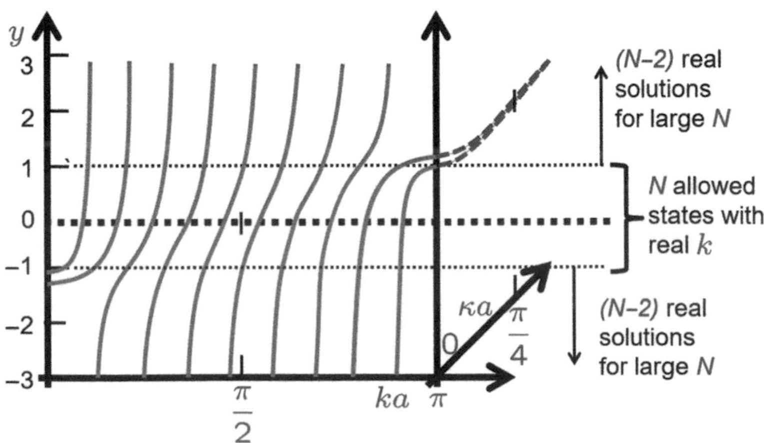

Figure 5.3: Normalized energy y as a function of ka for a linear of chain of 10 isotropic entities contributing a single state each.

$k = i\kappa$ implies a solution of the form

$$c_n = A\exp(-\kappa na) + B\exp(\kappa na), \tag{5.12}$$

for specific n of the mode for which this constraint arose, and corresponding to this is a change in the wavefunction solution of $|\psi(z)\rangle = \sum_n c_n|\phi(z - z_n)\rangle$. In the Bloch solution, this imaginary wavenumber leads to an attenuation away from $z = z_n$. The mode is confined near the locale—at the boundary where translational symmetry was broken, and it is a non-propagating confined mode. *It is a localized surface state.*

One may look at this argument mathematically. For $y > 1$, for the confined solution outside the propagating s band,

$$k = \frac{\pi}{a} + i\kappa, \tag{5.13}$$

with

$$y = \frac{-\sinh(N\kappa a) \pm \sinh(\kappa a)}{\sinh[(N-1)\kappa a]},$$
$$\therefore \lim_{N\to\infty} y \approx \exp(\kappa a). \tag{5.14}$$

This leads to the energy

$$E = E_0 - \alpha + 2\gamma\cosh(\kappa a) \approx E_0 - \alpha + 2\gamma\cosh(\ln y). \tag{5.15}$$

Since $y > 1$, these evanescent states have an energy that is larger than those of the bulk propagation states. They are also localized at the surface of this one-dimensional crystal, that is, to a point.

Since we have solved for these modes, we have now found all the coefficients that appear in our secular determinant equation set, since we now know their relationships and can normalize. In short, for an N-long array,

$$\begin{aligned} c_2 &= -c_1\exp(-\kappa a), \\ c_{N-1} &= -c_N\exp(-\kappa a), \\ c_{n+1} &= c_1(-1)^n\left[\exp(-n\kappa a) + (-1)^n\exp(-(N-n)\kappa a)\right] \\ &\qquad \text{for } 2 \le n \le N-1, \text{ and} \\ c_1 &= c_N \text{ by symmetry.} \end{aligned} \tag{5.16}$$

To go together with the picture of Figure 5.2, we now have the wavefunction solution in the form shown in Figure 5.4. Here, the form $\phi_n = \exp[-4(z - na)^2]$ is assumed with the solution drawn for $\kappa a = 0.5$. This corresponds to $y \approx 1.65$.

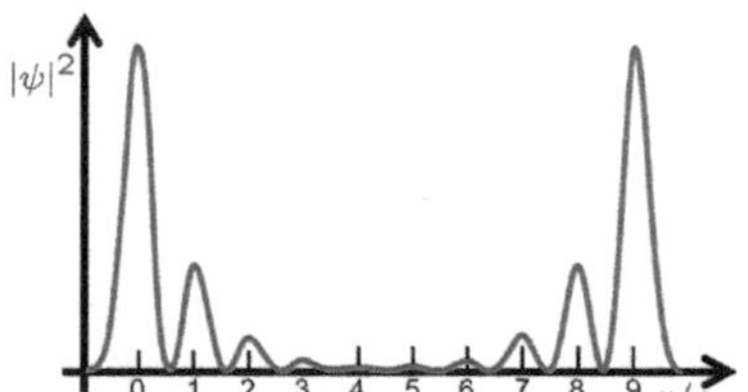

Figure 5.4: The wavefunction solution $\psi(z) = \sum_n c_n\phi(z - z_n)$ for the one-dimensional assembly of $N = 10$ of s-type states of Figure 5.3.

This picture can now be seen to lead to many of the observations that we generally see in semiconductors. But we need to make the situation a little more realistic, even for the one-dimensional toy model. Consider that each of these atoms has three energy states contributing. This is still a toy model, but now we have tried to include the possibility of different principal quantum numbers. When the basis states employed are very far apart, there is no degeneracy to worry about, since the states may be treated as asymptotically non-interacting. We have $3N$ states as the basis functions. As we bring them together, that is, decrease a—the lattice constant—degeneracy removal causes bands to form. Some of these modified states may eventually become non-propagating and confined to the surface. This is the picture, as shown in Figure 5.5 for $N = 8$. As lattice spacing is brought close together, three bands form, and we see non-propagating states at the bandedge. Upon continuing the decrease in lattice spacing, one begins to see mid-gap states, identified here as *ss*, which appear at a critical spacing. Also note that 2 states appeared, as in the previous example.

This figure is from William Shockley's treatment of this problem in the year 1939. See W. Shockley, "On the surface states associated with a periodic potential," Physical Review, **56**, 317 (1939).

These mid-gap surface states arose through the interaction between two orthogonal states. The conducting bands were also formed from these orthogonal states. This is essentially what we saw in the 10-atom example using *s*-states. There, we see these confined surface states appearing within the conducting band energy, but at $y > 1$. So, unusual mid-gap surface states—in the bandgap—appear now because of the mixing of bands.

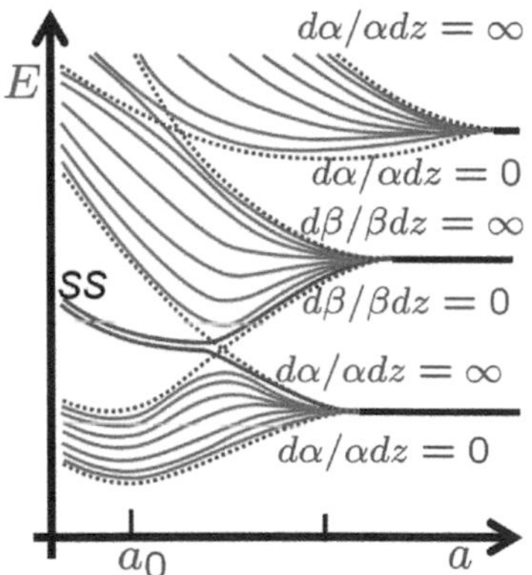

Figure 5.5: Energy of states as function of lattice spacing in a hypothetical one-dimensional crystal consisting of 8 atoms, each of which contributes 3 states. Electron states on the surface can be modeled through the basis functions, but we have freedom in the choice. Shockley modeled them using atomic functions. Tamm modeled them using the conducting band states. Either way, the end result is the same, although one sees them referred to as Shockley or Tamm states from time to time. Tamm states are those where the surface potential is modified. Shockley states are those where the periodic potential is interrupted.

Having looked at the nature of these surface states from the atomic function basis, and seeing the nature of band formation with surface states near the bandedge energies, it is apropos to also look at this picture from a nearly free electron perspective. This will make a correspondence to Chapter 3 and Chapter 4, where we made connections between a plane wave basis and a tight binding basis toward understanding bandstructure. The plane wave nearly free electron view instantly implies that the periodic potential is much less than the kinetic energy; that is, $V(z) \ll T$. We assume

$$V(z) = V_0 + V_K \exp(iKz), \tag{5.17}$$

where K is the reciprocal lattice basis vector. $V(z)$ is small and, as such, is treated as a perturbation. This is the form of periodic perturbation of the bulk state description reviewed earlier. The states at zone boundaries are nearly degenerate. One can therefore employ degenerate perturbation theory and, as for the Bragg reflection argument in the bandgap discussion, one can find the electron energy relationship to the wavevector with these raising and lowering square root terms of perturbation's contribution. We write the energy relationship in terms of the wavevector q, which is the deviation from the zone edge:

$$E(q) = V_0 + \epsilon_{\pi/a} + \epsilon_q \pm \left(4\epsilon_{\pi/a}\epsilon_q + |V_K|^2\right)^{1/2}$$
$$\text{for } q = \frac{\pi}{a} - k \ll \frac{\pi}{a}. \quad (5.18)$$

Here, the subscripted ϵ is the harmonic energy form, that is, $\epsilon_s = \hbar^2 s^2/2m$, where $s = \pi/a$ or q. This shows that the bandedge gap, the bandgap, appears at $q = 0$, that is, $k = \pi/a$.

There is a clear meaning embedded in this mathematical form. Any wavefunction with an imaginary wavevector decays exponentially spatially. In a crystal with perfect periodicity, this is unphysical; the wavefunction itself must be periodic. This means that the periodicity forces a zero imaginary component. No evanescent states shall exist in the bandgap. The periodic zone bandedge bandgap argument from Bragg reflection was but just one manifestation of it.

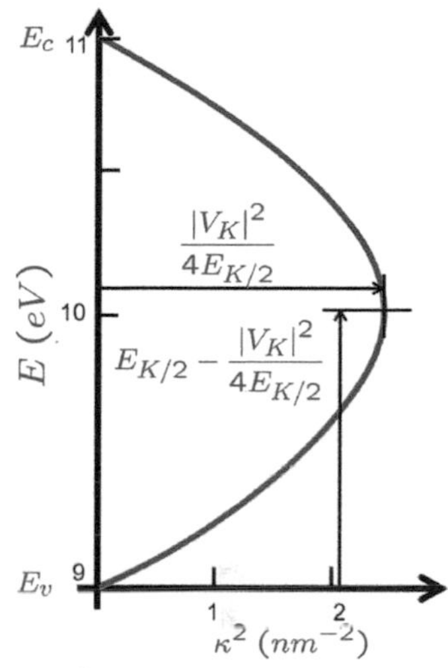

Figure 5.6: Energy of confined—evanescent—states with $q = -i\kappa_c$ in the free electron model.

At the surface, periodicity is broken, and it is permissible to have a decaying state. This $E(q)$, where q is the small deviation from the first Brillouin zone edge, is still the eigenenergy solution in this same mathematical form, but now with an imaginary $q = -i\kappa_c$ allowed. The energy $E(q)$ can exist as a continuous function in the complex plane where real and imaginary wavevectors are permitted to change. This energy, with $q = -i\kappa_c$, is

$$E(\kappa_c) = V_0 + \epsilon_{\pi/a} + \epsilon_{\kappa_c} \pm \left(|V_K|^2 - 4\epsilon_{\pi/a}\epsilon_{\kappa_c}\right)^{1/2}, \quad (5.19)$$

where the real energy solution exists for imaginary q for eigenenergy in the range $\epsilon_{\pi/a} - |V_K| < E < \epsilon_{\pi/a} + |V_K|$. This implies the maximum imaginary excursion allowed is $\kappa_c|_{max} = (m|V_K|^2/2\hbar^2\epsilon_{\pi/a})^{1/2}$. There exists a constant shift given by $V_0 + \epsilon_{\pi/a}$, and the κ_c-dependent confined states vary in energy as $\epsilon_{\kappa_c} \pm \left(|V_K|^2 - 4\epsilon_{\pi/a}\epsilon_{\kappa_c}\right)^{1/2}$. Figure 5.6 shows this surface state energy solution for $V_K = 1$ and $\epsilon_{K/2} = 10$ at the zone edge.

This attenuation wavevector takes the form shown in Figure 5.6 in this idealized free electron description. States in the center of the gap have the shortest extinction length. States near the bandedges are the longest. Mixing of the states—with similar wavefunction forms in the two bands—will emphasize the nearest band contribution more.

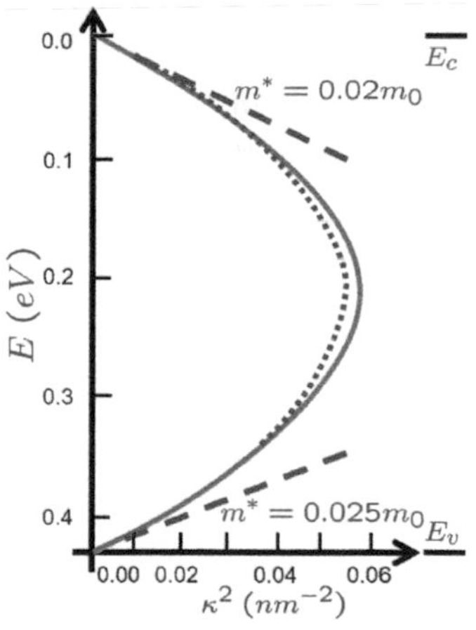

Figure 5.7: Surface state energies observed by surface tunneling measurements in *InAs*. The solid line is the free electron description of the energy of the states. The dotted line is the approximation to the tunneling-based experimental surface state energy.

Small bandgap semiconductors, such as *InAs* or *InSb*, come closest to a nearly free electron description. Figure 5.7 shows the observed surface state energy with this evanescent wavevector dependence in *InAs*. *InAs* is a small bandgap (~ 0.3 *eV*) semiconductor where the secondary valleys are far away. It also has a small effective mass and high mobility—characteristics of a nearly free electron behavior. The figure shows a very similar κ_c^2 dependence that the equation points to. A plane wave toy model comes pretty close to a realistic description!

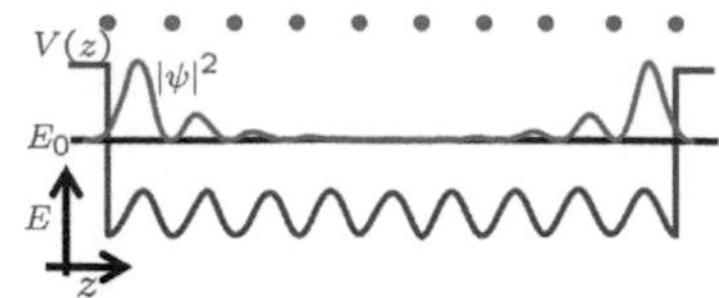

Figure 5.8: The toy free electron model with periodic potential and finite extent showing the probability per unit distance of finding an electron ($\propto |\psi|^2$).

This model leads us to the description of Figure 5.8, where a one-dimensional assembly of 10 atoms is shown. Propagating modes in the bulk, which reach out to the surface, given the boundary conditions from which the solution is built, are reflected in the low but equal probability across the large space of the sample—a wavefunction that is small and periodically changing. But $|\psi|^2$ peaks at the surface. There is a high probability of finding electrons at the surface—they are confined in the surface region. The wavefunction solution decays beyond the abrupt boundary exponentially. And there can be phase shifts in $\psi(z)$ w.r.t. the z_ns because of the finiteness. Here, we have *ad hoc* shown the exponential decay outside the chain while fixing the atomic positions and the boundary conditions. This deficiency can be addressed with more care. This correction here is related to the incorporation of boundary conditions at the surface from the nearly free electron model. This will bring the decay extent, the imaginary wavevector and the phase shifts together.

We will continue on this path of making small corrections to approximations that we find wanting, to understand implications and necessary corrections.

Now, in this nearly free electron model, we will try to probe the surface states as a result of the mixing of states built from the conduction and valence bands. This incorporates phase, wavevector and decay together a bit more self-consistently. The toy model picture of the potential perturbation is as in Figure 5.9.

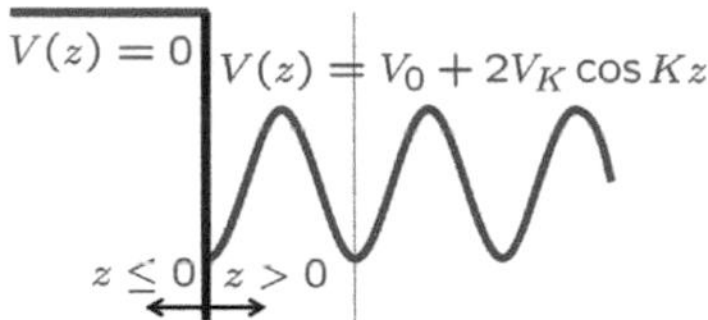

Figure 5.9: A nearly free harmonic potential perturbation picture at the surface in an approximation of the semiconductor where phase, wavevector and decay are all included together.

Let $\psi_s(z)$ be the wavefunction. At the interface $(z = 0)$, $\psi_s(z)$ and $\nabla\psi_s(z)$ are continuous because of the continuity of probability and probability current in finite energy conditions. This wavefunction must satisfy

$$\psi_s(z) = \begin{cases} \phi_v(z) \propto \exp(\kappa_v z) & \text{for } z \leq 0, \text{ and} \\ \phi_c(z) \propto \exp(-\kappa_c z)\cos[2(\pi/a)z + \delta] & \text{for } z > 0. \end{cases}$$

The rationale for writing this is as follows. When $z > 0$, inside the semiconductor, the wavefunction is evanescent and we wish to include a phase in this. The phase represents the consequence of matching and interference from forward and backward propagating Bloch functions. Together, this is represented in the exponential decay and the modulating cosine term. For $z \leq 0$, outside the semiconductor, this is an evanescent decay without any periodic propagation characteristics. Using the two boundary matching conditions, the phase follows as

$$\cos^{-2}\delta = \frac{V_0 + V_K}{\epsilon_{\pi/a}}. \tag{5.20}$$

If a solution for real energy exists, that is, for $E_{ss} = -\hbar^2\kappa_v^2/2m$, it exists with

$$\kappa_v = \frac{\pi}{a}\tan\delta - \kappa_c, \text{ and}$$
$$\kappa_c = \frac{mV_K a}{\pi\hbar^2 \sin^2\delta}. \tag{5.21}$$

This approach too gives a range of surface states of energy E_{ss} that will come about at the surface.

What we have discussed gives us a prescription for handling three dimensions. It will be more demanding since there are symmetry considerations to be included and the surface doesn't need to maintain an order that is identical to what it had within the crystal. By abruptly terminating, a surface would reconstruct to minimize its energy. The unsaturated dangling bonds will reconfigure, the atoms will move around and these changes will occur even deeper into the bulk, depending on the conditions. So, the electronic structure in general will be quite different from that of the bulk. But a surface will be periodic in its two-dimensional plane. The number of states contributing will determine the total number of states, and there will be two-dimensional bands that will form, since periodicity still exists in the plane. So, an advanced calculation of energetics will be needed where spatial order constraints will need to be relaxed. We will not do this. Our toy model has given us tools to speak to what kind of effects we might see, and that will suffice.

We will remark on these reconstructions but also discuss the surface and bulk modes of the movement of the atoms—phonons—presently.

5.4 *States at surfaces*

WE NOW EXTRAPOLATE the idealized behavior beyond the simple one-dimensional potential model, while still keeping it simple. Our interest is also in expanding our understanding of one dimension to three dimensions where the surface still has periodicity and therefore allows for propagation. We will use a sinusoidal potential in the crystal—the first Fourier term, for example, of a Kronig-Penney square potential. Figure 5.10 represents our approximation in the crystal extending onto the surface with the potential function as

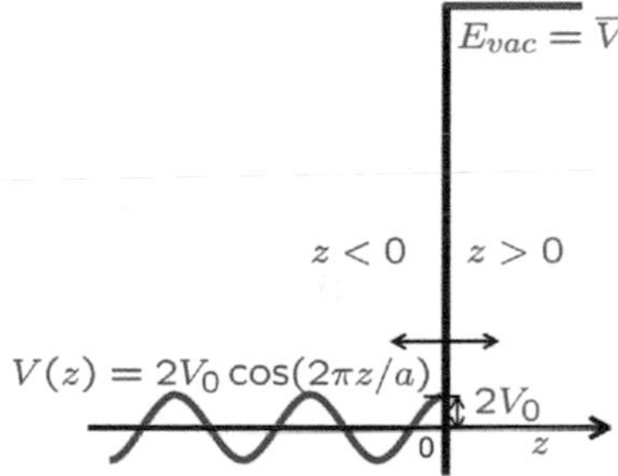

Figure 5.10: Potential $V(z)$ in a toy harmonic potential model extending to the surface.

$$\begin{aligned} V(z) &= V_0\left[\exp\left(i\frac{2\pi z}{a}\right) + \exp\left(-i\frac{2\pi z}{a}\right)\right] \quad \text{for } z < 0 \\ &= 2V_0\cos\left(\frac{2\pi z}{a}\right) \quad \text{for } z < 0, \text{ and} \\ E_{vac} &= \overline{V} \text{ for } z > 0 \end{aligned} \tag{5.22}$$

is the vacuum energy.

In bulk, the translational symmetry is

$$V(z) = V(z + na) \;\; \forall\, n = 0, \pm 1, \pm 2, \ldots. \tag{5.23}$$

With a Brillouin zone width of $K = 2\pi/a$, it is this periodicity that causes the perturbation-induced bandgap to appear at $k_\perp = \pm\pi/a = K/2$. Since the potential perturbation in the bulk is periodic harmonic, and the electrons are not too far from nearly free at the band minima, plane waves provide a good and efficient basis for the computation of the perturbation effect. The potential is a standing wave resulting from two counterpropagating electron crystal waves. Solving the Schrödinger equation, since the potential is a term that consists only of the first Fourier term of an expansion—the Bloch function, which is a Fourier function—now is relatively straightforward.

$$\psi(z) = A\exp(ik_\perp z) + B\exp\left[i\left(k_\perp - \frac{2\pi}{a}\right)|z|\right] \tag{5.24}$$

is the ansatz composed of the two-basis eigenfunctions for the bulk problem of $\hat{\mathscr{H}}|\psi(z)\rangle = E|\psi(z)\rangle$, which can be written more specifically as

$$\begin{aligned}&\left\{-\frac{\hbar^2}{2m_0}\frac{d^2}{dz^2} + V_0\left[\exp\left(i\frac{2\pi z}{a}\right) + \exp\left(-i\frac{2\pi z}{a}\right)\right]\right\}\\ &\quad\times\left\{A\exp(ik_\perp z) + B\exp\left[i\left(k_\perp - \frac{2\pi}{a}\right)|z|\right]\right\}\\ &\quad= E\left\{A\exp(ik_\perp z) + B\exp\left[i\left(k_\perp - \frac{2\pi}{a}\right)|z|\right]\right\}.\end{aligned} \tag{5.25}$$

For a solution to exist, the secular determinant—using multiplication by the complex conjugate of the two-basis eigenfunctions and integration over real space—must vanish, that is,

$$\begin{bmatrix}\frac{\hbar^2 k_\perp^2}{2m_0} - E(k_\perp) & V_0\\ V_0 & \frac{\hbar^2}{2m_0}\left(k_\perp - \frac{2\pi}{a}\right)^2 - E(k_\perp)\end{bmatrix}\begin{bmatrix}A\\ B\end{bmatrix} = 0. \tag{5.26}$$

The perturbation-induced bandgap that appears at the zone edge, that is, $k_\perp = \pm K/2 = \pm\pi/a$, where the two-basis states become asymptotically degenerate in energy, can be viewed through small changes around this point of symmetry, so we write

$$k_\perp = \frac{\pi}{a} + \Delta k, \tag{5.27}$$

and the solution is

$$\begin{aligned}E \approx\ &\frac{\hbar^2}{2m}\left(\frac{\pi}{a} + \Delta k\right)^2\\ &\pm |V_0|\left\{-\frac{\hbar^2\pi\,\Delta k}{ma|V_0|} + \left[\left(\frac{\hbar^2\pi\,\Delta k}{ma|V_0|}\right)^2 + 1\right]^{1/2}\right\},\end{aligned} \tag{5.28}$$

which shows the $\pm V_0$ splitting and the bandgap of $2V_0$ at the zone edge. So, we know the energy and the coefficients $A(\Delta k)$ and $B(\Delta k)$; therefore we know the eigenfunction

$$\psi(z) = C\left[\exp\left(i\frac{\pi z}{a}\right) + \frac{|V_0|}{V_0}\left\{-\frac{\hbar^2\pi\,\Delta k}{ma|V_0|} \pm \left[\left(\frac{\hbar^2\pi\,\Delta k}{ma|V_0|}\right)^2 + 1\right]^{1/2}\right\}\exp\left(-i\frac{\pi z}{a}\right)\right] \quad \text{for } z < 0. \tag{5.29}$$

Let us call this $\psi_i(z)$, to represent the wavefunction for $z < 0$.

What happens at the surface and beyond? The beyond is easy to answer. If $E < V_0$, this is a region where the wavefunction exponentially decays; that is,

$$\psi(z) \mapsto D\exp\left[-\sqrt{\frac{2m}{\hbar^2}(V_0 - E)}z\right] \quad \text{for } z \geq 0. \tag{5.30}$$

Let us call this wavefunction $\psi_o(z)$ for $z > 0$. As with our other discussions, the boundary conditions are continuity of ψ and $\partial\psi/\partial z$ at the interface. The former reflects the continuity of probability, and the latter of probability current (momentum)—both arising from finiteness of energy and the potential change at the interface. This form has no Δk-dependence even though inside the crystal it does. This is only possible if, for all the energies $E < V_0$, the Bloch function continuity prevails; that is,

$$\psi_o(z = 0^+) = \alpha\psi_i(z = 0^-, \Delta k) + \beta\psi_i(z = 0^-, -\Delta k). \tag{5.31}$$

Here, the two internal eigenfunctions are the two eigenfunctions from which the wavefunction comes about. The possible solution is a standing wave arising from the Bloch waves that match the exponentially decaying function across the boundary. And the standing wave that this represents arises from the Δk and $-\Delta k$ components. This matching condition must hold for all energies.

As with the earlier electron discussion, the bulk electron bandstructure exists up to the very surface with only some change. But we also obtain additional possible surface solutions that we can determine given the boundary constraints. Let

$$\Delta k = i\kappa, \tag{5.32}$$

that is, be imaginary, and, to simplify the writing,

$$\gamma = i\sin 2\delta = -i\frac{\hbar^2\pi\kappa}{maV_0}. \tag{5.33}$$

For the imaginary Δk, that is, confined modes, Equation 5.28 can remain real for a range. These have an energy of

$$E = \frac{\hbar^2}{2m}\left[\left(\pm\frac{\pi}{a}\right)^2 - \kappa^2\right] \pm |V_0|\left[1 - \left(\frac{\hbar^2\pi\kappa}{ma|V_0|}\right)^2\right]^{1/2} \quad (5.34)$$

with a wavefunction

$$\psi_i'(z) = F\exp(qz)\left\{\exp\left[i\left(\frac{\pi}{a}z \pm \delta\right)\right] \mp \exp\left[-i\left(\frac{\pi}{a}z \pm \delta\right)\right]\right\} \times \exp(\mp i\delta) \text{ for } z \leq 0, \quad (5.35)$$

and the evanescent exponential function of Equation 5.30 for $z \geq 0$. These are energy states within the bandgap. Figure 5.11 shows a conceptual view of the real part of conducting bulk states extending up to the surface, and the real part of localized surface states corresponding to $\Delta k = i\kappa$.

A commentary on workfunction is in order here. The precise definition of workfunction, as stated before, is the energy needed to remove the electron resident at the electrochemical potential energy to vacuum away from any interactions of any type with its surroundings. When we draw an energy diagram with a vacuum energy and the potentials and the bands, et cetera, simultaneously in a figure, it misses an important point, since this picture is very bulk-centric. All the description reflected in it is with the symmetries of the bulk. But an electron can only be extracted by taking out through a surface. *The workfunction contains both a bulk part and a surface part.* Effects at surfaces may not be ignored. And, here too, the traditional picture of field and image charges will have to be modified, given the importance of quantum-mechanical conditions as an electron passes through the surface.

As an electron passes through the surface, the picture one has to draw is not static. The concept of image charge must break down at quantum distance scales. This is reflected in the sequence shown in Figure 5.12 of the expectation of the electron charge density $q|\psi|^2$. An electron far away causes a minor perturbation in charge density

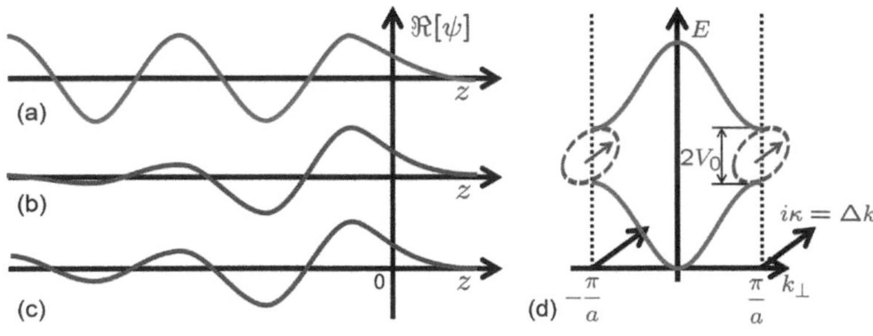

Figure 5.11: (a) A propagating state from the bulk extending up to the surface and decaying beyond. (b) A localized surface state that is in the bandgap. (c) Surface resonance is also possible as a superposition of confinement and propagation. (d) The E-**k** picture of the propagating bulk and the surface-confined states corresponding to (a) and (b).

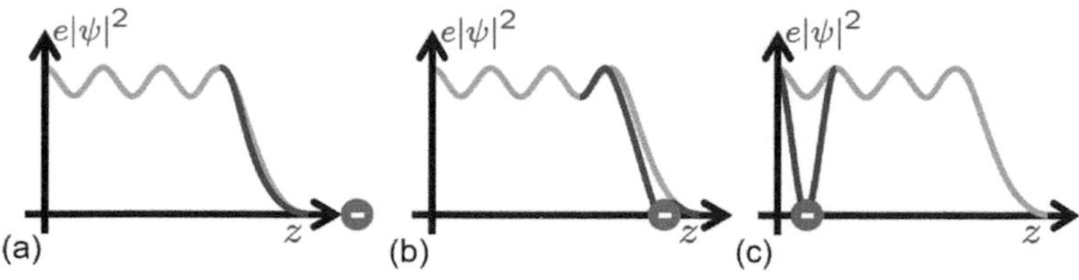

Figure 5.12: (a) An electron far away from the surface with minor perturbation of the surface. (b) An electron in the surface evanescence region pushing the resident charge into the material. (c) The electron within the bulk.

at the surface of the metal over the wavefunction evanescence scale. This is shown in Figure 5.12(a). This is the repulsion classically treated as a positive charge termination and modeled through the image charge due its identity in static solution in the vacuum region of interest. To us, this distributed repulsion is a distributed quasiparticle "hole" charge. As an electron enters this surface region, as in Figure 5.12(b), more of the charge in the region of the electron is repelled. But there is asymmetry here. There is less electron charge density toward the vacuum than in the metal. So, more of the electron charge is repelled in the metal, or, equivalently, a "hole" charge exists whose center is shifted toward the metal. *The net Coulomb energy has decreased.* Effectively, the existence of the electron in the surface region decreases the energy needed to pull it in, and, by reciprocity, in pushing it out.

In this "hole" one can see the equivalence between electric polarization and the "Coulomb hole" discussed in Chapter 1 together with the statement that an electron does not interact with itself.

The bulk bandedge picture that we draw differs. It is a picture of the long-range effect. Locally, there are changes taking place—the potential is rapidly changing—but the smooth lines describe the behavior as if all singular perturbations are smoothed out. This is as if a jellium smoothed out the high frequency or rapidly changing short-range perturbations.

Such modifications of surface are quite important. Adsorbates change workfunction. Cesiated sources—*Cs* being a very common workfunction-lowering alkali metal—are commonly employed in electron beam emission.

In the interior, the charge density approaches the bulk density. If we have a metal or a very heavily doped semiconductor, this is a degenerate material, so we know the length scale that is of import here. It is the Thomas Fermi wavelength of

$$\lambda_{TF} = \frac{2\pi}{k_F} = \frac{1}{2}\left(\frac{N}{a_B^3}\right)^{-1/6}, \tag{5.36}$$

where N is the carrier density, and a_B the Bohr radius. For Cu, with an $N \approx 8.5 \times 10^{22}\ cm^{-3}$, $\lambda_{TF} \approx 0.055\ nm$. This is miniscule—of the order of the atomic length scale—the screening is effective over an interatomic spacing as shown in Figure 5.12(c). But then, what might be a length scale of import at the surface? This is the length scale of evanescence with a workfunction barrier step of φ_w. So, the surface decay length can be characterized as

$$\Lambda = \left(\frac{\hbar^2}{2m\varphi_w}\right)^{1/2}, \tag{5.37}$$

In a nanoscale transistor—a small device—the number of dopants, for example, those that determine the threshold voltage, are limited in number. A continuum approximation of this that is executed classically will not capture the details of behavior, or the variation in device-to-device threshold voltage. Different dopant numbers and their distributions affect the local environment, and the bandedge picture does not capture it. So, nearly all properties of the transistor are affected by limited dopants in a small dimension, and the classical model ignores it. The classical model is a jellium approximation for the dopants. Effects are averaged out and drawn smoothly for their long-range behavior.

which is a *WKB* length scale approximation for the exponential decay. The Coulomb image charge lowering approximation for energy is

$$\varphi_{img} = -\frac{1}{4\pi\epsilon_0}\frac{e^2}{d} = -\frac{0.36}{d\ in\ nm}\ eV, \qquad (5.38)$$

where d is an effective distance. This image charge and the image force that it characterizes is an interaction between the electron and the surface. It is an excitation of a surface plasmon. While all this approximation through scales is limited by the quantum-mechanical-to-continuum span that it draws on, it does provide some insight into the magnitudes. The image lowering can be very significant, as Equation 5.38 indicates.

The charge density of the material will significantly affect this argument. Figure 5.13 shows this screening effect at two relative different densities. The lower density reflects a doped semiconductor, while the higher density is for a metal like *Cu*. Note that the low electron density leads to a larger amplitude perturbation within the material, and it extends deeper in.

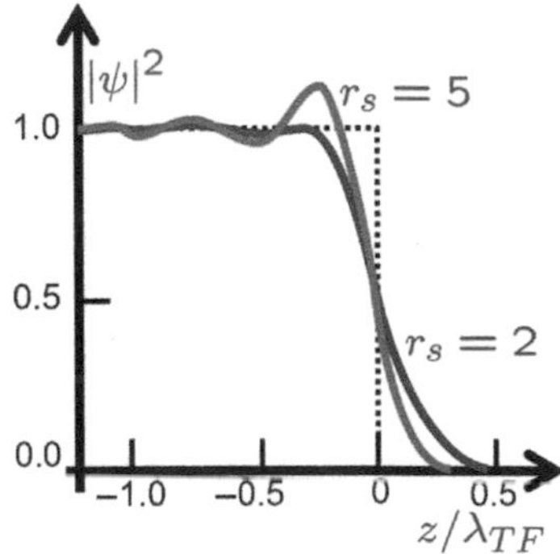

Figure 5.13: Normalized electron density distribution near the surface as the electron density changes. r_s is the Wigner-Seitz radius defined as the radius of the spherical volume per electron. Lower r_s therefore corresponds to higher electron density.

Because the screening is so short in metals—of atomic dimension—when two metals are brought together and shorted, the consequence is as described in Figure 5.14. When they are separated, we may draw them as shown in Figure 5.14, which reflects that an electron in a vacuum, independently of where it arose from, has the same reference energy. As one brings these very close and bring about thermal equilibrium by allowing the movement of particles, for example, by an electron-conducting contact separate from the interface shown, a dipolar layer appears between the two with an equilibration of the electrochemical potential. And when the two metals are in intimate contact, this dipole persists, with a contact potential given by the workfunction difference of the two materials. This is the built-in voltage such as that reflected in the conduction bandedge of Figure 5.14(c).

This rationalizes the use of workfunction differences in metal-to-metal contacts. But we will show that this picture bres down quite

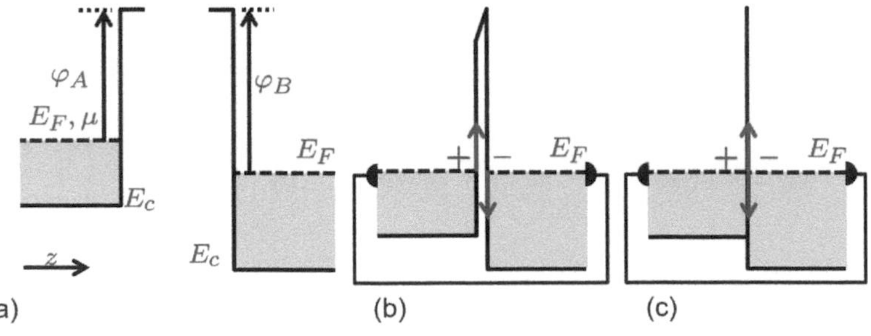

Figure 5.14: Part (a) shows two metals far apart, (b) shows them close together within atomic scales and (c) shows an intimate contact.

significantly when a general insulator and a semiconductor are in contact. The reason is that this workfunction description depended on this very dipolar—near subatomic scale—disturbance of charge at the high electron densities of metal. This is no longer true in semiconductors and insulators, and the induced states and their quantum-mechanical description will be essential. Before doing this, some comments on the energetics and order on a semiconductor surface are in order, to establish additional caveats for the idealized picture.

5.5 *Surface reconstruction*

WHAT HAPPENS TO THE ATOMIC arrangements at the surface as a result of the breaking of the symmetry of the bulk? We have expended some effort at understanding the electron states. The atomic arrangements—with the bonds and the states of the electrons, and the evolution that goes together with them—will also have an influence. We probe a little bit of this material perspective in this section.

As seen in Figure 5.15 in the unit cell of a zinc blende crystal, if one were to look at planes of different cuts, one would see a different ordering and a different set of bonds crossing the plane. If the symmetry is broken here, a lowest order inference one could draw from this bond breaking means that, first, a significant surface state contribution will arise from this perturbation, that is, the electron charge cloud realigning, and, second, that the energetics also demands that we relax the constancy constraint of the spatial ordering as one approaches the surface from inside the crystal. The electronic states are only a result of the equilibrium energetics of the

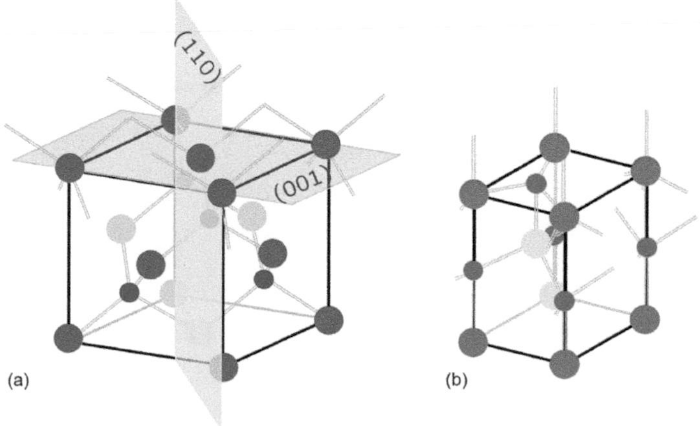

Figure 5.15: A pictorial view of the bonds across some of the cut surfaces in the two common semiconductor forms. Part (a) shows the zinc blende structure with two important cut planes: (110) and (001) and (b) shows the wurtzite form.

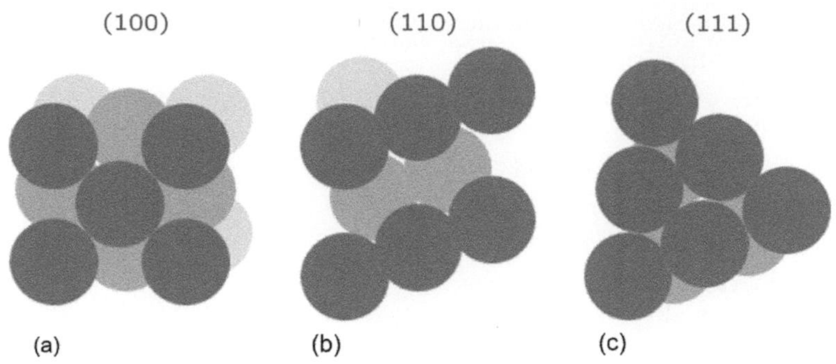

(100) (110) (111)

(a) (b) (c)

Figure 5.16: Planar cuts along the different planes of an *FCC* lattice. (100) is a face-centered square arrangement, (110) is a rectangular arrangement and (111) a hexagonal-packed arrangement—the one with the highest atomic density.

entire system of the atomic assembly, included within which are the electrons. Regarding the first consequence, we have already seen a toy approximation of the surface states. Regarding the second consequence, one reflection in our toy model was the raising of the potential at the surface, as illustrated in Figure 5.2, but a more accurate quantum "toy" calculation should also have included the energetics in a way where spatial variations were incorporated self-consistently.

The atomic arrangement across these planes for the *FCC* lattice is shown in Figure 5.16. One may see a different density of atoms, so each plane, in addition to the bonding, also exhibits different atomic density. The final arrangement of these different surfaces, depending on the conditions under that they are formed, will reorder themselves with atomic movement in plane and out of plane. (111), (110) and (100) is the order of the highest to lowest atomic densities, and on average, by symmetry in this three-dimensional arrangement, one must expect a similar order in the surface state densities due to unsatisfied bonds unless a specific reordering that satisfies bonding leads to an unexpected energy minimization. It doesn't seem to. But the surface reconstruction does make a difference in this density-of-atoms argument as well as in energetics through the surface state positioning and characteristics.

We explore the surface behavior and reconstruction through some classical and quantum thoughts.

Surface stress, which is the tension of the surface, is a consequence of the lowering of energy when the surface or interface is created. The macroscopic nature of the problem defines this surface/interface optimization in the continuum description of classical mechanics. One can mathematically follow through on this using thermodynamic arguments based on Gibbs free energy. In this description, which should work well for macroscopic conditions such as those in a water droplet on a surface, one needs to introduce some correction to account for the orientation dependences—anisotropy—that arise in the nature of a crystalline atomic structure. This surface energy, the surface tension and the details of the response of the interface will connect to this orientation dependence.

In the continuum description of elasticity, we may analyze through the stress and strain tensors. Let $\sigma_{ij} = \partial F_i/\partial a_j$ be the ijth stress tensor element arising in a differential force dF_i in the ith direction acting on an area element da_j that is oriented in the jth direction. Likewise, let $\varepsilon_{ij} = \partial u_i/\partial x_j$ be the ijth strain element with a du_i change in length in the ith direction for the volume element, and dx_j the position change in the jth direction that caused this length change to happen. In each of these cases, the other variables causing a change are kept constant. In the atomic arrangement at the surface, the electrons respond to the absence or change of arrangement by a change in distribution, bonding, et cetera, together with the repositioning of the atoms in the presence of forces that come about in this process of energy reduction at the surface or interface. The top layer—and the ones below, too—change, so the stress tensor will vary with position.

We account for this positional dependence of the stress σ, as reflected in the Figure 5.17 between the surface and the bulk, for the surface stress as a cumulative consequence from the bulk to the surface:

$$\sigma^s_{ij} = -\int_{-\infty}^{\infty} \left[\sigma_{ij}(z) - \sigma^b_{ij} \right] dz. \tag{5.39}$$

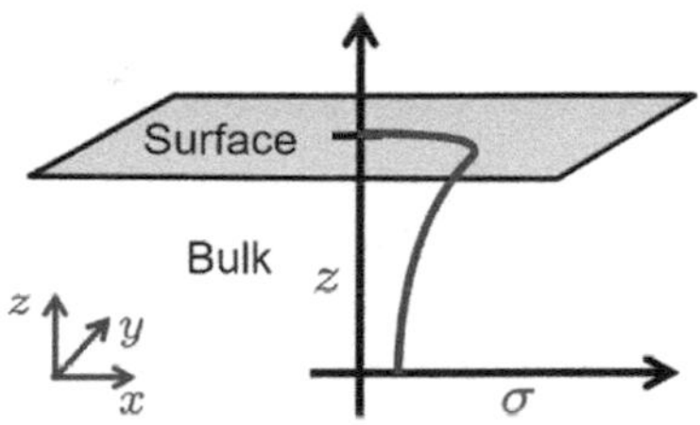

Figure 5.17: Building up of stress at a surface as a result of broken symmetry. Compressive stress that prevents the stretching out of the surface is shown here.

Note that surface stress is different in units from the bulk—it has an additional length arising in this accumulation. So, it has a different dimensionality, as is the case with other parameters that we think of when comparing volume versus surface. This picture reflects the following thinking. The surface-induced stress—a force per unit length—arises as an accumulation of stress—force per unit area—from the bulk. It exists in a thin region near the surface and disappears in the bulk. Figure 5.17 shows a negative stress condition. The surface wants to pull apart, and the stress exists to limit this expansion. This compressive stress is needed to regain a dimension similar to that of the bulk. A surface that contracts under its own stress has a positive stress. To stretch it, one needs to apply a tensile stress. The order of magnitude of these surface stresses are an N/m stretching out over a length scale of a nm, so 2 to 3 unit cells—a significant inversion-layer scale number.

Figure 5.18: A classical representation of a thin plate under planar shear stress.

The energy argument can now be drawn from this formulation. To place a plate, such as that shown in Figure 5.18, with an area A and thickness t, in a strained condition of ε_{ij}, the work needed to be performed is

$$\delta W = A \int_{-t/2}^{t/2} \left[\sum_{ij} \sigma_{ij}(z) \delta\varepsilon_{ij} \right] dz. \tag{5.40}$$

This can be broken up into the bulk and the surface parts of energy change:

$$\begin{aligned} \delta W &= \delta W^s + \delta W^b \\ &= 2A \sum_{ij} \sigma^s_{ij} \delta\varepsilon_{ij} + At \sum_{ij} \sigma^b_{ij} \delta\varepsilon_{ij} \\ &= \delta\mathcal{G}^s + \delta\mathcal{G}^b. \end{aligned} \tag{5.41}$$

If we define the areal surface energy in terms of a specific free energy γ—a free energy per unit area—one can write the surface Gibbs free energy term as

$$\delta\mathcal{G}^s = \delta(\gamma A) = \gamma\delta A + A\delta\gamma. \tag{5.42}$$

The surface energy change arises in these two causes. One part of the change is due to the surface area's energetic change effect. This is a part where, while the average area per atom is fixed with a specific free energy constant, the effective surface atom's numbers have changed. The second is due to the change in the surface atomic interactions causing change in their energy interaction contribution. This is the part due to surface reconstruction. So, both the change in surface atom numbers and their interatomic realignment lead to surface energy change and are included in $\delta\mathcal{G}^s$.

From Equation 5.41, where the first term is due to the surface and the second due to the bulk,

$$\delta W^s = 2A \sum_{ij} \sigma^s_{ij} \delta\varepsilon_{ij} = \gamma\delta A + A\delta\gamma, \quad \text{with } dA = A \sum_i d\varepsilon_{ii},$$

$$\therefore \quad \sigma^s_{ij} = \gamma + \frac{\partial\gamma}{\partial\varepsilon_{ij}}. \tag{5.43}$$

This is the Shuttleworth equation. In a liquid, because there is little resistance to atom flow at the surface or in the bulk, the second term vanishes. In a solid, how the atoms may move on the surface will depend on the specifics of the surface energy and surface stress in the proximity. What drives this movement is how this second term relates the stress tensor of the surface with the specific free energy and the strain. The driving force is

$$\frac{\partial\gamma}{\partial\varepsilon_{ij}} = \sigma^s_{ij} - \gamma \tag{5.44}$$

for the atom movement between the surface and the bulk. If $\sigma^s_{ij} - \gamma > 0$, atoms are driven to accumulate at a higher density at the surface. For the opposite sign, that is, $\sigma^s_{ij} - \gamma < 0$, atoms are less dense at the surface. γ—the specific surface free energy—is like

This equation is the simplest form describing the energetics for the purposes of this discussion. Take bubble formation in drinks. One immediately notices that bubbles do get seeded on the walls when beer is poured. But, this is not the complete story. Bubbles also appear in the liquid. How and where depends on the specifics, including those of the dissolved gases. Take a liquid oversaturated with gas (usually CO_2). Any asperity of surface lowers the energy at the gas-liquid interface, nucleating bubbles. A bubble-liquid interface, due to the surface tension, also gives rise to a pressure (the Laplace pressure) in the bubble. When bottled, the overpressure leads to the dissolution of the bubble. There exists a critical bubble radius and overpressure relationship. Uncorking, or pouring to create an interface, creates the bubble. Diffusion of CO_2 changes the bubble's size. Different sparkling wines—champagne, prosecco or cava—can have different enough bubble sizes that one notices them in the mouth. The growth of crystalline semiconductors proceeds via edges for this same energetics reason. If the growth occurs without the surface or edge templating, it loses crystallinity. The crystallization and amorphization used in optical disks also depend on this nucleation, which can be surface mediated or self-nucleated. These equations lead to taste and usefulness.

an excess free energy per unit area. It is the reversible work per unit area with other thermodynamic variables constant. *The surface exists in thermodynamic equilibrium because the work performed in creating the surface leads to an energy exchange between the surface component and the bulk component.*

In a liquid-solid interface, hydrophobicity, hydrophilicity or the specific shapes that appear are a consequence of the differences in this specific surface energy of the liquid and that of the liquid where it is in contact with a solid. Atoms are free to move around both at the surface and in the bulk, leading to the specifics of the shape. The specific surface free energy arises in the energy cost of creating additional surface while keeping the volume and the number of atoms constant. Bonds must be broken to expose new atoms at the surface. Surface defects, steps, et cetera all involve new surface areas and have energies associated with them. So, growth, defect generation during growth and three-dimensional, two-dimensional or one-dimensional growth will all relate to these considerations. And since atomic arrangements matter, orientations will matter. This is a simple description and will be subject to numerous complications. Polar surfaces with their electroenergy contributions, charge compensations therein, et cetera all will have a noticeable effect.

A direct small-dimensional manifestation of this surface-bulk tension is the nature of shapes of objects such as quantum dots. The equilibrium shape of a crystal is not necessarily a minimum surface area. It may be a complex shape. It is a shape where $\int_S \gamma(\hat{\mathbf{n}})\, d^2r$ is a minimum. Compound semiconductor growth often shows pyramidal forms because the (111) surface has the highest atomic density, as noted in Figure 5.16, and bonding linkages.

Figure 5.19: A scanning tunnel microscope observation showing the quantum corral when a closed circle was created by placement of *Fe* atoms on a (111) *Cu* surface. This figure is after M. F. Cromme, C. P. Lutz and D. M. Eigler, "Confinement of electrons to quantum corrals on a metal surface," Science, **262**, 218–220 (1993).

All the potential interactions will matter, and in complexity with wave mechanics central to it, this classical picture will break down. For example, the quantum corral (see Figure 5.19), created by placing *Fe* atoms on a (111) *Cu* surface, leads to a lowest mode wave pattern observation arising in the potential interaction of this electron-ion assembly on the surface. These are off-plane confined electron states on the surface that also form a standing wave pattern. So, they are confined states in the plane too, because of the bounding by *Fe*.

Reconstruction takes rather complicated forms. Even a simple cubic form—such as the salts—will show a number of features. Figure 5.20 shows a set of simple possibilities looking in a plane perpendicular to the surface through the unit cell surface crystal plane of a simple cubic arrangement. Atoms on the surface may relax to a position closer to the next plane of atoms, as in Figure 5.20(a). Or they may achieve a "Peierls"-like instability by

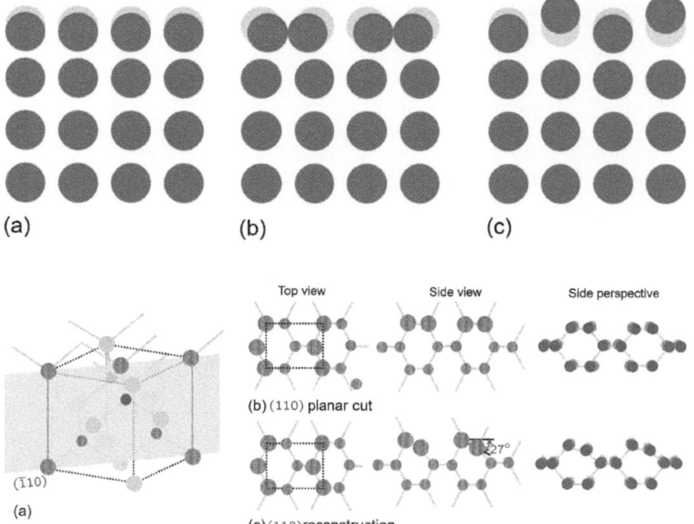

Figure 5.20: Example toy reconstructions of a cubic arrangement. Part (a) shows displacement where the top layer is pulled closer to the next plane, (b) shows where atoms shift forming pairs closer together with the pairs farther apart and (c) shows a reconstruction with alternating shifts of closer and farther for the atoms.

Figure 5.21: Surface reconstruction of (110) *GaAs*. The non-reconstructed view is on the left, and the relaxed (1×1) reconstruction is shown on the right. Part (a) shows the cut, (b) shows the top and side views and the side perspective and (c) shows the changes due to the (1×1) reconstruction.

forming closer pairs that are farther apart, as in Figure 5.20(b). Or they may reconstruct by moving atoms out of the plane and so have missing rows, as in Figure 5.20(c). It will depend on energetic conditions such as those of the quantum-mechanical boundary constraints at temperature T, thus reflecting thermal implications too. Adsorbates, if any, will also modify surface energies, so ambient conditions have an effect.

Much of what we discuss to emphasize the importance of the surface is based on observations in reproducible conditions of ultra high vacuum. So, we must keep this mind in drawing any conclusions regarding practical conditions.

Zinc blende and diamond crystals, such as many of the compound semiconductors, are considerably more complicated, with a variety of reconstructions under different conditions of temperature, ambient and surface preparation. *GaAs* cleaves quite easily in the (110) plane. So, historically, it has been among the easier surfaces to analyze. Figure 5.21 shows the surface reconstruction of (110) *GaAs*. The (1×1) mesh of the top view is identified by the dashed rectangle. Here, Figure 5.21(a) shows one of the (110) plane cuts of a zinc blende crystal so that one may visualize the atoms in the plane and those below the plane. This ideal cut view is shown for a normal view to it, a side view to it and a perspective view to it. Bond angles are easier to change than lengths. The reconstruction that follows is shown in the views of (c).

This angle-spacing argument is reflected in the exchange and repulsion argument. Carbon has both an sp^2 and an sp^3 hybridization in its crystalline form. Angular changes are easier. It is one of the reasons SiO_2 is a very low interface state density non-crystalline insulator in SiO_2/Si-grown arrangements.

The relaxed top view shows a pair of *Ga* and *As* atoms coming closer together. The averages must remain constant, so the next set is an another closer pair that is the same, with the next view set moving to a similar average distance. The real space periodicity on the surface is now twice as long in this direction, and so the Brillouin zone half as long. This is a broken symmetry of the same variety that causes "Peierls instability" leading to metal-insulator transitions. This is the (1×1) reconstruction of *GaAs*.

Silicon's (100) surface is the nearly ubiquitous form in use. As the surface undergoes quite a variety of reactions and treatments in the process of being put to use, a good understanding of it is very desirable. Figure 5.22 emphasizes the dimerization that is the principal effect. A (100) surface in the diamond and zinc blende system consists of parallel rows of atoms, where the two bonds reaching out of the surface are broken. Adjacent row atoms are close. So, the broken bonds from each adjacent atom dimerize together with a distortion which costs a deformation energy. The minimum energy surface is reflected in Figure 5.22(d).

The *Si* (111) surface is another surface of interest to emphasize the variety as well as complexity. When cleaved at room temperature, it reconstructs with a (2×1) pattern. The accepted reconstruction is shown in Figure 5.23. This is an example of deformation reaching down to at least the fourth atom layer. Figure 5.23(a) shows the top view where we see the $\pi/3$ rotational symmetry with different atoms in different planes. The (2×1) pattern on the surface reconstructs with a π-bonded chain, as shown in Figure 5.24. Atom pairs pull together on the surface within the (2×1) pattern; an sp^2 hybridization occurs in plane forming one bonded pair, and the p_z forms the second bonded

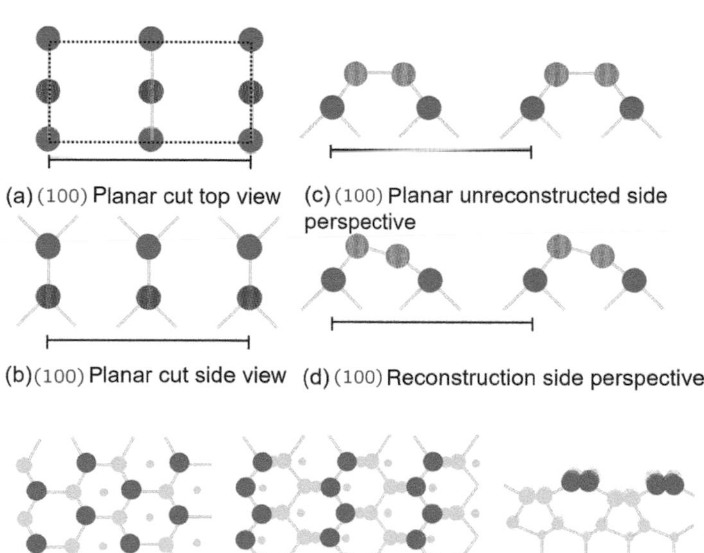

Figure 5.22: Surface reconstruction at room temperature in vacuum of the *Si* (100) surface. The unreconstructed views for the top and the side are shown in (a) and (b), respectively. Part (c) shows the planar side perspective of the unreconstructed view and (d) shows how the dimerization of the top layer atoms leads to changes in the bond lengths and angles. Note that this is a relaxed (2×1) surface unit.

(a) Top view (b) Side view (c) Side view perspective

Figure 5.23: Surface reconstruction at room temperature in vacuum of the *Si* (111) surface. The atomic arrangement in the planes orthogonal to the ⟨111⟩ is shown in (a). The size of the atoms is shown smaller deeper in from the surface. Part (b) shows a π-bonded chain causing the distortion on the surface, and (c) is the side view as atoms shift both in plane and out of plane.

pair. The distortion of the other sp^2 hybridized orbitals reach down into the plane below, with a change from the previous sp^3 hybridization. The distortion propagates multiple layers, as is shown in Figure 5.23 through the changes in size of atoms.

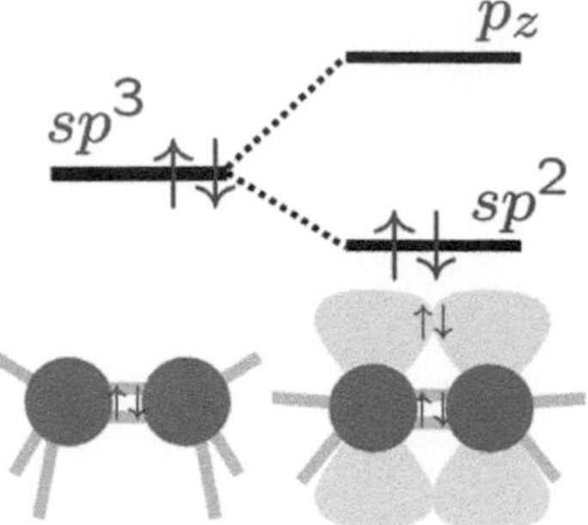

Figure 5.24: Surface reconstruction at room temperature in vacuum of the *Si* (111) surface through the change of extreme surface atom pairs to sp^2 hybridization with a p_z orbital. One of the sp^2 orbitals and one of p_z orbitals form the double bond between the surface atom pair.

If *Si* is annealed at an elevated temperature—650 *K* suffices—one observes a gold standard of reconstructions. It is a 7×7 reconstruction, which was tour de force in calculations in surface sciences since it involves movement as deep as four atom layers down and a large surface area. The primitive cell shown in Figure 5.25 now has 98 atoms. It consists of two triangular units, with a stacking fault in one of them. Figure 5.25(a) shows the placement of surface atoms, (b) shows view orthogonal to the surface plane a number of atomic layers down and (c) shows a scanning tunneling probe picture of the surface reconstruction.

It also was one of the first clear demonstrations of the power of the then new technique of surface tunneling microscopy, showing the mapping of surface tunnel current distribution that aligns with the charge distribution.

This discussion, classically of the surface forces, and of surface reconstruction as a manifestation of the energetics when symmetries are broken and the environmental conditions changed quite well, points out all the complexities and difficulties that the toy model sidesteps. But, they also point to us that we will see a rich set of behavioral outcomes in real situations.

Surface reconstruction is a place where we have now encountered a movement of the atoms and nuclei in a newer set of equilibrium positions.

But existence of finite temperature means that the nuclei, in the bulk and at the surface, also move around an equilibrium position. Electrons too move around, although the electron movement is a faster response compared to that of the nuclei. The electron movement will be more rapid since electrons are lighter and more responsive. Movement of the nuclei and the bound electrons—the core—creates an electron probability distribution that has a higher energy. The assembly forms a collection where there is a constant exchange of energy, whose lowest order approximation is that of a harmonic oscillator, where energy is being exchanged. The higher energy of the electron distribution is returned to that

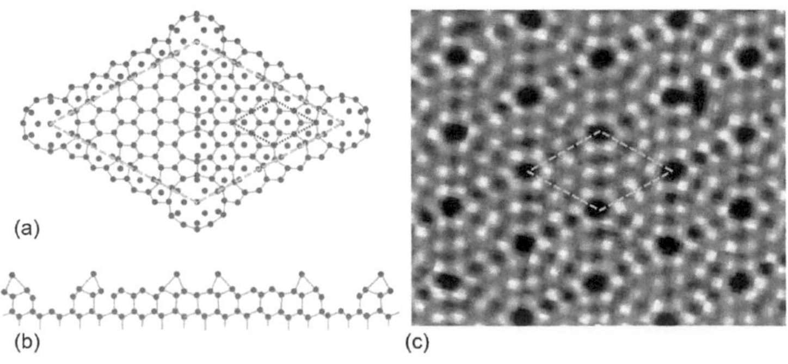

Figure 5.25: Surface reconstruction of (111) *Si*. Part (a) shows a top view of the placement of atoms of the smallest surface unit cell. It has a (7×7) arrangement of these ordered geometries with large and small rearrangements over a large number of lattice sites. Part (b) shows that the arrangement arises with the movement of many atoms perpendicular to the surface plane. Part (c) shows a scanning tunneling microscopic view where the diamond corresponds to the diamond drawn in (a).

of the nuclei/core when they return to their previous state—the equilibrium state. Neither the electron nor the nuclei core remain in their excited conditions. They shuffle back and forth.

In this description, the total electron energy is the potential for the movement of the nuclei/core.

A fast electronic movement compared to that of the nuclei can be interpreted as electron dynamics based on a static lattice where the nuclei/core, that is, a static nuclei/core, determines the potential for the electrons, and one may treat electron dynamics and nuclei/core dynamics as being separate and non-interacting.

This was our adiabatic approximation introduced in Chapter 1.

Adiabatic approximation is also the Born-Oppenheimer approximation which is employed in molecular description, where both the atomic nuclei and the electron charge cloud have to be tackled simultaneously. Adiabatic approximation can break in the solid state and needs to be corrected for as the description gets deeper. An example is electron scattering by the crystal—phonons, et cetera—where movement of the atoms causes perturbation that is the cause of scattering—a significant change in **k**, E or both for the electron.

To draw parallels with the electron picture at the surface, we will look at the behavior of this atomic movement. Crystal vibrations near the surface will be different than in the bulk. Surface phonons modes will be different from bulk phonon modes, in analogy with those of the electrons.

5.6 *Surface phonons*

As WITH ELECTRON MODES, crystal vibration modes will be different near the surface than in the bulk. Surface phonons will have specific properties distinguishable from their bulk brethren, and we are curious about these distinguishing, interesting characteristics. One reason is that electron transport is affected by energy loss through scattering via these modes; that is, the scattering from it. Another reason is that phonons are an important mechanism for heat transfer. Heat transfer is significantly affected by interfaces, just as electron current is significantly affected—and gainfully employed—in electron transport at interfaces. We will use a one-dimensional model to start with, to extract the essentials arising in breaking of the translational symmetry.

We take the three-dimensional picture of the surface with crystal vibrations allowed in three directions in Figure 5.26(a) and first

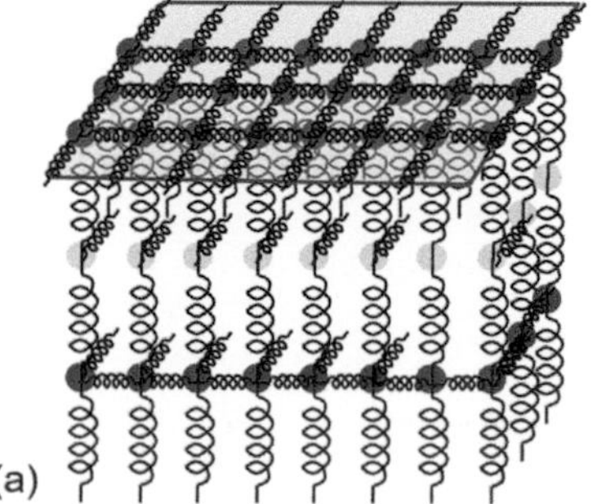

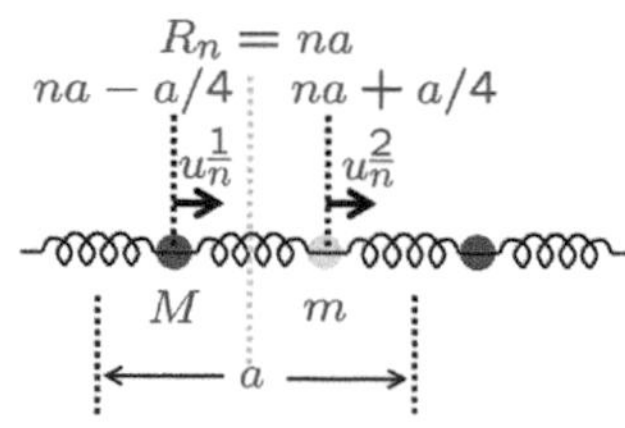

Figure 5.26: A surface of a two-basis crystal idealized to have coupling restricted to the orthogonal direction from the surface plane.

idealize it to the one-dimensional form of Figure 5.26(b). We have assumed that there is no coupling in plane, but we are interested in a two-atom basis. The infinite two-atom linear chain problem was solved in Equation 3.107 to show the physical basis for bulk infinite structures.

To solve for this surface instance, we employ an ansatz based on the bulk mode solution. With

$$u_n^1 = \sqrt{M}c_1 \exp\left\{i\left[qa\left(n-\frac{1}{4}\right)-\omega t\right]\right\}, \text{ and}$$
$$u_n^2 = \sqrt{m}c_2 \exp\left\{i\left[qa\left(n+\frac{1}{4}\right)-\omega t\right]\right\} \qquad (5.45)$$

as the atomic displacements, the solution of the underlying prescription of motion—harmonic and with only next neighbor interaction, as in Equation 3.105—is a solution form

$$-\omega^2\sqrt{M}c_1 = -k_s c_1 M^{-1/2} + 2k_s c_2 m^{-1/2}\cos(qa/2), \text{ and}$$
$$-\omega^2\sqrt{m}c_2 = -k_s c_2 m^{-1/2} + 2k_s c_1 M^{-1/2}\cos(qa/2). \qquad (5.46)$$

This also holds for the bulk solution, where we found that the continuous eigenenergy solution for the infinite chain (Equation 3.107) was

$$\omega_\pm^2 = \frac{k_s}{mM}\left\{(m+M) \pm \left[(m+M)^2 - 2mM(1-\cos qa)\right]^{1/2}\right\}. \qquad (5.47)$$

We are using this formulation and its solution now as the ansatz for the condition when the chain is terminated at a point—the surface.

We recognize that, far away from the terminated end, the solution must approach this infinite chain solution. If there is anharmonicity of interactions, they will have finite correlation lengths. This is a reason why at the surface we should expect solutions to be localized with amplitudes decaying away. This is to say that the wavevector $\mathbf{q}$ of the surface mode is complex. So, we start with the form

$$q = q_r + iq_i \qquad (5.48)$$

in our energy or frequency solution Equation 5.47 to find the implications. We find that the constraint for wavevectors is

$$\cos qa = \cos(q_r a)\cosh(q_i a) - i\sin(q_r a)\sinh(q_i a). \qquad (5.49)$$

The energy and frequency must be real. So, the imaginary part must vanish, that is,

$$\sin(q_r a)\sinh(q_i a) = 0 \qquad (5.50)$$

must hold.

The constraint of Equation 5.50 forces two possibilities. One is that $q_i = 0$. This is the bulk dispersion solution. *Bulk modes exist all the way to the surface, just as they did for the electrons.* The second possibility is that $q_r a = n\pi$ for $n = 0, \pm 1, \pm 2, \ldots$. This can be introduced into the solution of the form

$$\cos qa = \cos(n\pi)\cosh(q_i a) = (-1)^n \cosh(q_i a) \quad \text{for} \quad n = 0, 1. \tag{5.51}$$

For this condition, the eigenfrequency solution takes the form

$$\omega_{\pm}^2 = \frac{k_s}{mM}\left\{(m+M) \pm \left[(m+M)^2 - 2mM[1-(-1)^n \cosh q_i a]\right]^{1/2}\right\} \quad \text{for} \quad n = 0, 1. \tag{5.52}$$

For $n = 0$ here, $q_r = 0$. This is the Γ point of Brillouin zone. Under this condition, the frequency solution within the inside brackets within the root term can be simplified. The hyperbolic term that appears with $2mM$ simplifies to $(1 - \cosh q_i a)$, which is always negative ($\cosh() \geq 1$), so, for real frequency, one can take only positive roots. At $q_r = 0$, this reduces to the bulk optical branch. Now, if q_i is allowed to vary, one obtains an energy above. This is the zone center additional high energy branch with q_i varying shown in Figure 5.27. Note that there is no restriction on what q_i may be for this branch.

For $n = 1$ in Equation 5.52, that is, $q_r = \pm\pi/a$—the first Brillouin zone boundary—the condition for obtaining a real square root consistent with a real eigenfrequency is

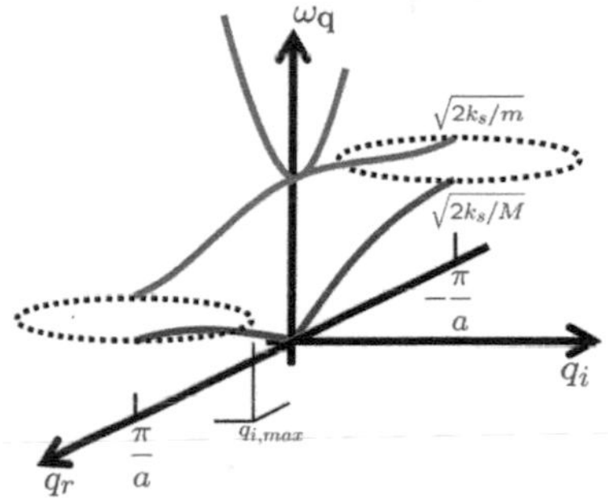

Figure 5.27: Phonon mode solutions for a two-atom basis linear chain. The bulk solutions are all real wavevectors shown in the acoustic and optical branch. Surface modes exist at $q_r = 0, \pi/a$, as shown.

$$|q_i| < \frac{1}{a}\operatorname{arccosh}\left(\frac{m^2+M^2}{2mM}\right) \equiv q_{i,max}. \tag{5.53}$$

This is an evanescent solution that exits at the zone boundaries, as shown in Figure 5.27, and the eigenfrequencies are

$$\omega_{\pm}^2\left(q_r = \frac{\pi}{a}, q_i\right) = \frac{k_s}{mM}\left\{(m+M) \pm \left[(m+M)^2 - 2mM(1-(-1)^n \cosh q_i a\right]^{1/2}\right\} \quad \text{for} \quad |q_i| < q_{i,max}. \tag{5.54}$$

At $|q_i| = q_{i,max}$,

$$\omega_{\pm}^2\left(q_r = \frac{\pi}{a}, q_{i,max}\right) = \left[k_s\left(\frac{1}{m} + \frac{1}{M}\right)\right]^{1/2}. \tag{5.55}$$

All these solutions are drawn in Figure 5.27. At the zone edge, there exist evanescent—confined to the surface—modes with a specific $q_{i,max}$ that is determined by the masses and the periodicity of the chain. At the zone center, there exist high energy modes—higher than the bulk optical frequency, where the q_i is unconstrained. Recall, though, that, the higher the q_i, the more rapidly it decays away from the surface. So, these are all confined modes. The only difference is that, at the zone edge, one has modes across the gap—akin to the surface states for electrons in the gap, and, at the zone center, there are surface modes that are unrestricted in energy above that of the bulk optical mode. So, to reemphasize, at the Γ point, surface modes exist above the maximum bulk phonon frequencies. And, at the zone edge, surface phonon modes exist within $|q_i| \leq q_{i,max} = (1/a)\text{arccosh}[(m^2 + M^2)/2mM]$. The surface modes are quite rich!

Having solved for eigenenergies, we may also write the displacement eigenfunctions. The displacements are

$$u_n^i = C_i \exp\left[i(qR_n^i - \omega t)\right], \quad \text{where}$$
$$R_n^i = \begin{cases} = a(n - 1/4) & \text{for } \underline{i} = 1 \text{ (atom 1), and} \\ = a(n + 1/4) & \text{for } \underline{i} = 2 \text{ (atom 2).} \end{cases} \tag{5.56}$$

These are all vibrations of the form

$$u_n^i \propto \exp\left(-q_i R_n^i\right) \exp(-i\omega t) \tag{5.57}$$

that decay away from the surface.

We can now outline how to generalize this by using what we learned from the electron problem. If surface bonds are weak, and, for most anisotropic materials, this weak parallel array description is a good approximation, we may look upon the behavior as a correlated response with vibrations in plane and out of plane. Since symmetry is not broken in plane, these vibration modes may be propagating. The different propagation modes, due to correlations, will be related through a phase difference. So, $\mathbf{q}_\parallel$ and $\mathbf{q}_\perp$ may be connected to each other, and one may write the displacements as

$$u_{\mathbf{q}}(\mathbf{r}) = A\hat{\mathbf{q}} \exp\left[i\left(\mathbf{q}_\parallel \cdot \mathbf{r}_\parallel + q_\perp z - \omega t\right)\right]. \tag{5.58}$$

Here, now parallel surface plane modes with real $\mathbf{q}_\parallel$ are possible. Also, perpendicular modes may be propagating (real $\mathbf{q}_\perp$) or evanescent (imaginary $\mathbf{q}_\perp$). This is captured in the displacement function

$$u_{\mathbf{q}_\parallel, q_\perp}(\mathbf{r}) = A\hat{\mathbf{q}} \exp(iq_{\perp,r} z) \exp(-q_{\perp,i} z) \exp\left[i\left(\mathbf{q}_\parallel \cdot \mathbf{r}_\parallel - \omega t\right)\right] \tag{5.59}$$

in a correlated response in primitive unit cells with both propagating and evanescent solutions included.

In the three-dimension picture, on the surface, one will have to describe these surface phonons in a two-dimensional Brillouin zone. This is the orthogonal cut of the plane. When we draw the bulk phonon modes, we show bands akin to the energy band across which electron energies are allowed in different bands, and we follow a path in directions represented by the symmetry points. Since surface phonon modes exist both in the gap and in the band, we will, in general, see a picture of allowed modes corresponding to this. The energies, the frequencies $\omega_{\mathbf{q}}$, the propagating wavevector $\mathbf{q}_{\parallel}$ and the evanescent wavevector perpendicular to the surface $\mathbf{q}_{\perp,i}$ are all related to each other. The solution is a consistent set from the "continuous" spectrum of possibilities between the surface and the bulk. Both the optical and acoustic modes and the evanescent modes exist in this midst. And, within this, the surface phonon—evanescent—modes have a two-dimensional spectrum.

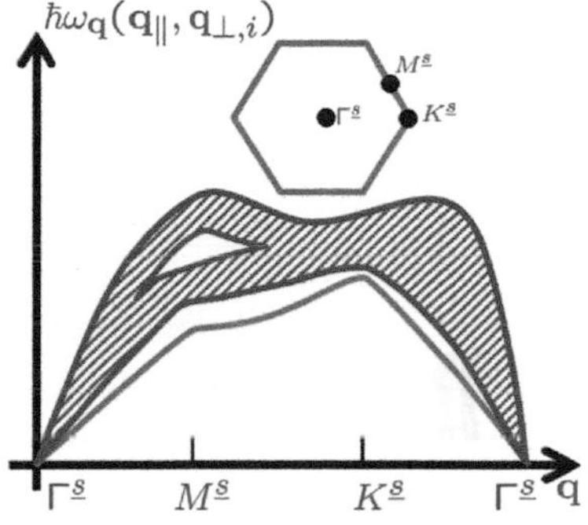

Figure 5.28: Bulk and surface phonon dispersion for an *FCC* crystal on the (111) surface.

Figure 5.28 shows an example of the nature of this mode dispersion for a (111) *FCC* bulk and surface. The surface modes exist, and so do bulk modes. If one took a line at any symmetry point, this shows the range of energies that are possible for all possible $\mathbf{q}_{\parallel}$ and $\mathbf{q}_{\perp,i}$ on that symmetry line. The phonon dispersion is drawn here as one follows the symmetry path $\Gamma^{\underline{s}} \rightarrow M^{\underline{s}} \rightarrow K^{\underline{s}} \rightarrow \Gamma^{\underline{s}}$. The two-dimensional surface Brillouin zone symmetry points can be identified through the s superscript and by viewing the appropriate cut of the three-dimensional 1st Brillouin zone. Figure 5.29 shows this picture as examples of different Brillouin zone forms for different surfaces of the *FCC* crystal.

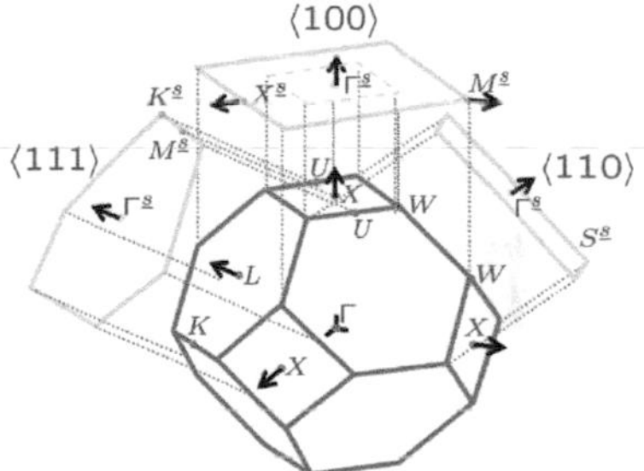

Figure 5.29: The first surface Brillouin zones for (100), (110) and (111) in an *FCC* crystal.

Real crystal dispersion curves are, of course, considerably complicated. This combination of surface and bulk projected to the surface phonon dispersion for (110) *InP* and *GaAs* is shown in Figure 5.30. The lines are the surface modes, while the hatched

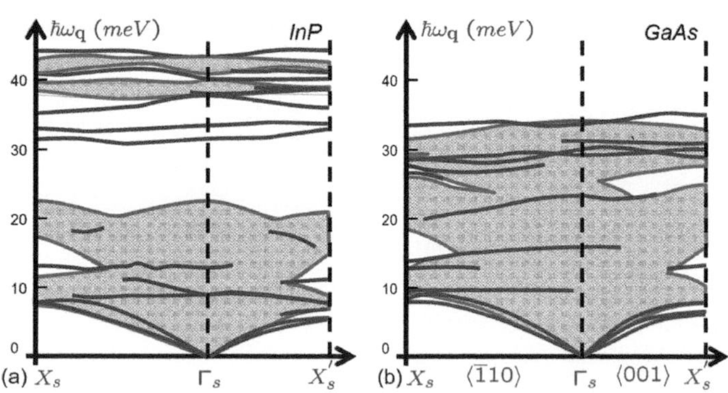

Figure 5.30: Surface phonon dispersion (with bulk phonons as shaded background) for (110) *InP* and *GaAs*. Modes with energies appear in the gap and in the optical bands. Adapted from J. Fritsch, P. Pavone and U. Schröder, "Ab initio calculation of the phonon dispersion in bulk InP and in the InP (110) surface," Physical Review B, **52** 11326-11334 (1995).

region shows the projections of the bulk mode. One can see the surface acoustic and optical branches around approximately the same region as the bulk modes. But, one can also see that there are a number of limitations to them. The surface acoustic branches certainly exist quite close to the bulk branches. There are also optical branches higher up in energy, but also restricted branches at lower energy.

As discussed in Chapter 3, we end up with $3N - 6$ modes for N atoms coupled in diatomic basis. These are the phonon modes that are represented in the dispersion relations. Longitudinal modes correspond to ones when the displacement is aligned with the axis, and transverse when the displacement is perpendicular. When the displacement phase is slowly varying, that is, successive displacements are relatively in phase, these are acoustic modes. When the displacement phase changes rapidly, such as when there are successive displacements opposite in sign, for example, then these are optical modes.

Having discussed, in a rather simple way, phonons at surfaces, we now return to understand the behavior at interfaces in Chapter 6, and in particular, the behavior of electrons at surfaces, given what we have observed in our "toy" models and the complexity of the real world.

5.7 *Summary*

Surfaces tell us much about black holes. Look up discussions of information, entropy and black holes, including controversies therein.

SURFACES ARE VERY DIFFERENT, and much happens in the surface region that evolves from and yet represents what happens in the bulk. Surfaces are essential to any coherent description of the nanoscale. Surfaces and interfaces are how we connect any microsystem to its surrounding—the environment—of the real world. This has been the focus of this chapter, where many essential points come across as a consequence of breaking of the symmetry of the bulk.

Workfunction and electron affinities are common measures employed in describing a material's electrochemical or limit energies of the conduction or valence bands. In conductors, the workfunction is the measure that one will see employed to remark on the built-in potential between materials, which is balanced by the opposite balancing term during a measurement of the potential difference. When the balancing junction and the measured junction are different in temperature, a voltage is measured that reflects the difference in temperature. This is the basis of the operation of thermocouples. It works, because minor errors arising in surfaces and interfaces balance out. *Workfunction or electron affinities are, however, strictly the properties of bulk.* And any such property defined by the energy required to remove an electron from a level in the material to vacuum also involves a removal through the surface. Surfaces matter. A good example of this error's consequence is the use and misuse of the metal-semiconductor workfunction difference in plain metal-semiconductors, metal-oxide-semiconductor systems or polysilicon-oxide-semiconductor systems, where it is a tool to add the correction term beyond the flatband voltage. Different surfaces will give rise to different threshold voltages because of this surface dependence, which a workfunction model will err at.

Even in the photoelectric effect—Einstein's Nobel—the photon dislodging the electron from the crystal is removing an electron from within a region that is adjacent to the surface and constrained by the photon's spatial extent of interaction and the electron achieving enough energy to be able to get to the surface, so with scattering and other interactions, and out.

The one-dimensional chain toy example showed how surface energies will be different from those within the chain—how Bloch states of an "infinite" crystal evolve as localized states on the surface. This one-dimensional model then gave us a means to expand the argument to show how assembly of such a crystal led to propagating states of the semiconductor—the bulk states—and the confined evanescent states at the surface. We could even show how the toy example comes quite close to showing the evanescent states' imaginary wavevector component in *InAs*, which comes quite close to the idealized free electron description of the model because the secondary valleys are far away.

For a more extended discussion of this problem of bulk/surface dichotomies, see S. Tiwari, "Device physics: Fundamentals of electronics and optoelectronics," Electroscience 2, Oxford University Press, ISBN 978-0-198-75984-3 (forthcoming). Barriers heights correlate, but are not directly related to bulk property differences. Silicon in different orientations has different barrier heights to SiO_2, and so metals to SiO_2.

Bulk electrons states extend to the surface, and the surface states can be seen as those arising from the mixing of bulk states due to the perturbation at the surface. So, we have propagating states extending all the way to the surface, as well as surface states that are localized orthogonally to the surface, with the total number of states arising from the valence of the constitutive atoms being conserved. Bulk electron bandstructure exists all the way to the surface with minor change, and localized states appear at the surface. Electrons in these surface states will have their own characteristic properties, such as of screening, as well as how they will cause the electron-hole interactions that we will discuss in Chapter 11.

The symmetry breaking of the surface also causes atomic bonding changes—at least, distortions and some breaking at the surface itself—and movement. Surface and bulk differences of stresses and strains interact in a few atom-thick regions. We took an energy-based approach to look at surface reconstruction and instabilities, to view some common and interesting examples.

The surface also has a pronounced effect on phonons. Phonon behavior is in many ways different from that of electrons. A two-atom model showed us that surface phonons have both localized and non-localized responses; that is, that there are allowed real (propagating) and imaginary (evanescent) wavevector solutions. Surface phonon modes at the zone center can be propagating. At the zone edge, one notes evanescent solutions. The surface modes at the zone center are above the bulk modes that exist all the way to the surface. So, surface phonons have a rich behavior, and, in nanoscale geometries, the way that these phonons cause local and nonlocal perturbation, especially when coupled to the electron cloud response, has important consequences.

5.8 *Concluding remarks and bibliographic notes*

SURFACE SCIENCE is a very important discipline with many interesting experimental tools of analysis that help understand the material and also verify a theory or tell us its incompleteness. Much of the literature on surfaces is experimental. The first major success of scanning tunneling microscopy was the observation of the atomic features at the surface, and the observation of many quantum-centric and reconstruction phenomena have made for major successes in the verification of theories, as well as pictures that spark imagination. A good introduction to semiconductor surface phenomena can be found in Balkanski and Wallis[1]. A very thorough study of surfaces is the book by Lannoo and Friedel[2]. This text, as implied by the title, covers the methods for modeling the electronic and phononic structures of surfaces; looks at the interesting aspects of transition metals and single element and compound semiconductors; and relates many of the observations to the metal-semiconductor system.

A caution needs to be stressed regarding pictures and imagination. There is much Photoshop subterfuge, dropping of data points, smoothing and introduction of imagination that cannot be condoned, especially when observations don't state it. This is the Millikan-ish dropping of data points in the oil drop experiment to get the electron charge consistent.

A reference—very semiconductor-specific—rich in parametric detail is the text by Mönch[3]. The book is particularly good at connecting the physics-based surface discussion to the practicalities of semiconductor consequences of interest to those utilizing semiconductors in devices.

For those interested in the materials science perspective—surface reconstruction, defects, steps, et cetera—there exist a number of good books that should be looked into to see the different insights of different authors. One good book is by Bechstedt[4], which discusses thermodynamics, diffusion, growth, reconstruction and defects. Another book, with a greater emphasis on experimental techniques, is that by Lüth[5]. A work complementary to this last example is the book by Ibach[6]. This book has an interesting chapter on surface magnetism in it.

[1] M. Balkanski and R. F. Wallis, "Semiconductor physics and applications," Oxford, ISBN 978-0-19-851740-5 (2007)

[2] M. Lannoo and P. Friedel, "Atomic and electronic structure of surfaces," Springer-Verlag, ISBN 978-3-642-08094-4 (1991)

[3] W. Mönch, "Semiconductor surfaces and interfaces," Springer, ISBN 978-3-642-08748-6 (2001)

[4] F. Bechstedt, "Principles of surface physics," Springer, ISBN 3-540-00635-4 (2003)

[5] H. Lüth, "Solid surfaces, interfaces and thin films," Springer, ISBN 978-3-642-13591-0 (2010)

[6] H. Ibach, "Physics of surfaces and interfaces," Springer, ISBN 13 978-3-540-34709-5 (2006)

We did not discuss electromagnetic phenomena at the surface—some of it appears for us through the plasmonic discussion in later chapters—but there is a considerable amount of Casimir force, dielectric function, tunneling and emission absorption phenomena that is very surface dependent. Sernelius[7] discusses modes, forces and interactions at the surface in considerable detail bearing on these aspects.

A discussion of Casimir interactions and their implications for nanoscale devices can be found in S. Tiwari, "Nanoscale device physics: Science and engineering fundamentals," Electroscience 4, Oxford University Press, ISBN 978-0-198-75987-4 (2017).

[7] B. E. Sernelius, "Surface modes in physics," Wiley-VCH, ISBN 3-527-40313-2 (2001)

5.9 Exercises

1. Calculate the thermionic current from a wire that is made of tungsten and is 3 cm long, has a radius of 0.1 cm and is at 2300 K. Tungsten has a workfunction of 4.5 eV. **[S]**

2. Surface state density tends to be highest near the bulk bandedges. Why? **[S]**

3. Surface states appear both within the bandgap and in the bands. We discussed at length that the states in the bandgap are evanescent modes that are localized at the surface. What about the surface states that appear in the band? Are they evanescent or can they be both evanescent and non-evanescent? **[S]**

4. Are surface phonons acoustic or optical? Or both? Argue your answer in short. **[S]**

6
Semiconductor interfaces and junctions

An understanding of the free-space-terminated semiconductor surface gives us some physical intuition for the changes wrought in electron states and phonon states at the surface. The physical underpinnings will help us understand the behavior when two different materials are brought together to form an interface. These are not two independent surfaces brought together, that is, material 1 to vacuum, and vacuum to material 2, with the vacuum removed in the transitive sense, since the two materials will interact in the interface region. Interactions don't necessarily lead to linear transformations. We have seen this with the Golden rule, when a perturbation that is not adiabatic is encountered. What we learned in Chapter 5 about the appearance of states in response to symmetry breaking and perturbation will be pertinent to our discussion. We will see in this chapter the role of the gap states, the differences between crystalline continuity versus its absence, and the nature of the materials—semiconductors or metals—that show up in the interface region where there may also be a junction. The changes will happen for electron states and phonon states.

By junction here, we mean the broader context of an interface together with surrounding regions where changes in composition of the materials, changes in the polarities or doping of the semiconductors, and others—inhomogeneities of various kinds—may exist. Together with the specifics of what happens at the interface, the details of what happens in the surrounding region will also matter. Both a metal-semiconductor junction and a semiconductor-semiconductor junction will behave differently, depending on the doping and the interface, and show a variety of different properties.

Such interfaces are of import in the metal-semiconductor junctions that are so often used for rectification and for ohmic transport; in insulator-semiconductor junctions, where one is either just isolating the surface or using the interface as a barrier for the creation of confined mobile charge layers; and in semiconductor-semiconductor junctions, where one often wishes to introduce changes in transport or electromagnetic interaction. So, this chapter will start with a brief introduction to interfaces and junctions and then proceed to the metal-semiconductor junction. This discussion builds the arguments toward understanding induced gap states—the states that appear in the bandgap (and above and below the bandedges)—when an interface exists with vacuum, as in Chapter 5, or with metal or with insulators. An understanding

Semiconductor Physics: Principles, Theory and Nanoscale. Sandip Tiwari.
 Published 2020 by Oxford University Press. DOI: 10.1093/oso/9780198759867.001.0001

of the nature of these states under the different circumstances in different materials lets us then look at the important crystalline interfaces between two semiconductors. This is the assembly that we call heterostructures. We will see how bands align due to the Bloch function changes that appear at the atomic scale in two semiconductors with crystalline continuity at an abrupt interface between the two. And we will then extend this discussion to where compositional mixing allows one to grade that interface. Both of these forms are ubiquitous in nanoscale devices.

6.1 *Interfaces and junctions*

AN IMPORTANT ILLUSTRATION of what happens due to the nature of bonding at material interfaces is the SiO_2/Si interface. Grown SiO_2 is an amorphous material that, in the process of the growth on Si from which it is formed, can reconfigure so that very few interface states are observed. This is illustrated in Figure 6.1(a) and (b), where the first shows the breaking of translation order and bond-driven termination of the silicon surface, and the second shows the interface trap density that is observed in different orientations. The (111) Si surface has the highest interface density. This is the orientation where each Si atom has either one or three bonds projecting out of the plane. It is also the orientation where multiple closely spaced Si planes are encountered as one goes from the corner of the cube, the atom along the diagonal, the atoms in the plane through the three adjacent corners of the cube and further on a mirror image. Accommodating this large and small bond number condition is not conducive to low interface state density. It is the angles of the bonds that are more flexible, but there are still energy consequences as angles distort. On the other hand, (100) Si has a

There was a saying in times past that silicon's success (in microelectronics) is not because of silicon, but because of SiO_2. Like many sayings, it has some truth to it.

In Chapter 7, we will talk a bit about point perturbations. Defects are one example of point perturbations. They matter in insulators too, where leakage current and reliability issues arise in them. SiO_2 is very robust because of a strong bond. SiO_2 bonds can flex, that is, they can change angles within limits. The (100) surface's spacing is quite right for being accommodative to this spacing leading to the low interface state density.

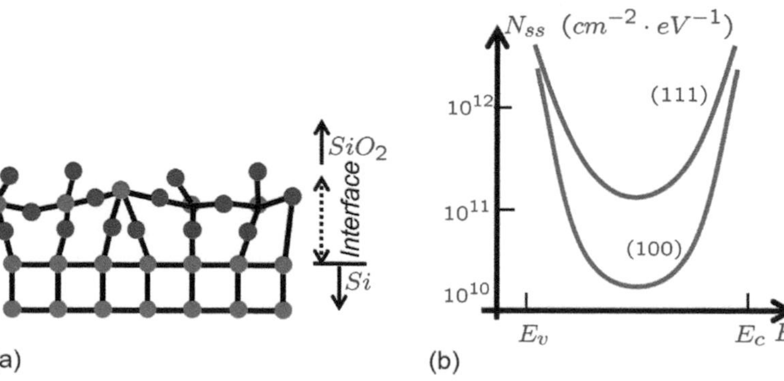

Figure 6.1: (a) A sketch of SiO_2/Si interface and (b) shows interface trap density in $cm^{-2} \cdot eV^{-1}$.

low interface density. The atomic density is low, it is an ordered surface of face-center and oxygen-bond termination that is more conducive. Adding to these thoughts, one would also have to see how the induced states arising from silicon and those arising from SiO_2 at the interface match. All in all, this is quite complicated. But, as Figure 6.1(a) indicates, this amorphous growth provides conditions for a large fraction of the atoms to bond, similar to the continuing crystalline growth by epitaxy of many compound semiconductors.

But, this is still not as simple as it sounds.

We are asking questions related to the electron state behavior as a function of position at the interface. We know that we can uniquely describe the $E(\mathbf{k})$ behavior in the bulk of silicon. We have also quantified under idealized conditions what this behavior is at the surface, but we have assumed that the spatial periodicity is still the same, and a very specific crystal potential $V(z)$ holds all the way to the surface. It is instructive to see how core shell energies probe the energy variation—observing them can show the nature of energy shifts at the interface. Figure 6.2 shows in an α-$Si/SiO_2/Si$ structure the oxygen K-edge intensity spectra with the step where the K-edge loss occurs for oxygen. In silicon, when deep enough, the step is absent, as it is in the amorphous silicon. In SiO_2, this step comes about, but note in this figure that it takes between 2 and 3 atomic distances before it comes about. In short, SiO_2 does not look like a bulk SiO_2 until ~ 0.5 nm into the oxide. The interface region is different. It has different atomic behavior. The orbital behavior is changing. The materials' characteristics are changing. If one were calculating tunneling through it, one would have to include these characteristics in it if we wished to be faithful. Fortunately, since the bandgap is very large, for the purposes of interface state discussion, this particular aspect of electronic structure turns out to be secondary. The highest density of induced states is close to the bandedges, and those regions are not usually accessed by electrons in device structures of interest.

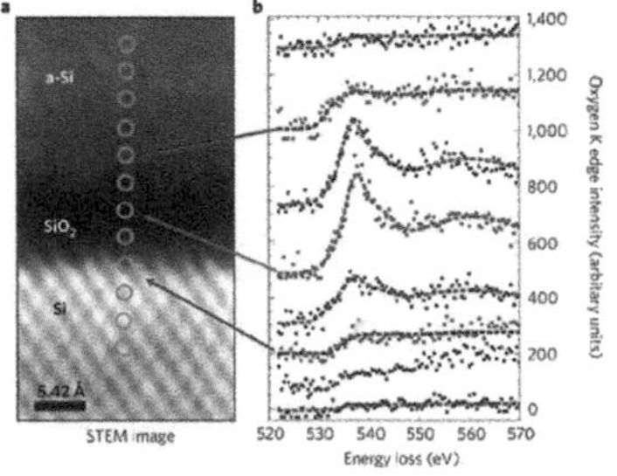

Figure 6.2: Energy loss spectra for an oxygen K-edge in an α-$Si/SiO_2/Si$ structure. From D. A. Muller, T. Sorsch, S. Moccio, F. H. Baumann, K. Evans-Lutterodt and G. Timp, "The electronic structure of the atomic scale of ultrathin gate oxides," Nature, ***399***, 758–761 (1999).

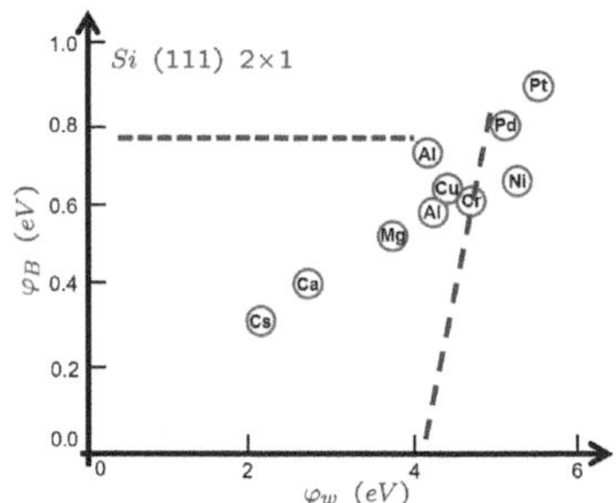

Figure 6.3: The barrier height of a metal-semiconductor junction with (111) Si of a 2×1 reconstructed surface versus the metal's workfunction.

Surface reconstruction is the form that the surface region takes as it minimizes energy—a new short-range order appears because of broken symmetry. Atoms realign and move up and down, and a lower energy comes about at the surface. We made some reconstruction-related observations in Chapter 5. We will discuss more shortly.

6.2 *Metal-semiconductor interface and junctions*

THE FIRST TOPIC we look at is the metal-semiconductor junction and what is observed as the interface manifestation. We remarked earlier that the workfunction difference type description should naturally be questioned. Figure 6.3 shows the barrier height of a metal to n-type (111) Si with (2×1) reconstruction. If the interface density was immense in numbers, one would see no correlation with the metal workfunction, since the interface locale in energy will define where the Fermi energy is at the surface pinned by the interfaces. And the metal electrochemical potential will align to this potential in the semiconductor. As the interface state density reduces, one would find increasing workfunction dependence. What one finds, as seen in Figure 6.3, is that there is weak correlation with the workfunction and the actual behavior is somewhere in-between. Schottky barrier height is a weak function of the metal workfunction. A 2 eV change in the metal workfunction causes about a 0.4 eV change in the Schottky barrier height. With *Cs*, the barrier height is low—a reflection of the lowering of the workfunction that we discussed in Chapter 5. For *Pt*, it is very high. *Pt* causes a large band bending in n-Si and, one finds, consistent with this workfunction, a much smaller band bending in p-Si. *Pt* is, at least from the electron transport viewpoint, more conducive for electron exchange with the valence band, that is, it is potentially a good p-type low resistance contact material.

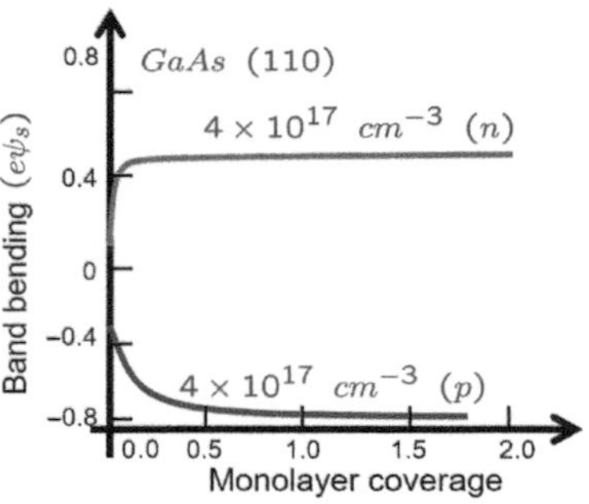

Figure 6.4: The change in band bending resulting from partial and complete coverage of (110) *GaAs* by *Sn* in n-type and p-type semiconductors. Only a very partial coverage is needed for band-bending changes for either. After M. Mattern-Klossen and H. Lüth, "The Schottky barrier of Sn on GaAs(110)," Surface Science, 162 610–616 (1985).

Sn on *GaAs* provides another contrast in Figure 6.4. A very fractional monolayer coverage, so only a partial coverage, is sufficient for metal-induced gap states resulting from *Sn*'s appearance to show their effect. The sum of the two bendings is about 1.3 eV and about 0.5–0.8 eV comes about for either of the two polarities. The figure also shows that a cleaved (110) surface of GaAs is relatively free of surface states in ultra high vacuum. For n-type material, the bands are relatively flat. For p-type, it is about 0.3 eV. This band bending corresponds to about 10^{12} cm^{-2} charges in surface states. Note that the surface atom density in crystals is of the order of $10^{22\times 2/3} \approx 10^{15}$ cm^{-2}. So, for both *Si* with the oxide and this cleaved-in-vacuum example, the surface states form only a very small fraction of the atomic concentration of the surface.

Figure 6.5 shows the energy states of intrinsic defects in a few of the compound semiconductors (*GaAs*, *GaSb* and *InP*), together with those for extrinsic chemisorbed species. In *GaAs*, these are all near mid-gap in about a 0.20 eV range, as is the surface Fermi level pinning that one observes. For *GaSb*, many of these are closer to valence bandedge, as is the surface pinning. And, in *InP*, these are closer to the conduction band, as is the surface pinning. This indicates some correlation—not causation—between energy levels of the induced states and the energies of intrinsic defects—missing bonding—in these structures. Indeed, for most Schottky barriers in compound semiconductors, it is the induced states—not the defect states—that are believed to be the predominant source of the barrier height observed.

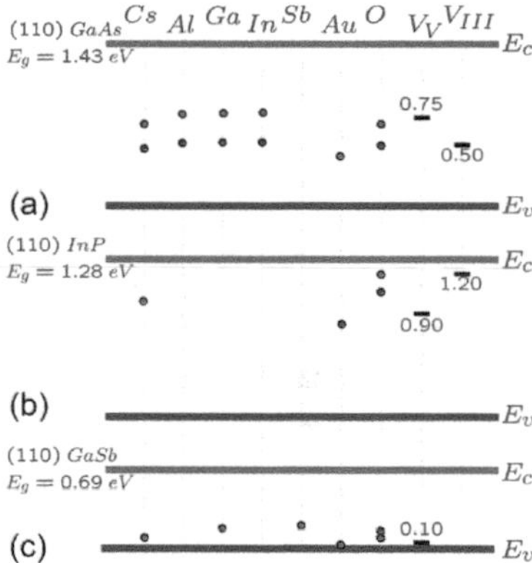

Figure 6.5: Energy of states—of intrinsic defects in the bulk, and of extrinsic chemisorbed species at the surface—in *GaAs*, *InP* and *GaSb* with a (110) surface. V_V is a vacancy of the group *V* element. V_{III} is the group *V* element on a group *III* site. After W. E. Spicer, I. Linday, P. Skeath and C. Y. Su, "Unified defect model and beyond," Journal of Vacuum Science and Technology, **17**, 1019–1027 (1980) and W. E. Spicer, R. Cao, K. Miyano, T. Kendelewicz, I. Lindau, E. Weber, Z. Liliental-Weber and N. Newman, "From synchrotron radiation to *I*-*V* measurements of *GaAs* Schottky barrier formation," Applied Surface Science, **41/42**, 1–16 (1989).

This collection of observations—from toy models that indicate the existence of confined interface states both in and outside the bandgap, with penetration of the electron Bloch state outside the crystal, to the nature of intrinsic defects where one will expect to have characteristics similar to those arising in bond breaking, to the way the band bending changes for cleaved surfaces when adsorbates are introduced—points to several lessons. First, the nature of interface is complex, can be very specific to the conditions and has a variety of origins with respect to how the phenomena will manifest when two materials are together. Second, the idealized wavefunction of one material will tail into the other and, no matter how complicated this wavefunction is, it must follow the continuity of probability and probability current.

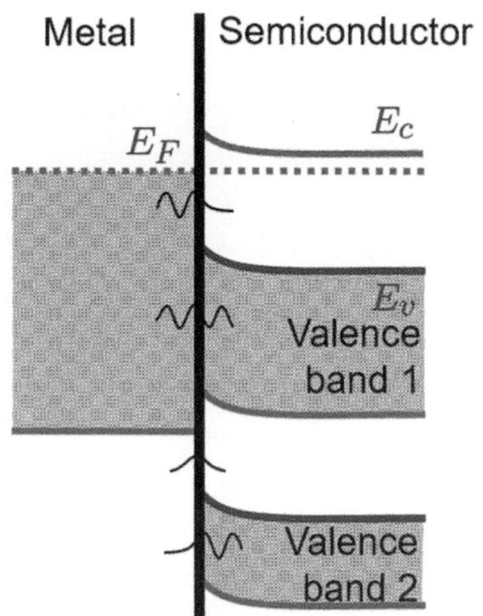

Figure 6.6: A representational view of the bands and waves at a metal-semiconductor interface. The wavefunctions of the propagating states of one penetrate into the other. Evanescent states are interface states and they are being induced by both the metal states and the semiconductor states.

If one placed a metal and a semiconductor together (see Figure 6.6), metal waves would penetrate the semiconductor and induce states in the forbidden gap. Metals also have a large number of states, since they are eVs up in the energy. The semiconductor itself, too, has an effect in this bandgap, due to confined states arising from the Bloch states within it. The nature of these states will not be identical to what they would have been with vacuum at the interface, even if we assume no atomic displacements near the interface. The states with imaginary wavevectors that we have introduced ($\mathbf{k} = \mathbf{k}^{\underline{r}} + i\mathbf{k}^{\underline{i}}$) will exist in all these instances. The only caveat is that when one places two semiconductors with similar translational symmetry together, so that there is a crystalline interface, the nature of the bonding that exists at the interface, which is an interfacial constraint on the Bloch states allowed, may provide a continuity of both the probability and the probability current density to be naturally obeyed without the large-scale appearance of $\mathbf{k}^{\underline{i}}$. We will see that this is the case for many heterointerfaces. In these situations, one will observe bandedge and bandgap discontinuities. Electrons in most situations don't see interface states. What they see is a change in the eigenenergies allowed. And the change in the lowest of those allowed appears as just a step in the conduction and valence bandedges, with the continuity conditions maintained. This sounds like a very difficult problem. It is. We have difficulty even reproducing $E(\mathbf{k})$ specifications without some input from experimental data that allows us to fix some of the parameters.

For some heterointerfaces, $GaSb/InAs$ for example, with crystalline continuity, there will still be states up in the conduction band and down in the valence band, but not in the bandgap.

Bandgap is a common parameter one fixes, but it could be effective mass, or others using data determined from precise optical measurements that couple states whose nature we understand in conduction and valence bands at various symmetry points. This is where the van Hove singularities are very useful markers.

We now consider what happens at the interface of two materials. An insulator/semiconductor interface such as SiO_2/Si is of particular interest, but so are metal-to-semiconductor interfaces and the heterostructures formed when two crystalline semiconductors form a continuous junction. Electromagnetic connectivity—such as in Poisson's relationship—and quantum-mechanical connectivity will have to be integrated in general. Our discussion with electrons in materials tells us that, for any perturbation, such as that of potential and of change in periodicity, will have consequences for states. The change in wavefunctions and probabilities that they correspond to will have charge consequences that will be represented in the energy. And this similar vein of discussion will apply to phonons, too, due to the change in translational symmetry, periodicity and the forces that keep the atoms around an averaged position during vibrations.

6.3 *Induced gap states and Fermi level pinning*

THE METAL-SEMICONDUCTOR JUNCTION in an idealized textbook model would have the following perspective.

The metal Bloch wavefunction tails into the semiconductor, overlapping with the forbidden states gap region of the semiconductor. Imaginary wavevector $\mathbf{k}^{\underline{i}}$ states due to broken symmetry exist. In a three-dimensional model, it fills this forbidden region with a distribution of states that are confined orthogonal to the plane and propagating parallel to the plane. These are metal-induced gap states (*MIGS*) described by an $E(\mathbf{k}^{\underline{i}})$ relationship. They extend up and down in energy in the gap, above it and below it since the states arose from the metal's conduction band. The states that are of interest to us are around the Fermi energy. These are in the bandgap region—usually forbidden—arising from the states that are spread out in the conduction band of the metal around the Fermi energy.

The semiconductor too induces states arising in the mixing of bulk states such as at heterostructures. These too can be higher up, lower down and spread in the forbidden gap. These are virtual induced gap states (*VIGS*), the details of which will depend on the complexity of the interface.

We will call these states, irrespective of their origin, induced gap states (*IGS*)—it is understood that, in general, these states arise from the perturbation at the interface. They represent the consequences of this perturbation and, in principle, one can outline a procedure to calculate them from the basis states of the two materials interfacing.

When a metal and a semiconductor are placed together, there are these interface modes due to both *MIGS* and *VIGS*. Bloch states of the metal—including the *MIGS* that arise from them—that match into the *VIGS*, that is, have energy and momentum matching (E and $\mathbf{k}_{\parallel}$, with $\mathbf{k}_{\perp}$ matching), will be filled. The states arising from the semiconductor, from the conduction band and the valence band, can be acceptor-like, that is, be neutral when empty, and negatively charged when filled, or donor-like:, that is, be charged when empty and neutral when filled. States close to the propagating conduction band energies are acceptor-like. The conduction band is the band which is largely empty and in which, when an electron is placed in it, the electron becomes propagating. It has a large number of such states that are empty and capable of receiving an electron. And the state is neutral in charge if not occupied. This is acceptor-like. The complement to this is that states closer to the valence band are donor-like.

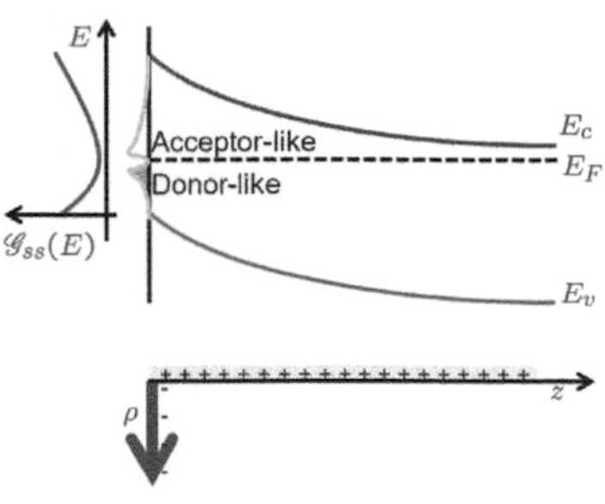

Figure 6.7: Band bending and Fermi level pinning on the surface of n-type *GaAs*. The figure shows a schematic of the density of states ($\mathscr{G}_{ss}$ distribution of the donor- and acceptor-like surface states). A similar figure—a mirror of this—may be drawn for p-type *GaAs*.

Take, for example, the case of a surface such as of *GaAs*, as shown in Figure 6.7. This illustrates the density of state distribution of the donor- and acceptor-like surface states. Donor states are largely filled, and acceptor states are largely empty. These are neutral. But, with the Fermi energy slightly enhancing the filling acceptor state numbers so that they are in excess of the donor state numbers, a small negative charge exists at the surface. The Fermi energy at the surface is in thermal equilibrium with that of the bulk by the filling of acceptor-like states in excess of the donor-like states. As *GaAs* is direct gap semiconductor with similar symmetry for the Γ point of the Brillouin zone for both the conduction and the valence states, the density of states for the surface is fairly symmetric. Mid-gap—a little closer to the valence band, which has a larger concentration of propagating states—is where the Fermi level pins and the electron-filled surface state density of few $10^{12}\ cm^{-2}$ is balanced by about a similar charge in the depletion region in the bulk. Away from thermal equilibrium, one may not make a larger excursion in this Fermi level pinning at the surface, since the density of the interface states rises rapidly in energy. Typically, one sees $<200\ meV$ of positional change in this surface pinning for *GaAs*.

So, the argument for free surfaces and interfaces is that there exists an approximate energy which separates donor-like and acceptor-like states. This is a neutrality level. In a symmetric system, it exists in the middle as shown in the toy model picture of Figure 6.8, but, in general, it doesn't have to be. Any deviation of Fermi energy from this energy causes charge build-up. If the excursion is toward donor-like states, it is positive, and if it is toward acceptor-like states, it is negative. Such an excursion cannot be very pronounced because of the high density of interface states. And this is what will prescribe the properties of the interface, such as the Schottky barrier height with a semiconductor, as well as other extrapolations, such as bandedge discontinuities.

Figure 6.8(a) shows the origin of the neutral level where the donor-like and acceptor-like states below and above are equal

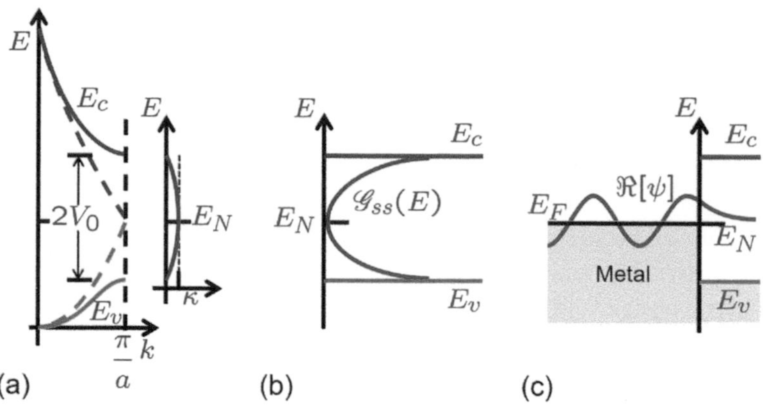

Figure 6.8: A conceptual outline of the neutral level E_N. (a) shows the toy model picture of breaking of degeneracy that causes the bandgap to appear with E_N as the energy where the evanescent states have maximum evanescent wavevectors. This is in the middle because of the symmetry of the toy model, and the density of surface states rapidly rises around it, as shown in (b). When a metal-semiconductor contact is formed, in thermal equilibrium, the neutral level of the composite system's states aligns to the Fermi energy in the metal.

in number. It is symmetric in this toy model. And one can see its origin in the introduction of the perturbation that causes the bandgap to appear. Figure 6.8(b) shows the density of these surface states, which is in the middle of the bandgap due to the symmetry of the model. When a metal and this semiconductor are brought in contact, as in Figure 6.8(c), and thermal equilibrium established, a neutrality level that arises in the confined states at the interface will determine where the Fermi energy in the metal will align. If these states are identical or dominated by what existed at the semiconductor surface alone, then it is undisturbed and the Fermi energy of the metal aligns to it. However, if the semiconductor did not have a large $\mathscr{G}_{ss}$ to begin with, more *Si*-like than *GaAs*-like, for example, then the neutral level will shift, since both the metal and the semiconductor are giving rise to confined modes of the surface.

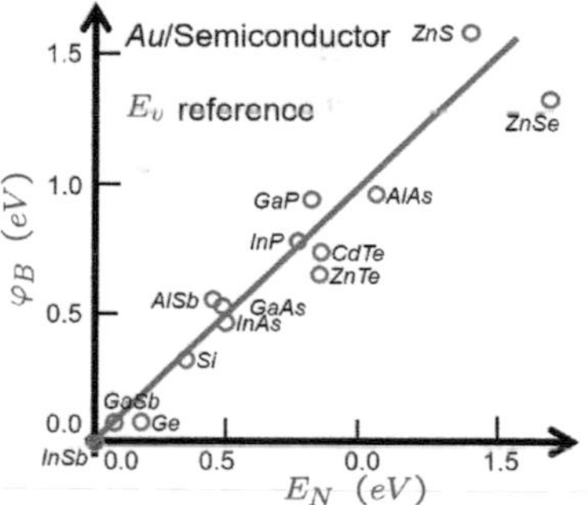

Figure 6.9: Barrier height of the metal *Au* for various semiconductors as a function of their neutral level referenced to the valence bandedge.

This description of the neutral level arising in the virtual and metal-induced gap states works pretty well with Schottky barrier models. Figure 6.9 shows the barrier height for *Au* contacts for a variety of semiconductors, from small to large bandgap, as a function of the neutral level energy, with all energies referenced to the valence bandedge. This level does not have to be in the forbidden gap. *InAs*, for example, is a small bandgap ($\sim$0.36 eV) semiconductor. The neutral level energy is higher than this. It is well into the conduction band. Metals make rather good ohmic contacts with n-type *InAs* for this reason. Our prior discussion of *GaAs*, *InP* and *Si* is also aligned with what is shown here. *Si*, for example, will have different neutral level with different metals—a workfunction dependence—since *VIGS* states and *MIGS* states both matter, unlike the *GaAs* case. Another interesting observation here is the very low height for the *Sb*-based systems *InSb* and *GaSb*. The former has quite a small bandgap, but the latter's is quite reasonable (∼0.7 eV). Here, forming ohmic contacts to p-type material is easier.

Semiconductors can have both direct and indirect bandgaps. The indirect conduction band minimum arises, as we have discussed, through mixing of $|s\rangle$ and $|p\rangle$ states. These have high density. An example of this is the nearly factor-of-10-lower effective conduction bandedge density of states of $GaAs$ compared to Si. Since induced states arise from mixing these basis states, one would expect a dependence on indirect bandgap energy. This is shown in Figure 6.10. The reason $InAs$, a direct gap material, has the large barrier height plotted here w.r.t. the valence bandedge and, we claim, has a neutral level far up in the conduction band, is that its direct conduction band's density of states is small—this material has a very large electron mobility because its effective mass is so small—and it is the indirect conduction band minimum—about an eV up in the valence band—that determines where the neutral level lies. The pinning and barrier height arising in the occupation of these surface states force it into the conduction band. None of the semiconductors have a neutral level in the valence band, since its maximum is always at the zone center, and the mixing will pull the level toward the conduction band energy.

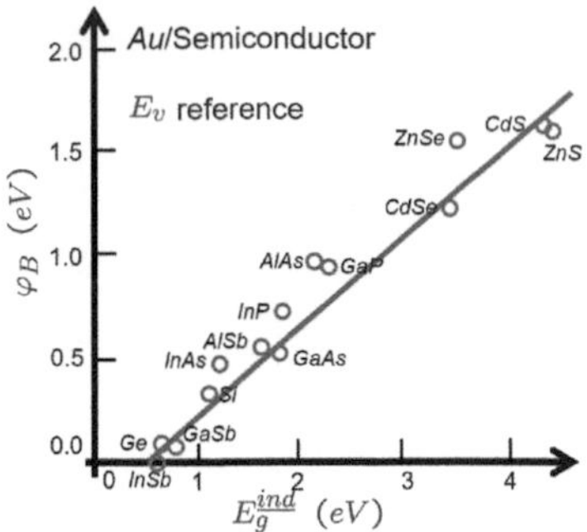

Figure 6.10: Barrier height of the metal Au for various semiconductors as a function of their indirect bandgap energy.

This argument of induced states describes the behavior of a semiconductor-semiconductor crystalline system too, and one may describe it with the same rationale. Our argument is summarized in Figure 6.11(a) and (b). The first panel shows our summary of the metal-semiconductor junction. Fermi energy in the metal aligns with the neutral level that shows the balancing within the interface at the junction. These states—here described as induced gap states (IGS) due to both the metal and the semiconductor—result in the alignment. Figure 6.11(b) shows the semiconductor-semiconductor situation. The mutually induced states arise from

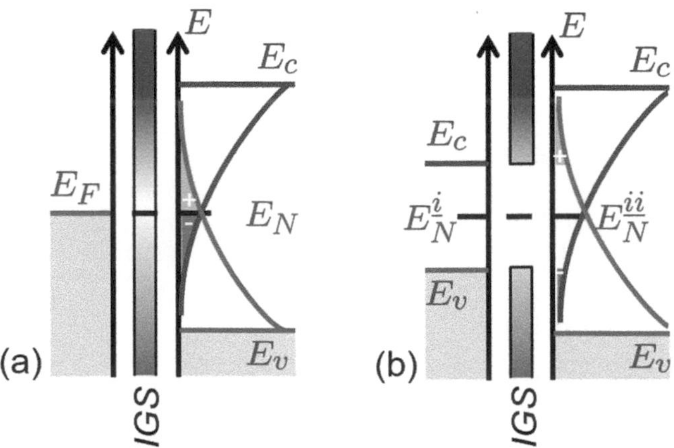

Figure 6.11: Interpretation of the metal-semiconductor and semiconductor-semiconductor heterostructure picture of energy levels alignment based on virtual induced gap states ($VIGS$). We denote these interface gaps states—regardless of their origin—as IGS here.

both semiconductors. If the symmetry of these states is similar, so that the interface is a continuation of bonding with the change in translations, then the matching of probability and probability current is quite conducive and induced states exist where Bloch states of one evanescences into the forbidden band of the other. In the forbidden gap region at the interface, where no propagating states exist on either side, there exist no induced states. So, the neutrality level of both semiconductors, to the lowest order, match.

6.4 Heterostructure

IN MATERIAL COMBINATIONS that are crystallinally continuous but electronically quite different—*InAs*/*GaSb* being one example—there may well be states extending in the energy region forbidden on either side. Figure 6.11(b) shows a situation where the wavefunction matching is quite exact at the interface. Where there are propagating states on both sides, propagation may proceed, although one may see quantum-mechanical reflections in the propagation, due to changes in the kinetic energy of the particles. In Figure 6.11(b), with the ease of probability and probability current matching, one sees a very localized filling and emptying of the induced states in the overlap region, with the neutral level of the two semiconductors *i* and *ii* aligning. There is now a balancing charge in the conduction state and the valence state, but it is entirely localized at the interface—an effect arising in $q|\psi|^2$ perturbation. This entire effect due to the interface dipole, highly localized as it is, appears in the conduction band and the valence band as the discontinuity. And charge neutrality in the *IGS*, with matching of the neutral levels, defines it.

6.5 Abrupt heterostructures

LET US START FIRST WITH THE PHYSICAL MEANING of the band alignment at the interfaces. The effective mass theorem gives us the interpretation for what is happening to the bandedge. The use of the effective mass theorem argument also implies, in turn, that we have a sufficiently wide region of multiple unit cells on each side of the interface so that the bandstructure picture applies. A bandedge energy as a function of position certainly is in contradiction with the uncertainty notion that states at bandedge are dependent on momentum. The resolution here is that we are really working with position-dependent potential as one crosses from one material to another. In this, there is no contradiction, and the effective mass theorem lets us write a Schrödinger equation form (the effective mass equation) that applies. Band discontinuities now become a set of parameters that may be applied as an interface condition.

Semiconductor heterostructures between two dissimilar materials are possible for a variety of semiconductors. *GaAs*, for example, forms heterojunctions with *Ge* and even *ZnSe*, with which it can be closely lattice matched. But there are technical issues with such junctions. First, depending on the nature of the chemical bonding and the Bloch function change from one semiconductor to the other, there may very well be a large number of localized interface states. *InAs*/*GaSb* is a classic example of this, where the lattice matching is quite good but interface states and discontinuity both exist. Another problem, particularly acute for the *GaAs* examples mentioned, is that atoms of the semiconductor on either side are also dopants on the other side. Any interdiffusion of atoms then makes electrical control tenuous. Most heterostructure deployment therefore has been in systems that are chemically matched.

In general, in semiconductors, heterostructures appear with the three different types of alignment that are shown in Figure 6.12. Type *I*—also called straddling gap alignment—is the most common, where a large bandgap semiconductor and a small bandgap semiconductor interface, with a barrier to bandedge electrons, and holes for transit from the small gap to the large gap. Type *II*—also called staggered gap alignment—places barriers for opposite polarity carriers, with the semiconductors exchanged. Type *III*—broken gap alignment—accentuates Type *II* to a point where one of the discontinuities is even larger than the bandgap of one of the semiconductors. Type *I* appears in $Ga_xAl_{1-x}As/GaAs$, in *InAs* containing multimolar composites and in numerous others. Strained *SiGe* grown on *Si*, too, has this Type *I* alignment, where the strain has an essential role to play in how this alignment appears. Type *II* alignment is seen at the *AlSb*/*InSb* interface. Type *III* alignment is encountered in the $InAs/Ga_{1-x}Al_xSb$ combination.

We discuss strain effects separately in Chapter 17.

In molar composites of compound semiconductors, for example, $Ga_xAl_{1-x}As$, which is a mix of an x molar fraction of *GaAs* and

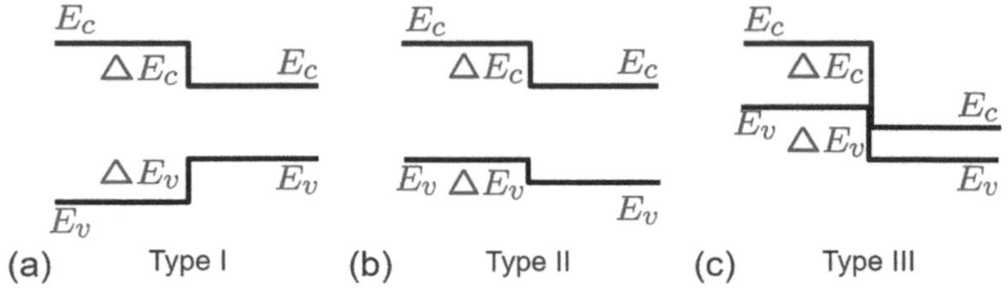

Figure 6.12: Three common bandedge alignments in heterostructures. Part (a) is Type I alignment, which is also called straddling gap alignment, (b) is Type II alignment, which is also called staggered gap alignment and (c) is Type III alignment, which is also called broken gap alignment.

$1 - x$ of *AlAs*, we will see changes in which band becomes a minimum as a function of the molefraction. *GaAs* is direct, with a Γ-L-X sequence for next conduction minima in increasing energy. *AlAs* has X as its lowest conduction energy. As the molefraction of *AlAs* is changed in the mix, many changes will come about, since the chemical potential is changing. The direct-to-indirect bandgap transition in this mixed system occurs at $x \approx 0.43$, as shown in Figure 6.13, where the valley minima are shown using the *Au* Schottky diode in these materials as a reference for the neutral level. The rationale in drawing this is that, in our metal-semiconductor picture, E_F in the metal aligns to the neutral level. So, a lowest order description of changes in the semiconductor is to use this *Au* barrier as a reference. In fact, in this system, at this crossover, all the different valleys are quite close in energy. This is a situation given to numerous interaction effects as a consequence of the mixing of states.

This approach of using an *Au* reference can be generalized, and Figure 6.14 draws the conduction and valence bandedges for many of the semiconductors of interest by using *Au* as the neutral level reference. If strain is absent, this figure gives quite an accurate representation of the conduction and valence band discontinuities that one would obtain in abrupt heterostructures and for various

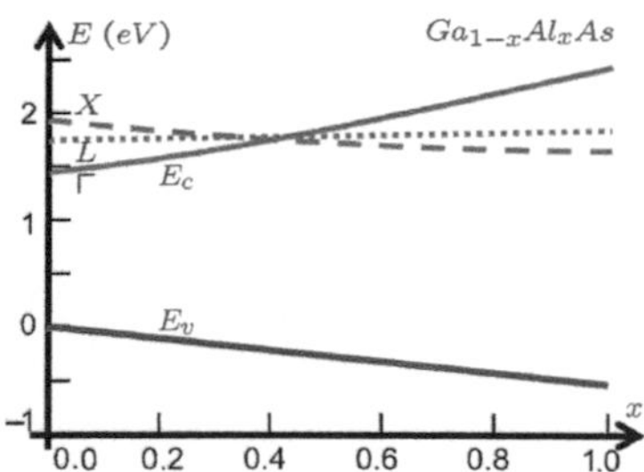

Figure 6.13: The change in conduction band minimum at Γ, L and X, and the valence band maximum in the $Ga_xAl_{1-x}As$ compound semiconductor system.

Another very consequential one is that of the behavior of dopants. *Si* in *GaAs* is a shallow donor—one that fits well to the hydrogenic model—or is it? Which conduction states does it couple to, and is there any local deformation? This will show up in the *AlAs* compositional mix. It forms deep donors—*DX* centers—and these are most cleanly observed in the behavior of the material at molefractions near and above this $x \approx 0.43$ molefraction. In fact, it shows multiple donor energy

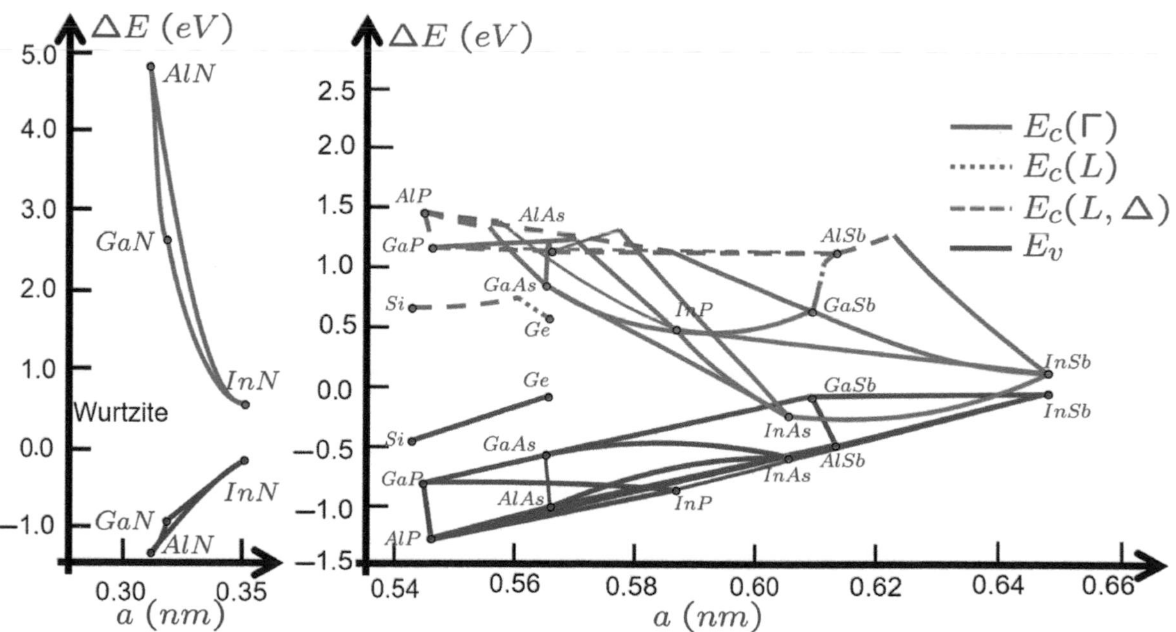

Figure 6.14: A "neutrality"-referenced alignment of conduction and valence bandedges. On the left is the alignment under unstrained conditions for *IIIV* nitrides in their wurtzite form. On the right is the *Au*-referenced alignment for many of the zinc blende and diamond systems.

compositions. One can directly observe here the changed alignment of *GaSb/AlSb* as being of the Type *I* type, but of *GaSb/InAs*, which too is quite reasonably lattice matched as being of the Type *III* type. *AlSb/InAs*, on the other hand, should be expected to be in-between—Type *II*, the staggered alignment. When strain is present, a variety of effects (strain energy, degeneracy splitting, etc.) will be present, and this figure becomes more inaccurate, but is still useful as a guide.

levels that are related to the environment that this *Si* sees. The group *III* atoms that it will see in next-nearest-neighbor surroundings may be all *Ga*, one *Al* and the rest *Ga*, or two *Al* and the rest *Ga*, and so on. The energy of this donor state will have to be different. There will be more on this in the discussion of point perturbations in Chapter 7.

6.6 *Abrupt heterojunctions in equilibrium*

WITH DISSIMILAR MATERIALS on either side of the heterojunctions, with the discontinuities of ΔE_c and ΔE_v, the equilibration of the electrochemical potential requires us to account for the different chemical environment of either side of the junctions (the chemical potential), together with the electrical potential. The energy bandedge profile and the Fermi energy give us the tools to visualize this equilibrium and the disturbance to it when we apply external potentials. The electrostatic potential $\psi(z)$ depends on the net charge $\rho(z)$. In thermal equilibrium, the electrostatic potential change is more straightforward to calculate since the charges—mobile and immobile—can be accounted from the materials' description and the statistical constraints. When off equilibrium, this is more complicated, since it depends on the dynamics of the transport of the mobile carriers. The thermal equilibrium description of the junction will follow from the boundary and interface conditions coupled with that of the charge description in the semiconductors. Heterostructures often form degenerate regions. So, we outline here a simple procedure, with degeneracy, to see how one would draw the band profile of heterojunctions in equilibrium.

The Fermi energy, the electrostatic potential and the bandedge energies are related through the state distribution in the bands, and, for our simplified example, the carrier density in degenerate conditions can be written as

$$n(z) = \mathcal{N}_c \frac{2}{\sqrt{\pi}} \mathcal{F}_{1/2}\left(-\frac{E_c(z) - E_F(z)}{k_B T}\right), \text{ and}$$

$$p(z) = \mathcal{N}_v \frac{2}{\sqrt{\pi}} \mathcal{F}_{1/2}\left(-\frac{E_F(z) - E_v(z)}{k_B T}\right), \tag{6.1}$$

where $\mathcal{N}_c = 2[2\pi m_c^*(z) k_B T/h^2]^{3/2}$ and $\mathcal{N}_v = 2[2\pi m_v^*(z) k_B T/h^2]^{3/2}$, are the effective densities of states for the conduction and valence

Here, we are building on the discussion of devices in S. Tiwari, "Device physics: Fundamentals of electronics and optoelectronics," Electroscience 2, Oxford University Press, ISBN 978-0-198-75984-3 (forthcoming). The reader should refer to any microelectronics text and find the origins of these relationships and their assumptions.

bands, respectively, E_F is the Fermi energy and $\mathcal{F}_{1/2}$ are the Fermi integrals of order 1/2. Since the materials on the two sides are different, writing the equations referenced to the Fermi energy is more convenient than writing those referenced to the intrinsic energy E_i which will be changing as the material changes. The ionized charges on either side also follow from the impurities' degeneracies (g_D and g_A) as

A Fermi integral of order ν is

$$\mathcal{F}_\nu(\eta_c) = \int_0^\infty \frac{\eta^\nu}{1+\exp(\eta-\eta_c)} d\eta.$$

η here is a normalized energy.

$$N_D^+(z) = N_D\left(1 + g_D \exp\{-\left[E_D(z) - E_F(z)\right]/k_BT\}\right)^{-1}, \text{ and}$$

$$N_A^-(z) = N_A\left(1 + g_A \exp\{-\left[E_F(z) - E_A(z)\right]/k_BT\}\right)^{-1}. \quad (6.2)$$

With the total charge as

$$\rho(z) = e[p(z) - n(z) + N_D^+(z) - N_A^-(z)], \quad (6.3)$$

and Poisson's equation of

$$\frac{d}{dz}\epsilon(z)\frac{d\psi}{dz} = \rho(z), \quad (6.4)$$

we have a description of continuity of displacement. In thermal equilibrium, since electrochemical potential equilibrates, $E_F(z)$ is a constant. If the junction is long enough so that electric fields disappear farther away from junctions, that is, the charges balance, a set of boundary conditions will be set for either side. At the interface, the band discontinuity prescribes the change in the bandedges. This is a complete description and gives the behavior of the heterojunction whether the polarities are the same (an isotype junction) or are opposite (an anisotype junction). Exercise 1 in this chapter shows examples of such junctions.

Off equilibrium, with current flow, the nature of dynamics of transport will matter. If the transport process is such that transport to and away from the interface is not a rate-limiting step, then it is the interface that will limit the transport.

S. Tiwari, "Device physics: Fundamentals of electronics and optoelectronics," Electroscience 2, Oxford University Press, ISBN 978-0-198-75984-3 (forthcoming), at an introductory level, and S. Tiwari, "Nanoscale device physics: Science and engineering fundamentals," Electroscience 4, Oxford University Press, ISBN 978-0-198-75987-4 (2017), at an advanced level, tackle the behavior of such junctions depending on the rate-limiting behavior.

This algorithm for the electronic description of a heterojunction has been in a generalized, position-dependent form, so even if the interface is not abrupt, the approach applies, with the parameters of the materials positionally changing.

6.7 *Graded heterostructures*

WHEN A HETEROJUNCTION IS CONSTRUCTED out of two materials that can be continually varied as a solid solution—*GaAs* and *AlAs* being classic examples, but also *InAs*, *InP* and other mixes, where strain may limit the range—then the chemical transition can be gradual. This is a graded junction. In $Ga_{1-x}Al_xAs$,

$0 < x < 1$ is this continuum, where the changing x is changing the local electrochemical description of the semiconductor. One can incorporate these changes as an alloy potential whose consequences are reflected in the total potential change in the semiconductor. Such heterojunctions will have tunability of properties that are determined by the distance over which grading occurs: from an abrupt to a near-adiabatic change.

This grading, and thus changes in the semiconductor's electrochemistry, are included in semiconductor parameters such as the effective density of states—$\mathcal{N}_c$ and $\mathcal{N}_v$—since, together, they describe the nature of the material. Thus, the equations outlined in Section 6.6 suffice to describe graded junctions too.

6.8 Polarized heterojunctions

NITRIDE SEMICONDUCTORS—*GaN*, *AlN* and *InN*, for example, in their wurtzite form—exhibit both spontaneous polarization and a piezoelectric effect. Spontaneous polarization ($\mathbf{P}_{sp}$) is the existence of a stable dipole in the unit cell due to phase transition, which itself is a thermodynamic equilibrium consequence arising in a lowering of the free energy by a small displacement of the constituent atoms w.r.t. each other, with this displacement leading to an electric dipole. This is a stable state, and a stacking of such dipoles means that a charge density $\rho = \nabla \cdot \mathbf{D} = \nabla \cdot (\epsilon_0 \boldsymbol{\mathcal{E}} + \mathbf{P})$ will arise from it on the faces, while, within the crystal, the dipole charge neutralizes at the unit cell interfaces. Because of the lack of symmetry and ionicity a strain-induced polarization ($\mathbf{P}_{pz}$) will also exist. In periodic atomic vibrations, it is this strain-induced polarization that causes a piezoelectric scattering through its dynamic interaction. In crystals, if it exists, due to the presence of mechanical deformation, it will also cause a static charge/polarization-induced field.

See S. Tiwari, "Nanoscale device physics: Science and engineering fundamentals," Electroscience 4, Oxford University Press, ISBN 978-0-198-75987-4 (2017) for a detailed discussion of spontaneous polarization, piezoelectricity and their uses. Piezoelectricity will appear in crystals that lack inversion symmetry. Zinc blende semiconductors have a piezoelectric effect, while diamond structure semiconductors do not. Piezoelectricity's most pronounced effects appear there either in mobility due to scattering, when other scattering considerations are eliminated, or in the charge fields that will exist in the strained regions of fabricated structures. Zinc blende and diamond forms will not have spontaneous polarization. When spontaneous polarization exists, the electrostatic charge densities that are produced in the material will be compensated by charges that the surface will accumulate from the environment as well as surface reconstructions. If one were to place two metal plates and shorted them, the charge displacement between the metals will compensate the polarization charge that arose from the medium in-between.

For the wurtzite form, Figure 6.15 shows such a spontaneous polarization. Group *IIIV* nitrides are semiconductors that show spontaneous polarization, with an increase as one proceeds from *GaN*, to *InN*, to *AlN*. It happens because the c-directed bond increases in length. Simultaneously, the in-plane distances decrease. So, the wurtzite cell becomes longer in the c direction. This spontaneous polarization also has a negative sign due to the asymmetry of the wurtzite bonding. The polarization vector points between the N atom and the group *III* atom along the [0001] direction. Therefore, crystals with the group *III* face have their polarization directed away from the face into the crystal. If the face is of N, it is directed toward the face.

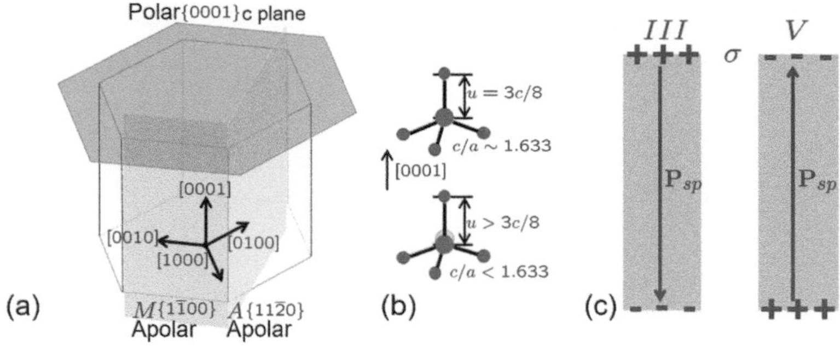

Figure 6.15: Polarization in wurtzite crystals of group *IIIV* nitrides. Part (a) shows that the {0001} is polar, while there also exists faces that are apolar (*M* and *A* plane) and (b) shows the origin of spontaneous polarization due to the relative displacement of a group *III* atom from the group *V* atom in the *HCP* assembly. The *c*-directed bond should be $3c/8$ in length, and $c/a \approx 1.633$ for *HCP*. In the presence of spontaneous polarization, the *c*-directed bond increases in length, pushing the group *III* atom here further down. There now exists a net dipole in the assembly, as shown in (c), depending on the face. Note the negative sign of the polarization due to the asymmetry of the crystal.

This spontaneous polarization of the crystal happens together with any strain-induced polarization ($\mathbf{P}_{pz}$); that is, piezoelectric polarization. Therefore,

$$\mathbf{P} = \mathbf{P}_{sp} + \mathbf{P}_{pz}. \tag{6.5}$$

Here, the piezoelectric polarization is

$$P_{pz} = e_{33}\varepsilon_{zz} + e_{31}(\varepsilon_{xx} + \varepsilon_{yy}), \tag{6.6}$$

where the *e*s are piezoelectric constants with units of C/m^2, and the εs are the strain tensor elements. When combinations of films of different materials are grown, the lattice constant differences that one can see in Figure 6.15 will also cause this additional strain-induced polarization, and it will be different in different directions. This makes for rather interesting consequences in polarization, the changes in the conduction bandedge at interfaces, and because of $\rho = \nabla \cdot \mathbf{D}$, interesting carrier polarity manipulation at group *IIIV* nitride semiconductor interfaces. As an example, consider a $Ga_{1-x}Al_xN$ layer grown on a relaxed *GaN*, that is in its natural wurtzite dimensions, in the polar [0001] direction. This $Ga_{1-x}Al_xN$ layer will be tensile strained. The spontaneous polarization at the interface can be found from, in the lowest order, the linear extrapolation for that composition between its two limits of *GaN* and *AlN*. Simultaneously, one will also have to find the piezoelectric polarization, since the $Ga_{1-x}Al_xN$ has been deformed, and this will depend on the elastic deformation properties. The net effect will be the actual polarization at the interface. Now, if, instead, a compositional mixing of *GaN* and *InN* was employed for the elastic film, the film will be compressive, and the piezoelectric consequence opposite. But this will have to be considered together with the spontaneous polarization that will also exist in these nitrides.

Note $\sigma_{ij} = \sum c_{ijkl}\varepsilon_{kl}$. Stress and strain are symmetric, so the nature of the elastic constants, which are fourth-order tensors, depends on the symmetry of the crystal. c_{1111} is customarily written in contracted notation as c_{11}, c_{2323} relating σ_{23} and ε_{23}—a shear term—as c_{44}. Strain will be dealt with in more detail in Chapter 17. The piezoelectric effect exists in zinc blende crystals too but occurs from off-diagonal components. For wurtzite nitrides, because the $e_{33}\varepsilon_{zz}$ term is in line, the piezoelectric polarization is strongly in line. e_{33} is 1.46 C/m^2 for *AlN*, 0.73 C/m^2 for *GaN* and 0.97 C/m^2 for *InN*.

We will tackle strain-induced consequences in Chapter 17. Since the deformation is elastic, the anisotropic stress-strain relationships will determine the deformation, and these can be calculated in principle. It is secondary to our discussion here, where we are stressing the unusuality of polarization-induced interface effects.

We illustrate this combination of different polarizations on the bandedge diagram in Figure 6.16, for a superlattice (quantization-sized wells and barriers) of crystalline *GaN* and *AlN* grown on a [0001]-oriented *AlN* surface. The electric fields at the interfaces reverse! Also note that, in this structure, the discontinuities are quite significant, since it is a nitride system.

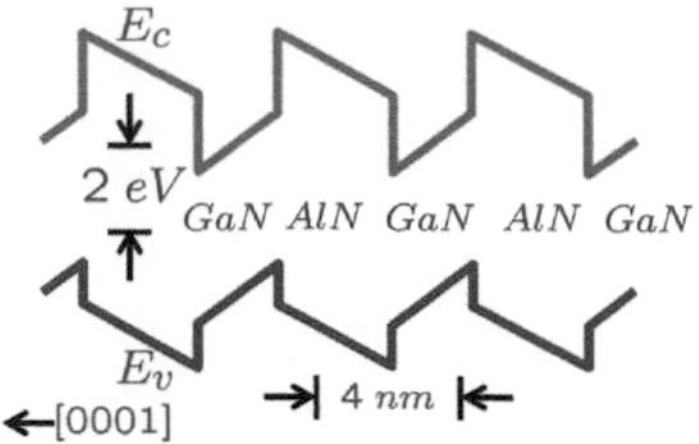

Figure 6.16: The bandedge diagram of *GaN*(2 *nm*)/*AlN*(3 *nm*) crystalline quantization-sized films grown on an *AlN* substrate in the [0001] direction. Note the unusual electric field reversal at the interface due to the combination of spontaneous polarization and strain-induced piezoelectric polarization without violating Gauss's law.

6.9 *Summary*

INTERFACES AND JUNCTIONS are inevitable in any system connected to the environment for exchange of energy and particles. These are regions of broken symmetry and of a variety of perturbations. The nanoscale utilizes small dimensions, and various properties that appear at the interface matter even more so because of the comparable dimensions of the interface region and the device region. This chapter built on Chapter 5, so that one could understand what happens at interfaces and junctions for a variety of materials—non-crystalline and crystalline, conducting, insulating or semiconducting—that one encounters with semiconductors.

Some of the salient points are the following.

When one looks at the correspondence between the workfunctions of metals and semiconductors (e.g., *Si* and *GaAs*) and barrier heights, the change in Fermi energy at the interface in *Si* is significant, while *GaAs* appears largely pinned, independent of the workfunctions. When the metal and the semiconductor surfaces come together, the assembly has interface states—similar to the states for a symmetry-breaking surface—and induced gap states exist. In *GaAs*, a direct gap semiconductor, these are large, even as their density decreases toward the middle of the gap, because of the nature of their creation from the propagating states of the bulk to confined states at the surface. In *Si*, this state density is small, with a prominent reason for this being the indirect bandgap and asymmetry of the Bloch conducting states. These states on the surface are virtual induced gap states. When a metal is placed, bringing its own states interacting in the surface region, the metal also induces states. These are the metal-induced gap states. Together with the low interface state density observed with the giant bandgap material SiO_2, this postulates a neutral level as a consequence of the various virtual induced gap states that correspond to the metal-semiconductor barrier height.

This idea of the neutral level also suggests the realignment of bandedges at the interface of crystalline semiconductors, where the Bloch functions conforming to two different semiconductors transition over a very short atomic-scale length scale. These structures—heterostructures—have abrupt discontinuity, with most forms where anion or cation species are common on the two sides existing without an interface state density in the gap. But it does not preclude low interface state density in the gaps—farther up in energy—due to Bloch state mixing of the induced variety. Different heterostructures are observed with different types of staggering in the conduction and the valence bandedges. When the molar composition of compound mixtures is varied, the bandstructure changes, and one observes consequences of the changes in the bandedges reflected in the alignment.

Composition at the crystalline interface region can be gradually varied instead of abruptly changed. In such a case, the discontinuities can be made to vanish, but the changes in the conduction bandedge and the valence bandedge must still conform to the electrochemical equilibration constraints. The nature of the bandedge variations therefore arise in electrostatic and chemical (electrical charge and alloy species) energies that must be reconciled.

6.10 *Concluding remarks and bibliographic notes*

As for Chapter 5, understanding surfaces through the surface techniques and how they relate to the important window through which one connects to the semiconductor of interest for particle and energy exchange, that is, interfaces, was the focus of this chapter. This connection is how we use devices. So, the literature and publications stretch very far back. For example, Schottky's theory of metal-semiconductor junctions goes back to the 1950s. With *Si*'s technological importance, interface state densities received much attention early on. With the relative immunity of the barrier height of *GaAs*, and related surface effects in it, there was also early attention to the Fermi level pinning of compound semiconductors.

Much of this early work related to interfaces such as SiO_2/Si or metals with compound semiconductors focused on characterization through electrical, optical and energy measurements, in high vacuum, of the interface states, as in the former, or of the role of defect states, as metals were introduced. Metal-induced gap states—a continuum of states within the semiconductor bandgap permeating into a few layers of the semiconductor—lead to Fermi level pinning, which gives local charge neutrality. A small number of such states suffices. The barrier height arises in a short-range part related to the surface dipole, electronegativity differences and any bonding changes, and an additional part due to metallic screening by the metal-induced gap states. The nearest band in energy tends to determine which gap states prevail. This is why valence band tends to have a stronger weight in determining charge neutrality.

Mönch's book[1] discusses the induced gap states view of interfaces of metal-semiconductor systems. Another early book, with an electrical engineering perspective, is that by Rhoderick and Williams[2].

A comprehensive text from the surface physics perspective is by Lüth[3]. This book too takes a surface science view to the experimental probing techniques and combines it with a principles-based discussion.

A more general understanding of the interfaces came into its own only after heterostructures became ubiquitous and the surface studies advanced enough to probe surface phenomena minutely.

Some of the earliest work exploring heterostructures was by Frensley and Kroemer[4] as the methods of heterostructure growth came into their own and layering of a variety of semiconductors pseudomorphically became possible.

The completeness of the understanding of the relationship between metal-semiconductor barrier heights, the positing of a neutrality level, and the transformation of this to an understanding of band lineups is due to Tersoff. Three publications are worth reading through. The first[5] discusses Schottky barriers and band structures. This paper highlights the importance of pinning strength and barrier heights in what one observes in the range of semiconductors. The second[6] emphasizes *Au*'s behavior on semiconductor surfaces and shows the correlation of defect levels with barrier heights. The third[7] integrates this learning. The band lineup of Figure 6.14[8] shows the resulting variations based on a semi-empirical interpretation that is consistent with the bands of the semiconductors.

A good review of spontaneous and piezoelectric polarization in the nitride systems is from Yu et al.[9].

[1] W. Mönch, "Semiconductor surfaces and interfaces," Springer, ISBN 978-3-642-08748-6 (2001)

[2] E. H. Rhoderick and R. H. Williams, "Metal-semiconductor contacts," Oxford, ISBN 019 859335 (1998)

[3] H.H. Lüth, "Solid surfaces, interfaces and thin films," Springer, ISBN 1868-4513 (2010)

[4] W. R. Frensley and H. Kroemer, "Theory of the energy-band lineup at an abrupt semiconductor heterojunction," Physical Review B, **16**, 2642–2652 (1977)

[5] J. Tersoff, "Schottky barriers and semiconductor bandstructures," Physical Review B, **32**, 6968–6971 (1985)

[6] J. Tersoff, "Reference levels for heterojunctions and Schottky barriers," Physical Review Letters, **56**, 675 (1986)

[7] J. Tersoff, "Transition-metal impurities in semiconductors: Their connection with band lineups and Schottky barriers," Physical Review Letters, **58**, 2367–2370 (1987)

[8] S. Tiwari and D. Frank, "An empirical fit to band discontinuities and barrier heights in *III–V* alloy systems," Applied Physics Letters, **60**, 630–632 (1992)

[9] E. T. Yu, X. Z. Dang, P. M. Asbeck, S. S. Lau and G. J. Sullivan, "Spontaneous and piezoelectric polarization effects in *IIIV* nitride heterostructures," Journal of Vacuum Science and Technology B, **17**, 1742–1749 (1999)

6.11 Exercises

1. Figure 6.17 shows $Ga_{1-x}Al_xAs/GaAs$ heterojunctions for three cases—the *n-p* junction, the *n-n* junction and the *p-p* junction for thermal equilibrium ((a), (b) and (c)) and forward bias ((d), (e) and (f), respectively). Here, forward bias is defined as the case with higher current flow. While the Fermi energy in thermal equilibrium is quite straightforward, given its definition, the quasi-Fermi levels in forward bias reflect the transport dynamics as discussed in w.r.t. the metal-semiconductor case. Are Figure 6.17(d), (e) and (f) drawn correctly? Also, indicate what the quasi-Fermi levels will look like throughout the structure, and why. **[S]**

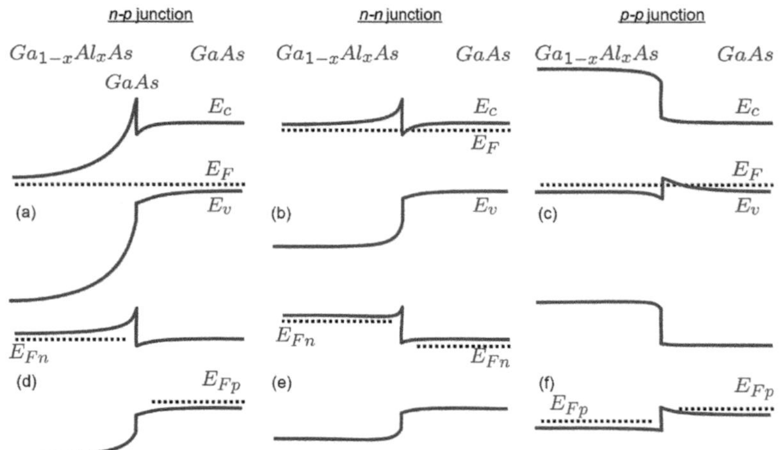

Figure 6.17: Different abrupt junctions formed in the $Ga_{1-x}Al_xAs/GaAs$ combination.

2. In crystalline heterostructures, the conduction band and valence band extrema are drawn as changing discontinuously at the interface. Can you argue a physical reason for this? [S]

3. *InAs* has surface Fermi level pinning in the conduction band. This certainly means that ohmic contacts are easy to make to n-tyoe *InAs*. How could one then make an ohmic contact to p-type *InAs* in order to make any device using it? [S]

4. Figure 6.18 shows a Type *III* heterostructure (also called a broken gap) in thermal equilibrium. The left side is p-type, and the right side is n-type. When we disturb the equilibrium, will it conduct by tunneling and be non-rectifying or will it behave more like a normal p/n junction that is rectifying? [S]

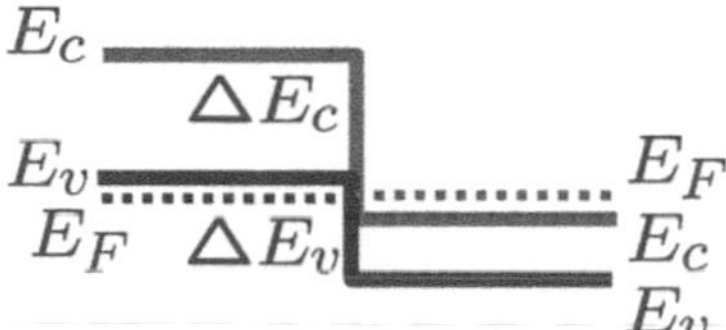

Figure 6.18: A broken-gap Type *III* lineup together with a change in doping from p to n.

7
Point perturbations

A GOOD-SIZED CRYSTAL CAN NEVER BE PERFECT. It is built at a finite temperature, so there is a thermodynamic propensity for defects: bonds that did not form, misplaced atoms and, of course, a variety of phenomena at the surface, where the crystal interfaces with its environment, as we have discussed. A missing atom is a defect. This is a vacancy. A substitution of a shallow hydrogenic impurity—*P* or *B* in *Si*—in the atomic site allows one to provide an excess electron or take it away (a hole) from the crystal. A vacancy may or may not be useful; an impurity may or may not. Vacancies have broken the symmetries that existed in the crystal by breaking the bond and may even cause a little shifting in the surrounding atoms. States different from the bandstructure picture will appear that may be shallow, or deep, or maybe even in the bands, just as what happened at the surface and interface. Both are examples of a point perturbation. *P* or *B* in *Si*, properly substituted in the crystal, are useful. But if it is a transition element such as *Ti*, *Cu* or *Au* that had been substituted, most likely, its consequences will be quite deleterious. The energy states arising in the presence of these elements are often a distraction, unless one is interested in killing the carrier lifetime of the semiconductor. A perfect crystal gains its usefulness by the introduction of imperfections. Functional devices are created through the controlled introduction of point defects and interfaces.

Take graphene, a two-dimensional material. If one plucks a *C* atom out of the crystal matrix, there will be a rearranging to optimize the energetics. In *Si*, it is the tetrahedral bonding that disappeared. The two will have quite different behaviors. In fact, much of the discussion of stresses at surfaces in Chapter 5 doesn't apply to monolayer materials. Continuum approximation and the stress-strain equations will be quite inappropriate.

Adding a dopant to a crystal to make it of one polarity or another is also an "imperfection," a desired imperfection that will be somewhat randomly distributed in the crystal. A shallow donor—a dopant that easily contributes, for example by thermal excitation, an excess electron to the conduction band—and a shallow acceptor—a dopant that picks up an electron from the valence band—are energetically shallow states. Their wavefunctions are spatially extended and are composed of basis states over a small

Semiconductor Physics: Principles, Theory and Nanoscale. Sandip Tiwari.
 Published 2020 by Oxford University Press. DOI: 10.1093/oso/9780198759867.001.0001

spread in **k**. So, they can couple efficiently to either the conduction band or the valence band for the two cases. States can also come about—from dopants such as many of the transition elements—that are highly spatially localized. Such states then are composed of basis states from the entire Brillouin zone. We have developed the tools that let us analyze both of these situations. But such a distinction can be too simple for other point defects. For example, crystals may distort locally because of a defect, and then there can be strong dependence of the bonding structure on the state. In compound semiconductors, the departure from covalent bonding can lead to a strong dependence of the bonding structure on the charge state, so that both shallow and deep electronic states may be associated with the same impurity. We will also discuss such defects.

Composition, even if quite precise on the macroscale, can be randomly fluctuating at the nanoscale. If one makes a ternary crystal, $Ga_{1-x}Al_xAs$, for example, then even the distribution of Ga and Al species in this crystal will have some randomness. In materials such as $Ga_{1-x}In_xAs$, which are of useful bandgap, the mobility improvements are limited by the scattering of electrons that arises from the compositional randomness, a process that is called alloy scattering.

There are a number of different types of such deviation from perfection that exist in the crystal, some useful and some not, that we discuss in this chapter. Often, these are called defects. Defect is perhaps a poor moniker for these deviations, since they would then be always deleterious. A crack (a dislocation, in materials science) or a hole (a vacancy, in materials science) is both a reason for concern and yet may be useful for certain situations. So, the appearance of impurities can either be a defect or have a desired consequence. Intrinsic phenomena, that is, those that arise from within, such as the appearance of a vacancy, are defects. But, as an extrinsic phenomenon, the appearance of Si as a substitutional dopant may be desirable for doping. On the other hand, such a substitution also comes together with complexity that is important for ionization, metastability and dependences on temperature and light. So, there is much ambiguity here. The surfaces and interfaces of the crystal have even many more varieties of these deviations, some inevitable because of the breaking of translation symmetry.

Reading any of the classic texts on solids will give a perspective on intrinsic defects such as point defects, antisite defects, their combinations or extended defects such as dislocations, edge, screw and mixed dislocations, their extension to grain boundaries, their consequences for diffusion and conductivity, et cetera. Our interest is in those deviations from perfections that have unusual properties—largely with electrical or optical activity—that need to be looked at carefully. Accumulation of impurities at grain boundaries can give hardness. The strength of Damascus steel draws on the careful introduction of carbon—fullerenes can be seen here—randomly in the iron matrix. The same comment holds true for substitutions at vacancies. Alloys are a scaled form of these.

We will only focus on point sources of perturbation. They are quite important in the behavior of semiconductors and the transport that takes within them. We will also focus on only a select few of the deviations that are important for understanding the behavior of semiconductors. The presence of carriers, their variety of properties, including those of transport, and the dependence of the properties on temperature, and even leakage, such as occurs in "insulating" dielectrics, depend on them. Ours will be a relatively simpler physical view in order to reflect on the semiconductor physics and draw on the discussion up to this point. But our discussion will include such deviations that exist even in amorphous materials that are of technical importance. So, we will include in this discussion SiO_2 and other dielectrics, such as high permittivity materials, that are of interest.

7.1 *Defects and perturbations*

VACANCIES AND IMPURITIES appearing as substitutions are point perturbations where one has created a disturbance around the locale. The creation of a vacancy leaves broken bonds. An impurity changes the nature of bonding in this locale. In either case, it is an energetic perturbation. A shallow substitutional impurity, which we can model in the hydrogenic and effective mass approximation, leads to an electron being donated to the crystal or an electron being plucked out of it. This is related to the nature of the interaction. Figure 7.1 shows an illustrative picture of what a vacancy versus what a hydrogenic shallow substitutional impurity potential picture may look like.

In our Hamiltonian description, the perturbation is a Coulomb interaction. In general, it may consist of a short-range part and a long-range part. In Figure 7.1(a), the vacancy is shown as having a localized perturbation, and the shallow hydrogenic impurity as having a spread. These two will have quite different consequences for how the perturbation and the energy state or states related to them will appear.

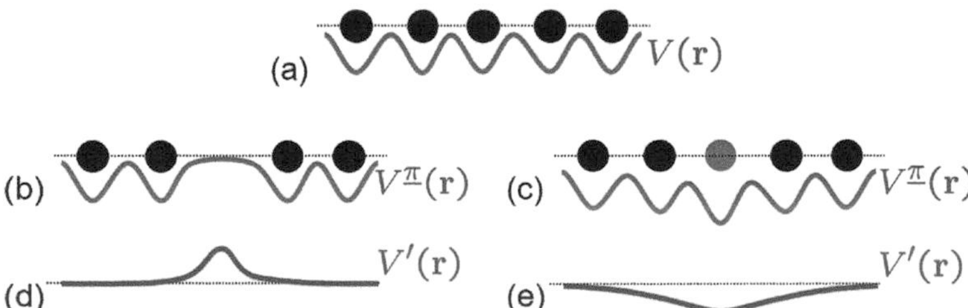

Figure 7.1: A sketch of the change in potential in a crystal as a result of a point perturbation (vacancy and hydrogenic substitutional impurity). Part (a) shows the periodic potential $V(\mathbf{r})$ consisting of potential contributions from core, Hartree and exchange correlation. Parts (b) and (c) show the potential when a vacancy and a hydrogenic impurity substitute for the atom. $\underline{\pi}$ is superscript to identify that this potential is perturbed from that of the ideal crystal. Parts (d) and (e) show the corresponding change between the potential in the presence of the point perturbation and the ideal crystal: (d) is a short-range perturbation and (f) is a long-range perturbation.

Our Wannier function-based analysis for effective mass (Section 4.7), and its use in understanding the ionization energy (E_i) of a shallow impurity, drew on the description of core levels. Effective mass description is useful if the variation of this core level description is slowly changing and therefore happens over a sufficient number of unit cells in the assembly for the description to become applicable. A case where there is a hydrogenic impurity in a semiconductor that is many unit cells long—so even in quantum-confined semiconductor structures—is an appropriate situation for using it. If the semiconductor is only a unit cell in width, this approximation will break. Since the electron now is spatially spread out, it couples to states closest in energy, which is why ionization energy becomes such a useful parameter. The short-range potential must have a minimum strength to be able to bind the electron to the state. If the kinetic energy is high enough, on average $\sim(3/2)k_BT$ in non-degenerate doping conditions, then if this ionization energy is smaller, the kinetic energy $T \geq E_i$, the electron will be unbound. On the other hand, take the example of vacancy in Figure 7.1(b) and (d). This is spatially local, and therefore it will be spread out in reciprocal space and allow interactions to occur with a broad range of states. We look at these cases individually to build a physical understanding.

In Chapter 20, we will discuss the usefulness of effective mass a little more to see how it is applicable in superlattices, where quite artificial bandstructure can be created through periodic layers. The electron needs to be spread out for this to work so that it feels the materials in its environment.

Take the quantum particle in a potential box problem. A confining potential of $V_0 > \hbar^2/8m^*a^2$ for a box of radius a leads to a bound state. Localization arising in this potential causes a non-vanishing momentum and kinetic energy through the Heisenberg uncertainty. So, the confinement must also exceed this expectation to prevent the electron from escaping. This argument is a 3-dimensional one. If the box is of 1 dimension, the bound states remain for all V_0. This arises from the simultaneous decrease of localization and leads to a fast-enough kinetic energy decrease. So, always be careful with models in 1 dimension. It will cause problems with deeper levels. 1 dimension also leads to interesting properties for the Ising model for spin that are grounded in this energy argument. And there are plenty of other similar examples that you can think up.

A vacancy is an unoccupied site of a crystal. In *Si*, a vacancy is quite straightforward: a missing *Si* atom site in the periodic arrangement. In the compound semiconductor *GaAs*, it may be at a *Ga* site (V_{Ga}) or it may be at an *As* site (V_{As}). If one has a foreign atom *X* at one of these sites, then we have X_{Ga} and X_{As} as the substitutional atoms. Si_{Ga} is a donor, and Si_{As} is an acceptor. *Si* in *GaAs* and other *IIIV* compounds is an amphoteric impurity: it can dope it either way, depending on which site it occupies. *S* in these same sites will become rather complicated. S_{Ga} will need a lot of thought, and since most of the semiconductors and dopants are group *III* or group *V* in a group *IV* crystal, or group *IV* in a group *IIIV* crystal, we will avoid this complexity, except for impurities such as the transition elements. In *III–V* crystals, there is one additional complexity: that of an antisite defects, where the opposite atom appears substitutionally. As_{Ga} is an *As* antisite defect. Ga_{As} is a *Ga* antisite defect.

Vacancies are unavoidable. Even nanotubes have them limiting their strength, and space elevators using nanotubes are a non-starter. The total energy for forming n vacancies in an assembly of N atoms is $\Delta\mathcal{G} - n\mathcal{H}_v - T\Delta S$. Considering only the entropy from atomic arrangements, $\Delta S = k_B \ln(N!/n!(N-n)!)$, where $\mathcal{H}$ is the enthalpy for formation of a vacancy. Minimizing the total energy using Stirling's formula gives

$$n = N\exp\left(-\frac{\mathcal{H}_v}{k_BT}\right).$$

Bond breaking is a few eV process, and substrates are made from boules grown near the melting point, so n will not be insignificant. For *Si*, $\mathcal{H}_v = 3.0\ eV$, and $T = 1687\ K$ for the melting point; with concentration frozen during cooling, $n \approx 5 \times 10^{12}\ cm^{-3}$.

If an impurity atom is not at a lattice site, then it is an interstitial atom. If an atom of the crystal goes interstitial, it is a self-interstitial. Crystals with reduced packing often exhibit such interstitials. Extrinsic impurities also quite often arise interstitially, since they cannot quite fit into the lattice site.

Vacancies, interstitials, antisite defects and substitutional impurities are all point defects. Some arise structurally and some are due to extrinsic impurities. An interstitial impurity is a compositional point perturbation, and these point perturbations arise in a variety of forms that are important for the semiconductor's behavior.

Point perturbations can also form complexes. A donor-acceptor pair can lower energy because of their opposite charges, so their appearance in several possible lattice spacings does occur stably. Multiple vacancies, particularly divacancies, are also possible, for thermodynamic reasons. A Frenkel defect is the movement of an atom from a lattice site to an interstitial site. In *Si*, one particular complex of importance is *H* bonding to an acceptor by giving it an electron. So, *p*-type *Si* has strong propensity for this. *GaAs*, being a compound and more weakly bonded, has plenty of examples of complexes. *DX* is one of these complexes. An *As* antisite defect is another example, leading to the *EL*2 center.

Indeed, one of the major reasons for instabilities and long term degradation is this *H* and *B* interaction. *H* is ubiquitous through the processes and, because of its small size, in SiO_2 and even in *Si*. A *B* acceptor bound to *H* becomes neutral. Doping has changed.

In the formation of any of these point perturbations, the atomic interactions are changing, so there will be new equilibrium positions, even if in some cases ever so slightly displaced in the vicinity of this perturbed region. This is lattice relaxation, which is not as easily observable. Only consequences such as changes in symmetry, or the appearance of a phenomenon, can be experimentally observed.

The name *DX* is a misnomer. It was initially believed to be a Si_{Ga} donor that was complexed with an unknown point perturbation. Evidence shows that it is a donor atom that is interstitial.

A symmetry breaking removes degeneracy. The average of an energy must remain the same, so levels must move up and down.

7.2 *Energetics of point perturbations*

To tackle the energetics of point perturbations, we need to tackle the bonding by valence electrons as well as the behavioral relationships arising in the core electrons. The high binding energy of non-shallow impurities can only arise in potential perturbation in which the core can have a meaningful role. This point goes back to our discussions of Chapter 1, and the treatment of independent and not-so-independent electrons in a periodic arrangement in the presence of the valence electrons. The perturbation arising from the point source is going to be the difference. The ideal potential of the crystal will include all those arising in atomic cores ($V_c^{\pi}(\mathbf{r})$), so the Hartree or Hartree-Fock potential ($V_H^{\pi}(\mathbf{r})$) with its antisymmetrization and the exchange correlation of the core electrons ($V_x^{\pi}(\mathbf{r})$). In Figure 7.1(b) and (c), then,

$$V^{\pi}(\mathbf{r}) = V_c^{\pi}(\mathbf{r}) + V_H^{\pi}(\mathbf{r}) + V_x^{\pi}(\mathbf{r}) \tag{7.1}$$

looks as shown and is the potential of the perturbed crystal (the fact that the crystal is perturbed is shown through the use of the superscript $\underline{\pi}$). And this potential is the input to the Hamiltonian equation of the crystal. In Figure 7.1, $V'(\mathbf{r})$ is the difference between the potential of the perturbed crystal ($V^{\underline{\pi}}(\mathbf{r})$) and the potential of the ideal crystal ($V(\mathbf{r})$).

Take the electron-core interaction term in the presence of perturbation:

$$V_c^{\underline{\pi}}(\mathbf{r}) = V_c(\mathbf{r}) + V_c'(\mathbf{r}) \tag{7.2}$$

is the sum of the core term of the ideal crystal and the change arising in the point perturbation. We consider only one point perturbation, and it is far away from others. The change $V_c'(\mathbf{r})$ then must asymptotically vanish far away, irrespective of whether it is a short-range or a long-range perturbation.

A substitutional impurity can have a variety of shell structures. These will have different consequences for how $V_c'(\mathbf{r})$ behaves over the short range and the long range. If a substitutional impurity has a core similar to that of the crystal, for *Si*, say, *P*, so that $1s^2 2s^2 2p^6$ is the core, then the charge of the core Ze, subscripted by i for the impurity, and h for the host, differ by the number of valence electrons. For *P*, or other such core impurities placed in place of the host, the largest change for potential energy of the valence electrons arises as

$$V_c'(\mathbf{r}) \approx -\frac{1}{4\pi\epsilon(0)} \frac{e^2(Z_i - Z_h)}{|\mathbf{r}|}, \tag{7.3}$$

an expression that loses accuracy the closer one gets to the impurity atom. This inaccuracy arises in that as one comes closer to the core, it will be necessary to account for non-point and additional quantum-mechanical constraints, including those of exchange and correlation. This argument also implies that if the cores of the host and the impurity are different, there will be additional energetic corrections that will be needed. In situations where Equation 7.3 is applicable, since it varies slowly over unit cell size scales, it is a long-range potential. So, impurities with identical cores—isocoric impurities—have a smooth and long-range potential. And because of its smoothness and long range, the effective mass approximation applies. This is the case for Figure 7.1(e).

In many of these problems, there is a need for some care w.r.t. the dielectric function. When a charge is spread out over a polarizable medium, this polarization must be accounted for in the screening. And care must be taken to identify static or high-frequency limits. The potential term for a valence electron—spread out—in a static environment then reflects a dielectric function of $\epsilon(0) = \epsilon_r(0)\epsilon_0$. When an electron is confined to the core or very near it, then it doesn't feel the polarization. The behavior is that of vacuum (a permittivity of ϵ_0).

If impurities do not have identical cores but have the same number of valence electrons, for example, *C* in an *Si* host, or similar substitutions for compound semiconductors, then the perturbation $V_c'(\mathbf{r})$ will reflect the consequences of differences in core electron charge distributions together with exchange and correlations. These decay rapidly; that is, are short range. This short-range consequence also arises for interstitials as well as for vacancies, as shown in Figure 7.1(d). For this latter case, the perturbation is approximately the negative of the potential of the missing host atom, that is, $V_c'(\mathbf{r})$.

Now, suppose the substituting atom has a different core charge and valence charge; for example, Cd : $[Kr]5s^2 4d^{10}$ in Si : $[Ne]3s^2 3p^2$. Cd has two electrons in the valence (a 2-fold charge in the core) and Si has four. For a compound semiconductor, a similar situation occurs with Sn : $[Kr]4d^{10}5s^2 5p^2$ at a Ga : $[Ar]3d^{10}4s^2 4p^1$ site. The core charges are different. Now, the perturbation potential must also include the screened Coulomb potential of Equation 7.3 to account for this change in core charge together with the other short-range potential encompassing any other differences of core such as the extent and screening's spatial dispersion, neither of which are in this equation. A stronger Coulomb potential is more effective in pulling electrons closer to the core. So, for a similar short-range potential, the perturbation consequence of a stronger Coulomb potential is larger.

To this, one must now add any changes to the Hartree $V_{\overline{H}}^{\pi}(\mathbf{r})$ and the exchange $V_{\overline{x}}^{\pi}(\mathbf{r})$ potentials due to the point perturbation. For the Hartree potential, this means a modification of Equation 1.69, which was written for the ith electron interacting with all the other electrons in the N-electron system. $o(\mathbf{z}')$ represented the summation over the occupation of states o_i for the ith electron. This point-perturbed crystal has states that include those in the bands which were the eigenenergies of the allowed unperturbed ideal crystal, and states whose eigenenergies are different as they arise from the perturbation source. These states may very well, and often are, in the gap. For an ideal crystal, the states would have been pure Bloch states. For the perturbed crystal, the former states are still Bloch states in the 0th order, or, more accurately, superpositions of Bloch states due to the perturbation. The latter states, for example, those arising from P introduced into the Si crystal, in the "forbidden" energy region, are spatially localized to the region of the point perturbation. The former are extended states, and the latter are localized. Let superscripting $\underline{xtd}$ identify extended states, and $\underline{loc}$ identify the local states, and let these states be denoted by the quantum number ν.

These superposition Bloch states also extend spatially over the entire infinite crystal for most of the energies.

The perturbed crystal's Hartree potential arises in the extended and local contribution, so, for the ith electron,

$$V_H^{\pi}(\mathbf{r}_i) = V_H^{\underline{xtd}}(\mathbf{r}_i) + V_H^{\underline{loc}}(\mathbf{r}_i), \text{ with}$$

$$V_H^{\underline{xtd}}(\mathbf{r}_i) = \sum_k^{\underline{xtd}} \int \frac{1}{4\pi\epsilon_0} \frac{e^2}{\mathbf{r}_i - \mathbf{r}'} \langle u_{\nu k}(\mathbf{z}')|u_{\nu k}(\mathbf{z}')\rangle \, d\mathbf{z}'$$

for extended states $\nu_k^{\underline{xtd}}$, and

$$V_H^{\underline{loc}}(\mathbf{r}_i) = \sum_{k \neq i}^{\underline{loc}} \int \frac{1}{4\pi\epsilon_0} \frac{e^2}{\mathbf{r}_i - \mathbf{r}'} \langle u_{\nu k}(\mathbf{z}')|u_{\nu k}(\mathbf{z}')\rangle \, d\mathbf{z}', \qquad (7.4)$$

where we have subsumed $o(\mathbf{z})$ and o_i of Equation 1.69. The electrons at the perturbing center affect the localized Hartree part $V_H^{\underline{loc}}(\mathbf{r}_i)$ via their numbers, which will make the Hartree contribution repulsive, and via the wavefunction in the summation. The wavefunction is more localized for fewer electrons at the center and spreads out more if more electrons are added. So, depending on the number of electrons localized at the center, the wavefunction relaxes spatially. The extended part $V_H^{\underline{xtd}}(\mathbf{r}_i)$ is also affected because the number of extended electrons is changed by the number of electrons that localize at the center as well as by the relaxation of the wavefunction. The first part has an inverse dependence on the extended electron count and so is small. The second part, however, can be significant, since all the occupied extended states are affected. Spatially, close to the localized electron, the probability amplitude is suppressed by Coulomb repulsion. This positive excess charge around the localized electrons is screening the Coulomb potential of the localized electron. So, the localized Hartree perturbation potential part may be replaced by a screened potential $V'_H(\mathbf{r}_i)$ while simultaneously changing the extended-Hartree-perturbation-potential-screened potential $V_H^{\underline{xtd}}(\mathbf{r}_i)$ by the Hartree potential of the unperturbed crystal. So,

$$V_H^{\pi}(\mathbf{r}_i) = V_H^{\underline{loc}}(\mathbf{r}_i) + V_H^{\underline{xtd}}(\mathbf{r}_i) = V_H(\mathbf{r}_i) + V'_H(\mathbf{r}_i), \text{ with}$$

$$V'_H(\mathbf{r}_i) = \sum_{k \neq i}^{\underline{loc}} \int \frac{1}{4\pi\epsilon_0} \frac{e^2}{\xi(\mathbf{r}_i, \mathbf{r}')} \langle u_{\nu k}(\mathbf{z}')|u_{\nu k}(\mathbf{z}')\rangle \, d\mathbf{z}', \qquad (7.5)$$

where $\xi(\mathbf{r}_i, \mathbf{r}')$ is a screening function.

We also need to account for exchange potential. Following Chapter 1, exchange potential encapsulates the Coulomb interaction with the exchange hole that arises in the exclusion of two electrons of the same spin in the same space. The spatial uncertainty is smaller for localized electrons than for extended ones. So, following an approach identical to that of the Hartree argument, we split the exchange potential $V_x^{\pi}(\mathbf{r}_i)$ that the ith electron at the center feels to a localized part ($V_x^{\underline{loc}}$), where there is a fraction of the electrons, and an extended part ($V_x^{\underline{xtd}}$), where the rest of the electrons centered on the center are. The extended part then can be substituted by the exchange potential of the unperturbed crystal $V_x(\mathbf{r}_i)$ together with an effective exchange potential $V'_x(\mathbf{r}_i)$. So,

$$V_x^{\pi}(\mathbf{r}_i) = V_x^{loc}(\mathbf{r}_i) + V_H^{xtd}(\mathbf{r}_i) = V_x(\mathbf{r}_i) + V'_x(\mathbf{r}_i), \text{ with}$$

$$V'_x(\mathbf{r}_i)u_\nu(\mathbf{r}_i) = -\sum_{k\neq i,\sigma_k=\sigma_i}^{loc} \int \frac{1}{4\pi\epsilon_0}\frac{e^2}{|\mathbf{r}_i - \mathbf{r}'|}\langle u_{\nu k}(\mathbf{z}')|u_{\nu k}(\mathbf{z}')\rangle\, u_{nuk}(\mathbf{r}_i)d\mathbf{z}', \tag{7.6}$$

with the summation for particles of identical spin; that is, $\sigma_k = \sigma_i$.

We have now determined the three terms that constitute the entire perturbation due to the center in Equation 7.1, so that the governing equation for the ith electron reads

$$\left[-\frac{\hbar^2}{2m_0}\nabla_i^2 + V(\mathbf{r}_i) + V^{\pi}(\mathbf{r}_i)\right]|u_n\rangle = E_n|u_n\rangle, \tag{7.7}$$

with $V^{\pi}(\mathbf{r}_i)$ composed of Coulomb, Hartree and exchange parts. In the lowest order, the Coulomb term dominates, and the index i can be eliminated because all the electrons feel the core perturbation potential, so

$$\left[\hat{\mathscr{H}}_0 + V(\mathbf{r}) + V_c^{\pi}(\mathbf{r})\right]|u_n\rangle = E_n|u_n\rangle. \tag{7.8}$$

For a center that has only one electron localized, the Hartree and exchange perturbations vanish and this equation is precise.

For the Hartree potential of the electron i in the localized state, V_H^{loc} is the perturbation. We can use the position-dependent screening to evaluate the correction following the argument leading to Equation 7.5:

$$\langle u_{ni}|V'_H|u_{ni}\rangle = \sum_{k\neq i}^{loc} \int \frac{1}{4\pi\epsilon_0}\int d\mathbf{r}\int \frac{e^2}{\xi(\mathbf{r},\mathbf{r}')}\frac{|u_{\nu i}(\mathbf{z})|^2|u_{\nu k}(\mathbf{z}')|^2}{\xi(\mathbf{r}_i - \mathbf{r}')}\, d\mathbf{r}'. \tag{7.9}$$

The integration in $\mathbf{r}'$ can be simplified by replacing the slowly varying $1/\xi(\mathbf{r}_i - \mathbf{r}')|\mathbf{r}_i - \mathbf{r}'|$ to its value at a mean position $\bar{z}$. With this, since the integration over $\mathbf{r}'$ of the normalized wavefunction $u_{\nu k}(\mathbf{z}')$ is unity, this Hartree energy reduces to

$$U_{\nu i} = \frac{e^2}{4\pi\epsilon_0}\int \frac{1}{\xi(\mathbf{r},\bar{\mathbf{r}})|\mathbf{r} - \bar{\mathbf{r}}|}\, d\mathbf{r}. \tag{7.10}$$

$U_{\nu i}$ is the *Hubbard energy*.

The correction due to the third term, the exchange energy arising in $V'_x(\mathbf{r})$, is grounded in spin. In the Bloch states, there is a balance of spin-up and spin-down electrons under equilibrium. But, for electrons localized at the center, this need not be true. Let $n = n_\uparrow + n_\downarrow$ be the localized electron number at the center of the two spin orientations. So, the total spin projection is $M_s = (1/2)(n_\uparrow - n_\downarrow)$. Therefore,

Where the electron is localized can have a variety of strong consequences. In S. Tiwari, "Nanoscale device physics: Science and engineering fundamentals," Electroscience 4, Oxford University Press, ISBN 978-0-198-75987-4 (2017), we will encounter examples of metal-insulator transition due to this localization because of the Coulomb energy and kinetic energy changes that arise in the localization. It is really not that far from the truth to say that much of what we have been analyzing through these energetics-focused equations is the play between potential energy and kinetic energy. One leads to more localization, while the other leads to spreading out and moving around.

$$n_{\uparrow} = \frac{1}{2}(n + 2M_s), \text{ and } n_{\downarrow} = \frac{1}{2}(n - 2M_s). \tag{7.11}$$

Let σ_i denote the spin quantum number of the state ν_i. The exchange contribution is then

$$\begin{aligned}\langle u_{\nu i}|V_x'|u_{\nu i}\rangle &= -(n_{\sigma_i} - 1)\frac{e^2}{4\pi\epsilon_0}\int d\mathbf{r}' \int \frac{u^*_{\nu'}(\mathbf{r}')u^*_{\nu}(\mathbf{r})u_{\nu}(\mathbf{r}')u'_{\nu}(\mathbf{r})}{\mathbf{r}-\mathbf{r}'}\, d\mathbf{r} \\ &= -(n_{\sigma_i} - 1)J_{\nu_i} \text{ in general, with}\end{aligned}$$

$$J_{\nu} = \frac{e^2}{4\pi\epsilon_0}\int d\mathbf{r}' \int \frac{u^*_{\nu'}(\mathbf{r}')u^*_{\nu}(\mathbf{r})u_{\nu}(\mathbf{r}')u_{\nu'}(\mathbf{r})}{\mathbf{r}-\mathbf{r}'}\, d\mathbf{r} \tag{7.12}$$

as an exchange integral. In writing Equation 7.12, the orbital identity of the quantum state ν' has been left out.

This procedure outlines how, given a certain number of electrons at a center, one may determine the Coulombic, Hartree and exchange contributions of the perturbation that it causes.

7.3 *Electrons at the point perturbation center*

We now have a mathematical description that self-consistently describes the energetics of the point source, as well as the number of electrons in the localized state, because the Hamiltonian and the thermodynamic equilibrium constraint complete the constraining conditions. At $T = 0$ K, all the states with eigenenergies that are below the Fermi energy E_F are occupied, and this tells us the number of electrons that are localized. If you change the temperature, the Fermi energy changes, and therefore, in general, the number of electrons at the center will also change. So, depending on the conditions of the semiconductor system, there can be a change, and this change is not only due to temperature, that is, in thermal equilibrium conditions, but is also due to other external electrochemical stimuli when away from equilibrium. This latter does become important for some of the point perturbation centers that have eigenenergies deeper in the bandgap.

One can make some estimates of the number of electrons to consider in this point perturbation energetics based on physical arguments.

As a starting point assume that the process of creating the center did not change the charge state of the crystal. It is still neutral. An interstitial impurity, for example, a transition element in *Si*, was introduced in its entirety as an atom. The same, when a substitutional impurity, say, *P* in *Si*, is introduced. If the point perturbation center is a vacancy in *Si*, it is there because a neutral *Si* atom was plucked from the crystal.

Now consider the case of substitutional *P* in *Si*, so it is an atom with chemical bonding not unlike those of other *Si* atoms bonding to their surrounding *Si* atoms. *P*'s valence shell is $3s^2 3p^3$. The *s* and *p* states are involved in the formation of the valence band states of the crystal as well as the states that appear in the bandgap. Four of the *P* electrons occupy the valence band states, leaving one electron to occupy the localized state. For a neutral *P*-substitutional impurity, there is 1 electron localized. Had this been a substitutional *B* ($[He]2s^2 2p^1$) atom, the count would be -1. A hole can occupy a state localized at the center. For any such impurity atom, the electrons bonding have energies in the valence band. The rest of the electrons, a number that may be positive or negative, are the active electrons.

The bonding electrons determine the oxidation state of the atom. An oxidation state or number is the degree of oxidation; that is, loss or gain of electrons during the bonding.

Many transition atoms appear in crystals such as of *Si* both substitutionally and interstitially. The oxidation state is now different. For a substitutional atom, the oxidation state still comprises the electrons of the atom that are involved in the bonding. But the interstitial form needs additional considerations. Such atoms are weakly bonded, so the number of electrons from the atom in the valence band is vanishingly small. All the electrons are available for occupying the states in the bandgap. *Fe*: $[Ar]4s^2 3d^6$ in *Si* is a good example. Substitutional *Fe* has an oxidation state of 4+, while interstitial *Fe* is in an oxidation state of 0+. So, with 8 valence shell electrons, there are 4 electrons that can be at the center for substitutional conditions and 8 for interstitial conditions.

If the short-range potential dominates, then the solution to the Hamilton/Schrödinger equation (Equation 7.7) for the problem, under the enormous simplification that translational symmetry provides, is quite direct using perturbation theory. If the long-range potential is important, then this equation may be partitioned into two. One is for the periodic potential of the crystal, whose solution is our $E(\mathbf{k})$ description. And the other equation is for this perturbation potential, and this is precisely what we did in deriving the effective mass equation of Equation 4.69. Our caveat is that $V_{\frac{\pi}{c}}(\mathbf{r})$ in Equation 7.1 arises in the core, and this was excluded in the derivation of the effective mass equation. Coulomb point perturbations arising from the net charge imbalance of the core are fine, but those arising from within it due to the short range within the core are not. A useful way to distinguish this is to look back at the derivation of Equation 4.69 and note that the envelop functions are slowly varying over the atomic scale. Any perturbation that is more rapidly varying than that conflicts with the assumption of this slow change of this derivation.

We may now apply this learning to explore the behavior of point perturbations.

7.3.1 *Shallow dopants*

SHALLOW DONOR AND ACCEPTOR IMPURITIES, such as P or B in Si—substitutional impurities—have a long-range perturbation $V'(\mathbf{r})$. These are all examples of atoms that are near the row or column of the periodic table of the atom being replaced. Let $\Delta Z = Z_i - Z_h$ be the difference in the impurity and host core charge numbers. This point perturbation can be either positive or negative. Then,

$$V'_c = -\frac{e^2}{4\pi\epsilon|\mathbf{r}|}\Delta Z, \tag{7.13}$$

and to this one must add the Hartree potential $V'_H(\mathbf{r})$ and exchange potential $V'_x(\mathbf{r})$ due to other localized electrons of the center. *If the impurity atom is only one valence electron different from the host, then the Hartree and exchange perturbations vanish.* There is no electron-to-electron interaction to be included. So, V'_c is the entire perturbation. If, however, the magnitude of the valence charge difference exceeds this, S for example in Si, then the Hartree and exchange potentials will need to be included.

The effective mass equation suffices for this $\Delta Z = \pm 1$ case. If $\Delta Z = 1$, the states where the excess electron may dwell will appear close to the conduction band in the bandgap. The electrons involved in bonding all go to the valence band. Likewise, if $\Delta Z = -1$, the states will be in the bandgap near the valence bandedge.

First, consider the simplest case where there is one conduction band minimum and one valence band maximum at the Brillouin zone center, and $\Delta Z = \pm 1$. This implies

$$\begin{aligned} E_c(\mathbf{k}) &= E_g + \frac{\hbar^2}{2m_c^*}\mathbf{k}^2, \quad \text{and} \\ E_v(\mathbf{k}) &= -\frac{\hbar^2}{2m_v^*}\mathbf{k}^2. \end{aligned} \tag{7.14}$$

The effective mass equations with $\mathscr{H}' = V'_c$, for the states near the conduction and valence bands are

$$\begin{aligned} \left[E_g - \frac{\hbar^2}{2m_c^*}\nabla_{\mathbf{r}}^2 \mp \frac{e^2}{4\pi\epsilon|\mathbf{r}|}\Delta Z\right]\varphi_c(\mathbf{r}) &= E\varphi_c(\mathbf{r}) \quad \text{or} \\ \left[-\frac{\hbar^2}{2m_c^*}\nabla_{\mathbf{r}}^2 \mp \frac{e^2}{4\pi\epsilon|\mathbf{r}|}\Delta Z\right]\varphi_c(\mathbf{r}) &= (E - E_g)\varphi_c(\mathbf{r}), \quad \text{and} \\ \left[\frac{\hbar^2}{2m_v^*}\nabla_{\mathbf{r}}^2 \mp \frac{e^2}{4\pi\epsilon|\mathbf{r}|}\Delta Z\right]\varphi_v(\mathbf{r}) &= E\varphi_v(\mathbf{r}) \quad \text{or} \\ \left[-\frac{\hbar^2}{2m_v^*}\nabla_{\mathbf{r}}^2 \pm \frac{e^2}{4\pi\epsilon|\mathbf{r}|}\Delta Z\right]\varphi_v(\mathbf{r}) &= (-E)\varphi_v(\mathbf{r}). \end{aligned} \tag{7.15}$$

These are equations that appear similar to the Schrödinger equation for hydrogen with differences in the parameters, signs and references. An attractive potential will lead to discrete eigenenergies that are lower than the reference, and a continuum of them above it. For repulsive potentials, there exist positive eigenenergies. So, discrete energies appear below the conduction band and above the valence band within the bandgap.

For the conduction band, utilizing the hydrogen model solution, the eigenenergies are

$$\begin{aligned} E_n &= E_g - \frac{E_B}{n^2}, \text{ with} \\ E_B &= \frac{m_c^* e^4 |\Delta Z|^2}{2(4\pi\epsilon)^2\hbar^2} = \frac{m_c^*/m_0}{(\epsilon/\epsilon_0)^2}|\Delta Z|^2 E_R, \text{ with} \\ E_R &= \frac{m_0 e^4}{2(4\pi\epsilon_0)^2\hbar^2} \end{aligned} \tag{7.16}$$

as the Rydberg energy—the binding energy of the principal quantum number of 1—of the hydrogen atom. The solution wavefunction, again using the hydrogen analogy, is φ_{cnlm_l}, where c identifies the conduction band association, n the principal quantum number, l the orbital quantum number and m_l the azimuthal orbital quantum number. For $n=1$, $l=0$ and $m_l=0$, the wavefunction solution is

$$\begin{aligned} \varphi_{c100}(\mathbf{r}) &= \frac{1}{\sqrt{\pi a_B^{*3}}} \exp\left(-\frac{|\mathbf{r}|}{a_B^*}\right), \text{ with} \\ a_B^* &= \frac{4\pi\hbar^2\epsilon}{m_c^* e^2} = \frac{1}{|\Delta Z|}\frac{\epsilon/\epsilon_0}{m_c^*/m_0} a_B, \text{ where} \\ a_B &= \frac{4\pi\hbar^2\epsilon_0}{m_0 e^2}. \end{aligned} \tag{7.17}$$

a_B is the Bohr radius and a_B^* is the effective Bohr radius for this effective mass hydrogenic model of the point perturbation. For hydrogen, the lowest orbit's binding energy is the Rydberg energy of $E_R \approx 13.6\ eV$, and the maximum Bohr radius is $a_B = 0.05\ nm$. In the semiconductor, this hydrogenic point perturbation, with $m_c^*/m_0 \approx 0.26$, $\epsilon/\epsilon_0 \approx 11.9$, and $|\Delta Z| = 1$, $a_B^* \approx 2.42\ nm$ and $E_B \approx 25\ meV$. The electron cloud has spread out over nearly 5 unit cells and therefore is extended. This binding energy for $n=1$ is a very small fraction of the bandgap, of the order of room temperature thermal energy, and the states with higher n will be even closer. The wavefunction decays with the effective Bohr radius length scale. It is a localized state, with a weak attraction between the impurity atom and the electron but spread out enough to satisfy the constraints that we placed in our analysis following Figure 7.1(c) and (e).

Only a small spread in **k** contributes, and this can be seen through the following argument. For $\varphi_{c100}(\mathbf{r})$, the Fourier transform varies inversely as $[1 + (2a_B^* k)^2]$. At the edge of the first Brillouin zone, $a_B^* k \approx a_B^* \pi / a$, which is of the order of 10. So, the Fourier transform factor falls by nearly 10^4 between the center and the edge.

So, this extended state is a shallow state. Since $Z_i = Z_h + 1$, the impurity has one excess electron over the host atom. It formed bound states in the bandgap. But it also leads to eigenvalues in the continuum of energy eigenvalues. The number of states has not changed; only, an excess electron has appeared, and potential perturbation has been introduced. But since bounded states appear in the bandgap, the density of states in the nearly continuous part of the energy has changed, while the net number of states has remained constant. The number of states in the valence band remains the same and, except for the additional electron of the impurity, all the rest of the valence electrons dwell in the valence band. So, all the host states and the states of the impurity, except for this one, appear in the valence band. At $T = 0\ K$, this excess electron will go into the lowest energy bound state; that is, the $n = 1$ level below the conduction band. Raising the temperature excites the electron from this shallow level into the conduction band, and now it is not localized anymore. It has become a nearly free electron that is mobile. This level is therefore a donor level, and the impurity is a donor impurity.

This discussion of states tied closer to the conduction band also holds true for a valence band with a negative point perturbation. The binding energy and the effective Bohr radius now are

$$E_n = \frac{E_B}{n^2} = \frac{m_v^* e^4 |\Delta Z|^2}{2(4\pi\epsilon)^2 \hbar^2} = \frac{m_v^*/m_0}{(\epsilon/\epsilon_0)^2} |\Delta Z|^2 E_R, \text{ and}$$
$$a_B^* = \frac{4\pi\hbar^2\epsilon}{m_v^* e^2} = \frac{1}{|\Delta Z|} \frac{\epsilon/\epsilon_0}{m_v^*/m_0} a_B. \tag{7.18}$$

At $T = 0\ K$, the $n = 1$ level will be empty. With an increase in temperature, an electron from the valence band can be excited to this impurity level, leaving a hole behind. Since the level accepts an electron, it is an acceptor level, and the impurity is an acceptor.

This entire discussion was based on an isotropic single conduction and a single valence band at the zone center. Bands can be anisotropic, band minima can be away from the zone center, and even if they are located at zone center, there can be band degeneracy, as in the case of a valence band with light- and heavy-hole bands.

A shifting of the band extremum does not affect binding energies. But anisotropy and valley degeneracy will. This requires a multiband effective mass equation, whose discussion we will skip, but whose conclusions we will comment on. Our $\mathbf{k} \cdot \mathbf{p}$ discussion (Section 4.6) and its subsequent use for valence bands (Section 4.8) showed the use of the diagonalization of the Hamiltonian matrix employing a basis set. These matrix elements are linear or quadratic functions of projections of $\mathbf{k}$. For degenerate bands, this implies a multi-component effective mass equation. The one-component envelope function now becomes a multi-component envelope function. And recall that these envelope functions were developed from the Bloch functions, which now have the complexities of interaction arising in the degeneracy or the off-center multiple valleys. For multiple valley minima, intervalley matrix elements become important. In the case of anisotropy, unlike the hydrogen-like model, now one needs to account for the anisotropy in the reciprocal space. A simple approximation is to replace the mass, for example $1/m_c^*$ of *Si*, by $(1/m_l^* + 2/m_t^*)$ as a direction-averaged reciprocal effective mass. For *P* in *Si*, Figure 7.2 shows the results of the hydrogen model and the changes due to the inclusion of effective mass anisotropy. The intervalley matrix element leads to a 3-fold splitting of the *P* donor ground state that we have simply described as a single energy level. It is 3-fold, because of the 6 minima along the three orthogonal coordinate axes.

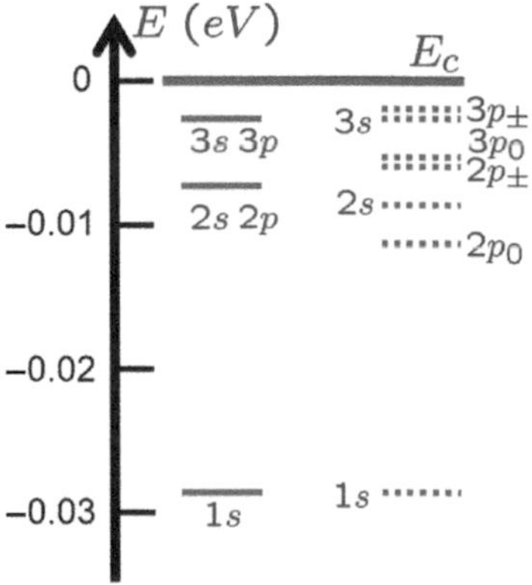

Figure 7.2: Ground and higher energy states of a substitutional *P* donor impurity in *Si*. (a) The results using the hydrogenic model. (b) The changes arising in anisotropy.

Although our description of these ionization or binding energies has been somewhat detailed, it is still incomplete. Fortunately, since most of these binding energies are small, the consequences of any inaccuracies are minor. Table 7.1 show an example of one additional factor: that of chemical shift. The table lists the binding energies of different shallow donors and acceptors. *Si*'s show a fair spread, but *Ge*'s do not, nor do those of *GaAs*. In the case of *GaAs*, the large difference between the donor and acceptor binding energies arises in the effective masses and the degeneracy of the valence band. In *Si*, the donors and acceptors have large binding energies, with *In* as quite an outlier. The table also shows that the binding energy has a dependence on the chemical nature of the impurity atom; that is, there are chemical shifts also beyond the perturbation $V'(\mathbf{r})$ of our model.

When a substitutional impurity has a $|\Delta Z| > 1$ while having an identical core, then these excess electrons or holes that are only

	Donor	E_B (meV)	Acceptor	E_B (meV)
Si	*P*	45	*B*	45
	As	54	*Al*	68
	Sb	43	*Ga*	74
	Bi	71	*In*	153
Ge	*P*	13	*B*	11
	As	14	*Al*	11
	Sb	10	*Ga*	11
	Bi	13	*In*	12
GaAs	C_{Ga}	5.9	Be_{Ga}	28
	Si_{Ga}	5.8	Mg_{Ga}	29
	Ge_{Ga}	5.9	Zn_{Ga}	31
	S_{Ga}	5.9	C_{As}	27
	Se_{Ga}	5.8	Si_{Ga}	35

Table 7.1: Experimentally measured binding energies of different shallow donors and acceptors in *Si*, *Ge* and *GaAs*.

weakly bound to the impurity atom can be thermally excited to the nearby band, as with the singly ionizable impurity, and they leave behind an ionized impurity that could be singly or multiply ionized. These carriers bound on the center are describable through the Hubbard energy of Equation 7.10. This Hubbard correction shifts the energy higher. The magnitude of this shift depends on the number of carriers bound on the center. For *S* substitutional in *Si*, the 2 excess electrons of *S* can appear with up and down spin at the level, and the Hubbard correction term will be given by

$$U = \frac{e^2}{4\pi\epsilon_0} \int_\Omega d\mathbf{r} \int_\Omega \frac{|\varphi_{c100}(\mathbf{r})|^2 |\varphi_{c100}(\mathbf{r}')|^2}{\xi(\mathbf{r}, \bar{\mathbf{r}})|\mathbf{r} - \bar{\mathbf{r}}|} \, d\mathbf{r}'. \tag{7.19}$$

This is the energy by which the *S* level (a 1*s* level, in hydrogenic notation) is shifted up. When a neutral *S*, that is, S^0 ionizes and becomes S^+, this S^+'s level (again, a 1*s* level, in hydrogenic notation) does not have this Hubbard energy shift, since it has only one electron. So, this level of S^+ is shifted down by U from the level of S^0. S^0 is a donor at 0.31 *eV*. S^+ is a donor at 0.59 *eV*. These are both very large energies compared to an estimate of $4 \times 30 = 120$ *meV* for a hydrogenic model.

S in *Si* is now not a shallow donor but a deep donor and a deep center.

We have stressed that a three-dimensional confinement a barrier $V_0 > \pi^2\hbar^2/2m_0a^2$, where V_0 is the confining potential and a is the size of the confinement box, is necessary for a bound state. There will be more on this in Chapter 20. So, a narrow quantum well may have no allowed confined state subject to the narrowness-to-confining potential constraints. Minimum potential confinement—strength—is necessary for any magnitude of the short range. But this confinement argument is dimensionality dependent. A one-dimensional potential well will always have bound states. The confining potential depth must exceed the expectation value of kinetic energy to keep the particle confined. The average kinetic energy $\propto \langle \mathbf{p}^2 \rangle / 2m_0$ decreases faster than the localization due to potential under Heisenberg uncertainty, which relates the uncertainty in momentum and position as a linear product. Kinetic energy is in the second order of momentum. So, these binding behaviors of shallow or deep centers should be expected to be dependent on dimensionality.

7.4 *Deep centers*

SHORT-RANGE AND LONG-RANGE potential perturbations have a major difference in their implication for binding. Shallow levels have long-range perturbation—a confining long-range Coulomb potential—and their short-range potential is inadequate for strong binding. Spatially, the eigenfunctions stretch out at length scales of a_B^*.

A deep center has a strong-enough confining short-range potential to create strongly bound states. So long as this binding energy is larger than that due to the long-range Coulomb potential, the short-range potential will dominate, and the behavior of such a center will be significantly different from that of the Coulomb-potential-limited shallow center. This is what characterizes deep centers or deep levels. Comparing the energy difference between the valence s level energies of impurities with those of host atoms for donors, and of the valence p level energies for acceptors, serves as a reasonable indicator of whether an impurity is likely to be shallow or deep, due to the stronger perturbation potential when the impurity is incorporated in the crystal. Given the centrality of the short-range potential confinement, one may not employ the effective mass approach. Bloch functions in a short spread in **k** are insufficient for constructing the wavefunction solution. Using atomic orbitals, that is, tight binding, is more suitable for understanding the deep levels.

It is a mischaracterization to think of deep in the sense of energy, that is, of an eigenenergy quite separated from the bandedge energies. This separation deeper into the bandgap is often true, but there are also these "deep" centers that may be very close to bandedge or even in the bands. *It is the dominance of short-range potential and the spatial confinement of electrons to the atomic size scale that characterizes them as "deep."*

Deep levels can be donors and acceptors or both. While shallow levels in our discussion have been singly ionized (S doubly ionized became deeper due to the Hubbard energy), the deep levels can be multiply ionized. Neutral and these multiply ionized states can continue to capture carriers from the bands.

Deep levels are poor at providing free carriers, but effective at removing them, precisely the opposite of what shallow levels do. Also, while shallow levels reach a steady state when equilibrium is disturbed through the transitions coupling the levels to the bands, and the thermal transitions for this process are entirely dominated by the nearest band, deep levels will have long dwell and empty times, even if the transition itself—as with shallow levels—happens rapidly. So, by coupling to both bands, they are effective generation and recombination centers affecting the lifetime of the material at moderate carrier concentrations. Shallow levels couple preferentially to one or the other band and hence do not directly affect lifetime at moderate carrier concentrations.

7.4.1 *Tight binding as a defect-molecule model*

WE WILL EMPLOY s^2p^2, our favorite semiconductor valence structure, as a toy model for understanding deep levels. But, in doing so, we will only outline the tight binding and use of atomic orbitals in understanding deep levels to complement the Hartree-exchange-correlation discussion of Section 7.2. There are other methods for treating highly localized states, for example, scattering theory with expansion via Wannier functions, that provide greater accuracy but with much greater computational effort. Tight binding is more intuitive. Employing it also elucidates for us the differences with the other methods for calculations in the semiconductors discussed in this text. The existence of a defect in the midst of order is the complement of the existence of a molecular order in the midst of ordered emptiness. So, a defect in a crystal is like a molecule in reverse, and this gives us quite interesting insights into understanding defects that the analytic formulation of Section 7.2

does not. So, ours will be simple toy models again, where we will ignore the formation of bands because of the interactions within the crystal. This is dealing with local atomic orbital interaction as if in a molecule. For this reason, this simple model is a defect-molecule model.

Take *Si* with its $[Ne]3s^2 3p^2$ electron structure. Tight binding takes the eigenfunctions of this valence as the basis set for the solid. There are one *s* orbital ($|s\rangle$) and three *p* orbitals ($|p_x\rangle$, $|p_y\rangle$ and $|p_z\rangle$) on each atom that exists at the fixed lattice site. So, starting from the problem that each electron obeys $\hat{\mathscr{H}}|\psi_k\rangle = E_k|\psi_k\rangle$ for the *k*th energy level, with $|\psi_k\rangle$ as the corresponding eigenvector, we build a wavefunction from the free atom basis, and here identifying by *i* the specific atom from which the basis arises, and α its state,

$$|\psi_k\rangle = \sum_{i,\alpha} c^k_{i\alpha}|u_{i\alpha}\rangle. \tag{7.20}$$

This is now a linear combination of the atomic orbitals. If we bring two atoms, *i* and *j*, together, the solution can be described in terms of the matrix elements $\mathscr{H}_{i\alpha,j\beta} = \langle u_{i\alpha}|\mathscr{H}|u_{j\beta}\rangle$. α and β are, in general, the two different states. Since the atomic orbitals on different atoms do not have to be orthogonal, we must include the interatomic overlap $S_{i\alpha,j\beta} = \langle u_{i\alpha}|u_{j\beta}\rangle$ in this two-center approximation. The solution then follows from the secular determinant vanishing, that is, $\det|\mathscr{H}_{i\alpha,j\beta} - ES_{i\alpha,j\beta}| = 0$. Chapter 1 used precisely this approach for the 2-electron, 2-atom system in Section 1.5. In the *LCAO* method, if the overlap is large and varies slowly spatially, the matrix elements may not converge rapidly. In tight binding, if one ignores overlap integrals, this makes it useful as a tool for understanding rather than a detailed analysis.

Now it should be more understandable that these methods may be useful for molecules. We applied this approach to the 6-carbon-atom ring—a molecule—and then discussed it for semiconductor bandstructures, emphasizing the valence band. For crystals, with their multitudes of interactions, they are good toy tools, and just that.

We build the wavefuncton for this *sp* basis so that we can employ it to build a covalently bonded solid. The expansion of Equation 7.20 for a one-atom $|s\rangle$ and $|p\rangle$ orbital basis problem is

$$|\psi_k\rangle = \sum_i c^k_{is}|u_{is}\rangle + c^k_{ip_x}|u_{ip_x}\rangle + c^k_{ip_y}|u_{ip_y}\rangle + c^k_{ip_z}|u_{ip_z}\rangle. \tag{7.21}$$

Diagonalization gives us the coefficient *c*s. A more useful basis that gives better insight is one where the interactions can be ordered in strength, with a leading term being true covalency. Hybrids are this basis.

Hybrids are an important tool for analysis in chemistry and very useful for covalent systems. See any good introductory chemistry text to understand them. They represent a convenient change in basis, with direct intuition for bonding. A *p* orbital has equal probability along its principle axis in mirror symmetry. *s* is radially symmetric. A mixing of these states in the bonding of a three-dimensional solid is conveniently seen through the hybrid for the sp^3. This creates basis sets where now the probabilities change in two opposite directions along the major axis. Of course, to form the bond or, equivalently, the change to these hybrids, basis, requires the energy exchanging of the reaction. *Si* crystal formation requires going to high temperatures for the chemical reaction that lets *Si* atoms attach themselves in the ordered way of the crystal.

sp hybrid construction employs the angular properties of the *p* orbitals: so, $\hat{\mathbf{x}} \cdot \mathbf{r}/r$, together with the $\hat{\mathbf{y}}$ and $\hat{\mathbf{z}}$ dot products supplying us with the angular part. Since any orthonormal combination of the *p* orbitals gives another equivalent combination set in $|u_{p\alpha}\rangle$, where α is a set of axes, this new set too has an angular part defined by $\hat{\boldsymbol{\alpha}} \cdot \mathbf{r}/r$. Figure 7.3 is a representational view of one of these orthonormal $|u_{p\alpha}\rangle$ orbitals centered on the atom i. Let the atom at position i have n equivalent bonds with neighbors j. Figure 7.3 is a representational view of one of these orthonormal $|u_{p\alpha}\rangle$ orbitals centered on the atom i.

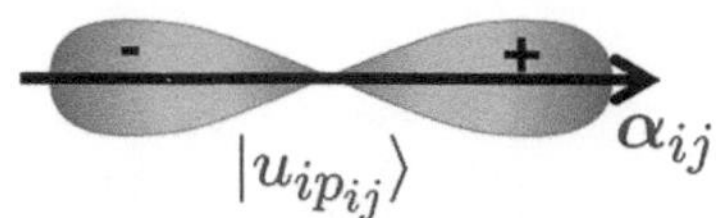

Figure 7.3: The angular dependence of a p orbital. Here it is identified as $|u_{ip_{ij}}\rangle$, centered on the indexed location of an atom i and along a direction $\boldsymbol{\alpha}_{ij}$ that passes through i and j.

When s and p orbitals are used to build a new orthonormal set, it is the form of bonding that will decide for us the new hybrid set of convenience. It may be sp, sp^2 or sp^3. The first exists in ethyne (also called acetylene (C_2H_2)) and in its polymer chains, the second exists in graphite, graphene and fullerenes, and the third is the stable state for group *IV Si* and *Ge* crystals. These n equivalent sp^n hybrids ($|u_{ij}\rangle$) have the form

$$|u_{ij}\rangle = \frac{1}{\sqrt{1+\lambda^2}}\left[|u_{is}\rangle + \lambda|u_{ip_{ij}}\rangle\right]. \tag{7.22}$$

The prefactor here is assuring normalization. For different $|u_{ij}\rangle$s to be orthogonal, the λ is constrained by

$$1+\lambda^2\cos\theta = 0, \tag{7.23}$$

where θ is the angle between the principal axes. Since the n bonds are spatially equivalent, $\sum_j \hat{\boldsymbol{\alpha}}_{ij} = 0$. So, this identity projected on any of the $\hat{\boldsymbol{\alpha}}_{ij}$s forces $1+(n-1)\cos\theta = 0$. $\cos\theta = -1/(n-1)$, and $\lambda = \sqrt{n-1}$. For the hybrids of interest to us here, $\lambda = 1, \sqrt{2}$, and $\sqrt{3}$. The first is an sp hybridization, the second is an sp^2 hybridization and the third is an sp^3 hybridization. For the sp^3,

$$\begin{aligned}
|u_{i111}\rangle &= \frac{1}{2}\left(|u_{is}\rangle + |u_{ip_x}\rangle + |u_{ip_y}\rangle + |u_{ip_z}\rangle\right), \text{ for } \hat{\boldsymbol{\alpha}}_{ij} = [111]/\sqrt{3},\\
|u_{i1\bar{1}\bar{1}}\rangle &= \frac{1}{2}\left(|u_{is}\rangle + |u_{ip_x}\rangle - |u_{ip_y}\rangle - |u_{ip_z}\rangle\right), \text{ for } \hat{\boldsymbol{\alpha}}_{ij} = [1\bar{1}\bar{1}]/\sqrt{3},\\
|u_{i\bar{1}1\bar{1}}\rangle &= \frac{1}{2}\left(|u_{is}\rangle - |u_{ip_x}\rangle + |u_{ip_y}\rangle - |u_{ip_z}\rangle\right), \text{ for } \hat{\boldsymbol{\alpha}}_{ij} = [\bar{1}1\bar{1}]/\sqrt{3}, \text{ and}\\
|u_{i\bar{1}\bar{1}1}\rangle &= \frac{1}{2}\left(|u_{is}\rangle - |u_{ip_x}\rangle - |u_{ip_y}\rangle + |u_{ip_z}\rangle\right), \text{ for } \hat{\boldsymbol{\alpha}}_{ij} = [\bar{1}\bar{1}1]/\sqrt{3}.
\end{aligned} \tag{7.24}$$

Figure 7.4 is a representational view of one of these orthonormal $|u_{p\alpha}\rangle$ orbitals centered on the atom i. This set represents four different possibilities for the angles of $\boldsymbol{\alpha}$; these are the tetrahedral geometric angles. Note now how one gets an easier view to the ordering in the bonding through the larger probability.

Figure 7.4: The angular dependence of an sp hybrid orbital. Here it is identified as $|u_{i\alpha}\rangle$, centered on indexed location of an atom i and along a direction $\boldsymbol{\alpha}$ that satisfies the hybridization condition $\lambda = \sqrt{3}$, $\sqrt{2}$ or 1.

Using the original orbitals,

$$\langle u_{is}|\hat{\mathscr{H}}|u_{is}\rangle = E_s, \text{ and } \langle u_{ip}|\hat{\mathscr{H}}|u_{ip}\rangle = E_p, \tag{7.25}$$

for the three different $|p\rangle$ orientations. These are the s orbital and p orbital energies in the solid, and these are not too far from that of the free atom. This accuracy is particularly true for the separation

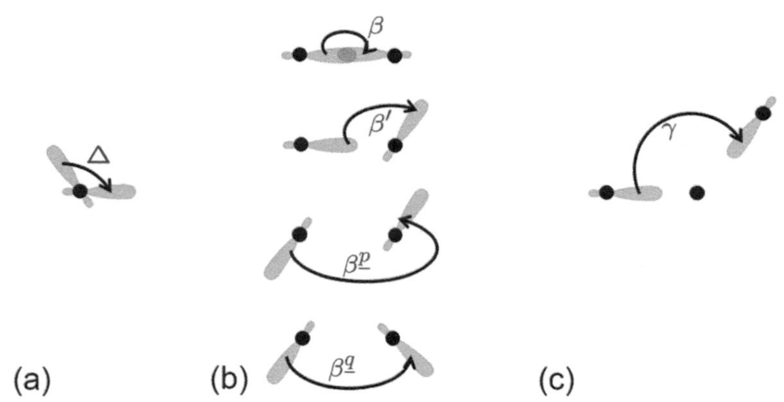

Figure 7.5: An illustration of the interactions of the hybrid orbitals. Part (a) shows an interaction with the same atom, (b) shows the interaction with first neighbors and (c) shows the interaction with second neighbors. The figure also notes the different angles that these project at.

$E_p - E_s$ of the free atom. This $E_p - E_s$ is the energy to move an electron from an s state (a ground state) to a p state (here to be thought of as an excited state). An atom was originally in an s^2p^2 ground state. The valence sp^3 state that favors covalent bonding requires the movement of an s state to a p state, and this is an excited state. In semiconductors of interest to us, this energy is 6–8 eV. In the covalent bonding favoring the sp^3 basis, now the matrix elements are

$$\begin{aligned} \langle u_{i\alpha}|\hat{\mathscr{H}}|u_{i\beta}\rangle &= \frac{1}{4}\left(E_s + 3E_p\right) = \langle E\rangle, \text{ when } \alpha = \beta, \text{ and} \\ \langle u_{i\alpha}|\hat{\mathscr{H}}|u_{i\beta}\rangle &= \frac{1}{4}\left(E_s - E_p\right) = \Delta, \text{ when } \alpha \neq \beta. \end{aligned} \tag{7.26}$$

β here is the angle of another allowed direction for the four possibilities of α. The energy $\langle E\rangle$ is the sp^3 bond's average energy. The energy Δ is negative. What really matters to us are the differences, given that we are going to ignore all other interactions beyond those between two atoms, as also the breaking of degeneracies and the formation of bands. $\langle E\rangle$ is therefore just a reference for us. The rank ordering of interactions between the hybrid orbitals within the semiconductor will arise in the same atom, the nearest neighbor, the next (second) neighbors, and so on. Figure 7.5 shows the 0th, 1st and 2nd neighbor possibilities. One will have to consider more than the nearest neighbor in an analysis, no matter how approximate, to at least be able to analyze the nature of a vacancy.

Figure 7.5(a) shows a self-interaction that is the energy cost of the hybridization. The top illustration of Figure 7.5(b) is for covalent bonding, and the rest is for other angles of the adjacent neighbor, with Figure 7.5(c) for the 2nd neighbor. The energies associated with the two-center approximation for the case of Si are listed in Table 7.2. If a vacancy exists, it is the (b) part that vanishes, and one does need to account (even in the lowest order) the next neighbor beyond that, as is represented here in (c). Estimates of the magnitudes of the energies for the interactions—identified through the angle—are in Table 7.2. Note that the highest energy decrease—through the bonding—arises as β, the direct link in the top part of Figure 7.5(b). Note also that anti-alignment (the net from the bottom of Figure 7.5(b)), when the wavefunctions point opposite, leads to a positive energy ($\beta^{\underline{p}}$) and that Figure 7.5(c) lowers the energy γ.

	$\langle u_{i\alpha}\lvert\hat{\mathscr{H}}\rvert u_{j\beta}\rangle$ (eV)
Δ (self)	−1.12
β (1st neighbor)	−3.75
β' (1st neighbor)	−0.51
$\beta^{\underline{p}}$ (1st neighbor)	+0.22
$\beta^{\underline{q}}$ (1st neighbor)	−0.33
γ (2nd neighbor)	−0.25

Table 7.2: Tight binding energy parameters for self, nearest and 2nd nearest neighbor for *Si*.

Although inaccurate, one should see the band formation through the lifting of degeneracy in the simple tight binding calculation that we explored in Chapter 4. Consider here only the nearest neighbor bonding interactions (characterized by β), where only the orbital interactions pointed toward each other matter. The diagonal matrix elements vanish, and the off-diagonal elements are $\langle u_{ij}|\hat{\mathscr{H}}|u_{ji}\rangle = \beta$. This problem has the eigenvalue solution of

$$E_A = -\beta, \quad \text{and} \quad E_B = \beta, \quad \text{and}$$
$$|u_{Aij}\rangle = \frac{1}{\sqrt{2}}\left[|u_{ij}\rangle - |u_{ji}\rangle\right], \quad \text{and} \quad |u_{bij}\rangle = \frac{1}{\sqrt{2}}\left[|u_{ij}\rangle + |u_{ji}\rangle\right]. \qquad (7.27)$$

A identifies the antibonding state. B is the bonding state. So far, this is similar to the 2-electron, 2-atom problem (hydrogen!) in Chapter 1. Our picture with only this interaction is of a solid of diatomic molecules. There are N atoms, $2N$ bonds between nearest neighbors and two energies, E_A (antibonding) and E_B (bonding), that are $2N$ degenerate. At $T = 0\ K$, E_A can have $4N$ electrons once one accounts for the spin degeneracy ($g_s = 2$).

More than our inorganic semiconductors, the organic semiconductors that usually consist of polymer chains come close to this bonding-dominated energy picture. E_A are the highest occupied molecular orbitals (*HOMO*), and E_B are the lowest unoccupied molecular orbitals (*LUMO*).

This was our molecular description. We can introduce broadening to this by starting to include the interactions that were ignored. Let us include only Δ arising in the $|s\rangle \mapsto |p\rangle$ excitation. With Δ included, we rebuild the wavefunction as

$$|\psi\rangle = \sum_{ij} c_{ij}|u_{ij}\rangle. \qquad (7.28)$$

This one additional interaction is not so bad at describing to the lowest order what happens in amorphous semiconductors.

The solution, with the usual use of projection (multiplying by the conjugate (bra) and spatial integration),

$$\begin{aligned} Ec_{ij} &= \Delta \sum_{j'\neq j} c_{ij'} + \beta c_{ji}, \\ \therefore\quad (E+\Delta)c_{ij} &= \Delta S_i + \beta c_{ji}, \quad \text{and} \\ Ec_{ji} &= \Delta \sum_{i'\neq j} c_{ji'} + \beta c_{ij}, \\ \therefore\quad (E+\Delta)c_{ji} &= \Delta S_j + \beta c_{ij}. \end{aligned} \qquad (7.29)$$

Here, S_i is the sum over all the c_{ij}s connecting the ith atom with the other four nearest j neighbors. Writing $\sum_j S_j = \delta S_i$, where

$$\delta = \frac{1}{\beta\Delta}\left[(E-\Delta)^2 - (\beta^2 + 4\Delta^2)\right], \tag{7.30}$$

we have now reduced the problem to that of one band. δ is the normalized averaging of interaction energy to nearest neighbor interaction, being treated as unity. $-4 < \delta < 4$ because of the 4 nearest neighbors, following Equation 7.25. This neighbor energy should not change because a basis has been changed. The solution to Equation 7.30 is the formation of two bands given by

$$E = \Delta \pm (\beta^2 + 4\Delta^2 + \beta\Delta\delta)^{1/2}. \tag{7.31}$$

If $S_i = 0$, the solution is $E = -\Delta \pm \beta$; that is, two flat bands of discrete energy showing the assumed non-degeneracy because of the absence of any interaction. Otherwise, we now have a spread of states in two energy bands. The Hamiltonian form and the presence of interaction (crystalline, amorphous, etc.) caused the formation of the band. Figure 7.6 shows a sketch of the states' energy change—the broadening—resulting from the normalized averaged interaction δ as a solution of Equation 7.30. The larger the δ interaction parameter is, the larger is the broadening.

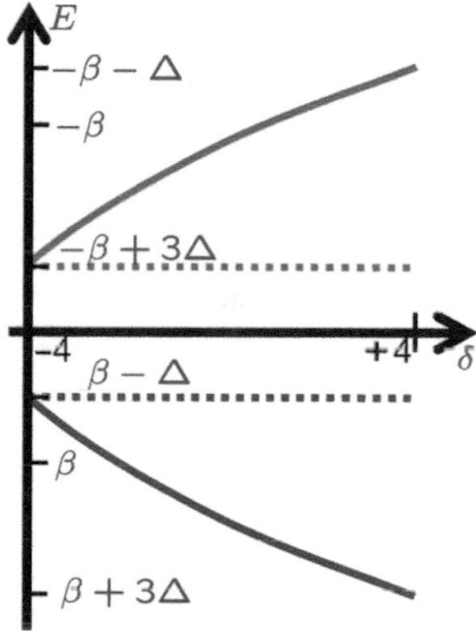

Figure 7.6: Bandstructure formation as a result of broadening due to neighborly interaction in the tight binding-based molecular model.

The width of the band is $4|\Delta|$. So long as the bands do not cross each other—bands can cross if $\beta < 4|\Delta|$—one sees the semiconductor-like band formation. This is the situation of covalency. Note also that energies shift down. Interaction between the bonding and antibonding states lowers the minimum energy to below β. One could incorporate all the other interactions of Figure 7.5(b) and (c), and this will change the bandstructure quantitatively, but it will not change the qualitative description. They cause further broadening. So, now we have a picture of the molecule and an extended picture of the formation of bands through tight binding. And the picture of the molecule will give us an ability to look at the complement as defect molecule in a crystal.

7.4.2 *Vacancy*

THIS MOLECULAR MODEL gives us a tool to qualitatively view the behavior of a vacancy, again through a toy model. Take a linear chain in the three-dimensional system, and first consider only one *s* and one *p* hybrid forming a bond along the chain, so two valence electrons per atom. There also exist *p* orbitals perpendicular to the chain, and they are not of interest in the bonding. The hybrid bases are

Graphite and graphene have a p_z orbital out of the plane, and sp^2 bonds in the plane. Ours here is an even more simple model, with polyethyne as a linear chain analog.

$$|u_{i,i\pm1}\rangle = \frac{1}{\sqrt{2}}(|u_{is}\rangle + |u_{ip_x}\rangle), \tag{7.32}$$

with β as the nearest neighbor interaction energy. $\Delta = (E_s - E_p)/2$, corresponding to the $|s\rangle \mapsto |p\rangle$ excitation. $|\delta| \leq 2$, since there are two neighbor interactions, and it doesn't matter which basis we consider. This linear chain will, using the same procedure to incorporate interactions, exhibit a formation of bands. Figure 7.7 shows the evolution of density from the limit $\Delta/\beta = 0$, that is, a coupled set of *sp* orbitals leading to a bonding state at energy β and an antibonding state at $-\beta$ in (a) (a set of delta functions as density of states) to the formation of bands due to the interactions Δ.

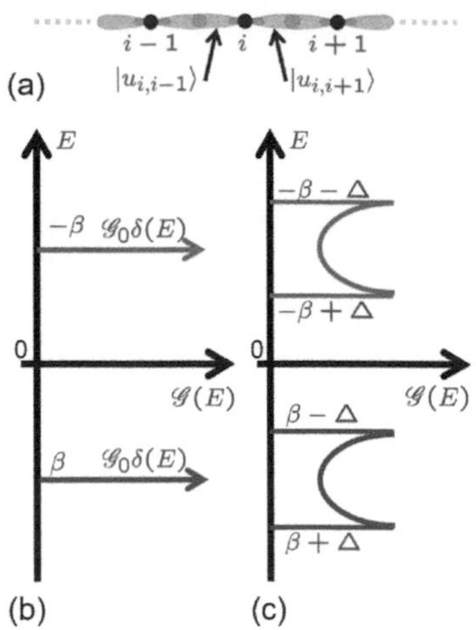

Figure 7.7: A linear covalent bonded chain with *sp* hybrid orbitals. Part (a) shows covalent bonding between successive atoms, (b) shows the density of states (a Dirac delta function) without the self-interaction Δ and (c) shows the broadening due to interactions within the approximations of the molecular model.

Now introduce a vacancy at the site $i = 0$ in the chain, as shown in Figure 7.8(a). In our molecular model, we have kept interactions very limited. This picture that we have drawn is now really two identical semi-infinite chains interacting through the site, indexed as $i = 0$. Given the restricted interactions being considered, we can look at only one chain, terminated, and assign the level a degeneracy of 2. This is a "surface state" for this linear chain in this approximation. Figure 7.8(b) shows the bonding and antibonding energies in the molecular limit ($\Delta = 0$), and (c) shows when broadening happens due to $\Delta = (E_s - E_p)/2$. $(E_s + E_p)/2$ is the origin of the ordinate in the energy figures. The level at $E = 0$ is the localized state at the vacancy. It has two-fold degeneracy, with the eigenvectors $|u_{10}\rangle$ and $|u_{-10}\rangle$. These are dangling orbitals forming a mid-gap deep-level state. For finite Δ, the bonding and antibonding degenerate levels broaden. At $E = 0$, the dangling orbital $|u_{10}\rangle$ interacts with the neighboring bonding and antibonding states. So does $|u_{-10}\rangle$. Both interaction energies, following Equation 7.32, are $\Delta/\sqrt{2}$. So, the situation is still symmetric, and $E = 0$ remains the solution for this state.

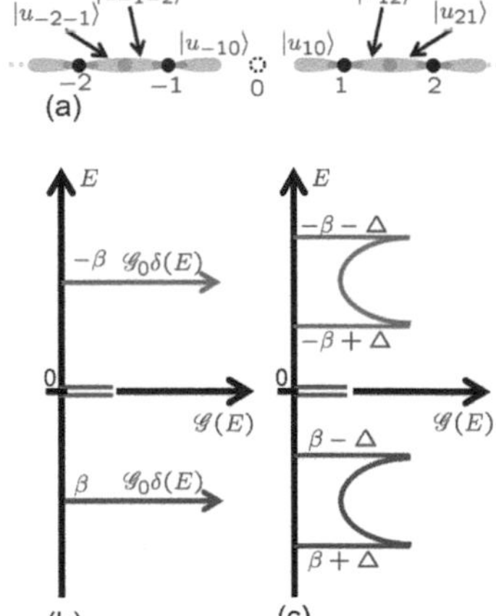

Figure 7.8: An *sp*-hybridized linear co-valent bonded chain with a vacancy. Part (a) shows a vacancy at $i = 0$, (b) shows the density of states, including the states arising in the vacancy and (c) shows the same under broadening. This figure can be compared to Figure 7.7, where there was no vacancy. A doubly degenerate surface state has appeared.

We can now write the wavefunction for this state too. Since

$$\begin{aligned} Ec_{10} &= \Delta c_{12} \text{ and} \\ Ec_{12} &= \Delta c_{10} + \beta c_{21} \text{ for } i < 2, \text{ and} \\ Ec_{i,i-1} &= \Delta c_{i,i+1} + \beta c_{i-1,i} \text{ and} \\ Ec_{i,i+1} &= \Delta c_{i,i-1} + \beta c_{i+1,i} \text{ for } i \geq 2, \end{aligned} \quad (7.33)$$

for $E = 0$,

$$\begin{aligned} c_{i,i+1} &= 0 \ \forall \ i \text{ and} \\ c_{i+1,i} &= -\frac{\Delta}{\beta} c_{i,i-1} = \left(-\frac{\Delta}{\beta}\right)^i c_{10}. \end{aligned} \quad (7.34)$$

The wavefunction of the vacancy deep state, therefore, is

$$|\psi_{vac}\rangle = \left(1 - \frac{\Delta^2}{\beta^2}\right)^{1/2} \sum_{i=0}^{\infty} \left(-\frac{\Delta}{\beta}\right)^i |u_{i+1,i}\rangle. \quad (7.35)$$

The amplitude of the wavefunction decreases rapidly so long as $\Delta/\beta < 1$. The dangling orbital has a contribution of $|u_{10}|\psi_{vac}\rangle|^2 = 1 - \Delta^2/\beta^2$. At perfect covalence, this will be the entire contribution since $\Delta/\beta \rightarrow 0$.

This procedure, with a lot more complexity, can be applied to the sp^3-hybridized covalent crystal in the molecular model approximation. Figure 7.9 formulates a description of our problem and the meaning of these equations. Four dangling orbitals will exist at energy $E=0$. The origin of energy, which was $(E_s + E_p)/2$ in the sp hybrid, will now be $(E_s + 3E_p)/4$, and the other bonding happens at an energy of $\pm\beta$. There is now a 4-fold degenerate bound state localized on the four dangling bonds pointing at the vacancy. Now, to this we add the interaction Δ. This arrangement is a Bethe lattice, that is, a cycle-free graph with a 4-fold connection here, and because of this, $E = 0$ still holds true for the vacancy state. The reason that splitting of the localized level is not adequately described in this description is that important direct interactions such as in γ between pairs of dangling orbitals such as of Figure 7.5(c) have not been incorporated.

We will only summarize the results with γ incorporated. Including the γ—which is an off-diagonal term in the interaction matrix, since the interaction arises between different atoms' dangling orbitals—one may build the new basis states. These eigenfunctions, $|v\rangle$, which is symmetric, and $|t_x\rangle$, $|t_y\rangle$ and $|t_z\rangle$, which are $|p\rangle$-like, are identified by their irreducible representations: A_1 for the first, and T_2 for the latter three. For the $|s\rangle$ and $|p\rangle$ states, a lower non-degenerate energy level at 3γ and a 3-fold degenerate energy level at $-\gamma$ result, interactions connecting the bonding and antibonding orbitals to the valence and the conduction bands. A third interaction here is the coupling of the eigenstate of the defect molecule that results from the dangling orbitals and the rest of the crystal. It is this interaction that leads to the delocalization of the bound-state wavefunction.

Note that these are just like the singlet (bonding) and triplet (antibonding) states that we have encountered before.

The A and T symmetry, and we will see another one (E), notations are related to group theory and are Mullicken symbols for a point group, indicating its symmetry operations and irreducible representations. A indicates symmetry w.r.t. the rotation of the principle axis. T indicates that the group is triply degenerate. E is doubly degenerate. These group-theoretic notations arose in the description of molecular symmetry and are tertiary to our interests.

The molecular description of this problem—in parallel with the linear chain approach—is to write a localized state description of the molecular defect—any defect, not just the vacancy—in the form

$$|\psi\rangle = c_d|u_d\rangle + \sum_{\eta} c_\eta|u_\eta\rangle, \tag{7.36}$$

an equation that, on the right side, consists of of $|u_d\rangle$ as an eigenfunction of the defect molecule, and $|u_\eta\rangle$s as the eigenfunction of the remaining crystal; that is, $|u_\eta\rangle$s have captured all the transformations that we had to make to go from $|s\rangle$ and $|p\rangle$ orbitals to the eigenfunctions with the atoms in the crystal. This leads to the coefficients being related through

$$
\begin{aligned}
(E - E_d)c_d &= \sum_\eta V_{d\eta} c_\eta, \quad \text{and} \\
(E - E_\eta)c_\eta &= V_{\eta d} c_d, \\
\therefore \quad E &= E_d + \sum_\eta \frac{|V_{d\eta}|^2}{E - E_\eta}.
\end{aligned}
\tag{7.37}
$$

with E_d and E_η as the eigenenergies of the composing basis, and V being the matrix element of the coupling of the defect "molecule" to the environment. E_η is either the collection of valence band states or the collection of conduction band states. Equation 7.37 can have solutions within the bandgap. These will be the localized solutions and, for these, the defect-molecule eigenfunction provides a sound basis to start from. These are the T_2 states shown Figure 7.9. However, the A_1 states are generally in the valence band and can be propagating. One can see here an analogous picture—because of this molecular atomic orbital localization emphasis—a behavior of propagating and non-propagating states that we found on the surface of a three-dimensional solid in Chapter 5. How much is the localization or delocalization of this defect state? This can be extracted from the coefficient part of Equation 7.37. Following normalization, one gets

If the interaction energy $V_{d\eta}$ is small, the defect energy is close to E_d, and it looks like a localized defect state that has not been very perturbed from the $|u_d\rangle$ eigenfunction of the defect molecule.

$$
|c_d|^2 = \left(1 + \sum_\eta \frac{|V_{d\eta}|^2}{|E - E_\eta|^2}\right)^{-1}. \tag{7.38}
$$

The bound defect state does have delocalization, since $|c_d|^2 < 1$. The higher the strength of the coupling perturbation $V_{d\eta}$ is, the more delocalization there is. Note also that it increases with decreasing $E - E_\eta$ also. So, this defect-molecule approach needs to be approached cautiously. If there is significant delocalization, that is, a large perturbation or a smaller gap, then the use of localization as the foundation of the defect molecule is in contradiction.

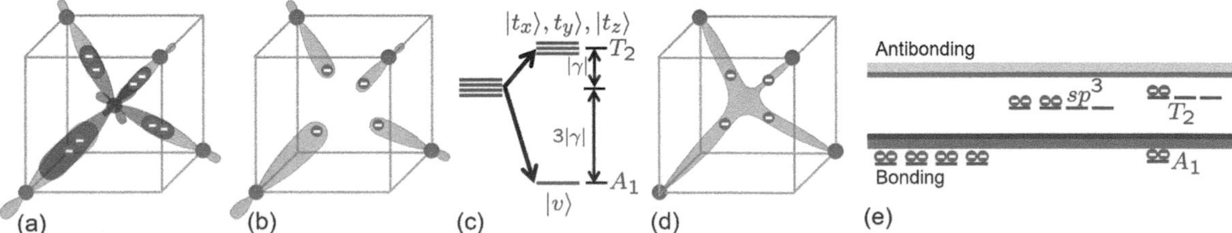

Figure 7.9: (a) The covalent bonding in the Si tetrahedral arrangement. Part (b) shows the appearance of vacancy in the molecular model picture, (c) shows the vacancy energy level splitting due to the interaction energy γ arising in the dangling orbitals of different atoms, (d) shows the wavefunction arising in the interaction and (e) is a representational view of the appearance of the states due to the vacancy defect within the distribution in energy of the propagating states that are bonding (valence) and antibonding (conduction) states of the crystal.

However, what we have now seen is that our one-dimensional picture does extend to the three-dimensional form, also under constraints, and shows that T_2-type states appear within the bandgap.

7.4.3 *Interstitials*

THIS MOLECULAR MODEL, with a bit more of a reduction in accuracy, is also instructive for seeing what happens when one has an interstitial. And this lets us see the differences from the vacancy case. Take the example of the interstitial atom identified as I and inserted into the diamond unit cell of a semiconductor such as Si, as shown in Figure 7.10. Again, we will only summarize results, since the calculation will require the inclusion of the interstitial's interaction with four other atoms, and each of those atoms has the four sp^3 hybrids. For the vacancy, we had to handle just four hybrid orbitals, for which we built a defect eigenstate and then calculated the interaction with the rest of the crystal. The approach remains quite similar.

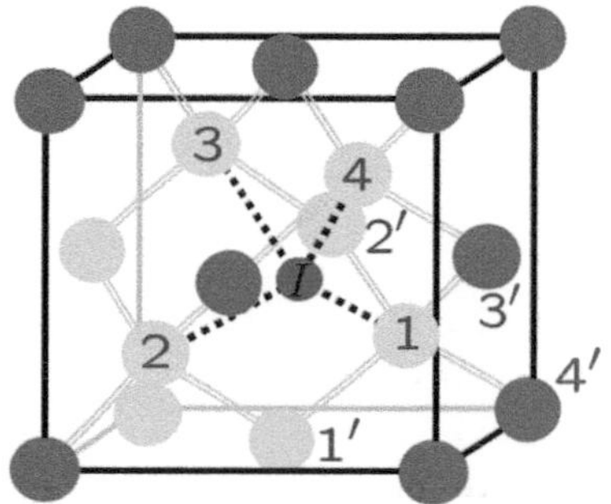

Figure 7.10: An interstitial marked as I in a diamond unit cell, and the neighbors with which the interactions must be considered. Numbers 1 through 4 are the first-nearest neighbors. These are the atoms along the diagonals within the unit cell, with 1 and 2 twisted perpendicular to 3 and 4, with the interstitial I in-between them. 1′ through 4′ are the next-nearest neighbors, and these are either on the face center or in the corner.

Start with the interstitial I, and its hybrid $|u_{Ii}\rangle$ coupled to another sp^3 hybrid $|u_{iI}\rangle$, where these hybrids need to be created from the sp^3 hybrids of the perfect crystal, viz. $|u_{ij}\rangle$, and these are not orthogonal to them either. The four $|u_{Ii}\rangle$ of the interstitials couple to the normal sp^3 hybrids of the neighbors $|u_{ij}\rangle$. Take the coupling of $|u_{I1}\rangle$ with the set $|u_{1j}\rangle$ for the different js. Just as with Equation 7.36, we work through the coupling of the defect $|u_d\rangle$ with the eigenstates of the crystal, which are the bonding and antibonding states built from $|u_{ij}\rangle$s. For example, we get

$$|\psi_{B1j}\rangle = \frac{1}{\sqrt{2}}(|u_{1j} + |u_{j1}\rangle), \text{ and}$$
$$|\psi_{A1j}\rangle = \frac{1}{\sqrt{2}}(|u_{1j} - |u_{j1}\rangle), \; j = 1', \ldots, 4', \tag{7.39}$$

and these bonding and antibonding states have energies $\pm\beta$. The energy perturbation term

$$\sum_{\eta} \frac{|V_{d\eta}|^2}{E - E_\eta} = \sum_{j} \left(\frac{|\langle\psi_{B1j}|u_{I1}\rangle|^2}{E - \beta} + \frac{|\langle\psi_{A1j}|u_{I1}\rangle|^2}{E + \beta} \right)$$
$$= \frac{E}{E^2 - \beta^2} \langle u_{I1}| \sum_{j} |u_{1j}\rangle\langle u_{1j}|u_{1I}\rangle. \tag{7.40}$$

$|u_{1j}\rangle\langle u_{1j}|$ doesn't change under any basis change of atom 1. So, an sp^3 hybrid $|u_{1I}\rangle$ is chosen toward the interstitial from atom 1, and three independent combinations are extracted from $|u_{11'}\rangle$, $|u_{12'}\rangle$, $|u_{13'}\rangle$ and $|u_{14'}\rangle$. $\langle u_{01}|u_{10}\rangle = \beta$, so

$$\sum_{\eta} \frac{|V_{d\eta}|^2}{E - E_\eta} = \frac{E\beta^2}{E^2 - \beta^2}. \tag{7.41}$$

Equation 7.37 says that, with $E_d = 0$, the eigenvalues are solutions of

$$E\left(1 - \frac{\beta^2}{E^2 - \beta^2}\right) = 0, \tag{7.42}$$

with $E = 0$, and $E = \pm\beta\sqrt{2}$, as the two solutions. Since the same solution holds for $|u_{I2}\rangle$, as well as for atoms 3 and 4's pointed hybrids, these energy levels are 4-fold degenerate.

A 4-fold degenerate state has arisen at $E = 0$. This is not different from that for the vacancy. Four orbitals of the interstitial led to this 4-fold degeneracy. The new feature is the 4-fold degenerate state at $E = \beta\sqrt{2}$. Since the valence band level is at β in this molecular model, this new feature has appeared below the valence band. Similarly, another 4-fold degenerate state appeared at $E = -\beta\sqrt{2}$, above the conduction level of $E = \beta$.

As with vacancy, we determine $|c_d|^2$ to determine the degree of localization:

$$|c_d|^2 = \left\{1 + \frac{\beta^2}{2}\left[\frac{1}{(E+\beta)^2 + (E-\beta)^2}\right]\right\}^{-1}, \tag{7.43}$$

which is different from what we found in Equation 7.38 for the vacancy. When $E = 0$, $|c_d|^2 = 1/2$. The wavefunction is only half localized on the interstitial. For a vacancy, the defect wavefunction is entirely localized. A similar calculation shows that the other states below the valence and above the conduction level are a quarter localized. Of course, now, if we include additional interactions, the degeneracies will break for all these three solutions. A lower non-degenerate A_1 and a higher three-fold-degenerate T_2 component will arise. The energy Δ for the $|s\rangle \mapsto |p\rangle$ excitation also needs to be included. And once we do that, for the $E = 0$ localized state, the A_1 component falls into the valence band, and T_2 falls into the conduction band. For the $E = \sqrt{2}\beta$ states, the A_1 component a appears as a localized state under the valence band, and T_2 is raised to being in the valence band. Complementary behavior arises for the $E = -\sqrt{2}\beta$ states.

This discussion shows us that, although only a rough approximation, the molecular model also shows us the features of states appearing across the energies, including potentially in the bandgap

Ti has the electronic structure $[Ar]3d^2 4s^2$. *Cu* is $[Ar]3d^{10}4s^1$. *d* mixing will be considerably more complicated. These electronic structures and their great difference from the *ss* and *ps* of the normal semiconductor will mean that these impurities will not be hydrogenic, and since they mostly appear as interstitial, with some being substitutional, quite a rich deep-level behavior should be expected. They both kill lifetime, and *Si* particularly worries about *Cu* since it is a very common interconnect metal in electronics and a fast diffuser, as most interstitials will be. With other orbitals from the introduced species, there will be a rich diversity in behavior. Both localized and conducting states should be expected, together with the ability of the atom to be in multiple ionization states.

depending on the energies of interaction. While we considered an interstitial that had the host-atom's valence structure, its presence in the crystal did bring in states that are quite illustrative. Other impurities, transition elements, et cetera, will provide even richer behavior.

7.4.4 *Substitutional impurities*

BASED ON THIS LEARNING from the molecular model, a few interesting implications are useful for understanding the nature of substitutional impurities: whether they are shallow hydrogenic or deep. In our simple tight binding approach, the impurity can be described again by Hamiltonian matrix elements that are different from that of the crystal in the impurity atom's neighborhood. The s and p energies of the impurity are different. For the formation of the sp^3 hybrid, there will be an energy needed that is different from that of the host. This is equal to the shift of an average energy (U), and a Δ' for the s to p excitation. The impurity atom and its nearest neighbors will have an interaction parameter β' different from that of the host. The unperturbed bonding and antibonding states are β and $-\beta$, and these become different for the bonds that exist between the substitutional impurity and its 4 neighbors. If we build $|u_{si}\rangle$ and $|u_{is}\rangle$, with $i = 1,\ldots,4$ indexing the host atoms as the basis, then the matrix elements are

$$\begin{bmatrix} U & \beta' \\ \beta' & 0 \end{bmatrix}, \tag{7.44}$$

with the eigenenergy solution as

$$E = \frac{U}{2} \pm \left[\left(\frac{U}{2}\right)^2 + \beta'^2 \right]^{1/2} \tag{7.45}$$

for the antibonding and the bonding states. Now, β' is quite close to β, based as they are on similar hybrids. This equation therefore shows how the energies will change from the $U = 0$ and $\beta' = \beta$ limit of the host and the impurity to the effect of the energy U arising in the average shift. And this should be seen through a normalized ratio such as E/β that characterizes how the impurity introduction changes the bonding and antibonding eigenenergies. Figure 7.11 shows this parameter versus U.

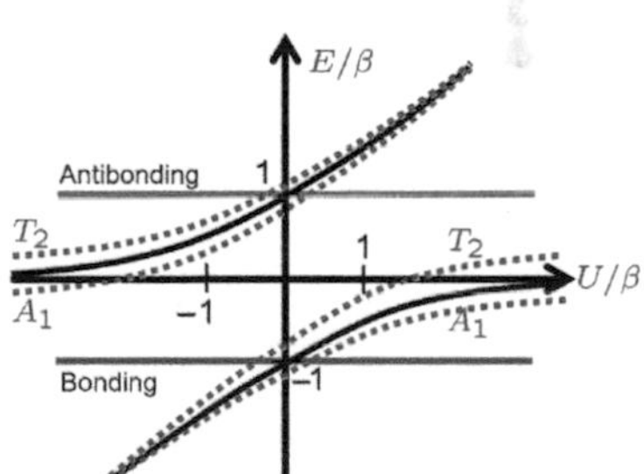

Figure 7.11: Change in the bonding and antibonding energy of a substitutional sp^3-hybridized impurity in an sp^3 host normalized to the bonding energy parameter β. For Si, $\beta = -3.75$ eV. The abscissa is the average shift in the sp^3 energy arising in the impurity vis-à-vis the host. The figure also shows the changes in the T_2 and A_1 splitting and its variation.

The energy U—the difference in the sp^3 energies of the host and impurity atom—are close to the energies of the free atom. Table 7.3 gives some relevant numbers across columns in order to compare

		sp^3 energy		
(eV)	(eV)	(eV)	(eV)	(eV)
	$B : 9.7$	$C : 12.9$	$N : 16$	$O : 20.3$
	$Al : 7.3$	$Si : 9.5$	$P : 11.9$	$Se : 13.4$
$Zn : 6.0$	$Ga : 7.6$	$Ge : 9.5$	$As : 11.3$	$Te : 12.0$
$Cd : 5.8$	$In : 7.3$	$Sn : 8.3$	$Sb : 10.8$	$S : 14.0$

Table 7.3: sp^3 energies of some atoms of interest across columns and rows of the periodic table around *Si* and *Ge*.

them to that of *Si* or *Ge*. A negative *U*, as with a group *V* or group *VI* impurity—impurities with extra electrons in outer shell—will donate single or multiple electrons. Here, the antibonding molecular state will be lowered. So, a donor state will form. Four antibonding molecular states are lowered into the average gap. Eight electrons occupy. And if *U* is positive, four bonding states will be raised up. Again, 8 electrons can occupy, but since there are fewer electrons in the molecule, there will be missing electron(s) or hole(s).

When energy differences are small, and this holds true for *B*, *P* or *As* in *Si*, which has a $\beta = -3.75\,eV$, the energy perturbation is small, and a shallow level arises in practice. *Al*, *In* or *N* do not work out well. Move one more column over, and none of these impurities are reasonable shallow dopants. We referred to the example of *S* in the discussion of energetics (Section 7.2) and in our discussion of the energies of shallow hydrogenic dopants (Subsection 7.2) with the implications of the multi-electron Hubbard energy consequences. We have arrived at a similar conclusion from the molecular defect perspective. This table also points out a few other interesting attributes. *N* is so vastly different in energy that this hybridization is quite discouraged. Si_3N_4 is the preferred compound, and if it is crystalline, it is normally trigonal or hexagonal. If we look at the zinc blende crystals from the *III*-*V* combination, one can now see that *Se* and *Te* will be a good *n*-type dopants, *S* marginally so, and the group *II* atoms *Zn* and *Cd* are reasonable, although technologically their fast diffusion makes them difficult to use. Group *IV* impurities in the group *III* or group *V* site work best. Also note from this table why it is so difficult to dope *N*- or *O*-containing compounds such as *GaN* or *ZnO*.

When the energy *U* is strongly attractive or repulsive, the impurity levels are pinned to energies that are close in energy and correspond to those of the vacancy levels. In *Si*, this is the T_2 level in the bandgap. For zinc blende structures, similar implications hold, except that one will now have two sets of A_1 and T_2 levels.

With this background, we can now summarize some illustrative and important perturbation defects. First, we will discuss behavior of transition atoms.

7.5 *Transition metal impurities*

WHAT HAPPENS WHEN d ORBITALS are also involved? Ti : $[Ar]3d^24s^2$, Cr : $[Ar]4s^13d^5$, Fe : $[Ar]4s^23d^6$ and Cu : $[Ar]3d^{10}4s^1$ are transition elements that are not uncommon as impurities in semiconductors. They all have occupied d orbital states, and they appear as deep levels with various ionization states. As we have remarked, they affect lifetime and reduce carrier concentration. Indeed, this can be useful in getting semi-insulating semiconductors with a bit more than an eV bandgap. These metals are transition elements from the $3d$ row and, being small in size, they predominantly appear interstitially in Si. $5d$ transition metals prefer substitutional sites and, in the case of III-V, the cation sites. $4d$ metals show both interstitial and substitutional behaviors. Most show energy states deep in the bandgap in common semiconductors.

In the early days of the use of *GaAs* in high-frequency transistors, *Cr* served to strip out the electrons contributed by residual shallow donors. Another way was to use intrinsic defects within *GaAs*—the various *EL* centers being the common ones—through technical skullduggery during crystal growth.

Since the transition element can have several states of ionization, that is, with different number of electrons centered on them, have both excess electrons and room for more, they can appear both donor-like and acceptor-like. If an electron localized at the center can transition, from it to the conduction band, this is a donor transition and the complement of this—hole transfer to the valence band (or, equivalently, electron capture from the valence band)—is an acceptor transition. It is the ability to do both that makes these centers generation-recombination centers. A neutral donor may undergo a transition $D^0 \mapsto D^+ + e^-$. This D^+ state may still be a deep-level state, and $D^+ \mapsto D^{++} + e^-$ may also be allowed. The same can happen with acceptor states. An acceptor state A^- may appear by capturing an electron from the valence band (or, equivalently, emitting a hole). If this $A^- \equiv D^-$, it can undergo a $D^- \mapsto D^0 + e^-$ transition. Through this two-step process, an electron from the valence band has been bumped up to the conduction band. An electron-hole pair has been generated. Also, in such multistate possibilities, because there are different numbers of electrons that are possible at the center, there will also be configuration interaction contributions.

Incidentally, transition elements also form compounds with II-VI semiconductors; $(Zn, Mn)Te$ and $(Cd, Mn)Te$ are semiconductors. And because Mn : $[Ar]4s^23d^5$ is magnetic, unusual useful properties can be found below the Curie point. S. Tiwari, "Nanoscale device physics: Science and engineering fundamentals," Electroscience 4, Oxford University Press, ISBN 978-0-198-75987-4 (2017) discusses phase-transition-based device-oriented usage of transition metal compounds in a variety of their forms.

Figure 7.12 shows a toy defect model picture of a transition metal impurity on a substitutional site. The transition element has a $4s^13d^5$ configuration—Cr substituted on the Ga cation site being our prototype—and the four sp^3 orbitals of the surrounding cation atoms point toward the transition metal atom, which, as we have seen, results in states with A_1 symmetry and T_2 symmetry. The s orbital of the transition atom deforms into an orbital of A_1

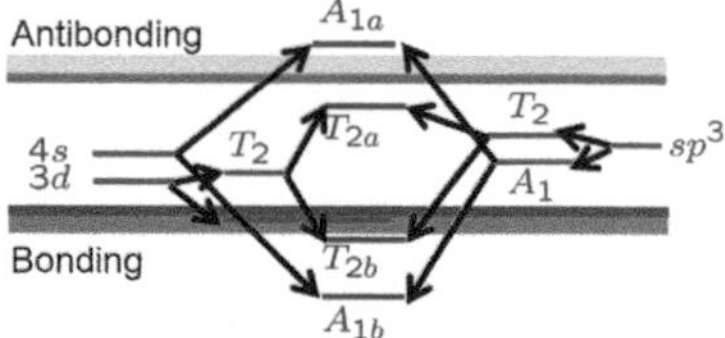

Figure 7.12: A defect-molecule toy model for a transition atom impurity in a tetrahedral semiconductor with *Cr* in *GaAs* as a prototypical example, *Cr* being substituted on the *Ga* site.

symmetry. The d states—5 of them—break into three states with T_2 symmetry, and two with an E symmetry. These E symmetry centers are highly localized. They are just linear combinations of d orbitals. The bonding T_2 symmetry states mainly arise in the d states. The antibonding state is dominantly from the sp^3 hybrids of the host. This becomes a deep center. This deep center has a wavefunction that spreads out to the surrounding host. It is quite host-like and not transition-atom-like anymore.

We have avoided a discussion of symmetry and its role. For our intuitive understanding in the simple molecule picture, they are distractions. Suffice to say that symmetries are very important, as we see here, where s and d ended up creating states where the original symmetries were crucial. The same happened for the s and p orbital decomposition.

7.6 Complexes

BESIDES THE POINT DEFECTS that we have considered up to this point—vacancy, interstitial and substitutional—perturbations can also arise in slightly more elaborate form while still being a point perturbation. A common cause for this lies in the energy that can be stored in lattice distortion. This makes it possible for metastable states to arise where distortion and accompanying energetics keep the system stable against small perturbations. An early example was the *EL*2 double donor in *GaAs*. This center is possibly As_{Ga}, or close to it, that is, is a mispositioning of the *As* on a *Ga* site. Another interesting example is that of the *DX* center. This arises in a group *IV* impurity, which, while in the normal course it would be expected to be a shallow center, becomes deep and the donor atom is singly negatively charged with a positioning on an interstitial site. We discuss these next because of their technical and usage importance.

7.6.1 The As antisite defect in GaAs ($\mathbf{As_{Ga}}$)

THERE EXISTS A CENTER, most likely a complex, called *EL*2, which appears as a double donor. A judicious use of it turns *p*-type *GaAs* semi-insulating, just as *Cr* incorporation turns *n*-type *GaAs* semi-insulating. However, it also affects lifetime and thus affects light emitting usage. The *EL*2 story is not quite completely understood. The *As* antisite defect (As_{Ga}), however, is believed to be either a part of the *EL*2 complex or its entirety. We focus here on As_{Ga} as a devilish play on our substitutional intrinsic defect twist in the molecular model.

This is a complex in the sense that it exists together with some other set of distortions and perturbations.

Consider first this case being that of an sp^3 bonding in a tetrahedral geometry as a defective substitution of *As*. We should expect A_1 bonding and antibonding levels. We have noted that the bonding level usually is in the valence band. The A_1 antibonding level here is in the bandgap. Likewise, the antibonding T_2 level is shifted and in the conduction band. The defect molecule has 10 electrons (2 in excess, since *As* appeared at a *Ga* site). Eight of these belong in the bonding states of the valence band. The rest (2 electrons) can populate the A_1 level, which is the next higher energy state available. So, this level is a double donor, because these electrons can now transition to the antibonding conduction states. But there is an energy cost to this level populating.

Figure 7.13 shows in (a) the spatial configuration that we just considered and in (b) a small displacement of the substitutional *As* along the [111] direction—a displacement away from one of the *As* neighbors—that is metastable. There is a local energy minimum that is about 0.24 eV above the minimum of the substitutional *As*. This *As* is now in an interstitial position about 0.02 nm displaced from the plane of the other 3 *As* atoms. The stable state is separated from the metastable state by a 0.6 eV energy barrier. And, in the reverse direction, this barrier is 0.36 eV. If one excites out one or both of the donor electrons from the substitutional center, the barrier decreases, and this makes the transition to the metastable state easier. This is high-enough energy that an optical excitation is needed. With this change, the center cannot capture back the optically excited electron that now exists in the conduction band. As a result, persistent photoconductivity happens. To recover to the substitutional state from the metastable state, one has to warm up the semiconductor so that enough energy exists for the *As* to return to the substitutional state.

We see here a 0.6 eV, 0.36 eV barrier to transition, and a metastable level that is about 0.24 eV above the double donor. These energies are about $10\times$ that of k_BT at room temperature. That is a confinement energy large enough for observation of persistence in the photoconductivity and the difficulty of recovery of the substitutional state. Yet, it is a small energy on the scale of bond energies, and it is the bonding that is being distorted when *As* is displaced. The reason for this small energy difference between the

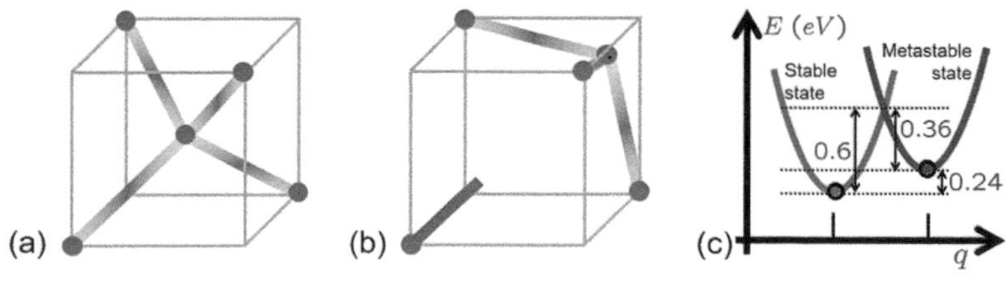

Figure 7.13: (a) An *As* atom substituting for a *Ga* site in the tetrahedral arrangement. (b) An energetically favorable interstitial displacement of this substitute *As*. (c) The configuration coordinate diagram of this stable-metastable system due to lattice distortion.

substitutional and the interstitial locations is that the interstitial *As* forms quite strong sp^2 bonds, which are akin to those in graphene, to the *As* atoms. The leftover p orbital of the *As* atom, which is high energetically, however, does not increase the energy of the interstitial state, since it remains largely unoccupied. The 2 leftover electrons largely reside with the 4th *As* atom from which this sp^2 assembly parted ways.

This energetics can be seen best through configuration diagrams that we first encountered in the discussion related to Figure 1.7 for the Franck–Condon shift. Figure 7.13(c) is a similar picture, where the metastable state lies above the stable state, but the transitions between the two require surmounting of a barrier whose energy is different in different directions. And this energy may be supplied by thermal or optical means. Optical processes will require higher energy for the stable-to-metastable transition, since the excitation will be higher up in energy, with little momentum available from the photon. Optical phonons, on the other hand, have a broad distribution, and the thermal excitation back from the metastable state requires a raising of temperature, although not on the same scale of energy, with the transitions determined by a Boltzmann transition rate.

7.6.2 *DX centers*

THE *DX* CENTER is another example of an unusual point perturbation center—As_{Ga}-like—that arises in the incorporation of impurity atoms in *GaAs* and $Ga_{1-x}Al_xAs$, and a few other compound semiconductors. A group *IV* impurity in a cation site, and a group *VI* impurity in an anion site, both forming shallow donors, exhibit metastable behavior and under certain conditions—a 0.22–0.35 molefraction of *AlAs* in the $Ga_{1-x}Al_xAs$ case—even makes these centers appear deep. They also exhibit the persistent photoconductivity that arises in the behavior of the $As_{Ga}/EL2$ center.

The name *DX*, for deep unknown *X* donor, is a misnomer. It took quite some time and persistence to unravel its origins. All that one knew at the beginning was that it arose because of the presence of substitutional donor impurities that are normally shallow hydrogenic donors. The far-infrared optical transitions between shallow donor levels in *GaAs* provided beautiful evidence for the validity of the hydrogenic model consistent with Figure 7.2 and Table 7.1. There was little evidence to suspect that the same center could give rise to a lattice-relaxed highly localized state.

A *DX* center is a singly negatively charged—again, metastable, as with *EL2*, but not a double donor—state that appears pronouncedly as an interstitial under particular conditions in the compound semiconductors. If it does not, then the system continues to have a shallow substitutional donor as its stable state for the impurity. The relative shift of the energy minimum of the *DX* center is of the order of 0.2 eV above the shallow donor in $Ga_{1-x}Al_xAs$, which is the only case we discuss. This energy difference is the net energy cost of populating the antibonding A_1 level of the shallow donor. When $Ga_{1-x}Al_xAs$ is heavily doped, electron occupation causes it to become singly negatively charged. And as with *EL2*, it can happen by optical excitation and subsequent loss of excess energy to the crystal to appear in this metastable state. Figure 7.13(a) and (b) are still representative for this system, with the substitutional donor binding to the host atoms—*Si* in a cation site—in (a) and distorting in (b), where it is now sp^2 bonded to *As* atoms. When in the metastable state, this *Si* atom now has 2 electrons. This raises the energy cost even more. The non-bonding p orbital of *Si* remains largely unoccupied, while the dangling hybrid orbital of the fourth *As* atom becomes occupied. This is a deeper-lying state.

The energies involved will depend on the surroundings—nearest, next nearest, et cetera—and these energies will depend on the host. For any change in the molefraction of *AlAs*, there is a change in the next nearest neighbors, and this means that this center will have a molefraction dependence. Figure 7.14 shows this molefraction dependence of the energy of the *DX* center as the conduction bandedge energy changes in this $Ga_{1-x}Al_xAs$ system. A corresponding configuration coordinate diagram sketching the details in $Ga_{1-x}Al_xAs$ is shown in Figure 7.15 for a molefraction of 0.32.

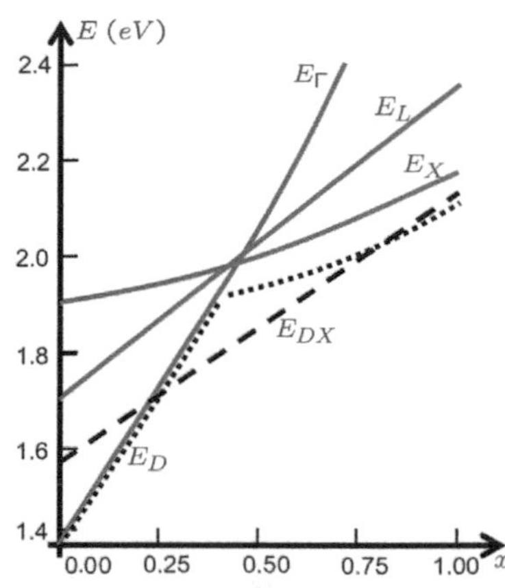

Figure 7.14: The variations in energy of the conduction band minima, shallow hydrogenic *Si* and the metastable *DX* state as a function of the molefraction of *AlAs* in $Ga_{1-x}Al_xAs$.

When the electron is at the *DX* center (q_{DX} state), one may optically excite the electron into the conduction band, making the center neutral. This requires E_{opt} of energy. This transfer changes the *DX* center into a shallow donor ground state (q_D). The electron, following the loss of excess energy to the crystal, is mobile and remains so persistently. It can stay in this mobile state, with the substitutional shallow donor stable, since there exists a barrier of energy E_φ if the electron is *L*-like, and now it cannot easily return to the q_{DX} configuration with electron binding in the *DX* center form. To get to this *DX* metastable state, thermal excitation by raising the temperature is needed.

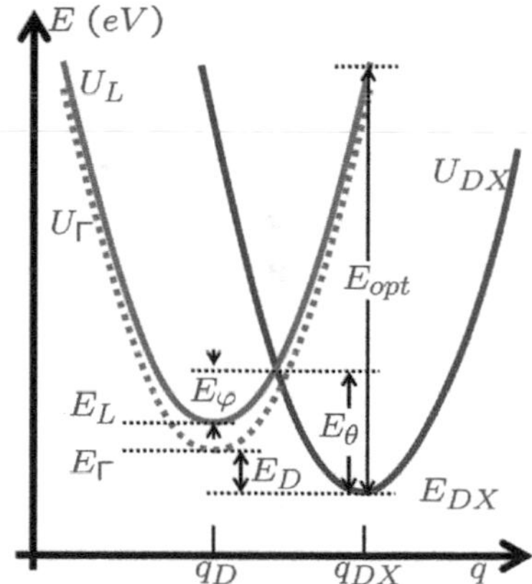

Figure 7.15: The configuration diagram of $Ga_{1-x}Al_xAs$ at a molefraction of 0.32 for *Si* as the donor. The electronic and mechanical energy (*U*) is shown together with the energies of the bandedge minima and the *Si* center state (shallow and deep). q_D identifies the shallow donor form with the electron in the conduction band (in either the Γ or the *L* valley, which both become quite accessible at $x - 0.32$. q_{DX} identifies configuration when the electron is bound to the *DX* metastable form.

So, this *DX* center exhibits a deep-level character but is quite different from the other point perturbations we have discussed. It couples to the conduction band, not the valence band, so it is not like the transition-metal-based deep levels. It has metastability, where this lattice energy exchange is very central. It makes the deployment of doping in compound semiconductor structures difficult with photoconductive and thermal instabilities, particularly when ionization is employed to transfer charge to smaller bandgap regions from selectively doped large bandgap regions.

Structural metastability, similar to these *DX* and *EL2* forms, since they arise in two different hybridizations (sp^3 and sp^2), when there is multiple electron storage, and when the crystal's elastic energy

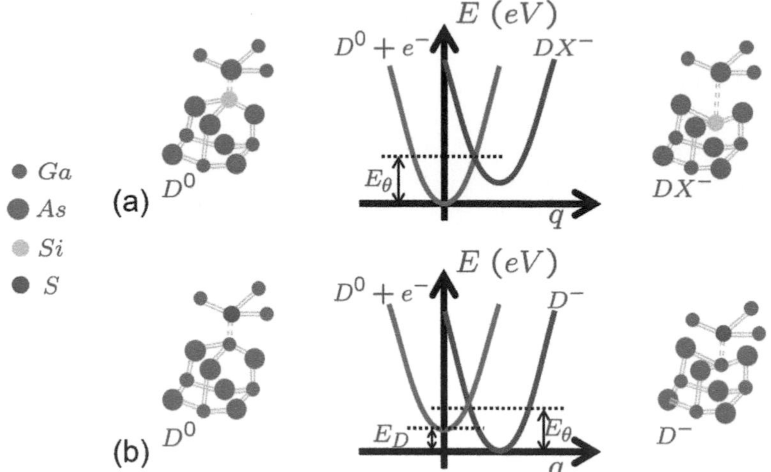

Figure 7.16: The configuration view of *Si* as a donor in *GaAs* is shown in (a). Part (b) shows a similar configuration view for *S* in *GaAs*. The deep-donor-like behavior of *Si* becomes pronounced and quite observable in many measurements only when the crystal is compositionally mixed and the configuration picture looks more like that in (b). In defect physics, negative *U* refers to a center that can lower its total energy by capturing an additional electron even as Coulomb energy increases due to repulsion. The net energy, however, reduces when lattice relaxation more than compensates the Coulomb repulsion. This is the case with (b).

causes lower total energy to be bound in a configuration, appears in many other point perturbations. The stability of such a state, of course, depends on the specifics of the semiconductor itself.

We can now draw intuitive connections between substitutional impurities such as *S*, which has a deep state due to the Hubbard contribution, versus the *DX* center with, say, *Si* as the substitutional impurity. Both can be incorporated into *GaAs*, and Figure 7.16 shows the configuration picture difference between the two—the lattice distortion in the case of *S* is exaggerated here for illustrative convenience—with the free energy changes when the electron is localized on the donor versus when it is not. In Figure 7.16(a), of the *DX* center, in *GaAs*, an ionized donor has lower energy, and if an electron is trapped, that is, with the *Si* and the electron in a DX^- state, then thermal energy is needed to detrap it. The *Si* has distorted away from its normal position in the crystal structure. If the donor is *S*, then it is the electron at the donor (the D^- state) that is the lower energy state of the system, due to the Hubbard energy. On the other hand, for *Si* in a crystal with *AlAs* compositional mixing into *GaAs*, a similar configuration picture (similar to that of Figure 7.15) can also appear for the *DX* center at higher alloy concentrations.

Following this discussion, we turn to a few other interesting defects of importance with implications for semiconductors.

7.7 *Interface and bulk defects in dielectrics*

DEFECTS IN DIELECTRICS AND INSULATORS, and their interface with semiconductors, are also important. Thin dielectrics abound in devices, and defects provide the means for electrons and atoms to move around, in turn causing instabilities. Defects can be electrically active or optically active and, even if inactive, they could be a conduit for diffusion and other problems that appear in the reliability of devices under use with the energy flows in them. We look summarily at some of the important ones to bring about some completeness to the discussion up to this point. Some of these are point perturbations, and some are more elaborate complexes.

7.7.1 *Pb and other centers in SiO_2 and Si*

DEFECTS ARE HARD TO OBSERVE since they are rare and largely sub-nanoscale. The tools for observations tend to be measurements of their effects in response to stimulation. Pass a current, and electrically active defects may become visible. Optically excite, and optically active defects may become observable. A very useful and common method is electron spin resonance, also called electron paramagnetic resonance, a technique of high sensitivity, which can show properties of paramagnetically active centers. So, mostly one can observe their consequences and find the most consistent and complete description that fits with the observations.

Electron spin resonance exploits the excitation of spin of unpaired electrons. It is similar to nuclear magnetic resonance where the nuclear spin is excited. A broken bond—a defect—will exhibit such a paramagnetic signature if an unbounded electron exists in it. In response to an oscillating magnetic flux B, the electron, because of its spin $s = 1/2$ with the secondary spin quantum number $m_s = \pm 1/2$ can move between two energies, up and down, that are separated in energy by $E = g\mu_B B$, where $g = 2.00231$ is the g factor of the electron, and μ_B is the Bohr magneton. When a spin flips, that is, $m_s = 1/2 \mapsto -1/2$, then a photon of this energy is radiated.

SiO_2, being amorphous in this example of our interest, has plenty of distortions of bonds. If SiO_2 were an isolated molecule, that is, molecular SiO_2 similar to CO_2, it would be a linear structure. In its solid-state form, when produced by reacting Si, it is mostly amorphous except when very close to the interface (a sub-nm region). Two basic forming units exist under normal conditions. The SiO_4 unit has tetrahedral coordination, and the Si_2O unit exists with a bond angle that varies from $4\pi/3$ to π. The Si-O spacing in the tetrahedral configuration varies (0.15 to 0.17 nm; the angle doesn't change much). The varying angles indicate the diversity of bonding. For sp^3 hybrids of Si bonding with p orbitals of O, the energetically favored angle is $\pi/2$, which is quite different from the normal tetrahedral sp^3 angles we are used to in zinc blende and diamond structures.

In CF_4 the interatomic distance is 0.136 nm, while the sum of atomic radii per Pauling's self-consistent radii definition is 0.141 nm, so it is not that far off. But, in SiF_4, the measured bond length is 0.154 nm, while the radii add up to 0.181 nm. The bonding reduces the separation, but there can be much variation. In SiO_2, the separation is about 0.162 nm, even if the sum is 0.183 nm. Si has occupiable $3d$ orbitals available for bonding with O's p orbitals. And these orbitals reach further out.

Figure 7.17 shows some example defects—a few of which are particularly important—that are paramagnetically active in the SiO_2/Si system. Several of these defects are electrically active. It is the paramagnetic point defect—the P_b center—at the interface that is of the most import for us. It is the dominant defect of the interface. The P_b center is an Si atom bonded to three other atoms with a dangling bond. So, different surfaces and roughness of surfaces will reflect this orientation of the orbital away from the

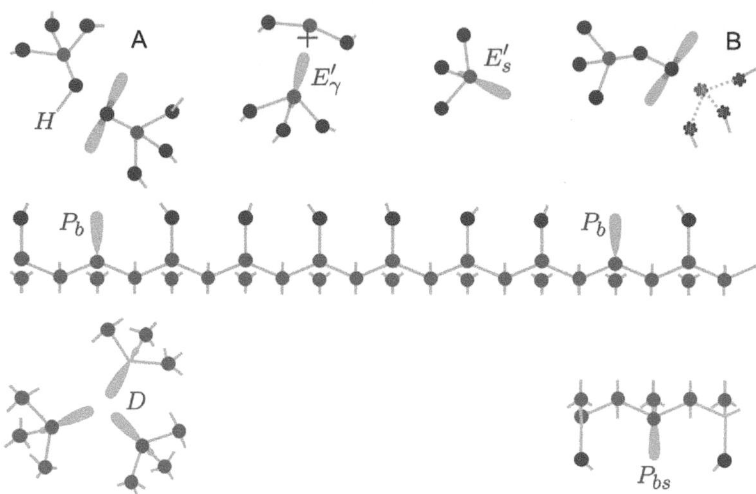

Figure 7.17: Some of the defects that are observed in SiO_2 and Si and their interfaces.

bonding. The random fluctuations in the periodic potential at the surface arising in the P_b center, and the other centers near the interface, comprise a major factor affecting the mobility of carriers along the surface.

In addition to the P_b, a few other centers of import are the following. The E' center arises in a broken Si bond in the amorphous region, and it occurs in a variety of forms with other neighboring arrangements. These are centers that arise with a deficiency of O and an excess of Si. The figure shows two of these: E'_γ, with a positively charged unit near, and the more common E'_s center, which is an isolated center. Non-stoichiometry enhances the appearance of these centers.

Thin thermal oxide, on the other hand, show centers arising in excess oxygen. One of these, the A center, shown in Figure 7.17, is the excess oxygen center. It is a non-bridging-oxygen hole center in the form. Another is the center marked B, which is the peroxyl radical center. Thin thermal oxides are susceptible to these defects. Within the crystalline Si, near the interface, one observes an increased number of defects arising in the processes used to build the SiO_2, which often involve diffusing the O in the variety of ways. Two, similar to the oxide defects, are the D and the P_{bs} center. If nitrogen-oxygen based gases are employed, or enhancement employs H or H containing compounds, other defects will also appear. Si device processing also employs implantation and incorporation of dopants at high densities. These too will create additional defect generation mechanisms because high-energy processes break bonds that may not entirely recover and increased incorporation of foreign atoms will also create defects. Our earlier point perturbation discussion pointed to some examples of these defects. Vacancies, for example, exist in implanted and annealed semiconductors.

7.7.2 F centers

DEFECTS CAN BE OPTICALLY ACTIVE, and one implication of this is the color that one often observes in what are normally transparent dielectrics. This is particularly an important feature of crystalline ionic solids. Ionic solids—high permittivity dielectrics, for example—have point defects—a missing atom leaving broken bonds—that become optically active via electrons or holes. A low energy ground state can then get excited to a higher energy state. This lost frequency in the broadband of light means that the light leaving will be depleted of this color. On the other hand, when the charged carrier drops back from the excited state to the ground state, it will emit photons at this frequency with some linewidth. These are color centers. An example is the *F* center, where an electron is trapped at an anion vacancy. This is a system akin to the hydrogen system in that there is a negatively charged electron together with a positive charge as a quantum system. There will exist a distribution of excited energy states in the system between which and the ground state transitions will take place when we deviate from thermal equilibrium. Al_2O_3 is a dielectric often used. Rubies are crystals of Al_2O_3. Cr^{3+} occupying Al^{3+}, so a Cr_{Al} defect, gives ruby its red color. The energy degeneracy of the three $3d$ electrons of Cr's outer shell interacting with the surrounding O^{2-} ions is broken into two groups in the crystal field. Three new energy levels appear above the ground state, as shown in Figure 7.18, and none of these were present before.

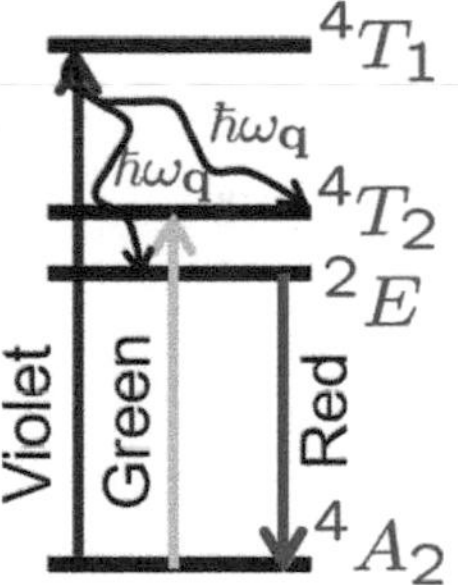

Figure 7.18: Absorption and emission of light in ruby (Cr-doped) Al_2O_3. The absorption spectrum is broad, with excitation from the ground state ($^4A_2$) to $^4T_1$ (violet) and $^4T_2$ (green to yellow). This means that red (with a undercurrent of blue-purple) is selectively transmitted. Electrons in the ^2E state arise more from the other excited state, with excess energy lost to the crystal through phonons. The emission of this light at red also enhances the color.

Upon shining light, the $^4T_1$ (violet) $^4T_2$ (green to yellow) are preferentially occupied by electrons. The transition to ^2E is relatively weak. All these transitions have a reverse process too, but the first two levels have a dominant transition to the ^2E level, with the excess energy lost to the crystal. So, in absorption, it is the shorter wavelengths that are absorbed, with much less at red, giving ruby its color. In good-quality rubies, one will also see the color enhanced by the red optical emission from the ^2E energy. The earliest solid-state lasers employed this capability to fill the ^2E energy level indirectly to achieve the laser breakthrough.

7.7.3 Poole-Frenkel conduction

THE POOLE-FRENKEL EFFECT is the lowering of barrier energy at a defect in a dielectric, for example, at a positively charged defect that can capture an electron, and has related Coulomb effects. See Figure 7.19 for an illustrative schematic of the essentials of the process and those under leakage mechanisms at work at the defect site. The effect has parallels with image force lowering, with the major difference being the absence of the conducting plane.

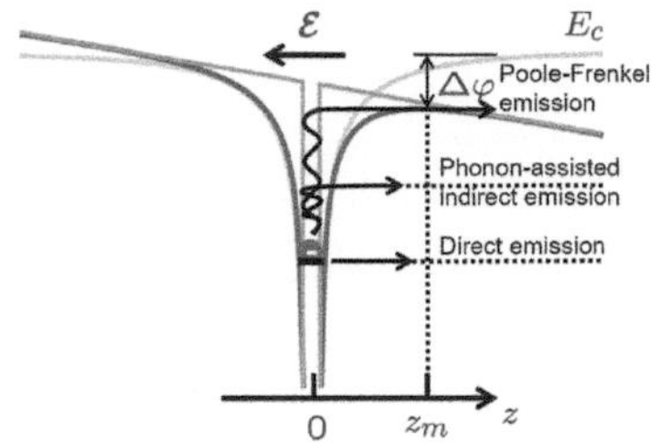

Figure 7.19: Schematic of the Poole-Frenkel process. Three tunneling processes are shown. Direct tunneling and phonon-assisted indirect tunneling are the two additional processes via which electrons pass through. Poole-Frenkel emission is barrier lowering without the image charge.

The Poole-Frenkel effect is a lowering of the Coulomb potential energy barrier in the presence of an electric field. A trap that is uncharged when filled and positively charged when empty, for example, will have this type of barrier lowering. Since it arises from the Coulomb interaction between charge on the defect and the electron emitted, it can exist for multiply charged states too. It is important in materials with defects, since the lowering of barrier enhances the current that can flow. The Poole-Frenkel process takes place, for example, in a dielectric such as the gate oxide, leading to enhanced current. It also occurs in semiconductors that are highly defected under high fields.

Consider the simple case of a singly charged defect and an electron in the presence of an electric field $\mathcal{E}$. At any position z, the change in energy is

$$\Delta U = e\mathcal{E}z + \frac{1}{4\pi\epsilon}\frac{e^2}{z}. \tag{7.46}$$

The maximum occurs at

$$e\mathcal{E} = \frac{1}{4\pi\epsilon}\frac{e^2}{z_m^2} \quad \therefore \quad z_m = \left(\frac{1}{4\pi}\frac{e}{\epsilon\mathcal{E}}\right)^{1/2}. \tag{7.47}$$

Therefore, the barrier height lowering is

$$\Delta\varphi = \Delta U|_{max} = 2e\mathcal{E}z_m = 2e\mathcal{E}\left(\frac{1}{4\pi}\frac{e}{\epsilon\mathcal{E}}\right)^{1/2} = \left(\frac{e^3\mathcal{E}}{\pi\epsilon}\right)^{1/2}. \tag{7.48}$$

Such a barrier lowering will cause excess conduction as electrons hop from defect to defect.

7.8 Summary

ATOMS ARE LONG-LIVED stable entities conforming to quantum mechanics and relativity as nature's principles, with electrons' wavefunctions quite cleanly definable. Our crystals are stable, bonded arrangements of these atoms. We have seen the states of the valence electrons of the atom evolve to bonding and antibonding states, and, with a large collection of them together, formation of valence bands and conduction bands. The valence bands have their origin in bonding, and the conduction bands in antibonding. Thermodynamics, however, also teaches us the likelihood of appearance of defects and other possible arrangements that are different from the primary one, depending on statics (and dynamics) and related to the energy of their formation. Take the example of *Si*. The $3s^2 3p^2$ atomic configuration evolves to the sp^3 hybrid in the formation of the diamond crystalline form of *Si*. This crystalline structure will have sites where the *Si* atom may be missing, or a *Si* atom is replaced by another atom, say, *P*, which has one excess electron, and can still form the sp^3 hybrid. It may be that there exists a different impurity, say, *Ti*, and this atom exists in the crystal substitutionally, as did the *P*, or interstitially, and, in each of these cases, there may be some bonding to the other surrounding *Si* atoms, as well as broken bonds. For that matter, there may even be lattice distortion. A compound semiconductor $Ga_{1-x}Al_xAs$, for example, may have all these complexities, and even more, since it is a compound. These are all point perturbations arising from a locale of the crystal, where the expected symmetries of the crystalline structure run amok.

Some of these perturbations will have a short-range effect, and some will have a long-range effect. This chapter developed the formalism for understanding several important examples of such perturbations. The case of *P* substitutional at a *Si* site was the example of a hydrogenic impurity that has a long-range perturbation. States arising in the valence of *P*, even though spatially localized to the equilibrium position of *P*, so a Wannier function approach could be applied, have a long reach, and therefore effective mass can become a suitable tool for determining the binding energy. On the other hand, if there exists a vacancy at a *Si* site, this is a localized perturbation. Bonds to the surrounding *Si* are absent, and the up and down electron spin filling of the bonding now has the contributions of the missing *Si* missing. The local arrangements will change, even though farther away from this arrangement the Bloch function picture is still valid. This defect has many of the attributes of what we saw when symmetry broke at the surface. An *Si* vacancy will give rise to states that are deep in energy, that is, we now have a deep level. *Ti* substitutional, or interstitial too, will have consequences similar to this, where now we need to think about the evolution of the states contributed by the *Ti*.

We developed a number of techniques to understand the variety of point perturbations and, in doing so, drew on the learning of the earlier chapters related to Hartree potential, exchange correlation and mixing of states in perturbation. We could analyze shallow dopants as long-range perturbations with Hartree and exchange potentials absent due to the one-electron difference. The result is a Schrödinger-like equation where multiple binding energy levels, all quite small—of the order of room temperature thermal energy—are obtained. This was the case with one excess or deficient electron. If it is more than one, then one needs to include the Hubbard correction term—a Hartree energy—which we found through position-dependent screening. *S*, for example, is a deep center.

Deep centers—strongly bound states arising from strong confining short-range potential—appear from a variety of causes, extended defects, transition metal impurities and so on. We approached these through tight binding thinking, where one viewed it as a defect molecule and employed the orbitals as the basis. This approach let us analyze vacancies, where the simplest toy instance is the appearance of a defect in the center of the gap due to the termination of a chain. The defect-molecule approach is very useful in showing how the bonding and antibonding approach to molecules can be seen to lead to states in the bands and in the bandgap in semiconductors. One could apply this to both interstitial and substitutional defects.

Recall that tight binding works particularly well with covalent molecules, and that is where it was developed first. It also happens to be useful in many situations of semiconductors.

We also looked at defects that we called complexes, since they arise in more than one simultaneous important change. As_{Ga} was our prototypical example. *As* has 5 electrons in the outermost orbit, and it forms the bond by sharing it with *Ga*, which has 3 electrons in the outermost orbit. As As_{Ga}, a metastable state is displacement of *As* along the cube's main diagonal bonded to three other *As* (an sp^2 hybridization), and the excess electron with the fifth *As* of the cube available for conduction. There is also a stable state with *As* tetrahedrally arranged. But, since the lattice distorts, there exists a barrier for change from each of this configurations to the other, and one can illustrate this behavior through the configuration coordinate, where both the electron and the lattice energy can be brought in together by using a generalized coordinate.

Another example of a very interesting complex is the *DX* center in compound semiconductors, an sp^3 hybrid of a substitutional donor such as *Si* with metastable behavior akin to that of As_{Ga} where it distorts out and is now sp^2 hybridized. The stable-metastable state transitions are barrier modulated but are now also related to the electron population and the barrier itself, which depends on the composition of the mixed compound such as

$Ga_{1-x}Al_xAs$, so one observes this deep character, but it also has unusual persistence photoconductivity and thermal dependence.

A few other defects that we touched on are the *Pb* center of SiO_2, which appears at the interface with *Si*, and others that appear in the bulk of SiO_2. These, and excess oxygen defects, do have a significant role in the noise and reliability behavior of small *Si* structures. An optically active center, the *F* center, was also discussed, since it has relevance to high permittivity oxides that we discuss in Chapter 18. And we concluded with a short discussion of the energetics of conduction around point defects through the Poole-Frenkel mechanism.

7.9 *Concluding remarks and bibliographic notes*

DEFECTS VIEWED from a mechanical perspective, where strength and other interface-centric properties assume importance, determine strength and a variety of other properties of materials. For us, the perspective was from physics with a focus on electronic and optical implications. So, our discussion has been the limited to these and their origins in the quantum nature of defects.

For understanding hybrids, and a chemistry-centric view of the deployment of quantum mechanics, I have regularly found Atkin's book[1] to be a very convenient, compact and readable source.

The Hartree, exchange and correlation view of the electron states help us understand the nature of the electron state, whether it is spread out like a Bloch function—long-range–or as a localized deep defect with short-range perturbation. An early and detailed discussion is by Lannoo[2]. The techniques are developed comprehensively for deep levels in many of the important semiconductors, with particular attention to transition elements in them, by Kikoin and Fleurov[3]. The book by Enderlein[4] is a very readable source for understanding the detailed analytic analysis of shallow and deep intrinsic and extrinsic defects of semiconductors.

A material and thermodynamic view of defects is the subject of a book by Pichler[5]. This book tackles the subject in silicon and, being materials-focused, also handles diffusion, which is quite affected by defects, comprehensively. A broadening of this approach to more semiconductors is by Seebauer and Kratzer[6].

The experimental techniques of studying defects use a variety of interaction and signaling phenomena that allow one to measure low energy signals. It is an art in itself, utilizing a variety of resonance phenomena and signal modalities. Readers interested in exploring

[1] P. W. Atkins, "Quanta," Oxford, ISBN 0-19-855572-2 (1991)

[2] M. Lannoo and J. Bourgoin, "Point defects in semiconductors I," Springer-Verlag, ISBN 13 978-3-642-81576-8 (1981)

[3] K. A. Kikoin and V. N. Fleurov, "Transition metal impurities in semiconductors," World Scientific, ISBN 981-02-1883-4 (1994)

[4] R. Enderlein and N. J. M. Horing, "Fundamentals of semiconductor physics and devices," World Scientific, ISBN 981-02-2387-0 (1997)

[5] P. Pichler, "Intrinsic point defects, impurities and their diffusion in silicon," Springer-Verlag, ISBN 978-3-7091-7204-9 (2004)

[6] E. G. Seebauer and M. C. Kratzer, "Charged semiconductor defects," Springer, ISBN 978-1-84882-058-6 (2009)

this range of techniques will find the book edited by Stavola stimulating[7].

SiO_2/Si being a major semiconductor assembly of industrial technical interest, there are numerous books that summarize the latest understanding. The ability of SiO_2 to contort—it forms a solid but doesn't really exist the way CO_2 does as a molecule and a gas—as a solid is essential to the success of this SiO_2/Si system. A beautiful paper[8] discussing the bulk electronic structure of SiO_2 by Laughlin, Joannopuoulus and Chadi, three folks who made a variety of other major contributions to condensed matter, is very worthwhile reading. Equally of significance is an early paper by Helms and Poindexter[9]. This paper—well before its time—points out numerous reliability and other issues that will arise due to defects, as oxides were made thin well before they acquired significance in technology.

A comprehensive introduction to the subject of defects in the SiO_2 systems is in the edited volume of Pacchioni, Skua and Griscom[10]. The book discusses measurement techniques, defects connected to different technologies and processing procedures, and defects in bulk and at interfaces.

For understanding the DX center, a few papers are suggested. A thorough review of the center is by Pat Mooney[11]. She and Tom Theis undertook the early comprehensive experimental studies when $Ga_{1-x}Al_xAs$ unveiled a variety of its interesting electronic signatures. A recent review of such defects is by Alkauskas et al.[12] A very stimulating set of reading, including on the process to scientific clarity arrived through debate, are the papers by Chadi, who was referred to earlier, including one of the author's responses. These papers are excellent examples of how to write, debate and elucidate, and how experiments drive theory, and theory drives experiments, in search of clarity.[13,14,15]

7 M. Stavola (ed.), "Identification of defects in semiconductors," **51A**, Semiconductors and Semimetals, Academic, ISBN 0-12-752159-3 (1998)

8 R. B. Laughlin, J. D. Joannopoulos and D. J. Chadi, "Bulk electronic structure of SiO_2," Physical Review B, **20**, 5228–5237 (1979)

9 C. R. Helms and E. H. Poindexter, "The silicon-silicon-dioxide system: Its microstructure and imperfections," Report on Progress in Physics, **57**, 791–852 (1994)

10 G. Pacchioni, L. Skuja and D. L. Griscom (eds), "Defects in SiO_2 and related dielectrics: Science and technology," Springer, ISBN 978-0-7923-6686-7 (2000)

11 P. M. Mooney, "Deep donor levels (*DX* centers) in *III-V* semiconductors," Journal of Applied Physics, **67**, R1–R26 (1990)

12 A. Alkauskas, M. D. McCluskey and C. G. Van de Walle, "Tutorial: Defects in semiconductors—Combining experiment and theory," Journal of Applied Physics, **119**, 181101 (2016)

13 D. J. Chadi and K. J. Chang, "Theory of the atomic and electronic structure of *DX* centers in *GaAs* and $Al_xGa_{1-x}As$ alloys," Physical Review Letters, **61**, 873–876 (1988)

14 D. J. Chadi and K. J. Chang, "Energetics of *DX*-center formation in *GaAs* and $Al_{1-x}Ga_xAs$ alloys," Physical Review B, **39**, 10063–10074 (1989)

15 D. J. Chadi, K. J. Chang and W. Walukiewicz, "Reply to Maude, Eaves, Foster and Portal," Physical Review Letters, **62**, 1923 (1989)

7.10 Exercises

1. Let us try to use the δ function as an approximation for defect potential. The point imperfection is a potential of $-V_0$ of a square well of radius a.
   - Find the normalized ground state wavefunction.
   - Let the bound state have an energy of $0.5\ eV$, let the effective mass be free electron mass, that is, $m^* = m_0$, and let $a = 0.01\ nm$. What is V_0?

- What is the extent of the bound electron in the ground state?
- Estimate the number of Si atoms and valence electrons that will be enclosed in the sphere of the extent of the previous bound electron. **[M]**

2. As^+ and In^- form a complex defect pair in Si. Si has a lattice constant of 0.543 nm, a static dielectric constant of 11.8 and a bandgap of 1.1 eV. Calculate and plot as a function of spacing the ionization energies of the nearest, second-nearest and third-nearest pairs. The levels can be assumed to shift by the same amount. Are any of these energy states in the bandgap? **[M]**

3. A donor-acceptor complex exists in a one-dimensional crystal that has a force constant of 10 N/m. It has a Franck-Condon shift of 0.1 eV. Estimate the spacing shift in the complex for an electron in the ground state versus the excited state. **[S]**

8
Transport and evolution of classical and quantum ensembles

UNDERSTANDING THE BEHAVIOR OF ENSEMBLES with or without perturbation, that is, with or without an external stimulus, is essential to predicting macroscopic observable properties of any semiconductor or, more generally, condensed matter devices. Particles—treated classically or quantum-mechanically—maintain equilibrium and respond to a stimulus under constraints placed by the environment. We have stressed that, in any of the analysis, it is critical to determine what part of the behavior should be treated quantum-mechanically, since that is the level of detail that is necessary to capture the essence, but that any such quantum-mechanically constrained part is surrounded by a classical environment to which it connects and through which one observes. And any such predictions of observable properties based on quantum-mechanical properties should gracefully asymptote to the classical observations upon which most of our intuition is built. This correspondence property must come about when one views a collection:, that is, an ensemble, evolving as a classical or quantum-mechanical collection of particles.

An electron, a photon or a phonon is both a particle and a wave. And as such, how their response evolves under stimulus is of immense interest. Particles interact with each other. These interactions lead to the appearance of thermal equilibrium in the absence of any stimulus except that of an energy and particle-exchanging contact with an environment at temperature T. When the equilibrium is disturbed, the collection evolves. Classically, the response that one observes from a system—in equilibrium or away—is one of statistical expectations of various moments. Noise, for example, is an observation of fluctuations and is a direct descendant of one of the moments. As a system response to stimulus, the

Semiconductor Physics: Principles, Theory and Nanoscale. Sandip Tiwari.
 Published 2020 by Oxford University Press. DOI: 10.1093/oso/9780198759867.001.0001

expectation and the noise in expectations are connected to each other through this ensemble's evolution. Diffusion is a spreading out of a distribution of an ensemble grounded in random walk. Drift is the expectation of the net motion of this distribution caused by the stimulus. These must be related to each other.

For example, a resistor has its resistance representing the expectation of the impeding that takes place to the momentum, that is the current, of the charge carriers of an ensemble. The Johnson–Nyquist noise represents a fluctuation that is related to the resistance.

In semiconductors, we have particularly stressed the behavior of electrons. Many semiconductor properties are an ensemble response of these carriers. For electrons, movement of charge, that is current, movement of energy, that is amplification based on conversion in desired energy forms or heat flow, et cetera, are all of interest with semiconductors. But, an electron can be viewed as both a particle and a wave. The $E(\mathbf{r}, \mathbf{k})$ of this electron—and its ensemble with a distribution of the different $E(\mathbf{r}, \mathbf{k})$ states it occupies within even an independent electron picture—is a quantum-mechanical consequence. Properties arising in the response of this electron charge cloud will have plenty of subtleties arising in the classical and quantum-mechanical considerations.

We are writing the energy dispersion relationship as $E(\mathbf{r}, \mathbf{k})$ since there are going to be positional dependences in a general problem. Electrostatic potential changes cause bandedge energy— usually our reference when we write $E(\mathbf{k})$ to change at any $\mathbf{r}$.

We will see in this discussion the correspondences and differences between the classical behavior and the quantum-mechanical behavior. The ensemble response may be represented in an expectation of a parameter of interest, but the response also reflects the interactions that occur within the ensemble as a collection of the particles or waves. Indeed, the ensemble is a probability distribution that is evolving in time, whose consequences are reflected in the numerous parameters of interest. There are numerous ways to describe these. The Liouville equation, the Langevin equation, the Fokker Planck equation, Kolmogorov's forward and backward equations, and Markov chains relate to such statistical evolution and are relevant to a wide variety of observed natural phenomena. We will make connections from the classical to the quantum through the Liouville equation and density matrices, and also remark on the others' relevance to quantum-mechanical nonequilibrium, but concentrate on Liouville's Boltzmann transport equation reformulation. The Liouville equation describes particle flow in phase space, and its quantum-mechanical formulation describes the evolution of mixed states through the density matrix. The Boltzmann transport equation becomes a particularly powerful tool for describing semiconductor behavior, where quantum considerations can be suitably brought in. In Chapter 13, we will concentrate on a Green's function approach to look at the evolution through a quite different formulation—an integrated functional response rather than a differential form—to remark on classical-quantum subtleties.

We start with a simpler problem first.

What happens when particles are treated as independent and move with little or no interaction with each other, that is, with minimal scattering?

8.1 *Transport with vanishing scattering*

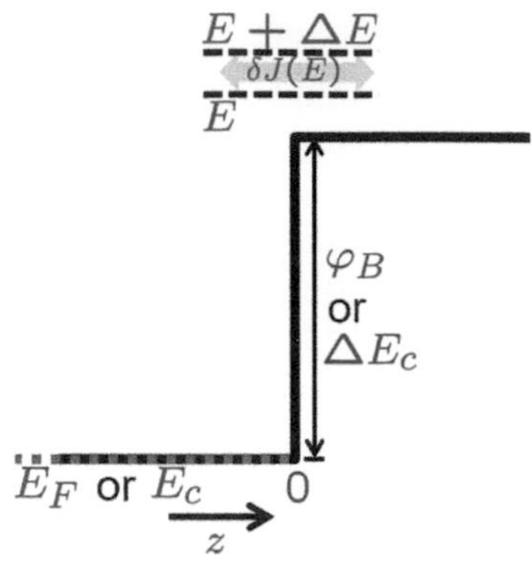

Figure 8.1: A conceptual view of the energy-position diagram at a semiconductor heterostructure or metal-semiconductor interface. For metal, the energy edge on left reflects the electrochemical potential or Fermi energy. For the semiconductor, it is a representation of the conduction bandedge.

TAKE THE EXAMPLE OF THERMIONIC EMISSION. Consider a heterostructure interface between two crystalline semiconductors or a metal/semiconductor interface, as shown in Figure 8.1. A conduction bandedge discontinuity (ΔE_c) or a barrier height ($\Delta\varphi_B$) describes a region at the interface where states and electrons may exist on the left, but states do not exist on the right. Let the rate-limiting step be transport at the interface, that is, arising in electrons moving to the right or the left, restricted by the considerations at the interface. Copious supply and extraction capability exists for the electron transport away from the interface. In the band of energy of $E(\mathbf{r}, \mathbf{k})$ states shown there will exist populated extended states that propagate toward the left as well as toward right. At thermal equilibrium, each and every transition, including propagation, is balanced in detail. In the presence of a stimulus, a net effect arises as one disturbs this equilibrium. For simplicity of discussion, we assume isotropic simple bandstructure, an abrupt barrier and no scattering or perturbational interaction such as quantum-mechanical tunneling at the interface.

Copious supply to the interface means that the rate-limiting step is determined by the constraints from the interface, and therefore that is the only region that we need to rigorously analyze. The more general considerations of this important problem of transport in devices is in S. Tiwari, "Nanoscale device physics: Science and engineering fundamentals," Electroscience 4, Oxford University Press, ISBN 978-0-198-75987-4 (2017). Here, we are only interested in extracting a salient feature of this transport at the interface when no other consideration of interaction except for the propagation of a state applies.

Conservation of energy—the total energy consisting of potential and kinetic—and momentum must hold, since scattering through other sources in the environment has been discounted. Only carriers—occupied states—that can surmount the barrier and have the properly directed momentum in motion may transmit. For motion from left to right, the possibly lowest energy electron is one that has momentum toward the right (positive z or $\perp$ direction), and all this energy is associated with this $\hbar\mathbf{k}_\perp$, to overcome the barrier and be in a propagating step on the right. The lowest $k_\perp$ state on the right is for $k_\perp = 0$. But only an electron with $\hbar^2 k_\perp^{*2}/2m^* = \varphi_B$ or ΔE_c, where $\mathbf{k}_\perp^*$ is the wavevector at $z = 0^-$, has the minimum total energy with the proper direction for this transport. If the momentum is less than $\hbar\mathbf{k}_\perp^*$, no matter how much momentum exists parallel to the interface ($\hbar\mathbf{k}_\parallel$, and energy corresponding to that motion $E_\parallel = \hbar^2\mathbf{k}_\parallel^2/2m^*$), it shall not pass. It must also have the matching momentum, but, for a metal, with its nearly free electrons, and a fairly filled Brillouin zone, this should not be an issue. For a semiconductor in the $z > 0$ region, this may very well be an issue, but we will ignore it, since it doesn't distract from the central result

For semiconductor/semiconductor systems, this momentum matching will be quite a bit more complicated. Resonant tunneling, conduction currents across heterostructure barriers where the symmetry of the barrier states is in contrast to that of injecting, and their dependence reflected in tunneling and barrier widths and presence of sequential tunneling, that is, tunneling with scattering and changes in momentum and energy, show this.

we want to derive. So, we assume that energy and occupation of injecting states, and properly directed momentum, are the major considerations. What we have just said is captured in Figure 8.2 as a region that is blocked, and a region beyond where energy and sufficient momentum exist for injection from left to right.

For electrons that can transit left to right in the sliver of energy δE at E, there exists a current that can be written as a product of charge, velocity and the number of electrons, that is, the number of states that are occupied. This current, a thermionic injection current in this sliver, is

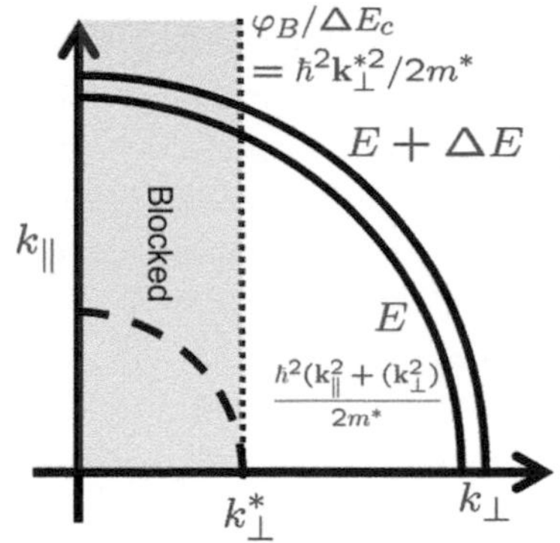

Figure 8.2: Energy and momentum, reflecting regions where transport is allowed or blocked for injection from left to right of the interface shown in Figure 8.1.

$$dJ = e\frac{\hbar k_\perp}{m^*}2\frac{d^3k}{(2\pi)^3}\mathscr{F}\left(\frac{E_\parallel + E_\perp}{k_B T}\right)$$

$$\therefore \quad \frac{dJ}{dE_\perp} = \frac{2e}{h}\frac{d^2k_\parallel}{(2\pi)^2}\mathscr{F}\left(\frac{E_\parallel + E_\perp}{k_B T}\right), \tag{8.1}$$

where $\hbar k_\perp/m^*$ is the velocity, with a spin degeneracy of $g_s = 2$ for states that are occupied given by the product of the density of states and the Fermi-Dirac occupation function. The result, rewritten, is a product of $2e/h$, a conductance, and the density of occupied states. For each state available for this ballistic conduction, there exists a conductance $2e/h$ or $-2q/h$ ($q = -e$) for change in current per change in energy.

Each conductance channel provides, in bias voltage dependence, a $2e^2/h$ or $2q^2/h$ of conductance in current/voltage units. We have arrived at a significant result. *For ballistic conduction of an electron—a quantum wavefunction motion—the quantum conductance associated with each channel is a nature's constant,* $2e^2/h$. The factor 2 here arose in each channel being occupied by two electrons, each with a separate secondary spin quantum number.

We can now integrate Equation 8.1, knowing that $E_\perp$ can vary φ_B and higher and that $d^2k_\parallel = 2\pi k_\parallel dk_\parallel = (2\pi m^*/\hbar^2)dE_\parallel$. This complete expression is

$$\begin{aligned} J &= \frac{2e}{h}\frac{m^*}{(2\pi)\hbar^2}\int_{\varphi_B}^{\infty} dE_\perp \int_{E_\parallel} dE_\parallel \mathscr{F}\left(\frac{E_\parallel + E_\perp - E_F}{k_B T}\right) \\ &= \frac{2em^* k_B T}{(2\pi)^2\hbar^3}\int_{\varphi_B}^{\infty} dE_\perp \ln\left[1 + \exp\left(-\frac{E_\perp - E_F}{k_B T}\right)\right], \end{aligned} \tag{8.2}$$

which, in the Boltzmann limit, reduces to a thermionic current of

$$J_{te} = \frac{2em^* k_B^2}{(2\pi)^2\hbar^3}T^2 \exp\left(-\frac{\varphi_B - E_F}{k_B T}\right) = \mathscr{A}^* T^2 \exp\left(-\frac{\varphi_B - E_F}{k_B T}\right). \tag{8.3}$$

This equation is the Richardson equation, with $\mathscr{A}^* = 2em^* k_B^2/(2\pi)^2\hbar^3$ as the Richardson constant. An exponential energy factor—bias voltage dependence—arose in the Boltzmann occupation probability.

The prefactor is related to the motion in the occupied channels:, that is, a quantum conductance and a temperature dependence that arose as a prefactor from the integration of the occupation probability of the channels. This is thermionic emission.

This e^2/h, usually written in the inverse form, h/e^2, appears as nature's parameter often in conduction. Superconductivity, single electron tunneling, quantum Hall effect and others all have an invariant parameter that is related to h/e^2.

Thermionic emission is a direct reflection of quantum conductance arising in ballistic—unscattered—conduction. This was current from left to right. There is current flowing from right to left. Thermal equilibrium forces one of these conditions, and one gets from this the $\exp(qV/k_BT) - 1$ dependence. In a metal/n-type semiconductor junction, it is the electron transport from the right to the left that will dominate. And the assumption that we made that there are states available for filled states to couple to empty states is certainly very valid. The left side has many unoccupied states.

Mesoscopic transport literature is surfeit with h/e^2s, including in noise, since the occupation of channels is quantized, thus affecting fluctuations. Supremely high mobility materials, for which a mobility of $10^6\ cm^2/V\cdot s$ is quite common at low temperatures in many high quality compound semiconductor heterostructure interfaces, can show ballistic transport for dimensions of many μm. So, ballistic conduction—mesoscopic transport—where the electron is both a charge particle that can be bent by electric and magnetic fields, and a wave that shows interference and other effects, is quite observable at relatively large dimensions. There is nothing necessarily nanoscale about it. See S. Tiwari, "Nanoscale device physics: Science and engineering fundamentals," Electroscience 4, Oxford University Press, ISBN 978-0-198-75987-4 (2017) for many such device implications.

The emphasis in this discussion is to bring out the importance of the $2e^2/h$ factor. It arises through that intimacy of the momentum-velocity relationship in this quantum wave-particle correspondence. If scattering events are absent, or limited in numbers, one sees it in transport.

This wave-particle (charged) interlink can be seen through a calculation of conduction in channels (see Figure 8.3). Consider current arising from electrons in an energy interval dE of the nth band with propagation in the y direction. The x and z direction are confined and have no propagation. As a particle, one can view the current subscripted by this band as arising from dN_n electrons in that energy spread of dE. Let the channel be of length L. Then $dJ_n = edN_n/(L/v)$, which is looking at the current density passing through any cross-section at velocity v with a density of dN_n of such electrons in the L-long channel. The velocity is $\mathbf{v}_{n\mathbf{k}} = (1/\hbar)\nabla_{\mathbf{k}}E_{n\mathbf{k}}$. Or, for this problem, $v = (1/\hbar)\partial E_{n\mathbf{k}}/\partial k_y$. The k_y direction has an equidistant spacing of states given by $2\pi/L$. So, up to an energy E, the number of filled electron states is

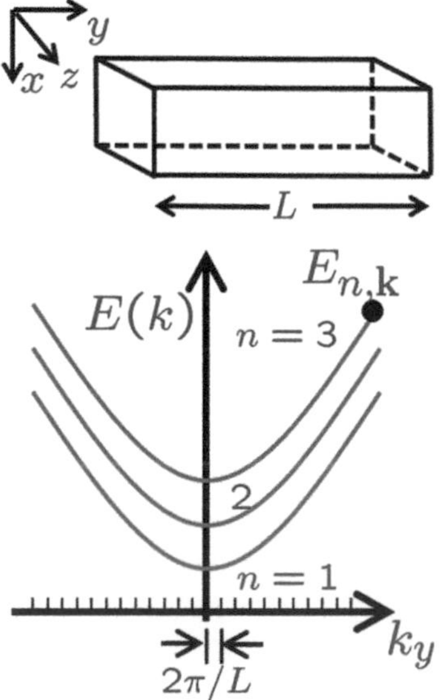

Figure 8.3: Quantum conductance arising from electron transport as a particle or propagation as a wave through a conducting channel in the y direction.

$$N_n = g_s \frac{L}{2\pi} k(E) = 2\frac{L}{2\pi} k(E). \tag{8.4}$$

Hence,

$$\frac{dN_n}{dE} = 2\frac{L}{2\pi}\left(\frac{\partial E(n,k)}{\partial k}\right)^{-1} = 2\frac{L}{2\pi}\frac{1}{\hbar v}. \tag{8.5}$$

It follows therefore that

$$dJ_n = \frac{2e}{h}dE = \frac{2e^2}{h}dV, \text{ or } g_q = \frac{dJ_n}{dV} = \frac{2e^2}{h}. \tag{8.6}$$

The conductance per open channel populated with an up spin and a down spin electron is $g_q = 2e^2/h \approx 80\mu S$. This is the ballistic conductance due to the motion of the charged particle as a wave.

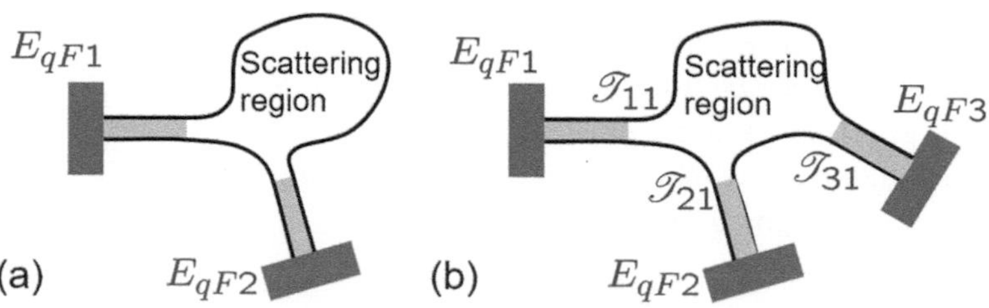

Figure 8.4: Quantum conductance arising from electron transport as a particle or propagation as a wave through a conducting channel in the y direction.

If channels are filled up to the quasi-Fermi energy E_{qF1}, and if it is a low temperature so that we do not worry about the tail of occupation, and transport takes place to a region where the quasi-Fermi energy is E_{qF2}, the current will be

$$I_j = \frac{2e}{h}(E_{qF1} - E_{qF2}) = \frac{2e}{h}\Delta E_{qF}. \qquad (8.7)$$

This is shown in Figure 8.4(a). If there is some scattering, and multiple contacts to the environment, as in Figure 8.4(b), the generalization of this is

$$I_{mn} = \frac{2e}{h}\mathscr{T}_{mn}(E_{qF1} - E_{qF2}), \qquad (8.8)$$

where m and n identify two of these injector/receiver regions, and $\mathscr{T}_{mn}$ is the transmission coefficient. This is a direct use of electromagnetic scattering approach to this semiconductor conduction problem, with the electron as a charge-carrying wave. The multi-port scattering theory directly translates to this limited-scattering problem.

We now turn to a discussion of ensembles where interactions will be allowed to take place. First, take classical particles. Liouville's theorem describes for us this evolution.

8.2 *Classical Liouville's theorem*

Canonical, as an English word, means connected or by law, so something that follows a rule or is orthodox. It comes from the Latin word *canonicalis*, which means of an ecclesiastical edict. Add to it the interpretation that, as a result, a standard behavior or attribute can be tied. $\mathbf{q}$ as position—Cartesian or polar—can be connected to $\mathbf{p}$—linear momentum or angular momentum—through the rule that the time-dependent change is related to the partial derivative of the Hamiltonian of the conjugate coordinate. It doesn't matter which coordinate description one chooses.

HAMILTONIAN MECHANICS CONVENIENTLY DESCRIBES the behavior of particles through two canonical coordinates $(\mathbf{q}, \mathbf{p})$, where $\mathbf{q}$ is a generalized position coordinate and $\mathbf{p}$ is the momentum coordinate. Let us identify the ith particle by subscripting with i. Hamiltonian dynamics is convenient since it allows us to write first order equations via

$$\dot{q}_i = \partial\mathscr{H}/\partial p_i, \text{ and } \dot{p}_i = -\partial\mathscr{H}/\partial q_i. \qquad (8.9)$$

Each particle's state is represented by the two coordinates $(\mathbf{q}, \mathbf{p})$. The continuous space of these coordinates is the phase space. Each particle at any instant of time can be described by the $(\mathbf{q}, \mathbf{p})$ point. For one particle, it is a two-dimensional vector space. n particles

The vector is represented by three projections.

can be represented by a $2n$-dimensional phase space of coordinates, and one can write $2n$ first order equations to describe the trajectory of all the particles. At any point in this phase space, only one trajectory passes through for a given Hamiltonian.

Figure 8.5 specifically identifies trajectories of three particles as they evolve in time from t to t' in phase space.

Liouville's theorem asserts that, in classical motion, with the Hamiltonian complete, and no other energy input or output into the system, the phase space of any particle or their collection will be conserved. *The classical phase space is incompressible.* This is to say that, for the three particles shown, the volumes $\Omega = \Omega'$. Consider a displacement of δt in time. We have transformed coordinates

$$q_i' = q_i'(q_i, p_i) = q_i(t+\delta t) = q_i(t) + \dot{q}_i \delta t = q_i(t) + \frac{\partial \mathscr{H}}{\partial p_i}\delta t$$

$$p_i' = p_i'(q_i, p_i) = p_i(t+\delta t) = p_i(t) + \dot{p}_i \delta t = p_i(t) - \frac{\partial \mathscr{H}}{\partial q_i}\delta t. \qquad (8.10)$$

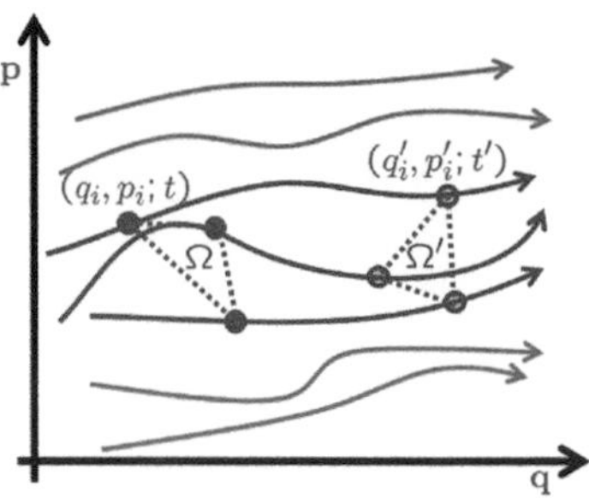

Figure 8.5: An example of the evolution of trajectories of classical particles in the phase space. The phase volume (Ω) of three particles at time t is shown here to evolve to Ω' at time t'. This volume will be conserved for classical particles.

To determine the transformation for the volume, we need to determine the Jacobian, which is

$$J(q_i', p_i'; q_i, p_i) = \frac{\partial(q_i', p_i')}{\partial(q_i, p_i)}. \qquad (8.11)$$

This is

$$\begin{vmatrix} \frac{\partial q_1'}{\partial q_1} & \frac{\partial q_1'}{\partial q_2} & \cdots & \frac{\partial q_1'}{\partial p_1} & \frac{\partial q_1'}{\partial p_2} & \cdots \\ \frac{\partial q_2'}{\partial q_1} & \frac{\partial q_2'}{\partial q_2} & \cdots & \frac{\partial q_2'}{\partial p_1} & \frac{\partial q_2'}{\partial p_2} & \cdots \\ \cdots & \cdots & \cdots & \cdots & \cdots & \cdots \\ \frac{\partial p_1'}{\partial q_1} & \frac{\partial p_1'}{\partial q_2} & \cdots & \frac{\partial p_1'}{\partial p_1} & \frac{\partial p_1'}{\partial p_2} & \cdots \\ \frac{\partial p_2'}{\partial q_1} & \frac{\partial p_2'}{\partial q_2} & \cdots & \frac{\partial p_2'}{\partial p_1} & \frac{\partial p_2'}{\partial p_2} & \cdots \\ \cdots & \cdots & \cdots & \cdots & \cdots & \cdots \end{vmatrix}$$

$$= \begin{vmatrix} 1 + \frac{\partial}{\partial q_1}\frac{\partial \mathscr{H}}{\partial p_1}\delta t & \frac{\partial}{\partial q_2}\frac{\partial \mathscr{H}}{\partial p_1}\delta t & \cdots & \frac{\partial}{\partial p_1}\frac{\partial \mathscr{H}}{\partial p_1}\delta t & \frac{\partial}{\partial p_2}\frac{\partial \mathscr{H}}{\partial p_1}\delta t & \cdots \\ \frac{\partial}{\partial q_1}\frac{\partial \mathscr{H}}{\partial p_2}\delta t & 1 + \frac{\partial}{\partial q_2}\frac{\partial \mathscr{H}}{\partial p_2}\delta t & \cdots & \frac{\partial}{\partial p_1}\frac{\partial \mathscr{H}}{\partial p_2}\delta t & \frac{\partial}{\partial p_2}\frac{\partial \mathscr{H}}{\partial p_2}\delta t & \cdots \\ \cdots & \cdots & \cdots & \cdots & \cdots & \cdots \\ -\frac{\partial}{\partial q_1}\frac{\partial \mathscr{H}}{\partial q_1}\delta t & \frac{\partial}{\partial q_2}\frac{\partial \mathscr{H}}{\partial q_1}\delta t & \cdots & 1 - \frac{\partial}{\partial p_1}\frac{\partial \mathscr{H}}{\partial q_1}\delta t & -\frac{\partial}{\partial p_2}\frac{\partial \mathscr{H}}{\partial q_1}\delta t & \cdots \\ -\frac{\partial}{\partial q_1}\frac{\partial \mathscr{H}}{\partial q_2}\delta t & -\frac{\partial}{\partial q_2}\frac{\partial \mathscr{H}}{\partial q_2}\delta t & \cdots & -\frac{\partial}{\partial p_1}\frac{\partial \mathscr{H}}{\partial q_2}\delta t & 1 - \frac{\partial}{\partial p_2}\frac{\partial \mathscr{H}}{\partial q_2}\delta t & \cdots \\ \cdots & \cdots & \cdots & \cdots & \cdots & \cdots \end{vmatrix} \qquad (8.12)$$

Since the primary term in δt comes from the diagonal, look at the product of the diagonal term

$$\prod_D [J(q_i', p_i'; q_i, p_i)] = \left(1 + \frac{\partial}{\partial q_1}\frac{\partial \mathscr{H}}{\partial p_1}\delta t\right)\left(1 + \frac{\partial}{\partial q_2}\frac{\partial \mathscr{H}}{\partial p_2}\delta t\right)\cdots$$
$$\times \left(1 - \frac{\partial}{\partial p_1}\frac{\partial \mathscr{H}}{\partial q_1}\delta t\right)\left(1 - \frac{\partial}{\partial p_2}\frac{\partial \mathscr{H}}{\partial q_2}\delta t\right)\cdots \qquad (8.13)$$

It is a product of 1s and terms in δt^2. Of the latter, terms of opposite pairs cancel, since the order of partial derivatives is immaterial. The off-diagonal terms are also in the 2nd power of δt. No term exists in linear dependence. So, the derivative of this Jacobian vanishes. Hence, phase-space volume is an invariant.

Stated in the form of particle density (ρ) in phase space,

$$\frac{d\rho}{dt} = \sum_i \left(\frac{\partial \rho}{\partial q_i} \dot{q}_i + \frac{\partial \rho}{\partial p_i} \dot{p}_i \right) + \frac{\partial \rho}{\partial t} = [\mathscr{H}, \rho]_{\mathbf{q},\mathbf{p}} + \frac{\partial \rho}{\partial t} = 0, \qquad (8.14)$$

where we have also explicitly incorporated other time dependences ($\partial \rho/\partial t$) that are not part of the Hamiltonian and the classical particle evolution that we employed in the Jacobian argument. This $\partial \rho/\partial t$ may arise from many causes. It is an energetic interaction that leads to loss of particle energy to the environment that is not accounted for in the Hamiltonian, such as friction of the environment. Sugar dropped into water dissolves, with the particle "annihilated." Multiple particles, if they were charged, as they are in atmospheric phenomena, will be subject to motional discrepancies, since there are other explicit time dependences not accounted for. In a semiconductor, in the drift-diffusion equation—fluid flow equations—electrons and holes annihilate each other. All of these are explicit and separate $\partial \rho/\partial t$ dependences. With this incorporation,

$$\begin{aligned} \frac{d\rho}{dt} &= 0 \\ \therefore \quad \frac{\partial \rho}{\partial t} &= -[\mathscr{H}, \rho]_{\mathbf{q},\mathbf{p}} \\ &= -\sum_i \left(\frac{\partial \rho}{\partial q_i} \dot{q}_i + \frac{\partial \rho}{\partial p_i} \dot{p}_i \right) \\ &= -\sum_i \left(\frac{\partial \rho}{\partial \mathscr{H}} \frac{\partial \mathscr{H}}{\partial q_i} \frac{\partial \mathscr{H}}{\partial p_i} - \frac{\partial \rho}{\partial \mathscr{H}} \frac{\partial \mathscr{H}}{\partial p_i} \frac{\partial \mathscr{H}}{\partial q_i} \right). \end{aligned} \qquad (8.15)$$

Here [,] is the Poisson bracket. When may $\partial \rho/\partial t = 0$ hold true? When ρ is only a function of energy and all energetics are included is one such instance. This is the equivalent of being a stationary state in phase space. The phase-space density now remains a constant at each locale in phase space. Thermal equilibrium also implies an independence from time. So, in thermal equilibrium too, this holds. Disturb the thermal equilibrium, and ρ can have an explicit time dependence. This disturbance from thermal equilibrium was a change of the energetics where the cause of the disturbance has not been included in the Hamiltonian. Include it in the Hamiltonian, and let us say that it doesn't have time dependence, and one will find a steady-state solution, so long as there is no other explicit $\partial \rho/\partial t$ dependence that is not accounted for.

This classical form of Liouville equation describes the complete evolution of each and every classical particle, given complete

In Maxwell's equations of thermodynamics, $\partial(T,S)/\partial(p,V) = 1$, which embodies the various thermodynamic interrelationships between temperature T, entropy S, pressure p and volume V and makes the statement that, in any cyclic process, work performed—the area enclosed in the relational curve of T and S, or p and V—is equal to the heat exchanged, has a very direct underlying connection to this incompressibility argument. $\mathbf{q}$ and $\mathbf{p}$ as canonic variables tie in to the thermodynamic variables.

In Galileo's experiment of dropping a stone and a feather in the air from the leaning tower of Pisa, with the Hamiltonian determined by the gravitational attraction, this phase-space volume changed, with the feather losing much of its potential momentum to the friction with the air.

See S. Tiwari, "Quantum, statistical and information mechanics: A unified introduction," Electroscience 1, Oxford University Press, ISBN 978-0-198-75985-0 (forthcoming), where we use the Poisson bracket to describe and extract a variety of quantum-mechanical relationships. The Poisson bracket is a convenient notion that, of course, dates back to the classical science period. Poisson's name appears, like that of many other French scientists from the classical science development period, in various places, the Poisson equation being the most common. Poisson is also associated with discoveries in probability. His name, Siméon Denis Poisson, is inscribed on the Eiffel tower in the company of 71 other great French scientists, mathematicians and engineers, including Pierre-Simon Laplace, whose equation and probability thoughts too show up so often in our explorations.

One should interpret a system in equilibrium as one whose extensive and intensive variables are time independent. When this system is isolated, all these variables must still not change. Thermal equilibrium is the condition of such a system in an environment at temperature T. Balance must hold in all transactions occurring in the system in every detail for equilibrium to hold. Off-equilibrium specification of the $(\mathbf{q}, \mathbf{p})$ space, and it describes a collection of them through its rewriting in a phase-space density form. This is a total of $2n$ vector equations for n particles. Trajectories do not cross, that is, any point in phase space can only be associated with one and only one particle. Many of the real problems of interest to us, such as describing electron transport, arise in a very large number—closer to Avogadro than thousands—of such electrons moving around. And these are also quantum conditions.

or nonequilibrium is any condition where the system is not in thermodynamic equilibrium. Nonequilibrium makes exchanges of energy leading to work—energy exchanged to some desired form—possible.

So, one can see two particular concerns. Particles may have a distribution form that comes close to the statistical distribution of the ensemble, but not every particle can be specified via its position q_i or momentum p_i. The second is that an electron may be describable at one level—depending on the size, properties of interest and, therefore, the depth of the details of the description—as a classical particle, but it is also a quantum particle—electrons very near other electrons will have to be viewed quantum-mechanically, and we know that the Poisson bracket must be subjected to uncertainty constraints through the $i/\hbar$ factor and that there are also quantum-scale interactions, such as scattering, that will occur during the course of the electrons' travel. So, probability distribution and the probabilistic nature of quantum mechanics both enter. Classical equations will have analogs or can be transformed into quantum-mechanical forms where the action quantum and quantum nature can be incorporated, and one can have these evolution equations as equations in statistical probability distributions forms both classically and quantum-mechanically. The quantum Liouville equation is the form that the classical Liouville equation takes. The probabilistic evolution classically and quantum-mechanically can both be written, through the Fokker Planck and the Kolmogorov master equation forms.

It is a master equation in the sense that a general form can, with a suitable choice of parameters, which may be fitting parameters, suffice to describe the relationships that the equation captures.

8.3 *Quantum Liouville equation*

Evolution of a collection of particles—classical or quantum-mechanical—is conveniently describable through probability distributions without a significant loss of accuracy for determination of most macroscopic parameters of interest. The classical form of the Liouville equation, with its incompressibility of volume, describes such an evolution in the phase-space density form where the classical aspect is embedded in the non-crossing of paths and the uniqueness of each point in the phase space as a characteristic of one particle only at any instant of time. The probability distribution in phase space gives us the underlying measure for the classical statistical ensemble. A quantum particle is described as a superposition state built from the different eigenstates that it may be found in. This possibility of finding the particle, upon observation, in one of the eigenstates is a probability—a quantum version different from the statistical one—at one level.

How we tackle many particles quantum-mechanically brings out a number of these quandaries and notions of imprecision in quantitatively formulatable form from which one can learn a bit about how to interpret the classical observations. At the simplest level, we may write the statefunction of a single particle as $|\Psi(q,p)\rangle$. For a multiple (N) particle ensemble, one may write this state function as $|\Psi(q_1,\cdots,q_N)\equiv|\Psi(\mathbb{Q})\rangle$. For classical systems, such as, in classical thermodynamics, the argument for building the Boltzmann measure of entropy ($S=k_B\ln\Omega$, where S is entropy, and Ω the number of possible microstates) or its equivalent Gibb's formulation ($S=-k_B\sum_i \mathfrak{p}_i\ln\mathfrak{p}_i$, where $\mathfrak{p}_i$ is the probability of the ith microstate), which are formally equivalent, since in random distribution all microstates are equally likely, one finds that the most likely possibility is the distribution with the largest number of microstates and thereon rapidly decreasing in large ensembles. Equivalent configurations arise in different assemblies of the random particles in their classical states. The collection is a mixed state arising from different particles, that is, particles identifiable with different (q_i,p_i), or a collective mixed state $(\mathbf{q},\mathbf{p})$. A Gaussian distribution is a distribution of a large number of classical particles at thermal equilibrium with decreasing distribution density of arrangements—the numbers contributing to the mixed states—around the central maximum with the highest number of such arrangements. The quantum ensemble has its analog to this in the form $|\Psi(\mathbb{Q})\rangle$, which is our short version for $|\Psi(\mathbb{Q},\mathbb{P})\rangle$.

Probability as a statistical notion for the quantum superposition should be clearly distinguished from the classical statistical probability. Often, they coincide. But as Bell showed, which was discussed in some depth in S. Tiwari, "Nanoscale device physics: Science and engineering fundamentals," Electroscience 4, Oxford University Press, ISBN 978-0-198-75987-4 (2017), one can show conditions where they will not be the same, and experiments confirm this thesis. This idea—as an ensemble with multiple particles—appears in entropy and other places. If there is not an explicitly captured connection between particles, that is, the two together are not really independent, then the classical independence-based randomness has to break down when compared to the quantum situation. Entangled photons, two simultaneously generated photons of connected polarization, are two photons physically, but have this link still there through the property of polarization. Tackle one's and the other's is revealed immediately. No independence here in polarization. Entropy in the Boltzmann sense is not the only form in which incompleteness of knowledge exists. π or Euler's constants may be transcendental, whose ever-longer writing shows equidistribution with seemingly random appearance of the digits in equal 1/10th measure in digital form, but they are also intimately connected through Euler's equation $\exp(i\pi)+1=0$. There exists an internal link that this equation specifies. And there is an entropy related to this: the algorithmic or Kolmogorov entropy, which is measured through the smallest program that can determine that quantity, even if the complete specification of that quantity is not possible because infinity is only a notion, not something precise.

There is an additional complexity. Because of the uncertainty principle, states are not points in phase space. There exists a spread. And, for a single particle, which is therefore an independent-particle quantum system, the constituting particle has a statefunction $\Psi(q,p)$, which itself is built from eigenfunctions that comprise the solution to the Hamiltonian. A mix of particles will have two direct consequences. The probability density that now exists in the quantum ensemble is spread over various configurations. A definite state $\Psi(\mathbb{Q})$ has a probability spread among its various configurations of $|\Psi(\mathbb{Q})|^2$. This probability in itself is different from the quantum probability arising in the various eigenfunction possibilities of a quantum particle. We now have a hierarchy of probabilities.

An example of a two-particle system should illustrate this. Take two photons. This two-photon collection can be in a variety of mixed states. $|\Psi_1\rangle = \frac{1}{\sqrt{2}}\left[\begin{pmatrix}\leftrightarrow\\ \leftrightarrow\end{pmatrix} + \begin{pmatrix}\updownarrow\\ \updownarrow\end{pmatrix}\right]$ is an entangled state where the polarization of one is tied to the polarization of the second. An observation of the system will find either one of these constituting particles, where the property of one will be tied to the other, in the form of horizontal or vertical polarization. On the other hand, $|\Psi_2\rangle = \begin{pmatrix}\leftrightarrow\\ \leftrightarrow\end{pmatrix}$ is a horizontally polarized pair. $|\Psi_3\rangle = \begin{pmatrix}\updownarrow\\ \updownarrow\end{pmatrix}$ is a vertically polarized pair. And we could have built a fourth, $|\Psi_4\rangle$, where the two particles are entirely decoupled, each observable independently in their composing states of horizontal or vertical polarization. These are all various mixed states. The ensemble expectation of any observable $\mathscr{A}$ for the quantum assembly then is

$$\langle \mathscr{A} \rangle = \sum_n \mathfrak{p}_n \langle \Psi_n(\mathbb{Q}) | \hat{\mathscr{A}} | \Psi_n(\mathbb{Q}) \rangle. \qquad (8.16)$$

These Ψ_ns are not orthonormal in general, as the two-photon example illustrates. We can rework this in an orthonormal basis. Let $|\Phi_\alpha\rangle$ be an orthonormal basis, so that $\mathbb{I} = \sum_\alpha \langle \Phi_\alpha \rangle \langle \Phi_\alpha |$. So, working through the expectation for the observable,

The technique for achieving orthonormality is discussed in S. Tiwari, "Quantum, statistical and information mechanics: A unified introduction," Electroscience 1, Oxford University Press, ISBN 978-0-198-75985-0 (forthcoming) and encompasses taking inner products of the vector functions, which gives the degenerate part, and then subtracting and normalizing. With vectors, the equivalent is finding a dot product, and working out from it.

$$\begin{aligned}
\langle \mathscr{A} \rangle &= \sum_n \mathfrak{p}_n \langle \Psi_n | \left(\sum_\alpha |\Phi_\alpha\rangle\langle\Phi_\alpha| \right) \hat{\mathscr{A}} |\Psi_n\rangle \\
&= \sum_n \mathfrak{p}_n \sum_\alpha \langle \Phi_\alpha | \hat{\mathscr{A}} \Psi_n \rangle \langle \Psi_n | \Phi_\alpha \rangle \\
&= \sum_\alpha \langle \Phi_\alpha \hat{\mathscr{A}} | \left(\sum_n \mathfrak{p}_n |\Psi_n\rangle\langle\Psi_n| \right) |\Phi_\alpha\rangle \\
&= \mathrm{Tr}(\hat{\mathscr{A}} \rho), \\
\text{where } \rho &= \sum_n \mathfrak{p}_n |\Psi_n\rangle\langle\Psi_n|
\end{aligned} \qquad (8.17)$$

is the density matrix. The trace of the density matrix (the sum of the terms along the diagonal of the density matrix) is

$$\mathrm{Tr}(\rho) = \mathrm{Tr}\left(\sum_n \mathfrak{p}_n |\Psi_n\rangle\langle\Psi_n| \right) = \sum_n \mathfrak{p}_n \mathrm{Tr}(|\Psi_n\rangle\langle\Psi_n|) = \sum_n \mathfrak{p}_n = 1. \quad (8.18)$$

So, all possibilities of mixed states are included, and the density matrix suffices to know everything about an observable using its operator.

We can now work from the density matrix. The evolution in it due to changes in time maps for us the evolution of expectations of observables. We know, from the time evolution of the statefunction arising in the Hamiltonian, $-(\hbar/i)\partial|\Psi_n\rangle/\partial t = \hat{\mathscr{H}}|\Psi_n\rangle$. Therefore, the Hamiltonian dynamics gives a time dependence to the density matrix of

$$\frac{\partial \rho}{\partial t} = \sum_n \mathfrak{p}_n \left(\frac{\partial |\Psi_n\rangle}{\partial t} \langle \Psi_n| + |\Psi_n\rangle \frac{\partial \langle \Psi_n|}{\partial t} \right)$$
$$= -\frac{i}{\hbar}[\hat{\mathscr{H}}, \rho]. \tag{8.19}$$

Add to this the time dependences not included in this dynamics, extrinsic to the Hamiltonian, and we have, in general, with energy flowing in and out of the system,

$$\frac{d\rho}{dt} = -\frac{i}{\hbar}[\hat{\mathscr{H}}, \rho] + \frac{\partial \rho}{\partial t}. \tag{8.20}$$

This is the quantum Liouville equation. The quantum uncertainty is reflected in the $i/\hbar$ factor that we have seen before when quantum action was incorporated in the Poisson formulation of Hamiltonian dynamics. This equation form in other ways looks very much like the classical Liouville equation.

See S. Tiwari, "Quantum, statistical and information mechanics: A unified introduction," Electroscience 1, Oxford University Press, ISBN 978-0-198-75985-0 (forthcoming) for the Poisson bracket's action-based Heisenberg reformulation of the quantum-mechanics equation to complement the wave-based Schrödinger formulation.

The uncertainty and the quantum have deeper implications reflected in this Liouville equation. We illustrate this through the unraveling of the difference between the interpretation of the wave-packet as a localized representation in wave dynamics for a particle and a classical particle. For this, we use the Heisenberg picture, where operators are time dependent and the state vector is time independent. For any operator $\hat{\mathscr{A}}(t)$ for an observable, we have

$$\frac{d}{dt}\hat{\mathscr{A}}(t) = \frac{i}{\hbar}[\hat{\mathscr{H}}, \hat{\mathscr{A}}(t)] + \left.\frac{\partial \hat{\mathscr{A}}}{\partial t}\right|_{\mathscr{H}}. \tag{8.21}$$

In the Heisenberg notation, the equations of motion, following the above, are

$$\frac{d}{dt}\hat{q} = \frac{\hat{p}}{m}, \text{ and}$$
$$\frac{d}{dt}\hat{p} = \frac{i}{\hbar}[\hat{\mathscr{H}}, \hat{p}(t)] = -\frac{dV(\hat{q})}{d\hat{q}}. \tag{8.22}$$

Here, $V(q)$ is the potential, with the corresponding operator as $\hat{V}(q)$ in the Hamiltonian $\hat{\mathscr{H}} = (\hat{p}^2/2m) + \hat{V}(q)$. Let the potential change slowly—an adiabatic change—so that

$$\left\langle \frac{dV(\hat{q})}{d\hat{q}} \right\rangle \approx \frac{dV(q)}{dq}. \tag{8.23}$$

The wavepacket movement corresponds quite well to the particle movement, since adiabaticity removes quantum reflection considerations. This is also another example of the correspondence principle at work. Now, let us deconvolve the consequences of this potential and its change in position through series expansion so that higher moment effects on the wavepacket motion can be explored. So, we remove the adiabaticity constraint and write

$$V(\hat{q}) = V(q) + (\hat{q} - q)\frac{dV(q)}{dq} + \frac{1}{2}(\hat{q} - q)^2\frac{d^2V(q)}{dq^2} + \cdots$$

$$\therefore \quad \frac{dV(\hat{q})}{d\hat{q}} = \frac{dV(q)}{dq} + (\hat{q} - q)\frac{d^2V(q)}{dq^2} + \frac{1}{2}(\hat{q} - q)^2\frac{d^3V(q)}{dq^3} + \cdots. \quad (8.24)$$

Looking at the moments,

$$\langle \hat{q} - q \rangle = 0, \quad (8.25)$$

which states that the expectation of position on the wavepacket and the particle are identical—no surprise here. For the next higher moment,

$$\langle (\hat{q} - q)^2 \rangle = (\Delta\hat{q})^2. \quad (8.26)$$

The wavepacket disperses. Its width changes and it spreads out. *A particle doesn't do that. What we are seeing is the uncertainty-induced fuzziness.* The expectation, in detail, now is

$$-\langle \frac{dV(\hat{q})}{d\hat{q}} \rangle = -\frac{dV(q)}{dq} - \frac{1}{2}\frac{d^3V(q)}{dq^3}(\Delta\hat{q})^2 - \cdots. \quad (8.27)$$

The wavepacket is being distorted via these higher-order terms, a domino of perturbations on perturbations.

Now, to see the incompressibility of the Liouville equation and the fuzziness of the quantum-Liouville equation, consider the three wavepackets ψ, ψ' and ψ''. These form a triangle enclosing an area as shown in Figure 8.5; consider the motion under only a single degree of freedom. The equations of motion for one of these wavepackets (ψ') are

This wavepacket-particle difference is one facet of uncertainty, and the other facet is the Fisher information propagation limit that we noted in Chapter 2. Wavefunction, particle, uncertainty and information complete their cycle through this simple analysis.

$$\frac{d}{dt}(q' - q) = (p' - p), \text{ and}$$

$$\frac{d}{dt}(p' - p) = \langle \psi', \frac{d\hat{V}(\hat{q})}{d\hat{q}}\psi' \rangle + \langle \psi, \frac{d\hat{V}(\hat{q})}{d\hat{q}}\psi \rangle$$

$$= -\left[\frac{d^2V(q)}{dq^2} + \frac{1}{2}\frac{d^4V(q)}{dq^4}(\Delta\hat{q})^2\right](q' - q)$$

$$-\frac{1}{2}\frac{d^3V(q)}{dq^3}\left[(\Delta'\hat{q})^2 - (\Delta\hat{q})^2\right]. \quad (8.28)$$

The second term—a negative term—in this last expression causes quantum distortion. Area is not preserved anymore. And this change in phase-space area can be seen through its rate of change,

$$\frac{d}{dt}\frac{(q' - q)(p'' - p) - (q'' - q)(p'' - p)}{2}$$

$$= \frac{d^3V(q)}{dq^3}$$

$$\times \frac{(q'' - q)[(\Delta'\hat{q})^2 - (\Delta\hat{q})^2] - (q' - q)[(\Delta''\hat{q})^2 - (\Delta\hat{q})^2]}{4}. \quad (8.29)$$

It is a non-vanishing term if the size of the wavepacket and that of the triangle constituting the means of these three wavepackets are of the same order of magnitude.

In the classical picture, the points that form the boundary of the domain in phase space of Figure 8.5 move according to Hamilton's formulation, with the enclosed volume invariant. Quantum action ($\hbar$) constraint distorts the domain. It causes a projection of longer and thinner filaments with a volume of $\hbar^{-N}$, where N is the number of the quantum particles. Localized particles may be viewed more conveniently through Wannier functions, and the smoothening of them in a statefunction of the collective ensemble leads to a quantum dynamical evolution that appears to be fuzzier.

Recall our resorting to Wannier functions for the understanding of effective mass, and other situations where localization flourishes, such as shallow hydrogenic dopants.

8.4 *Fokker-Planck equation*

THIS PICTURE OF THE CLASSICAL-TO-QUANTUM CORRESPONDENCE and their dynamic evolution is a very multi-faceted story with elements related to quantum action and elements related to probabilities, as well as our description of mixed states in this soup. The convenient edifice of describing probabilities in ensembles through distributions, and the description of distributions and their evolutions through moments, appear in many different forms in the classical and quantum domains. The Fokker-Planck equation is one such form that goes back a hundred plus years, and recourse to it and to the Langevin equation is common when describing ensemble dynamics. We will return to the Fokker-Planck and Langevin equations later (Section 8.10) but make some remarks here on the Fokker-Planck equation, since it stresses probability evolution and the various places it shows up quite beautifully.

See the first part of Appendix F in S. Tiwari, "Nanoscale device physics: Science and engineering fundamentals," Electroscience 4, Oxford University Press, ISBN 978-0-198-75987-4 (2017) for an exploration of probabilities, distribution functions and moments and their interrelationships.

An ensemble responds to the forces through an evolution as a collective ensemble undergoing particle-particle and particle-environment interactions. The 0th—lowest—order measure of the response is the mean of the evolution. This is a system response that appears as a net response, such as current, for example, if a voltage is applied to which an electron charge responds as in-between two biased conducting plates. But the particle-particle dynamics is also present as this mean response unfolds. Take any plane of many-electron charge, and it spreads out evolving in time and space. This is diffusion. If one measures the current, one will see fluctuations in the current of what may be otherwise a constant mean electron density in between the plates. This is noise. Noise is also the response of the charge ensemble. So, on the one hand, drift of the charge cloud and the diffusion in the charge cloud are connected. On the other hand, the resistance of the charge cloud and the Johnson-Nyquist noise of the charge cloud are also connected. They reflect an underlying physical interaction in the ensemble and the response of the particles and the ensemble through its interactions with the external environment. Under perturbations and interactions, the flow of the distribution, which is a representation of a random matrix with overarching higher-order constraints, as a function shows macroscopic characteristics like those of the ensemble. The distribution functions take on a dynamic evolution that reflects the statistical consequences for the ensemble under constraints (such as interactions with the environment—canonical, microcanonical and macrocanonical ensembles—and energy and particle exchange). The Fokker-Planck equation captures this, and it will appear in many forms as a master equation.

One form of the Fokker-Planck equation, applicable to our interests, is the one-dimensional form, with $\mathcal{D}_1(z)$ and $\mathcal{D}_2(z)$ as functions representing the ensemble dynamics in play,

$$\frac{\partial f}{\partial t} = \left[-\frac{\partial}{\partial z}\mathcal{D}_1(z) + \frac{\partial^2}{\partial z^2}\mathcal{D}_2(z) \right] f, \tag{8.30}$$

where $f \equiv f(\mathbf{r}, \mathbf{p} \text{ or } \mathbf{k})$ is the phase-space distribution—a Liouville density—and, in its simplest form for us for semiconductors, the Boltzmann distribution function, which holds for classical-like non-degenerate conditions of a semiconductor. The first term in Equation 8.30 is a drift term proportional to the positional derivative, and the second is a diffusion term that is proportional to a second derivative. This operator relationship with first and second order derivatives operating on a function product with a distribution density is the lowest order linearized off-equilibrium response in the presence of interactions within the ensemble. This equation is just as useful for describing the "not-so-free financial" market response as it is for the Brownian motion phenomenon observed classically in fluid flow, and as it is for describing electron motion in the drift-diffusion conditions in a semiconductor.

The Fokker-Planck description is particularly useful for describing continuous macroscopic variables in conditions that are not too far from equilibrium.

8.5 Boltzmann transport equation

WE NOW RETURN to the problem of central interest to us: the description of an ensemble of various particles—electrons and phonons—responding to stimulus in the semiconductor. The

Boltzmann transport equation is a simplification of the Liouville equation, where one suitably mixes classical and quantum, so a semi-classical version, to arrive at a form that is very convenient for exploring transport dynamics with careful accounting for the quantum-defined nature of scattering, as well as the quantum-mechanical constraints reflecting in the state distributions, statistics of occupation, and much else where the quantum nature of electron, photon and phonon behavior will matter. This becomes possible if one can ignore particle-particle interaction and uncertainty implications—the fuzziness aspect of quantum Liouville—and it is in this sense that the particles are being treated as independent classical particles.

The Hamiltonian of the electrons in the crystal reflects a variety of energetic interactions, as discussed in different contexts in Chapters 1 through 7. A simple interpretation is that this Hamiltonian has a form $\mathscr{H} = \mathscr{H}_e + \mathscr{H}_c + \mathscr{H}'$, where the first term is the Hamiltonian of electrons in a perfect crystal, the second represents the rest of the crystal, and the third term is a perturbation that is to be tackled separately to explore and quantitatively evaluate the consequences of an interaction. These first two terms represent the consequences of the presence of electrons in the "perfect" crystal and the periodic potential of the crystal. But then what we do is partition—atoms are made fixed in position—and their movement, for example, is included through a perturbation. This is equivalent to saying that we have split the eigenstates into a product representing the electron and the crystal, that is $|\mathbf{k}, c\rangle = |\mathbf{k}\rangle|c\rangle$. This is the Born-Oppenheimer or adiabatic approximation.

Consider now a system of particles—electrons—described by a distribution function $f(\mathbf{r}, \mathbf{k}; t)$ evolving in time under the influence of external forces and internal particle interactions. So the forces cause a change in time, and eventually, at some point in time, the system reaches a steady state. The Liouville equation tells us that

$$\frac{d\rho}{dt} = \sum_i \left(\frac{\partial \rho}{\partial q_i} \dot{q}_i + \frac{\partial \rho}{\partial p_i} \dot{p}_i \right) + \frac{\partial \rho}{\partial t}, \tag{8.31}$$

where ρ represents a particle density in phase space.

The transformation of this equation into the Boltzmann transport form—a form in terms of the distribution function—can be understood through Figure 8.6. The charged particle—electron or hole (also phonon)—system can be given a quasi-classical interpretation. The E-$\mathbf{k}$ description provides each mobile particle properties from their assignment to the band—a Bloch band, with implications beyond just the assignment of energy—and a wavevector $\mathbf{k}$, which represents the particle's momentum in the crystal of $\hbar\mathbf{k}$. We also

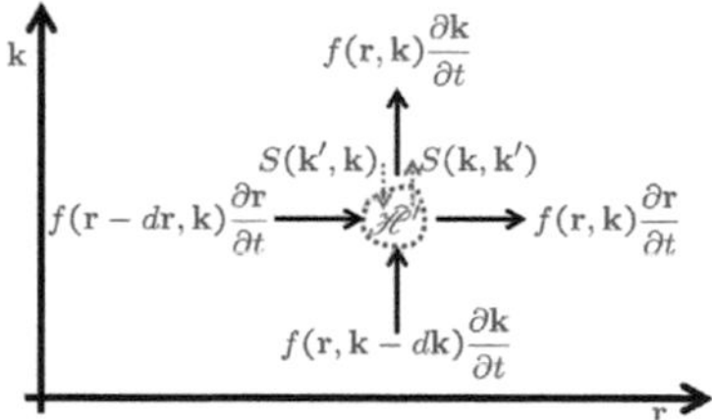

Figure 8.6: Continuity of a distribution function in phase space.

This Liouville and Boltzmann transport equation form can also be written for phonons. We usually do not, and only introduce it in our discussion of remote and off-equilibrium processes (Chapter 19) to see the interesting electron-phonon interaction consequences in thermoelectric transport. For now, our lowest-order approximation is that the phonon distribution doesn't stray far enough from equilibrium.

have a position of the particle. The state of the system is specified by the distribution function of these variables.

Since the crystal momentum is $\hbar\mathbf{k}$, $(\mathbf{p},\mathbf{q}) \equiv (\mathbf{k},\mathbf{r})$, and, from now on, we will mostly discuss in terms of the wavevector and position coordinate. The continuity equation that the Liouville equation is (classically and quantum-mechanically) can now be seen as the evolution of the distribution function $f(\mathbf{r},\mathbf{k};t)$, in the form

$$\begin{aligned}\frac{df}{dt} &= -\frac{\partial f}{\partial \mathbf{k}}\cdot\dot{\mathbf{k}} - \frac{\partial f}{\partial \mathbf{r}}\cdot\dot{\mathbf{r}} + \frac{\partial f}{\partial t}\\ &= -\dot{\mathbf{k}}\cdot\nabla_{\mathbf{k}}f - \dot{\mathbf{r}}\cdot\nabla_{\mathbf{r}}f + \frac{\partial f}{\partial t}, \text{ which is}\\ &= 0 \text{ in steady state.}\end{aligned} \tag{8.32}$$

Slowing down in momentum or real space increases the distribution function. The distribution $f(\mathbf{r},\mathbf{k};t)$ at time t arises in changes in momentum (or wavevector) and changes in position. The changes in momentum due to external forces are from the distribution function $f(\mathbf{r},\mathbf{k}-\dot{\mathbf{k}}\Delta t)$. These were the carriers from Δt time earlier that drifted into the distribution function of interest. Similarly, it is the carriers that were at $(\mathbf{r}-\dot{\mathbf{r}}\Delta t,\mathbf{k})$ that diffused into the distribution function of interest. We need to add to this time-dependent changes arising in the perturbations that are not explicitly included in the dynamics of evolution. In Figure 8.6, this $\partial f/\partial t$ dependence of change is shown as arising in the perturbation Hamiltonian $\hat{\mathscr{H}}$, and occurring due to interactions in states—$S(\mathbf{k}',\mathbf{k})$ represents scattering into the $(\mathbf{r},\mathbf{k})$-state from all others, that is, $(\mathbf{r},\mathbf{k}')$, and $S(\mathbf{k},\mathbf{k}')$ is the complementary process. This is the Boltzmann transport equation.

df/dt is the net rate of change of distribution function. In evolving conditions, it will be non-zero. In steady state, it vanishes. The first two terms of the equation's right-hand side are the consequence of changes arising in momentum space and in real space. And the last term is due to interactions with time-dependent consequences that are not included in the Hamiltonian. This is the rate of change in phase space. The real space consequence is through the real space motion (velocity of the particle and any real space particle density changes, that is, those that are due to real space diffusion). The momentum-space consequence is through the momentum change, which arises from forces at work, and any phase-space particle density changes, that is, those that are due to momentum space diffusion.

Since f represents particle density, that is, a probability density, this Boltzmann transport equation represents a continuity equation of probability density. In steady state, it is a constant. In conditions where time dependence exists, such as in a transient, it gives us

It is important to stress here that position and momentum are canonical coordinates. Forces cause change in momentum. Classically, velocity, the time-dependent position evolution in real space, is related to the momentum through the inertial mass. Quantum-mechanically, velocity ($\dot{\mathbf{r}}$) is a function of momentum. For semiconductors, this velocity is $\dot{\mathbf{r}} = (1/\hbar)\nabla_{\mathbf{k}}E_{n\mathbf{k}}$, where $E_{n\mathbf{k}}$ is the $E-\mathbf{k}$ description in the bandstructure. Quantum-mechanically, there also happens to be the useful concept of effective mass related through $(1/\hbar^2)\nabla^2_{\mathbf{k}}E_{n\mathbf{k}}$, which allows quite a bit of the inertial motion to be understood in the crystal. But, it is the former relationship that is the explicit functional connection of velocity to momentum for an electron that happens to have an energy E and the quantum number $\mathbf{k}$ for the wavevector and the band index n in the crystal. Momentum-dependent changes are what we classically call drift. Position-dependent changes are what we classically call diffusion.

the probability change in time. The probability change in time is happening because of real space motion, because of momentum-space motion, and because of phase-space changes arising due to other time-dependent phenomena whose consequences have not been included in the first two terms. This last term arises in the interaction of this distribution with the surroundings, or even within itself, if that was not explicitly included in the first two terms, which came from the Poisson bracket with the Hamiltonian.

An illustration of how these distribution functions relate to our description of semiconductor dynamics is in Figure 8.7. Consider an isotropic distribution. In thermal equilibrium, at any position $\mathbf{r} = \mathbf{r}'$, this distribution function in the momentum space ($f_0(\mathbf{r}',\mathbf{k})$) has a zero mean. There is no net momentum for the distribution even if there is a spread. When a force is applied, such as a field in the opposite direction for electrons, this distribution function, still at $\mathbf{r} = \mathbf{r}'$, is shifted. Being centered at a different wavevector, here at a $\mathbf{k}_{\|} \neq 0$, there is a net momentum. We will see this as a net drift velocity of electrons arising from the force. Note, though, that we still have a distribution function with a momentum spread, only that the expectation of this momentum spread has shifted from the origin, where it was at thermal equilibrium.

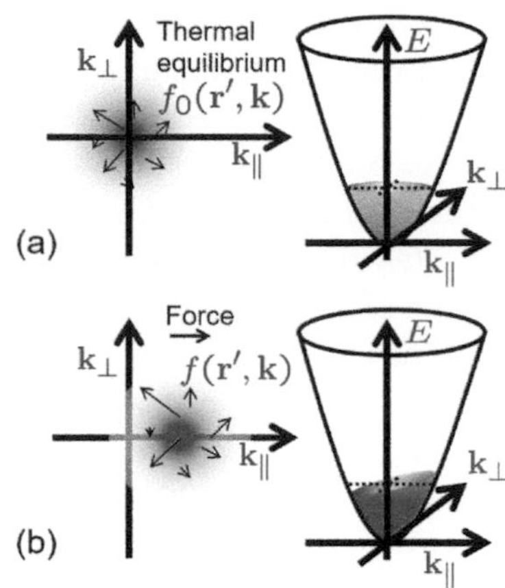

Figure 8.7: (a) An isotropic distribution function in momentum space at thermal equilibrium. This is the distribution function $f_0(\mathbf{r}',\mathbf{k})$ at $\mathbf{r}=\mathbf{r}'$ at a specific position. (b) Nonequilibrium distribution under a force, for example, an electric field whose direction is opposite to that of the force for electrons. The distribution function now is $f(\mathbf{r}',\mathbf{k})$ at the same position in space of $\mathbf{r} = \mathbf{r}'$.

This spread is the thermal spread. Electrons have momentum and energy. It is thermally distributed. And it balances antipodally.

8.5.1 Scattering

SCATTERING IS THE MOST IMPORTANT FORM through which $\partial f/\partial t$'s consequence appears for us. For steady state, whose quantum-mechanical analog can be thought of as a stationary state, we have

$$\left.\frac{\partial f}{\partial t}\right|_{scat} = \dot{\mathbf{k}} \cdot \nabla_{\mathbf{k}} f + \dot{\mathbf{r}} \cdot \nabla_{\mathbf{r}} f. \tag{8.33}$$

In the semiconductor, the scattering, which occurs at some real space position with its associated quantum uncertainty, arises from interactions that are in momentum space. Electrons at momentum $\mathbf{k}$ (an occupied state at wavevector $\mathbf{k}$, which is associated with a crystal momentum $\hbar\mathbf{k}$) may scatter to an empty state at wavevector $\mathbf{k}'$. We show this as the perturbation-dependent time-dependent change in Figure 8.6, separate from the known Hamiltonian-induced phase-space trajectory. Electrons from an occupied state at a wavevector $\mathbf{k}'$ may also scatter to an unoccupied state at $\mathbf{k}$. If we are interested in what happens at $\mathbf{k}$, then we need to include all the possible $\mathbf{k}'$ states with which this interaction is possible. This is

$$\left.\frac{\partial f}{\partial t}\right|_{scat} = \int \{S(\mathbf{k}',\mathbf{k}) f(\mathbf{k}') \left[1 - f(\mathbf{k})\right] - S(\mathbf{k},\mathbf{k}') f(\mathbf{k}) \left[1 - f(\mathbf{k}')\right]\} \, d\mathbf{k}', \tag{8.34}$$

where $S(\mathbf{k},\mathbf{k}')$ is the scattering rate—a probability per unit time of a transition from initial state $\mathbf{k}$ to a final state $\mathbf{k}'$ in a unity volume of the $\mathbf{k}$-space. A similar definition holds for $S(\mathbf{k}',\mathbf{k})$. In Equation 8.34, the net partial time dependence of the distribution function is due to scattering into $\mathbf{k}$ and scattering out from $\mathbf{k}$. Scattering in is proportion to $S(\mathbf{k}',\mathbf{k})$, which is probability per unit time of the transition, the supply function, which is proportional to the occupation of the $\mathbf{k}'$ state, and the receiving function, which indicates whether a state at $\mathbf{k}$ is unoccupied, and therefore available. This supply function is just the distribution function $f(\mathbf{k}')$. The unoccupied state density defines the receiving function as $(1-f(\mathbf{k}))$. The integral accounts for all possible $f(\mathbf{k}')$ states that interact with the $f(\mathbf{k})$ state. The second term is for the opposite process. In thermal equilibrium, these two terms balance. And this balance must be in detail for all $\mathbf{k},\mathbf{k}'$ pairs.

Equation 8.34 is the scattering or collision integral. To gain more insight, consider what it reflects. From an initial occupied state $|\mathbf{k}'\rangle$ an electron may scatter to unoccupied $|\mathbf{k}\rangle$ states. Likewise, from an initial occupied $|\mathbf{k}\rangle$ state, an electron may scatter to unoccupied $|\mathbf{k}'\rangle$ states. There is a total of N states in the system. fN reflects the occupied states, and $(1-f)N$ the unoccupied states. If there are N_s scattering centers, our interpretation is that

$$\left.\frac{\partial f}{\partial t}\right|_{scat} = N_s \sum_{\mathbf{k}',N-1} \left\{S(\mathbf{k}',\mathbf{k})f(\mathbf{k}')\left[1-f(\mathbf{k})\right] - S(\mathbf{k},\mathbf{k}')f(\mathbf{k})\left[1-f(\mathbf{k}')\right]\right\}. \quad (8.35)$$

The integral represents scattering from one center, while the above equation reflects the consequence from many. If the system has many states, that is, $N \gg 1$, then $N-1 \approx N$, and the summation can be reduced over all the states. The volume associated with each $|\mathbf{k}'\rangle$ state is $\Omega_{\mathbf{k}'} = (2\pi)^3/\Omega$, where Ω is the total volume. Therefore,

$$\left.\frac{\partial f}{\partial t}\right|_{scat} = \frac{N_s\Omega}{(2\pi)^3}\int_{\Omega_{\mathbf{k}'}} \left\{S(\mathbf{k}',\mathbf{k})f(\mathbf{k}')\left[1-f(\mathbf{k})\right] - S(\mathbf{k},\mathbf{k}')f(\mathbf{k})\left[1-f(\mathbf{k}')\right]\right\} d\mathbf{k}'. \quad (8.36)$$

In thermal equilibrium, the distribution function is a constant; therefore, $\partial f/\partial t = 0$ in thermal equilibrium. This implies

$$S(\mathbf{k}',\mathbf{k}) = S(\mathbf{k},\mathbf{k}')\frac{f_0(\mathbf{k})\left[1-f_0(\mathbf{k})\right]}{f_0(\mathbf{k}')\left[1-f_0(\mathbf{k}')\right]}. \quad (8.37)$$

If conditions are non-degenerate, then the occupation probability is very small, that is, $f(\mathbf{k}), f(\mathbf{k}') \ll 1$. This, in turn, means that

$$S(\mathbf{k}',\mathbf{k}) \approx S(\mathbf{k},\mathbf{k}')\exp\left[\frac{E(\mathbf{k}')-E(\mathbf{k})}{k_BT}\right]. \quad (8.38)$$

Under non-degeneracy, we may also write

$$\left.\frac{\partial f}{\partial t}\right|_{scat} = \int \left[f(\mathbf{k}')S(\mathbf{k}',\mathbf{k}) - f(\mathbf{k})S(\mathbf{k},\mathbf{k}') \right] d\mathbf{k}'. \tag{8.39}$$

This scattering rate relationship implies that, for $S(\mathbf{k}',\mathbf{k}) = S(\mathbf{k},\mathbf{k}')$, $E(\mathbf{k}') = E(\mathbf{k})$. The scattering rate from state $|\mathbf{k}\rangle$ to $|\mathbf{k}'\rangle$ and its inverse are equal only for elastic collisions. The energy of carriers before and after must remain the same. If the scattering is restricted to a parabolic approximation region, this in turn means that $|\mathbf{k}'| = |\mathbf{k}|$.

Equation 8.39, or the more general form (Equation 8.34), suffices to describe the dynamics in Boltzmann transport equation. So long as one can describe the scattering rate, and we saw an example of its calculation for Coulomb potential in Chapter 1, one can calculate the distribution function's evolution and therefore various properties of interest.

For many macroscopic conditions such as those in larger dimension devices, this equation can be considerably simplified through a master-equation-like parameter of a time constant.

8.6 *Relaxation time approximation*

WE CLAIM THAT CHANGES IN THE DISTRIBUTION FUNCTION due to perturbations, under certain conditions, may be written as

$$\left.\frac{\partial f}{\partial t}\right|_{scat} = -\frac{f(\mathbf{k}) - f_0(\mathbf{k})}{\tau(\mathbf{k})} = -\frac{\Delta f}{\tau_{\mathbf{k}}}. \tag{8.40}$$

It is useful to note here that Equation 8.40 is not one of general validity. Equation 8.39 has broader validity and Equation 8.36 even more so. But being able to write a relaxation time that connects the perturbation in the distribution to its rate of change as a time constant that only depends on the change makes equations convenient and intuition easier. Solving using scattering rates between states through Equation 8.39 or Equation 8.36 will need computation and is usually pursued through a Monte Carlo simulation.

This is the *relaxation time approximation*. And if this time-constant approximation is valid, then the Boltzmann transport equation can be explored conveniently analytically for a variety of dynamics of transport. And it is also quite applicable for a variety of semiconductor situations. This time constant $\tau(\mathbf{k})$—a momentum relaxation time—will variously be written as $\tau_{\mathbf{k}}$ and, when applied to expectations, as $\langle\tau(\mathbf{k})\rangle = \langle\tau_{\mathbf{k}}\rangle$. This relaxation time is a function of $\mathbf{k}$; its expectation too is a function of $\mathbf{k}$, but it does not have a dependence on the form that $f(\mathbf{k})$ takes. So, it does not depend on the distribution function as it appears under various perturbation and real space geometries and other conditions under which a real use of a semiconductor in a device appears. This is what makes it powerful. If the relaxation time approximation is not valid, one cannot find the distribution function in an explicit analytic form using this semi-classical equation of motion that is the Boltzmann transport equation.

The drift-diffusion equation, and its even more extreme form of just plain drift (as in conductors) or diffusion (as in injection in materials with low carrier densities), or the Drude near-classical approximation, are examples of the use of relaxation time approximation to its extremes.

Now let us explore when this relaxation time approximation may be applicable. In thermal equilibrium,

$$\left.\frac{\partial f}{\partial t}\right|_{scat,0} = \int \left[f_0(\mathbf{k}')S(\mathbf{k}',\mathbf{k}) - f_0(\mathbf{k})S(\mathbf{k},\mathbf{k}') \right] d\mathbf{k}' = 0. \tag{8.41}$$

Off-equilibrium, that is, with perturbation, $f(\mathbf{k}) = f_0(\mathbf{k}) + \Delta f(\mathbf{k})$, and, using the thermal equilibrium equality,

$$\left.\frac{\partial f}{\partial t}\right|_{scat} = \int \Delta f(\mathbf{k}')S(\mathbf{k}',\mathbf{k})\,d\mathbf{k}' - \Delta f(\mathbf{k}) \int S(\mathbf{k},\mathbf{k}')\,d\mathbf{k}'. \tag{8.42}$$

Perturbation changes the distribution function. A change in distribution function in phase space (real and momentum) is directly related to the flow of carriers, which relates to current, that is, momentum flow; energy flow, that is, how carriers transport energy from one place to another; and any other parameters we may be interested in. This form of equation also tells us when the relaxation time approximation might hold. If the first term of Equation 8.42 vanishes, one will be able to write

$$\begin{aligned} \left.\frac{\partial f}{\partial t}\right|_{scat} &= -\Delta f(\mathbf{k}) \int S(\mathbf{k},\mathbf{k}')\,d\mathbf{k}' \\ &= -\frac{\Delta f(\mathbf{k})}{\tau(\mathbf{k})} = -\frac{f(\mathbf{k}) - f_0(\mathbf{k})}{\tau(\mathbf{k})}, \quad \text{where} \\ \tau(\mathbf{k}) &= \frac{1}{\int S(\mathbf{k},\mathbf{k}')\,d\mathbf{k}'}. \end{aligned} \tag{8.43}$$

Δf is an odd function of $\mathbf{k}$. It is a perturbation that caused the disturbance from thermal equilibrium. In thermal equilibrium, $f_0(\mathbf{k})$ axiomatically is an even function. Perturbation, in order to have a net effect, has to have an odd power dependence on $\mathbf{k}$. Current, for example, is a result of Δf in real space and momentum space. At any position, the current arises in the momentum of the particles. It is proportional to it. So, current, which is proportional to Δf, being an odd function of momentum, also implies that Δf is an odd function of $\mathbf{k}$. The first term of Equation 8.42 therefore vanishes when $S(\mathbf{k}',\mathbf{k})$ is an even function.

So, a sufficient condition for the validity of relaxation time approximation is that the first term of the integral—the scattering or collision integral—the term related to the change in distribution function due to scattering to $\mathbf{k}$, vanishes. $\Delta f(\mathbf{k}')$ is odd, so $S(\mathbf{k},\mathbf{k}')$ must be even.

There are many conditions of scattering in semiconductors under which $S(\mathbf{k},\mathbf{k}')$ is an even function.

One example is if all combinations of $\mathbf{k}$s and their antipodes lead to the same scattering rates (see Figure 8.8), that is, if

$$S(\mathbf{k},\mathbf{k}') = S(\mathbf{k},\mathbf{k}'^{\vdash}) = S(\mathbf{k}^{\vdash},\mathbf{k}') = S(\mathbf{k}^{\vdash},\mathbf{k}'^{\vdash}). \tag{8.44}$$

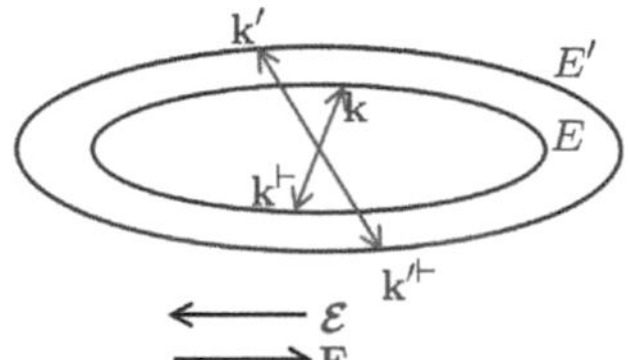

Figure 8.8: Scattering in the anisotropic conditions of ellipsoidal equienergy surfaces at energy E and E' when a perturbational field $\boldsymbol{\mathcal{E}}$ is present.

In the presence of the electric field $\boldsymbol{\mathcal{E}}$ or, equivalently, the force $\mathbf{F}$, the electron charge cloud responds with an increase in population at $\mathbf{k}'^{\vdash}$, and a decrease in population at $\mathbf{k}'$. That is, $\Delta f(\mathbf{k}'^{\vdash}) > 0$, and $\Delta f(\mathbf{k}') < 0$. So, on the equienergy surface of energy E',

$\Delta f(\mathbf{k}') = -\Delta f(\mathbf{k}'^{\vdash})$. If the $\mathbf{k}'^{\vdash}$ states have more carriers, that increases the transition rate from $\mathbf{k}'^{\vdash}$ to $\mathbf{k}$. But there is an equal decrease in transition rate from $\mathbf{k}'$ to $\mathbf{k}$. However, $S(\mathbf{k}',\mathbf{k}) = S(\mathbf{k}'^{\vdash},\mathbf{k})$ from this equienergy surface. Therefore, the sum of the transition rates from $\mathbf{k}'^{\vdash}$ and $\mathbf{k}'$ remains unchanged for small fields. All the states can be paired. Therefore, the transition rate into $\mathbf{k}$ is unchanged from the thermal equilibrium conditions. The only change in the distribution function at $\mathbf{k}$ therefore arises in transitions from $\mathbf{k}$; that is, scattering out of $\mathbf{k}$. So, for random scattering processes, that is, processes subject to conditions of Equation 8.44, the relaxation time approximation is valid, and the relaxation time will be given by Equation 8.43.

There are other non-random scattering processes too, where relaxation time approximation holds. A particularly common example for relaxation time approximation is elastic scattering, that is, processes in which the loss of kinetic energy vanishes. By extension, in the lowest order, this also applies to quasielastic scattering, that is scattering processes where the change in the electron energy is quite small compared to thermal energy ($\Delta E \ll k_B T$). Another one is when the scattering probability is dependent on the angle between the incoming and outgoing momentum.

Consider elastic scattering. The scattering probability normalized to unit area of the constant energy surface, from $\mathbf{k}'$ to $\mathbf{k}$ or its reverse, must balance so that there is no net power flow in an elastic process. This implies that

$$S_{\hat{\mathbf{n}}}(\mathbf{k}',\mathbf{k}) = S_{\hat{\mathbf{n}}}(\mathbf{k},\mathbf{k}'), \tag{8.45}$$

where $S_{\hat{\mathbf{n}}}$ is a real scattering rate that is normalized to account for energy flow, that is, through a normalization to $\hat{\mathbf{n}}d^2r$. At thermal equilibrium,

$$\left.\frac{\partial f}{\partial t}\right|_{scatt,0} = \int_A [f_0(\mathbf{k}')S_{\hat{\mathbf{n}}}(\mathbf{k}',\mathbf{k}) - f_0(\mathbf{k})S_{\hat{\mathbf{n}}}(\mathbf{k},\mathbf{k}')]\, d^2k' = 0. \tag{8.46}$$

If Equations 8.45 and 8.46 are true, then $f_0(\mathbf{k}') = f_0(\mathbf{k})$. Therefore, off-equilibrium,

$$\begin{aligned}
\left.\frac{\partial f}{\partial t}\right|_{scat} &= \int_A \{[f_0(\mathbf{k}') + \Delta f(\mathbf{k}')]\, S_{\hat{\mathbf{n}}}(\mathbf{k}',\mathbf{k}) \\
&\quad - [f_0(\mathbf{k}) + \Delta f(\mathbf{k})]\, S_{\hat{\mathbf{n}}}(\mathbf{k},\mathbf{k}')\}\, d^2k' \\
&= -\Delta f(\mathbf{k}) \int_A S_{\hat{\mathbf{n}}}(\mathbf{k},\mathbf{k}') \left[1 - \frac{\Delta f(\mathbf{k}')}{\Delta f(\mathbf{k})}\right] d^2k' \\
&= -\frac{\Delta f(\mathbf{k})}{\tau_{\mathbf{k}}},
\end{aligned}$$

$$\text{with } \frac{1}{\tau_{\mathbf{k}}} = \int_A S_{\hat{\mathbf{n}}}(\mathbf{k},\mathbf{k}') \left[1 - \frac{\Delta f(\mathbf{k}')}{\Delta f(\mathbf{k})}\right] d^2k', \tag{8.47}$$

which is the momentum-dependent time constant for elastic and, approximately by extension, quasielastic scattering.

Now consider scattering processes where the scattering rate depends on the deflection θ of the outgoing wave $\mathbf{k}'$. The Coulomb scattering of Figure 1.6 is one example that we have already seen quite early on. Another is acoustic phonon scattering in isotropic valleys (see Figure 8.9). Let there be a small electric field $\boldsymbol{\mathcal{E}}$ and let the scattering be quasielastic; then, the distribution function in this small field can be written as $f = f_0 + g\boldsymbol{\mathcal{E}} \cdot \mathbf{k}$:

$$\left.\frac{\partial f}{\partial t}\right|_{scat} = \int_A \left(g\boldsymbol{\mathcal{E}} \cdot \mathbf{k}' - g\boldsymbol{\mathcal{E}} \cdot \mathbf{k}\right) S_{\hat{\mathbf{n}}}(\theta)\, d^2\mathbf{k}'. \tag{8.48}$$

Figure 8.9: Quasielastic angle-dependent scattering such as with acoustic phonons in valleys with spherical constant energy surfaces or by impurity ions.

Let $\hat{\mathbf{z}}$ be the incident direction; then,

$$d^2\mathbf{k}' = (\mathbf{k}'\, d\theta) \cdot (\mathbf{k}' \sin\theta\, d\phi) = \mathbf{k}'^2 \sin\theta\, d\theta\, d\phi. \tag{8.49}$$

The consequences of azimuthal dependence will vanish, since they are harmonic in the azimuthal angle, and the integral is over 2π. So,

$$\left.\frac{\partial f}{\partial t}\right|_{scat} = g\boldsymbol{\mathcal{E}} \cdot \mathbf{k} \int_A (1 - \cos\theta)\, S_{\hat{\mathbf{n}}}(\theta)\, d^2\mathbf{k}' = -\frac{\Delta f}{\tau_{\mathbf{k}}}, \text{ where}$$

$$\frac{1}{\tau_{\mathbf{k}}} = \int_A (1 - \cos\theta)\, S_{\hat{\mathbf{n}}}(\theta)\, d^2\mathbf{k}'. \tag{8.50}$$

The semi-classical analog of this equation is

$$\frac{1}{\tau_{\mathbf{k}}} = N_s v \int_{-\pi}^{\pi} \sigma(\theta)(1 - \cos\theta) d\theta,$$

where N_s is the number of scattering centers, v is the velocity of carriers, and $\sigma(\theta)$ is a cross section of the scattering through angle θ. Note the equivalence with velocity, capture cross section and the deviation angle corresponding to the scattering rate through a momentum area as it sweeps through momentum space. It is a statement of the reduced effect on the current flow when carriers scatter through a small angle. Ionized impurities and acoustic phonons both cause small-angle scattering.

Since the scattering rate defines meantime between the scattering events as

$$\frac{1}{\tau_c} = \int_A S_{\hat{\mathbf{n}}}(\theta)\, d^2\mathbf{k}', \tag{8.51}$$

the relaxation time varies as

$$\tau_{\mathbf{k}} = \frac{\tau_c}{\langle 1 - \cos\theta \rangle}. \tag{8.52}$$

This scattering angle dependence of the relaxation time says that if all angle deflections are equally likely, then $\langle \cos\theta \rangle = 0$, and $\tau_{\mathbf{k}} = \tau_c$ is the average time between scattering events. However, if small changes in directions are favored, that is, $\langle \cos\theta \rangle \to 1$, then $\tau_{\mathbf{k}} \gg \tau_c$. When this happens, then the relaxation time is the same over the entirety of the constant energy surface. So, $\tau_{\mathbf{k}}$ is now only a function of energy, a case that is quite distinct from when the scattering randomizes momenta.

We will discuss scattering for various crystalline semiconductor-specific important processes in Chapter 10 but here make a few remarks on the efficacy of the relaxation time approximation. It is convenient in that it allows one to make analytic calculations. But one could just as well have employed the transport equation directly by keeping track of scattering events, incorporating them as they occur in a computation randomly at the expected

Broad category	Perturbation	Scattering
Lattice	Optical phonon	Polar
		Non-polar
	Acoustic phonon	Piezoelectric
		Deformation potential
Inter-carrier scattering	Coulomb	Electron-electron (majority)
		Electron-hole (minority)
Defect		Crystal
		Impurity
		Alloy
		Interface

Table 8.1: Some of the significant scattering mechanisms in a semiconductor.

probabilities, such as via Monte Carlo techniques. The analytic approximations give us insights, and, as we will presently see, and the undergraduate device courses employ, a pretty acceptable description of semiconductor device behavior.

In a semiconductor, one will encounter numerous perturbations that disturb particle flow. Table 8.1 breaks these into three categories: (a) those arising in electrons interacting with the solid's motion termed lattice and encompassing the phonon interactions; (b) carrier-carrier scattering and other coupled scattering mechanisms, where electrons interact with other coupled excitations; and (c) scattering arising in interactions with randomness of unintentional and intentional imperfectness of the crystal, such as impurities or alloys. Some of these will be dealt with in Chapter 10 because of their importance or their interesting nature.

The perturbation, being a finite-duration perturbation, the scattering rate in the presence of energy gain or loss in scattering can be written using Golden rule as

$$S(\mathbf{k},\mathbf{k}') = \frac{2\pi}{\hbar}\int_{\Omega} \psi^*_{\mathbf{k}'}\hat{\mathscr{H}}'\psi_{\mathbf{k}}\delta\left[E(\mathbf{k}) - E(\mathbf{k}') - E_{scat}\right] d\mathbf{r}. \qquad (8.53)$$

When the departure of the distribution function from thermal equilibrium due to the perturbation is small, that is, $\Delta f \ll f_0$, one may rework the scattering-related perturbation term in Equation 8.34 as

$$\left.\frac{\partial f}{\partial t}\right|_{scat} = \int S(\mathbf{k}',\mathbf{k})f_0(E)\left[1 - f_0(E')\right]\left[\phi(\mathbf{k}') - \phi(\mathbf{k})\right] d\mathbf{k}'. \qquad (8.54)$$

$\Delta f = -\phi(\mathbf{k},\mathbf{r})\partial f_0/\partial E$ represents the linear term of the small perturbation, and, because the perturbation is small, we have assumed that $\partial f/\partial E = \partial f/\partial E|_0$. The interpretation of Figure 8.7 now should be clear. For an electric field $\boldsymbol{\mathcal{E}}$ pointed in the $-\hat{\mathbf{k}}_{\parallel}$ direction, the shift in distribution is $\partial f/\partial k_{\parallel} \approx \partial f/\partial k_{\parallel}|_0$. The drift of the distribution due to the force arises in $\hbar\dot{\mathbf{k}} = -e\boldsymbol{\mathcal{E}}$, and, in steady state, the consequence is

$$\frac{e\boldsymbol{\mathcal{E}}}{\hbar} \cdot \nabla_{\mathbf{k}} f = \frac{\Delta f}{\tau_{\mathbf{k}}}, \tag{8.55}$$

so that

$$\Delta f = \frac{e\tau_{\mathbf{k}}\mathcal{E}}{\hbar} \left. \frac{\partial f}{\partial k_{\parallel}} \right|_0 . \tag{8.56}$$

This is equivalent to saying that the distribution function is

$$f = f_0 + \frac{e\tau_{\mathbf{k}}\mathcal{E}}{\hbar} \left. \frac{\partial f}{\partial k_{\parallel}} \right|_0 = f_0 \left(k_{\perp}, k_{\parallel} - \frac{e\tau_{\mathbf{k}}\mathcal{E}}{\hbar}, \mathbf{r}; t \right). \tag{8.57}$$

A rigid shift took place in the distribution. This new distribution is the distribution arising from $(k_{\perp}, k_{\parallel} - e\tau_{\mathbf{k}}\mathcal{E}/\hbar, \mathbf{r}; t)$. Recall Figure 8.7; this shows the rigid shift and the meaning reflected in it. The momentum shift can also be written in terms of the effective mass m^* and the drift velocity, which is the expectation velocity—the net velocity—of this distribution. $v_d = \hbar\langle\Delta k\rangle/m^*$, so

$$v_d = \frac{e\langle\tau_{\mathbf{k}}\rangle}{m^*}\mathcal{E} \quad \therefore \mu = \frac{\partial v}{\partial \mathcal{E}} = \frac{e\langle\tau_{\mathbf{k}}\rangle}{m^*}, \tag{8.58}$$

where μ is the mobility. This also implies that the expectation of the momentum and its shift from thermal equilibrium is $\langle\hbar\mathbf{k}\rangle = \langle\hbar\Delta\mathbf{k}\rangle = e\boldsymbol{\mathcal{E}}\langle\tau_{\mathbf{k}}\rangle$, and the expectation of the carrier energy change is $\langle\Delta E\rangle = e\mathcal{E}v_d\langle\tau_{\mathbf{k}}\rangle$.

For now, we remain focused on the relaxation time approximation and relate it to overall features of transport as well as relationships within. The approximation is quite valid, as we just saw for spherical energy surfaces and low-loss energy processes. So, in direct gap compound semiconductors, it works well for acoustic phonon and impurity scattering. It works well for randomizing scattering. So, it is an effective approximation in non-polar semiconductors—the ones where conduction minima are generally not at the zone center, and constant energy surfaces not spherical—for optical phonon scattering. This is so since optical phonons have a near-constant distribution over all wavevectors. So, the total momentum and energy conservation constraints can be satisfied in all orientations. However, in non-polar semiconductors, it now does not work for low energy or elastic scattering processes, so it does not work for ionized impurity scattering or acoustic scattering. Likewise, in polar semiconductors—many of the compound semiconductors—it does not work for optical phonon scattering.

If the probability of occurrence of different scattering events is independent, and they cause only a moderate change in energy of the carrier, then the scattering rates of the different mechanisms add up. As an extension of this reasoning, if the "average" scattering rates add up, and these scattering events are also subject to the relaxation time approximation, then one can see that

If the energy of the carrier falls below the optical phonon energy threshold, optical phonon emission is suppressed, so one can see here one scattering process affecting the possibility of another, thus breaking the independence criterion.

$$\frac{1}{\langle\tau_{\mathbf{k}}\rangle} \approx \sum_i \frac{1}{\langle\tau_{\mathbf{k}}\rangle_i} \quad \therefore \quad \frac{1}{\mu} \approx \sum_i \frac{1}{\mu_i}. \tag{8.59}$$

Equation 8.59 is known as Matthiessen's rule. It is a semi-empirical relationship useful in understanding the mobilities of semiconductors as parameters such as temperature, doping, et cetera, are changed and one is interested in finding general trends.

8.7 *Conservation equations from Boltzmann transport*

THE DISTRIBUTION FUNCTION, by describing the particle density in canonical coordinates, gives us a semi-classical description from which properties we desire may be obtained. Among the properties that we are usually interested in—both at and away from thermal equilibrium—are the particle densities, which in thermal equilibrium we generally understand well but away from thermal equilibrium can be quite beholden to the spatial and time dependences in transport; the current, which is the charge flux and therefore proportional to momentum; the energy that is carried around by these particles as they flow; and others that we may be interested in, for example, spin currents if transport can be made spin dependent, or the exchange between the energy being carried by different particles, for example, electrons and phonons. All the information related to these is contained in the $f(\mathbf{r}, \mathbf{k})$ of the particles and the state description of the particles contained in the E-$\mathbf{k}$ description.

The Boltzmann transport equation allows us to derive a set of conservation equations that determine these parameters. If we know the distribution function, we can multiply it by the relationship for the property associated with the particle, so turning it into the moment, and from this determine the equation for expectation. We now have the moment equation associated with that property while accounting for the distribution function. These moment equations—statistically defined—are similar to moments of functions. We will see that there is a correspondence here. Moments now mean the accumulated magnitude of the variable whose expectation is being determined from the equation of distribution function evolution. Of these moment equations, the 0th of which will be the continuity equation for the particle density, the 1st, which will be the continuity equation for current density, and the 2nd, which will give the continuity equation for energy density, are particularly important.

For a probability distribution function, a mean is the first moment, a variance is the second moment, and so on. Moment is like a lever. The consequence of a parameter is bootstrapped by the lever. In a transistor, the gate voltage is levering through the oxide capacitance. A carrier higher up in energy is carrying more energy and has that lever in energy transport. A moment characterizes this leverage.

Take a generalized variable $\varphi(\mathbf{k})$ that characterizes a property, and we would like to find its continuity relationship in real space. So, φf gives the distribution function of the property φf in phase space. We integrate over the $\mathbf{k}$-space, and obtain the continuity relationship of $\langle\varphi\rangle$, the expectation value of φf in real space at any position **r**, where $\langle\rangle$ reflects the ensemble averaging over all of the **k**-space. So,

$$\begin{aligned}\frac{d}{dt}\langle\varphi f\rangle &= -\langle\dot{\mathbf{k}}\cdot\nabla_{\mathbf{k}}\varphi f\rangle - \langle\dot{\mathbf{r}}\cdot\nabla_{\mathbf{r}}\varphi f\rangle + \langle\left.\frac{\partial\varphi f}{\partial t}\right|_{scat}\rangle \\ &= -\langle\dot{\mathbf{k}}\cdot\nabla_{\mathbf{k}}\varphi f\rangle - \nabla_{\mathbf{r}}\cdot\langle\dot{\mathbf{r}}\varphi f\rangle + \langle\left.\frac{\partial\varphi f}{\partial t}\right|_{scat}\rangle \\ \because\ \nabla_{\mathbf{r}}\cdot(\dot{\mathbf{r}}\varphi f) &= \varphi f\nabla_{\mathbf{r}}\cdot\dot{\mathbf{r}} + \dot{\mathbf{r}}\cdot\nabla_{\mathbf{r}}\varphi f, \text{ where} \\ \varphi f\nabla_{\mathbf{r}}\cdot\dot{\mathbf{r}} &= 0, \end{aligned} \tag{8.60}$$

since $\dot{\mathbf{r}}$ is only a function of **k**. $\langle\varphi f\rangle$ is now only a function of position **r** and time t.

Recall our comment that $\partial f/\partial t$ are all the time dependences of interactions whose energies are not accounted for in the Hamiltonian. Generally, in semiconductors, we work with electrons and holes, where holes are the quasiantiparticles of electrons, representing vacant electron states in an otherwise filled valence band. When one includes interactions within bands within our description, and include electrons and holes, the most significant one for particle continuity is of the generation and recombination of electrons and holes. The most common of these is the generation of electron and hole pairs, if the population is depleted to below that of thermal equilibrium ($np < n_i^2$, where n_i is the intrinsic carrier concentration), and recombination of electron and hole pairs when it exceeds that of thermal equilibrium ($np > n_i^2$).

In Chapter 11, we will look at particle generation and recombination processes in some detail.

Some caution is needed here, since there can certainly be multiparticle interactions processes, and they are processes that can span both short and long time scales. Defect-assisted processes are slow, while direct band-to-band processes can be fast. We can take care of this by bringing in a net recombination term

$$\mathcal{U} = \mathcal{R} - \mathcal{G}, \tag{8.61}$$

where $\mathcal{R}$ is the recombination rate, and $\mathcal{G}$ is the generation rate in units of per unit volume and time. We can even give these $\mathcal{U}$s, $\mathcal{R}$s and $\mathcal{G}$s generalized meanings for the generalized variable $\varphi(\mathbf{k})$. It could be for particle recombination and generation, current recombination and generation, and even kinetic energy recombination and generation and so on. For all these too, the flows will have to arise in these separate particles in different bands for them to interact in this way. Include this in the time-based dependence, and in steady state we have

$$\frac{d}{dt}\langle\varphi f\rangle = -\langle\dot{\mathbf{k}}\cdot\nabla_{\mathbf{k}}\varphi f\rangle - \nabla_{\mathbf{r}}\cdot\langle\dot{\mathbf{r}}\varphi f\rangle - \frac{\langle\varphi f\rangle - \langle\varphi f\rangle_0}{\tau_\varphi}. \tag{8.62}$$

The term $(\langle\varphi f\rangle - \langle\varphi f\rangle_0)/\tau_\varphi = \mathcal{U}$ is a generalization of the recombination-generation of various moments, and τ_φ is a time constant characterizing the rate. Note that if only one type of band is considered, only one particle type interacting—an example being just looking at particle density by making $\varphi = 1$—then $\tau_\varphi \to \infty$, since there can be no loss of carriers in a single band. If there are particle and antiparticle bands to be considered, then τ_φ is a relaxation time connecting that interaction.

A lifetime is a relaxation time. It happens to be a time constant associated with the entropic drive of relaxation toward equilibrium via the annihilation or generation of particles that have been disturbed from their equilibrium statistical distribution.

For $\varphi(\mathbf{k}) = 1$, the implication is that the expectation value is

$$\langle \varphi f(\mathbf{r}, \mathbf{k}; t)\rangle = \int f(\mathbf{r}, \mathbf{k}; t)\, d\mathbf{k} = n(\mathbf{r}, t), \tag{8.63}$$

the carrier concentration. This is the mean of the distribution function—the expectation value of particle density per unit volume. For $\varphi(\mathbf{k}) = (\hbar\mathbf{k}/m^*)(-e)$, a velocity times the charge, that is, an electron charge flux,

$$\langle \varphi f(\mathbf{r}, \mathbf{k}; t)\rangle = \int \frac{e\hbar\mathbf{k}}{m^*} f(\mathbf{r}, \mathbf{k}; t)\, d\mathbf{k} = \mathbf{J}(\mathbf{r}, t), \tag{8.64}$$

the current density. For $\varphi(\mathbf{k}) = \hbar^2\mathbf{k}^2/2m^*$, the kinetic energy of the particle,

$$\langle \varphi f(\mathbf{r}, \mathbf{k}; t)\rangle = \int \frac{\hbar^2\mathbf{k}^2}{2m^*} f(\mathbf{r}, \mathbf{k}; t)\, d\mathbf{k} = W(\mathbf{r}, t), \tag{8.65}$$

the kinetic energy density of the carrier ensemble. $W/n = w$ is the kinetic energy per carrier.

The continuity equations of the various moments now follow. For the 0th moment, with $\varphi(\mathbf{k}) = 1$, that is, $\langle\varphi f\rangle = n$:

$$\begin{aligned}
\langle \dot{\mathbf{k}} \cdot \nabla_{\mathbf{k}} \varphi f\rangle &= 0, \\
\nabla_{\mathbf{r}} \cdot \langle \dot{\mathbf{r}} \varphi f\rangle &= -\frac{1}{e} \nabla_{\mathbf{r}} \cdot \mathbf{J}, \\
\therefore \quad \frac{dn}{dt} &= -\frac{1}{e} \nabla_{\mathbf{r}} \cdot \mathbf{J} + \mathcal{G} - \mathcal{R}, \\
\text{or} \quad \frac{dn}{dt} &= \frac{1}{q} \nabla_{\mathbf{r}} \cdot \mathbf{J} - \frac{\Delta n (= n - n_0)}{\tau_n},
\end{aligned} \tag{8.66}$$

the particle continuity equation, or the particle conservation equation. The first term on the right describes the real space divergence in current, that is, the accumulation or depletion of carriers due to the current flow, and the second term is the particle density change arising in particle creation and annihilation processes due to interactions between bands. The quantum analog of this equation is the probability current equation.

For the 1st moment with $\varphi(\mathbf{k}) = -e\hbar\mathbf{k}/m^*$, that is, $\langle\varphi f\rangle = \mathbf{J}$:

$$\langle\dot{\mathbf{k}}\cdot\nabla_{\mathbf{k}}\varphi f\rangle = -\frac{en}{m^*}\hbar\dot{\mathbf{k}} = -\frac{en}{m^*}\mathbf{F},$$

$$\nabla_{\mathbf{r}}\cdot\langle\dot{\mathbf{r}}\varphi f\rangle = -e\nabla_{\mathbf{r}}\cdot\left(\frac{\hbar\mathbf{k}}{m^*}\frac{\hbar\mathbf{k}}{m^*}f\right) = -\frac{2e}{m^*}\nabla_{\mathbf{r}}W$$

$$\therefore\ \frac{d\mathbf{J}}{dt} = \frac{en}{m^*}\mathbf{F} + \frac{2e}{m^*}\nabla_{\mathbf{r}}W - \frac{\mathbf{J}}{\tau_{\mathbf{k}}},$$

$$\text{or}\ \frac{d\mathbf{J}}{dt} = -\frac{qn}{m^*}\mathbf{F} - \frac{2q}{m^*}\nabla_{\mathbf{r}}W - \frac{\mathbf{J}}{\tau_{\mathbf{k}}}, \tag{8.67}$$

the current continuity equation, or the current conservation equation. Here, in its last form, we substituted $q = -e$, to make it more general and work for electrons and holes. The first term on the right arises in momentum change. It is what we think of classically as drift. The second term on the right is the diffusive flow. A gradient in energy density is due to particle density and their energy content in the distribution in momentum space. Integrated over $\mathbf{k}$, this represents the energy density. The last term is the perturbation—scattering—precipitated momentum changes. The term has been simplified through a momentum relaxation time.

For the 2nd moment, with $\varphi(\mathbf{k}) = \hbar^2\mathbf{k}^2/2m^*$, that is, $\langle\varphi f\rangle = W$:

$$\langle\dot{\mathbf{k}}\cdot\nabla_{\mathbf{k}}\varphi f\rangle = \langle\frac{\mathbf{F}}{\hbar}\cdot\left(2\frac{\hbar^2\mathbf{k}}{2m^*} + \frac{\hbar^2\mathbf{k}^2}{2m^*}\nabla_{\mathbf{k}}f\right)\rangle = -\langle\frac{\mathbf{F}}{e}\cdot(e\hbar\mathbf{k}f)\rangle$$

$$= -\frac{\mathbf{F}}{e}\cdot\mathbf{J}$$

$$\nabla_{\mathbf{r}}\cdot\langle\dot{\mathbf{r}}\varphi f\rangle = \nabla_{\mathbf{r}}\cdot\langle\dot{\mathbf{r}}\frac{\hbar^2k^2}{2m^*}\rangle$$

$$\therefore\ \frac{dW}{dt} = \frac{1}{e}\mathbf{F}\cdot\mathbf{J} - \nabla_{\mathbf{r}}\cdot\langle\dot{\mathbf{r}}\frac{\hbar^2k^2}{2m^*}\rangle - \frac{W - W_0}{\tau_w}$$

$$\text{or}\ \frac{dW}{dt} = -\frac{1}{q}\mathbf{F}\cdot\mathbf{J} - \nabla_{\mathbf{r}}\cdot\langle\dot{\mathbf{r}}\frac{\hbar^2k^2}{2m^*}\rangle - \frac{W - W_0}{\tau_w}, \tag{8.68}$$

which is the energy continuity equation, or the energy conservation equation. The first term is the flow of energy in momentum space—a force pushing the electrons along in energy in momentum space; the second is the real space flow of energy, that is, carriers with energy moving with a velocity; and the third is the scattering-related loss of energy. The third term is the entropic drive toward equilibrium. Note that it has a time constant τ_w, different from $\tau_{\mathbf{k}}$, since this is related to the changes in $f \times \hbar^2\mathbf{k}^2/2m^*$. τ_w is the energy relaxation time. We will have different relaxation time constants for each one of these different moment equations, but because they arise in the relaxation of the distribution (and the variable $\varphi(\mathbf{k})$), they will be dependent on the momentum relaxation time $\tau_{\mathbf{k}}$.

One can see the commonalities in the equations. The conservation or continuity of the different parameters is related to changes taking place due to the time-dependent motion of the parameter in momentum space, its time-dependent motion in real space, and its entropic propensity toward thermal equilibrium captured in a relaxation time. For particle density, if only one band exists, then the relaxation time is infinity. If multiple bands exist, we will have to calculate it, and it will be different. This parameter is the lifetime of the particle. We will see several unusual features related to this in Chapter 11. If this interaction happens radiatively between electrons and holes, it is the radiative lifetime. If it happens due to defects in the material—a phonon-assisted process—it is the Hall-Shockley-Ridley non-radiative lifetime. If it happens due to multiple charge particles interacting with each other, for example, an electron at high energy losing some of its energy to the creation of an electron-hole pair and the rest possibly in phonons, then it is an Auger generation process. An impact ionization process is such a process. A recombination event, which often occurs when the bandgap is small or when doping is high is when an electron and a hole recombine and give their energy to another particle in the band. So, Auger processes come in a large variety, often occur at high doping and are common to small bandgap materials and high fields.

These moment equations describe the continuity of various moments of the distribution function. For semiconductor electronics, the correspondence of these is to particle, current and energy continuity. In fairly large-sized devices, it is common to use the drift-diffusion equation—a Fokker-Planck-like equation—consisting of an explicit drift and diffusion terms. It is of quite limited validity; Equations 8.66–8.68 are more precise, although they are restricted to relaxation time approximation if $\partial f/\partial t$ is not explicitly managed.

The drift-diffusion equation (which works pretty reasonably for large devices), however, is better than the Drude equation, which is cruder and, at best, is a good example of finding a relationship through matching of units and the good luck of matching of errors.

It is useful to see drift diffusion as a limit case from these more precise equations. Take the 1st moment equation of current continuity (Equation 8.67). First, assume non-degenerate conditions, so Maxwell-Boltzmann distribution conditions, where particles may be treated classically with $k_BT/2$ of energy per motional degree of freedom. So, $W = (3/2)nk_BT$ at temperature T for the electron distribution and, in this case, the crystal too. Let the electric field be $\boldsymbol{\mathcal{E}}$. We can write

$$\tau_{\mathbf{k}}\frac{d\mathbf{J}}{dt} = \frac{e^2\tau_{\mathbf{k}}}{m^*}\boldsymbol{\mathcal{E}}n - \mathbf{J} + \frac{e^2\tau_{\mathbf{k}}}{m^*}\nabla_{\mathbf{r}}nk_BT. \qquad (8.69)$$

The time scales of interest in the use of these equations, as in devices, are much larger than $\tau_{\mathbf{k}}$. So,

$$
\begin{aligned}
0 &= e\frac{e\tau_{\mathbf{k}}}{m^*}n\mathcal{E} - \mathbf{J} + \frac{e\tau_{\mathbf{k}}}{m^*}k_BT\nabla_{\mathbf{r}}n, \quad \text{or} \\
\mathbf{J} &= e\mu_n n\mathcal{E} + e\mathcal{D}_n\nabla_{\mathbf{r}}n, \quad \text{where} \\
\mu_n &= \frac{e\tau_{\mathbf{k}}}{m^*}, \quad \text{and} \\
\mathcal{D}_n &= \frac{k_BT\tau_{\mathbf{k}}}{m^*} = \frac{k_BT}{e}\mu_n.
\end{aligned} \tag{8.70}
$$

Momentum relaxation times are of the order of sub-*ps*, while device times of greater than *ps* are common. The relaxation times are, of course, dependent on the scattering process and are temperature dependent, but an accumulation of these will tend to be such that the drift-diffusion equation should be applicable in large geometries where plenty of scattering events occur.

The second equation in this set is the drift-diffusion equation, a consequence of current continuity in relaxation time approximation, and the assumptions that a classical Maxwell-Boltzmann distribution is valid for the transport, that the carriers and the crystal are at the same temperature T, that there are a large number of these randomizing scattering events over the time scales it is employed and that the length scales of spatial variations (fields and any other parameters that will affect transport) are much larger than the length scale between scattering, that is, the momentum mean free path ($\lambda_{\mathbf{k}}$). Note also that a direct connection exists between the mobility μ_n and the diffusion coefficient $\mathcal{D}_n$ through the thermal voltage k_BT/e. This is the Einstein relationship, written here in classical Boltzmann conditions as $\mathcal{D}_n/\mu_n = k_BT/e$. Einstein first derived a relationship between the system response and random events in the analysis for Brownian motion. In general, for electrons and holes, under these assumptions, the drift-diffusion relationship may be written as

The Einstein description, circa 1905, shows that both fast (the random scatterings) and slow (the system response to an external stimulus) are at play in Brownian motion and that these are connected to each other.

$$
\begin{aligned}
\mathbf{J}_n &= en\mu_n\mathcal{E} + e\mathcal{D}_n\nabla_{\mathbf{r}}n, \quad \text{and} \\
\mathbf{J}_p &= ep\mu_p\mathcal{E} - e\mathcal{D}_p\nabla_{\mathbf{r}}p.
\end{aligned} \tag{8.71}
$$

The drift-diffusion equation describes Brownian motion, under a potential field, of a classical particle movement—a fluid movement. As written, particularly with this assumption that the temperature of the particles is the same as that of the lattice, that is, $T_e = T_l = T$, one could write an equivalent classical continuity equation for energy, in the form

$$W = \frac{1}{e}\tau_w\mathbf{F}\cdot\mathbf{J} - \tau_w\nabla_{\mathbf{r}}\cdot\langle\dot{\mathbf{r}}\frac{\hbar^2k^2}{2m^*}\rangle + W_0, \tag{8.72}$$

where W_0 is the energy in the particles at thermal equilibrium, that is, $W_0 = (3/2)nk_BT$.

It should now be clear why this path in phase space, from Liouville, quantum-Liouville, to Boltzmann transport is so powerful. One equation sufficed in projecting a variety of flux parameters that are of interest in flow. Configuration space would have been messier. You will see in Chapter 9, as we discuss the interactions arising in axial and polar fields, that this phase-space formulation gives a direct route to all kinds of calculations where the inter-field interactions, the electric causing an in-line force and the magnetic causing a transverse cross term, can be subsumed gracefully.

Electrons, as the particles in motion in the field, are losing their energy to the crystal, for example, through phonon emission. So, using a little more general form, we can write the relationship with the electron particle temperature and the phonon particle temperature separate. With this separation of energy form—electrons and phonons, where the phonon form is really a heat in a broadband—Equation 8.68 can be recast as

$$\begin{aligned} \frac{dW}{dt} &= \frac{1}{e}\mathbf{F}\cdot\mathbf{J} - \nabla_{\mathbf{r}}\cdot\langle\dot{\mathbf{r}}\frac{\hbar^2k^2}{2m^*}\rangle - \nabla_{\mathbf{r}}\cdot\mathscr{Q} - \frac{W-W_0}{\tau_w}, \\ \text{or}\quad \frac{dW}{dt} &= -\frac{1}{q}\mathbf{F}\cdot\mathbf{J} - \nabla_{\mathbf{r}}\cdot\langle\dot{\mathbf{r}}\frac{\hbar^2k^2}{2m^*}\rangle - \nabla_{\mathbf{r}}\cdot\mathscr{Q} - \frac{W-W_0}{\tau_w}, \end{aligned} \tag{8.73}$$

where $\mathscr{Q}$ is the heat energy flux density.

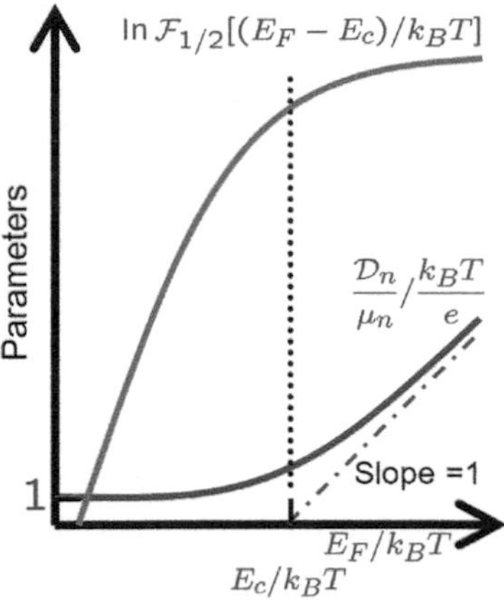

Figure 8.10: Ratio of the diffusivity to mobility and logarithm of the Fermi 1/2 integral as Fermi energy is swept from non-degenerate to degenerate conditions.

In this drift (momentum space), diffusion (real space) and perturbational loss reduced to a drift-diffusion relationship for current, mobility is a parameter reflective of friction to movement. Diffusivity is movement based on real space differences in concentration of species. Both arise in the fast scattering that $\tau_{\mathbf{k}}$ characterizes. And it is this fast scattering, and its power spectrum, that is, its ability to exchange energy with the environment in which the particle is flowing, that determine the friction and the diffusivity of the particles. It can also be generalized to degenerate conditions, where Fermi-Dirac occupation statistics must be employed. It still holds that connection between drift and diffusion (or slow and fast or systemic response and fluctuation noise) as a variation on this relationship there. The occupation of states, or, equivalently, the carrier densities, are now up in the band. The occupation probability rapidly rises to approximately unity $\sim k_BT/e$ below the quasi-Fermi energy. The carrier populations are related to energies (Fermi and bandedge or some other bandstructure energy standard) through the Fermi 1/2 integral—the same integral that we encountered in transport equation at the interface in Section 8.1. Figure 8.10 shows the change as the Fermi energy is swept and the carrier density changes.

Under non-degenerate conditions, the ratio is the thermal voltage (thermal energy divided by the electron charge magnitude), but, under degenerate conditions, it increases. That is, diffusivity becomes stronger with Fermi energy in the conduction band. Filled states surrounded in proximity in the $(E, \mathbf{k})$ are also filled, and very little interaction may occur between the states. There are no empty states to scatter into. Near the Fermi energy, this occupation and emptiness changes and interaction between a particle and its surrounding states becomes possible, although a particle losing energy has only states near the Fermi energy but not too far below it, for transitions. The scattering happens on the surface of this filled $(E, \mathbf{k})$ region. The non-degenerate condition behaves more like a classical particle interaction condition; the degenerate Fermi gas does not.

This Einstein relationship has within it deeper meanings related to the short-range and long-range interactions—interactions over short range and scattering resulting in a wave on a liquid surface with long range—leading to scattering as its consequence embedded in it. It is a general relationship that is seen in a variety of processes where fast interactions happen together with slower long scale interactions. So, fast and slow together lead to a collective movement. It also has implications for fundamental measurements, fluctuation-dissipation and other places, some of which we will come to in due course.

One significance of this fast-and-slow, fluctuation-and-systemic connection is that, even in the conceptual limit of absolute zero temperature ($T = 0\,K$), the zero point motion—a quantum-mechanical randomness in uncertainty of the canonic variables—shows up as a diffusion.

An understanding of the elementary real space view of particle motion and this phase-space statistical view is illuminating. If one viewed these particles as just moving independently, one would write the current as $\mathbf{J} = -e\sum_{\mathbf{k}} n_{\mathbf{k}}\mathbf{v}_{\mathbf{k}} = -e\sum_{\mathbf{k}} n_{\mathbf{k}}\dot{\mathbf{r}}_{\mathbf{k}}$ for the electrons. Here, $n_{\mathbf{k}}$ is the number of electrons in the state $|\mathbf{k}\rangle$, and $\mathbf{v}_{\mathbf{k}}$ their velocity. Concentration gradients didn't appear. The consequence of the force was velocity that moved the particles. An equation for energy flux—equivalent to that for the charge flux—is $\mathbf{W} = \sum_{\mathbf{k}} w_{\mathbf{k}} n_{\mathbf{k}}\mathbf{v}_{\mathbf{k}} = \sum_{\mathbf{k}} w_{\mathbf{k}} n_{\mathbf{k}}\dot{\mathbf{r}}_{\mathbf{k}}$, where $w_{\mathbf{k}}$ now is viewable as the energy carried by each electron in the state $|\mathbf{k}\rangle$, and a summation over all the occupied states. The velocity in these equations has subsumed momentum space, real space and particle interactions—all of them—in this one velocity parameter. If one knows it, this is useful, but mostly it is not useful or rigorous, since it lumps far too many of the underlying canonic connections. This formulation is useful in order to get some intuition. An example is in understanding the nature of the hole as the semiconductor electron's quasiparticle. Valence band or bands are all nearly filled. Empty states in this electron-filled band now appear in the opposite direction of the $\mathbf{k}$-space filling by electrons in the presence of a field. So, the response of this band can be viewed as movement of the emptying states with a positive charge—opposite to that of the filling of states by the electrons' charge—in response to the force. The $(E, \mathbf{k})$ of the valence band still determines the mass of these electrons of the band. So, it does for the collective response, represented by the empty states. The hole is a quasiparticle—an excitation response—of electrons that fill the valence band.

In semiconductor devices, when carriers are moving with a saturated velocity in steady state in a region with the velocity saturation arising in the energy gain (from field) and losing (from scattering) balancing, we sometimes write $\mathbf{J} = qn\mathbf{v}$. Here, it is more true, since $\nabla_{\mathbf{r}} n$ is a vanishing number. But when we use this relation in, say, the transport region of the inversion layer of a *MOSFET*, we must recognize that there is a $\nabla_{\mathbf{r}} n$ that is not of vanishing significance. In gradual channel approximation, it is the diffusive $\nabla_{\mathbf{r}} n$ term that is maintaining current continuity as one approaches the pinched-off region of the channel. Such equations are of limited utility and should be used with care lest some phase-space interaction of significance is lost in this translation.

8.8 *Brownian motion*

Since Brownian motion as a fundamental constraint is so important, and connects to power, energy, memory of events, ensembles, probabilities and other ideas, it is worthwhile discussing it for the underlying fundamental ideas. Brownian motion is the irregular motion of macroscopic objects that one can observe in an optical microscope, such as in liquids, but also in gases.

When an external force is applied, the Brownian particle responds, in the presence of this scattering—fast—interaction with the surrounding medium. The motion therefore suffers friction, arising in these random encounters of redirection. The response to the force has two parts. One is a systematic part of friction. The force and this friction define the net response under external force. But there is also a random part of the response due to the scattering-based fluctuations. This is noise in electric circuits, the bouncing around of the colloidal particles, or others in different systems, but with this common slow-and-fast characteristic. Since the friction and the random fast scattering are related, the net motion under force and the diffusive flow must be related. This is what the Boltzmann transport equation leads to and the Einstein relationship shows.

This internal relationship between the systematic and the random parts of microscopic forces is a general feature. And we connect these two through the *fluctuation-dissipation theorem*, which relates the dissipation of a system with a correlation power spectrum. The fluctuation-dissipation theorem is founded on Nyquist's theoretical work backing Johnson's thermal noise observations. Nyquist showed that the power spectrum of thermal noise in a resistive circuit is proportional to the absolute temperature with a proportionality constant that is a function of the resistance at each frequency. Resistance is quite recognizable as a dissipation element. But, this systemic-power spectrum relationship is more general and is the fluctuation dissipation theorem. We will look at it in this general context (Chapter 14) when discussing the Kramers-Kronig relations that quantify how the imaginary and real parts of a general conductivity response—and many other more general linear system responses, such as susceptibility for example—are related. For now, suffice it to say that the theorem provides us with a general relationship between the response of a system to external stimulus and the internal fluctuations that exist in the absence of that disturbance. It also therefore provides us with a means of ascertaining the fluctuations-bounded limit to observational measurements. The internal fluctuations here are characterized by a correlation function or a fluctuation spectrum of physical observables that are fluctuating. In an electrical measurement, the impedance or admittance is the systemic response, and noise is a fluctuation correlation function. A scanning probe's measurement capability, for example, is limited by the fluctuations from the environment and from within.

An ink blob dropped into water spreads out. The edge regions of this spreading are quite irregular if observed at a resolution slightly better than that of the human eye, and, over time, the ink spreads out and fills out uniformly, at least as observable through the color. This macroscopic uniformity is its equilibrium arising in uniformity of chemical potential with the density of "ink particles" becoming uniform and its most random form of assembly. But, even then, if one were to observe at sufficient resolution, one would see these particles showing abrupt irregularity in motion. The spreading of the ink is a slow process in the chemical disequilibrium, but, even when in equilibrium, there is still thermal motion at work, with the particles interacting with the molecules of their surroundings. There is this scattering taking place between the molecules of the liquid—the environment—and the particle. The spreading out of the ink—a diffusion—occurs and is related to the scattering occurring within this system. A thermal equilibrium exists because of the scattering. And a steady state arises under stimulus because of the scattering.

Human thinking too, as Daniel Kahneman and Amos Tversky point out well, has a slow-and-fast part. The slow part is the analytic response—system 2 thinking—worked through in the brain, and the fast part is the heuristics-based response—system 1 thinking—drawing on parallel experiences. In Ithaca, in September, when new students arrive on the campus with their large vehicles and suburban driving experiences, it is prudent to resort to system 2 thinking when in a hurry and crossing streets. The Kahneman book, *Thinking, Fast and Slow*, is really interesting reading and speaks much to human foibles arising in our hardware-programmed thinking. Fluctuation-dissipation is quite relevant to societal behavior. See S. Tiwari, "Nanoscale device physics: Science and engineering fundamentals," Electroscience 4, Oxford University Press, ISBN 978-0-198-75987-4 (2017) for the application of fluctuation-dissipation in scanning probe measurements. Johnson-Nyquist noise is related to the resistance R—a systemic response—as

Brownian motion is the observation of sudden bursts in the motion of particles in different directions while they are undergoing a net flow in an environment where other particles are also in motion.

Consider the Brownian motion example of Figure 8.11: a particle in a field, moving while also undergoing scattering. The particle acquires a constant drift velocity v_d, acquiring energy in the potential field, and losing energy via the random scattering. Let γ be a friction coefficient relating these together with m–the mass's inertial characteristic in this response—to build

$$v_d = -\frac{1}{m\gamma}\frac{dV}{dz}. \tag{8.74}$$

For a concentration $n(z)$ of these particles, the net particle flux is

$$J(z) = v_d n(z) - \mathcal{D}\frac{\partial n(z)}{\partial z}. \tag{8.75}$$

For current to vanish in thermal equilibrium, $\mathcal{D} = k_B T/m\gamma$. Since classical particles satisfy the Maxwell-Boltzmann distribution, with an exponentially decreasing probability of occupation, the carrier density has the form $n(z) \propto \exp(-V/k_B T)$, reflecting the consequence of the potential in an environment of temperature T, which, by itself, separately, also has a residual thermal equilibrium kinetic energy $(3/2)k_B T$, or, more precisely—in a semiconductor, where the number of states vanish at bandedge—as $\sqrt{8k_B T/\pi m}$. Drift in this equation is the systemic part arising from forces on the system; diffusion here is the part arising in the fluctuations caused by the energy- and momentum-exchanging scattering events. This diffusion reflects the changes in particles' position over time. So, we take an ensemble average,

$$\mathcal{D} = \lim_{t\to\infty} \frac{1}{2t}\langle [z(t) - z(0)]^2\rangle, \text{ where}$$

$$z(t) - z(0) = \int_0^t v_d(\tau)\, d\tau. \tag{8.76}$$

So, using the tricks of rereferencing variables in time for the double integration,

$$\begin{aligned} \mathcal{D} &= \lim_{t\to\infty} \frac{1}{2t}\int_0^t dt_1 \int_0^t \langle v_d(t_1)v_d(t_2)\rangle\, dt_2 \\ &= \lim_{t\to\infty} \frac{1}{t}\int_0^t dt_1 \int_0^{t-t_1} d\langle v_d(t_1)v_d(t_1+\tau)\rangle\tau \\ &= \int_0^\infty \langle v_d(t_0)v_d(t_0+t)\rangle\, dt, \end{aligned} \tag{8.77}$$

where we assumed that the correlation is lost in infinite time, that is, $\lim_{t\to\infty}\langle v_d(t_0)v_d(t_0+t)\rangle = 0$. Since we started off from $\mathcal{D} = k_B T/m\gamma$, this leads to

$4k_B TRB$, where B is the bandwidth over which the power spectrum is spread. An equivalent expression holds for mechanical movement.

Brownian motion is so named after Robert Brown, a Scottish botanist, who observed the rapid fluctuating movement of pollen in liquids under an optical microscope. The first observation of this phenomenon seems to have been made by the Dutch physicist Jan Ingenhousz nearly forty years earlier. This is not surprising: lenses and optical instrumentation are major Dutch creations and, through Galileo, and other astronomers, their creation is one of the major catalyzing events in the birth of modern science.

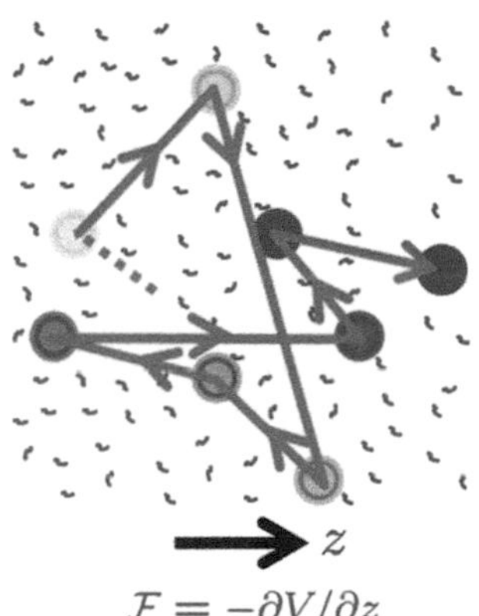

Figure 8.11: Brownian motion of a classical particle as it undergoes random scattering while under the influence of a force, that is, a potential gradient.

The exponential probability is the consequence of entropy $S = k_B \ln \Omega$, and perturbation of the distribution from the equilibrium as derived in S. Tiwari, "Quantum, statistical and information mechanics: A unified introduction," Electroscience 1, Oxford University Press, ISBN 978-0-198-75985-0 (forthcoming).

$$\mu = \frac{1}{m\gamma} = \frac{\mathcal{D}}{k_BT} = \frac{1}{k_BT}\int_0^\infty \langle v_d(t_0)v_d(t_0+t)\rangle\, dt. \qquad (8.78)$$

There exists a direct relationship between the systemic response—the mobility—to external stimulus, and the fluctuations represented through the diffusion coefficient, which is an ensemble average of expectations of squares of separations—a parameter proportional to variance of a distribution—per time. This is the Einstein relationship in a more general form. The drift-diffusion equation describes the Brownian approximation of the motion of electrons as semi-classical particles in a semiconductor. The diffusion is the real space consequence in motion due to the fluctuations. Repeated scattering events with surroundings—similar and dissimilar particles—act as if there were a fluctuating force—arising in fast scattering events. The mobility of the electron is a friction coefficient representing the momentum space consequence of the stimulus causing the net motion.

Diffusion is a random walk—a wandering. In a regular walk, distance traveled is proportional to time t. Take N steps, and the object is N steps away, building up toward infinity as $t \to \infty$. The reason velocity—drift velocity—is so much of interest is that it is an invariant quantity that describes this regular walk. The ratio N/t, or z/t for continuous variables, is a fixed number characterizing the behavior of the system of interest. In a random walk, this invariant is $\langle z^2\rangle/t$—wandering in equally likely opposite directions. The expectation for distance traveled in N steps is $\sqrt{N}$. So, the interesting quantity—the invariant—is $\langle z^2\rangle/t$, which characterizes the diffusion. This is precisely what Equation 8.77 is calculating.

8.9 *Randomness and stochasticity*

RANDOMNESS IS QUITE FUNDAMENTALLY PERVASIVE in nature. By this, we mean that determinism—a form of absoluteness—is a unique notion of axiomatic constructs such as Boolean logic, or natural numbers.

Uncertainty principle is a quantum-mechanical randomness—a fluctuation. It is also a noise. This quantum noise can be viewed as an internal noise. It is intrinsic to the system. An example of an external or extrinsic noise is the thermal noise. It arises from the system being in an environment. It is the interaction with the environment that brings about the temperature T.

Classical randomness, such as in a fair coin or fair dice toss, is a pseudo-randomness based on incompleteness of the description of the problem. All one needs to know are the initial conditions and the equation of the dynamics, and, classically, one should be able to predict. It may take a lot of calculating, by which time the toss may have reached its steady state. The toss problem is not even chaotic, but a classical calculation requires the observation of a lot of variables of the starting conditions and the interactions as the motion of the coin or dice evolves classically. So, only in the mathematically constrained problems do we arrive at determinism.

The natural world's classical observation arises in their statistical limits—an ensemble expectation—of the underlying quantum-mechanical reality. To tackle randomness in physical phenomena, the use of stochastic approaches abound. We have mentioned the Fokker-Planck equation, Markov chains and the Langevin equation and taken a first look at Brownian motion. These are all pertinent to what we call stochastic processes.

Physically, stochasticity is the appearance of the underlying randomness in the observed behavior. Mathematically, by a stochastic process, we mean a family of random variables, say, $\{x_i\}$, that are defined on a probability space. Scattering events happen, but the occurrence of different types of them (phonon, ionized impurity, electron-electron or, for that matter, within-phonon-based, acoustic, optical, emission, absorption, longitudinal and transverse are just a few of these), even if they may be quantifiable by an average, as in the results of a dice toss, are random in time. There is a chain of such scattering events. Some are really random. Some maintain characteristics with buried mutual information. Take, for example, absorption of acoustic phonons during the scattering of electrons in silicon. Lowest valleys, near the X point, are sixfold degenerate and ellipsoidal. Acoustic phonons have an occupation probability that decreases with energy following Bose-Einstein occupation probability. So, absorption of low energy phonons, which exist in larger numbers, is preferred. But conservation of energy and momentum must hold for the phonon-electron system. The wavevector is stretched out along the [100] direction but shrunk along the transverse directions, on the equienergy surface. Preferred directions have to exist, since different directions are not equivalent in momentum. The scattering is not random for an $(\mathbf{r}_i, \mathbf{k}_i) \mapsto (\mathbf{r}_i', \mathbf{k}_i')$ sequence in scattering. The chain of events will matter. This is a Markov chain.

Markov-chain-based techniques are used in communications, stock market or disease or weather forecasting, in artificial learning and in nearly every other place. If it is not being used now in any problem with an analysis and prediction theme, someday it is likely to be, since incompleteness and connections in sequence of events are tackled through this branch of probability theory that A. A. Markov, a Russian mathematician, founded. A fan of Pushkin, while reading *Eugene Onegin*, a novel in verse, Markov developed the approach to analyze patterns of vowels and consonants in Pushkin's writings in 1913. The proof showed that a Markov chain must asymptotically settle to a stable configuration that corresponds to the long term average behavior of the system. The appreciation of Pushkin didn't change because of this, but one now had a new technique by which one could make qualitative predictions of whether a writing was Pushkin's. The methodology of tackling chains of linked events—the future state's dependence on the current state—makes probability go far beyond dice tossing. Together with the Bayesian approach and the frequentist probability confidence measures, these are three of the most powerful mathematical techniques for the real world.

A Markov process is one where all history is lost during the change of the state. The past does not matter anymore. Markov-chain-based methods let us explore these sequential links. In the quantum Liouville equation, the changes that occur to the set of $\mathbf{r}, \mathbf{k}$ are stochastic, arising in scattering events defined by probability functions and a superset of many such probability functions. We will only probe this underlying stochasticity for its most important themes relevant to semiconductors. It is a very rich field. We connect this probabilistic stream through related equations with semiconductor bearings.

8.10 The Langevin and Fokker-Planck equations redux

Paul Langevin is an early 20th century French physicist who, along being known for the development of stochastic dynamics, is also known for tying electron spin to para- and diamagnetism, so for the early developments of quantum mechanics. Langevin was a strong anti-fascist who lost his teaching position in Nazi-occupied France. The story also goes that the Nobel organizers asked Marie Curie to delay her receipt of the second Nobel Prize till a duel between Langevin, apparently Curie's beau post Pierre, and the press was fought and taken care of. Marie refused this as an irrelevant sideshow. Marie stood tall, and society has still not caught up nearly hundred years later.

THE LANGEVIN EQUATION, like the Fokker-Planck equation mentioned earlier, is a way to view the dynamics through the stochastic theme and the probability evolution. The Langevin equation is a master equation form pulling together fluctuations and dissipation in a stochastic rendering:

$$m\frac{d\mathbf{v}}{dt} + m\gamma\mathbf{v} = \mathcal{F}(t) + \mathfrak{F}(t), \tag{8.79}$$

a first-order differential equation, where the net force $\mathbf{F}$ has been split into a macroscopic force $\mathcal{F} = -\nabla\mathscr{H}$, which is a slow force, and $\mathfrak{F}$, which is a fast and randomly fluctuating force, both of which act on the particles. γ is the same Einsteinian phenomenological friction coefficient. $\mathfrak{F}$ arises as a force in fluctuations of a Gaussian variable of zero mean and short-range correlations. Let $\mathfrak{F}(t) \equiv \mathfrak{F}_i(t)$ represent this fluctuation variable. This fast force appears sporadically indexed as instants i. $\langle \mathfrak{F}_i(t) \rangle = 0$. The variance arises in the correlations of these fast scattering events, where

$$\left\langle \mathfrak{F}_i(t)\mathfrak{F}_{i'}(t') \right\rangle = \alpha \delta_{ii'}(t - t'), \tag{8.80}$$

with $\alpha > 0$. α is the strength of the fluctuation. For simplicity, assume that it is also a constant. Referring back to the Brownian discussion, γ and α are related to each other. Any position solution of Equation 8.79 can only be expressed in a nondeterministic form in terms of probability distribution.

The fluctuation-dissipation connection can now be seen in the unperturbed situation when there exists no external macroscopic force $\mathcal{F} = 0$. This force is the potential field $-\partial V/\partial z$ in Figure 8.11. The Fourier transform of the velocity is

$$\begin{aligned} \mathbf{v}(\omega) &= \int \exp(-i\omega t)\mathbf{v}(t)\, dt \\ \therefore \quad &= \frac{1}{m(-i\omega + \gamma)} \mathfrak{F}(\omega). \end{aligned} \tag{8.81}$$

The velocity $\mathbf{v}(t)$ is now also a random variable as a result of these fast random scattering events arising in $\mathfrak{F}_i(t)$. The first moment $\langle v_i(t) \rangle = 0$. The second moment is

$$\langle v_i^2(t) \rangle = \frac{\alpha}{m^2} \frac{1}{2\pi} \int_{-\infty}^{\infty} \frac{1}{\omega^2 + \gamma^2}\, d\omega = \frac{\alpha}{2m^2\gamma}. \tag{8.82}$$

But since $m\langle v_i^2 \rangle/2 = (1/2)k_B T$, $\alpha = 2m\gamma k_B T$, which is an Einstein relation. This relates the variance of fluctuating fast forces to the slow dissipative frictional forces and the temperature. Note also the use of Fourier transformation in this extraction. We will see this connection appear in response functions later.

In thermal equilibrium, the dynamics of individual particles is diffusive. Upon application of an external force, there is a drift motion, which, instead of being ballistic, that is, unimpeded acceleration, is still subject to the scattering hindrance. Scattering brings about a diffusive term that is dependent on the density gradient ∇f—this is Fick's or Fourier's law—and the application of the external force creates a drift ($d\mathbf{v}/dt = 0$ in Equation 8.79); the two are related.

It should now be clear that fluctuations, dissipation, drift and diffusion have neighborly connections. The Fokker-Planck equation that we first encountered in Section 8.4 looks at these same connections, which exist in the evolution of $\mathbf{v}$ arising in stochasticity in the Langevin equation, through probability. Let $\mathfrak{p}(\mathbf{v}, t)$ be the probability distribution for particle velocity at any time t at any position $\mathbf{r}$. $\mathfrak{p}(\mathbf{v},t|\mathbf{v}_0,t_0)$ is the conditional probability of observing velocity $\mathbf{v}$ at time t, when starting with velocity $\mathbf{v}_0$ at some time $t_0 < t$. For any change from some intermediate configuration $(\mathbf{v}',t')$ subject to this slow fluctuating force, so long as this evolution process does not depend on how the configuration $(\mathbf{v}',t')$ itself was arrived at, that is, there is no dependence on prehistory, one can write

$$\mathfrak{p}(\mathbf{v},t|\mathbf{v}_0,t_0) = \int \mathfrak{p}(\mathbf{v},t|\mathbf{v}',t')\mathfrak{p}(\mathbf{v}',t'|\mathbf{v}_0,t_0)\,d\mathbf{v}' \text{ for } t' \in [t_0,t]. \quad (8.83)$$

We will see equations of evolution similar to this in our discussion of Green's functions. It represents chaining of events. When there exists independence from prehistory, we call it a Markov process. Only one-step probabilities matter. All the rest of the past does not. And this is because the events (of scattering) are random.

One might even argue that drift-diffusion-fluctuation-dissipation is such a chain too. Drift is connected to diffusion, diffusion to fluctuations, and fluctuations to dissipation, and dissipation and drift are interlinked too. It is now a circular chain.

We simplify notation at this point. First, the initial condition of $(\mathbf{v}_0,t_0)$ is understood without being written every time. Second, we also choose to write arguments reflecting increments so that series expansion is easier to see.

In society, a lot of problems arise that have this Markovian process character. When one forgets history, the sum of histories and how one arrives at a certain societal point in time, it becomes much easier to assign blame and commit terrible acts. Grays become black and white.

The incremental probability of configuration evolution to $(\mathbf{v}+\Delta\mathbf{v}, t+\Delta t)$ from $(\mathbf{v},t)$—a scattering—is

$$S(\mathbf{v},t;\Delta\mathbf{v},\Delta t) = \mathfrak{p}(\mathbf{v}+\Delta\mathbf{v},t+\Delta t|\mathbf{v},t). \quad (8.84)$$

The equation of probability evolution (Equation 8.83) then becomes

$$\mathfrak{p}(\mathbf{v},t+\delta t) = \int S(\mathbf{v}-\delta\mathbf{v},t;\delta\mathbf{v},\delta t)\mathfrak{p}(\mathbf{v}-\delta\mathbf{v},t)\,d\delta\mathbf{v}, \quad (8.85)$$

with the final time, previously t, now as $t+\delta t$, and intermediate time, previously t', as t. When these fluctuation processes are significant, small time increments δt also correspond to small velocity changes $\delta\mathbf{v}$. One may then expand in $\delta\mathbf{v}$. This expansion is only valid when this short-range correspondence with one-step transitions in short distances is valid. The expansion is

Small velocity change with small time increments will not be the case for the mesoscopic conduction discussed earlier. Scattering is important here in limiting these incremental changes.

$$\begin{aligned}\mathfrak{p}(\mathbf{v},t+\delta t) &= [\mathscr{I}^{\underline{0}}\mathfrak{p}](\mathbf{v},t) - \frac{\partial}{\partial v_i}[\mathscr{I}^{\underline{1}}_i\mathfrak{p}](\mathbf{v},t)\\ &\quad + \frac{1}{2}\frac{\partial^2}{\partial v_i \partial v_j}[\mathscr{I}^{\underline{2}}_{ij}\mathfrak{p}](\mathbf{v},t) + \cdots. \end{aligned} \quad (8.86)$$

Here, the marginals of the perturbations are

$$\begin{aligned}\mathscr{I}^{\underline{0}}(\mathbf{v},t) &= \int S(\mathbf{v},t;\delta\mathbf{v},\delta t)\,d\delta v,\\ \mathscr{I}^{\underline{1}}_i(\mathbf{v},t) &= \int \delta v_i S(\mathbf{v},t;\delta\mathbf{v},\delta t)\,d\delta v,\\ \mathscr{I}^{\underline{2}}_{ij}(\mathbf{v},t) &= \int \delta v_i \delta v_j S(\mathbf{v},t;\delta\mathbf{v},\delta t)\,d\delta v, \text{ and so on.}\end{aligned} \quad (8.87)$$

If δt is small enough, Equation 8.79 can be reduced to a simpler form—the consequences of slow forces disappearing and linear expansion terms prevailing—to

$$m\frac{\mathbf{v}(t+\delta t)-\mathbf{v}}{\delta t}+m\gamma\mathbf{v}\approx\mathfrak{F}(t)$$
$$\therefore\quad \mathbf{v}(t+\delta t)\approx(1-\gamma\delta t)\mathbf{v}+\frac{\mathfrak{F}(t)}{m}\delta t. \tag{8.88}$$

If $\mathbf{v}(t)=\mathbf{v}$, then, for the random variable $\mathbf{v}(t+\delta t)$, the effects of $\mathfrak{F}(t)$ are in the other terms of this equation. They represent the effect of transition probabilities via

$$\begin{aligned} S(\mathbf{v},t;\delta\mathbf{v},\delta t) &= \mathfrak{p}(\mathbf{v}+\delta\mathbf{v},t+\delta t|\mathbf{v},t)\\ &= \langle\delta[\mathbf{v}+\delta\mathbf{v}-\mathbf{v}(t+\delta t)]\rangle_{\mathfrak{F}}\\ &= \langle\delta[\delta\mathbf{v}+\gamma\mathbf{v}\delta t-\delta t\frac{\mathfrak{F}(t)}{m}]\rangle_{\mathfrak{F}}. \end{aligned} \tag{8.89}$$

Equation 8.80 leads to the distribution of the fluctuating fast force in the form

$$\begin{aligned} \mathfrak{p}(|\mathfrak{F}|) &= A\int\mathcal{D}\mathfrak{F}\exp\left(-\frac{1}{2\alpha}\int|\mathfrak{F}(\tau)|^2d\tau\right)dt'\\ &\equiv A\int\mathcal{D}\mathfrak{F}\exp\left(-\frac{\delta t'}{2\alpha}\sum_{\tau}|\mathfrak{F}(\tau)|^2\right)dt', \end{aligned} \tag{8.90}$$

where A is a normalization constant, and the second expression is discretized in time. The fluctuating force, therefore, has a distribution function that appears Gaussian:

$$\mathfrak{p}[\mathfrak{F}(t)]=A\exp\left[-\frac{|\mathfrak{F}(t)|^2}{2\alpha}\delta t\right]. \tag{8.91}$$

This implies that, for the scattering process,

$$\begin{aligned} S(\mathbf{v},t;\delta\mathbf{v},\delta t) &= \int\mathfrak{p}(\mathfrak{F})\,\delta[\delta\mathbf{v}+\gamma\mathbf{v}\,\delta t-\frac{\mathfrak{F}(t)}{m}]\rangle_{\mathfrak{F}}\,d\mathfrak{F}\\ &= A\frac{m}{\delta t}\exp\left(-\frac{m^2}{2\alpha\delta t}|\delta\mathbf{v}+\gamma\mathbf{v}\,\delta t|^2\right). \end{aligned} \tag{8.92}$$

This gives the marginals

$$\begin{aligned} &\mathscr{I}^{\underline{0}}(\mathbf{v},t)=1,\\ &\mathscr{I}_i^{\underline{1}}(\mathbf{v},t)=-\gamma v_i\,\delta t,\ \text{ and}\\ &\mathscr{I}_{ij}^{\underline{2}}(\mathbf{v},t)=\frac{2\alpha}{m}\delta t\,\delta_{ij}+v_iv_j(\gamma\,\delta t)^2. \end{aligned} \tag{8.93}$$

With $\delta t\to 0$, we ignore the higher marginals of $\mathscr{I}^{\underline{\geq 2}}$, which are in $\mathscr{O}(\delta t^2)$, with the expansion in first order becoming

$$\left(\frac{\partial}{\partial t}-\frac{\partial}{\partial v_i}\gamma v_i-\frac{\partial^2}{\partial v_i\partial v_j}\mathcal{D}_v\right)\mathfrak{p}(\mathbf{v},t)=0. \tag{8.94}$$

Here, $\mathcal{D}_v = \alpha/2m^2$ is the diffusivity of velocity. In equilibrium, $\mathcal{D}_v = \gamma T/m$. $\mathcal{D}$, our macroscopic diffusion constant for particle movement, is $\mathcal{D} = \mathcal{D}_v/\gamma^2$. Equation 8.94 is the formal form of the Fokker-Planck equation.

The Fokker-Planck equation is a partial second-order differential equation of probability distribution that can be solved given any initial distribution $\mathfrak{p}(\mathbf{v}, t_0)$. It describes the evolution of the probability distribution. The first term is the linear change in time. The second term is the drift term. The slow force $\mathcal{F}$, if included in our analysis, leads to an $\mathcal{F}/m$ shift in the $\mathscr{I}_i^1(\mathbf{v}, t)$ contribution of probability. If one includes and finds the expectation of $\mathbf{v}$ by multiplying the modified Fokker-Planck equation with v_i and integrating, it will result in the equation

$$\frac{d}{dt}\langle \mathbf{v} \rangle + \gamma \langle \mathbf{v} \rangle - \frac{\mathbf{F}}{m} = 0, \tag{8.95}$$

which is the drift-diffusion equation. It is also what one would get from the Langevin equation by integrating out the noise. $\langle \mathbf{v} \rangle = \mathbf{F}/m\gamma$ is drift.

The third term in the Fokker-Planck equation is the diffusion term. A drift cloud characterized by the probability function $\mathfrak{p}(\mathbf{v}) \propto \exp[-(\gamma/2\mathcal{D}_v)|\mathbf{v} - \mathbf{F}/m\gamma|^2]$ travels centered at a drift trajectory. One also sees in this relationship, at thermal equilibrium, the familiar Maxwell-Boltzmann relationship of $\mathfrak{p}(\mathbf{v}) \propto \exp(-m\mathbf{v}^2/2k_BT)$. The second term and the third term are related to each other. The equation represents the Markovian chain of connections arising in the scattering events. At long time spans, the second and third terms prevail, that is, drift and diffusion prevail. In any evolution trajectory, the second term shows the consequence as a net motion, while the third term shows the consequences of the fluctuations that are observable through noise. When no stimulus exists, that is, there is thermal equilibrium, they work together, balancing each other so that the system asymptotically ends in thermal equilibrium.

8.11 Markov process and Kolmogorov equation

WE NOW EXPLORE THE CHAIN OF PROBABILISTIC INTERACTIONS: THE MARKOV CHAIN. A Markov process has no memory of history. This randomness gives tremendous analytic power, as we have seen in the relaxation time approximation. Is there a way to optimize the descriptive accuracy by accounting for memory and analytic tractability? This is what a Markov process approach attempts to do, and this is what makes it useful across disciplines. Random events may be analytically much more tractable, but they also will not display interesting dynamics through correlations. So, first, let us look at how the notion of stochastic process and a chain of events can be connected to each other through probabilities.

Let $\{s_i, t_i\}$ be a sequence where $i = 1, \ldots, n$ tags discrete observations of the state. When the sequence of observations, i followed by $i+1$, has some randomness to it, the process is a stochastic process. A joint probability $\mathfrak{p}(s_n, t_n; \ldots; s_1, t_1)$ identifies each of the collection of all such possible sequences. This stochastic process is stationary if the probability does not change under simultaneous time translation for each i. This is translational invariance. Since

$$\begin{aligned}\mathfrak{p}(s_n, t_n; \ldots; s_1, t_1) &= \mathfrak{p}(s_n, t_n | s_{n-1}, t_{n-1}; \ldots; s_1, t_1) \\ &\times \mathfrak{p}(s_{n-1}, t_{n-1} | s_{n-2}, t_{n-2}; \ldots; s_1, t_1) \\ &\times \cdots \times \mathfrak{p}(s_2, t_2 | s_1, t_1), \end{aligned} \tag{8.96}$$

is a completely random process—absent correlations and all these probability terms independent of each other—we have

$$\mathfrak{p}(s_n, t_n; \ldots; s_1, t_1) = \prod_i^n \mathfrak{p}(s_i, t_i). \tag{8.97}$$

If the probability terms are dependent only on the previous state, but not on how the previous state was arrived at, then

$$\mathfrak{p}(s_n, t_n; \ldots; s_1, t_1) = \mathfrak{p}(s_n, t_n | s_{n-1}, t_{n-1}). \tag{8.98}$$

This is an example of a Markov process. There exists no memory, since there is no dependence on the history of how (s_{n-1}, t_{n-1}) came to be. This one-step connection is a powerful interregnum between pure randomness and detailed tracking. By choosing the time step of updating such that short term memory is lost even as the important characteristics of the dynamics are kept, one finds an effective compromise that becomes tractable. For Brownian motion, this is achieved by making the time step larger than the fast scattering force's interaction time duration. The process has now been made an effective Markovian process, which will be approximate, but very useful. The right choice of the coordinate is important to succeed at achieving the Markovian form for a process. In Brownian motion, the velocity evolves in a Markovian process, but the position as the integral of velocity over time does not. Position has long-term memory and is not Markovian.

If one chooses equal time steps, when position coordinate separation $|q_j - q_{j-1}|$ is large, then $|q_{j+1} - q_j|$ is also likely to be large since the first arose in a large velocity. The premise that $\mathfrak{p}(q_j, t_j | q_{j-1}, t_{j-1}; \cdots; q_1, t_1) = \mathfrak{p}(q_j, t_j | q_{j-1}, t_{j-1})$ has been broken. Velocity has less memory since changes in it arise in the interaction event itself immediately. One could have made this argument in phase space instead of configuration space.

The Markov process can be transplanted into a useful alternative relationship in continuous form. A state (s, t), starting from an initial condition of (s_0, t_0), comes about through an intermediate event at (s', t'). The transition arises as an integrative consequence over all possibilities of the intermediate event, that is,

$$\begin{aligned}\mathfrak{p}(s,t;s_0,t_0) &= \int \mathfrak{p}(s,t;s',t';s_0,t_0)\,ds' \\ &= \int \mathfrak{p}(s,t|s',t';s_0,t_0) \times \mathfrak{p}(s',t'|s_0,t_0)\,ds'. \end{aligned} \tag{8.99}$$

If the process is Markovian, $\mathfrak{p}(s,t|s',t';s_0,t_0) = \mathfrak{p}(s,t|s',t')$, and we have

$$\mathfrak{p}(s,t;s_0,t_0) = \int \mathfrak{p}(s,t|s',t') \times \mathfrak{p}(s',t'|s_0,t_0)\,ds', \tag{8.100}$$

where the terms are only starting and ending state dependent, that is, maintain locality. This equation is the Kolmogorov equation. Like other integrative relationships, this can also be written in a summation through matrix form for discrete processes. It factorizes the description. In the probability of ending up at (s,t) starting at (s_0,t_0), consider the role of the intermediate (s',t'), with $|t-t'=\delta t| \ll |t - t_0|$; that is, a very short duration final hop. For small δt, $\mathfrak{p}(s,t|s',t) = \delta t \approx \delta(s - s') + \mathcal{O}(\delta t)$, the process is stationary. $\mathcal{O}(\delta t)$ is due to transitions out of s' except those to s, and the transitions into s. Let $s_i \mapsto s_j$ in time δt be written as a scattering term $S(s_j|s_i)\delta t$; then,

$$\mathfrak{p}(s,t|s',t-\delta t) = \left[1 - \delta t \int S(s''|s')\,ds''\right] \delta(s''|s') + S(s|s')\,\delta t. \tag{8.101}$$

Substituting Equation 8.101 into Equation 8.100, with $\delta t \to 0$, we have

$$\frac{\partial}{\partial t}\mathfrak{p}(s,t) = \int \left[S(s|s')\mathfrak{p}(s',t) - S(s'|s)\mathfrak{p}(s,t)\right] ds', \tag{8.102}$$

a master equation that is very similar, together with related correspondences, to the scattering equation forms of Equations 8.34 and 8.39.

This master equation—as did the scattering equations—has embedded in it the principle of detailed balance. At large time scales, time-independent equilibrium $\mathfrak{p}_0(s)$ arises as a ratio of the transition rates, that is, $S(s|s')/S(s') = \mathfrak{p}_0(s)/\mathfrak{p}_0(s')$.

8.12 *Drude equation*

A discussion of carrier transport in matter will not be complete without some remarks on the Drude equation—an equation circa 1900—that is a projection of the classical Newtonian form of kinetics in the solid as a bouncing motion among impediments, so the form

$$\frac{d}{dt}\langle \mathbf{p} \rangle = -e\boldsymbol{\mathcal{E}} - \frac{\langle \mathbf{p} \rangle}{\tau}, \tag{8.103}$$

with a current $\mathbf{J} = (ne^2\tau/m)\boldsymbol{\mathcal{E}}$. This current expresses semi-heuristically Ohm's law. Paul Drude wrote this equation for metals—where there are lots of electrons that could be free to move between the impediments in the medium—and it is a form that precedes the quantum formulations. It worked for metals and gained traction, with Sommerfeld later on deploying it in his hybrid formulations. But it is an equation that seems to work only because a number of errors largely balance out for some of the problems of interest. Metals are not describable as a gas of classical particles. A large number of states in the conduction band are filled, and transport-related interactions and scattering are largely localized to the Fermi surface. What happens around the Fermi energy prevails. These are but two of its serious problems. That electrons carry along energy became a source of serious discrepancy laid bare in the thermal capacity of metals. These shortcomings are very stark. It is often amusing to see the Drude thinking still being applied in modern problems where quantum-mechanical interactions are crucial; examples are graphene and optoelectronics. One has to be really careful when resorting to it lest some imbalance in errors crops up and is lost in the quantitative details.

Albert Einstein, before the fame and while still a student transitioning to patent examiner and looking for a faculty position, strongly disliked Drude's cut-and-paste formulation. One can see in this subject his interest in exploring Brownian motion as well as thermoelectricity, that is, the notion that particles also carried along energy with them, which was his first attempt at a doctorate degree with H. F. Weber at ETH Zürich. Einstein wrote to Paul Drude his criticism, the reply came, and Einstein wrote to his friend Jost Winteler, "To two pertinent objections which I raised about one of his theories and which demonstrate a direct defect in his conclusions, he responds by pointing out that another (infallible) colleague of his shares his opinion". Einstein to Jost Winteler, July 8, 1901; quoted in A. Douglas Stone, "Einstein and the quantum: The quest of the valiant Swabian," Princeton, ISBN 978-0-691-16856-2 (2013), 44. By the way, Einstein's photoelectric effect equation too ignores much that happens by way of scattering and interactions in the surface region, where the photon is interacting and dislodging an electron out of the crystal. The electron has to make its way to the surface and out.

This completes our description of the evolution of states. We are now left with understanding the physical properties of the transport under stimulus simultaneously present from different forms and the details of the transitions in the semiconductors. This we tackle in the next set of chapters.

8.13 *Summary*

THIS CHAPTER IS ONE where we make our first foray into describing time evolution—dynamic changes—under perturbation in a solid as one burrows in from a classical description to a quantum description due to the change in the length scale of the problem, with the correspondence principle expectation intact. Quantum-mechanical and statistical-mechanical constraints all enter, so the chapter spanned a vanishing scattering view in quantum form, which is the analog of a projectile ballistic motion, the classical Liouville description of particles and their ensembles, the evolution of the classical to the quantum Liouville description, the Fokker-Planck equation and on through a Boltzmann transport semi-classical description to the nature of scattering and its modeling. This latter part let us take a broader view of Brownian motion, randomness, stochasticity and how these are dealt with in Markov processes and through the Kolmogorov equation. Between these ends, we made our first explorations of scattering processes and their quantum and semi-classical descriptions. So, it is a chapter tackling cause and chance in the classical-quantum interregnum.

Our discussion of quantum transport, absent scattering, showed a quantum conductance of $2e^2/h$, with two electrons (up and down) occupying a channel. Open and close such channels with a change in the energy slit controlling the transport, and the conductance will change proportional to the number of channels opened or closed. This is the equivalent of modes of electromagnetics that give rise to a corresponding wave impedance. This led us to a view for the mesoscale—a scale where the quantum manifests itself under limited scattering, and through the other quantum properties of the electron particle—from the one end of the spectrum of no scattering and quantum behavior.

In the classical description of a particle or ensembles, two canonical coordinates—position and momentum in the Hamiltonian description—suffice to describe the evolution. The phase space in this multi-dimensional space composed of the two canonic coordinates for each of the particles is incompressible. The Liouville equation describes the evolution of this ensemble, and if one knows the Hamiltonian, all the time dependences of interaction that are not in the Hamiltonian, such as when more than one set of particles are involved, and the state at some initial condition, then the evolution is completely describable in forward time and backward time. Quantum mechanics through uncertainty prescribes a spread, so the evolution of states has spreads. The quantum version of the Liouville equation describes this spread, and it brings in probabilities in our description because quantum mechanics is a statistical theory. The quantum Liouville equation did away with incompressibility in phase space.

Writing the evolution in terms of probabilities is a natural progression when incorporating cause and chance into the evolutionary discussion. The quantum Liouville equation in terms of the expectations on an observable $\langle \mathcal{A} \rangle$ is one form. The Fokker-Planck equation summarizing drift and diffusion over a distribution is one such form. The Langevin equation is another such master equation, where fast and slow forces are assimilated together in a particle response so that one can see the connection between drift and diffusion in response to forces, and their connections to correlations, which too are connected to the variety of moments, one of which is thermal noise. Probabilities, memory and history are pulled together in the discussion of Markov process and the Kolmogorov equation, where the evolution of the state under scattering—with randomization and without randomization—can be brought together.

The Boltzmann transport equation is the form of this evolution of state—captured through the distribution function—that we concentrated on. We could interpret changes in real space (position canonic coordinate) and momentum space (wavevector $\mathbf{k}$ representing the crystal momentum $\hbar\mathbf{k}$ canonic coordinate). Scattering representing changes in the momentum of particles constituting the distribution could be incorporated, with the scattering specific to the various interactions that are simultaneously present in the crystal. This is the general form that can be implemented computationally. But we found it very meaningful, particularly to gain intuition, to employ a relaxation time approximation where a single momentum—the canonic coordinate—relaxation time constant could be used. We noted conditions under which this approximation would be quite valid—elastic scattering, angular scattering and others—where the linear evolution of the distribution function acquired very specific meanings, such as mobility and diffusion. These mobilities and diffusions are precisely what Einstein postulated for Brownian motion, which is observed for small particles in fluids. These are connected to fluctuation-dissipation, noise and other points of discussion of the chapter. The Boltzmann transport equation also showed us how the moments of the equation give rise to an understanding of particle continuity, current continuity, energy continuity and other higher moments, all of which are representative of the evolution of probabilities and the consequences of correlations.

The Boltzmann transport equation applies to single particles as well as their collection, such as in the distribution function. So, the equation also works to show how transport happens without scattering, so with a quantum conductance of $2e^2/h$, since the bandstructure characteristics can be embedded in it, as it does in the presence of scattering. This Boltzmann transport equation therefore will be one of our basic tools for describing the behavior of semiconductors under stimulus in the following chapters.

8.14 *Concluding remarks and bibliographic notes*

THIS CHAPTER WAS OUR WAY OF SETTING UP the equations of evolution as collections of particles undergo change under stimulus with interactions within and with the environment. Brownian motion, as an example of fluctuation-dissipation, is a classic example of this evolution, where mobility and diffusivity are related and with the fast process of scattering providing the fluctuation and dissipation in the force field. Ensembles treated as statistical distributions undergoing interactions subject to different probabilities can be viewed through probabilistic equations. And this is what the Fokker-Planck, Langevin, Markov chain, Kolmogorov, Liouville, quantum Liouville and Boltzmann forms that we concentrated on represent.

To get an appreciation for this subject and its history, Brownian motion provides a very important starting point. A paper by Kramers[1] is a very rich and generous starting point. Essential concepts of particle distribution dynamics come together here with chemical reactions, where the interaction is represented through two potential wells between which there is the barrier energy related to the chemical reaction. This is very different from the evolution of the distributions that we are interested in, but one can immediately see the fundamental similarities. At the other end, a comprehensive development of mesoscopic transport using quantum Landauer approaches is provided in the book by Nazarov and Blanter[2]. Another good text is by Datta[3].

The Fokker-Planck, Langevin, Markov chain and Kolmagorov equations, as ways to view stochastic processes, are also applied in many branches of sciences. One either studies individual trajectories or studies the evolution of the probability in position and time. As such, there are a large number of books on the subject. Two books, both by Friedman[4,5], provide an advanced quantitative introduction to the subject. These books are very mathematical. An advanced approach, and a standard for physics-centric books, is the text by Altland and Simons[6]. Its chapter on classical nonequilibrium builds the thread from Fokker-Planck/Langevin through Boltzmann and Einstein to Markov and Kolomogorov majestically with completeness.

The Liouville equations and the Boltzmann transport equations appear in nearly all the semiconductor physics texts, with slightly different twists. Jacoboni's[7] is one of the most modern forms, and it integrates the development of these equations physically and analytically in a very appealing way. A comprehensive text[8] by Markowich et al. discusses the kinetic theories of transport, including Boltzmann, quantum Liouville and other transport phenomena such as tunneling, that we did not discuss.

[1] H. A. Kramers, "Brownian motion in a field of force and the diffusion model of chemical reactions," Physica, **VII**, 284–304 (1940)

[2] Y. N. Nazarov and Y. M. Blanter, "Quantum transport," Cambridge, ISBN 13 978-0-521-83246-5 (2009)

[3] S. Datta, "Electronic transport in mesoscopic systems," Cambridge, ISBN 0-521-41604-3 (1999)

[4] A. Friedman, "Stochastic differential equations," **1**, Academic, ISBN 9781483217871 (1975)

[5] A. Friedman, "Stochastic differential equations," **2**, Academic, ISBN 9781483217888 (1975)

[6] A. Altland and B. Simons, "Condensed matter field theory," Cambridge, ISBN 13 978-0-521-76975-4 (2010)

[7] C. Jacoboni, "Theory of electron transport in semiconductors," Springer, ISBN 978-3-642-10585-2 (2010)

[8] P. A. Markowich, C. A., Ringhofer and C. Schmeiser, "Semiconductor equations," Springer-Verlag, ISBN 0-387-82157-0 (1990)

8.15 Exercises

1. Quantum conductance—$2e^2/h$ in units of S for *Siemens*, or $2e/h$ in units of S per electron charge—appeared when there was unscattered transport in an occupiable channel with a spin up

and a spin down electron. Can you give a simple explanation of why this conductance is a constant irrespective of the length of the channel? **[S]**

2. What is a mixed state? Are there classical mixed states? If yes, what does that physically mean? **[S]**

3. In thermal equilibrium, detailed balance holds. For a conductor, one implication is that the current that one will measure in the shorting wire that assures that there exists no external electromagnetic interaction is also zero—no net flow of particles or energy. How can one tell that there is motion of electrons in the conductor then? Or are they standing still? **[S]**

4. Away from equilibrium, why is the change in distribution function an odd function of momentum? Is there an argument more general than our argument using current? **[S]**

5. Should a perfect crystal (i.e., one with no defects, identical atoms at the lattice sites, etc.) have perfect thermal conductivity? Phonons don't disappear and the wavepacket consists of the linear combination of eigenstates, so it seems phonon modes in **q**-space persist and would give thermal superconductivity. Provide a short argument for or against, please. **[S]**

6. Fick's first and second laws for diffusion, written for a one-dimensional system, are

$$J = -\mathcal{D}\frac{\partial c}{\partial z}, \text{ and } \frac{\partial c}{\partial t} = \frac{\partial}{\partial z}\left(\mathcal{D}\frac{\partial c}{\partial z}\right).$$

J here is a flux, $\mathcal{D} = (1/2)\lambda^2\nu$, where $\mathcal{D}$ is a diffusion coefficient, λ is a "jump" length scale, ν a frequency of the jumps, and t is time. If diffusion is a random walk, write down an equation for this jump frequency and suggest a meaning. **[S]**

7. In many ways, the Boltzmann transport equation is a remarkable result. It predicts irreversibility. The Liouville equation from which it came, however, is reversible. Speculate on what may be going on for this to come about. **[S]**

8. Assuming Boltzmann and relaxation time approximations, and a small departure from thermal equilibrium, show, using the Boltzmann transport approach, that a gradient of carrier density leads to diffusion whose diffusion coefficient is $\lambda v_\theta/3$, where λ is the mean free path. **[S]**

9. The third moment equation from the Boltzmann transport equation, the energy conservation equation, has as its last term $\nabla_{\mathbf{r}} \cdot \langle \dot{\mathbf{r}} \cdot \hbar^2\mathbf{k}^2/2m^* \rangle$. What exactly is this saying? **[S]**

10. Why is the relaxation time $\tau(\mathbf{k})$, which we have also written as $\tau_{\mathbf{k}}$, a function of $\mathbf{k}$? **[S]**

11. Suppose the thermal equilibrium distribution, f_0, is approximated by a Boltzmann distribution function, that is,

$$f_{\mathbf{k}}^0 = C\exp\left(-\frac{E_{\mathbf{k}}}{k_B T}\right).$$

For free carriers located in a spherical band with effective mass m^*, the electron energy is given by $E = (1/2)m^* v_{\mathbf{k}}^2$. The resultant carrier distribution is the Maxwell-Boltzmann distribution. Show that, in the presence of a weak external electric field $\boldsymbol{\mathcal{E}}$, the distribution function is approximated by

$$f_{\mathbf{k}} = C\exp\left(-\frac{m^*|\mathbf{v}_{\mathbf{k}} + \mathbf{v}_d|^2}{2k_B T}\right),$$

where $\mathbf{v}_d$ is the drift velocity. What is its magnitude in terms of parameters that characterize the semiconductor? The interpretation of this result is that the external field causes the carrier velocities to increase uniformly by an amount equal to $\mathbf{v}_d$ while leaving the distribution function unchanged. The resulting distribution function is therefore known as a drifted (or displaced) Maxwell-Boltzmann distribution function. **[S]**

12. Using the Maxwell-Boltzmann distribution function at equilibrium, show that the average kinetic energy of an electron is given by $3k_B T/2$. Show that, for the displaced Maxwell-Boltzmann distribution, the average kinetic energy is given by

$$\frac{1}{2}m^* v_d^2 + \frac{3}{2}k_B T.$$ **[S]**

13. Apply the Boltzmann transport equation in a relaxation time approximation for a semiconductor where both electrons and holes are present, and show that, in homogeneous conditions, the Boltzmann gas of electrons and holes can be viewed as a charge gas with a conductivity of $\sigma = e(n\mu_n + p\mu_p)$, where the μ_n and μ_p are drift mobilities given as

$$\mu_n = \frac{e}{m_n^*}\frac{\langle v_n^2 \tau_n(v_n)\rangle}{\langle v_n^2\rangle}.$$

Here, subscripting with n identifies the characteristic with that of an electron. An analogous expression holds for holes. $\tau_n(v_n)$ and $\tau_p(v_p)$ are the relaxation times for electrons and holes, respectively. **[S]**

14. Prove that

$$\frac{d}{dt}\langle\psi|\mathscr{A}|\psi\rangle = \frac{i}{\hbar}\langle\psi|[\mathscr{H},\mathscr{A}]|\psi\rangle + \langle\psi|\frac{\partial\mathscr{A}}{\partial t}|\psi\rangle,$$

where $|\psi\rangle$ satisfies

$$-\frac{\hbar}{i}\frac{\partial|\psi\rangle}{\partial t} = \mathscr{H}|\psi\rangle.$$

If $\mathscr{A}$ has no explicit dependence on time, then $\partial\mathscr{A}/\partial t = 0$. Employing

$$\mathscr{H} = \frac{p^2}{2m} + V(z,t),$$

show that

$$\frac{d\langle z\rangle}{dt} = \frac{\langle p\rangle}{m}, \quad \text{and} \quad \frac{d\langle p\rangle}{dt} = \langle -\frac{\partial V}{\partial z}\rangle,$$

using $\langle\mathscr{A}\rangle \equiv \langle\psi|\mathscr{A}|\psi\rangle$. The expectation values follow classical equations of motion. This is Ehrenfest's theorem. [S]

9
Scattering-constrained dynamics

THE UTILITY OF SEMICONDUCTORS comes from their ability to efficiently transform energy forms. This effectiveness is particularly suited to the information enterprise through the processing, communicating and sensing that these transformations enable. Optoelectronic conversion such as in solar cells, digital cameras, lasers, transistor-based computing and communication, environment and health care sensors, including those that perform ultrasensitive magnetic probing of the brain, et cetera, are all based on the transformations of energy, which itself is based on the flow of particles, electromagnetic interactions, with robustness, reliability and sensitivity, while being perturbed by the environment. An example of a perturbation to the particles' evolution is scattering. By this, when considering electrons and holes, we mean all interactions that change the charge carriers' states, be they transitions between bands, within bands, with photons or with phonons. Understanding the flow and the processes under these different perturbations with the specifics of semiconductors taken into account is important and physically quite rich.

In this chapter, we will develop the analytic description of this particle dynamics, while the particles undergo scattering, that is, interact with the surroundings under various influences of the electrical, magnetic, optical and thermal variety. This will influence the transport and energy conversion of various forms that can be viewed through the distribution function that describes particle ensemble in the canonical coordinates developed in Chapter 8. We will use the Boltzmann transport description of this dynamics where multiple influences may be simultaneously at work.

Semiconductor Physics: Principles, Theory and Nanoscale. Sandip Tiwari.
 Published 2020 by Oxford University Press. DOI: 10.1093/oso/9780198759867.001.0001

9.1 *Thermal equilibrium*

THERMAL EQUILIBRIUM IS MUCH EASIER TO DESCRIBE PHYSICALLY. A closed system where energy exchange is allowed with its surroundings, with no external energy stimulus except thermal through the existence at temperature T, achieves this state, a steady state, where the entropy is maximum and detailed balance—in detail, for every process, an equal and opposite rate—holds, and if one were to disconnect the system from its surroundings, there the system would remain. The system is in the universe's cavity, immersed in a blackbody radiation bath at temperature T, with no other stimulus. The particle distribution function remains the same, and if one were to make a measurement for any physical observable, it would remain the same. This is to say that an operator operating on this distribution function gives us the observable, physically and mathematically.

Boltzmann, through the phase-space description, provided a statistical foundation. Interactions within the system increase the entropy. Boltzmann's H-theorem sets the analytic prescription for how this happens. For a collection of independent classical particles, Boltzmann's H-factor, which gives the name to Boltzmann's H-theorem, is given by

Boltzmann had considerable difficulty in convincing others of his statistical views, even though Maxwell and Gibbs too were also questioning the pervasive dogma of determinism. This period was the end of the 19th century. While Euler's collected works may be the largest, Boltzmann's two-volume *Vorlesungen über Gastheorie* is the sharpest small collection.

$$H(t) = \int_0^\infty f(E,t)\left[\ln\left(\frac{f(E,t)}{E^{1/2}}\right) - 1\right] dE, \tag{9.1}$$

where $f(E,t)$ is the particle energy distribution in time. In an isolated system, this H-factor is at a minimum, when the particles obey the Maxwell-Boltzmann distribution. For any other distribution, for the same total kinetic energy E, this H-factor will be higher. When scattering is allowed, any starting distribution of classical particles approaches the minimum of H and the Maxwell-Boltzmann distribution.

This H-theorem is a statement of the inevitability of irreversibility. It is a theorem that exemplifies the 2nd law of thermodynamics, which itself is really a postulate.

From reversible processes comes the irreversibility.

A small change from this thermal equilibrium, consequently, the changes in the distribution, can be managed and understood as a perturbation from this thermal equilibrium. So, we may apply electric fields or magnetic fields, shine light or impose temperature gradients, and, in principle, the change can be found, and one may determine the characteristics of the system arising in the transport and the interactions.

9.2 *Transport in generalized form*

A SMALL CHANGE FROM THERMAL EQUILIBRIUM, under electric, magnetic, optical or thermal stimulation, can be viewed through the first order terms of linear change. Where important, one may include even higher order terms selectively. The macroscopic properties that these distribution functions change correspond to what can be viewed directly through macroscopic expectations in Onsager's approach. This we will tackle separately in Chapter 15. If, on the other hand, the changes are significant, we will tackle these too later through Onsager's approach. Thermal equilibrium means for us no net change in macroscopic observables in time. There must be no net change in the observable's expectation, even if there may be fluctuations in measurements due to fluctuations such as noise. This must require, in detail, a balance so that the variety of specific interactions do not become a conduit to a net transformation. The net rate of any detailed state-to-state transition must vanish in thermal equilibrium. If there is a transition rate from a state $|i\rangle$ to state $|f\rangle$, then there must be an equal rate for transition from state $|f\rangle$ to state $|i\rangle$. In situations of interest to us in the Boltzmann transport picture, for the particles in $|\mathbf{k}\rangle$, with an occupation given by $f(\mathbf{k})$, and in $|\mathbf{k}'\rangle$, with an occupation given by $f(\mathbf{k}')$, $f(\mathbf{k})S_{\mathbf{k}\mathbf{k}'}[1-f(\mathbf{k}')]$ must balance with $f(\mathbf{k}')S_{\mathbf{k}'\mathbf{k}}[1-f(\mathbf{k})]$ in thermal equilibrium. Only then can no net transport in charge or energy, et cetera, can be ensured in thermal equilibrium. If we now stimulate this system, from many sources simultaneously, we will need to find the nonequilibrium distribution function in order to determine the characteristics arising from the transport and interactions of carriers engendered by the stimulations. In the linearized form, in principle, this determination will follow a quite simple algorithm based on the rate of change, even if the nature of excitation by polar and axial vector forces has divergence and curls as the operating mechanism, which will complicate the forms of the results.

In writing the Boltzmann transport equation as

$$\begin{aligned}\frac{df}{dt} &= -\dot{\mathbf{k}}\cdot\nabla_{\mathbf{k}}f - \dot{\mathbf{r}}\cdot\nabla_{\mathbf{r}}f + \left.\frac{\partial f}{\partial t}\right|_{scat} \\ &= -\dot{\mathbf{k}}\cdot\nabla_{\mathbf{k}}f - \mathbf{v}\cdot\nabla_{\mathbf{r}}f + \left.\frac{\partial f}{\partial t}\right|_{scat},\end{aligned} \tag{9.2}$$

the first term arises in momentum-space changes, and the second in those of real space. If there exists a force $\mathbf{F}$, due to the stimulus applied, we can rewrite the evolution of the distribution function as

$$\frac{df}{dt} = -\frac{1}{\hbar}\mathbf{F}\cdot\nabla_{\mathbf{k}} f - \mathbf{v}\cdot\nabla_{\mathbf{r}} f + \left.\frac{\partial f}{\partial t}\right|_{scat}. \tag{9.3}$$

While we will employ quantum-mechanical arguments in the quantum-specific subsystem, where it is critical to deploy them, for example, in the scattering itself, we will keep in mind that *the observations are in a classical environment*. Following our arguments, therefore, for us, here $df/dt = 0$, as a direct consequence of Liouville's equation and phase-space incompressibility. We also adopt relaxation time approximation to make solutions analytic; to some limited level, tractable; and, to a larger extent, useful for gaining intuition. $df/dt = 0$, in Equation 9.3, with relaxation time approximation, leads to

$$\frac{df}{dt} = -\frac{1}{\hbar}\mathbf{F}\cdot\nabla_{\mathbf{k}} f - \mathbf{v}\cdot\nabla_{\mathbf{r}} f - \frac{f - f_0}{\tau_{\mathbf{k}}} = 0, \tag{9.4}$$

whose solution, away from the equilibrium, has the form

$$\begin{aligned} f &= f_0 - \frac{\tau_{\mathbf{k}}}{\hbar}\mathbf{F}\cdot\nabla_{\mathbf{k}} f - \tau_{\mathbf{k}}\mathbf{v}\cdot\nabla_{\mathbf{r}} f \\ &= f_0 - \frac{\tau_{\mathbf{k}}}{\hbar}\mathbf{F}\cdot\frac{\partial f}{\partial E}\nabla_{\mathbf{k}} E - \tau_{\mathbf{k}}\frac{1}{\hbar}\nabla_{\mathbf{k}} E\cdot\nabla_{\mathbf{r}} f \\ &= f_0 - \frac{\tau_{\mathbf{k}}}{\hbar}\nabla_{\mathbf{k}} E\cdot\left(\frac{\partial f}{\partial E}\mathbf{F} + \nabla_{\mathbf{r}} f\right) \\ &= f_0 - \tau_{\mathbf{k}}\mathbf{v}\cdot\left(\frac{\partial f}{\partial E}\mathbf{F} + \nabla_{\mathbf{r}} f\right). \end{aligned} \tag{9.5}$$

These last two equation forms are meaningful. A nonequilibrium in the distribution occurs because of the stimulation and is also mediated by the scattering process, a flow in through the group velocity, and changes arising in two terms contributing to a flux: the force that is modulated through an energy dependence (the **k**-space effect with energy marginality), and the real space effect of the concentration gradient. The quantum-mechanical connection of states through the bandstructure is built in. The $\tau_{\mathbf{k}}$ as a prefactor shows the relaxation effect of scattering, and the velocity **v** shows the linear dependence on the flow of carriers under the bandstructure constraints. And this relaxing and moving in works with the consequences of forces in **k**-space that cause changes in energy and the consequences of real space concentration changes.

But Equation 9.5 is a nonlinear equation. The simplest example of a force of interest is in the form of an electric field. Let there be no other force, and let the material be homogeneous with a vanishing concentration gradient that may be ignored. Then, the distribution function, in the presence of the electric field $\boldsymbol{\mathcal{E}}$, is

$$f = f_0 + q\tau_{\mathbf{k}}\frac{\partial f}{\partial E}\mathbf{v}\cdot\boldsymbol{\mathcal{E}}. \tag{9.6}$$

For electrons, $q = -e$ here. For holes, $q = +e$. We can linearize it. Changes in distribution function are small. So, write $\partial f/\partial E \approx \partial f_0/\partial E$, which says that while the occupation at any state E did change, it was a small change (a redrawn Figure 8.7 with $E \propto \mathbf{k}^2$ is a small change for any $\mathbf{k}^2$ contour), and it pretty much looks similar in shape—its marginal propensity is maintained—even if the magnitude of the distribution function is different. So, now

$$f = f_0 + q\tau_{\mathbf{k}} \frac{\partial f_0}{\partial E} \mathbf{v} \cdot \boldsymbol{\mathcal{E}}$$

$$\text{or the form } f = f_0 + \frac{\partial f_0}{\partial E} \mathbf{v} \cdot q\tau_{\mathbf{k}} \boldsymbol{\mathcal{E}}, \tag{9.7}$$

which is explicitly linear in an electric field. It also tells us that marginal energy causing a change in the distribution function is the product term $\mathbf{v} \cdot q\tau_{\mathbf{k}}\boldsymbol{\mathcal{E}}$. $q\boldsymbol{\mathcal{E}}$ is a force, and $q\tau_{\mathbf{k}}\boldsymbol{\mathcal{E}}$ is the impulse over the relevant time scale of $\tau_{\mathbf{k}}$ between scattering events. The distribution function evolves in the field under the influence of these impulses appearing on the $\tau_{\mathbf{k}}$ time scales. The spatial extent of this force acting in time is determined by the velocity $\mathbf{v}$ with which the flow occurs. Equivalently, the force $q\boldsymbol{\mathcal{E}}$ is acting over the distance $\mathbf{v}\tau_{\mathbf{k}}$ and imparting the energy that results in the change in the distribution function.

So, we have found a linearized solution, under these "fast" impulse events, with the assumption of relaxation time and the distribution function not distorting excessively as it changes.

Now, we generalize by using our learning from this. Let there be all the three stimuli of interest to us—an electric field, a magnetic field and thermal gradients—present simultaneously. The starting form of Equation 9.5 implies

$$f = f_0 - \frac{\tau_{\mathbf{k}}}{\hbar} \left[q(\boldsymbol{\mathcal{E}} + \mathbf{v} \times \mathbf{B}) \right] \cdot \nabla_{\mathbf{k}} f - \tau_{\mathbf{k}} \mathbf{v} \cdot \nabla_{\mathbf{r}} f. \tag{9.8}$$

With Equation 9.8, the force has changed from just $q\boldsymbol{\mathcal{E}}$, and we are also interested in incorporating the effects of the concentration changes from the $\nabla_{\mathbf{r}} f$ term. Thermal effects too will appear in here since we are going to employ a general formalism. Our ansatz is that the linearized form of the solution remains as the lowest-order effect; that is, that the solution is of the form

$$f = f_0 + \frac{\partial f_0}{\partial E} \mathbf{v} \cdot \mathbf{G}, \tag{9.9}$$

where $\mathbf{G}$ is a generalized impulse that we need to determine. Take the dot product term of the square-bracketed force and gradient of the distribution in Equation 9.8, and substitute this ansatz:

$$q\left(\mathcal{E}+\mathbf{v}\times\mathbf{B}\right)\cdot\nabla_{\mathbf{k}}f=q\left[\mathcal{E}\cdot\nabla_{\mathbf{k}}f_0+\left(\mathbf{v}\times\mathbf{B}\right)\cdot\nabla_{\mathbf{k}}f_0\right.$$
$$\left.+\,\mathcal{E}\cdot\nabla_{\mathbf{k}}\left(\frac{\partial f_0}{\partial E}\mathbf{v}\cdot\mathbf{G}\right)+\left(\mathbf{v}\times\mathbf{B}\right)\cdot\nabla_{\mathbf{k}}\left(\frac{\partial f_0}{\partial E}\mathbf{v}\cdot\mathbf{G}\right)\right]. \tag{9.10}$$

In the first term, $\nabla_{\mathbf{k}}f_0=(\partial f_0/\partial E)\nabla_{\mathbf{k}}E=(\partial f_0/\partial E)\hbar\mathbf{v}$. The second term too contains this same gradient. The impulse $\mathbf{G}$ in the third term is also a function of the electric field, so the dot product is second order in the electric field. With the system in small perturbation, we neglect the second order term. For the fourth term,

$$\begin{aligned}
&\left(\mathbf{v}\times\mathbf{B}\right)\cdot\nabla_{\mathbf{k}}\left(\frac{\partial f_0}{\partial E}\mathbf{v}\cdot\mathbf{G}\right)\\
&=\frac{\partial f_0}{\partial E}\left(\mathbf{v}\times\mathbf{B}\right)\cdot\nabla_{\mathbf{k}}\left(\mathbf{v}\cdot\mathbf{G}\right)+\left(\mathbf{v}\cdot\mathbf{G}\right)\left(\mathbf{v}\times\mathbf{B}\right)\cdot\nabla_{\mathbf{k}}\frac{\partial f_0}{\partial E}\\
&=\frac{\partial f_0}{\partial E}\left(\mathbf{v}\times\mathbf{B}\right)\cdot\nabla_{\mathbf{k}}\left(\mathbf{v}\cdot\mathbf{G}\right)
\end{aligned} \tag{9.11}$$

because $\nabla_{\mathbf{k}}(\partial f_0/\partial E)=(\partial^2 f_0/\partial E^2)\nabla_{\mathbf{k}}E=(\partial^2 f_0/\partial E^2)\hbar\mathbf{v}$, which results in the second term of this expansion vanishing, since $\mathbf{v}\times\mathbf{B}$ with which it has a dot product is orthogonal to it. So, we have

$$\begin{aligned}
q\left(\mathcal{E}+\mathbf{v}\times\mathbf{B}\right)\cdot\nabla_{\mathbf{k}}f\approx{}& q\hbar\frac{\partial f_0}{\partial E}\mathbf{v}\cdot\mathcal{E}\\
&+\frac{q}{\hbar}\frac{\partial f_0}{\partial E}\mathbf{v}\cdot\left[\mathbf{B}\times\left(\mathbf{G}\cdot\nabla_{\mathbf{k}}\right)\nabla_{\mathbf{k}}E\right].
\end{aligned} \tag{9.12}$$

Now, we tackle the last term of Equation 9.8,

$$\mathbf{v}\cdot\nabla_{\mathbf{r}}f=\mathbf{v}\cdot\nabla_{\mathbf{r}}f_0+\mathbf{v}\cdot\nabla_{\mathbf{r}}\left(\frac{\partial f_0}{\partial E}\mathbf{v}\cdot\mathbf{G}\right), \tag{9.13}$$

where the second term is a spatial dependence in the impulse effect, while the first one is the primary spatial dependence of the distribution function. The second term is a spatial dependence of the perturbation, is of higher order and can be neglected. So,

$$\begin{aligned}
\mathbf{v}\cdot\nabla_{\mathbf{r}}f&\approx\mathbf{v}\cdot\nabla_{\mathbf{r}}f_0\\
&=\frac{\partial f_0}{\partial\left[(E-\mu)/k_BT\right]}\mathbf{v}\cdot\nabla_{\mathbf{r}}\left(\frac{E-\mu}{k_BT}\right)\\
&=k_BT\frac{\partial f_0}{\partial E}\mathbf{v}\cdot\nabla_{\mathbf{r}}\left(\frac{E-\mu}{k_BT}\right),
\end{aligned} \tag{9.14}$$

where μ is the chemical potential, that is, the non-electrical part of electrochemical potential ($E_{qF}=\mu-q\psi$, where ψ is the electrostatic potential). We pool these approximations together, and compare them to the impulse-based ansatz to look for further simplifications:

$$\frac{1}{\tau_{\mathbf{k}}}\frac{\partial f_0}{\partial E}\mathbf{v}\cdot\mathbf{G} = q\frac{\partial f_0}{\partial E}\mathbf{v}\cdot\boldsymbol{\mathcal{E}} + \frac{q}{\hbar^2}\frac{\partial f_0}{\partial E}\mathbf{v}\cdot[\mathbf{B}\times(\mathbf{G}\cdot\nabla_{\mathbf{k}})\nabla_{\mathbf{k}}E] - k_BT\frac{\partial f_0}{\partial E}\mathbf{v}\cdot\nabla_{\mathbf{r}}\left(\frac{E-\mu}{k_BT}\right) \tag{9.15}$$

has the $\partial f_0/\partial E$ multiplier in common; therefore,

$$\begin{aligned}\frac{\mathbf{G}}{\tau_{\mathbf{k}}} &= q\boldsymbol{\mathcal{E}} + \frac{q}{\hbar^2}[\mathbf{B}\times(\mathbf{G}\cdot\nabla_{\mathbf{k}})\nabla_{\mathbf{k}}E] - k_BT\nabla_{\mathbf{r}}\left(\frac{E-\mu}{k_BT}\right)\\ &= q\mathcal{F} + \frac{q}{\hbar^2}[\mathbf{B}\times(\mathbf{G}\cdot\nabla_{\mathbf{k}})\nabla_{\mathbf{k}}E],\\ \text{where } q\mathcal{F} &= q\boldsymbol{\mathcal{E}} - k_BT\nabla_{\mathbf{r}}\left(\frac{E-\mu}{k_BT}\right)\end{aligned} \tag{9.16}$$

is what we will call a *force* ($\mathcal{F}$ is an electrothermal field), where the electrical force and the concentration and thermal gradient forces can be lumped. These forces arise in causes that are polar in nature. In the laboratory frame of reference, one can view the electric field as tying electric charges, the entropic concentration forces tie the chemical-centric concentration of species, and the thermal forces tie differences in temperature. All these can be added vectorially. The magnetic field, on the other hand, causes a rather convoluted effect through the cross-product of an axial vector. It also brings in the complication of the need to determine the impulse **G** self-consistently, since it is in an implicit equation.

So, our solution for the general Boltzmann transport problem is through a generalized impulse function.

$$\mathbf{G} = q\tau_{\mathbf{k}}\mathcal{F} + \frac{q\tau_{\mathbf{k}}}{\hbar^2}[\mathbf{B}\times(\mathbf{G}\cdot\nabla_{\mathbf{k}})\nabla_{\mathbf{k}}E], \tag{9.17}$$

which represents electrical, magnetic and thermal stimuli, leading to a linearized small-perturbation change in the distribution function of

$$f = f_0 + \frac{\partial f_0}{\partial E}\mathbf{v}\cdot\mathbf{G}. \tag{9.18}$$

Calculating the entire impulse **G** from the implicit Equation 9.17 is the challenge to us. Through $\nabla_{\mathbf{k}}E$ as well as $(\mathbf{G}\cdot\nabla_{\mathbf{k}})$ operating on it, so through the bandstructure-defined allowed states, the propagational properties of these states and then the interaction of the generalized impulse of this propagational response, one must find it. The flux of energy is $\mathbf{v}\cdot\mathbf{G}$, and it has the primary effect on the change in the distribution function.

Although it is not quite explicit, we now have the algorithm for finding the response. One can apply it to the various circumstances one finds in a semiconductor. Many of these circumstances are interesting. In electronics, one may only have electric fields. This is the simplest problem, and Chapter 8 had examples of it together, simultaneously, with concentration gradients. The $\dot{\mathbf{k}}$ and the $\dot{\mathbf{r}}$ resolved those. But, even in electric devices, there may be temperature changes across it, since devices dissipate. This is an additional complication, which, again, the electrothermal force $q\boldsymbol{\mathcal{F}}$ will tackle through the linearized Boltzmann-distribution function representation.

Our first simplification toward analyticity comes from bandstructure. Carriers are largely at bandedges when the disturbance is limited from thermal equilibrium. They may be in isotropic, anisotropic or even multiple bands, as in the case of holes. The $E(\mathbf{k})$ approximation for this is subsumed, here written for the electron case, in

$$E = E_c + \frac{\hbar^2}{2}\mathbf{k}\cdot\frac{1}{\mathbb{M}^*}\cdot\mathbf{k}, \text{ where } \frac{1}{\mathbb{M}^*} = \begin{bmatrix} 1/m_1^* & 0 & 0 \\ 0 & 1/m_2^* & 0 \\ 0 & 0 & 1/m_3^* \end{bmatrix} \quad (9.19)$$

is the effective mass tensor. This implies

See Appendix J for an expanded discussion of the effective mass tensor.

$$(\mathbf{G}\cdot\nabla_{\mathbf{k}})\,\nabla_{\mathbf{k}}E = \hbar^2\frac{1}{\mathbb{M}^*}\cdot\mathbf{G} \;\;\therefore\;\; \mathbf{G} = q\tau_{\mathbf{k}}\boldsymbol{\mathcal{F}} + q\tau_{\mathbf{k}}\mathbf{B}\times\frac{1}{\mathbb{M}^*}\cdot\mathbf{G}. \quad (9.20)$$

The solution for the impulse is

$$\mathbf{G} = q\tau_{\mathbf{k}}\frac{\boldsymbol{\mathcal{F}} - q\tau_{\mathbf{k}}(1/\mathbb{M}^*)\cdot(\boldsymbol{\mathcal{F}}\times\mathbf{B}) + (q\tau_{\mathbf{k}})^2\det[1/\mathbb{M}^*]\boldsymbol{\mathcal{F}}\cdot\mathbf{B}[(1/\mathbb{M}^*)^{-1}\cdot\mathbf{B}]}{1 + (q\tau_{\mathbf{k}})^2\det[1/\mathbb{M}^*][(1/\mathbb{M}^*)^{-1}\cdot\mathbf{B}]\cdot\mathbf{B}}. \quad (9.21)$$

In the case of spherical constant energy surfaces, the distribution function is

$$f = f_0 + \frac{\partial f_0}{\partial E}q\tau_{\mathbf{k}}\mathbf{v}\cdot\left[\frac{\boldsymbol{\mathcal{F}} - (q\tau_{\mathbf{k}}/m^*)(\boldsymbol{\mathcal{F}}\times\mathbf{B}) + \left(q\tau_{\mathbf{k}}/m^*\right)^2(\boldsymbol{\mathcal{F}}\cdot\mathbf{B})\mathbf{B}}{1 + \left(q\tau_{\mathbf{k}}/m^*\right)^2\mathbf{B}\cdot\mathbf{B}}\right]. \quad (9.22)$$

This solution is instructive by showing the entangled relationships arising in the different fields—polar and axial—and their divergence and curl operation on the independent electron or hole. The denominator shows that while undergoing scattering during the duration of each impulse, the particles are also being forced into circular orbits by the magnetic flux. This is a cyclotron resonance term, where the carrier exhibits an angular frequency consequence of $\omega_c = q|\mathbf{B}|/m^*$ while undergoing scattering at the rate $1/\tau_{\mathbf{k}}$. The impulse's energetic consequences are being degraded by this path increase, and scattering, even as a field—**B**—does not provide energy, since the motion is perpendicular to the field. The first term in the numerator shows the direct polar electrothermal consequences. An electric field, a thermal field or a concentration field all behave similarly and co-act through exchanges between the energy forms. Electrical conductivity, thermal conductivity and electrothermal effects (Seebeck, Peltier and Thomson) are consequences that can be analyzed through the distribution function in the linearized lowest order approximation of the polar field's consequences. The second term in the numerator is the first order consequence of the interaction between the axial and the polar fields. Deviation from axial paths will lead to voltage, temperature and concentration consequences arising in the magnetic field. The Hall effect is one consequence showing the magnetic-electric interaction, but there are others, with thermal involvement (Ettinghausen, Nernst and Righi-Leduc). The third term in the numerator is the next order consequence in an electrothermal-magnetic field interaction that is a magnetoresistive effect. The distribution function, while explicitly showing the momentum relaxation time $\tau_{\mathbf{k}}$ representing interaction of states, also has a complexity that arises in the energy flow and impulse consequences that the second term represents. The time constants—relaxation times—for any parameter of interest under the different excitations will have to be evaluated under those circumstances. So, the various effects just mentioned will have complicated expectation relationships of the various flow consequences involved and, through that, dependences on $\tau_{\mathbf{k}}$ and energy.

Following on an earlier margin note, $\mathbf{J} = qn\langle\mathbf{v}\rangle$, with $\mathbf{v}$ subsuming the phase-space momentum and real space components, will have to be evaluated using this nonequilibrium f, and similarly for the energy flow density, $\mathbf{W} = n\langle w\mathbf{v}\rangle$. One can see in the product $w\mathbf{v}$ a verity of energy dependences in a relaxation time.

Consequences of light stimulation are incorporated into the Equation 9.22 relationship implicitly. Electromagnetic fields affect carrier transport through interaction. Lasers, because they are coherent, will bunch particles through electromagnetic interaction at the wavelength scale of the light in the medium. Light's interaction also creates particles, which is a change in the distribution function. The former will have to be modeled through the electrical and magnetic terms, which will not be an easy task. The latter—the creation of particles under small intensity and so small field consequences in the medium—directly affects the chemical potential. And because both carrier types may be created, one would have to resort to two distribution functions, and therefore two Boltzmann transport equations will have to be solved.

9.2.1 *Expectations and time constants off-equilibrium*

We start by resolving the expectations of parameters of interest—how to calculate them—and with the development of some general relationships of use when energy forms, exchange and the relaxation expectation relationships get enmeshed. First, take the velocity $\langle\mathbf{v}\rangle$ in our formulation. Let us look at the simplest of situations, $\mathbf{B} = 0$, T a constant, and the material homogeneous:

$$\langle \mathbf{v} \rangle = \frac{\int_{-\infty}^{\infty} \mathbf{v} f_0 \, d\mathbf{v} + q \int_{-\infty}^{\infty} \tau_{\mathbf{k}} (\partial f_0 / \partial E) \mathbf{v} (\mathbf{v} \cdot \boldsymbol{\mathcal{E}}) \, d\mathbf{v}}{\int_{-\infty}^{\infty} f_0 \, d\mathbf{v} + q \int_{-\infty}^{\infty} \tau_{\mathbf{k}} (\partial f_0 / \partial E) (\mathbf{v} \cdot \boldsymbol{\mathcal{E}}) \, d\mathbf{v}}$$

$$\approx \frac{q \int_{-\infty}^{\infty} \tau_{\mathbf{k}} (\partial f_0 / \partial E) \mathbf{v} (\mathbf{v} \cdot \boldsymbol{\mathcal{E}}) \, d\mathbf{v}}{\int_{-\infty}^{\infty} f_0 \, d\mathbf{v}}, \qquad (9.23)$$

since the expectation of velocity on the thermal equilibrium distribution vanishes, and the second term in the denominator is small under the small-perturbation assumption under which we have executed these derivations. The principle in determining any expectation is to employ the distribution function as the weighting function. *The distribution function is the statistical model of the probability for the various possibilities of $(\mathbf{r}, \mathbf{k})$ of the particles.* Again, take the simplest example of isotropic single band with effective mass m^*. Since $E - E_c = (1/2) m^* v^2$, the elemental velocity space volume is $|d\mathbf{v}| = 4\pi v^2 dv$, and, in an energy basis,

$$\langle \mathbf{v} \rangle = \frac{q \int_{E_c}^{\infty} \tau_{\mathbf{k}} (\partial f_0 / \partial E) \mathbf{v} (\mathbf{v} \cdot \boldsymbol{\mathcal{E}}) (E - E_c)^{1/2} \, dE}{\int_{E_c}^{\infty} f_0 (E - E_c)^{1/2} \, dE}. \qquad (9.24)$$

The three real space directions are equivalent. In thermal equilibrium, $\langle \mathbf{v} \rangle_0 = 0$. But carriers still move, since the system is in an environment at temperature T and there is a distribution function that varies with the energy that is related to the velocity. The mean may vanish in the thermal equilibrium, but the $\langle \mathbf{v}^2 \rangle$ will not. So, one can speak to net velocity—which vanished—but also the most probable speed, which will be the speed without regard to the direction with highest probability, to the root mean square velocity, or even to the mean of the speed. The thermal velocity in this collection of velocities is a measure of the thermal motion, so a velocity associated with the form the distribution takes with temperature.

Thermal velocity ($\langle v \rangle_\theta = v_\theta$, with v a scalar), being a measure of a velocity associated with the distribution at the temperature T, takes a variety of definitions. It is the root mean square of the velocity, but also the most probable speed, or the mean of the magnitude of the velocity. The root mean square of the total velocity, and the mean of the magnitude of the velocity, are the most common accepted forms.

The disturbance from equilibrium is small, with a net noticeable effect $\langle \mathbf{v} \rangle$ in response to the electric field. For this isotropic case, we can handle this easily by choosing the direction of the field as one of the directions. Let this be the y direction. So, $\boldsymbol{\mathcal{E}} = \mathcal{E} \hat{\mathbf{y}}$. A small disequilibrium, that is $\langle \mathbf{v} \rangle \ll \langle \mathbf{v}^2 \rangle^{1/2}$, implies that $|\mathbf{v}|^2 \approx 3 v_y^2$. We calculate this using Equation 9.24:

$$\langle v_y \rangle = \frac{2}{3} \frac{q \mathcal{E}_y}{m^*} \frac{q \int_{E_c}^{\infty} \tau_{\mathbf{k}} (\partial f_0 / \partial E) (E - E_c)^{3/2} \, dE}{\int_{E_c}^{\infty} f_0 (E - E_c)^{1/2} \, dE}. \qquad (9.25)$$

In our discussion of the meaning of the distribution function and its drifted form, with the model example in Equation 8.58, we encountered $q\tau_{\mathbf{k}}/m^*$ as a proportionality constant that we interpreted as a mobility. We can now give this parameter much more specificity given that the distribution function provides a clear phase-space dependence of where the particles are. So, if we write by equivalence $v_d|_y = (q\mathcal{E}_y/m^*)\langle\tau_{\mathbf{k}}\rangle$ as the response, then the momentum relaxation time to be used must be the expectation value over all the energies possible, that is,

$$\langle\tau_{\mathbf{k}}\rangle = \frac{2}{3}\frac{\int_0^\infty \tau_{\mathbf{k}}\left(-\partial f_0/\partial E\right)\eta^{3/2}\,d\eta}{\int_0^\infty f_0\eta^{1/2}\,d\eta}, \text{ where } \eta = \frac{E - E_c}{k_B T}. \tag{9.26}$$

η is a normalized kinetic energy parameter that we have employed before. The form of averaging of relaxation time for tackling the drift velocity of a distribution function is prescribed by this form and arose in the linearized Boltzmann transport equation solution. The mobility associated with this drift is the conductivity mobility. In general, then,

Particularly of note here is that the drifted distribution's flow effect has appeared within the averaging, and one must incorporate the state density and marginality in it. It involves an energy dependence prescribed by the distribution function change. If the mass is anisotropic, this will get relatively complicated.

$$\mathbf{v} = \mu_c\boldsymbol{\mathcal{E}}; \tag{9.27}$$

not accounting for the sign of the charge of the particle, the current is

$$\mathbf{J} = \frac{q^2 n\langle\tau_{\mathbf{k}}\rangle}{m^*}\boldsymbol{\mathcal{E}} = \sigma\boldsymbol{\mathcal{E}}, \tag{9.28}$$

where $\sigma = q^2 n\langle\tau_{\mathbf{k}}\rangle/m^* = qn\mu_c$ is the conductivity.

The expectation $\langle\tau_{\mathbf{k}}\rangle$ appeared in a very specific form for the purposes of determining the drift velocity in Equation 9.26. In general, we will expect the averaging to be considerably more complex; after all, there is a $\partial f_0/\partial E\mathbf{v}\cdot\mathbf{G}$ in the perturbation of the distribution function, with $\mathbf{G}$ a prescribed mix of interaction between electrothermal and magnetic fields coupled through an energy-dependent momentum relaxation time $\tau_{\mathbf{k}}$. And then there are energy dependences in the quantity whose expectation value may be desired.

Let the momentum relaxation be in the form $\tau_{\mathbf{k}}(\eta) = \tau_0\eta^r$, where τ_0 is energy independent. For a general expectation, consider $\langle\tau_{\mathbf{k}}^s\eta^t\rangle$, a power s of the specific energy-dependent relaxation time value, and an additional power t in normalized energy arising in the observable of interest. For drift velocity, we had $s=1$ and $t=0$, but, for others, say, $\langle w\mathbf{v}\rangle$, which is of particular interest, it will not. So, we have three different powers: r arising in the energy dependence of the momentum relaxation time, s for the power dependence on the momentum relaxation time and t in the explicit energy dependence from arising in the quantity whose expectation is sought. For this general case,

$$\langle \tau_{\mathbf{k}}^s \eta^t \rangle = \frac{2}{3}\tau_0^s \frac{\int_0^\infty (-\partial f_0/\partial E)\eta^{sr+t+3/2}\,d\eta}{\int_0^\infty f_0 \eta^{1/2}\,d\eta}$$
$$= \frac{4}{3\sqrt{\pi}} \times \left(sr + t + \frac{3}{2}\right)!\tau_0^s \frac{\mathcal{F}_{sr+t+1/2}(\zeta)}{\mathcal{F}_{1/2}(\zeta)},$$
$$\text{for } \tau_{\mathbf{k}} = \tau_0 \eta^r, \tag{9.29}$$

where $\mathcal{F}_j(\zeta)$ are Fermi integrals of order j, and ζ is a specific parameter, such as the normalized Fermi energy ($\eta_F = (E - E_F)/k_B T$), that may be a part of the conditions that the distribution function describes.

This Fermi integral representation, analytically expressed, is

$$\mathcal{F}_j(\zeta) = \frac{1}{j!}\int_0^\infty f_0(\zeta)\eta^j\,d\eta.$$

ζ represents a parameter that expresses the natural constraints on the distribution function. For example, in thermal equilibrium, there exists an electrochemical equilibrium. And E_F, the Fermi energy, together with the state distribution, tells us about carrier concentration as a macroscopic parameter. If we define $\eta_F = (E - E_F)/k_B T$ as a normalized electrochemical energy parameter, then $\zeta = \eta_F$ gives us the energy-connected dependences of the various parameters through the Fermi integral. The carrier concentration n, and, equivalently, p, are expressible through the $\mathcal{F}_{1/2}$ integral in terms of the effective density of states from non-degenerate to degenerate conditions, that is, with a broader validity than just the exponential, under the constraint of the validity of effective density of states assumptions:

$$n = \mathcal{N}_c \frac{2}{\sqrt{\pi}}\mathcal{F}_{1/2}(\eta_F),$$

where $\eta_F = (E_F - E_c)/k_B T$, and $\mathcal{N}_c$ is the effective density of states. Note $(1/2)! = \sqrt{\pi}/2$, which follows from gamma functions. The 1/2 is the power of the energy dependence of the density of states in three-dimensional conditions.

The expectation of the momentum relaxation time, appropriate to determining the drift velocity, then is

$$\langle \tau_{\mathbf{k}} \rangle = \frac{4}{3\sqrt{\pi}} \times \left(r + \frac{3}{2}\right)!\tau_0 \frac{\mathcal{F}_{r+1/2}(\zeta)}{\mathcal{F}_{1/2}(\zeta)}. \tag{9.30}$$

Now, complicate this problem by adding a magnetic field as a perturbation in addition to the electric field, while still in the spherical conduction minimum approximation, with homogeneity and no thermal field. From Equation 9.21, we have

$$\mathbf{G} = q\tau_{\mathbf{k}} \frac{\boldsymbol{\mathcal{E}} - (q\tau_{\mathbf{k}}/m^*)(\boldsymbol{\mathcal{E}} \times \mathbf{B}) + (q\tau_{\mathbf{k}}/m^*)^2\mathbf{B}\,(\boldsymbol{\mathcal{E}} \cdot \mathbf{B})}{1 + (q\tau_{\mathbf{k}}/m^*)^2\mathbf{B}\cdot\mathbf{B}}. \tag{9.31}$$

If we are interested in the drift velocity and current in the presence of both these fields,

$$\mathbf{J} = qn\mathbf{v}_d = qn\langle \mathbf{v} \rangle = qn\frac{2}{3m^*}\frac{\int_0^\infty \mathbf{G}(-\partial f_0/\partial E)\eta^{3/2}\,d\eta}{\int_0^\infty f_0\eta^{1/2}\,d\eta}$$
$$= \frac{q^2 n}{m^*}\langle \frac{\tau_{\mathbf{k}}}{1 + (\omega_c\tau_{\mathbf{k}})^2}\rangle \boldsymbol{\mathcal{E}} - \frac{q^3 n}{m^{*2}}\langle \frac{\tau_{\mathbf{k}}^2}{1 + (\omega_c\tau_{\mathbf{k}})^2}\rangle (\boldsymbol{\mathcal{E}} \times \mathbf{B})$$
$$+ \frac{q^4 n}{m^{*3}}\langle \frac{\tau_{\mathbf{k}}^3}{1 + (\omega_c\tau_{\mathbf{k}})^2}\rangle \mathbf{B}(\boldsymbol{\mathcal{E}} \cdot \mathbf{B}). \tag{9.32}$$

In this latter form, the first term is the ohmic term, now in the presence of both these fields. It is a variation on the conductivity term we obtained in the presence of $\boldsymbol{\mathcal{E}}$ alone. Conductivity has been degraded by the presence of larger path lengths arising due to magnetic field, and this magnetic field coupling—a magnetoresistance contribution—has a square power dependence on the momentum relaxation time. The averaging needed is of $\langle \tau_{\mathbf{k}}/(1 + \omega_c^2\tau_{\mathbf{k}}^2)\rangle$. The second term is the Hall effect, a cross-coupling of the electric and magnetic fields, so an introduction of orthogonality, and it too has a magnetoresistance in it. The numerator now has a second power dependence on the momentum relaxation time. The third term is another magnetoresistance that is now in the direction of the magnetic field, but with a second order dependence in the magnitude. If the magnetic field is small, this last term, as in its complementary appearance in Equation 9.22, will be excluded as a small irritant.

The simultaneous use of electric and magnetic fields is one of the simplest tools used to probe semiconductor properties. Resonance-relaxation interaction is probable through the magnetic field when looking at the conduction response. But the most convenient tool this approach provides is the Hall measurement for mobility, carrier type and carrier concentration. At low magnetic fields, $(\omega_c\tau_{\mathbf{k}})^2 \ll 1$, and the third term of the equation is immaterial. This sets up a relationship between the current density, the electric field and the magnetic field—a vector relationship—and is therefore a tool to use orientational measurement for extracting the properties of the material in which the flow happens:

$$\mathbf{J} = \frac{q^2 n}{m^*}\langle\tau_{\mathbf{k}}\rangle\boldsymbol{\mathcal{E}} - \frac{q^3 n}{m^{*2}}\langle\tau_{\mathbf{k}}^2\rangle(\boldsymbol{\mathcal{E}} \times \mathbf{B}). \tag{9.33}$$

First, set the magnetic field direction as the z orientation in Cartesian coordinate. So, with $\mathbf{B} = B_z\hat{\mathbf{z}}$, the currents and the fields are related as

$$\begin{aligned} J_x &= \frac{q^2 n}{m^*}\langle\tau_{\mathbf{k}}\rangle\mathcal{E}_x - \frac{q^3 n}{m^{*2}}\langle\tau_{\mathbf{k}}^2\rangle\mathcal{E}_y B_z, \\ J_y &= \frac{q^2 n}{m^*}\langle\tau_{\mathbf{k}}\rangle\mathcal{E}_y + \frac{q^3 n}{m^{*2}}\langle\tau_{\mathbf{k}}^2\rangle\mathcal{E}_x B_z, \text{ and} \\ J_z &= \frac{q^2 n}{m^*}\langle\tau_{\mathbf{k}}\rangle\mathcal{E}_z. \end{aligned} \tag{9.34}$$

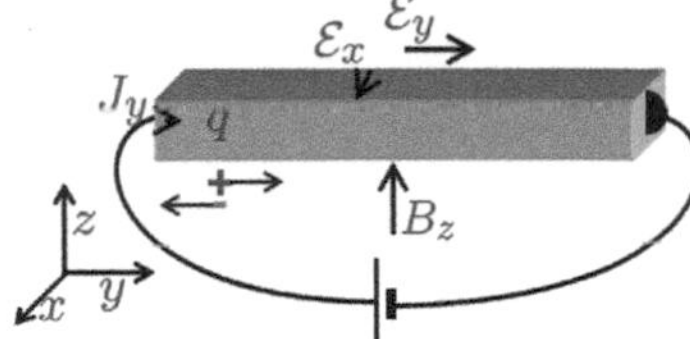

Figure 9.1: Excitation conditions for standard Hall measurement.

Now, we force conditions as shown in Figure 9.1. Current is forced to flow in the y direction, and, in steady, state there is no current flow in the x direction or the z direction. We will set up a measurement of voltage in the x direction so that the electric field's consequences are ascertained, but this is through a high impedance contact. Under these constraints,

$$\begin{aligned} J_z &= \frac{q^2 n}{m^*}\langle\tau_{\mathbf{k}}\rangle\mathcal{E}_z = 0 \quad \therefore \quad \mathcal{E}_z = 0, \text{ and} \\ J_x &= \frac{q^2 n}{m^*}\langle\tau_{\mathbf{k}}\rangle\mathcal{E}_x - \frac{q^3 n}{m^{*2}}\langle\tau_{\mathbf{k}}^2\rangle\mathcal{E}_y B_z = 0 \\ \therefore \quad \mathcal{E}_x &= \frac{qB_z}{m^*}\frac{\langle\tau_{\mathbf{k}}^2\rangle}{\langle\tau_{\mathbf{k}}\rangle}\mathcal{E}_y. \end{aligned} \tag{9.35}$$

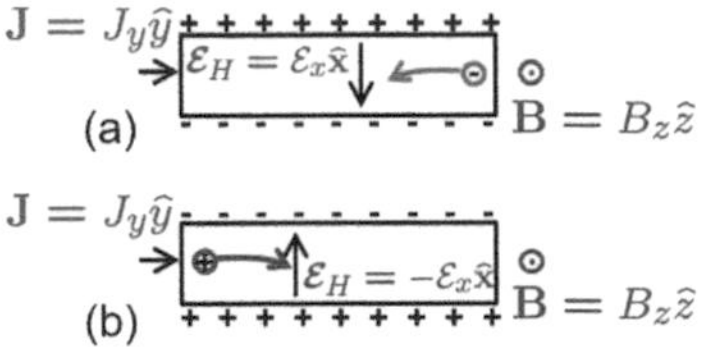

Figure 9.2: The orientation of the electric field and the build-up of Hall voltage when a magnetic field is out of the plane of the sample, and current flows along the plane. In (a), the Lorentz force $q(\mathbf{v} \times \mathbf{B})$ causes electrons to bend as shown, with the excess charge building the transverse electric field that balances the two different field-induced currents, with a net current $J_x = 0$. Part (b) shows the same for holes, with the electric field in the opposite orientation.

In all these relationships, the charge q has a sign, and these field magnitudes are magnitudes for the orientation in which they are specified. Electrons and holes will cause a sign dependence, as in the last field relationship, since the sign of q reverses, as reflected in Figure 9.2. If the flowing charge is a hole, a field develops in the $\hat{\mathbf{x}}$ direction. A positive voltage—a Hall voltage V_H—will be measured in the opposite direction to the field ($V_H = -\int \mathcal{E}_x\, dx$).

For electrons, it will be the opposite, so measuring the voltage in the plane, orthogonal to the flow of current and the magnetic field, provides the polarity of the charge.

The appearance of this field in a direction orthogonal to the imposed electric and magnetic field, under the constraint of no current flow in the other directions, is the *Hall effect*. The measurement also gives us a tool for understanding the relationship between the time constants of energy coupling. Here, $\langle \tau_{\mathbf{k}} \rangle$ arose in the flow of current, and the $\langle \tau_{\mathbf{k}}^2 \rangle$ arose in the magnetic field interaction with this electric field-induced flow. Since the longitudinal electric field $\mathcal{E}_y$ is related to the transverse field $\mathcal{E}_x$, the relationship between the current flow, the magnetic field and the Hall-voltage-associated field follows as

$$\begin{aligned} J_y &= \frac{q^2 n}{m^*} \frac{m^*}{qB_z} \frac{\langle \tau_{\mathbf{k}} \rangle}{\langle \tau_{\mathbf{k}}^2 \rangle} \mathcal{E}_x + \frac{q^3 n}{m^{*2}} \langle \tau_{\mathbf{k}}^2 \rangle \mathcal{E}_x B_z \\ &= \frac{qn}{B_z} \left(\frac{\langle \tau_{\mathbf{k}}^2 \rangle}{\langle \tau_{\mathbf{k}} \rangle^2} + \frac{q^2}{m^*} \langle \tau_{\mathbf{k}}^2 \rangle B_z^2 \right) \mathcal{E}_x. \end{aligned} \tag{9.36}$$

We define two parameters, a Hall constant R_H and a Hall factor r_H, as follows:

$$\begin{aligned} R_H &\equiv \frac{\mathcal{E}_x}{J_y B_z} = \frac{1}{qn} \left(\frac{\langle \tau_{\mathbf{k}}^2 \rangle}{\langle \tau_{\mathbf{k}} \rangle^2} + \frac{q^2}{m^*} \langle \tau_{\mathbf{k}}^2 \rangle B_z^2 \right) \\ &\approx \frac{1}{qn} \frac{\langle \tau_{\mathbf{k}}^2 \rangle}{\langle \tau_{\mathbf{k}} \rangle^2}, \quad \text{and} \\ r_H &\equiv \frac{\langle \tau_{\mathbf{k}}^2 \rangle}{\langle \tau_{\mathbf{k}} \rangle^2}, \quad \text{so } R_H \frac{1}{qn} r_H. \end{aligned} \tag{9.37}$$

In the Hall constant equation, we have ignored, again, the second order term in magnetic flux in our useful approximation. The Hall factor is a property of the material for each polarity type, and the Hall constant is specifically related to the charge polarity, the doping and the Hall factor. The expectation values for the time constants follow from the procedures we have described. For different doped semiconductors, these are known. Recall that $\tau_{\mathbf{k}} = \tau_0 \eta^r$ is a good approximation, with r determined by the dominant scattering mechanism. The prefactor τ_0 cancels out, and a good estimate for the Hall factor is

$$r_H = \frac{3\sqrt{\pi}}{4} \frac{\left(2r + \frac{3}{2}\right)!}{\left[\left(r + \frac{3}{2}\right)!\right]^2}. \tag{9.38}$$

One can find these Hall factor in tables of classic texts. Since polarity is known from the sign of the Hall voltage, the Hall constant measurement gives the carrier concentration as

$$n \text{ or } p = \frac{1}{q}\frac{r_H}{R_H}. \tag{9.39}$$

One can associate a mobility with the motional characteristics that the Hall measurement provides. We have found the conductivity $\sigma = q^2 n \langle \tau_{\mathbf{k}} \rangle / m^* = qn\mu_c$. In correspondence to this, one can define a Hall mobility, that is, a mobility measured when the electric field and the magnetic field are simultaneously present and the setting up of the Hall field by the transverse force induced by the magnetic field lets us evaluate motional characteristics as

$$\mu_H = R_H \sigma = \mu_c \frac{\langle \tau_{\mathbf{k}}^2 \rangle}{\langle \tau_{\mathbf{k}} \rangle^2} = r_H \mu_c. \tag{9.40}$$

During drift, it is for the motion of the carriers themselves in the applied electric field that one is measuring this drift mobility characteristic—the frictional response while scattering. It is not the same as finding the mobility from a transverse field that arose because a magnetic field bends the path of charge carriers. This transverse field builds up since no net current is flowing in the transverse direction. So, the transverse flux from magnetic causes is balanced by the transverse flux from the field, which itself arose as charge accumulated and depleted in the transverse direction. Hall measurement therefore gives a mobility that is somewhat different from the conductivity or drift mobility. The two are related to each other through the Hall factor.

Mobilities are an important parameter reflective of the friction arising in scattering under the external stimulation in transport. The movement of carriers that it reflects will be a function of how the motion takes place and what scattering it undergoes. Drift or conduction mobility—under only an electric field—will be different from Hall mobility—under electric and magnetic fields—and it is not necessarily obvious whether one will be higher than the other, in general, because of the additional path length arising in magnetic field. After all, the relaxation rate under such motion may change, since the states being occupied in the two different circumstances are slightly different. The net rate is a product of occupation of a state, availability of the state to which transition happens, the scattering matrix element that couples and an integration of all the possibilities. Magnetic field changes this. The energy and momentum matching may change under these constraints. Figure 9.3 shows an example of the difference between Hall and drift (conduction) mobility in *GaAs*. While they come very close at moderate to high dopings, the Hall mobility is higher in less-doped material. Suppression of ionized impurity scattering leads to $\langle \tau_{\mathbf{k}}^2 \rangle \gtrsim \langle \tau_{\mathbf{k}} \rangle^2$. The higher the mobility, therefore, the higher is the velocity due to acceleration in the impressed electric field

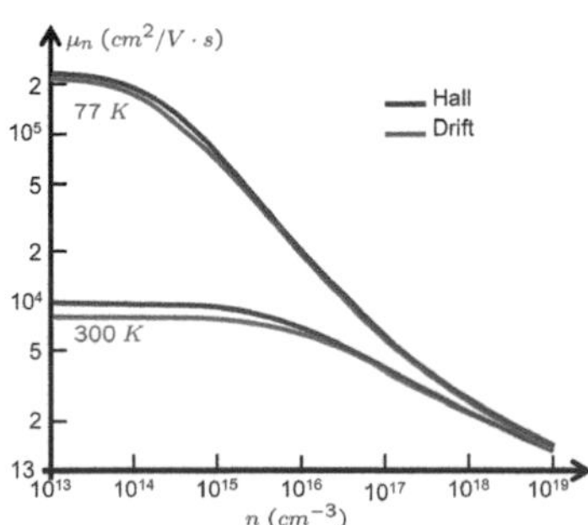

Figure 9.3: The Hall and drift (conduction) mobilities in n-type *GaAs*.

and the less is the change in angle in the scattering, although the higher is the angle due to the Lorentz force: a higher build-up in the transverse field, and a higher effective Hall factor. *GaAs*, with its isotropic bottom of the conduction band, ends up with a higher Hall mobility when ionized impurity is concentration is low.

An interesting contrast to this is the behavior of the electron as a minority carrier. Compare electron conduction in similarly doped n-type and p-type conditions. Most of the holes are heavy and so not too far off from their Coulomb energy perturbation. Should the mobility of an electron as a majority carrier be very similar to that of an electron as a minority carrier? As with Hall mobility versus conduction, one now encounters a variety of dependencies where these expectations of various powers of energy and, indirectly, directionalities enter. Figure 9.4 shows this important example. Now, the consequences are larger at the higher doping end, where the Coulomb scattering due to the type of species causing it matters.

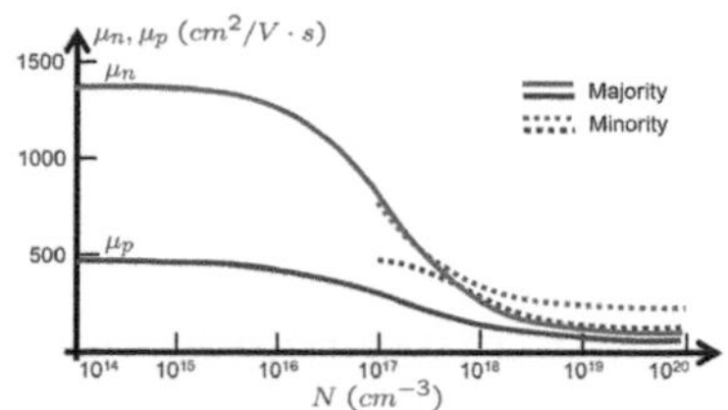

Figure 9.4: Mobility of electrons and holes as majority and minority carriers in *Si*.

This question of minority carrier transport is somewhat complicated by the fact that, to maintain quasi-neutrality, there exists a built-in field, since $\nabla_{\mathbf{r}} n_p = \nabla_{\mathbf{r}} p_p$ under low level injection, which aids electron motion in conditions such as in an *npn* bipolar transistor's base. A discussion of this can be found in S. Tiwari, "Device physics: Fundamentals of electronics and optoelectronics," Electroscience 2, Oxford University Press, ISBN 978-0-198-75984-3 (forthcoming).

The Einstein relationship discussed in Chapter 8—in drift diffusion and Brownian motion—can now be revisited in light of this discussion of mobility. Our electrothermal field—in the absence of a magnetic field—arises in the polar electric field, together with a polar concentration field and a polar thermal field. Concentration changes are reflected in the chemical potential μ change, and the net is our electrothermal field (Equation 9.16)

$$\mathcal{F} = \mathcal{E} - \frac{k_B T}{q} \nabla_{\mathbf{r}} \left(\frac{E - \mu}{k_B T} \right). \tag{9.41}$$

Recall, $E_{qFn} = \mu + q\psi$, with $q = -e$, is the electrochemical potential. ψ is related to the conduction bandedge that determines the lowest potential energy, implying $\nabla_{\mathbf{r}} E_c = -e\nabla_{\mathbf{r}}\psi$ for the electrons as the particle. And there is a corresponding relationship in the case of holes.

$\mathcal{E} = \nabla_{\mathbf{r}} E_c / q$, with $q = -e$, so

$$\mathcal{F} = -\frac{k_B T}{q} \nabla_{\mathbf{r}} \left(\frac{E - E_c - \mu}{k_B T} \right) = \frac{k_B T}{q} \nabla_{\mathbf{r}} \xi. \tag{9.42}$$

ξ is a generalized and normalized electrothermochemical potential.

Let there be no temperature gradient, just the electrical and concentration effects. Then,

$$\mathbf{G} = q\tau_{\mathbf{k}} \mathcal{E} + \tau_{\mathbf{k}} \nabla_{\mathbf{r}} \mu = \tau_{\mathbf{k}} \nabla_{\mathbf{r}} \xi; \tag{9.43}$$

therefore, current density is

$$\begin{aligned}
\mathbf{J} = \frac{qn}{m^*} \langle \mathbf{G} \rangle &= \frac{q^2 n}{m^*} \langle \tau_{\mathbf{k}} \rangle \mathcal{E} + \frac{qn}{m^*} \langle \tau_{\mathbf{k}} \rangle \nabla_{\mathbf{r}} \mu \\
&= qn\mu_n \mathcal{E} + n\mu_n \nabla_{\mathbf{r}} \mu, \quad \text{where} \\
\nabla_{\mathbf{r}} \mu &= \frac{\partial \zeta}{\partial n} \frac{\partial \mu}{\partial \zeta} \nabla_{\mathbf{r}} n \\
&= \left[\frac{\partial}{\partial \zeta} \mathcal{N}_c \frac{2}{\sqrt{\pi}} \mathcal{F}_{1/2}(\zeta) \right]^{-1} k_B T \nabla_{\mathbf{r}} n
\end{aligned}$$

$$= \left[\mathcal{N}_c \frac{2}{\sqrt{\pi}} \mathcal{F}_{-1/2}(\zeta) \right]^{-1} k_B T \nabla_{\mathbf{r}} n,$$

$$\therefore \; \mathbf{J} = qn\mu_n \mathcal{E} + k_B T \mu_n \frac{\mathcal{F}_{-1/2}(\eta_F)}{\mathcal{F}_{-1/2}(\eta_F)} \nabla_{\mathbf{r}} n, \tag{9.44}$$

a form that extends the non-degenerate form to the degenerate form. So, we again obtain, under constant temperature, but with electrical- and concentration-driven conditions,

$$\frac{\mathcal{D}_n}{\mu_n} = \frac{k_B T}{|q|} \frac{\mathcal{F}_{1/2}(\eta_F)}{\mathcal{F}_{-1/2}(\eta_F)}, \tag{9.45}$$

a relationship now valid for both electrons and holes. This is the Einstein relationship when one writes current in the simplest of Fokker-Planck forms of drift diffusion, and writable for both electrons, as here, and holes.

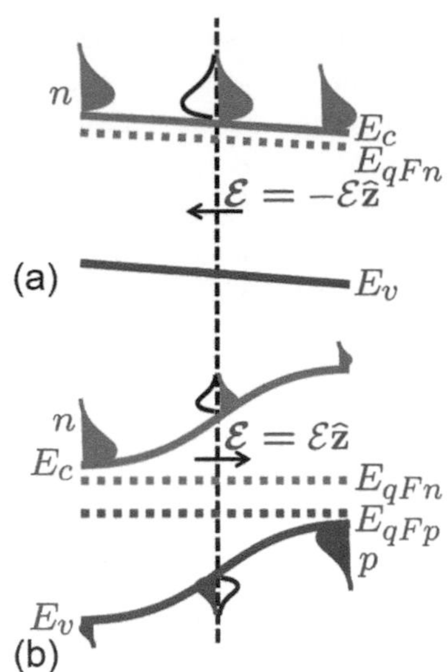

Figure 9.5: Carrier transport in two common situations in semiconductor problems. In (a), which shows a long sample, an electric field causes the majority carriers' electrons to drift under an applied bias, with carrier concentrations only marginally changed. In (b), in an *n*-*p* junction, electrons flow in a forward-biased junction. The electric field in (b) is opposite to that in (a). In (a), momentum-space change dominates. In (b), real space change dominates. But this is the case when the two fields oppose. Is the mobility the same in the two cases when the field is identical in magnitude?

But, this equation reflects a linear small disturbance in distribution function from thermal equilibrium. In its use in Figure 9.5(a), it implies that the distribution functions—shown here for forward and reverse momenta in a central region—are quite close to each other and have an expectation in energy of $(3/2)k_BT$, with the forward flux slightly in excess. The forward and reverse flows are close in balance in the midst of the randomizing scattering and the field's propensity to increase the forward flux of electrons. If there is no scattering, the motion will be ballistic. If there is scattering, it diffuses the motion. The near balance of forward and reverse flux, if existent, indicates both sufficient scattering and a small-enough stimulation. So, a back flux and a forward flux come in near balance as the field causes limited change in the momentum for the forward flux. State occupation is exponentially tailing, and k_BT of energy change during the impulse is a representational energy for this. In Figure 9.5(a), this argument means that the mobility and diffusivity found in low field or thermal-equilibrium conditions should be used. If the field becomes high, or changes suddenly, the argument is violated—an example is in the drain end of channel of a *MOSFET*—where the channel-drain junction is reverse biased and potential change happens rapidly over a short distance. So, carriers acquire kinetic energy over a short distance, with scattering events insufficient in number during the transit. Another place where this argument will break is during a rapid time transient upon a sudden application of bias, even if the asymptotic steady state itself is in accord with the constraints placed by the argument. When time periods are in *pss*, there are changes happening at relaxation time scales, and these change regions are propagating in the structure, sending a rapid field change front with a flow that is considerably off-equilibrium.

In S. Tiwari, "Nanoscale device physics: Science and engineering fundamentals," Electroscience 4, Oxford University Press, ISBN 978-0-198-75987-4 (2017), we discuss the k_BT argument as a Bethe condition while exploring thermionic emission and ballistic motion. Here, energy gain and loss during motion between scattering events should be limited for the near-equilibrium description to be applicable.

The more significant consequence of this discussion is the situation in Figure 9.5(b) of a forward biased *n-p* junction, where the field now opposes the net flux of the particles. Even if high, this field does not raise the kinetic energy of the distribution, that is, make it hotter. Carriers still conform to thermal equilibrium distribution and have $(3/2)k_BT$ of energy. What it has is a large concentration gradient. One can then look at the distribution function, again split between a forward flux and a reverse flux. Carrier concentration is decreasing proportional to $\exp(-q\Delta\psi/k_BT)$, where $\Delta\psi$ is the electrostatic potential change with position. In any small infinitesimally thin cross-sectional region of this junction, as the scattering takes place during the transit, there is also a change in concentration, a change that did not exist in the case of Figure 9.5(a). The velocity of electrons as carriers moving backward is higher as they see an accelerating field. The velocity of carriers moving forward is not. They are largely diffusing up the barrier. The velocity that the carriers can have, with scattering and with our assumption that we have not strayed too far from thermal equilibrium still intact, is limited to $\langle v\rangle \approx v_\theta$, the thermal velocity. So, the mobility is not the low field mobility, but a high field mobility that is constrained by the field to

$$\mu_n(\mathcal{E}) = \frac{\mu_n(0)}{1+\mu_n(0)\mathcal{E}/v_\theta}. \tag{9.46}$$

Diffusivity and this high field mobility are still related through the Einstein relationship—this follows directly from the thermal equilibrium and near-thermal equilibrium argument—its just that their magnitudes are different in case (b) versus case (a). We will tackle steady-state scattering-constrained high field transport in Chapter 10.

Our argument for discussing Equation 9.29, which helps one determine the expectation of general energy-modified and energy-dependent power of time constants, was that the powers of r, s and t arise in the relaxation times' energy dependence, the power of the relaxation time, and the energy dependence of the parameter of interest. In looking at energy flux, a relationship we are interested in is

$$\langle w\mathbf{v}\rangle = \frac{\int_{-\infty}^{\infty} w\mathbf{v}f d\mathbf{v}}{\int_{-\infty}^{\infty} f d\mathbf{v}}, \tag{9.47}$$

which, in the example of Figure 9.5(b), provides us with how the energy flow takes place at each and every cross-section, so at a specific $\mathbf{r}$. In thermal equilibrium, it will vanish. w has some form that depends on what the excitation conditions are and in what form is

See S. Tiwari, "Quantum, statistical and information mechanics: A unified introduction," Electroscience 1, Oxford University Press, ISBN 978-0-198-75985-0 (forthcoming) for a discussion of the different energy is being carried. In thermal equilibrium, $w = w_0 = (3/2)k_BT$. But, off-equilibrium, one would have to determine w specifically. When the field is general, we may define energy flux as

$$\mathbf{W} = -\frac{n}{m^*}\langle w\mathbf{G}\rangle, \tag{9.48}$$

where, again, one has a $\langle w\mathbf{G}\rangle$ expectation as a product of energy and impulse averaged over the distribution function. This expression is the generalization of the current density expression—now for energy—and for electrons, for energy and with current flow in the opposite direction.

So, what precisely is w, the energy carried by each carrier, and what does a temperature T represent? To resolve this, one has to go back to our understanding of free energies, and the form that we need to worry about is Helmholtz energy. It is the temperature and entropy that is changing in the exchange of energy in all their forms in our problem. Let there be n charge particles in the distribution; then,

$$w = \left.\frac{\partial(TS)}{\partial n}\right|_{T,V} = \left.\frac{\partial U}{\partial n}\right|_{T,V} - \left.\frac{\partial \mathcal{F}}{\partial n}\right|_{T,V} = E - \mu, \tag{9.49}$$

where the derivative is meant to extract the change arising through a change of one particle to the system of free energy U and Helmholtz energy $\mathcal{F}$, with the temperature as T and entropy as S. In the derivatives, the first term in this relationship tells us free energy per electron; that is, electrochemical potential per electron. The second term tells us the chemical potential per electron. Chemical potential per electron is reflected in the concentrations. Free energy is reflected in the electrical and chemical potentials, so the charged kinetics and the concentration. If concentration is not changing, then, in the single particle nearly free electron picture, this energy has been reduced to the classical equipartition energy of $(3/2)k_BT$. But this is, not so in general.

As for the second question we asked, what is the temperature T here? The temperature of any system is defined thermodynamically as equal to the temperature of a large reservoir with which it is in contact and in equilibrium. Temperature can then be defined thermodynamically ($1/T = dS/dU$), since the reservoir is large, with a large number of particles and a large volume. Temperature, according to this interpretation, is an invariant result of an equilibration with the reservoir. When one has a very small system, say, a two-state system such as that in Rabi oscillations, if the perturbation is taken away, the system will be found in either the higher energy state or the lower energy state. This does not mean that it has a

conditions that expose to us the exchangeable parts of free energy. Energies represent the thermodynamic potential of a system. Work performed and heat generated are forms of energy, so the energies of interest, free energies, occur in a variety of forms that can be converted into each other. The internal energy of the system, U, can be thought of as the energy required to create a system in the absence of temperature or volume changes. So, it includes the various forms that may undergo change, for example, electrostatic or electromagnetic energies, except those due to temperature, that is, thermal energy and the occupation of a volume, that is, mechanical movement. If the environment has a temperature T, then some energy is spontaneously transferred from the environment to the system, with a concomitant change in the entropy of the system. The system, which is now in a state of higher disorder than before, has less energy to lose. This is the Helmholtz free energy $\mathcal{F} = U - TS$. The system occupies a volume V in the environment at temperature T by requiring the additional work of PV. This energy now is the Gibbs free energy $\mathcal{G} = U - TS + PV$. The mechanical energy form PV is of important concern to chemistry, but not as much to us where pressures and volume changes are quite small. It is of importance in semiconductor problems only when there is phase change and deformation (strain), because volume may change significantly. Enthalpy, the fourth thermodynamic potential, is an energy measure that is useful when systems release energy, such as in an exothermic reaction. Enthalpy, $\mathcal{H} = U + PV$, is the energy change associated with internal energy and the work done by the system. So, we limit ourselves to only those changes in Helmholtz energy that occur as a result of changes in internal energy, temperature and entropy.

Charles Kittel's opinion column titled "Temperature fluctuation: An oxymoron," in the May 1988 edition of *Physics Today* (**41**, 93), is highly recommended reading for understanding temperature and asking yourself, "What is the temperature of an atom?"

higher temperature or a lower temperature. *Tying it to the reservoir is a necessity in the definition of the temperature.* Absent perturbation, the two-state system can have fluctuations, but not fluctuations of temperature.

We can now tackle the energy form exchange through the general impulse approach. Again, let the disturbance from thermal equilibrium be small. And let there be excitation through $\boldsymbol{\mathcal{E}}, \mathbf{B}, \nabla_{\mathbf{r}}T$ and $\nabla_{\mathbf{r}}n$. The current density and energy flux density are, respectively,

This argument leads to the observation that nanoscale and quantum systems have temperature, but one need not prescribe temperature fluctuations to them. A single isolated system has no temperature. If one knew its state, the entropy vanishes. If its energy is known, the fluctuation vanishes. Temperature is the invariant of an equilibration with a large reservoir. Our use of temperature assignment to particle distribution (T_e) or to lattice (T_l) then is merely a numerical expedient of fitting. A classical harmonic oscillator, with the virial implication, has a $(1/2)k_BT$ internal energy assignment for every quadratic term in momenta (more generally, $\langle E_i\rangle = k_BT/m$ for every term of $\mathscr{H}_i$ in mth powers of position and momentum). But this is not true for quantum oscillators.

$$\mathbf{J} = \frac{qn}{m^*}\langle \mathbf{G}\rangle = \frac{q^2n}{m^*}\left[\langle \tau_{\mathbf{k}}\boldsymbol{\mathcal{F}}\rangle - \frac{q}{m^*}\langle \tau_{\mathbf{k}}^2(\boldsymbol{\mathcal{F}}\times\mathbf{B})\rangle\right], \text{ and}$$

$$\mathbf{W} = -\frac{n}{m^*}\langle w\mathbf{G}\rangle = -\frac{qn}{m^*}\left[\langle \tau_{\mathbf{k}}w\boldsymbol{\mathcal{F}}\rangle - \frac{q}{m^*}\langle \tau_{\mathbf{k}}^2 w(\boldsymbol{\mathcal{F}}\times\mathbf{B})\rangle\right], \tag{9.50}$$

where

$$\boldsymbol{\mathcal{F}} = \boldsymbol{\mathcal{E}} - \frac{1}{q}\nabla_{\mathbf{r}}w + \frac{w}{qT}\nabla_{\mathbf{r}}T. \tag{9.51}$$

We have now rewritten the electrochemothermal field in a broader form. The first term is the electric field. The second arises in the energy content of the particle sans the chemical component. The third term is due to thermal fields toward equilibration. This is a complete representation of the near-equilibrium transport, including scattering and the exchanges taking place during transit.

We now discuss several of the consequences that this general relationship form of Equation 9.50 leads to.

9.2.2 *Thermal conductivity due to carriers*

THE THERMAL CONDUCTIVITY CONTRIBUTION FROM CARRIERS (κ_c, a parameter to relate heat flux—energy transported—to temperature change through $\mathbf{W} = -\kappa_c\nabla_{\mathbf{r}}T$, when no other stimulations except temperature change exists) can now be calculated. Take $\mathbf{J} = 0$ and $\mathbf{W} = 0$; then, the current density from Equation 9.50 gives

$$\mathbf{J} = \frac{q^2n}{m^*}\left[\langle \tau_{\mathbf{k}}\boldsymbol{\mathcal{F}}\rangle - \frac{q}{m^*}\langle \tau_{\mathbf{k}}^2(\boldsymbol{\mathcal{F}}\times\mathbf{B})\rangle\right]$$

$$\therefore\ 0 = \langle \tau_{\mathbf{k}}\rangle\left(\boldsymbol{\mathcal{E}} + \frac{1}{q}\nabla_{\mathbf{r}}\mu\right) + \frac{1}{qT}\langle \tau_{\mathbf{k}}w\rangle\nabla_{\mathbf{r}}T, \text{ and}$$

$$\mathbf{W} = -\frac{qn}{m^*}\left[\langle \tau_{\mathbf{k}}w\rangle\left(\boldsymbol{\mathcal{E}} + \frac{1}{q}\nabla_{\mathbf{r}}\mu\right) + \frac{1}{qT}\langle \tau_{\mathbf{k}}w^2\rangle\nabla_{\mathbf{r}}T\right]. \tag{9.52}$$

The null equation relates $\boldsymbol{\mathcal{E}} + (1/q)\nabla_{\mathbf{r}}\mu = -(1/qT)\langle \tau_{\mathbf{k}}w\rangle\nabla_{\mathbf{r}}T$, so that

$$\mathbf{W} = -\frac{n}{m^*T}\left[\langle \tau_{\mathbf{k}}w^2\rangle + \frac{\langle \tau_{\mathbf{k}}w\rangle^2}{\langle \tau_{\mathbf{k}}\rangle}\right]\nabla_{\mathbf{r}}T, \text{ and}$$

$$\therefore\ \kappa_c = \frac{n}{m^*T}\left[\langle \tau_{\mathbf{k}}w^2\rangle + \frac{\langle \tau_{\mathbf{k}}w\rangle^2}{\langle \tau_{\mathbf{k}}\rangle}\right]. \tag{9.53}$$

The particle-transport-mediated thermal conductivity is directly proportional to particle density, the energy content per particle and momentum relaxation time and inversely proportional to effective mass and temperature. More particles carry more energy, more energy content per particle is an increase in energy flow, and more relaxation time is less motional loss in the direction of flow. If effective mass increases, the particles drag, and if temperature is lower, the carriers stay closest to the lowest energy states allowed and so carry less energy along.

Table 9.1 summarizes some of the room temperature thermal conductivities in materials of interest. Some are dominated by carrier conduction, and some by phonons. Take, for example, the difference between pure *Si* and doped *Si*. Figure 9.6 draws out the temperature dependence of some of these thermal conductivities in example semiconductors and contrasts them with those of diamond and *Cu*. Thermal conductivity in semiconductors tends to peak at low temperature. The peaking is due to the increase in Umklapp processes at higher temperature.

	κ $W/cm \cdot K$
Al	2.37
Cu	4.01
Au	3.17
Si (pure)	1.48
Si (10^{19} cm^{-3})	1.11

Table 9.1: Thermal conductivity of example solids at room temperature. Note the decrease in thermal conductivity of *Si* when doped due to scattering.

So, absent current flow and no external stimulation, except due to temperature, one finds the particle flow contribution to thermal conductivity as a function of several randomizing relaxation times. This calculation does not include phonon contribution. It is not unreasonable for metals with a near-Avogadro's number of carriers, thermal conduction through carriers dominating. For semiconductors, it may or may not. In many large devices, although not in all, phonons may dominate in taking away heat. In power devices, or devices with scales and geometries where heat spreading is an important consideration, if different regions have different levels of dissipation, and temperature differences affect the carrier dynamics, then both the carrier and the phonon heat conduction become important. Silicon, if it isotopically pure, has much improved thermal conductivity due to improved phonon transport.

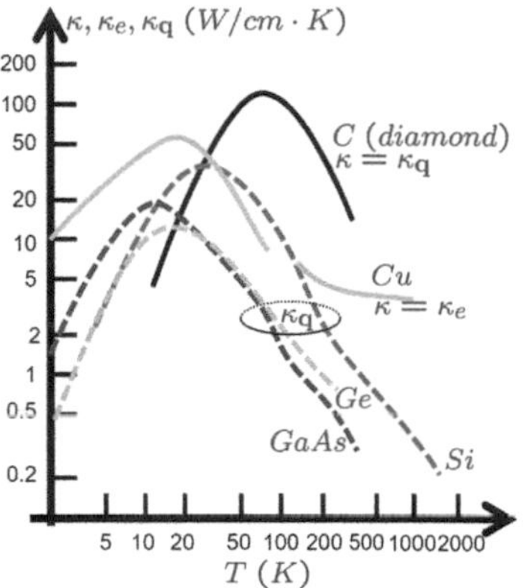

Figure 9.6: Thermal conductivity of *Si*, *Ge* and *GaAs* semiconductors and a metal (*Cu*) as a function of temperature. $\kappa_c = \kappa_e$ is the thermal conductivity due to electrons. κ_q is the thermal conductivity due to phonons. Lattice, that is phonon-limited, thermal conductivity, which dominates for semiconductors, is shown here. *Cu*'s is dominated by electrical thermal conductivity. A high thermal conductivity insulator (diamond) is also shown. In this case, the thermal conductivity will be entirely due to the phonons. With temperature change, the initial rise arises in the increase in increased energy content of the phonons and the electron population. The drop immediately after the peak is due to Umklapp processes. Isotropically pure semiconductors can have very large improvements in the peak thermal conductivity.

This discussion shows the importance of the constraints under which an energy exchange is being studied. We take this thought further by looking at thermal-electric energy exchange under a variety of circumstances next.

9.2.3 *Thermoelectric effects*

If no current flows, and a thermal gradient exists, thermal and electric current must balance everywhere. This is part of the origin of the electric field in Equation 9.52. The other part is due to the concentration gradient. An electric field results in a potential change over distance. So, when one writes

$$0 = \langle\tau_{\mathbf{k}}\rangle \left(\boldsymbol{\mathcal{E}} + \frac{1}{q}\nabla_{\mathbf{r}}\mu \right) + \frac{1}{qT}\langle\tau_{\mathbf{k}} w\rangle \nabla_{\mathbf{r}} T, \tag{9.54}$$

the diffusion of electrons down the temperature gradient is being opposed by the electric field that comes about from the carrier population change that arises when the thermal equilibrium was disturbed. This electric field establishes its direction from the net positive change to the net negative charge. For electrons, it causes a force to balance the thermally induced diffusive flow. This is the *Seebeck effect* or the *thermoelectric effect*. Our derivation says that an electric field arises in the presence of a temperature gradient, and it is given as

$$\boldsymbol{\mathcal{E}} = -\frac{1}{q}\nabla_{\mathbf{r}}\mu - \frac{1}{q\langle\tau_{\mathbf{k}}\rangle T}\langle\tau_{\mathbf{k}} w\rangle \nabla_{\mathbf{r}} T. \tag{9.55}$$

The first term in this equation—the chemical potential term—is due to concentration inhomogeneity, and the second term is due to temperature inhomogeneity. Since $\nabla_{\mathbf{r}}\mu = \partial\mu/\partial T \nabla_{\mathbf{r}} T$, the electric field can be written as

$$\begin{aligned} \boldsymbol{\mathcal{E}} &= -\frac{1}{qT}\left(T\frac{\partial\mu}{\partial T} + \frac{\langle\tau_{\mathbf{k}} w\rangle}{\langle\tau_{\mathbf{k}}\rangle} \right) \nabla_{\mathbf{r}} T \\ &\approx T\frac{d}{dT}\left(\frac{\langle\tau_{\mathbf{k}} w\rangle}{qT\langle\tau_{\mathbf{k}}\rangle} \right) \nabla_{\mathbf{r}} T = \Upsilon \nabla_{\mathbf{r}} T, \\ \text{where } \Upsilon &= T\frac{d}{dT}\left(\frac{\langle\tau_{\mathbf{k}} w\rangle}{qT\langle\tau_{\mathbf{k}}\rangle} \right) = -T\frac{d\mathcal{P}}{dT}, \\ \text{with } \mathcal{P} &= \frac{\langle\tau_{\mathbf{k}} w\rangle}{qT\langle\tau_{\mathbf{k}}\rangle}. \end{aligned} \tag{9.56}$$

Here, Υ is the *Thompson coefficient* and $\mathcal{P}$ is *thermoelectric power*. A common measure of thermoelectricity is the ratio of the electric potential differential produced when no current is flowing and the temperature differential. This is the Seebeck coefficient.

The Thompson coefficient can be quite vanishing, and can be positive or negative depending on the scattering mechanism through which the temperature dependence arises in the ratio $\langle\tau_{\mathbf{k}} w\rangle / qT\langle\tau_{\mathbf{k}}\rangle$. For n-type semiconductors, the electric field and the temperature gradient oppose each other. So, electrons have a negative Thompson coefficient. For holes, the electric field and the temperature gradient reinforce each other, and the Thompson coefficient is positive. The thermoelectric power is in the opposite direction for electrons and holes.

Consider an n-type semiconductor under a temperature gradient, as shown in Figure 9.7. $T_2 > T_1$, and two contacts are used to

The Seebeck effect is the appearance of a thermoelectric voltage under open circuit conditions. If two materials A and B are connected in a loop, and the loop is broken to measure the voltage ΔV with the two A-B junctions separated in temperatures ΔT, then $S_{AB} = \lim_{\Delta T \to 0} \Delta V / \Delta T$ gives the thermoelectric power of the couple. $\Delta V = \int_{T_1}^{T_2} (S_B - S_A)\, dT$, with $S_{AB} = S_B - S_A$. S is, in general, a tensor, but, for cubic crystals, one can look upon it as a scalar. This is the Seebeck coefficient. In any material, the thermoelectric power flow in the opposite direction means that if one created a temperature gradient and looked at the direction of current through the polarity, one could tell whether the semiconductor was n type or p type. This is the basis for a traditional simple technique—the hot-cold probe technique—to identify the polarity of a material.

measure—under high impedance conditions, so vanishing current—the induced bias voltage that establishes itself across the semiconductor. This measuring point where the meter is connected to the wire is at temperature T_0:

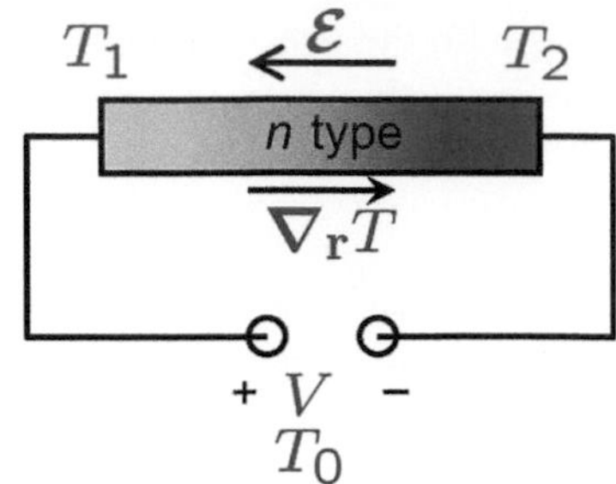

Figure 9.7: An n-type semiconductor under thermal gradient with an induced voltage being measured for $\mathbf{J} =$ and $T_0 < T_1 < T_2$.

$$\begin{aligned} V &= -\oint \mathcal{E} \cdot d\mathbf{r} = \oint \Upsilon \nabla_{\mathbf{r}} T \cdot d\mathbf{r} \\ &= \int_{T_0}^{T_1} \Upsilon_m \, dT + \int_{T_1}^{T_2} \Upsilon_s \, dT + \int_{T_2}^{T_0} \Upsilon_m \, dT \\ &= \int_{T_1}^{T_2} (\Upsilon_s - \Upsilon_m) \, dT. \end{aligned} \tag{9.57}$$

Here, Υ_m is the Thompson coefficient of the metal wire connecting to the semiconductor, and Υ_s is the Thompson coefficient of the semiconductor. The result is that the induced voltage is an integrated difference of the coefficients over the temperature. Now, if the metal is chosen to have a Thompson coefficient close to vanishing, the induced voltage is directly proportional to the difference of temperature.

Thermocouples use this approach to measure temperature using contacts using two different metals of known coefficients.

When one lets the current flow—so allowing thermoelectric power transmission—one gets numerous components to the energy of heat flow. Take a situation with electrical, concentration and thermal fields but no magnetic field ($\mathbf{B} = 0$). We then have no $\boldsymbol{\mathcal{F}} \times \mathbf{B}$, so

$$\begin{aligned} \mathbf{J} &= \frac{q^2 n}{m^*} \left[\langle \tau_{\mathbf{k}} \boldsymbol{\mathcal{F}} \rangle - \frac{q}{m^*} \langle \tau_{\mathbf{k}}^2 (\boldsymbol{\mathcal{F}} \times \mathbf{B}) \rangle \right] \\ \therefore \quad \mathcal{E} + \frac{1}{q} \nabla_{\mathbf{r}} \mu &= \frac{m^*}{q^2 n \langle \tau_{\mathbf{k}} \rangle} \mathbf{J} - \frac{1}{qT} \frac{\langle \tau_{\mathbf{k}} w \rangle}{\langle \tau_{\mathbf{k}} \rangle} \nabla_{\mathbf{r}} T, \quad \text{and} \\ \mathbf{W} &= -\frac{\langle \tau_{\mathbf{k}} w \rangle}{q \langle \tau_{\mathbf{k}} \rangle} \mathbf{J} - \frac{n}{m^* T} \left(\langle \tau_{\mathbf{k}} w^2 \rangle - \frac{\langle \tau_{\mathbf{k}} w \rangle^2}{\langle \tau_{\mathbf{k}} \rangle} \right) \nabla_{\mathbf{r}} T \\ &= T \mathcal{P} \mathbf{J} - \kappa_e \nabla_{\mathbf{r}} T = \Pi \mathbf{J} - \kappa_e \nabla_{\mathbf{r}} T, \end{aligned} \tag{9.58}$$

where Π, a product of temperature and thermoelectric power,

$$\Pi = T \mathcal{P} = -\frac{\langle \tau_{\mathbf{k}} w \rangle}{q \langle \tau_{\mathbf{k}} \rangle}, \tag{9.59}$$

is the Peltier coefficient. Equation 9.58 now describes energy density flux, with the current present, as the consequence of the additional component of energy arising in the externally induced flow of current and that due to the movement of energy by thermal conductivity. This is the *Peltier effect*, where current carriers heat, which reinforces or, depending on the sign of the carrier charge, opposes the heat flow arising in the temperature gradient. The temperature gradient had induced a real space diffusive flow. Current has now provided an additional contribution arising in momentum space.

The flow current also generates heat through dissipation. The heat generated is

$$P_\theta = \mathbf{J} \cdot \boldsymbol{\mathcal{E}} - \nabla_{\mathbf{r}} \cdot \mathbf{W}, \tag{9.60}$$

which is the dissipation term subtracted by the flux of heat taken away by the carriers. This is the heat generated per unit volume that is being moved to the environment. Since

$$\begin{aligned} \boldsymbol{\mathcal{E}} &= \frac{1}{\sigma}\mathbf{J} + \Upsilon\nabla_{\mathbf{r}}, \text{ and} \\ \mathbf{W} &= \Pi\mathbf{J} - \kappa_e\nabla_{\mathbf{r}}T, \\ P_\theta &= \frac{\mathbf{J}\cdot\mathbf{J}}{\sigma} + \Upsilon\mathbf{J}\cdot\nabla_{\mathbf{r}}T - \Pi\nabla_{\mathbf{r}}\cdot\mathbf{J} + \kappa_e\nabla_{\mathbf{r}}^2 T. \end{aligned} \tag{9.61}$$

In Equation 9.61, an equation for the heat generated per unit volume in the presence of electric, concentration and thermal fields, the first term is the Joule heat—Brownian dissipation—the second term is the Thompson heat term—the thermal-electric interaction—the third term is the divergence arising in heat carried away by current, and the fourth term is the current due to the diffusive flow of heat. The heat generated per unit volume is not simply the result of a product of current and voltage drop. If there are gradients in temperature, if current divergence is significant and if temperature divergence is significant, then numerous consequences will arise in temperature differentials and thermal generation across a device.

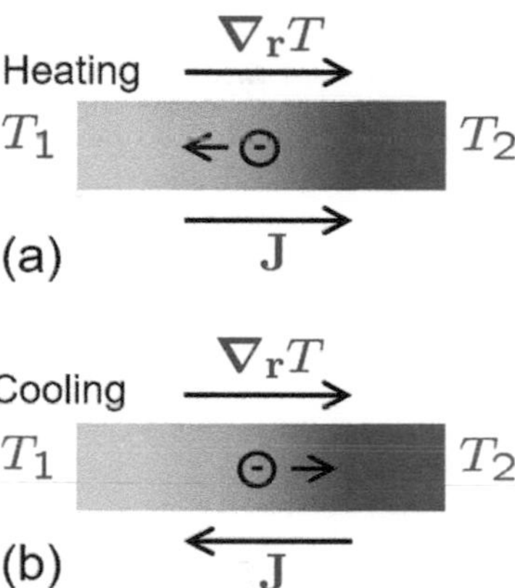

Figure 9.8: An n-type semiconductor under thermoelectric stimulated conditions, where the temperature gradient and current are in the same direction in (a) and opposite in (b).

While terms such as the ohmic dissipation of $\mathbf{J}\cdot\mathbf{J}/\sigma$ are largely an irritant, except when heating is a desired result, this thermal-electrical exchange can be used gainfully. One of these is through the deployment of the Thompson heat ($\Upsilon\mathbf{J}\cdot\nabla_{\mathbf{r}}T$) Peltier term to cooling and heating. In Figure 9.8(a) and (b), two conditions are shown for the flow of current in the presence of a gradient in temperature in an n-type semiconductor. In (a), The flow of electrons, due to the current, is from the hot temperature to lower temperature, with the current in the same direction as the temperature gradient. Hotter carriers arrive at the lower temperature T_1. Electrons arriving from higher temperature to lower temperature give this heat energy to the lattice. This happens to be also true for the last term of the equation, except that both the diffusive carrier flow and the electrically stimulated carrier flow—largely a drift flow, since the material is n-type and so not far from thermal equilibrium—are bringing heat to the crystal at the cold end. This statement of identical polarity thermal and current-driven energy flow is the same as saying that Υ is negative. It causes the net heat generated per unit volume at the cold end to increase, which is what Equation 9.61 tells us. The opposite is true for Figure 9.8(b).

For an n-type semiconductor, when the temperature gradient and the current oppose, the cold end is cooled further by the Thompson heat term—the electrons are now taking heat from the crystal as they go from the lower temperature region to the higher temperature region—even as the thermal conduction—arising in thermal diffusion—heats it. This picture is exactly the reverse when Υ is positive, as for holes and p-type semiconductors.

The n- and p-materials' current-direction- and temperature-gradient-direction-induced heat flow interaction can be employed for cooling and heating. Figure 9.9 shows the biasing under which current flows in opposite directions in n-type and p-type semiconductors, with the current in the n-type semiconductor flowing toward-a surface that is being cooled as a result of the current flow. Here, both the n and the p arms of what is generally a stack of such fins carry heat away from the stage to the heat sink. This arrangement, without the battery but with the gradients of the temperature, is now also a thermoelectric generator. Although quite inefficient, it is a useful low electric power source where heat is available and otherwise being wasted.

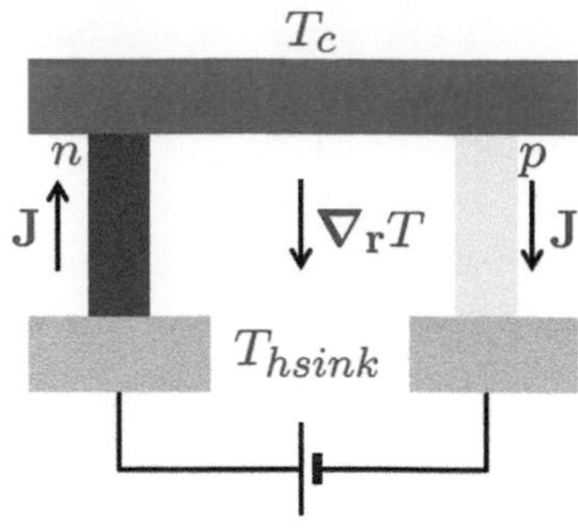

Figure 9.9: A thermoelectric cooler employing n-type and p-type semiconductors to cool a stage by extracting heat to a heat sink, using Thompson heat flow and the opposite polarity of the Thompson coefficient for electrons and holes.

Thermoelectric cooling is also called Peltier cooling. It is an inexpensive way to use small-sized container cooling as a substitute for a large, expensive refrigerator. These coolers are also employed to maintain the stage temperatures of precision single-mode lasers. There, the objective is to stabilize the wavelength to the desired magnitude by using the temperature upon which it depends through the delicate lasing energy states undergoing stimulated recombination.

Since electric fields also arise in concentration gradients when quasineutrality is broken, an equivalent thermoelectric effect may be obtained for electrons and holes through chemical potential change. Assume $\nabla_{\mathbf{r}}T = 0$ for n material, so Equation 9.60 describes the dissipation under heat flow. The electric field follows from the degeneracy-generalized drift-diffusion Equation 9.44, where current and the concentration gradient lead to an electric field of

$$\mathcal{E} = \frac{\mathbf{J}}{\sigma} - \frac{k_B T}{q\mathcal{N}_c\mathcal{F}_{-1/2}(\eta_F)}\nabla_{\mathbf{r}} n, \tag{9.62}$$

and therefore a heat per unit volume of

$$P_\theta = \frac{\mathbf{J}\cdot\mathbf{J}}{\sigma} - \frac{k_B T}{q\mathcal{N}_c\mathcal{F}_{-1/2}(\eta_F)}\mathbf{J}\cdot\nabla_{\mathbf{r}} n - \Pi\nabla_{\mathbf{r}}\cdot\mathbf{J}. \tag{9.63}$$

The thermoelectric power source has no moving mechanical parts or moving ions, unlike most common sources of electric power.

Note that the drift-diffusion equation has built into it the $\nabla_{\mathbf{r}}T = 0$ assumption. It only considers near-equilibrium behavior under randomized scattering.

If the current density $\mathbf{J}$ and the concentration gradient $\nabla_{\mathbf{r}} n$ are in the same direction, electrons will take heat away from the crystal as they diffuse from a high concentration region to a low concentration region. Note that, for electrons, both the gradient of concentration and the current have the same sign. In this situation, cooling will happen.

Since thermoelectric power is of scientific and engineering interest, a figure of merit that accounts for the ability to generate electric power under temperature differential is useful in making comparisons. A larger thermal gradient is possible with lower thermal conductivity, so this figure of merit zT is

$$zT = \frac{\sigma S^2 T}{\kappa}. \tag{9.64}$$

Note that, in our calculations, we have not determined the phonon component of thermal conductivity κ. S is the Seebeck coefficient. κ is in the denominator, since a lower thermal conductivity allows much of the energy flow to appear in electrical form. A higher temperature for this ability is desirable, so the temperature T is in the numerator.

9.2.4 *Thermoelectromagnetic effects*

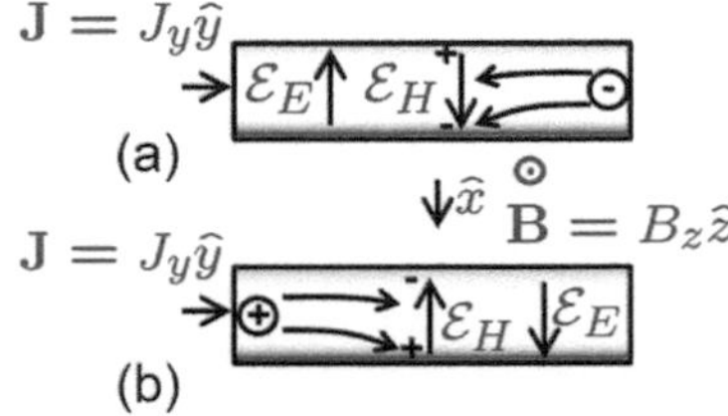

Figure 9.10: The Hall effect (electrical-magnetic interaction) and the Ettinghausen effect (thermal-electrical-magnetic interaction) in n-type and p-type semiconductors. Note that the conditions for Hall measurement (current and magnetic field) are identical to those of Figure 9.2.

THE HALL EFFECT, WITH WHICH WE STARTED the discussion of energy exchanges, is an electromagnetic effect, and, with this previous discussion of thermoelectric effect, it should be quite straightforward to see the thermal—diffusive—consequences as a variation on the classical Hall effect. An integrative illustration of this is in Figure 9.10, where, in addition to the bias-stimulated electric current, there also exists a thermal gradient—in this case, in the x direction. So, compare this illustration to that of Figure 9.2. The Hall effect is shown in Figure 9.10, in accord with the earlier example for identical conditions. Hall measurement is under isothermal conditions, with the Hall voltage measured in the orthogonal direction under condition of no current flow. The force from the Hall field formed balances the Lorentz force, due to the absence of current in that orthogonal direction. Slower charge carriers deflect less, faster charge carriers deflect more, and an averaging over the velocities determines the Hall field. But we have seen that momentum relaxation time increases with energy, that is, $r > 0$ in $\tau_{\mathbf{k}} = \tau_0 \eta^r$. The faster electrons deflect even more, since they do not undergo as much relaxation. This deflected direction accumulates more of the hotter carriers and gets hotter. A temperature gradient arises. But then there is the thermal diffusive flow from hot regions to the cooler regions. Since there is no current allowed in this direction, an electric field must arise that balances the thermal diffusive flow. This appearance of an electric field due to thermal causes, in the presence of a magnetic field and a current flow, is the *Ettinghausen effect*.

The precise conditions for Ettinghausen effect measurement are thermal isolation in addition to the current isolation of the sample in the orthogonal direction, and no temperature gradient in the current flow direction. In Figure 9.10's coordinate choices, this is $J_x = 0$, $\partial T/\partial x = 0$, as is $W_y = 0$. The Ettinghausen coefficient is

$$C_E \equiv -\frac{\partial T/\partial x}{J_y B_z} = \frac{\mu_c}{q\kappa_e}\left(\frac{\langle \tau_{\mathbf{k}}^2 E\rangle}{\langle \tau_{\mathbf{k}}\rangle^2} - \frac{\langle \tau_{\mathbf{k}}^2\rangle\langle \tau_{\mathbf{k}} E\rangle}{\langle \tau_{\mathbf{k}}\rangle^3}\right). \tag{9.65}$$

Note that the energy dependence of the momentum relaxation time matters, as well as the sign of charge. Figure 9.10 illustrates the case for electrons and holes for $r \geq 0$. Since there are scattering mechanisms where $r < 0$, for example, with deformation potential scattering by acoustic or optical phonons, the behavior of electrons and phonons can reverse where these effects are dominant. We will see that this is possible at low temperature and high purity in non-polar semiconductors, so under conditions of low impurity scattering.

9.2.5 *Thermomagnetic effects*

THERMAL STIMULATION CAUSES CARRIER FLOW via diffusive transport. It causes flow analogous to the flow arising in electrical stimulation, except that the details of the mechanism are different. So, just as Hall and Ettinghausen effects appeared under electrical and magnetic stimulation, respectively, and thermal changes were a consequence in Ettinghausen effect, if one has thermal and magnetic stimulation, one will see complementary consequences. The *Nernst effect* is the analog of the Hall effect for thermomagnetic conditions. The *Righi-Leduc effect* is the analog of the Ettinghausen effect for thermomagnetic conditions. These are illustrated in Figure 9.11 for n and p semiconductors.

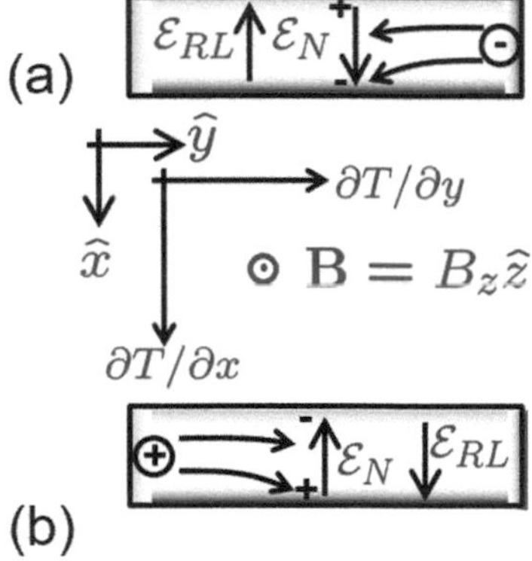

Figure 9.11: The Nernst effect and the Righi-Leduc effect in semiconductors. Part (a) shows the effect in an n-type semiconductor and (b) shows the effect in a p-type semiconductor.

In the Nernst condition, temperature different at the contacts provides the thermal excitation; instead of the electrical bias, no current flows (so, $J_y = J_x = 0$) and the orthogonal measurement conditions are isothermal (so $\partial T/\partial x = 0$ at the contacts but not in the sample). A transverse electric field is produced because thermally driven carriers—cold and hot—deflect differently under Lorentz force, making one surface hotter than the other, and no current is allowed to flow in that direction. The Nernst coefficient ($cm^2/s \cdot K$) is

$$C_N \equiv \frac{\mathcal{E}_x}{B_z \partial T/\partial y} = \frac{\mu_c}{qT}\left(\frac{\langle \tau_{\mathbf{k}}^2 E\rangle}{\langle \tau_{\mathbf{k}}\rangle^2} - \frac{\langle \tau_{\mathbf{k}}^2\rangle\langle \tau_{\mathbf{k}} E\rangle}{\langle \tau_{\mathbf{k}}\rangle^3}\right) = \frac{\kappa_e}{T} C_E, \tag{9.66}$$

where the kinetic energy E of the carrier appears together with the relaxation time dependences of conductivity.

In the Righi-Leduc measurement, conditions are of no current flow, as in this last example, that is, $J_y = 0$ and $J_x = 0$, but also that the boundary conditions are isothermal transversally, that is, $W_x = 0$. Energy was allowed to flow in the Nernst measurement—those conditions were adiabatic. Now, the Righi-Leduc coefficient (again $cm^2/s \cdot K$) is

Effect	Definition	Conditions	Form
Hall	$\mathcal{E}_y/J_x$	$J_y = \nabla T_x = \nabla T_y = 0$	$-\rho_{xy}$
Ettinghausen	$(\nabla T_y)/J_x$	$J_y = \nabla T_x = W_y = 0$	$-\Pi_{xy}/\kappa_{exx}$
Transverse Magnetoresistance	$\mathcal{E}_x/J_x$	$J_y = \nabla T_x = \nabla T_y = 0$	ρ_{xx}
Nernst	$(\nabla T_x)/J_x$	$J_y = W_x = \nabla T_y = 0$	Π_{xx}/κ_{exx}
Longitudinal Magnetoresistance	$\mathcal{E}_z/J_z$	$\nabla T_z = 0$	ρ_{zz}
Righi-Leduc	$(\nabla T_y)/(\nabla T_x)$	$J_x = J_y = W_y = 0$	$\kappa_{exy}/\kappa_{exx}$
Nernst	$\mathcal{E}_y/\nabla T_y$	$J_x = J_y = \nabla T_y = 0$	$-\mathcal{Q}_{xy}$
Transverse thermal conductivity	$-W_x/\nabla T_x$	$J_x = J_y = \nabla T_y = 0$	κ_{exx}
Ettinghausen-Nernst	$\mathcal{E}_x/\nabla T_x$	$J_x = J_y = \nabla T_y = 0$	$\mathcal{Q}_{xx}$
Longitudinal thermal conductivity	$-W_z/\nabla T_z$	$J_z = 0$	κ_{ezz}
Seebeck	$\mathcal{E}_z/\nabla T_z$	$J_z = 0$	$\mathcal{Q}_{zz}$
Peltier	W_z/J_z	$\nabla T_z = 0$	Π_{zz}

Table 9.2: Some of the major electromagnetothermal effects and their measurement conditions with the resulting characterization.

$$\begin{aligned} C_{RL} &\equiv \frac{\partial T/\partial x}{B_z \partial T/\partial y} \\ &= \frac{n\mu_c^2}{q\kappa_e T}\left(\frac{\langle\tau_{\mathbf{k}}^2 E^2\rangle}{\langle\tau_{\mathbf{k}}^2\rangle} + \frac{\langle\tau_{\mathbf{k}}^2\rangle\langle\tau_{\mathbf{k}} E\rangle^2}{\langle\tau_{\mathbf{k}}\rangle^4} - \frac{\langle\tau_{\mathbf{k}}^2 E\rangle\langle\tau_{\mathbf{k}} E\rangle}{\langle\tau_{\mathbf{k}}\rangle^3}\right). \end{aligned} \tag{9.67}$$

So, there are numerous effects in these electrical-magnetic-thermal exchanges of energy. The Ettinghausen effect has its coefficient (C_E, in units of $cm^3K/W \cdot s$) given by $(\partial T/\partial x)/J_y B_x$. The conditions of observations of these effects—adiabatic, that is, no exchange, and isothermal, that is, temperature gradient absent—matter. We did not stress these in the discussion, nor did we discuss the wider numbers of them, but the Table 9.2 summarizes these together with the main consequences of interest.

9.3 *Frequency dependence*

Our derivation of Boltzmann transport in the relaxation time approximation is predicated on the validity of the deployment of $\tau_{\mathbf{k}}$ as a valid parameter for relaxation time. From this, we have extracted valuable averaged properties of interest that also let us evaluate current and energy flow. One consequence of the underlying assumption is that there are time scales of validity.

It stands to reason to consider this to be of the order of several $\tau_{\mathbf{k}}$ for the purposes of the averaged calculations. Since momentum relaxation times are a few 10s of *fs* to sub-*ps*, the frequencies up to which we may employ the derivative equations from Boltzmann transport is certainly in the 100s of *GHz*, so long as the right properties of the material at that frequency are employed.

It should be noted that the foundation is rigorous. If the relaxation time is made large, the Boltzmann transport equation, while including the $E(\mathbf{k})$ description of what is happening in the semiconductor, is just as valid for mesoscopic transport. What it does not account for is the partitioning that arises in the state occupation of the channels and the associated correlation. The $\mathbf{k}$ term tells us the flow and the group velocity argument that we employed to find $g_q = 2e^2/h$ must follow from it. It is just not a useful way to approach mesoscopic problem. Boltzmann transport is much more useful where scattering is taking place, and there are all the spatial and momentum effects that need to be accounted for.

Our interest is still in a steady-state solution, so $df/dt = 0$. Let the electric field $\boldsymbol{\mathcal{E}} = \boldsymbol{\mathcal{E}}_0 \exp(-i\omega t)$ be the only stimulus. Then, our impulse-based form of solution is

$$f = f_0 + q\tau_{\mathbf{k}} \frac{\partial f_0}{\partial E} \mathbf{v} \cdot \boldsymbol{\mathcal{E}}_0 \exp(-i\omega t), \tag{9.68}$$

which lets us find the time dependence:

$$\begin{aligned}
\frac{df}{dt} &= q\tau_{\mathbf{k}} \frac{\partial f_0}{\partial E} \left(-i\omega \mathbf{v} + \frac{d\mathbf{v}}{dt} \right) \cdot \boldsymbol{\mathcal{E}}_0 \exp(-\omega t) \\
&\approx -iq\tau_{\mathbf{k}} \frac{\partial f_0}{\partial E} \omega \mathbf{v} \cdot \boldsymbol{\mathcal{E}}_0 \exp(-\omega t) \\
&= -i\omega q\tau_{\mathbf{k}} \frac{\partial f_0}{\partial E} \mathbf{v} \cdot \boldsymbol{\mathcal{E}}_0 \exp(-i\omega t) \\
&= -i\omega (f - f_0),
\end{aligned} \tag{9.69}$$

where the approximation is one of neglecting higher order terms. Since the velocity follows from

$$\frac{d\mathbf{v}}{dt} = -q \frac{1}{\mathbb{M}} \cdot \boldsymbol{\mathcal{E}}_0 \exp(-i\omega t),$$

we should expect a solution that exhibits consequences of non-linearity: second order terms in the electric field amplitude, and at twice the frequency. As before, we maintain fields small for linear approximation to be valid. Since the distribution satisfies Equation 9.4,

$$\frac{df}{dt} = -\mathbf{v} \cdot \left(\frac{\partial f}{\partial E} \mathbf{F} + \nabla_{\mathbf{r}} f \right) - \frac{f - f_0}{\tau_{\mathbf{k}}}, \tag{9.70}$$

it follows that

$$\begin{aligned}
-i\omega (f - f_0) &= -\mathbf{v} \cdot \left(\frac{\partial f}{\partial E} \mathbf{F} + \nabla_{\mathbf{r}} f \right) - \frac{f - f_0}{\tau_{\mathbf{k}}} \\
(f - f_0) \left(\frac{1}{\tau_{\mathbf{k}}} - i\omega \right) &= -\mathbf{v} \cdot \left(\frac{\partial f}{\partial E} \mathbf{F} + \nabla_{\mathbf{r}} f \right) \quad \text{or} \\
f - f_0 &= -\frac{\tau_{\mathbf{k}}}{1 - i\omega\tau_{\mathbf{k}}} \mathbf{v} \cdot \left(\frac{\partial f}{\partial E} \mathbf{F} + \nabla_{\mathbf{r}} f \right).
\end{aligned} \tag{9.71}$$

This solution is similar in form to Equation 9.5, except that the form of the relaxation time has changed from a real term to a complex term. We now have $\tau_{\mathbf{k}} \mapsto \tau_{\mathbf{k}}/(1 - i\omega\tau_{\mathbf{k}})$. Define a complex relaxation time $\tau_{\mathbf{k}}^{c}$,

$$\tau_{\mathbf{k}}^{c} = \frac{\tau_{\mathbf{k}}}{1 - i\omega\tau_{\mathbf{k}}}, \tag{9.72}$$

and all of our derivations are applicable. The complex relaxation time will mean that one will observe lag effects. The distribution function itself becomes a complex function. While at low frequencies the time dependence can be followed, higher frequencies will have lag—a phase delay—reflected in the imaginary parts. In will precipitate attenuation and dissipation, but this frequency dependence—as it does with cyclotron resonance where magnetic field causes a radial response that then makes material characteristics and properties characterizable—will also let us extract properties of materials.

9.4 Summary

THE RESPONSE OF THE STATISTICAL ENSEMBLE of electrons in the semiconductor under various stimuli—single or multiple present simultaneously—is at the heart of the usage of semiconductors. This response is the confluence of these particles being perturbed by each other and by the environment they are in, as well as responding to the external stimuli. The Boltzmann transport equation, with scattering captured through a momentum relaxation time $\tau_{\mathbf{k}}$ under the assumption that the scattering is randomizing, was our semi-classical tool—the quantum nature is embedded in the description of the particle motion and their interaction—to predicting the response. In order to make the tool general, we derived a generalized impulse $\mathbf{G}$ through which one could see the response under electric, magnetic and thermal linearized perturbation from the stimuli. The distribution function varies as a product of $\partial f_0/\partial E$, which is the linear variation of the distribution function under energy changes in the crystal, and $\mathbf{v} \cdot \mathbf{G}$, which is the amount of work done—the energy change—through this interaction with the stimuli. The reason why the Fermi surface is key to understanding transport properties is that $\partial f_0/\partial E$ has its largest changes at the Fermi energy.

We considered only the Boltzmann transport equation for electrons. We could have written one for phonons too. We did not, assuming that phonons are pretty much in equilibrium for the conditions of interest to us. At the mesoscale, this may very well not be a good assumption.

The chemical potential changes, the electrical potential changes and the vagaries of the bandstructure that are reflected in the response behavior of the electron to forces in the presence of the crystal energetics are all thus included in this response under the assumption of randomization, scattering dominated transport and linearization.

We found how the effective mass tensor appears in this generalized response, even though, to simplify the description, we subsequently replaced it by an effective mass.

If only electrical forces are in play, one sees the classical electrical response. If one has electrical and magnetic forces present, one sees the Hall response. If electrical and thermal forces are present, one sees an electrothermal response. If magnetic and thermal forces are present, then one sees a magnetothermal response, which has correspondences to the electrothermal response.

But underlying all this was the incorporation of the randomizing momentum relaxation time $\tau_{\mathbf{k}}$ that is a function of energy. Slow, low energy electrons will respond quite differently than fast, higher energy electrons. One can see that for a magnetic field, where Lorentz forces will have different angular consequences. Thermally, too, this will be different, since population characteristics are different. This will be through the way the chemical potential changed. And scattering time should change with energy, since scattering processes themselves have those energy dependences as transitions take place between states with different momentum. And so will the response through expectations on the powers of the relaxation times and energy.

For electrical force alone, we could show how mobilities change, how majority and minority carriers can behave differently, how diffusivity and mobility can be related to each other and how these responses change as the field increases and therefore the energy of the carriers increases.

Including magnetic field in this description showed us the development of Hall voltage and how this allows us to determine Hall mobility and carrier concentrations and carrier types of the material.

When thermal gradients exist, there is now the flow of particles, with their charge and their energy in the system. Like magnetic forces, where, although energy is not imparted but the motion is bent orthogonally, there are going to be equivalent electrothermal interactions. So, we found Thompson, Seebeck and Peltier coefficients with corresponding phenomena that reflect external conditions forced on the system, which are of relevance to thermoelectric conversion.

When thermal and magnetic stimuli are present, we noted the equivalents of the thermoelectric effects—somewhat complicated in their equations because of the way magnetic field interacts with particle motion—and could again see the appearance of fields. If one were to have electrical, thermal and magnetic stimuli present, this becomes a little more complicated, but, in all these cases, our impulse-based approach could make reasonable predictions so long as the underlying assumptions of the derivation held.

When the stimulus is time dependent, this approach can be extended through the introduction of a complex relaxation time where the static relaxation time and the signal's time constant couple. The frequency dependence appears in the response through a phase lag, assuming that the applied signal is slower than the relaxation time of the system.

9.5 *Concluding remarks and bibliographic notes*

SEMICONDUCTORS, METALS AND INSULATORS have many similarities and many differences. And since particle distribution, the number of electrons or the nature of dominance of electrons or phonons in the stimuli response can be different, one will see quite a bit of difference between their properties and their response to stimuli. Our focus in this chapter has been on semiconductor specifics. Here, electrons (or their antiparticle hole) are certainly important for many of the observed stimulation responses. But when thermal considerations are important, phonons will enter too.

A starting point—good for the fundamentals and intuition—is provided by two old books, one by Ehrenberg[1] and the other by Ziman[2]. Both books are excellent sources for understanding how interactions affect the transport of a charge cloud as the variety of interactions happen. From the same time period is the book by Smith[3], which brings out the flow interdependences due to multiple stimuli.

A to-the-point treatment of this transport under multiple interactions is the text by Wolfe et al.[4]. This book is a wonderful balance between brevity and analytic completeness of an argument. Another treatment, at a similar level, is by Balkanski and Wallis[5]. A more advanced discussion is by Ridley[6].

The Boltzmann transport equation, written with $\partial f/\partial t$ to account for all events—scattering specifically—not included in the explicit time dependence in spatial and momentum dependence, is an accurate semi-classical description, even if scattering is not randomizing, and even if only a few scattering events take place, or even if there are only a few carriers. One just has to follow the entire ensemble with all particles accounted for individually and the scattering events accounted for. This is an important approach

[1] W. Ehrenberg, "Electric conduction in semiconductors and metals," Oxford (1958)

[2] J. M. Ziman, "Electrons and phonons," Oxford (1960)

[3] R. A. Smith, "Wave mechanics of crystalline solids," Wiley, (1961)

[4] C. M. Wolfe, N. Holonyak and G. E. Stillman, "Physical properties of semiconductors," Prentice Hall, ISBN 0-13-669961-8 (1989)

[5] M. Balkanski and R. F. Wallis, "Semiconductor physics and applications," Oxford, ISBN 978-0-19-851740-5 (2007)

[6] B. K. Ridley, "Quantum processes in semiconductors," Oxford, ISBN 0-19-850-580-9 (1999)

to understanding the response where neither no scattering nor a lot of scattering suffices, and quantum consequences are significant. It requires Monte Carlo techniques where various random events—even if not randomizing—are accounted for properly. Peter Price's seminal paper[7] is the classic reference for this. A proper quantum transport formalism starting from the semi-classical is in the book by Fischetti and Vandenberghe[8]. The discussion in this text of open boundary conditions, which is important for understanding nanoscale devices, along with the other discussions, is highly recommended reading.

Marder[9] discusses the transport phenomena, particularly with attention to Fermi liquids, so high carrier concentration systems. The discussion includes electric, magnetic and thermal effects. A good text for understanding the nature of thermoelectric power in metals—which have very filled conduction bands but considerably different scattering than semiconductors—is by Blatt et al.[10]. This book is a good source for understanding the nature of scattering in metals as well as the consequences of magnetic impurities and other scattering for thermoelectric power at low temperatures, where metals are the most efficient.

[7] P. J. Price, "Monte Carlo calculation of electron transport in solids," Semiconductors and Semimetals, **14**, Academic, ISBN 0-12-752114-3 (1979)

[8] M. V. Fischetti and W. G. Vandenberghe, "Advanced physics of electron transport in semiconductors and nanostructures," Springer, ISBN 978-3-319-01100-4 (2016)

[9] M. P. Marder, "Condensed matter physics," Wiley, ISBN 978-0-470-61798-4 (2010)

[10] F. J. Blatt, P. A. Schroeder, C. L. Foiles and D. Greig, "Thermoelectric power of metals," Plenum, ISBN 13 978-1-4613-4268-7 (1976)

9.6 Exercises

1. We will work on understanding the conductivity tensor for silicon in this problem. For electrons, there are 6 constant energy surfaces just above the conduction bandedge. These surfaces are ellipsoidal and are described by two effective masses: m_t for the transverse mass, and m_l for the longitudinal mass. For example, for the ellipsoid along the z axis, the surface of constant energy is described by

$$E(\mathbf{k}) = E_c + \frac{\hbar^2}{2}\left[\frac{(k_x - k_{x0})^2}{m_t} + \frac{(k_y - k_{y0})^2}{m_t} + \frac{(k_z - k_{z0})^2}{m_l}\right].$$

   - The acceleration $\mathbf{a} = \ddot{\mathbf{r}}$ response to an external driving force $\mathbf{F}$ is given in terms of the inverse of the effective mass tensor $(1/\mathbb{M})\mathbf{F}$. Show that the total acceleration responds isotropically to the force, that is, $\mathbf{a}_{total} = M^{-1}\mathbf{F}$, where M is a scalar. Calculate the contributions from all six ellipsoids and find M explicitly.
   - What is the numerical value of this effective mass ratio M/m_0 for silicon?

We tackled, through superposition arguments, the isotropicity of response of the electron conductivity in the class. Now, we do this by direct manipulation.

- Refer to the first part of this question and write out the total effective mass tensor for the conductivity of electrons in *Si* explicitly.

- Do the same for the holes, assuming that there are two hole bands with different effective masses. Also give a numerical value for the effective (conductivity) mass ratio for the holes in *Si*. **[S]**

2. What should determine the shortest time scale (or highest frequency scale) for determining the applicability of the Boltzmann transport equation? **[S]**

3. When a metal is subjected to a magnetic field $\mathbf{B} = B_z\hat{\mathbf{z}}$, the wavevector $\mathbf{k}$ of an electron state traces an orbit in the $\mathbf{k}$-space, with $\mathbf{k}_z$ and energy E constant. We now include ϕ as the angle swept by the wavevector,

$$d\phi = \frac{1}{m^* v_\perp}\hbar dk_\parallel,$$

where $\perp$ and $\parallel$ indicate respectively components in directions perpendicular to $\mathbf{B}$ and parallel to the path. Set up the Boltzmann transport equation in terms of E, k_z and ϕ for a small electric field. Keeping only first order terms of the electric field, find the current density and the conductivity tensor. From this, show that when the orbits are closed in the k_x-k_y plane, the magnetoresistance saturates at high fields. Hence, the orbits are open. Show that it grows as the second power in a magnetic field. **[M]**

4. Hall measurements are performed on a sample in the configuration shown in Figure 9.12 with $I_x = 5\ \ mA$, $B_z = 0.1\ \ T$ and $T = 77\ K$. Measured values are $V_x = 10\ V$, $V_y = -5\ V$. Assuming that $m_e^* = 0.1m_0$, $m_h^* = 0.5m_0$ and $\tau_\mathbf{k} = \tau_0\eta^{3/2}$ due to impurities, determine the following:
   - the polarity type of the material
   - the Hall constant
   - the carrier concentration
   - the conductivity
   - the conductivity mobility
   - the Hall mobility **[M]**

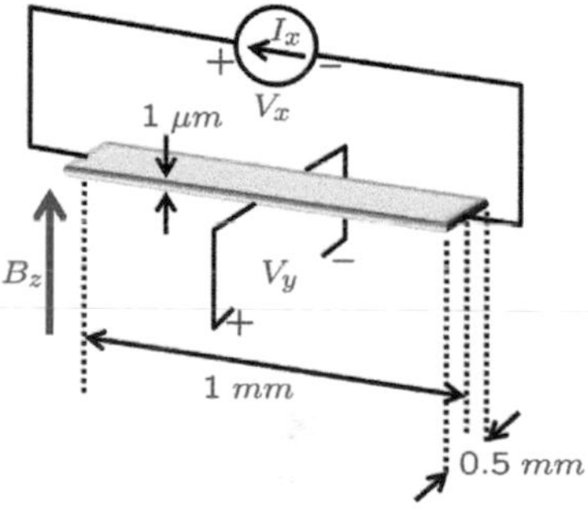

Figure 9.12: Hall measurement in a semiconductor with the conditions for measurement.

Note the coordinate orientations.

5. Determine an expression for the Hall factor r_H, assuming Fermi statistics and $\tau_\mathbf{k} = \tau\eta^r$. Plot r_H versus η ($\eta = (E - E_c)/k_BT$) for $-4 \leq \eta \leq 10$, with $r = -1/2$ and $r = 3/2$. **[S]**

6. We wrote an expression for $\langle\tau(\mathbf{k})\rangle$ in terms of the distribution function and $\tau(\mathbf{k})$ by working through the expectation of velocity

$$\langle\mathbf{v}\rangle = \frac{q\int_{-\infty}^{\infty}\tau_{\mathbf{k}}(\partial f_0/\partial E)\mathbf{v}(\mathbf{v}\cdot\boldsymbol{\mathcal{E}})\,d\mathbf{v}}{\int_{-\infty}^{\infty} f_0\,d\mathbf{v}},$$

which ended up as

$$\langle\tau_{\mathbf{k}}\rangle = \frac{4}{3\sqrt{\pi}}\left(r+\frac{3}{2}\right)!\tau_0\frac{\mathcal{F}_{r+1/2}(\zeta)}{\mathcal{F}_{1/2}(\zeta)}$$

for $\tau_{\mathbf{k}} = \tau_0\eta^r$. Write a form for $\langle\tau_w(\mathbf{k})\rangle$—an expectation for the time constant for energy—in terms of $\tau(\mathbf{k})$ and the distribution function. Or at least point to how you might derive it with a starting point for an expression. **[S]**

7. Should energy relax faster or slower than momentum? **[S]**

8. Plot the distribution function in energy for electrons in *GaAs*, $f(E)$, when a field of 0.5, 1.0 or 2.0 kV/cm is applied. Assume that the relaxation time is 10^{-12} *s* and that the electron gas is non-degenerate. **[S]**

9. An acoustic wave of the form $\mathcal{A}\exp[i(\mathbf{q}\cdot\mathbf{r}-\omega t)]$ propagates through an n-type semiconductor with a parabolic band, where it produces a variation in energy of the electrons

$$E = \overline{E}_1\exp[i(\mathbf{q}\cdot\mathbf{r}-\omega t)].$$

Since the force on an electron is $\mathbf{F} = -\nabla_{\mathbf{r}}E$, show that, in the relaxation time approximation, a good relationship for the electron distribution is

$$f = f_0 + \frac{\partial f_0}{\partial E}\frac{i\tau_{\mathbf{k}}\mathbf{v}\cdot\mathbf{q}E}{1+i\tau_{\mathbf{k}}\mathbf{v}\cdot\mathbf{q}}.$$

Does this distribution provide conduction? **[M]**

10. Magnetic breakdown is a phenomenon that can occur when an electron can transition across a bandgap under the application of a magnetic field. Metals with small energy gaps at zone boundaries exhibit this. Show that the condition for this magnetic breakdown is

$$\hbar\omega_c E_F > E_g^2,$$

where the symbols have their usual meaning. **[S]**

11. A semiconductor has both electrons and holes in reasonable concentration. Such a condition is called *ambipolar*. Find an expression for the Hall constant under $\omega_c^2\tau^2 \ll 1$. What does the Hall constant expression reduce to when the mean free paths are independent of velocity? **[S]**

12. Show that, in the presence of a concentration gradient in an isotropic semiconductor, a diffusion flux $-\langle\lambda_{\mathbf{k}}\rangle\langle v\rangle_\theta/3$ times the concentration gradient comes about. The $\langle v\rangle_\theta$ is the expectation of the thermal velocity magnitude. If the fractional variation of concentration over the mean free path of $\langle\lambda_{\mathbf{k}}\rangle$ is small, and the mean free path is independent of velocity, then the distribution function can be written in the form $n(\mathbf{r})f(\mathbf{v})$, with $n(\mathbf{r})$ as a concentration at any position, and $f(\mathbf{v})$ a the velocity distribution which is not dependent on the position $\mathbf{r}$. **[S]**

13. Find a differential equation that can describe the ambipolar drift and diffusion of a distribution of excess electron and hole pairs in a semiconductor whose thermal equilibrium concentrations are n_0 and p_0 and comparable to the excess populations and where approximate electric neutrality prevails across the crystal. **[M]**

14. Interestingly, following Exercise 13, a distribution of excess electron-hole collection drifts in the direction of the field, despite electrons of the n-type semiconductor moving the other way. Consider a one-dimensional semiconductor. Take the motion of an excess carrier density pulse, where $\Delta n = \Delta p$ within the pulse and vanishing outside, and explain physically why such a behavior is observed. **[M]**

15. Could one make a thermoelectric cooler using the Peltier effect, using metals only, that is, no semiconductors? The complement of this phenomenon—the Seebeck effect—is employed in thermocouples without resorting to semiconductors. Provide a short reason, please. **[S]**

16. A classic experiment for measuring the drift mobility of minority carriers is the Haynes-Shockley experiment of Figure 9.13. It measures the time needed for a distribution of excess carriers injected at an emitter point contact to drift to a collector point contact in the presence of an electric field $\mathcal{E}_0$. A simple interpretation would be that carriers drift with a velocity $\mu^*\mathcal{E}_0$, so the time T to move a distance d will be $T = d/\mu^*\mathcal{E}_0$, and hence the mobility. Consider a one-dimensional diffusion with a delta function pulse of excess carriers injected at $z = 0$ and $t = 0$. Include diffusion and recombination, and show that the maximum collector signal appears at $z = d$ at a time T_0 of

$$T_0 = \frac{\left[1 + (4\alpha d^2/\mathcal{D}^*)\right]^{1/2} - 1}{4\alpha}, \text{ where}$$

$$\alpha = \frac{\mu^{*2}\mathcal{E}_0^2}{4\mathcal{D}^*} + \frac{1}{\tau} = \frac{e\mu_p\mathcal{E}_0^2}{4k_BT}\frac{b(n-p)^2}{(n+p)(bn+p)} + \frac{1}{\tau}.$$

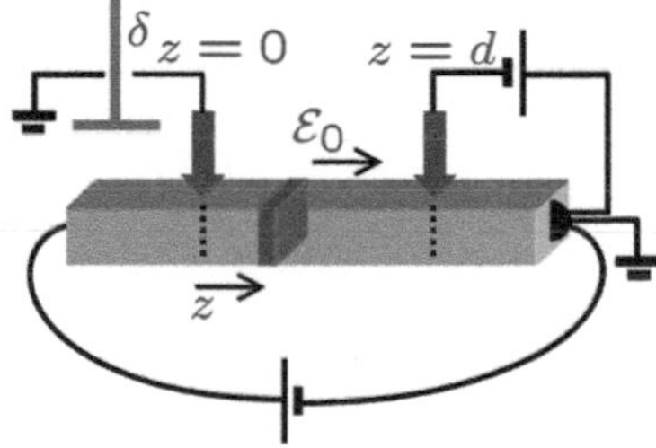

Figure 9.13: The Haynes-Shockley experiment to determine drift mobility. An excess carrier pulse is injected into a background electric field of $\mathcal{E}_0$ at the emitter, and its response appears at the collector following drift of the excess carrier pulse.

A superscript of $*$ is meant to denote that this is a measurement based on an excitation and therefore a small local perturbation. μ and $\mathcal{D}$ are mobility and diffusivity, respectively, in thermal equilibrium.

The mobility is related as

$$\mu^* = \frac{d}{\mathcal{E}_0 T_0}\left[(1+x^2)^2 - x\right], \text{ where}$$
$$x = \frac{2k_B T}{e\mathcal{E}_0 d}\left(\frac{T_0}{t} + \frac{1}{2}\right)\left(\frac{n+p}{n-p}\right). \quad \textbf{[M]}$$

17. The Ettinghausen, Nernst and Righi-Leduc effects are effects of magnetic induction that are equivalent to what happens with electric field. It is appropriate to explore these a little more, so that they are not forgotten, even if one usually doesn't encounter them. Identify and show which one of the following list of effects can be canceled in a Hall experiment by making four measurements with reversal in **B** and **J** during a Hall measurement experiment: (a) the Etinghausen effect, (b) the Nernst effect, (c) the Righi-Leduc effect, (d) the thermoelectric effect and (e) the ohmic (*IR*) voltage drop, such as that due to contacts. **[S]**

We might have to study them carefully if magnetic fields and spin-related phenomena become important in electronic devices

10
Major scattering processes

How SCATTERING—as a quantum-mechanical or classical transition—arises in perturbation interactions—in thermal equilibrium or off it—was one of the starting thoughts in Chapter 1, where an ionized impurity's Coulomb potential perturbation was tied to the scattering that the carrier undergoes through the interaction. The Golden rule was the tool that let us calculate the scattering matrix element, given the from and to states' quantum descriptions. In the case of an ionized impurity interacting with a nearly free electron, the electron as a charge wave undergoes a change of momentum during a short interaction—short, but not so short as to break the Golden rule's assumptions—whose perturbation Hamiltonian is now known, and hence the scattering matrix element can be calculated. We noted the correspondence between scattering and momentum quantum-mechanically and captured cross-section and velocity classically in Chapter 8. For use in Boltzmann transport, by taking all the from states, which the distribution function represents, and all the available states where the electrons can potentially change to, and integrating, one has a way to determine the time-exclusive dependence due to this interaction. If it fits into a relaxation time approximation, and we have found that this interaction has an energy dependence since the fast electrons deviate less from their path, the Boltzmann transport equation is more analytically tractable. Classically, this same interaction can be evaluated. It is a problem with an electron of an initial momentum undergoing a change in momentum due to a Coulomb attractive or repulsive force. So, there also exists a quantum and classical correspondence here, with the correspondence principle connecting.

Ionized impurity, however, is not the only form of scattering that charge particles will undergo in the variety of semiconductor constructs one employs. There are interfaces; temperature causes an

Semiconductor Physics: Principles, Theory and Nanoscale. Sandip Tiwari.
 Published 2020 by Oxford University Press. DOI: 10.1093/oso/9780198759867.001.0001

atomic motion of a multiple phonon-abstracted variety, which is a form of deformation as well as a charge effect dependent on ionicity viewable through the Born-Oppenheimer approximation lens; two-dimensional materials will have exposed surfaces with physisorbed species, and so on.

Any occurrence that is not periodic, or not accounted for in the $(E, \mathbf{k})$ description of the electron, is a perturbation with an interaction effect that is a scattering.

Table 10.1 is a limited summary of the major semiconductor scattering processes broken across three categories that identify a broader categorization of the source mechanism: crystal vibrations, the crystal medium origins and those arising from mobile particle interactions with each other. Crystal vibrations, modeled through the phonons, lead to interactions with acoustic and optical branches. The longitudinal and transverse branches may even have different interactions, since momentum must also be conserved.

Deformation is the movement of atoms—a form of tensile and compressive straining of the crystal—whose primary effect can be seen as a small perturbation in the $E(\mathbf{k})$ itself. If the atoms have an effective charge, then one must account for the piezoelectric consequence. Non-polar and polar semiconductors will also show different optical phonon interactions, where presence and absence

Scattering	Cause	Interaction	Form	Where
Crystal vibrations	Phonon	Carrier-acoustic	Deformation	
			Piezoelectric	
		Carrier-optical	Non-polar	
			Polar	
Crystal medium	Charged impurity	Carrier-impurity	Coulomb	
	Neutral impurity	Carrier-impurity	Perturbation	Low temperature
	Surface/interface	Carrier-boundary	Phonons, roughness and fluctuations	
	Alloy	Carrier-alloy	Fluctuations	
	Defects	Carrier-defect	Crystal defect	
	Dislocations	Carrier-dislocation	Crystal defect	
Mobile particles	Carrier-carrier	Electron-electron	Coulomb	High concentration and inversion
		Hole-hole	Coulomb	High concentration and inversion
		Electron-hole	Coulomb	Minority transport
	Coupled particles	Polariton-surrounding	Various	Surfaces and coupled situations

Table 10.1: Scattering mechanisms, interactions and their formal origins and locales.

of charge matters. The crystal itself has random distribution of various causes. Impurities and ternary and higher-order compositions are non-periodically distributed. Interfaces have potential steps, material adjacent to even an ideal interface with non-periodic potential perturbations. SiO_2 is amorphous, and the ternaries and crystalline materials will also have doping. Defects—charged and neutral—will exist both in bulk and at surfaces. Mobile carriers also interact with other moving particles: electrons with electrons, electrons with holes if conditions are bipolar, and if photons or other quasiparticles such as excitons, or other polaritons, are present, then there are interactions associated with these. Each one of these can be quite complex. Take, for example, an ionized impurity. A charge is screened. So, even an ionized impurity will have screening effects that need to be incorporated.

Our focus of this chapter is to show that our generalized algorithm can be applied to all these circumstances to various levels of detail—it is only a matter of detailed exercising. So, we will discuss the salient points, and summarize conclusions for a select set.

A polariton, as noted earlier, is our quasiparticle representation when two energy excitations interact because of energy and momentum matching, and the degeneracy is lifted. The resulting state representation is through polaritons. Electron-photon, electron-phonon and other interactions can create such coupled-particle assemblies, where the coupling of one a form of "dressing" by the other. An exciton is not usually called a polariton but is another one of these coupled-particle assemblies.

10.1 *General comments on scattering*

SCATTERING IS A RESPONSE that is a transition of the state of a particle undergoing scattering as a result of a perturbation-induced interaction. Electrons deflecting due to interaction with atomic motion is a scattering, but so is an electron and a hole recombining—an electron changing its state from the conduction band to the valence band—producing a photon and scattering. We just tend to visualize these a little differently, but the approaches one uses to calculate the rates of these interactions are quite the same. When in thermal equilibrium, all the transition rates state to state must balance for each connecting process. No energy form and no individual interaction then may transfer energy to another form in net. There will be fluctuations that one may measure—as in noise—but, on average, the system remains in a macrostate where observables maintain their average properties. Entropy then is the maximum. So, when one disturbs the system from equilibrium, these scattering process rates that connect states will change. Equilibrium is disturbed. There is an excess of one process over another. In a *p-n* junction, in forward bias, there is excess recombination in net and in the individual state couplings (electrons and holes recombining both in the transition regions and in the quasineutral regions), and, in reverse bias, there is excess generation in net and

The study of scattering has a long history. Rutherford's fame by opening up our understanding of atoms is from scattering experiments by beams of He^{2+} through a gold foil, thus showing the existence of the nucleus. Nuclear fission is an extreme consequence of such a scattering event. In all this, the energies and species involved matter. U^{235} or Pu^{239} are extreme examples for fission via neutrons: 4 *kg* of Pu^{239} suffices to create >300 *TJ* of energy. And a thin *Be* foil suffices via scattering of the neutrons to suppress the criticality. Oppenheimer, following his graduation from Harvard, and having been declined at Cambridge by Rutherford, became a student of J. J. Thompson. He was put to the task of making proper *Be* films—a highly toxic material. Oppenheimer detested this to the extent that he put some of the material in the apple of his tutor, Patrick Blackett. Unlike today's absolutism and zero tolerance that doesn't seem to recognize the development of a mind and body's chemistry with age, Cambridge was foresighted enough to send him off to Gottingen to do theory with Max Born. *Be* remains important in much of the nuclear saga. A number of science workers died in the Manhattan Project due to radioactivity and especially in one particular accident where criticality did happen. The federal apparatus had to work hard to suppress reports and compensation. Even Fermi, who was aware of radioactivity's consequences, Marie Curie, having preceded in 1934 from her work, and Oppenheimer himself were victims of cancer. Science and engineering has quite a bit of this element of foolishness precipitated by an exciting sandbox. Richard Feynman in the individual state couplings. A near-ideal *p-n* diode has only a small amount of this, and it makes for quite a nonlinear and useful device. But, so is a light-emitting diode—a more complex implementation of the *p-n* form—where the forward-biased excess recombination becomes useful as a source of light because it is dominated by a direct photon-emitting process in a suitable material.

is the only scientist involved with the Manhattan Project who seems to have stayed away from defending his involvement in it.

Scattering is also the means by which any disturbance from thermal equilibrium is being pushed back toward equilibrium. This is an entropic force, an attempt toward maximization of "lack of knowledge," which writing $\partial f/\partial t = -(f - f_0)/\tau_{\mathbf{k}}$ in the relaxation time approximation of the Boltzmann transport equation states. So, in this process of attempting to restore equilibrium, different scattering processes will have different levels of effectiveness.

We have explored ionized impurity scattering as an example of the use of the Golden rule in Chapter 1. A carrier moving with high momentum deviates less from its path than a slower carrier does. The interaction time and the impulse of the interaction in the former is smaller. There is now an angular dependence that is a function of the magnitude of momentum in the quantum picture and of velocity in the classical picture. If scattering is a small-angle scattering, it lets the carriers keep their momentum longer, so it is not as effective at randomization. This is what our writing of Equation 8.52 implied. Scattering by phonons, impurities, other electrons, defects such as dislocations, surface states and other randomness there—each will have some form of angular dependence. Some will have a strong dependence, and some will have none at all. And which of these forms should be of most concern to us will depend on the conditions. High doping will make impurities important. Operation at room temperature will keep phonon interactions strong. If one has low impurities but high carrier densities, such as in inversion regions, interparticle scattering will be important.

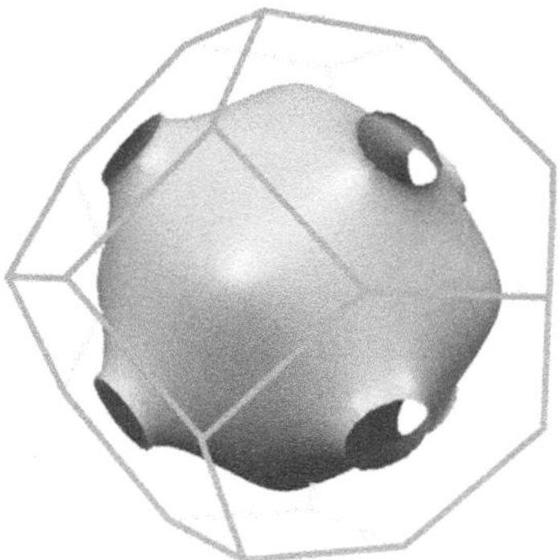

Figure 10.1: The Fermi surface of copper, together with first Brillouin zone boundaries.

Metals have their conduction bands quite filled. Nearly half of the Brillouin zone is filled. Figure 10.1 shows the Fermi surface of copper. Since filled and empty states are simultaneously present within a few k_BT of the Fermi energy, and energy conservation and momentum conservation must happen during the scattering event, a large change in $\mathbf{k}$ is likely. Optical phonons are spread out nearly evenly in momentum across the entire Brillouin zone. This implies that $\langle\theta\rangle \approx \pi/2$ in metals. In semiconductors, with very few carriers compared to the number of states, the Brillouin zone is barely occupied, whether at the center Γ point in direct gap materials or off it, as in indirect materials, and $\langle\theta\rangle \ll \pi/2$. So, semiconductors maintain closer alignment for the momentum during scattering, while metals orthoganalize it. Scattering in metal is more

This argument regarding semiconductors requires some care. Electrons will scatter between the different non-zone-center valleys; however, the probability of staying within the valley is large, that of scattering to nearest set of symmetric valleys is less, and that of scattering farther out to the mirror valley is even less.

randomizing, and therefore, a time constant $\tau_{\mathbf{k}} \approx \tau_c$ independent of energy seems to work out. This time τ_c can be viewed, classically, as a consequence of N scattering centers with a scattering cross-section σ interacting with electrons moving at a speed v. If an electron exists within a volume σv, it interacts with the scattering center, so there are N such elemental interaction volumes undergoing scattering. Therefore, $\tau_c \approx 1/N\sigma v$ in this classical picture of scattering. $N\sigma v$ is the rate at which this scattering interaction is taking place anywhere. We now place this scattering cross-section on a firmer quantum-mechanical footing.

But this is true only for matters related to conductivity, and this is a Drudeian thought.

10.1.1 *Scattering cross-section*

THE SCATTERING CROSS-SECTION, which we have now encountered numerous times, represents an areal parameter—momentum- and energy dependent—that represents the areal projection of the region of influence affecting an incident particle wave and causing it to scatter. It can be viewed both classically and quantum-mechanically. Classically, it is an area of obstruction. The obstruction's shape affects how particles incident will get deflected. Quantum-mechanically, it is a parameter with areal units reflecting the matrix element of the interaction, the matrix element being the coupling term between an incident and a scattered state due to a potential perturbation.

We illustrate this cross-section notion through elastic scattering. Assume an arbitrary field—our generalized impulse arose in such a collection of fields that affected motion divergently or curly—and we wish to relate the matrix element and momentum relaxation time to this perturbation. There exists a scattering center at the origin of Figure 10.2 that causes a carrier of wavevector $\mathbf{k}$ to scatter to wavevector $\mathbf{k}'$ in the presence of this impulse $\mathbf{G}$. We have chosen a coordinate system so that the initial state is pointed in $\hat{\mathbf{z}}$ and $\mathbf{G}$ is in the $k_x k_z$ plane. Off-equilibrium, the distribution function, because $\mathbf{v} = \hbar\mathbf{k}/m^*$, is

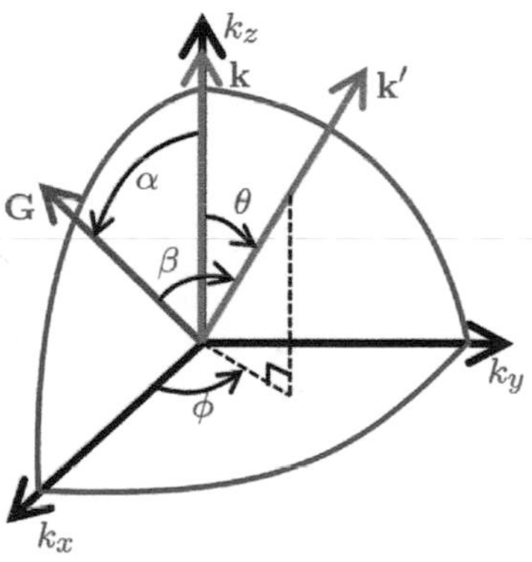

Figure 10.2: An electron traveling along the $\mathbf{k}_z$ direction, undergoing scattering due to a center at the origin in the presence of an impulse $\mathbf{G}$.

$$f(\mathbf{k}) = f_0(\mathbf{k}) + \left.\frac{\partial f(\mathbf{k})}{\partial E}\right|_0 \frac{\hbar}{m^*}\mathbf{k}\cdot\mathbf{G}, \qquad (10.1)$$

with $\mathbf{k}\cdot\mathbf{G} = kG\cos\alpha$, $\mathbf{k}'\cdot\mathbf{G} = k'G\cos\beta$ and $k = k'$. Note that we have again assumed that the marginal distribution function change with energy is small, and the thermal equilibrium marginal suffices:

$$\begin{aligned} & f(\mathbf{k}')\left[1-f(\mathbf{k})\right] - f(\mathbf{k})\left[1-f(\mathbf{k}')\right] \\ & = \left.\frac{\partial f(\mathbf{k})}{\partial E}\right|_0 \frac{\hbar}{m^*} kG\left[\cos\beta(1-\cos\alpha) - \cos\alpha(1-\cos\beta)\right] \\ & = \left.\frac{\partial f(\mathbf{k})}{\partial E}\right|_0 \frac{\hbar}{m^*} kG\left[\sin\alpha\sin\theta\cos\phi - \cos\alpha(1-\cos\theta)\right]. \qquad (10.2) \end{aligned}$$

Therefore,

$$\left.\frac{\partial f}{\partial t}\right|_{scat} = -\left.\frac{\partial f(\mathbf{k})}{\partial E}\right|_0 \frac{\hbar}{m^*} G \frac{N_s \Omega}{(2\pi)^3} \int_{\Omega_{\mathbf{k}'}} S(\mathbf{k}', \mathbf{k}) \left[\cos\alpha (1 - \cos\theta) - \sin\alpha \sin\theta \cos\phi\right] k \, d\mathbf{k}'. \quad (10.3)$$

Here, $|d\mathbf{k}| = k^2 \sin\theta d\theta d\phi \, dk$, and the azimuthal angle ϕ spans 0 to 2π. From the choice of the coordinate system and the impulse,

$$\frac{\partial f_0(\mathbf{k})}{\partial E} \frac{\hbar}{m^*} k G \cos\alpha = -\left[f(\mathbf{k}) - f_0(\mathbf{k})\right]. \quad (10.4)$$

Here, recall our first pass at discussing scattering with relation to the Boltzmann transport equation (Subsection 8.5.1).

So, if N_s are the number of scattering centers, occupying a volume Ω, then, with elastic scattering, the relaxation time can be written as

$$\frac{1}{\tau_{\mathbf{k}}} = \frac{N_s \Omega}{(2\pi)^2 k} \int_k \int_0^{\pi} S(\mathbf{k}, \mathbf{k}') \sin\theta (1 - \cos\theta) \, d\theta k^3 \, dk. \quad (10.5)$$

Details of what $S(\mathbf{k}, \mathbf{k}')$ looks like will depend on the scattering process that is causing this elastic scattering,

$$S(\mathbf{k}, \mathbf{k}') = \frac{2\pi}{\hbar} |\mathscr{H}_{\mathbf{k}\mathbf{k}'}|^2 \delta(E_{\mathbf{k}} - E_{\mathbf{k}'}), \quad (10.6)$$

and the matrix element $\mathscr{H}_{\mathbf{k},\mathbf{k}'}$ will depend on the real space perturbation potential $U(\mathbf{r})$ causing the scattering. For parabolic bands, this relaxation time—under elastic scattering conditions—reduces to

$$\frac{1}{\tau_{\mathbf{k}}} = \frac{N_s \Omega m^{*2} v}{2\pi \hbar^4} \int_0^{\pi} |\mathscr{H}_{\mathbf{k}\mathbf{k}'}|^2 \sin\theta (1 - \cos\theta) \, d\theta. \quad (10.7)$$

One can give this form a classical interpretation. We have N_s/Ω of scattering centers per unit volume. Let $\sigma_{\mathbf{k}}$ be the scattering cross-section that is momentum dependent. Classically, this means that

$$\frac{1}{\tau_{\mathbf{k}}} = \frac{N_s v}{\Omega} \sigma_{\mathbf{k}}. \quad (10.8)$$

Any carrier found within a cross-section area of $(N_s/\Omega)\sigma_{\mathbf{k}}$ surrounding the scattering centers in a distance $v\tau_{\mathbf{k}}$ is scattered. And this determines the time constant for the scattering. In Equation 10.7, one can see that an electron that is scattered with the solid angle of (θ, ϕ) loses $(1 - \cos\theta)$ of its initial momentum. The cosine factor determines the loss from the initial direction. To find a classical cross-section, one must account for all these dependencies of angles, so

$$\sigma_{\mathbf{k}} = \int_{\phi=0}^{2\pi} \int_{\theta=0}^{\pi} \sigma(\theta) \sin(\theta)(1 - \cos\theta) \, d\theta \, d\phi, \text{ where}$$

$$\sigma(\theta) = \left(\frac{\Omega m^*}{\pi \hbar^2} |\mathscr{H}_{\mathbf{k}\mathbf{k}'}|\right)^2 \quad (10.9)$$

is a differential cross-section describing the consequence of the matrix element of the perturbation. With this, one gets

$$\frac{1}{\tau_{\mathbf{k}}} = \frac{2\pi N_s v}{\Omega} \int_0^{\pi} \sigma(\theta) \sin\theta (1 - \cos\theta)\, d\theta. \qquad (10.10)$$

This is the classical form of the scattering integral. In elastic scattering, there exist these angular relationships of the change in direction of the particle undergoing the elastic scattering; the differential cross-section captures the perturbation's momentum-dependent consequence, and the momentum's relaxation time is an angle-modified integral, where all the angular dependences—of the differential cross-section and the change in direction—must be accounted for.

The scattering integral is also often called the collision integral. Classically, this is collision.

10.1.2 *Matrix element calculation*

ALL THESE SCATTERING CALCULATIONS require the calculation of the matrix element from the perturbation. In practice, this is a laborious process, which quite often requires approximations and assumptions to obtain analytic forms. We will mostly summarize these later but describe here the Fourier-based rationale of the calculation. Let $\Delta U(\mathbf{r})$ be a potential perturbation. $\hat{\mathscr{H}}' = \Delta\hat{U}(\mathbf{r})$ for each of the N centers is Hermitian, and the matrix element is

$$\mathscr{H}_{\mathbf{kk}'} = \int_{\Omega} \psi^*_{\mathbf{k}} \hat{\mathscr{H}}' \psi_{\mathbf{k}}\, d^3\mathbf{r} = \frac{1}{N} \int_{\Omega} \psi^*_{\mathbf{k}} \Delta U \psi_{\mathbf{k}'}\, d\mathbf{r}. \qquad (10.11)$$

To calculate it, we first expand the scattering potential in a Fourier series:

$$\Delta U(\mathbf{r}) = \sum_g \mathcal{A}_g \exp(i\mathbf{g}\cdot\mathbf{r}), \quad \text{where}$$
$$\mathcal{A}_g = \frac{1}{\Omega} \int_{\Omega} \Delta U(\mathbf{r}) \exp(-i\mathbf{g}\cdot\mathbf{r})\, d^3\mathbf{r}. \qquad (10.12)$$

For Bloch wavefunctions, $\psi_{\mathbf{k}}(\mathbf{r}) = \exp(i\mathbf{k}\cdot\mathbf{r}) u_{\mathbf{k}}(\mathbf{r})$, this gives

$$\mathscr{H}_{\mathbf{kk}'} = \frac{1}{N} \sum_g \int_{\Omega} \exp(-i\mathbf{k}\cdot\mathbf{r}) u^*_{\mathbf{k}} \mathcal{A}_g \exp(i\mathbf{k}'\cdot\mathbf{r}) u_{\mathbf{k}'}(\mathbf{r})\, d^3\mathbf{r}, \qquad (10.13)$$

an integral that vanishes for all wavevectors, except when $\mathbf{g} = \mathbf{k} - \mathbf{k}'$. So, the matrix element is

$$\mathscr{H}_{\mathbf{kk}'} = \frac{1}{N} \mathcal{A}_{\mathbf{k}-\mathbf{k}'} \int_{\Omega} u^*_{\mathbf{k}}(\mathbf{r}) u_{\mathbf{k}'}(\mathbf{r})\, d^3\mathbf{r}. \qquad (10.14)$$

$\mathcal{A}_{\mathbf{k}-\mathbf{k}'}$ are the Fourier coefficients of $\Delta U(\mathbf{r})$, and the $u(\mathbf{r})$s are the modulation functions of the propagating particle wavepackets. *This is a general result.*

Now consider parabolic bands, $u_{\mathbf{k}}(\mathbf{r}) = u_{\mathbf{k}'}(\mathbf{r})$ and

$$\mathscr{H}_{\mathbf{kk}'} = \mathcal{A}_{\mathbf{k}-\mathbf{k}'} = \frac{1}{\Omega} \int_{\Omega} \Delta U(\mathbf{r}) \exp[-i(\mathbf{k}' - \mathbf{k})\cdot\mathbf{r}]\, d^3\mathbf{r}. \qquad (10.15)$$

The matrix element is the Fourier coefficient of the perturbation potential itself. The scattering is most efficient when the periodicity between the potential perturbation's Fourier expansion and the incident and the scattered wavefunctions matches. The scattered wavefunction emphasizes the interaction of these components, which makes sense.

We saw bandgaps arising for this same reason.

10.1.3 *Matrix element for ionized impurity scattering*

SCREENED AND UNSCREENED IONIZED IMPURITY SCATTERING provide a convenient example for illustrating the use of this matrix element calculation approach. The screening here is the dressing of a charge perturbation, such as of an ionized scattering, by the mobile carriers. In non-degenerate semiconductors, as we have discussed before, this is the Debye screening, with a length scale of λ_D. For degenerate materials, it is Thomas-Fermi screening, with a length scale of λ_{TF}. First, consider an unscreened potential scatterer: $\Delta U(\mathbf{r}) = \pm Ze^2/4\pi\epsilon(0)|\mathbf{r}|$, where Ze is the charge of the scatterer. The Fourier coefficients for this potential perturbation are

We discussed Debye and Thomas-Fermi screening in Chapter 1, and it finds its usage extensively in S. Tiwari, "Device physics: Fundamentals of electronics and optoelectronics," Electroscience 2, Oxford University Press, ISBN 978-0-198-75984-3 (forthcoming). Carriers, being mobile, respond to local fields and cause local charge density changes that reduce the perturbation. Carriers of opposite polarity pull in.

$$\begin{aligned} \mathcal{A}_g &= \pm\frac{Ze^2}{4\pi\epsilon(0)\Omega}\int_\Omega \exp(-i\mathbf{g}\cdot\mathbf{r})\frac{1}{|\mathbf{r}|}\,d^3\mathbf{r} \\ &= \pm\frac{Ze^2}{\epsilon(0)\Omega}\int_0^\infty r\exp(-igr)\,dr \\ &= \pm\frac{Ze^2}{\epsilon(0)\Omega}\frac{1}{|\mathbf{g}|^2} = \pm\frac{Ze^2}{\epsilon(0)\Omega}\frac{1}{|\mathbf{k}-\mathbf{k}'|^2}, \end{aligned} \tag{10.16}$$

where we have used the volume element $d^3\mathbf{r} = r^2\sin\theta\,d\theta\,d\phi\,dr$. This means that the unscreened ionized impurity matrix element is

$$\mathcal{H}_{\mathbf{kk}'} = \pm\frac{Ze^2}{\epsilon(0)\Omega}\frac{1}{|\mathbf{g}|^2} = \pm\frac{Ze^2}{\epsilon(0)\Omega}\frac{1}{|\mathbf{k}-\mathbf{k}'|^2}, \tag{10.17}$$

which decays by the square of the separation in wavevectors of the incident and scattered waves.

Now consider the case with screening where the potential varies as

$$\Delta U(\mathbf{r}) = \frac{Ze^2}{4\pi\epsilon(0)}\exp\left(-\frac{|\mathbf{r}|}{\lambda}\right), \tag{10.18}$$

so the Fourier coefficients are

$$\begin{aligned} \mathcal{A}_g &= \frac{Ze^2}{4\pi\epsilon(0)\Omega}\int_\Omega \exp\left(-\frac{r}{\lambda}\right)\exp(-i\mathbf{g}\cdot\mathbf{r})\frac{1}{|\mathbf{r}|}\,d^3\mathbf{r} \\ &= \frac{Ze^2}{\epsilon(0)\Omega}\frac{1}{|\mathbf{g}|^2+1/\lambda^2}, \end{aligned} \tag{10.19}$$

and therefore the matrix element is

$$\mathscr{H}'_{\mathbf{kk}'} = \frac{Ze^2}{\epsilon(0)\Omega} \frac{1}{|\mathbf{k}-\mathbf{k}'|^2 + 1/\lambda^2}. \tag{10.20}$$

Comparing the unscreened (Equation 10.17) and the screened potentials (Equation 10.20), we see that there is a factor

$$\Lambda = \frac{|\mathbf{k}-\mathbf{k}'|^2}{|\mathbf{k}-\mathbf{k}'|^2 + 1/\lambda^2} \tag{10.21}$$

that relates screened and unscreened potentials. So, we now have here, starting from the potential perturbation, the scattering element for ionized impurity scattering.

In induced two-dimensional electron gases, this screening, without introducing new impurities, allows one to improve mobility.

10.2 *Scattering by phonons*

PHONONS EXIST AS acoustic—long-range phase correlation in atomic displacement—and optical—anti-correlated atomic displacement—and a scattering process may generate them or absorb them. For these phonons of energy $\hbar\omega_{\mathbf{q}}$ and momentum $\hbar\mathbf{q}$, conservation with a state change indexed as i for initial, and f for final (see Figure 10.3), requires conservation of energy,

$$E_i - E_f = \pm\hbar\omega_{\mathbf{q}} = \pm\frac{\hbar^2}{2m^*}\left(k_i^2 - k_f^2\right), \tag{10.22}$$

and momentum,

$$\mathbf{k}_i - \mathbf{k}_f = \pm\mathbf{q}. \tag{10.23}$$

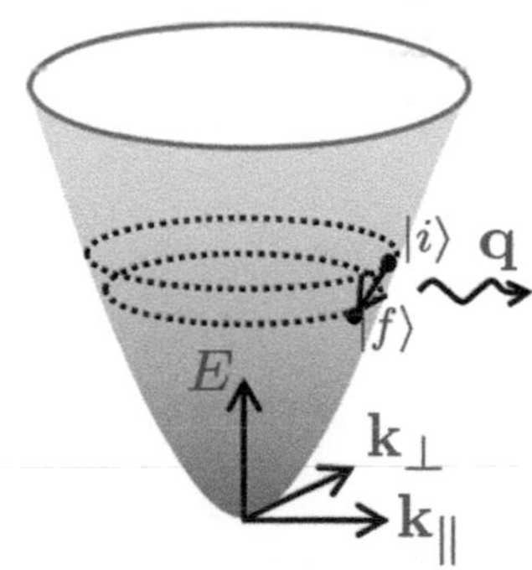

Figure 10.3: An electron undergoing scattering while emitting a phonon. The f in E_f indexes and identifies the final state, not a state at Fermi energy, where we use the capital form F.

The rate of a transition depends on the existence of the particle—an electron here—in the initial state, which is characterized by the distribution function $f(\mathbf{r},\mathbf{k})$. The state it is transitioning to must be empty; availability of the final state is described by a distribution $1 - f(\mathbf{r},\mathbf{k})$. And the probability of a transition taking place is the scattering rate $S_{\mathbf{k}_i\mathbf{k}_f} = S(\mathbf{k}_i,\mathbf{k}_f)$. This transition probability is tied to the strength of the interaction that is embodied in the matrix element.

The density of available final states—most of which are empty under degenerate and near-degenerate conditions—is just the density of states at E_f. This is

$$\mathscr{G}(E_f) = \frac{(2m^*)^{3/2}E_f^{1/2}}{2\pi^2\hbar^3} = \frac{(2m^*)^{3/2}\left(E_i \pm \hbar\omega_q\right)^{1/2}}{2\pi^2\hbar^3} \tag{10.24}$$

in the isotropic, or force-fitted anisotropic, band at the bandedge density of state approximation. Let $S_{\mathbf{k}_i\mathbf{k}_f} \equiv G(\mathbf{q})$ be the electron-phonon coupling. The probability of a phonon at wavevector $\mathbf{q}$—its occupation density—is

$$n(\mathbf{q}) = \frac{1}{\exp(\hbar\omega_{\mathbf{q}}/k_BT) - 1}. \qquad (10.25)$$

Absorption of the phonon is proportional to this source function $n(\mathbf{q})$. *This is spontaneous absorption.*

The probability of emission, however, can be both spontaneous and stimulated. The net rate contributing to the change in the distribution function—a product of the number of initial states occupied (it exists and is occupied, so 1), the density of final states that are available for occupation ($\mathcal{G}(E_f)$, since nearly all are empty), and the probability of the phonon absorption process and the phonon emission process—can now be written, where the coupling $G(\mathbf{q})$ that connects the electron-to-phonon perturbation enters. We have not yet discussed the details of allowed, weak, strong and disallowed electrical interaction with the acoustic or optical phonons that have their own short-range, long-range and polarization characteristics. The general form of this rate is

Emission can be both spontaneous and stimulated. Stimulated here means that the existence of another phonon at $\mathbf{q}$ leads to a sympathetic emission. So, emission has a proportionality to $1+n(\mathbf{q})$. This is the second term in the bracket in Equation 10.26. This stimulated term can be thought of as a driven oscillation of the phonon. The phonon at $\mathbf{q}$ is interacting with the electron with a specific energy difference, associated with the oscillation frequency of the phonon. If a phonon exists at this frequency, energy coincides, as does the momentum, a tight coupling exists and it causes an emission—a stimulated emission—of an identical phonon. This is a unique and, at first glance, surprising result. But the Einstein A and B coefficients that are discussed in Appendix K and are grounded in the detailed balanced argument at thermal equilibrium underly this stimulation. And lasers—a very useful invention (perhaps not a precisely correct use of the noun "invention," since galaxies have shown lasing at microwave wavelengths)—are an important byproduct.

$$\begin{aligned}
\frac{1}{\tau_c} &\approx \frac{(2m^*)^{3/2}}{4\pi^2\hbar^3}\sum_{\mathbf{q}} G(\mathbf{q})\left[\frac{(E_i+\hbar\omega_{\mathbf{q}})^{1/2}}{\exp\left(\frac{\hbar\omega_{\mathbf{q}}}{k_BT}\right)-1} + \frac{(E_i-\hbar\omega_{\mathbf{q}})^{1/2}}{1-\exp\left(-\frac{\hbar\omega_{\mathbf{q}}}{k_BT}\right)}\right] \\
&= \mathcal{G}(E_f)\begin{bmatrix} n(\mathbf{q}) \\ 1+n(\mathbf{q}) \end{bmatrix} G(\mathbf{q}), \\
\therefore \quad \frac{1}{\tau_{\mathbf{k}}} &= \left\langle\frac{1-\cos\theta}{\tau_c}\right\rangle \\
&= \left\langle(1-\cos\theta)\mathcal{G}(E_f)\begin{bmatrix} n(\mathbf{q}) \\ 1+n(\mathbf{q}) \end{bmatrix} G(\mathbf{q})\right\rangle, \qquad (10.26)
\end{aligned}$$

a notation form where the top term of the column is for absorption of a phonon, and the bottom is for emission of a phonon. In the first set of terms, $1/\tau_c$ is the rate associated with a state at energy E_i interacting with phonons at wavevector $\mathbf{q}$. Different momenta and energies will have different angular dependence determined by the conservation equations. So, the net rate—the difference of scattering out and in—involving absorption or emission of phonons is to be averaged over all possibilities of states coupling to the specific $\mathbf{k}$. This averaging will involve θ, the density of states at E_f, the occupation probability of phonons of a $\mathbf{q}$ wavevector, and the electron-phonon coupling term.

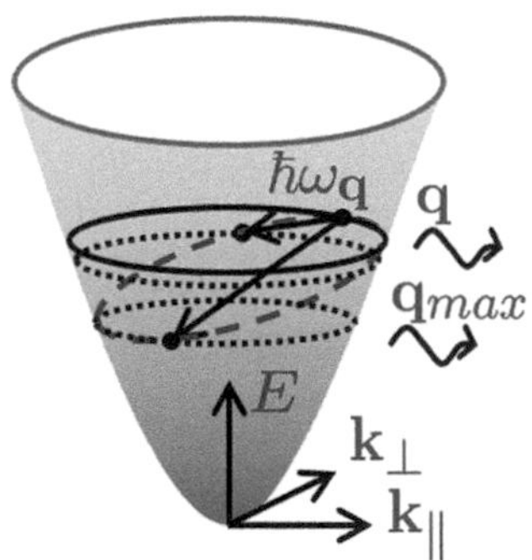

Figure 10.4: Acoustic phonon emission during a scattering event between states of a parabolic band. Both a spread of phonon energy and phonon momentum exist, where the transition takes place with energy and momentum conservation. But, with a limit to the maximum wavevector change in the band, there will exist a limit to the highest phonon wavevector. Energy changes are small. Acoustic scattering is quasielastic.

With acoustic phonons, the momentum spread of the allowed states of carriers limits events to low energy. This is illustrated in Figure 10.4 for a parabolic band. Carriers are spread out over a few k_BT of energy. Two emission events illustrated here show that ranging from zone center phonons to a maximum phonon

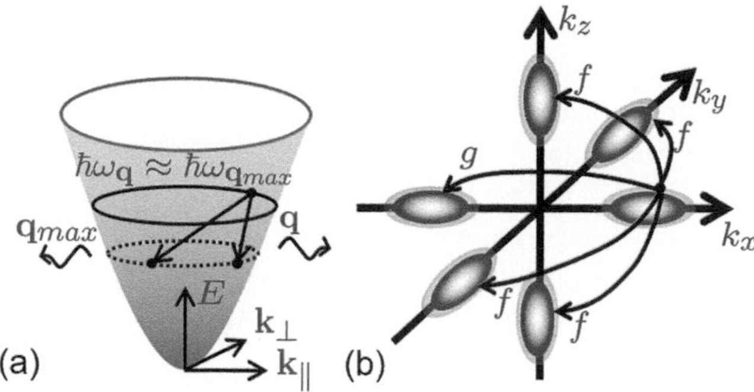

Figure 10.5: Optical phonon emission in a direct gap semiconductor in (a) and an indirect semiconductor (*Si*) in (b). Optical phonons have nearly a constant occupancy across wavevectors and nearly a constant energy. So, a spread of **q**s up to a maximum of $\mathbf{q}_{max}$ exists within the same valley. This is true for the case in (b) too, where the illustration is in the form of constant energy surfaces. Part (b) shows that scattering may also occur with the four degenerate valleys in the orthogonal plane (f scattering) and with the mirror symmetric valley (g scattering). Both emission and absorption will occur, the latter if a sufficient number of phonons are available for the absorption process.

wavevector $\mathbf{q}_{max}$ exists, limited by the energy bandstructure of the semiconductor. The energy loss is limited by this maximum phonon wavevector.

Optical phonons have larger energies—10s of meV and higher. Figure 10.5 shows two examples: one in a direct semiconductor, for example, *GaAs* in (a), and one in an indirect material, in this case *Si* in (b). For scattering within the band, a spread of **q**s—again limited by the states $\hbar\omega_{\mathbf{q}}$ away in energy—and limited in momentum matching to $\mathbf{q}_{max}$ is allowed. Only a small region at the zone center of the Brillouin zone is active in scattering. In indirect materials, *Si* here, the large phonon wavevectors matter. In the emission process illustrated in Figure 10.5(b), scattering may now take place with states in the four valleys in the orthogonal plane and one valley that is mirror symmetric. The former is f scattering, here an f emission. The latter is g scattering, here a g emission.

The limits of energy change of the carrier and its correspondence to the range of momenta of phonons and their availability can now be understood. Acoustic phonons allow energy exchange to their vanishing energy at the zone center. The maximum is defined by $\hbar\omega_{\mathbf{q}_{max}}$. This will correspond to the largest displacement possible for the atoms. So, in principle, the maximum ΔE will correspond to the largest acoustic phonon energy defined by the momentum change allowed by the $E(\mathbf{k})$ description. Optical phonons have a fairly small spread around the $\hbar\omega_{\mathbf{q}}$ for all the branches. That is the energy exchangeable so long as the momentum exchange is allowed. But if $E_i < \hbar\omega_{\mathbf{q}}$—optical phonons are typically k_BT (room temperature T) in energy—then, emission will not occur but absorption is still allowed. This is illustrated in Figure 10.6. If no state exists in the bandstructure $\hbar\omega_{\mathbf{q}}$ below the emitting state, then the process is not allowed. This energy-based consideration places restrictions on phonon processes both as a function of energy and as a function of temperatures. $E_i \ll \hbar\omega_{\mathbf{q}}$ is very easy to achieve at low temperatures.

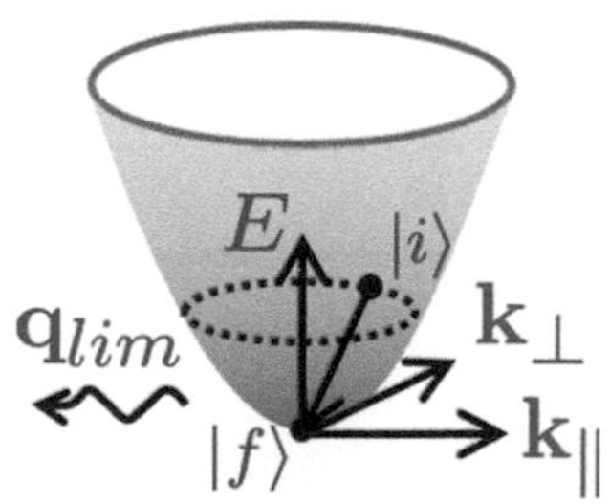

Figure 10.6: A limiting case for optical phonon emission for a transition from a parabolic band state. With all optical phonons in the energy range of $\hbar\omega_{\mathbf{q}}$, ranging from a few 10s to almost 100 meV of energy, the lowest energy of an initial state $E_i = \hbar\omega_{\mathbf{q}}$ for an optical phonon emission to a state at the band's edge.

Table 10.2 shows the occupation factor for phonon states at three different temperatures for *GaAs*, which has a longitudinal (and not too far from transverse) optical phonon energy of $\hbar\omega_{\mathbf{q}} \approx 36\ meV$. Optical phonons get fewer as temperature is lowered, nearly vanishing near liquid *He* temperatures. At low temperatures, acoustic phonons, which are available even at low temperatures, will dominate. And if the charge particles acquire high enough energy due to external stimulus, optical phonon emission too will become important. So, in high fields, carrier transport can still saturate in static conditions in semiconductors that are more than a few scattering lengths long, because optical phonon emission will release the energy acquired during the acceleration.

T (K)	n ($\mathbf{q}$)
300	0.3
77	0.004
4	10^{-45}

Table 10.2: Phonon occupation—probability of of occupation of a phonon state $E(\mathbf{q}) = \hbar\omega_{\mathbf{q}}$ as a function of temperature for optical phonons of *GaAs*, with $\hbar\omega_q = 36\ meV$.

Now consider the nature of the behavior of phonons, longitudinal and transverse, as well as acoustic and optical, and their mechanisms of causing a perturbation. Figure 10.7 shows a representative view of longitudinal and transverse acoustic phonons in (a) and (b) for zone center and zone edge phonons, and a similar view for optical phonons in (c) and (d).

The acoustic phonons have long-range, neighboring atoms that have a gradual change in displacement, that is, the phase of the displacement $\mathbf{u} = \breve{\mathbf{u}} \exp[i(\mathbf{q} \cdot \mathbf{r} - \omega t)]$ shifts gradually. Since these displacements are only gradually changing, whether longitudinally or transversely, they appear as a wave of compression and rarefaction traveling in space. So, the function $\nabla u(\mathbf{r})$ is phase shifted by $\pi/2$ from the displacement $u(\mathbf{r})$.

We have used the breve symbol to identify the parameter as the amplitude of the time and space harmonic.

Optical phonons have a very rapid change—out of phase from one unit cell to next—as shown in Figure 10.7(c) and (d). The zone center phonons in (c) are shown here with the maximum displacement for both longitudinal and transverse modes. These are $\pi/2$ out of phase from one unit cell to the net. The zone edge phonons too have large displacement. But now, as in Figure 10.7(d), one atom of the basis may be used as the reference, and the other atom—the light one here in (d)—undergoes maximum displacement for both longitudinal and acoustic modes. This figure could also equivalently be seen with the smaller atom as the reference, and then the larger one undergoes maximum displacement.

How an electrical (also called polar) or electromagnetic interaction will happen with these phonons will depend on how well a coupling of the two excitations takes place. In this, the orientation of the polarizations and their magnitude, or the crystal's bandstructure consequences from the deformation waves, will all matter. So, the perturbation Hamiltonian can take a multitude of forms depending on the circumstances. Acoustic and optical branches will behave differently. So can transverse and longitudinal modes.

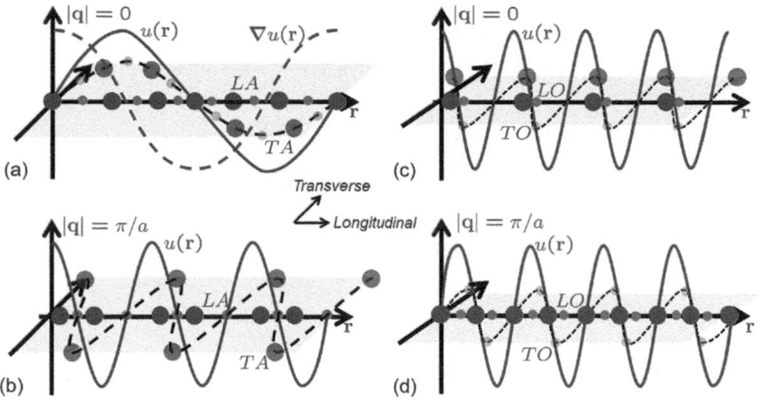

Figure 10.7: A pictorial view of the nature of acoustic and optical, longitudinal and transverse, phonons in a two-atom basis crystal. $u(\mathbf{r})$ is a positional sketch of the displacement for both the longitudinal and the transverse branches. The heavier atom is shown as the larger of the two. Parts (a) and (b) show the acoustic mode displacements at the zone center and the zone edge, (c) and (d) show the optical mode displacements at the zone center and the zone edge. Part (a) also shows the gradient of displacement for the acoustic branch—most clearly viewable in the atoms' displacement in the *LA* mode. In (b), which is the zone edge condition for acoustic phonons, the choice of reference phase is such that only the bigger atoms are displaced, and the smaller atoms not at all. In (d), which is the zone edge condition for optical phonons, the choice of reference phase is such that only the smaller atoms are displaced, and the bigger atoms not at all.

10.2.1 *Acoustic phonon interactions*

IN DETERMINING THE $E(\mathbf{k})$ STATE VIEW of independent electrons in the crystal, we employed the adiabatic/Born-Oppenheimer approximation in eliminating the interaction term in the crystal Hamiltonian. We have observed two properties for the acoustic phonons in Figure 10.7(a) and (b). The first is that, over a long distance scale, there is a deformation—at its simplest, a slow change in the unit cell size—that is propagating. The *LA* phonon causes longitudinal deformation while the *TA* phonon causes lateral deformation. Both perturbations to periodicity will modulate the bandstructure. Deformation can be viewed as a slow change of the energy of the state that an electron is occupying as an acoustic phonon passes by. This is an energy perturbation, and we largely concern ourselves with the changes in the bandedge energies of E_c and E_v, where the carriers are. This is the *deformation coupling* arising in the perturbation of the bands. In the coupling and consequent scattering of carriers, symmetries will matter through momentum matching. In n-type semiconductors, with electrons at the zone center, the *LA* phonons, with their more pronounced in-line deformation, will be stronger. But, in non-zone center semiconductors such as *Si* and *Ge*, the transverse acoustic phonons will couple strongest deformationally, even if the effect is still weak.

This discussion ignored any charge consequences. Charge distribution was assumed to be following the atomic motion. And only the bandstructure effect on the electron state due to the atomic motion was evaluated. But if the crystal is polar—partially ionic with charge transfer from one constituent atom onto another—even with the charge distribution assumed following the atomic motion, as in group *III-V* and group *II-VI* semiconductors—there will exist a dipole arising in some effective charge e^* spaced $\langle \mathbf{z} \rangle$ apart, so

We will discuss the dipole again with optical phonons, where, instead of this compression-rarefaction acoustic deformation, we will have oscillating dipoles of neighboring atoms. This will be a strong electromagnetic interaction, but one that is fundamentally very different from the piezoelectric interaction of the acoustic phonon example. Acoustic frequencies are in *kHz*, and optical phonon frequencies are decades of *THz*. One is slow and spatially long range, while the other is fast and spatially short range.

a dipole moment of $\mathbf{p} = e^* \langle \mathbf{z} \rangle$ in neighboring bonds. Over a unit cell, this integrates to zero. Under acoustic motion, this dipole is being perturbed by the mechanical motion, and the perturbation is changing over the neighboring atom distance scale. If the crystal lacks inversion symmetry, this perturbation of the dipole will be accumulative over $\lambda/2$ of displacement, and depletive over the next half cycle. So, mechanical motion causes a polarization; that is, a piezoelectric polarization comes about due to the acoustic wave. The acoustic oscillation has caused a slow piezoelectric perturbation that the carriers Coulomb couple to, resulting in *piezoelectric coupling*. So, acoustic phonons in ionic and non-centrosymmetric crystals will give rise to piezoelectric interaction.

Si is both centrosymmetric and non-ionic. It will have acoustic-induced deformation scattering, but not piezoelectric scattering. *GaAs* is non-centrosymmetric and partially ionic. It will have deformation and piezoelectric scattering due to acoustic phonons.

10.2.2 *Optical phonon interactions*

THE DISPLACEMENT OF INDIVIDUAL ATOMS in the optical phonon branch is very rapid; Figure 10.7 parts (c) and (d) show, for the zone center and the zone edge, these rapid changes, with consecutive phases of opposite sign. This is a high frequency. Optical phonons have high energy. They are spread out at about this constant energy over the whole zone but are of the order of a few 10s of *meV*. Atomic motion again will cause *deformation interaction* in the same spirit as for acoustic phonons. But if there is dipolar polarization, oscillating dipole moments, so oscillating fields, will couple electromagnetically to the charged carriers. This is *electromagnetic coupling* arising in the polar nature of the optical phonons that are interacting with the charged particle. Since dipole moments are coupling, longitudinal dipole moments will be more pronounced in causing scattering. In *III-V* and *II-VI* compound semiconductors, the effect will be pronounced and, since it is electromagnetic in nature, if large carrier populations are present, interactions will also arise through collective excitations of the carrier ensemble; that is, with plasmons. In devices where one has large carrier populations at interfaces, and field termination is usually employed using dielectrics or high permittivity materials that are quite ionic, strong electromagnetic coupling via optical phonons will arise.

In *GaAs*, the optical phonons are about 36 *meV* in energy; in *Si* they are about 54 *eV*; and in SiO_2, a strong mode exists at 112 *meV*. 50 *meV* is a frequency in far infrared, about 10 *PHz*. Only the zone edge acoustic phonon reaches this energy and frequency.

We have now distinguished electromagnetic coupling from piezoelectric coupling. The latter arises in mechanical deformation. The electrical coupling is piezoelectricity based. Electromagnetic coupling, on the other hand, should be viewed as coupling arising in an oscillating field arising from oscillating dipoles, a forte of

Phonon	Mechanism	Constraint	Example
LA	Deformation		Si, GaAs
	Piezoelectric	Non-centrosymmetry Ionicity	GaAs
LO	Deformation		Si, GaAs
	Polar	Ionicity	GaAs
TO	Deformation		Si, GaAs

Table 10.3: Acoustic and optical phonons, the coupling and perturbation mechanism and its scattering consequence.

optical phonons of a polar semiconductor. Table 10.3 is a summary view of this short discussion of the nature of the phonon-carrier interaction and its relevance to different semiconductors.

We now set up the perturbation problem so that we may describe the approach to calculating the scattering rate, or its implication for us, a time constant that will have energy dependence and other dependences on the parameters of the problem.

10.2.3 *Deformation interaction (LA, LO and TO)*

To set up the incorporation of deformation, we start with the displacement of the atoms around their mean position. Let $\hat{\mathbf{n}}$ be a unit vector in the direction of displacement, which will be longitudinal or transverse for our calculations. This displacement is

$$\mathbf{u} = \breve{\mathbf{u}} \exp[i(\mathbf{q} \cdot \mathbf{r} - \omega_q t)] = \breve{u} \hat{\mathbf{n}} \exp[i(\mathbf{q} \cdot \mathbf{r} - \omega_q t)]. \qquad (10.27)$$

Here, $\mathbf{r}$ is a spatial coordinate around the mean position. For acoustic phonons, the strain arises in the graded accumulation of the displacements. It is maximum at the node of the wave and it should also be harmonic in the linear limit. Since the relative volume change $\Delta\Omega/\Omega = \boldsymbol{\nabla} \cdot \mathbf{u}$, the strain follows a dependence in the gradient $\boldsymbol{\nabla} \cdot \mathbf{u}(\mathbf{r}, t)$:

In detail, deformations in different directions will be different and so will the shear effect. Symmetry makes it possible to express what would be six different deformation potential constants in terms of two. We will not dwell on the details of this since our interest is in understanding the physical nature.

$$\boldsymbol{\nabla} \cdot \mathbf{u}(\mathbf{r}, t) = i\hat{\mathbf{n}} \cdot \mathbf{q} \breve{u} \exp[i(\mathbf{q} \cdot \mathbf{r} - \omega_q t)]. \qquad (10.28)$$

Strain vanishes for transverse modes; so, acoustic transverse modes do not cause deformation scattering. To describe the potential perturbation arising in the strain, we introduce a parameter Ξ_a—a deformation potential constant—relating the bandedge energy deformation arising in the dilation. For the longitudinal acoustic mode,

$$\Delta U = \frac{\partial E}{\partial \Omega} \Delta\Omega = \Omega \frac{\partial E}{\partial \Omega} \frac{\Delta\Omega}{\Omega} = \Xi_a \boldsymbol{\nabla} \cdot \mathbf{u}(\mathbf{r}, t). \qquad (10.29)$$

Ξ_a is a deformation potential constant for acoustic phonons. It is in units of energy. Since this is a dot product of the displacement, only longitudinal acoustic phonons cause deformation interaction. But shear strain that arises in the longitudinal deformation will also affect the energy states. It quantifies the susceptibility of the conduction band minimum, or valence band maximum, to the dilation of the crystal. An *LA* phonon causes a rise and then a fall in the band extrema, which in turn cause the deformation scattering. When the carrier randomly perturbs due to an encounter with such a series of compressions and rarefactions, if the angle between its travel and the normal to the acoustic phonon's plane (its direction of travel) is θ, the angle $\pi - 2\theta$ is the scattered particle wave's angle with highest probability. In such a scattering, the minimum phonon momentum is $\mathbf{k}_i = \mathbf{q}$ (with $\mathbf{k}_f = 0$ in Equation 10.23); therefore, if the electron has an energy of the order of $k_B T$, then the momentum change must be large. So, this scattering involves a large change in momentum at modest temperatures and is isotropic, being quasielastic. Only at the lowest temperatures does it favor forward directions.

The propagation velocity of acoustic phonons is approximately the speed of sound—strictly true at $q = 0$, and vanishing at zone edge due to interference. Since this sound velocity is much smaller than electron velocity ($E(\mathbf{q}) = \hbar v_s q$), the energy is then much smaller than the electron energy for nominal energy electrons. Momentum matching means that the acoustic phonon's low energy will be reflected in the energy change even as the momentum change is large. Take a minimum energy electron undergoing an *LA* phonon deformation scattering, where it loses the energy: $E_i \approx k_B T$; therefore, $k_i = \sqrt{k_B T 2m^*/\hbar^2}$. Momentum matching says that, for this minimum condition, $q = k_i$. So, the energy of the phonon is $E(q) = \hbar v_s q = \sqrt{k_B T}\sqrt{m^* v_s^2}$. The speed of electrons—thermal velocity is a good measure—is of the order of $10^5\, m/s$. $k_B T \approx m^* v_\theta^2$. Silicon has a sound velocity $v_s \approx 5800\, m/s$. This is about $0.058 \times k_B T$. So, the phonon energy is only 5.8 % of the $k_B T$ energy. The electron doesn't lose much energy.

Since $\hbar\omega_{\mathbf{q}} \ll k_B T$ and $\hbar\omega_{\mathbf{q}} \ll E$ for acoustic phonons, only the low $\mathbf{q}$ part of acoustic branch is active. Since $G(\mathbf{q}) \propto \mathbf{q}$ and $\omega_{\mathbf{q}} \propto \mathbf{q}$, the acoustic phonon statistics have a contribution of

$$\begin{aligned} \frac{1}{\exp(\hbar\omega_{\mathbf{q}}/k_B T) - 1} &\approx \frac{1}{1 + (\hbar\omega_{\mathbf{q}}/k_B T) - 1} \\ &\approx \frac{k_B T}{\hbar\omega_{\mathbf{q}}} \propto \frac{k_B T}{|\mathbf{q}|} \quad \text{for absorption, and} \\ \frac{1}{1 - \exp(\hbar\omega_{\mathbf{q}}/k_B T)} &\approx \frac{1}{1 - [1 - (\hbar\omega_{\mathbf{q}}/k_B T)]} \\ &\approx \frac{k_B T}{\hbar\omega_{\mathbf{q}}} \propto \frac{k_B T}{|\mathbf{q}|} \quad \text{for emission.} \end{aligned} \tag{10.30}$$

This implies that the product of $G(\mathbf{q})$ and the statistical factor is independent of $\mathbf{q}$. The resulting scattering will then have only an energy dependence:

$$\left.\frac{1}{\langle \tau_{\mathbf{k}} \rangle}\right|_{\text{acoustic deformation}} \approx m^{*3/2} E^{1/2} k_B T. \tag{10.31}$$

Lower mass, lower energy of the carrier and lower temperature will increase the relaxation time arising in acoustic deformation scattering.

LO phonons too have this compression-rarefaction effect, except it is at a much shorter wavelength. The relative displacement between atoms is

The longitudinal mode has the larger effect compared to transverse mode. So, we only consider *LO* phonons.

$$\delta\mathbf{u}(\mathbf{r}, t) = \mathbf{u}_1(\mathbf{r}, t) - \mathbf{u}_2(\mathbf{r}, t) = \hat{\mathbf{n}}\delta u(\mathbf{r}, t), \tag{10.32}$$

where $\mathbf{u}_1 = \breve{\mathbf{u}}_1 \exp[i(\mathbf{q}\cdot\mathbf{r} - \omega_q t)]$, with $\mathbf{u}_2$ expressed similarly. The potential perturbation arises from the relative displacement, so

$$\Delta U(\mathbf{r},t) = \Xi_d \delta\mathbf{u}(\mathbf{r},t), \tag{10.33}$$

where the deformation parameter Ξ_d is the direct relationship between bandedge perturbation and dilation; that is, units of energy per length scale.

Note the difference in units and proportionality for acoustic phonons (to strain, which is unitless) and for optical phonons (to displacement, which has units of length).

10.2.4 *Piezoelectric interaction (LA)*

THE STRAIN PROPAGATING WITH *LA* PHONONS, in crystals lacking inversion symmetry (non-centrosymmetry), polarizes and the polarization accumulates over unit cells across half wavelengths. This results in internal electric fields varying in space and time, which will Coulomb interact with charge carriers. This is piezoelectric scattering. The local scattering potential, if $\Delta U(\mathbf{r},t) = -q\psi(\mathbf{r},t)$, where $\psi(\mathbf{r},t)$ is the internal-field-induced electrostatic potential, is

$$\Delta U(\mathbf{r},t) = -q\psi(\mathbf{r},t) = q\int \boldsymbol{\mathcal{E}}(\mathbf{r},t)\cdot d\mathbf{r}, \tag{10.34}$$

expressed in terms of the local field. The semiconductor here has—absent mechanical strain, such as at absolute zero—no piezoelectricity. It acquires a position- and time-varying piezoelectricity, due to the weak propagating "piezoelectric" wave arising from the *LA* phonon in presence of non-centrosymmetry:

$$\mathbf{D}(\omega) = \epsilon(\omega)\boldsymbol{\mathcal{E}} = \epsilon_0\boldsymbol{\mathcal{E}} + \mathbf{P}(\omega). \tag{10.35}$$

For static conditions, the convention has been to write

$$\mathbf{D}(0) = \epsilon(0)\boldsymbol{\mathcal{E}} = \epsilon_0\boldsymbol{\mathcal{E}} + \mathbf{P}(0), \tag{10.36}$$

where $\mathbf{P}(0)$ is the polarization consequence of charges that are associated with the atomic core and the ionic dipole, that is, the presence of the crystal. If a spontaneous polarization exists, examples of which are in several of the perovskites, then that too must be included. For most semiconductors of the group *IV*, group *III-V* and group *II-VI* semiconductors, we can write the displacement as

$$\mathbf{D}(0) = \epsilon(0)\boldsymbol{\mathcal{E}} + e_{pz}\nabla u(\mathbf{r},t) = \epsilon_0\boldsymbol{\mathcal{E}} + \mathbf{P}(0) + e_{pz}\nabla u(\mathbf{r},t), \tag{10.37}$$

with e_{pz} as a piezoelectric constant (C/m^2), where we have again ignored the anisotropic; that is, tensor, nature of strain effects. Since the sources of the fields are the atomic polarization, the ionic polarization and the piezoelectricity, the static displacement leads to

$$\boldsymbol{\mathcal{E}}(\mathbf{r},t) = -\frac{e_{pz}}{\epsilon(0)}\nabla u(\mathbf{r},t), \tag{10.38}$$

Perovskites, discussed in Chapter 3, are the ABO_3 class of materials, where *A* and *B* are transition elements. Such materials have a variety of interesting properties, including ferroelectricity that arises from spontaneous polarization. $PbTiO_3$ is a particularly popular example. These materials can also be piezoelectric. $LiNbO_3$ is a particularly popular example. They can also have moderate bandgaps, which makes them semiconducting, albeit, with quite different characteristics than those discussed here. $SrTiO_3$ is one example of this, where electrons and holes transport not unreasonably.

and hence, using Equation 10.34, there exists a scattering potential of

$$\Delta U(\mathbf{r},t) = -\frac{q e_{pz}}{\epsilon(0)} u(\mathbf{r},t) = i\frac{q e_{pz}}{\epsilon(0) q_s} \nabla \cdot \mathbf{u}(\mathbf{r},t), \qquad (10.39)$$

where q_s is the phonon wavevector.

q being also the symbol for charge, we write the s subscript for phonons from time to time.

Note from Equation 10.29 for deformation perturbation, and Equation 10.39 for piezoelectric perturbation (both for acoustic phonons and for *LO* phonons), there exists a $\pi/2$ phase difference. *The strain maximum and the piezoelectricity maximum are $\pi/2$ apart of the acoustic wave at any instant in time. They operate independently.*

10.2.5 *Polar mode interaction (LO)*

THE IONIC POLARIZATION AND THE SHORT LENGTH RANGE of the optical phonons lead to internal electric field perturbations within the unit cell. The interaction between polar optical phonons and electrons is the Fröhlich interaction. Again, as with piezoelectricity, and the use of Equation 10.34, we need to find the internal field. The ionic contribution can be found by looking at the difference in dielectric response function at low frequency and very high frequency. The low-frequency response includes both the atomic and the ionic responses. And we are going to include the free electron response separately. The high-frequency (asymptotically infinite) response includes only the atomic response. So, atomic and ionic polarization is reflected in the $\omega = 0$ response, and the atomic in the $\omega \rightarrow \infty$ response. Let $\mathbf{P}_i$ be the ionic polarization contribution. We can conclude from

For acoustic phonons, the ionic consequence was piezoelectric—long length range—and only in non-centrosymmetric crystal. For optical phonons, the ionic charge consequence is at the unit cell length scale and exists whether there is inversion symmetry or not.

Some of these comments and reflections are also buried in the discussion of Section 3.11, where we introduced phonons, and longitudinal, transverse, acoustic and optic phonon notions through the toy model.

$$\begin{aligned} \mathbf{D}(\infty) &= \epsilon(\infty)\boldsymbol{\mathcal{E}} = \epsilon_0\boldsymbol{\mathcal{E}} + \mathbf{P}(\infty) \text{ that} \\ \mathbf{D}(0) &= \epsilon(0)\boldsymbol{\mathcal{E}} = \epsilon_0\boldsymbol{\mathcal{E}} + \mathbf{P}(0), \\ &= \epsilon_0\boldsymbol{\mathcal{E}} + \mathbf{P}(\infty) + \mathbf{P}_i \\ &= \epsilon(\infty)\boldsymbol{\mathcal{E}} + \mathbf{P}_i. \end{aligned} \qquad (10.40)$$

The polarization of a unit cell depends on the relative displacement of ions ($\delta\mathbf{u}(\mathbf{r},t)$) within it and an effective charge e^*—a Born effective charge—that we have encountered before. The effective internal polarization, over the unit cell volume, is

$$\mathbf{P}_i = \frac{e^*}{\Omega_0'} \delta\mathbf{u}(\mathbf{r},t), \qquad (10.41)$$

where Ω_0' is the basis normalized volume, that is, $\Omega_0' = \Omega_0/\nu$, where ν is the number of bases of the primitive unit cell.

The ionizability is embedded in the difference between $\epsilon(\infty)$ and $\epsilon(0)$. The Born effective charge reflecting this ionizability, and appearing in the *LO* mode, follows as

	TA	LA	TO	LO
Si				
Conduction	Deformation	Deformation	—	—
Valence	Deformation	Deformation	Deformation	Deformation
GaAs				
Conduction	Piezoelectric	Deformation and piezoelectric	—	Fröhlich
Valence	Deformation and piezoelectric	Deformation and piezoelectric	Deformation	Deformation and Fröhlich

Table 10.4: Electron-phonon interactions in *Si* and *GaAs*.

$$e^* = \Omega_0' \omega_{LO} \epsilon(\infty) \rho^{1/2} \left[\frac{1}{\epsilon(\infty)} - \frac{1}{\epsilon(0)} \right]^{1/2}. \qquad (10.42)$$

ρ here is the density of the semiconductor. We have written this equation without proof here. We will find this charge equation in terms of $\epsilon(\infty)$ and $\epsilon(0)$ during the discussion of electromagnetic-semiconductor interaction (Chapter 12). The internal field due to the ionicity arises entirely from internal polarization (external consequences for Equation 10.40 vanish and displacement is absent); therefore,

$$\boldsymbol{\mathcal{E}}(\mathbf{r}, t) = -\frac{e^*}{\Omega_0' \epsilon(\infty)} \delta \mathbf{u}(\mathbf{r}, t), \qquad (10.43)$$

and hence the perturbation potential is

$$\Delta U(\mathbf{r}, t) = -\frac{q e^*}{\Omega_0' \epsilon(\infty)} \int \delta \mathbf{u}(\mathbf{r}, t) \cdot \mathbf{r} = i \frac{q e^*}{\Omega_0' \epsilon(\infty) q_s} \delta u(\mathbf{r}, t). \qquad (10.44)$$

Note again that the polar perturbation of Equation 10.44 is $\pi/2$ out of phase from the optical deformation perturbation of Equation 10.33. Optical deformation and polar perturbation—just like acoustic deformation and piezoelectric perturbation—act independently of each other.

We have now discussed a number of electron-phonon interactions based on the transverse and longitudinal modes. A summary in a tabular form (Table 10.4) for the centrosymmetric *Si* and non-centrosymmetric *GaAs* helps to integrate and deconvolve them in context.

One can now see correspondences between scattering caused by acoustic (long length range) and optical (short length range) phonons. Both cause deformation perturbation interaction and scattering, and the electromagnetic form of scattering. For acoustic ones, it appears through the strain in a piezoelectric form that builds over a longer length scale. For optical ones, it appears in a polar form that appears at the unit length scale. Piezoelectric perturbation needs a non-centrosymmetric crystal. Polar perturbation only appears in polar semiconductors.

10.3 *Umklapp processes*

PHONON PROCESSES that we have discussed so far are all examples of a *normal process*, normal in the sense that all the initial and final states are being considered within the Brillouin zone. It is, however, possible to have interactions where a new state is in the next Brillouin zone while satisfying the energy and momentum conservation. We illustrate these processes, called *Umklapp processes*, through the example of phonon annihilation. So, instead of a carrier and a phonon, as we have discussed so far, we now have two phonons interacting with each other. The interaction ends with one phonon resulting through a merger. The conservation equations in their normal form state

$$\hbar\omega_{q1} + \hbar\omega_{q2} = \hbar\omega_{q3}, \text{ and}$$
$$\hbar\mathbf{q}_1 + \hbar\mathbf{q}_2 = \hbar\mathbf{q}_3. \tag{10.45}$$

Here, the direction of the energy flow is determined by the direction of $\mathbf{q}_3$, which is a vector sum of the incident phonons. Since normally energy flows in the same direction, and has the same total magnitude, phonons interacting with each other do not cause any change in thermal resistance in this normal process. In fact, if thermal resistance arose only through such normal process phonon interaction, the material would be thermally superconducting. But, with $\mathbf{K}$ as the reciprocal lattice vector, the momentum conservation also holds for

$$\mathbf{q}_1 + \mathbf{q}_2 = \mathbf{q}_3 + \mathbf{K}. \tag{10.46}$$

Phonons of wavevector $\mathbf{q}_3$ and of $\mathbf{q}_3 + \mathbf{K}$ are indistinguishable in the periodic crystal. This was the basis for our building up the reduced zone representation and was one of the properties that directly followed from translational symmetry operations in the Bloch function discussion. This folding back into the first zone is the Umklapp process, and an illustration of this is in Figure 10.8.

The Umklapp process has changed the direction of the energy flow. It has also destroyed the conservation of crystal momentum. The meaning of the incorporation of $\mathbf{K}$ here is that the vibration has jumped the mode by one unit cell. This is possible since the mode states are arising across the entire periodicity of the structure (see Figure 3.15 for a related electron state discussion).

Through the Umklapp process, both a thermal resistance to phonon flow and a mechanism for thermalization of phonon distribution have appeared. While we discussed this w.r.t. phonon-phonon processes, this change of wavevector through interaction to outside the first Brillouin zone, and its appearance through the reciprocal space wavevector incorporation inside the first Brillouin zone, but with very significant property changes, will also appear in phonon-electron processes. Extending beyond the zone boundaries is a necessity for this process, so the interacting states need to be significantly farther away from zone center, and closer to the zone boundaries.

Umklappen in German means "flip over." *Umklapprozesse*, coined by Rudolf Peierls, is an insightful observation by Peierls, who also predicted the Peierls instability, which we discuss in S. Tiwari, "Nanoscale device physics: Science and engineering fundamentals," Electroscience 4, Oxford University Press, ISBN 978-0-198-75987-4 (2017) and which has a similar fundamental observational flavor. The Umklapp process, along with what happens at defects and interfaces, is a major mechanism affecting thermal conductivity in near-ideal crystals at high temperatures. It is also a reason why thermal superconductivity cannot exist.

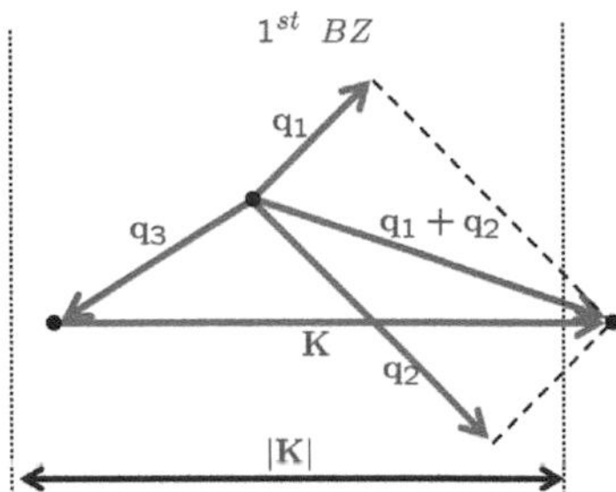

Figure 10.8: Two phonons, of wavevector $\mathbf{q}_1$ and $\mathbf{q}_2$, interacting and collapsing into one phonon of wavevector $\mathbf{q}_3$ through the Umklapp process, while still conserving crystal momentum. $\mathbf{K}$ is the reciprocal lattice vector. The figure also shows the first Brillouin zone (*BZ*) boundary, where, while $\mathbf{q}_1 + \mathbf{q}_2$ are outside the Brillouin zone, the reciprocal lattice vector incorporation leads to a final wavevector $\mathbf{q}_3$ that is in a direction away from the direction of $\mathbf{q}_1 + \mathbf{q}_2$.

An electron traveling and accelerating in a vanishing scattering material under a field will also be subject to this folding over. This is Bloch oscillation. It is easier to imagine this Bloch oscillation in superlattices with their artificial bandstructure of long unit cells and short Brillouin zones. As of 2018, I have not yet seen an incontrovertible proof of Bloch oscillation. Leo Esaki, who received the Nobel Prize for tunneling in the tunnel diode, and Leroy Chang invented superlattices, with the aim of finding Bloch oscillations. It is an invention that has been very useful in many other places, as inventions are wont to do.

This Umklapp process is also possible in carrier-phonon scattering, since Equation 10.46—momentum conservation—and energy conservation may have different contributing terms but still exist in similar forms. One now has for the wavevector

$$\mathbf{k}_i + \mathbf{q} = \mathbf{k}_f + \mathbf{K}, \tag{10.47}$$

which can be represented in a similar form as shown in Figure 10.8. We can view the Umklapp process as one where a phonon is either created or destroyed with a simultaneous Bragg reflection. The Bragg reflection—the change through $\mathbf{K}$—involves transfer of momentum and energy to the crystal.

By this point, it should be clear that the only reason we have accorded the Umklapp process a distinguishable mention is its use of the reduced zone picture of reciprocal space. Had we used the repeated zone picture, this process would be a normal process.

10.4 *Scattering potentials, matrix elements and scattering time constants*

CLEARLY, THE CALCULATION OF THE CONTRIBUTIONS TO TRANSPORT arising in the various mechanisms is of enormous importance, since that is where many of a device's limitations, for example, in speed, lie. We are going to summarize here the characteristics of the breadth of these scattering mechanisms.

We are limiting the scope of our analysis here. This is a subject with much published literature and books, some of which are pointed out in Section 10.10. We have even confined our discussion to three-dimensional conditions, although two-dimensional considerations appear because of the ubiquitous presence of interfaces. Atomically thin semiconductors will bring out even more complexity than that discussed as would nanowires. Use of parameter fitting behooves a strong caution. As von Neumann remarked, "With four parameters I can fit an elephant, and with five I can make him wiggle his trunk". (F. Dyson, "A meeting with Enrico Fermi," Nature, **427**, 297 (2004). Vladimir Vapnik, who did much of the foundational work of statistical learning theory, ascribes a similar quote to Lev Landau (V. Vapnik, "Estimation of dependences based on empirical data," Springer, ISBN 10 0-287-30865-2 (2006) 476) in the midst of a discussion of non-inductive methods of inference. Landau did not trust physical theories that combined more than a few factors.

The scattering potential and matrix elements of the major scattering processes that we have mentioned, and some that we have discussed in detail, are in Table 10.5.

Knowing the scattering potential and the matrix elements—both with approximations—one can calculate the momentum relaxation time that we determined in Equation 10.7. Table 10.6 summarizes some of the important—and often quite approximate—analytic relationships.

Table 10.5: Scattering mechanisms, their scattering potential and the matrix element.

Scattering mechanism	Potential (ΔU)	Matrix element ($\mathscr{H}_{\mathbf{kk}'}$)
Impurity		
Ionized	$\frac{Ze^2}{4\pi\epsilon(0)r}$	$\frac{Ze^2}{\epsilon(0)\Omega}\frac{1}{\lvert\mathbf{k}-\mathbf{k}'\rvert^2}$
Screened	$\frac{Ze^2}{4\pi\epsilon(0)}\exp\left(-\frac{\lvert\mathbf{r}\rvert}{\lambda}\right)$	$\frac{Ze^2}{\epsilon(0)\Omega}\frac{1}{\lvert\mathbf{k}-\mathbf{k}'\rvert^2 1/\lambda_2}$
Neutral	$\frac{\hbar^2}{m^*}\left(\frac{a_B^*}{r^5}\right)^{1/2}$	$\frac{2\pi\hbar^2}{m^*\Omega_0}\left(\frac{Z_0 a_B}{k}\right)^{1/2}$
Acoustic phonons		
Deformation	$\Xi_a \boldsymbol{\nabla}\cdot\mathbf{u}$	$\Xi_a\left(\frac{\hbar}{2\Omega_0\rho\omega_s}\right)^{1/2}\hat{\mathbf{a}}\cdot\mathbf{q}_s\left(n_q+\frac{1}{2}\pm\frac{1}{2}\right)^{1/2}$
Piezoelectric	$i\frac{ee_{pz}}{\epsilon(0)q_s}\boldsymbol{\nabla}\cdot\mathbf{u}$	$\frac{ee_{pz}}{\epsilon(0)}\left(\frac{\hbar}{2\Omega_0\rho\omega_s}\right)^{1/2}\left(n_q+\frac{1}{2}\pm\frac{1}{2}\right)^{1/2}$
Optical phonons		
Deformation	$\Xi_d\delta\mathbf{u}(\mathbf{r},t)$	$\Xi_d\left(\frac{\hbar}{2\Omega_0\rho\omega_{LO}}\right)^{1/2}\left(n_q+\frac{1}{2}\pm\frac{1}{2}\right)^{1/2}$
Piezoelectric	$i\frac{qe^*}{\Omega_0\epsilon(\infty)q_s}$	$\frac{ee^*}{\Omega\epsilon(\infty)q_s}\left(\frac{\hbar}{2\Omega_0\rho\omega_{LO}}\right)^{1/2}\left(n_q+\frac{1}{2}\pm\frac{1}{2}\right)^{1/2}$

$e^* = \Omega_0'\omega_{LO}\epsilon(\infty)\rho^{1/2}[1/\epsilon(\infty) - 1/\epsilon(0)]^{1/2}$

That relaxation time approximation should be used with caution is illustrated in Figure 10.9, which shows the dependence of the parameter r (the power of normalized energy in the relaxation time relationship) fitted for mobility, thermoelectric power and Hall constant calculation. Near the region where $\hbar\omega_{\mathbf{q}}$ is of the order of 1–$10k_BT$, the factor diverges and changes sign for the dependence in mobility, while reversing sign for the thermoelectric power. When these energies are about equal, the nature of optical phonon interaction changes, from a positive energy dependence for the inverse relaxation time to a negative one. A higher temperature rapidly lowers the relaxation time, since polar phonon scattering becomes quite efficient. At low temperatures, it gradually becomes independent of temperature.

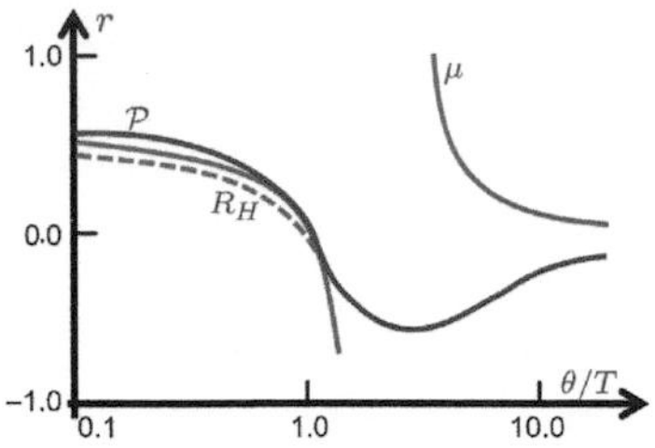

Figure 10.9: The use of relaxation time approximation for polar scattering is of limited validity. This variational calculation by H. Ehrenreich shows energy power r's dependence for variational solution when applied to mobility, thermoelectric power and the Hall constant.

10.5 *Different scattering mechanisms simultaneously*

WE HAVE ASSUMED—AN ASSUMPTION THAT IS VIOLATED ONLY UNDER A LIMITED NUMBER OF CIRCUMSTANCES—THAT THE DIFFERENT SCATTERING MECHANISMS are both random and independent. Under this assumption, the net scattering probability is a sum arising from the different mechanisms, that is,

$$\frac{1}{\tau_{\mathbf{k}}(\eta)} = \sum_i \frac{1}{\tau_{i,\mathbf{k}}(\eta)}, \quad \text{and} \quad \frac{1}{\langle\tau_{\mathbf{k}}(\eta)\rangle} = \sum_i \frac{1}{\langle\tau_{i,\mathbf{k}}(\eta)\rangle}. \qquad (10.48)$$

Scattering mechanism	Relationship $1/\langle\tau_{\mathbf{k}}\rangle$	Comment
Impurity		
Ionized	$\frac{2.41Z^2N_I}{\epsilon_r^2(0)T^{3/2}}g(n^*,T,\eta)\left(\frac{m_0}{m^*}\right)^{1/2}\eta^{-3/2}$	$r \approx 3/2$, isotropic bands
Neutral	$1.22\times10^{-7}\epsilon_r(0)N_N\left(\frac{m_0}{m^*}\right)^2$	$r = 0$
Acoustic		
Deformation	$\frac{4.17\times10^{19}\,\Xi_a^2T^{3/2}}{c_l}\left(\frac{m^*}{m_0}\right)^{3/2}\eta^{1/2}$	$r = -1/2$, $c_l = \frac{1}{5}(3c_{11}+2c_{12}+4c_{44})$
Piezoelectric	$1.05\times10^7h_{14}^2\left(\frac{3}{c_l}+\frac{4}{c_t}\right)T^{1/2}\left(\frac{m^*}{m}\right)^{1/2}\eta^{-1/2}$	$r = 1/2$
Optical		
Deformation	$\frac{2.07\times10^{19}\,\Xi_d^2T^{1/2}\theta}{c_l[\exp(\theta_q/T)-1]}\left(\frac{m^*}{m}\right)^{3/2}$ $\times\left[\left(\eta+\frac{\theta_q}{T}\right)^{1/2}-\exp\left(\frac{\theta_q}{T}\right)\left(\eta-\frac{\theta_q}{T}\right)^{1/2}\right]$	$r = -1/2$, $\theta_q = \hbar\omega_{LO}/k_B$
Polar	$\frac{1.04\times10^{14}[\epsilon_r(0)-\epsilon_r(\infty)]\theta_q^{1/2}(\theta_q/T)^r}{\epsilon_r(0)\epsilon_r(\infty)\left[\exp\left(\frac{\theta}{T}\right)-1\right]}\left(\frac{m^*}{m_0}\right)^{1/2}\eta^{-r}$	Limited validity, $r(\theta_q/T)$ variational $T \ll \hbar\omega_{LO}/k_B$ and $T \gg \hbar\omega_{LO}/k_B$

$g(\zeta) = \ln(1+\zeta) - \frac{\zeta}{1+\zeta}$, where $\zeta = b = 4.31\times10^{13}\frac{\epsilon_r(0)T^2}{n^*}\left(\frac{m^*}{m_0}\right)\eta$

$n^* = n + [(n+N_A)(N_D-N_A-n)]/N$ in the Brooks-Herring approximation.

Table 10.6: Scattering mechanisms, their scattering potential and the matrix element.

For different transport relaxation times of interest, that is, $\langle \tau_{\mathbf{k}}^{s} \eta^{t} \rangle$, which is an integral relationship (Equation 9.29), one may perform this task numerically. As an aside, in the Monte Carlo solution of the Boltzmann transport equation, since each of the relaxation times are known by evaluation, one has an expected probability of each scattering type; therefore, in simulating many particles to get a statistically meaningful result, one may use an adequately small time step and then have each of the particles, whose position and momentum are known, undergo scattering conforming to the probabilistic expectation by using random number generation to simulate the possibility of the different scattering events or the absence of them. If, in the calculation of these times, there are slow functions involved, for example, the $g(n^*, T, \eta)$ in ionized impurity scattering, one may use its mean at a constant energy, and evaluate the integrals. This inverse relaxation summation also leads to Matthiessen's rule—again, under independent random approximation—that the mobility will follow

The Monte Carlo technique has its origins in the need to calculate nuclear fission rates (the scattering of the high energy variety of fission) with a bound necessary for the correctness of the estimate. It is a powerful approach for statistical inferences. It is an invention of Stanislaw Ulam, who is also credited with the invention of the detonation technique for hydrogen bomb. Lwov (now Lviv) was and is a major center of mathematics, with the famous Stefan Banach as the leader. Banach, like Sartre and de Beauvoir in Paris, did much of his work and inspiring others in cafes. The Poland-Prussia-Russia-Ukraine borders have moved around with each war. Lwov used to be in Poland and is now in Ukraine.

$$\frac{1}{\mu} = \sum_i \frac{1}{\mu_i}, \tag{10.49}$$

where i indexes the different scattering processes.

10.6 Mobilities of semiconductors

OUR ANALYTIC DESCRIPTION OF SCATTERING is now adequate to discuss observed mobilities—therefore, the momentum relaxation defined property—of semiconductors. In these, we can see the various attributes and limits placed by the scattering mechanisms. The behavior of the electron, as through the E–$\mathbf{k}$ description, where effective mass appears as a very key parameter, also enters in this description.

First, let us look at the mobility as a function of temperature under common doping conditions. Figure 10.10 shows the mobility in *Si* and *GaAs*, two classical and widely used semiconductors. Silicon is non-polar, so optical phonon scattering is non-polar, arising in deformation and electromagnetic interaction that is not piezoelectric. *GaAs*, however, is polar, and one sees the signature of polar phonon scattering at the high temperatures. We will see this *GaAs* behavior shortly; this figure is a general description under varying doping to show the effect of impurities with temperatures while the other scattering mechanisms are also present. At room temperatures, in both these materials, phonon scattering

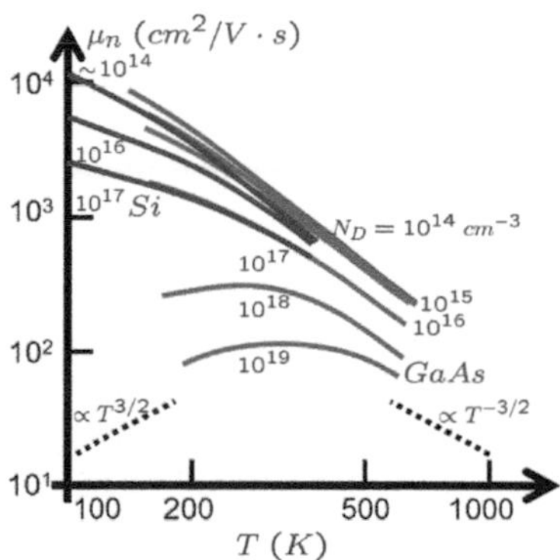

Figure 10.10: Mobility of *Si* and *GaAs* as a function of temperature for various doping conditions.

is important, but one also sees the contribution arising in ionized impurity scattering. The temperature signature of ionized impurity scattering is positive higher power; that is, higher temperature and higher velocity leads to reduced ionized impurity influence on mobility. And increased temperature causes more phonon scattering and reduction in the net velocity, hence the mobility.

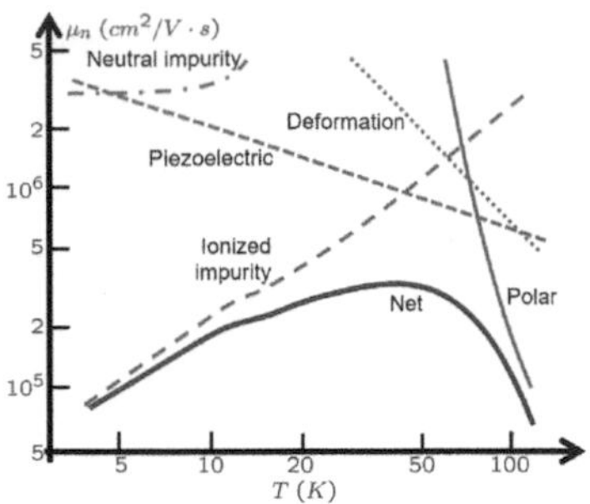

Figure 10.11: A detailed view of the different mechanisms affecting the mobility of a moderately doped *GaAs*.

GaAs and several other compound semiconductors are interesting because of the larger mobility arising in lower effective mass, even as the scattering mechanisms are largely similar in their characteristics and relative effect. *GaAs* lets us see the consequences of the various mechanisms. Figure 10.11 shows the different mechanisms' mobility limits, together with an observed and combined Matthiessen-relationship-dictated mobility behavior. Consider lower temperature first. Neutral impurity scattering is the weakest. Neutral impurities cause a very small perturbation. Piezoelectric scattering is next, and this arises in acoustic phonons that can be active at low temperatures since their $\hbar\omega_{\mathbf{q}}$ extends down to nothing. At temperatures still below room temperature, first deformation potential and then polar scattering take over. The polar scattering is dominated by optical phonons, but the deformation potential scattering is due to acoustic phonons, with an increasing contribution from optical phonons as the temperature rises. At lowest temperatures, therefore, it is the ionized impurity that dominates, and, at the highest temperatures, it is the polar scattering that dominates.

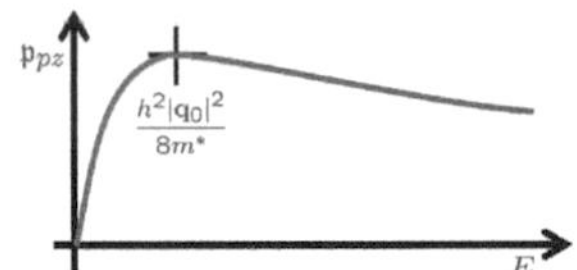

Figure 10.12: Piezoelectric scattering probability as a function of the electron's energy.

The piezoelectric scattering's carrier energy dependence—and the carrier energy in the small off-equilibrium is tied to the temperature—is that of a rapid rise followed by a slow decay, as shown in Figure 10.12. Piezoelectric scattering, which arises in acoustic phonons and therefore couples to vanishing wavevector phonons, rises rapidly as more electron states become available for scattering, reaching a maximum before slowly decreasing again. At the larger energy range, the coupling and the state-statistics product leads to this relative energy independence.

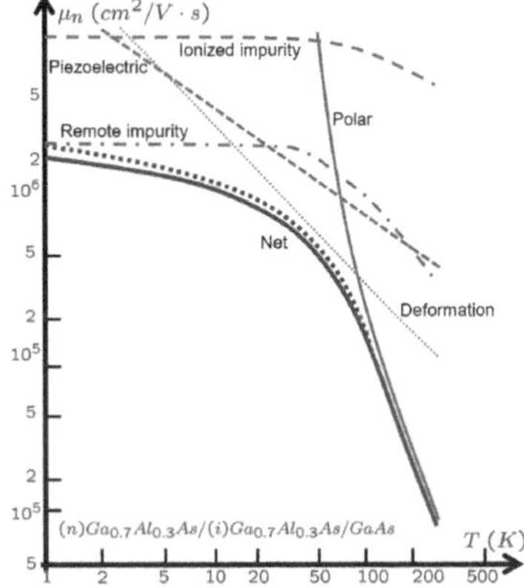

Figure 10.13: A detailed view of the different mechanisms affecting the mobility of a high mobility two-dimensional electron gas in *GaAs* at the interface with $Ga_{0.7}Al_{0.3}As$.

One may reduce this ionized impurity scattering even as the carriers remain in heterojunctions, by using remotely doped heterostructures where the ionized dopants remain in the higher conduction bandedge material while electrons remain in the lower conduction bandedge semiconductor at the heterostructure interface. The mobility behavior is shown in Figure 10.13. Now the local ionized impurity has been removed, and remote ionized impurity cause the scattering limitation at low temperatures. Phonon scattering at the interface will be mildly affected by the presence of the interface and the coupling of modes of $Ga_{0.7}Al_{0.3}As$—the higher bandgap and conduction discontinuity barrier used here—with those of *GaAs*. But note that mobilities at the low temperatures, where the removal of the ionized impurity had its effect, now exceeds $10^6\ cm^2 \cdot V/s$. At low temperatures, in such structures, the screening by mobile carriers suppresses scattering by ionized impurities, whether they are local or remote. Once past about 100 K, there is not that much difference between the mobility in bulk *GaAs* with moderate doping and this heterostructure with no intentional doping. It is all dominated by phonon scattering.

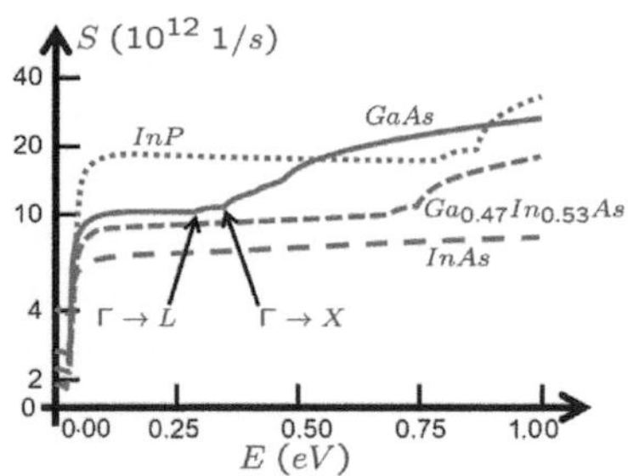

Figure 10.14: Scattering rate of electrons, for example, compound semiconductors (*GaAs*, *InP*, $Ga_{0.47}In_{0.53}As$ and *InP*), at room temperature. The first jump in scattering rate occurs when a significant electron population reaches the optical phonon emission threshold. Higher up, the thresholds for onset that occur with an increase in density of states when additional band minima are encountered is marked for *GaAs*. Where these secondary minimum exist is important for scattering rate, as carriers pick energy. *InAs* has a secondary valley nearly an eV up from the band minimum.

The scattering rates are a function of the matrix element, the occupation of states to scatter from and the states to scatter to. The number of states at the bandedge vanishes. Scattering of a low energy carrier therefore must vanish. A carrier that is high up in energy—either as part of the distribution function by chance in the Boltzmann tail or because one is considerably off-equilibrium by applying a large electric field (the mobility plots drawn up to now are for vanishingly small electric fields)—will usually see a larger number of states, and because bands are complex—not just the simplistic parabolic—there will be unusual sudden rises and falls in the density of states (see, e.g., Figure 4.13 for *Si* in Chapter 4), and therefore of scattering. This is illustrated in Figure 10.14 for a few different compound semiconductors in the form of scattering rates and the threshold for increase in density of states for *GaAs*. Where these thresholds are matters when carriers pick up energy off-equilibrium. *InAs*, even though a small bandgap material, has a very high mobility because of the lower effective mass, also has a lower scattering rate because of it (lower density of states) but, additionally, no secondary minimum is encountered up to $1\,eV$ up in energy. That is much larger than the bandgap of *InAs* itself. These carriers can go quite up in energy; that is, get "hot." Whenever the lattice temperature, so, the temperature associated with phonon distribution, is different from that of the electron, we have these hot electrons. We can also have a localized region of the semiconductor at a higher temperature than elsewhere for the phonon distribution. So, phonon distribution can also get hot. Our discussion will be restricted to conditions where the phonon distribution has a temperature that is the same as that of the reservoir surrounding the semiconductor. It is uniform and characterized by a temperature $T_l = T$. On the other hand, electron distribution may be associated with a temperature T_e or T_l that is not equal to T.

Recall the f and g scattering between the different X valleys of Si as sketched in Figure 10.5(b). The involvement of these long wavevector—short wavelength—phonons, mostly optical, is important for transport in silicon. Relaxation time, or scattering rate as a function of energy, again is useful and important for how the hot carriers exist in silicon in samples where distance and time scales allow these transfers to take place. Figure 10.15 shows the emission and scattering rates for these. Note that g emission is the most efficient. It is easy to launch a longitudinal phonon oscillation. Compared to f absorption, the g absorption too is more efficient. But f emission has a higher scattering rate. This is not surprising. Energetic electrons find it easier to emit phonons and lose energy rather than the fortuitous encountering of an existing phonon for absorption. Note also in this figure the very high rate of total scattering. A few 10^{13} s^{-1} rate is about a relaxation time constant of fractions of ps.

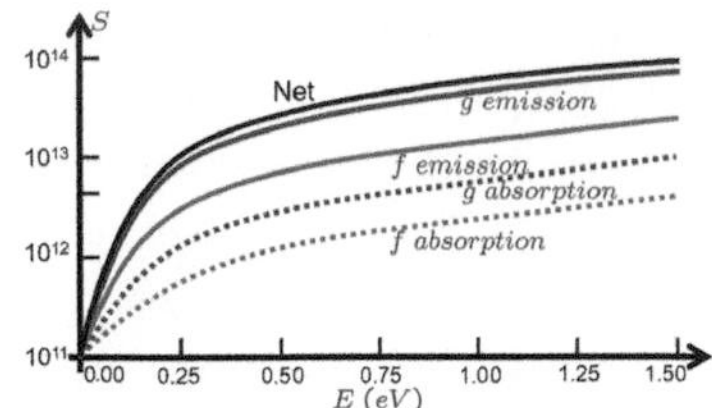

Figure 10.15: f and g emission and capture rates in Si at 300 K. g scattering is scattering to the antipodal X valley. f scattering is to orthogonal valleys.

We can also see in the relaxation times the scattering origins of the processes that lead to negative differential velocity. In Figure 10.16, in the low energy region, one can see the importance of both the acoustic phonons and the optical phonons. The energies here are larger than the optical phonon emission threshold. The order of magnitudes of these time constants are sub-ps and up. When the ~ 0.3 eV threshold of the secondary valley (X calculated here) is reached, this scattering to large crystal wavevector electrons occurs via zone-edge phonons; that is, our Fröhlich process. This relaxation time is much smaller, so this scattering is very efficient so long as enough electric field exists that can push these electrons up to these energies despite the acoustic and polar phonon losses.

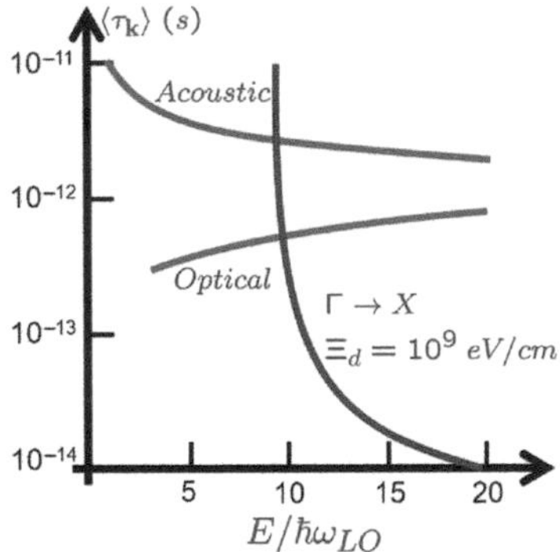

Figure 10.16: Relaxation times in $GaAs$ as a function of energy. Acoustic phonon relaxation time decreases (scattering rate increases) with energy. Polar optical phonon scattering by zone edge phonons is needed for intervalley scattering.

This scattering behavior changes with acquiring of energy, due to changes as one goes further off-equilibrium. The higher the energy is the higher the scattering rate (usually) of losing this energy. But so is the acquisition of the energy from the field. This has a direct consequence in the velocity-field behavior. There are two aspects of this. When the region over which this energy is acquired is short—of the order of few mean free paths over which there are a limited number of scattering events, and these are often non-randomizing—one sees carriers going further up in the $E(\mathbf{k})$ states, and velocity overshoot will happen. When the region is long—many mean free paths and a sufficient number of randomizing scattering events—then one sees a local balancing in the acquisition and loss of energy.

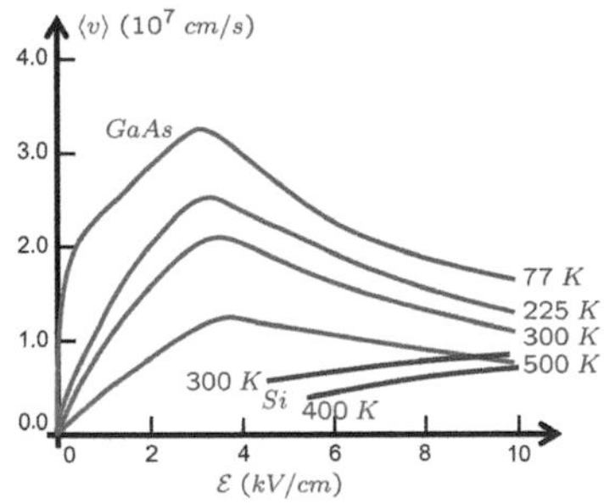

Figure 10.17: Velocity-electric field behavior for electrons in long sections of uniform field of $GaAs$ and Si at a few different temperatures at moderate doping. $GaAs$ has a secondary valley in which scattering may occur at high electric fields. Si does not. $GaAs$ goes from having a relatively low scattering rate, compared to that of Si to having comparable rates at high electric fields.

This imbalance is what leads to the nonlinearities of the velocity-field curves, whose examples are shown in Figure 10.17 for $GaAs$ and Si, which show two prototypically different forms. $GaAs$ shows a peak in velocity and then a decrease for doping conditions where ionized impurity doesn't dominate, so with not excessively high doping concentrations (less than mid-10^{18} cm^{-3}). There exists a region where applying a higher field leads to a lower velocity as carriers transfer over to the secondary valley. This is a negative differential velocity, and when properly employed at the right size scales, the negative differential characteristic will even show up as a negative differential resistance in current-voltage characteristics that can be employed gainfully. Silicon doesn't show such a negative differential velocity. Many compound semiconductors do. But, again, because of the gently increasing scattering rate, *InAs* does not. Its secondary valleys are too high.

Gunn diodes are example of this negative differential resistance appearing at the output terminals of a device. This can be employed for generating high frequencies in *GHz* range. Other design—doping and length dependent—may create a region of accelerating and piling carriers, that is, charge domains, that may stay pinned or that may also travel. These too can be used for high-frequency generation. Some of these uses are low power, but some can be used for quite high power generation.

10.7 *Frequency effects*

The secondary valley in *InAs* may be high and not allow negative differential velocity. But it has a profound effect. As we discussed, the Fermi level pinning on the *InAs* surface arises in this very high, but high mass, secondary valley. It pulls the surface states up.

THE OFF-EQUILIBRIUM DISTRIBUTION FUNCTION of Equation 9.21 becomes approximate when the relaxation time approximation becomes quite invalid. We have seen this to be the case under certain scattering circumstances. It is also invalid when forces are changing suddenly over regions of length of the order of the mean free path, that is, where insufficient scattering occurs, or even in time durations of the order of the relaxation time.

What happens in frequency if the rest of the constraints of validity are met? The description is certainly quite adequate under quasistatic conditions. What about at frequencies higher than $1/\tau_{\mathbf{k}}$? The answer is yes, so long as the approximations in our response discussion—polarization, ionicity, et cetera—are adequately accounted for. In a semiconductor, if an oscillating electric field is applied with a vanishing average, all the carriers have a force one way and then the other, and the average position of the distribution must remain the same in a linear response system. The movement stays centered, even if the shape of the distribution function changes. For electric field, carriers move one way in one half of the cycle, and then equally the other way in the other half of the cycle. And it samples accumulatively the space many times over, and scattering as it does so.

It is the response at higher magnetic field that is quite interesting in this response because of the denominator of Equation 9.21. Write a frequency

$$\omega_c^2 \equiv q^2(\det|1/\mathbb{M}|)[(1/\mathbb{M})^{-1} \cdot \mathbf{B}] \cdot \mathbf{B}; \tag{10.50}$$

ω_c here is a cyclotron frequency indicating that the charged particle likes to cycle around the axis of the magnetic field. The denominator now, with the frequency response included (following Section 9.3), is

$$1 + \left(\omega_c \tau_{\mathbf{k}}^*\right)^2 = 1 + \frac{(\omega_c \tau_{\mathbf{k}})^2}{(1 - i\omega\tau_{\mathbf{k}})^2} = \frac{\left(\omega_c^2 - \omega^2\right)\tau_{\mathbf{k}}^2 - 2i\omega\tau_{\mathbf{k}} + 1}{(1 - i\omega\tau_{\mathbf{k}})^2}. \tag{10.51}$$

For $\omega\tau_{\mathbf{k}} \gg 1$, this denominator has a sharp minimum at $\omega = \omega_c$. The distribution function shows a damped resonance. All properties, including current, thermal response, et cetera, will show a resonant peak when the applied frequency is this cyclotron frequency. And since it does, it is a means of sampling the mass tensor, because of the cyclotron frequency's dependence on it through Equation 10.50. This is now another parameterized mass, called the cyclotron mass, to join the list of density of states, conductivity, et cetera.

10.8 *Carrier response in high electric fields*

As one increases the electric field across a long sample, carriers do acquire more energy. The expectation value of the energy over the ensemble rises, and the distribution function initially just shifts—this was our shifted distribution function of Figure 8.7. We derived a mobility—a constant mobility—in this linear response condition. However, as fields increase, scattering rates change—usually increasing at a faster rate than the electric field's increase—this mobility is now a less meaningful parameter. It is a fit to how the distribution function is evolving, where the scattering rates and the $\tau_{\mathbf{k}}$ under the relaxation time approximation are also changing. After all, these are functions of the energy (E) or its normalized form ($\eta = E/k_BT$). In detail, the tail of the distribution will change, and the expectation of energy will change as a balance between the acquisition and the loss of energy that exists in the steady state.

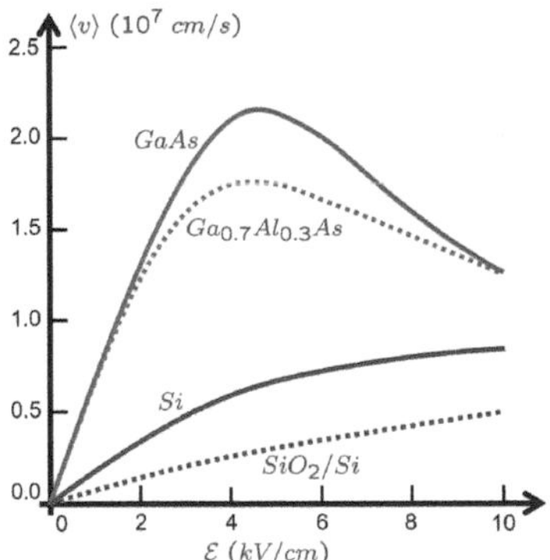

Figure 10.18: Velocity-field characteristics for exemplar semiconductor systems at 300 K under moderate doping. Note the decrease in velocity and the low field mobility in the SiO_2/Si system.

Mobility is now a fitting parameter—variational, and with internal parameter dependences such as η on $\mathcal{E}$ and T and doping, et cetera. Such an approach is still useful—and, in limits, quite insightful. Figure 10.18 shows a few example velocity-field characteristics of electrons under low doping conditions, so, in the limit of low ionized impurity scattering, which we have seen becomes important at moderate to high doping levels and even more so at low temperatures as the thermal velocities decrease.

This figure also shows the drop in mobility at the SiO_2/Si interface. We have not discussed the origins of this. Its origin is in the mechanisms that arise at interfaces. SiO_2 is amorphous, and the interface oxide is not a crystalline bulk oxide as we discussed previously, so there is a random potential fluctuation coupling to the electrons of the inversion layer, even if the interface is atomically smooth. The roughness of this interface—with step heights in it—also adds to the scattering. The randomness leads to a random redirection of momentum away from the SiO_2, but confined within the inversion layer in its two-dimensional conditions, and

In semiconductor structures, the SiO_2 will never be crystalline. That is only possible if quartz is bonded onto Si, but then the interface will have plenty of other technical issues.

leads to near halving of the mobility from that of bulk *Si*. We have looked at the $\mathcal{E} \rightarrow 0$ limit in quite some detail in this discussion of scattering and consequences of multiple stimulations at work. Now consider the $\mathcal{E} \rightarrow \infty$ limit, but under conditions where the carrier population is being conserved and there are no conduction-valence band interactions such as generation and recombination.

To separate the distribution function's "temperature" of the electron from that of its environment—the crystal lattice—we ascribe a temperature T_e to the first, and T_l to the second, as discussed in the margin of Section 9.2 on transport under generalized forces. The distribution drifts, and we had concluded that the expectation velocity of this is $\langle v \rangle = \mu(= q\langle \tau_{\mathbf{k}} \rangle / m^*)\boldsymbol{\mathcal{E}}$. The velocity is now lower than the linear change with increases in scattering rate, or, equivalently, its inverse in the relaxation time. The electron charge distribution is higher up in the E-$\mathbf{k}$ space, even as there are cold electrons. Since exponentials still relate the probabilities in the scattering rate, and these are a function of energy of electrons, one may model the distribution function, which is now drifted and distorted, by introducing this temperature T_e; that is, the Fermi-Dirac occupation function at any spatial position is

We employ these gas- and plasma-centric terms, such as electron cloud, here and there, since early investigations of many of these behaviors of field, scattering and interaction effect was in plasmas of electrons in vacuum. Vacuum tubes and much of the high density technology of accelerators, as well as plasma etching technology, have these same phenomena at work. In an etching apparatus, glowing under these plasma conditions, it does look a bit like a cloud.

$$f = \frac{1}{1 + \exp[(E - E_{qF})/k_B T_e]}. \tag{10.52}$$

Concurrent with this, application of the electric field $\boldsymbol{\mathcal{E}}$ leads to a drift velocity

$$\mathbf{v}_d = \langle \mathbf{v} \rangle = \mu(T_e)\boldsymbol{\mathcal{E}}. \tag{10.53}$$

The mobility parameter is now being modeled as changing, and it is a function of the electron cloud temperature T_e. To find $\mu(T_e)$, we determine $\langle \tau_{\mathbf{k}}(T_e) \rangle$:

$$\begin{aligned}
\langle \tau_{\mathbf{k}}(T_e) \rangle &= \frac{\int_{-\infty}^{\infty} \tau_{\mathbf{k}}(T_e) f \, d\mathbf{v}}{\int_{-\infty}^{\infty} f \, d\mathbf{v}} \\
&= \tau_0(T_e) \frac{\int_0^{\infty} f \eta_e^{r+1/2} \, d\eta_e}{\int_0^{\infty} f \eta_e^{1/2} \, d\eta_e} \\
&= \frac{2}{\sqrt{\pi}} \times \left(r + \frac{1}{2} \right)! \tau_0(T_e) \frac{\mathcal{F}_{r+1/2}(\eta_{Fe})}{\mathcal{F}_{1/2}(\eta_{Fe})},
\end{aligned} \tag{10.54}$$

where $\eta_e = (E - E_c)/k_B T_e$, and $\eta_{Fe} = (E_{qF} - E_c)/k_B T_e$. We have noted that, at low fields, the momentum relaxation time has a dependence on energy. The proportionality factor changes with temperature depending on increasing or lowering the strength of the mechanism, even as the energy dependence remains the same. So, we extend from a low field $\tau_{\mathbf{k}} = \tau_0 \eta^r = \mathcal{K} T_e^u \eta^r$ dependence. The relationship at higher fields is

With an increase in temperature, phonon scattering must increase, even if the energy dependence is the same, since the phonon population is changing. Similarly, with ionized impurity scattering, lower temperatures cause more scattering, since the velocity of carriers, on average, is reduced.

$$\begin{aligned}\tau_{\mathbf{k}}(T_e) &= \tau_0(T_e)\eta^r = \mathcal{K}T_e^u\eta^r \\ &= \tau_0\left(\frac{T_e}{T_0}\right)^u \eta_e^r \end{aligned} \tag{10.55}$$

and hence

$$\begin{aligned}\mu(T_e) &= \frac{q\tau_0}{m^*}\left(\frac{T_e}{T}\right)^u \frac{2}{\sqrt{\pi}} \times \left(r+\frac{1}{2}\right)! \frac{\mathcal{F}_{r+1/2}(\eta_{Fe})}{\mathcal{F}_{1/2}(\eta_{Fe})} \\ &= \mu_0\left(\frac{T_e}{T}\right)^u. \end{aligned} \tag{10.56}$$

We have now related the asymptotic low field mobility to a mobility parameter at higher fields through the temperature T_e. All we need to do is relate this temperature to the electric field, and we have a velocity-field relationship that is nonlinear and a function of energy. Steady state means all balances must hold. So, we invoke the conservation of energy in the Boltzmann transport equation. Starting with the Boltzmann transport equation describing the time-dependent response to the application of an electric field,

$$\frac{df}{dt} = \frac{e}{\hbar}\boldsymbol{\mathcal{E}} \cdot \nabla_{\mathbf{k}} f - \mathbf{v} \cdot \nabla_{\mathbf{r}} f - \frac{f - f_0}{\tau_{\mathbf{k}}}, \tag{10.57}$$

we multiply by energy and average over the distribution, so

$$\begin{aligned} &\frac{\int_{-\infty}^{\infty} E(df/dt)\, d\mathbf{v}}{\int_{-\infty}^{\infty} f\, d\mathbf{v}} \\ &= \frac{(e/\hbar)\boldsymbol{\mathcal{E}} \cdot \int_{-\infty}^{\infty} E\nabla_{\mathbf{k}} f\, d\mathbf{v}}{\int_{-\infty}^{\infty} f\, d\mathbf{v}} \\ &\quad - \frac{\int_{-\infty}^{\infty} E\mathbf{v} \cdot \nabla_{\mathbf{r}} f\, d\mathbf{v}}{\int_{-\infty}^{\infty} f\, d\mathbf{v}} - \frac{\int_{-\infty}^{\infty} E[(f - f_0)/\tau_{\mathbf{k}}]\, d\mathbf{v}}{\int_{-\infty}^{\infty} f\, d\mathbf{v}}. \end{aligned} \tag{10.58}$$

The left side of the equation the expectation of the time dependence of energy, that is, the evolution of the change in it. The first term on the right is the first order term in field dependence. This is one of the gains in energy as a result of the field performing work resulting in a change in the momentum. The second term we neglect. This is the real space diffusive flow. Concentration gradients cannot be large when the field flow has been enhanced by the high fields. The third term is the loss term from scattering. So, we have a gain in energy from the field in the first term, and a loss of this gained energy to scattering in the last term. We can write this form spatially integrated as

$$\frac{d}{dt}\langle W\rangle = en\mathbf{v}_d \cdot \boldsymbol{\mathcal{E}} - \frac{\langle W - W_0\rangle}{\tau_w}. \tag{10.59}$$

The left side of the equation the net power to the electron distribution as a difference, shown on the right, of the power supplied by the electric field and the power lost in scattering. We found this equation by looking at the energy flow in the Boltzmann transport equation. Now, this τ_w is the energy relaxation time that characterizes the relaxation of the hot electron distribution. This is the time constant one will see when the $\boldsymbol{\mathcal{E}}$ is turned off and the electron distribution relaxes toward thermal equilibrium:

$$\frac{\langle W - W_0 \rangle}{\tau_w} = \frac{\langle W \rangle - \langle W_0 \rangle}{\tau_w} = \frac{\int_{-\infty}^{\infty} E[(f - f_0)/\tau_{\mathbf{k}}]\, d\mathbf{v}\, d\mathbf{r}}{\int_{-\infty}^{\infty} f\, d\mathbf{v}\, d\mathbf{r}} \tag{10.60}$$

relates this energy relaxation time. The equivalent relationship for the momentum relaxation time is

$$\langle \tau_{\mathbf{k}} \rangle = \frac{2}{3} \frac{\int_0^{\infty} \tau_{\mathbf{k}} (-\,\partial f/\partial E|_0) \eta^{3/2}\, d\eta}{\int_0^{\infty} f_0 \eta^{1/2}\, d\eta}, \tag{10.61}$$

which we had already found. The energy dependences in these relationships come from the scattering process, but the two are not the same in general, that is, $\tau_{\mathbf{k}} \neq \tau_w$.

In steady state, the left side vanishes, and we have

$$\frac{\langle W - W_0 \rangle}{\tau_w} = \frac{\langle W \rangle - \langle W_0 \rangle}{\tau_w} = en\mathbf{v}_d \cdot \boldsymbol{\mathcal{E}}. \tag{10.62}$$

Here, normalized by n,

$$\langle w \rangle = \left\langle \frac{W}{n} \right\rangle = k_B T_e \langle \eta_e \rangle = \frac{3}{2} k_B T_e \frac{\mathcal{F}_{3/2}(\eta_e)}{\mathcal{F}_{1/2}(\eta_e)}, \text{ and}$$

$$\langle w_0 \rangle = n k_B T \langle \eta_e|_0 \rangle = \frac{3}{2} k_B T \frac{\mathcal{F}_{3/2}(\eta)}{\mathcal{F}_{1/2}(\eta)}. \tag{10.63}$$

Figure 10.19 shows these relaxation times in thermal equilibrium for *GaAs* in (a). Figure 10.19(b) shows how the application of an electric fields leads to changes in the expectation energy and expectation velocity and its asymptote toward steady state in a few *ps* when an electric field is applied. Note how the energy relaxation time is larger than the momentum relaxation time across the energy span, with both higher at 77 *K* due to lower scattering. Also note how, as

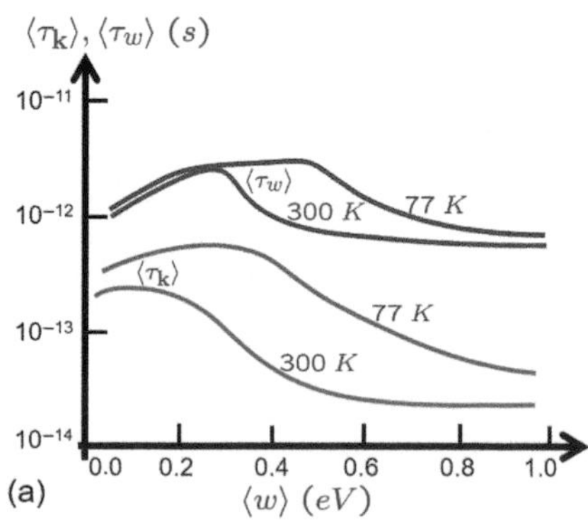

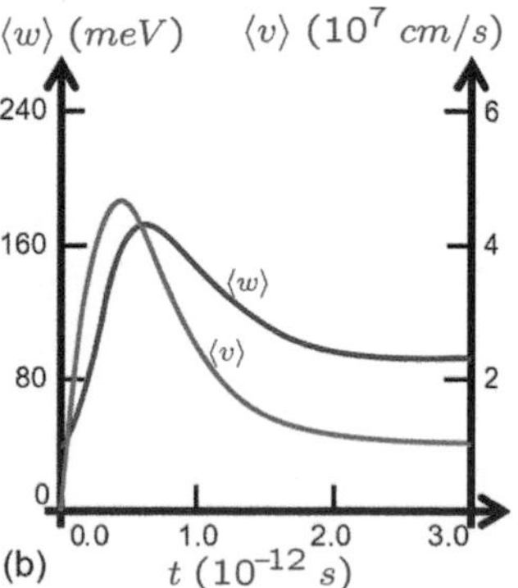

Figure 10.19: The momentum and energy relaxation times of *GaAs* as a function of energy at 300 and 77 *K* is shown in (a). The solution of the Boltzmann transport equation for energy response and velocity response when an electric field is applied is shown in (b).

a result, the velocity relaxes faster than the energy, with a few *ps* needed for the time transient to subside. Also note that the velocity reaches as high as $\sim 4 \times 10^7$ *cm/s* before settling to the $\sim 10^7$ *cm/s*. So, the application of the field causes the thermal distribution to first be disturbed farther up the $E(\mathbf{k})$—in small time periods where enough scattering has not taken place due to times being of the order of the relaxation times—and then it subsides to an off-equilibrium distribution, which, in our approximate description, has a temperature of T_e.

The relationship

$$\frac{\langle w \rangle - \langle w_0 \rangle}{\tau_w} = \frac{\int_{-\infty}^{\infty} E[(f - f_0)/\tau_{\mathbf{k}}]\, d\mathbf{v}}{\int_{-\infty}^{\infty} f\, d\mathbf{v}} \tag{10.64}$$

leads to

$$\begin{aligned} \frac{\langle w \rangle}{\tau_w} &= \langle \frac{w}{\tau_{\mathbf{k}}} \rangle = k_B T_e \langle \frac{\eta_e}{\tau_{\mathbf{k}}} \rangle \\ &= \frac{2}{\sqrt{\pi}} \times \left(\frac{3}{2} - r\right)! \frac{k_B T_e}{\tau_0} \left(\frac{T}{T_e}\right)^u \frac{\mathcal{F}_{3/2-r}(\eta_e)}{\mathcal{F}_{1/2}(\eta_e)}, \text{ and} \\ T_e &= \frac{3\sqrt{\pi}}{4} \frac{\tau_0}{[(3/2) - r]!} \left(\frac{T_e}{T}\right)^u \frac{\mathcal{F}_{3/2}(\eta_e)}{\mathcal{F}_{3/2-r}(\eta_e)}. \end{aligned} \tag{10.65}$$

For non-degenerate conditions, so with simplification of the Fermi integrals, this also becomes a direct analytic relationship between T_e and energy E of

$$\begin{aligned} \left(\frac{3}{2} - r\right)! \frac{k_B}{\tau_0} \left(\frac{T}{T_e}\right)^u (T_e - T) &= \left(r + \frac{1}{2}\right)! \frac{q^2 \tau_0}{m^*} \left(\frac{T_e}{T}\right)^u E^2, \text{ or} \\ \left(\frac{T_e}{T} - 1\right) \left(\frac{T}{T_e}\right)^{2u} &= \frac{q^2 \tau_0^2}{m^* k_B T} \frac{[r + (1/2)]!}{[(3/2) - r]!} E^2 \\ &= (\beta E)^2, \end{aligned} \tag{10.66}$$

where we have introduced β as a parameter. When electrons become quite hot, that is, $T_e \gg T$, we can simplify all these different parameter relationships to

$$\begin{aligned} \frac{T_e}{T} &= (\beta E)^{-2/(2u-1)}, \\ \mu(T_e) &= \mu_0 \left(\frac{T_e}{T}\right)^u, \text{ or} \\ \mu(E) &= \mu_0 (\beta E)^{-2/(2u-1)}. \end{aligned} \tag{10.67}$$

We have now connected the high field mobility to energy through the scattering process modeled in the relaxation time approximation. The high field mobility decreases from the low field mobility value and asymptotically vanishes. This latter implies that the velocity saturates. Our model has many limitations. A gradual

When a charge cloud accelerates as the result of an applied electric field, the cloud and the crystal may be out of equilibrium. If the field-induced energy rise is high enough, this leads to $T_e \neq T_l$. But if in an electric field it does not accelerate, for example, the drift-diffusion in the *p-n* junction, then it maintains $T_e = T_l$. So, electric fields can exist without this energy-associated consequence. The field must be providing the energy to the electrons for the temperature to change. This underlay the discussion of which mobility to use in our discussion of *p-n* junctions in the marginalia of Section 9.2.

decrease in mobility, while staying positive, means that the differential negative behavior discussed earlier is not captured. One can see the reason for this in the fact that our relaxation time relations are all monotonic functions, so that the onset of new scattering processes, such as to the secondary valleys, will produce changes in the energy dependence parameter r, which our formalism did not include.

The velocity saturation lends itself to a reasonably simple explanation. If carriers have higher energy, emission of optical phonons—the longitudinal variety—becomes the dominant mechanism. Each of these emissions takes out a fair amount of energy. Longitudinal variety is preferred because it is the direction most conducive to momentum matching in the scattering process. Steady state implies that carriers gain energy and lose energy at an equal rate, so

$$\frac{\langle w\rangle - \langle w_0\rangle}{\tau_w} \approx \frac{\hbar\omega_{LO}}{\tau_w} \approx e\mathbf{v}_d \cdot \boldsymbol{\mathcal{E}}. \tag{10.68}$$

With energy relaxation time a good approximation for how the steady state is being achieved, on average the picked-up energy in time τ_w is all lost, as $\hbar\omega_{\mathbf{q}}$ on average in each scattering event, so

$$\tau_w = \frac{(\hbar\omega_{LO} m^*)^{1/2}}{e\mathcal{E}}. \tag{10.69}$$

Also, the excess energy on average is $(1/2)m^*\langle v^2\rangle$, so

$$v_d = v_{sat}, v_{pk} \approx \sqrt{\langle v^2\rangle} = \left(\frac{\hbar\omega_{LO}}{m^*}\right)^{1/2} = f(m^*, \omega_{op}, T). \tag{10.70}$$

So, the saturated velocity of semiconductors is quite closely related to the effective mass and the longitudinal optical phonon energy. Figure 10.20 illustrates this by showing a normalized parameter for several semiconductors.

Figure 10.21 illustrates the build-up temperature tails. It illustrates the carrier distribution function in a small cross-section of the drain depletion high field region of a transistor at drain biases in the current saturation region. The *Si* semiconductor is at 300 *K*. At the onset of saturation, the field are gradual across the entire inversion channel, and the distribution function is quite Maxwell-Boltzmann-like and can be fitted with a $T_e = 300$ *K*. Raise the drain bias further, with most of this bias increase appearing in a small region at the drain end to which the inversion region couples, and one sees a long tail forming and the peak shifting slightly. Note that there are electrons, albeit in very small numbers—the plot is on a logarithmic scale—and one sees temperatures in a tail region up to 2880 *K* here. This is a hot electron tail. A major fraction of the electrons are not too far from $T_e = 300$ *K*. These are the electrons

The electron accelerates between scattering. The average is lower than the peak. A more accurate development of this argument is the following. At the velocity of interest v_d, the energy balance is $ev_d\mathcal{E} = (3\hbar\omega_{\mathbf{q}}/2)[\exp(\hbar\omega_{\mathbf{q}}/k_BT) - 1]$, where optical phonons dominate. The exponential accounts for emission and absorption. When velocity doesn't rise any further (peak or saturated), T_e is large, the momentum balance leads to $e\mathcal{E} = 2m^*v_d[\exp(\hbar\omega_{\mathbf{q}}/k_BT) + 1]$ for non-polar optical phonons. So,

$$v_{sat}, v_{pk} = \left[\frac{2\hbar\omega_{op}}{4m^*}\tanh\left(\frac{\hbar\omega_{op}}{2k_BT}\right)\right]^{1/2} = g(m^*, \omega_{op}, T).$$

This equation, or others with different mechanisms dominating at velocity peaking, are largely within a few percent of each other.

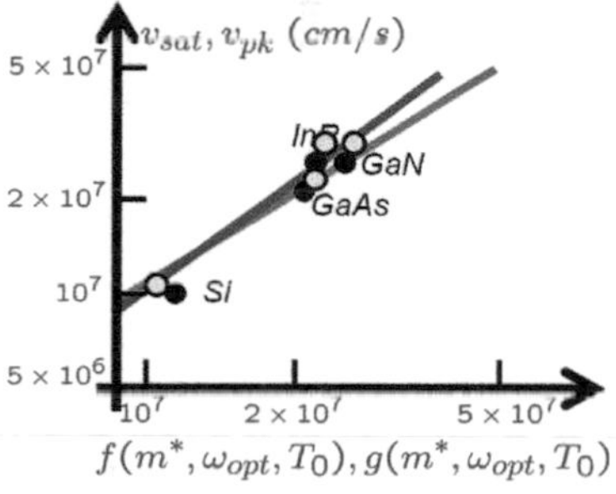

Figure 10.20: The correlation between high field velocity and the ratio of the longitudinal optical phonon energy and effective mass, shown through a plot across several semiconductors. Two different estimations based on the simple optical phonon emission argument ($f(m^*, \omega_{op}, T)$) and the non-polar emission and absorption mechanism dominating ($g(m^*, \omega_{op}, T)$) are shown.

spread in the region of 0.0–0.6 eV. These electrons remained cold since they underwent sufficient scattering. But a fraction of the electrons injected from the inversion region did not scatter—after all, it is a probabilistic process, the region is narrow and the time it takes to cross this region small—and it is these electrons that appear in this second high energy tail.

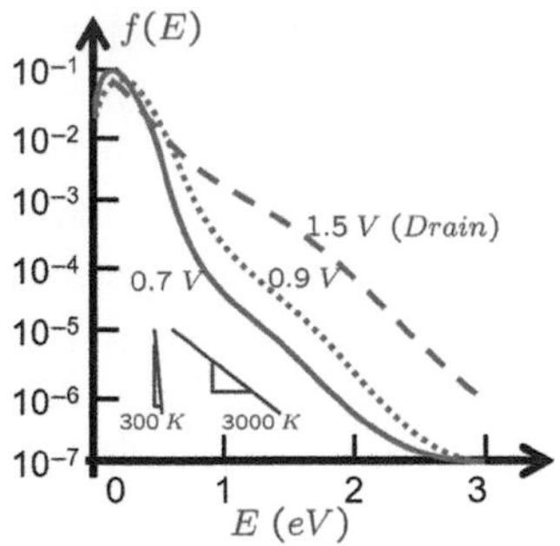

Figure 10.21: Electron distribution along the channel of a *MOSFET*. The 300 K distribution is in a region with low electrostatic potential change. The others are for increasing electrostatic potential change. The electrostatic potential change, after saturation of the current, occurs at the drain end. Adapted from S. E. Laux and M. V. Fischetti, "Issues in modeling small devices," International Electron Devices Meeting 1999. Technical Digest (Cat. No. 99CH36318), IEEE, 523–526 (1999).

We can also see in Figure 10.22 how the energy changes when electrons show negative differential velocity as they undergo transfer to the secondary valleys—first L and then X—when sufficient electric field is applied in *GaAs*. At low fields, the average energy remains close to the $(3/2)k_BT$ that one expects close to thermal distribution. Beyond a kV/cm, the field is high enough for carriers to start acquiring enough energy while the scattering is still moderate. The average energy rises. But, once it reaches that $\sim 0.3\ eV$ threshold for the transfer to the L valley, the rise saturates to quite below 1 eV, even as the field reaches $10^5\ V/cm$. The valley population in the central Γ valley drops as first the L valley and then the X valley start being scattered into. These have a large density of states, polar optical scattering is efficient, and carriers now move slower in the high mass higher valleys. One goes from high mobility and high velocity to a lower mobility and lower velocity.

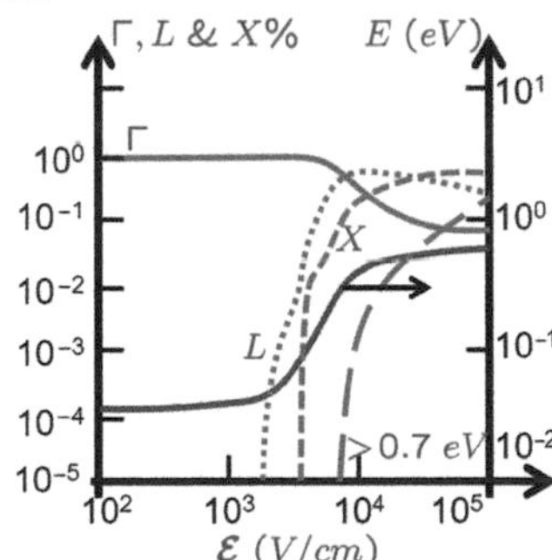

Figure 10.22: Average energy and valley occupation in steady state for an electron in (100) *GaAs* under a constant electric field. The electric field ranges from low to high magnitudes and passes through the negative differential velocity region. Also shown is the fraction of electrons at an energy of 0.7 eV. Adapted from S. E. Laux and M. V. Fischetti, "Issues in modeling small devices," International Electron Devices Meeting 1999. Technical Digest (Cat. No. 99CH36318), 523–526 (1999).

10.9 *Summary*

We stressed the variety of scattering mechanisms of import in common semiconductors of interest and related their properties: scattering rates, relaxation times, cross-sections and perturbation potentials and, through this, derived some of the field-mediated transport properties. Many points flow through our arguments, since there are many possible sources of perturbation in a semiconductor, and, depending on the choice of conditions, a group of these need to be considered simultaneously.

The scattering cross-section has a clear classical meaning. Semi-classically, one can also assign a meaning that corresponds to an effective area that the moving electrons feel and get scattered from or to. This reflects a parameter with areal units reflecting the matrix element, and a relaxation time tied to the scattering rate can be related quite straightforwardly for elastic processes. The matrix element for scattering between $|\mathbf{k}\rangle$ and $|\mathbf{k}'\rangle$ can be written in terms of the perturbation potential of the scattering process. We derived this for ionized impurity scattering while including screening.

In a metal, the Fermi energy is very high into the conduction bands, so much of the electron state change is on the Fermi surface with large wavevector phonons

Scattering by phonons is a dominant theme in semiconductors. Acoustic phonons are atomic oscillation modes with correlated displacement, and optical phonons are atomic oscillation modes with anti-correlation. Both oscillations can be longitudinal and transverse. Acoustic mode energies reach zero at vanishing wavevector. Optical mode energies remain finite. So, these modes have quite different temperature signatures. Absorption requires occupation of these modes. So, as temperature decreases, optical absorption vanishes, and if measurements involve only low kinetic energy, then optical phonon emission too gets suppressed.

dominating. Semiconductor bands are largely empty for typical carrier concentrations, and numerous scattering processes can cause a change.

Longitudinal and transverse acoustic modes will modulate the bandstructure through this dilation and compression of the crystal. This shows up as perturbation in the potential energy—the bandedge energy—and causes deformation coupling and scattering. *LA* phonons cause pronounced in-line deformation in direct gap semiconductors. For indirect bandgap semiconductors, it is the *TA* phonons that cause the strongest deformation coupling. Charge also matters. Semiconductors where the inversion symmetry is absent accumulate the dipole charge over half a wavelength, and this leads to piezoelectric scattering due to acoustic phonons.

Optical phonons—both *LO* and *TO*—too cause deformation interaction. But if charge is present, then there will be a dipolar coupling effect. This we call electromagnetic coupling, and it matters for *LO* phonons. Electrons undergo scattering by optical phonons dominated by the polar optical scattering in crystals with ionicity, such as *GaAs*. This electromagnetic coupling will also become important in photon interactions.

Because the dipolar effect arises in ionic polarization of the crystal, there also exists frequency dependence. The polarization part of the response is reflected in the difference in the inverse dielectric function at static conditions and at high frequency, between which the polarization freezes out. So, the polarization effect can be extracted from this. Also, one notes from this that when one is considering the dipole interaction with electromagnetic waves—photon processes—then the frequency dependence will appear. So, it will appear in plasmonic responses where electron charge clouds' frequency response couples. All these will appear in our consideration in discussions of optical response (Chapter 12) and of high permittivity materials (Chapter 18) and of remote processes (Chapter 19).

Phonon processes can involve large wavevectors, since Brillouin-zone-size phonons exist. It is possible for two phonons to coalesce, leading to a wavevector in the next Brillouin zone. This is equivalent in this periodic structure to a phonon wavevector in the opposite direction. The energy carried by the phonons is now moving in the opposite direction. This is an Umklapp process. Thermal properties have become poorer, and phonons become thermalized in this way. This same behavior can happen with electron-photon interaction and for electron-electron interaction, albeit with much smaller likelihood.

When different scattering mechanisms exist, are important simultaneously and are independent of each other, then the net scattering is a sum of the contributing scattering mechanisms, and one can therefore reduce the net effect geometrically in the mobility. This is Matthiessen's rule.

The practicalities of these different scattering mechanisms were observed through their temperature and energy dependences, where we noted how semiconductors such as *GaAs* achieve very high mobilities and improve further by improved screening and reduced ionized impurity scattering in two-dimensional electron gases. We also noted how the multiple valleys of *Si* near the *X* point lead to idiosyncratic effects there. As one raises the energy of electrons, scattering to secondary valleys may enter. So, *GaAs* shows a negative differential velocity in its velocity-electric field behavior as $\Gamma \rightarrow L$ scattering prevails. On the other hand, the extremely high secondary valley in the small effective mass and small bandgap semiconductor *InAs* leads to high mobilities and velocities further up in energy. By employing the Boltzmann transport formalism here, one observed the behavior by which these off-equilibrium electrons relax and how this relaxation in momentum and energy is reflected in the relevant relaxation time constants that can all be written in terms of the momentum relaxation time and the various scattering processes' energy dependences. We also noted here that multiple mechanisms and multiple constraints can lead to a distribution function that may have multiple exponential tails—our example was of two—corresponding to two electron temperatures present in the distribution function when a high field existed over a short length scale. One of the temperature tails corresponded to the crystal temperature, and the other corresponded to the hot electrons that acquired higher energy because they were lucky to not undergo scattering at the high fields.

10.10 Concluding remarks and bibliographic notes

SCATTERING IN SEMICONDUCTORS, with the multiple mechanisms, different semiconductors with different mechanisms of importance, the dependence on temperature of all these, and the suppression of scattering in two-dimensional electron gases with their own unusual properties, has meant a tremendous amount of research interest spanning many decades. There is a vast literature, and we point out only a few of the many significant publications that have the flavor of a review, of intuition or of a comprehensive summary.

The early books by Ziman[1] and Peierls[2] are still just as lucid and fundamentally clear in the discussion of scattering and interactions in matter as any writing since then. Ziman, in particular, tackles electron-phonon interaction, screening effects and lattice scattering through a number of chapters.

Hamaguchi[3] specifically analytically explores electron-electron, electron-phonon and plasmon scattering comprehensively. Another intermediate text is by Wolfe, Holonyak and Stillman[4].

Jacoboni's text[5] discusses scattering at a more advanced level, is also more complete, and also discusses alloy scattering in mixed ternary and quaternary semiconductors as well as holes. Readers of this chapter should be able to follow the arguments of this text. Another set of comparable discussions is in the text by Fischetti and Vandenberghe[6], and that by Ridley[7]. Finally, as mentioned in Chapter 4, the book series "Handbook on Semiconductors" is a very thoughtful source[8] that gives insights into the thinking process that led to the consensus.

High field, the modeling of optical phonon scattering in high field conditions, and the concept of electron temperature are treated by Conwell in her classic review published in the "Semiconductors and Semimetals" series from Academic Press[9].

[1] J. M. Ziman, "Electrons and phonons," Oxford (1960)

[2] R. E. Peierls, "Quantum theory of solids," Oxford, ISBN 0-19-850781 (1960)

[3] C. Hamaguchi, "Basic semiconductor physics," Springer, ISBN 978-3-642-03302-5 (2010)

[4] C. M. Wolfe, N. Holonyak and G. E. Stillman, "Physical properties of semiconductors," Prentice Hall, ISBN 0-13-669961-8 (1989)

[5] C. Jacoboni, "Theory of electron transport in semiconductors," Springer, ISBN 978-3-642-10585-2 (2010)

[6] M. Fischeti and W. G. Vandenberghe, "Advanced physics of electron transport in semiconductors and nanostructures," Springer, ISBN 978-3-319-01100-4 (2016)

[7] B. K. Ridley, "Quantum processes in semiconductors," Oxford, ISBN 0-19-850-5809 (1999)

[8] W. Paul (ed.), "Band theory and transport properties," 1, North-Holland, ISBN 0-444-85346-4 (1982)

[9] E. M. Conwell, "High field transport in semiconductors," Academic (1967)

10.11 *Exercises*

1. In a semiconductor at a low temperature, how likely is the optical phonon absorption process compared to the acoustic phonon absorption process? **[S]**

2. How important is the screening in ionized impurity screening at low doping concentrations compared to high? **[S]**

3. Why would a neutral impurity cause scattering? **[S]**

4. Is piezoelectric scattering a significant scattering mechanism due to phonons in silicon? Note silicon has inversion symmetry. Which kind of phonons—acoustic or optical—may be relevant? **[S]**

5. Why does momentum appear in determining the coupling arising in oscillator strength? **[S]**

6. Why are transverse optical phonons unaffected by electric fields while longitudinal ones are? **[S]**

7. Take Si with a conduction effective mass $m^* = 0.26m_0$. What is the energy, momentum, wavevector and de Broglie wavelength of a conduction electron at thermal velocity at 300 K? Si also has a sound velocity of $\sim$5800 m/s. Using conservation of energy and crystal momentum, calculate the highest percent change in energy of an electron undergoing an LA phonon deformation scattering. **[S]**

8. During scattering by an acoustic phonon, an electron with an initial velocity v_i will either lose or gain energy. Show that, for a sound velocity of c, this loss of gain in energy is bound, with

$$\Delta E \leq \frac{4c}{v_i} - 4\left(\frac{c}{v_i}\right)^2.$$ **[S]**

9. Does electron-electron scattering randomize the energy distribution of hot electrons in 3-dimensional, 2-dimensional and 1-dimensional systems? Please explain concisely. **[S]**

10. At high fields, electrons emit optical phonons at a very high rate, causing "hot phonon" effects. Estimate the optical phonon generation rate in $GaAs$ at electron energies of 200 meV. If optical phonons cannot dissipate rapidly, the phonon occupancy $(n(\omega_{\mathbf{q}}))$ becomes large. What effect will this have on hot carrier relaxation? **[S]**

11
Particle generation and recombination

The 0th moment of the Boltzmann transport equation leads to the continuity equation for particles. In this equation, the time dependence not included in the particle Hamiltonian—$\partial f/\partial t$—also contributes to the evolution of the distribution. In single particle situations, a large population of electrons in quite filled conduction bands of metals, for example, where one doesn't need to consider any appearance or disappearance of them, the term vanishes if the distribution function lumps occupation across all the conduction bands. In semiconductors, with very partial occupation of bands, electrons interact across bands of different indexes and, in this, the interaction between the conduction and valence bands, and others, can be quite significant. These two distributions cannot really be lumped together. Their charges are different: one is a mostly empty band, and one is mostly filled. And the interaction between them will take a multitude of forms. Even at the simplest, this 0th moment equation needs to be written separately for particles in the conduction band and particles in the valence band. The interaction between them leads to a $\partial f/\partial t$ term in each of the equations. The simplest approximation to this change term, upon disturbance from thermal equilibrium, follows as

Looking for a hole in metals would be like looking for aliens arriving in *UFOs*. It is highly unlikely that they would be found, since filled bands are far up and away—as are inhabitable planets far away in space and time.

$$\frac{\partial f}{\partial t} = -\frac{f - f_0}{\tau} \quad \therefore \quad \frac{\partial \langle f\varphi \rangle}{\partial t} = -\frac{\langle f\varphi \rangle - \langle f_0\varphi \rangle}{\tau_\varphi}, \tag{11.1}$$

where $\varphi = 1$ for particle density calculation. This time constant $\tau_{\varphi=1}$ has the meaning of a particle lifetime—τ_n for electrons, or τ_p for holes—that characterizes the generation or recombination of particles, since Equation 11.1 is a particle continuity equation. One could look upon these time constants as the equivalents of relaxation times, with the randomizing part of the approximation where the relaxation is of the particle density disturbed from thermal equilibrium. These time constants of approximations of the various moment equations—particle relaxation,

Semiconductor Physics: Principles, Theory and Nanoscale. Sandip Tiwari.
 Published 2020 by Oxford University Press. DOI: 10.1093/oso/9780198759867.001.0001

momentum relaxation, energy relaxation—all reflect the entropic urge toward thermal equilibrium.

For particles, when a single band picture is employed, for example, in many conditions of transport with high majority carrier concentration and limited minority carrier concentration change, using a *constant* single time constant is quite a reasonable model, because the number of carriers is only limitedly perturbed. Many circumstances, however, involve multiple bands and this time-constant fitting can become quite complex, just as it did for the momentum relaxation time. The high field drain-channel region of a *MOSFET* is also a region of short extent. Carriers are well out of equilibrium and acquire high kinetic energy, and such a high energy carrier, say, an electron in an n-channel *MOSFET*, may dislodge an electron from an occupied state in the valence band into the conduction band while losing sufficient energy to make that process happen. This is impact ionization. In a highly doped small bandgap material, this high initial energy triggered process may even occur without any significant field, because energy states at a bandgap order of magnitude become more likely to be occupied. This is Auger generation. At least two bands are involved, particle count is not conserved and the time constant now has an energy dependence. In regions of bipolar transport, as in a p-n junction or the bipolar transistor, the electrons and holes recombine (and generate) by again an interaction either directly across the bandgap or through intermediate states in the band. So, there are numerous circumstances where this τ_φ for the 0th moment equation is finite and has significant implications. And one would have to employ a Boltzmann transport equation for each one of these particles, with their different distribution functions, to describe the physics. The electron or hole lifetime is this relaxation time. And it arises in many different processes. For some, a simple classical reasoning suffices, but some require quantum-mechanical rigor to get reasonable accuracy and honorable insight. Lasers, for example, arising in radiative recombination, would not exist were it not for quantum mechanics at work.

There is also nothing stopping us, except for the complexity, of even writing a Boltzmann transport equation for phonons. It is just immensely hard, since degenerate bands, Bose-Einstein statistics and a broadband distribution of optical phonons exist. We will attempt this in Chapter 19 to explore thermoelectric phenomenon at nanoscale.

In this chapter, we will explore classical and quantum-mechanical approaches to understanding these lifetimes for non-radiative and radiative processes—some only classically, but some in both ways to see the correspondence and the details. The non-radiative processes are emphasized through the Hall-Shockley-Read process discussion for a single charged state and through one specific Auger process that is quite suitable for quantum-mechanical discussion, and which in turn gives us a means of understanding impact ionization process. The radiative recombination process is employed to understand the basics that let us demystify the general approach to these calculations.

11.1 *Radiative recombination and generation*

RADIATIVE PROCESSES are, of course, quantum-mechanical processes. Particles are an emergent view to compact energy excitations. Two different particles—electrons and holes—combining to produce a third—a photon—with very different properties is certainly not trivially expected in a classical view. But, quantum-mechanically, from the energy and the wave-particle view, it is all quite rational. Lasing depends on stimulated emission—the creation of another photon identical to an existing photon—which too is only possible to visualize quantum-mechanically. This stimulation is at the heart of the A and B coefficients, and this basic notion of spontaneous and stimulated emission is summarized in Appendix K, together with the implications for electromagnetic-matter interactions where photons are generated or annihilated. Our discussion here is semi-classical and parameterized, taking recourse to detailed balance as the initial matching condition. The parameterization buries in it the quantum-mechanical origins. A more substantive treatment based on scattering involving phonons is a task for us in Chapter 12.

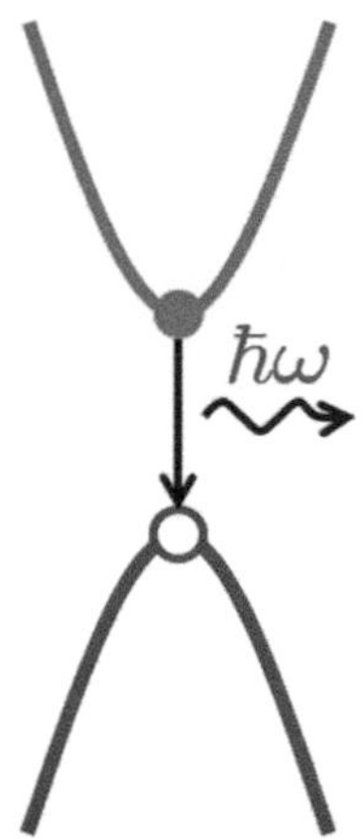

Figure 11.1: A classical view of the annihilation of an electron-hole pair generating a photon in a direct transition. An electron occupying a state in the conduction band transitions to occupying a state in the valence band. This valence band state must be empty prior to the transition and is the state of the quasiantiparticle hole.

Take $GaAs$ with a bandgap of $1.4\,eV$. A photon of this energy will have a momentum $\hbar k = E\overline{n}/c \approx 1.4 \times 3/3 \times 10^8\, eV \cdot s/m$, which is about $2.24 \times 10^{-27}\, kg \cdot m/s$. An electron at the thermal energy of k_BT—about the energy of the peak in the distribution—has, because of $E = \hbar^2k^2/2m^*$, a momentum in the crystal of $\hbar k = \sqrt{E2m^*}$, which is about $2.2 \times 10^{-26}\, kg \cdot m/s$. The momentum at the first Brillouin zone edge is $\hbar k = \hbar\pi/a$, about $6.3 \times 10^{-25}\, kg \cdot m/s$. The electrons occupy about 3 % of the length expanse of the reciprocal space in $GaAs$. Photons are in another factor-of-10-shrunk region of the order of 0.3 %. This span difference has a variety of implications, which differentiate the photon from the phonon—the other boson—even if they are both "charge neutral." Both have electromagnetic effects: a photon is an electromagnetic bundle of energy, while a phonon has charge effects through deformation, piezoelectricity and polar behavior. For photons, the small momentum makes classical direct calculations not unreasonable for several problems. In this, the approach is similar to that of the use of capture cross section in ionized impurity scattering analysis.

In Figure 11.1, an electron-hole pair recombination leads to the generation of a photon of energy $\hbar\omega$. The electron, in transitioning from a state in the conduction band to a state in the valence band, loses its energy in the form of the photon. Photons have a very high velocity c modified by the index of refraction, so the energy of a photon $E = \hbar(c/\overline{n})k$, where $\overline{n}$ is an averaged index of refraction and k the wavevector of a photon, involves a very small k. Photons have a very small momentum compared to that of the electron.

The approach to such semi-classical conditions that we have seen before and will see again is to use thermal equilibrium—detailed balance—to relate parameters that then lead to a parameterized equation dependent on the carrier population, and preferably in useful limits, directly on the excess carrier population.

We may write the recombination rate $\mathcal{R}_r$ as

$$\mathcal{R}_r = c_r np, \tag{11.2}$$

where c_r is a rate constant for capture in units of cm^3/s. One electron and one hole recombine, so the proportionality is to the product of electron density—the number of electrons available—and hole density—the number of holes available, which is the number of empty states in the valence band—through some proportionality that determines the time-dependent rate. This is the c_r here. And normalizing to working with unit volumes being the norm, Equation 11.2 is the proper form, based on dimensionality considerations. In thermal equilibrium, using the 0 subscript to identify this condition, as we have done throughout the text, we have

$$\mathcal{R}_{r0} = c_r n_0 p_0 = c_r n_i^2; \tag{11.3}$$

therefore, off-equilibrium,

$$\mathcal{R}_r = \mathcal{R}_{r0} \frac{np}{n_i^2}, \tag{11.4}$$

where $n = n_0 + \Delta n$, and $p = p_0 + \Delta p$, with Δn and Δp as the excess population from equilibrium. In general, these can be positive or negative. The change in the recombination rate is

As before, our analysis will include the lowest order perturbation terms.

$$\Delta\mathcal{R}_r = \mathcal{R}_r - \mathcal{R}_{r0} = \frac{(n_0 + \Delta n)(p_0 + \Delta p)}{n_0 p_0} \mathcal{R}_{r0} - \mathcal{R}_{r0}$$
$$= \left(\frac{\Delta n}{n_0} + \frac{\Delta p}{p_0} \right) \mathcal{R}_{r0}. \tag{11.5}$$

In thermal equilibrium, the change in the recombination rate vanishes. Indeed, in thermal equilibrium, recombination and generation balance. They balance in detail for each individual pair of state interactions, and therefore for the collective ensemble too. It is the recombination process that leads to the radiative emission and therefore is of interest to us.

The time constant for the 0th moment equation—the particle conservation time constant—is what we call lifetime. It is a measure of the relaxation time in the entropic drive toward maximum entropy of thermal equilibrium. For radiative processes, this time constant is the radiative lifetime

$\mathcal{B}$ is called the bimolecular radiative constant. It was in molecules that the radiative rate was first calculated and employed. It is related to the matrix element of radiative processes and is related to the Einstein B coefficient. They are both related to the matrix element for radiative processes.

$$\tau_r = \frac{\Delta n}{\Delta\mathcal{R}_r} = \frac{\Delta n}{\Delta n \times p_0 + \Delta p \times n_0} \frac{n_i^2}{\mathcal{R}_{r0}}. \tag{11.6}$$

Since one electron and one hole recombine, or get generated from photon absorption, in the radiative process, $\Delta n = \Delta p$, and we have

$$\tau_r = \frac{\Delta n}{\Delta\mathcal{R}} = \frac{n_i^2}{n_0 + p_0} \frac{1}{\mathcal{R}_{r0}} = \frac{1}{n_0 + p_0} \frac{1}{\mathcal{B}}. \tag{11.7}$$

$\mathcal{B} = \mathcal{R}_{r0}/n_i^2$ is a radiative constant. The constant $\mathcal{B}$ is directly drawn from the matrix element coupling the states and is therefore related to how well matched the electron and hole states are coupled through the transition dipole matrix (see Appendix K, and we follow through in detail in Chapter 12). Table 11.1 summarizes bimolecular recombination coefficients for some of the semiconductors. Note how, for direct bandgap semiconductors (*GaN*, *GaAs*, *InP*, *InAs*), it is much larger, compared to the values for indirect

	$\mathcal{B}$ (cm^3/s)
AlAs	7.5×10^{-11}
GaAs	1.0×10^{-10}
GaN	1.1×10^{-8}
GaP	3.0×10^{-15}
InAs	2.1×10^{-11}
InP	6.0×10^{-11}
4H SiC	1.5×10^{-12}
Si	1.1×10^{-14}

Table 11.1: Bimolecular recombination coefficients for some semiconductors. The numbers are for room temperature.

gap semiconductors, where the matrix element will be small from symmetry arguments.

The radiative lifetime in Equation 11.7 also has implications. If the semiconductor is n-type or p-type, the thermal equilibrium carrier density of the opposite polarity is low. So, in the denominator, the larger carrier density dominates. The proportionality then is also to the inverse of this density. It is the excess lower carrier density that is being radiatively removed in a noticeable way by the radiative process and constitutes the rate-limiting step. For example, if $n_0 \gg p_0$ (an n-type material), then $\tau_r \propto 1/n_0$.

11.2 *Non-radiative processes: Hall-Shockley-Read*

DEFECTS IN SEMICONDUCTORS arise in many forms, and we tackled their quantum-mechanical origins in our discussion of point perturbations (Chapter 7). Intrinsic defects, such as vacancies, dislocations, surface states and other crystal imperfection complexes, arise in the thermodynamic nature of the material preparation and existence. Extrinsic defects, unintentional and intentional, exist because of impurities incorporated during the process of preparing the material or devices. Our discussion of shallow donors and acceptors—as hydrogenic impurities—as sources of additional electrons or holes in the semiconductor was predicated on their forms being similar to a semiconductor species, but for an excess or deficiency of electrons. Such an impurity then can be seen as hydrogenic impurity, mildly perturbative in the crystal's periodic potential, leading to an electron that was localized over a much larger space (many unit cells and $\times 100$ or more than the orbital localization). Since the confining energy is very small comparable to thermal energy at room temperature, and because of the statistics of occupation when there are few dopant states and a large number of states in the band, it occupies states in the conduction band. These conduction band states are propagating states. Spreading in real space is localization in reciprocal space. So, hydrogenic donors or acceptors are to viewed as being closely coupled in the momentum space of the E-$\mathbf{k}$ picture and as a long-range perturbation spatially. At low temperatures, the electron may stay quite localized, which is what we call freeze-out. This description is simplistic; of course, higher doping means degeneracy as dopants are closer together—a Mott description—and the E-$\mathbf{k}$ description needs modification.

In contrast to the hydrogenic impurities, with their low ionization energy of carriers to band extrema, defects—intrinsic and extrinsic—also arise as a short-range spatial perturbation in the

We are calling the defect-mediated recombination process the Hall-Shockley-Read process. Much literature refers to this as the Shockley-Read-Hall process. I suspect that there is a curious story behind the experiments and theory leading to this point here. Hall's papers, "Germanium rectifier characteristics," Physical Review, **83**, 228 (1951) and "Electron-hole recombination in germanium," Physical Review, **87**, 387 (1952), describe the equations close to the form we see. Shockley and Read have a more generalized theory in "Statistics of the recombinations of holes and electrons," Physical Review, **87**, 835–842 (1952). There are curious footnotes in these publications. What should the defect-mediated process be called? It depends on who knew what when, when it was published and under what constraints. These credit ideas are pervasive in science, especially since, as Koestler in *The Act of Creation* has argued, outbreaks of creativity happen as many bring the relevant thoughts together from many perspectives, and the moment is ripe. The modern scientific process arose as Copernicus, Kepler, Brahe, Lippershey, Bruno, Galileo and so many others stirred the broth in the midst of the Renaissance. Exclusive credit is an oxymoron.

See S. Tiwari, "Quantum, statistical and information mechanics: A unified introduction," Electroscience **1**, Oxford University Press, ISBN 978-0-198-75985-0 (forthcoming) for the elementary description of localization and transitions, and S. Tiwari, "Nanoscale device physics: Science and engineering fundamentals," Electroscience **4**, Oxford University Press, ISBN 978-0-198-75987-4 (2017), for an advanced discussion with application to devices.

potential, both in the bulk and at the surfaces. These are deep levels—which can be quite complex in their behavior—that are sometimes also called traps. Instead of a hydrogenic substitutional impurity such as *P* or *As* as a donor for an *n*-type semiconductor, consider a transition metal atom—atoms with *d* or *f* orbitals with electrons—substituting in the *Si* crystal. *P* and *As* had an excess electron in the *p* orbital, which has a symmetry similar to that of the *p* orbital of the s^2p^2 of *Si* that hybridized to sp^3. So, *P* and *As* can also undergo such a wavefunction restructuring easily, except that there exists an excess electron. Transition elements such as *Cu*, *Ti* and *Au* have many more states in the *d* outer orbital, a number of electrons partially filled in them and a symmetry that is very different from that of the atoms from which the crystal assembly is formed. The consequence is that many states will appear for different levels of ionization of the impurity; many of these are in the bandgap, but many could be in the band. So, states exist in many states of ionization (*Au* could be Au^0, i.e., neutral, as well as several states of ionization, Au^+, Au^{2+}, Au^{+3}, etc., with different energies in the bandgap and possibly in the bands). Since the outer orbital involved in the interaction is highly localized, it is spread out in **k**. Electrons of all different types of envelope symmetry will interact with them, albeit much, much more slowly, that is, emission and capture of electrons (or holes, i.e., coupling to the valence band) will happen much more slowly. Hydrogenic donors interact with the conduction band; we only consider their valence band interactions under very specific conditions, such as when a photon has an energy that can cause an electromagnetic coupling. Deep impurities can couple with both bands. So, they are mediators for electron-hole interactions, even with thermal energy as the catalyst. Thermal energy is not such a catalyst for hydrogenic impurities, since the barrier energy is the entire bandgap for the opposite band. Deep levels have a much lower energy barrier to such a interaction. Deep levels can be both donor-like and acceptor-like. Such deep levels will also arise in intrinsic defects, which too have short spatial ranges.

The complementary argument is that *B* will have one less electron, which leads to the hole.

Many common traps are singly ionized with an energy in the bandgap. Oxygen is an example of a donor-like trap. Chromium is an example of acceptor-like trap. The intrinsic deep levels come in a complex variety, and often we may be able to measure the presence and characteristics of the trap by thermally activated measurements, although we can only give verisimilitude to its origin. *GaAs*, for example, has a deep acceptor *EL3* at 0.58 *eV*, an electron trap *EB3* at 0.91 *eV*, and another electron trap *EB6* at 0.41 *eV*, where the energies are referenced to the conduction bandedge.

A deep donor, that is, a donor of electrons, is one where, when the electrons exist with the impurity, it is charge neutral and, when the electrons detach, the deep impurity is charge positive. An acceptor-like deep trap is one that accepts an electron, turning charge negative.

This hydrogenic and deep-level behavior was looked at in detail in Chapter 7 from the energetics viewpoint. We are now interested in the dynamic effect of the deep levels in the semiconductor.

We will take a simple example—a deep donor with one ionization state—to understand the capture and emission mechanics and its consequences. The model is the Hall-Shockley-Read model, and a simple illustration of it is in Figure 11.2. Figure 11.2(a) here illustrates the emission of an electron from a neutral trap to the conduction band, which leaves the trap in a positively charged state. This emission process is characterized by e_e as the volume-normalized emission time constant, so units of $cm^{-3} \cdot s^{-1}$. In the inverse of this process (Figure 11.2(b)), an electron is captured by a charged trap with the capture time constant c_e. A similar set of parameters exists for hole capture. We write in terms of the holes to be unambiguous about the interaction being between the trap state and the valence band states. It could have been viewed as an electron transition process to the valence band, which is mostly filled with electrons. All these four processes are possible, so if an electron is captured by a trap (Figure 11.2(b)), and subsequently a hole is captured by a trap (Figure 11.2(c)), an electron and a hole have disappeared. This is a recombination. If an electron emission ((Figure 11.2(a)) and a hole emission happen (Figure 11.2(d)), we now have a generation process. In thermal equilibrium, detailed balance holds, so (a) and (b) balance, and (c) and (d) balance separately—not just to the band, but for each and every trap state to each and every band state.

Figure 11.2 is the simplest singly charged trap example of the Hall-Shockley-Read (*HSR*) generation-recombination process. The *HSR* process is non-radiative. The excess energy is being supplied to the crystal (recombination) or extracted from the crystal (generation). So, phonons must also be a part of this interaction. Our semi-classical parameterized treatment sidesteps the conservation requirements by taking recourse to parameters and the balancing at thermal equilibrium.

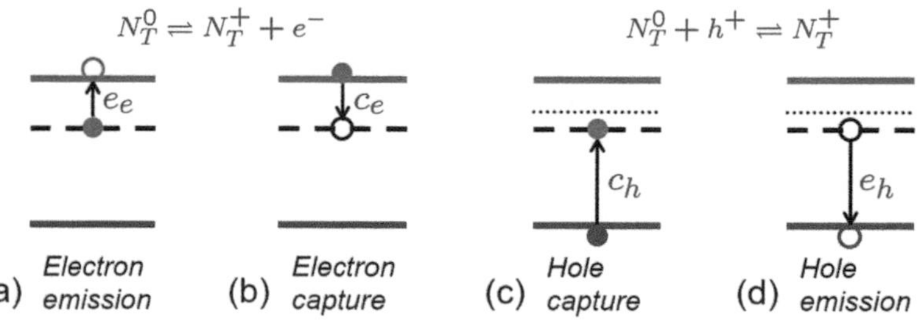

Figure 11.2: Part (a) shows the emission of an electron from a deep level ($N_T^0 \rightarrow N_T^+ + e^-$), (b) shows the capture of an electron to the deep level ($N_T^+ + e^- \rightarrow N_T^0$), (c) shows the same for a hole capture ($N_T^0 + h^+ \rightarrow N_T^+$) and (d) shows the emission of a hole from the deep level ($N_T^+ \rightarrow N_T^0 + h^+$). Note that the hole process could be equivalently viewed as an electron process coupling to the valence band. Part (c) would then be an electron emission to an empty state in the valence band, and (d) an electron capture from a filled state in the valence band.

Let N_T be the trap density of the donor-like trap. It exists either as a positively charged trap of density of N_T^+, that is, donor-like and singly ionized, or as a neutral trap of density N_T^0. In thermal equilibrium,

$$N_T = N_{T0}^+ + N_{T0}^0, \tag{11.8}$$

and, away from thermal equilibrium,

$$N_T = N_T^+ + N_T^0. \tag{11.9}$$

The rate of capture of electrons from the conduction band is

$$\mathcal{R}_{ce} = c_e n N_T^+. \tag{11.10}$$

As with the radiative recombination discussion, this relationship can be seen as the proportionality, and therefrom equality, in the recombination process being proportional to the number of carriers available for capture (n) and the available sites for capture (N_T^+). Similarly, the rate of emission of electrons to the conduction band—the generation rate of electrons in the conduction band—is

$$\mathcal{G}_{ce} = e_e N_T^0 (\mathcal{N}_c - n). \tag{11.11}$$

Note $(\mathcal{N}_c - n)$, because the effective density of states is the density of electron states parameterized to the edge of the band and n of these are occupied. At thermal equilibrium, detailed balance forces

$$\mathcal{G}_{ce0} = \mathcal{R}_{ce0},$$

$$\text{or } e_e \overline{N}_T^0 (\mathcal{N}_c - n_0) = c_e n_0 N_{T0}^+$$

$$\therefore\ e_e = c_e \frac{n_0}{\mathcal{N}_c - n_0} \frac{N_{T0}^+}{N_{T0}^0}, \tag{11.12}$$

where we have now related the emission proportionality constant to the capture proportionality constant. This is precisely the same approach for determining parameter relationships as in the radiative recombination discussion. Electron emission and capture constants are related, and the same is true for the other transitions between states, that of hole capture and emission. For holes, for capture,

$$\mathcal{R}_{vh} = c_h p N_T^0, \tag{11.13}$$

and, for emission,

$$\mathcal{G}_{vh} = e_h N_T^+ (\mathcal{N}_v - p). \tag{11.14}$$

And the thermal equilibrium constraint translates to

$$\mathcal{G}_{vh0} = \mathcal{R}_{vh0},$$

$$\text{or } e_h N_{T0}^+ (\mathcal{N}_v - p_0) = c_h p_0 N_{T0}^0$$

$$\therefore\ e_h = c_h \frac{p_0}{\mathcal{N}_v - p_0} \frac{N_{T0}^0}{N_{T0}^+}. \tag{11.15}$$

The hole emission constant (e_h) has now been related to the hole capture constant (c_h).

These constants are related to the ability of a trap to capture or to emit. We tackled a similar problem for ionized impurity scattering when the $\cos\theta$ dependence and the interpretation of scattering cross-section was provided a quantum-classical interpretation (Subsection 10.1.1). Figure 11.3 shows a few ionized traps of trap density N_T^+, with their capture range viewed through cross-section σ, in the presence of a density of n electrons moving around thermally at a velocity v_θ. As this is a positively charged trap, electron capture is possible. The fractional extent of the capturing volume that the electron is likely to see per unit time is $\sigma v_\theta N_T^+$. This is the capture rate, so

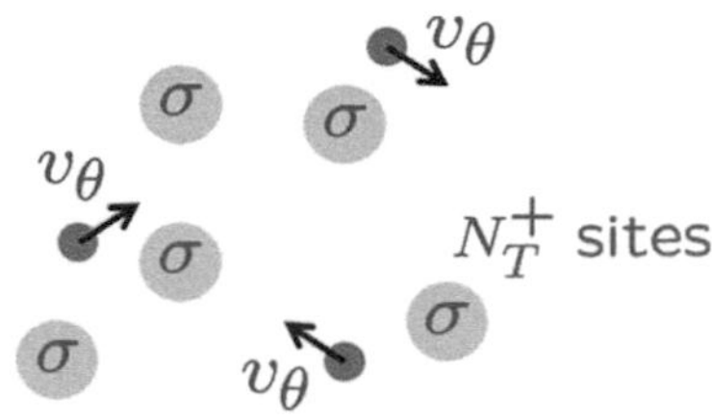

Figure 11.3: An illustration of a density of N_T^+ charged traps of cross-section areas σ in a sea of n electron density moving with thermal velocity v_θ.

$$c_e = \sigma v_\theta N_T^+. \tag{11.16}$$

With $n_0 \ll \mathcal{N}_c$, using the Boltzmann approximation,

$$\begin{aligned} e_e &= c_e \frac{n_0}{\mathcal{N}_c - n_0} \frac{N_{T0}^+}{N_{T0}^0} \\ \therefore \quad &= c_e \frac{n_0}{\mathcal{N}_c} \frac{N_{T0}^+}{N_{T0}^0}, \quad \text{with } n_0 = \mathcal{N}_c \exp\left(\frac{E_F - E_c}{k_B T}\right). \end{aligned} \tag{11.17}$$

Similarly,

$$\begin{aligned} e_h &= c_h \frac{p_0}{\mathcal{N}_v - p_0} \frac{N_{T0}^0}{N_{T0}^+} \\ \therefore \quad &= c_h \frac{p_0}{\mathcal{N}_v} \frac{N_{T0}^0}{N_{T0}^+}, \quad \text{with } p_0 = \mathcal{N}_v \exp\left(\frac{E_v - E_F}{k_B T}\right). \end{aligned} \tag{11.18}$$

An additional constraint, from thermal equilibrium, is $n_0 p_0 = n_i^2$ by the law of mass action.

A remark is in order regarding the law of mass action. Its antecedents are equilibrium in chemical reaction. Equilibrium demands forward and reverse processes to balance and also relate to the reactants and products in dilute conditions. $n_0 p_0 = n_i^2$ is true in dilute conditions when only n (electrons) and p (holes) are the reacting species. Electrons in the conduction band arise from electrons jumping from a valence band, leaving a hole behind. All are dilute. If we have non-degenerate doping, this still holds. But if degeneracy exists, conditions are not dilute; additional states have been introduced by the dopants. Now, $n_0 p_0 \neq n_i^2$.

Let Δn and Δp be the density of excess electrons and holes, respectively. Their change in time arises in the recombination and generation rate. So,

$$\begin{aligned} -\frac{d\Delta n}{dt} &= \mathcal{R}_{ce} - \mathcal{G}_{ce} = c_e n N_T^+ - c_e \frac{n_0}{\mathcal{N}_c} \frac{N_{T0}^+}{N_{T0}^0} N_T^0 \mathcal{N}_c \\ &= c_e \left(n N_T^+ - n_0 N_T^0 \frac{N_{T0}^+}{N_{T0}^0} \right) \\ &= c_e \left[n N_T^+ - \mathcal{N}_c \exp\left(-\frac{E_c - E_T}{kT}\right) N_T^0 \right], \end{aligned} \tag{11.19}$$

and, similarly,

$$\begin{aligned} -\frac{d\Delta p}{dt} &= \mathcal{R}_{vh} - \mathcal{G}_{vh} = c_h \left(p N_T^0 - p_0 N_T^+ \frac{N_{T0}^0}{N_{T0}^+} \right) \\ &= c_h \left[n N_T^0 - \mathcal{N}_v \exp\left(-\frac{E_T - E_v}{kT}\right) N_T^+ \right]. \end{aligned} \tag{11.20}$$

In thermal equilibrium,

$$\Delta n = \Delta p = 0, \tag{11.21}$$

and, in steady state,

$$\frac{d\Delta n}{dt} = \frac{d\Delta p}{dt}, \tag{11.22}$$

since the carriers are allowed to only interact with each other via the *HSR* process. Therefore,

$$N_T^+ = \frac{N_T\left[c_e\mathcal{N}_c \exp\left(-\frac{E_c-E_T}{k_BT}\right) + c_h p\right]}{\left[n+\mathcal{N}_c \exp\left(-\frac{E_c-E_T}{k_BT}\right)\right]c_e + \left[p+\mathcal{N}_v \exp\left(-\frac{E_T-E_v}{k_BT}\right)\right]c_h} \tag{11.23}$$

and

$$N_T^0 = \frac{N_T\left[c_e n + c_h\mathcal{N}_v \exp\left(-\frac{E_T-E_v}{k_BT}\right)\right]}{\left[n+\mathcal{N}_c \exp\left(-\frac{E_c-E_T}{k_BT}\right)\right]c_e + \left[p+\mathcal{N}_v \exp\left(-\frac{E_T-E_v}{k_BT}\right)\right]c_h}. \tag{11.24}$$

For known emission and capture characteristics—these are a function of the *HSR* centers in play—one can determine precisely the ionized and the neutral trap densities. The expressions look somewhat complex, but one can see quite clear sense in them. If the electron and hole capture rate was vanishingly small, the density is only determined by the background doping, and therefore the position of Fermi energy. Say, the material is n-type and the left side of the denominator dominates as $c_e, c_h \to 0$. If $N_T \ll n$, then

In Exercise 6 of this chapter, we tackle *Au* with its multiply charged states; this same statistical treatment will lead to further complexity.

$$N_T^+ \approx N_T\frac{\mathcal{N}_c}{n}\exp\left(-\frac{E_c-E_T}{k_BT}\right) = N_T\exp\left(-\frac{E_F-E_T}{k_BT}\right), \tag{11.25}$$

the expectation based on Fermi energy, trap energy position and the Maxwell-Boltzmann statistics of occupation.

The net rate of decrease of excess carrier density can be written as

$$\begin{aligned} \mathcal{U} &= -\frac{d\Delta n}{dt} = -\frac{d\Delta p}{dt} = \mathcal{R} - \mathcal{G} \\ &= \frac{np - n_i^2}{\left[n+\mathcal{N}_c \exp\left(-\frac{E_c-E_T}{k_BT}\right)\right]\frac{1}{c_hN_T} + \left[p+\mathcal{N}_v \exp\left(-\frac{E_T-E_v}{k_BT}\right)\right]\frac{1}{c_eN_T}}. \end{aligned} \tag{11.26}$$

This equation too has a direct meaning. When $np > n_i^2$, that is, there are excess carriers, then there is net recombination. If it is less, there is net generation. Both are entropically driven. This equation can be written in a simpler and more meaningful form by parameterizing with trap-determined time constants for capture of holes (τ_{p0}) and electrons (τ_{n0}) as

$$\tau_{p0} = \frac{1}{c_h N_T}, \text{ and } \tau_{n0} = \frac{1}{c_e N_T}. \tag{11.27}$$

These two constants are related to the properties of the deep levels and how many of them there are. We can now write

$$\mathcal{U} = -\frac{d\Delta n}{dt} = -\frac{d\Delta p}{dt} = \frac{np - n_i^2}{(n + n_T)\,\tau_{p0} + (p + p_T)\,\tau_{n0}}, \tag{11.28}$$

where we have used the additional interpretation of

$$n_T = \mathcal{N}_c \exp\left(-\frac{E_c - E_T}{k_B T}\right) \text{ and}$$

$$p_T = \mathcal{N}_v \exp\left(-\frac{E_T - E_v}{k_B T}\right), \tag{11.29}$$

as a density of carriers—connected to traps—so, as if the Fermi energy is at the trap level.

Equation 11.28 is an equation from which much can be explored in many of the common conditions in devices off-equilibrium, and the carrier interactions therein. This includes *p*-*n* junction regions, transport in quasineutral regions such as in bipolar transport, and a variety of phenomena in silicon-on-insulator transistors, and the back-depletion regions of *MOSFETs*. τ_{p0} and τ_{n0} were lifetime parameters associated with the trap's capturing characteristics and the trap concentration. Equation 11.28 also tells us that if a material is *n*-type, the first term of the denominator will usually be large $n \gg p, n_T, p_T$, and if this were the case, what we will mostly be concerned with is the parameter τ_{p0}. In other words, in a quasineutral *n*-type material, hole concentration is small and the amounts of excess electrons and holes are of the same magnitude, so what matters is what happens to the excess holes, which are the minority carriers. The rate-limiting step is these excess holes. In general, though, such as in *p*-*n* junction transition regions, both will matter, since carrier concentrations are low for both, and even switch as one traverses the region.

For an *n*-type semiconductor, the lifetime we are particularly concerned with is the hole lifetime,

$$\begin{aligned}\tau_p &= -\frac{\Delta p}{d\Delta p/dt} \\ &= \frac{\Delta p}{np - n_i^2}\left\{\left[n + \mathcal{N}_c \exp\left(-\frac{E_c - E_T}{k_B T}\right)\right]\tau_{p0}\right. \\ &\quad \left. + \left[p + \mathcal{N}_v \exp\left(-\frac{E_T - E_v}{k_B T}\right)\right]\tau_{n0}\right\} \\ &= \frac{\Delta p}{np - n_i^2}\left[(n + n_T)\,\tau_{p0} + (p + p_T)\,\tau_{n0}\right]. \end{aligned} \tag{11.30}$$

For a p-type semiconductor, the lifetime we are particularly concerned with is the electron lifetime,

$$\begin{aligned}\tau_n &= -\frac{\Delta n}{d\Delta n/dt} \\ &= \frac{\Delta n}{np - n_i^2}\left\{\left[n + \mathcal{N}_c \exp\left(-\frac{E_c - E_T}{k_B T}\right)\right]\tau_{p0}\right. \\ &\quad \left. + \left[p + \mathcal{N}_v \exp\left(-\frac{E_T - E_v}{k_B T}\right)\right]\tau_{n0}\right\} \\ &= \frac{\Delta n}{np - n_i^2}\left[(n + n_T)\,\tau_{p0} + (p + p_T)\,\tau_{n0}\right]. \end{aligned} \tag{11.31}$$

In quasineutral regions, the excess population is the same, so these lifetimes are the same if all the rest of the parameters are. In quasineutral regions, the rate-limiting step is via the excess minority carriers, so it is the minority carrier lifetime that we mostly concern ourselves with in quasineutral conditions. But, in general, both matter.

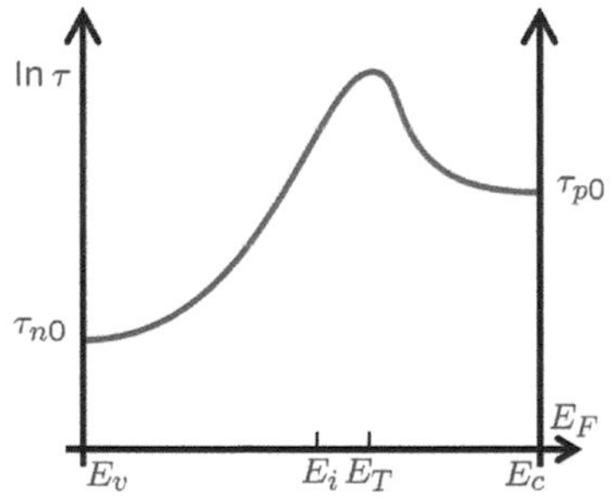

Figure 11.4: A sketch of lifetime as the Fermi energy is swept from p-type to n-type for a singly charged deep level.

Figure 11.4 shows how the lifetime changes as the Fermi energy is swept in a quasineutral material. If it is quite p-type, the minority carrier is n-type, and it is τ_{n0} that matters. This is what matters in the quasineutral base region of n-p-n bipolar transistors and the p-type region of long p-n diodes. The complement of this is what happens in an n-type material when the Fermi energy is closer to the conduction bandedge. When the Fermi energy is close to the middle of the gap, both carrier populations are small, and the lifetime actually peaks. The carrier and the trap terms in each of the denominator terms of Equation 11.28 are comparable and small, peaking the lifetime. A trap close to the middle of the gap tends to be most efficient at generation and recombination. Such a trap can communicate effectively with both the conduction band and the valence band, and the energy impediment to the transition balances to about half the bandgap for both carriers. Of course, in stating this we have assumed that capture or emission rate constants (the cs and the es) for electrons and holes are quite similar, in which case the exponential term of the transition will dominate.

From Equation 11.4, one can also extract some limiting values for low, moderate and high levels of excess carriers. For small values of Δn and Δp,

$$\tau = \tau_{low} = \tau_{p0}\frac{n_0 + n_T}{n_0 + p_0} + \tau_{n0}\frac{p_0 + p_T}{n_0 + p_0}. \qquad (11.32)$$

For large values of Δn and Δp,

$$\tau = \tau_{high} = \tau_{p0} + \tau_{n0}, \qquad (11.33)$$

and, for intermediate values of Δn and Δp,

$$\tau = \tau_0 \frac{1 + [(\tau_{p0} + \tau_{n0})\Delta p]/[(n_0 + n_T)\tau_{p0} + (p_0 + p_T)\tau_{n0}]}{1 + \Delta p/(n_0 + p_0)}. \qquad (11.34)$$

This discussion stressed the consequences of intentional or unintentional interaction centers that allow efficient capture and emission with both the conduction band and the valence band and thus limit the lifetime—the relaxation time for the 0th moment equation—for electrons and holes. It is a "defect"-induced process. It is a non-radiative process where energy and momentum matching will involve interaction of the crystal, and hence phonons. We did not tackle this problem quantum-mechanically, sweeping those details into a cross-section, the constants *c*s and *e*s or the lifetime parameters τ_{p0} and τ_{n0}. It is not that instructive, except to show the power of quantum mechanics. The radiative recombination and generation discussion before this, again without recourse to quantum mechanics, which we will address in Chapter 12, was a discussion of direct interaction between bands.

I hope that, by this time, this enormous power of quantum mechanics is front and center in your imagination. Classical observations can always be reached through the correspondence principle from quantum mechanics. But there is so much richness and detail that this approach is enlightening.

There is an additional extremely significant way by which carriers interact between bands, and through other states, which becomes important under a variety of conditions, particularly high carrier concentration, low bandgap, et cetera. This is the phenomena of Auger recombination and generation. Auger interactions in semiconductors and their states—not the atomic states—is the subject of this upcoming section.

We will pursue this discussion both classically and quantum-mechanically, to stress the fundamentals.

Pierre Auger was among the first, together with Lise Meitner, who had a major part in the discovery of uranium fission, to show how, in atomic systems, electrons changing state from a higher energy orbit to a lower energy orbit can give up this energy and kick another electron up in energy. Auger electron spectroscopy at this atomic level is a useful approach in the analysis of materials.

11.3 *Non-radiative processes: Auger*

When the excess energy of a generation or recombination process is coupled through another particle—usually one additional particle—the process is called an Auger process. The origin of this name is its initial observation in atomic systems. It is of considerable importance in semiconductors, since high doping conditions, high injection conditions, such as of bipolar lasers, and small bandgap materials are quite susceptible to it. Indeed, impact ionization, where an electron gaining energy in an electric field parts with it by generating an electron-hole pair is an electric-field-activated Auger generation process. Figure 11.5 shows the simplest

If you look at the literature, you will find a particularly intriguingly titled paper "The first 70 Auger processes" (P. T. Landsberg and D. J. Robbins. "The first 70 semiconductor Auger processes," Solid-State Electronics, **21**, 1289–1294 (1978)). It is a pretty good analytic paper.

description—akin to the radiative recombination description—for an example Auger process. This one is an idealized example of a simple, isotropic, parabolic, band-to-band recombination. In general, Auger processes may happen between band-to-shallow levels and involve excitons, multiple bands, combinations of carriers and phonons, and so on.

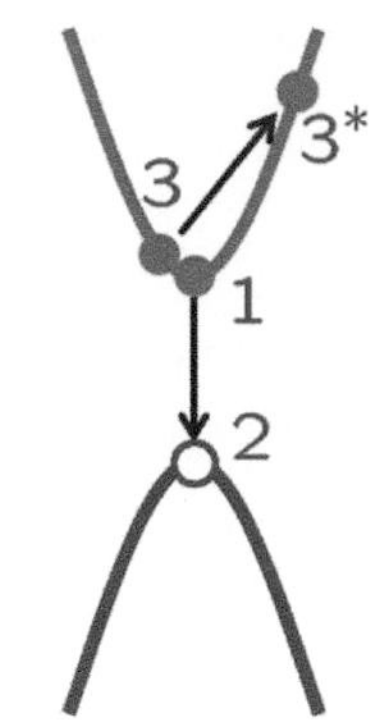

Figure 11.5: An illustration of a 2-electron and 1-hole Auger process. Electron 1 combines with hole 2, and imparts excess energy to electron 3, which gets excited to a higher energy state, 3*.

We tackle the process of Figure 11.5 first, using our semi-classical picture. Two electrons and one hole are involved, and all need to exist for this recombination process to take place, so

$$\mathcal{R}_{Ae} \propto n^2 p \tag{11.35}$$

is a compact description of the Auger recombination rate. We could have placed a constant in the relationship to make it an equality. In thermal equilibrium,

$$\mathcal{R}_{Ae0} \propto n_0^2 p_0. \tag{11.36}$$

For Auger generation, all one needs is a sufficiently hot electron that can lose its energy to bounce an electron from the filled valence band to the nearly empty conduction band. The rate-limiting condition here is the presence of this high energy electron. So,

$$\mathcal{G}_{Ae} \propto n, \tag{11.37}$$

off-equilibrium, and

$$\mathcal{G}_{Ae0} \propto n_0 \tag{11.38}$$

at thermal equilibrium. In thermal equilibrium, the recombination and the generation match, that is, $\mathcal{R}_{Ae0} = \mathcal{G}_{Ae0}$, and therefore

$$\mathcal{R}_{Ae} = \mathcal{R}_{Ae0}\frac{n^2 p}{n_0^2 p_0}, \text{ and}$$
$$\mathcal{G}_{Ae} = \mathcal{G}_{Ae0}\frac{n}{n_0}. \tag{11.39}$$

So, the net recombination—decrease—rate is

$$\mathcal{R}_{Ae} - \mathcal{G}_{Ae} = \mathcal{G}_{Ae0}\left(\frac{n^2 p}{n_0^2 p_0} - \frac{n}{n_0}\right)$$
$$= \gamma_n n(np - n_i^2), \tag{11.40}$$

where $\gamma_n = \mathcal{G}_{Ae0}/n_0^2 p_0$ is a material- and Auger-coupling-related parameter, similar to our deployment of c_e and c_h or their equivalent e_e and e_h for the *HSR* process. These are parameters that can be characterized through measurements. An equation similar to Equation 11.40 for a hole-initiated Auger process is

$$\mathcal{R}_{Ah} - \mathcal{G}_{Ah} = \gamma_p p(np - n_i^2). \tag{11.41}$$

And the net Auger recombination rate due to electron-initiated and hole-initiated processes is

$$\begin{aligned} \mathcal{U}_A &= \mathcal{R}_{Ae} - \mathcal{G}_{Ae} + \mathcal{R}_{Ah} - \mathcal{G}_{Ah} \\ &= (\gamma_n n + \gamma_p p)(np - n_i^2). \end{aligned} \tag{11.42}$$

Now consider a heavily doped p-type material such as the quasineutral base of a n-p-n bipolar transistor. We have Δn and Δp, as the excess carriers, about equal. Here

$$\mathcal{U}_A \approx \left[\gamma_n(n_0 + \Delta n) + \gamma_p(N_A + \Delta p)\right]\left[(n_0 + \Delta n)(N_A + \Delta p) - n_i^2\right]. \tag{11.43}$$

The first part of the first term of the product on the right-hand side is small, since electrons are the minority carrier, and the size of the excess population is limited. In the second part of the first term of the product, $\gamma_p N_A$ is much larger than $\gamma_p \Delta p$. So, we get

$$\mathcal{U}_A \approx \gamma_p N_A (n_0 N_A + \Delta n \times N_A + n_0 \Delta p + \Delta n \Delta p - n_i^2), \tag{11.44}$$

where the dominant term arises from N_A, and since $\Delta n \gg n_0$, we can reduce this to

$$\mathcal{U}_A \approx \gamma_p N_A (\Delta n N_A) = \gamma_p N_A^2 \Delta n. \tag{11.45}$$

The net Auger rate is a product of the hole-initiated process, so a term that will be second power in hole concentration ($\sim N_A$), and the recombination of electrons, so a term that is first power in excess electron concentration. Hence, the Auger lifetime has the form

$$\tau_A = \frac{1}{\gamma_p N_A^2} \tag{11.46}$$

to fit $\mathcal{U}_A = \Delta n / \tau_A$. *The Auger lifetime has a second power dependence on background doping.*

Figure 11.6 shows the lifetime of *Si* and *GaAs*. Note that, at moderate doping, it has a first inverse power dependence and, for both, at higher power, it has a second power dependence. In both cases, 10^{19} cm^{-3} is a doping threshold where this slope change takes place. Note also how the lifetime in silicon is actually larger than that of *GaAs*, even though the latter is a larger bandgap material. At low doping levels, the *HSR*-dominated lifetime is superior in *Si*. It is purer, more ideal, compared to a compound semiconductor with a lower melting point but whose stoichiometry as well as residual deep-level content is worse. *GaAs* is also a direct, bandgap material, with a more efficient radiative process also contributing to lifetime reduction.

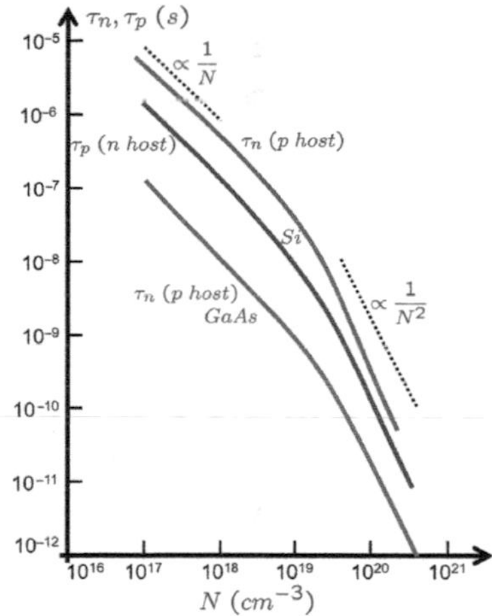

Figure 11.6: Lifetime as a function of background concentration in *Si* and *GaAs*. Both n- and p-type concentrations are shown. The lifetime follows an approximate inverse-doping and inverse-doping-squared dependence as the doping is increased. The concentration spans a large range, and the lifetime therefore has dependence arising in non-radiative/radiative recombination at low concentrations, and Auger recombination at high concentrations. Note, the significantly smaller lifetime of *GaAs* compared to that of *Si*.

11.3.1 Quantum treatment of the Auger process

THE AUGER PROCESS in a semiconductor, at its simplest involving only three charge particles and phonons, is quantum-mechanical in its very nature of coupling at this scale, and as such, an ideal means for introducing how such complex calculations that then become relevant classically, as in the earlier discussion, can be performed. Figure 11.7 shows the variety of Auger generation processes—these are electron activated—that can be seen. The complements to these are the reverse processes of recombination, and another ten of such processes that are hole activated.

This progression from quantum-mechanical to classical was also reflected in the discussions of capture cross section. These are all scattering processes that are inherently quantum-mechanical in this condensed matter.

Quantum-mechanical treatment of all these processes is non-trivial, particularly since phonons have to be accounted for. But we will explore the processes of Figure 11.7(a) and (b)—direct and indirect bandgap multiparticle interaction—to bring out the salient underpinnings.

By the way, there is no reason why there cannot be even more particle interactions, such as, for example, in Figure 11.7(a), two electrons gaining energy. It is just that their likelihood gets very rapidly smaller. Note that the proportionality of inverse-square dependence on doping in lifetime behavior is only approximate.

In the direct gap Auger process (Figure 11.7(a)), no phonon energy or momentum is involved. An electron-electron scattering ends up with one of them recombining with a hole in the valence band, and the other electron rising in energy. We view this process as two electrons interacting in the conduction band, with one transitioning to an empty state in the valence band and providing its lost energy to the other electron in conduction band, which rises up in energy and stays in the conduction band. Only the same index band is considered here. But, in practice, there can be complexities, since conduction bandstructure is really an $E(n, \mathbf{k})$ description, and the valence band has a heavy-hole band, a light-hole band and a split-off band. Restricting ourselves to the simplest case of 2 electrons, 2 spins and the energy and momentum conservation in the processes, one can see four possibilities in Figure 11.8, with the particle before and after an interaction being identified with the two quantum numbers of relevance: momentum $\mathbf{k}$ and spin ($\uparrow$ or $\downarrow$). Note that the change of band index (conduction band to valence band) is being kept implicitly. The electron-electron scattering is taken into account through the Golden rule, so

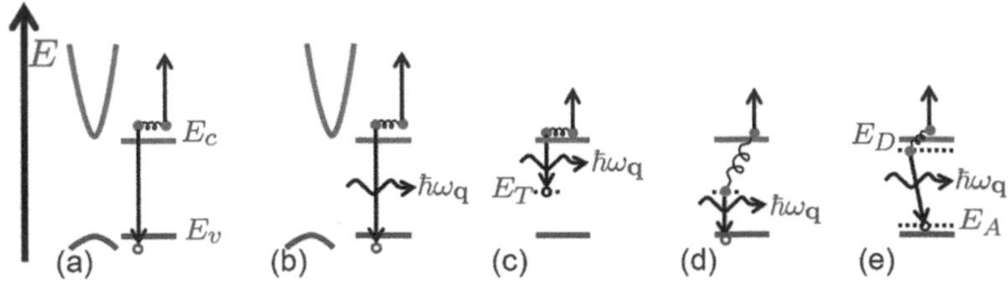

Figure 11.7: Five electron-initiated Auger processes. Parts (a) and (b) show the electron-electron-hole process in direct and indirect semiconductors, respectively. The latter necessarily requires phonon involvement. Part (c) is a localized electron-electron-trap Auger process, (d) is an electron-trap-hole-phonon-based localized Auger process and (e) involves an electron at a donor state, with an empty acceptor state and a free electron.

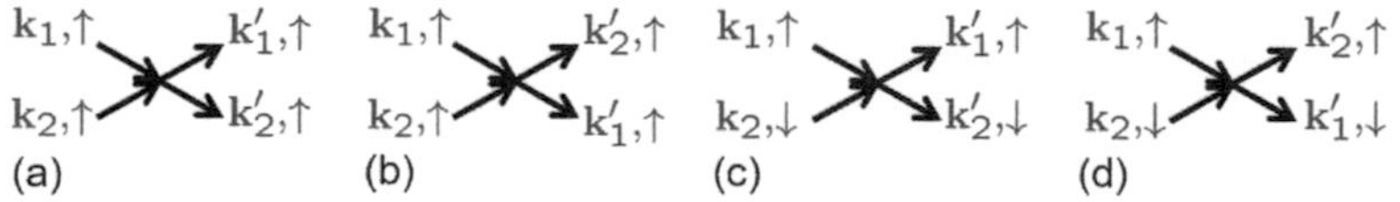

$\mathbf{k}_1,\uparrow$ $\mathbf{k}'_1,\uparrow$ $\mathbf{k}_2,\uparrow$ $\mathbf{k}'_2,\uparrow$ (a)
$\mathbf{k}_1,\uparrow$ $\mathbf{k}'_2,\uparrow$ $\mathbf{k}_2,\uparrow$ $\mathbf{k}'_1,\uparrow$ (b)
$\mathbf{k}_1,\uparrow$ $\mathbf{k}'_1,\uparrow$ $\mathbf{k}_2,\downarrow$ $\mathbf{k}'_2,\downarrow$ (c)
$\mathbf{k}_1,\uparrow$ $\mathbf{k}'_2,\uparrow$ $\mathbf{k}_2,\downarrow$ $\mathbf{k}'_1,\downarrow$ (d)

Figure 11.8: Two electrons interacting, leading to a change in their states during the direct gap Auger process. Part (a) shows identical spins remaining with the two particles. (b) shows identical spins remaining with particle exchange, (c) shows no spin exchange of opposite spins between particles and (d) shows the interaction with spin exchange. Parts (a) and (b) are indistinguishable, (c) and (d) are distinguishable and (a) and (b) cause interference.

$$S_{\mathbf{kk}'} = \int \frac{2\pi}{\hbar} |\mathscr{H}'_{\mathbf{kk}'}|^2 \delta(E_f - E_i)\, d\mathbf{k}' \tag{11.47}$$

The two processes of Figure 11.7(a) and (b) are indistinguishable: spin and momentum (even with the change of band) leave two particles that have identical quantum numbers. These will interfere; that is, the degeneracy is removed via a reconstruction of the wavefunctions under interaction. The interaction being screened Coulomb,

This is the Slater determinant again!

$$\begin{aligned}\mathscr{H}'_{12} = \frac{1}{\Omega^2} \int & u^*_{v\mathbf{k}'_1}(\mathbf{r}_1) \exp(-i\mathbf{k}'_1 \cdot \mathbf{r}_1) u^*_{c\mathbf{k}'_2}(\mathbf{r}_2) \exp(-i\mathbf{k}'_2 \cdot \mathbf{r}_2) \\ & \times \frac{e^2 \exp(-|\mathbf{r}_1 - \mathbf{r}_2|/\lambda)}{4\pi\epsilon|\mathbf{r}_1 - \mathbf{r}_2|} \\ & \times u_{c\mathbf{k}_1}(\mathbf{r}_1) \exp(i\mathbf{k}_1 \cdot \mathbf{r}_1) u_{c\mathbf{k}_2}(\mathbf{r}_2) \exp(i\mathbf{k}_2 \cdot \mathbf{r}_2)\, d^3\mathbf{r}_1\, d^3\mathbf{r}_2.\end{aligned} \tag{11.48}$$

To simplify—as in other multibody calculations, to remove the static term—we use a center-of-mass frame. Conservation of momentum specifies

$$\begin{aligned} & \mathbf{k}_1 + \mathbf{k}_2 - \mathbf{k}'_1 - \mathbf{k}'_2 = 0 \\ \therefore\ & \frac{1}{2}(\mathbf{k}'_1 - \mathbf{k}'_2) - \frac{1}{2}(\mathbf{k}_1 - \mathbf{k}_2) = \mathbf{k}'_1 - \mathbf{k}_1.\end{aligned} \tag{11.49}$$

For normal scattering, that is, for Figure 11.7(c) and (d), this integral can be simplified to

$$\begin{aligned}\mathscr{H}'_{12} &= \frac{e^2}{4\pi\epsilon\Omega} \frac{\mathcal{O}(\mathbf{k}_1,\mathbf{k}'_1)\mathcal{O}(\mathbf{k}_2,\mathbf{k}'_2)}{|\mathbf{k}'_1 - \mathbf{k}_1|^2 + 1/\lambda^2}, \text{ where} \\ \mathcal{O}(\mathbf{k}_1,\mathbf{k}'_1) &= \int_{\Omega_0} u^*_{v\mathbf{k}'_1}(\mathbf{r}_1) u_{c\mathbf{k}_1}(\mathbf{r}_1)\, d^3\mathbf{r}_1, \text{ and} \\ \mathcal{O}(\mathbf{k}_2,\mathbf{k}'_2) &= \int_{\Omega_0} u^*_{c\mathbf{k}'_2}(\mathbf{r}_2) u_{c\mathbf{k}_2}(\mathbf{r}_2)\, d^3\mathbf{r}_2.\end{aligned} \tag{11.50}$$

What do we do about the indistinguishable processes of Figure 11.7(a) and (b), where spins are identical, and interference follows? The rates here must depend on $\mathscr{H}'_{12} - \mathscr{H}'_{21}$ to make the wavefunction asymmetric under the exchange of electrons. The distinguishable processes depend on $\mathscr{H}'_{12}$ and $\mathscr{H}'_{21}$, respectively. So, we have

This Hamiltonian function is asymmetric. See Chapter 1 for a discussion of wavefunction construction for manybody conditions through the Slater determinant, and the underlying notions.

$$|\mathscr{H}'|^2 = |\mathscr{H}'_{12} - \mathscr{H}'_{21}|^2 + |\mathscr{H}'_{12}|^2 + |\mathscr{H}'_{21}|^2. \tag{11.51}$$

To calculate, we also need to know the state occupation probabilities and the state vacancy probabilities for the heating electron and the energy losing electron transitions. These are

$$f(\mathbf{k}_2) = \frac{n}{\mathcal{N}_c} \exp\left(-\frac{E_{c\mathbf{k}_2}}{k_B T}\right), \text{ and}$$
$$1 - f(\mathbf{k}_1') = \frac{p}{\mathcal{N}_v} \exp\left(-\frac{E_{v\mathbf{k}_1'}}{k_B T}\right), \tag{11.52}$$

where the energies are referenced to the respective bandedges. The joint probabilities of events can be viewed through a weighting factor,

$$p(\mathbf{k}_2, \mathbf{k}_1') = \frac{np}{\mathcal{N}_c \mathcal{N}_v} \exp\left(-\frac{E_{c\mathbf{k}_2} + E_{v\mathbf{k}_1'}}{k_B T}\right)$$
$$= \exp\left(-\frac{E_g + E_{c\mathbf{k}_2} + E_{v\mathbf{k}_1'}}{k_B T}\right). \tag{11.53}$$

The source function and the sink function that define the existence of conditions for the events to have a probability of occurrence are captured through a source-sink factor $ss(\mathbf{k}_1, \mathbf{k}_2, \mathbf{k}_1')$ that accounts for the probability of having an electron state $|\mathbf{k}_1\rangle$, an electron state $|\mathbf{k}_2\rangle$ and an empty electron state (a hole state) $|\mathbf{k}_1'\rangle$. These are all describable through the distribution function.

We now have the wherewithal to calculate the integral. For a direct gap and parabolic band semiconductor, with conservation of energy and momentum, one finds the solution to be

$$E_{c\mathbf{k}_1} = E_{c\mathbf{k}_2} = \mu E_{v\mathbf{k}_1'}$$
$$= \frac{\mu^2}{1 + 3\mu + 2\mu^2} E_g \text{ with } \mu = \frac{m_c^*}{m_v^*}. \tag{11.54}$$

The source-sink factor is

$$ss(\mathbf{k}_1, \mathbf{k}_2, \mathbf{k}_1') = \frac{n}{\mathcal{N}_c} \exp\left(-\frac{1 + 2\mu}{1 + \mu} \frac{E_g}{k_B T}\right), \tag{11.55}$$

which implies that when the mismatch in effective mass is larger, that is, the band states have different energetic dependences of different modulation functions, and when the bandgap is larger, the factor decreases exponentially. Simultaneously, the energy of the electron kicked up in energy is

$$E_{c\mathbf{k}_2'} = \frac{1 + 2\mu}{1 + \mu} E_g. \tag{11.56}$$

When $\mu = 1$, that is, there are identical band states,

$$E_{c\mathbf{k}_1} = E_{c\mathbf{k}_2} = \mu E_{v\mathbf{k}_1'} = \frac{1}{6} E_g,$$
$$\text{and } E_{c\mathbf{k}_2'} = \frac{3}{2} E_g. \tag{11.57}$$

So, for an idealized symmetric direct bandgap semiconductor, where electrons and holes look very similar, the energy that the hot electron ends up with during recombination is $(3/2)E_g$, of which $E_g/6$ comes from the initial electron's conduction band energy, together with the kicked electron's conduction band energy, the state that the first electron jumps to and the bandgap energy.

This $(3/2)E_g$ energy is the threshold energy for the reverse of this process—the Auger generation process—and so will correspond to the impact ionization process, which arises in the electron acquiring this energy from the electric field. This is an idealized calculation for one of the Auger processes in direct gap semiconductors when $\mu = 1$ and the parabolic bands are entirely symmetric.

In *GaAs*, $\mu \approx 0.1$, and so

$$E_{c\mathbf{k}_1} = E_{c\mathbf{k}_2} = \mu E_{v\mathbf{k}'_1} \approx 0.01E_g,$$
$$\text{and } E_{c\mathbf{k}'_2} \approx 1.1E_g, \qquad (11.58)$$

that is, in recombination, the hot electron achieves an energy of about the bandgap, and this is the threshold for the hot electron to create an electron-hole pair. It is not a very large energy; at room temperature, it is of the order of $40k_BT$. In devices, application of an electrostatic potential of this order of magnitude will increase the likelihood of such electrons, and so long as this energy is very rapidly acquired during transit in times of the order of relaxation time, impact ionization will become quite important. Likewise, in lasers, where one creates an inversion condition by creating a large carrier population, by necessity, they occupy higher states (if they exist) and Auger recombination becomes significant. The same is true at high dopings (higher carrier concentrations), for the same reason.

Auger recombination is certainly an important efficiency and performance issue for bipolar lasers. The optical efficiency drops at high injection in the larger bandgap materials such as $Ga_xIn_{1-x}N$ and other *GaN*-based visible diodes, and, for smaller bandgap materials, it is important throughout the span of operation.

Consider now what properties of interaction of states affects the direct Auger transitions. Transitions involve matching of states through the overlap functions ($\mathcal{O}(\mathbf{k}_1, \mathbf{k}'_2)$ and $\mathcal{O}(\mathbf{k}_2, \mathbf{k}'_1)$) and the weighting factor reflecting the existence of the occupied state for coupling of energy to and the unoccupied state for transition to and coupling from ($p(\mathbf{k}_2, \mathbf{k}'_1)$). Recall that states within the same band have very slowly changing modulation functions, that is, $u_{c\mathbf{k}}$s and $u_{v\mathbf{k}}$s are each slowly varying as $\mathbf{k}$ changes. Overlap integrals vary slowly:

This slow change was the basis for the $\mathbf{k} \cdot \mathbf{p}$ method for calculating bandstates.

$$\mathcal{O}(\mathbf{k}_1, \mathbf{k}'_1) = \mathcal{O}(\mathbf{k}_2, \mathbf{k}'_1), \text{ and}$$
$$\mathcal{O}(\mathbf{k}_2, \mathbf{k}'_2) = \mathcal{O}(\mathbf{k}_1, \mathbf{k}'_2). \qquad (11.59)$$

This leads to the conclusion that $k_1 \approx k_2$, and $\mathscr{H}'_{12} \approx \mathscr{H}'_{21}$. So, we can ignore the spin-indistinguishable processes under the perturbation

Hamiltonian. $\mathscr{H}'_{12} - \mathscr{H}'_{21} \approx 0$. The interaction Hamiltonian for the direct Auger process then is $|\mathscr{H}'|^2 \approx 2|\mathscr{H}'_{12}|^2$. The electron is close to the bandedge—relatively speaking, compared to the bandgap energy E_g—under these conditions. Therefore, the Auger scattering rate can be written as

$$S_{Ar} = \frac{4\pi}{\hbar}\left[\frac{e^2}{(2\pi)^3\epsilon}\right]^2 \frac{np}{\mathcal{N}_c\mathcal{N}_v} \int \frac{\left|\mathcal{O}_{\mathbf{k}_1\mathbf{k}'_1}\mathcal{O}_{\mathbf{k}_2\mathbf{k}'_2}\right|^2 \exp\left[-\frac{E_{c\mathbf{k}_2}+E_{v\mathbf{k}'_1}}{k_BT}\right]}{\left(|\mathbf{k}'_1-\mathbf{k}_1|^2+1/\lambda^2\right)^2}$$
$$\times\,\delta(E_{c\mathbf{k}'_2} - E_g - E_{v\mathbf{k}'_1} - E_{c\mathbf{k}_1} - E_{c\mathbf{k}_2})\,d\mathbf{k}_2\,d\mathbf{k}'_1,$$

integrated over the spins.

The overlap integrals also depend on whether the overlap being calculated is in the same band or a different band. In different bands, the modulation functions changes: $u_{c\mathbf{k}}(\mathbf{r})$ and $u_{v\mathbf{k}}(\mathbf{r})$ are orthogonal for the same $\mathbf{k}$. This just states that the electron—as a fermion—has a different quantum number of the band index. This sameness in the same band and separateness in different bands implies

$$\mathcal{O}(\mathbf{k}_2,\mathbf{k}'_2) \approx 1, \text{ and}$$
$$\mathcal{O}(\mathbf{k}_1,\mathbf{k}'_1) \approx \frac{\hbar}{mE_g}(\mathbf{k}_1\cdot\langle v|\hat{\mathbf{p}}|c\rangle - \mathbf{k}'_1\cdot\langle v|\hat{\mathbf{p}}^*|c\rangle) \approx 0, \qquad (11.60)$$

where we again have the momentum matrix element. Heavy holes—anisotropic and quite different with respect to E-$\mathbf{k}$ characteristics than the conduction electrons—are hard to tackle. We will just use the f sum rule to calculate this second function of Equation 11.60. Although small, the overlap can be calculated to be

Because of interactions, again, we are returning here to oscillator strength and the coupling of energy through it. Called out before, Appendix I is highly recommended reading.

$$|\mathcal{O}(\mathbf{k}_1,\mathbf{k}'_1)|^2 \approx \frac{\hbar^2\langle v|\hat{\mathbf{p}}|c\rangle^2}{m_0^2E_g}|\mathbf{k}_1-\mathbf{k}'_1|^2$$
$$= \frac{\hbar^2}{2E_g}\left(\frac{1}{m_0}+\frac{1}{m_v^*}\right)|\mathbf{k}_1-\mathbf{k}'_1|^2. \qquad (11.61)$$

If we ignore screening, that is, $\lambda \to 0$, the Auger scattering rate is

$$S_{Ar} = \frac{2\pi}{\hbar}\left(\frac{e^2}{8\pi^3\epsilon}\right)^2\frac{\hbar^2}{E_g}\left(\frac{1}{m_0}+\frac{1}{m_v^*}\right)$$
$$\times\,\frac{np}{\mathcal{N}_c\mathcal{N}_v}\int\frac{\exp[-(E_{c\mathbf{k}_2}+E_{v\mathbf{k}'_1})/k_BT]}{|\mathbf{k}_1-\mathbf{k}'_1|^2}$$
$$\times\,\delta(E_{c\mathbf{k}'_2} - E_g - E_{v\mathbf{k}'_1} - E_{c\mathbf{k}_1} - E_{c\mathbf{k}_2})\,d\mathbf{k}_2\,d\mathbf{k}'_1. \qquad (11.62)$$

We need to find when the delta function peaks. If we assume that $m_c^* \ll m_v^*$ and $\mathbf{k}'_1 \gg \mathbf{k}_1$, then, following Figure 11.9 for momentum matching, the delta function energy matching condition can be written as

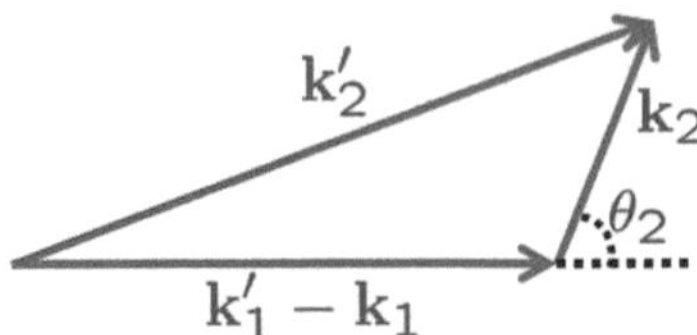

Figure 11.9: Momentum matching during the direct Auger interaction.

$$\delta(\cdots) = \frac{\hbar^2|\mathbf{k}_1 - \mathbf{k}_1'|^2}{2m_c^*} - E_g - \frac{\hbar^2\mathbf{k}_1'^2}{2m_v^*} - \frac{\hbar^2\mathbf{k}_1^2}{2m_c^*} + \frac{\hbar^2\mathbf{k}_2 \cdot (\mathbf{k}_1' - \mathbf{k}_1)}{m_c^*}\cos\theta_2. \quad (11.63)$$

Integrating over $\mathbf{k}_2$ reduces the recombination integral to

$$S_{Ar} = \frac{2\pi m_c^{*2} k_B T}{\hbar^4} \int \frac{1}{|\mathbf{k}_1' - \mathbf{k}_1|^3} \exp\left\{-\left[E_{v\mathbf{k}_1'} + \frac{m_c^*}{2\hbar^2}\frac{1}{|\mathbf{k}_1' - \mathbf{k}_1|^2} \times \left(E_g + \frac{\hbar^2 k_1'^2}{2m_v^*} + \frac{\hbar^2 k_1^2}{2m_c^*} - \frac{\hbar^2}{2m_c^*}|\mathbf{k}_1' - \mathbf{k}_1|^2\right)\right]/k_BT\right\} d\mathbf{k}_1'. \quad (11.64)$$

Since $\mathbf{k}_1' \gg \mathbf{k}_1$,

$$S_{Ar} = \frac{8\pi^2 m_c^{*2} k_B T}{\hbar^4} \exp\left[\frac{(1-\mu)E_g}{2k_BT}\right] \times \mathscr{K}_0\left[\frac{(1+2\mu+4\mu^2)^{1/2}E_g}{2k_BT}\right], \quad (11.65)$$

where $\mathscr{K}_0$ is a zero order modified Bessel function. This can be simplified in limits to

$$S_{Ar} = \frac{8\pi^{5/2} m_c^*(k_BT)^{3/2}}{\hbar^4(1+\mu)^{1/2}E_g^{1/2}} \exp\left(-\frac{\mu E_g}{k_BT}\right) \quad \text{for} \quad \frac{E_g}{k_BT} \gg 1. \quad (11.66)$$

And we end up with a direct Auger scattering rate of

$$S_{Ar} = \frac{e^4 m_c^*(k_BT)^{3/2}(m_c^*/m_0 + \mu)}{4\pi^{5/2}\epsilon^2\hbar^3(1+\mu)^{1/2}E_g^{3/2}} \exp\left[-\frac{(1+\mu)E_g}{k_BT}\right]. \quad (11.67)$$

For *GaAs*, which has a bandgap of 1.43 eV, the Auger recombination rate is $S_{Ar} = 5 \times 10^{-17}\ s^{-1}$. For *InSb* with a bandgap of 0.18 eV, a little less than a factor of 10 smaller, it is $8 \times 10^6\ s^{-1}$. *InSb* is dominated by this direct Auger recombination along with the other Auger processes that we have mentioned, while, for *GaAs*, while mostly *HSR* recombination will prevail, Auger recombination will become important at carrier densities exceeding $10^{19} cm^{-3}$.

Phonon-assisted processes become important in indirect gap materials—electrons and holes have larger momentum differences—so, *Si* and even more so *Ge* have high recombination rates. Regardless of the carrier concentrations, Auger processes need to be considered carefully in bipolar conditions in semiconductors at bandgaps of 0.75 eV and lower. But, as remarked, in devices, lasers and base and emitter regions of bipolar transistors, the high carrier concentration will make Auger processes' consequence significant even in materials of bandgap as large as 3 eV.

It is useful after this to see what happens to recombination lifetime in some of the useful materials when dopings are changed.

Figure 11.6 showed the lifetime dependence for *Si* and *GaAs*. At doping concentrations of $10^{19}\ cm^{-3}$ or more both show the change in slope of lifetime from an inverse first power to an inverse second power dependence. This is precisely what our recombination discussion indicates when rising in *HSR* processes at lower doping and Auger processes at higher doping.

We should also note the significance of the Auger generation process. For a direct gap generation, this is when an electron high up in energy (identified by occupation of the state $|\mathbf{k}_1\rangle$) couples to a hole state (or an empty electron state) $|\mathbf{k}_2\rangle$; the final states $|\mathbf{k}_1'\rangle$ and $|\mathbf{k}_2'\rangle$ are in a largely empty band. Occupation factors therefore drop out of the calculations. We have

$$\begin{aligned}\mathcal{O}(\mathbf{k}_1,\mathbf{k}_1') &= \mathcal{O}(\mathbf{k}_2,\mathbf{k}_1'), \text{ and}\\ \mathcal{O}(\mathbf{k}_2,\mathbf{k}_2') &= \mathcal{O}(\mathbf{k}_1,\mathbf{k}_2').\end{aligned} \tag{11.68}$$

Again, ignoring the spin-indistinguishable processes because of the vanishing perturbation Hamiltonian, $k_1 \approx k_2$, $\mathcal{H}_{12}' \approx \mathcal{H}_{21}'$, and $|\mathcal{H}'|^2 \approx 2|\mathcal{H}_{12}'|^2$. If electrons not too far off-equilibrium, that is, if this is generation due to electric-field-stimulated hot electrons, then

$$S_{Ag} = \frac{4\pi}{\hbar}\left[\frac{e^2}{(2\pi)^3\epsilon}\right]^2 \iint \frac{\mathcal{O}(\mathbf{k}_1,\mathbf{k}_1')\mathcal{O}(\mathbf{k}_2,\mathbf{k}_2')}{|\mathbf{k}_1'-\mathbf{k}_1|^4}\delta(E_f - E_i)\, d\mathbf{k}_2\, d\mathbf{k}_1'. \tag{11.69}$$

Now consider the threshold that we have discussed for recombination, but now in impact ionization generation conditions. Carriers accelerate and generate, and, in a direct isotropic material, this will happen with directions aligned. So, the conservation conditions for energy and momentum are, respectively,

$$\begin{aligned}E_1 = E_{AiiT} &= E_2 + E_1' + E_2' + E_g, \text{ and}\\ k_1 &= -k_2 + k_1' + k_2'.\end{aligned} \tag{11.70}$$

E_{AiiT} is a threshold energy for Auger impact ionization. The matrix element is nearly a constant at threshold. With bands isotropic and parabolic, using $\mathcal{O}_c = \mathcal{O}(\mathbf{k}_1,\mathbf{k}_1')$ and $\mathcal{O}_v = \mathcal{O}(\mathbf{k}_2,\mathbf{k}_2')$,

$$\begin{aligned}S_{Aii} &= \frac{2}{\hbar}\left(\frac{e^2}{\epsilon}\right)^2 \frac{1}{(2\pi)^5}\frac{\hbar^4}{(2m_c^*)^2}\frac{\mathcal{O}_c^2\mathcal{O}_v^2}{E_{AiiT}}\left(\frac{1+2\mu}{1+\mu}\right)^4 \iint \delta(E_f - E_i)\, d\mathbf{k}_2\, d\mathbf{k}_1'\\ &= \frac{2}{\hbar}\left(\frac{e^2}{\epsilon}\right)^2 \frac{1}{(2\pi)^5}\frac{\hbar^4}{(2m_c^*)^2}\frac{\mathcal{O}_c^2\mathcal{O}_v^2}{E_{AiiT}}\left(\frac{1+2\mu}{1+\mu}\right)^4\\ &\quad\times\left(\frac{2m_c^*}{\hbar^2}\right)^3\frac{\pi^3}{2}\frac{(1+\mu)^2}{(1+2\mu)^{7/2}}(E_1 - E_{AiiT})^2\\ &= S_0\left(\frac{\epsilon_0}{\epsilon}\right)^2\frac{m_c^*}{m_0}\frac{\mathcal{O}_c^2\mathcal{O}_v^2}{(1+2\mu)^{3/2}}\left(\frac{E_1 - E_{AiiT}}{E_g}\right)^2,\end{aligned} \tag{11.71}$$

where

$$S_0 = \left(\frac{e^2}{4\pi\epsilon_0}\right)^2 \frac{m_0}{\hbar^3} \approx 4.14 \times 10^{16}\ s^{-1}. \qquad (11.72)$$

This is the Auger generation rate at impact ionization threshold.

For *GaAs*, electrons have an $\mathcal{O}_c \approx 1$. Heavy holes, although high in density, have a poor momentum match, and therefore the matrix element is small and, for heavy holes, $\mathcal{O}_v \approx 0$. For light holes, $\mathcal{O}_v \approx 0.91$, using a $\mathbf{k} \cdot \mathbf{p}$ calculation with the f sum rule. This leads to, after accounting for the density of states of the light holes ($\mathscr{G}_{lh}$), $\mathcal{O}_v^2 \approx 0.06$. *GaAs* therefore has

$$S_{Aii} \approx 5 \times 10^{11} \left(\frac{E_1 - E_{iiT}}{E_g}\right)^2\ s^{-1}, \qquad (11.73)$$

where $E_1 = E_2 + E_1' + E_2' + E_g$. This relationship shows the rate increasing in second power with energy. It is a hard threshold, since the relationship only holds for $E_1 \geq E_{iiT}$. If phonon processes (emission) are important, the threshold will be soft, since phonons provide a soft means for increasing or decreasing the initiating electron's energy.

Figure 11.10 shows electron- and hole-initiated impact ionization rates (probability of generation per unit length of travel) for a few semiconductors. Note the slope change at about $3 \times 10^6\ cm/V$ (inverse electric field). This reflects changes in the leading cause of the generation process, among the different processes we have mentioned.

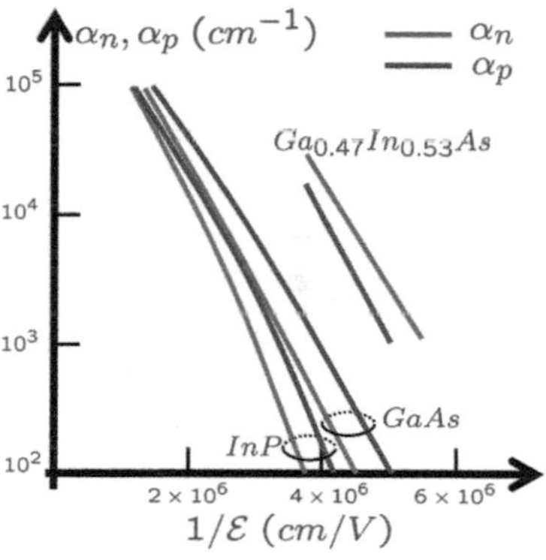

Figure 11.10: Impact ionization coefficient as a function of the inverse of electric field for a few semiconductors. Both electron-initiated and hole-initiated impact ionization processes are shown.

11.4 *Surface recombination*

Early in the text, we started with an extensive discussion of surface states and their significant role in Fermi level pinning and other interactions arising when interfaces and surfaces have to be passed through or exist as boundaries of regions of interest. Surface states are surface-confined states—donor-like or acceptor-like, which can be both in the bandgap and outside the bandgap—so they can have the wherewithal to be intermediaries for communication between the valence band and the conduction band, as well as cause carrier degeneracy in the surface region. The major difference is that these states are localized at the surface—per unit area instead of unit volume is the relevant measure—and that, unlike the transport in three dimensions, where carriers are moving every which way, here the "classical" capture cross-section will need an interpretation that looks different in an orthogonal direction to the surface than in a lateral direction. Also, since the local condition is different from that in the bulk, for example, Fermi level pinning forces changes to the Fermi energy position at surface compared to the bulk, these changes and their consequences to transport must be accounted for. The result is that we also think more in terms of surface recombination velocity, which characterizes the velocity of carrier arriving and annihilating at the surface, so, an electron-hole mating process. If this is the rate-limiting step, it is easier to analyze, if it is not, one needs to look at this a bit more rigorously. We will address select important considerations of these phenomena, and leave some for self-reading through the bibliographic notes.

When we assign velocities to transport across interfaces and junctions—through forward momentum or the Richardson velocity—we are describing something quite complementary: a particle flux, which can be thought through with velocity or current density. The reader may wish to ponder why there is no surface generation velocity being discussed.

11.4.1 *Neumann boundary conditions*

Besides ohmic and rectifying interfaces, another significant boundary for semiconductors is in-between the semiconductor and an insulator. No or negligible current flows through the insulators. Gauss' law requires that, for the electric fields normal to the interface $\hat{\mathbf{n}} \cdot \boldsymbol{\mathcal{E}}$ in the semiconductor and in the insulator,

$$\epsilon_{sem}\hat{\mathbf{n}} \cdot \boldsymbol{\mathcal{E}}\big|_{sem} - \epsilon_{ins}\hat{\mathbf{n}} \cdot \boldsymbol{\mathcal{E}}\big|_{ins} = Q_{surf}, \qquad (11.74)$$

where Q_{surf} is the interface charge density. If the insulator can be assumed to be infinitely thick, the field in the insulator can be ignored, and one has

$$\epsilon_{sem}\hat{\mathbf{n}} \cdot \boldsymbol{\mathcal{E}}\big|_{sem} = Q_{surf}. \qquad (11.75)$$

If one further assumes that the surface is charge-free, then

$$\epsilon\hat{\mathbf{n}} \cdot \boldsymbol{\mathcal{E}}\big|_{sem} = 0, \qquad (11.76)$$

which is often referred to as the Neumann boundary condition. In compound semiconductors, there is significant surface state density and hence quite often a significant surface charge. The Neumann boundary condition is rarely valid. Fermi level pinning at the surface, which results from this surface charge, may be incorporated in two ways. We may make the *ad hoc* assumption that the surface potential is pinned due to this Fermi level pinning; this is akin to an electrostatic potential Dirichlet boundary condition. Alternately, we may consider a large surface state density of donor and acceptor traps that do not allow any significant Fermi level excursion because of charge imbalance. In this case, we may use charge neutrality, including these interface charges as well as Gauss's law at the interface.

11.4.2 *Surface recombination*

IN THE DISCUSSION OF PARTICLE INTERACTION and at various boundaries, particularly important is the non-conservation arising in the recombination and generation of excess carriers. In this discussion of the behavior of excess carriers, a particularly important one both at surfaces and in treatments of certain boundary conditions is that of recombination at surfaces and also, by extension, at interfaces. Compound semiconductor surfaces usually occur with considerable numbers of states in the forbidden gap. Generally, there is a distribution of states, and the surface is quite often pinned because of the large number of states. Sometimes this pinning may occur in the bands; for example, in *InAs* it occurs in conduction band while in *GaSb* it occurs in the valence band. When these states occur in the forbidden gap, recombination transitions occur through the non-radiative *HSR* process at the surface. Carriers within a few diffusion lengths can readily recombine by drift diffusion to the surface, leading to a net flow of current to the surface that we will call surface recombination current.

The treatment of surface recombination should be actually quite complex if in detail and without parameterization. Simple relations, however, can be derived for low level injection conditions. When excess carriers exist in the bulk, we can derive the diffusive current toward the surface for a uniformly doped sample. It is this diffusive flux that supplies the surface recombination current under low level injection conditions. We will treat this first in a simple way, to show the simplified equations that are used in the Dirichlet boundary conditions and that readily lead to the concept of surface recombination velocity. Following that, we will discuss where this simplified treatment will break down.

Let ϱ_s be the reflection coefficient at the surface representing the probability that a particle returns to the bulk without recombining at the surface. Similarly, let ϱ_b be the reflection coefficient representing the probability that a carrier headed toward the bulk will show up at the surface. We assume, for the simple analysis, that these are independent of current density and carrier concentrations. Following Figure 11.11, the total flux $\mathcal{F}_s$ to the surface from the bulk is given by

$$\mathcal{F}_s = \mathcal{F}_i + \varrho_b \mathcal{F}_b, \tag{11.77}$$

where $\mathcal{F}_i$ is the incident flux, and $\mathcal{F}_b$ is the total reverse flux.

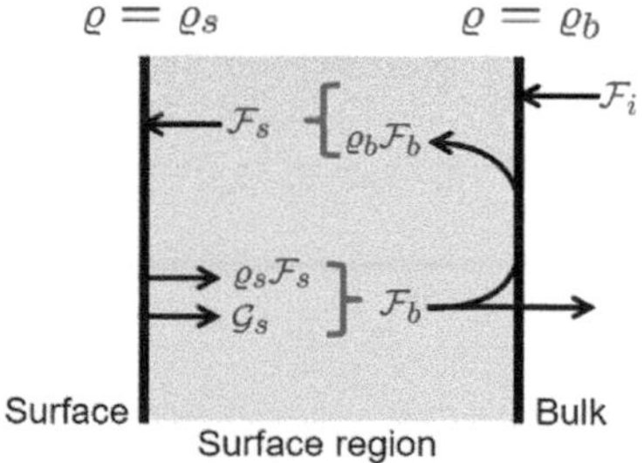

Figure 11.11: Fluxes of carriers representing transport processes taking place in the surface region of semiconductors during surface recombination.

Likewise, we may write the total flux from the surface to the bulk as

$$\mathcal{F}_b = \mathcal{G}_s + \varrho_s \mathcal{F}_s, \tag{11.78}$$

where $\mathcal{G}_s$ is the generation rate at the surface. The total fluxes, in terms of the reflection parameters, generation rate and incident flux, are

$$\mathcal{F}_s = \frac{\mathcal{F}_i + \varrho_b \mathcal{G}_s}{1 - \varrho_s \varrho_b}, \quad \text{and}$$
$$\mathcal{F}_b = \frac{\mathcal{G}_s + \varrho_s \mathcal{F}_i}{1 - \varrho_s \varrho_b}. \tag{11.79}$$

For a classical gas distribution, that is, for non-degenerate materials, the fluxes $\mathcal{F}_s$ and $\mathcal{F}_b$ are related to the thermal velocity and the free carrier concentration as $\sim n v_\theta/4$, which reduces to $\sim n_0 v_\theta/4$ in thermal equilibrium. This follows from the argument that, at any given instant of time, half of the carriers are directed toward the surface, and since their direction is random, they have an average velocity of $\sim v_\theta/2$ in the orthogonal direction. This allows us to write the generation rate and the incident flux $\mathcal{F}_{i0}$, at thermal equilibrium, at the surface, as

$$\mathcal{G}_{s0} = \frac{n_0 v_\theta}{4}(1 - \varrho_s), \quad \text{and}$$
$$\mathcal{F}_{i0} = \frac{n_0 v_\theta}{4}(1 - \varrho_b). \tag{11.80}$$

Note that the generation rate remains constant; this follows using similar arguments as for the bulk *HSR* recombination mechanism. Departure from equilibrium results in a disparity between the net flux to and from the surface, the difference of which is the recombination flux. The method applied here is similar to the one sometimes used in the discussion of metal-semiconductor junctions. The flux directed toward the surface originates from a distance that is, on average, equal to the mean free path. Thus, the fluxes to and from the surface at any position are

$$\mathcal{F}_s = \frac{v_\theta}{4}\left(n - \gamma\frac{dn}{dz}\right), \quad \text{and}$$
$$\mathcal{F}_b = \frac{v_\theta}{4}\left(n + \gamma\frac{dn}{dz}\right), \tag{11.81}$$

where γ is a proportionality constant. The sum of the surface and bulk fluxes is $n_s v_\theta/4$ at the surface. Our equations may now be used to determine the incident flux as

$$\mathcal{F}_i = \frac{n_s v_\theta}{2}\frac{1 - \varrho_s \varrho_b}{1 + \varrho_s} - \frac{n_0 v_\theta}{4}(1 + \varrho_b)\frac{1 - \varrho_s}{1 + \varrho_s}, \tag{11.82}$$

and hence the net flux of carriers, assuming only diffusive transport of carriers, is

$$\mathcal{F}|_{surf} = \mathcal{F}_s - \mathcal{F}_b = \frac{n_s - n_0}{2} v_\theta \frac{1-\varrho_s}{1+\varrho_s} = -\mathcal{D}_n \left.\frac{d\Delta n}{dz}\right|_{surf}. \tag{11.83}$$

This is simply written as

$$-\mathcal{D}_n \left.\frac{d\Delta n}{dz}\right|_s = S \; \Delta n|_s, \tag{11.84}$$

with S, the surface recombination velocity, as

$$S = \frac{v_\theta}{2} \frac{1-\varrho_s}{1+\varrho_s}. \tag{11.85}$$

Note that if all carriers incident at the surface recombine, then the surface recombination velocity is $v_\theta/2$, its largest value.

Assuming an infinite surface recombination velocity, as is common for many boundaries encountered in device modeling, is tantamount to assuming that the excess carrier concentration at the surface is zero. For many practical cases, this assumption is justified. The function ϱ_s represents the statistics of recombination at the surface, because it represents the probability that a carrier will return. Thus, $1 - \varrho_s$ is proportional to the recombination rate at the surface.

We have discussed the statistics of *HSR* recombination; these also hold for most surfaces, because the recombination occurs through deep traps. The recombination rate at the surface can be represented by similar expressions to those used for a single level. Assuming that there exists a single dominating trap level, the surface recombination rate is

$$\mathcal{R}_s = \sigma_n \sigma_p v_\theta N_{Ts} \frac{n_s p_s - n_i^2}{(n_s + n_{Ts})\sigma_n + (p_s + p_{Ts})\sigma_p}, \tag{11.86}$$

where n_s and p_s are the surface carrier concentrations, and n_{Ts} and p_{Ts} are the surface carrier concentrations if the Fermi level were at the trap level.

Consider an example that is a simplification of Equation 11.86. For a trap with equal hole and electron capture cross-sections, if the electron is a minority carrier, hence if hole concentrations are large, then the recombination rate is

$$\mathcal{R}_s \approx \sigma v_\theta N_{Ts} n_s. \tag{11.87}$$

The rate of recombination of electrons, per unit time, per unit area, is given by this expression. The reflection coefficient is related to this rate since it expresses the probability of not recombining. The recombination rate is proportional to the carrier concentration; it is provided by the difference of incident and reflected flux, and the constant of proportionality $\sigma v_\theta N_{Ts}$ is the surface recombination velocity in the absence of any surface space charge, with identical conditions at the surface as in the bulk. For low interface state density oxide-silicon interfaces, $\sigma \approx 10^{-16}\ cm^2$, and $N_{Ts} \approx 10^{10}\ cm^{-2}$, resulting in a surface recombination velocity on the order of $\sim 10\ cm \cdot s^{-1}$. In compound semiconductors such as *GaAs*, even if we ignore the effect of surface space charge and other effects which we will soon discuss, $N_{Ts} \approx 10^{14} \cdot cm^{-2}$, $\sigma \approx 10^{-15}\ cm^{-2}$ and, hence, surface recombination velocity is of the order of $10^6\ cm \cdot s^{-1}$. Strictly speaking, the recombination rate characterizes the surface effectively for most purposes. Surface recombination velocity is a concept introduced because it characterizes a meaningful constant in some situations. If, in an *ad hoc* manner, we defined it as a parameter that related the recombination rate to the excess carrier concentration, then it would vary as a function of biasing condition, et cetera, because this is only a simple derivation from the more complicated *HSR* expression.

11.4.3 *Surface recombination with Fermi level pinning*

WHEN A CHARGE REGION IS PRESENT AT THE SURFACE, the surface recombination velocity takes an even more complicated form. It now depends on the surface state density, the characteristics of the surface states, surface charge, et cetera—parameters that determine the surface carrier concentrations. We can see from Equation 11.86 that the surface recombination will actually go through a maximum when $\sigma_n (n_s + n_{Ts}) + \sigma_p (p_s + p_{Ts})$ goes through a minimum. This will occur when both electron capture and hole capture processes are equally active, that is, when the quasi-Fermi levels straddle the trap level.

Since the surface space charge situation is important to most compound semiconductors, we will look at it in more detail. We consider variations introduced on our simple model due to the presence of band bending from Fermi level pinning. For low level injection (see Figure 11.12), we may again derive a relationship that relates surface carrier concentrations to the bulk.

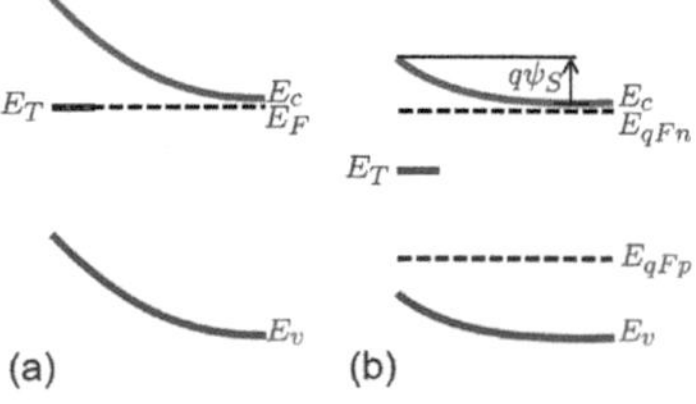

Figure 11.12: Band diagram for estimating surface recombination at a trap, in the presence of Fermi level pinning, at (a) thermal equilibrium and (b) low level injection conditions.

The presence of surface depletion changes the thermal equilibrium concentration at the surface. In the case of Figure 11.12, this would mean changing from the bulk thermal equilibrium magnitude by the Boltzmann exponential related to the total band bending at the surface. The total recombination rate is a more general case of the above, where we again ignore the n_{Ts} and p_{Ts} terms and the intrinsic carrier density terms, so

$$\mathcal{R}_s = \sigma_n \sigma_p v_{\theta_n} v_{\theta_p} N_{Ts} \frac{n_s p_s}{n_s \sigma_n v_{\theta_n} + p_s \sigma_p v_{\theta_p}}. \quad (11.88)$$

Let the bulk value of carrier concentrations be n_0 and p_0 under thermal equilibrium, and let n and p be the bulk values under the low level injection condition. If the recombination is the rate-limiting step, then the low level injection condition implies flat quasi-Fermi levels between the bulk and the surface. If the total band bending is ψ_s from the bulk, the carrier concentration at the surface is

$$n_s = n \exp\left(-\frac{q\psi_s}{k_B T}\right), \text{ and}$$
$$p_s = p \exp\left(\frac{q\psi_s}{k_B T}\right), \quad (11.89)$$

with

$$np = n_i^2 \exp\left(\frac{E_{qFn} - E_{qFp}}{k_B T}\right), \quad (11.90)$$

at both the surface and in the bulk. The surface recombination rate can then be written as

$$\mathcal{R}_s = \frac{(\sigma_n \sigma_p v_{\theta_n} v_{\theta_p})^{1/2} N_{Ts} (n_s p_s)^{1/2}}{(n_s \sigma_n v_{\theta_n} / p_s \sigma_p v_{\theta_p})^{1/2} + (p_s \sigma_p v_{\theta_p} / n_s \sigma_n v_{\theta_n})^{1/2}}. \quad (11.91)$$

The occupation probability f of the trap N_{Ts} at the surface is given by

$$f = \frac{\sigma_n v_{\theta_n} n_s + \sigma_p v_{\theta_p} p_{Ts}}{\sigma_n v_{\theta_n} (n_s + n_{Ts}) + \sigma_p v_{\theta_p} (p + p_{Ts})}. \quad (11.92)$$

Since, at thermal equilibrium, the Fermi level is pinned at the trap level E_T, the occupation probability is very close to 1/2. Away from thermal equilibrium, the resultant surface charge density is given by

$$\mathcal{Q}_s = qN_{Ts}\left(f - \frac{1}{2}\right) = \frac{qN_{Ts}}{2} \frac{\sigma_n v_{\theta_n} n_s - \sigma_p v_{\theta_p} p_s}{\sigma_n v_{\theta_n} n_s + \sigma_p v_{\theta_p} p_s}, \quad (11.93)$$

where we again ignore the insignificant terms away from thermal equilibrium. For large trap densities that lead to Fermi level pinning, this charge is insignificant compared to the charge qN_{Ts} if all the traps were ionized or, equivalently, only a very small deviation from the 1/2 occupation probability occurs. This, however, implies that

$$\frac{n_s}{p_s} = \frac{\sigma_p v_{\theta_p}}{\sigma_n v_{\theta_n}}. \quad (11.94)$$

The ratio of carrier concentrations at the surface is a constant, and, in a large trap density, and low level injection limit, the recombination rate is

$$\mathcal{R}_s = \frac{(\sigma_n \sigma_p v_{\theta_n} v_{\theta_p})^{1/2} N_{Ts}}{2} (n_s p_s)^{1/2}. \quad (11.95)$$

The recombination rate is not proportional to the minority carrier concentration anymore; it is proportional to the square root of the electron and hole concentration at the surface, and since the quasi-Fermi levels are flat, it is also proportional to the square root of the electron and hole concentration in the bulk. A consequence of this is an $\exp(eV/2k_BT)$ dependence of surface recombination current for low level injection bias conditions where this analysis applies. This square root dependence also appears for recombination in a p–n junction space charge region and is responsible for a similar $\exp(eV/2k_BT)$ dependence there.

Surface recombination velocity, defined as a prefactor to excess minority concentration during recombination calculations, is no longer a constant but a function of bias conditions. Arguments have been forwarded that we should define surface recombination velocity in an alternate form that preserves a constancy. In the above, one may introduce S_0 as an intrinsic surface recombination velocity that follows the recombination relation

$$\mathcal{R}_s = S_0\sqrt{n_s p_s}. \tag{11.96}$$

This intrinsic recombination velocity is related to the conventional definition of surface recombination velocity in terms of excess carrier concentration (e.g., one useful in Dirichlet boundary conditions) via

$$S = S_0\sqrt{\frac{n_s}{p_s}}. \tag{11.97}$$

The intrinsic surface recombination velocity is the surface recombination velocity that occurs when the electron and hole densities at the surface are equal. Rigorous calculations of these parameters show that significant deviations occur at high level injection conditions, where our theory is not valid. Differences also occur when the surface recombination rate is the rate-limiting step. Under these conditions, the band diagram and the quasi-Fermi levels look as in Figure 11.13, indicating that the carrier flux needed for recombination occurs via both drift and diffusion to the surface.

Figure 11.13: The quasi-Fermi levels and conduction and valence band edges in the presence of a high surface recombination rate.

11.5 *Summary*

THE 0TH MOMENT OF THE BOLTZMANN transport equation describes particle conservation. Electrons (or holes) exist as the particles in the semiconductors while undergoing interactions. If these particles can appear or disappear, then this time dependence, which is not accounted for in the momentum change and positional change terms, must be accounted for. This is the time dependence in the distribution function that needs to appear separately to take care of the totality. Electrons and holes can be generated and can recombine. So, in principle, and especially when this rate is significant, such as when both populations exist in meaningful numbers or when other processes change them—rapid potential changes that lend kinetic energy sufficient to cause band-to-band processes, or sufficient carriers in high energy tails even at low fields, or radiation-induced change—then these need to be accounted for. We looked at a few of such situations for the bulk and surface.

Radiative recombination is a process we will return to later since it is an important part of understanding electromagnetic-dipole interaction in materials. But we employed it here to introduce the basic notions of how to incorporate a time constant akin to the scattering time for the direct electron-hole interaction, which is the changing of an electron's state from one band to another. We could view this through capture and emission rates that are a classical description of transition probability between occupied and unoccupied states. Since in thermal equilibrium, all processes must balance with a reverse process in detail, one could use linearization from this condition to write a lifetime—the radiative lifetime—for the process. This time constant characterizes the time dependence of change from equilibrium, and it is related to the thermal equilibrium concentration of carriers, since they are involved in capture and emission through the occupation of states and a material parameter $\mathcal{B}$, which is the radiative constant that follows from the matrix element coupling these states. Appendix K discusses this parameter and $\mathcal{A}$, the spontaneous parameter, both of which are tied to the matrix element connecting the states through the electromagnetic potential and the dipole interaction.

The Hall-Shockley-Read process is a non-radiative process of electron-hole interactions through states in the bandgap that both electrons and holes can interact with. These are interactions requiring phonon mediation. Again, starting with thermal equilibrium and disturbing it, one could write net equations for recombination and generation processes through these states and tie them to classical emission and capture constants that connect occupied and unoccupied states through particle emission and capture. The capture, for example here, can be viewed semi-classically as a carrier encountering such a state arising in an intrinsic or extrinsic defect of a certain cross-section where the defect state has an influence, and when moving carriers encounter such a region, they are captured. And from this captured state they may be re-emitted or they may capture another carrier of opposite polarity. The existence of this in the bandgap, with its smaller energy separation, as compared to that of the bandgap, makes such Hall-Shockley-Read centers efficient as mediating sites for electron-hole generation and recombination. The net rates will depend very much on the properties of such centers as well as the presence of carriers of both types for the process to proceed.

As an example of another important non-radiative process, we considered the Auger process, a process involving multiple carriers. Impact ionization, with an energetic electron or hole losing its kinetic energy in creating an electron-hole pair, is an Auger generation process. But this process also has significance as a recombination process, since small bandgaps or high carrier concentration conditions such as in lasers are very conducive to this recombination. Because of multiple carrier involvement, the lifetimes arising in this process have a strong background doping dependence. We used the Auger process to explore the quantum basis, and how it also can be mapped to the classical interpretations. So, our discussion of scattering, where multiple electrons—in both the conduction band and the valence band—interact, with the involvement of phonons, conserving energy and momentum and accounting for their fermion nature, led us to writing the perturbation Hamiltonian. One could calculate this under constrained conditions and, as a result, note the appearance of the minimum energy dependences, and the dependence on the nature of the electron and hole bands. From this, one could see how even large gap materials such as *Si* and *GaAs* show an Auger non-radiative lifetime dependence at high doping conditions. Impact ionization as an Auger generation process under conditions of high fields could also be seen through the quantum-mechanical relationships that we derived.

At the nanoscale, and even larger scales in bipolarity conditions in many *IIIV* semiconductors, surface recombination can be seen, since the surface states are deep levels where electrons and holes can be active with each other. It is an important aspect where the recombination is now tied to the flux of carriers that need to come to the surface, a region where Fermi level pinning and other constraints may exist. One could look at this variety of behavior through arguments based on charge conservation, flux and the rates of interaction in the surface states, to come up with a measure—the surface recombination velocity—that quantifies the net effect.

11.6 *Concluding remarks and bibliographic notes*

THIS CHAPTER put together aspects of point perturbations, surface states, Boltzmann transport, bandstructure and electromagnetic interactions in matter to explore how electrons and holes generate and recombine in a semiconductor. An initial introduction to radiative processes, Hall-Shockley-Read processes, Auger processes and processes at the surfaces was the range for our discussion.

Casey and Panish[1] discuss the A and B coefficients, their spontaneous and stimulated origins and the radiative generation and recombination processes over a number of direct bandgap semiconductor materials. Many other books on optical processes are also useful for understanding these concepts, whose origins go back to Einstein and his exploration of Planck's quantum postulate and the nature of radiation process and related statistics.

[1] H. C. Casey and M. B. Panish, "Heterostructure lasers," **1**, Academic, ISBN 978-0323157698 (1978)

Blakemore's text[2] comprehensively and semi-classically explores recombination processes in semiconductors, although it does not dwell on surface effects. For surface phenomena, a good introduction is in Tiwari's text[3] devoted to compound semiconductor devices. Surface recombination is a particularly strong detriment in bipolar structures because of the vagaries of compound semiconductor surfaces. Surface recombination also makes an appearance in McKelvey's text[4], as does a classical treatment of the Hall-Shockely-Read recombination. For surface recombination velocity, a discussion worth following is the argument by Rees for a velocity in an alternative form that preserves an invariance[5], and the comment by De Visschere[6].

[2] J. S. Blakemore, "Semiconductor statistics," Pergamon, Library of Congress 61-12443 (1962)

[3] S. Tiwari, "Compound semiconductor device physics," Academic, ISBN 13 978-0126917406 (1992)

[4] J. P. McKelvey, "Solid state and semiconductor physics," Krieger ISBN 13 978-0898743968 (1982)

[5] G. J. Rees, "Surface recombination velocity—a useful concept?," Solid-State Electronics, **28**, 517–519(1985)

[6] P. De Visschere, "Comment on G. J. Rees' 'Surface recombination velocity—a useful concept?'," Solid-State Electronics, **29**, 1161–1165(1986)

Ridley[7] is quite comprehensive in his discussion of Auger processes, including impact ionization. The great Soviet physicist Keldysh has much to say about many things semiconductor, and his name is synonymous with many effects. Keldysh's treatment of impact ionization (also Auger processes)[8] are very worthwhile following through.

[7] B. K. Ridley, "Quantum processes in semiconductors," Oxford, ISBN 0-19-850-580-9 (1999)

[8] L. V. Keldysh, "Kinetic theory of impact ionization in semiconductors," Soviet Physics JETP, **10**, 509–524 (1960)

11.7 *Exercises*

1. Let E_D and g_D be the donor energies in a semiconductor sample that is n-type but compensated. Show that the carrier concentration is related as

$$n\left(n+N_A-\frac{n_i^2}{n}\right)=\frac{\mathcal{N}_c}{g_d}\left(N_D-N_A-n+\frac{n_i^2}{n}\right)\exp\left(-\frac{E_c-E_D}{k_BT}\right).$$

How does the activation energy behave in a plot of carrier concentration with temperature in different temperature ranges? **[A]**

2. In this problem, we will evaluate statistics of deep trap energy levels arising from independent defects. Consider a semiconductor with two defects associated with energy levels E_1 and E_2.
   - What is the occupation probability of the two levels?
   - If the Fermi energy is between the two levels, with $E_2 - E_F \gg k_B T$, what do the occupation probabilities reduce to? **[S]**

3. Consider a semi-infinite n-type sample of $GaAs$ doped to $10^{16}\ cm^{-3}$ and with a mid-gap trap that has identical electron and hole capture cross-sections of $10^{-15}\ cm^{-2}$. Derive, using justifiable simplifications, the expression for decay of excess carriers in both the low injection limit and the high injection limit. What are the short time and long time limits if $10^{15}\ cm^{-3}$ carriers were created? What are the short and long time limits if $10^{17}\ cm^{-3}$ carriers were created? **[S]**

4. If $\tau_{n0} = \tau_{p0} = \tau_0$, show that the maximum lifetime occurs when the intrinsic and Fermi energy coincide. Show that this lifetime is

$$\tau = \tau_0 \left[1 + \cosh \left(\frac{E_T - E_i}{k_B T} \right) \right]. \qquad \text{([S])}$$

5. We now extend the previous problem to an analysis of recombination in a semiconductor when the energy levels arise from the same impurity. A technologically relevant example of this is gold, a transition element, as a substitutional impurity in silicon. In $GaAs$, multiple levels occur due to chromium. For gold, a closed d shell is inert. A positively charged state can accept an electron, while a negatively charged state can donate an electron. The higher negatively charged states occur in the valence band. So, the substitutional impurity in Si causes a lower energy level E_D, a donor level, and a higher energy level E_A, an acceptor level.
   - Calculate the equilibrium density of Au atoms in the positively charged state (N_T^+), the density of Au atoms in the neutral charge state (N_T^0), the density of Au atoms in the negatively charged state (N_T^-), and the total charge density on Au. Consider the degeneracies as g_A, and g_D.
   - What are the rate equations? Let σ_n^+ represent the capture cross-section for the electron capture process on the donor site, leading to a capture rate of c_n^+. Let σ_n^0 represent the capture cross-section for the capture of an electron on an acceptor,

Such traps are discussed as deep levels in Chapter 7.

leading to the capture rate of c_n^0. Let σ_p^0 represent the capture cross-section for the hole capture process on the donor site, leading to a capture rate of c_p^0. Let σ_p^- represent the capture cross-section for the capture of a hole on an acceptor, leading to the capture rate of c_p^-. Let e_n^0, e_n^-, e_p^+ and e_p^0 identify the electron and hole emission rates from donor and acceptor sites.

- Invoke the principle of detailed balance to evaluate the emission rates in terms of the capture rates at thermal equilibrium.
- What is the low level neutral region recombination rate?
- Calculate the low level neutral region minority carrier lifetime for trap concentrations significantly smaller than the shallow donor concentration.
- In high-level injection conditions, what is the limiting form of the lifetime?
- Capture cross-sections for Au in Si, at 300 K, are as follows:
  - $\sigma_n^+ = 3.5 \times 10^{-15}$ cm²
  - $\sigma_n^0 = 5.0 \times 10^{-16}$ cm²
  - $\sigma_p^- = 1.0 \times 10^{-15}$ cm²
  - $\sigma_p^0 = 3.0 \times 10^{-16}$ cm².

  Assuming that recombination occurs only due to Au in Si, what are the low level and majority carrier lifetimes for 10^{16} cm^{-3} Au doped Si at 300 K for
  - intrinsic material,
  - 10^{16} cm^{-3} n-type material and
  - 10^{16} cm^{-3} p-type material? **[A]**

6. We will develop the recombination statistics for a 2-level system. An accepted picture of Au in Si is that of a monovalent impurity (closed d shell that can be considered inert). If gold is substitutional, gold can be conjectured to be in the following bonded charge states: (a) Au^+, (b) Au^0, (c) Au^-, (d) Au^{2-} and (e) Au^{3-}. The bonded state is shown in Figure 11.14. The Au^{2-} and Au^{3-} presumably occur outside the bandgap. Adding an electron to the Au^+ state occurs at the energy E_D. The second electron in gold goes on at energy E_A. The degeneracy factors of these levels are g_D and g_A. Also, $E_D < E_A$.

   - Using the grand partition function, calculate for the equilibrium case:

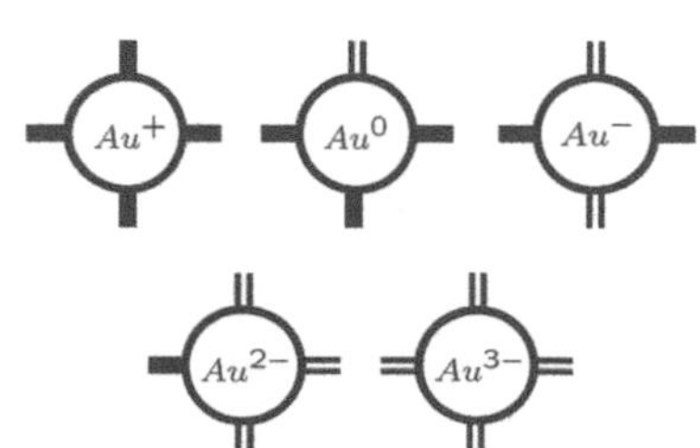

Figure 11.14: The different bonded states of gold.

- N_T^+: the average density of gold atoms in the +-charged state
- N_T^0: the average density of gold atoms in the neutral state
- N_T^-: the average density of gold atoms in the −-charged state
- ρ: the total charge density on the gold atoms.

Express your answer in terms of $N_T, E_D, E_A, E_F, g_D, g_A$ and k_BT.

- Suppose we really had two independent levels at E_A and E_D, with equal concentrations. Consider these to be independent and arriving from different particles. Calculate the above factors in this case. Comment on the feasibility of distinguishing these two cases experimentally.
- Write the rate equation for case (a) using the following notation:
  - electron capture on the donor: $\sigma_n^+ v_\theta = c_n^+$
  - electron capture on the acceptor: $\sigma_n^0 v_\theta = c_n^0$
  - hole capture on the donor: $\sigma_p^0 v_\theta = c_p^0$
  - hole capture on the acceptor: $\sigma_p^- v_\theta = c_n^-$
  - electron emission from the donor: e_n^0
  - electron emission from the acceptor: e_p^-
  - hole emission from the donor: e_p^+
  - hole emission from the acceptor: e_p^0.

  The signs here refer to the charge state prior to the corresponding process.
- Invoke detailed balance to find es in terms of cs.
- Calculate the low level ($\Delta n \approx \Delta p \approx 0$) neutral region recombination rate.
- Calculate the low level neutral region minority carrier lifetime for the case when N_T is much less than the shallow dopant density.
- What is the limiting life time in high level injection conditions, that is, when $\Delta n, \Delta p \gg n_0, p_0, n_{TA}, p_{TD}$?
- Measured values of capture cross-sections of gold in silicon at $T = 300$ K are $\sigma_n^+ = 3.5 \times 10^{-15}\ cm^2, \sigma_n^0 = 5 \times 10^{-16}\ cm^2$, $\sigma_p^- = 1.0 \times 10^{-15}\ cm^2$ and $\sigma_p^0 = 3.0 \times 10^{-16}\ cm^2$. A sample of Si has $10^{16}\ cm^{-3}$ gold atoms. Assume recombination solely through the gold atoms. Find the low level minority and majority carrier lifetime when

- $n_0 = 1.0 \times 10^{16}\ cm^{-3}$, and $p_0 = 2.2 \times 10^{4}\ cm^{-3}$,
- $n_0 = 1.0 \times 10^{15}\ cm^{-3}$, and $p_0 = 2.2 \times 10^{5}\ cm^{-3}$,
- $n_0 = 1.5 \times 10^{10}\ cm^{-3}$, and $p_0 = 1.5 \times 10^{10}\ cm^{-3}$,
- $n_0 = 2.2 \times 10^{5}\ cm^{-3}$, and $p_0 = 1.0 \times 10^{15}\ cm^{-3}$, and
- $n_0 = 2.2 \times 10^{4}\ cm^{-3}$, and $p_0 = 1.0 \times 10^{16}\ cm^{-3}$. **[A]**

7. Why is there a threshold in energy for a carrier to trigger an Auger generation process? And why is it close to that of the bandgap usually? **[S]**

8. Show that, under low-level injection, the expression for the Auger recombination rate reduces to a direct proportionality to the excess carrier concentration, with the proportionality constant dependent on the square of the dopant concentration. **[S]**

9. Argue how the recombination rate at the surface is related to the reflection coefficient at the surface. Cast the relationship in a mathematical form. **[S]**

10. Show that if the Fermi level is pinned at the surface, that is, if the sheet density of traps substantially exceeds the sheet charge density in the surface depletion region, then the occupation probability of a trap level is close to 1/2. Under what conditions may this break down? **[S]**

11. If the flux of carriers in any direction is given as $nv_\theta/4$, how can the surface recombination velocity exceed $v_\theta/4$? **[S]**

12. An n-type semiconductor crystal is cut to form a thin plate of thickness $2t$. The length and width are large enough that we ignore edge effects. Let S be the surface recombination velocity at the two surfaces. When this crystal is illuminated with light near the absorption edge, electron-hole pairs are generated. Since the wavelength is long enough, carriers are generated without significant absorption of the light. Find the steady-state excess carrier concentration at all points in the crystal, and describe its photoconductive response in steady state. **[M]**

13. Using the conditions given in Exercise 12, show that the result when S attains its maximum of $v_\theta/2$ and the result when S is taken to infinity are essentially the same when $\mathcal{L}_p \gg \lambda_{\mathbf{k}}$. **[S]**

12
Light interactions with semiconductors

LIGHT IS AN ELECTROMAGNETIC WAVE in the Maxwell view of electromagnetic unification and also a particle of energy excitation in the Planck and Einstein views, from where this led to the quantum mechanics wave-particle unification. If we go hundreds of years further back, it is also representable as a ray in the direction of energy flow for geometric optics, which does not tell us anything about phase, but which Huygens' approach of wavefront does. These last two notions are very limited but quite useful. The first two represent to us the wave-particle duality that pervades throughout this text. Convenience will decide our use of either the wave view or the particle view of light-matter interactions in this chapter. For example, the electromagnetic view tells us that there will be electrically and magnetically induced interactions of many varieties with charged particles, oscillating nuclei with charge clouds around them and charged donors, acceptors or traps, with all the coupled excitations represented by newer quasiparticles. Employing the photon exclusively as a neutrally charged particle for such a discussion leads to too difficult a road.

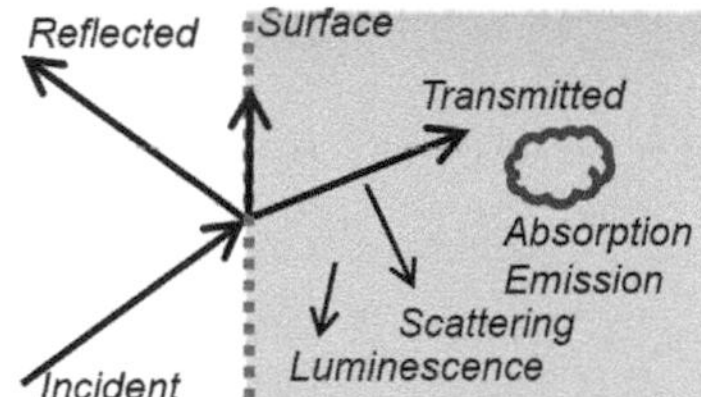

Figure 12.1: An illustration of some of the major phenomena of light's behavior in matter. It is reflected, some of it can be channeled along the surface by plasmonic interactions, and some of it is transmitted into the medium. This transmitted light can be absorbed but may re-emit, with the photon possibly breaking into two. Some of the light is scattered, and some of it leads to luminescence of various varieties, including fluorescence. Strictly, though, luminescence must be considered to be a cold body radiation arising in a spontaneous process.

Concerning light incident on matter, we consider only the moderate energy light (few μeV to several eVs) incident on a semiconductor (see Figure 12.1). Some of it gets reflected; if there exist mobile electrons, some of it even interacts with the plasmon excitation of these electrons and get channeled along the surface, and some gets transmitted into the material. The transmitted light can undergo absorption through a variety of processes that we will discuss; in some quite restricted conditions, the photon may even split, forming an entangled pair of photons; some of the photons will scatter off, and some may be absorbed, only to be re-emitted later through a reconversion into light of the absorbed energy. Scattering is usually the weakest of these effects in a semiconductor because of its quality and will appear only under artificially created

Semiconductor Physics: Principles, Theory and Nanoscale. Sandip Tiwari.
 Published 2020 by Oxford University Press. DOI: 10.1093/oso/9780198759867.001.0001

circumstances. Scattering occurs because the wavelength of the photon and the size of a discernible object within the material (e.g., nanocrystal, with different light propagation properties) are of similar dimensions. Surface channeling through coupled plasmon interactions is of interest since it can be significant in conducting materials and makes many useful applications possible. Absorption is of interest and employed very gainfully in such device structures as solar cells. We will also remark on luminescence and emission, but not much on reflection.

Our treatment of plasmons is rather limited. S. Tiwari, "Nanoscale device physics: Science and engineering fundamentals," Electroscience 4, Oxford University Press, ISBN 978-0-198-75987-4 (2017), discusses it in much more detail for nanoscale applications.

As a preliminary, consider just the consequence of the boundary condition and a very narrow region around it, without details of what may happen at the interface in view of several of the surface consequences we have talked about and their implications through interaction with the material. Figure 12.2 shows a uniform semi-infinite semiconductor on the right and a vacuum on the left, with a transverse electromagnetic (*TEM*) wave incident. The electric fields and magnetic fields have a relationship with each other through the divergence and curl connection of polar and axial fields—the consequence of which is the propagating wave—so, presence of sources, and the nature of the permittivity (ϵ) and permeability (μ), enter here. We have no sources. But permittivity and permeability are the consequences of the nature of the material itself and the field responses from what constitutes the matter. Electric polarization in all its forms from various causes affects the permittivity. And the spin—an angular momentum attribute—and currents affect the permeability. These can be complex in general. Since $c = 1/\sqrt{\epsilon\mu}$, the Maxwell relations will have various complex consequences floating around in the behavior of the model.

Recall that only *TEM* modes are allowed in free space. This follows from the Maxwell relationships that are summarized in the glossary.

In Figure 12.2, this is shown through the index of refraction ($n^{\underline{c}} = n^{\underline{r}} + in^{\underline{i}}$) as the sum of a real part and an imaginary part arising in the complex dielectric function ϵ. We will ignore permeability, assuming that the semiconductor is non-magnetic. Figure 12.2 also shows at the interface, with x being the axis along the interface, the

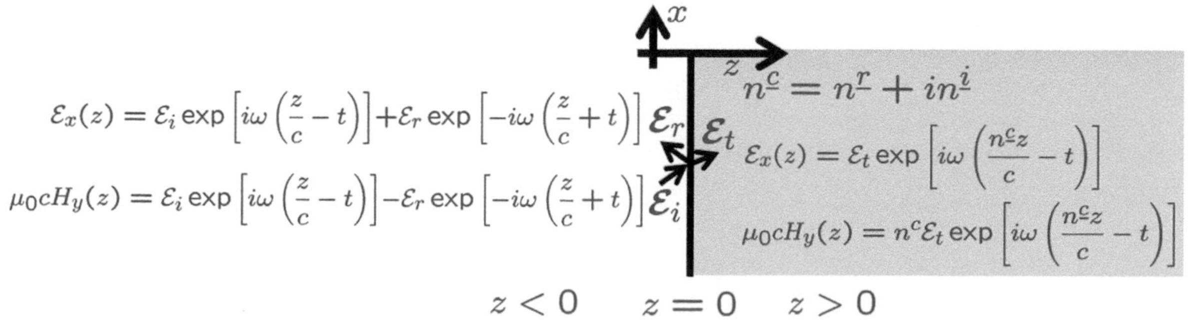

x, z, $\mathcal{E}_r$, $\mathcal{E}_t$, $\mathcal{E}_i$

$$\mathcal{E}_x(z) = \mathcal{E}_i \exp\left[i\omega\left(\frac{z}{c} - t\right)\right] + \mathcal{E}_r \exp\left[-i\omega\left(\frac{z}{c} + t\right)\right]$$

$$\mu_0 c H_y(z) = \mathcal{E}_i \exp\left[i\omega\left(\frac{z}{c} - t\right)\right] - \mathcal{E}_r \exp\left[-i\omega\left(\frac{z}{c} + t\right)\right]$$

$$n^{\underline{c}} = n^{\underline{r}} + in^{\underline{i}}$$

$$\mathcal{E}_x(z) = \mathcal{E}_t \exp\left[i\omega\left(\frac{n^{\underline{c}}z}{c} - t\right)\right]$$

$$\mu_0 c H_y(z) = n^c \mathcal{E}_t \exp\left[i\omega\left(\frac{n^{\underline{c}}z}{c} - t\right)\right]$$

$z < 0 \quad z = 0 \quad z > 0$

Figure 12.2: An illustration of transmission and reflection at a vacuum/semiconductor boundary.

form for the fields, with subscripts i, r and t for incident, reflected and transmitted, respectively. If no charge exists at the surface, the boundary conditions are

$$\begin{aligned} \mathcal{E}_t &= \mathcal{E}_i + \mathcal{E}_r, \text{ and} \\ H_t &= H_i - H_r, \text{ which is equivalent to} \\ n^{\underline{c}}\mathcal{E}_t &= \mathcal{E}_i - \mathcal{E}_r, \end{aligned} \tag{12.1}$$

when the relative permeability of the medium is unity. As an example of the dependence on this complex index of refraction, the reflectivity is

$$\varrho = \left(\frac{1 - n^{\underline{c}}}{1 + n^{\underline{c}}}\right)^2 = \frac{(n^{\underline{r}} - 1)^2 + (n^{\underline{i}})^2}{(n^{\underline{r}} + 1)^2 + (n^{\underline{i}})^2}. \tag{12.2}$$

Metals have high reflectivity, since the conductivity and the boundary conditions force the transverse electric field to vanish. Skin depth arises as the electromagnetic wave evanesces in the interaction with the electron cloud. And the index of refraction is large and imaginary. So, reflectivity is close to unity. Semiconductors, on the other hand, tend to be all gray if thick, as nearly 66 % of light passes and gets absorbed during passage through the medium.

So, reflection and transmission happen at the boundary because the two materials have different propagation properties, and because the electric field is polar, the magnetic field is axial and they are constrained through the relationships of Maxwell. This places constraints on the transverse and normal fields' continuity at the interface affecting both. If charge exists at the interface, it places constraints on the polar electric field, some of which terminates on these charges. If the charge is mobile, it suppresses the lateral electric field at the interface, since it can move. If electric fields are changing, the magnetic fields need to also, bound as they are through Maxwell's equations. In metals, reflection is significant because of the mobile charge. So is the decay of the electric field away from the interface in the metal. But, along with this reflection, there is some transmission of energy along the metal interface. The propagating wave will interact with the material—its charges, its phonons and its other mechanisms for energy exchange—which will lead to loss of the photons, which we call absorption.

As an introduction to the subject of this chapter, Figure 12.3 illustrates a rough view of absorption, measured as an absorption coefficient (α) in units of per unit length—the percentage fraction of photons, for example, or its equivalent in decay in intensity or energy flow per unit cross-section—in a semiconductor from a meV to an eV, so from very low energies—smaller than those of ionization or of optical phonons—to bandgap energies. Two

Maxwell's relations are a beautiful creation. A divergence-based equation speaks of a decline of a parameter. As one moves away from a charge, the potential effect of that charge decreases inversely with distance, vanishing asymptotically. A curl-based equation speaks to an attraction that keeps something going round and round—forever, if there is no dissipation term in the equation. A free space solution to the coupling of divergence and curl with no dissipation then is a propagating wave. The divergence is being curled to keep propagating. Another significant result within this is that even if light does not dissipate in idealized free space, there is still a propagation impedance because the fields must be related to each other, and even if the modes are continuous in free space, the nature of propagation, planar or spherical, will change the relationship and change the impedance. Impedance here does not mean dissipation, just as it does not mean dissipation for the quantum conductance. The interesting properties of divergence and curls are brought out in Appendix L,

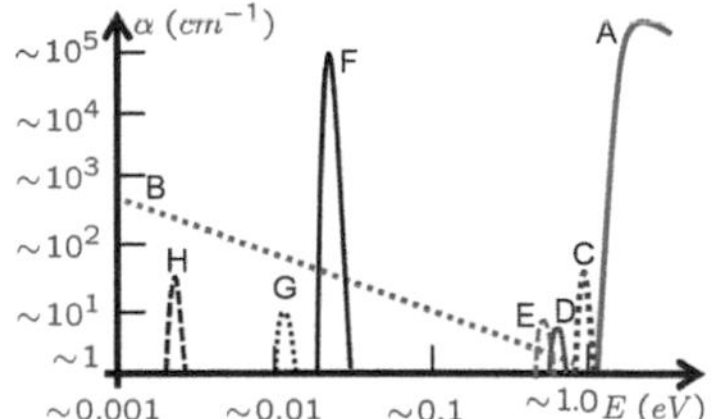

Figure 12.3: A sketch of photon absorption coefficients in a semiconductor as a function of energy due to some of the major mechanisms. Curve A shows band-to-band absorption, curve B is free carrier absorption, curve C is exciton absorption, curve D is valence-band-to-donor absorption, curve E is acceptor-to-conduction band absorption, curve G is optical phonon absorption, curve H is valence-band-to-acceptor absorption, and curve I is donor-to-conduction-band absorption. Also shown here is a curve F, where efficient photon absorption happens between deep levels. This happens in some rare examples. Zn and O form such a center in the indirect bandgap GaP.

major effects one sees are the band-to-band absorption close to the bandgap when electrons can transition from the valence band to the conduction band, and the smaller, gently rising toward low energy free carrier absorption of the photon. Smaller energy photons are more efficiently absorbed by the free carriers in a semiconductor. Between these two prominent features, one additional major recombination arises at optical phonon frequencies, but this is in the narrow bandwidth of energy of these phonons. And then, in addition, one has recombination processes where donors, acceptors and bands have transitions. Additionally, excitons (electron-hole pair) have absorption just below the bandedge, and acoustic phonons too couple to photons.

where the Helmholtz theorem and vector splitting are discussed. A discussion of the modes and their coupling, including enhancements to the coupling such as through the Purcell effect, is in Appendix M.

We are mostly concerned with what happens inside the semiconductor. So, the variety of transitions arising in photon interactions with phonons, directly, through impurities, the longitudinal and transverse constraints in this, and the coupling of a variety of particles, including their coupled incarnations, so oscillator strengths, since the coupling of excitations requires prudent matching of these oscillators, will be the focus of this chapter. We will understand recombination and the nature of the dielectric function, which captures the polarization induced by fields from this, leading us to the linear regions of the theory. The real and imaginary parts of response functions are related through the Kramers-Kronig relationship, which we will particularly concentrate on in a more general way in Chapter 14.

Oscillation, as the smallest stable representation with potential and kinetic energy exchange, constitutes a fundamental process. Harmonic oscillators are fundamental and ubiquitous. Bound electrons on atoms oscillate since the core and the electron are dipolar. These are at optical frequencies. Atoms in an assembly vibrate with the accompanying charge. These are vibrational oscillators at infrared frequencies. Free electrons oscillate, although these are primarily important at very high concentrations in highly doped semiconductors, inversion layers and, of course, metals. As the frequency of stimulus is raised, these dipoles progressively become inactive, and therefore the electromagnetic interaction with the semiconductor changes, with the dielectric function as a principal outcome of interest to us.

12.1 *Electron-photon interactions across the bandgap*

WE START BY DEVELOPING THE INTERACTION HAMILTONIAN for an electron-photon system. To distinguish an electron of wavevector **k** from that of a photon, which too has a wavevector, we will employ **q**—the same notation as for phonons—as the notation for the wavevector of a photon. Let **A** be the vector potential of the photon,

$$\mathbf{A} = \frac{1}{2}A\hat{\mathbf{a}}\exp[i(\mathbf{q}\cdot\mathbf{r}-\omega t)] + \frac{1}{2}A\hat{\mathbf{a}}\exp[-i(\mathbf{q}\cdot\mathbf{r}-\omega t)], \tag{12.3}$$

which is a photon of a transverse electromagnetic wave, whose vector potential is in the orientation $\hat{\mathbf{a}}$ and of energy $E = \hbar\omega$, and, of course, because it is transverse electromagnetic, $\hat{\mathbf{a}}\cdot\mathbf{q} = 0$. The wave also satisfies

$$|\mathbf{q}| = \frac{\omega \underline{n}}{c}, \tag{12.4}$$

where $\underline{n}$ is the index of refraction.

A is the magnetic vector potential. Since a magnetic field is an axial vector, its potential must be a vector too. The electric field has, as we have employed throughout, an electric scalar potential. The electrostatic potential acquires its meaning through the electrical field via the derivative. The vector potential does the same for the magnetic field through the curl. Adding a constant to the scalar or the vector potential—a change of reference—does not change the fields. For the photon, **A** plays the same role as the wavefunction ψ. See Appendix N for a discussion of vector and scalar potentials.

The Hamiltonian for this electron-photon assembly is

$$\hat{\mathscr{H}} = \frac{1}{2m_0}(\hbar\mathbf{k} - q\mathbf{A})^2 = \frac{1}{2m_0}\left(\hbar^2\mathbf{k}^2 - \hbar q\mathbf{k}\cdot\mathbf{A} - \hbar q\mathbf{A}\cdot\mathbf{k} + q^2\mathbf{A}^2\right). \quad (12.5)$$

$\mathbf{k} = (\hbar/i)\nabla_{\mathbf{r}}$, and $\nabla_{\mathbf{r}}\cdot\mathbf{A} = 0$; therefore, $\mathbf{k}\cdot\mathbf{A} = 0$, so $\mathbf{A}\cdot\mathbf{k} = 0$, and

$$\begin{aligned}\hat{\mathscr{H}} &= \frac{1}{2m_0}\left(-\hbar^2\nabla_{\mathbf{r}}^2 + iq\hbar\mathbf{A}\cdot\nabla_{\mathbf{r}} + q^2\mathbf{A}\right)\\ &\approx -\frac{\hbar^2}{2m_0}\nabla_{\mathbf{r}}^2 + \frac{iq\hbar}{m_0}\mathbf{A}\cdot\nabla_{\mathbf{r}}\\ &= \hat{\mathscr{H}}_0 + \hat{\mathscr{H}}',\end{aligned} \quad (12.6)$$

the rationale for which is that a $q^2\mathbf{A}$ term can be neglected at small light intensity compared to the term right before it. In Equation 12.6, the first term is our unperturbed electron Hamiltonian $\hat{\mathscr{H}}_0$, and the second ($\hat{\mathscr{H}}'$) is the perturbation term for the electron-photon interaction. In Section 1.2, we derived Equation 1.41 for the time dependence under perturbation for a two-level system started in one state. Our analysis is quite the same for a multilevel system appropriate to the case of an electron-photon interaction. Our governing equation is

$$-\frac{\hbar}{i}\frac{\partial\psi}{\partial t} = \hat{\mathscr{H}}\psi = \left(\hat{\mathscr{H}}_0 + \hat{\mathscr{H}}'\right)\psi, \quad (12.7)$$

and the solution is composed of a new wavefunction built from the eigenfunctions of the unperturbed system as

$$\psi(\mathbf{r},t) = \sum_n c_n(t)\psi_n(\mathbf{r})\exp\left(-\frac{iE_n t}{\hbar}\right), \quad (12.8)$$

leading to the time dependence of coefficients through

$$i\hbar\dot{c}_m(t) = \sum_{n=1}^{N} c_n(t)\mathscr{H}'_{mn}(t)\exp\left[\frac{i(E_m - E_n)t}{\hbar}\right], \text{ where}$$

$$\int \psi_m^*(\mathbf{r})\psi_n(\mathbf{r})\,d\mathbf{r} = \delta_{nm}. \quad (12.9)$$

The problem is of photon interaction in a crystal, where there are N primitive unit cells. We also write the vector potential, which is a real harmonic quantity, through Equation 12.3, where the first term gives rise to stimulated absorption, and the second leads to stimulated emission. The matrix element for electron transition from state $|\mathbf{k}\rangle$ at energy $E_{\mathbf{k}}$ to state $|\mathbf{k}'\rangle$ at energy $E_{\mathbf{k}'}$—a two-level-like problem—can be written as

$$\mathscr{H}'_{\mathbf{k}\mathbf{k}'}(t) = \frac{1}{N}\int \psi_{\mathbf{k}'}^*(\mathbf{r})\hat{\mathscr{H}}'\psi_{\mathbf{k}'}(\mathbf{r})\,d\mathbf{r}, \quad (12.10)$$

with the perturbation, as in Section 1.2, containing a time dependence through $\exp(i\omega t)$.

If the system started in a state $|\mathbf{k}\rangle$ at energy $E_{\mathbf{k}}$, its transition probability to state $|\mathbf{k}'\rangle$ is $\left|c_{\mathbf{k}'}(t)\right|^2$, found via

$$c_{\mathbf{k}'}(t) \approx \frac{1}{i\hbar}\int_0^t \mathscr{H}'_{\mathbf{kk}'}(t')\exp\left[\frac{i\left(E_{\mathbf{k}'}-E_{\mathbf{k}}\right)t'}{\hbar}\right]dt', \text{ with}$$

$$\begin{aligned}\mathscr{H}'_{\mathbf{kk}'}(t) &= \frac{iq\hbar}{2m_0 N}\int_\Omega \psi^*_{\mathbf{k}'}\left(\mathbf{A}\cdot\nabla_{\mathbf{r}}\right)\psi_{\mathbf{k}}\,d\mathbf{r} \\ &= \frac{iq\hbar A}{2m_0 N}\exp(-i\omega t)\int_\Omega \psi^*_{\mathbf{k}'}\exp(i\mathbf{q}\cdot\mathbf{r})(\hat{\mathbf{a}}\cdot\nabla_{\mathbf{r}})\psi_{\mathbf{k}}\,d\mathbf{r}. \end{aligned} \quad (12.11)$$

The time-independent part of this matrix element ($\mathscr{H}'_{\mathbf{kk}'}$) is the prefactor and the integral, and the time dependence is in the $\exp(i\omega t)$, so that $\mathscr{H}'_{\mathbf{kk}'}(t) = \mathscr{H}'_{\mathbf{kk}'}\exp(-i\omega t)$, where $\mathscr{H}'_{\mathbf{kk}'}$ is a magnitude, and we may write the probability of transition in time duration of t as

$$\left|c_{\mathbf{k}'}(t)\right|^2 = \frac{2\pi\left|\mathscr{H}'_{\mathbf{kk}'}\right|^2 t}{\hbar}\delta\left(E_{\mathbf{k}'}-E_{\mathbf{k}}-\hbar\omega\right), \quad (12.12)$$

which associates the transition energy conservation condition of $E_{\mathbf{k}'} - E_{\mathbf{k}} = \hbar\omega$ as a selection condition and states that the probability of transition increases with time t. Note that this is the transition for the electron—in the crystal—undergoing a change to another state, again in the crystal. The interaction is with the electron—local—and the transition is for the electron to move to a state defined by the crystal. The first condition is reflected in the mass being the free electron mass m_0 in the prefactor in Equation 12.11. The second—of the states—is reflected in the Bloch function, whose one emergent property is the effective mass. So, this equation reflects transitions such as that shown in Figure 12.4, where the electron changes its state, which means a change in the wavefunction through its modulation function via the electron-photon interaction. No crystal momentum change took place.

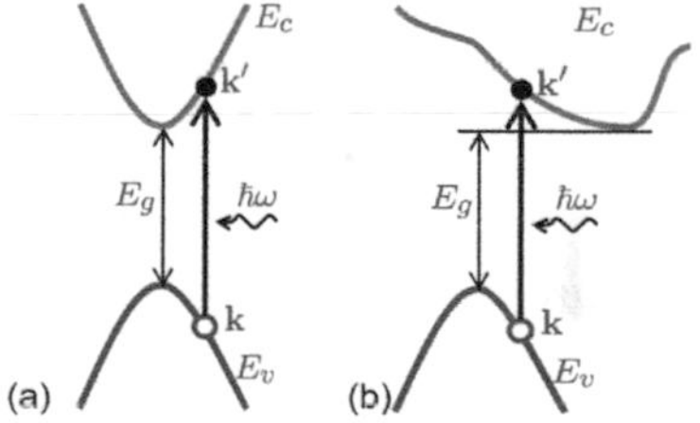

Figure 12.4: An electron-photon-interaction-induced transition from the valence band to the conduction band, as reflected through a direct band-to-band absorption. Part (a) shows this for a direct bandgap semiconductor, and (b) for an indirect semiconductor. The matrix coupling elements can be very different, since the Bloch functions in the two situations will be very different.

The $|\mathbf{k}\rangle$ is the Bloch function ($\psi_{\mathbf{k}} = u_{\mathbf{k}}(\mathbf{r})\exp(i\mathbf{k}\cdot\mathbf{r})$), so Equation 12.11 may now be recast as

$$\begin{aligned}\mathscr{H}'_{\mathbf{kk}'} &= \frac{iq\hbar A}{2m_0 N}\int_\Omega \exp\left[i\left(\mathbf{k}-\mathbf{k}'+\mathbf{q}\right)\cdot\mathbf{r}\right] \\ &\quad\times u^*_{\mathbf{k}'}\left[\hat{\mathbf{a}}\cdot\nabla_{\mathbf{r}}u_{\mathbf{k}} + i\left(\hat{\mathbf{a}}\cdot\mathbf{k}\right)u_{\mathbf{k}}\right]d\mathbf{r} \\ &= \frac{iq\hbar A}{2m_0 N}\sum_{\mathbf{R}_{mn}}\exp\left[i\left(\mathbf{k}-\mathbf{k}'+\mathbf{q}\right)\cdot\mathbf{R}_{mn}\right] \\ &\quad\times\int_{\Omega_0} u^*_{\mathbf{k}'}\left[\hat{\mathbf{a}}\cdot\nabla_{\mathbf{r}}u_{\mathbf{k}} + i\left(\hat{\mathbf{a}}\cdot\mathbf{k}\right)u_{\mathbf{k}}\right]d\mathbf{r} \\ &= \frac{iq\hbar A}{2m_0}\int_{\Omega_0} u^*_{\mathbf{k}'}\left[\hat{\mathbf{a}}\cdot\nabla_{\mathbf{r}}u_{\mathbf{k}} + i\left(\hat{\mathbf{a}}\cdot\mathbf{k}\right)u_{\mathbf{k}}\right]d\mathbf{r}. \end{aligned} \quad (12.13)$$

In the rewritten form of Equation 12.13, we have used two properties. The first step was to employ the periodicity of the Bloch functions to reduce the integral from the entire volume to a unit cell. The second was the use of wavevector (momentum) conservation, that is, $\mathbf{k} - \mathbf{k}' + \mathbf{q} = 0$, with N unit cells each contributing through the exponential result of 1.

This last form of Equation 12.13 has two terms, both dependent on the modulation function, inherent within which is the symmetry of the state. The first term leads to what we call *allowed transitions*, and the second leads to what we call *forbidden transitions*. The allowed transitions reflect the fact that it is a term which does have a finite meaningful value, since $u_{\mathbf{k}}$ does have a spatial variation in all directions, including that of the vector potential of light. The "forbidden" transitions are weaker electron-photon momentum conserving transitions *without any other involvement such as that of phonons. Bloch functions between two different bands, that is different index quantum numbers, are orthogonal. So, for* $\mathbf{k} = \mathbf{k}'$, *terms of this integral contribution vanish.* Photons, as we have discussed, have small wavevectors. Overall, the term contributes, even if it is small.

The forbidden transition is really not absent; it just has a small magnitude compared to the allowed transitions. The origin of the use of the word is from spectral transitions, where certain lines are forbidden by selection rule but become allowed when the approximation leading to that rule is not made. I have not been able to find who introduced this nomenclature. It has stuck. Many people like a deterministic coda. Quantum mechanics allows much through its uncertainty. The Forbidden City is also not forbidden, and very worthy of a visit. These are words used by a system to specify a rule that is then meant to be broken. It is kind of like Chekhov's rifle, which is on the wall in the first chapter and then, as the book is finishing, must be taken down in the last chapter. Forbidden transitions should not be confused with transitions where conservation of momentum involves phonons. This latter is an important source of transitions—light absorption—in silicon, which is so useful in photovoltaics, and is an indirect transition.

12.1.1 *Allowed transitions*

THE MATRIX ELEMENT FOR ALLOWED TRANSITIONS is

$$\mathscr{H}'_{\mathbf{kk}'}\big|_{allowed} = \frac{iq\hbar A}{2m_0} \int_{\Omega_0} u^*_{\mathbf{k}'} \left(\hat{\mathbf{a}} \cdot \nabla_{\mathbf{r}} u_{\mathbf{k}}\right) d\mathbf{r}. \tag{12.14}$$

Since $\mathbf{p} = (\hbar/i)\nabla_{\mathbf{r}}$, we define a momentum matrix element,

$$\mathbf{p}_{\mathbf{kk}'} = \frac{\hbar}{i} \int_{\Omega_0} u^*_{\mathbf{k}'} \nabla_{\mathbf{r}} u_{\mathbf{k}} \, d\mathbf{r}, \tag{12.15}$$

leading to

$$\mathscr{H}'_{\mathbf{kk}'}\big|_{allowed} = -\frac{qA}{2m_0} \left(\hat{\mathbf{a}} \cdot \mathbf{p}_{\mathbf{kk}'}\right). \tag{12.16}$$

We have now again related a transition of the change of state of the electron from one index to another—$|\mathbf{k}\rangle \mapsto |\mathbf{k}'\rangle$, valence to conduction—in terms of the momentum, and therefore a momentum matrix element. Why is this so? The interaction with light, a source of energy with a very small momentum, causes a Bloch state spread out in real space to change to another Bloch state spread out in space upon the photon's absorption. Its crystal momentum remained quite unchanged, but its modulation function changed, which is reflected in the E-$\mathbf{k}$ relationship of the band, and whose immediate view to us is through the effective mass. A less mobile electron in the valence band—a sea of electrons—became a more mobile electron, so it is now being viewed as an independent, nearly free electron. This is a transition from a liquid to a gas. It is the modulation function that represents the wavepacket-like extent

Note that the matrix element involves m_0, which is the free electron mass representing this local interaction. But the change of state from the valence to the conduction now makes the electron move in the crystal under different E-$\mathbf{k}$ constraints represented in the effective mass.

of the propagating electron's reach and its responsiveness to forces. Momentum change is the response. And therefore it stands to reason that one sees the momentum matrix element as a reflection of allowed absorption.

The probability of a $\mathbf{k} \mapsto \mathbf{k}'$ transition in the crystal is

$$\left|c_{\mathbf{k}'}(t)\right|^2 = \frac{2\pi t}{\hbar}\left(\frac{qA}{2m_0}\right)^2 (\hat{\mathbf{a}} \cdot \mathbf{p}_{\mathbf{k}\mathbf{k}'})^2 \delta\left(E_{\mathbf{k}'} - E_{\mathbf{k}} - \hbar\omega\right), \tag{12.17}$$

with the transition rate

$$\frac{\left|c_{\mathbf{k}'}(t)\right|^2}{t} = \frac{2\pi}{\hbar}\left(\frac{qA}{2m_0}\right)^2 (\hat{\mathbf{a}} \cdot \mathbf{p}_{\mathbf{k}\mathbf{k}'})^2 \delta\left(E_{\mathbf{k}'} - E_{\mathbf{k}} - \hbar\omega\right). \tag{12.18}$$

To calculate the probability of band-to-band transitions, we need to know this transition rate of Equation 12.18, the probability of occupation of the valence state transitioning (the Fermi-Dirac distribution function $f = f_{FD}(E, E_F, T) = 1/\{1 + \exp[(E - E_F)/k_BT]\}$, the probability that the state being transitioned to is unoccupied in the conduction band $(1 - f_{FD})$, the volume in reciprocal space occupied for each wavevector quantum number $((2\pi)^3/\Omega)$, and the spin degeneracy $g_s = 2$, so the joint probability is

We will return to the implications of selection rules for the transitions in Chapter 20. These become significant in confined conditions, affecting particularly the polarization of light from lasers. In bulk, for now, the specific heavy-hole, light-hole, split-off-hole and electron conduction band interactions will be left more general, which is good enough. In Section 20.7, you will find reference back to the selection rules for bulk conditions too. The section should be readable without the intervening fertile fields.

$$\mathfrak{p}_{\mathbf{k}\mathbf{k}'} = \frac{2\Omega}{(2\pi)^3}\int_{\Omega_{\mathbf{k}}} \left|c_{\mathbf{k}'}(t)\right|^2 f_0(1 - f_0)\, d\mathbf{k}, \tag{12.19}$$

which is an integration over the whole reciprocal space.

The transition per unit volume—the recombination rate (here, for the allowed band-to-band transitions)—is the transition probability per unit volume per unit time and so, assuming little disturbance from thermal equilibrium, and an incident wave of energy $\hbar\omega$ per photon, is

$$\begin{aligned} r|_{allowed} &= \frac{2}{(2\pi)^3}\int_{\Omega_{\mathbf{k}}} \frac{\left|c_{\mathbf{k}'}(t)\right|^2}{t} f_0(1 - f_0)\, d\mathbf{k} \\ &= \frac{2}{4\pi^2\hbar}\left(\frac{qA}{2m_0}\right)^2 \int_{\Omega_{\mathbf{k}}} (\hat{\mathbf{a}} \cdot \mathbf{p}_{\mathbf{k}\mathbf{k}'})^2 \delta\left(E_{\mathbf{k}'} - E_{\mathbf{k}} - \hbar\omega\right) \\ &\quad \times f_0(1 - f_0)\, d\mathbf{k}. \end{aligned} \tag{12.20}$$

The energy difference of the states transitioning is

$$\begin{aligned} E_{\mathbf{k}'} - E_{\mathbf{k}} &= E_g + \frac{\hbar^2 k'^2}{2m_e^*} + \frac{\hbar^2 k^2}{2m_h^*} \\ &= E_g + \frac{\hbar^2 k^2}{2m_r^*}, \end{aligned} \tag{12.21}$$

where we again use the reduced mass approach of two-body problems with

$$m_r^* = \frac{m_e^* m_h^*}{m_e^* + m_h^*}. \quad (12.22)$$

m_r^* is the reduced effective mass for this electron-hole pair that is recombining. Since the Dirac delta is the rapidly changing function in the transition rate equation, we can write

$$r|_{allowed} = \frac{q^2 A^2 \omega f_{osc}}{16\pi^2 m_0} f_0(1 - f_0) \int_{\Omega_{\mathbf{k}}} \delta\left(E_g + \frac{\hbar^2 k^2}{2m_r^*} - \hbar\omega\right) 4\pi k^2\, dk. \quad (12.23)$$

f_{osc} here is the oscillator strength

$$f_{osc} = \frac{2(\hat{\mathbf{a}} \cdot \mathbf{p}_{\mathbf{k}\mathbf{k}'})^2}{\hbar m_0 \omega} \approx 1 + \frac{m^*}{m_h^*}, \quad (12.24)$$

a dimensionless parameter that captures the strength of the interaction through the momentum matrix element connecting the two states between which the transition is taking place.

See Appendix I for a discussion of oscillator strength. Matrix elements for calculating transition rates involve $\mathbf{p}_{\mathbf{k}\mathbf{k}}$ and so are called momentum matrix elements. Through $\hat{\mathbf{a}} \cdot \mathbf{p}_{\mathbf{k}\mathbf{k}'}$, they are also dipole matrix elements. Here, with light, it arose in the proportionality $\mathbf{A} \cdot \mathbf{p}$. In the $\mathbf{k} \cdot \mathbf{p}$ discussion, it arose again through the matrix term $\langle \mathbf{k}' | \mathbf{p} | \mathbf{k} \rangle$ in the coupling between the states. It is convenient to describe this strength of coupling between two states through this oscillator strength. Multiple state coupling follows through the oscillator sum rule, which is usually valid.

In Equation 12.23, the integral is quite simplifiable:

$$\begin{aligned} &\int_{\Omega_{\mathbf{k}}} \delta\left(E_g + \frac{\hbar^2 k^2}{2m_r^*} - \hbar\omega\right) 4\pi k^2 dk \\ &= \frac{4\pi}{3}\left(\frac{2m_r^*}{\hbar^2}\right)^{3/2} \frac{d}{d(\hbar\omega)} (\hbar\omega - E_g)^{3/2} \\ &= \frac{2\pi}{\hbar^3} (2m_r^*)^{3/2} (\hbar\omega - E_g)^{1/2}, \end{aligned} \quad (12.25)$$

which leads to the dependence

$$r|_{allowed} = \frac{q^2 A^2 \omega f_{osc} (2m_r^*)^{3/2}}{8\pi\hbar^3 m_0} f_0(1 - f_0)(\hbar\omega - E_g)^{1/2}, \quad (12.26)$$

a square root of energy dependence on energy for the allowed transitions.

The parameter that one is usually interested in is the absorption per unit length ($\alpha(E)$),

$$\alpha = \frac{r}{\Phi}, \quad (12.27)$$

where r is the transition probability per unit volume per unit time, and Φ is the quantum flux, that is, the number of photons per unit area per unit time:

$$\Phi = \frac{\langle \mathbf{S} \rangle}{\hbar\omega}, \quad (12.28)$$

where $\mathbf{S}$ is the Poynting vector ($\mathbf{S} = \boldsymbol{\mathcal{E}} \times \mathbf{H}$, which is energy incident per unit area per unit time). With the vector potential as $\mathbf{A}$ of Equation 12.3, the fields are $\boldsymbol{\mathcal{E}} = A\omega\hat{\mathbf{a}} \sin(\mathbf{q} \cdot \mathbf{r} - \omega t)$, and $\mathbf{H} = -(A/\mu)(\mathbf{q} \times \hat{\mathbf{a}}) \sin(\mathbf{q} \cdot \mathbf{r} - \omega t)$, so that

$$\langle \mathbf{S} \rangle = \frac{1}{2} \frac{\omega A^2}{\mu} = \frac{1}{2} n \epsilon_0 c \omega^2 A^2, \quad (12.29)$$

from which the absorption coefficient for allowed band-to-band transitions follows as

$$\alpha|_{allowed} = \frac{q^2 A^2 (2m_r^*)^{3/2} f_{osc}}{4\pi\epsilon_0 ncm_0\hbar^2} f_0(1-f_0)(\hbar\omega - E_g)^{1/2}. \quad (12.30)$$

In the idealized condition of a filled valence band and an entirely empty conduction band,

$$\alpha|_{allowed} = 2.7\times 10^5 \left(\frac{2m_r^*}{m_0}\right)^{3/2} \frac{f_{osc}}{n} (\hbar\omega - E_g)^{1/2}, \quad (12.31)$$

where n is the index of refraction. The smaller the reduced effective mass—usually controlled by the conduction band—the higher is the absorption.

12.1.2 *Forbidden transitions*

THE MATRIX ELEMENT FOR THE FORBIDDEN TRANSITIONS is

$$\mathscr{H}'_{\mathbf{kk}'} = -\frac{q\hbar A}{2m_0} (\hat{\mathbf{a}}\cdot\mathbf{k}) \int_\Omega u_{\mathbf{k}'}^* u_{\mathbf{k}}\, d\mathbf{r}, \quad (12.32)$$

which we arrived at through the second part of Equation 12.13, and is dependent on the overlap of the modulation functions of the two bands. This is going to be small, because of the orthonormality of the Hermitian solutions. We also have

$$\frac{|c_{\mathbf{k}'}|^2}{t} = \frac{2\pi}{\hbar}\left(\frac{q\hbar A}{2m_0}\right)^2 |\hat{\mathbf{a}}\cdot\mathbf{k}|^2 \mathcal{O}_{cv}\delta\left(E_{\mathbf{k}'} - E_{\mathbf{k}} - \hbar\omega\right), \quad (12.33)$$

where $\mathcal{O}_{cv}$ is the overlap integral. With different bands and a finite but small photon wavevector, $0 < \mathcal{O}_{cv} \ll 1$. The term $|\hat{\mathbf{a}}\cdot\mathbf{k}|^2$ has a dependence on $\mathbf{k}$. This is an average arising in the volume of the source transition state that couples under the conservation constraints. We approximate this by the reciprocal volume of $k^2/3$. So, the transition rate follows as

$$r|_{forbidden} = \frac{q^2 A^2 \mathcal{O}_{cv} (2m_r^*)^{5/2}}{12\pi m_0^2 \hbar^4} f_0(1-f_0)(\hbar\omega - E_g)^{3/2}, \quad (12.34)$$

and the forbidden absorption coefficient follows as

$$\alpha|_{forbidden} = \frac{q^2 (2m_r^*)^{5/2} \mathcal{O}_{cv}}{6\pi\epsilon_0 ncm_0^2\hbar^2} f_0(1-f_0)\frac{(\hbar\omega - E_g)^{3/2}}{\hbar\omega}, \quad (12.35)$$

and, under the idealized simplification of a filled valence band and an empty conduction band, as

$$\alpha|_{forbidden} = 1.8\times 10^5 \left(\frac{2m_r^*}{m_0}\right)^{5/2} \frac{\mathcal{O}_{cv}}{n} \frac{(\hbar\omega - E_g)^{3/2}}{\hbar\omega}. \quad (12.36)$$

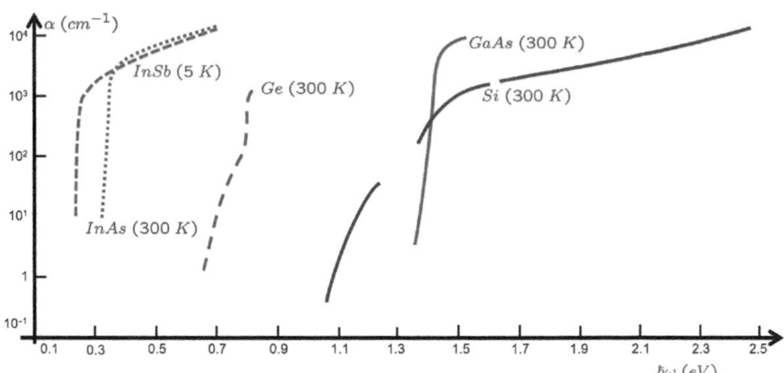

Figure 12.5: Absorption coefficients at bandedge for direct and indirect semiconductors illustrating the allowed and forbidden transition consequences.

There exists now an inverse frequency tail in the expansion for the forbidden transition compared to the allowed transition. And since $\mathcal{O}_{cv}$ is small, the forbidden transition is significantly weaker.

Figure 12.5 show some of these characteristics for example direct and indirect bandgap semiconductors. In such characteristics, when plotted near the bandedge, where the change is most pronounced, the region of rapid change is known as the *absorption edge*. It is close to the bandgap. In direct bandgap materials, this edge is very pronounced. In indirect materials, where phonon-assisted processes are very important, it is not so pronounced, as Figure 12.5 shows.

12.1.3 *Phonon-assisted indirect transitions*

PHOTON ABSORPTION WITH ELECTRON transition from one band to another will also occur through the participation of a phonon. Although their population density or available state density is temperature and phonon energy dependent, the phonons provide a means for momentum conservation. At room temperature, the density is high, whether of acoustic or optical phonons. Phonon involvement makes transitions involving wavevector changes possible, and while it is a second-order process, since another particle is involved, there are many more states that are coupled, making this process not inefficient.

In Figure 12.6, one can see how a spread of the phonon energy and the wavevector allows a band of valence electrons to transition to a band of conduction band states in an indirect bandgap material. Panel (a) shows the state at the valence band maximum coupling to a number of bandgap states in the conduction band by virtue of the spread of phonon energies as a function of the wavevectors available. We have employed the s subscript to distinguish the phonon parameters from the

"Not inefficient" is stated with caution. Solar cells based on *Si* are ubiquitous, even if its bandgap is indirect, with the zone center energy spread between conduction and the valence bandedge nearly 3.0 eV in a 1.1 eV bandgap material. But, then, silicon does not make lasers, while the direct gap materials with strong allowed transitions do.

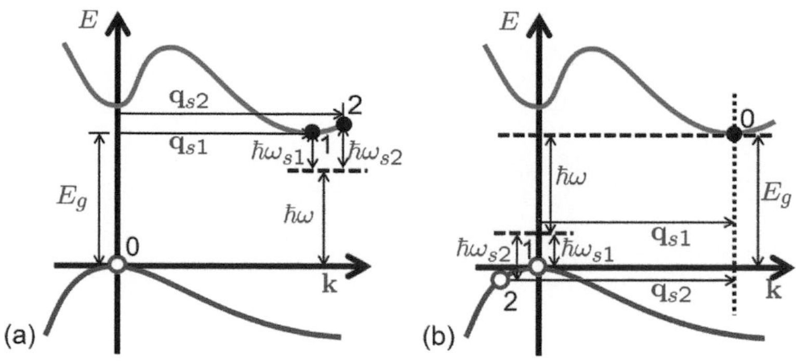

Figure 12.6: Phonon-absorption-initiated absorption, where significant momentum and energy are supplied by the phonon. Part (a) shows the absorption from the valence bandedge to several conduction bandstates and (b) shows that many states from the valence band also participate, and here the processes to the conduction bandedge state are shown.

photon parameters. Panel (b) shows the same number of states in the valence band coupling to the conduction bandedge state. In these instances, the transition from state 0 in Figure 12.6(a) to 1 or 2 requires a phonon of wavevector $\mathbf{q}_{s1}$ and energy $\hbar\omega_{s1}$, for coupling to the state identified by 1, or a phonon of wavevector $\mathbf{q}_{s2}$ and energy $\hbar\omega_{s2}$, for coupling to the state identified by 2. In both of these, the energy of the photon involved is still the same $\hbar\omega$. Both of the phonons are absorbed here. Processes involving the emission of phonons should also be possible. In all these cases, conservation of energy and momentum holds through the participation of the phonon. An important energy consequence is that the indirect band-to-band absorption has a minimum at an energy lower than the bandgap energy. $\hbar\omega \geq E_g - \hbar\omega_s$.

Analysis of such indirect transitions requires 2nd order perturbation theory. One can view the electron transition from the valence band to conduction band as arising in a direct transition to a virtual state in the conduction band with the simultaneous emission of a phonon. Likewise, one could also have a virtual transition to a valence band state with the emission of a phonon and a direct transition. The virtual state is a very short-lived state describable through the uncertainty relationship. And there will be allowed transitions and forbidden transitions for this too. All we have done is incorporated a new particle in the dynamics. The electromagnetic wave still leads to a perturbation term with two components.

We will not dwell on the derivations but just state the main results.

12.1.3.1 *Allowed phonon-assisted indirect transitions*

THE ALLOWED TRANSITIONS in an indirect semiconductor—phonon-assisted and through the conduction band—as depicted in Figure 12.7, have an absorption coefficient of

$$\alpha_c(\pm\omega_s)|_{ai} = \frac{g_c q^2 m_h^{*3/2} \omega_s f_{cosc} E_0 (\hbar\omega \pm \hbar\omega_s - E_g)^2}{32\pi\epsilon_0 ncm_0 m_e^{*1/2} \hbar\omega\lambda_c k_B T (E_0 - \hbar\omega)^2} \frac{\pm 1}{\exp\left(\pm\frac{\hbar\omega_s}{k_B T}\right) - 1}, \tag{12.37}$$

where the $+$ sign is for the phonon absorption process, and the $-$ sign is for phonon emission.

For the allowed process through the valence band, the absorption coefficient is

$$\alpha_v(\pm\omega_s)|_{ai} = \frac{g_c q^2 m_e^{*3/2} \omega_s f_{vosc} E_1 (\hbar\omega \pm \hbar\omega_s - E_g)^2}{32\pi\epsilon_0 ncm_0 m_h^{*1/2} \hbar\omega\lambda_v k_B T (E_1 - \hbar\omega)^2} \frac{\pm 1}{\exp\left(\pm\frac{\hbar\omega_s}{k_B T}\right) - 1}, \tag{12.38}$$

with the total absorption given by the sum of all such processes (allowed and indirect), that is,

$$\alpha|_{ai} = \alpha_c(+\omega_s) + \alpha_c(-\omega_s) + \alpha_v(+\omega_s) + \alpha_v(-\omega_s). \tag{12.39}$$

In these sets of equations, g_c and g_v are the degeneracies of the minima, for example, for the X valley of the *Si* conduction band, $g_c = 6$. f_{cosc} and f_{vosc} are the oscillator strengths for the conduction band and the valence band, respectively, and λ_c and λ_v are the mean free paths for electron scattering.

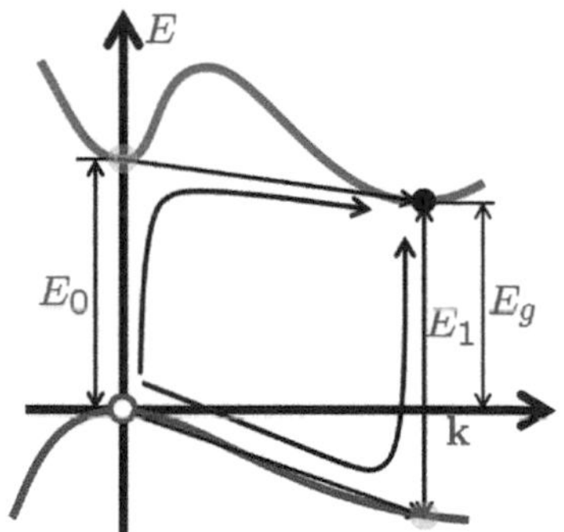

Figure 12.7: Band-to-band transitions in indirect bandgap and phonon assistance, modeled through virtual states either in the conduction band or in the valence band.

12.1.3.2 *Forbidden phonon-assisted indirect transitions*

For the "forbidden" transitions, again phonon assisted and in indirect materials, treated similarly through virtual states, the absorption coefficient is given by

$$\alpha_c(\pm\omega_s)|_{fi} = \frac{g_c q^2 m_h^{*5/2} \omega_s \mathcal{O}_c' (\hbar\omega \pm \hbar\omega_s - E_g)^3}{48\pi\epsilon_0 ncm_0 m_e^{*1/2} \hbar\omega\lambda_c k_B T (E_0 - \hbar\omega)^2} \frac{\pm 1}{\exp\left(\pm\frac{\hbar\omega_s}{k_B T}\right) - 1} \tag{12.40}$$

for processes through the conduction band and

$$\alpha_v(\pm\omega_s)|_{fi} = \frac{g_v q^2 m_e^{*5/2} \omega_s \mathcal{O}_c' (\hbar\omega \pm \hbar\omega_s - E_g)^3}{48\pi\epsilon_0 ncm_0 m_h^{*1/2} \hbar\omega\lambda_v k_B T (E_1 - \hbar\omega)^2} \frac{\pm 1}{\exp\left(\pm\frac{\hbar\omega_s}{k_B T}\right) - 1} \tag{12.41}$$

for processes through the valence band. E_0 and E_1 here are the direct gaps in the indirect semiconductor at the valence band maximum and the conduction band minimum, respectively, as seen in Figure 12.7.

The conclusion of this discussion is that there are a large number of recombination processes for photons, phonon-free and phonon-assisted for direct and indirect semiconductors. Each then shows a different absorption edge shape. And the significant change in the photon energy dependence is as summarized in Table 12.1.

Type	Dependence
Direct	
"allowed"	$\propto (\hbar\omega - E_g)^{1/2}$
"forbidden"	$\propto (\hbar\omega - E_g)^{3/2}$
Indirect	
"allowed"	$\propto (\hbar\omega \pm \hbar\omega_s - E_g)^2$
"forbidden"	$\propto (\hbar\omega \pm \hbar\omega_s - E_g)^3$

Table 12.1: Energy dependence of various conduction-valence absorption processes in direct and indirect semiconductors.

Since both phonon emission and absorption are possible when the photon energy exceeds the bandgap by the phonon energy magnitude, $\alpha = \alpha_a + \alpha_e$ for $\hbar\omega > E_g + \hbar\omega_q$. If the temperature is low, the phonon density is low, so α_a will become small. This provides us with a tool for understanding the semiconductor's optical interaction through its behavior near the bandgap. Figure 12.8 shows the absorption coefficient in phonon-dominated conditions at the bandedge. At moderate and high temperatures, one will see a tail in the absorption curve with a $\sqrt{\alpha_a}$ linear dependence on the photon energy. And this will have a fairly noticeable cut-off at a phonon energy below the bandgap. When the temperature is lowered, this phonon absorption will be reduced, and because the bandgap is increased, the absorption characteristics will shift to higher energy.

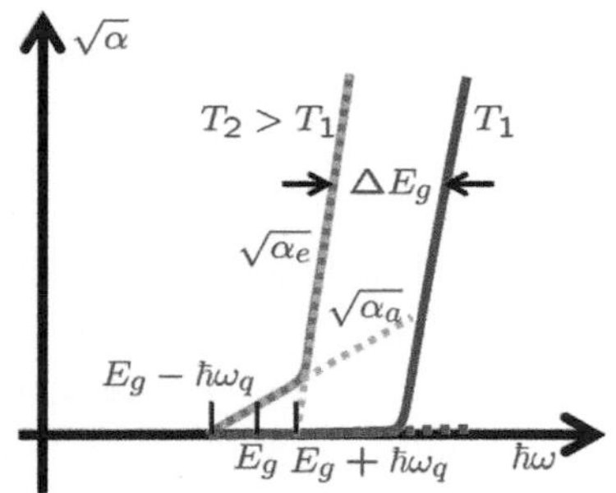

Figure 12.8: The square roots of absorption coefficients in phonon-dominated absorption conditions such as in an indirect semiconductor of low doping. As temperature is lowered, bandgap increases and phonon absorption disappears.

12.1.3.3 *Doping consequences in band-to-band transitions*

DOPING MATTERS in band-to-band transitions through the availability of states to transition from and to. If a material is heavily doped, say, n-type, then the conduction bandedge states are filled, so a band-to-band absorption cannot take place to these states, and absorption must shift to higher energy, where the states are available. This is near the Fermi energy, where filled states and unfilled states are both available. From the Fermi-Dirac distribution, one also sees that the exponential in the denominator implies that, at energies a few ($\sim 4k_BT$) below the Fermi energy, unoccupied states become available for transitions. These conditions are sketched in Figure 12.9. At an energy identified here as $E_{F'} \approx E_F - 4k_BT$, where the wavevector is $k_{F'}$, about 2 % of states are unoccupied, and a band-to-band absorption to these states becomes noticeable. This condition corresponds to a photon energy of

$$\hbar\omega \approx E_g + \frac{\hbar^2 k_{F'}^2}{2}\left(\frac{1}{m_e^*} + \frac{1}{m_h^*}\right). \tag{12.42}$$

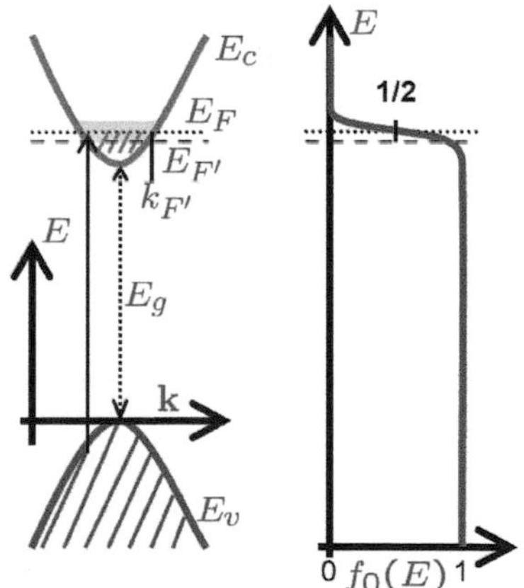

Figure 12.9: Occupation of states in n-type high doping conditions. In the conduction band, noticeable empty states only appear at energies close to $E_{F'} \approx E_F - 4k_BT$, where the Fermi-Dirac distribution is $f_0(E_{F'}) = 0.982$; that is, about 2 % of states unoccupied.

The absorption edge has shifted, a shift known as a *Burstein-Moss shift*. There is also a significant temperature effect in high doping conditions with low impurity density. The phonon population is decreasing, so, for phonon-assisted processes, emission dominates. If one had a pure semiconductor, as we have seen, this will mean an $E_g + \hbar\omega_q$ intercept edge in the $\sqrt{\alpha_e}$ for the absorption coefficient plot, such as in Figure 12.8. When the doping is higher, but bandgap shrinkage and other effects are negligible, then the absorption intercept will shift to a higher photon energy.

Doping effects manifest themselves in multitudinous ways. Heavy doping causes band tails, an effective bandgap shrinkage and a change in state behavior, including distortion of the Bloch function from the idealized representation in the parabolic relationship of Equation 12.42. In indirect semiconductors, with heavy doping, momentum conservation due to electron-electron scattering will also become significant. We will visit the carrier absorption in more detail in the discussion of free carrier absorption (Section 12.2). But, in heavy doping, at these near-bandgap energy conditions, the absorption process can be aided by the momentum change for conservation provided by the density of carriers, so proportionally. This absorption coefficient is proportional to the square of energy and the population, so $\alpha \propto n(\hbar\omega - E_g - E_F)^2$. A plot of $\sqrt{\alpha}$ is linear with the incident photon energy, and the plot shifts proportional to the square root of doping, being a plot of $\sqrt{\alpha(E)}$ with E. See, for example, the doping-dependent behavior of Figure 12.10 for the absorption coefficient for *As*-doped *Ge*.

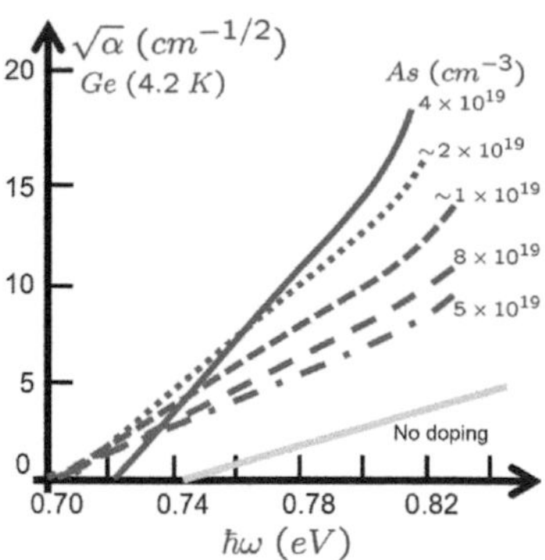

Figure 12.10: Pure, low and high doping absorption edges in a few semiconductors.

12.1.3.4 *Field dependence of absorption*

FIELDS TOO CAUSE A CHANGE IN ABSORPTION behavior. An electric field causes change, since it is a positional band change due to change in the electrostatic potential, and this will affect overlap functions. Magnetic field changes it because the Lorentz-force-induced cyclotron motion is a spatial confinement—a dimensional reduction—and it will cause subbands to form and allow energies of states to change.

First, consider the electric field effect, as shown in Figure 12.11. Since there is an electric field, the Bloch functions that are evanescent in the bandgap increasingly overlap with the field, and this leads to the tunneling between the valence band and the conduction band. Such tunneling occurs through an approximate triangular energy barrier that the electron tunneling from the valence band to the conduction band sees, as depicted by the dashed lines in the structure. Such tunneling becomes important when the ~ 1 *eV* barrier region becomes many *nm*s thick, that is, when there is an electric field of the order of a *MV/cm*. In SiO_2, with its larger barrier energy, it is larger than this. But there is an interesting overlap consequence shown through the upper representation of a conduction band propagating Bloch function. Tunneling with only field involved depends on the overlap shown along the lower representation of the Bloch functions. Now, if a photon can participate, it can provide an excess of $\hbar\omega$ of energy, and the barrier energy that is involved in such a tunneling process is $\sim(E_g - \hbar\omega)$. In the spatial illustration of Figure 12.11, it is a larger overlap with the conduction band Bloch function(s) shown in the

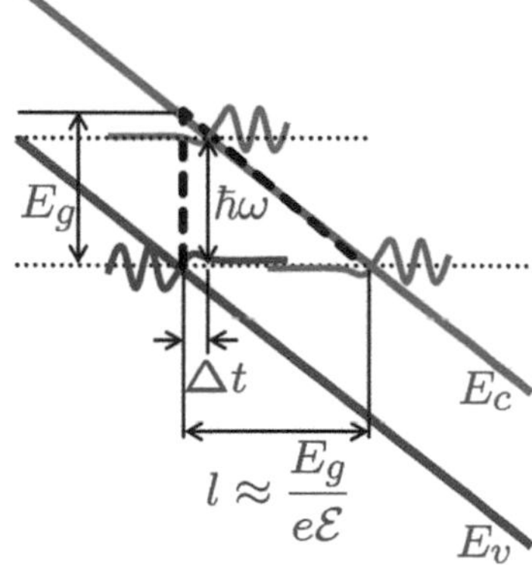

Figure 12.11: Illustration of the Franz-Keldysh effect. In the presence of an electric field, the overlap between the evanescing valence and conduction bands is over a distance that decreases approximately by the electric field and the bandgap. With the participation of a photon, the overlap can be with a spatially closer but higher energy evanescing conduction band wavefunction enhancing the absorption; that is, the transitioning of a valence electron to a conduction band state.

For tunneling, triangular barriers and their various manifestations, including in quantization in the inversion layer of a *MOSFET*, see S. Tiwari, "Device physics: Fundamentals of electronics and optoelectronics," Electroscience 2, Oxford University Press, ISBN 978-0-198-75984-3 (forthcoming).

top region of this figure. From a width of $l \approx E_g/e\mathcal{E}$ for electric-field-based tunneling, over which the evanescing wavefunctions overlap, the region has now been reduced to $\Delta t \approx (E_g - \hbar\omega)/e\mathcal{E}$ in the presence of this field. This can now be much larger.

Tunneling of an electron from the valence band to the conduction band is a transition, and we normally think of it spatially. But it is a transition from one band to another, and therefore also equivalent to an absorption process. An electron from the valence band transitioned to the conduction band. What we have seen with the photon is that this transition was made more efficient. This is the *Franz-Keldysh effect*. It is a photon-assisted tunneling through the energy gap and an enhancement of the absorption process. Higher fields will increase it. Figure 12.12 shows the absorption coefficient in *GaAs* under an electric field, illustrating this Franz-Keldysh effect.

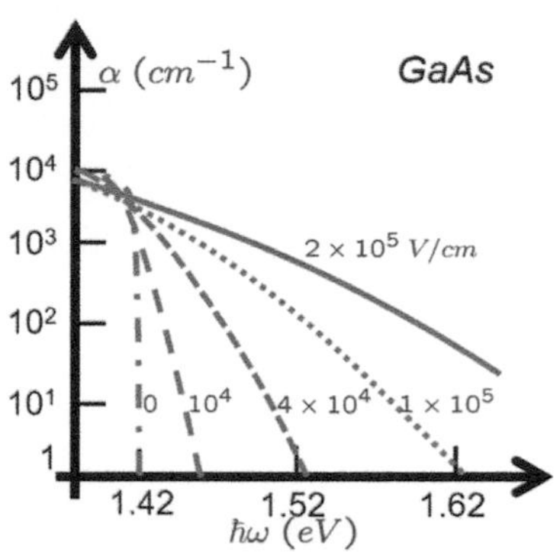

Figure 12.12: The dependence of absorption coefficient on photon energy as a function of the electric field in *GaAs*.

Since the tunneling still occurs across an approximately triangular barrier, the solutions follow from the use of Airy functions. The electroabsorption coefficient is approximately

$$\alpha(E)|_{\mathcal{E}} \approx 1.0 \times 10^4 \frac{f_{osc}}{n} \left(\frac{2m_r^*}{m_0} \right)^{1/3} \int_{\beta}^{\infty} |\text{Ai}(z)|^2 \, dz, \text{ where}$$

$$\beta = 1.1 \times 10^4 \left(\frac{2m_r^*}{m_0} \right)^{1/3} \frac{E_g - \hbar\omega}{\mathcal{E}^{2/3}}. \quad (12.43)$$

Note that, in an indirect semiconductor, the symmetry of the wavefunctions will be different, and, just by the photon participation, the overlap will remain quite small. After all, the momentum matching will be quite poor. Just as for indirect gap absorption, a phonon process will need to be involved, and now this has become a three-carrier process, involving the photon, the phonon and the electron from the valence band. This will have to be inefficient. So, indirect bandgap materials show quite weak electric field dependence.

There may be weak field dependence for the Franz-Keldysh effect in indirect gap materials, but normal photon-free tunneling is quite similar for indirect and direct gap materials. This is because the distances over which this interaction takes place in tunneling is large (l of Figure 12.11), and phonons may participate in the time scales involved. The tunneling is incoherent tunneling.

Now consider the consequence of magnetic field. If scattering were absent, the electron, still a free particle in a nearly free electron gas, has a time response dictated by the Lorentz force of

$$m^* \frac{d\mathbf{v}}{dt} = q\mathbf{v} \times \mathbf{B}, \quad (12.44)$$

where we have assumed an isotropic mass in parabolic approximation. So, the electron orbits around the field with the cyclotron frequency of $\omega_c = eB_z/m^*$ in a radius of $r_c = v/\omega_c$. We encountered this frequency in our Boltzmann transport equation discussion (e.g., see, Equation 9.22). It can be written more generally in terms of the mass tensor. Classically, no limits have been placed on what velocity, momentum or position this electron may have. However, quantum-mechanically, these are prescribed through coherence—integral wavelengths of circumference of cyclotron orbit—that is,

$$2\pi r_c = \frac{nh}{p} = \frac{nh}{m^* v} \quad \forall\, n = 1, 2, \ldots . \tag{12.45}$$

This relates the velocity and the radius via $v = r_c \omega_c = n\hbar\omega_c / m^* v$ for the electron's motion characteristics. The kinetic energy, by extension, is dictated to have discrete possibilities of

$$\frac{m^* v^2}{2} = n \frac{\hbar\omega_c}{2}. \tag{12.46}$$

So, the presence of magnetic field then changes the energies of the allowed states under this quantum-mechanical prescription with its zero point uncertainty to

$$E_n - E_0 = \left(n + \frac{1}{2}\right)\hbar\omega_c = \left(n + \frac{1}{2}\right)\hbar\frac{eB_z}{m^*}. \tag{12.47}$$

This says that the energy bands now form a ladder of subbands. Figure 12.13 shows these Landau levels in the presence of a magnetic field, with all the allowed energies of states shifting according to this equation.

This Landau subband consequence is the first of a series of hierarchy of consequences, including entirely new collective responses, that are discussed in S. Tiwari, "Nanoscale device physics: Science and engineering fundamentals," Electroscience 4, Oxford University Press, ISBN 978-0-198-75987-4 (2017).

The new energy subbands can be written as

$$\begin{aligned} E &= E_{c0} + \frac{\hbar^2 k^2}{2m_e^*} + \hbar\omega_{ce}\left(n + \frac{1}{2}\right) \quad \text{for electrons, and} \\ E &= E_{v0} - \frac{\hbar^2 k^2}{2m_h^*} + \hbar\omega_{vh}\left(n + \frac{1}{2}\right) \quad \text{for holes,} \end{aligned} \tag{12.48}$$

where $n = 0, 1, \ldots$. The effective bandgap has now changed to

$$E_{g,eff}(B) = E_g(0) + \frac{e\hbar B}{2m_{cr}^*}, \tag{12.49}$$

where the reduced cyclotron effective mass m_{cr}^* is the geometric mean of the cyclotron conduction and hole masses that sample the reciprocal space during motion and may be different from the parabolic mass under the more general semiconductor conditions. The absorption edge will shift higher in energy as a result of the application of magnetic field.

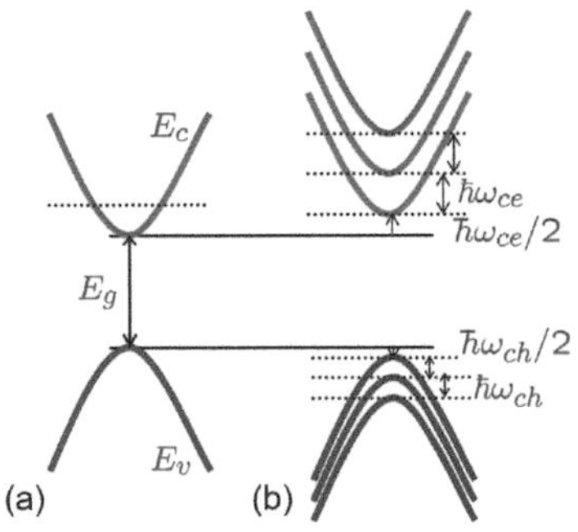

Figure 12.13: Formation of Landau subbands in the presence of a magnetic field. Part (a) shows the E-$\mathbf{k}$ (parabolic bands) without field, and (b) shows the Landau subbands. The Landau level energy is the bandedge energy of these subbands.

12.1.3.5 *Temperature dependence of absorption*

Bandgap changes with temperature. When the heavy doping effects, which we have remarked on before, are absent, so the semiconductor has moderate to low doping, a primary effect is the shrinkage of the lattice when the temperature is lowered. Usually, the consequence of this is an increase in the bandgap, since the state interactions have increased. This is certainly so in group *IV* and *IIIV* compounds. The lower the temperature is, the larger the bandgap will be and hence, again, the absorption edge will shift higher in energy. However, this is not necessarily a universal behavior, In several other semiconductor forms, for example, those consisting

Some of the more important semiconductor properties that we often employ, including some of the temperature dependences, can be found at the end of the glossary.

of the significantly more ionic group *VI* elements, the bandgap can actually shrink with lowering of temperature.

12.2 *Free carrier absorption*

ABSORPTION DOES NOT ENTIRELY DISAPPEAR past the absorption edge at low energies. There are numerous additional absorption mechanisms that will still occur, as noted in Figure 12.3, many due to the lower energy of the interacting mechanism, such as from optically active traps, deep levels, the electron-hole pair that we call an exciton, and so on. But there is one particularly significant one arising in just the conducting nearly free electron still interacting with the photon—an electromagnetic coupling—and a free electron undergoing a kinetic energy change across the energy swath. In the process, the photon energy is absorbed by the electron and the phonon. This is illustrated in Figure 12.14, which shows the electron being accelerated or raised in energy through photon absorption in a phonon-coupled process.

Skin depth, the region over which an electromagnetic signal extinguishes, is precisely because of this absorption, as well as the electromagnetic rearranging at the interface due to the interaction between the electromagnetic wave and the conducting electron cloud under the boundary constraint.

We have explored the frequency-dependent behavior of electrons (Section 10.7) and observed that there is both a real part and an imaginary part in the response. The imaginary part reflects the phase consequences. With electromagnetic waves, we will see similar consequences, because both an electric field and a magnetic field exist simultaneously. Here, one has a static conductivity of $\sigma = e^2 n\langle\tau_{\mathbf{k}}\rangle/m^*$, and its dynamic counterpart $\sigma^* = e^2 n\langle\tau^*_{\mathbf{k}}\rangle/m^*$, with

$$\tau^*_{\mathbf{k}} = \frac{\tau_{\mathbf{k}}}{1 - i\omega\tau_{\mathbf{k}}}. \tag{12.50}$$

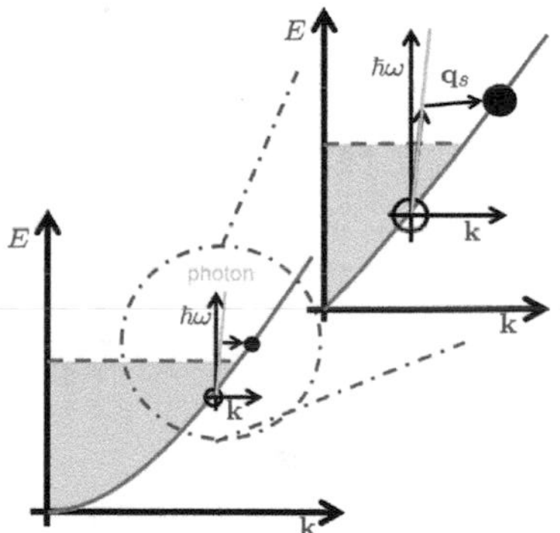

Figure 12.14: Part (a) illustrates the energy-wavevector relative relationship for a photon, with its high group velocity, interacting with an electron in the conducting gas in the conduction band. Part (b) expands this to emphasize the importance of the need for a phonon to make momentum matching possible together with the energy matching.

The expectation over all the variations is then

$$\langle\tau^*_{\mathbf{k}}\rangle = \langle\frac{\tau^*_{\mathbf{k}}}{1+\omega^2\tau^{*2}_{\mathbf{k}}}\rangle + i\langle\frac{\omega\tau^{*2}_{\mathbf{k}}}{1+\omega^2\tau^{*2}_{\mathbf{k}}}\rangle. \tag{12.51}$$

The electromagnetic frequencies that we normally are interested in are in the terahertz-to-petahertz range, that is, the far infrared-to-visible range—extremely high compared to the electronic device frequencies that we were interested in there—so, $\omega\tau_{\mathbf{k}} \gg 1$ is a good assumption, and this leads to

$$\langle\tau^*_{\mathbf{k}}\rangle = \langle\frac{1}{\omega^2\tau^*_{\mathbf{k}}}\rangle + i\frac{1}{\omega}. \tag{12.52}$$

In this analysis, this relaxation process has subsumed in it the phonon involvement necessary for the scattering of the electron to the higher energy when a photon gets absorbed. So, phonons in energy and momentum matching do not appear explicitly.

The conductivity then is

$$\begin{aligned}\sigma^* &= \frac{e^2 n}{m^*\omega^2}\langle\frac{1}{\tau^*_{\mathbf{k}}}\rangle + i\frac{e^2 n}{m^*\omega} \\ &= \sigma^r + i\sigma^i.\end{aligned} \tag{12.53}$$

If we eliminate **H** in Maxwell's equations under conducting conditions, but with no source, we obtain a propagating solution through the electric field equation

$$\nabla^2 \boldsymbol{\mathcal{E}} = \epsilon\mu_0 \frac{\partial^2 \boldsymbol{\mathcal{E}}}{\partial t^2} + \sigma^* \mu_0 \frac{\partial \boldsymbol{\mathcal{E}}}{\partial t}, \tag{12.54}$$

with the traveling wave solution $\boldsymbol{\mathcal{E}} = \hat{\mathbf{a}}\mathcal{E}_0 \exp[i(\mathbf{q}\cdot\mathbf{r} - \omega t)]$.

Note that we are using **q** as the wavevector for electromagnetic waves (as also for phonons, where we sometimes use s subscript selectively to identify phonons, separate from other interfering symbols, **k** being the preferred usage for electrons).

Here,

$$q = \frac{\omega}{c}\left(\epsilon_r + i\frac{\sigma^*}{\epsilon_0\omega}\right)^{1/2}, \tag{12.55}$$

with $c = 1/\sqrt{\epsilon_0\mu_0}$ as the free space speed of light. In free space, the wavevector $q = \omega/c = 2\pi/\lambda$, with λ as the wavelength in free space. But, as Equation 12.55 says, it propagates at a different wavevector and simultaneously extinguishes. This change arises in the square root term. Let us write this as a complex index of refraction,

$$n^{\underline{c}} = \left(\epsilon_r + i\frac{\sigma^*}{\epsilon_0\omega}\right)^{1/2} = n^{\underline{r}} + i n^{\underline{i}}. \tag{12.56}$$

The complex index of refraction leads to the following connections between the real and the imaginary parts:

$$\begin{aligned} n^{\underline{r}2} - n^{\underline{i}2} &= \epsilon_r - \frac{\sigma^{\underline{i}}}{\epsilon_0\omega}, \text{ and} \\ 2n^{\underline{r}}n^{\underline{i}} &= \epsilon_r - \frac{\sigma^{\underline{r}}}{\epsilon_0\omega}. \end{aligned} \tag{12.57}$$

For a *TEM* wave, propagating in the z direction with the electric field oriented in the x direction, the solution is

$$\boldsymbol{\mathcal{E}} = \hat{\mathbf{x}}\mathcal{E}_0 \exp\left(-\frac{\omega n^{\underline{i}} z}{c}\right) \exp\left[i\left(\frac{\omega n^{\underline{r}} z}{c} - \omega t\right)\right], \tag{12.58}$$

where one can explicitly see the attenuation of the amplitude and the change in the wavevector or, equivalently, the wavelength. The free space wavelength has been scaled to $\lambda/n^{\underline{r}}$. The conductivity and the attenuation factor are related through

$$\frac{\omega n^{\underline{i}}}{c} = \frac{\sigma_r}{2n^{\underline{r}}c\epsilon_0} \tag{12.59}$$

and therefore the absorption coefficient, the per unit length measure of loss of energy/flux/number of photons (quantities proportional to the square of the amplitude), is

$$\alpha = \frac{\sigma_r}{n^{\underline{r}}c\epsilon_0}. \tag{12.60}$$

In terms of the momentum relaxation, which is related to the conductivity, we can write the absorption coefficient as

$$\alpha = \frac{e^2 n \lambda^2}{4\pi^2 \underline{n}^r c^3 \epsilon_0 m^*} \langle \frac{1}{\tau_{\mathbf{k}}} \rangle. \tag{12.61}$$

The longer the wavelength is, the more absorption there is. It is easier to absorb a low energy photon through the free carrier absorption.

So, with conducting electrons, we have found that the index of refraction is a complex number $\underline{n}^c$, and the real and the imaginary parts are related through the equation

$$\underline{n}^{r2} = \epsilon_r - \frac{e^2 n}{\epsilon_0 m^*} \omega^2 + \underline{n}^{i2}, \tag{12.62}$$

and herein is the electron cloud that causes the dissipation, and as we had seen before, provides the oscillatory response as a collective ensemble that we called the plasma frequency,

$$\omega_p^2 \equiv e^2 n / \epsilon m^*. \tag{12.63}$$

The electron charge cloud has a dipolar oscillatory response that has appeared for us in absorption because of the scattering that takes place. And the charge cloud will show up in responses in any situation where there is any electromagnetic interaction involved, including, as we have mentioned, in the electromagnetic propagation at the surface through the surface plasmon polariton. We can also write these index relationships through the plasma and the relaxation time properties as

Indeed, what we see here is that semiconductors have this unusual characteristic: the optical properties will have a metal-like behavior at low frequencies and an insulator-like behavior at high frequencies.

$$\underline{n}^{r2} = \epsilon_r \left[1 - \left(\frac{\omega_p}{\omega} \right)^2 \right] + \frac{\epsilon_r^2}{4 \underline{n}^{r2}} \left(\frac{\omega_p}{\omega} \right)^4 \langle \frac{1}{\omega \tau_{\mathbf{k}}} \rangle^2. \tag{12.64}$$

When $\omega \tau_{\mathbf{k}} \gg 1$,

$$\begin{aligned} \underline{n}^{r2} &\approx \epsilon_r \left[1 - \left(\frac{\omega_p}{\omega} \right)^2 \right] \\ &= \epsilon_r - \frac{e^2 n \lambda^2}{4\pi^2 \epsilon_0 m^* c^2}. \end{aligned} \tag{12.65}$$

The real part of the index of refraction is lowered with doping, and hence *heavily doped regions do guide light weakly, even as they attenuate due to the free carrier absorption.*

We had started this chapter with a discussion of the index of refraction as a complex parameter, among others, reflective of the electromagnetic interactions with matter, where electron plasma is one of the causes of the interaction. Our expression of reflectivity (see Equation 12.2) subsumed this. Figure 12.15 shows how this reflectivity appears at the absorption edge, here in *InSb* due to the interaction with the electron plasma. At higher doping shifts, the edge and the longer wavelength photons, that is, the ones with lower energy, interact much more strongly with the electron plasma.

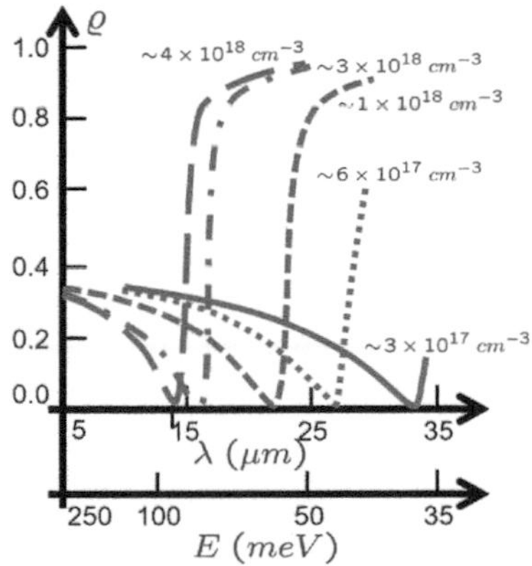

Figure 12.15: Reflectivity in *InSb* for various dopings. *InSb* is a very small bandgap material. Lower energy photons interact strongly with the electron plasma, causing a large change in the index of refraction and increasing reflectivity well beyond the ~30 % of most semiconductors.

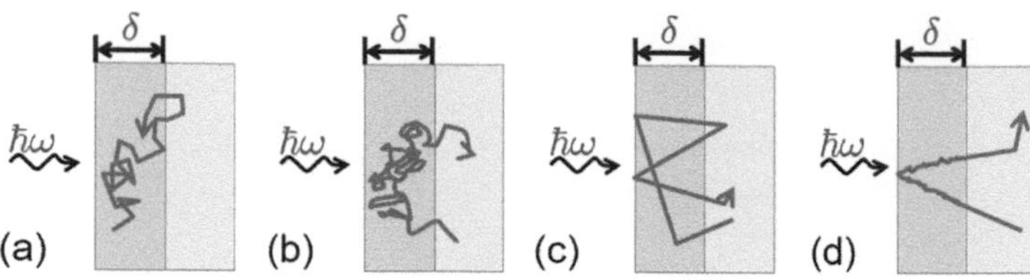

Figure 12.16: Four distinctly different regimes in electromagnetic-electron interaction in the skin layer. Scattering of electrons is shown through sudden changes in the path directions. For a mean free path $\langle\lambda_{\mathbf{k}}\rangle$, a skin depth δ and an electron distance travel in a fraction of an electromagnetic cycle $\ell = \langle v\rangle/\omega$, (a) shows the classical skin effect, such as at infrared and longer wavelengths, including at the low microwave frequencies ($\langle\lambda_{\mathbf{k}}\rangle \ll \delta$, and $\langle\lambda_{\mathbf{k}}\rangle \ll \ell$), (b) shows the conditions where an electron can relax within the skin depth, but, being at higher frequencies, the inertial effects of electrons become important ($\ell \ll \langle\lambda_{\mathbf{k}}\rangle \ll \delta$), (c) shows an anomalous skin depth effect when the mean free paths are very large ($\delta \ll \langle\lambda_{\mathbf{k}}\rangle$, and $\delta \ll \ell$) and (d) shows an extreme anomalous skin effect when $\ell \ll \delta \ll \langle\lambda_{\mathbf{k}}\rangle$.

The phenomenon of skin depth needs a few additional remarks in light of what we have learned about electron dynamics and the interaction of light with the electrons.

Figure 12.16 shows four different situations for metals or even high doped or inversion layers where electrons and not crystal oscillations dominate. The description is based on the scaling dimensions of the mean free path $\langle\lambda_{\mathbf{k}}\rangle$, the skin depth δ and the distance traveled by the electron in a $1/2\pi$ fraction of the cycle ($\ell = \langle v\rangle/\omega$).

Figure 12.16(a) reflects what one observes in metals at infrared and longer wavelengths, such as the short frequencies of microwaves in *GHz*s. This is when $\langle\lambda_{\mathbf{k}}\rangle \ll \delta$ and $\langle\lambda_{\mathbf{k}}\rangle \ll \ell$, the electron is largely moving around with scattering-dominated motion, and many scattering events happen within one cycle of light in the skin depth region. It is a local and a near instantaneous response between the fields and the currents. As the frequency is increased further, as in Figure 12.16(b), the electron's inertia becomes important. The electron is undergoing collisions, but the light wave is also rapidly oscillating, causing damping. It is a classic case of fluctuation-dissipation in the electromagnetic response. The electron response lags behind that of the light, with the phase lag approaching $\pi/2$ in the case of extreme relaxation. Electron scattering is not as important, and the electron responds to the oscillating field, undergoing occasional collisions. Electrons now screen the field. This is why the reflection coefficient becomes high. Absorption drops because of the phase lag. Figure 12.16(c) is an anomalous region where the mean free path is high. Examples in metals are the use of low temperature and in semiconductors the high mobility inversion layers. Figure 12.16(d) is a region of anomalous reflection. This is a bit like the case in Figure 12.16(b), but the scattering is less. Now the surface scattering dominates.

12.3 *Excitons, and absorption by excitons*

THE NEARLY FREE ELECTRON PICTURE is an excellent model for many purposes. But there is a large variety of conditions where one observes consequences of two seemingly different particles as carrier of energies, coupling with energies and momentum quite close, and so behaving together in a form that is much easier to think of as a new quasiparticle. An electromagnetic field coupled to an electron has just been treated as two different forms of particles as carriers of energies. But, on an interface, such as a dielectric with a high density of electrons, the boundary, the boundary's compulsions reflected in the constraints on Maxwell's equations, and the consequences of a conductive electron cloud make it possible for a surface channel to form that can also propagate, with the electromagnetic wave/photon and the electron cloud acting together in unison. This is an example of a *plasmon polariton*. All the different forms in which energy is carried by particles, when looked at in sufficient detail, will show conditions where the energy carried by the particle couples to other energy forms to make a new quasiparticle, which is a good way of representing the response. We will see some of these later.

When, in band-to-band absorption, an electron state in the valence band is emptied and one in the conduction band is filled, one now has made a transition in energy, a change of state, while the localization of the particles is still determined by uncertainty. An electron left behind a hole. How does this electron interact with this hole or the sea of electrons that has settled in the valence band? We had made remarks regarding correlation holes in Section 1.2, and Coulomb holes in Section 1.7, during our discussion of Hartree and Hartree-Fock methods. These are very important in high carrier concentration systems, since exchange, with electrons as fermions, is very central to the filled states of a Fermi liquid or in atomic orbitals. In a Fermi gas, to the first order, just considering Coulomb interaction between an electron and a hole suffices. The electron now appears as one electron in a sea, where the sea's effect is a periodic perturbation. And this Coulomb interaction will depend on the permittivity. For semiconductors, $\epsilon_r \approx 10$, but, for organic semiconductors, $\epsilon_r \approx 4$ and, for high permittivity materials, $\epsilon_r \gtrsim 10$; each material has a different attractive force between electrons and holes. The hole, in this view, is a valence band particle. Only when both holes and electrons were in large numbers did we have to worry about it, but, given the extent of the spatial distances and these attractions, one does need to treat it as a higher-order effect.

How shall we treat many-body effects, such as an electron and its local environment, when the electron is generated by the photon absorption process? The electron has a zone surrounding it in this interacting system that reduces the probability of finding other electrons within the vicinity and thus has created this hole that is attracted to it. So, Coulomb interactions and exchange both matter. For the Fermi gas picture, just considering the electron-hole pair as a correlated quasiparticle is sufficient for our interests. This is the *exciton*.

Attraction between the electron and the hole make their motion correlated, as is represented in this combined assembly of the exciton. We have seen that the effective Bohr radius is of the order of a few *nms* to 10 *nm* in inorganic semiconductors. When confined in a quantum well, we increase the attraction, through pushing the particles together, of the interacting pair—the exciton. This picture is somewhat similar to that of an electron and a donor, where the ionization energy will change due to the confining, except that the exciton is free to move.

An exciton itself is uncharged. *It diffuses; it does not drift.* It is a quasiparticle with a center-of-mass motion and relative motion within it. The exciton as an electron-hole pair has an energy lower than that of the bandgap by the Coulomb attraction energy. And a photon can cause it to recombine, so one sees an excitonic recombination, as was illustrated in Figure 12.3. It can also be broken apart by perturbation. A measure of the strength of the exciton is the comparison between the lowering of energy in its formation compared to thermal energy. If $k_BT < E_{xn}$, the exciton energy then will be important. If strong forced confinement of electrons and holes exists, then it too will be important.

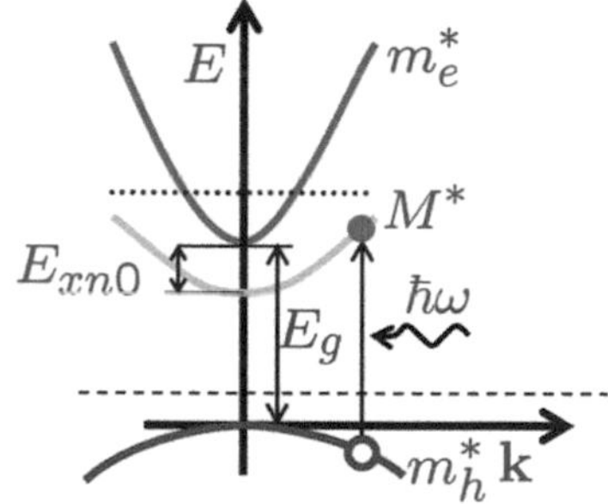

Figure 12.17: An exciton being created as a result of recombination.

Similar to the donor-electron hydrogenic calculation, one may view the ionization energy of such a pair scaled from the Rydberg energy as

$$E_{xn0} = -13.6\left(\frac{1}{n\epsilon_r}\right)^2 \frac{m_r^*}{m_0}\ eV \qquad (12.66)$$

arising in an effective radius of electron-hole separation of $r_{xn} = 0.053n^2\epsilon_r m_0/m_r^*\ nm$. Here, the relative effective mass of the two-body system (m_r^*) is the geometric mean of the effective electron mass and the hole mass. Figure 12.17 illustrates its nature within this electron-hole framework. This exciton is free to move in the crystal, with a dispersion relationship of

$$E_{xn}(\mathbf{k}) = E_{xn} = \frac{\hbar^2 k^2}{2M^*}, \qquad (12.67)$$

where the energy reference is the conduction bandedge, and $M^* = m_e^* + m_h^*$.

See S. Tiwari, "Quantum, statistical and information mechanics: A unified introduction," Electroscience 1, Oxford University Press, ISBN 978-0-198-75985-0 (forthcoming), for detailed remarks on hydrogenic donor or acceptor energies, as well as our discussion of shallow hydrogenic dopants and Wannier functions in Chapters 4 and 7.

This picture suffices for understanding the two-particle picture of a photon interacting with excitons, but now we need to introduce the collective behavior aspect of the photon-exciton coupled state, because the transformation resulting in a photon interacting in the semiconductor creating an exciton needs to be described with energy and momentum being conserved. Conservation of energy means that the creation of an exciton from a photon must occur where the dispersion curves of the photon and the exciton intersect. The two are uncharged—but not quite, since, in perturbation, it is the higher-order interaction arising in the photon as an electromagnetic particle, and the exciton as an electric particle that is polarizable that leads to a weak interaction. And we have seen how degeneracy is lifted in weak interaction. This new creation is the exciton polariton, and Figure 12.18 shows this degeneracy lifting. Recall that the photon has a high velocity, so the dispersion curve is linear and sharply rising ($E = \hbar c\mathbf{q}$), and the exciton is much broader.

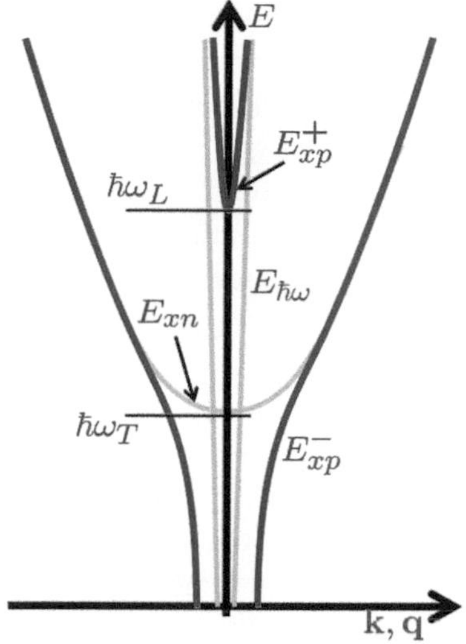

Figure 12.18: The dispersion curves for a photon and an exciton, that is, as bare particles, and the creation of the two branches of the exciton polariton through the bare particles' interaction.

Polaritons are made up of an electromagnetic wave coupled to some form of polarization (excitons, here) to form a new quasiparticle, and the dispersion curve shows how this new quasiparticle's energies E_{xp}^{+} and E_{xp}^{-} change with the wavevector. The degeneracy of the photon and the exciton has now been broken. We also see this electromagnetic wave coupling to other polarizations, for example, those with mechanical oscillations of the phonons.

Figure 12.19 shows the absorption edge of *GaAs* as the temperature is lowered. Note the shift to higher bandgap, as expected from the discussion in Subsection 12.1.3, and the appearance of the exciton absorption peak modifying the absorption edge. The absorption coefficient for allowed transitions is modified in the presence of exciton absorption to

$$\alpha \approx 2.7 \times 10^5 \left(\frac{2m_r^{**}}{m_0}\right)^{3/2} \frac{f_{osc}}{n_r^r} \left(\hbar\omega - E_g - E_{xn}\right)^{1/2} \; cm^{-1}, \qquad (12.68)$$

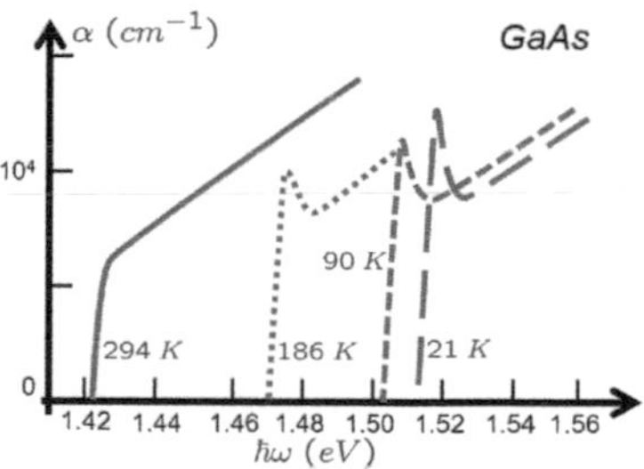

Figure 12.19: The absorption coefficient of *GaAs* as temperature is lowered. This increases the bandgap, shifting the absorption edge but also showing the peaking due to excitonic absorption.

where m_r^{**} is now the geometric mean of the exciton effective mass and the hole effective mass. In detail, one will observe in the absorption behavior consequences due to phonon participation also. These usually appear with thresholds.

12.4 *Absorption by crystal vibrations*

PHOTONS CAN ALSO DIRECTLY COUPLE to the crystal's vibrations—phonons—because atoms are both ionic and polarizable. Such a coupling clearly will be stronger with increasing ionicity. *Also, since the electromagnetic waves are TEM, they will strongly couple to transverse optical phonons, not longitudinal optical phonons, even though both these forms have polarization.* The consequence of this is that the dielectric function, or other forms of electromagnetic characteristics, will show interesting features between the transverse and the longitudinal optical phonon frequencies, in particular, a high reflectivity. Equation 12.55 is the wavevector of the solution of the transmitting wave in a medium characterized by a complex dielectric function, a complex index of refraction or other, similar, interdependent functions.

We consider a two-atom basis primitive cell. Let the displacement around a mean position be of the form

$$\delta\mathbf{u} = \hat{\mathbf{b}}\delta\breve{u}_{1,2}\exp\left[i(\mathbf{q}_s\cdot\mathbf{r}-\omega_s t)\right] \tag{12.69}$$

for the vibrations. Conservation of energy implies $\omega_s = \omega$; conservation of energy means $q_s = k \approx 0$, since photons have a very sharp dispersion curve. This means that zone center phonons are of interest. Let k_s be the "spring" coupling coefficient between the ions. The relative displacement equation of motion (recall our phonon dispersion solution in Section 3.11) can be written as

We are using $\delta\mathbf{u}$ to emphasize the movement around the mean with the amplitude magnitude $\delta\breve{u}_{1,2}$. We have brought in the s subscript again to distinguish charge from phonons. The photons are subscript-less.

$$-\omega^2\delta\mathbf{u} = -\frac{k_s}{2M_r}\delta\mathbf{u} + \frac{e^*}{M_r}\boldsymbol{\mathcal{E}}'. \tag{12.70}$$

Here, M_r is the geometric mean of the two atomic masses, and e^* is the effective ionic charge. $\boldsymbol{\mathcal{E}}'$ is a local electric field in the crystal when the incident photon electric field is $\boldsymbol{\mathcal{E}}$. This local field must be different, since the crystal is ionic and has a polarizing environment. The solution for displacement is

$$\delta\mathbf{u} = \frac{e^*/M_r}{\omega_0^2-\omega^2}\boldsymbol{\mathcal{E}}', \text{ with } \omega_0^2 \equiv \frac{k_s}{2M_r}. \tag{12.71}$$

The photon-free phonon response, which was harmonic, in the presence of an electromagnetic field now appears as a forced response oscillating at the photon frequency but with an amplitude that has a Lorentzian lineshape. The more the difference between ω, the electromagnetic frequency, and ω_0, which is related to the phonon frequency, the less is the displacement perturbation from the electromagnetic field. $\omega = \omega_0 = \sqrt{k_s/2M_r}$ defines a resonance condition where the coupling is the strongest and the electromagnetic field causes the largest changes in displacement. The dipolar field arising in the polarization of the crystal will interact with the electromagnetic field around this resonance condition. So, what this simple relational analysis shows is that when electromagnetic fields have frequencies in the range of the crystal vibrational frequency (30 $meV/\hbar$, i.e., fractions of a PHz, which is far infrared), strong interaction should be expected.

From the Maxwellian electromagnetic viewpoint, the difference between vacuum and matter is that matter contributes electric and magnetic polarization because of its composition. As electrons, nuclei, charge and spin respond to the electromagnetic field in the material, the material has a different dielectric function or permeability function than that of vacuum. Ionic polarization is because of the ionic dipole due to interatomic charge transfer. Electronic polarization is because of the nearly free electron charge cloud. Atomic polarization is because of the atomic distortion arising in the dressed nuclear dipole.

Now consider the polarization contribution from the ions,

$$\mathbf{P}_i = \frac{Ne^*}{\Omega}\delta\mathbf{u} = \frac{Ne^*/M_r\Omega}{\omega_0^2 - \omega^2}\boldsymbol{\mathcal{E}}'. \tag{12.72}$$

N here is the number of unit cells in the volume Ω. The ionic polarization arises in the transfer of charge from one atom to another as part of chemical bonding, leading to an oscillating dipole as the ions oscillate. The ion itself is composed of the nucleus and the surrounding core electrons, what we call atomic cores. The dielectric function, that is, the connection between the displacement and the electric field, arises in this and other polarization interactions arising in the material. The displacement is

$$\mathbf{D}(\omega) = \epsilon(\omega)\boldsymbol{\mathcal{E}} = \epsilon_0\boldsymbol{\mathcal{E}} + \mathbf{P}(\omega). \tag{12.73}$$

At low frequencies, $\epsilon(0) = \epsilon_r(0)\epsilon_0$, where both atomic motion and ionic motion respond in inorganic semiconductors (as do electrons, which we ignore for now). At high frequencies, $\epsilon(\infty) = \epsilon_r(\infty)\epsilon_0$, and the ionic response will vanish, since the displacement response to the electromagnetic field vanishes. But the atomic polarization, the part arising in the displacement of the dressed nuclear dipole—the atomic polarizability—still remains. We refer to this remnant polarization at high frequencies $\mathbf{P}(\infty)$ as the background. We can write polarization at any frequency as

So, even at the highest frequencies, the material does not appear as vacuum; it still has atomic polarization. Light will still move at a lower speed than it does in vacuum, as it interacts with this polarization as it propagates.

$$\mathbf{P}(\omega) = \mathbf{P}_i(\omega) + \mathbf{P}(\infty), \tag{12.74}$$

and the polarization at the very high frequencies as

$$\mathbf{P}(\infty) = [\epsilon(\infty) - \epsilon_0]\boldsymbol{\mathcal{E}}. \tag{12.75}$$

Now, we may write the local field $\boldsymbol{\mathcal{E}}_{local}$, which is being modified by the polarization of the crystal. This electric field is the sum of the external field together with the field contribution arising in the material's dipoles. This calculation is not trivial, but we can make a suitable approximation with justification that the dipoles are all parallel and arranged on a cubic lattice. As shown in Figure 12.20, we treat a small region around the locale as being local, and the region outside as the rest consisting of uniform polarization. The sphere needs to be large enough to allow the averaging. The sum of the dipole field inside the sphere can now be treated as being at its center, and one can calculate the effect of the outside. Since the volume varies as $(4/3)\pi r^3$ and the surface area as $4\pi r^2$, the local field is

$$\boldsymbol{\mathcal{E}}_{local} = \boldsymbol{\mathcal{E}} + \frac{1}{3\epsilon_0}\mathbf{P}. \tag{12.76}$$

If we write each atom as contributing χ_a of susceptibility, that is, each atom's polarization is related as $\mathbf{p} = \epsilon_0\chi_a\boldsymbol{\mathcal{E}}_{local}$, with N atoms per unit volume, then

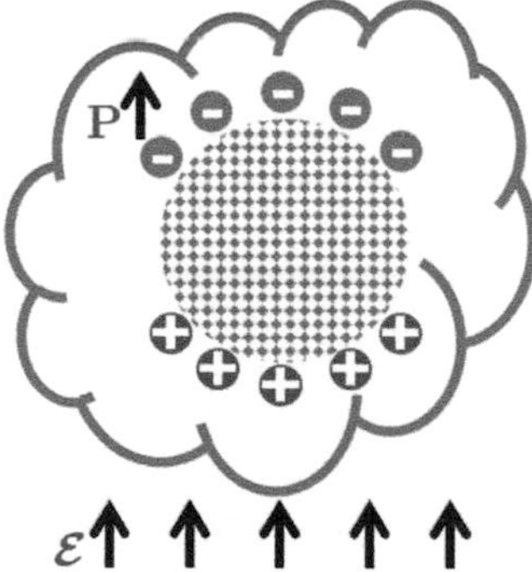

Figure 12.20: The estimation of field at any locale due to the surrounding dipoles. One may draw a sphere large enough to calculate the consequence of nearby dipoles treated as those inside the sphere, and farther dipoles outside the sphere. This outside region is treated as being of uniform polarization.

$$\mathbf{P} = N\epsilon_0\chi_a\mathcal{E}_{local} = N\epsilon_0\chi_a\left(\mathcal{E} + \frac{1}{3\epsilon_0}\mathbf{P}\right) = (\epsilon_r - 1)\epsilon_0\mathcal{E}. \qquad (12.77)$$

Rearranged, the relationship between the relative dielectric constant and the atomic polarizability is

$$\frac{\epsilon_r - 1}{\epsilon_r + 2} = \frac{1}{3}N\chi_a. \qquad (12.78)$$

This is the *Claussius-Mossotti relationship.*

From Equation 12.74, it follows that

$$\epsilon(\omega)\mathcal{E} = \frac{Ne^*/M_r\Omega}{\omega_0^2 - \omega^2}\mathcal{E} + \epsilon(\infty)\mathcal{E}, \qquad (12.79)$$

with

$$\epsilon_r(\omega) = \epsilon_r(\infty)\left(1 + \frac{\omega_1^2}{\omega_0^2 - \omega^2}\right), \text{ and}$$
$$\epsilon_r(0) = \epsilon_r(\infty)\left(\frac{\omega_0^2 + \omega_1^2}{\omega_0^2}\right), \qquad (12.80)$$

where

$$\omega_1^2 = \frac{Ne^{*2}}{M_r\Omega\epsilon(\infty)} \qquad (12.81)$$

and $\omega_0^2 = k_s/2M_r$, as derived earlier.

The interaction between the electromagnetic field and the crystal's vibrations leads to a dielectric response function that has a resonance part arising from frequency matching between the crystal's natural oscillations and a frequency that is connected to the ionicity, the masses and the atomic polarizability.

Now consider transverse optical phonons, the ones that couple efficiently for *TEM* waves,

$$\omega_{TO}^2\delta\mathbf{u}_{TO} = \omega_0^2\delta\mathbf{u}_{TO} - \frac{e^*}{M_r}\mathcal{E}. \qquad (12.82)$$

The curl of a transverse displacement $\nabla \times \delta\mathbf{u}_{TO} = i\mathbf{q} \times \mathbf{u}_{TO} \neq 0$, since the transverse displacement is perpendicular to the propagation direction. But $\nabla \times \mathcal{E} = 0$. Therefore, $\omega_{TO}^2 = \omega_0^2$. *The transverse phonons are not affected by the electric field.*

Let us now calculate the electromagnetic coupling to the longitudinal oscillations. We have

$$\omega_{LO}^2\delta\mathbf{u}_{LO} = \omega_0^2\delta\mathbf{u}_{LO} - \frac{e^*}{M_r}\mathcal{E}$$
$$\therefore\ \omega_{LO}^2\nabla\cdot\delta\mathbf{u}_{LO} = \omega_0^2\nabla\cdot\delta\mathbf{u}_{LO} - \frac{e^*}{M_r}\nabla\cdot\mathcal{E}$$
$$= \omega_0^2\nabla\cdot\delta\mathbf{u}_{LO} - \frac{e^*}{M_r}\times\left[-\frac{Ne^{*2}}{\Omega\epsilon(\infty)}\right]\nabla\cdot\delta\mathcal{E}, \qquad (12.83)$$

and since the divergence $\nabla\cdot\delta\mathbf{u}_{LO} = i\mathbf{q}\cdot\mathbf{u}_{LO} \neq 0$, because displacement and propagation are dimensionally aligned, and $\nabla\cdot\mathbf{D} = 0$, since there is no spontaneous polarization or external charge, we can simplify. We derived the displacement in Equation 12.71, which we employ for

$$\begin{aligned}\mathbf{D} &= \frac{Ne^*}{\Omega}\delta\mathbf{u}_{LO} + \epsilon(\infty)\boldsymbol{\mathcal{E}}\\ \therefore\quad \nabla\cdot\boldsymbol{\mathcal{E}} &= \nabla\cdot\left[\frac{\mathbf{D}}{\epsilon(\infty)} - \frac{Ne^*}{\Omega\epsilon(\infty)}\delta\mathbf{u}_{LO}\right]\\ &= -\frac{Ne^*}{\Omega\epsilon(\infty)}\nabla\cdot\delta\mathbf{u}_{LO}.\end{aligned} \tag{12.84}$$

Equation 12.83 has now been reduced to

$$\omega_{LO}^2 = \omega_0^2 + \omega_1^2. \tag{12.85}$$

While the TO oscillation frequency is unchanged at ω_0^2, the LO oscillation frequency is higher. This gap between the two is related to the effective ionic charge, which is related to ionic polarizability, and residual atomic polarization. The dielectric functions are now relatable as

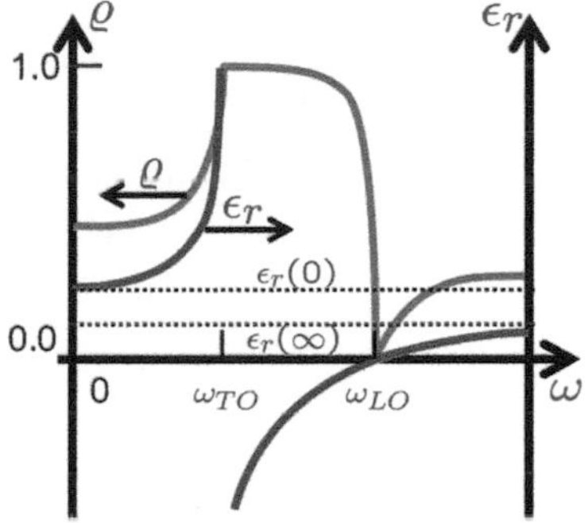

Figure 12.21: Electromagnetic frequency dependence of the relative dielectric constant (red lines) and the reflectivity (blue line) of semiconductors around the optical phonon frequencies.

$$\begin{aligned}\frac{\epsilon_r(0)}{\epsilon_r(\infty)} &= \left(\frac{\omega_{LO}}{\omega_{TO}}\right)^2, \text{ and}\\ \epsilon_r(\omega) &= \epsilon_r(\infty)\left(\frac{\omega_{LO}^2 - \omega^2}{\omega_{TO}^2 - \omega^2}\right).\end{aligned} \tag{12.86}$$

Figure 12.21 shows the dielectric and reflectivity responses as a function of electromagnetic frequency around the optical phonon frequencies. Note that, around the ω_{TO}, the permittivity diverges and changes sign. This is the natural response of resonance where the phase changes sign, and it reflects Equation 12.86's frequency dependence. The permittivity is negative for $\omega_{TO} < \omega < \omega_{LO}$, and this is a region of high reflectivity, following Equation 12.2. Beyond ω_{LO} frequencies, phonon interaction vanishes and only atomic polarization effects remain. This region is known as *restrahlen*, or more properly, reststrahlen, that is, residual radiation in German.

By the way, in a metal, at low frequencies, the dielectric constant is complex, with a high imaginary part due to the plasmon resonance. At high frequencies, the dielectric constant usually becomes negative. When negative, just as in the $\omega_{TO} < \omega < \omega_{LO}$ range of semiconductors, the reflectivity becomes very high. At even higher frequencies, the dielectric constant becomes positive, and metals look like a "dielectric."

These analytic relationships were derived without any damping. One would expect some—nothing is lossless—and there are plenty of anharmonicities and other couplings involve. If included, a useful phenomenological relationship is

$$\epsilon(\omega) = \epsilon(\infty) + \frac{\epsilon(0) - \epsilon(\infty)}{1 - (\omega^2/\omega_{TO}^2) - i(\gamma\omega/\omega_{TO}^2)}, \tag{12.87}$$

where γ is a damping factor and, again, $\omega_{TO} = \omega_0$.

Silicon is covalent, therefore bereft of ionicity, and hence $e^* = 0$. So, Si does not show a similar restrahlen and high reflectivity window behavior, unlike compound semiconductors, which have varying degrees of ionicity. A practical example of this behavior,

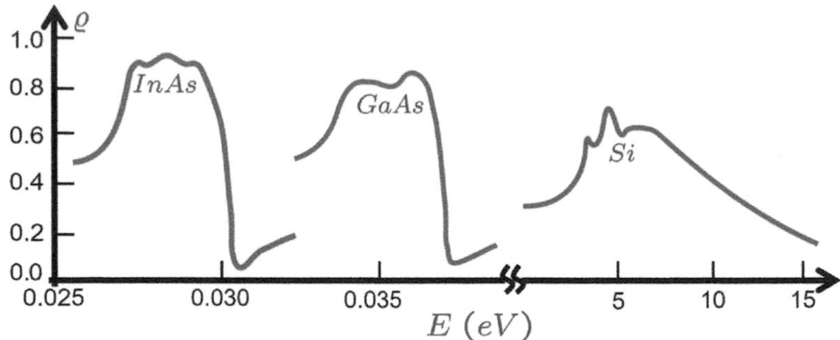

Figure 12.22: Reflectivity of Si, together with that of two compound semiconductors ($InAs$ and $GaAs$).

comparing Si and two compound semiconductors, is shown in Figure 12.22.

Light that is not reflected is transmitted into the semiconductor and undergoes absorption. This will now have coupling to the various atomic modes. We will not dwell on it; suffice to say that absorption will show signatures of efficient or poor coupling to these modes and will therefore be dependent on the wavevector. Figure 12.23 shows the absorption coefficient of pure Si at low energies.

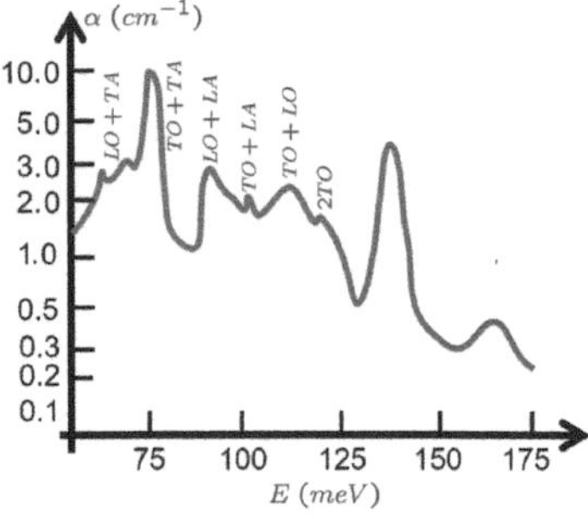

Figure 12.23: Absorption coefficients due to crystal vibrations in pure Si at low energies.

We conclude this section with an important parameter extraction. How do we determine the effective ionic charge e^*? The reflectivity measurement accurately determines ω_{TO} and ω_{LO}, at least for the compound semiconductors, and the difference of the squares of these frequencies is proportional to the effective ion charge squared, that is,

$$\omega_1^2 = \frac{Ne^{*2}}{M_r\Omega\epsilon(\infty)} = \omega_{LO}^2 - \omega_{TO}^2, \tag{12.88}$$

and since $\omega_{LO}^2/\omega_{TO}^2 = \epsilon_r(0)/\epsilon(\infty)$, it follows that

$$e^{*2} = \frac{M_r\Omega}{N}\omega_{LO}^2\epsilon^2(\infty)\left[\frac{1}{\epsilon(\infty)} - \frac{1}{\epsilon(0)}\right]. \tag{12.89}$$

$M_r\Omega/N = (M_r/\Omega_0)(\Omega/N)\Omega_0 = \rho(\Omega/N)\Omega_0$, and therefore

$$\begin{aligned} e^* &= \Omega_0\epsilon(\infty)\omega_{LO}\rho^{1/2}\left[\frac{1}{\epsilon(\infty)} - \frac{1}{\epsilon(0)}\right]^{1/2} \\ &= \Omega_0\omega_{TO}\rho^{1/2}[\epsilon(0) - \epsilon(\infty)]^{1/2}. \end{aligned} \tag{12.90}$$

Knowing the low-frequency and high-frequency permittivity, density and one of the optical phonon frequencies, it is possible to derive the effective ionic charge for any semiconductor whose primitive cell volume is known. We have encountered this relationship before (although without proof), in Equation 10.42, and we will encounter this relationship again, since this effective charge

Unusual situations also arise in semiconductors through defect complexes. GaP is an indirect bandgap material. But when doped with oxygen for n-type doping, and zinc for p-type doping, Zn-O forms next site complexes that are decently luminescent centers. Some of the earliest light-emitting diodes were based on such luminescence—an indirect material luminescing! Of course, they had poor lifetime, since light of large-enough energy also has the propensity to create defects. This is an important degradation phenomena in lasers at surfaces, where defects are easier to generate and light of high intensity is undergoing interactions. Getting light from indirect material is almost—but not quite—like an attempt at making perpetual machines. There are plenty of interesting stories of hope regarding luminescence from porous silicon and of hot electron germanium lasers.

has important implications for charge transport at interfaces and other places in nanoscale devices.

12.5 *Absorption by impurity states*

TRANSITIONS BETWEEN IMPURITY STATES AND PROPAGATING STATES via photon-mediated electromagnetic coupling are also sources of absorption, albeit with weak and low energies. They are weak since the states have a small matrix element with quite strong dissimilarities in the wavefunctions and the localization of the impurity states. Nevertheless, particularly at low temperatures, and with careful observations, donor states, acceptor states and propagating conduction band and valence band states can be seen to be present. For impurity transitions, occupation of states, availability of empty states and the matrix element coupling the two, integrated over all possibilities is again the measure of absorption coefficient. It has the form

$$\alpha_{if} \propto N_I f_i (1 - f_f) \frac{(\hbar\omega - \Delta E_{fi})^{1/2}}{\hbar\omega}, \tag{12.91}$$

where N_I is the state density that is occupied, f_i is the occupation factor from which the transition happens, and f_f is the occupation factor of the states being transitioned to. One is assuming in writing this relationship that one of the two—occupied or unoccupied—is the rate-limiting step of the transition, together with the coupling matrix element. The transition is proportional, in a way similar to that for allowed direct transitions, to the square root power in energy, with ΔE_{if} replacing the bandgap E_g. Consider the valence band transition to the empty donor state—a large energy change transition—in an n-type semiconductor, following Equation 12.91. The valence band is largely filled, so $f_i \approx 1$, and the donor states are largely empty, so $N_I(1-f_f) = N_D^+$; therefore, the absorption coefficient for this situation is

$$\alpha_{vd} \propto \frac{N_D}{1 + g_D \exp[(E_F - E_D)/k_B T]} \frac{[\hbar\omega - (E_D - E_v)]^{1/2}}{\hbar\omega}. \tag{12.92}$$

Figure 12.24 shows, at low temperatures, several of these transitions in phosphorous-doped *Si*.

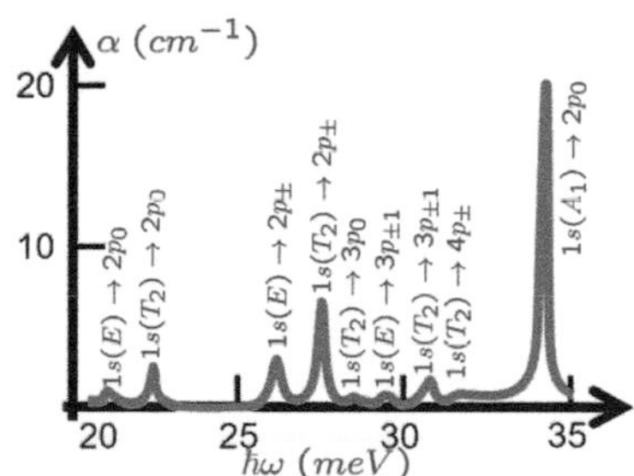

Figure 12.24: Absorption between conduction band and phosphorous donor states in *Si* that is doped low. Note that one observes not one state but several states of the donor.

12.6 *Luminescence*

LUMINESCENCE IS THE RADIATIVE EMISSION from solids. For semiconductors, there are two particular forms: *electroluminescence*, which is the emission due to electrical injection of carriers, and *photoluminescence*, which is the re-emission of light following absorption of light at higher energy. Since emission is an energy relaxation mechanism, it is more complicated than absorption, because many relaxation mechanisms are possible simultaneously. The spectrum of emission will be affected by the source function, how the carriers are distributed in energy—thermal or otherwise—the emission rates, et cetera. We have already discussed numerous non-radiative recombination processes in Chapter 11. These will decrease the fraction of recombination that is radiative. Our discussion in Appendix K also shows that the Einstein A coefficient, which will be related to the radiative lifetime, is also proportional to the Einstein B coefficient. This means that absorption and emission probabilities are related. Transitions that efficiently absorb also efficiently emit, except the source and sink functions determine which of these is preponderant off-equilibrium.

In photoluminescence, the injected carriers generated by above-bandgap light spread in energy and recombine through the various energy relaxation pathways via direct band-to-band emission, but also through some of the complements of the other absorption processes we have discussed. As such, photoluminescence is a good tool for understanding the quality of a material, and the radiative and non-radiative processes that exist in it.

In electroluminescence, efficient introduction of carriers that can then efficiently radiatively recombine is of immense interest. So, all these luminescence discussions naturally gravitate toward population densities, carrier distribution, competing non-radiative processes and efficient ways to make luminescence happen in devices. This is a subject by itself, so we will not dwell further on luminescence. But it should be clear by this point that luminescence in all its forms will be more efficient in direct gap materials, where electrons and holes can couple efficiently, compared to indirect gap materials, where phonons become necessary.

12.7 *Summary*

Light's interaction with matter is through the electromagnetic interaction with the oscillating ionic dipole or other free charge (electronic). The verity of this is in the dielectric function, which, as a function of frequency, shows the consequences of polarization and plasmonic interaction in the electromagnetic interaction response.

In the interaction with and between bands, the transfer of an electron from one state to another, the perturbation Hamiltonian is a term proportional to $(iq\hbar/m_0)\mathbf{A}\cdot\nabla_\mathbf{r}$. Conservation of energy and momentum must still hold, and since the photon has a very low wavevector, the crystal momentum of the initial and final states are pretty identical. Note here the mass is m_0, for obvious reasons. An electron just morphs from one state to another, due to a rapid, photon-coupling-induced transition and a change in the modulation function. It is the $|k\rangle$ of the electron that reflects its presence in the crystal. And when changing to its new state, it doesn't move around in the crystal. No crystal momentum change takes place. This form, due to the $\mathbf{A}\cdot\nabla_\mathbf{r}$ term, when coupling a state $|\mathbf{k}\rangle$ to another $|\mathbf{k}'\rangle$, leads to a term that is aligned with the vector potential and the gradient of the periodic modulation function of the Bloch state, and another term that is out of phase and aligns with the electron wavevector. Even though both these terms will cause transitions, the first term is, for historic reasons, called an "allowed" transition, and the second is called a "forbidden" transition. The allowed transition is a dipole-vector potential coupling and is caused by a momentum matrix element dependent on the gradient of the periodic modulation function. The forbidden transition element arises in the overlap of the periodic modulation functions. In direct bandgap semiconductors, the first term is large and has a recombination rate that varies as the square root of the photon energy separation from the bandgap, and a number of other parameters, including oscillator strength. The forbidden transition has a 3/2 power dependence, together with the overlap integral, and no oscillator strength, since this transition is not mediated by the momentum matrix element. The nature of transitions can be surmised from the bandedge absorption characteristics.

Transitions also happen with phonon assistance, and these too have allowed and forbidden behaviors. Phonon assistance is an additional energy and momentum conservation constraint in a process where virtual states in the conduction band or the valence band participate in enhancing the rate. For indirect semiconductors such as *Si*, this is a significant mechanism. Doping affects band-to-band transitions, since it affects filled and unfilled state distribution. Now, Fermi energy rather than just the bandedge matters for the energies around which transitions can couple. The electric field also influences absorption, since tunneling between bands is influenced by the field-determined barrier between the conduction bandedge and the valence bandedge. Absorption now is a photon-assisted tunneling process. The magnetic field, through the formation of Landau subbands, also affects absorption. The effect of temperature is largely through the change in bandstructure, particularly the bandedge energies.

An important absorption mechanism is through the free carriers. Electrons respond to the electromagnetic field through Coulomb interaction and Lorentz interaction. Metals reflect light because of the mediation by large electron density, and it is reflected in both the real and the imaginary responses of the conductivity and the dielectric function. The imaginary part of the dielectric function reflects a non-propagating effect; that is, absorption. For electromagnetic waves, the coupling response, together with the presence of the electron charge cloud, results in absorption connected to conductivity that is proportional to charge carrier density and the momentum relaxation time of the electrons $\tau_{\mathbf{k}}$. Free carrier absorption will show up along the entire energy spectrum but is particularly noteworthy throughout the bandgap, where other mechanisms may be suppressed, even as lower energy photons make the process of free carrier absorption more efficient. This absorption also shows up pronouncedly in smaller bandgap semiconductors.

Excitons—electron-hole quasistable coupling—lowers the energy below the bandgap. So, near the bandedges, in the bandgap energy region, one also observes absorption peaks related to excitonic processes. In quantum wells, where electrons and holes can be pushed closer together, such an absorption acquires additional significance. Another, similar, coupling is through impurity states, where again the energy of shallow impurities will be reflected in selective absorption in the bandgap region of the semiconductors.

Phonons also mediate absorption, as in the indirect materials, but it is the electromagnetic-phonon coupling that is a rich source of information regarding the material, because of the specifics of the phonon-photon coupling. An electromagnetic wave, when it is transverse, couples selectively with phonons, where the interaction is grounded in the polarization of the crystal. Optical modes, with their out-of-phase motion, have a significant enough dipole. So, the photon frequency, the longitudinal and transverse phonon responses, and semiconductor properties can all be related to each other through the dielectric function. This is again another means of characterizing semiconductor parameters.

We ended with a short set of comments on luminescence—radiative emission from the solids—which happens through electrical injection of carriers (electroluminescence) but also as re-emission due to photons at another higher energy (photoluminescence).

12.8 Concluding remarks and bibliographic notes

ELECTRON-PHOTON INTERACTIONS IN A SOLID, particularly a semiconductor, lie at the heart of many important technologies, but two, those of lasers and of solar cells, clearly stand out. The first has exacting needs for an understanding of the interaction, and, for the second, a good general description and physical intuition is necessary. There is considerable literature for both these subjects. Here, we refer to a few works through which the reader may gain additional understanding.

A classic source in the study of the optical properties of semiconductors is by Pankove[1]. Add to this the two volumes of semiconductor laser-specific texts from their early era by Casey and Panish[2].

One modern book, an advanced text, that develops the quantum theory of the optical and electronic properties of semiconductors is by Haug and Koch[3]. This text covers many of the subjects, including electromagnetic interactions in confined structures, and others from an optical view, thoroughly. Frank Wooten's book[4] is also a good source. The Wolfe, Holonyak and Stillman book[5] is a very trusted source for the analytic foundations of the absorption processes. At a similar level is the discussion in the text by Balkanski and Wallis[6].

Rosencher and Vinter[7] discuss electron-photon dipolar interaction and the optical properties of semiconductors quite comprehensively.

Excitons appear in all kind of materials, including semiconductors—after all, electrons and holes have a Coulomb attraction. Weak and strong binding, and the nature of the state description—based on the Bloch states of the crystalline description—all enter in, making it quite a difficult undertaking. For a broader view of excitons, see the book by Knox[8]. It tackles them in weak and tight binding and relates their properties to their optical absorption effects. For semiconductors, one of the places where their technological implications is the strongest is in confined structures of small bandgap semiconductors. There are claims of their usefulness in photovoltaics, where multicarrier generation becomes possible from light. A book for looking over these implications of

[1] J. I. Pankove, "Optical processes in semiconductors," Dover, ISBN 978-0486602752 (1971)

[2] H. C. Casey and M. B. Panish, "Heterostructure lasers, Part A and Part B" Academic, ISBN 13 978-0124334571 and ISBN-13 978-0121631024 (1978)

[3] H. Haug and S. W. Koch, "Quantum theory of the optical and electronic properties of semiconductors," World Scientific, ISBN 981-2238-609-2 (2004)

[4] F. Wooten, "Optical properties of solids," Academic, Library of Congress Catalog 72-187257 (1972)

[5] C. Wolfe, N. Holonyak and G. E. Stillman, "Physical properties of semiconductors," Prentice Hall, ISBN 0-13-669961-8 (1989)

[6] M. Balkanski and R. F. Wallis, "Semiconductor physics and applications," Oxford, ISBN 978-0-19-851740-5 (2007)

[7] E. Rosencher and B. Vinter, "Optoelectronics," Cambridge, ISBN 0-521-77129-3 (2004)

[8] R. S. Knox, "Theory of electrons," Academic, Library of Congress Catalog LC 63-22334 (1963)

excitons is by Kilina and Habenicht[9], concentrating particularly on quantum dots of *II-VI* compounds and carbon monolayer structures.

A very readable text on optical properties is by Fox[10]. Interband absorption, excitons, free electrons and luminescence quantum confinement are all treated at an intermediate level in this text. For the Franz-Keldysh effect, see the original publications[11,12]. Tharmalingham's paper[13] provides an integrated picture with the analytic formulation of the Franz-Keldysh relations.

[9] S. V. Kilina and B. F. Habenicht, "Excitonic and vibrational dynamics in nanotechnology," Pan Stanford, ISBN 13 978-981-4241-20-4 (2009)

[10] M. Fox, "Optical properties of solids," Oxford, 978-0-19-957336-3 (2010)

[11] W. Franz, "Einfluß eines elektrischen feldes auf eine optische absorptionskante," Zeitschrift für Naturforschung, **13a**, 484–489 (1958)

[12] L. V. Keldysh, "The effect of a strong electric field on the optical properties of insulating crystal," Soviet Physics JETP, **7**, 788–790 (1958)

[13] K. Tharmalingham, "Optical absorption in the presence of a uniform field," Physical Review, **130**, 2204–2206 (1963)

12.9 *Exercises*

1. Take a polar dielectric that has a single relaxation time. An alternating voltage of constant amplitude is applied to a capacitor made using this dielectric. Find an expression for the heat dissipation as a function of frequency. Show that the maximum in ϵ_r^i occurs at a frequency where the heat dissipation is half of its maximum value. Does this make this approach straightforward for determining relaxation frequency? **[S]**

2. Would it be possible for thermal equilibrium to be achieved between an absorber and thermal radiation if emission did not depend on the radiation density embodied in Planck's radiation law? Why or why not, and what does this mean for stimulated emission? **[S]**

3. What is this oscillator strength parameter? Why is it important, and why does one use free electron mass to determine the momentum-based coupling of the states being connected in the transition? **[S]**

4. Consider a semiconductor in a uniform electric field, with atoms on a cubic lattice and dipoles all pointing along the direction of the external field ($\boldsymbol{\mathcal{E}}$).
   - Referencing Figure 12.20, show, using the dipole field relationship, that the electric field generated at the center of the sphere is
   $$\mathcal{E}_{sph} = \frac{1}{4\pi\epsilon_0} p_i \sum_i \frac{3z_i^2 - r_i^2}{r_i^5},$$
   where p_i is the ith site's dipole moment, and z_i is its position in the field's direction.
   - Show that, when all p_is are aligned and the semiconductor is homogeneous, $\mathcal{E}_{sph} = 0$.

- Now, assume that, outside the sphere, the environment is of uniform polarization **P** and parallel to the electric field. Show that the surface charge density on the sphere at any angle θ from the z axis is $-P\cos\theta$.

- Next, show that it therefore follows that the polarization outside the spherical surface generates a field of $-\mathbf{P}/3\epsilon_0$ at its center. **[S]**

5. Calculate the frequency of the longitudinal plasma oscillations of an electron gas with a jellium, positively charged background. How does this result change when one includes quantum-mechanical constraints? **[S]**

6. A metal's conductivity is independent of the thickness t unless the electron mean free path becomes comparable to t. Assume random surface scattering, a bulk conductivity of σ_0 and mean free path of λ_0. Show that this scattering terminating the scattering-free path—the mean free path—at the surface will change the conductivity and the mean free path as

$$\frac{\sigma}{\sigma_0} = \frac{\lambda_{\mathbf{k}}}{\lambda_0} = \frac{3t}{4\lambda_0} + \frac{t}{2\lambda_0}\ln\left(\frac{\lambda_0}{t}\right)$$

[M]

7. Take a metal-vacuum interface at $z=0$, with the metal in the $z \geq 0$ half space and the vacuum in $z<0$. At long wavelengths, in the metal, the dielectric function is

$$\epsilon_M(\omega) = \epsilon_0\left(1 - \frac{\omega_p^2}{\omega^2}\right).$$

For potential ψ, in the metal where $\nabla^2\psi = 0$, a solution is

$$\psi_M(x,z) = A\cos(kx)\exp(-kz),$$

with the electric field components being $\mathcal{E}_{Mz} = kA\cos(kx)\exp(-ikz)$ and $\mathcal{E}_{Mx} = kA\sin(kx)\exp(-ikz)$.

- Show that, at $z<0$, in vacuum, the potential

$$\psi_0(x,z) = A\cos(kx)\exp(ikz)$$

satisfies the boundary condition of the tangential electric field continuity. What then is $\mathcal{E}_{0x}$?

- The displacements are $\mathbf{D}_M = \epsilon_M(\omega)\boldsymbol{\mathcal{E}}_M$, and $\mathbf{D}_0 = \epsilon_0\boldsymbol{\mathcal{E}}_0$. The normal component of displacement is continuous at the boundary; so, show that this implies that

$$\epsilon_M(\omega) = -\epsilon_0$$

and that the interface plasma oscillates at a frequency of

$$\omega_s = \frac{1}{\sqrt{2}}\omega_p. \qquad \textbf{[S]}$$

8. An n-type $InSb$ crystal has an electron concentration of $10^{18}\,cm^{-3}$. Assume that the semiconductor is nearly isotropic, with an electron effective mass of $0.015m_0$. Calculate the plasma frequency and the wavelength at which there is a minimum in reflectivity by considering electrons to be lossless. The crystal has a dielectric constant of 16. **[S]**

9. Consider a direct gap cubic zinc blende semiconductor, and use the $\mathbf{k}\cdot\mathbf{p}$ approach in a two-band model (conduction, which is s-like, and valence, which is triply degenerate—no spin-orbit—and p-like) to obtain the power series expansion in k around $\mathbf{k}=0$ for the conduction band to determine the effective mass that will contribute to the free carrier term of the dielectric constant ($\Delta\epsilon_r = -Ne^2/\omega^2 m^*$). This is an "optical" effective mass. In which semiconductor(s) should this neglecting of spin-orbit be a reasonable assumption? **[M]**

10. Take a hypothetical semiconductor whose relevant parameters are shown in Figure 12.25. We show here the process of photon absorption and a density of states for the conduction band and the valence band. There are no restrictions from the k selection rules for generation or recombination transitions.
    - At low temperature, we pump a thin sample of this material and measure its transmission in order to determine the absorption coefficient $\alpha = \alpha(E) = \alpha(\hbar\omega)$. The magnitude of the absorption depends on filling probabilities, a coupling constant linking valence and conduction band states and the joint density of the upper and lower states separated by $\hbar\omega$. What is the variation of α with E like?
    - If the pump intensity I_0 is increased to very high levels, how does the absorption of the thin sample change with increasing I_0 and fixed E? Justify your answer. **[M]**

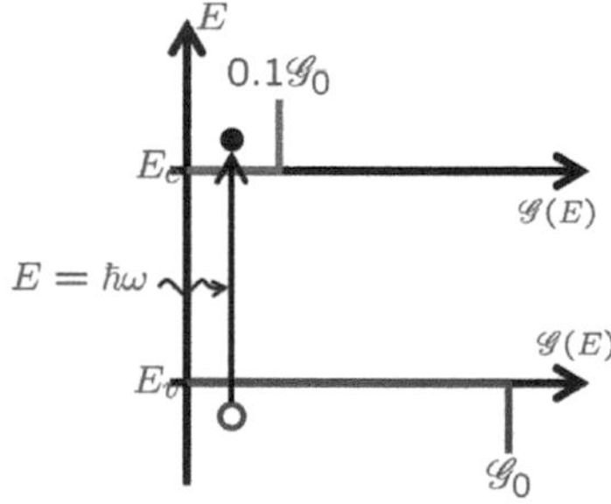

Figure 12.25: Photon absorption in a hypothetical semiconductor.

11. We will now find the energy states and eigenfunctions of an exciton within the effective mass approximation. This exciton is the bound electron-hole quasiparticle due to Coulomb interaction. Within this effective mass approximation, the time-independent Schrödinger equation for the exciton wavepacket ψ_x is written as

$$\left(\frac{\mathbf{p}^2}{2m_n^*} + \frac{\mathbf{p}^2}{2m_p^*} - \frac{e^2}{4\pi\epsilon|\mathbf{r}_n - \mathbf{r}_p|}\right)\psi_x = E\psi_x.$$

By introducing the new coordinates

$$\mathbf{r} = \mathbf{r}_n - \mathbf{r}_p$$

and

$$\rho = \frac{m_n^* \mathbf{r}_n + m_p^* \mathbf{r}_p}{m_n^* + m_p^*},$$

we can separate the variables as $\psi_X(\mathbf{r}_n, \mathbf{r}_p) = X(\mathbf{r})Y(\rho)$.

- Rewrite the time-independent Schrödinger equation in separate forms in terms of $\mathbf{r}$ and ρ and a reduced effective mass of

$$m_x^* = \frac{m_n^* m_p^*}{m_n^* + m_p^*}.$$

- Find the eigenvalues of the center-of-mass equation. What kind of eigenstates do they correspond to?

- Find the eigenvalues of the difference equation. What kind of eigenstates do these correspond to?

- Using these results, find the exciton's effective Bohr radius in bulk *GaAs*. Assume $m_n^* = 0.063 m_0$ and that the static permittivity is $\epsilon = 12.9\epsilon_0$. How does the estimate of the effective Bohr radius compare to the unit cell dimension of *GaAs*? Is the use of the effective mass approximation justified?

12. A *Ge*-like semiconductor has a bandgap E_g at the Brillouin zone boundary that is isotropic and optically allowed. It leads to the strongest optical absorption. The dielectric constant is in the near infrared, so $\epsilon_r(\infty) \approx 12$, and the cubic lattice constant is 0.542 *nm*. Calculate E_g. **[S]**

13. Using the $\mathbf{k} \cdot \mathbf{p}$ method, determine the variation in the optical effective mass of *InSb* electrons as a function of concentration. Ignore the spin-orbit splitting since it is far away. Also calculate, at 10^{18} cm^{-3}, the plasma edge λ_p, where the reflectivity changes to unity, and the wavelength of the reflectivity minimum, assuming lossless electrons. The conduction band minimum effective mass is $m^* = 0.15 m_0$, the relative dielectric constant $\epsilon_r = 16$ and the bandgap $E_g = 0.24$ eV. How does this result compare with that of Exercise 8? **[S]**

14. Find the dependence of the matrix element of $\mathbf{p}$ for direct interband transitions near $\mathbf{k} = 0$ using the two-band model given in Exercise 9. **[S]**

15. Let M_0 be the minimum critical point for direct interband transitions. Take a parabolic expansion of the density of states around M_0, and assume validity for all energies. Calculate the shape of the corresponding real part of the dielectric constant (ϵ_r^r), assuming constant matrix elements of $\mathbf{p}$. What is the shape of ϵ_r^r for other types of van Hove critical point? **[M]**

13 Causality and Green's functions

Powerful words mellow and are melded by time, tipping the balance toward a preferred point of view. Causality is interpreted by philosophers and scientists in different ways. Sciences prefer precision and unambiguity: symbolic manipulation and objective precision with a reductive style. Philosophers are integrative. Science and engineering methods—except for the rare revolutionary jumps—work as band-aid applications—tackling a known problem or inconsistency—and losing sight of the long term symptom. Society, for example, doesn't know how to rapidly respond and create mechanisms for mitigating generational consequences of sciences—nuclear, plastics, global warming from energy transformations, and population via agriculture and medicine—while philosophy helps with understanding our being via probing questions and dialog, with the meaning of being human in nature front and center. This is the central conundrum that phenomenology deals with a little more to my satisfaction. A reflection on this incompleteness comes in the form of the following joke: what is the difference between a mathematician and a philosopher? Philosophers use pencil and paper; mathematicians use pencil, paper and a wastepaper basket. Heisenberg's uncertainty principle led to nearly two decades of philosophical writings on "free will." Cause and chance have much to say on this.

CAUSE AND CHANCE both appear to play meaningful roles in our daily experiences, as well as in the behavior of the materials we are studying in this text. Stimulus is a cause. The movement of the electron interacting with its surroundings—scattering—is often a matter of chance. But even if these terms are employed across science and humanities, they are dissatisfyingly vague, have quite different meanings and are used for different implications.

A common notion/understanding is that cause implies a necessity of connection and relation between a sequence of events, while chance is a randomness. Necessity and accident both appear to guide events. An accident is not really entirely arbitrary. Bayes' theorem is a powerful example of a law for probabilities. It is our quantitative means of tackling chance. Having a cause also doesn't make predicting the future with certainty possible, since complete knowledge of both the present and the past is not quite available, except in the circumscribed and idealized textbook problems. We remove contradictions and arbitrariness through observation and development of a predictive theory. It is in this sense that science and humanities, particularly philosophy or theology, are very different.

We tackle the concepts of chance—probabilities—and the probabilities' relationships elsewhere and focus on the cause-effect connection as represented in the concept of causality, the mathematical tool to practice it in the form of Green's functions, and their use in orthogonal space for properties of materials in this chapter. In the following chapter (Chapter 14), we will employ the implications of these ideas to develop the dispersion relationships in properties of materials, and, more broadly, the Kramers-Kronig relationships for linear systems.

Probabilities' implications in information mechanics are the subject of the last volume of this series. The first volume does the same for the development of our understanding of ensembles, such as the implications of statistical mechanics for condensed matter, and thermostatics and thermodynamics in general. In quantum mechanics, it is the stochastic nature of the quantum-mechanical measurement process.

Semiconductor Physics: Principles, Theory and Nanoscale. Sandip Tiwari.
 Published 2020 by Oxford University Press. DOI: 10.1093/oso/9780198759867.001.0001

13.1 *Causality, determinism and correlations*

CAUSALITY IS WEAKLY RELATED to determinism—the unique evolution of states—and even more weakly to correlation—a quantitative evaluation of a degree of correspondence (pairwise and higher order) between the simultaneity of two events in space and time—but is quite distinguishably different. What relates these terms is the idea of dependence. If two concepts are being connected, and the concepts can be symbolically represented, then a function allows us to express it. A logical dependence is such a function.

When the connection is being probed between two different sets, the verifiable observational and predictability criterion has a larger hill to climb. This caution is also necessary for inference by induction. Induction—a powerful tool of philosophy—allows one to generalize a number of observations. Inference by induction is useful in daily life (the sun rises in the morning, winter follows summer, leaves turn green in spring, etc.), although there is no clear criterion for the validity of induction—it is largely intuition—but it is not an inference by causality. Another similar inferential approach is that of using correlations—direct and higher order—but these too are not inference by causality. So, drawing absolute conclusions is quite non-trivial.

Consider another example. When it is raining or the path I take to work is not cleared of snow, I take Bus No. 30. According to the timetable, the bus will arrive at my stop at 10 minutes past the hour, and 40 minutes past the hour. This law of the timetable is deterministic. I can predict future events from it. Asking the question "Why?" is not appropriate. However, the law of the timetable is not immutable: the bus operating authority may change the schedule, a bus may break down or a driver may call in sick. Determinism should be viewed as the ability, through rules, to predict the occurrence of event B (bus timing) from the knowledge of A (the timetable) but without a physical, timeless and spaceless link between all things in set A and all things in set B. *Causality, that B is caused by A, is to claim that the phenomenon B can be entirely traced to A for all variations of conditions in time and space.* If you take away B, then A must also be absent. This definition is easier to understand in time—a detailed discussion follows—but the spatial invariance is a little more challenging. When the state of one particle of an entangled pair is observed, the state of the other is also simultaneously fixed, no matter how far apart they may be. Neither the immutability of the speed of light as a limit

If all degrees of freedom were followed in the sense of expectations, then the system will be entirely deterministic. Chance appears in classical treatments via incompleteness of information. Even the outcome of a fair dice, given all the information and time to calculate, can be determined.

Applied mathematicians view causality in the general form that physical laws—expressible as mathematical equations—connecting continuous variables are such that, for any finite number of parameters, some variable or set of variables that appears in the equations is uniquely determined in terms of the others.

Take connecting constellations to human fate—two very different spaces of nature—as astrology claims. Verifiable repeatable observations and experimentable predictions are the sine qua non of science. Astrology fails this test. Physical nature and living nature are too different to be considered identical space.

There is one other approach that I have sometimes seen being practiced in the task of providing proofs, which I call "proof by intimidation." It may get a teacher out of a tough spot in the short term but is sure to be a failure in the long term.

Even if absolute ideals may be unreachable, approaching ideals is worthy. To reject experience in the absence of logical proof is foolish. Being anti-vaccination, applauding boxing or American football—a violent university business—as worthy sports and believing in astrology are also foolish.

A key question is "How does one know that no other parameter than the ones stated are needed, that is, are there any hidden variables?" Another hidden variable aside is as follows: an "entity" judging everybody's actions as right or wrong is ethically unacceptable. Let each action of ours have 3 possibilities: good, indifferent and bad. And let us keep s as the smallest unit of time, so that a 75-year life has $I \approx 4 \times 10^7$ actions. For an individual, this is 3^I instances over life. If one keeps track of $N = 10^7$

speed prescribed by relativity nor the independence from space is violated.

To clarify and distinguish causality from other engineering inferential thoughts, consider the following event spaces:

- if A then B (in English)
- $A \Rightarrow B$ (in logic),
- $B \vee \bar{A}$ (Boolean).

These are all limit statements. But they are not statements of A causes B.

Statistically, cocks crow (event A) at dawn (event B). So, the probability $\mathfrak{p}(B|A) \lesssim 1$, but we can be quite sure that cocks do not cause dawn.

A statement of implication (cock crowing $\Rightarrow$ dawn) is statistically fine. If one makes more observations from this sample space, one will continue to see $B \vee \bar{A}$. Causation is a far, far stronger statement. If A causes B, then even if an intervention changes the distribution of the observations, $B \vee \bar{A}$ will still be true in the new distribution. *In a direct cause-and-effect relationship, that is, a relationship of a single cause and a direct effect arising from it, removing the effect also removes the cause.*

Another important point is that absolute independence does not follow from pairwise independence. Let $\mathfrak{p}(x_1, x_2) = \mathfrak{p}(x_1)\mathfrak{p}(x_2)$, and $\mathfrak{p}(x_2, x_3) = \mathfrak{p}(x_2)\mathfrak{p}(x_3)$. But this does not imply the independence of x_1 and x_3, that is, $\mathfrak{p}(x_1, x_3) \neq \mathfrak{p}(x_1)\mathfrak{p}(x_3)$ in general. This is an example where events have correlation but not causality. Most relationships, however, involve multiple factors. And, as we noted in the example discussed here, $\mathfrak{p}(x_1, x_3) \neq \mathfrak{p}(x_1)\mathfrak{p}(x_3)$, with pairwise interactions, generally. This makes the non-direct case hard. In more complex situations, each statement about causation cannot be proved, only disproved. When repeatedly tested and not disproved, we raise our estimate of the reliability of the hypothesis, using the Bayesian rule. Having multiple low probability factors in a series or high probability factors in parallel makes causal inference very, very tricky.

Here are a few examples to illustrate the difficulty in determining causation.

First, take the following vernacular statement: "I know Joseph Fourier was born in 1768 since Louis *XVI* was executed in 1793." "Since" is attached to knowing the execution year, not Fourier's birth year. The execution did not cause the birth of Fourier, nor does it explain the birth year. It does, however, allow one to guess the birth year.

individuals, we have $N3^I$ actions kept in memory. Each action connects to other people—is affected by them or affects them, and, because this happens in time, is thus a sum of history—which then corresponds to another $N\sum_{i=1}^{I}\sum_{i=1}^{i} i(N-1)$ instances to track. This is

$$M = N3^I + \frac{1}{6}N(N-1)I(I+1)(I+2)$$

of memory. The first term—the individual's self-action—dominates and is $\sim 10^{3\times10^7}$. The estimate of the total number of particles in the universe is 10^{78}. And we have not even accounted for the action's consequences in time, where a judgment may need be modified later after seeing consequences, as with the creation of the atomic bomb. The universe, and by recursion, any external entity, does not have the resources for such a task—an *np*-impossible Maxwell's demon–like task. The really bright side of this conclusion is that we can always look forward to the beauty of creative insights in science and literature, and new Bruchs, Schuberts and Kumar Gandharvas.

Muffling the cock will not stop the dawn.

Joseph Fourier, a man of humble birth, was orphaned at the age of 10. Progressive education in the hands of Benedictine monks, an early discovery and passion for mathematics, and hard work led to his graduation at the age of 14. Legendre seems to have been his mentor. College graduates at that time could either go into the army or go into the Church. His progress was obstructed —in a sign of the times—by the minister of war, who said "*Fourier étant pas noble ne pourrait entrer dans l'artillerie, quand même il serait un second Newton,*" that is, "Were he even a second Newton, Fourier could not enter artillery since he is not noble" (F. Arago, "Joseph Fourier" (French edn), CreateSpace Independent Publishing Platform, ISBN 1533677298 (2016), p. 9). Fourier had to choose the Church, when, just three days before he would have had to take the vow, in 1789, the National Assembly suspended the requirement—a Poissonian stroke of luck that was of immense benefit

Second, take the mechanical statement $\mathbf{F} = m\mathbf{a}$, that is, $\mathbf{F}(t) = m\mathbf{a}(t)$, which says that a force at any instant of time is proportional to the acceleration at that time. The two vectors—$\mathbf{F}$ and $m\mathbf{a}$—are equal. But this does not necessarily mean that $\mathbf{F}(-t) = m\mathbf{a}(-t)$. Thermodynamic laws are not equalities. Time-reversal symmetry is not invariant for many situations, the expansion of the universe being the most prominent example. This $\mathbf{F} = m\mathbf{a}$ relationship is a statement of equality, not causality. Forces can lead to situations where acceleration exists, but not the other way around. But determining this requires other observations on the system. There are situations where acceleration can lead to a force too (and not the other way around). This too requires other observations on the system. F and ma coexist. They are an equality, and you cannot have one without the other.

to mankind, since he did so much afterwards, including his studies and teaching at École Normale and École Polytechnique. The Fourier heat equation is from this period. Fourier was an ardent believer in *liberté, égalité, fraternitè*. The entanglement of states is reflected in his appearance in Victor Hugo's *Les Misérables*, which will not be quoted here, to be charitable. It is best not to derive any causality inferences here. Life and French history, both of the revolution and of science, is surfeit with interactions among states.

Third, examine the RS(Reset-Set) flip-flop in Figure 13.1, which can be viewed as a network. It has a bit of memory built in, since the past and the present matter in its operation. When one applies a 0 pulse to $\overline{S}$, it sets $A = 1$ and $B = 0$, following the $1 \mapsto 0$ transition. When $\overline{S}$ is brought back to $\overline{S} = 1$ by the $0 \mapsto 1$ part of the $\overline{S}$ pulse, no change occurs. $(A, B) = (1, 0)$ is remembered. To make $(A, B) = (0, 1)$, one must do the complement operation: apply a 0 to $\overline{R}$. Again, the change happens at the $1 \mapsto 0$ transition and is remembered in the $0 \mapsto 1$ part. Observe the changes over a period of time, or through a pencil-and-paper experiment on changes, and one can identify causative factors—but not the cause. This flip-flop cannot look back in the past uniquely and therefore, for even this constrained problem, fails in causality.

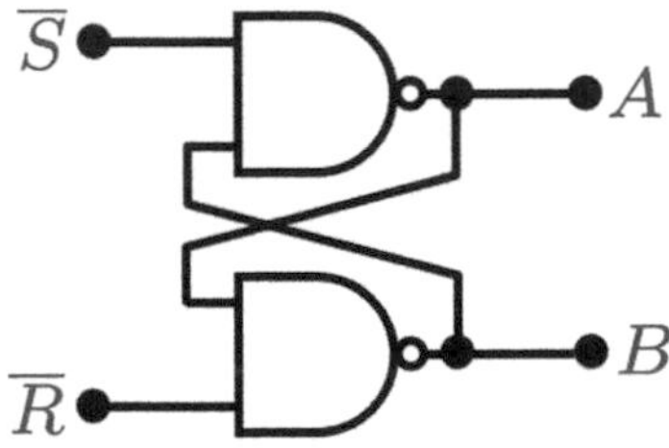

Figure 13.1: The reset-set flip-flop.

Fourth, in a feedback loop, all elements become causes and have effect as the signal propagates through and continues to do so, with consequences locally and delayed. Cause cannot be distinguished from effect here.

13.2 *Causality, and time and space immutability*

Our statement on the causality of time and space immutability requires more serious thinking. Space immutability, in particular, is a more challenging idea and harder to grasp. In classical mechanics, gravitational and electromagnetic forces act across space, that is, are not contiguous, and relativity establishes the constraint of the speed of light. In quantum mechanics, chance—as in randomness—appears through fluctuations, that is, the canonical variables, that cannot both be precisely known simultaneously. So, how does causality play out? We illustrate our conclusion through the Aharonov-Rohrlich paradox.

Figure 13.2 shows a long cylinder containing a particle and a frictionless piston to which a box with two open end faces is connected. This allows a ball, which is external to the cylinder, to pass through the box while bouncing off one or two walls that are parallel to the plunger wall. The particle is far away from the plunger in the cylinder, where the wall and the piston are a distance L apart. When the ball, arriving at this assembly, strikes the box wall as shown, it moves the plunger in. The ball bounces elastically twice, leaves the box and moves the piston a distance ΔL at the completion of this process.

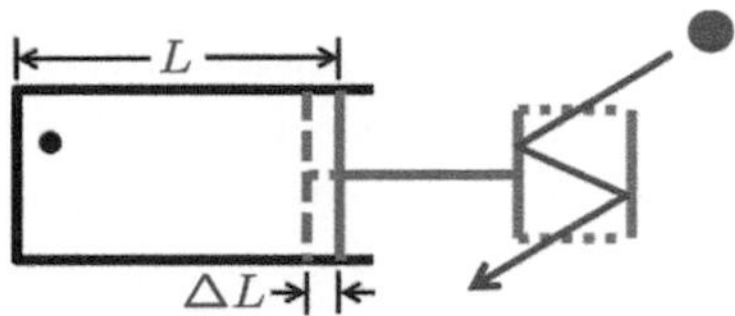

Figure 13.2: A particle in a cylinder with a frictionless piston attached to an open box through which a particle can enter and bounce out.

The question we ask is, does the particle in the cylinder affect the ball's trajectory?

Viewed classically, the first bounce of the ball on the wall causes the piston to move in. Upon bouncing the second time, without the particle and the piston coming in contact, the ball recovers its energy and momentum. The piston is back to its original position of rest. If the particle and piston had made contact during the transit time period between the two wall collisions of the ball, then the ball would not be able to recover all the energy and momentum, because the piston-particle interaction has its own energy and momentum exchange.

Viewed quantum-mechanically, consider the two limits of time duration (long and short) for the inter-wall transit time of the ball. Long here means that the particle has many of its own collisions with the piston and the opposite cylinder wall during this time period. We can employ adiabatic approximation here. The particle's energy increases at the expense of the ball. The ball has been affected by the particle. So far, so good. This is quite understandable and not really that far from what classical mechanics will tell us for a moving particle undergoing collisions with the piston during the ball's transit.

The particle is in a superposition of eigen states $|u_n\rangle$, with energies $E_n = n^2\pi^2\hbar^2/2mL^2$, where m is mass of the particle. With L decreasing, E_n increases. Decrease in L is slow, so the probability amplitude of the eigenstates remains unchanged, and the expectation value of the energy of the particle increases.

Consider now the slow limit indeed, one so slow that the particle doesn't reach the piston during this transit period. Let the particle be in the state $|\psi(0)\rangle$ at $t = 0$. This is a wavepacket of low Δp in momentum and large Δz in position. Let $L \gg \Delta z \gg \Delta L$. If the particle's expectation for velocity is $\langle v\rangle$, then, in the absence of a ball, this particle makes one round trip in a time of $T = 2L/|\langle v\rangle|$. The wavepacket keeps its shape but will spread out ever so slightly. If $\langle v\rangle$ is large, we will have

$$\left(\frac{\Delta p}{p}\right)^2 \ll \left(\frac{\Delta z}{z}\right)^2. \tag{13.1}$$

This latter point of detail we will not dwell on since it is not relevant to the present argument. One example to see this is that the barriers confining the particle are not infinite barriers, so leakage and some shape change must follow.

For a Gaussian wavepacket of spread $\Delta z \approx a/\sqrt{2}$,

$$\left|\int_{-\infty}^{\infty} \psi^*(z,t)\psi(z - \frac{\hbar k t}{m}, 0)\, dz\right|^2 = \left(1 + \frac{\hbar^2 t^2}{4m^2a^4}\right)^{-1/2}.$$

Equation 13.1 then follows.

For the matrix element $\langle\psi(0)|\exp(-i\hat{\mathscr{H}}T/\hbar)|\psi(0)\rangle$, the expectation value at $t=0$ is $\exp(-i\hat{\mathscr{H}}T/\hbar)$ (its real and imaginary parts are the observables); yet, the state of the particle at $t=T$ is

$$|\psi(T)\rangle = \exp(-i\hat{\mathscr{H}}T/\hbar)|\psi(0)\rangle. \tag{13.2}$$

With wavepacket spreading absent, we may choose L so that the phase factor is unity, that is, $|\psi(T)\rangle = |\psi(0)\rangle$, so that the expectation value is unity. This, our first case, is for when the ball has not struck the box wall anytime before $t=T$.

Now, consider a second case where the ball strikes the box wall, and therefore the piston, at a reference time chosen to be $t=0$, causing the piston to move inwards, and then again at time $t \ll T/2$. $|\psi(0)\rangle$ is vanishingly small in the piston. But the movement of the piston is reflected in the Hamiltonian $\hat{\mathscr{H}}$. The operator $\exp(-i\hat{\mathscr{H}}T/\hbar)$ evolves $|\psi(0)\rangle$, with the expectation of $|\psi(T)\rangle$ now peaking a little further inside in the cylinder (a displacement of about $2|\Delta L|$ from $|\psi(0)\rangle$). This is a translation under the operation $\exp(i\hat{p}2|\Delta L|/\hbar)$. So,

$$|\psi(T)\rangle = \exp(i\hat{p}2|\Delta L|/\hbar)|\psi(0)\rangle. \tag{13.3}$$

If $\Delta z \gg |\Delta L|$, then while the overlap $\langle|\psi(T)|\psi(0)\rangle$ is still ~ 1, the phase has changed by about $\exp(2i\langle p\rangle T/\hbar)$, with $\langle p\rangle = \langle\psi(0)|\hat{p}|\psi(0)\rangle$. The ball has changed the expectation value of $\exp(-i\hat{\mathscr{H}}T/\hbar)$ at time $t=0$ without any apparent regard for T. The change in the expectation value of the particle is about $-2\langle p\rangle|\Delta L|/T$. A particle far away from the piston has had its energy changed due to the ball hitting the box wall/piston. If its energy changed, the ball's energy too has changed. Measuring the change in the ball's energy tells us the change in the particle's energy—this, despite any large spatial separation. And we know that there is a particle in the cylinder if the ball's energy changed. This is an example of action at a distance, and a spatial immutability—a nonlocality—that complements the time immutability of causality.

To summarize this discussion, both causality and chance play a role in both the classical and the quantum-mechanical—Copenhagen—view of the world. Correlations and determinism are weak offshoots with relaxed conditions and are more easily digestible for the human. A systematic use of both of these approaches is necessary in understanding what we observe. In the subject area of this text—the interaction and response in semiconductors during an interaction—chance, through scattering during transport, and causality, through orthogonal linkages in properties of materials such as the frequency dependence of the susceptibility arising in wave-matter interactions, appear. Both need to be employed. Using of Green's functions is one of the

A particularly appealing theory of life on earth is that the ferment of billions of years led to creation of the simple and complex molecules a billion years ago—by chance that was enhanced by the prevailing conditions—and these molecules were qualitatively different. It was at this point that the matter produced began to reproduce at the expense of the surroundings. *The process was now less susceptible to chance and more dependent on causal transformations.* Think of chance and causality as two complementary—perhaps opposite—sides leading to a transformation. If motion is observed for a particle in the process of diffusion—chance—one sees $\Delta p\Delta z = constant$, since $(z-\langle z\rangle)^2 = a\Delta t$, where a is acceleration. Under conditions where quantum mechanics must be employed, this is replaced by the Heisenberg uncertainty of $\Delta p\Delta z \geq \hbar/2$.

most useful techniques for tackling causal problems. Solutions of polynomial equations with initial value (time) or boundary value (space) can be powerfully found through it.

13.3 *Green's functions*

SINCE WE ARE INTERESTED IN ANALYZING AND PREDICTING what happens to systems under some form of stimulation—such as by an electromagnetic wave—or an interaction within a system—disorder, defects, doping, et cetera—a general way of analysis is of interest. More generally, what this says is that we are interested in understanding what happens to a system as an interaction strength is dialed up or turned on from zero to some magnitude. An external perturbation—a photon/electromagnetic wave interaction, for example—may appear, and one wants to know the system's response. Using Green's functions is one of the ways to examine the change of the system for which a causal relationship exists, that is, one where if you take away the source, the effect disappears.

In pharmaceuticals, double blind studies allow one to determine a correlative measure of the chances of usefulness of the medicine. But this does not imply any causality. Diseases are complex, with many interlinkages. And an experiment in reverse, where even the notion of what any disease is in a precisely definable way is, is a nonstarter.

Consider the functional relationship of electrostatic interaction. A point charge e at $\mathbf{r}_0$ causes an electrostatic potential ψ at $\mathbf{r}$ of

$$\psi(\mathbf{r}) = \frac{1}{4\pi\epsilon}\frac{e}{|\mathbf{r}-\mathbf{r}_0|}. \tag{13.4}$$

This followed from the function—the Poisson function—written as

$$\nabla^2\psi(\mathbf{r}) = -\frac{\rho(\mathbf{r})}{\epsilon}, \tag{13.5}$$

where, in general, there is a charge distribution of $\rho(\mathbf{r})$ in space. The solution for this distribution is also known to us as

$$\psi(\mathbf{r}) = \frac{1}{4\pi\epsilon}\frac{\rho(\mathbf{r}')}{|\mathbf{r}-\mathbf{r}'|}\,d\mathbf{r}'. \tag{13.6}$$

This was a simple boundary value problem, where we have specified the boundary conditions of charge. The charge was our stimulus. The system responds to an external stimulus in its own characteristic way causally. The response of the system is our response function, which we will probe more thoroughly in a subsequent chapter (Chapter 14).

For any function f of the spatial ($\mathbf{r}$) and time (t) coordinates of interest to us, one may write the response as

This discussion and the discussion of Chapter 14 are limited to linear response. But the approach is just as useful, even if more complex, for nonlinear response.

$$Z(\mathbf{r},t|\mathbf{r}',t') = \int_{-\infty}^{\infty}\int_{-\infty}^{\infty} G(\mathbf{r},t|\mathbf{r}',t')f(\mathbf{r}',t')\,d\mathbf{r}'\,dt'. \tag{13.7}$$

The function $G(\mathbf{r},t|\mathbf{r}',t')$ is the Green's function here, a response function that describes the response $Z(\mathbf{r},t|\mathbf{r}',t')$ at $(\mathbf{r},t)$ due to the stimulus $f(\mathbf{r}',t')$, with the boundary and initial value $(\mathbf{r}',t')$ spanning space and time. This stimulus may be of any kind—the most common being an electromagnetic wave—but it could as well be a force applied to a mass, an excitation to a string, an electron's scattering or fluid flow at a turn. Using a Green's function is an alternative powerful technique to evaluate the response that the function f has embedded in it that just as well applies to the quantum-mechanical domain.

A remark on the notation using " |." We have used this with probability to indicate a prior. The probability of event B knowing that A has occurred is $\mathfrak{p}(A|B)$. We use this same notation to indicate the causal prior for the Green's function. The response operator for determining what happens at $\mathbf{r},t$, given that there is a cause at $\mathbf{r}',t'$, is $G(\mathbf{r},t|\mathbf{r}',t')$. If this operator relationship is a function of the spatial separation $\mathbf{r}-\mathbf{r}'$ and the time separation $t-t'$, then one can also write it as $G(\mathbf{r},t|\mathbf{r}',t')=G(\mathbf{r}-\mathbf{r}',t-t')$.

Comparing Equations 13.6 and 13.7, it will follow that, for the time-independent Laplace problem, the Green's function is

$$G(\mathbf{r}|\mathbf{r}') = \frac{1}{4\pi\epsilon}\frac{1}{|\mathbf{r}-\mathbf{r}'|}. \tag{13.8}$$

For this problem, this Green's function—a response function—acts on the function $f = \rho$ (ρ being the stimulus) to generate the potential response consequence.

This simple problem will illustrate for us a number of points and the methodology for Green's functions. The Green's function in this problem is the Coulomb potential due to a unit point charge at $\mathbf{r}=\mathbf{r}'$. For any $\mathbf{r}\neq\mathbf{r}'$, this charge is zero. Over the entire space, $\int \rho(\mathbf{r})\,d\mathbf{r}=1$. This implies that

$$\rho(\mathbf{r})\,d\mathbf{r} = \delta(\mathbf{r}-\mathbf{r}') \tag{13.9}$$

is the Dirac delta function as our stimulus, with a boundary value at $\mathbf{r} = \mathbf{r}'$. We wrote this relationship because we know this solution from our past using other procedures. Now, let us view this through Green's function methods. For the Green's function to satisfy this problem,

The Dirac delta function is of incredible utility and is in widespread use, not just in these Green's function calculations. It should be very clearly distinguished from the Kronecker delta. We have used them both in the text. See Appendix B for the Dirac delta function and other functions that appear in our mathematical practices in this text.

$$\nabla^2 G(\mathbf{r}-\mathbf{r}') = -\frac{\delta(\mathbf{r}-\mathbf{r}')}{\epsilon} \tag{13.10}$$

must hold, and the boundary conditions need to be satisfied. The potential vanishes as $r\to\infty$. Let $\mathbf{R}=\mathbf{r}-\mathbf{r}'$, so we are in search of $G(\mathbf{r}-\mathbf{r}')=G(\mathbf{R})$. We use a Fourier transform and write

$$G(\mathbf{k}) = \frac{1}{(2\pi)^{3/2}}\int \exp(-i\mathbf{k}\cdot\mathbf{R})G(\mathbf{R})\,d^3\mathbf{R} \tag{13.11}$$

under the transform constraint of

$$\int \exp(-i\mathbf{k}\cdot\mathbf{R})G(\mathbf{R})\delta(\mathbf{R})\,d^3\mathbf{R} = 1. \tag{13.12}$$

Equation 13.10 therefore leads to

$$-k^2 G(\mathbf{k}) = \frac{1}{(2\pi)^{3/2}\epsilon}. \tag{13.13}$$

$G(\mathbf{k}) = 1/\epsilon k^2$ transformed back gives

$$G(\mathbf{R}) = \frac{1}{(2\pi)^{3/2}} \int \exp(i\mathbf{k}\cdot\mathbf{R}) \left[\frac{1}{(2\pi)^{3/2}\epsilon k^2}\right] d^3\mathbf{k} = \frac{1}{4\pi\epsilon R}. \qquad (13.14)$$

$G(\mathbf{R}) = G(\mathbf{r}-\mathbf{r}')$ satisfies the Poisson equation for a point unit charge.

Having found the Green's function, we can use it to determine the potential for $\rho(\mathbf{r}')$ as a general problem. The response function equation (Equation 13.7) tells us to write this electrostatic potential as

$$\psi(\mathbf{r}) = \int G(\mathbf{r}-\mathbf{r}')\rho(\mathbf{r}')\, d^3\mathbf{r}'. \qquad (13.15)$$

Therefore, employing Equation 13.10, we get

$$\begin{aligned}\nabla^2\psi(\mathbf{r}) &= \int \nabla^2 G(\mathbf{r}-\mathbf{r}')\rho(\mathbf{r}')\, d^3\mathbf{r}' \\ &= -\int \frac{\delta(\mathbf{r}-\mathbf{r}')}{\epsilon}\rho(\mathbf{r}')\, d^3\mathbf{r}' \\ &= -\frac{\rho(\mathbf{r})}{\epsilon}. \qquad (13.16)\end{aligned}$$

Our method was to employ a unit point stimulus/source, hence the delta function, to determine the Green's function as a solution to the Poisson equation through Fourier transformation techniques. It is a unique solution, where the boundary condition of potential vanishing has been incorporated in the Fourier transform (Equation 13.12) employed. This Green's function now becomes our tool to solve a more general problem.

Now, let us tackle the dynamic cousin of this problem: electromagnetic propagation, where the causal evolution and boundary/initial conditions of space and time must be considered. The vector potential $\mathbf{A}$ and the scalar potential ψ satisfy

This will also become the basis of tackling scattering, a problem where Green's function techniques are indispensable.

$$\begin{aligned}&\left(\nabla^2 - \frac{1}{c^2}\frac{\partial^2}{\partial t^2}\right)\mathbf{A}(\mathbf{r},t) = -\mu\mathbf{J}(\mathbf{r},t), \text{ and} \\ &\left(\nabla^2 - \frac{1}{c^2}\frac{\partial^2}{\partial t^2}\right)\psi(\mathbf{r},t) = -\frac{\rho(\mathbf{r},t)}{\epsilon}. \qquad (13.17)\end{aligned}$$

The equations are very similar—the only difference is that one is a scalar relationship while the other is a vector relationship—so our point function approach to developing Green's function is applicable in the same way as before. For this Dirac δ function—now as independent functions in two coordinates—we can write

$$\begin{aligned}\left(\nabla^2 - \frac{1}{c^2}\frac{\partial^2}{\partial t^2}\right) G(\mathbf{r}-\mathbf{r}', t-t') &= -\delta(\mathbf{r}-\mathbf{r}', t-t') \\ &= -\delta(\mathbf{R})\delta(s), \qquad (13.18)\end{aligned}$$

where we have used $s = t - t'$ for the time translation, in analogy with our spatial translation variable **R**. The spatial Fourier transform satisfies

$$\frac{1}{c^2}\frac{\partial^2 G(\mathbf{k},s)}{\partial s^2} + k^2 G(\mathbf{k},s) = \delta(s). \tag{13.19}$$

This is a second-order harmonic equation. The right-hand side can be interpreted to mean an impulse excitation (at $s = 0$), with k^2 a restoring force constant, for example, a spring constant, and $1/c^2 = \mu\epsilon$ an inertia term, for example, a mass, for a displacement G or a velocity $\partial G/\partial s$. So, this is a general equation that appears in a variety of problems. The boundary condition of the problem and the nature of the problem tell us which Green's function out of many choices must be used. So, this choice of Green's function for the homogeneous part relates to both the nature of the problem and the boundary condition. G is a causal function—a response to a unit impulse—as a finite response to a finite stimulus. It is continuous for all ss and **R**s, as is $\partial G/\partial s$. To see this mathematically, integrate under the curve in the intervals $s = 0 - \epsilon$ and $s = 0 + \epsilon$ and take the limit,

$$\frac{1}{c^2}\left.\frac{\partial G}{\partial s}\right|_{-\epsilon}^{\epsilon} + k^2 \int_{-\epsilon}^{\epsilon} G(\mathbf{k},s)\, ds = 1$$

$$\therefore \lim_{\epsilon\to 0} \frac{\partial G}{\partial s} = c^2 \tag{13.20}$$

with the integral contribution vanishing

Equation 13.19 under the initial value conditions of $\partial G/\partial s = c^2$ at $s = 0$ and $G = 0$ for $s \leq 0$ is

$$G(\mathbf{k},s) = \frac{c}{k}\sin(kcs)\Theta(s), \tag{13.21}$$

where $\Theta(s)$ is the Heaviside function. Inverse spatial transformation of this equation gives

$$\begin{aligned} G(\mathbf{R},s) &= \frac{c\Theta(s)}{(2\pi)^3}\int \exp(i\mathbf{k}\cdot\mathbf{R})\frac{\sin(kcs)}{k}\, d^3k \\ &= \frac{c\Theta(s)}{2\pi^2 R}\int_0^\infty \sin(kR)\sin(kcs)\, dk \\ &= \frac{c\Theta(s)}{8\pi^2 R}\int_0^\infty \{\exp[ik(R-cs)] - \exp[ik(R+cs)]\}\, dk \\ &= \frac{c\Theta(s)}{4\pi R}\left[\delta(R-cs) - \delta(R+cs)\right]. \end{aligned} \tag{13.22}$$

$\delta(R + cs) = 0$ for positive R and s, so

$$G(\mathbf{R},s) = \frac{c}{4\pi R}\delta(R-cs), \tag{13.23}$$

The harmonic oscillator equation is a truly basic equation of nature. It is the lowest-order equation defining stability. Restoring force appears for any change: positive or negative, stretching or straining, tensile or compressive, et cetera. In quantum mechanics, we start with the harmonic oscillator as the basic constituent of the edifice. It applies to photons, phonons, blackbody radiation, field quantization, and so on, and this lowest form for stability is at the heart of it.

This is the area under the curve. This is where one can sympathize for arguments against calling the Dirac δ a good function.

This argument is identical to the quantum-mechanical argument we made in S. Tiwari, "Quantum, statistical and information mechanics: A unified introduction," Electroscience 1, Oxford University Press, ISBN 978-0-198-75985-0 (forthcoming), for continuity of $\partial\psi/m\partial z$ and of the wavefunction at a potential boundary. There, the equation is the linear Schrödinger equation. A discontinuous change means infinite energy perturbation, which doesn't exist. This same reasoning is true here. Here, too, it is the probability density and the probability density current (a momentum)—of different characteristics of the system—that must be continuous.

The Heaviside function (see Appendix B), which is also called the unit step function, appears in nearly all problems where one turns on a stimulus at some instant and looks at the response in time. Teaching of transformations—and, in electrosciences, transformations are critical to solving physical or signal problems—cannot be complete without it. The Heaviside function is defined by

$$\Theta(s) = \begin{cases} 1 & \text{for } s \leq 0, \text{ and} \\ 0 & \text{for } s = 0. \end{cases}$$

Note that $d\Theta(s)/ds = \delta(s)$.

that is, a change at $\mathbf{r}'$ causes a change at $\mathbf{r}$, a time c/s later. This is a *retarded interaction*. So, we can now find the solution for the general dynamic problem of electrical potential arising in $\rho(\mathbf{r}', t')$ as

$$\begin{aligned}\psi(\mathbf{r}, t) &= \frac{1}{\epsilon} \int G(\mathbf{r} - \mathbf{r}')\rho(\mathbf{r}', t')\, d^3\mathbf{r}'\, dt' \\ &= \frac{c}{4\pi\epsilon} \int d^3\mathbf{r}' \int \frac{\delta(R - cs)}{R} \rho(\mathbf{r}', t')\, d^3 r' \\ &= \frac{c}{4\pi\epsilon} \int \frac{1}{R} \rho\left(\mathbf{r}', t - \frac{R}{c}\right) d^3\mathbf{r}'. \end{aligned} \tag{13.24}$$

This is the general spatial- and time-dependent form of the retarded potential. It satisfies Equation 13.17. One could find a similar solution for the vector potential arising in the current.

A few comments are in order from this exposition. In linear systems, the response to the perturbation can be expressed using Green's functions. These Green's functions, developed using unit impulse perturbation, hold true independent of the form the actual perturbation takes. Many Green's functions can satisfy the requirement. The Green's function becomes unique when the boundary and initial value conditions are specified. So, the Green's function developed in this algorithm is unique and specific to the boundary value and initial value under the mathematical functional relationship prescribing the physical problem. The solution itself is then found as an integral, with the Green's function operating on the stimulus. Since an integral superposes for a linear system, the solution holds generally for linear systems. Two separate causes will give an effect that is the sum of the effects from the causes separately. A caution is warranted. We ensured here that, spatially, the Green's function vanishes as $\mathbf{r} \to \infty$ by using its spatial Fourier transform. For a temporal initial value condition, we have to walk more gingerly.

We have seen now this approach applied in classical static (a static Poisson's equation problem) and dynamic conditions (a wave propagation problem). When an electron travels and undergoes scattering, it is a situation where transport is undergoing scattering. It may very well be that transport may be well described by a near-classical description, such as a Drude model with $\tau_{\mathbf{k}}$ capturing an expectation of scattering $\langle\tau_{\mathbf{k}}\rangle$, which must be dealt with quantum-mechanically to be accurate. The Green's function approach is useful in both these conditions. And transport is one of the very essential properties of semiconductors that is of interest to us. The evolution of the dynamics takes place in both time and space. So, let us look at its deployment in these conditions.

13.4 *Green's functions in classical and quantum evolution under scattering*

TAKE THE MOTION OF nearly free electrons of effective mass m^* in the presence of scattering with an expectation of a momentum relaxation time of $\langle\tau_{\mathbf{k}}\rangle$ and an electric field $\boldsymbol{\mathcal{E}}$. The perturbation is the force $-e\boldsymbol{\mathcal{E}}$ arising in this field. Let us determine the Green's function by just viewing this problem as a response problem. In a time dt', this electron—absent scattering and starting from rest—acquires a momentum $-e\boldsymbol{\mathcal{E}}dt'$ and a velocity $\boldsymbol{\mathcal{E}}dt'/m^*$. Should this field be present only for this time duration of dt', with scattering absent, the electron maintains this acquired velocity and momentum for all the later times. Now let us introduce scattering at the rate of $1/\tau_{\mathbf{k}}$. In time interval dt, the scattering probability is $dt/\tau_{\mathbf{k}}$. For a cloud of electrons, the number of electrons that will not undergo scattering between $t = 0$ at a population of $n(0)$ and time $t = t$, with the electron population that did not undergo scattering written as $n(t)$, is

$$n(t) = n(0)\exp\left(-\frac{t}{\langle\tau_{\mathbf{k}}\rangle}\right) \tag{13.25}$$

The expectation of the velocity at time t follows from all the contributions to the velocity over all the increments dt' during the course of the time evolution from $t = 0$ to $t = t$. This is

$$\langle\mathbf{v}\rangle = -\int_{-\infty}^{t}\left[-\exp\left(-\frac{t-t'}{\langle\tau_{\mathbf{k}}\rangle}\right)\right]\frac{e}{m^*}\boldsymbol{\mathcal{E}}\,dt' = -\frac{e\boldsymbol{\mathcal{E}}\langle\tau_{\mathbf{k}}\rangle}{m^*}. \tag{13.26}$$

This is as expected. The change in velocity can be viewed as a response to an electric field acting for time dt' at time t' with a response function $G(t-t')$. This Green's function is

$$G(t-t') = -\frac{e}{m^*}\exp\left(-\frac{t-t'}{\langle\tau_{\mathbf{k}}\rangle}\right). \tag{13.27}$$

We have captured the response of the past through the Green's function, and we may now integrate this entire past history to get the expected velocity at time t. Both the effect of the field causing energy and momentum gain, and the randomization due to scattering, have been incorporated. This interpretation casts the Green's function acting on the source function under the boundary and initial constraints leading to the evolution of the response. Our source function is the force arising in the field and causing a change in velocity. The Green's function as the response function incorporated into it the scattering in the environment that is always present.

The sum of history as incorporated in the Feynman diagrams is another way of incorporating this history. Julian Schwinger and Richard Feynman both showed quantum electrodynamics at work through their part. One of their earliest successes was the correction terms of the Lamb shift, which is a consequence of the interaction between vacuum energy fluctuations and the hydrogen electron. Schwinger brought a relativistically covariant form to quantum electrodynamics, and the idea of renormalization, which removes self-energy infinities. He was successful at using Green's functions in a relativistic invariant way, drawing on his use of Green's functions in solving electromagnetic problems for radar development at the *MIT* Radiation Laboratory during the Second World War. Sin-Itiro Tomanaga had developed Schwinger-like space-time ideas in imperial Japan earlier. Feynman used his namesake diagrams. Freeman Dyson showed how these two seemingly very different approaches map to each other, using the *S*-matrix.

We now explore this for the quantum-mechanical problem. Causality here again is a reflection of looking into the evolution due to a cause in space (in a place) and time (in the past). The time-dependent Schrödinger equation tells us that, for a Hamiltonian $\hat{\mathscr{H}} = \hat{\mathscr{H}}_0 + \hat{\mathscr{H}}'$, where $\hat{\mathscr{H}}_0$ is the steady state, and $\hat{\mathscr{H}}'$ is the perturbation, the time evolution is given by

$$-\frac{\hbar}{i}\frac{\partial\psi(z,t)}{\partial t} = \hat{\mathscr{H}}\psi(z,t) = \left[\hat{\mathscr{H}}_0 + \hat{\mathscr{H}}'(z,t)\right]\psi(z,t). \qquad (13.28)$$

For the simplest of problems—no potential energy—$\hat{\mathscr{H}}_0 = -(\hbar^2/2m)\nabla^2$. The wavefunction of the particle of mass m evolves as a result of a perturbation $\hat{\mathscr{H}}'$ governed by this equation. If the system has more than one particle—say, 2—this equation will still function through the use of a reduced mass.

The differential Equation 13.28 is of first order in time and prescribes how the evolution of the wavefunction takes place in time. This evolution solution is a phase change given by $\exp(-i\langle\hat{\mathscr{H}}'\rangle t/\hbar)$. In the presence of perturbation, the state of the system evolves in a more complex way. If the wavefunction is known at time t_0, that is at $t = t_0\ \forall\ z$, then the wavefunction can be determined for all times *before and after*, that is, for $t \leq t_0$ and for $t \geq t_0$. *The observation for this deterministic wavefunction evolution will still be nondeterministic.*

Quantum mechanics is a non-deterministic theory. The adverb "deterministic" here implies that we know how ψ evolves. When an observation is performed at a future moment, the eigenstate observed may be different if the wavefunction is not an eigenfunction. Only when the wavefunction is an eigenfunction will we observe just the change in phase, which is equivalent to finding the system at the same eigenenergy.

Equation 13.28 is also linear in ψ. Superposition holds. Since solutions are superposable linearly, the relationship between the wavefunction $\psi(z,t)$ at time t and $\psi(z,t_0)$ at time $t=t_0$ must be linear. So, the wavefunction evolution satisfies a linear homogeneous form:

$$\psi(z,t) = i\int_{\Omega\to\infty} G(z,t|z',t')\psi(z',t')\,d^3z', \qquad (13.29)$$

which prescribes the Green function $G(z,t|z',t')$ for the Hamiltonian $\hat{\mathscr{H}}$.

What is different here is the agnosticism to the arrow of time. Quantum mechanics is linear and reversible.

We will be able to distinguish forward and reverse time by how we define the Green's functions. The *retarded Green's function* ($G^+(z,t|z',t')$) operates forward in time, and the *advancing Green's function* ($G^-(z,t|z',t')$) operates backward in time. We also call these functions propagators—an operator that performs propagation. The Green's function, or propagator, is a sum of these two, and our definition of these—consistent with the above comments—is

$$G^+(z,t|z',t') = \begin{cases} G(z,t|z',t') & \text{for } t > t', \\ 0 & \text{for } t < t', \end{cases} \quad \text{and}$$

$$G^-(z,t|z',t') = \begin{cases} -G(z,t|z',t') & \text{for } t < t', \\ 0 & \text{for } t > t'. \end{cases} \qquad (13.30)$$

Take the Heaviside function $\Theta(\tau)$; then,

$$\Theta(t-t')\psi(z,t) = i\int_{\Omega\to\infty} G^{+}(z,t|z',t')\psi(z',t')\,d^3z', \quad \text{for } t > t'. \tag{13.31}$$

This equation is still valid for $t < t'$. It reduces to an identity of $t - t' = 0$. So, the Heaviside operator operating on the wavefunction—here, the wavefunction at forward time—is given by the retarded propagator operating on the prior—a wavefunction known to us. This is precisely causal evolution. Now take the advancing Green's function, which lets us determine the past wavefunction $\psi(z,t)$ from the prior $\psi(z',t')$, where $t<t'$. Now, we operate with the Heaviside function, and get

$$\Theta(t'-t)\psi(z,t) = -i\int_{\Omega\to\infty} G^{-}(z,t|z',t')\psi(z',t')\,d^3z' \quad \text{for } t < t'. \tag{13.32}$$

Again, for $t > t'$, we get an identity relationship, and this description remains valid.

What this has accomplished is that the Green's function for the time-reversible, linear, quantum-mechanical description has been broken up into a part that describes forward time (the retarded propagator) and reverse time (the advancing propagator). The sum of the two is still the complete Green's function. The Heaviside function gives us the means to separate forward time from backward time.

Green's function's linear response properties can now be explored, and, having satisfied ourselves of them, one should be able to see the evolution as the system responds, even as it undergoes different interactions. We will take scattering as one of the important examples that is of special significance for us. But first consider a few properties.

Consider an instant of time t_1 such that $t > t_1 > t'$. So, we choose an instant somewhere in the forward time and look beyond.
Claim:

$$G^{+}(r,t|r',t') = i\int_{\Omega\to\infty} G^{+}(r,t|r_1,t_1)G^{+}(r_1,t_1;r',t')\,d^3r_1 \tag{13.33}$$

for $t > t_1 > t'$. This says that we can choose arbitrary intervening times in the future and that the Green's function operator cumulates. The end result is independent of the intervening interval's partitioning choice for a linear system.
Proof: Since $t > t'$,

$$\psi(r,t) = i\int_{\Omega\to\infty} G^{+}(r,t|r',t')\psi(r',t')\,d^3r'. \tag{13.34}$$

This is propagation into the future. So long as $t > t'$, one may start with $\psi(r',t')$ at any time t':

$$\begin{aligned}\psi(r,t) &= i\int_{\Omega\to\infty} G^{+}(r,t|r_1,t_1)\psi(r_1,t_1)\,d^3r_1\\ &= i\int_{\Omega\to\infty} G^{+}(r,t|x_1,t_1)\,d^3r_1\\ &\quad\times\; i\int_{\Omega\to\infty} G^{+}(x_1,t_1|r',t')\psi(r',t')\,d^3r'\\ &= i\int_{\Omega\to\infty} d^3r' i\int_{\Omega\to\infty} G^{+}(r,t|x_1,t_1)\\ &\quad\times\; G^{+}(r_1,t_1|r',t')\psi(r',t')\,d^3r_1.\end{aligned} \tag{13.35}$$

The retarding operator operating on $\psi(r',t')$ is therefore $G^{+}(r,t|r',t')$.

By symmetry, it also follows that

$$G^{-}(r,t|r',t') = -i\int_{\Omega\to\infty} G^{-}(r,t|x_1,t_1)G^{-}(x_1,t_1|r',t')\,d^3r_1,\quad \text{for } t < t_1 < t, \tag{13.36}$$

for going backward in time. So, the independence of the result on the choice of any intervening time also holds for going backward in time.

Claim: If we choose t_1 as some instant of time for the reference, then, in three-dimensional space,

$$\delta(r'-r) = \int_{\Omega\to\infty} G^{+}(r,t'|r_1,t_1)G^{-}(r_1,t_1|r',t')\,d^3r_1,\quad \text{for } t' > t_1. \tag{13.37}$$

Proof:

$$\begin{aligned}\psi(r,t') &= i\int_{\Omega\to\infty} G^{+}(r,t'|r_1,t_1)\psi(x_1,t_1)\,d^3r_1\\ &= i\int_{\Omega\to\infty} G^{+}(r,t'|r_1,t_1)\,d^3r_1\\ &\quad\times\;(-i)\int_{\Omega\to\infty} G^{-}(r_1,t_1;r',t')\psi(r',t')\,d^3r'\\ &= \int_{\Omega\to\infty} d^3r'\int_{\Omega\to\infty} G^{+}(r,t';x_1,t_1)\\ &\quad\times\; G^{-}(r_1,t_1;r',t')\psi(r',t')\,d^3r_1.\end{aligned} \tag{13.38}$$

But if $t' > t_1$, then

$$\psi(r,t') = \int_{\Omega\to\infty} \delta(r'-r)\psi(r',t')\,d^3r'; \tag{13.39}$$

therefore, our claim stands proven.

This also implies by symmetry that

$$\delta(r'-r) = \int_{\Omega\to\infty} G^{-}(r,t'|r_1,t_1)G^{+}(r_1,t_1|r',t')\,d^3r_1,\quad \text{for } t' < t_1. \tag{13.40}$$

This proof shows that the linear response just means that the propagation is a continuation of events. In Green's function form, the operators multiply and cumulate through the integration. One may even go back in time and come back, and, so long as the Green's functions hold—that is, the causal effect is the same—one gets an identical result.

So, whether one goes forward in time, backward in time, chooses other instances of time in-between as intermediate steps—for that matter, instants of time in the past or future as intermediate steps—for the linear response and for quantum-mechanical determination, with its in-built reversibility, one arrives at the same effect by using the properly chosen Green's function representing the causal response. It reflects the sum of histories and causality in this reversible environment. This propagator method allows one to proceed "physically" and "intuitively," and write Equation 13.29 with the proper choice of $G(z,t|z',t')$ commensurate with the physics of the stimulus and with the boundary and the initial conditions. We now proceed from this so that we can tackle scattering.

Let $\{|u(z)\rangle\}$ be the orthonormal eigensolution set from which we compose the wavefunction $|\psi(z)\rangle$. First, consider a time-independent perturbation for a zero-potential problem. The solution to the zero potential Hamiltonian is a wave, that is, free flight. For an eigenfunction that has eigenenergy E, the Schrödinger equation is

$$\left[\hat{\mathscr{H}}_0 + \hat{\mathscr{H}}'(z)\right]|u_E(z)\rangle = \left[-\frac{\hbar^2}{2m}\nabla^2 + \hat{\mathscr{H}}'(z)\right]|u_E(z)\rangle = E|u_E(z)\rangle. \tag{13.41}$$

If the solution is known for any time t', the time-dependent Schrödinger equation

$$-\frac{\hbar}{i}\frac{\partial}{\partial t}|\psi(z,t)\rangle = \left[-\frac{\hbar^2}{2m}\nabla^2 + \hat{\mathscr{H}}'(z)\right]|\psi(z,t)\rangle \tag{13.42}$$

gives formally the solution for any time t when one knows the solution at $t = t'$. In general, the wavefunction solution is

$$|\psi(z,t)\rangle = \sum_E c_E(t)|u_E(z)\rangle, \tag{13.43}$$

where the coefficients

$$c_E(t) = \int_{\Omega\to\infty} u_E^*(z)\psi(z,t)\,d^3z \tag{13.44}$$

are time dependent. Substituting this into Equation 13.42, we obtain

$$-\frac{\hbar}{i}\sum_E \frac{dc_E(t)}{dt}|u_E(z)\rangle = \sum_E c_E(t)E|u_E(z)\rangle, \tag{13.45}$$

where we are subscripting by E the different eigenfunctions with different eigenenergies. Because of orthonormality, we may multiply by the bra $\langle u_E(z)|$ and get

$$-\frac{\hbar}{i}\frac{dc_E(t)}{dt} = Ec_E(t)$$
$$\therefore\quad c_E(t) = c_E(t')\exp\left[-\frac{iE(t-t'))}{\hbar}\right]. \qquad (13.46)$$

Recall that, in this procedure, which was often used in S. Tiwari, "Quantum, statistical and information mechanics: A unified introduction," Electroscience 1, Oxford University Press, ISBN 978-0-198-75985-0 (forthcoming), orthonormality allows one to extract coefficients by multiplying with the transpose, resulting in only one term on the left and one term on the right. This is the way to get the diagonal terms of the matrix representing the set of equations that these relationships represent. This procedure also appears in the extraction of the Golden rule for transitions. These c_Es represent the probability of finding the system in that eigenfunction state of $|u_E\rangle$ if an observation is made.

The probability of finding the state to be $|u_E(z)\rangle$ with an energy E is unchanged, with $|c_E(t)|^2 = |c_E(t')|^2$. We have now found

$$\begin{aligned}\psi(z,t) &= \sum_E c_E(t)|u_E(z)\rangle \\ &= \sum_E c_E(t')\exp\left[-\frac{iE(t-t')}{\hbar}\right]|u_E(z)\rangle \\ &= \int_{\Omega\to\infty}\left[\sum_E \langle u_E(z)|u_E(z)\rangle\right] \\ &\quad\times\exp\left[-\frac{iE(l-l')}{\hbar}\right]|\psi(x,t)\rangle\, d^3z'. \qquad (13.47)\end{aligned}$$

So, we now have our Green's function:

$$G(z,t|z,t) = -i\left[\sum_E \langle u_E(z)|u_E(z)\rangle\right]\exp\left[-\frac{iE(t-t')}{\hbar}\right]. \qquad (13.48)$$

If $t = t'$, the right-hand side of this equation reduces to $-i\delta(z-z')$, as it should. This Green's function is complete.

We derived this Green's function in the presence of a perturbation $\hat{\mathscr{H}}'(z)$, which was independent of time. It is also valid when there exists no perturbation ($\hat{\mathscr{H}}'(z) = 0\ \forall t$). Absent any perturbation, this Green's function is called the free Green's function. This free Green's function then leads to

$$u_E(z) = \frac{1}{(2\pi/\hbar)^{3/2}}\exp(-i\mathbf{k}\cdot\mathbf{z}), \qquad (13.49)$$

where $\mathbf{k} = \mathbf{p}/\hbar$, and $E = \mathbf{p}^2/2m$, a solution for a free wave.

Let us look at some of the properties of this free wave arising under the condition that $\hat{V} = 0$ and $\hat{\mathscr{H}}'(z) = 0$ for all t:

$$\begin{aligned}G_0(z,t|z',t') &= -i\frac{1}{(2\pi/\hbar)^{3/2}} \\ &\quad\times\int\exp[i\mathbf{k}\cdot(\mathbf{z}-\mathbf{z}')]\exp\left[-\frac{iE(t-t')}{\hbar}\right]d^3p \\ &= -i\frac{1}{(2\pi/\hbar)^{3/2}}\end{aligned}$$

$$\times \int \exp\left\{\frac{i}{\hbar}\left[p_x(x-x') + p_y(y-y') + p_z(z-z')\right]\right.$$
$$\left. - \frac{i}{\hbar}(t-t')\right\} dp_x\, dp_y\, dp_z. \qquad (13.50)$$

Subscript 0 identifies it as a Green's function for a free wave. This is an explicitly solvable analytic problem. We show this through one of the three symmetric Cartesian-directed components:

Note that the term inside the integral has a "Gaussian" form where one may employ

$$\int_{-\infty}^{\infty} \exp(-i\alpha z^2)\, dz = \left(\frac{\pi}{i\alpha}\right)^{1/2}.$$

$$\begin{aligned}
&\frac{i}{\hbar}p_x(x-x') - \frac{i}{\hbar}\frac{p_x^2(t-t')}{2m} \\
&= -\frac{i}{\hbar}\left[\frac{p_x\sqrt{t-t'}}{\sqrt{2m}} - \frac{\sqrt{2m}(x-x')}{2\sqrt{t-t'}}\right]^2 + \frac{i}{\hbar}\frac{m(x-x')^2}{2(t-t')} \\
&= -\frac{i}{\hbar}\frac{t-t'}{2m}\varpi^2 + \frac{i}{\hbar}\frac{m(x-x')^2}{2(t-t')},
\end{aligned} \qquad (13.51)$$

where

$$\varpi = p_x - \frac{m(x-x')}{t-t'}, \qquad (13.52)$$

which lets us manipulate using Gaussian integration employing $\alpha = t - t'/2m\hbar$. For this one-of-three symmetric term,

$$\begin{aligned}
&\int_{-\infty}^{\infty} dp_x \exp\left\{\frac{i}{\hbar}\left[p_x(x-x') - \frac{p_x^2(t-t')}{2m}\right]\right\} \\
&= \exp\left[\frac{i}{\hbar}\frac{m(x-x')^2}{2(t-t')}\right]\int_{-\infty}^{\infty} \exp\left[-\frac{i}{\hbar}\frac{t-td\varpi}{2m}\varpi^2\right] d\varpi \\
&= \exp\left[\frac{i}{\hbar}\frac{m(x-x')^2}{2(t-t')}\right]\sqrt{\frac{2\pi m\hbar}{i(t-t')}},
\end{aligned} \qquad (13.53)$$

and therefore the Green's function solution can be written as

$$\begin{aligned}
G_0(x,t|x',t') &= -\frac{i}{(2\pi\hbar)^3}\left[\frac{2\pi m\hbar}{i(t-t')}\right]^{3/2} \exp\left[i\frac{|x-x'|^2 2m\hbar}{4\hbar^2(t-t')}\right] \\
&= -i\left[\frac{m}{2\pi i\hbar(t-t')}\right]^{3/2} \exp\left[\frac{im|x-x'|^2}{2\hbar(t-t')}\right].
\end{aligned} \qquad (13.54)$$

We can split this Green's function to forms for forward propagation (retarded function) and backward propagation (advancing function) by using the Heaviside approach:

$$\begin{aligned}
G_0^+(z,t|z,t) &= G_0(z,t|z',t')\Theta(t-t') \\
&= -i\left[\frac{m}{2\pi i\hbar(t-t')}\right]^{3/2} \exp\left[\frac{im|z-z'|^2}{2\hbar(t-t')}\right]\Theta(t-t'), \text{ and} \\
G_0^-(z,t|z',t') &= -G_0(z,t|z',t')\Theta(t'-t) \\
&= +i\left[\frac{m}{2\pi i\hbar(t-t')}\right]^{3/2} \exp\left[\frac{im|z-z'|^2}{2\hbar(t-t')}\right]\Theta(t'-t).
\end{aligned} \qquad (13.55)$$

Note that

$$G_0^+(z,t|z',t') = G_0^{-*}(z',t'|z,t), \tag{13.56}$$

that is, they are complex conjugates.

When a perturbation $\hat{\mathscr{H}}'(z,t)$ is present, we need to modify the free Green's function. This perturbation may be a scattering event arising in an energetic interaction that turns on at t_1 for a short time Δt_1. Let $\hat{\mathscr{H}}'(z_1,t_1)$ be the perturbation during this time Δt_1. For $t < t_1$, the wavefunction is that of a free particle (a free wave), and its propagation is guided by $G_0(z,t|z',t')$. At $t = t_1$, $\hat{\mathscr{H}}'(z_1,t_1)$ turns on and causes this free wave to scatter. Writing ϕ as the free wave solution, and rewriting the time-dependent Schrödinger equation as

Now you can see why we looked at Green's function through an intermediate time. If no perturbation exists at that intermediate time, the Green's function's operation accumulates.

$$\left[-\frac{\hbar}{i}\frac{\partial}{\partial t_1} - \hat{\mathscr{H}}_0\right]\psi(z_1,t_1) = \hat{\mathscr{H}}'(z_1,t_1)\psi(z_1,t_1), \tag{13.57}$$

where $\hat{\mathscr{H}}'(z_1,t_1)$ is active for $t_1 < t < t_1 + \Delta t_1$, the result of the interaction is

$$\psi(z_1,t_1) = \phi(z_1,t_1) + \Delta\psi(z_1,t_1). \tag{13.58}$$

Here, we have written in a form to indicate that $\Delta\psi(z_1,t_1)$ is a change in the wavefunction resulting from the interaction, with

$$\left[-\frac{\hbar}{i}\frac{\partial}{\partial t_1} - \hat{\mathscr{H}}_0\right]\phi(z_1,t_1) = 0, \tag{13.59}$$

and $\Delta\psi(z_1,t_1) = 0$, for $t < t_1$. So,

$$\left[-\frac{\hbar}{i}\frac{\partial}{\partial t_1} - \hat{\mathscr{H}}_0\right]\Delta\psi(z_1,t_1) = \hat{\mathscr{H}}'(z_1,t_1)\left[\phi(z_1,t_1) + \Delta\psi(z_1,t_1)\right], \tag{13.60}$$

where, after discounting $\hat{\mathscr{H}}'(z_1,t_1)\Delta\psi$ as a second order perturbation term, we write

$$\left[-\frac{\hbar}{i}\frac{\partial}{\partial t_1} - \hat{\mathscr{H}}_0\right]\Delta\psi(z_1,t_1) = \hat{\mathscr{H}}'(z_1,t_1)\phi(z_1,t_1), \tag{13.61}$$

whose solution with a limited time of interaction is the integral

$$\Delta\psi(z_1,t_1+\Delta t_1) = -\frac{i}{\hbar}\int_{t_1}^{t_1+\Delta t_1}\left[\hat{\mathscr{H}}_0\Delta\psi(z_1,t') + \hat{\mathscr{H}}'(z_1,t')\phi(z_1,t')\right]dt'. \tag{13.62}$$

The first term here is small, a second-order energy change in the perturbation that arose. The second term is the primary contribution of the perturbation on the incident free wave. This is

$$\Delta\psi(z_1,t_1+\Delta t_1) = -\frac{i}{\hbar}\hat{\mathscr{H}}'(z_1,t_1)\phi(z_1,t_1)\Delta t_1. \tag{13.63}$$

When the perturbation disappears after time Δt_1, the scattered wave again propagates freely. For this scattered wave,

$$\begin{aligned}\Delta\psi(z,t) &= i\int G_0(z,t|z',t')\Delta\psi(z_1,t_1)\,d^3z_1\\ &= \int G_0(z',t'|z,t)\frac{1}{\hbar}\hat{\mathscr{H}}'(z_1,t_1)\phi(z_1,t_1)\Delta t_1\,d^3z_1. \end{aligned} \quad (13.64)$$

The free wave Green's function operator operating on the perturbation term of the wavefunction, which is the cumulative response over the interval Δt_1 during which the perturbation operates, gives the change in wavefunction of the scattered free wave.

We have now developed an analytic description, and a figurative picture as shown in Figure 13.3, of the evolution and interaction in space-time through the Green's function technique. For the free wave, the Green's function describes, in space-time, how the description at a time (z,t) can be connected to a time (z',t'), as shown in panel (a) of Figure 13.3. In the presence of a single interaction at time t_1, the interaction leads to a scattered wave. This figure describes the wavefunction evolution as

$$\begin{aligned}\psi(z,t) &= \phi(z,t) + \Delta\psi(z,t)\\ &= \phi(z,t) + \int G_0(z,t|z_1,t_1)\frac{1}{\hbar}\hat{\mathscr{H}}'(z_1,t_1)\phi(z_1,t_1)\Delta t_1\,d^3z_1\\ &= i\int\Big[G_0(z,t|z',t')\\ &\quad + \int G_0(z,t|z_1,t_1)\frac{1}{\hbar}\hat{\mathscr{H}}'(z_1,t_1)\Delta t_1 G_0(z_1,t_1|z',t')\,d^3z_1\Big]\\ &\quad \times \phi(z',t')\,d^3z'. \end{aligned} \quad (13.65)$$

Figure 13.3 maps to the last of these analytic forms. The first term is the Green's function propagator evolving the incident free wave over the interval $(z',t') \mapsto (z,t)$. The second term is the net effect of the free wave propagation (the last term captured through $G_0(z_1,t_1|z',t')$), its perturbation captured through the interaction with the perturbation Hamiltonian and, finally, the free wave

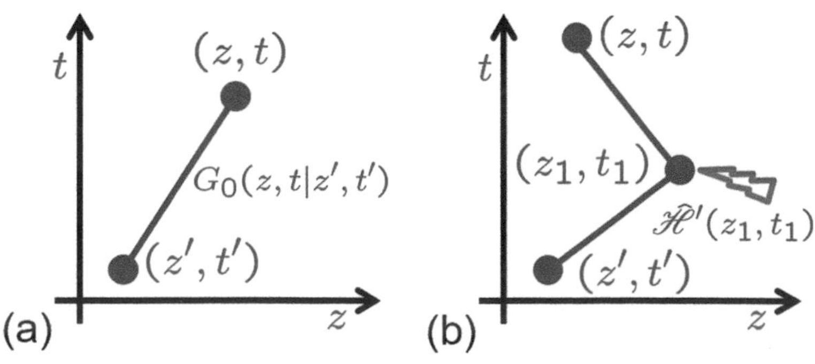

Figure 13.3: Part (a) shows the propagation of a free wave with the associated propagator $G_0(z,t|z',t')$ acting on the prior. Part (b) shows a free wave undergoing scattering at t_1 due to a perturbation $\hat{\mathscr{H}}'(z_1,t_1)$. $G_0(z,t|z',t')$ is still the propagator during the free flight.

propagation of this perturbation (captured through $\int G_0(z,t|z_1,t_1)$). So, the causal sequence of events have been partitioned into a free wave evolution over $(z',t') \mapsto (z,t)$, and a part related to the perturbation, which is built through evolution over $(z',t') \mapsto (z,t)$, at (z_1,t_1), and then $(z_1,t_1) \mapsto (z,t)$.

So, the Green's function for this example of single scattering can be summarized as

$$G(z,t|z',t') =$$
$$G_0(z,t;z',t') + \int G_0(z,t|z_1,t_1)\frac{1}{\hbar}\hat{\mathscr{H}}'(z_1,t_1)G_0(z_1,t_1|z',t')\Delta t_1\, d^3z_1. \quad (13.66)$$

This propagator is the simplest of examples involving a perturbation: a free propagation together with a short perturbation in time spread out in space. It sets up our description of *a single scattering event*.

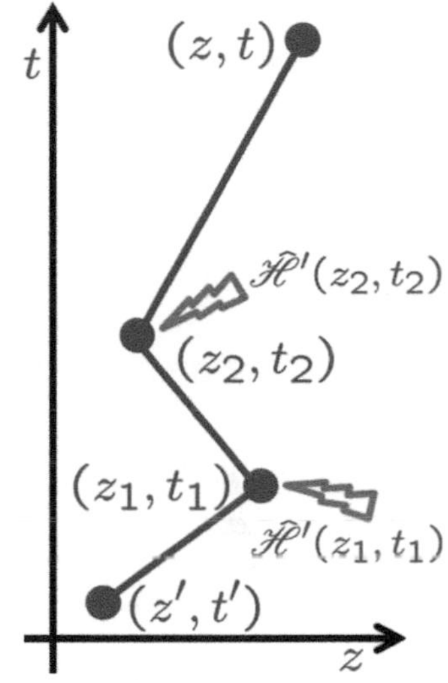

Figure 13.4: Space-time depiction of two interactions of a free wave in free flight.

The evolution description for multiple scattering events can now be written as a recursive iteration algorithm. What we need to do is incorporate any of the events through the perturbation effect $(1/\hbar)\hat{\mathscr{H}}'(z_i,t_i)\Delta t_i$ for the ith event occurring over the time duration Δt_i by integrating over space before the next free propagation. If two scattering events take place, their perturbations, which are represented by viewing Figure 13.4 as showing a sequence of causal events, lets us write

$$\begin{aligned}
\Delta\psi(z) &= G_0(z|z_2)\frac{1}{\hbar}\hat{\mathscr{H}}'(z_2)\psi(z_2)\Delta t_2 \int d^3z_2 \\
&= i\int d^3z'\, d^3z_2\, \Delta t_2 G_0(z|z_2)\frac{1}{\hbar}\hat{\mathscr{H}}'(z_2) \\
&\quad \times \left[G_0(z_2|z) + \int d^3z_1\, \Delta t_1 G_0(z_2|z_1)\right. \\
&\quad \left. \times \frac{1}{\hbar}\hat{\mathscr{H}}'(z_1)G_0(z_1|z')\right]\phi(z',t'), \quad (13.67)
\end{aligned}$$

where the time parameter has not been shown within the Green's function, for brevity. It can be seen from the context as corresponding to the subscript of the spatial parameter. In Equation 13.67, the first equation shows the evolution following the scattering event at (z_2,t_2). The longer second equation breaks this up into the propagation from the initial reference (z',t') through both of the scatterings while still undergoing free flight. This second form shows that the sequence of the two events leads to a term that has as its propagation and perturbation consequence the integration of

$$G(z,z_2,z_1,z') = G_0(z|z_2)\frac{1}{\hbar}\hat{\mathscr{H}}'(z_2)G_0(z_2|z_1)\frac{1}{\hbar}\hat{\mathscr{H}}'(z_1)G_0(z|z_1)\phi(z'). \quad (13.68)$$

The propagator is just an accumulation of the free propagation and a short perturbation in time over the entire space through the sum of their histories. The wavefunction change that arose is

$$\begin{aligned}\Delta\psi(z) = \phi(z) &+ \int d^3z_1\, \Delta t_1 G_0(z|x_1)\frac{1}{\hbar}\hat{\mathscr{H}}'(z_1)\phi(z_1) \\ &+ \int d^3z_2\Delta t_2 G_0(z|z_2)\frac{1}{\hbar}\hat{\mathscr{H}}'(z_2)\phi(z_2) \\ &+ \int d^3z_1\Delta t_1 \int d^3z_2\Delta t_2 G_0(z|z_2)\frac{1}{\hbar}\hat{\mathscr{H}}'(z_2)G_0(z_2|z_1) \\ &\times \frac{1}{\hbar}\hat{\mathscr{H}}'(z_1)\phi(z_1). \end{aligned} \tag{13.69}$$

Note how this change arose due to the perturbation first at (z_1, t_1), then free propagation, then the perturbation (z_2, t_2) followed by the free propagation onto (z, t). The pattern is clear. One is integrating over the spatial coordinates and summing over the time of events; that is, ending up with calculations in four dimensions. The order of time is strict in this expansion. It also matches the retarded Green's function $G_0^+(x, t|x', t')$ in that it only acts forward. So, with multiple scattering events, one can write this retarded Green's function as

$$\begin{aligned}G^+(z|z') = G_0^+(z|z') &+ \int d^4z_1\, G_0^+(z|z_1)\frac{1}{\hbar}\hat{\mathscr{H}}'(z_1)G_0^+(z_1|z') \\ &+ \int d^4z_1\, d^4z_2\, G_0^+(z|z_1)\frac{1}{\hbar}\hat{\mathscr{H}}'(z_1)G_0^+(z_1|z_2) \\ &\times \frac{1}{\hbar}\hat{\mathscr{H}}'(z_2)G_0^+(z_2|z') + \cdots, \end{aligned} \tag{13.70}$$

with $d^4z_i \equiv d^3z_i\, dt_i$. Setting up the equation in this way with the sequencing of the order starting at the initial/boundary value at z', t' lets us write in a recursive algorithmic form and thus solve with the initial and boundary values employed consistently.

For a finite time-period interaction, that is, $\lim_{t\to-\infty} \mathscr{H}'(z, t) = 0$, and also $\lim_{t\to\infty} \mathscr{H}'(z, t) = 0$, which perturbs an incoming wave, that is, $\lim_{t\to-\infty} \psi(z, t) = \phi(z, t)$, using the procedure developed, one can immediately write

$$\begin{aligned}\psi^{(+)}(z, t) &= \lim_{t\to-\infty} i \int d^3z'\, G^+(z, t|z', t')\phi(z', t') \\ &= \phi(z) + \int d^4z_1\, G_0^+(z|z_1)\frac{1}{\hbar}\mathscr{H}'(z_1)\psi^{(+)}(z_1). \end{aligned} \tag{13.71}$$

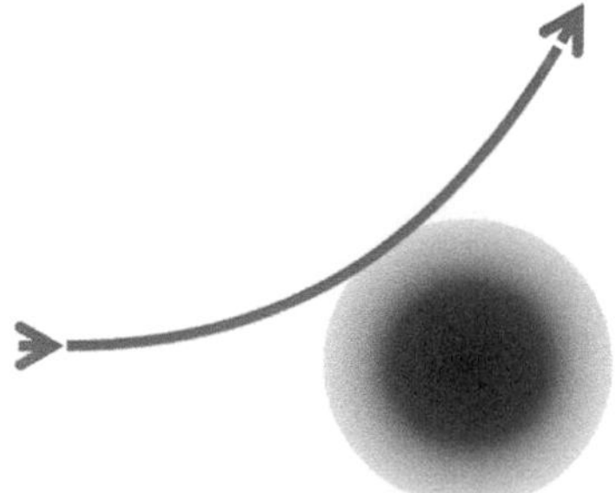

Figure 13.5: A wave undergoing an interaction that leads to a scattered wave propagating into the future.

The second integral term is the scattered wave propagating into the future, as shown in Figure 13.5.

If the scattering is adiabatic, that is, the interaction turns on much more slowly than do the kinetic excitation energies of the system, then the state change evolves slowly. The wavefunction continues to satisfy the time-independent Schrödinger equation in time. A classic example is the one-dimensional problem of a wave encountering a planar barrier. If the potential change is sharp—the length scale of the potential change is much shorter than that of the wavelength $\lambda = 2\pi/k$ of the wave (where the kinetic energy is $\hbar^2k^2/2m$)—then reflections happen and transmitting and reflecting waves appear. The wave breaks up into two. If the interaction is adiabatic, that is, there is a slow change in the barrier, then reflection asymptotically disappears. The time counterpart to this spatial example is a perturbation $\mathscr{H}'(x,t) \mapsto \mathscr{H}'(x,t)\exp(-\alpha|t|)$, where $\alpha > 0$, with α suitably chosen to preserve the essentials of the interaction. So, in this adiabatic perturbation, the plane wave in the past history evolves to a plane wave in the future history. And scattering has occurred, and one has the quantitative form for it. This Green's function usage in quantum-mechanical conditions is formally the equivalent of the Green's function usage in classical conditions. They are both examples of usage in linear response theory.

The Green's function solving the Schrödinger equation, $G(\mathbf{r},t|\mathbf{r}',t')$, gives the conditional probability—the Bayesian notion—of finding the particle at $(\mathbf{r},t)$ as $|G(\mathbf{r},t|\mathbf{r}',t')|^2\,d^3\mathbf{r}$, given that it was at $(\mathbf{r}',t')$. This quantum-mechanical mapping of the earlier classical discussion can also be interpreted to tell us that we can find the Green's function—as we did for classical conditions—by solving the time-dependent Schrödinger equation

$$-\frac{\hbar}{i}\frac{\partial G}{\partial t} - \hat{\mathscr{H}}G = \delta(t-t')\delta(\mathbf{r}-\mathbf{r}'). \tag{13.72}$$

Starting precisely at $(\mathbf{r}',t')$ within the Dirac delta view, the Green's function evolves in time and space under the conservation dynamics dictated by the Schrödinger equation. This equation form also tells us how the Green's function evolves at $t = t'$. As in the earlier flux analogy, we choose a time interval $(t'-\Delta t, t'+\Delta t)$ strapping the Dirac delta function and shrink it:

$$\lim_{\Delta t\to 0}\int_{t'-\Delta t}^{t'+\Delta t}\left[-\frac{\hbar}{i}\frac{\partial G}{\partial t} - \hat{\mathscr{H}}G = \delta(t-t')\delta(\mathbf{r}-\mathbf{r}')\right]dt \tag{13.73}$$

then gives

$$G(t=t'^{+}) - G(t=t'^{-}) = -\frac{i}{\hbar}\delta(\mathbf{r}-\mathbf{r}'). \tag{13.74}$$

The Green's function discontinuously changes at $t = t'$. The propagator-retarding Green's function is the first term here. It vanishes in the past and prescribes evolution in time. The advanced Green's function is the second term. It vanishes for the future and prescribes going back in time. Building on plane wave eigenfunctions, the solution is

$$G^{+}(\mathbf{r},t|\mathbf{r}',t') = \sum_n \lim_{\varepsilon\to 0}\int_{-\infty}^{\infty}\frac{d\omega}{2\pi}\exp[-i\omega(t-t')]\frac{\psi_n(\mathbf{r})\psi_n^*(\mathbf{r}')}{\hbar\omega - E_n + i\varepsilon}$$
$$= -\frac{i}{\hbar}\sum_n \exp\left[-\frac{iE_n(t-t')}{\hbar}\right]\psi_n(\mathbf{r})\psi_n^*(\mathbf{r}')\Theta(t-t') \quad (13.75)$$

for the retarding Green's function, and

$$G^{-}(\mathbf{r},t;\mathbf{r}',t') = \sum_n \lim_{\varepsilon\to 0}\int_{-\infty}^{\infty}\frac{d\omega}{2\pi}\exp[-i\omega(t-t')]\frac{\psi_n(\mathbf{r})\psi_n^*(\mathbf{r}')}{\hbar\omega - E_n - i\varepsilon} \quad (13.76)$$

for the propagator-advancing Green's function. The $\varepsilon \to 0$ is the mathematical subterfuge for tackling infinities in working through the Dirac delta function under the integral. When a system can be described by a finite number of quantum states $|n\rangle$, these equations are a conducive way to tackle the interaction problem. Semiconductor scattering problems are in this class of problems. The Green's function can then be viewed as a vector matrix in space-time, analogous to the Dirac vector formulation versus the Schrödinger wavefunction formulation. The Dirac delta function then is a unit matrix ($\mathbb{I}$) function.

We define the inverse of the Green's operator describing this evolution as follows:

$$K^{\pm}G^{\pm} = 1$$
$$\therefore\ K^{\pm} = \lim_{\varepsilon\to 0}\left(-\frac{\hbar}{i}\frac{\partial}{\partial t} - \hat{\mathscr{H}} \pm i\varepsilon\right), \text{ and}$$
$$K(\mathbf{r},t;\mathbf{r}',t') = K(\mathbf{r},t)\delta(\mathbf{r}-\mathbf{r}')\delta(t-t'). \quad (13.77)$$

The matrix product here can be viewed as an integration over all intermediate coordinates in space-time.

When the perturbation is weak, that is, $\hat{\mathscr{H}} = \hat{\mathscr{H}}_0 + \hat{\mathscr{H}}'$, with $\hat{\mathscr{H}}' \ll \hat{\mathscr{H}}_0$, the retarding Green's function is

$$G^{+} = [K^{+}]^{-1}$$
$$= [K_0^{+}[1 - [K_0^{+}]^{-1}\mathscr{H}']]^{-1}$$
$$= (1 - [K_0^{+}]^{-1}\mathscr{H}')^{-1}[K_0^{+}]^{-1}$$
$$= [1 + G_0^{+}\mathscr{H}' + G_0^{+}\mathscr{H}'G_0^{+}\mathscr{H}' + \cdots]G_0^{+}$$
$$= G_0^{+} + G_0^{+}\mathscr{H}'G_0^{+} + G_0^{+}\mathscr{H}'G_0^{+}\mathscr{H}'G_0^{+} + \cdots. \quad (13.78)$$

So,

$$K_0^{+} = \lim_{\varepsilon\to 0}\left(-\frac{\hbar}{i}\frac{\partial}{\partial t} - \hat{\mathscr{H}}_0 + i\varepsilon\right) \quad (13.79)$$

and

$$G_0 = K_0{}^{-1}. \tag{13.80}$$

We discussed Green's functions here in real space and time and formed a reciprocal Green's function. One could also just as well have written these two functions in terms of the reciprocals of real space and time; that is, wavevector and frequency. So, one can map the $(\mathbf{r}, t) \leftrightarrow (\mathbf{k}, \omega)$ equivalence to a $G(\mathbf{r}, t) \leftrightarrow G(\mathbf{k}, \omega)$ equivalence. In characterizing materials, our recourse is observation of the spectral response rather than the time response, which largely works only for device and transport response. This spectral response in linear systems and the use of the Kramers-Kronig relationship that ties in the real and the imaginary parts of susceptibility will be discussed in Chapter 14. For now, we make a few remarks on the time-domain response by generalizing the scattering response discussed. If sequential scattering events on one wave, which can be viewed as arriving from one port and being redirected elsewhere, that is, other ports, can be calculated, so can multiple wave streams and a generalized scattering matrix. The equivalence here is that if a wavefunction $\lim_{t\to\infty} |\psi_i^+(r,t)\rangle$ arises and, absent scattering, it would have been $\lim_{t\to\infty} |\phi_f(x,t)\rangle$, then

$$S_{if} = \lim_{t\to\infty} \langle \phi_f(r,t) | \psi_i^+(r,t) \rangle \tag{13.81}$$

One can view this as the quantum mechanical equivalent of the scattering matrix formalism useful in electromagnetic; that is, wave, and specifically microwave, theory.

is the scattering matrix element—the Heisenberg scattering matrix, or the S-matrix. This S-matrix will have symmetries that reflect the symmetries of the corresponding Hamiltonian.

It now becomes possible to look upon the device spatial and temporal response in a way similar to the port theory of microwaves, as shown in Figure 13.6. One can have contacts, in which case their Green's functions must be prescribed. Or one can have open boundary conditions. An open boundary condition is one where an incident wave must be directly specified and an exiting wave is one that exits never to return.

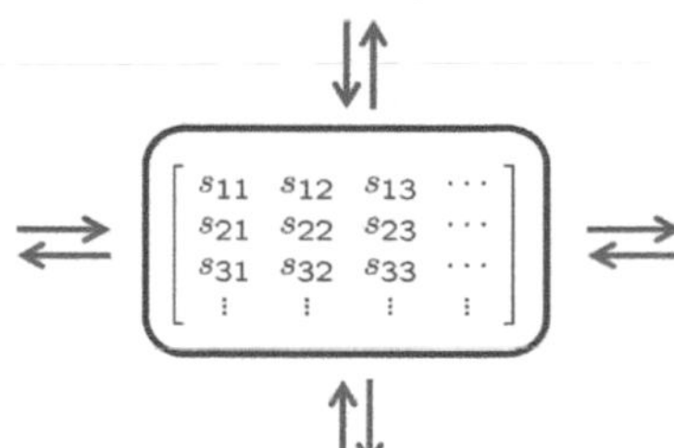

Figure 13.6: A multiport representation of the Heisenberg S-matrix for wavefunction interactions.

This description up to this point is for thermal equilibrium conditions, that is, conditions where, while the system is at temperature T, no other external stimulus is present. When external stimuli are present (bias voltages, light, magnetic field, etc.), then, to evaluate the causal consequences, one has to resort to nonequilibrium Green's functions. These are useful in both closed and open systems. Non-equilibrium Green's functions also work with strong external fields and can work with them nonperturbatively. In such a situation, electron-electron interactions can be treated as an infinite summation in a series. And these calculations can be performed while keeping the system approximately constrained to conservation of the number of particles in it, linear momentum, angular momentum, et cetera. Dissipation, that is, loss of energy from the particles, their wavefunctions of interest, memory, that is, a signature of the past in the present behavior, et cetera, all appear naturally and can be analyzed.

13.5 *Summary*

The Green's functions approach is a powerful technique for modeling a system's response while accounting for both a stimulus causation and a stochastic environmental interaction. Development of a Green's function, with known initial or boundary conditions, then gives a powerful way of determining the system's evolution. This has been the focus of this chapter, where we employed this alternative approach to analyze a range of semiconductor response behaviors, using classical, semi-classical and quantum-mechanical approaches. Green's function works with all these. But the necessity of connection in causal behavior, and chance in random behavior, makes it incumbent on us to understand the strict differences between causality, determinism and correlations. Causality is the tracing of the observed phenomena entirely to the identified cause. In Kramers-Kronig relations, which will be subject of Chapter 14, the tie in between the real and imaginary parts of a linear response is causal, since they are orthogonal faces of the same response. Probabilities, and correlations, let us tackle chance objectively—although not completely—and this underlies much of the Bayesian view of assigning probabilities to hypothesis and their refinement with observations. Using the quantum paradox due to Aharonov and Rohrlich, we could illustrate the space and time nature of the immutability embedded in causality of the quantum view versus the classical view.

This obfuscation between causality, determinism and correlation underlies much of polity, economics, business and health-related undertakings. Sometimes it is because of foolishness, and sometimes because of ignorance, but the worst is when it is due to perfidy.

With this background, we noted how the response can be written as a Green's function—an impulse response—acting on a function in accumulation over space and time. We could illustrate this with known and previously encountered problems of perturbation. Coulomb interaction with a point charge (a delta function) gives us a way to develop the Green's function, and this in turn lets us solve the spatial problem of distributed charge. The same approach, in space and time, led us to the Green's function for electromagnetic propagation.

George Green was a miller and a mathematician. His work, besides this Green's function and theorem, included early work on electricity and magnetism. He was largely self-taught and was quite a reluctant miller since the constant trimming of the sails of the mill interfered with other passions.

Julian Schwinger used Green's function in much of his work on radar and the early fundamental work on quantum electrodynamics. A common objection to Green's function is the elaborate mathematical edifice in which intuition can be lost. Not so with Feynman diagrams, which this Schwinger-Feynman simultaneous exploration reemphasized.

Using this foundation, one could explore the classical and quantum-mechanical evolution of the response of a system as it undergoes scattering while being under stimulus. Classically, this was very much a Drude-like response, but, for the quantum-mechanical time-dependent evolution of the state function, it gave a powerful way of analyzing the evolution—with linearity and reversibility of quantum mechanics—using advancing and retarding Green's functions. Different perturbations, spatially and temporally distributed, can be accounted for. This is the way one would be able to capture the consequences of single and multiple scattering events and, through them, the evolution of the wavefunction. The scattering matrix is a representation of the scattering response.

13.6 *Concluding remarks and bibliographic notes*

THE GREEN'S FUNCTIONS TECHNIQUE goes back to the year 1828 and influenced Maxwell, and many others, even though the work was not quite recognized during Green's lifetime. Using Green's functions is now a standard technique for solving inhomogeneous linear differential equations.

The Aharonov-Rohrlich paradox discussed here is from their book[1] on paradoxes. For seriously imbibing quantum mechanics, resolving paradoxes is one of the most powerful and convincing ways. This book is very instructional in this way. Also, its cover is itself another reason for finding it in the library or getting a copy of one's own.

A general physics-oriented introduction to Green's functions is in the book by Duffy[2]. This book tackles wave equations, heat equation, Helmholtz equation, et cetera, and is a good introduction to the application of the techniques. A particular favorite is the book by Melnikov[3].

For the quantum-oriented usage of Green's functions, a very readable and lucid starting source is by Datta[4]. The book develops the use of equilibrium and nonequilibrium Green's functions techniques in nanoscale semiconductors. Walter Greiner has written a number of books on a range of physics subjects, all very thorough and complete. In his book on quantum electrodynamics[5], he tackles the techniques of using propagators. For a more advanced understanding of Green's functions in quantum mechanics, a number of books are good references. The first is the book[6] by Economou. Another advanced treatment is by Haug and Jauho[7]. For those interested in farther forays in solid state, such as correlated systems, the book by Doniach and Sondheimer[8] is a good source.

[1] Y. Aharonov and D. Rohrlich, "Quantum paradoxes," Wiley-VCH, ISBN 13 978-3-527-40391-2 (2005)

[2] D. G. Duffy, "Green's functions with applications," Chapman & Hall, ISBN 1-58488-110-0 (2001)

[3] Y. A. Melnikov, "Green's functions and infinite products," Birkhäuser, ISBN 978-0-8176-8279-8 (2011)

[4] S. Datta, "Quantum transport," Cambridge, ISBN 13 978-0-521-63145-7 (2005)

[5] W. Greiner, "Quantum electrodynamics," ISBN 13 978-3540875604 (2009)

[6] E. N. Economou, "Green's functions in quantum physics," Springer, ISBN 10-3-540-28838-4 (2006)

[7] H. J. W. Haug and A.-P. Jauho, "Quantum kinetics in transport and optics of semiconductors," Springer, ISBN 978-3-540-73561-8 (2008)

[8] S. Doniach and E. H. Sondeheimer, "Green's functions for solid state physicists," Imperial, ISBN 1-86094-078-1 (1998)

13.7 Exercises

1. Consider random fluctuations in classical conditions. An example is a smoke particle undergoing rapid collisions in still conditions; that is, zero mean velocity. Show that

$$\langle \frac{\Delta z^2}{\Delta t} \rangle^{1/2} = a^{1/2} (\Delta t)^{1/2},$$

where a is acceleration and therefore $\Delta p \Delta z = ma = constant.$ **[S]**

2. Show that

$$\int_a^b \tau \delta(t - \tau)\, d\tau = t[\Theta(t - a) - \Theta(t - b)],$$

where $\Theta()$ is the Heaviside step function. **[S]**

3. Find the transfer and Green's functions for the system obeying

$$\frac{d^2 y}{dt^2} - 3\frac{dy}{dt} + 2y = f(t),$$

with $y(0) = 0$, and $dy/dt = 0$, at $t = 0$. **[S]**

4. Show that electromagnetic propagation specified by the Maxwell's equations (as captured in Equation 13.17) is satisfied by the Green's function generated in Equation 13.23. **[S]**

5. Since damped harmonic oscillators are encountered so often, find the Green's function for it. So, find Green's function for the differential equation

$$m\frac{d^2 y}{dt^2} + \Gamma\frac{dy}{dt} + ky = f(t),$$

where m is a mass, Γ is a damping coefficient, k is a restoring constant, and $f(t)$ is a forcing function. **[S]**

6. We encountered the Fokker-Planck equation in Chapter 8. Find its Green's function. This requires one to solve the boundary value problem

$$\frac{d}{dz}\left(z^2\frac{dg}{dz}\right) - a\frac{d}{dz}(zg) - bg + \lambda g = -\delta(z - \zeta), \text{ for } 0 < x, \zeta < \infty.$$ **[S]**

14
Quantum to macroscale and linear response

SCHRÖDINGER'S AND MAXWELL'S EQUATIONS, upon which much of our discussion up to now has been built, are linear equations. Reversibility holds in linear transformations. The world, however, is irreversible to the extent that time always seems to march forward. This irreversibility is something we have remarked on earlier in the discussion of probabilities, probability sequences, the Golden rule, uncertainty as chance, et cetera, which arises as ensembles build complexity and new emergent properties, and rules appear as a more suitable form for representing the underlying physical laws. Maxwell's equations are reversible, except when one brings into them consequences from materials, that is, of ensembles, in forms that break the electrical-magnetic field and energy exchange—by storing away or providing additional sources—which do not respond linearly.

There is no clearer indicator, but not proof, of irreversibility than that we get older and nobody has been observed to live forever. But caution is needed. Time is not an observable. It is a parameter that can be treated relatively. There is a sense of before, present and future—but no now. Now is gone no sooner than the mind's inkling for it. Relativity throws its own wrenches into this cauldron. An interpretation regarding time in which I find very good food for thought is that these "nows" are being continuously generated. It is a dimension with some very special characteristics. Past in this view is "classical" that we know. The future is "quantum-mechanical" with many different possibilities that may unfold.

When matter undergoes spontaneous polarization, the energy gets stored in the polarization associated with the minute rearrangement of the lattice. Ferroelectrics, materials that result from spontaneous polarization, are nonlinear, even if there are conditions where they may appear as linear.

In the electromagnetic analysis of linear systems, we learn about reciprocity, which simply says that a medium is reciprocal if when it is stimulated a certain way at a multitude of places, and its responses is observed, impressing the response will lead to the appearance of a signal identical to that of the earlier stimulation. In other words, a medium as a bounded system is reciprocal if is stimulated at some input ports, and an output response signal is observed at output ports, reversing the experiment by applying a signal identical to the response signal to the output ports will bring about a signal identical to that of the original at the input ports.

Semiconductor Physics: Principles, Theory and Nanoscale. Sandip Tiwari.
 Published 2020 by Oxford University Press. DOI: 10.1093/oso/9780198759867.001.0001

Microwave networks, if they don't contain nonlinear elements including the source, are reciprocal. Resistors, capacitors and inductors are linear elements that we introduce as low frequency—small, compared to the wavelength of the signals being used—elements, and any network built out of these is reciprocal because these are linear elements. This is precisely what we mean when vectors of current, voltage and their derived parameters for various ports are used to build a matrix connecting them—impedance, admittance, hybrid, S, et cetera—which can all be inverted and transformed into each other, with the vector on the left meaning the stimulus, and the vector on the right meaning response. Reversibility leads to reciprocity between the stimulus and the response.

A source is a non-reciprocal element. It stores and supplies energy, by transforming the energy form. Batteries are electrochemical elements. A nuclear battery is a radioactivity-based source.

This argument of reversibility breaks as soon as one has a gain element; that is, the introduction of a separate energy source akin to the spontaneous polarization as a latent stored energy source that can be sprung. It is this introduction or removal of energy forms from the form that the equation is modeling—the electromagnetic from—that makes the network non-reciprocal. And it is a good thing too, since we use such gain elements—transistors, lasers, et cetera—very gainfully.

Spontaneous polarization is a property that arises in phase transitions when complexity is built into the matter, but underlying it is still the linear Schrödinger's equation or the Hamiltonian form in all its complexity. For understanding the scale-induced complexity of matter, we do have to then resort to statistical mechanics and thermodynamics, which take us toward irreversibility. Phase transitions and the Navier-Stokes equation are nonlinear, and yet they both are the culmination of quantum and microscale processes.

Nonlinearity may also arise when a system is pushed into extremes where nonlinear transformations may occur. A high magnetic field in a very pure high mobility material brings new forms of Hall effects, including a fractional quantum Hall effect that is entirely a collective response of the system. The change is nonlinear and collective, and, in many of these instances, renormalizations are essential to conquer singularities. So, much of what we observe and model linearly can be pushed to nonlinearity. Much that is reversible becomes irreversible in the transformation toward the classical world from the quantum foundations. Reciprocity and reversibility, in a certain sense, can hold true at the macroscale when grounded in linearity, but they also may not. This brief summary of the range of topics in this area (see Figure 14.1), many of which we have mentioned in passing in previous discussions, is to reinforce the importance of linearity and nonlinearity, reversibility and irreversibility, and reciprocity and non-reciprocity, which we see all around us and which make studying this area so interesting.

See S. Tiwari, "Nanoscale device physics: Science and engineering fundamentals," Electroscience 4, Oxford University Press, ISBN 978-0-198-75987-4 (2017), for related discussion.

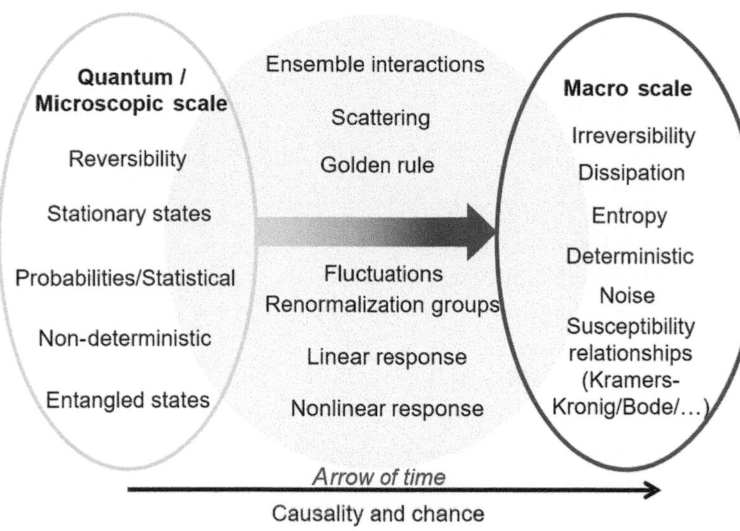

Figure 14.1: The evolution of the classical world and its observations from the quantum and the microscale, with attributes transforming under the complexity of interactions that the assemblage undergoes. Most of these we have now discussed in one form or other in understanding the properties of semiconductors.

Even noise, as a measure of fluctuations in an ensemble, can be visualized through some of these ideas, where measures of the spread of fluctuations are connected to the expectation of the system's response, with dissipation appearing as a parameter in linear response. This is a broader subject by itself, and we will tackle it in its entirety in Chapter 16.

In this chapter, we will focus in particular on the properties associated with linear response in light of this discussion. A primary property of linearity is the existence of the Kramers-Kronig relationship as a general behavior of linear response, even as we will employ the dielectric function to illustrate it primarily here. This will also let us relate some attributes of the dielectric response of semiconductors that we noted were so essential in Chapter 12. Another important topic we will touch on are the Onsager relationships (Chapter 15), going beyond the laws of thermodynamics, which give us a convenient means for transforming energy forms through the development of canonical conjugate fluxes and forces and which show us how some of the properties we have discussed can be arrived at far more simply.

14.1 *Causality's implication in linear response*

In Chapter 13, we dwelled at length on the notion of causality—that a claim that an effect B is caused by the occurrence of A is to say that B in entirety is due to A, for all variations in space and time—remove B, then A must be absent—and introduced the approach of Green's functions. As an example, we showed the application of Green's functions in solving a problem of broader interest to us: how do the nearly free electrons move in the presence of a field and what are the chance events of scattering. We found $\langle v \rangle = -e\mathcal{E}\langle \tau_{\mathbf{k}} \rangle / m^*$ and a Green's function $G(t-t') = -(e/m^*)\exp[-(t-t')/\langle \tau_{\mathbf{k}} \rangle]$ to describe the build-up of this response. The present gets connected to the past through a function that defines the response evolution. And all one does is integrate this entire past history.

We have discussed the dielectric function and the various contributions to it from the interactions between an electromagnetic wave and matter. It turned out to be complex, taking quite a few forms, all complex, even if the real term dominated at select frequencies, and there was a lot of difference between pure insulators, semiconductors and metals across the frequency range. Are the real and the imaginary parts of this dielectric function independent or tied to each other? Or, reframing the question, are the real and the imaginary parts of the index of refraction $\underline{n}^c = \underline{n}^r + i\underline{n}^i$ independent?

This index of refraction is the ratio $\sqrt{\epsilon\mu/\epsilon_0\mu_0}$, the scaling of the speed of light in the medium. Further to this, we have assumed that the material is non-magnetic, that is, $\mu = \mu_0$. So, $\underline{n}^c = \sqrt{\epsilon_r}$ and is related quite explicitly, such as in Equation 12.57, for conductivity arising in free carriers through this.

They cannot be independent, since they arose in the same cause. This is the essence of what the Kramers-Kronig relationship, the interrelationship of the imaginary and the real parts of a response of a causal process in a linear system, says. Light propagation as a linear system—transmitting with absorption—serves as our case study.

Light is absorbed, with the extinction arising in the imaginary part of the response, and the propagation behavior arising in the real part. The solution for the transmitted portion may be written as

$$\begin{aligned} \boldsymbol{\mathcal{E}} &= \hat{\mathbf{x}}\mathcal{E}_0 \exp\left(-\frac{\omega \underline{n}^i z}{c}\right) \exp\left[i\left(\frac{\omega \underline{n}^r z}{c} - \omega t\right)\right] \\ &= \hat{\mathbf{x}}\mathcal{E}_0 \exp\left(-\frac{\alpha z}{2}\right) \exp\left[i\left(\frac{\omega \underline{n}^r z}{c} - \omega t\right)\right] \end{aligned} \quad (14.1)$$

for a *TEM* incident wave transmitting in the z direction with an x-directed electric field and a suitable choice of phase. $\alpha = 2\omega\underline{n}^i/c$ is the absorption coefficient.

That the real part and the imaginary part are related in such a propagation, with causality underlying it, can be illustrated through a gendanken experiment. If the real and the imaginary parts, or the extinction and the propagation, are unrelated to each other, then it should be possible to have a medium with ϵ or $\underline{n}^c$ such that one may have a notch in the transmission at ω_0; that is, that light transmits across all frequencies except at ω_0. Take the case of light passing through a δ thick material, as shown in Figure 14.2. The *TEM* electromagnetic wave, normally incident, is

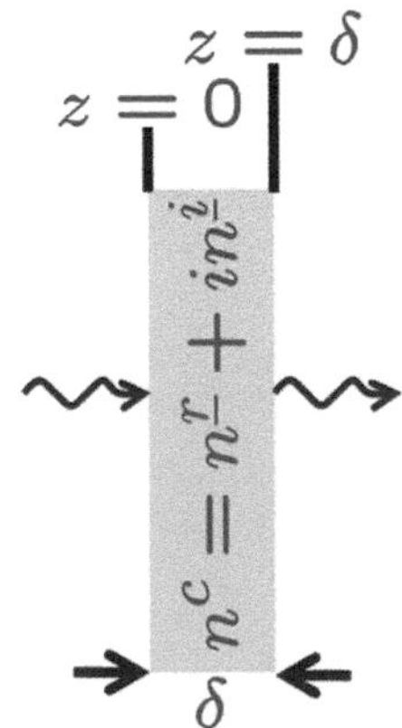

Figure 14.2: Light passing orthogonally through a slab of thickness δ.

$$\mathcal{E}_{in}(z,t) = \mathcal{E}_{in}(0)\exp\left[-i\omega\left(t - \frac{z}{c}\right)\right], \tag{14.2}$$

and the transmitted wave, following passage through the δ thick slab, is

$$\begin{aligned}\mathcal{E}_{out}(\delta,t) &= \mathcal{E}_{out}(\delta)\exp(-i\omega t)\\ &= \exp\left(i\frac{n^{\mathfrak{c}}\omega}{c}\right)\mathcal{E}_{in}(0,t)\\ &= \mathcal{E}_{in}(0)\exp\left[-i\omega\left(t - \frac{n^{\mathfrak{c}}\delta}{c}\right)\right],\end{aligned} \tag{14.3}$$

which, with time dependence ignored, is $\mathcal{E}_{out}(\delta) = \mathcal{E}_{in}(0)\exp(i\omega n^{\mathfrak{c}}\delta/c)$. By making δ arbitrarily small, we can ignore transmission attenuation. Reflection is ignored, since that is just a normalization of the amplitude. We use the Green's function

$$G(\delta,\tau) = \frac{1}{2\pi}\int_{-\infty}^{\infty}\exp\left(i\frac{\omega n^{\mathfrak{c}}}{c}\delta\right)\exp(-i\omega\tau)\,d\omega \tag{14.4}$$

so that

$$\mathcal{E}_{out}(\delta,t) = \int_{-\infty}^{\infty} G(\delta,t-t')\mathcal{E}_{in}(0,t')\,dt'. \tag{14.5}$$

Since the fields are real, the Green's function must be a real function too. So, the conjugation property is $n^{\mathfrak{c}}(-\omega) = [n^{\mathfrak{c}}(\omega)]^*$; $n^{\mathfrak{r}}(-\omega) = n^{\mathfrak{r}}(\omega)$; and $\alpha(-\omega) = \alpha(\omega)$.

Causality enters for us through the speed of light constraint, which determines how much time it takes for light to pass through the slab. $\mathcal{E}_{out}(\delta,t)$ depends on $\mathcal{E}_{in}(0,t')$ for $t' < t - \delta/c$. So, $G(\delta,\tau) = 0$ for $\tau < \delta/c$, and

$$\begin{aligned}\exp\left(i\frac{\omega n^{\mathfrak{c}}}{c}\delta\right) &= \int_{\delta/c}^{\infty} G(\delta,\tau)\exp(i\omega\tau)d\tau\\ \therefore\ \exp\left\{i\frac{\omega}{c}\left[n^{\mathfrak{c}}(\omega)-1\right]\delta\right\} &= \int_{\delta/c}^{\infty} G\left(\delta, t+\frac{\delta}{c}\right)\exp(i\omega t)dt.\end{aligned} \tag{14.6}$$

The function must have analytic continuation, with δ being arbitrary, even if small.

We may now apply these results. Let light arrive at the slab at $t = 0$, with a sharp front. This is the wave from which the ω_0 frequency component will be sharply removed. Let the slab of thickness δ be arbitrarily thin. Figure 14.3(a) shows the light consisting of a continuous distribution of frequency components that arrived. Figure 14.3(b) shows the component at frequency ω_0, which is removed upon passage. And what this means, with this component arbitrarily removed in the frequency space, is that the real space output wave will be spread out across all times, including for $t < 0$. This breaks causality. The output exists in pre-time or, to have it not exist in pre-time, the canceling signal must

Relativistic considerations of causality, although not important in this example, are profound. Time, like entropy, is an emergent notion. There is no precise time, only relative time, and a human interpretation of before and after, which relativity plays wonders with as the first introduction to a deep idea of science when we are young. I have always thought that, instead of using *BC* or *AD* for specifying time, or the modern archaeologist notation of *CE* or *BCE*, where *CE*, by the way, stands for "common era" or "current era" and, in the process, only compounds the polarizing questions it raises, should be replaced by "before Heisenberg" and "after Heisenberg." This usage captures the notion of time and its ambiguity and honors a revolutionary scientist without entering religious wars.

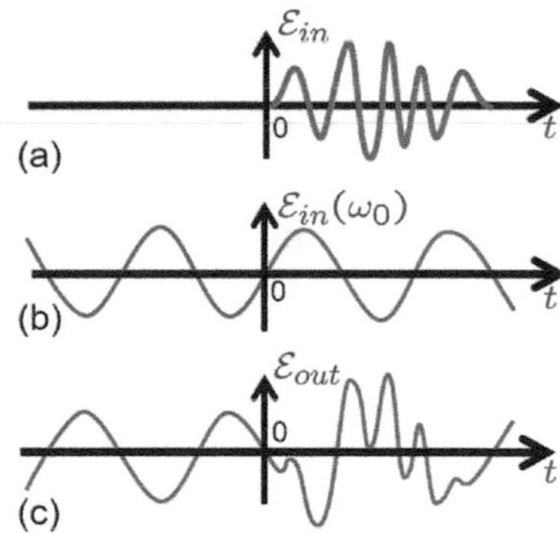

Figure 14.3: Part (a) shows an incident light front consisting of a continuous distribution of frequencies onto an arbitrarily thin slab. Since it vanishes for $t < 0$, but is built from a continuous distribution in frequencies, it includes the wave (b) at ω_0, which is the only component the slab is supposed to absorb in this gedanken experiment. But, to have destructive interference for $t < 0$, the output will need to be as shown in (c) where the ω_0 component must cancel out for all pre-time ($t < 0$). That is, the output exists at ω_0 for $t < 0$. Causality is violated.

exist in pre-time. Both break causality, and therefore an arbitrarily sharp notch may not exist at ω_0.

If the real and the imaginary parts of a dielectric function or an index of refraction were arbitrary, it would be possible to build a system where the absorption explodes and the transmission vanishes at only ω_0 and nowhere else. This gedanken experiment has refuted this possibility.

The absorption of a frequency, that is, the term dependent on the imaginary part of the index, needs to be accompanied by a compensating phase shift at all other frequencies, which depends on the real part of the index in order for the causality not to be violated. This phase shift then destructively interferes for all $t < 0$, and the output signal does not show up before the input has arrived, including the speed of light limitation. This also means that the real part of the index and the imaginary part of the index are related. A frequency-dependent relationship, that is, a dispersion relationship, relates the real and the imaginary components in the linear response analysis. This is the Kramers-Kronig relationship. This is a general result of linear systems and not restricted just to dielectric function.

Since inductors, capacitors and resistors form linear networks, Kramers-Kronig holds for the linear system. This is what Bode plots are. It also applies to mechanical systems. You should think of other places that this holds for, once you have read Section 14.2.

14.2 *Linear response theory and dielectric function*

We should precisely state what linear response means. For a single input function x, if the response is some function $f(x)$, then, in a linear response, for two independent inputs of x and y, the response will be $f(x+y)$. This is easily seen to be true for capacitors, inductors, resistors, Schrödinger's equation, Maxwell's equation in free space and dielectrics, the drift-diffusion equation and many others, but not for the Navier-Stokes equation or the Boltzmann transport equation, in the more generalized way when both an electric field and a magnetic field, as two independent inputs, are included.

WHAT DOES A SYSTEM DO when perturbed from its equilibrium? The first term in any response's expansion is linear. And, for a range of interactions, since the starting equations are linear, there is a significant range of input and output where the response remains linear. Two cases where we have already seen in some detail this linear response are fluctuation-dissipation and Brownian motion (see Sections 8.7 and 8.8). We will see in the following section, in our toy model view of the damped oscillator response (Section 14.3), again these parametric connection between real and imaginary (as opposed to expectation and fluctuation spread in noise). Fundamental to much of this entanglement in the response is the intimate connection between the transformations taking place between the interacting states and their close relationships. This gives rise to reversibility. Chance and the possibility of states being transformed to being many lead down the path to irreversibility. But when states are tightly coupled at the level of the original interactions, key transformation physical rules must hold: there is conservation of energy and momentum across the transition's participating entities, and the energy transformations are connected, even if they move from one form to another. The thermoelectric

This intimacy of state connection is essential for efficiency; otherwise, another damping activity leaks away the energy, leading to inefficiency. Transformers and power generators need this very intimate electric and magnetic energy stream coupling to be efficient. The largest hydroelectric turbine, as of 2017, made by Voith for the Three Gorges Dam, generates 784 *MW*. It had better be efficient, pretty close to 100 %, or it will get really hot.

effect, for example, is the connection of one energy flux type to another. This holds true for hurricanes, too. And even if these conditions are non-equilibrium, this intimate connection results in Onsager relationships, which reflect another form of linear transformation.

We will employ causality and the Green's function approach to develop our ideas. The induced response to an external stimulus in space and time can be written as

$$Z(\mathbf{r},t) = \int_{-\infty}^{\infty} G(\mathbf{r},t|\mathbf{r}',t') f(\mathbf{r}',t')\, d\mathbf{r}'\, dt', \tag{14.7}$$

where Z is the response at the position $(\mathbf{r},t)$, and G is the response function at $(\mathbf{r},t)$ to the stimulus $f(\mathbf{r}',t')$ at $(\mathbf{r}',t')$. This is a causal function, whereby the response to stimulus only depends on the past. We will also embed linearity in it; that is, employ it in situations where linearity holds. In general, the Green's function will be complex and will be deterministic, with chance—therefore, fluctuation and its consequence, dissipation—embedded in it. It is just a function that sums the histories.

We have already seen one example, the response of electrons undergoing causal motion under the stimulus of an electric field and chance of scattering, which was real. In general, though, functions and relationships will be complex. It is an unfortunate matter that we call functions with imaginary components complex. Imaginary is certainly an incorrect term. It represents orthogonality, not something magically brought out of thin air. When two parts are manifestations due to causations from two separate parameter spaces that both need to be described for completeness, they are not independent but orthogonal. Using the imaginary axis is just a way to represent this notion mathematically. I have always wished that this simple notion were taught properly in early schooling. We unnecessarily turn off bright minds by rules, edicts and obtuse nomenclature. It is somewhat similar to telling people in an early class where power supplies, inductors, capacitors, resistors, that is, some of our electric elements first appear, that you may not suddenly short a power supply to a capacitor. Of course you can. It is the same as why should we not take the square root of negative numbers. New ideas arise and thinking develops by asking such questions. The battery-to-capacitor example leads to an understanding of the displacement current as a real current, and dissipation in capacitors through radiation and other forms. Imagine mathematics and all our engineering learning without the idea of i to represent an orthogonal variable's axis.

Now consider the general connection between the electric displacement and the electric field. In the Green's function form, it can be written convolutionally as

$$\mathbf{D}(\mathbf{r},t) = g_0\boldsymbol{\mathcal{E}}(\mathbf{r},t) + \int_{-\infty}^{\infty}\int_{-\infty}^{t} g(\mathbf{r}-\boldsymbol{\xi},t-\tau)\boldsymbol{\mathcal{E}}(\boldsymbol{\xi},\tau)\, d\tau\, d\boldsymbol{\xi}, \tag{14.8}$$

that is, that the displacement response in a linear medium, such as a dielectric, is a function of some linear transformation of the local field in time, and the sum of its history and nonlocal consequences. There is an instantaneous part (g_0 is a constant, which turns out to be the vacuum dielectric constant), and the nonlocal space and time part g, which, in general, may be a tensor for anisotropic media.

What Equation 14.8 says is that displacement, in general, depends on the fields at that position at that time, and also on neighboring positions and previous instants of time, through the Green's function g.

We will limit ourselves to just temporal dependence, with no spatial, so that the displacement is spatially local. The Green's function then is

$$g(\mathbf{r}-\boldsymbol{\xi},t-\tau) = \delta(\mathbf{r}-\boldsymbol{\xi})g(t-\tau), \tag{14.9}$$

and the displacement may be written as

$$\mathbf{D}(t) = g_0\boldsymbol{\mathcal{E}}(t) + \int_{0}^{\infty} g(\tau)\boldsymbol{\mathcal{E}}(t-\tau)\, d\tau. \tag{14.10}$$

In the spatial reciprocal space, this local formulation, following Fourier transformation, is

$$\mathbf{D}(\mathbf{k},t) = g_0\boldsymbol{\mathcal{E}}(\mathbf{k},t) + \int_{-\infty}^{t} g(\mathbf{k},t-\tau)\boldsymbol{\mathcal{E}}(\mathbf{k},\tau)\,d\tau. \qquad (14.11)$$

Here, the time history is reflected in the integration limits, and the propagation may be viewed both in time and in space, although usually we don't have to worry about nonlocal propagational considerations. For simple media, we may ignore spatial considerations.

In general, again Fourier transforming Equation 14.11 from the time domain to the frequency domain,

$$\begin{aligned}\mathbf{D}(\mathbf{k},\omega) &= g_0\boldsymbol{\mathcal{E}}(\mathbf{k},\omega) + \boldsymbol{\mathcal{E}}(\mathbf{k},\omega)\int_{0}^{\infty} g(\mathbf{k},\tau)\exp(-i\omega\tau)\,d\tau \\ &= \boldsymbol{\mathcal{E}}(\mathbf{k},\omega)\left[g_0 + \int_{0}^{\infty} g(\mathbf{k},\tau)\exp(-i\omega\tau)\,d\tau\right]. \qquad (14.12)\end{aligned}$$

This we write as

$$\begin{aligned}\mathbf{D}(\mathbf{k},\omega) &= \boldsymbol{\mathcal{E}}(\mathbf{k},\omega)\epsilon = \boldsymbol{\mathcal{E}}(\mathbf{k},\omega)\epsilon_0\epsilon_r \\ &= \boldsymbol{\mathcal{E}}(\mathbf{k},\omega)\epsilon_0\left[1+\chi(\mathbf{k},\omega)\right]. \qquad (14.13)\end{aligned}$$

Equation 14.13 shows the few different ways we have employed the symbols for dielectric function or permittivity (ϵ), dielectric function or permittivity of free space (ϵ_0), the relative dielectric constant or relative permittivity (ϵ_r, also sometimes referred to as k or κ in the high permittivity engineering literature), and the susceptibility (χ) that is the consequence of the material beyond that of vacuum. In general, with the exception of ϵ_0, all are complex;

$$\chi(\mathbf{k},\omega) = \int_{0}^{\infty} g(\mathbf{k},\tau)\exp(-i\omega\tau)\,d\tau \qquad (14.14)$$

is susceptibility, and

$$\epsilon_r = 1 + \chi(\mathbf{k},\omega) \qquad (14.15)$$

is the relative dielectric function.

What does ϵ_0 mean? It is the linear response function of vacuum. It is the dielectric constant or permittivity that underlies the relationship of the electric field to the magnetic field, through the immutability of the speed of light. $\mathbf{D} = \epsilon_0\boldsymbol{\mathcal{E}}$ in vacuum. When looking at the response of a material, when $\omega \to \infty$, we expect and require that the material's susceptibility to the field vanishes, that is, that the material, although still there, is not responsive—that is to say, that $\chi(\mathbf{k},\omega)$ should be reflective of the material's extreme frequency response. The material still has atoms (electrons surrounding the nuclei in a stable but slightly deformable form),

Light's speed, the movement of electromagnetic energy, is an immutable constant. We resorted to this causality limitation in our first pass at a discussion of causality for the passage of light through a slab in Section 14.1. It is important for several problems, such as radiation from a high speed accelerating beam, even if it is only through bending, but not for our semiconductor problems, where the time and speed scales of interest are much lower.

Since the consequences of stimulus from past times are included in the Green's function, it is also called "the propagator." This propagational approach of the Green's function is also called the Keldysh approach, after the person who first described and employed it. Mstislav Keldysh is a name we have now encountered often. He was another exceptional Soviet mathematician and scientist. He was also a moving force in the space program, with interests in mechanics, and was often called the "chief theoretician" in the Soviet circles, somewhat akin to what Hans Bethe was called in the Manhattan Project.

and these will have consequences for how the ultra-high frequency electromagnetic wave will behave.

Equation 14.8, from which we started, has now been recast in a form that, for varying causes in space and time, leads to a complex function—the dielectric function $\epsilon(\mathbf{k},\omega)$—that gives us the response. For many situations, this can be simplified, but, for all, one may use a causal Green's function and propagator relationship to arrive at the response.

We brought in atomic polarization in the high frequency discussion of Chapter 12, but this is still not infinite frequency. We need to go deeper than X-rays, into a regime where these particle models that we have employed fail.

Whether we write the dielectric function,

$$\epsilon(\mathbf{k},\omega) = \epsilon^{\underline{r}}(\mathbf{k},\omega) + i\epsilon^{\underline{i}}(\mathbf{k},\omega), \tag{14.16}$$

or the susceptibility function,

$$\chi(\mathbf{k},\omega) = \chi^{\underline{r}}(\mathbf{k},\omega) + i\chi^{\underline{i}}(\mathbf{k},\omega), \tag{14.17}$$

these are causal relationships, and therefore the real and the imaginary parts are related. They have arisen conjoined from the propagator, accumulating in history through a causal relationship. And while we may have employed this causal approach to this specific problem, the approach is generalizable to all linear systems problems. The propagational relationship will hold true for a wide range of physical phenomena.

The propagator must be a physical function. This means it is a real function. This means that the real part of the functions of Equations 14.16 and 14.17 must be an even, that is, symmetric function, as a consequence of the Fourier transformation argument. Also, by complementarity, the imaginary part must be an odd, or asymmetric, function.

We write the Green's function as a sum of even and odd functions in time dependence, that is, $g(\mathbf{k},t) = g_e(\mathbf{k},t) + g_o(\mathbf{k},t)$:

$$g(\mathbf{k},t) = 2g_e(\mathbf{k},t) = 2g_o(\mathbf{k},t) \ \forall t > 0, \tag{14.18}$$

that is, the Green's function takes an even form for all times past the reference time. This means $|g_e(\mathbf{k},t)| = |g_o(\mathbf{k},t)|$. It also means that

$$g(\mathbf{k},t) = 0 \ \forall t < 0. \tag{14.19}$$

We can write these relationships in one equation as

$$g(\mathbf{k},t) = 2g_e(\mathbf{k},t)\Theta(t) = 2g_o(\mathbf{k},t)\Theta(t), \tag{14.20}$$

where $\Theta(t)$ is the Heaviside function, defined as

$$\Theta(t) = \begin{cases} 0 & \forall\ t < 0, \\ 1 & \forall\ t > 0 \end{cases} \equiv \lim_{\alpha\to 0} \begin{cases} 0 & \forall\ t < 0, \\ \exp(-\alpha t) & \forall\ t > 0. \end{cases} \tag{14.21}$$

The principal part of an integral, written here using $\mathscr{P}$ as the notation, is the value of the integral excluding the divergence. So, if the divergence is at b, then

$$\mathscr{P}\int_a^c f(x)\,dx = \lim_{\delta\to 0}\left[\int_a^{b-\delta} f(x)\,dx + \int_{b+\delta}^c f(x)\,dx\right].$$

Now, we may perform the Fourier transform integrations over the $-\infty$ and ∞ limits of time to obtain $\chi(\mathbf{k},\omega) \equiv \chi^{\underline{r}}(\mathbf{k},\omega) + i\chi^{\underline{i}}(\mathbf{k},\omega)$ from Equation 14.14.

In the integrals we are working with, the divergence is at $\omega = \omega_1$ on the real axis. With damping contributing to the oscillator, it will be in the lower half of the complex plane. For understanding the meaning of principal value, its importance to complex system analysis and its relationship to where the poles and zeros appear, see Appendix O.

The Fourier transform of the function $\exp(-\alpha t)$ is $(\alpha - i\omega)/(\alpha^2+\omega^2)$, and, taking its limits ($\alpha \rightarrow 0$), it is $\pi\delta(\omega)-(i/\omega)\mathscr{P}$. $\mathscr{P}$ here is to indicate that the principal value should be chosen when integrating. We write the transform of $g_e(\mathbf{k},t)$ as $\chi^{\underline{i}}(\mathbf{k},\omega)$, and the transform of $g_o(\mathbf{k},t)$ as $i\chi^{\underline{i}}(\mathbf{k},\omega)$, so that the result for the susceptibility is

$$\begin{aligned}\chi(\mathbf{k},\omega) &= \chi^{\underline{r}}(\mathbf{k},\omega) - \frac{i}{\pi}\mathscr{P}\int_{-\infty}^{\infty}\frac{\chi^{\underline{r}}(\mathbf{k},\omega_1)}{\omega-\omega_1}\,d\omega_1, \text{ or} \\ &= i\chi^{\underline{i}}(\mathbf{k},\omega) + \frac{1}{\pi}\mathscr{P}\int_{-\infty}^{\infty}\frac{\chi^{\underline{i}}(\mathbf{k},\omega_1)}{\omega-\omega_1}\,d\omega_1. \end{aligned} \qquad (14.22)$$

The implication of this final relationship and our even-odd Green's function approach is that

$$\begin{aligned}\chi^{\underline{i}}(\mathbf{k},\omega) &= -\frac{1}{\pi}\mathscr{P}\int_{-\infty}^{\infty}\frac{\chi^{\underline{r}}(\mathbf{k},\omega_1)}{\omega-\omega_1}\,d\omega_1, \text{ and} \\ \chi^{\underline{r}}(\mathbf{k},\omega) &= \frac{1}{\pi}\mathscr{P}\int_{-\infty}^{\infty}\frac{\chi^{\underline{i}}(\mathbf{k},\omega_1)}{\omega-\omega_1}\,d\omega_1, \end{aligned} \qquad (14.23)$$

which are the Kramers-Kronig relationships for susceptibility in a dielectric medium. One also sees these relationships written in the formally equivalent alternative forms

The reason to approach this Kramers-Kronig dielectric response through susceptibility is that the susceptibility falls off rapidly for large frequency, and this makes integrals of the form

$$\chi(\omega) = \frac{1}{2\pi}\int_{-\infty}^{\infty}\chi(\omega_1)\Theta(\omega-\omega_1)\,d\omega_1,$$

the susceptibility function in the Fourier transform space of time, with $\Theta(\omega-\omega_1)$ as the Fourier transform of the Heaviside step function, to converge fast. $\epsilon(\omega)$, on the other hand, will converge to ϵ_0.

$$\begin{aligned}\chi^{\underline{i}}(\mathbf{k},\omega) &= -\frac{2}{\pi}\int_0^{\infty}\int_0^{\infty}\chi^{\underline{r}}(\mathbf{k},\omega_1)\sin(\omega t)\cos(\omega_1 t)\,d\omega_1\,dt, \text{ and} \\ \chi^{\underline{r}}(\mathbf{k},\omega) &= -\frac{2}{\pi}\int_0^{\infty}\int_0^{\infty}\chi^{\underline{i}}(\mathbf{k},\omega_1)\cos(\omega t)\sin(\omega_1 t)\,d\omega_1\,dt, \end{aligned} \qquad (14.24)$$

as well as

$$\begin{aligned}\chi^{\underline{i}}(\mathbf{k},\omega) &= -\frac{2}{\pi}\mathscr{P}\int_0^{\infty}\frac{\omega\chi^{\underline{r}}(\mathbf{k},\omega_1)}{\omega_1^2-\omega^2}\,d\omega_1, \text{ and} \\ \chi^{\underline{r}}(\mathbf{k},\omega) &= \frac{2}{\pi}\mathscr{P}\int_0^{\infty}\frac{\omega_1\chi^{\underline{i}}(\mathbf{k},\omega_1)}{\omega_1^2-\omega^2}\,d\omega_1. \end{aligned} \qquad (14.25)$$

The Kramers-Kronig-to-Hilbert transform connection is discussed in Appendix O, together with analyticity.

The real and the imaginary parts of complex functions that have no poles in the lower plane or the upper plane are related through Hilbert transformations. These Hilbert transforms provide the formal mathematical underpinning of the connection between the real part and the imaginary part, since we are articulating this for linear systems undergoing linear transformations. Any physically acceptable causal function follows this relationship between the in-phase response and the out-of-phase response in their linear limits. In linear networks, the amplitude and phase relationships that we call Bode relations are instances of Kramers-Kronig relationships. In communications, the variety of linear system responses

The Kramers-Kronig relationship applies to susceptibility in paramagnetic systems, which are linear, but not to permeability, since the permeability does not vanish in the absence of stimulus, that is, it is a nonlinear, spontaneous-polarization-based energetic consequence. In mechanics, the attenuation and dispersion in ultrasonics too follow the Kramers-Kronig relationship.

subscribe to the Kramers-Kronig relationship. In paramagnetic systems, Kramers-Kronig relationships apply to the susceptibility.

The consequences of these Kramers-Kronig relationships can now be seen in the dielectric function behavior of semiconductors (see Figure 14.4). *GaAs* being ionic, our prior comments from Section 12.4 apply. ω_{TO} is unchanged, while the ω_{LO} oscillation mode responds to the applied signal's frequency. In *Si*, this effect is absent. The important point to note here is that if we know one of the two—the real or the imaginary—response curves in its entirety, then the other one can be calculated. This is quite useful, since sometimes one or the other of the responses is easier to measure. Phase changes, for example, are much easier to measure through lock-in techniques, while absolute response amplitudes are prone to higher error.

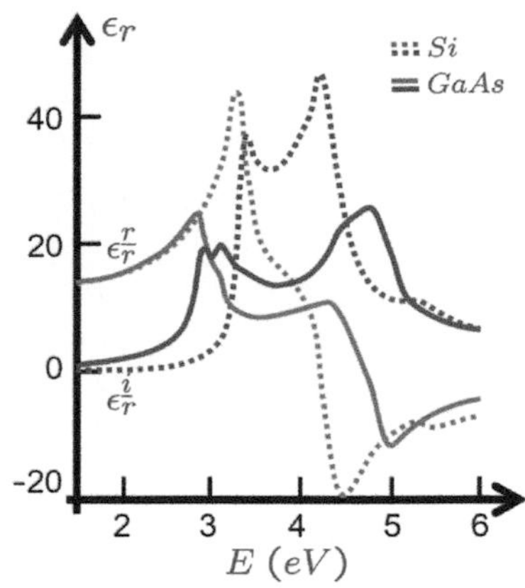

Figure 14.4: The real and imaginary parts of the dielectric function for *GaAs* and *Si*. Note that *GaAs* is partially ionic and *Si* is covalent. Adapted from D. E. Aspnes and A. A. Studna, "Dielectric functions and optical parameters of *Si*, *Ge*, *GaP*, *GaAs*, *GaSb*, *InP*, *InAs*, and *InSb* from 1.5 to 6.0 *eV*," Physical Review B, **27**, 985–1009 (1983).

If $\chi^{\underline{r}}=0$, then so should $\chi^{\underline{i}}=0$ for all frequencies, and $\epsilon=\epsilon_r\epsilon_0=(1+\chi)\epsilon_0=\epsilon_0$. A dispersionless medium, for example, vacuum, cannot be dissipative. As a corollary to this, *a dissipative medium cannot be dispersionless*. On the other hand, for the example of Section 14.1, a notch precisely at ω_0, that is, $\chi^{\underline{i}} = -\delta(\omega-\omega_0)$, will lead to $\chi^{\underline{r}} = (1/\pi)(\omega_0-\omega)^{-1}$. This implies that if there exists a resonance at this frequency ω_0, then the propagation characteristics change across a broader band. The phase shift changes sign at ω_0. At $\omega<\omega_0$, the response the phase is ahead, and, for $\omega > \omega_0$, it is opposite and lags. It is this precise change in propagation characteristics that made removing a frequency without affecting anything else a physical impossibility.

The lead-lag response change at resonance comes directly from Equation 14.25. Take the linear system analogy to an inductor of inductance L and a capacitor of capacitance C. The current-voltage response follows $\tilde{I} = -i\omega\tilde{V}/(\omega^2-\omega_0^2)$, where $\omega_0^2 = 1/LC$. Current is ahead in phase for $\omega<\omega_0$ and lags for $\omega > \omega_0$.

If a medium is anisotropic, the Kramers-Kronig relation still hold for $\mathbf{D}(\mathbf{k},\omega) = \epsilon(\mathbf{k},\omega)\boldsymbol{\mathcal{E}}(\mathbf{k},\omega)$. There has been nothing in our derivation, except the condition of linear response, that prevents different propagational features in different orientations. Now, $\epsilon(\mathbf{k},\omega)$ is a 3×3 tensor with 9 components ($\epsilon_{jl}(\mathbf{k},\omega)$, in general), which symmetry will reduce. For example, for cubic anisotropic systems, there are 6 components, and, for highly symmetric systems, there is only 1.

14.3 *Linear response of a damped oscillator*

THE KRAMERS-KRONIG REPRESENTATION has shown us that when two characteristics, for example, the real and the imaginary parts of a dielectric function, are equivalent when considered across the entire frequency range that is their overlapping domain, then they are really different views of the same function. This is *analytic continuity*, which has been extended to the broader domain of the complex plane, employing causality for a situation with stimulus, that is, perturbation. Refraction and absorption got connected. The analyticity of the dielectric function in the upper half plane made this Kramers-Kronig relationship possible, and we took care of the poles through the use of the principal value.

The existence of poles in a response is a manifestation of resonances in the linear system. In the dielectric function, we have seen resonances in phonon behavior and in the electron charge cloud behavior, and the Lorentz model was an integral part of this resonance behavior—with and without damping. We look at this next to link to the response we have noted.

14.3.1 Lorentz model

WE USE THE LORENTZ OSCILLATOR MODEL for the atom to show some of the general features. And we will follow this with a more causal discussion of linear response through electrons. The Lorentz model is the simplest of models, and, in Chapter 12, in our marginalia on oscillators, this was the first among many that we mentioned. The electron-nucleus assembly consists of a large mass and a small mass bound together as quantum strings under Coulomb interaction. Consider a one-electron, one-proton nucleus system. The general equation of motion—classical—for this is

The Lorentz model for physicists is the Cauchy model for the mathematicians.

At the highest of frequencies, the atomic polarization is the last man standing!

$$m_0\ddot{\mathbf{r}} + m_0\Gamma\dot{\mathbf{r}} + m_0\omega_0^2\mathbf{r} = -e\boldsymbol{\mathcal{E}}_{local}, \qquad (14.26)$$

where $\boldsymbol{\mathcal{E}}_{local} = \breve{\mathcal{E}}_{local}\hat{\mathbf{e}}\exp(-i\omega t)$, with the unit vector $\hat{\mathbf{e}}$ pointed in the local field's direction. We have assumed that the nucleus is extremely heavy so that the reduced mass is quite close to the electron mass. The local field $\boldsymbol{\mathcal{E}}_{local}$ is the field acting on the electron. The damping Γ incorporates energy losses, such as through radiation, scattering when the atom is in a solid, et cetera. The last term on the left is the restoring force. With fields varying time-harmonically through an $\exp(-i\omega t)$ form, the solution is

Why only the electric field as the external field? Only because $e\mathbf{v}\times\mathbf{B}$, where $\mathbf{B}$ is due to the electromagnetic field is very small and so is $\mathbf{v}$ in the outermost orbits.

$$\mathbf{r} = -\frac{e\boldsymbol{\mathcal{E}}_{local}/m_0}{(\omega_0^2-\omega^2) - i\Gamma\omega}, \qquad (14.27)$$

which has an induced dipole moment of

$$\mathbf{p} = -e\mathbf{r} = \frac{e^2\boldsymbol{\mathcal{E}}_{local}}{m_0}\frac{1}{(\omega_0^2-\omega^2) - i\Gamma\omega}. \qquad (14.28)$$

The atomic polarizability α is

$$\alpha = \frac{\mathbf{p}}{\boldsymbol{\mathcal{E}}_{local}} = \frac{e^2}{m_0}\frac{1}{(\omega_0^2-\omega^2) - i\Gamma\omega}. \qquad (14.29)$$

The macroscopic atomic polarizability with N atoms per unit volume is

$$\mathbf{P} = N\langle\mathbf{p}\rangle = N\alpha\langle\mathcal{E}_{local}\rangle = \chi_a\boldsymbol{\mathcal{E}}, \tag{14.30}$$

consistent with the symbols we have used up to this point. So, the complex dielectric function arising in atomic polarizability is

$$\epsilon_r = 1 + \frac{Ne^2}{m_0}\frac{1}{(\omega_0^2-\omega^2) - i\Gamma\omega}, \tag{14.31}$$

so that, with the index of refraction $n^{\underline{c}} = n^{\underline{r}} + in^{\underline{i}} = \sqrt{\epsilon_r}$ for unity relative permeability,

$$n^{\underline{r}2} - n^{\underline{i}2} = 1 + \frac{Ne^2}{m_0}\frac{\omega_0^2-\omega^2}{(\omega_0^2-\omega^2)^2 + \Gamma^2\omega^2}, \text{ and}$$

$$2n^{\underline{r}}n^{\underline{i}} = 1 + \frac{Ne^2}{m_0}\frac{\Gamma\omega}{(\omega_0^2-\omega^2)^2 + \Gamma^2\omega^2}, \tag{14.32}$$

a set of equations that may be compared to the equivalent expressions in Equation 12.57, which showed the consequences of electron conductivity.

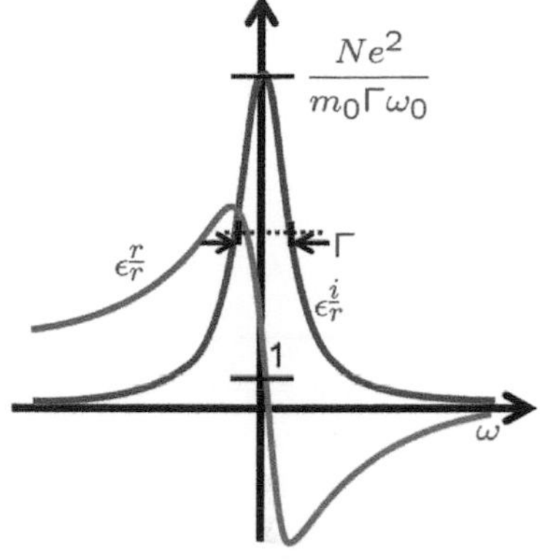

Figure 14.5: The Lorentzian lineshape and relative dielectric response in the Lorentz model. The real and imaginary parts of the relative dielectric function are shown.

The Lorentzian response behavior is shown in Figure 14.5. $\epsilon^{\underline{r}}$ increases with increasing frequency except in a narrow region defined by the damping near the resonance frequency. This decrease is an anomalous dispersion defined by the solution of

$$(\omega_0^2 - \omega_\pm^2)^2 = \pm\Gamma^2\omega_0^2. \tag{14.33}$$

The maximum in $\epsilon_r^{\underline{i}}$ is given by

$$\epsilon_r^{\underline{i}} = \frac{Ne^2m_0}{\Gamma\omega_0}, \tag{14.34}$$

so long as the damping is not excessive. $\epsilon_r^{\underline{i}}$'s full width at half maximum gives us the damping factor Γ.

Now if there are several states, the jth of which has N_j electrons per unit volume with a resonance frequency ω_j, then

$$\epsilon_r = 1 + \frac{e^2}{m_0}\sum_j \frac{N_j}{(\omega_j^2-\omega^2) - i\Gamma_j\omega}, \tag{14.35}$$

with $\sum_j N_j = N$. The corresponding quantum-mechanical form is

$$\epsilon_r = 1 + \frac{e^2}{m_0}\sum_j \frac{N_jf_j}{(\omega_j^2-\omega^2) - i\Gamma_j\omega}, \tag{14.36}$$

with ω_j being the transition frequency between two states of the atom, and f_j being the oscillator strength.

For free atoms, $\sum_j f_j = 1$, in equivalence to the classical $\sum_j N_j = N$.

At low frequencies, with $\omega \ll \omega_j$,

$$\Re[\epsilon_r(\omega)] \approx 1 + N\frac{e^2}{m_0}\sum_j \frac{f_j}{\omega_j^2}\left(1+\frac{\omega^2}{\omega_j^2}\right) \approx 1 + a + b\omega^2, \text{ and}$$

$$\Im[\epsilon_r(\omega)] \approx N\frac{e^2}{m_0}\sum_j \frac{f_j\Gamma_j}{\omega_j^4} \approx c\omega, \tag{14.37}$$

where a, b and c are specific constants. At the other extreme,

$$\lim_{\omega\to\infty} \epsilon_r(\omega) \approx 1 - N\frac{e^2}{m_0\omega^2}\sum_j f_j\left(1 - i\frac{\Gamma}{\omega}\right), \text{ with}$$

$$\Re[\epsilon_r(\omega)] \approx 1 - \frac{Ne^2}{m_0\omega^2}\sum_j f_j, \text{ and}$$

$$\Im[\epsilon_r(\omega)] \approx \overline{\Gamma}\frac{Ne^2}{m_0\omega^3}\sum_j f_j, \text{ with } \overline{\Gamma} = \frac{1}{\sum_j f_j}\sum_j f_j\Gamma_j \tag{14.38}$$

as a mean oscillator strength. So, we have succeeded in making connections between oscillator strengths, damping and collection of atomic assemblies to the dielectric function in a toy model. We can extend this by now including the electron charge cloud's resonance response too.

14.3.2 *Oscillating electron model*

THE LORENTZ MODEL'S APPROACH works reasonably well for many of the contributions to electronic polarizability. We illustrate this via the response of an electron as a particle in a charge plasma. Scattering in the plasma implies a change. Equation 14.26 still holds, although where damping comes from has changed, and the local field could as well just be called a field, so $\boldsymbol{\mathcal{E}}_{local} = \boldsymbol{\mathcal{E}}$. N is now the number of electrons. Since these electrons are in the semiconductor, we will use the effective mass m^*. The relative dielectric function can now be rewritten in a form that is slightly different,

$$\epsilon_r = 1 + N\alpha = 1 + \chi = 1 + \frac{\omega_p^2}{\omega_0^2 - i\Gamma\omega - \omega^2}, \text{ where}$$

$$\omega_p^2 = \frac{ne^2}{m^*} \tag{14.39}$$

is the plasma frequency, with n as the electron density.

The model says that the dielectric function is an analytic function with two poles in the lower half plane located at

$$\omega_\pm = -i\frac{\Gamma}{2} \pm \omega_0\left[1 - \left(\frac{\Gamma}{2\omega_0}\right)^2\right]^{1/2}. \tag{14.40}$$

With a weak damping, $\Gamma \ll \omega_0$; $\omega_\pm \approx -i(\Gamma/2) \pm \omega_0$; and the residues are $\pm\omega_p^2/2\omega_0$. Damping also means $\Gamma > 0$, so the poles are in the lower half plane. For a weak damping, there is a narrow peak around the resonant peak where the absorption will be. This is reflected in the imaginary part of the relative dielectric function of Figure 14.5. Spontaneously growing solutions are disallowed. The index of refraction slowly increases over much of the frequency range (a normal dispersion, with $\partial n_r^{}/\partial\omega > 0$) but, around the resonance, an anomalous dispersion comes about ($\partial n_r^{}/\partial\omega < 0$).

These poles in the lower half of the plane are direct consequences of causality. You should convince yourself of this through analytic argument. That only certain analytic functional forms will be allowed is a consequence of causality—and the reason why we emphasize it. I wish we stressed causality just as much in the real world of intra-earth matters.

We can now connect this causality discussion through the Green's function analysis with which we started. To evaluate the Green's function, we need to integrate along the contour shown in Figure 14.6,

$$\begin{aligned} G(t,0) &= \frac{1}{2\pi}\oint_C [\epsilon_r(\omega) - 1]\exp(-i\omega t)\, d\omega \\ &= -i(r_+\exp(-i\omega_+ t) + r_-\exp(-i\omega_- t)) \ \forall\, t > 0 \\ &= \omega_p^2 \exp\left(-\frac{\Gamma}{2}t\right)\frac{\sin\omega_0 t}{\omega_0}\Theta(t), \end{aligned} \quad (14.41)$$

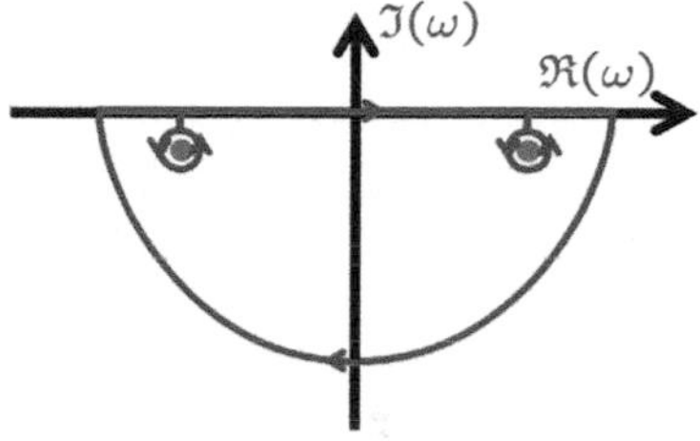

Figure 14.6: Green's function contour for the weakly damped single mode oscillator.

which follows from $-2\pi i$ times the sum of residues, and where causality exists.

The Green's function of Equation 13.7 lets us determine the response to a stimulus by integrating up to any time via the generalized Equation 13.7. The forced oscillator will have an initial period of change followed by settling down to some steady-state response. The upper limit of integration is time t; thus, only the stimuli at earlier times enters. This is causality. The effect cannot precede the cause. On the other hand, the Heaviside function lets us close the path of integration. This shows that analyticity and causality are connected, that the lower half plane allowed integration and that both are satisfied. The response vanishes at $t \leq 0$, initial growth is ω_0 guided and a decay exists with a $2/\Gamma$ time constant. When the applied signal is close to the natural frequency, and the damping small, many of the earlier periods reinforce the response at any $t > 0$, and this is through the sinc term, as it was for the Golden rule.

The curious student reader is encouraged to pursue Appendix O, where a few different strands of the connections from causality are emphasized.

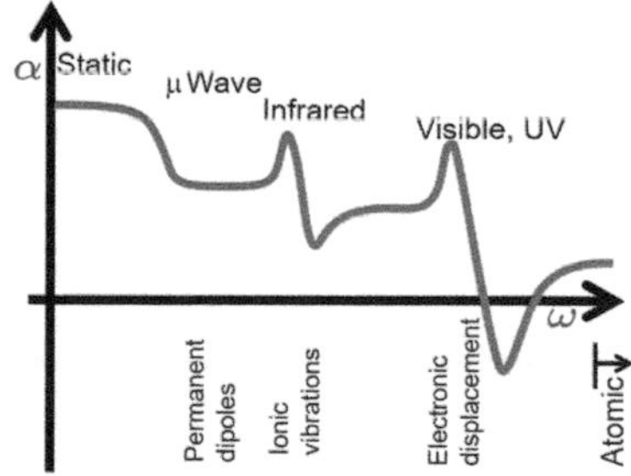

Figure 14.7: Polarization as a function of frequency, showing changes as one sweeps past and beyond ultraviolet (UV) frequencies.

We have now seen a number of oscillators that collectively affect the dielectric function through the polarization response. Figure 14.7 sketches some of the major ones that we see in semiconductors and which we have looked at in some detail in this text, all of which still leave the assembly in the linear response regime so long as the stimulation is small. First, there is some change from static conditions at near microwave frequencies, due to permanent dipoles. In infrared frequencies, one sees the ionic vibrations,

Note that this is not the spontaneous polarization of phase transitions, which is, by its very nature of energy transformation and statistical changes, nonlinear.

that is, phonon contributions, and then, in visible and ultraviolet frequencies, is the electron cloud response; farther on, one would see the consequences of atomic polarization.

14.4 *Quantum-statistical view of response*

IN CHAPTER 8, we explored the quantum-statistical view to evolution through density matrices. In this chapter, we have stressed the linear foundations of the foundational equations such as that of Schrödinger. In Chapter 8, we also looked at the density matrix ρ, where $\rho_{nm} = \overline{c_n^* c_m}$, with $c_{n,m}$s being the amplitudes in the state representation. $|\psi\rangle = \sum_n c_n|\phi_n\rangle$ lets us know the average of an expectation, given the state of knowledge. For an observable $\mathcal{A}$, this is $\overline{\mathcal{A}} = \sum_{n,m} \rho_{mn} A_{nm}$. $\overline{\mathcal{A}} = \mathrm{Tr}(\rho\mathcal{A})$. We also wrote, using the density matrix, that a system perturbed from equilibrium responds as

Schrödinger's is an equation of waves, describing how the music must be played. In Chapter 8, we also looked at the density matrix equation is the equation of realism, when, as in real life, the notes get a bit off, attention drifts, a string breaks or the cell phone of somebody in the audience goes off.

$$[\mathscr{H}', \rho] = -\frac{\hbar}{i}\left\{\frac{\partial}{\partial t}\rho(t) + \frac{i}{\hbar}[\mathscr{H}_0, \Delta\rho(t)]\right\}, \tag{14.42}$$

which is the quantum version of the Boltzmann transport equation. This all represents, both in the state function and in this evolution, the statistical nature of the theory. Since $\hat{\rho}|n\rangle = \mathfrak{p}_n|n\rangle$, with $\mathfrak{p}_n = \langle n|\hat{\rho}|n\rangle = (1/N)\sum_{i=1}^{n} |\langle i|n\rangle|^2$, $\hat{\rho}$ is an operator for quantum-mechanical probability distribution.

Response function is the parameter through which we tie the energetics of the interaction to the observables. Table 14.1 shows some examples of the response functions and their corresponding observables in linear systems examples of relevance for an understanding of the semiconductors.

So, the dielectric function, the magnetic susceptibility and the conductivities can all be viewed through the density matrix evolution equation (Equation 14.42). This equation leads to

$$-\frac{\hbar}{i}\frac{\partial}{\partial t}\Delta\rho(t) = \exp\left(\frac{i}{\hbar}\mathscr{H}_0 t\right)[\mathscr{H}'(t), \rho_0]\exp\left(-\frac{i}{\hbar}\mathscr{H}_0 t\right) \tag{14.43}$$

	Response function	$\mathscr{H}'$	Stimulus	Observable
ϵ	Dielectric function	$-\mathbf{p}\cdot\boldsymbol{\mathcal{E}}$	Electric field	Dielectric polarization
μ	Magnetic susceptibility	$-\mathbf{m}\cdot\mathbf{H}$	Magnetic field	Magnetic polarization
κ	Heat conductivity	$\mathbf{v}\cdot\nabla T$	Temperature gradient	Heat current density
σ	Electrical conductivity	$-\mathbf{J}\cdot\mathbf{A}$	Electric field	Electric current density

Table 14.1: Response function, stimulus and the observable, for a few examples of linear responses.

for evolution, and therefore the expectation for any observable $\mathcal{A}$ is

$$\langle \mathcal{A} \rangle_t = \text{Tr}[(\rho_0 + \Delta\rho(t)\mathcal{A}] = \text{Tr}(\rho_0\mathcal{A}) + \text{Tr}[\Delta\rho(t)\mathcal{A}], \qquad (14.44)$$

where the first term is the thermal equilibrium solution, and the second term gives the evolution in time. This equation is reducible in time using Fourier transformation. Also, one can see in this equation that the response function is determined by the thermal equilibrium through which the changes are connected when disturbed by an excitation.

14.5 *Summary*

THE PERPLEXING QUESTION OF how nonlinearities and irreversibility appear from linear equations such as those of Schrödinger and Maxwell has appeared again and again in this text, and so has a generous deployment of linear equations to describe numerous physical phenomena. The former we have quasi-resolved through the coupling from one to many states, wavefunction collapse, entropy and free energy exchanges such as in-phase transitions that bring spontaneous new properties to a system. The latter is for us an example of small changes around stability, in systems under stimulus. The electromagnetics of free space, and in a variety of material conditions, has a linear response. Only under high energy conditions, with material properties entering the nonlinear regime, or in many phase transitions, and others, do we start seeing the nonlinear response. Linear response is very much present in many of the physical phenomena. In electrical forays, the use of resistors, capacitors and inductors is evidence of their ubiquity. Only when external energy is somehow transplanted into this medium—as in gain that transforms static power to dynamic power—or when the free energy in the medium is removed by being changed into another form while still resident in the medium—as in polarization—do we see breakdown of linearity.

Causality has implications for the linear response. A simple example showed us, under the constraint of the speed of light's immutability, how phase shifts appear at all frequencies if absorption is concentrated at a specific frequency. The former is a real part response while the latter happens due to the imaginary part of the index of refraction. So, a linear response to a stimulus has a complex response, with the real and the imaginary parts of the response both due to the same causal stimulus, so embedded in them is information that is tied between them. The information may be spread out over a broad range such as of frequency. The

Kramers-Kronig relationship for the dielectric function tying the real and the imaginary parts of susceptibility (or of its equivalent, the index of refraction in this paragraph's illustration) in a linear response system is a statement of this embedded information. We used the Green's function approach to derive the Kramers-Kronig relationship for the susceptibility. The significance of this is that if one can make a measurement of one of these, one can predict what the other would be like. Absorption measurement can tell us about transmission measurement.

An oscillator is our unit of stability underlying the physical medium. It was therefore useful to look at the linear response of oscillators, with damping as a linear term in the response equation, to explore how atoms and electrons will respond in the medium. We noted a response that is an analog of the electrical system, phase changes being ahead or lagging depending on stimuli being slower or faster than the oscillator. Since a damping was introduced, one obtained both a real component and an imaginary component of the response in the dielectric function composed of such oscillators. And the real part and the imaginary part are related to each other. We found a similar behavior in the electron charge cloud in a material responding to an oscillating field. Both of these are examples of how the ionic and electronic contributors to polarization respond in a material, and this underlies the Kramers-Kronig response observed there. We then extended this linear response view to the quantum-statistical viewpoint through the density matrix and observables under stimulus, that is, a perturbation.

14.6 *Concluding remarks and bibliographic notes*

Just the question of "Is there free will?" or "Nature versus nurture?" will bring forth debates based on one's upbringing and all the baggage (biases?) we accumulate in the course of one's life. Philosophers and scientists are human too.

CAUSALITY IS A VERY RICH word with multifarious meanings based on one's inclinations. The advent of quantum mechanisms with the non-determinism and statistical nature embedded in them made it a rich ground for debates between philosophers and scientists. I particularly recommend Max Born's Waynflete lectures[1] for how he explores with a scientist's tools physical phenomena while showing how to avoid common contradictions of the everyday approach to looking at these. To this, I will also add David Bohm's book[2] as another writing requiring careful reading. The subject of correlation and its relationship to causality is dealt with by Kenny[3].

For understanding Kramers-Kronig relationships in depth, that is, the nature of dispersion with the subsumed information in it, see Nussenzveig[4]. Another, stretching out to the quantum-mechanical

[1] M. Born, "Natural philosophy of cause and chance," Oxford (1949)

[2] D. Bohm, "Causality and chance in modern physics," Routledge, ISBN 0-415-17440-6 (2005)

[3] D. A. Kenny, "Correlation and causality," Wiley, ISBN 978-0471024392 (1979)

[4] H. M. Nussenzveig, "Causality and dispersion relations," Academic, Library of Congress 72-7685 (1972)

view to response, is by Sethna[5], with a number of interesting exercises sprinkled throughout for computation-based exploration.

The linear response, as we will see in Chapter 16, also shows up in the connection between system response and the system's fluctuation response. Pécseli[6] explores noise, but also the Kramers-Kronig relationships from an electrical engineering perspective.

We had also remarked on the analytic function connection of these relationships. A good text for exploring analytic functions and these consequential relationships is by Kelly[7]. Cauchy principal value, the dispersion relations and other rich mathematical tools are spread out throughout this text.

For those interested in the broader implications of Kramers-Kronig relationships, Bode plots are another way of looking at them. Bode plots are in terms of magnitude and phase. The mechanical response of systems also shows Kramers-Kronig features, since they are linear responses. A good paper worth reading is by Bechoefer[8].

[5] J. P. Sethna, "Statistical mechanics: Entropy, order parameters, and complexity," Oxford, ISBN 13 978-0198566779 (2011)

[6] H. A. Pécseli, "Fluctuations in physical systems," Cambridge, ISBN 13 978-0521655927 (2000)

[7] J. J. Kelly, "Graduate mathematical physics," Wiley-VCH, ISBN 13 978-3-527-40637-1 (2006)

[8] J. Becchoefer, "Kramers-Kronig, Bode, and the meaning of zero," arXiv:1107.0071v1 (2011)

14.7 Exercises

1. Why are the real and the imaginary parts of the susceptibility—in all these linear response functions—related to each other? Please give a physically intuitive short reason. **[S]**

2. Our argument related to linear response theory, that is, the linear response to perturbations of the equilibrium state, has been that there are fluctuations—fast processes, and the system responds to external parameters in the presence of these fluctuations, leading to fluctuation-dissipation. A rapid change causes irreversibility in the presence of these fluctuations and hence the dissipation. The Kramers-Kronig relationship, fluctuation-dissipation, the second law of thermodynamics and Onsager relations are all manifestation of this dynamics. Is this true irrespective of the particular Hamiltonian describing the process of interest? Please provide a short argument. **[S]**

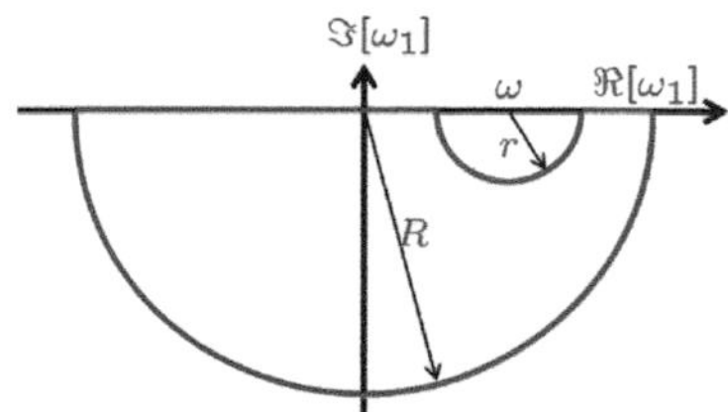

Figure 14.8: Integration path for the Kramers-Kronig relationship demonstration, with $R \to \infty$ and $r \to 0$.

3. Prove the Kramers-Kronig relation by integrating the susceptibility $\chi(\mathbf{k},\omega)/(\omega_1 - \omega)$ along the path shown in Figure 14.8 and parsing the real and the imaginary parts of the result. **[S]**

4. If one measures the absorption spectrum, can the refractive index be computed? **[S]**

5. The Kramers-Kronig relationship must hold also for all functions built out of $\underline{\epsilon}^r$ and $\underline{\epsilon}^i$. Find the relationship between $d\underline{\epsilon}^r/d\omega$ and $d\underline{\epsilon}^i/d\omega$. **[S]**

6. Derive the Kramers-Kronig relationship for the real and the imaginary parts of the complex index of refraction $\underline{n}^c$. **[S]**

7. A solid exhibits an absorption band at frequency ω_0 of width Δ, with an extinction coefficient of

$$\underline{n}^i = \begin{cases} n_0^i & \forall\ \omega_0 - \Delta/2 \leq \omega \leq \omega_0 + \Delta/2, \\ 0 & \text{otherwise.} \end{cases}$$

What is the refractive index at low frequencies if $\omega_0 \gg \Delta$? **[S]**

8. Fourier integrals let us write the general equation

$$\boldsymbol{\mathcal{E}}(\mathbf{r},t) = \int_{-\infty}^{\infty} \boldsymbol{\mathcal{E}}(\mathbf{r},\omega) \exp(i\omega t)\, d\omega,$$

where the fields are complex transforms in frequency or time of each other. Show that

- $\boldsymbol{\mathcal{E}}(\mathbf{r},-\omega) = \boldsymbol{\mathcal{E}}^*(\mathbf{r},\omega)$,

- in a linear isotropic medium, one may write $\sigma(\mathbf{r},\omega)$ and $\epsilon(\mathbf{r},\omega)$ such that

$$\mathbf{J}(\mathbf{r},t) + \frac{\partial}{\partial t}\mathbf{D}(\mathbf{r},t) = \int_{-\infty}^{\infty} [\sigma(\mathbf{r},\omega) + i\omega\epsilon(\mathbf{r},\omega)]\boldsymbol{\mathcal{E}}(\mathbf{r},\omega) \exp(i\omega t)\, d\omega,$$

and

- $\sigma(\mathbf{r},-\omega) = \sigma(\mathbf{r},\omega)$, and $\epsilon(\mathbf{r},-\omega) = \epsilon(\mathbf{r},\omega)$, which implies that the real part of $(\sigma + i\omega\epsilon)$ is an even function of frequency, and the imaginary part is an odd function of frequency. **[S]**

15
Onsager relationships

THERMAL EQUILIBRIUM ARISES IN THE ACCUMULATION OF ALL THE INTERACTIONS taking place in a system in the absence of any external stimulus, except thermal and particle exchange in contact with its large reservoir environment. The equilibrium state then is one where the macrostate with larger number of different configurational possibilities has a higher chance of existence than the more unique macrostates with only a few possibilities. In thermal equilibrium, in detail then, less is known about the precise configuration of the system, since the number of possibilities is higher, and although its macrostate provides expectations for observables, these observables will also fluctuate.

In the classroom, the possibility of all the molecules being confined to one corner instead of being evenly dispersed with some fluctuations around that distribution is infinitely smaller. It is possible to characterize the pressure or temperature of the room.

Entropy in this sense is a lack of information in detail about the system and, in thermal equilibrium, it is a maximum. But as we saw in the fluctuation-dissipation discussion through Brownian motion in Section 8.8, an important notion is that when macroscopic quantities deviate from equilibrium, the linear response, that is, the first-order response, can be reformed to connect the response through the random interactions taking place. Fluctuations are noise, fluctuations are drag and fluctuations are tied to the lack of pinning down of the precise description of the system, and therefore there is entropy. An important notion here is that since fluctuations are these random interactions, and the response to external forces are also beholden to them, response to any change in an external stimulus, taking place under these random interactions, will also show these characteristics. *The fluctuations follow the same decay rules as the macroscopic response and decay to the external forces.*

It is the microstates-to-microstates interaction where reversibility, including time reversibility, is applicable and where one may say that, in some sense, a local equilibrium exists, even though local equilibrium is absent at the macroscale through the exponential rise in one direction, the one corresponding to the response. So,

Semiconductor Physics: Principles, Theory and Nanoscale. Sandip Tiwari.
 Published 2020 by Oxford University Press. DOI: 10.1093/oso/9780198759867.001.0001

linear response in a way connects off-equilibrium to equilibrium, and fluctuation-dissipation is an outcome of it.

15.1 *Flux-flow and Onsager relationships as linear responses*

ANOTHER VERY IMPORTANT CONSEQUENCE is the existence of *Onsager relationships*, which express the equality of certain ratios between pairs of canonical flows and forces when thermodynamic systems are out of equilibrium.

Onsager relationships are fundamental and significant enough that many call their statement as the 4th law of thermodynamics. Nernst, who formulated the 3rd law, and called that law "my law," would have been unhappy about the appearance of a 4th one. His claim would be that there were three people (Rudolf Claussius, William Rankine and Germain Hess) associated with the first law, two people (Sadi Carnot and Rudolf Claussius) with the second, and only one—him, Walther Hermann Nernst—with the third. Ipso facto, there could be no more thermodynamic laws.

A few examples will illustrate this canonical flux-flow equality of ratios. It is quite palatable that heat flows from hot to cold, that is, from higher temperature to lower. Just some more thinking along the same lines, of numbers and of the rapidity of interactions, also leads us to the idea that, in the presence of pressure difference, particle flow will take place from higher to lower pressure, with concentration as the physical reason. But now consider what happens when temperature and pressure vary. Temperature difference at constant pressure causes particle flow. This is convective flow, as in water boiling on a stove. Pressure difference at constant temperature causes heat flow. We saw this in the concentration dependence of heat flow in semiconductors. What Onsager tells us is that heat flow per pressure difference, and density flow per temperature difference, are equal. This is not obvious but follows from the argument of local microscopic reversibility when deviating slightly from equilibrium. Tornadoes and hurricanes are an example of these connections and relationships. In the thermoelectric effect, this same connection holds true for these flows and forces, that is, conjugate variables. The relationships that connect the transport coefficients relating fluxes, which tend to restore a system to thermodynamic equilibrium so long as the system is not too far from this equilibrium, and the forces inducing such fluxes are symmetric and these are the Onsager relationships.

There exist a number of empirical laws that relate fluxes in terms of forces instantaneously. Table 15.1 illustrates a salient few. All these "laws" can be written in terms of the gradient of a canonical force ($\nabla_{\mathbf{r}}\Phi$). Their distillation, as well as their origin, is that one finds a parameter of response and a parameter of the forcing function that are empirically observed to be related linearly to the gradient, while maintaining all other conditions invariant. For example, for all the laws of Table 15.1, except Fourier's, temperature is kept constant. In all of these examples, energy in one form is converting to energy in another form, with potential and kinetic front and center.

These are empirical in their origin but placed on a firmer footing, where they may be seen as a good approximation in a restricted domain from fundamentals. Newton's, for example, is Ehrenfest's expectations from quantum at macroscale, or Ohm's from quantum Liouville through a string of approximations including that of Boltzmann transport to a Brownian situation.

The energy lost that is not accounted for—heat—is entropy increase in the off-equilibrium conditions. So, take the example of Figure 15.1, where passage of current in a conducting medium causes heat generation, as a restricted example of our thermoelectric discussion of Subsection 9.2.3. This is the Ohm's law situation, where the current density is

$$J_\phi = -\frac{1}{\rho}\frac{\partial \phi}{\partial y} = -\sigma\frac{\partial \phi}{\partial y}, \tag{15.1}$$

where ρ is the resistivity, and σ is the conductivity. The heat generation may be viewed in entropic terms. Entropy is being generated due the equilibrium being disturbed. This entropy generation density (s)—a flux—is

$$s = J_\phi\left[-\frac{1}{T}\frac{\partial \phi}{\partial y}\right] = \frac{1}{T}\rho J_\phi^2. \tag{15.2}$$

One can see meaning behind this equation. Heat is $\Delta Q = T\Delta S$, so an inverse temperature must appear in the denominator of an equation relating s. One can even assign meaning to the bracketed term of Equation 15.2. It is an electric force. In Equation 15.1, a flux—the current density—was caused by the forces applied, which can be viewed as a flux-force relationship. In Equation 15.2, we have a specific new instance of a canonical flux (of entropy) being generated by a canonical force (the electric force as defined here). This equation satisfies

$$\rho J_\phi^2 = sT, \tag{15.3}$$

where the left-hand side is Joule heating flux, which is the heat energy density flow per unit time. And it is in balance with the entropy density per unit time times temperature. Joule heating develops a temperature gradient and is caused by charge flow, and one gets a coupled transport of heat and charge that satisfies Equation 15.3. If a temperature gradient exists, there is also a flux arising in the thermal nonequilibrium. The entropy relations taking into account the consequence of external biasing causing current,

Frictional loss is loss to modes of vibrations, within atoms, of atoms, of electrons, and so on. Heat is, as we have seen, an energy form quite intimately tied to kinetic energy.

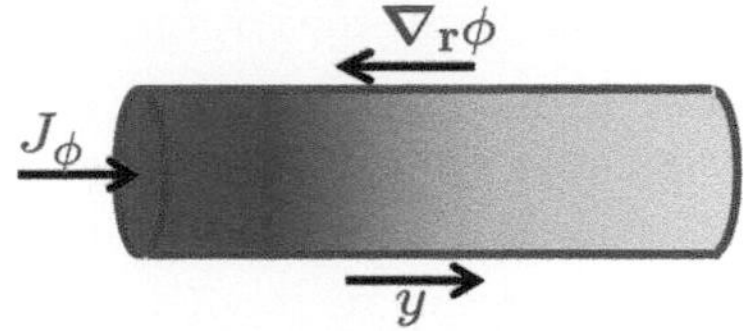

Figure 15.1: Heat generation and temperature and potential gradients when current is flowing in a conductive medium, and the only source of heat generation is in the friction of the particle flow.

Table 15.1: Flow and flux relationships for some classical empirical laws.

Law	Relationship	Definitions
Ohm's	$J_\phi = -\frac{1}{\rho}\frac{\partial \phi}{\partial y} = -\sigma\frac{\partial \phi}{\partial y}$	J_ϕ: Current density ρ: Resistivity σ: Conductivity
Fourier's	$J_T = -\kappa\frac{\partial T}{\partial y}$	κ: Thermal coefficient
Fick's	$J_\mu^1 = -\mathcal{D}\frac{\partial c_1}{\partial y}$	$\mathcal{D}$: Diffusion coefficient c_1: Concentration of the 1st phase
Darcy's	$J_Q = -KA\frac{\partial h}{\partial y}$	J_Q: Discharge flux KA: Conductivity times area h: Head of the porous column

which causes temperature changes, which cause a thermal flux, then says, in these linear conditions, that

$$s = J_{\mathfrak{T}} \frac{\partial}{\partial y}\left(\frac{1}{T}\right) + J_{\phi}\left[-\frac{1}{T}\frac{\partial \phi}{\partial y}\right]. \tag{15.4}$$

Here, s is still the entropy production rate ($J/m^3 \cdot s \cdot K$), with the first term—the new term—being a product of a thermal flux ($J_{\mathfrak{T}}$) and its canonical conjugate—a thermal force—given by $\partial/\partial y(1/T)$. Entropy generation is the energy in heat form ("the not-useful form"), so we can associate it with a thermal flux $J_{\mathfrak{T}}$. We have created two equations—one of electrical flux and one of thermal flux—in terms of two canonical forces—one electrical and one thermal—for which we now have a relational form.

This thermal dissipation under charge flow discussion for the example can be folded in as

$$\begin{aligned} J_{\phi} &= L_{\phi\phi}\left[-\frac{1}{T}\frac{\partial \phi}{\partial y}\right] + L_{\phi\mathfrak{T}}\frac{\partial}{\partial y}\left(\frac{1}{T}\right), \\ J_{\mathfrak{T}} &= L_{\mathfrak{T}\phi}\left[-\frac{1}{T}\frac{\partial \phi}{\partial y}\right] + L_{\mathfrak{T}\mathfrak{T}}\frac{\partial}{\partial y}\left(\frac{1}{T}\right). \end{aligned} \tag{15.5}$$

The first equation is for electric current flux. The second equation is for thermal current flux. They are writable through a linear combination of conjugate electric force and thermal force. Fluxes and forces are in a tensorial order. This was an example rooted in an electrical stimulation, exploring the energy exchange between the electrical from (a potential form) and the thermal form, together with the entropy.

The proof of Onsager relations stretch over 22 and 15 pages of *Physical Review*. We will not repeat them here and will keep our argument grounded in physical underpinnings.

Onsager tells us that each flux can be written as a linear combination of fluxes and forces as a linear perturbation from equilibrium. Energy exchanges take place in many forms, and these exchanges in different forms can be put together as a linear addition because of the linear response, while the equality of ratio holds between pairwise forms due to their microscopic origins. In the variety of situations we have considered, heat, electrostatic energy, concentration and free energy are the forms of energy that we have encountered most often. So, consider thermal (heat), charge (electrostatic energy), chemical energy (particle concentration) and free energy (Gibbs), which is exchanged between all these forms in the conditions we are interested in.

This is generalizable. One could have used other forms in which energy resides, for example, a magnetic form.

The entropy production is a sum of all conjugate *flux-force pairs*, such as in the example discussed. Taking the major forms, the entropy production is

$$s = J_{\mathfrak{T}} \frac{\partial}{\partial y}\left(\frac{1}{T}\right) + J_{\phi}\left[-\frac{1}{T}\frac{\partial \phi}{\partial y}\right] + \sum_i J_{\mu}^i \left[-\frac{1}{T}\frac{\partial \mu_T^i}{\partial y}\right] + r\left(-\frac{\Delta \mathscr{G}}{T}\right). \tag{15.6}$$

The first two terms are the same as the ones we discussed in our starting example. The third term describes the consequences of concentration changes—chemical potential changes—and has been generalized to multiple phases, for example, species (electrons and holes!). The last term is the Gibbs free energy force. It is the Gibbs free energy in these conditions that is being exchanged in different forms. Entropy, heat, electrical energy, chemical energy and Gibbs free energy are all the terms exchanging, and this equation balances them.

Next, the linear system formulation and the flux-force argument assert that the fluxes are linear homogeneous functions of all the forces of the same tensorial order:

$$\begin{aligned}
J_{\mathfrak{T}} &= L_{\mathfrak{T}\mathfrak{T}}\frac{\partial}{\partial y}\left(\frac{1}{T}\right) + L_{\mathfrak{T}\mu}\left[-\frac{1}{T}\frac{\partial\mu_{1,T}}{\partial y}\right] + L_{\mathfrak{T}\phi}\left(-\frac{1}{T}\frac{\partial\phi}{\partial y}\right), \\
J_{\phi} &= L_{\phi\mathfrak{T}}\frac{\partial}{\partial y}\left(\frac{1}{T}\right) + L_{\phi\mu}\left[-\frac{1}{T}\frac{\partial\mu_{1,T}}{\partial y}\right] + L_{\phi\phi}\left[-\frac{1}{T}\frac{\partial\phi}{\partial y}\right], \\
J_{\mu} &= L_{\mu\mathfrak{T}}\frac{\partial}{\partial y}\left(\frac{1}{T}\right) + L_{\mu\mu}\left[-\frac{1}{T}\frac{\partial\mu_{1,T}}{\partial y}\right] + L_{\mu\phi}\left[-\frac{1}{T}\frac{\partial\phi}{\partial y}\right], \text{ and} \\
r &= l\left(-\frac{\Delta\mathcal{G}}{T}\right),
\end{aligned} \tag{15.7}$$

where the first equation is for heat (thermal), the second is for electric, the third is for mass (chemical composition) and the last is for Gibbs free energy exchange. One phase is assumed here. The diagonal elements in these relationships are related to the various parameters of the classical laws: κ for the thermal coefficient, ρ for resistivity, $\mathcal{D}$ for diffusivity, et cetera.

Finally, Onsager relations connect the independent fluxes and forces:

$$L_{\mathfrak{T}\mu} = L_{\mu\mathfrak{T}}, \;\; L_{q\phi} = L_{\phi q}, \;\text{ and }\; L_{\phi\mu} = L_{\mu\phi}. \tag{15.8}$$

The physical argument to justify these forms is the following. The Boltzmann entropy describes for this thermodynamic off-equilibrium situation the fluctuations that exist at equilibrium. Microscopic reversibility also holds, and when the system is disturbed, the macroscopic laws describe the relaxation of fluctuations to equilibrium through the slow and fast forces at work in the midst of all the scattering processes. If one looks at time scales where sufficient number of the microscopic events take place leading to a proper expectation of the macroscopic picture, then, given independent forcing functions—our variables—the responsiveness of the system for any specific canonical force i is

$$\alpha_i(t+\tau) = \alpha_i(t) + \tau\sum_j L_{ij}X_j, \tag{15.9}$$

where the sum term describes the fluctuation events, such as scattering, that are connected microscopically and that change the state of the system from what it was at time t to what it will be at time $t + \tau$, the microscopic events having taken place under microscopic reversibility for the time duration of τ. This is the basis for the linearity connecting the macroscale with its long time response to the microscale and its small time response.

The fluxes are linear and homogeneous functions of the forces. The coefficients are functions of the state variables (the macroscopic parameters). In a practical situation, one may make observations and connect empirical transport coefficients to the Onsager coefficients, which are symmetric.

In Equation 15.7, which is, of course, writable in a matrix form, the coefficients along the diagonal are the dissipative elements. They are self-interaction terms, not energy exchange interaction terms. The off-diagonal terms represent work through reversible processes. If off-diagonal terms are large, that is, the exchange coupling is efficient, the dissipation is small and the entropy production is low. These connections make practical measurements useful in the design and prediction off-equilibrium of complex situations, because the observations are a manifestation of Onsager off-equilibrium linearity.

15.2 *Examples of Onsager consequences*

WE CAN NOW ILLUSTRATE some of the implications of Onsager's linear response relationships.

Fick's law, in Onsager terms, is

$$J_{\mu}^{1} = -D\frac{\partial c_1}{\partial y} = -L_{\mu\mu}\frac{1}{T}\frac{\partial \mu_{1,T}}{\partial y} = -L_{\mu\mu}\frac{R}{c_1}\frac{\partial \mu_{1,T}}{\partial c_1}\frac{\partial c_1}{\partial y}, \tag{15.10}$$

where R is the molar gas constant, and the canonical chemical force is $(1/T)\partial\mu_{1,T}/\partial y$, with T subscripting indicating a constant temperature condition. Fourier's law, in Onsager terms, is

$$J_{\mathfrak{T}} = -\kappa\frac{\partial T}{\partial y} = L_{\mathfrak{T}\mathfrak{T}}\frac{\partial}{\partial y}\left(\frac{1}{T}\right) = -\frac{L_{\mathfrak{T}\mathfrak{T}}}{T^2}\frac{\partial T}{\partial y}. \tag{15.11}$$

So, $L_{\mu\mu}$ and $L_{\mathfrak{T}\mathfrak{T}}$ are the Onsager coefficients for chemical force and thermal force, respectively. And $L_{\mu\mathfrak{T}} = L_{\mathfrak{T}\mu}$, because of the connection between independent fluxes and forces.

These connections between independent fluxes and forces can be seen through a set of examples that one doesn't normally discuss but which involve chemical and thermal forces. The *Soret effect* is the phenomenon of thermal diffusion in hydrodynamics whereby thermal gradients can be employed to separate small species from large species. The *Dufour effect* is the complement of the Soret effect. Diffusion now causes the thermal gradient. These can be seen through the following:

$$J_{\mathfrak{T}} = L_{\mathfrak{T}\mathfrak{T}} \frac{\partial}{\partial y}\left(\frac{1}{T}\right) + L_{\mathfrak{T}\mu}\left(-\frac{1}{T}\frac{\partial \mu}{\partial y}\right), \text{ and}$$

$$J_{\mu} = L_{\mu\mathfrak{T}} \frac{\partial}{\partial y}\left(\frac{1}{T}\right) + L_{\mu\mu}\left(-\frac{1}{T}\frac{\partial \mu}{\partial y}\right)$$

$$\therefore\ J_{\mathfrak{T}} = -\left(L_{\mathfrak{T}\mathfrak{T}} - \frac{L_{\mu\mathfrak{T}}L_{\mathfrak{T}\mu}}{L_{\mu\mu}}\right)\frac{1}{T^2}\frac{\partial T}{\partial y} + \frac{L_{\mathfrak{T}\mu}}{L_{\mu\mu}}J_{\mu}. \qquad (15.12)$$

This equation shows the power of this Onsager approach. The presence of a chemical gradient force reduces the thermal conductivity, as the first term's bracketed coefficient shows. In addition, as a result of the coupling, the second term says that there is a convective heat current arising from the coupling.

This is not quite relevant to solid state, but, in a cell membrane, with the salty watery environment and temperature differentials, the mass flow and heat flow are beholden to these linear response interrelationships. Neuronal response too is subject to the Onsager relationships together with the other intricacies of the dynamics there.

We complete this discussion by returning to thermoelectrics, for which we derived a number of energy exchange effects. For this example, we now also include possibilities as a source of energy through energy exchange (see Figure 15.2). The relevant relationships are

$$s = J_{\mathfrak{T}} \frac{\partial}{\partial y}\left(\frac{1}{T}\right) + J_{\phi}\left[-\frac{1}{T}\frac{\partial \phi}{\partial y}\right],$$

$$J_{\mathfrak{T}} = L_{\mathfrak{T}\mathfrak{T}} \frac{\partial}{\partial y}\left(\frac{1}{T}\right) - L_{\mathfrak{T}\phi}\left(\frac{1}{T}\frac{\partial \phi}{\partial y}\right), \text{ and}$$

$$J_{\phi} = L_{\phi\mathfrak{T}} \frac{\partial}{\partial y}\left(\frac{1}{T}\right) - L_{\phi\phi}\left(\frac{1}{T}\frac{\partial \phi}{\partial y}\right). \qquad (15.13)$$

Conduction and charge transfer superpose, and we can rewrite this as

$$J_{\mathfrak{T}} = -\frac{1}{T^2}\left(L_{\mathfrak{T}\mathfrak{T}} - \frac{L_{\mathfrak{T}\phi}L_{\phi\mathfrak{T}}}{L_{\phi\phi}}\right)\frac{\partial T}{\partial y} + \frac{L_{\mathfrak{T}\phi}}{L_{\phi\phi}}J_{\phi}. \qquad (15.14)$$

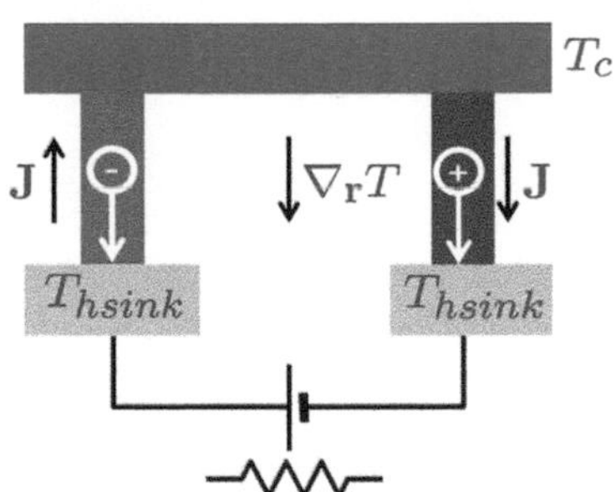

Figure 15.2: A thermoelectric arrangement for cooling and voltage generations.

This equation describes the reversible heat transport in the conductor. In this equation, the prefactor of the second term shows that coupling leads to heat transport via electric current, which is a convection. This is the Peltier heating term. The first bracketed term now says that the thermal conductivity of this stationary state is lower than that of a homogeneous conductor. This equation also tells us what the Peltier coefficient is. By definition,

$$\Pi = \left.\frac{J_{\mathfrak{T}}}{J_{\phi}}\right|_{dT=0} = \frac{L_{\mathfrak{T}\phi}}{L_{\phi\phi}}. \qquad (15.15)$$

To calculate the voltage generation from the temperature gradient when there is a particle current,

$$\Delta\phi = \int_L \frac{\partial \phi}{\partial y}\, dy = \int_L \left[\frac{\Pi}{T}\frac{\partial T}{\partial y} - \frac{T}{L_{\phi\phi}} J_\phi \right] dy. \tag{15.16}$$

Here, the first term in the integrand is work arising in the temperature gradient, and the second term gives the ohmic loss in the presence of the temperature gradient. The Onsager equations gave us quite straightforwardly the positional accumulation of potential in the presence of transport.

To find out the maximum thermoelectric potential, we employ

$$\left.\frac{\Delta\phi}{\Delta T}\right|_{J_\phi=0} = -\left.\frac{\Pi}{T}\right|_{dT=0}. \tag{15.17}$$

The Seebeck coefficient and the Peltier coefficient can be related through the Onsager relationships. The Seebeck coefficient is easier to measure, since it is measurement of a voltage with no current flowing. And, from that, the Peltier coefficient can be determined.

We now show the entropic connection in this example:

$$\left.\frac{J_{\mathfrak{T}}}{J_\phi}\right|_{dT=0} = \frac{L_{\mathfrak{T}\phi}}{L_{\phi\phi}} = \frac{\Pi}{\mathcal{F}} = \frac{1}{\mathcal{F}} TS^*. \tag{15.18}$$

The entropy is being transported by electrons and holes, and if we deconvolve the potential contributions,

$$\begin{aligned}
\Delta\phi &= \int_0^L \frac{\partial \phi_p}{\partial y}\, dy + \int_L^0 \frac{\partial \phi_n}{\partial y}\, dy = \int_0^L \frac{\partial \phi_p}{\partial y}\, dy - \int_0^L \frac{\partial \phi_n}{\partial y}\, dy \\
&= -\int_0^L \left(S_p^* \frac{1}{\mathcal{F}} \frac{\partial T}{\partial y} - r_p j \right) dy - \int_0^L \left(S_n^* \frac{1}{\mathcal{F}} \frac{\partial T}{\partial y} - r_n j \right) dy \\
&= -\frac{1}{\mathcal{F}} (S_p^* - S_n^*) \Delta T + (r_p + r_n) j \Delta y,
\end{aligned} \tag{15.19}$$

which says that the entropic contribution generates and that there is a potential drop across the resistances.

If used at the idealization of no current, this thermoelectric generator has

$$\left.\frac{\Delta\phi}{\Delta T}\right|_{j=0} = -\frac{1}{\mathcal{F}} (S_p^* - S_n^*). \tag{15.20}$$

So, while the thermoelectric discussion of Chapter 9 was very instructive in understanding the implications of fluctuation-dissipation in the midst of multimodal stimulation, it generated equations that became increasingly harder to interpret unless followed through carefully. Here, Onsager's linear off-equilibrium flux-force argument leads to the same conclusion in a more general and intuitive way.

15.3 *Summary*

THE ONSAGER RECIPROCITY RELATIONS encapsulate the most basic description of how out-of-equilibrium systems increase in entropy due to the fluxes of matter and energy that the thermodynamic forces—the affinities—generate. They are statements of linear relationships in the off-equilibrium response of the ensemble. Onsager's reciprocity relations relate linear response coefficients between flux densities and thermodynamic forces to one another. And these also correspond to fluctuations and dissipation, as we noted the connections in our discussion of Brownian motion and other types of fluctuation-dissipation.

The relationships let us make calculations at a system-scale quite conveniently, and we saw examples of this in the thermoelectric response. So far, in this text, we have found numerous such correspondence relationships. These various relationships—Kramers-Kronig, Brownian motion, fluctuation-dissipation and Onsager—are tied to each other in a deep way. In the next chapter, we will see one more connection, that of noise, which is due to the fluctuation-dissipation correspondence.

One power of Onsager relationships that we did not discuss is that they allow one to tackle surfaces, edges and other dimensionality and scale changes within a common framework, unlike the detailed scattering-based formulations, where one had to tackle the problem in detail for the scattering-dominated flow and calculate the parameters accordingly.

15.4 *Concluding remarks and bibliographic notes*

ONSAGER RELATIONS provide another example of linear response and the adequacy of the thermodynamic description for irreversible phenomena.

A good starting point for seeing the connections between classical thermodynamics, equilibrium and the Onsager description in off-equilibrium with conjugate forces is the paper by Miller[1] that explores electrokinetics, thermoelectricity, isothermal diffusion, thermomagnetism and galvanomagnetism. A more complete discussion of linear response theory, straddling its various manifestations, including Onsager relations, is the open internet archive lecture resource[2] from Hertel.

Kubo's contribution[3] is a complete and advanced thesis on irreversible processes and the centrality of Onsager relationships in them.

[1] D. G. Miller, "Thermodynamics of irreversible processes," Chemical Reviews, **60**, 15–37 (1960)

[2] P. Hertel, "Linear response theory," available at https://archive.org/details/Peter_Hertel___Linear_Response_Theory (2001)

[3] R. Kubo, "Some aspects of the statistical-mechanical theory of irreversible processes," in *Lectures in theoretical physics*, Interscience, Library of Congress 59-13034, 120–203 (1959)

A text-based source[4] for a more chemistry-centric discussion is by Kjelstrup and Bedeaux. For heat transfer, with a special focus on the nanoscale, is the book[5] edited by Volz.

[4] S. Kjelstrup and D. Bedeaux, "Non-equilbrium thermodynamics of heterogeneous systems," World Scientific, ISBN 13 978-981-277-913-7 (2008)

[5] S. Volz (ed.), "Microscale and nanoscale heat transfer," Springer, ISBN 13 978-3-540-36056-8 (2007)

15.5 *Exercises*

1. Why should off-equilibrium conditions of energy exchange have linear relationships? **[S]**

2. We will probe the thermoelectric effect for its Onsager reciprocity manifestation. The thermal current and the electric current are the two manifestations under temperature and potential, respectively, as the stimuli. Using entropy change as the basis, explore and show that the correspondence between fluctuations in kinetic energy and macroscopic currents leads to the current relationships that Onsager relationships show. **[M]**

16
Noise

THIS CHAPTER IS ABOUT CAUSALITY AND CHANCE COMING TOGETHER AS NOISE and also of integrating the various facets of fluctuation-dissipation into noise. Our approach will be physical and conceptual, with an interest in attributes and consequences of interest for devices where this approach is quite relevant.

For those interested in more general, fundamental and mathematical probing from a statistical viewpoint, the appendix on noise in S. Tiwari, "Nanoscale device physics: Science and engineering fundamentals," Electroscience 4, Oxford University Press, ISBN 978-0-198-75987-4 (2017), is recommended as a complement to this discussion.

Noise has appeared again and again as a fluctuation, that is, a form of randomness, in the characteristics observed. It may be a seemingly random spread that exists over multiple measurements, even as there is a predictable expectation of a system's response. Or it may be a seemingly random fluctuation that is observed in a sequence of time. So, this noise to us is a random deviation from expectation in any measurement. If we measure a voltage across a resistor by passing a certain current through it, $V = IR$ would be valid in classical limits as really $\langle V \rangle = \langle IR \rangle$, where one may perform the statistical averaging in time or over resistors of identical R. Within the resistor, there exist numerous scattering events—fluctuations—that lead to the dissipation represented by R but which are also manifested in one measurement from another in that the precise voltage measured will not be the same, even if, after many measurements in time, that $\langle V \rangle$ asymptotes to a precise value for a precise current I impressed on this system. Sometimes it is larger, some it is smaller, and the first statistical measure of this difference would be that $\langle V^2 \rangle \neq \langle V \rangle^2$; that is, the voltage signal has a variance. We employ this or equivalent measures of the random deviation over a collection of such measurements that we will call noise.

We must make a clear distinction between classical statistical randomness and quantum randomness. This difference between classical and quantum randomness is very important (see S. Tiwari, "Nanoscale device physics: Science and engineering fundamentals," Electroscience 4, Oxford University Press, ISBN 978-0-198-75987-4 (2017), in the discussion of Bell's inequalities). Zero point fluctuations are also random, and therefore a noise, but a quantum noise. We will stick here with the noise that appears in a system consisting of classical or quantum objects that are assembled in numbers but are not entangled and therefore have a distribution, when observations are made, that is bereft of entanglement consequences. An electron moving in a semiconductor acquires noisiness through its movement in the midst of random interactions. A quantum harmonic oscillator has the noise as a natural outlet of uncertainty in the precise determination of the canonical coordinates. Engineers do tend to have a tendency to play loose. This is a characteristic that is quite useful sometimes but is also irritating and

In this chapter, we start with the understanding of a few important concepts—important here and elsewhere too where we have used related ideas and terms—and follow it with the development of an understanding of noise through a treatment of

Semiconductor Physics: Principles, Theory and Nanoscale. Sandip Tiwari.
 Published 2020 by Oxford University Press. DOI: 10.1093/oso/9780198759867.001.0001

electrons in a solid structure. This approach will directly draw on our semiconductor discussions.

a source of failure sometimes. Cross-talk is often lumped in with noise. And there are books with such titles. Cross-talk is deterministic, knowing enough about the pulses coupling and the system, particularly the system's capacitances, cross-talk's effect can be undone. True randomness cannot be undone.

16.1 *Characterization of signals and their randomness*

NOISE IS A STOCHASTIC PROCESS representing a random variation in a function in time and space. Most of our concern will be with time. As a random process, it needs a statistical characterization. For any system, the averaging may be accomplished over a time (or space) interval, or it could be done over many identical systems, as sketched in Figure 16.1. The former is time averaging, and the latter is ensemble averaging.

A stochastic process is the collection of random variables that take on values from a common set in a common probability space. We encountered this state set and its evolution in the Markov discussion of Section 8.9.

For time averaging, and ensemble averaging, one can define some of the important characteristics of this stochastic function $z(t)$, as summarized in Table 16.1.

When time averages and ensemble averages are equal, the ensemble is an ergodic ensemble. Not all ensembles are ergodic. The autocorrelation is a measure of how the nature of the signal changes with time delays, and its equivalent measure—covariance—is a measure of the difference of the nature of two different signals. If signals are discrete, one may look at these measures discretely through summations, and if they are continuous, one resorts to the equivalent integration approach. We have $\mathfrak{p}_2$ as a second-order probability distribution function; that is, it looks at how two different signal streams appear under the time-shifting operation that lead to the integrated covariance measure of the two streams. In general, one may write a kth order probability density function as

A number of these characteristics are related to the Markovian and Kalmogorovian probabilistic discussions of sequence of events. The functions we are discussing are a sequence of events in the game of causality and chance. The chance part we still have to define more clearly. Ergodic's etymological origin is the Greek word $\varepsilon\rho\gamma o\nu$ (work) and $\acute{o}\delta\acute{o}\varsigma$ (way). Ergodic processes, with time evolution, lose memory of the initial state. In thermodynamics, the ergodicity hypothesis is the matching of the behavior of a system in time with that over the phase space. It implies that, given enough time, all the points of the probability space will be reached. If Liouville's theorem applies, then the starting states in phase space of the system determine its trajectory, and the phase-space volume is a constant. And not all the phase space can be sampled in finite time. Phase transitions—symmetry breakdown—are examples of nonergodicity. If one moved through a Curie temperature, the average order parameter should have remained unchanged.

$$\begin{aligned} & \mathfrak{p}_k(z(t_0), z(t_1), \ldots, z(t_k); t_0, t_1, \ldots, t_k) \\ &= \mathfrak{p}_k(z(t_0), z(t_1), \ldots, z(t_k); t_0 + \delta t, t_1 + \delta t, \ldots, t_k + \delta t). \end{aligned} \tag{16.1}$$

If this kth order probability density function is invariant with this time shifting, and $k+1$ is as well, then this process is *stationary in the order, k*. If a stochastic process remains stationary in any order, then it is *strictly stationary*. When the mean and the mean square

Recall that "stationary," in quantum-mechanical terms, is that the probability density remains invariant in time, so this usage is equivalent.

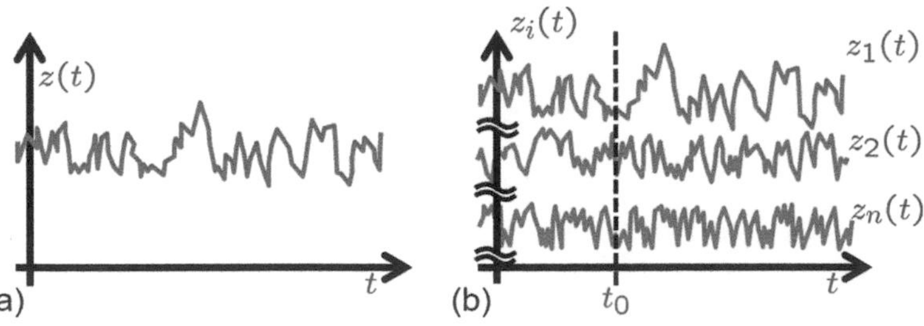

Figure 16.1: Part (a) shows a random signal $z(t)$ over time t and (b) shows the random signal $z_i(t)$ over instances of n identical systems. Time averaging is over the time of one system. Ensemble averaging is over multiple systems at an identical instance of time, which is shown here as $t = t_0$.

Parameter	Relationship
Time averages	
Mean	$\langle z(t)\rangle = \lim_{T\to\infty} \frac{1}{T}\int_{-T/2}^{T/2} z(t)\,dt$
Mean square	$\langle z^2(t)\rangle = \lim_{T\to\infty} \frac{1}{T}\int_{-T/2}^{T/2} z^2(t)\,dt$
Variance	$\sigma_z^2 = \langle z^2(t)\rangle - \langle z(t)\rangle^2$
Autocorrelation	$\phi_z(t) = \langle z(t)z(t+t')\rangle$ $= \lim_{T\to\infty} \frac{1}{T}\int_{-T/2}^{T/2} z(t)z(t+t')\,dt'$
Ensemble averages	
Mean	$\langle z(t_0)\rangle = \lim_{N\to\infty} \sum_{i=1}^{N} z_i(t_0)$ $= \int_{-\infty}^{\infty} z(t_0)\mathfrak{p}_1(z(t_0),t_0)\,dz(t_0)$
Mean square	$\langle z(t_0)\rangle = \lim_{N\to\infty} \sum_{i=1}^{N} z_i^2(t_0)$ $= \int_{-\infty}^{\infty} z^2(t_0)\mathfrak{p}_1(z(t_0),t_0)\,dz(t_0)$
Variance	$\sigma_z^2 = \langle z^2(t_0)\rangle - \langle z(t_0)\rangle^2$
Covariance	$\langle z(t_0)z(t_1)\rangle = \lim_{N\to\infty} \sum_{i=1}^{N} z_i(t_0)z_i(t_2)$ $= \int_{-\infty}^{\infty} z(t_0)z(t_1)\mathfrak{p}_2(z(t_0),z(t_1);t_0,t_1)\,dz(t_0)\,dz(t_1)$

$\mathfrak{p}_1(z_0,t_0),\ldots$ are first-order probability density functions; that is, the probability of finding $z(t)$ in the range of z_0 and z_0+dz_0 at time t_0.

$\mathfrak{p}_2(z(t_0),z(t_1);t_0,t_1),\ldots$ are second-order probability density functions; that is, the joint probability of finding $z(t)$ in the range of z_0 and z_0+dz_0 at time t_0, as well as in the range z_1 and z_1+dz_1 at time t_1.

Table 16.1: A few useful parameters to characterize time- and ensemble-averaged signals. $z(t)$ is the signal in time. $\langle\rangle$ is being used for time and ensemble averaging and can be understood from context. i is a subscript to identify the ith event.

are independent of time t_0, that is, are constants, and $\langle z(t_0)z(t_1)\rangle$ is independent of the choice of t_0 and t_1 but depends on the difference $\tau = t_1 - t_0$ then it is *weakly stationary*, sometimes called wide-sense stationary.

The Wiener-Khintchin theorem tells us that the autocorrelation function of a wide-sense stationary process can be spectrally decomposed via the power spectrum of the process. Appendix A discusses a number of attributes that can be observed between the direct and reciprocal domains through Fourier transformations. It spans a discussion of Perseval's theorem, convolution, correlation, the Wiener-Khintchin theorem and Carson's theorem. In noise, we will see this view of the frequency spread of a energetic fluctuation phenomena and its memory persistence as an important reciprocal space manifestation.

The power spectral density ($S_z(\omega) = \lim_{T\to\infty} 2\langle|Z(\omega)|\rangle^2/T$), where $Z(\omega)$ is the Fourier transform of $z(t)$, is a powerful way to analyze and understand signals; spectral density is quite measurable and also often easier to analyze. As an example, take a statistically stationary noisy waveform such as the one in Figure 16.1(a), with the autocorrelation

$$\phi_z(t') = \phi_z(0)\exp\left(-\frac{|t'|}{\tau_0}\right), \tag{16.2}$$

where $\phi_z(0) = \langle z^2\rangle$, and τ_0 is some relaxation constant. The power spectral density is

$$S_z(\omega) = 4\phi_z(0)\frac{\tau_0}{1+\omega^2\tau_0^2}, \tag{16.3}$$

The Hall-Shockley-Read process (Chapter 11), with its longer capture and emission times, is one pathway for such a relaxation time constant to appear as noise atop the signal from the semiconductor.

that is, a Lorentzian spectrum has appeared with a cut-off frequency defined by the relaxation time, and a low-frequency spectral power density in noise of $S_z(0) = 4\phi_z(0)\tau_0$. *Autocorrelation in this noisy signal can now be seen, as also the power spectral response, and one can see that the memory of the past is lost on a time scale of the order of* τ_0.

Figure 16.2 shows the autocorrelation and spectral density for this statistically stationary process. In a time shift of the order of t'/τ_0, the correlation is down to $1/e$ of its peak. The power spectrum shows that, at radial frequencies above τ_0, the spectrum rolls off at 6 *dB/octave*.

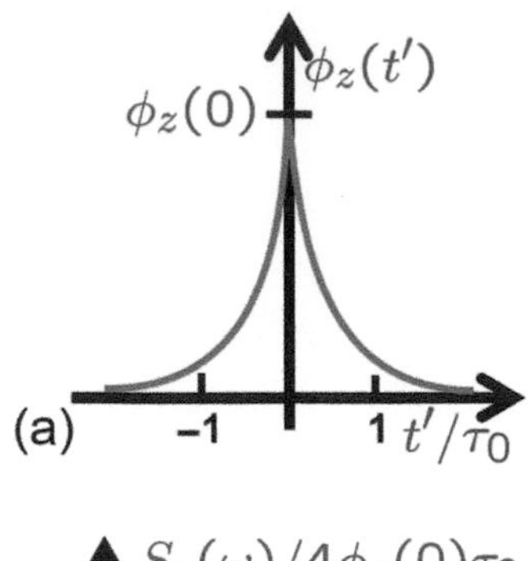

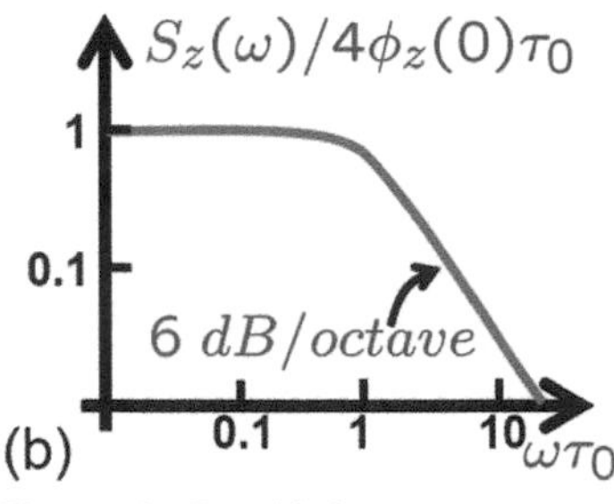

Figure 16.2: Part (a) shows autocorrelation and (b) shows the spectral power density function for a stationary process with $\phi_z(t') = \phi_z(0)\exp(-|t'|/\tau_0)$, which has a Lorentzian spectrum.

To get a peek at ergodicity, consider the signal $z(t) = \cos(\omega t + \theta)$, where θ is a phase noise in an oscillator. If θ is uniform, that is, $\mathfrak{p}(\theta) = 1/2\pi$, then its time average is 0, and an ensemble average at any time for $z_i(t)$ also vanishes. The process is ergodic in the mean. It is also ergodic in the autocorrelation $(\phi_z(t') = \langle z(t)z(t+t')\rangle)$. But, for any other non-uniform choice of $\mathfrak{p}(\theta)$, this process is not stationary and not ergodic.

The signals that one observes are of finite time duration, either because of the signal's time duration or because of the finiteness of the observation time. Strictly, we must approach these situations through a gated function that is finite and generally non-zero over an interval, and vanishes otherwise. Such a gate function will be identified with a subscript T. A time-integrated function of a statistically stationary process is statistically non-stationary. It has a deterministic evolution of expectation. An example is the random walk diffusion of $z(t')$. Consider

$$\zeta_T(t) = \begin{cases} \int_0^t z(t')\,dt' & \forall\ 0 \le t \le T, \text{ and} \\ 0 & \text{otherwise,} \end{cases} \tag{16.4}$$

which is a ramping signal with noise atop it. It is a Wiener-Levy process, which is statistically non-stationary. To evaluate the memory of the process, we determine the covariance—a power spectrum form—using the Wiener-Khintchin theorem, and since the process is assumed ergodic, employing autocorrelation ($T \rightarrow \infty$), this covariance is

See Appendix A for a discussion of the Wiener-Khintchin theorem.

$$\begin{aligned}
\langle\zeta(t)\zeta(t+t')\rangle &= \int_0^t \int_0^{t+t'} \langle z(t')z(t'')\rangle\, dt'\, dt'' \\
&= \int_0^t \int_0^{t+t'} dt'\, dt'' \frac{1}{2\pi} \int_0^\infty S_z(\omega)\cos(\omega t')\, d\omega \\
&= \frac{1}{2\pi} \int_0^\infty S_z(\omega) \int_0^t \int_0^{t+t'} \cos[\omega(t'-t'')]\, dt'\, dt'' \\
&= \frac{1}{2\pi} \int_0^\infty S_z(\omega) \frac{1}{\omega^2} \\
&\quad \times \left\{1 + \cos(\omega t') - \cos(\omega t) - \cos[\omega(t+t')]\right\} d\omega \\
\therefore\ \langle\zeta^2(t)\rangle &= \frac{1}{\pi} \int_0^\infty S_z(\omega) \frac{1}{\omega^2}(1 - \cos\omega t)\, d\omega.
\end{aligned} \tag{16.5}$$

If memory is vanishingly short, the power spectral density stretches out in frequency, that is, this noise is *white noise*, and the mean square $\langle\zeta^2(t)\rangle$ reduces to

$$\langle \zeta^2(t) \rangle = \frac{S_z(0)}{2} t. \tag{16.6}$$

One may define a diffusion coefficient $\mathcal{D}_\zeta$,

$$\mathcal{D}_\zeta = \frac{1}{4} S_z(0), \tag{16.7}$$

which is related to the zero frequency value of the power spectral density of the signal $z(t)$. The corresponding autocorrelation function is

$$\begin{aligned} \phi_z(t', T) &= \frac{1}{T} \int_0^{T-|t'|} \langle \zeta(t+t')\zeta(t) \rangle \, dt \\ &= T\left(1 - \frac{|t'|}{T}\right)^2 \mathcal{D}_\zeta. \end{aligned} \tag{16.8}$$

$\zeta(t)$ is a cumulative process. And if it is memoryless, then

$$\begin{aligned} \langle \zeta(t+t')\zeta(t) \rangle &= \langle [\zeta(t) + \Delta\zeta(t')]\zeta(t) \rangle \\ \therefore \ \langle \zeta^2(t) \rangle &= 2\mathcal{D}_z t, \end{aligned} \tag{16.9}$$

where $\langle \zeta(t)\Delta\zeta(t') \rangle \rightarrow 0$ since correlation time has been assumed to vanish. In Equation 16.9, one now sees the length scale features that arise in a random process—such as of diffusion arising in random walk—that we have seen reflected in diffusion lengths with its finite correlation times. We have again seen the fluctuation-dissipation at work in this application of the Wiener-Khintchin theorem.

In this discussion of the signal- and randomness-based properties, which are independent of the origin, we have sidestepped a number of important mathematical notions, particularly those of moments, characteristic functions and the behavior of various probabilistic distribution functions. But one notion we do need to emphasize, since it has been used with abandon throughout this text, is that of randomness.

For those interested in at least an introduction to these connections between measures of randomness and signals, the moments, the probabilistic distributions and the generative characteristic functions, the appendices, particularly Appendix F, of S. Tiwari, "Nanoscale device physics: Science and engineering fundamentals," Electroscience 4, Oxford University Press, ISBN 978-0-198-75987-4 (2017), provide a good starting point.

16.2 *Randomness*

NOISE AND RANDOMNESS are quite interlinked. And we have also used the noun "fluctuations" often. In the discussion of diffusion in the Wiener-Levy process, we have seen how the same conclusion arises through the different perspectives. Randomness, however, needs some emphasis to discourage its abuse.

Randomness is often projected to unpredictability. But this notion of unpredictability—from a human perspective—is subjective. What one observer cannot predict in a sequence of events—a representation of observations (data) on a signal—does not mean that it is also unobvious and unpredictable to another observer who has more *a priori* information.

A signal varying in time is deterministic, that is, is predictable, so long as it is differentiable infinite times and known *a priori* to infinitesimally narrow time intervals in instants of time. Taylor expression, in principle, permits arbitrary accuracy at arbitrary times later or in the past. This corresponds to our statements on analytic functions in Appendix O. Unpredictability is related to discontinuities in functional values or in time derivatives at some order. It is this discontinuity-induced unpredictability that results in failure of the expansion approach.

In semiconductors, we often encounter physical events that do not have a precise mathematical description because of their random evolution, and this random development may be because of a single variable or from many. Scattering, as we saw, arises in a multitude of different causes. We have also noted that these random phenomena are coupled into systemic responses. Diffusivity in this recent example ending Section 16.1 was a systemic response from noise. For semiconductors, this noise would have arisen in fluctuations due to scattering with a finite non-zero memory time.

In electrical engineering, anything that is not an information-bearing signal of the information sought is sometimes called noise. Interference effects arising in couplings et cetera are not noise. They can be deterministically tackled, just as chaos-caused interferences. See S. Tiwari, "Nanoscale device physics: Science and engineering fundamentals," Electroscience 4, Oxford University Press, ISBN 978-0-198-75987-4 (2017).

Classical electrical noise arises from fluctuations in the motion of particles and in the number of particles traversing volumes and boundaries. To this, one must add uncertainty-related quantum fluctuations such as those of spontaneous processes.

This range of random processes may have memory, that is, the occurrence of a random event in the past may be remembered for finite time, or such a memory may be entirely lacking. In this latter case, the knowledge of the actual state of the system suffices to describe the distribution of the future states.

This separation of the future from the past by the present, the description of an average future independent of the past and solely from the present, is the Markovian process discussed in Section 8.11 that makes tackling of this subject easier.

Regarding this memory it is interesting that we are all born with some hardwiring but also memory that has to be short term—a *TV* memory. And then as time progresses, and we get older, we all acquire different extents of the Markov chain!

For a large class of problems, information is available only to a certain—and limited—extent, even in classical conditions. Quantum mechanics, by its nature, makes behavior nondeterministic. There are quantum-mechanical statistical probabilities of the measurement of eigenvalues even in a solvable stationary state.

Even if we assume that the forces at the smallest scale of interest are exactly known, there may be so many degrees of freedom that it is impossible to obtain the entire initial condition required to solve the dynamics.

For systems with few degrees of freedom, the initial conditions are known only within a certain accuracy. Predicted temporal changes may get modified by minute changes of initial conditions because of nonlinearity. This is the case for chaos, which is a deterministic nonlinear response.

So, dynamic evolution needs to be described in statistical terms. In statistical mechanics, we employ statistical probabilities of all states as equal for equal energies irrespective of the microstate simply because *we do not, and cannot, have information on all microstates.* This is the "insufficient knowledge hypothesis." The insufficient knowledge hypothesis expresses the absence of *a priori* information as an equality in *a priori* probability. But mathematical techniques can only derive the probabilities of outcomes of experiments on trials from *a priori* probability densities or distributions.

The fellowship of probability, clearly a very important one, like religions, has multiple schisms. The main one is subjective versus objective that we referred to in Chapter 2. This has taken on a very important modern context since machine learning techniques are very much a play on finding most-probable solutions fast. And *a priori* guesses, or not making a guess and being agnostic, is important to speed and correctness. Matters of belief begin to enter in this discussion and schism. And beliefs are biases. Sometimes they help to a solution quickly. Sometimes they are a path to tragedies.

Probability densities and distributions comprise an important aspect, but only one aspect, of what we have to deal with in analyzing what we cannot predict. Many times, symmetry arguments can give us probability distributions.

Symmetry arguments can give us probability distributions as for the roll of a die. We assign equal probabilities for all the face eigenvalues of a die, and call it fair. But, can any real-world die be truly fair? If it were, as a classical object, given all the initial conditions and interactions at work, one would be able to predict the outcome. Both the intractability of such a calculation and the minute imbalances in an object, of Avogadro number scale, make the die unfair, although fair enough. Symmetry arguments often have this problem of being convoluted and non-intuitive. We will tackle this using one example. But, before that, here is a short set of remarks on probabilities. Classical probabilities are positive definite. Quantum probabilities, such as when entanglement exists, are complex, and, like classical probabilities, analog. Probabilities are normalizable, although this condition is not axiomatically necessary. Probabilities do not necessarily have a mathematical requirement that averages should exist, unless axiomatically demanded. It is, however, hard to see the absence of averages in physical processes, except through mathematical counters.

For a more definitive discussion of probabilities, see the appendix on probabilities and the Bayesian approach in S. Tiwari, "Nanoscale device physics: Science and engineering fundamentals," Electroscience 4, Oxford University Press, ISBN 978-0-198-75987-4 (2017). Kolmogorov placed the theory of probability on its firm axiomatic foundations and developed a rigorous objective view.

Here is an example of a mathematical counter regarding averages. Take the probability density

$$\mathfrak{p}_n(z) = n\frac{\exp(-z)}{z} I_n(z),$$

where $I_n(z)$ is the modified Bessel function. This probability density satisfies

$$\int_0^\infty \mathfrak{p}_n(z)\,dz = 1 \;\; \forall\, n = 1, 2, \ldots .$$

However, $\int_0^\infty z\mathfrak{p}_n(z)\,dz$ diverges. An average doesn't exist, but there is no good reason to argue that such a physical process cannot exist.

16.2.1 *Bertrand's paradox*

To explore the conundrum of symmetry and its resolution in establishing a probability distribution, we look at a classic paradox problem: Bertrand's paradox. The question posed is: "Consider an equilateral triangle inscribed in a circle. Suppose a chord of the circle, as in Figure 16.3, is chosen at random. What is that probability that the chord is longer than a side of the triangle?" So, we are being asked what the probability is that a chord on this circle is longer than $\sqrt{3}R$.

Symmetry arguments lead to at least three ways of determining the probability:

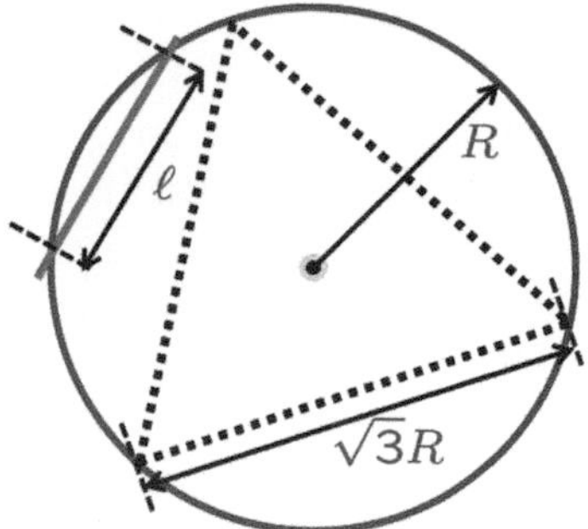

Figure 16.3: A graphical view of Bertrand's paradox with a chord of length ℓ randomly chosen in a circle of radius R. What is the probability that $\ell > \sqrt{3}R$?

Random end point: On any fixed point on the circle, for all the lines passing through the point with an angle θ to the tangent, consider a θ that is uniformly distributed over $(0, \pi)$. All the lines pass through the circle except those that are tangent. Chords longer than $\sqrt{3}R$ require the angular constraint $\pi/3 < \theta < 2\pi/3$; therefore, the probability is $(\pi/3)/\pi = 1/3$.

Random diameter: Take a line through the diameter and consider all the lines orthogonal to it. The chord is longer than $\sqrt{3}R$ when the point of intersection on the line lies in the middle half of the diameter. Assuming that these points are uniformly distributed, the probability is $1/2$.

Random midpoint: For any chord to be longer than $\sqrt{3}R$, the center must lie at a distance less than $R/2$ from the center. This area is 1/4th of the circle's area. The chord centers are uniformly distributed over the circle; therefore, the probability is $1/4$.

In these three cases, all have an *a priori* assumption of uniform distribution. But the answers are $1/3$, $1/2$ and $1/4$, respectively. We employed statistical reasoning, employed assumptions at a lower level of description—end-point angles, orthogonality to a line through the center, and centers of line—and arrived at the consequence of these assumptions leading to certain statistical properties.

One way to test the validity of the result is to test the hypothesis against measurement.

The other way to test and convince one's correctness is to state and formulate the problem unambiguously. Is the uniform distribution assumption in the three different cases unambiguous, and *truly* a random distribution that is uniform? Did we employ information that is not part of the statement of the problem? We have not been told anything about the position or the size of the circle. The solution must not depend on this choice. The solution must be both *translation and scale invariant.* Since the distribution of chords should be independent of the size and the position of the circle, so should the probability. So, if we choose a smaller circle, as in Figure 16.4, and move it around in the bigger circle, the distribution of the chords and the probability must not change.

The last case—random midpoint—fails because it is not translationally invariant. The first case—random end points—fails on both the translation and the scale invariance count. A point on the diameter and its angle are being chosen, not chords dropped on the circle.

The no-overt-or-covert-information procedure is the use of the random diameter. By choosing all possible angles through the center, and then all lines orthogonal to it, all possibilities of all chords are covered randomly and uniformly, and the choice subscribes to the invariants of the system. Therefore, $1/2$ is the probability for a random chord being longer than the side of an inscribed equilateral triangle.

So, maximum ignorance must be exercised in being random, and for symmetry arguments to not fail us.

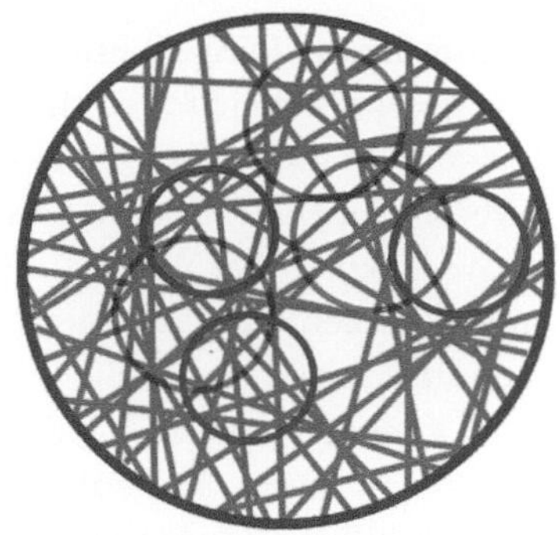

Figure 16.4: Bertrand' paradox as a problem with chords randomly dropped, and the criteria of invariance with size and position being tested through circles of different sizes and placements.

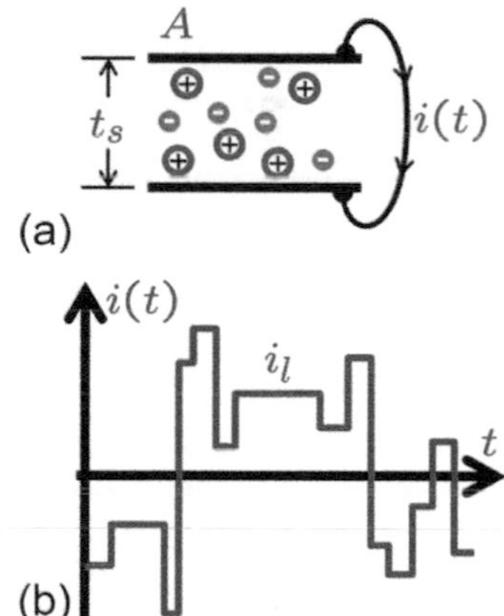

Figure 16.5: A semiconductor-based capacitor with stationary ions and mobile charge. Part (a) shows the geometry and (b) shows time-dependent current flowing in a loop.

16.3 *Fluctuations and noise*

FLUCTUATION-DISSIPATION AND BROWNIAN MOTION have been important themes in understanding transport in semiconductors. We will set up our noise discussion drawing on these themes and expanding them to placing associated noise on a quantitative and analytic footing that connects causal transport, fluctuations, dissipation and the lumped elements of inductor, capacitor and resistor in perspective. In an earlier discussion, the Einstein relationship as a systemic feature connecting the slow response—inherent in the Wiener-Levy process as we have now seen—connects to the fast fluctuation and dissipation.

Figure 16.5 show an idealized capacitor, shorted through idealized ohmic plates, with a conducting (n-type here and quasineutral, with immobile donors and mobile electrons) semiconductor medium in-between. The electron and the shallow donor ion that contributes excess electrons will respond in the environment, subject to the constraints, such as temperature T and the short, placed by it. Numerous different scattering processes—energy and momentum exchange processes—will exist, contributing to the response. Thermal motion—a consequence of kinetic energy, temperature as a consequence of the exchange of energy connecting particles to the environment, and the spread of the motional properties of the particles—will exist for all the particles. We will consider only the motion of electrons and treat the dopants as largely fluctuating around their mean position. So, we will look at the dopants as a

Induces charge here is to say that theelectric field—a polar field—terminates on these conducting plates. To show that the static solution of field is identical if an image charge is considered, and the solution restricted to one half, one is introduced to image charge and image force approach in undergraduate years. There are a number of subtleties in this image charge picture that come out when the charge is moving. There are symmetry considerations that need to be thought through. See Appendix K in S. Tiwari, "Nanoscale device physics: Science and engineering fundamentals," Electroscience 4, Oxford University Press, ISBN 978-0-198-75987-4 (2017).

jellium maintaining quasineutrality. The electrons move around. A charged particle induces charge on the conducting plates. The plates are shorted, and this moving charge induces fluctuating current through this field termination through the shorted electrodes.

We assume that the separation of plates $t_s \gg \lambda_{\mathbf{k}}$, so the electrons undergo plenty of scattering. The scattering event takes a vanishingly small time, while the free flight in-between takes place over a longer time scale. Scattering is the fast event, and the flight is the slow event. The current in the circuit reflects the electron motion in-between the plates. Continuity of current means that, in any cross-section in-between the plates, the same current (the particle motion current and the displacement current arising in the changing of the displacement) exists. Scattering causes the current to change as the carrier gets redirected. So, the current—inside or outside—has discontinuities at these scattering moments and remains constant in-between. Figure 16.5(b) shows this step-changing current in time in the outside short. Its average is zero, but it has a variance. Another assumption in this description is that the number of electrons reaching the plate is a rare event and will be ignored. This is, and we will find it to be important, *shot noise.* It will be discussed separately.

The simultaneity of particle flow current and displacement current, both of which are real currents, can be handled through Maxwell's equations, which carry them together. The consequences on signal delays are important in high-frequency devices, where these coexisting forms affect the speed. We will think through this conceptually.

See S. Tiwari, "Device physics: Fundamentals of electronics and optoelectronics," Electroscience 2, Oxford University Press, ISBN 978-0-198-75984-3 (forthcoming) for a discussion of Ramo's theorem and signal delay in the base-collector region of a bipolar transistor. A carrier does not need to arrive at a quasineutral region before its presence is felt because of the field termination of moving charges.

If the circuit were open, the work W performed to induce a charge q on the plates with a potential difference of V_c must equal the energy U of a charge particle moving through the potential difference that it sees. An electron moving a distance z has an energy change $U = eV = eV_c(z/t_s)$. As seen externally, the work $W = qV_c$. The electron—localized charge—as it moves this distance z between the plates that are t_s apart induces a proportional charge when shifting from one plate to the other. An electron close to one plate, for example, where it starts, has a nearly equal and opposite polarization charge distributed over the near plate where the field terminates. A small field also extends out to the distant plate, where there is a minute, opposite terminating polarization charge. As the electron moves between the plates, this terminating charge of opposite polarity shifts from one plate to the other, and when the electron passes into the plate, all the polarization charge associated with this electron disappears. While the charge was moving, the polarization charge on the plates was changing and there was current in between the plates, but there was also current in the outside short circuit, and current continuity was being maintained. Ramo's theorem helps analyze this, or its frequency-dependent form, self-consistently. Consider the polarization charge on the distant plate as the electron travels a distance z in one-dimensional form. The induced charge is $q = ez/t_s$. The current in the plate is $i = dq/dt = ev_z/t_s$. If electrons travel at a constant velocity, such as in a linear region, then the current due to the electron's travel appears as a ramp in time. If this situation were open circuit, there would be no outside current, and, everywhere, the particle flow current and the displacement current would balance each other. In a short circuit, the current flows in the connecting wire.

Stating that this is in one-dimensional form is just to say that current flows only one-dimensionally. There is a current density and a sheet charge flows between the plates.

So, the effect of the flow of the electron is gradually building and present at the plate through the field termination, or what we have called the induced polarization charge. When it reaches the plate, real charge and induced charge neutralize and the current disappears. The ramp of current exists for the duration of the electron flight. Ramo's theorem is a more rigorous way of tackling this, and there is a bit of unusuality there that this discussion has skirted.

In Figure 16.5(b), we didn't show any ramps, since the picture represents an accumulated consequence of many electrons traveling, some in the $+z$ direction, and some in the $-z$ direction, with the steps happening when any one of these electrons changes direction. Let $\mathcal{T}$ be a time duration such that electron removal through the plates is ignorable, even as the steps in current $i(t)$ of Figure 16.5(b) arise from the scattering events taking place for bouncing electrons.

We expand the current in a Fourier series:

$$\begin{aligned} i(t) &= I_0 \Re\left[\sum_{n=0}^{\infty} \sqrt{2} I_n \exp\left(\frac{2\pi n t}{\mathcal{T}}\right)\right] = \frac{1}{\mathcal{T}} \sum_{l=1}^{N} i_l, \quad \text{where} \\ I_0 &= \lim_{\mathcal{T}\to\infty} \frac{1}{\mathcal{T}} \int_0^{\mathcal{T}} i(t)dt, \quad \text{and} \\ I_n &= \lim_{\mathcal{T}\to\infty} \frac{\sqrt{2}}{\mathcal{T}} \int_0^{\mathcal{T}} i(t) \exp\left(-i\frac{2\pi n t}{\mathcal{T}}\right) dt \end{aligned} \qquad (16.10)$$

are the amplitudes of the Fourier components. i_l is an individual step in the current waveform, l is the index for the steps in current in time and n is the index for the Fourier expansion.

The Fourier component I_n in terms of the individual event steps is

$$
\begin{aligned}
I_n &= \frac{\sqrt{2}}{\mathcal{T}} \left[\int_0^{t_1} i_1 \exp\left(-i\frac{2\pi nt}{\mathcal{T}}\right) dt + \cdots + \int_{t_{l-1}}^{t_l} i_l \exp\left(-i\frac{2\pi nt}{\mathcal{T}}\right) dt \right. \\
&\quad \left. + \cdots + \int_{t_{N-1}}^{\mathcal{T}} i_N \exp\left(-i\frac{2\pi nt}{\mathcal{T}}\right) dt \right] \\
&= \frac{\sqrt{2}}{\mathcal{T}} \sum_{l=1}^{N} i_l f_l \exp\left(-i\frac{2\pi n t_l}{\mathcal{T}}\right), \quad \text{where} \\
f_l &= \frac{1 - \exp[-i2\pi n(t_{l-1} - t_l)/\mathcal{T}]}{i2\pi n/\mathcal{T}}.
\end{aligned} \tag{16.11}
$$

f_l represents the modulation function of the Fourier expansion arising in the different steps of the current.

For an ensemble of such systems, the average of all the Fourier indices must vanish, since they occur randomly, and detailed balance and thermodynamics' second law apply in thermal equilibrium. So, $\langle I_n \rangle = 0 \ \ \forall \ n$. We can also write the square of the Fourier amplitude,

$$
|I_n^2| = \frac{2}{\mathcal{T}^2} \sum_{l=1}^{N} \sum_{m=1}^{N} i_l i_m f_l f_m^* \exp\left[-\frac{2\pi n(t_l - t_m)}{\mathcal{T}} \right], \tag{16.12}
$$

which captures the collective effect of statistically varying quantities that are tied together. What this says is that the number of collisions N in the time $\mathcal{T}$ varies, as do the time instant t_l of the lth scattering instance, and the current i_l. These independent random events can be put together in the form of this autocorrelation, with $t' \to 0$.

We pick an ensemble out of the observed set, where N is a constant. The plate-directed velocity v_z, after scattering, is assumed to have no memory, that is, it is truly random. More important in the way we have developed this analysis is that the angle of motion, post scattering, varies randomly. This assures that the plate-directed component has randomness. The current i_m—the mth step—is independent of the number of scattering events before. Therefore,

This independent randomness assumption is only approximately true. We have seen examples of correlations in our scattering discussion of Chapter 10.

$$
\langle |I_n^2| \rangle = \frac{2}{\mathcal{T}^2} \sum_{l=1}^{N} \sum_{m=1}^{N} \langle i_l i_m \rangle \langle f_l f_m^* \exp\left[-\frac{2\pi n(t_l - t_m)}{\mathcal{T}} \right] \rangle. \tag{16.13}
$$

The currents in these different intervals of scattering events are independent, so $\langle i_l i_m \rangle = \langle i_m^2 \rangle \delta_{l,m}$. Equation 16.13 then reduces to

$$
\langle |I_n^2| \rangle = \frac{2}{\mathcal{T}^2} \sum_{l=1}^{N} \langle i_l^2 \rangle \langle f_l f_l^* \rangle. \tag{16.14}
$$

Note that the modulation function f_l is complex, while the current i_l is real.

We are now ready to put our statistical implications in here. We know the velocity distribution in thermal equilibrium. For non-degenerate conditions, the Maxwell-Boltzmann distribution tells us the z-directed velocity has a probability related through

$$\mathfrak{p}(v_z)d(v_z) = \left(\frac{m^* v_z^2}{2\pi k_B T}\right)^{1/2} \exp\left(-\frac{m^* v_z^2}{2k_B T}\right) dv_z. \tag{16.15}$$

Current $i = ev_z/t_s$ therefore follows

$$\begin{aligned}\langle i_l^2 \rangle = \langle i^2 \rangle &= \int_{-\infty}^{\infty} i^2 \left(\frac{m^*}{2\pi k_B T}\right)^{1/2} \frac{t_s}{e} \exp\left(-\frac{m^* t_s^2 i^2}{2e^2 k_B T}\right) di \\ &= \frac{e^2}{m^* t_s^2} k_B T, \\ \text{with } \langle i \rangle = 0. \end{aligned} \tag{16.16}$$

This says that the process is stationary. The system is statistically time stationary.

For calculating the scattering implications, we need a model of scattering. Classically, scattering time is related to the velocity, so scattering cross-section is inversely related to velocity, as we saw in Chapters 10 and 11. This is a fairly applicable assumption for many processes with neutral species. We assume that there is no persistence of velocity. The scattered velocity is independent of history, that is, this motion with scattering is a Markov process, and at worst we make a Markov approximation, where we ignore any statistical bearings.

We use a simple independent randomizing scattering model.

The probability of scattering in any time interval Δt is $\mathfrak{p}(\Delta t) = \nu \Delta t$ where ν is a scattering rate independent of velocity. This says that the scattering cross-section is inversely proportional to the velocity. In a Markov process, in the time up to $t + \Delta t$, splitting the time period to one of duration t followed by another of Δt, and using $\mathfrak{q}(t + \Delta t)$ to write the no-scattering probability,

We employed a similar approach in finding Green's function for carrier velocity expectation in Section 13.3.

$$\begin{aligned} \mathfrak{q}(t + \Delta t) &= \mathfrak{q}(t)\mathfrak{q}(\Delta t) \text{ and} \\ \mathfrak{q}(\Delta t) &= 1 - \nu \Delta t; \\ \therefore\ \frac{1}{\Delta t}\left[\mathfrak{q}(t + \Delta t) - \mathfrak{q}(t)\right] &= -\nu \mathfrak{q}(\Delta t) \\ \therefore\ \frac{d\mathfrak{q}(t)}{dt} &= -\nu \mathfrak{q}(t), \text{ and} \\ \mathfrak{q}(t) &= \exp(-\nu t), \end{aligned} \tag{16.17}$$

as the probability of no scattering during the duration of time t. Scattering events are occasional events. There are durations of flights with moments of scattering. So, this a low average probability for an event, and we use Poisson statistics to describe the probability. For K scattering events in time duration t,

$$\mathfrak{p}(K,t) = \frac{(\nu t)^K}{K!}\exp(-\nu t), \text{ and}$$

$$\langle K\rangle = \sum_{K=0}^{\infty} K\frac{(\nu t)^K}{K!}\exp(-\nu \mathcal{T}) = \nu\mathcal{T}, \text{ with}$$

$$\int_0^\infty t\exp(-\nu t)\nu\, dt = \frac{1}{\nu} \tag{16.18}$$

for the average time between scattering. Equation 16.11 tells us the modulation factor, whose expectation value once we choose the random approximation and $\mathcal{T} \gg 0$ is

$$\begin{aligned}\langle|f_l|^2\rangle &= \int_0^{\mathcal{T}} \frac{2[1-\cos(2\pi nt/\mathcal{T})]}{4\pi^2 n^2/\mathcal{T}}\nu\exp(-\nu t)\, dt \\ &= \frac{2}{\nu^2 + (2\pi n/\mathcal{T})^2},\end{aligned} \tag{16.19}$$

an expression independent of l—any specific instant of scattering. This simplifies the expression for $\langle|I_n|^2\rangle$ (the Fourier amplitudes of Equation 16.14) into a simple summation, with

$$\langle|I_n|^2\rangle = 4\frac{N}{\mathcal{T}^2}k_B T\frac{e^2}{m^* t_s^2}\frac{1}{\nu^2 + (2\pi n/\mathcal{T})^2}. \tag{16.20}$$

Since scattering follows Poisson statistics, we know the average number of scattering events in time interval $\mathcal{T}$ is $\langle N\rangle = \nu\mathcal{T}$. Since $\omega = 2\pi n/\mathcal{T}$,

$$\langle|I_n|^2\rangle = 4k_B T\frac{e^2}{m^* t_s^2}\frac{\nu/\mathcal{T}}{\nu^2+\omega^2}. \tag{16.21}$$

This expression was found using transformations with $\omega > 0$, since $0 < n < \infty$. Mathematically, we could have employed $-\infty < n < \infty$, in which case an additional factor of 2 would appear in Equation 16.21 because of the opposite polarity rotation in the complex plane.

We have extracted mean square Fourier coefficients of current. These are inversely related to the length of the time series. Equivalently, the frequency resolution—its density—is proportional to the time interval $\mathcal{T}$, consistent with our expectations.

We now define an effective current using Parseval's theorem,

$$I_{eff}^2 = \frac{1}{\mathcal{T}}\int_0^{\mathcal{T}} i^2(t)\, dt = \sum_n |I_n|^2\mathcal{T}. \tag{16.22}$$

Parseval's theorem is another mathematical consequence of integral transforms such as Fourier. See Appendix A.

Now, to find the state density dependence, consider the radial frequency span $\Delta\omega = (\omega;\omega+\Delta\omega)$. There are $\Delta\omega\mathcal{T}/2\pi$ states:

$$I^2_{eff}(\omega;\omega+\Delta\omega) = \sum_{n=T\omega/2\pi}^{n=T(\omega+\Delta\omega)/2\pi} \langle |I_{eff}|^2 \rangle$$

$$\approx \frac{2}{\pi} k_B T \frac{e^2}{m^* t_s^2} \frac{\nu}{\nu^2+\omega^2} \Delta\omega, \quad \text{or}$$

$$I^2_{eff}(f;f+\Delta f) = 4k_B T \frac{e^2}{m^* t_s^2} \frac{\nu}{\nu^2+(2\pi f)^2} \Delta f. \tag{16.23}$$

The system response and the fluctuations are now relatable through these extractions.

We distinguish the two frequencies: ν for scattering, where the scattering events are measurable as steps, and f for the real frequency as an analog signal measure here. In the rest of the text, ν is the cyclic frequency.

An ideal capacitor—an insulating dielectric clad with two perfectly conducting plates—has an admittance of $Y = i\omega C$. We have a semiconductor with mobile electrons in it. The admittance must change. To determine the admittance for our structure—a nonideal capacitor for which we have determined a current response in the presence of fluctuations under reasonable approximations—we apply a time-dependent bias voltage of $V(t) - \Re[V \exp(i\omega t)]$ and remove the short. The system is now off-equilibrium. There exists an electric field, with homogeneity assumed, of $\mathcal{E}(t) = \Re[(V/t_s)\exp(i\omega t)]$. Scattering modulates the electron's velocity. If a scattering event happened at $t = t_1$, then, following the scattering ($t > t_1$) but before the next scattering,

$$v_z(t) = v_z(t_1) - \Re\left\{\frac{eV}{im^* t_s \omega}\left[\exp(i\omega t) - \exp(i\omega t_1)\right]\right\}. \tag{16.24}$$

The current due to particle flow and the displacement is

$$i(t) = \frac{e v_z(t_1)}{t_s} - \Re\left\{\frac{e^2 V}{i\omega m^* t_s^2 \omega}\left[1 - \exp(-i\omega(t-t_1)\right]\exp(i\omega t)\right\} \tag{16.25}$$

until the next scattering event. Both of these are statistically varying. This statistical distribution in time is Poissonian for all scattering events, that is, for the lth event,

$$\Delta t = t - t_l = \exp(-\nu\Delta t)\nu\, d\Delta t$$

$$\therefore \quad \langle i(t) \rangle = \Re\left[\frac{e^2 V}{\omega m t_s^2}\frac{\exp(i\omega t)}{\nu + i\omega}\right]. \tag{16.26}$$

The integral

$$\int_0^\infty \left[1 - \exp(-i\omega t)\right]\exp(-\nu t)\nu\, d\tau = \frac{i\omega}{\nu + i\omega}, \tag{16.27}$$

so the impedance—including both the particle flow and the displacement contribution—is

$$Z'(\omega) = \nu \frac{m^* t_s^2}{e^2} + i\omega \frac{m^* t_s^2}{e^2}, \tag{16.28}$$

composed of a dissipative and a storing component (fluctuations(!) in flow together with integrated displacement as the fields change due to flow). To this we add the dielectric displacement component—a linear contribution—arising from the field change due to voltage change at the plates, which exists even in the absence of electron flow (so, $Z = Z' \parallel 1/i\omega C$). Figure 16.6 shows the system response model of the capacitor that we have now arrived at.

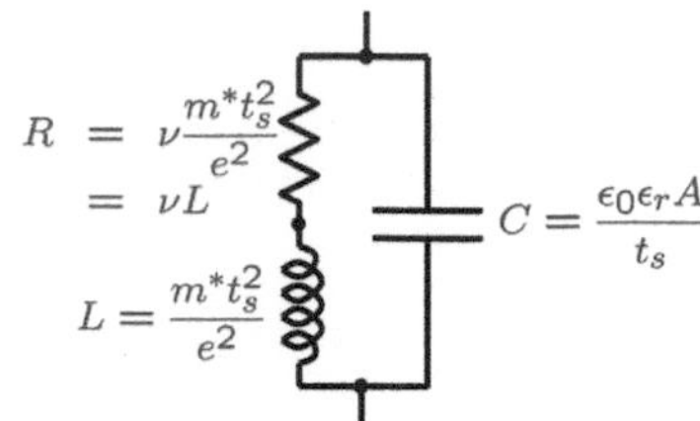

Figure 16.6: Impedance model of the capacitor with randomly scattering electron flow. ϵ_r is the relative dielectric constant of the semiconductor.

From the admittance

$$Y(\omega) = \frac{\nu}{\nu^2 + \omega^2} \frac{e^2}{m^* t_s^2} + i \left(\omega C - \frac{\omega}{\nu^2 + \omega^2} \frac{e^2}{m^* t_s^2} \right), \qquad (16.29)$$

one can now rewrite the effective current in a band (Equation 16.23) as

$$I_{eff}^2(\omega; \omega + \Delta\omega) = \frac{2}{\pi} k_B T \frac{e^2}{m^* t_s^2} \frac{\nu}{\nu^2 + \omega^2} \Delta\omega$$

$$= \frac{2}{\pi} k_B T \Re[Y(\omega)] \Delta\omega,$$

$$\text{or } I_{eff}^2(f; f + \Delta f) = 4 k_B T \Re[Y(\omega)] \Delta f. \qquad (16.30)$$

An important note is that, while in thermal equilibrium, Nyquist relationship in the form here or its other versions, are independent of the material or how the structure came to be, whether in Silicon Valley or in Tikamgarh or Hong Kong, but that once the equilibrium is disturbed and current passed, the noise characteristics may change quite substantially. We will see examples of this through the Hooge's parameter later.

Equation 16.30 is a special case of Nyquist's theorem relating the resistance—fluctuations—to the noisiness of current, and an example of the fluctuation-dissipation theme at play. More importantly, this equation says that, for the circuit drawn in Figure 16.6, the noisy current source goes in parallel with the noise-free circuit elements. The fluctuations in current originating in the microscopic scattering processes lead to a macroscopic response that is a directly measurable average response property of the system. There is a dissipative element—the resistor—that appears as a circuit element tied to the associated thermal fluctuations. The electrons also hold kinetic energy, and this shows up as the inductance. And the capacitor arises in the potential energy directly related to the bias voltage applied. One could have equivalently made a Thévenin noise voltage source equivalent with

$$V_{eff}^2(f; f + \Delta f) = 4 k_B T \Re \left[\frac{1}{Y(\omega)} \right] \Delta f. \qquad (16.31)$$

The two equivalent representations with ideal noiseless lumped elements for the semiconductor with cladding plates is shown in Figure 16.7.

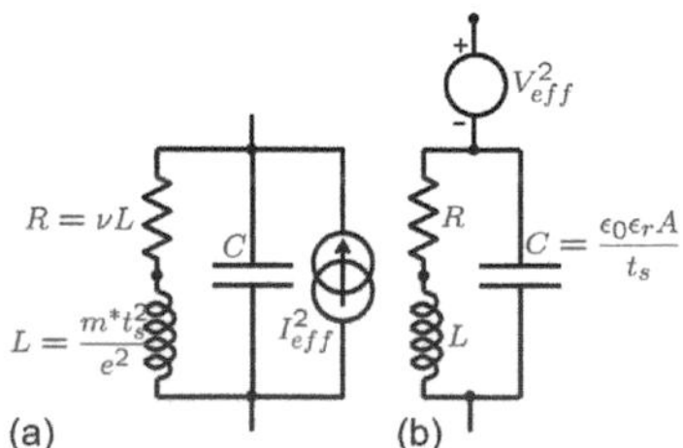

Figure 16.7: Norton and Thévenin equivalent models with current (in (a)) and voltage (in (b)) noise sources together with noiseless lumped elements.

Our derivation employed following the response of one electron in time and accumulating an ensemble of identical systems. But we have also stated that the electrons are independent and that the processes are random. The generalization for many electrons then follows rather straightforwardly, with most relations changing to scale for the change in numbers. The current, assuming N electrons, is

$$i(t) = i_1(t) + i_2(t) + \cdots + i_N(t), \tag{16.32}$$

with the current expectation due to uncorrelated and statistically independent electrons being

$$\begin{aligned} \langle i_i(t) i_j(t) \rangle &= \langle i_i(t) \rangle \langle i_j(t) \rangle = 0 \quad \forall\, i \neq j \\ \therefore \quad \langle i^2(t) \rangle &= \langle i_1^2(t) \rangle + \langle i_2^2(t) \rangle + \cdots + \langle i_N^2(t) \rangle \\ &= N \langle i_k^2(t) \rangle. \end{aligned} \tag{16.33}$$

The electron density is $n = N/At_s$, where A is the cross-section area, so

$$\begin{aligned} I_{eff}^2(\omega; \omega + \Delta\omega) &= \frac{2}{\pi} k_B T \frac{A}{t_s} \frac{ne^2}{m^*} \frac{\nu}{\nu^2 + \omega^2} \Delta\omega \\ \therefore \quad I_{eff}^2(\omega; \omega + \Delta\omega) &= \frac{2}{\pi} k_B T C \frac{ne^2}{\epsilon m^*} \frac{\nu}{\nu^2 + \omega^2} \Delta\omega, \end{aligned} \tag{16.34}$$

where we have introduced the capacitance C of the structure as a parameter.

So far in this analysis, we have used ν as the scattering rate parameter arising in the various scattering mechanisms that may be present and which are all assumed to be independent and random. We also have a cloud of electrons which are interacting with each other and moving around thermally or with stimulation if a bias voltage is applied. This electron charge cloud responds to the signal. Naturally, our earlier description of such a plasma is relevant to this description. The recapitulation and specific application of that description, in short, is as follows. In the presence of the electric field $\boldsymbol{\mathcal{E}}$, the movement of electrons under scattering—fluctuations—is damped, and this displacement kinetics can be written in our one-dimensional description as

$$\langle \ddot{\mathbf{z}} \rangle + \nu \langle \dot{\mathbf{z}} \rangle = \frac{e}{m^*} \Re \left[\boldsymbol{\mathcal{E}} \exp(i\omega t) \right], \tag{16.35}$$

whose solution is

$$\langle \mathbf{z} \rangle = \frac{e}{m^*} \Re \left[\frac{\boldsymbol{\mathcal{E}}}{i\omega} \frac{\exp(i\omega t)}{\nu + i\omega} \right]; \tag{16.36}$$

therefore, polarization is

$$\mathbf{P} = en \langle \mathbf{z} \rangle = \frac{ne^2}{i\omega m^*} \boldsymbol{\mathcal{E}} \frac{\exp(i\omega t)}{\nu + i\omega} \equiv \epsilon_0 \epsilon_r \boldsymbol{\mathcal{E}} \exp(i\omega t). \tag{16.37}$$

We have made a simple Newtonian connection through one scattering rate parameter to the plasmonic response, and it has appeared as the relative dielectric constant. Since

$$\frac{\mathbf{P}}{\epsilon_0 \mathcal{E}} = \epsilon_r - 1,$$

it follows that

$$\begin{aligned}
\epsilon_r(\omega) &\equiv \epsilon_r^r(\omega) + i\epsilon_r^i(\omega) = 1 - \frac{i\omega_p^2}{\omega(\nu + i\omega)}, \text{ with} \\
\epsilon_r^r(\omega) &= 1 - \frac{\omega_p^2}{\nu^2 + i\omega^2}, \text{ and} \\
\epsilon_r^i(\omega) &= -\frac{\nu\omega_p^2}{\omega(\nu^2 + i\omega^2)},
\end{aligned} \tag{16.38}$$

where $\omega_p^2 = ne^2/m^*$. These relationships must subscribe to the Kramers-Kronig relationship, due to causality and the linearity of Equation 16.35.

Returning now to the fluctuation response, when the structure is shorted, it is more useful to look at the fluctuations in current, that is,

$$I_{eff}^2(\omega; \omega + \Delta\omega) = \frac{2}{\pi} k_B T C \frac{\nu\omega_p^2}{\nu^2 + \omega^2} \Delta\omega, \tag{16.39}$$

and if it is open, the voltage fluctuations, that is,

$$V_{eff}^2(\omega; \omega + \Delta\omega) = \frac{2}{\pi} k_B T \frac{1}{C} \frac{\nu\omega_p^2}{\left(\omega_p^2 - \omega^2\right)^2 + \nu^2\omega^2} \Delta\omega, \tag{16.40}$$

where we have deployed the electron cloud's frequency characteristics through ω_p. With this parameterization, our noise model for the structure is illustrated in Figure 16.8.

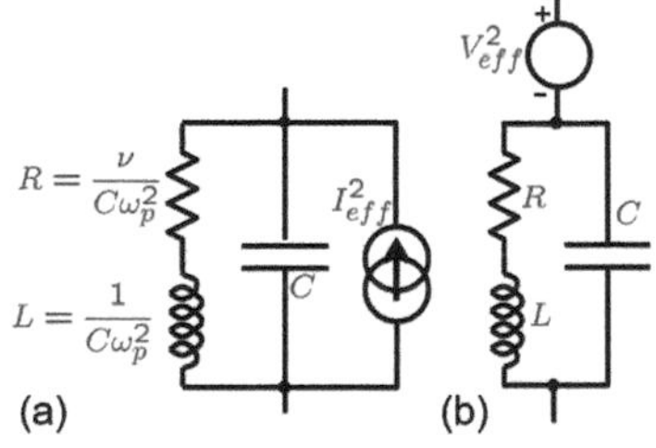

Figure 16.8: Norton and Thévenin equivalent models with current (in (a)) and voltage (in (b)) noise sources together with noiseless lumped elements, written in terms of plasma frequency to reflect the electron scattering modulated response to applied stimulus.

This electron transport system acts a generator of noise, with an internal frequency-dependent impedance that is subsumed in these equations.

At small scattering rate, the limits are

$$\lim_{\nu \to 0} I_{eff}^2(\omega; \omega + \Delta\omega) = 2k_B T C \omega_p^2 \delta(\omega) \Delta\omega, \tag{16.41}$$

which is like a shot noise arising in shot pulses at low frequencies. Note that $R = \nu/C\omega_p^2$, and we have seen a similar oscillator strength response for atomic vibrations, and others. The open circuit voltage noise is

$$\begin{aligned}
\lim_{\nu \to 0} V_{eff}^2(\omega; \omega + \Delta\omega) &= 2\frac{k_B T}{C}\left(\frac{\omega_p}{\omega}\right)^2 \delta\left[1 - \left(\frac{\omega_p}{\omega}\right)^2\right] \Delta\omega \\
&= 2\frac{k_B T}{C} \delta\left[1 - \left(\frac{\omega_p}{\omega}\right)^2\right] \Delta\omega.
\end{aligned} \tag{16.42}$$

These equations represent the thermal fluctuations around the plasma frequency where resonance will have the strongest effect.

When the scattering rate is large, the equivalent expressions for noise sources are

$$\lim_{\nu \gg \omega} I^2_{eff}(\omega; \omega + \Delta\omega) = \frac{2}{\pi} k_B T C \frac{\omega_p^2}{\nu} \Delta\omega, \text{ and}$$

$$\lim_{\nu \gg \omega} V^2_{eff}(\omega; \omega + \Delta\omega) = 2 \frac{k_B T}{C} \frac{1}{\nu} \left(\frac{\omega_p}{\omega} \right)^2 \Delta\omega. \tag{16.43}$$

As an aside, we should remark on the admittance that would be present, even if the particle flow were absent, for example, in a non-conducting dielectric. The applied voltage and current follow the relationships

$$v = \Re[t_s \mathcal{E} \exp(i\omega t)], \text{ and}$$

$$i = A\Re[\dot{\mathbf{D}}] = \Re[i\omega\epsilon_0(\epsilon_r^r + i\epsilon_r^i)\mathcal{E} \exp(i\omega t)], \tag{16.44}$$

so the admittance is

$$Y(\omega) = \frac{i}{v} = -\frac{A}{t_s}\epsilon_0\epsilon_r^i\omega + i\frac{A}{t_s}\epsilon_0\epsilon_r^r\omega. \tag{16.45}$$

The admittance still has a real dissipative part arising in all the contributors to the polarization response through the different dipolar or other oscillator interactions. The imaginary part will cause a dissipative conductance term. Vacuum may not have this dissipation form, but all materials will, even if it is vanishingly small. The dissipation is dependent on the imaginary part of the dielectric function, and the average dissipated power is

$$\frac{\Re[vi]}{At_s} = \frac{1}{4}(Vi^* + V^*i) = -\frac{1}{2}\omega\epsilon_0\epsilon_r^i|\mathcal{E}|^2. \tag{16.46}$$

The presence of the flow of electrons increases this dissipation through the electron cloud's conductive response.

16.3.1 *Quantum and thermodynamic connection to resonance*

WE HAVE NOW BUILT A MODEL consisting of an *RLC* (*R* for resistor, *L* for inductor, and *C* for capacitor) network to capture the essence of fluctuation-dissipation in a system response of a linear system. The resistor here is our dissipative element that introduces randomization, and the inductor and the capacitor are storage elements, the first for the kinetic form of energy, and the second for the potential. And it is *RLC*, since this combination—an electrical engineering basic lumped element model (so a long-wavelength graphical tool is used for representing energy interactions)—is sufficient to describe a linear system response. It is also very pertinent, since a simple combination of it represents for us resonance (a parallel LC), and the resonator represents to us a fundamental building block—classical and quantum-mechanical—of stability and change such as through the damped polarization interaction response of the dielectric function or the undamped electrical-magnetic resonance of an electromagnetic field. To see the noise aspect of the quantum and the thermodynamic—that is, the quantum-statistical link—we take a diversion to probe the resonance from a quantum-mechanical view.

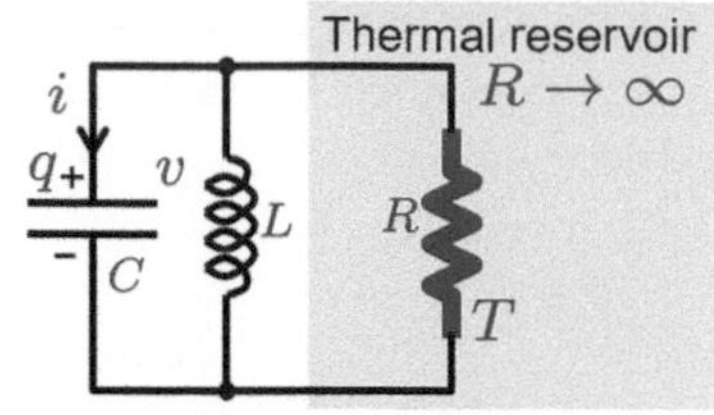

Figure 16.9: An RLC network as a dissipative resonator. R here represents the loss of energy to the environment in untracked and untrackable degrees of freedom represented as a thermal reservoir. An ideal dissipationless reservoir is $R \to \infty$, but where the environment thermodynamically still leads to fluctuations in the capacitor and the inductor.

Figure 16.9 shows a parallel RLC network that one can obtain as a graphical representation of any linear differential equation. Here, the resistor represents the connection to the environment—also the large number of unaccounted degrees of freedom where energy is lost—as a representation of the environment of the thermodynamic system. The inductor and capacitor form our resonance and have energy in the second power of a canonic variable. They are quantizable.

The networks that we draw are really graphs of information flow showing the linkages (connectors) between nodes (vertexes), the connectors representing the prescription for information flow along that path. See the chapter on information mechanics of S. Tiwari, "Nanoscale device physics: Science and engineering fundamentals," Electroscience 4, Oxford University Press, ISBN 978-0-198-75987-4 (2017).

With charge on the plates as $\pm q$, voltage as v, and current i flowing through it in the limit of $R \to \infty$, one may write

$$i = \dot{q},\ q = Cv,\ v = -L\dot{i},\ \text{and}\ \mathscr{H} = \frac{1}{2}Cv^2 + \frac{1}{2}Li^2. \tag{16.47}$$

In the limit, with no dissipative elements, the energy is conserved, and we may determine a set of canonical conjugate coordinates. We will choose charge q as one, satisfying

$$\begin{aligned} \frac{dq}{dt} &= \frac{\partial \mathscr{H}}{\partial p},\ \text{and}\ \frac{dp}{dt} = -\frac{\partial \mathscr{H}}{\partial q}, \\ \therefore\ p &= Li,\ \text{with} \\ \mathscr{H} &= \frac{q^2}{2C} + \frac{p^2}{2L} \end{aligned} \tag{16.48}$$

satisfying the equation of motion

$$\ddot{q} + \frac{1}{LC}q = 0. \tag{16.49}$$

This resonant circuit is in thermal equilibrium, with the reservoir at temperature T. $R \to \infty$ implies that no particles are exchanged, but energy is exchanged with the reservoir. We use Boltzmann statistics (see Appendix F) to determine the probability of finding the system in the volume $(q, q + dq; p, p + dp)$,

$$\mathfrak{p}(q,p)\,dq\,dp = \frac{\exp[-\mathscr{H}(q,p)/k_BT]\,dq\,dp}{\iint_{-\infty}^{\infty} \exp[-\mathscr{H}(q,p)/k_BT]\,dq\,dp}. \tag{16.50}$$

This distribution is a statement that all the microstates are equally probable when $\mathscr{H}(q,p)$ is the same. The denominator integrates to $2\pi k_BT\sqrt{LC}$. This is classical. Had we assumed that the energy is not continuously divisible, that is, is quantized, we would have ended up with a Bose-Einstein distribution. Here, we have just picked up the classical limit of the distribution and cast the system description in canonical coordinates.

Using this probability, we have

$$\begin{aligned}
\langle v^2 \rangle &= \iint_{-\infty}^{\infty} v^2 \mathfrak{p}(v,i)\, di\, dv = \frac{k_B T}{C}, \\
\langle i^2 \rangle &= \iint_{-\infty}^{\infty} i^2 \mathfrak{p}(v,i)\, di\, dv = \frac{k_B T}{L}, \text{ and} \\
\langle \mathscr{H} \rangle &= k_B T. \qquad (16.51)
\end{aligned}$$

These equations represent the equipartition of energy in classical ensemble systems and are a fundamental result for classical systems under the harmonic dependence. $(1/2)k_BT$ of energy is distributed in each of the harmonic variables (q,p), and therefore associated with their "particle"—the circuit elements here—the capacitor and the inductor. A thermal equilibrium leads to this energy distribution consequences. The reservoir deployed here as a resistor $R \to \infty$ resistance at temperature T has a very weak-vanishing interaction with the inductor-capacitor circuit that retains its identity. $R \gg \sqrt{L/C}$ suffices as a condition, since it means that the damping time (RC, L/R) is much larger than the time associated with the inverse resonance frequency $\sqrt{LC}$. The heat reservoir has many times larger degrees of freedom than the circuit. The circuit here has only two degrees of freedom. The capacitor is noisy, and the inductor is noisy, and this noise is associated with the necessity that any observation of the resonator requires one to connect it to the reservoir with which it exchanges energy.

16.3.2 *Thermal noise in linear systems*

WE CAN NOW SEE THE STATISTICAL PROPERTIES of the system. It has a fluctuating current given by

$$i(t) = I_0 + \Re\left[\sum_{n=1}^{\infty} \sqrt{2} I_n \exp\left(\frac{2\pi n t}{T}\right)\right], \qquad (16.52)$$

and a fluctuating voltage given by

$$v(t) = V_0 + \Re\left[\sum_{n=1}^{\infty} \sqrt{2} V_n \exp\left(\frac{2\pi n t}{T}\right)\right]. \qquad (16.53)$$

Since it is a linear circuit, I_ns and V_ns are related through

$$V_n = \frac{i\omega_n/\Omega}{1 - (\omega_n/\Omega)^2 + i(\omega_n/\Omega)\mathcal{D}} Z_0 I_n, \qquad (16.54)$$

where $Z_0 = \sqrt{L/C}$ is the characteristic impedance, $\mathcal{D} = Z_0/R \equiv \mathcal{Q}$ relate the dissipation and quality factors, $\Omega = 1/\sqrt{LC}$ is the natural frequency, and $\omega_n = 2\pi n/\mathcal{T}$ is the Fourier frequency component.

Parseval's theorem lets us write the Fourier amplitude relationship

$$\langle v^2 \rangle = \sum_{n=1}^{\infty} \langle V_n^2 \rangle,$$

$$\therefore \quad \frac{k_B T}{C} = \sum_{n=1}^{\infty} \frac{(\omega_n/\Omega)^2}{\left[1-(\omega_n/\Omega)^2\right]^2 + (\omega_n/\Omega)^2 \mathcal{D}^2} Z_0^2 \langle |I_n|^2 \rangle. \qquad (16.55)$$

The number of each of the components in the summation in the frequency interval $\Delta\omega$ is $\Delta\omega/(2\pi/\mathcal{T})$, which increases linearly with time. For a sufficiently large density of these frequency components, we replace the summation by integration:

$$\frac{k_B T}{C} = \int_0^{\infty} \frac{(\omega/\Omega)^2 Z_0^2 \langle |I_n|^2 \rangle}{\left[1-(\omega/\Omega)^2\right]^2 + (\omega/\Omega)^2 \mathcal{D}^2} \frac{\mathcal{T}}{2\pi} \, d\omega. \qquad (16.56)$$

When the capacitor's dissipation factor is small, that is, the quality factor is large, because this integrand peaks sharply at the resonant frequency Ω, and the current component factors vary slowly, the current can be pulled outside the integral. The use of the *LC* network resonance has allowed us to pick up the noise Fourier component. With this extraction,

$$\begin{aligned} \frac{k_B T}{C} &= \frac{\mathcal{T}}{2\pi} Z_0^2 \langle |I_n|^2 \rangle \Omega \int_0^{\infty} \frac{\eta^2}{(1-\eta^2)^2 + \eta^2 \mathcal{D}^2} \, d\eta \\ &= \frac{\mathcal{T}}{2\pi} Z_0^2 \langle |I_n|^2 \rangle \Omega \frac{\pi}{2\mathcal{D}} \\ &= \frac{\mathcal{T}}{4} \langle |I_n|^2 \rangle \frac{R}{C}, \\ \therefore \quad \langle |I_n|^2 \rangle &= 4 k_B T \frac{1}{R} \frac{1}{\mathcal{T}}. \end{aligned} \qquad (16.57)$$

From this relationship, Nyquist's theorem follows. *Nyquist's theorem* states that the thermal noise of a fluctuating resistor is viewable as a noise-free resistor of resistance R in parallel with a fluctuating current source $I_{eff}^2(f, f+df)$ or in series with a fluctuating voltage source $V_{eff}^2(f, f+df)$, given by

For now, we are using f for frequency and ν has been employed for scattering rate.

$$\begin{aligned} I_{eff}^2(f, f+df) &= 4 k_B T (1/R) df, \quad \text{and} \\ V_{eff}^2(f, f+df) &= 4 k_B T R df. \end{aligned} \qquad (16.58)$$

The noise is a *thermal noise*, since it exists in thermal equilibrium when the statistical observations are made on the system by connecting it to a reservoir at temperature T and placing it in thermal equilibrium.

We have derived this under assumptions that the classical rules apply. If quantization is included, so energies/frequencies are quantized in the resonator analysis, and employing Bose-Einstein statistics to account for the thermal modes,

$$\langle \mathscr{H} \rangle = \frac{\sum_{n=0}^{\infty} nh\omega \exp(-n\hbar\omega/k_B T)}{\sum_{n=0}^{\infty} \exp(-n\hbar\omega/k_B T)} = \frac{\hbar\omega}{\exp(\hbar\omega/k_B T) - 1}, \tag{16.59}$$

with zero point energy ignored. With this, we get the voltage fluctuations as

$$V_{eff}^2(f, f + df) = 4R \frac{hf}{\exp(hf/k_B T) - 1} df. \tag{16.60}$$

In the limit $\hbar \to 0$, this relationship reduces to the classical equations of Equation 16.58. If $k_B T \ll \hbar\omega$, so quantization becomes important since the energy separation of the states is larger than thermal energy, then Equation 16.60 becomes the defining Nyquist relationship. We worry about such quantum noise for photons in the semiconductor systems.

Thermal noise is the noise associated with the relaxation time of the thermal fluctuations in a statistical system. For resistors, one may view it as the voltage fluctuations induced by the thermal motion of electrons, which then relax back due the "resistance" that is reflective of energy loss through the scattering. For capacitors and inductors, it arises in the fluctuations that will exist when the normally storing element is attached to the environment for energy and particle exchange at temperature T. For a resistor, this noise is

See S. Tiwari, "Nanoscale device physics: Science and engineering fundamentals," Electroscience 4, Oxford University Press, ISBN 978-0-198-75987-4 (2017) Appendix F for the distribution-centered analysis of noise.

$$\langle V_{noise}^2 \rangle = 4k_B T R \Delta f \tag{16.61}$$

in the frequency band of Δf. As written here, it is a white noise—also Gaussian, an aspect we did not cover—with the tailing occurring at frequencies of the order of $1/\langle \tau_{\mathbf{k}} \rangle$. Thermal noise is also referred to as Johnson noise and Nyquist noise in the literature. We can write the thermal noise in terms of its power spectral distribution. For voltage, this is

We have stayed with the acronym of the noise as thermal noise to identify its physical origin. John Bertrand Johnson's publication dates to 1928 with a conference talk in 1926. Harry Nyquist's theory is from 1928. Both were at Bell Laboratories where electrical systems were the focus. There are many connections here between Brownian, fluctuation-dissipation, linear system themes that several equivalent relationships exist in a variety of literature in various disciplines. So, the Nyquist relationship is general in nature and not just restricted to thermal noise.

$$S_v = \frac{\langle V_{noise}^2 \rangle}{\Delta f} = 4k_B T R. \tag{16.62}$$

We can now also relate this back to the mathematical discussion of randomness-measuring statistical parameters with which we started this chapter.

Figure 16.10 shows the thermal distribution of the velocity (or, equivalently, momentum), which represents the flux of charge and thus the current in (a), its autocorrelation for current in (b) and the consequent spectral power distribution of current in (c). These all conform to Equations 16.2 and 16.3, with $\tau_0 = \tau_{\mathbf{k}}$. Note that, in thermal equilibrium, we start with a Gaussian distribution for the Maxwell-Boltzmann function, but it is skewed when

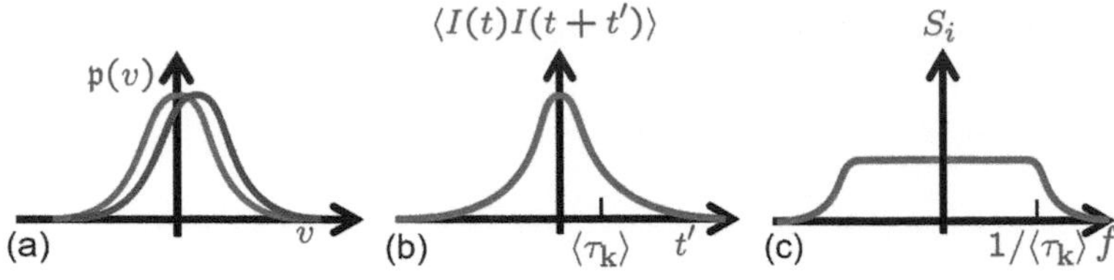

Figure 16.10: Part (a) shows velocity distribution in thermal equilibrium, and off-equilibrium under the influence of a field Part. (b) shows the autocorrelation in current, which is proportional to the velocity or momentum distribution. Part (c) shows the power spectral density in current.

off-equilibrium. We have seen electron temperature changes and multiple distributions superposing off-equilibrium. So, the thermal noise relationship is quite accurate in the thermal equilibrium and up to the roll-off point at $1/\langle\tau_{\mathbf{k}}\rangle$ or ν.

The current is the sum of all the currents from each particle $I = \sum_i I_i$, and the fluctuation correlation with a one-dimensional calculation in the z direction for a quasineutral n-type region is

$$\begin{aligned} \langle\delta I^2\rangle &= q^2 n \langle v^2\rangle \frac{dx\,dy}{dz} \\ \therefore\ \ S_i &= \langle\delta I^2\rangle \frac{4\tau_{\mathbf{k}}}{1+\omega^2\tau_{\mathbf{k}}^2} \\ &= 4q^2 n \frac{k_B T}{m^*}\tau_{\mathbf{k}} \\ &= 4k_B T q\mu n \frac{A}{L} = 4q^2 \mathcal{D} n \frac{A}{L} \\ &= 4k_B T \frac{1}{R} \end{aligned} \tag{16.63}$$

and is the current power spectral density ($\equiv S_v = 4k_B TR$), where the autocorrelation conforms, with z representing current, to

$$\begin{aligned} \phi_z(t') &= \lim_{T\to\infty} \frac{1}{T} \int_{-T/2}^{T/2} \langle z(t) z(t+t')\rangle\, dt \\ &= \phi_z(0) \exp\left(-\frac{|t'|}{\tau_0}\right) = \langle z^2\rangle \exp\left(-\frac{|t'|}{\tau_0}\right). \end{aligned} \tag{16.64}$$

This correlation has a Lorentzian lineshape; that is,

$$S_z(\omega) = 4\int_0^\infty \phi_z \cos(\omega t')dt' = \frac{4\phi_z(0)\tau_0}{1+\omega^2\tau_0^2}. \tag{16.65}$$

Though with confinement and quantization consequences, there will be the implications of the relative magnitude of $k_B T$ and $\hbar\omega = E$ considerations to know when to worry about thermal noise and when to worry about quantum noise.

When $\tau_0 \equiv \langle\tau_{\mathbf{k}}\rangle \to 0$, that is, absent memory, this spectrum is white. With $\tau_{\mathbf{k}} \leq 10^{-12}$ s, it is white for most electronic device purposes, since $2\pi/\tau_{\mathbf{k}}$ is in hundreds of *GHz*.

16.3.3 Partition thermal noise

How does this thermal noise change when the number of channels is limited, and non-classical correlated quantum-occupation fermionic constraints become important? In the spirit of

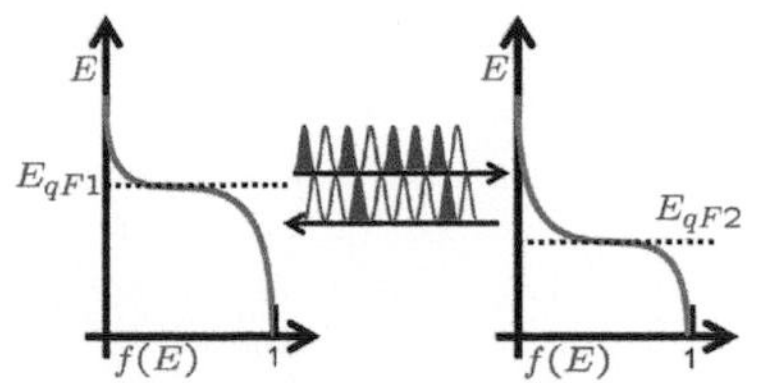

Figure 16.11: Current flow under bias in between two ports connected with limited number of channels and limited scattering.

Figure 8.4 from the quantum conductance discussion of Chapter 8, we probe this through the current that flows between two ports (see Figure 16.11). As discussed in Section 8.1, the current in these conditions is

$$i = \frac{2e}{h}(E_{qF2} - E_{qF1}) = \frac{g_q}{e}(E_{qF2} - E_{qF1}). \tag{16.66}$$

With finite scattering events, the conductance is

$$g = \frac{2e^2}{h}\sum_{j=1}^{J}\mathcal{T}_j = g_q\sum_{j=1}^{J}\mathcal{T}_j, \tag{16.67}$$

the sum of transmission coefficients over all the channels, with signals potentially traveling in both directions and undergoing some scattering.

Take near-static conditions of $\omega \approx 0$. The electron injection is stochastic, the occupation of channels is under fermionic constraints, and the events are independent. The emission probability is definable through the occupation probability arising in the Fermi-Dirac distribution function. And uncertainty is manifested through time intervals between wave packets. The uncertainty-constrained time interval is of the order of

$$\Delta t \approx \frac{h}{E_{qF2} - E_{qF1}}. \tag{16.68}$$

The number of channels occupiable is in the range defined by the quasi-Fermi energies. If $E_{qF1} > E_{qF2}$, then more electrons travel from the first port to the second, even though there are electrons with directed momenta in both directions. This totality defines the current. The current from the reservoir j, using an overbar to denote averaging, is

$$\overline{i_j(E)}\,dE = \frac{g_q}{e}f_j(E)\,dE. \tag{16.69}$$

The filled states couple to empty states for the conduction to take place. So, the current is proportional to $f(1-f)$. This dependence on both a fill factor and an empty factor is partitioning, so the stochasticity of state occupation and its reflection in the noise is now modified, with a choice being made between two paths. The fluctuations of this are a type of noise. This choice—stochasticity—is one of the important noises in lasers, where it is called mode partition noise. The current's noise power from the reservoir is

$$\begin{aligned} S_j(E)\,dE &= 2e\overline{i_j(E)}\,dE(1 - f_j(E))\,dE \\ &= \frac{4e^2}{h}f_j(E)[1 - f_j(E)]\,dE \quad \text{for } j = 1, 2. \end{aligned} \tag{16.70}$$

The total current noise power from a reservoir is

$$\begin{aligned} S_j(\omega \approx 0) &= \frac{4e^2}{h} \int_0^\infty f_j(E)[1 - f_j(E)]\, dE \\ &= 2g_q k_B T \int_{\zeta(0)}^\infty \frac{1}{\zeta^2}\, d\zeta, \\ &\quad \text{where } \zeta(E) = \exp\left[(E - E_{qFj})/k_B T) + 1\right], \\ &= 2g_q k_B T \frac{1}{\exp(-E_{qFj}/k_B T) + 1} \approx 2g_q k_B T, \\ &\quad \text{for } E_{qFj} \gg k_B T. \end{aligned} \tag{16.71}$$

Both of the reservoirs in this example are independent sources of noise, so noise adds. The current noise in these quantized channels therefore can be written as

Both are Gaussian, so variances add.

$$S(\omega \approx 0) = 4g_q k_B T, \tag{16.72}$$

which is a form very similar to what we found for thermal noise in classical conditions.

Thermal noise under classical is therefore quite similar to that under mesoscopic quantum-constrained conditions. The other important point of note is that with the passage of current off-equilibrium, the carrier population and the channel occupancy change, and the change in the occupation is reflected in a change in the temperature. We have not discussed the off-equilibrium conditions. If the changes are not too far from equilibrium, the classical behavior is largely unchanged, but the quantum-constrained systems do show changes arising in the constraints placed by the correlation properties of occupation—any channel of a specific energy can only have an up- and down-spin electron, so if one is present, the choice of the other is of opposite spin, and nothing else. So, there is now a suppression of noise. These differences can be seen in Table 16.2, which shows thermal noise in equilibrium

	Thermal noise equilibrium $eV \ll k_BT$	Spectral density off-equilibrium $eV \gg k_BT$
Mesoscopic		
Ballistic	$4k_BTg_q$	$4k_BTg_q$
Elastic scattering	$4k_BTg_q\mathscr{T}$	$2q\bar{i}(1-\mathscr{T})$
Distributed elastic scattering	$4k_BTG$	$\frac{1}{3}(2q\bar{i})$
Macroscopic		
Distributed inelastic scattering	$4k_BTG$	$4k_BTG$

Table 16.2: Equilibrium and nonequilibrium thermal noise in mesoscale and classical macroscale semiconductors.

and nonequilibrium systems under mesoscopic and macroscopic classical conditions, both in equilibrium and off it.

16.4 *Shot noise in linear systems*

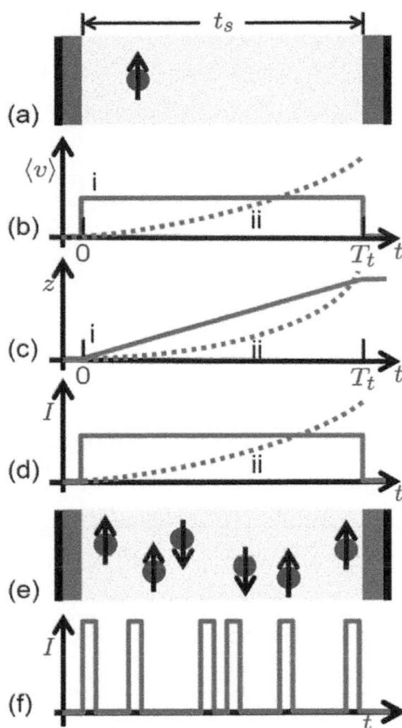

Figure 16.12: Part (a) shows the transit of one electron between two high conductivity regions. Case (i) is when the carrier moves with constant velocity, and case (ii) is for constant field motion. Part (b) shows the position of the movement of the electron as it traverses for case (i) and (ii). Part (c) shows the current at any cross-section including the current during the electron's motion for case (i) and case (ii). On a longer time scale with an ensemble of electrons, (d) shows the presence and movement of multiple electrons and (e) the corresponding observed current as the discrete number of electrons traverse the region in between the highly conducting contacts. Electrons are injected and collected into the external environment at these highly conducting boundaries.

ELECTRONS ARE DISCRETE PARTICLES—or a localized wavepacket, in the quantum description—traversing the semiconductor. Current exists in the circuit during the motion of the electron. When the number of electrons traversing is small, the current is small, but so is the number of electrons, and therefore the fluctuations in them are more noticeable. Since, for the electron, passing the boundary to the environment for measuring the current takes it out of the measurement, the passage across the boundary suddenly removes the current contribution of that electron. It is like a shot—an impulse. And the fluctuations arising in the discreteness of the electrons contributing to the current is the *shot noise.*

Figure 16.12 shows a few sketches to describe the basics of the shot noise phenomenon. Electron emission into the rate-limiting conduction region is a Poisson point process. The emission and the transport are independent in the sense that the injection phenomena at the highly conducting boundary—a contact plate or a highly conducting region—and the scattering phenomena in the transport region are mutually independent. As the carriers travel from one boundary to the other under scattering mediation, as shown in Figure 16.12(a) and (b), it is observable through the current in (c) in the external circuit. Current continuity and our earlier discussion of displacement and field termination at the highly conducting boundary apply. When many such electrons are flowing across the rate-limiting region, on a longer time scale one sees shots of such pulses of time widths bounded by electron entry and exit at the boundaries. The stream is of independent pulse; again, one may view the variance consequences through correlations of $\langle I(t)I(t+t')\rangle$, with the transit time $T_t = t_s/v$, where t_s is the length, and v is the velocity, as the time constant for the relevant pole for spectral noise power density. For thermal noise, this was $\langle \tau_{\mathbf{k}} \rangle$, which represented the averaged time constant of motion between scattering events. What we have here is a "scattering" at the boundaries for detection. Since the process is Poissonian, the spectral density will have the Lorentzian signature, and one will see a noise power spectral density in current $S_i \propto I/(1+\omega^2 T_t^2)$. It will be proportional to the current, since that is proportional to how many electrons are transiting, and it will be white with a pole high up in frequency—as it was with thermal noise—determined by the amount of time it

We will return to this theme of different manifestations in noise—fluctuations—a little later. Literature treats thermal noise and shot noise as two entirely independent phenomena. They are not. They are manifestation of the fluctuation in any observation where the "scattering" in transport regions and at the boundaries takes to transit. As with thermal noise, if mesoscopic transport—correlated transport—is present, the independence condition is broken, Poisson approximation breaks and this shot noise will change.

determines the observation. In this sense, the random telegraph phenomena and the other $1/f$ noise manifestations of fast-and-slow interactions, too, are such manifestations. These noise manifestations are quite physically interlinked.

16.4.1 *Transit, surface charge and shot noise analysis*

CONSIDER AN ELECTRON injected at one electrode and transiting to the other electrode through the rate-limiting transport region of Figure 16.12(a). The injection is a Poisson process—small numbers injected from a large density—and, during the scattering, the transport happens with a multitude of scattering, whose systemic response is the net velocity v as a Wiener-Levy process. The current is $i(t) = qv/t_S$. Assume a short circuit (thermal equilibrium). The energy gained per unit time by an electron in *in-between scattering events* is $q\mathcal{E}v$. The energy supplied by the electrodes is $Vi(t)$. These balance. This is one statement of Ramo's theorem.

If there is an external circuit with a series resistance of R_s and the electron transit time $T_t = t_S/v \ll R_sC$, where C is the capacitance of the semiconductor transit region, then our argument for what happens during the electron transit is the following. Let the power supply apply a bias voltage of V. Transit is being approximated as being instantaneous, and the circuit's response time R_sC is the dominant time for the circuit dynamic response. As soon as the electron shoots into the collecting electrode (positive; the anode), the voltage of the anode electrode is $V_{ael} = V - q/C$ w.r.t. the cathode electrode, since an electron has been removed. Current continuity implies

$$\frac{1}{R_s}\left[V - V_{ael}(t)\right] = \frac{d}{dt}\left[CV_{ael}\right] \tag{16.73}$$

with the solution

$$\begin{aligned} V_{ael} &= V - \frac{q}{C}\exp\left(-\frac{t}{R_sC}\right), \text{ and} \\ i(t) &= \frac{1}{R_s}(V - V_{ael}) = \frac{q}{R_sC}\exp\left(-\frac{t}{R_sC}\right). \end{aligned} \tag{16.74}$$

Each electron, following its removal, leaves behind its signature as an exponential decay.

Now consider the surface charge on the electrode regions as one electron traverses, and we will take three situations: from very long transit time to impulsive.

First, let us look at the consequence of the transit of the electron on the polarization charge on the electrodes in the following three cases:

For thermal noise, in our discussion related to Figure 16.5, the steps were due to the scattering taking place within the semiconductor, and electrons changing directions. We ignored the shot process, when an electron reaches and transits into the electrode. The two processes—different scatterings—are being assumed to be independent.

(i) Constant velocity with $R_sC \ll T_t$, $R_s \to 0$:
This is the example with which we started this section in Figure 16.12. Before the injection of the electron into the semiconductor, the surface charge is CV. An electron of $-e$ charge, when injected, induces a $+e$ charge when it is infinitesimally close to the cathode—the injecting electrode. During the transit, this charge movement and the field termination on the electrodes is being compensated externally from the circuit. The cathode charge is

$$Q_{cel} = -CV - e - \int_0^t i(t')\,dt', \quad \text{with the solution}$$
$$= \begin{cases} -CV - e(1 - vt/t_s), & \text{for } 0 < t < t_s/v, \\ -CV & \text{otherwise.} \end{cases} \tag{16.75}$$

So, the surface charge on the anode, CV pre-injection, changes by $+e$ during electron flight as a result of the electron current. The transiting electron is initially dominantly field connected to the cathode and finally—toward the end of the flight—to the anode. So, the anode charge is

$$Q_{ael} = CV + \int_0^t i(t')\,dt', \quad \text{with the solution}$$
$$= \begin{cases} CV - evt/t_s, & \text{for } 0 < t < t_s/v, \\ CV & \text{otherwise.} \end{cases} \tag{16.76}$$

Current from the supply is instantaneous, in accord with the electron transit, and the voltage between the electrodes has remained V, even as the surface charge of the electrode changes during the transit.

(ii) Zero velocity start with $R_sC \ll T_t$, $R_s \to 0$:
We assume that the electron starts from rest and accelerates due to the applied field, yet the region is short enough that scattering is vanishingly small. The velocity, under these conditions, as we have seen in Equation 13.26, is of the form $\langle v = dz/dt \rangle = -e\mathcal{E}t/m^*$. This means that $T_t = \sqrt{2m^*t_s^2/eV}$, and the current $i(t) = ev/t_s = (e^2V/m^*t_s^2)t$. The cathode charge is

$$Q_{cel} = -CV - e - \int_0^t i(t')\,dt', \quad \text{with the solution}$$
$$= \begin{cases} -CV - e[1 + (eV/2m^*t_s^2)t^2] & \\ \quad = -CV - e[1 - v(t)t/2t_s], & \text{for } 0 < t < t_s/v, \\ -CV & \text{otherwise.} \end{cases} \tag{16.77}$$

Again, the surface charge increases to $CV + e$ over T_t via the bias current from CV. And it is compensated by the change at the anode of

$$Q_{ael} = CV + \int_0^t i(t')\,dt', \text{ with the solution}$$
$$= \begin{cases} CV + (e^2V/2m^*t_s^2)t^2 \\ \quad = CV + ev(t)t/2t_s & \text{for } 0 < t < t_s/v, \\ CV & \text{otherwise.} \end{cases} \tag{16.78}$$

Again, the voltage change across the semiconductor is V and a constant.

(iii) Impulsive transit with $R_sC \gg T_t$:
Since the electron spends vanishing time in the semiconductor, the current response is related only to the decay following the electron transit following on our starting discussion of this section. The anode charge is

$$Q_{ael} = CV_{ael} = \begin{cases} CV + e\exp\left(-\frac{t}{R_sC}\right), & \text{for } t > 0, \\ CV, & \text{for } t < 0, \end{cases} \tag{16.79}$$

and the cathode charge is

$$Q_{cel} = -CV_{ael} = \begin{cases} -CV - e\exp\left(-\frac{t}{R_sC}\right), & \text{for } t > 0, \\ -CV, & \text{for } t < 0, \end{cases} \tag{16.80}$$

all with the R_sC charging time constant in the response.

Having understood the charge and the voltage response, we can now look at the noise power spectrum for the three different cases under Poisson emission as an independent process:

(i) Shot noise with constant velocity, $R_sC \ll T_t$, and $R_s \rightarrow 0$:
Let all electrons have the identical pulse shape, $f(t)$. The current with K electrons in the semiconductor is $i(t) = \sum_{k=1}^{K} c_k f(t - t_k)$. Let ν be the rate of arrival; then, the spectral density is

$$S(\omega) = 2\nu a_k^2 F^2(\omega) + 4\pi\left[\nu a_k \int_{-\infty}^{\infty} f(t)\,dt\right]^2 \delta(\omega), \text{ with}$$
$$f(t) = \begin{cases} -ev/t_s, & \text{for } 0 < t < t_s/v, \\ 0 & \text{otherwise, and} \end{cases}$$
$$F(\omega) = \int_{-\infty}^{\infty} f(t)\exp(-i\omega t)\,dt = \int_0^{t_s/v} \frac{ev}{t_s}\exp(-i\omega t)\,dt$$
$$= \frac{ev}{t_s}\frac{1 - \exp(i\omega t_s/v)}{i\omega}$$
$$= e\exp\left(-i\omega\frac{t_s}{2v}\right)\frac{\sin(\omega t_s/2v)}{\omega t_s/2v}. \tag{16.81}$$

If ν is the average rate of electron arrival, $I = q\nu$, therefore substituting it into the spectral expression, we obtain

$$
\begin{aligned}
S_i(\omega) &= 2\nu e^2 \frac{\sin^2(\omega t_s/v)}{(\omega t_s/v)^2} + 4\pi v^2 e^2 \delta(\omega) \\
&= 2eI[\mathrm{sinc}(\omega t_s/v)]^2 + 4\pi I^2 \delta(\omega) \\
&= 2eI \text{ for } \omega \ll 2v/t_s.
\end{aligned} \tag{16.82}
$$

The shot noise spectral noise density is $S_i = 2eI$ up to frequencies that are a fraction of the inverse of transit times.

(ii) Shot noise with zero velocity start with $R_sC \ll T_t$, $R_s \to 0$:
Now we have $a = dv/dt = e^2V/t_s^2m^*$, and $f(t) = t$, for $0 < t < T_t$, and vanishing elsewhere. So,

$$
\begin{aligned}
I = \langle i(t) \rangle &= a\frac{1}{T_t}\int_0^{T_t} f(t')\,dt' \\
&= a\nu\frac{T_t^2}{2} = e\nu, \text{ and} \\
F(\omega) &= \int_{-\infty}^{\infty} f(t)\exp(-i\omega t)\,dt = \int_0^{T_t} t\exp(-i\omega t)\,dt \\
&= iT_t\frac{\exp(-i\omega T_t)}{\omega} - \frac{1-\exp(-i\omega T_t)}{\omega^2}, \\
\therefore \quad F^2(\omega) &= \frac{2+\omega^2T_t^2 - 2\omega T_t\sin(\omega T_t) - 2\cos(\omega T_t)}{\omega^4}.
\end{aligned} \tag{16.83}
$$

The noise spectral density, which is

$$
S_i(\omega) = 2\nu\left(\frac{e^2V}{t_s^2m^*}\right)^2 F^2(\omega) + 4\pi\nu^2e^2\delta(\omega), \tag{16.84}
$$

can be simplified with approximations to the harmonic functions in $F(\omega)$,

$$
\begin{aligned}
\sin(\omega T_t) &= \omega T_t - \frac{1}{3!}(\omega T_t)^3 + \mathcal{O}(\omega^5), \text{ and} \\
\cos(\omega T_t) &= 1 + \frac{1}{2!}(\omega T_t)^2 + \frac{1}{4!}(\omega T_t)^4 + \mathcal{O}(\omega^6),
\end{aligned} \tag{16.85}
$$

to

$$
\begin{aligned}
S_i(\omega) &= 2\nu\left(\frac{e^2V}{t_s^2m^*}\right)^2\left[\frac{2}{3!}T_t^4 - \frac{2}{4!}T_t^4 + \mathcal{O}(\omega^5)\right] + 4\pi\nu^2e^2\delta(\omega) \\
&= 2eI + 4\pi I^2\delta(\omega), \text{ neglecting } \mathcal{O}(\omega^5).
\end{aligned} \tag{16.86}
$$

For $0 < \omega < 1/T_t$, the current noise power spectral density is

$$
S_i = 2eI. \tag{16.87}
$$

(iii) Shot noise for impulsive transit with $R_sC \gg T_t$:
Now we have $f(t) = (e/R_sC)\exp(-t/R_sC)$, for $t \geq 0$, and $f(t) = 0$, for $t < 0$. The Fourier transform is

$$
F(\omega) = \int_{-\infty}^{\infty} f(t)\exp(-i\omega t)\,dt = \frac{e}{1+i\omega R_sC}, \tag{16.88}
$$

so the noise power spectral density is

$$\begin{aligned} S_i(\omega) &= 2\nu \frac{e^2}{1 + i\omega^2 R_s^2 C^2} + 4\pi \nu^2 e^2 \delta(\omega) \\ &= 2eI \frac{1}{1 + i\omega^2 R_s^2 C^2} + 4\pi I^2 \delta(\omega) \\ &= 2eI, \text{ for } \omega \ll 1/R_s C, \end{aligned} \tag{16.89}$$

using Carson's theorem.

So, we have found that the shot noise spectral density under a variety of different conditions has the same magnitude, $2eI$, just as we found the thermal noise spectral density $4k_BT(1/R)$.

Shot noise is also suppressed by correlations. Our derivation depended on electron emission being independent of the transport process. If $T_t \gg R_sC$, and the velocity is constant, so $T_t = t_s/v$—the first situation analyzed—if channels occupation is limited due to correlation such as in mesoscopic conditions, then an electron emission of identical spin is suppressed into a channel where an electron already exists and is transmitting in the channel. This is also over time scales for which electrode voltages are at their initial values. So, the electron mission rate ν must be less than $1/T_t$. If, on the other hand, $R_sC \gg T_t$, then the electrode voltages are different, in which case electron emission will be dependent on this inter-electrode voltage. Significant time ($\gg R_sC$) must elapse before statistical independence reappears. So, independence will only happen if $\nu \ll R_sC$. So, there are constraints to the shot noise relationship of $S_i = 2eI$ related to limited scattering, electrode potentials and time scales. Besides mesoscopic conditions, if a space charge effect exists in the transport region, then presence of charge affects potentials and thus transport. If there are situations where a memory effect exists in external circuits, there too the emission will be regulated by the electrode voltages, and the independence approximation will be invalidated.

Space-charge-limited current exists in many situations. Semi-insulating or poorly conducting regions are common to devices. Silicon-on-insulator or even bulk *MOSFET* needs non-conducting regions under the inversion layer; *III-V* transistors employ such regions along with heterostructures. All low doped regions will show consequences from the ideal expectation in minor or major ways.

16.5 *Low-frequency noise*

THE POISSONIAN CHARACTERISTIC of emission or capture at an electrode appears in multiple ways in materials. Take for example, the behavior of traps or deep levels. They capture or emit carriers as a fast process, but the captured or emitted state persists for long periods of times. Such a trapping effect—Hall-Shockley-Read or others—appears in many places from a variety of intrinsic or extrinsic sources. An SiO_2/Si interface will have some interface states, and some of these will act as traps, including those in regions near the silicon interface, where their local effect will be present. Even in any region that is substantially conducting, trapping and detrapping has local consequences—in screening and scattering—whose effects will be present in the conduction. So, fast capture or emission with a long persistence is a random fluctuation whose signature will appear at very low frequencies. This is often called *flicker* noise, $1/f$, although the signatures can have a variety of power of $1/f$ noise, generation-recombination noise, random telegraph noise, noise due to imperfect contacts or other types of noise, based on their source or history. So, noise with a noise power signature of the form

$$S_z(\omega) \propto \omega^{-\alpha}, \;\; 0.5 \lesssim \alpha \lesssim 2, \tag{16.90}$$

is low frequency noise.

We are calling it "low-frequency noise" to identify the symptom, and not the cause. This is somewhat dissatisfying, but even as the causes are different, one also often finds that there are situations where there are multiple causes, and one cannot be confident of the attribution of the source of noise. So, use the term low-frequency noise with the ambiguity and disdain that it deserves.

16.5.1 *Noise from trapping-detrapping*

IF THE CARRIER POPULATION FLUCTUATES due to trapping and detrapping, such as with largely empty or largely filled bands, with both, as in the generation-recombination process or with other circumstances, the trapping and detrapping events are fast and the persistence of the trapped or detrapped state is long. These persistent times are Poissonian—with low averages—just as we found the electrode emission process to be usually approximable by Poisson statistics. Let the trapped state have a time of τ_1 and the detrapped state a time of τ_2, both of which obey Poisson statistics; these then become important.

We first tackle the physical basis of single band or bipolar band interaction. Single band interaction, that is, a single carrier interacting with a single band, leads to random telegraph noise or signal. The approach to tackling bipolarity of bands and carriers is quite similar. Figure 16.13 shows an isolated defect capturing and then later reemitting a carrier. A capture process eliminates a carrier available for conduction or, equivalently, screens the region in its neighborhood, so that the total charge available for conduction is reduced. The current is now smaller. An emission process, in contrast, raises the current. A random telegraph signal, as in Figure 16.13(b), shows discreteness because the times in the captured or the emitted states are significant—larger than scattering or transit times.

In S. Tiwari, "Nanoscale device physics: Science and engineering fundamentals," Electroscience 4, Oxford University Press, ISBN 978-0-198-75987-4 (2017), you will see a more detailed discussion of random telegraph signal arising in trapping and detrapping. Even if rare, Poisson processes and this random fluctuation will have a strong effect—strong because if the geometry has a small signal but a fluctuation exists due to a single electron, it has a relatively higher signal and consequence. Small transistors, unless made carefully, have a stronger random telegraph signal from carrier trapping and detrapping near interface regions.

A single defect can provide plenty of complexity. The defect can be in a multitude of charged states. It may couple to other physically proximate states. The behavior of this multitude of charge states may be influenced by the state of the spatially local

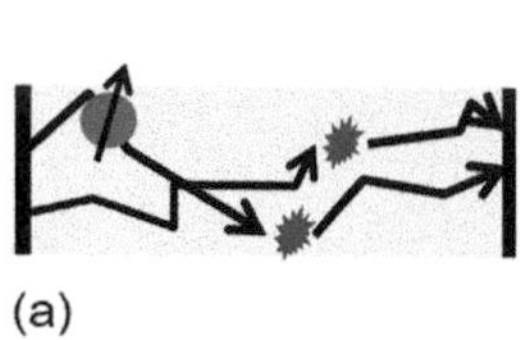

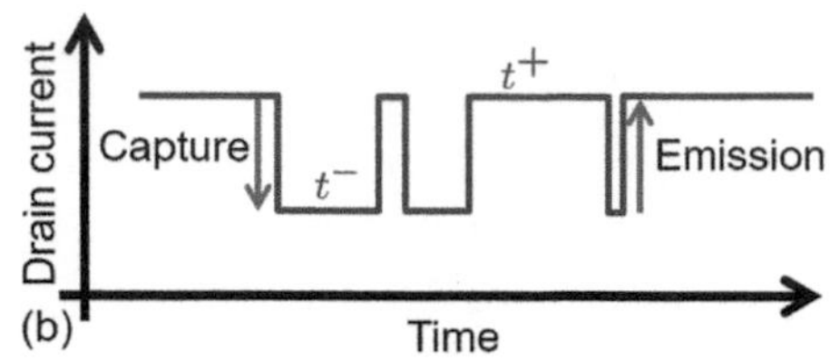

Figure 16.13: Capture and emission of electrons during semiconductor traversal, and a random telegraph signal-like effect in current resulting from it. This process can happen in bulk, but particularly so at interfaces such as those of insulator/semiconductors. Generation-recombination is a more complex form of this effect arising in both bands—so, electrons and holes—interacting.

defect states. We will keep the problem simple, to extract basic physical features. We will consider an isolated, singly charged state, random and independent, to understand the physical character. Defects, being small in number, have a Poisson distribution, and they capture and emit, playing a role not unlike that of electrodes in the appearance of shot noise in this respect. This slow process of capturing and emitting, in this respect, is a shot noise that now shows up as low frequency $1/f$ noise in small structures.

This marriage of fast and slow is why I have often mused that we really have two-and-a-half noises in the macroscale electronic conditions: thermal noise and one-and-a-half shot noise, which arise either from the reservoir boundaries or from the nature of fast-and-slow interaction inside the material itself. The low-frequency noise is another consequence of fast and slow, for example, slow trapping and fast emission in the random telegraph noise.

This emission or capture in time is a discrete event of low probability, while the mean is finite. In a symmetric system, the system is, on average, in each of these states half of the time. Poisson statistics describes the event distribution of this process. The event probability is

$$\mathfrak{p}(k, \nu T) = \frac{(\nu T)^k}{k!} \exp(-\nu T), \tag{16.91}$$

where ν is the mean rate of transitions per second, T is the time interval and νT is the net average time in the state. If τ_+ and τ_- are the average times spent in the emitted and captured states, respectively, the probability distribution of the dwell time in the emitted state (t^+) or the captured state (t^-) is

$$\mathfrak{p}(t^{\pm}) = \frac{1}{\tau^{\pm}} \exp\left(-\frac{t^{\pm}}{\tau^{\pm}}\right). \tag{16.92}$$

We will call the high current state—the "emitted" state—e, and the low current state—the "captured" state—c. The probability per unit time of transition from state e to state c through the process of capture is given by $1/\overline{\tau}_e$, and, for the reverse process of emission, the probability is given by $1/\overline{\tau}_c$. The transitions are assumed to be instantaneous; $\mathfrak{p}_e(t)\,dt$ is the probability that the state does not transition from e to c during the time interval $(0, t)$ but does so in $(t, t + dt)$. This implies that

$$\mathfrak{p}_e(t) = \frac{A(t)}{\overline{\tau}_e}, \tag{16.93}$$

where $A(t)$ is the probability that the state has not made a transition during $(0, t)$, and $1/\overline{\tau}_e$ is the probability of the transition per unit time at time t:

$$A(t+dt) = A(t)\left(1 - \frac{dt}{\overline{\tau}_e}\right), \tag{16.94}$$

which states that the product of not making a transition during $(0,t)$ and $(t, t+dt)$ is the probability of not making a transition during the time interval $(0, t+dt)$. This equation can be rewritten in the simple form

$$\frac{d}{dt}A(t) = -\frac{A(t)}{\overline{\tau}_e}, \tag{16.95}$$

whose solution is

$$A(t) = \exp\left(-\frac{1}{\overline{\tau}_e}\right), \tag{16.96}$$

with $A(0) = 1$. Then,

$$\mathfrak{p}_e(t) = \frac{1}{\overline{\tau}_e}\exp\left(-\frac{t}{\overline{\tau}_e}\right), \tag{16.97}$$

with

$$\int_0^\infty \mathfrak{p}_e(t)\,dt = 1. \tag{16.98}$$

The equivalent expression for the captured state c is

$$\mathfrak{p}_c(t) = \frac{1}{\overline{\tau}_c}\exp\left(-\frac{t}{\overline{\tau}_c}\right). \tag{16.99}$$

When the emission and capture transitions happen at a single characteristic attempt rate, the times should be exponentially distributed. As a result, the mean time in state e is

$$\int_0^\infty t\mathfrak{p}_e(t)\,dt = \overline{\tau}_e, \tag{16.100}$$

and the standard deviation is

$$\left[\int_0^\infty t^2\mathfrak{p}_e(t)\,dt - \overline{\tau}_e^2\right]^{1/2} = \overline{\tau}_e. \tag{16.101}$$

A similar expression applies for the captured state.

The standard deviation of the mean time spent in any state is the same as the mean time spent in the state for this simple capture-emission model. But trapping and energetics can be quite complicated and, experimentally, a variety of random telegraph signals are observed. Multiple pathways may open or close. If trapping at a defect causes local deformation, the energy is in an elastic and Coulomb form. This configuration allows multiple metastable states that may be reflected in different time constants and signal intensities during different periods. Coupled charge states of defects may be active, resulting in the turning on and off of different behaviors. Grain boundary defects, interface defects and others will have their own idiosyncrasies.

The power spectral density—the spectral correlations—measures the randomness and the net energy reflected in the phenomena behind the measured parameter. The power spectral density is

An increase in the spectral density, the noise energy arising from the fluctuations, causes a higher probability of errors in digital devices, and a higher noise floor in analog and high-frequency-use devices. So, these are very important for the nanoscale.

$$S(\omega) = \int_{-\infty}^{\infty} \exp(-\omega\tau) \langle z(t+t')z(t) \rangle \, dt'. \tag{16.102}$$

A random process with the characteristic time τ, such as the one discussed, has a Lorentz spectrum:

$$S(\omega) = \frac{\tau}{1+\omega^2\tau^2}. \tag{16.103}$$

This approach can be extended, for example, when there is a distribution of characteristic times, or when the time arises from thermally activated defects:

$$S(\omega) = \int \frac{\tau}{1+\omega^2\tau^2} \mathscr{C}(\tau) \, d\tau, \tag{16.104}$$

for example, extends this behavior to a distribution of traps. If this distribution function $\mathscr{C}(\tau) \propto 1/\tau$, for $\tau_1 \leq \tau \leq \tau_2$, then $S(\omega) \propto \omega^{-1}$, for $\tau_2^{-1} \leq \tau \leq \tau_1^{-1}$.

This power spectral dependence reduces to a $1/f$ dependence only for specific conditions of the distribution being constant for $k_B T \ln(\tau_1/\tau_0) \leq ET \ln(\tau_2/\tau_0)$, where E is the activation energy. For a distribution of traps with a spread of the characteristic time τ,

$$S(\omega) = \int_0^{\infty} \frac{2\tau}{1+\omega^2\tau^2} \mathfrak{p}(\tau) \, d\tau. \tag{16.105}$$

It the distribution is flat, that is, $\mathfrak{p}(\tau) = 1/\tau$, this Lorentzian behavior of distribution of sites will lead to $S(\omega) \propto 1/\omega$. What this implies is that a variety of frequency dependences may be observed in the spectrum, all increasing as the frequency is lowered, depending on the nature of the trapping processes that are active. If the characteristic time is thermally activated, that is, $\tau = \tau_0 \exp(E/k_B T)$, then the frequency dependence of the spectrum is more complex than that for a simple single defect with a constant characteristic time. For example, an exponential dependence in the characteristic time leads to a $1/\omega^\alpha$ dependence, with α varying as mentioned at the beginning of this section.

16.5.2 Hooge parameters and mélange

CONDUCTORS—RESISTORS, METALS WITH GRAIN BOUNDARIES, amorphous materials, or structures with different electrodes, whether quite conductive or resistive, also show a pronounced increase in the spectral power of noise with lowering of frequency.

This is not necessarily a large fluctuation, but it is nevertheless important, and it arises from a variety of sources. The observed behavior tends to follow

$$\frac{\langle \Delta R^2 \rangle}{R^2} = \alpha \frac{1}{f} N \, df, \tag{16.106}$$

where α is a parameter called the Hooge parameter, N is the number of electrons, and $\alpha = 10^{-3}$–10^{-6} for most materials. One mechanism responsible for this is conductivity fluctuation, that is, $\Delta\sigma = q\mu\Delta n + qn\Delta\mu$, where both Δn and $\Delta\mu$—effects in carrier concentration and mobility, respectively—arise in charge fluctuations. Poorer quality materials have poorer Hooge parameters. It is lower when the mobility of the material is lower.

16.6 Summary

THIS CHAPTER DISCUSSED at length the interconnections between chance and causality coming together as noise and as a manifestation of fluctuation-dissipation. It is the randomness of the fluctuations that appears as noise, but this noise also has a systemic connection through the features of dissipation that fluctuations cause. The statistical nature of randomness in classical conditions does not have to be identical to that of quantum conditions. The classical randomness represents the distribution in canonic variables that an ensemble of particle has. When the quantum nature is very central, this is considerably different. Uncertainty is noise. If there is entanglement in the system, it forms a nonlocal link, and Bell's inequalities tell us the differences one will see in the statistics of the observations. Randomness, however, does underly noise, and, in this sense, any deterministic phenomena that is too hard to calculate is not to be thought of as noise.

Much of this physical behavior and manifestation can be observed in our natural world too. In our society, on an individual scale, what we do—random or deliberate—has a local spatial and temporal consequence whose correlative effect is lost at higher size scales. But, where society heads and the dissipative way it heads toward any direction are affected through this fluctuation-dissipation from each one of us. What we do matters, even if it may be only in an ever-so-small way.

Our approach to understanding noise took a more conceptual and physical approach, although we started with an introduction of how mathematically one may statistically characterize a signal in time, or signals from many different similar systems in time. This let us introduce mean, mean square, variance, autocorrelation and covariance as some of the measures that are important for understanding noise, as well as stationarity and its order as a means of characterizing signals. We used this to establish power spectral density as a means of characterizing the correlations in a system response and, through this, noise and the systemic dissipative response.

Our exploration of thermal noise was through understanding the response of a semiconductor-based capacitor through which electrons flowed while undergoing scattering. This allowed us to connect a physical understanding of the scattering-constrained transport to the appearance of thermal noise in both slow (free flight) and fast (scattering) processes, relate this charge movement to particle and displacement current and the appearance of inductors and capacitors as kinetic and potential storage elements, respectively, and of the resistor as a dissipation element. A suitable model for our example was the storage and dissipation elements together with a noise source. This noise source could be written either in the Norton form with a current source or in the Thévenin form with a voltage source. *Inductors are noisy, and so are capacitors.* This point was emphasized via highlighting the quantum and thermodynamic connections extracted through the inductor-capacitor resonance. The act of observation requires connecting the system to the environment at temperature, T. Inductors and capacitors then will show fluctuations, following thermodynamic and statistical arguments. We also tied this to the quantum implications, when energy is not continuously distributed and where the distribution function may be different, such as in photon systems, or where quantum-exclusion applies, such as in quantized channels. Partition thermal noise is an example of this.

Shot noise arises in the fast disappearance of moving particles across a system's boundary. An electron moving between two plates is observable through the current in the external circuit. And when this electron passes through the reservoir boundaries, the associated current disappears like a shot. Since the flow of current here is the source, the noise is proportional to the current, with a dependence not unlike that of thermal noise. Shot noise will also exist for photons.

The last example of noise was low-frequency noise arising in a variety of causes, with all of them showing an increase in the power spectrum at lower frequencies. We chose trapping-detrapping as an example to show that this noise has a "Lorentzian" spectrum. Surfaces give rise to a pronounced increase in this noise, since surfaces have higher number of states with a propensity to trap carriers for a pronounced period of time. But it also happens in the bulk, where Hall-Shockley-Read trap states will also capture and emit carriers. The time that a carrier stays captured or emitted is long, whether it is through a process involving a single band or one involving both the conduction and the valence bands, and it is this slow process that leads to the inverse frequency dependences. Local environment changes in this process also lead to conductivity changes where both carrier densities and carrier mobilities are affected. Particularly appropriate for poor transport materials—amorphous and polycrystalline semiconductors—is the Hooge parameter, through which one can characterize resistance fluctuations.

16.7 Concluding remarks and bibliographic notes

NOISE IS PERVASIVE once one brings together an ensemble of particles at a temperature T. The system will have a statistical distribution, so a measurement that gives a snapshot of a certain property of the system at a certain time will not be identical to the thermodynamically averaged property or the property measured at another instant. This is noise, with its origin in the random phenomena that occur in the system, whether viewed classically or quantum-mechanically. The beauty is that this random phenomena—fluctuations—also manifests as a dissipative response of the system. So, in linear systems, one sees a very direct link between fluctuation and dissipation, and fluctuation's noise incarnation.

When semiconductors are used, this noise is important, as it places a variety of constraints. A signal needs to be ascertained with sufficient accuracy and fidelity. Noise places a floor, and the signal-to-noise ratio is a common parameter important in analog and high-frequency usage and precision measurements. Noise in threshold voltage fluctuations arising in fluctuations such as those of dopants in a device is an important constraint for transistors. Noise places constraints on the energy and voltages that one may use to operate a deterministic logic gate in order to maintain logical validity. This chapter brought together several of the considerations specifically important for semiconductors as well as for their usage in devices. But noise is an esoteric domain at the intersection of statistics and mechanics in matter, and often an afterthought for the engineer. As energy and precision continues to become increasingly important, the understanding of noise does too, even though there is not enough of a literature exploring the subject at the scale it deserves.

An important source for seeing the transit of understanding through the Brownian, probability, stochastic, random walk, Fokker-Planck and Langevin path is the mongraph by Mazo[1]. It is a very well-written source that makes an easy transfer between history, mechanics, physics, mathematics and the applications. Electrical connections exist in harmony with hydrodynamics, colloidal motion and other systems where random processes lead to noise. A similar book, but with a more engineering bent, is by Pécseli[2]. The text

[1] R. M. Mazo, "Brownian motion: Fluctuations, dynamics, and applications," Clarendon Press, ISBN 0-19-851567-7 (2002)

[2] H. L. Péscseli, "Fluctuations in physical systems," Cambridge, ISBN 978-0521655927 (2000)

also draws the connections between fluctuation-dissipation and the Kramers-Kronig relations of a linear system.

An important comprehensive discussion is by Kogan[3]. It tackles in depth correlation functions, ergodicity and Markov processes. It also explores fluctuation-dissipation in both equilibrium and off-equilibrium conditions, ballistic systems, disordered systems and other conditions for the general domain of solids, with a number of chapters devoted to semiconductors.

[3] S. Kogan, "Electronic noise and fluctuations in solids," Cambridge, ISBN 0-521-46034-4 (1996)

MacDonald's book[4] is one of the older books, but it is still a gem. It discusses the various aspects of fluctuations in electrical circuits as seen through the vacuum tube view very well. The ideas discussed are quite relevant to the semiconductor environment too.

[4] D. K. MacDonald, "Noise and fluctuations", John Wiley, Library of Congress 62-16153 (1962)

For understanding the power spectrum, convolutions, reciprocal domains, the Wiener-Khintchin theorem and the treatment of electrical systems and linear systems, Champeney's text[5] is particularly noteworthy.

[5] D. C. Champeney, "Fourier transforms and their applications," Academic, ISBN 0-12-167450-9 (1973)

For those interested in shot noise and mesoscopic systems, a particularly interesting reference is by de Jong[6]. It explores shot noise, as well as its suppression due to quantum constraints, when one employs mesoscopic systems with ballistic transport across channels.

[6] M. J. M. de Jong, "Shot noise and electrical conduction in mesoscopic systems," Copynomie Veldhoven, ISBN 90-74445-19-5 (1995). This is a Ph.D. thesis from Leiden University

Mode partition noise in semiconductor lasers, together with its underlying analysis using a Master equation written in the Langevin fluctuation form, is treated by Pao-Lo Liu for steady-state and dynamic conditions in the book edited by Y. Yamamoto[7].

[7] Y. Yamamoto, "Coherence, amplification, and quantum effects in semiconductor lasers," John Wiley, ISBN 0-471-51249-4 (1991)

Electrical engineering has its own style—drawn from the implications for the operation of circuits—that the reader may want to also peek at. We studied here the underlying phenomena, but what an engineer who puts a device or circuit to use is interested in are measures such as the signal-to-noise ratio, the noise figure, effects in a multiport environment when circuits are chained, and noise as it exists due to various causes in different devices. A good source for such an understanding is the book by Gabriel Vasilescu[8]. It discusses how noise is treated in device modeling and how it is incorporated in the design of circuits, and it takes a side diversion to cross-talk interference modeling in circuits.

[8] G. Vasilescu, "Electronic noise and interfering signals: Principles and applications," Springer, ISBN 3-540-40741-3 (2005)

Low-frequency noise is of particular importance in field-effect devices such as the *MOSFET*. A book that dwells entirely on this $1/f$ noise is by von Haartman and Mikael Östling[9]. The book integrates the engineering operation of these devices with the noise phenomena as observed and understood in practical structures. So, it incorporates the variety of mobility and field effects that occur over generations of bulk and silicon-on-insulator devices.

[9] M. von Haartman and M. Östling, "Low-frequency noise in advanced *MOS* devices," Springer, ISBN 978-1-4020-5909-4 (2007)

16.8 Exercises

1. What is the theoretical maximum thermal noise power that can be delivered to an ideal noiseless load from a resistor R at temperature T? **[S]**

2. Will photons exhibit thermal noise? Please provide a reason. **[S]**

As an aside, think too about photon entropy.

3. Do processes that cause noise have to be random? Yes or no, and a short argument, please. **[S]**

4. We brought out the thermal noise and shot noise by considering an electron's transport in-between electrodes. Thermal noise depends on temperature. Does shot noise too? If not, can you suggest why not? **[S]**

5. Could one use the Boltzmann transport picture for analyzing scattering, noise and other phenomena in mesoscopic conditions where quantized conductance is important and few channels available for transport? Why or why not? **[S]**

6. We typically do not write Maxwell's equations using a fluctuation term, although all statistical effects are within its instantaneous validity. When we form noise models showing fluctuations, we write ideal capacitors, resistors and inductors together with noise sources to show the fluctuation effect—the fluctuation of charge and its consequences—and this is useful. We also typically do not write Maxwell's equations in reciprocal space and time. Show that, by using a tilde (˜) as an overscript in the conventional notation to represent the inclusion of fluctuations, one may write Maxwell's equations as follows:
In the space and time domain:

$$\begin{aligned}
\nabla \times \mathbf{H}(\mathbf{r},t) &= \tilde{\mathbf{J}}(\mathbf{r},t) + \frac{\partial \mathbf{D}(\mathbf{r},t)}{\partial t}, \\
\nabla \times \boldsymbol{\mathcal{E}}(\mathbf{r},t) &= -\frac{\partial \mathbf{B}(\mathbf{r},t)}{\partial t}, \\
\nabla \cdot \mathbf{D}(\mathbf{r},t) &= \tilde{\rho}(\mathbf{r},t) \text{ and} \\
\nabla \cdot \mathbf{B}(\mathbf{r},t) &= 0,
\end{aligned}$$

with

$$\frac{\partial \tilde{\rho}(\mathbf{r},t)}{\partial t} + \nabla \cdot \tilde{\mathbf{J}} = 0.$$

In the reciprocal space and time domain:

$$\begin{aligned}
i\mathbf{k} \times \mathbf{H}(\mathbf{k},\omega) &= \tilde{\mathbf{J}}(\mathbf{k},\omega) - i\omega \mathbf{D}(\mathbf{k},\omega), \\
i\mathbf{k} \times \boldsymbol{\mathcal{E}}(\mathbf{k},\omega) &= i\omega \mathbf{B}(\mathbf{k},\omega),
\end{aligned}$$

$$i\mathbf{k}\cdot\mathbf{D}(\mathbf{k},\omega) = \tilde{\rho}(\mathbf{k},\omega), \quad \text{and}$$
$$i\mathbf{k}\cdot\mathbf{B}(\mathbf{k},\omega) = 0,$$

with

$$i\omega\tilde{\rho}(\mathbf{k},\omega) - i\mathbf{k}\cdot\tilde{\mathbf{J}}(\mathbf{k},\omega) = 0.$$

Using the complex dielectric function ($\epsilon(\mathbf{k},\omega) = \epsilon^{\underline{r}}(\mathbf{k},\omega) + i\epsilon^{\underline{i}}(\mathbf{k},\omega)$), this latter set can be rewritten as

$$\begin{aligned}
\mathbf{k}\times\mathbf{H}(\mathbf{k},\omega) &= -i\tilde{\mathbf{J}}(\mathbf{k},\omega) - \omega\epsilon_0\epsilon(\mathbf{k},\omega)\boldsymbol{\mathcal{E}}(\mathbf{k},\omega),\\
\mathbf{k}\times\boldsymbol{\mathcal{E}}(\mathbf{k},\omega) &= \omega\mathbf{B}(\mathbf{k},\omega),\\
\mathbf{k}\cdot\boldsymbol{\mathcal{E}}(\mathbf{k},\omega)\epsilon(\mathbf{k},\omega) &= \frac{i}{\omega\epsilon_0}\mathbf{k}\cdot\tilde{\mathbf{J}}, \quad \text{and}\\
\mathbf{k}\cdot\mathbf{B}(\mathbf{k},\omega) &= 0,
\end{aligned}$$

with the direct connection shown. This representation is formally the equivalent of separating a generator to account for the fluctuations. Now we have a deterministic dielectric function and fluctuating charges and currents. **[S]**

7. The integrated value of an energy spectrum over all frequencies is proportional to the total energy transfer of a signal, that is, if we define the energy spectrum $S_f(y)$ of a function $f(x)$ as

$$S_f(y) = F(y)F^*(y),$$

where

$$F(y) = \int_{-\infty}^{\infty} f(x)\exp(-ixy)\,dx,$$

which is consistent with

$$\int_{-\infty}^{\infty} S_f(y)\,dy = \int_{-\infty}^{\infty} F(y)F^*(y)\,dy = 2\pi\int_{-\infty}^{\infty} f(x)f^*(x)\,dx.$$

Show that the integrated value of the power spectrum should match the mean rate of energy transfer in the signal, that is,

$$\langle|f(x)|^2\rangle = \frac{1}{2\pi}\int_{-\infty}^{\infty} P_f(y)\,dy,$$

where

$$P_f(y) = \lim_{X\to\infty}\frac{1}{2X}F_X(y)F_X^*(y).$$

[S]

8. Two resistors—R_1 at temperature T_1, and R_2 at temperature T_2—are connected either in series or in parallel. What is the root mean square noise voltage at the output? **[S]**

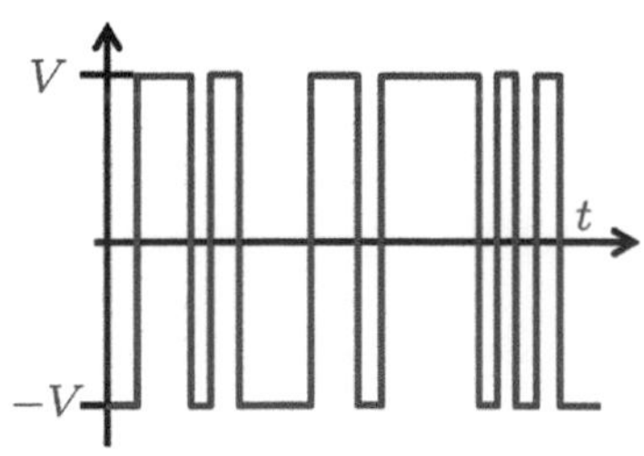

Figure 16.14: A random telegraph signal fluctuating between $\pm V$.

9. Derive an expression for the autocorrelation function of the random telegraph signal shown in Figure 16.14, assuming that the time instants of zero crossings follow a Poisson distribution. In the process of deriving, show that the product $z(t)z(t-\tau)$ with values of $\pm V^2$ depends on the number of changes k. Thus, the spectral power density can be found. Sketch it, and comment on the implications of the magnitude of τ vis-à-vis T. **[S]**

The result should be clear from the text's discussion. Here, just consider the kth transition probability as $\mathfrak{p}(k, \nu T) = [(\nu T)^k/\nu!]\exp(-\nu T)$ and derive the result through the series of the probability of events.

17
Stress and strain effects

How may one change a semiconductor's transport or other properties arising in interfacial and bulk changes? For devices at nanoscale, with their poor scaling of parasitics' consequences for desired operational scaling, this is an important question. Can one reduce scattering, emphasize one characteristic over another or change the potential constraints of an interface so that the operational characteristics—frequency limits or gain, for example—still continue to improve? In the case of optoelectronics, there may be other operational characteristics where device size may not be as central as it is for electronic devices. For lasers, for example, these may be efficiency, temperature span of operation or reliability. Introducing strain is one way of changing electronic and optoelectronic interactions. Strain is easier to accomplish in thin layers where boundaries—a planar boundary such as a substrate defining a grown layer's periodicity, or an edge boundary that pushes in or pulls out a small region—can be introduced. These modifications introduce changes in bandstructures and interfacial bandedges and can also change scattering through changing the states' characteristics and their occupation. The use of the nanoscale also provides additional strain-related possibilities that do not exist at larger scales. In this chapter, we extend our discussion of bandstructure, band discontinuities and transport—much of the text up to this point—to a manipulation of them through strain. The use of nanoscale in this also brings a few additional physical phenomena, because of their increased importance, into the discussion. These are matters related to long-range interactions and their fluctuation consequences, the increasing importance of plasmons, and the phonon interactions at interfaces. We will extend some of these discussions in the next chapter (Chapter 18) to high relative dielectric constant materials—used as gate insulators and elsewhere—to tie many of the consequential scattering effects in

Semiconductor Physics: Principles, Theory and Nanoscale. Sandip Tiwari.
 Published 2020 by Oxford University Press. DOI: 10.1093/oso/9780198759867.001.0001

the presence of the stronger Coulomb coupling that high dielectric constants permittivity brings.

17.1 Strained layers

GALLIUM ARSENIDE, ALUMINUM ARSENIDE AND THEIR COMPOSITIONAL MIXES, as we discussed in Chapter 6, form a pristine interface when two compositions are changed from one to the other, with a precisely predictable conduction band and valence band discontinuity, and a reproducible undisturbed bulk-like bandstructure nearly up to the interface. This is because the lattice constant of the two materials is quite close—$\leq 0.13\,\%$ at room temperature—and their thermal expansion coefficients are also insignificant, leading to a nearly bulk-to-bulk unstrained interface. *Si*, on the other hand, has a lattice constant of $0.543095\,nm$, and *Ge* has a lattice constant of $0.564613\,nm$ at room temperature, which is an in-plane strain of $3.81\,\%$, which is considerably larger. The discontinuity will depend on the strain, and the bulk bandstructure will also change in the strained region near the interface. If one can accommodate the elastic strain energy without the crystal coming apart by creating defects of various types, through the temperatures employed and thickness of material grown, even minutely thin layers—on the order of a few *nms* thick—can be grown. This thickness is too small for any effective device usage, but a compositional mix of $Si_{1-x}Ge_x$, with $0 < x < 0.4$, can be achieved in thick-enough layer form ($\sim 30\,nm$) to be useful for many interesting device applications. So, strain allows a multitude of semiconductors to be grown on others. The strain also changes the properties of these materials at the interface, as well as away from it. In the $GaAs/Ga_{1-x}Al_xAs$ system, the strain does not play much of a role, but the interface does in a significant way. And creating unorthodox strain patterns through stress at chosen boundaries provides a way to manipulate the local bandstructure artificially.

The strain at the *GaAs/AlAs* interface is very, very slightly tensile and observable in a red shift in the optical response. *AlAs* has a slightly larger lattice constant ($0.56605\,nm$) compared to *GaAs*, whose lattice constant is $0.56533\,nm$. *Al* is smaller but *AlAs* is also more ionic. The thermal expansion coefficients are $6.86 \times 10^{-6}\,K^{-1}$ for *GaAs*, and $5.2 \times 10^{-6}\,K^{-1}$ for *AlAs*.

Lower temperature implies lower energy for disturbing the stability through the thermodynamic processes, and lower thickness means that the accumulated elastic energy contained in the strain is kept limited.

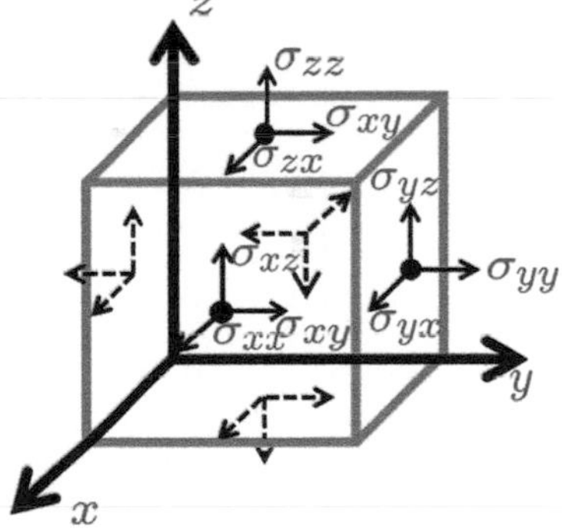

Figure 17.1: Nomenclature for forces and stresses. Forces distributed on a surface are stress. The normal component and shear components of the stress can be noted through the subscript. Shear has two components parallel to the plane of shear. The figure shows the direction of normal and shear stresses, with solid lines for the visible faces and dotted lines for the opposite. Normal stress causing tension is positive by convention. For shear, the positive component is aligned in the direction that causes tensile stress if it aligns along that coordinate direction.

First, consider what happens in a continuum media in elastic theory. The convention and nomenclature for stresses is shown in Figure 17.1. Stress is force distributed over a surface. Pressure on a surface exercises stress. The force per unit area can be resolved into three spatial degrees of freedom, so three coordinate components. The normal component of stress uses the symbol σ with a double subscript (or because it is duplicated, a single subscript) identifying the direction normal to the plane. It is positive if it is tensile; that is, stretches the body in that direction. The in-plane stress, also denoted by σ, can be identified through the subscripts as a shear stress. It is subscripted by the two coordinate components of its plane. The positive direction of shear stress is the positive direction where tensile stress on the same side would have a positive direction on the corresponding axis. The stresses on the six faces of a cubic element can be described by three normal stresses (σ_{xx}, σ_{yy} and σ_{zz}) and six shear stresses (σ_{xy}, σ_{yx}, σ_{xz}, σ_{zx}, σ_{yz} and σ_{zy}). Considering an infinitesimally small element, the shear forces on any face and equilibrium will lead to a balance between complementary stresses. For example, $\sigma_{xy}\,dx\,dy\,dz = \sigma_{yx}\,dx\,dy\,dz$ on the x-y face implies equality of σ_{xy} and σ_{yx}. Therefore, there exist three independent normal stresses and three independent shear stresses.

Treated as a continuum, Hooke's law experimentally establishes the relationship between strains and stresses. A generalization of Hooke's law, that is, under elastic and linear deformation conditions, $\sigma_{ij} = C_{ijkl}\varepsilon_{kl}$, connecting all directions of stress to all directions of strain. C_{ijkl} is the elastic stiffness tensor. It is of the fourth order, since two different directions in three-dimensional space are being connected. The symmetry of cubic semiconductors forces the stiffness tensor to have only four independent components relating stress and strain through

Robert Hooke was one of the early great English scientists. He was a contemporary of van Leeuwenhoek, who is credited with the microscope and a decade older than Newton. Hooke was a curator of experiments for the Royal Society, and the story is—or the claim goes—that it was Hooke who suggested that Newton look at gravity and Kepler's observations. Hooke's law does have correspondences to Newton's. Note that all this analysis is in a continuum picture. It will break for two-dimensional material. The treatment of sharing of an atom has to be a quantum-mechanical undertaking.

$$\begin{bmatrix} \sigma_{xx} \\ \sigma_{yy} \\ \sigma_{zz} \\ \sigma_{yz} \\ \sigma_{xz} \\ \sigma_{xy} \end{bmatrix} = \begin{bmatrix} c_{11} & c_{12} & c_{12} & 0 & 0 & 0 \\ c_{12} & c_{11} & c_{12} & 0 & 0 & 0 \\ c_{12} & c_{12} & c_{11} & 0 & 0 & 0 \\ 0 & 0 & 0 & c_{44} & 0 & 0 \\ 0 & 0 & 0 & 0 & c_{44} & 0 \\ 0 & 0 & 0 & 0 & 0 & c_{44} \end{bmatrix} \begin{bmatrix} \varepsilon_{xx} \\ \varepsilon_{yy} \\ \varepsilon_{zz} \\ 2\varepsilon_{yz} \\ 2\varepsilon_{xz} \\ 2\varepsilon_{xy} \end{bmatrix}. \quad (17.1)$$

The stress and strain relationship in *Si* and *Ge*, for example, can be characterized by the stiffness constants given in Table 17.1. Note that when the subscript terms are identical, so *xx*, *yy* and *zz*, then the stress is normal, and if they are different, then it is a shear stress.

	c_{11} *GPa*	c_{12} *GPa*	c_{44} *GPa*
Si	166.0	64.0	79.6
Ge	126.0	44.0	67.7

Table 17.1: Stiffness constants at room temperature for *Si* and *Ge*.

The implication of these independent components can be illustrated through the consequence of strain in different directions. Take a hypothetical *Si* pseudomorphic film; that is, one that is laterally lattice aligned to a larger $Si_{1-x}Ge_x$ substrate. If the substrate surface is (001), then shear strains (ε_{xy}, ε_{yz} and ε_{xz}, which are all of identical magnitude) must vanish. So, as shown in Figure 17.2, there are different normal strains for *Ge* and *Si* under uniaxial stress in the [001] direction, but shear components vanish. However, if one had uniaxial stress in the [110] and [111] directions of *Si* and *Ge*, then the shear components are non-zero.

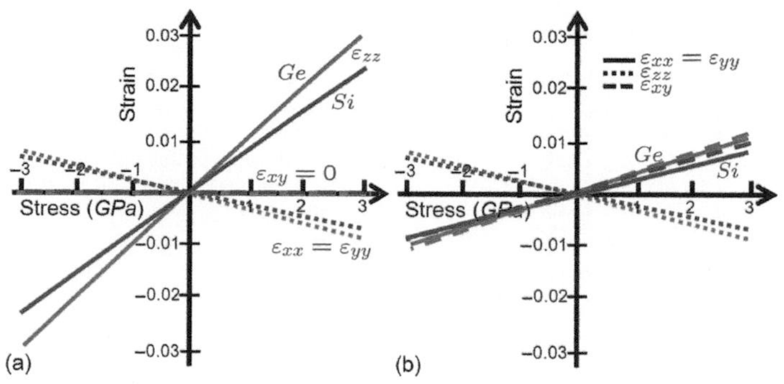

Figure 17.2: Strain response on (001) *Ge* and *Si*. These figures are a function of applied stress in *GPa*. Part (a) shows the strain response for [100] uniaxial stress, so along the conventional choice of a non-primitive cell's directions. Shear strain components are absent. Part (b) shows the strain response when a uniaxial stress is applied along the [110] direction. Both normal and shear strains exist.

For an isotropic material, a normal stress will not produce any distortion in the angles of the element, and $\varepsilon = \sigma/Y$, where Y is the modulus of elasticity (Young's modulus), defines the linear elastic relationship. Extension of an element is accompanied by lateral contraction, so while $\varepsilon_{xx} = \sigma_{xx}/Y$ for a normal stress applied in the x direction, $\epsilon_{yy} = -\nu\sigma_{xx}/Y$, and $\epsilon_{zz} = -\nu\sigma_{xc}/Y$, also result for the isotropic elastic material. ν here is Poisson's ratio. If there exist three normal stresses (σ_x, σ_y and σ_z), then the strains will have both the normal extension and the lateral contraction components, so

$$\begin{aligned}
\varepsilon_x &= \frac{1}{Y}\left[\sigma_x - \nu(\sigma_y + \sigma_z)\right], \\
\varepsilon_y &= \frac{1}{Y}\left[\sigma_y - \nu(\sigma_x + \sigma_z)\right], \text{ and} \\
\varepsilon_z &= \frac{1}{Y}\left[\sigma_z - \nu(\sigma_x + \sigma_y)\right] \qquad (17.2)
\end{aligned}$$

will result in the linear limit by superposition.

For us, for our cubic lattices, the in-plane strain of $\varepsilon_{\parallel}$ and the out-of-plane strain ($\varepsilon_{\perp}$) can be related as

$$\frac{\varepsilon_{\perp}}{\varepsilon_{\parallel}} = -\frac{2\nu}{1-\nu}. \qquad (17.3)$$

The strain distorts the cubic cell to a tetragonal cell. Table 17.2 gives the measured Poisson's ratio of some semiconductors of interest.

Semiconductor	Poisson's ratio (ν)
GaAs	0.312
GaP	0.305
Si	0.279
Ge	0.270

Table 17.2: Room temperature Poisson's ratio for some of the semiconductors.

If one were to form a strained layer on top of a substrate, as in Figure 17.3, for example, by growth, and the strained layer takes on the in-plane lattice periodicity of the substrate by being elastically strained—known as pseudomorphic approximation—then the strained crystal has an elastic energy of

$$E = 2\mu\frac{1+\nu}{1-\nu}\varepsilon_{\parallel}^2, \qquad (17.4)$$

where μ is the shear modulus $\mu = Y/2(1+\nu)$.

Any strain tensor can be decomposed into the sum of three separate tensors:

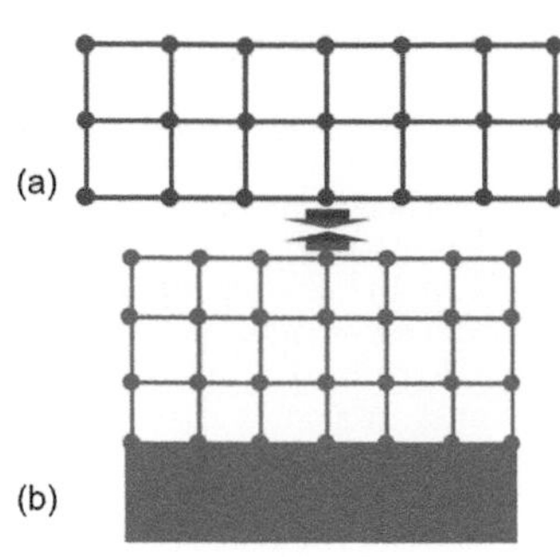

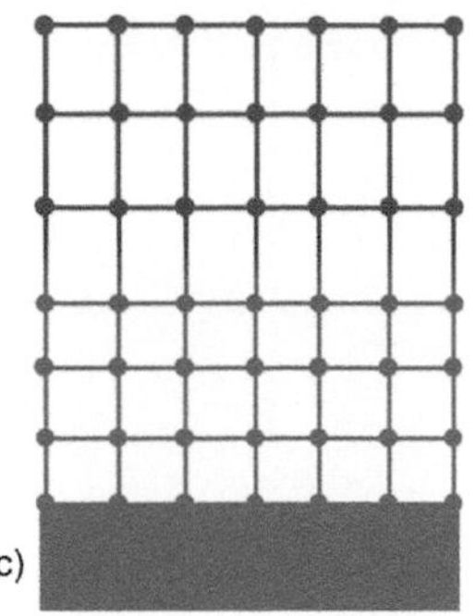

Figure 17.3: Part (a) shows a toy model of a large lattice constant crystal and (b) shows a smaller lattice constant substrate on which the semiconductor of (a) is formed as a continuing—morphic—layer that takes the substrate's in-plane periodicity, as shown in (c).

$$\begin{bmatrix} \varepsilon_{xx} & \varepsilon_{xy} & \varepsilon_{xz} \\ \varepsilon_{yx} & \varepsilon_{yy} & \varepsilon_{yz} \\ \varepsilon_{zx} & \varepsilon_{zy} & \varepsilon_{zz} \end{bmatrix}$$
$$= \frac{1}{3}\begin{bmatrix} \varepsilon_{xx}+\varepsilon_{yy}+\varepsilon_{zz} & 0 & 0 \\ 0 & \varepsilon_{xx}+\varepsilon_{yy}+\varepsilon_{zz} & 0 \\ 0 & 0 & \varepsilon_{xx}+\varepsilon_{yy}+\varepsilon_{zz} \end{bmatrix}$$
$$+ \frac{1}{3}\begin{bmatrix} 2\varepsilon_{xx}-(\varepsilon_{yy}+\varepsilon_{zz}) & 0 & 0 \\ 0 & 2\varepsilon_{yy}-(\varepsilon_{xx}+\varepsilon_{zz}) & 0 \\ 0 & 0 & 2\varepsilon_{zz}-(\varepsilon_{xx}+\varepsilon_{yy}) \end{bmatrix}$$
$$+ \begin{bmatrix} 0 & \varepsilon_{xy} & \varepsilon_{xz} \\ \varepsilon_{yx} & 0 & \varepsilon_{yz} \\ \varepsilon_{zx} & \varepsilon_{zy} & 0 \end{bmatrix}. \qquad (17.5)$$

The first term has the diagonal terms identical—it is a constant tensor where the terms are one third of the trace—and it reflects a volume change. It is a hydrostatic response. The other two terms arise in shear strain and involve a shape change. These represent the effect of a shear stress. The first of these—the second term in this strain tensor expansion—is due to changes in lengths along the three axes, and the second—the last term with no diagonal elements—is a rotational distortion term. This second term with diagonal elements represents a uniaxial stress along one of the cube's coordinate axes; that is, along the ⟨100⟩ direction.

The response is hydrostatic—a constant stress applied from all directions on the cube—as if it was placed in oil and pressure was applied with a plunger for compressive stress. This is also how many hydrostatic experiments are performed.

Uniaxial stress has caused the three dimensions of the cubic crystal to change. The last term contributes when stress exists along the [110] or [111] directions; that is, directions other than along the cube's axes. When a cube is under hydrostatic strain, its shape is unchanged. When stress is applied along the [100] axes, the cube distorts to an orthorhombus. And when the stress is along the [110] or [111] directions, the shape becomes triclinic, which is the least symmetric of the 14 Bravais lattices. The axes are now of unequal lengths and are now non-orthogonally inclined w.r.t. each other. This is the general response for cubic materials. Note the fractional factors in front, though. The response to a compressive uniaxial stress along [001] is not identical to the response when a biaxial tensile stress of identical magnitude is applied along [100] and [010]. Decompose this stress tensor, and one sees that the last two terms are identical. But the first term—the hydrostatic strain—will differ by a factor of 2.

What we are interested in is the specific strain response to stress that can be practically obtained during semiconductor usage.

The example of Figure 17.3—a grown crystalline layer on a thicker substrate where the substrate has the stronger influence on defining the in-plane periodicity—is an example of biaxial stress that arose in the layer being compressed to close to the substrate's dimensions. Strain will exist at the interface and its adjacent region on both sides of the interface. But the layer will show most of the consequence, since the substrate is thick and the film will have to store the elastic energy arising in the deformation. In Figure 17.3, the layer has become tetragonal by expanding in the out-of-plane direction. This new shape can be seen as a sum of hydrostatic compression and a uniaxial tension, as can also be seen from Equation 17.5. The reverse of this case is when the substrate has a larger lattice constant than the grown film does. An example of this is growing a sufficiently thick $Si_{1-x}Ge_x$ film so that it has relaxed to its natural periodicity and then growing a Si film on it. The Si periodic spacing will now be stretched in the plane and will shrink out of the plane. Figure 17.4 shows the resulting strain in the Si layer on a (001) surface. So, the [001] direction is shrunk while the [100] and [010] directions are stretched. Note, following Equation 17.1, the shear components vanish.

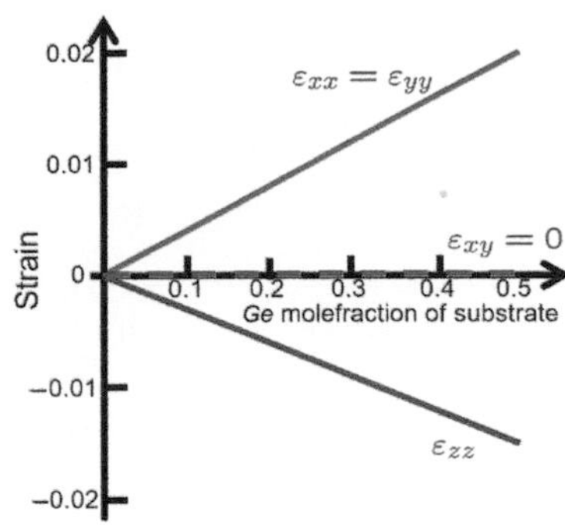

Figure 17.4: When pseudomorphic Si is grown on a relaxed crystal of $(001)Si_{1-x}Ge_x$, it is stretched in the plane and shrunk out of the plane. The strain is shown here for a (001) surface for a silicon thickness that can accommodate the elastic strain. Because the crystal is cubic, note that $\varepsilon_{xy} = \varepsilon_{yz} = \varepsilon_{zx} = 0$. The net strain is the sum of the hydrostatic compression and a uniaxial strain in the out-of-plane direction.

The strained configuration of important technological use is that of $Si_{1-x}Ge_x$ grown on Si. For small strains, a grown film can accommodate the strain energy elastically through the tetragonal distortion. As the film is made thicker, misfit dislocations arise in the interface region. Our discussion of surfaces in Chapter 5 explored some of these stress issues and was used as the argument for the reconstitution of the surface region. A discussion of where misfit dislocations will come about in order to accommodate the stress energy—how much of it is accommodated by strain and how much by the dislocations—is beyond our interests here and certainly will require plenty of assumptions given, the nonequilibrium nature of a growth process. Suffice it to say that properties of materials and the conditions and nature of the growth will define thicknesses that can accommodate the strain without the creation of dislocations. Figure 17.5 shows, for $Si/Si_{1-x}Ge_x$ and $GaAs/Ga_{1-x}In_xAs$ systems, an approximate thickness for films where

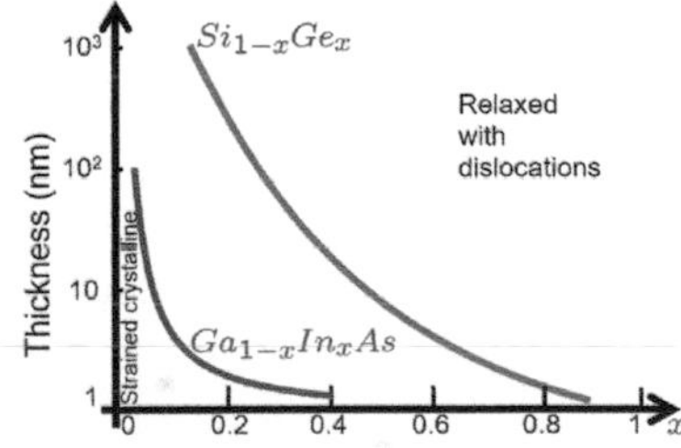

Figure 17.5: The pseudomorphic thickness, that is, the limit thickness up to which elastic strain can be accommodated by the film taking the in-plane lattice periodicity of the substrate, for $Si_{1-x}Ge_x$ grown on Si, and $Ga_{1-x}In_xAs$ grown on a $GaAs$ substrate.

The $Si_{1-x}Ge_x$ film—thin—is of enormous interest in the development of very high-frequency bipolar transistors for mobile communications. Bipolar transistors have an additional beneficial attribute of low noise, since much of the operation happens without the involvement of surfaces—a frequent source of defect interaction with $1/f$ features—and the base region is, by necessity, desired to be very thin. See S. Tiwari, "Device physics: Fundamentals of electronics and optoelectronics," Electroscience 2, Oxford University Press, ISBN 978-0-198-75984-3 (forthcoming) for

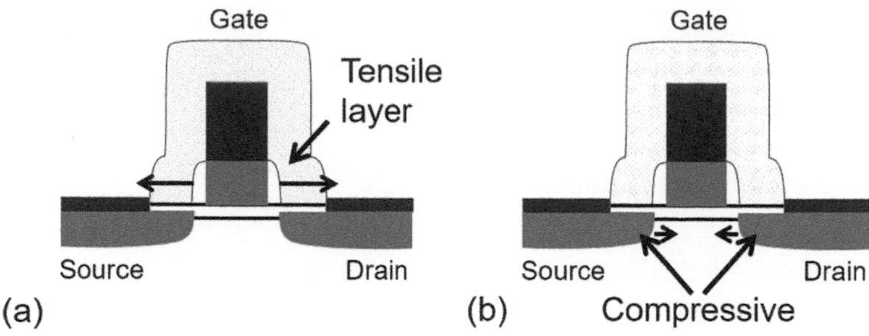

Figure 17.6: When the aspect ratio is large, processing can be employed to cause predominantly one direction of strain in small geometries. Films on the gate can be used to stretch the semiconductor underneath. Regions cladding the channel, such as the doped source and drain regions, may be used to cause compressive strain.

strain can be accommodated as a compressive film containing *Ge* and grown on a silicon substrate.

more discussion of the importance of strain and the resulting band changes in transistors that enable the frequency limits to be extended.

The other possibility for introducing strain in semiconductors is to take regions of a device—for a transistor, the gate or even the confined transistor or channel area, which have a relatively distorted aspect ratio—and apply a strain along the narrower direction via processing trickery. A tensile film, as shown in Figure 17.6(a), can cause a stretching laterally by pulling the gate out along its length, or a compression laterally, from the doped contacting region—an example is the incorporation of *Ge* as $Si_{1-x}Ge_x$ in this region—or from the isolation region, while keeping the channel still in *Si*.

In these uniaxial stress situations—uniaxial because the stress is in one dimension (usually width, which is much larger than length in this scenario)—Equation 17.1 can again be simplified. Figure 17.2 shows the strain response of *Ge* and *Si* under uniaxial strain in the [001] and [110] directions.

In semiconductors, the transistors are usually made in the [110] direction. The top view of the deformation looks as shown in Figure 17.7 and has two dominant independent components in the stress-strain relationship of the type that we have already discussed. This is also what Figure 17.2(b) indicates.

Historically, the reason was that wafers cleaved along this edge, so square dies resulted. The choice of the (001) surface is due to its lower density of interface states, which led to better control in the early years. The [110] orientation for gates is also partly from the higher electron mobility that one observes in this direction. The [100] direction band curvature leads to the higher longitudinal mass.

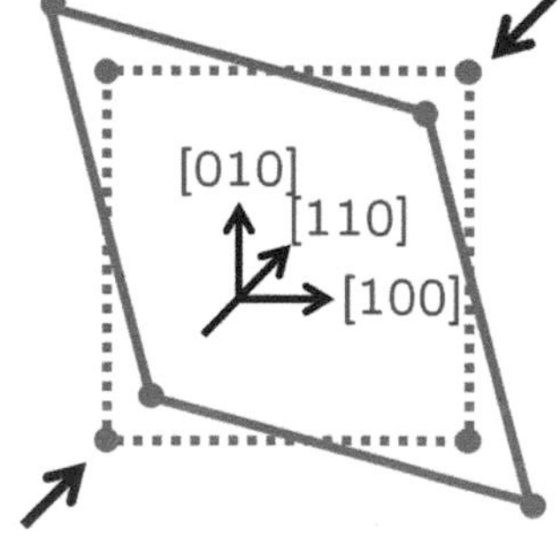

Figure 17.7: Uniaxial strain in *Si*, here compressive, in the [110] direction. The (110) plane of *Si* is usually aligned along the width direction of the field-effect transistor's gate.

Both the biaxial and the uniaxial strains are of technological interest. We have seen that biaxial stresses can be decomposed into hydrostatic and uniaxial forms. And the application of uniaxial stress will result in an in-plane strain without the out-of-plane effect that biaxial stress has.

The consequences of the different strains for electrons and for holes is the next subject of interest to us.

17.2 *Band alignment and bandstructure consequences*

A CHANGE IN LATTICE PERIODICITY will lead to bandstructure change and will also have consequences at the interface, as discussed in Chapters 4 and 5 on band alignment. So, both the "bulk" environment within a thin film and the "interface" environment are going to change. And these consequences will be for both electron states and hole states, a change in the anharmonicity as well as their shifting in energy, and for the transport properties arising in the bandstructure deformation as well as due to scattering.

Since a general stress can be decomposed into a hydrostatic component and a shear component, we need to look at what forms these take for biaxial stress, as occurs when a layer is grown, and uniaxial strain, as occurs when pattern-based techniques are employed. We approach this by first looking at the more general trends of the changes.

17.2.1 *Effect of hydrostatic stress*

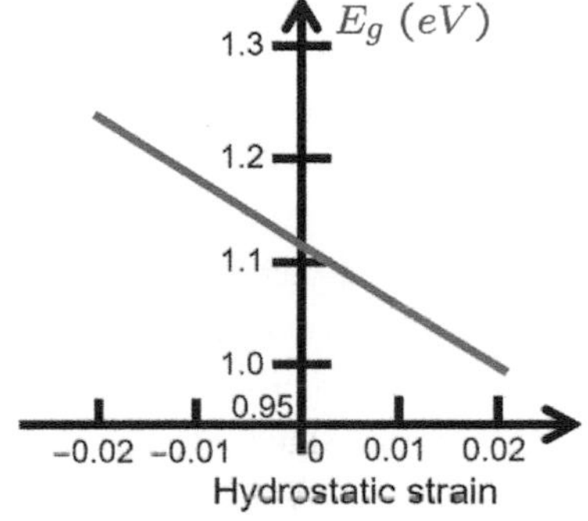

Figure 17.8: The room temperature change in the bandgap of *Si* under hydrostatic strain.

SINCE, IN CUBIC STRUCTURES, A HYDROSTATIC STRAIN causes identical deformation in all directions, the primary consequence then is a change in the bandgap. Symmetries do not change, so the degeneracies of the band minima should not change. But since the separation between atoms is changing, the interatomic interactions will change. If atoms are closer together—a compressive hydrostatic strain—the bandgap should increase, since the interaction perturbation has become larger. Figure 17.8 shows the bandgap change in *Si* under hydrostatic conditions. Separation of the bands is a function of the hydrostatic strain. *Ge* has a unit cell of $\sim 0.566\ nm$, and *Si* one of $\sim 0.543\ nm$, so if one could manage to obtain a sub-pseudomorphic thickness film of *Si* on a *Ge* surface, it will have an in-plane strain of the order of 0.04. An extension of this graph says that a nearly $-0.2\,eV$ decrease in the bandgap would arise due to hydrostatic tension. In reality, this is a very high strain, but strains of the order of 0.02 are quite feasible for useful layers under elastic strain without defect generation during all processing and usage.

Bipolar transistors employing strain will also compositionally change the strained layer. The strained base layer of $Si_{1-x}Ge_x$ on an *Si* collector will be compressive, and the molefraction x will change so that one may preferentially introduce an electric field through the bandgap changes to sweep injected electrons faster toward the collector.

17.2.2 *Effect of shear stress*

SHEAR STRESSES break crystal symmetry. In the last two tensors of Equation 17.5 arising in shear, the first—the diagonal form—is a change in length along the three directions of $\langle 100\rangle$. We remarked that a uniaxial strain along these directions will result in this term, while the second—a rotational tensor—will vanish. Figure 17.7 showed this rotation and distortion consequence for a uniaxial stress along $\langle 110\rangle$ directions. So, uniaxial stress along $\langle 100\rangle$ and along $\langle 110\rangle$ have different effects because of the change in symmetry. The biaxial strain in our example illustrates the uniaxial consequence in the growth direction of [001] with a change in periodicity. It made the crystal tetragonal. The pattern approach illustrates the uniaxial consequence from the [110] direction, and here the in-plane structure distorted from a square to a rhombus. The cubic symmetry is now changed to a tetragonal symmetry.

The real space symmetry change has consequences for the reciprocal space. Consider first the valence band. Recall that the light-hole, heavy-hole and split-off bands have their origins in the spin-orbit interaction. The states at the top of the valence band, following our discussion in Chapter 4, are $|3/2, 3/2\rangle$, $|3/2, -3/2\rangle$, $|3/2, 1/2\rangle$ and $|3/2, -1/2\rangle$, and $|1/2, 1/2\rangle$ and $|1/2, -1/2\rangle$, due to the spin-orbit interaction. The former, degenerate set of 4 zone center states are at the top of the valence band, and the second set in the split-off valence band. When the Hamiltonian has inversion symmetry, these fourfold states split into a set of twofold degenerate bands corresponding to the direction of the wavevector. The $|3/2, \pm 3/2\rangle$ states are the higher states of the heavy-hole band, and the $|3/2, \pm 1/2\rangle$ states are those of the light-hole bands. Deformation of the crystal breaks the zone center degeneracy and causes further change. Figure 17.9 shows the consequences for *Si* under biaxial tensile and uniaxial compressive stress. This behavior for valence band distortion and splitting is quite similar for most of the zinc blende and the diamond lattice semiconductors.

For conduction bands, unlike valence bands, where the bandedge behavior is quite similar for the zinc blende and diamond lattice semiconductors, the bandedge behavior is quite different. As elaborated in Chapter 4, the differences in covalent and polar interactions lead to conduction band minima at different points of the Brillouin zone. *GaAs* has a minimum at the zone center, *Si* has it near the X point along the Δ direction, and *Ge* has it at the L point. Strain effects will vary for these different types. The *GaAs* minimum being non-degenerate, there is no splitting. *Si* has sixfold degeneracy due to the sixfold symmetry of this minimum point near X and along the $\langle 100\rangle$ directions. The *Ge* minimum, being at the zone edge L point, should be seen as having fourfold degeneracy.

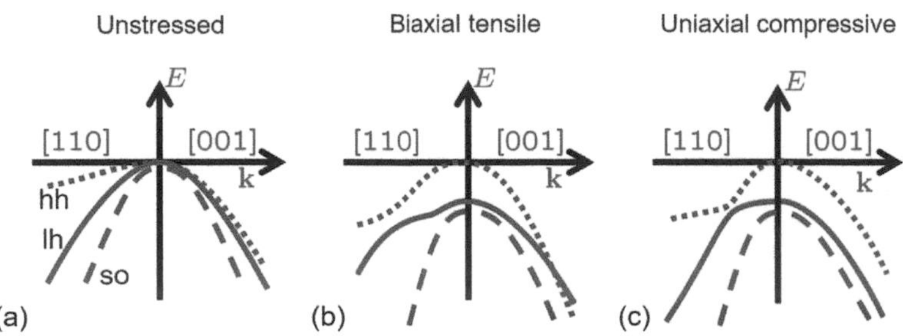

Figure 17.9: Light-hole (lh) and heavy-hole (hh) band deformation of *Si* under only shear stress. Part (a) shows the near valence bandedge unstressed $(E, \mathbf{k})$ behavior, (b) shows the consequence of shear under biaxial tensile stress, so compression in the [001] direction, and (c) shows the consequence of [110] uniaxial compressive stress. The split-off band is also shown in these figures.

Symmetry arguments tell us how these degeneracies for *Si* and *Ge* should be expected to change.

For *Si*, the biaxial and [110] uniaxial stresses distort the [100] and the [010] directions—the in-plane directions—similarly and the [001] direction—the out-of-plane direction—differently. So, one gets a splitting into fourfold and twofold degeneracy of the conduction band. Therefore, we will call the former Δ_4, and the latter Δ_2, since Δ is the direction we have associated with [001].

For *Ge*, for biaxial stress, with its identical [100] and [010] consequences, all the *L* valleys respond similarly—they are in the $\langle 111 \rangle$ orientation—leading to no degeneracy lifting. But the projection of these $\langle 111 \rangle$ valleys for the [110] uniaxial stress is different—two of the projections are along [110], and two are along $[\bar{1}10]$—with the valley degeneracy splitting into two twofold degenerate groups.

Uniaxial stress, on the other hand, being an applied force in only one direction in real space, affects bandstructure very differently. Now, the interaction of states contributing to the bands in that specific orientation changes in a significant way. There is now asymmetry, and degeneracies that existed in the cubic arrangement are lifted. Dimensions have changed. So, the conduction band and the valence band will reflect this change in symmetry and hence degeneracy. Consider [110] uniaxial stress. Band degeneracy will split. In *Si*, where the conduction band minima are in from the *X* point ($\langle 100 \rangle$), the out-of-plane direction will show one behavior, and the in-plane directions will show another behavior. In the valence band, the light-hole states and the heavy-hole states, with their different characters, will also separate differently, and the degeneracy at the zone minimum will be broken.

This change in band behavior—particularly the splitting in the conduction and valence states—is shown in Figure 17.10 for *Si*, *GaAs* and *Ge*, with their three different electron bandedge state symmetries.

Now consider what happens to the occupied states. The conduction bandedge states are easier to think through, and we will discuss *Si* here. Equienergy surfaces are ellipsoidal, and there are six different directions along which they are centered with a longitudinal mass along that direction. For biaxial tensile strain, the result of the degeneracy splitting and the change in the bandedge energy in the state occupation is as shown in Figure 17.11(a), which shows the description important to in-plane motion of these thin films, and Figure 17.11(b) shows the occupation. The transverse mass is smaller, and the twofold degenerate valley along the $\langle 001 \rangle$ axes are now the lower bandedge energy valley. It has a smaller mass, so it stretches more broadly in the transverse direction of

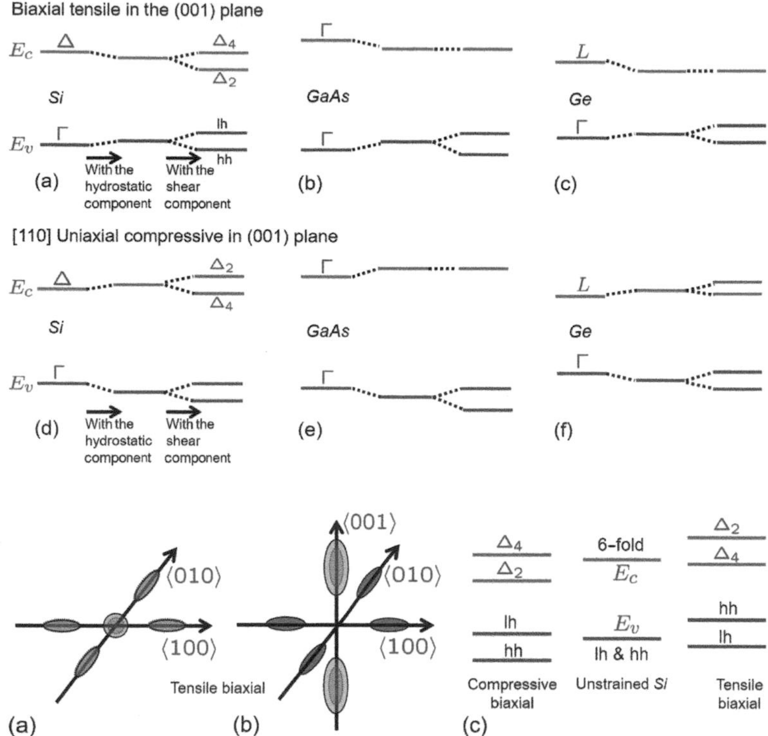

Figure 17.10: Changes in the bandedge state energies for *Si*, *GaAs* and *Ge* under biaxial tensile stress in a (100) plane in (a) through (c), and [110] uniaxial compressive stress, also in a (100) plane, in (d) through (f). The splittings have been decomposed into the hydrostatic and the shear contributions. The valence band heavy-hole (hh) and light-hole (lh) behavior is quite similar at the zone center—a splitting of heavy- and light-hole band degeneracy—although considerable differences will arise as one looks at higher hole energies and different wavevectors. The conduction bandedge state response is quite different for the different symmetries of those states.

Figure 17.11: Part (a) shows the constant energy contour of conduction band states on a (001) surface under biaxial tensile stress. Dashed lines are for unstrained conditions. The six ellipsoids with fourfold and twofold degeneracy are shown in (b) for the biaxial tensile stress conditions. The lower transverse mass and the lowering of the energy makes the twofold ellipsoids larger. Part (c) shows the degeneracy breaking and bandedge alignment changes under a polarity change of this biaxial stress; hh, heavy hole; lh, light hole.

motion in the (001) plane. It should also be understandable that tensile stress and compressive stress will have the opposite effect, as sketched in 17.11(c). Note that the energies of the states in the twofold and fourfold valleys are moving in opposite directions. The twofold valley minima also shift at about twice the rate of the fourfold valley minima.

17.2.3 *Band warping*

THE OTHER IMPORTANT CHANGE due to stress is that of band warping. The equienergy surfaces in the conduction band were ellipsoidal for the following reason. The *X*-point directed axis—a Δ axis—connects to a square face of the *FCC* crystal. The conduction valleys must follow this symmetry, since they are along this same direction. The symmetry forces the transverse contours to be circles. Along the axis, it is elongated, since there was little perturbation from states of any bands nearby. In the presence of stress, if this square symmetry in the perpendicular plane is reduced, the energy symmetry in this plane will also be lost. So, the ⟨110⟩ uniaxial stress, which leads to the in-plane rhombus distortion of a square, will distort the circle to an ellipse, as shown in Figure 17.12, with major and minor axes along the ⟨110⟩ directions.

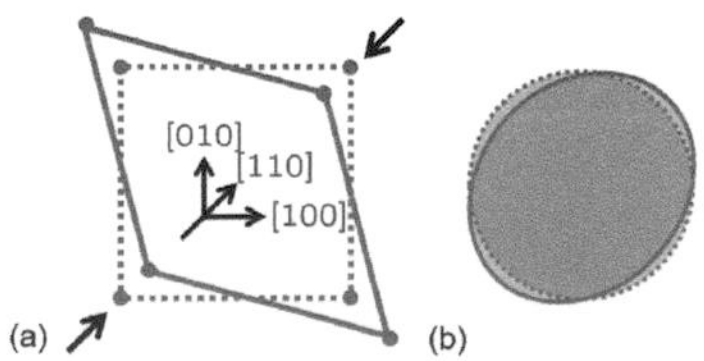

Figure 17.12: The symmetry consequence in band warping arising from the [110] uniaxial stress for (001) silicon. The two out-of-plane valleys will have constant energy contour distort from the circle, as shown in (b).

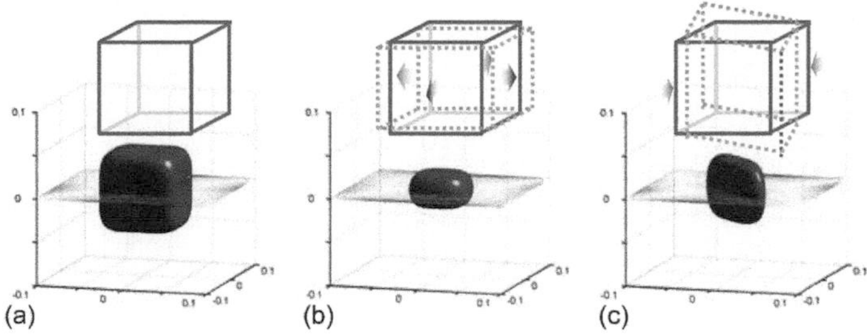

Figure 17.13: Very low constant energy valence surface in *GaAs*. Part (a) shows unstrained conditions with its cubic symmetries, (b) shows the case under biaxial tensile stress and (c) shows the case under [110] uniaxial stress. Note that these symmetries reflect the real space symmetries because of the Γ center maximum, and very low energy with small state interactions between light and heavy holes, due to degeneracy lifting and because the split-off band is farther away. The stress in the (b) and (c) cases is 1 *GPa*. Adapted from Y. Sun, S. E. Thompson and T. Nishida, "Strain effects in semiconductors," Springer, ISBN 978-1-4419-0551-2 (2010).

Band warping is the distortion of the wavevector for the same constant energy in different directions. Warping is much more significant in the valence band because of the interaction of states, spin and the symmetry consequences. The valence bandedge is at the Γ point. Because of this central locale, the reciprocal space energy surfaces will follow all the changes in symmetry of the real space under stress. *Si* valence band warping is going to be relatively complicated because the split-off band is so close to the light-hole and heavy-hole bands. The split-off band is separated by only 0.044 eV for *Si* at the zone center.

So, first consider the valence band in *GaAs* where the split-off band is nearly 0.34 eV below, so at least state interactions with split-off are small and degeneracy lifting under strain makes matters easier. Figure 17.13 shows the unstressed condition and the two stressed conditions that we have been interested in. The unstressed *GaAs* valence band has the symmetry of a cube. A biaxial tensile stress keeps the in-plane directions symmetric—elongated—and the out-of-plane direction symmetry is broken and that direction shrunk. So, the equienergy surface near the top of the valence band distorts the cube to a square cuboid. Uniaxial stress along [110] causes the cubic symmetry to distort to a tetragonal symmetry. Since this is a much more significant symmetry change, the warping is much more significant. The rhombus in-plane real space symmetry is reflected, at very low energies, in the axis of revolution turning to the [110] direction. The symmetry of the rhombus is maintained in the plane of the equienergy surface in the plane. If one looks at equienergy surfaces at higher energy, the heavy-hole, particularly, and the split-off states' interactions become important. Note that, at low energies, with the splitting of the bands, and being at the center of the zone, the interband coupling will be weaker. Indeed, the bands become more parabolic.

Now consider the consequence of the interaction of the states or of band mixing. The heavy-hole states are characterized by $|3/2, \pm 3/2\rangle$, and the light-hole states by $|3/2, \pm 1/2\rangle$. The angular momenta are related to rotational invariance. In a cubic crystal, it is the fourfold rotational symmetry that leads to the angular momentum L being 1. Include spin $S = 1/2$, and this led to the eigenstate description of the light hole and the heavy hole. Even though these bands are degenerate at the Γ point, the heavy- and light-hole states do not mix with each other, due to symmetry. Under biaxial stress, stretching or straining in plane, and its complement orthogonally, symmetry change occurs only in two orthogonal directions. The out-of-plane mixing between heavy-hole and light-hole states is relatively unchanged. So, the bands and states may rise or fall, depending on the polarity of the stress; degeneracy will lift, but band warping will be small along $\langle 001 \rangle$. In the case of [110] uniaxial stress, the out-of-plane rotational symmetry is now 2 and not 4. Angular momentum differing by 2 units of action cannot be distinguished. So, states such as $|3/2, -1/2\rangle$ and $|3/2, 3/2\rangle$ can couple. Light-hole and heavy-hole states can interact. So, with uniaxial $\langle 110 \rangle$ stress, degeneracy will lift, state mixing will be significant and band warping will be pronounced.

In quantum wells, this warping, by its consequences for confined quantum-well bandedge shifts, can lead to interesting instances of anti-crossing: two energy levels that cross near the Γ point curve away when they approach each other. This is a case of wave mixing resulting from symmetry reduction as in the [110] uniaxial stress instance.

These consequences arising in valence band state mixing will be much stronger in *Si*, where the split-off band is only few *meV*s away in energy at the zone center. Figure 17.14—to be compared with the *GaAs* result of Figure 17.13—shows the significant consequences for the uniaxial condition. Note also the tremendous warping that exists due to the split-off state interaction in the unstressed condition. *Si*'s valence bandstructure has quite consequential spin-orbit effects that show up particularly strongly under strain. Note also the correspondence of these equienergy surfaces with the E-$\mathbf{k}$ diagrams for unstressed and stressed conditions in Figure 17.9. It is the light-hole band that couples most strongly to the split-off band in the unstressed condition. When stress is introduced, degeneracy lifts, but the heavy-hole band still remains relatively unaffected by the split-off band. But the coupling between the light-hole band and the split-off band causes the two to shift in opposite directions. Under biaxial tensile strain, the light-hole mass increases with an inverse dependence on the stress through the mixing with the split-off band, with the heavy hole remaining immune.

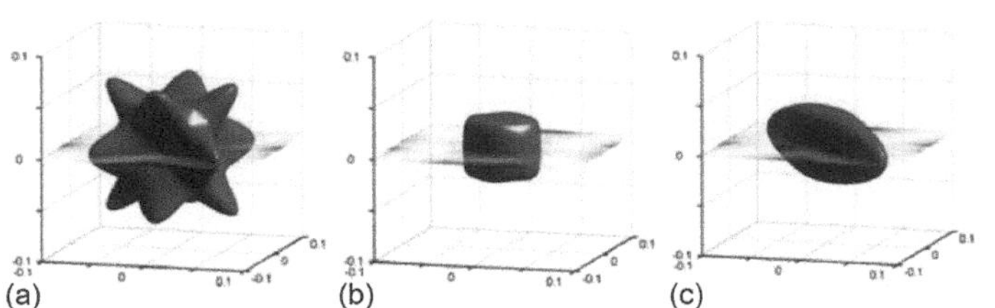

Figure 17.14: The constant energy valence band surface at very low energies in *Si*. Part (a) shows unstrained conditions with its cubic symmetries, (b) shows the case under biaxial tensile stress and (c) shows the case under [110] uniaxial stress. The stress in the (b) and (c) cases is 1 *GPa*. Adapted from Y. Sun, S. E. Thompson and T. Nishida, "Strain effects in semiconductors," Springer, ISBN 978-1-4419-0551-2 (2010).

The tight binding picture also gives us a perspective on this valence band shifting. This is illustrated in Figure 17.15.

Reduced distances between atoms increases interatomic interaction via increased wavefunction overlap, which is why the bandgap widens between bonding states and antibonding states. This was our reasoning for hydrostatic stress's effect. With shear stress, the consequence on bands will be related to symmetries. s states have little effect arising in warping from shear. The more pronounced effect arises in the hydrostatic stress through the wavefunction overlap. Warping arises in interband coupling. And, for s states rotated by shear w.r.t. one another, if it leaves the separation the same, no energetic difference should be expected. However, if angles change—as with oriented states, shown as the hybridized tetrahedral symmetry bonds of Figure 17.15—then overlap parameters are modified. This is a strong effect for p states. In the case of biaxial strain (Figure 17.15(a)), the four bonds rotate either toward or away from the plane of strain. In-plane orbital coupling increases, and out-of-plane coupling decreases. The weight of p_x and p_y in the bond increases, and that of p_z decreases. So, the [001] direction has an increased overlap of in-plane orbitals. The heavy-hole band is lowered. A decreased out-of-plane overlap of the p_zs leads to the light-hole band being raised. In the in-plane directions, the light-hole band is lowered (less overlap of p_x and p_y), leaving the heavy-hole band higher. The conduction band along the X direction is primarily composed of antibonding p states. So, the two valleys along the out-of-plane direction drop in energy under biaxial tension, from decreased overlap. The four valleys in the in-plane direction rise.

Note that the bands are not pure heavy-hole or light hole-bands. In biaxial tension, in the out-of-plane direction, the top valence band states are light-hole-like. In the in-plane directions, these states are heavy-hole-like. See the curvatures in Figure 17.15.

Under [110] compressive uniaxial stress, the in-plane shear changes symmetry, as seen in Figure 17.15(b), where two of the bonds are pulling in and two are separating out. With change in bond length and angles, the symmetry in the plane has changed. The atoms along the [110] direction are pulling in, and those

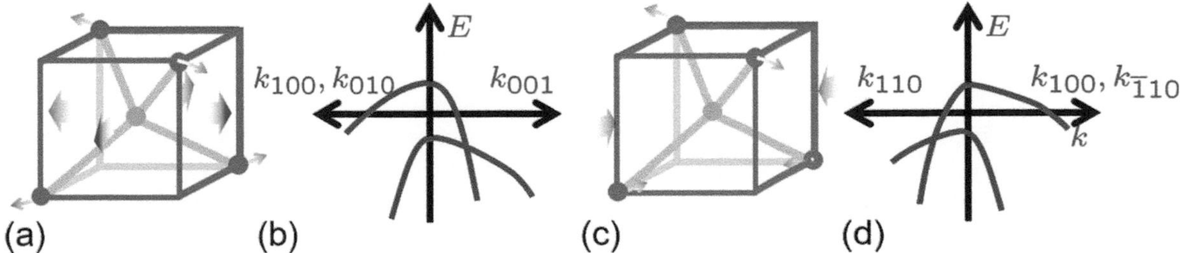

Figure 17.15: Parts (a) and (b) show the core effect under biaxial strain, and (c) and (d) show the core effects under [110] uniaxial strain. Parts (a) and (c) show the corresponding bond rotation, while (b) and (d) show the consequence in valence band shifting and its warping.

along [$\bar{1}$10] are separating. So, two quite opposite interactions are happening for the atoms in the lower and the upper planes. A bond length change also happens with the angular change. Rotation has the more significant consequence. The weight of the out-of-plane orbitals is increased, and that of the parallel ones decreases. Along the [110] direction, the heavy-hole states are lowered, and the light-hole states are raised. A complementary situation exists for the other two atoms and their bonding. Pulling apart increases the parallel projection and decreases the perpendicular projection. Band splitting is a result of the two types of shear in Equation 17.5. The states along the [100] and [$\bar{1}$10] directions are heavy-hole band states, and those along [110] are light-hole states for the lowest energy states of the valence band.

The conduction band too will have a similar descending of the two out-of-plane valleys and a rising of the four in-plane valleys.

We may now relate these states-related consequences to the changes in bandgap, that is, the consequences for the lowest energies in the conduction band and the valence band.

17.2.4 *Bandgap changes*

THE CHANGES IN BANDEDGE STATES in the conduction band and the valence band, as we have now seen, arise from a multitude of interactions. In *Si*, the valence bandedge states are changing, with split-off band interactions, and behaving as light or heavy holes in different directions. Likewise, the conduction band states, being off the X point along the Δ axis, also show different consequences arising in the changes in changes in symmetry taking place under different stresses.

The two stress situations, because of their practical importance, are the biaxial tensile and the [110] uniaxial compressive. It is useful to keep their consequences for the bandgap in mind, given the differences in the symmetry of the three semiconductors—*Si*, *GaAs* and *Ge*—of interest to us with their different band picture.

First, the consequence of strain in the three cases is shown in Figure 17.16 for both biaxial and uniaxial conditions under compressive and uniaxial stress, respectively. Compression results in less of a bandgap change than tension does. And the degenerate conduction valleys of *Si* and *Ge* result in a reduction in the bandgap unlike that for *GaAs*, where the conduction minimum is at the zone center. More than holes, it is the electron states that have a larger consequence for the bandgap.

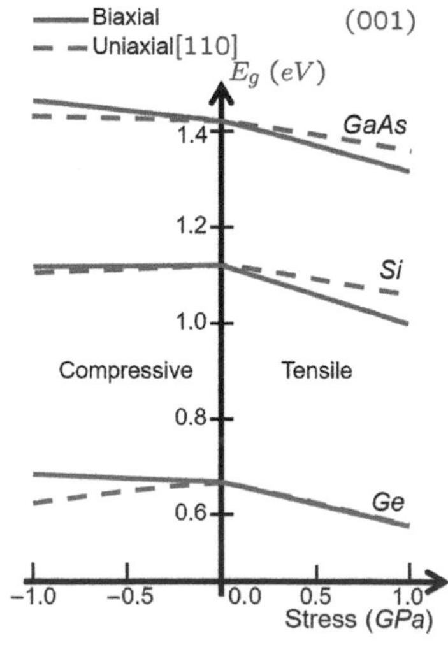

Figure 17.16: Bandgap as a function of biaxial and uniaxial strain for *Si*, *Ge* and *GaAs*. The surface is (001), and the uniaxial stress is in the [110] direction.

This is illustrated in Figure 17.17, which shows the nature of the case of biaxial strain for *Si* and *GaAs*. For *Si*, the twofold degenerate conduction valleys move faster as a function of the stress, while the fourfold valleys are relatively unperturbed. And the valence band states—light-hole states—have a significant shift arising in the split-off band interaction. This picture, to a lesser extent, also holds true for *GaAs*.

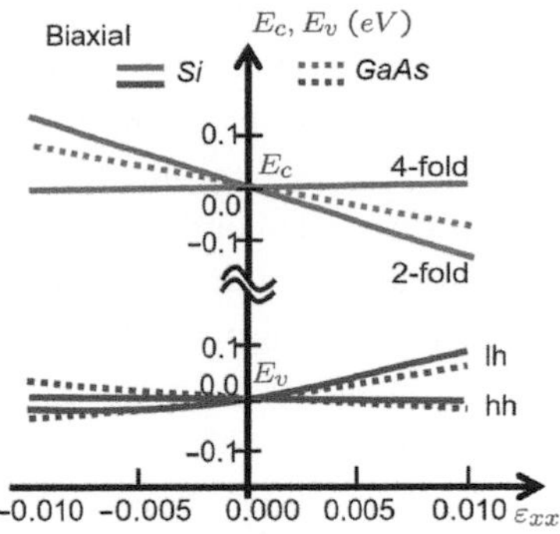

Figure 17.17: Bandedge change as a function of biaxial strain, for *Si* and *GaAs*; hh, heavy hole; lh, light hole.

See S. Tiwari, "Device physics: Fundamentals of electronics and optoelectronics," Electroscience 2, Oxford University Press, ISBN 978-0-198-75984-3 (forthcoming), for the role of inversion ladders in field-effect transistors, and a more advanced discussion in S. Tiwari, "Nanoscale device physics: Science and engineering fundamentals," Electroscience 4, Oxford University Press, ISBN 978-0-198-75987-4 (2017).

17.3 *Transport and confinement*

WE CAN NOW INTEGRATE this discussion of the changes in the nature of the occupiable $E(\mathbf{k})$ states in the conduction and valence bands and relate that to the consequences for transport. Figure 17.10 showed the changes that take place in the biaxial tensile and the [110] uniaxial compressive conditions. Transport as characterized through mobility will be affected by the changes in effective mass and by the scattering. Our scattering discussion in Chapter 10 has already addressed occupation, the perturbation matrix element and the states available for occupation into the scattering rates and, for some of the processes, the relaxation time approximation. So, without dwelling on the details of the changes in scattering parameters, we will now discuss some of the salient consequences of strain on transport. In addition, we will discuss the consequences of confinement, which often appears in strain-utilizing field-effect transistors.

Degeneracy	$m_\parallel$	$m_\perp$	Ladder
2	m_t	m_l	Unprimed
4	m_l	m_t	Primed

Table 17.3: Bandedge effective masses for subband ladders of a (001) *Si* surface.

If confinement is present, as in field-effect transistors, where it is due to the inversion along the plane, it will be reflected in a quantization of energy for out of plane momentum. So, the first effect of this in devices is reflected in the out-of-plane direction confinement in the interface inversion layer. Table 17.3 is a summary of the relevant parametric characteristics of the valleys. The Δ_2 valleys that are oriented out of plane have a heavy mass ($m_l = 0.916m_0$) normal to the interface, and a lower mass ($m_t = 0.19m_0$) in the plane. The heavier mass makes the subband confinement ladder of these valleys lower, with lower separation. The other four valleys have the lower mass affecting the raising of the lowest subband energy and the subband separations. Electron states in these subbands will have a higher conductivity mass parallel to the interface.

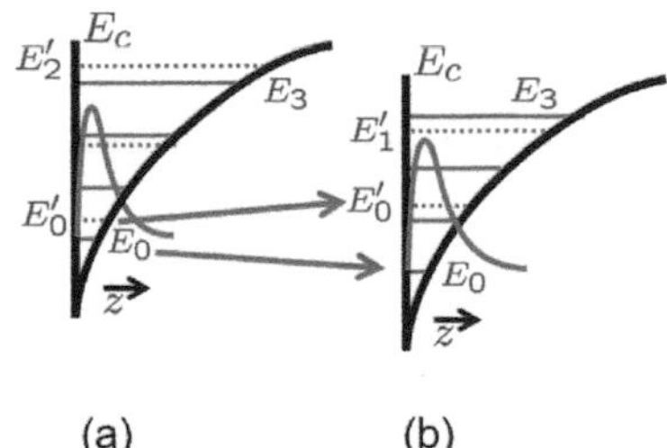

Figure 17.18: Part (a) shows the subband ladders of the Δ_2 and Δ_4 conduction valleys in unstrained *Si* and (b) shows the shifts in them as a result of biaxial tensile strain.

The ladder representation (unprimed letters correspond to Δ_2 valleys, while primed ones correspond to Δ_4 valleys) of this confinement in unstrained conditions is shown in Figure 17.18 for the unstrained and tensile biaxial strained conditions in *Si*. The twofold out-of-plane valleys have higher occupancy in the tensile biaxial stress condition. These are the states that will be first occupied in an n-type material or where electrons are brought in through a field. The conductivity mass is $0.19m_0$ for electrons in the Δ_2 valleys, and the charge envelope is also closer to the interface. The conductivity mass for the Δ_4 valleys is $0.315m_0$, and the charge envelope is relatively farther away as the confinement energies are higher. Biaxial tensile strain lowers Δ_2 valleys even more (see Figure 17.10), so the lower conductivity mass subband is lowered, and the higher conductivity mass subband is raised. For Si, for biaxial tensile strain on $Si_{0.75}Ge_{0.25}$, the separation of the unprimed and primed lowest subband energies is of the order of 0.28 eV, compared to $\sim$0.10 eV for relaxed conditions at sheet carrier concentrations of 10^{13} cm^{-2}. This difference in energy is nearly a change of $\exp(0.18/0.026) \approx 1000$ in occupancy. The mass consequence therefore has a very significant effect, and the mobility enhancements are significant—reaching as much as a factor of 2—at high electric fields and sheet charge concentrations. The electron mobilities can improve by nearly a factor of 2 with this lowering of inter- and intraband scattering and preferred occupation of lower conductivity mass states.

We now look at the valence band with its complexity of warping and the three different bands close together by looking at equienergy surfaces in the plane of motion. Figure 17.19 shows the heavy-hole, light-hole and split-off bands under unstrained conditions, a 1 % compressive strain and a 1 % tensile strain in the presence of a field of 10^6 V/cm in the insulator (about 2×10^{11} cm^{-2}

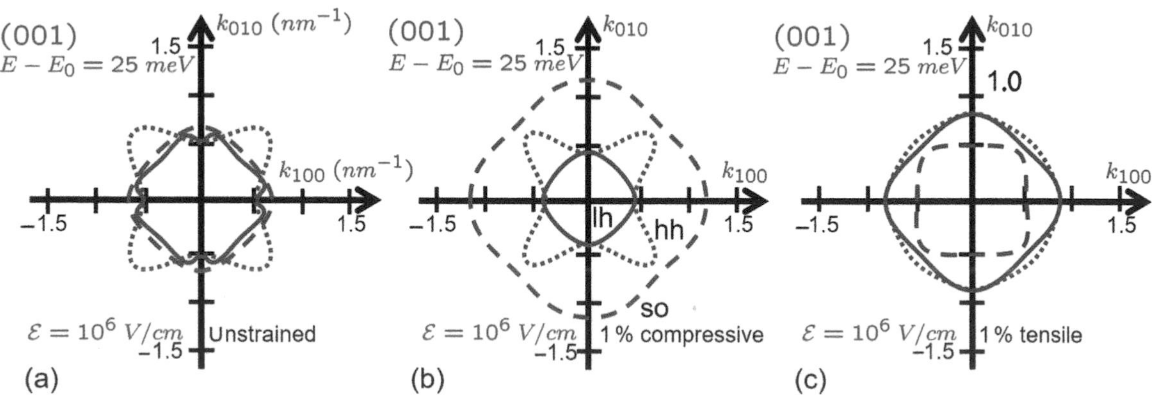

Figure 17.19: Equienergy surfaces for valence bands on a (001) surface 25 meV below the subband edge under unstrained (a), biaxial 1 % compressive and 1 % tensile stress at an electric field of 10^6 V/cm in the oxide at the surface. This corresponds to a charge density of 2×10^{12} cm^{-2}; hh, heavy hole; lh, light hole. Adapted from M. V. Fischetti, Z. Ren, P. M. Solomon, M. Yang and K. Rim, "Six-band $k \cdot p$ calculation of hole mobility in silicon inversion layers: Dependence on surface orientation, strain and silicon thickness," Journal of Applied Physics, **94**, 1079–1095 (2003).

of charge) in the inversion region at the surface at an energy 25 *meV* below the lowest subband energy.

As in Figure 17.14, one can see the high anisotropy and the ⟨110⟩-directed large mass of the heavy-hole band in Figure 17.19(a). This figure also shows that, with the lowest subband 25 *meV* below this energy, light-hole states and split-off states are also occupied. The light hole also shows the coupling consequence that exists in the ⟨100⟩-directed momentum states. Under compressive conditions, the technologically preferred condition for improvement of electron mobility, much of the symmetry is still maintained, the split-off band has become wider and the light and heavy holes have shrunk as their energy got raised. In the biaxial tensile conditions shown in Figure 17.14(c), all these three bands stretch out to about a similar length, with the split-off a bit smaller, and the effective mass is reduced. Biaxial tensile conditions are therefore preferable for hole mobility. Hole mobility improves for compressive conditions too, but, for tensile conditions, this improvement is considerably more significant. It can be as much as a factor of 2 to 3 times higher than for unstrained conditions.

Now consider the [110]-oriented uniaxial strain. If it is compressive, the Δ_2 valleys respond in a manner opposite to that for biaxial tensile stress, since the separation of the out-of plane direction responds in an opposite manner. This is not too conducive to mobility improvement, but the breaking of degeneracy and splitting is helpful. The improvement is not significant. Uniaxial compressive stress does improve characteristics significantly. This is seen in Figure 17.20, where panel (a) shows the unstrained equienergy valence surface on (001) silicon—heavy hole, light hole and split-off included—while panel (b) shows the changes when a uniaxial stress of 2 *GPa* is applied. The four-corner star pattern has now been foreshortened in the direction of carrier travel. The conductivity mass is also lower. The mobility will increase, especially with the

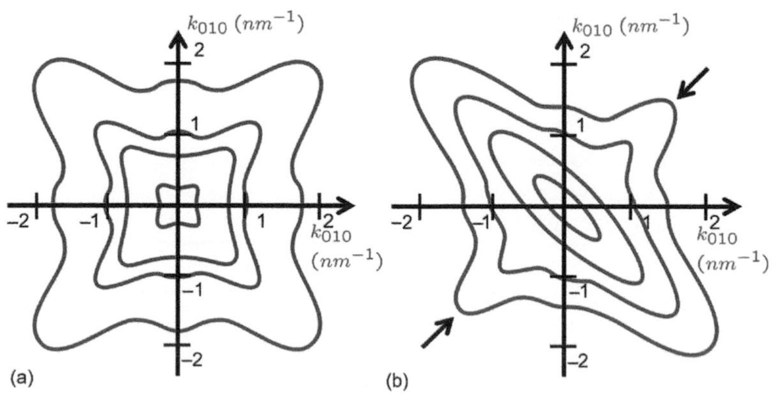

Figure 17.20: Part (a) shows the unstrained constant valence energy contours in the (001) plane. Part (b) shows changes arising as a result of a 1 *GPa* uniaxial compressive stress.

decreased scattering due to degeneracy lifting. A strain of 0.01 is sufficient to enhance hole mobility by nearly a factor of 3 over unstrained conditions.

We now put all this discussion together for electrons and holes for the two stress conditions that we have emphasized. An approximate description with the correct trends of mobility is shown in Figure 17.21 for *Si*. It is approximate, since mobility calculations are subject to all the deformation and other parameters assumed in any calculation of scattering and for experimental data being subject to technological artifacts. One can see in this that hole mobility improvement—the characteristic most desired, since holes are slow—is quite significantly affected by compressive uniaxial stress.

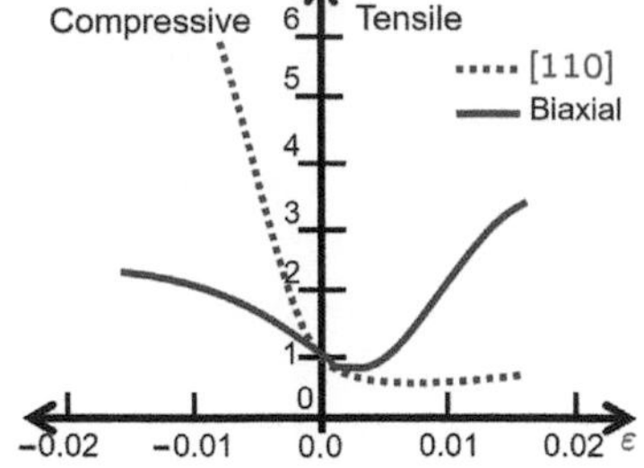

Figure 17.21: The fractional increase and decrease in mobility for holes under biaxial and [110] uniaxial stress for (001) *Si* as a function of strain.

17.4 *Strain with compositional consequences*

High-frequency bipolar transistors employ vertical transport and, to achieve high frequency, employ quasifields selectively affecting the transport of electrons. This is achieved through compositional change, particularly in the short base region, where high *p*-base doping and fast electron transport is desired. We therefore discuss the [001]-directed transport in the presence of strain and compositional change. It is the transport of a minority carrier (electron) in a sea made up of a heavily doped base with opposite polarity.

The reader should follow the discussion of heterostructure bipolar transistors—*SiGe* being of interest here—in S. Tiwari, "Device physics: Fundamentals of electronics and optoelectronics," Electroscience 2, Oxford University Press, ISBN 978-0-198-75984-3 (forthcoming), for an in-depth understanding of the dependence of the behavior of the device on the transport in the film as also that of the junction when heterostructures are employed. These are intimately connected.

First, consider the consequences when an abrupt junction is made between *Si* and *Ge*. Strain will exist, so the description given in Chapters 4 and 6 (see Figures 6.5 and 6.13) must change to account for the strain. *Si* and *Ge* have quite different forms spin-orbit splitting, which affects the valence bandstructure, and their conduction valley minima are of quite different symmetries. The nearly ideal description of the $Ga_{1-x}Al_xAs/GaAs$ junction where strain is nearly non-existent does not hold. Figure 17.22 shows the discontinuities for (001) surfaces when one or the other material is strained. It is a very contrasting picture. When *Ge* is strained and on unstrained *Si*, its Δ_4 minima nearly coincide with the *L* minimum. We have already discussed why this *L* valley degeneracy is not broken. Also, since *Ge* is now compressive strained, the Δ_4 valleys are lowered, and the Δ_2 valleys raised—quite like what happens with *Si* when it is compressed. This follows from the overlap and symmetry arguments already considered. The bandgap of *Ge* has shrunk well beyond its unstrained value. The valence bandedge discontinuity is much larger than the conduction bandedge discontinuity. Of course,

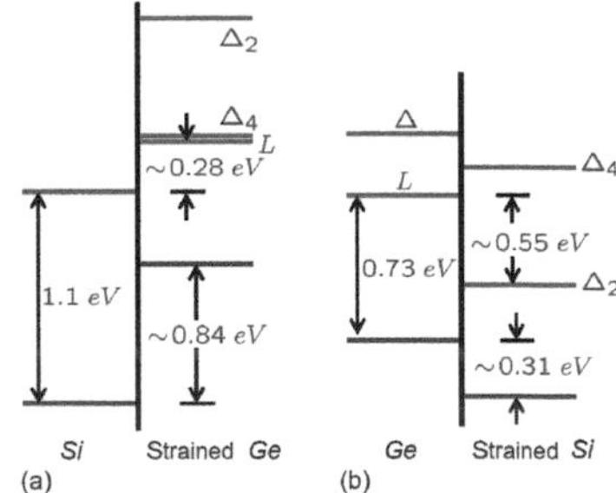

Figure 17.22: Band discontinuities in the presence of strain at the *Si*/*Ge* interface. This is a theoretical picture showing the result of changes in symmetries of the reciprocal space, the spin-orbit splitting and the consequences of immense strain in a system with a nearly 4 % strain at the interface.

the structure has extreme strain and, looking at Figure 17.5, one sees that it is quite beyond the limits of applicability of the theory of metastability. So, consider the discussion as a gedanken exercise for understanding the consequences of strain on bandstructure in materials with very different symmetries in reciprocal space. When strained *Si* exists on *Ge*, now the conduction band valley shifts are reversed. Both the conduction bandedge and the valence bandedge in *Si* are lower than in *Ge*, and the bandgap of strained *Si* is actually lower than that of *Ge*. These abrupt changes at the interface when the composition changes are sufficient to create a variety of interface confinement effects in the $Si_{1-x}Ge_x/Si$ or $Si_{1-x}Ge_x/Ge$ system. However, these properties are not significantly improved over that of insulator structures, since sub-eV discontinuities are too small to limit transverse conductivity in field-effect transistors. The strain in a film, on the other hand, provides much more significant improvements.

This change in the location and character of different valleys is shown in Figure 17.23 across the range of molefractions of the compositional mix on unstrained *Si* and unstrained *Ge*. The bandgap of $Si_{1-x}Ge_x$ may be varied all the way from the $\sim 1.1\ eV$ bandgap of *Si* to $\sim 0.5\ eV$, again in *Si*. The first is unstrained. The second is extremely tensile biaxial strained on *Ge*.

For the pseudomorphic thicknesses shown in Figure 17.5, for $Si_{1-x}Ge_x$, which can handle the elastic strain, these two figures—Figure 17.22, which shows the discontinuity, and Figure 17.23, which shows the change in the bandgap—demonstrate what becomes possible in a bipolar transistor employing heterostructures, a.k.a. compositional changes. In bipolar transistors, the need for a high base doping in a thin base for low base resistance and short charging times while also maintaining gain by suppressing the carrier injection from the base into the emitter as well as

For a detailed discussion of the design, attributes and relationships of various changes that one can achieve in a bipolar transistor using strain, see S. Tiwari, "Device physics: Fundamentals of electronics and optoelectronics," Electroscience 2, Oxford University Press, ISBN 978-0-198-75984-3 (forthcoming).

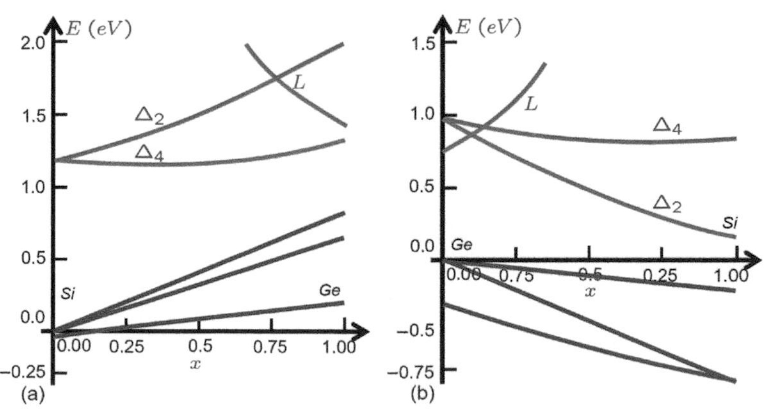

Figure 17.23: The bandedge changes for the conduction band and valence bands in the (*Si*, *Ge*) system for a (001) surface. In (a), a strained $Si_{1-x}Ge_x$ is located on unstrained *Si*. In (b), strained $Si_{1-x}Ge_x$ is located on unstrained *Ge*. After C. G. Van de Walle and R. M. Martin, "Theoretical calculations of heterojunction discontinuities in the *Si*/*Ge* system," Physical Review B, **34**, 5621–5634 (1986).

maintaining gradual changes that avoid carrier pile-up anywhere, as well as having useful breakdown voltages, requires a judicious use of compositional changes and choices of compositions in the transition regions of the emitter-base junction and the base-collector junction, as well as across the base. The practical constraints of thickness, doping and the compositional changes of $Si_{1-x}Ge_x$ allow considerable improvements in the bipolar device behavior because of the properties that we have discussed.

17.5 *Summary*

THE NANOSCALE, WITH ITS DOMINANCE OF SURFACES AND INTERFACES—in the case of devices, hordes of them—will have strain built in. It will also be non-uniform if caused by heterogeneity of materials. Even with very close matching of lattice constants in the *AlAs/GaAs* interface (or a mixed composition of $Ga_{1-x}Al_xAs$ instead of *AlAs* as is usual), there will be a very minute stress and strain. *AlAs* has a minutely larger lattice constant at 300 *K*. If the materials are very different, with even amorphous or polycrystalline interfaces to a single crystal, then there will be stress resulting from the propensity toward pseudomorphism for deposited material, the different energetics for deposition and growth, and the expansion coefficients as temperatures are changed. Changes in the unit cell dimensions of a crystalline structure will lead to bandstructure changes. Bandstructure changes lead to the changes in inherent velocities associated with the Bloch states in different directions, and also to changes in scattering, since occupied states, unoccupied states and the coupling between them will change. Particularly in *Si*, intentional introduction of strain has been an important transistor design tool. So it has been for many varieties of compact lasers where desired wavelengths become possible, and the non-radiative Auger recombination mechanisms suppressed.

We established—in the continuum assumption—the stress-strain description of semiconductors, where four parameters suffice. In general, the strain tensor can be split into three components. The first of these is a diagonal response, so it is uniform in the three directions and is hydrostatic. The second and the third are due to shear stress; the second has diagonal terms—is uniaxial—and is a dilation for positive stress, while the third term has no diagonal elements and is a rotational distortion. Biaxial strain, as occurs when a film is grown in lattice conformity on a substrate, can be split into a hydrostatic strain and a uniaxial strain. The third term of our expansion is absent. When a uniaxial stress is present—which is possible in transistors through a change in the gate length direction, which is usually much shorter than the gate width direction—it distorts the crystal with an in-plane strain whose behavior will depend on the orientation of the stress. Using $Si_{1-x}Ge_x$, pattern-induced stress and growth of pseudomorphic films for both the compound and the single element semiconductors are common ways of introducing strain.

Hydrostatic strain, by changing the lattice uniformly in all directions, has a direct effect on modulating the bandgap. Compressive strain will reduce the lattice parameter and increase the bandgap. Shear stress changes symmetry. In general, by splitting it into the two tensor forms (diagonal and off-diagonal), depending on the form of the stress, one will observe different consequences. A uniaxial stress along the cube's axes doesn't cause the rotation, but one along $\langle 110 \rangle$ does.

The conduction band effects can be seen through the changes in the bandgap and the changes in the band minimum. It is of considerable importance in *Si*, where there are six equivalent conduction band minima (along Δ), two of which (Δ_2)—orthogonal to the surface—will have one type of symmetry, with the other four (Δ_4) having another symmetry. If a biaxial stress exists—in a grown film—between the shift in the bandgap, with the valence bandedge as the reference, due to the hydrostatic component, and the shear consequence, the bandedges of the valleys will shift in opposite directions, due to the symmetries. For zone-centered *GaAs*, this complication is absent, as it is for *Ge* on a (001) surface, since the L valleys are equivalent. On top of this, if there is confinement, further shifting will take place in the conduction band energies. If a uniaxial stress exists, again there is a consequence in the distortion that reflects hydrostatic stress and shear stress, with a change in the bandgap and a splitting of the conduction bandedges for *Si*, but this time also for *Ge* if the stress is on the (001) plane.

The valence band too shows consequences, in many ways more pronounced, because the different valence bands have very different symmetries arising in their $|s\rangle$ and $|p\rangle$ constitution. Valence bands are very anisotropic, and the light- and heavy-hole bands are degenerate at the zone center in unstressed conditions. With stress, the band warping changes and the degeneracy is lifted. These changes, the response in the different directions, and the interactions between the bands need to account for the spin-orbit interaction together with the unit cell changes. Lighter holes appear at the bandedge without a simultaneous presence of heavy holes. In addition, confinement effects and the occupation of different band minima can change significantly from the unstressed conditions.

A further twist on these conduction and valence band energy changes is the introduction of compositional changes where the bandstructure itself is being changed by the atoms populating the distorted lattice. The conduction band minimum will also change from a Δ minimum to a Ł minimum in unstrained conditions. Now one has to account for compositional consequences. But if a strained $Si_{1-x}Ge_x$ is grown on *Si* and is still a single crystal, the *L* valley never appears as a minimum. If *Ge* is introduced into a *Si* crystal, Δ_2 valleys rise, and so do Δ_4 valleys but only after dropping initially up to about half mixing under the strained conditions.

Finally, the consequence of these electronic state changes, a next-order effect in the state changes due to confinement in inversion layers, as well as to the scattering interactions and the masses of relevance for conduction and state occupation, is the effect on the transport behavior of the electrons and holes. Strain generally improves transport, principally by suppressing scattering and improving the conduction mass. Compressive stress for in-plane hole transport is most conducive for improvement.

17.6 Concluding remarks and bibliographic notes

UNDERSTANDING STRAIN ASSUMED TECHNOLOGICAL importance with the arrival of nanoscale geometries and as transistor current drive capabilities' improvements became marginal with reduced dimensions due to scattering.

For those who have a preliminary understanding of group theory, an excellent early book, discussing strain starting from a discussion of symmetries, is by Bir and Pikus[1]. This book is an excellent source for approaching solid-state theory from a group theory perspective, group theory having been one of Wigner's and Weyl's important contributions. The book also is a good introduction to the theory of invariants, one of Luttinger's major contributions, and showing the consequences of strain is an excellent place to show the consequences of changes in symmetry and, from them, the invariances.

For strain discussion in compound semiconductors, an early reference is by Pearsall[2]. The chapter by Kasper and Schäffler is a good summary of the continuum mechanics relationship for determining strain in nanoscale and thicker films. A standard reference for the theory of elasticity is by Timoshenko and Goodier[3].

A set of good discussions related to *Si* and its implications for transistors is by Sverdlov[4] and by Sun, Thompson and Nishida[5].

[1] G. L. Bir and G. E. Pikus, "Symmetry and strain-induced effects in semiconductors," John Wiley, ISBN 0-7065-1367-3 (1974) (English translation from the Russian original published by Izdatel'stvo "Nauka")

[2] T. P. Pearsall (ed.), "Strained-layer superlattices: Materials science and technology," and semimetals, **33**, Academic, ISBN 0-12-752133-X (1991)

[3] S. Timoshenko and J. N. Goodier, "Theory of elasticity," McGraw-Hill (1951)

[4] V. Sverdlov, "Strain-induced effects in advanced *MOSFETs*," Springer-Verlag, ISBN 978-3-7091-0381-4 (2011)

[5] Y. Sun, S. E. Thompson and T. Nishida, "Strain effects in semiconductors," Springer, ISBN 978-1-4419-0551-2 (2010)

The former concentrates more on devices; the latter has a more robust discussion of bandstructure techniques.

An advanced book discussing the details of the bandstructure changes is by Fischetti and Vandenberghe[6]. It is particularly complete in its spin-orbit and strain the valence bandstructure, and the implications under confinement.

[6] M. Fischetti and W. G. Vandenberghe, "Advanced physics of electron transport in semiconductors and nanostructure," Springer, ISBN 978-3-319-01100-4 (2017)

17.7 *Exercise*

1. The split-off hole band in *Si* is only 0.044 eV below the valence band maximum, while that in *GaAs* is 0.34 eV below, a factor of ~ 10 different. Can you think of any intuitive reason of why this may happen? **[S]**

18
High permittivity dielectrics

HIGH PERMITTIVITY DIELECTRICS are of immense importance for many semiconductor devices: as a dielectric for field effect, as a capacitor in dynamic random access memories, in the "metamaterials," and even as semiconductors themselves. Permittivity is a characteristic reflective of the polarization of the material. We started a discussion of the dielectric function—permittivity—in Chapter 10, particularly in the sections on phonon interactions when the atomic motion is in a polar crystal. Fröhlich interaction—electron and polar optical phonon interaction—was a major consequence. Bound and free charges both respond to the electric field, and therefore the material has a polarization response that is characteristic of how this charge—the dipole—responds. Outer electrons and the ion—bound electrons with the nuclear core—respond differently. This has to be a function of frequency, since ions are slow in response, quasi-bound electrons faster and free electrons even faster. So, nearly free electrons, as in metals or, under conductive conditions, semiconductors—particularly in the high electron density inversion layer—exhibit the free electron response, as well as the electronic ionic dipole response embedded in the dielectric response modeled in Section 3.11 for phonon-field interaction. An insulator has an electronic response (the quasi-bound electrons of the valence band from the outer shell of the atoms) and an ionic response. The response can be quite complex, since interactions exist between multiple energy-storing excitations.

Polarization arising in mechanical strain leads to interesting consequences seen in piezoelectric materials, and the effect of spontaneous polarization is seen in ferroelectrics. Superconductivity is another feature in this multibody interaction. In organic materials, such as those composed of short or long molecules, the permittivity has additional frequency-dependent features that follow from the short- and long-range interactions of dipoles and radiation. We will

Permittivity is the proportionality relating electric field to electric displacement. It characterizes the tendency of the charge in the material, in the presence of an electric field, to distort; that is, for polarization to happen. A larger charge distortion—polarization—leads to higher permittivity. It is a measure of permitting less of the electrical field—so, not permittivity but negpermittivity, just as with entropy and negentropy à la Shannon. Permittivity $\epsilon = \epsilon_r \epsilon_0$. $\epsilon_0 = 8.854 \times 10^{-14}\ F/m$ is the permittivity of free space, where there exists no polarization. ϵ_r is the relative permittivity. It is a constant if the field is static, so a dielectric constant. A time-varying field causes ϵ to change with frequency or wavevector. So, in electrodynamics, permittivity and relative permittivity are now functions—hence the term "dielectric function" as a general description. Engineering literature also employs the symbol κ for ϵ_r. "Metamaterials" are physically sculpted—engineered—materials where interactions due to periodicity and at interfaces create unusual properties. Negative index of refraction, for example, in a frequency band, or bandgap for electromagnetic waves are two examples. A simple example is of quarter-wave stacks as mirrors through the destructive interference of normal incident electromagnetic waves. Other examples include those arising from charge-electromagnetic wave interaction. Semiconductors have a dielectric constant that is larger than that of SiO_2 and so, in quite a

Semiconductor Physics: Principles, Theory and Nanoscale. Sandip Tiwari.
 Published 2020 by Oxford University Press. DOI: 10.1093/oso/9780198759867.001.0001

restrict ourselves to inorganic high permittivity dielectrics as well as to semiconductors that too have a higher permittivity.

Use of high permittivity dielectrics with semiconductors became technologically of interest when the scaling of field-effect transistors reached a point where nitrided SiO_2 as a gate dielectric became thin enough that tunneling, as well as the variety of reliability consequences due to high electric fields and tunneling-induced defect generation, became strongly limiting. High permittivity dielectrics in such circumstances, even if their conduction and valence barriers may be lower compared to those of the SiO_2/Si interface, can be useful so long as carrier transport properties in the channel remain reasonable because they allow a thicker dielectric that restricts current while allowing for comparable mobile charge control.

The suppression of tunneling current while maintaining similar channel charge control follows from the relationship between displacement and charge as seen in Maxwell's first equation, and because currents by tunneling are exponentially related to the negative of the thickness through the accumulated wavevector of evanescing wavefunction. A higher permittivity dielectric permits a higher thickness and suppresses tunneling current, and yet the displacement can be maintained, so the mobile charge can be effectively controlled. These arguments have numerous caveats, since technologies and nature place constraints. One constraint, for example, is that SiO_2 is a very "chosen" oxide—stable, and preferred in equilibrium and kinetically when oxygen is present, due to favorable free energy—so there exists, along with high permittivity dielectrics, a small, interfacial, SiO_2-like film a film that is not quite bulk-like SiO_2 but which affects the charge control. Fortunately, one useful result of this natural preference is that the interface state density is lower than what it may have been had a high permittivity dielectric been atomically adjacent. Another is that it suppresses the interaction between the high permittivity dielectric's optical phonons with the electrons of the channel. A similar issue of compounding with favorable phase formation also arises between the gate, which is often replaced by a metal or its silicide to decrease resistance, and the high permittivity dielectric. We will ignore the technological complexity of these material combinations, which has a way of evolving over time, and focus on the fundamentals of the interactions in this chapter and the next.

In this chapter, this prelude specifically leads us to a discussion of the nature of these permittivities, and since they are the consequence of polarization, how the phonons of such materials and of

few respects, are high permittivity dielectrics. Materials where phase transitions due to some impressed energy changes properties such as of permittivity enormously—for example, the metal-insulator transition—are also of engineering significance. Ferroelectrics are another example where there is both high permittivity and hysteresis. See S. Tiwari, "Nanoscale device physics: Science and engineering fundamentals," Electroscience 4, Oxford University Press, ISBN 978-0-198-75987-4 (2017). Here, we stick with just a plain vanilla use and understanding of the higher permittivity with semiconductors.

A nitrided SiO_2 is a dominantly SiO_2 film grown from an Si substrate, but where a fraction of a percent of nitrogen is incorporated through the different processes that are employed to grow the oxide. It is a robust film that strongly barricades the diffusion of various species through the thin film. Among these diffusing species of interest are the B, H, variety of metals that are used in slicides, contacts, et cetera.

Si interact. In the following chapter (Chapter 19), we will continue on this theme by looking specifically at remote scattering processes and how they affect interface charge dynamics. We will also largely limit our discussion to electron transport, although, certainly, holes are just as important in usage. Holes have poorer mobility, and the scale of the effects that we discuss will be usually lower due to the large scattering already pre-existing and the heavier masses involved.

18.1 Permittivity and the material's related characteristics

WE START WITH A LOOK AT THE RELATIVE PERMITTIVITY, that is, the dielectric constant, of a variety of dielectrics that are potentially compatible with semiconductors or are themselves semiconductors. Our discussion is first to draw out the consequences of the ionicity arising in the quantum-mechanical nature of the bond, the polarization changes that result from the nature of the bonding, and, from these, how the electronic bandstructure changes and how all these different properties are correlated with each other. The broad outline of these connections is shown in Table 18.1, which includes some oxides and silicates, so *Si*- and *O*-containing, relatively stable compounds that have an inclination toward "silicon compatibility." We have also chosen dielectrics up to a relative permittivity that is useful in field-effect transistors.

We have noted the higher permittivity of the common semiconductors tackled in this text. $SrTiO_3$, $LaMnO_3$, their variety of mixed compounds, such as $SrRu_xTi_{1-x}O_3$ and $LaCu_xMn_{1-x}O_3$, and several others such as VO_2 exhibit both high permittivity and interesting electron correlation consequences. An insulator can become conducting through bandstructure change due to mechanical or electrical stress. These materials have a small-enough bandgap—semiconductor-like—and have a very unsemiconductor-like transport. See S. Tiwari, "Nanoscale device physics: Science and engineering fundamentals," Electroscience 4, Oxford University Press, ISBN 978-0-198-75987-4 (2017), for such phase transition examples.

Table 18.1 shows the cation electronegativity, the low-frequency relative permittivity, the bandgap and the low wavevector energy of the two optical phonon energies that are relevant. We have included the dioxide and nitride of *Si*—quite common dielectrics formed using *Si*—for comparison. The oxides, and another oxygen-containing set of compounds, the silicate and multi-element oxide, are included. The list is long, and could be longer, but, in practice, the high permittivity material that has practically worked the best—w.r.t. defects, sensitivity to processing, long-term interactions and general reliability—combining process technology and practical device needs, is HfO_2, and that is the one that our discussion will largely encircle. Table 18.1 is instructive, since it shows a number of correlations that are grounded in reasonable cause.

To explore the device argument as to why only a certain limited increase in permittivity is useful—this draws on scaling and on the electrical and physical dimensionality argument—see Section 2.3 in S. Tiwari, "Nanoscale device physics: Science and engineering fundamentals," Electroscience 4, Oxford University Press, ISBN 978-0-198-75987-4 (2017).

The electronegativity of an element—there are various measures of it—quantifies the element's ability to attract a shared electron, compared to other elements with which it forms a compound. The Gibbs free energy of formation listed here is an estimated value at 1000 *K*—typical high temperatures encountered during device fabrication—for a reaction toward SiO_2's formation when in contact with *Si*. The phonon energies are for the optical branch of the

	Electro-negativity of the cation (eV)	ΔG_f at 1000 K (eV)	Relative permittivity $\epsilon_r(0)$	$\epsilon_r(\infty)$	Bandgap E_g (eV)	Phonon energy $\hbar\omega_q$ (meV)
SiO_2	1.91		3.9	1.5	9.1	55.60 138.10
Si_3N_4	1.91		7.4		5.1	
Al_2O_3	2.75	1.61	∼9	3.4	8.8	48.81 71.41
$HfSiO_4$	Hf: 1.3		∼11		6.5	
Y_2O_3	1.22	5.06	∼15		6?	
HfO_2	1.3	2.06	∼25	4.0	5.8	12.40 48.35
ZrO_2	1.22	1.83	∼25	4.8	5.8	16.67 57.70
$BaZrO_3$	Ba: 0.89		∼26			
Ta_2O_5	1.5	−2.27	∼30	4.8	3.8–5.3	

Table 18.1: Pauling electronegativity of the cation, Gibbs free energy (per mole) at 1000 K for reduction to SiO_2 when in contact with Si, relative permittivity/dielectric constant of oxide, bandgap, and optical phonon energy of a select set of dielectrics.

transverse component and interact electromagnetically, as discussed in Subsection 10.2.2 for polar mode optical scattering.

Since electronegativity is the measure of an atom's propensity to attract electrons that are being shared, it is a measure of the tendency to move the electron density toward itself. It is affected by both the atomic number and how far the electrons are from the nucleus, as well as by the nature of the bond that arose—whether it is sharing among equals, as in a covalent bond, or due to a bond between atoms with unfilled or filled orbitals, which is tied to which column they are in the periodic table. *Si*, being from group *IV*, has a larger covalent tendency, *Al*, being from group *III*, will be more ionic, and *Ba*, being from group *II*, even more so. *Hf*, *Y*, *Zr* and *Ta* are all transition elements with an s^2d^i ground-state configuration with varying integer i for the number of outermost electrons in the d orbital, which is only partially filled.

Transition elements, because of d occupancy, and lanthanides and actinides, with f subshells—all of which spread out farther away from the nucleus—lead to many unusual properties as elements and compounds. There is a variety of phase transitions: ferroelectricity, piezoelectricity, ferromagnetism, superconductivity and others, including, quite likely, those not yet discovered, that arise in the subtle interplay of different energetic interactions at play in the condensed matter many-body state. Some of these are discussed in S. Tiwari, "Nanoscale device physics: Science and engineering fundamentals," Electroscience 4, Oxford University Press, ISBN 978-0-198-75987-4 (2017).

Electronegativity has many measures, Pauling's being the most common. Pauling's is a modification to the deviation of bond energy of a compound XY from a theoretical average of XX and YY, using 4 as the reference value for fluorine, which is the most electronegative element. Based on the difference between the electronegativity of the two atoms forming a bond, if the difference is below 0.5, the bond is usually treated by the chemistry community as non-polar covalent. If it is 0.5–2.0, it is polar covalent, and if it exceeds 2.0, it is ionic. This last case characterizes a large-scale transfer of the electronic charge from one atom to the other. O has an electronegativity of 3.44, and N has an electronegativity of

3.04. So, SiO_2 and Si_3N_4 are both polar covalent. But the transition element oxides and compounds are ionic. The pull of O, a highly electronegative element, is strong, making dipolar polarization significant. This is reflected in the change in the dielectric constant of these materials.

Another set of major consequences arise in the strength of the bonds. The bonds of SiO_2 and Si_3N_4, being more covalent, are stronger and have less polarization, but, also because of this bond strength, these materials have a larger bandgap, and their optical phonon energies are larger. Al_2O_3—a group *III-VI* compound—also has properties quite similar to their group *IV-VI* and *IV-V* counterparts in SiO_2 and Si_3N_4. In the transition element compounds, the ionic consequences become quite substantial.

The question of silicon compatibility is somewhat laden and charged. These high permittivity materials are deployed with an *Si* substrate and, because of their use with gates, often abut higher conductivity metal compounds (silicides, nitrides or perhaps even pure metals, with a discontinuity magnitude control through interface chemistry) that replace a very heavily doped polysilicon gate. This is achieved in the form of structures that undergo high temperature processes in a variety of ambients. In addition to changes within these films, there are reactions that become possible at the gate-side interface and the substrate-side interface. So, because of the presence of oxygen, there exists competition between the formation of SiO_2 and the stability of an oxygen-containing dielectric, as well as in the competition between the gate material—nitrides, silicides or others—and the gate dielectric, all involving interfaces with their enhanced propensity to a variety of reaction kinetics and changes in properties with implications for electronic control and electronic transport. The free energy in Table 18.1, if positive, indicates instability at the *Si*-side interface, that is, that the oxygen would prefer to bond with *Si*, and an interfacial oxide film is likely. The table implies that Ta_2O_5 is the only compound that has stability when in contact with *Si*. A similar free reaction energy evaluation is needed for the gate-side interface. It turns out, as mentioned earlier, that the substrate-side instability allows one to improve on interface states, and we will see in the discussion of remote scattering in Chapter 19, improvement in mobility, albeit at the cost of effective insulator thickness.

The relative permittivity listed here is the low-frequency value. In our discussion of atomic motion and Born-Oppenheimer approximation in Chapter 1, we had employed the adiabatic limit where the electron charge cloud could follow the residual atomic charge. But at high frequency, the mismatch in this motion leads to a change. This change in the polarizability at low frequency and high frequency was encountered in Chapter 10. We will return later to the connections between this frequency-dependent permittivity behavior, the phonon characteristics and its consequences for transport processes later in the chapter.

The next complexity of interest to us is the conduction and valence bandedge discontinuity, since it directly affects—exponentially, through the cumulative consequence of the tunneling electron's wavevector and the thickness—the current through these dielectric films. This detail of electronic bandstructure consequence can be seen in Figure 18.1, which shows the approximate conduction and valence bandedge discontinuities between these example materials and Si. These values should be treated as approximate for the following reason. The dielectrics are not in their crystalline stable phase, where much of the theory that we developed in Chapter 6 applies. And even if they were, they would be in a strained condition on the substrate. Materials deposited by various techniques will be amorphous or polycrystalline, with a variety of orientations. They will have plenty of defects, and, as they are thin, bulk-interface differences of what the electrons and holes see as their environment will also apply. Having said this, the predictive capability for what happens between these dielectrics and SiO_2 will be a bit more accurate because of the limited mixing of the conduction state and the valence band state at the interface at large bandgaps and large discontinuities. So, one may apply a transitivity relationship using this Si reference with some confidence. The magnitude of this discontinuity is best left as

See S. Tiwari, "Device physics: Fundamentals of electronics and optoelectronics," Electroscience 2, Oxford University Press, ISBN 978-0-198-75984-3 (forthcoming), and S. Tiwari, "Nanoscale device physics: Science and engineering fundamentals," Electroscience 4, Oxford University Press, ISBN 978-0-198-75987-4 (2017), for a discussion of quantum-mechanical tunneling and the nature of interfaces and of the electronic structure of the films.

Silicon, as we know, has a diamond lattice. The high permittivity oxides and oxide-containing compounds may be amorphous or polycrystalline. If crystalline, HfO_2 and ZrO_2, which form similar structures, appear in cubic, tetragonal, monoclinic or orthorhombic forms. Ta_2O_5 is orthorhombic. $HfSiO_4$ is tetragonal.

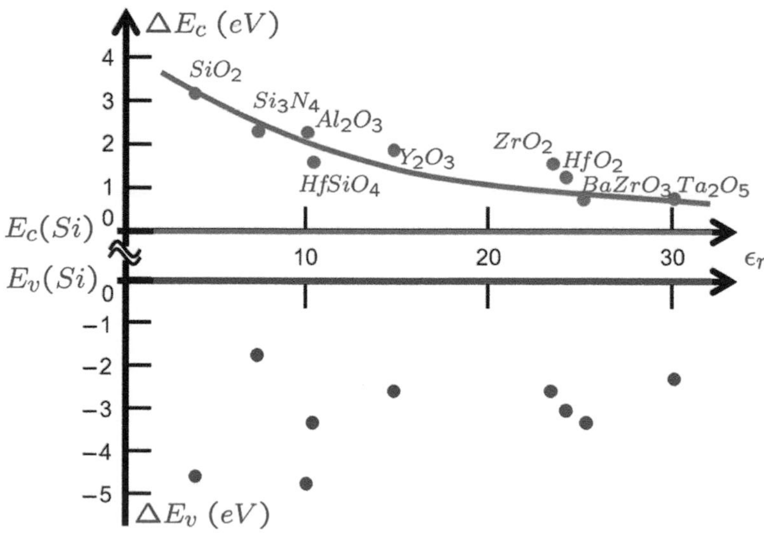

Figure 18.1: Example dielectrics of varying permittivity and their approximate conduction and valence bandedge discontinuities as a function of increasing permittivity. SiO_2 and Si_3N_4 are reference dielectrics. Al_2O_3 is the only other dielectric here that is s-p bonded.

an experimental parameter, given the shortcomings of the theory and the theory's ability to mimic a variable experimental condition.

For electrons, what is of interest is the variation one sees in the conduction band discontinuity that one sees as a function of permittivity. As the permittivity increases, the conduction discontinuity of the dielectric to Si decreases. For HfO_2/Si, it exceeds the bandgap of Si. For SiO_2/Si, it is about 3 times the bandgap. Discontinuity changed by about a factor of 3 and so did the permittivity. The tunneling current approximately relates exponentially through the integrated wavevector with position. Wavevector is also related as a square root of the tunneling electron's energy separation from the conduction band potential. So, thickness matters linearly, while energy separation is the square root in the exponential's input. The permittivity change allows a direct equivalent change in the dielectric's thickness without loss of charge control. So, if tunneling current is the property at stake—is excessive—then permittivity change is helpful. For gate insulator tunneling, this increase in thickness becomes useful. Via an increase in physical thickness through higher permittivity, it is possible to suppress tunneling current at equivalent displacement in the structures.

Often, one will find a fair contribution from defect-mediated current in thin dielectrics, which tends to worsen during usage because of electrical stress.

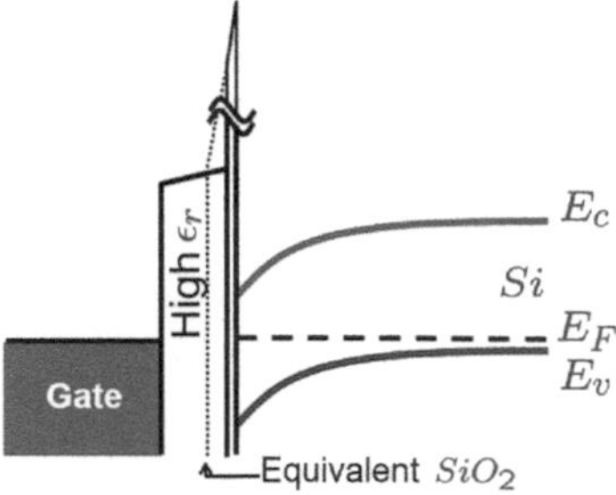

Figure 18.2: Band diagram at a metal/high permittivity/interface oxide/Si interface in thermal equilibrium that approximately mimics an HfO_2 high permittivity dielectric and a p-type substrate. The dotted line shows an approximate electrically equivalently thick SiO_2.

Figure 18.2 shows a cartoon of a metal/high permittivity/interface oxide/p-type Si band picture where HfO_2 is the representative dielectric and an interface SiO_2 is present. Also plotted is an equivalent oxide thickness for similar charge control electrostatics; that is, displacement. The oxide equivalent thickness is significantly smaller following the thickness-tunneling-displacement argument, and the higher permittivity is helpful in suppressing tunneling current for similar mobile charge control so long as the various circuit and usage-related needs of a device can be achieved.

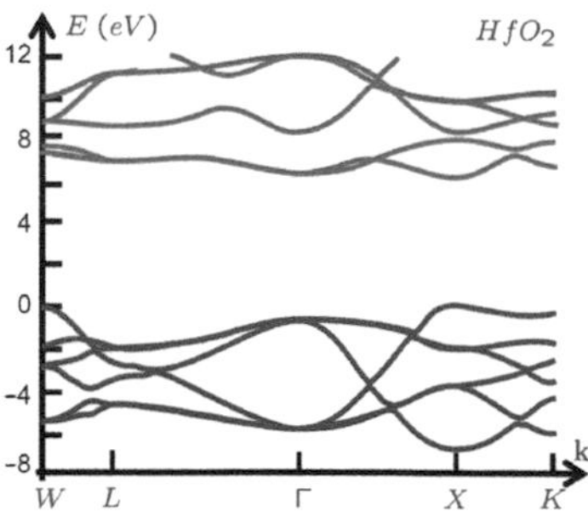

Figure 18.3: Bandstructure of HfO_2.

We illustrate the bandstructure of these dielectrics, whichs have to have an acceptably large bandgap for usefulness through that of HfO_2, which is the most common high permittivity dielectric. This is shown in Figure 18.3. ZrO_2 is a dielectric with quite similar permittivity and band discontinuity characteristics. Its bandstructure is also quite similar to that of HfO_2. Both have a bandgap a bit larger than 5 eV. The conduction band minimum is at the zone center—the Γ point—but, for the first time, we now encounter a valence band maximum at the X point. Recall our discussion in Chapter 4, where we used $|s\rangle$-state- and $|p\rangle$-state-mixing-based arguments to emphasize the ubiquity of the valence band maximum being at the zone center, albeit with a variety of light-hole, heavy-hole and split-off hole band consequences arising in their localization near the core.

The transition metal oxides are different because they are not *s-p* bonded; they have *s* and *d* orbitals as the outermost orbitals in the metal, and they are strongly ionic, with a large transfer of the electron charge density from the transition metal to the oxygen. In an ionic crystal, with a large-scale transfer of charge, and missing symmetric balancing over the volume, the valence band maximum can be away from the Brillouin zone center. $BaZrO_3$, on the other hand, does have a valence band maximum at the zone center with a very large mass. The valence band maximum doesn't really change much across the entire Brillouin zone, except for a small region around the X point.

So, valence band maxima do not have to be at the zone center. They are at the zone center in the group *III-V* and group *IV* compounds that we have discussed, due to the symmetries, including those of the basis states.

$BaZrO_3$ has a perovskite structure. Perovskites are extremely interesting as a result of the combination of an element with electrons farther away from the nucleus (as with *Zr*) placed together group *II* and group *VI* elements that have strong electronegativity. Perovskites show a variety of interesting phase transition properties that are discussed in S. Tiwari, "Nanoscale device physics: Science and engineering fundamentals," Electroscience 4, Oxford University Press, ISBN 978-0-198-75987-4 (2017).

We should make a few remarks regarding the high permittivity-gate interface to contrast that with the considerations of the high permittivity *Si* interface. Short length gates also place a large premium on the conductivity of the gate. Degenerately doped polysilicon, the common form of gate, when the gate length is many decades of nm, even at the highest dopings, will have a depletion region length of the order of a unit cell—half a nm—when the channel is inverted. So, the controlling potential is now further separated from the mobile charge. This depletion can be largely eliminated if one can employ much more conductive metals or their stable compounds for gates. Metals, for example, have typically an electron concentration of $10^{22}\ cm^{-3}$. This is to be compared to the mid-$10^{20}\ cm^{-3}$ that one may achieve with doped polysilicon. The difficulty is that such a gate must be able to achieve the threshold voltages desired, and it also must be immune to reactions with the high permittivity dielectric that one may employ. Practically, several nitrides—those of *Ti* and *Ta* being the most common—provide quite a stable metallic compound. But these metal gates are largely mid-gap materials, that is, that they have a Fermi energy that aligns to the middle of the *Si* bandgap. So, the interface of these materials needs to be tailored to permit the threshold voltage control through a suitable change in the barrier that the high permittivity dielectric has at the interface. We will not dwell on this subject; suffice it to say that technology at these dimensions and suitable harnessing of the reactions in the making and long-term usage of such structures is a difficult challenge.

18.2 Soft phonons

INSULATORS ARE "SEMICONDUCTORS" with a large bandgap. Permittivity arises from the polarization of the material ($\mathbf{D} = \epsilon_0 \boldsymbol{\mathcal{E}} + \mathbf{P} = \epsilon_0 \epsilon_r \boldsymbol{\mathcal{E}}$). For inorganic insulators, this relative permittivity arose from the ionic and electronic polarization. The electronic interaction is also reflected in the bandgap. This common origin shows up in the electronic polarization, averaged over the Brillouin zone, being inversely proportional to the bandgap squared.

Since a useful high permittivity dielectric is desired to be an insulator, there is a minimum requirement for the bandgap, so this electronic polarization needs to be small and is really not available as a tool to achieving high permittivity. Ionic polarizability is. So, this increase in permittivity is related to the effective charge on the ions. In our discussion of polar mode interaction (Section 10.2.5), we evaluated the Born effective charge e^* reflecting ionizability's consequence on the perturbation term for the interaction. This ionization polarizability is present under static electromagnetic conditions and absent at the high frequency, and it varied as a square root of $[1/\epsilon_r(\infty) - 1/\epsilon_r(0)]$ (Equation 10.42). The ionic polarizability's frequency-dependent consequence is reflected in the net polarization at high frequencies as

High frequency—written as ∞—is meant to reflect a frequency beyond the optical phonon frequencies, so *PHz*.

$$\mathbf{P}(\infty) = \left[1 - \frac{\epsilon(0)}{\epsilon(\infty)}\right]\mathbf{D} = \epsilon(0)\left[\frac{1}{\epsilon(0)} - \frac{1}{\epsilon(\infty)}\right]\mathbf{D}. \tag{18.1}$$

Ionic polarization arises in the longitudinal optical phonons, and if it is the only cause for change in polarization—if nearly free electrons were present, they too will have an effect, but we are assuming a perfect insulator—then the perturbation potential is caused by the ionic polarization. We initially explored ionic polarization in Section 3.11 with the expression of Equation 3.113 summarizing the dependence on Born effective charge. In our discussion of absorption by crystal vibrations (Section 12.4), in Equations 12.89 and 12.90, we directly expressed this charge in terms of the dielectric function in the form of a square-root dependence together with the radial frequency of phonons. Polarization is dependent through the Born effective charge in the square-root relationship of the difference in inverse of high-frequency and low-frequency permittivity. We know the density of these modes, and we may regard the expectation energy of the phonon modes as the zero point energy of those modes ($\hbar\omega_{LO}/2$). This led us to the perturbation energy of Equation 10.44. In these bulk conditions, for the phonon field, the unscreened Fröhlich field is

$$\mathcal{E}_q = \left\{\frac{\hbar\omega_{LO}}{2q^2}\left[\frac{1}{\epsilon(\infty)} - \frac{1}{\epsilon(0)}\right]\right\}^{1/2}. \tag{18.2}$$

The *LO* phonon in the bulk Fröhlich interaction has a scattering strength proportional to $\hbar\omega_{LO}\{[1/\epsilon(\infty)] - [1/\epsilon(0)]\}$. This bulk expression says that the bracket term is proportional to the square of the dipole field amplitude that arose in the ionic polarization.

In our high permittivity material, increasing the dielectric constant at the below-optical frequencies of electronics, it is the ionic polarization that is our tool. Larger ionic polarization requires more polarizability of the bond from the transition metal to oxygen. Such bonds are soft, and if the bonds are soft, then so are the optical phonons. This is to be contrasted with the case for SiO_2 and Si_3N_4, which are more covalent and have a stiffer bond. This difference in the nature of these bonds is reflected in the optical phonon energies shown in Table 18.1.

Extreme examples of ionic bonding are the group *I-VII* salts—*NaCl*, *KCl*, et cetera, which are all very polarizable and are soft enough that the bonds can be broken easily by dissolution in water

Mobility, as we have seen with semiconductors, is dominated by optical phonons at room temperature. The occupation of the phonon states at these energies is high enough so that absorption and emission are both relevant. For high permittivity materials, such as $La_{1-x}Sr_xTiO_3$ or $SrTiO_3$, where phase transitions are employed in using carrier transport, this mobility consequence will be of immediate relevance.

For devices based on metal-insulator phase transitions, see S. Tiwari, "Nanoscale device physics: Science and engineering fundamentals," Electroscience 4, Oxford University Press, ISBN 978-0-198-75987-4 (2017).

The question of interest to us is, what happens to the scattering as a result of this perturbation interaction when a dielectric is next to a semiconductor?

When a semiconductor, and we take Si as our prototype—with permittivity ϵ_{Si}—is adjacent to a dielectric with permittivity ϵ_{ins}, the coupling of the phonon modes in the two media via polar mediation involves surface optical modes. These surface optical modes, coupled transversely, have the frequency

$$\omega_{SO} = \omega_{TO}\left[\frac{\epsilon_{ins}(0) + \epsilon_{Si}(\infty)}{\epsilon_{ins}(\infty) + \epsilon_{Si}(\infty)}\right]^{1/2}. \tag{18.3}$$

This shows why TO modes become important. The scale length of the penetration of the coupling is related to permittivity through the charge-displacement relationship, and it has a square-root dependence through the relationship of the charge to the gradient of displacement. Surface optical modes (SO) have a coupling strength that is proportional to

$$\hbar\omega_{SO}\left[\frac{1}{\epsilon_{Si}(\infty) + \epsilon_{ins}(\infty)} + \frac{1}{\epsilon_{Si}(\infty) + \epsilon_{ins}(0)}\right], \tag{18.4}$$

a modification to the bulk Fröhlich expression. Equation 18.4 is applicable to the inversion layer of Si adjacent to SiO_2. It expresses a modification that arises in "image effect": that is, how the dipole field arising in the dielectric decays as it penetrates the inversion region at the interface.

Since the dielectric's properties are affecting the semiconductor's properties by being nearby, this is a nonlocal effect arising through coupling across the dipolar decay length scale. It is a "remote phonon scattering." In the SiO_2/Si system, this remote phonon scattering is vanishingly small for low energy carriers because of two reasons. The first reason is that ionic polarizability and the resulting coupling are quite small in SiO_2—its relative permittivity changes from 3.9 under static conditions to 1.5 at optical frequencies—since the bonds are hard. The second reason is related to the two optical phonons listed in Table 18.1. For phonons that are at 138 meV energy, the energy of the thermal electron is insufficient for phonon emission, and there are too few of these phonons to be absorbed by electrons. For phonons that have ~ 56 meV energy, it turns out that the oscillator strength is small. This latter is for the same reason that the permittivity of SiO_2 is low.

Si has an optical phonon energy of about 55 meV also, but it has a higher permittivity because of stronger oscillator strength.

For high permittivity insulators, the high-frequency dielectric response is via electronic polarization. At low frequencies, the dielectric response is largely via ionic polarization. So, while the high-frequency response of high permittivity insulators is not unlike that of SiO_2, the low-frequency response is very different. Since ionic strength prevails at low frequency, and a large permittivity exists, there is a large difference between low- and high-frequency magnitudes, and therefore a large scattering perturbation and a smaller *SO* phonon frequency. *The high permittivity goes together with a stronger ionic interaction, and it will have a stronger remote phonon scattering effect.*

We will discuss this consequence of high permittivity arising in ionic polarization, the soft phonons associated with them, and their resultant penetration farther away into the semiconductor in Chapter 19.

18.3 Summary

HIGH PERMITTIVITY DIELECTRICS of interest to semiconductors utilize ionic polarizability to achieve high permittivity. This ionic polarizability becomes possible mostly through the use of transition elements—elements with partially filled *d* orbitals—in the form of oxides. HfO_2 is a very common example. The *d* electrons stretch out farther from the nucleus, so the electron transfer to *O* is much greater, and the compound is much more ionic. It has a reasonable bandgap, so it works well as a insulating dielectric for field-effect control. The bandedge discontinuities for the conduction band and the valence band at an interface with *Si* also have magnitudes that are large enough to be useful. The high permittivity, combined with the presence of an interface oxide film, permits strong electrical control of the channel while managing tunneling current through the controlling film. We explored the variety of relevant properties, their origin and their implications for such dielectrics.

Since it is ionic polarization that is being employed, the bonds are soft. This means that the bandgaps are not large and the energy of the phonon modes is low: several 10s of *meV*; that is, of the order of thermal energy. So, these phonon modes can couple quite effectively. This coupling strength is related through the difference in the inverse of permittivity at high frequency versus that at lower frequency because of the dependence on polarizability or, equivalently, the Born effective charge characterizing the ionizability. The strength of the interaction is proportional to the *LO* phonon frequency and this inverse permittivity separation of low and optical frequencies. It is a strong interaction in high permittivity materials because of the large inverse permittivity separation. In covalent compounds, such as SiO_2, this interaction is very weak, because the permittivity doesn't change drastically and because either the phonon mode is very high in energy or it has low oscillator strength, as in the case of the second phonon mode. The high permittivity dielectric's strong ionic interaction connects perturbations over a longer length scale, that is, remotely, thus affecting carrier transport in *Si* that may be adjacent to the high permittivity material. With the presence of an interface oxide of *Si*, this remote effect is partially suppressed, and this allows a judicious use of the high permittivity dielectrics for scaling the thickness of the gate insulators in the transistors while limiting the rise in tunneling current flowing through the insulator.

In the following chapter, we will explore further the consequences of these perturbations and this and other local and remote interactions affecting electron transport.

18.4 Concluding remarks and bibliographic notes

THE PREPONDERANCE OF THE LITERATURE related to high permittivity dielectrics, and its use with *Si* is related to the practice and attributes of technology, and it becomes very subject to the techniques employed and their level of development. However, there are a number of references that the reader will find of relevance to explore the subject further.

A good primary reference is the book edited by Houssa[1]. A number of chapters in the text are of interest. J. Roberson and

[1] M. Houssa, "High-κ gate dielectrics," Institute of Physics, ISBN 0-7503-0906-7 (2004)

P. W. Peacock discuss the electronic structure and band offset. M. V. Fischetti, D. A. Neumayer and E. Cartier discuss the nature of electron mobility in high permittivity field-effect transistor systems. In addition, there are other articles that are related to the practical technology and device behavior, such as the issues related to reliability. These, however, I suspect are now dated because of their technological nature.

Another book is an edited collection by Howard Huff[2]. One reason for suggesting this reference is that there are a number of other perceptive peaks and predictions about the future from known people. It is always interesting to read what people thought in the past about the future, and especially why they thought that, and then compare it to what really happened.

[2] H. Huff, "Into the nano era," Springer, ISBN 978-3-540-74558-7 (2009)

Since this chapter has particularly stressed HfO_2 as a high permittivity dielectric, the review by Choi, Mao and Chang[3] is also suggested to those would like to obtain a practical materials-science-based discussion of the nature of the material, its interfaces and its compounding.

[3] J. H. Choi, Y. Mao and J. P. Chang, "Development of hafnium based high-*k* materials—A review," Materials Science and Engineering, **R 72**, 97–136 (2011)

We did not discuss conduction through the high permittivity insulators, although conduction through insulators was introduced summarily as a subject in Chapter 7. A good reference to explore it in metal oxide insulators is by Tsuda et al.[4], who discuss a broad range, from simple oxides to perovskites, and their variety of interesting properties, including that of conduction.

[4] N. Tsuda, K. Nasu, A. Fujimori and K. Siratori, "Electronic conduction in oxides," Springer, ISBN 3-540-66956-6 (2000)

18.5 Exercises

1. In high permittivity insulators with Si, such as HfO_2, why are surface optical phonons and their interaction with electron inversion layer so strong across the entire electron density range of interest in field-effect devices? **[S]**

2. It has been the LO polarization and not the TO polarization that we have stressed in exploring the high permittivity material's interaction with semiconductors. Why? **[S]**

3. An n-type semiconductor with an electron concentration n is terminated on a surface ($z = 0$) by an insulator spanning $z > 0$ of permittivity ϵ_{ins}. Take the long wavelength limit ($\mathbf{q} \rightarrow 0$) so that the dielectric function of the semiconductor can be written in terms of the static value as

$$\frac{\epsilon_{sem}(\mathbf{q}, \omega)}{\epsilon_r(0)\epsilon_0} = 1 - \frac{\omega_p^2}{\omega^2}.$$

At the interface, there will discontinuity in the orthogonal electric field, with

$$\epsilon_{sem}(\mathbf{q},\omega)\mathcal{E}_z(z=0^-) = \epsilon_{ins}\mathcal{E}_z(z=0^+).$$

With the electric filed as $\boldsymbol{\mathcal{E}} = \boldsymbol{\mathcal{E}}_0 \exp[i(\mathbf{q}\cdot\mathbf{r}+\omega t)]$, show that there is interface plasma oscillation with a frequency of

$$\omega_{SP} = \frac{\omega_p}{(1+\epsilon_{ins}/\epsilon_r(0)\epsilon_0)^{1/2}}. \qquad \textbf{[M]}$$

19
Remote processes

We separate the discussion of remote processes from that of high permittivity in order to emphasize the importance of interactions such as those arising in Coulomb, that is, charge-based, origins, and those arising in phonons, that is, coupling to modes of atomic vibrations, at the nano-dimensional device length scale. We recognize that fields penetrate and that there is a continuous exchange in the potential and kinetic energy forms of energy or of electrochemical energy across a system in thermodynamic equilibrium or off it, but, by and large, our relational description takes a local form. We also recognize that phonons—vibrations—also spread out. Optical phonons have a minimum energy, and electromagnetic modes couple to them through dipole interaction, and therefore field and dipole orientations matter. Acoustic phonons are essential to sound's propagation. Local fields, local carrier concentrations and local potentials define the nature of the excitation response—of electrons or phonons—in our quantitative description up to this point in this text, except in the short digressions on screening, correlations and plasmons. At the nanoscale, the spatial spread of these interactions, including that of phonons, becomes an issue that needs to be tackled separately and in itself. It is a major source of energy and momentum loss through interactions that arise remotely even as the local interactions remain present.

If phonons are not in equilibrium, they will strongly modify the electron distribution and be modified by the electron distribution. So, if distributions are off the equilibrium and in the limit where scattering events are also limited, the classical description of Chapter 9 will fail, and this phonon-electron coupling will very well play an important role. This is a drag on electron motion by phonons. With high permittivity dielectrics and their soft phonons of low energies that couple across from one region to another, this phonon drag will be of enormous consequence, since the mobility

Another set of places where this distribution change due to the limits on scattering has serious consequences is in quantum wells, including in solar cells of small bandgap materials (we discuss this as a multi-exciton problem in S. Tiwari, "Nanoscale device physics: Science and engineering fundamentals," Electroscience 4, Oxford University Press, ISBN 978-0-198-75987-4 (2017)) because narrow confined regions with interfaces for electrons and holes and phonons will have interfacial consequences in propagation. Such a description requires much more quantitative detail. Perhaps one even needs to resort to a Monte Carlo type of calculation since the calculation is not amenable to accurate-enough analytic solutions.

Semiconductor Physics: Principles, Theory and Nanoscale. Sandip Tiwari.
 Published 2020 by Oxford University Press. DOI: 10.1093/oso/9780198759867.001.0001

of the electrons in an inversion region is of primary interest. This is now a remote phonon drag process.

As another example, consider the transport in the inversion layer of a transistor. An electron entering a channel from the source reservoir or exiting a channel from the drain reservoir is a perturbation. It is introducing or removing energy from a region of interest. A long-range Coulomb interaction both during an electron's transit away from the reservoirs and during its entry and exit can then cause a plasma wave excitation in the degenerately doped reservoir regions. This collective excitation will eventually break into single particle excitation because of Landau damping.

Landau insightfully described the time decay of longitudinal space charge waves in a plasma. The electromagnetic pulse interacts: it accelerates the particles that have a velocity slower than the phase velocity and slows those that are faster. More particles gain energy from the wave than lose from the wave in the typical particle distribution function with velocity. So, the electromagnetic wave, in net, loses energy, no matter how low its amplitude. This is a reason why oscillatory instabilities get suppressed. It is applicable in many situations: from galactic dynamics—the electron gas of stars interacting with gravitation—to the channel electrons interacting with the electron sea of reservoirs. A surfer riding just a little slower than an ocean wave will gain energy from the wave. A stationary buoy just bobs up and down. A surfer moving faster than a wave will be pushing on the wave and lose his energy. This is when the surfer needs to do an aerial to avoid a wipeout.

Through this electron-plasma coupling, an exchange has happened between the momentum of the electron in the channel with that of the reservoir, even though they are separated from each other by longer than screening lengths. This is again a remote Coulombic process.

So, local and remote processes arising in charge excitations and phonon excitations are pervasive in nanoscale structures, and this is the focus of this chapter, with a particular emphasis on remote processes.

We will tackle three examples of remote effects that are all quite consequential. The first is of remote phonon scattering, as a continuation of the previous chapter's ionic polarization discussion. The second is of plasmonic scattering, where Coulomb interaction over a distance has a strong effect. The third is of phonon drag, where local phonons and the phonons of another region drag electrons. All these examples are of relevance to a miniature device, particularly those employing the field effect. In a field-effect transistor, here the first example is due to the use of high permittivity material. The second is due to the use of high carrier concentrations that exist in the source and the drain. And the remote part of the third is due to the coupling between the gate region and the channel region. All of these will affect the transport of electrons in the inversion region and other places where the electron cloud interacts with spatially remote (and local) perturbations that are energetically proximate.

19.1 *Remote phonon scattering*

Our description of the soft phonon's implication in electron-phonon interaction (Chapter 18) pursued the following argument. A dielectric's polarization field is proportional to the optical phonon amplitude. The interaction between two electrons in a medium is Coulomb mediated by an inverse proportionality to the permittivity. The polarizability of the medium is a function of frequency, since at the very least—for the two primary effects in an insulator—the electron cloud motion and the dressed nuclear motion have a mismatched response. The former is faster than the latter. Ionic response was the primary cause of high permittivity in the transition metal compounds of interest to us. $\epsilon(0)$ included both the electronic and the ionic polarizabilities. $\epsilon(\infty)$ included only electronic polarizability. The inverse difference of these $(1/\epsilon(\infty) - 1/\epsilon(0))$ is proportional to the changes arising from the removal of the ionic polarizability. It is directly related to the square of the amplitude of the dipole field, so to phonons and their frequency, which characterizes the ionic mass motion. So, we end up with an electron-longitudinal optical phonon scattering strength that varies as $\hbar\omega_{LO}[1/\epsilon(\infty) - 1/\epsilon(0)]$. The magnitude of $\hbar\omega_{LO}$ matters. The soft phonons of the high permittivity materials of interest to us have energies of the order of thermal energy (at room temperature), which is also an energy that characterizes the large population of electrons that is not too far from equilibrium. Phonons and electrons therefore have significant interaction. Traveling electrons transfer momentum to the crystal; that is, the traveling electrons are dragging the phonons along. Likewise, phonons are providing friction to the electron flow in this description. We will tackle this latter effect through the modifications to be discussed in Section 19.3.

When the high permittivity material is placed together with the semiconductor, as in use for field effect, these soft oscillations arising in soft bonds and the soft optical phonons will couple. This coupling takes place transversely. The coupling strength is now a modification of the inverse permittivity relationship where the permittivity of both the dielectric and the semiconductor matter, and this strength has a proportionality to the energy of the surface optical mode ($\hbar\omega_{so}$) that arises. These are phonon modes perpendicular to the interface, so they are related to ω_{TO} through the relationship of Equation 18.3 and a strength given by Equation 18.4. The insulator dipole fields, which are modified by image charge effects, decay in the semiconductor channel where electrons exist. But, what has happened is that soft phonons of the dielectric have a spatial reach into the semiconductor, where they are now causing additional scattering. This will reduce the mobility. The effect due to high permittivity is being felt in the semiconductor, which is spatially separated. This is *remote phonon scattering*. The surface optical modes are mediating remotely.

This remote phonon scattering effect is vanishingly absent in SiO_2 because of the large phonon energy, the poor oscillator

	$\epsilon_r(0)$	$\epsilon_r(\infty)$	$\hbar\omega_{TO1}$ (*meV*)	$\hbar\omega_{TO2}$ (*meV*)	E_g (*eV*)
SiO_2	3.90	2.5	55.60	138.10	∼9.0
Al_2O_3	12.53	3.20	48.18	71.41	8.8
AlN	9.14	4.80	81.40	88.55	
ZrO_2	24.0	4.00	16.67	57.70	5.8
HfO_2	22.00	5.03	12.40	48.35	5.8
$ZrSiO_4$	11.75	4.20	38.62	116.00	6.5*

* Bandgaps are experimental. $ZrSiO_4$'s is an estimate.

Table 19.1: Permittivity, transverse optical phonon frequencies and the bandgap of selected dielectrics. HfO_2 is of monoclinic crystalline structure.

strength of the weaker phonon collection that exists in SiO_2, and a poor coupling constant of the hard phonons arising in the stronger bonds, as discussed toward the end of Section 18.2.

Table 19.1 summarizes a few of the parameters for our discussion of phonon behavior for materials of interest here.

For both HfO_2 and ZrO_2 comparable dielectrics—which are useful high permittivity dielectrics for Si, there exists a giant change in their relative dielectric constant arising in the ionicity. Also, note that they both have transverse mode phonon energies that are lower than the room temperature thermal energy. Our arguments now need to be modified by at least two additional considerations. First, the coupling is between these surface phonon modes and what is usually an inversion region—an electron plasma of high electron density—and the equivalent of this situation at the gate/dielectric interface, where the gate too has a high electron density, although now three dimensionally. Second, the presence of an interface oxide, which although sub-*nm* and therefore not quite of the same characteristics as a good gate insulator oxide, will suppress the remote phonon effect. If the inversion region has a Thomas-Fermi wavelength of $\lambda_{TF} \propto 1/k_{TF}$, which varies as $1/\sqrt{n_s}$, that is, the square root of the sheet carrier concentration in the inversion region, then the remote scattering potential will decay through this screening. The presence of a large carrier concentration in the gate has the effect of increasing the screening of electron-interface optical modes so long as the physical thicknesses are small enough. So, the mobility will decrease, but the consequences will be mitigated by the presence of increased screening and phonon coupling suppression by an interfacial oxide film.

The reader should follow M. Fischetti's detailed analysis—referred to in the bibliographic notes—to explore how one may include interface films, two dimensionality and this remote phonon issue together in an analysis in the presence of multiple oscillation modes.

We will not dwell on the details of the analysis, since it is not of central interest to us. The physical nature of the problem is.

Figure 19.1 shows an approximate mobility consequence due to remote phonon scattering for HfO_2 compared to SiO_2. The caveat here is that the tunneling currents in the SiO_2 systems will be unbearable in any practice. The presence of the interface layer reduces the scattering strength of the SO modes by $\sim\exp(-2k_{TF}t_i)$, where t_i is the interface oxide thickness, and k_{TF} is the wavevector under this Thomas-Fermi screening condition. The interface oxide, while being of low permittivity, has an enormous effect through the suppression of the high permittivity dielectric's coupling mode. To this, one must also note as an addendum that this is the screening consequence in a structure of small dimensions. Thin films will also let the gate screen. So, this too will be an addition to the screening that arises in the large electron density in the inversion layer. The result is that, at high carrier concentrations, the mobilities in the different postulated examples are quite similar. Only in the thick SiO_2/Si case does this effect disappear and a high mobility come about at high carrier concentration.

Chapter 1 discussed the many-body based arguments regarding the Thomas-Fermi limit of the screening.

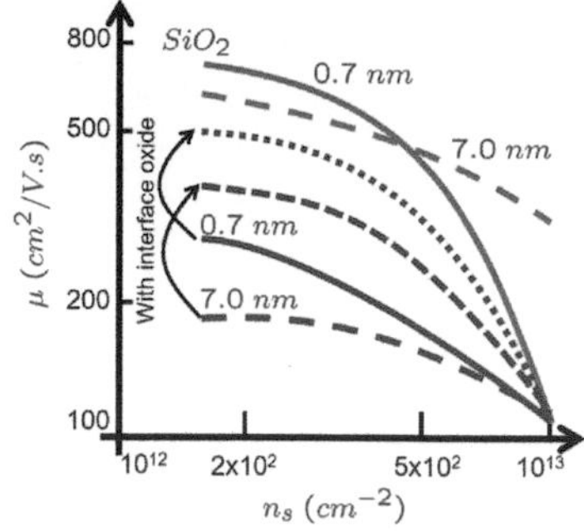

Figure 19.1: Calculated electron mobility in a silicon inversion layer for two equivalent oxide thicknesses: a small one of 0.7 nm, and a large one of 7.0 nm, together with mobilities when these same equivalent thicknesses are achieved using HfO_2. The mobilities in the latter are less than those in the former. But the presence of interface oxide—a monolayer—improves the mobility substantially.

This remote phonon scattering discussion was meant to illustrate the consequences of high permittivity arising in ionic polarization and how important a role it has at nanoscale. This discussion has included in it the changes that arose from high electron concentrations, and their electromagnetic coupling as seen in plasmons. The Coulomb interaction with large electron density is pervasive in inversion layers because of their reduced dimensionality. Electron transport in the channel will be subject to these plasmonic consequences. Our next section discusses the effects in small geometries arising in the interaction between the electrons of the inversion layer and the highly doped source and drain reservoirs of a field-effect transistor.

Strictly, Coulomb interaction—electromagnetic interaction is a more general and accurate term, since Coulomb is usually meant to imply "static" conditions—underlies all scattering. The bandstructure describes the electron's allowed state, and the Hamiltonian includes Coulomb interaction terms. So, this loose use of Coulomb interaction is meant to make a direct link to charge, which deformation, for example, does not. In this section, we are further restricting ourselves to the electron, and its collective response through the plasmon as its quasiparticle. The plasmon, to make this more tricky, is a boson. A plasmon is a hybridization of photons and the excess-depleted electron collective. Electrons are fermions, but the excitation is a paired excitation of two fermions, which is then a boson. A plasmon, as hybridization of two bosons, is a boson.

19.2 *Short- and long-range electron Coulomb effects*

In our discussion of mobility (Section 10.6), we explored the improvement in mobility that arose in a two-dimensional electron gas at the $Ga_{1-x}Al_xAs/GaAs$ interface when the dopants were separated from the electron gas. This was an example of improving mobility by reducing the scattering from ionized impurities by spacing them away. It is an example of a scattering directly arising in Coulomb interaction. In this section, we will be concerned with the remote electron's and the electron cloud's collective response. At nanoscale, similar to the case of the two-dimensional electron gas with separation of dopants, there are other high carrier density regions around. And these will have interactions, because the Coulomb interaction is a spatially remote interaction.

These electron interactions, for semiconductor conditions, can be broadly separated into two categories.

The first form of interaction is where we may treat the electron as a single particle interacting with its surroundings. This is under conditions where the electron may be treated as an individual particle, since the Coulomb potential energy arising in its interaction is small enough that the kinetic and potential effects can be considered separately. This single electron behavior—even in the presence of a large number of electrons—showed up in Equation 5.38, which quantified the image force interaction on a conducting surface. The energetics here leads to screening of the perturbation—a dopant ion being a common form—by the mobile charge. For electrons treated as independent (and far and few), the Coulomb energy varies proportional to $1/\epsilon|\mathbf{r} - \mathbf{r}'|$, where the $\mathbf{r}$s are the locales of the two independent charge particles. When the electron density is higher—so degeneracy prevails—it is the electrons around the Fermi energy that can respond. So, at low doping conditions, it is the Debye length scale ($\lambda_D = \sqrt{\epsilon k_B T/e^2 N}$, where N is the concentration of dopants), and at higher densities, the first-order correction is the Thomas-Fermi wavelength ($\lambda_{TF} = \sqrt{\epsilon/e^2 \mathscr{G}(E_F)}$, where $\mathscr{G}(E_F)$ is the density of states at the Fermi energy) that applies to the short length scale response.

One can look upon the screening expressions as representing the following characteristic. In thermal equilibrium, the screening is associated with a spread caused by the concentration N over $k_B T$ of energy, for non-degenerate conditions. For degenerate conditions, this spread is constrained by the allowed states' energy spread, where there are both occupied and unoccupied states in close proximity. So, this is related to the density of states.

The second form of interaction is when this electron density is large enough that this collection's interaction within itself cannot be ignored. Electrons are correlated—repulsively interacting—and this keeps them apart. This is the cause of skin depth and the reflection at metal surfaces and has numerous other implications. The plasmon represents an excitation of an interacting electron gas in a form where the plasmon frequency $\omega_p^2 = e^2 n/\epsilon m^*$ (Equation 12.63) is a long wavelength oscillation. The plasmon is the quantum of this oscillation with a low momentum and an energy of $\hbar\omega_p$. It is a boson. At electron densities of the Si inversion layer, $10^{19}\ cm^{-3}$, $\omega_p/2\pi = 2 \times 10^{13}\ Hz$ and, at those of the source and drain regions or the gate, $>10^{20} cm^{-3}$, $\omega_p/2\pi = 0.6 \times 10^{14}\ Hz$. These are energies in the 75–250 meV range for the various high electron densities in devices. The wavelength is of the order of 1000 nm, so the plasma frequency has a wavelength at the device length scale. The close proximity of high concentration regions will have long range, that is, remote Coulomb interactions: between the channel and the source and drain reservoirs, and between the channel and the gate. The collective oscillations will also have a limit at low wavelengths (smaller n) where they cut off, which will be determined by the

A simple way to view why there is this collective interaction is to imagine a high electron density, into which ensemble one places a hole. The hole's presence causes a perturbation. Electrons rush in, that is, the whole ensemble, because of its high density and strong coupling, pulls in. It overshoots because of the acquired kinetic energy. So, it then pushes out to compensate. And this process repeats. A small change has caused a long wavelength oscillation. This is the plasmon, whose primary equation we found in our discussion of free carrier absorption (Section 12.2).

S. Tiwari, "Nanoscale device physics: Science and engineering fundamentals," Electroscience 4, Oxford University Press, ISBN 978-0-198-75987-4 (2017), discusses at length the use of plasmons—in non-propagating and propagating conditions—for interesting nanoscale structures.

single electron perturbation equation's length scale in the presence of the n-electron concentration.

In the following, we will use Si as an example for the specifics, even though the conceptual description applies generally to all semiconductors.

We have observed that there is a continuum of Coulomb interactions that are mediated by quantum-mechanical constraints. We have discussed a variety of approximations of the Hamiltonian of the multitude of electrons in the crystal—Hartree, Hartree-Fock, Pauli exclusion, correlation, et cetera—in Chapter 1, and they will all be relevant here. The plasmon length scale will apply to longer range interaction at the high density. But, even for conditions of high density, there are short-range interactions that are simultaneously present. The size scale of nanoscale devices is in the 100 nm and below range, with insulators of a few nm, and source–drain distances of 10–100 nm as some of the other measures. The plasmon length scale at THz is 1000 nm in the semiconductor. The Thomas-Fermi length scale at 10^{19} cm^3 is of the order of 5.5 nm. And we are interested in evaluating how the transport of electrons in an inversion region gets affected under these conditions.

Take the example of inversion-to-gate region interaction as an example of short-range interaction. Electrons in the gate are around the depletion region of the gate. These are nearly at rest, so the gate will apply a drag on the momentum of the electrons in the inversion region. This is a *Coulomb drag*.

Another example of short-range interaction is what happens due to electron-electron scattering when an electron distribution encounters a rapid change in the electric field, such as at the channel-drain junction. The hot electrons, that is, electrons that are off-equilibrium with an excess distribution in the high energy tail, have a higher scattering rate in this high energy part through energy-losing processes such as phonon or ionized impurity, Auger generation, randomizing interface-induced scattering, et cetera. They lose this energy rapidly. Even though the high electron density in the drain does not by itself cause this momentum change, how these electrons came about to be there does.

The high electron density will cause momentum change through the plasmon oscillation process, but that is a separate long-range interaction process.

Now that we have a reasonable feel for the variety of interactions and their quantitative estimation in the semiconductor, we expand here the discussion of Chapter 1 to emphasize the Coulomb-kinetic energetic interaction. The Hartree potential of Equation 1.69 included the $1/|\mathbf{r} - \mathbf{r}'|$ interactions terms as well as a term of the electron interacting with itself. The spatial distribution of electrons and of the positive charge was considered uniformly homogeneous. If donors are considered discrete points and immobile, and the electron density low enough that donor-electron short-irange interaction is significant, then the screening has a $[1/4\pi\epsilon(\infty)r]\exp(-r/\lambda)$ dependence, where $1/\lambda = \sqrt{[e^2/\epsilon(\infty)]\partial n/\partial E_F}$. This is the Debye-Hückle result. For *Si*, this is quite a reasonable description at just below the degenerate conditions. If the electron density is low enough—at an electron concentration of 10^{18} cm^{-3}, electrons are, on average, 10^{-6} $cm \equiv 10$ nm apart, so that the miscounting arising in uniformity and screening—two conflicting assumptions—can be ignored, then the total kinetic energy is through an integration over all the occupied states from the bandstructure calculation of an undoped crystal. This is

$$t = \frac{T}{n}, \text{ with}$$
$$n = g \int f_{FD}[E_0(\mathbf{k})] \frac{1}{(2\pi)^3} d\mathbf{k}. \qquad (19.1)$$

E_0 is the electron energy, with a g-fold degeneracy. At near-degenerate conditions, uniform concentration begins to prevail. Each donor is screened in a volume with length scale $r_0 = (3/4\pi n)^{1/3}$. The energy associated with each dopant through the Coulomb interaction with the electron charge is reduced as

$$\delta U_{de} \approx - \int_0^{r_0} \frac{1}{\epsilon(\infty)} \frac{e^2}{(4/3)\pi r_0^3} r\, dr = -\frac{3}{2}\left(\frac{4\pi}{3}\right)^{1/3} \frac{e^2}{4\pi\epsilon(\infty)} n^{1/3}. \qquad (19.2)$$

The energy of each electron is raised by the Coulomb attraction as

$$\delta U_{ee} \approx \int_0^{r_0} \frac{e^2}{[(4/3)\pi r_0^3]^2} \frac{1}{4\pi\epsilon(\infty) r} \frac{4}{3}\pi r^3 4\pi r^2\, dr$$
$$= -\frac{3}{5}\left(\frac{4\pi}{3}\right)^{1/3} \frac{e^2}{4\pi\epsilon(\infty)} n^{1/3}. \qquad (19.3)$$

Since screening in the elemental volume makes donor-donor repulsion vanish, the net Coulomb attraction is

$$\delta U_C = \delta U_{de} + \delta U_{ee} = -\frac{9}{10}\left(\frac{4\pi}{3}\right)^{1/3} \frac{e^2}{4\pi\epsilon(\infty)} n^{1/3}. \qquad (19.4)$$

The magnitudes of the screening length λ of the Debye-Hückle result, the Thomas-Fermi screening length λ_{TF} and even the donor separation r_0 are quite similar at the degenerate conditions of interest to us. *The electron-phonon interactions and the random motion of electrons under these conditions leads to fluctuating potential in the thermally agitated environment.*

This change in energy δU_C is composed of a potential part δV_C and a kinetic part δT_C. The expectation for kinetic energy is

$$\langle T_C \rangle = \frac{1}{2} \sum_{i \neq j} \nabla_i V(\mathbf{r}_i - \mathbf{r}_j) \cdot \mathbf{r}_i = -\frac{1}{2} \langle V_C \rangle \qquad (19.5)$$

Solid-state textbooks introduce the Madelung constant as a dimensionless parameter that characterizes the energy per cell of point charges in a lattice in terms of the translation distance while including the neutralizing background. Ours is an equivalent calculation here while incorporating into it self-interactions.

by applying the Virial theorem to this Coulomb condition. Using ergodicity,

$$T_C = \frac{9}{10}\left(\frac{4\pi}{3}\right)^{1/3} \frac{e^2}{4\pi\epsilon(\infty)} n^{1/3}. \tag{19.6}$$

Since the electrons are interacting particles with Coulomb energy consequences, there is a shift in chemical potential that is not identical to the shift in electrochemical potential—Fermi energy—with the chemical potential still defined as $\mu = 1/\Omega \partial \mathcal{F}/\partial n$, following our discussion in Chapter 9. So,

$$\delta\mu = \frac{1}{\Omega}\frac{\partial \delta N_D \mathcal{U}_C}{\partial n} = -\frac{3}{10}\left(\frac{4\pi}{3}\right)^{1/3} \frac{e^2}{4\pi\epsilon(\infty)} n^{1/3}. \tag{19.7}$$

So the electron-phonon interaction is doing pretty much what we described was happening to the electron screening cloud when we dropped a hole into the assembly.

In calculating energies, the reference conduction band minimum has now shifted down by this magnitude in the presence of these electrons. *A bandgap narrowing has occurred.*

So far in this calculation, we have not accounted for the exchange energy that we noted in the Hartree-Fock approximation. The exchange energy is a reduction in the Coulomb repulsion energy because identical spin electrons cannot be spatially proximate due to Pauli exclusion. This exchange energy, in our notation from Chapter 1, for *Si*, is

$$\delta V_x = -\frac{3}{4}\frac{e^2}{4\pi^2\epsilon(\infty)}\left(\frac{m_l^*}{m_t^*}\right)^{1/3} \frac{\tan^{-1}\eta}{\eta}, \text{ where } \eta = \left(\frac{m_l^*}{m_t^*} - 1\right)^{1/2}. \tag{19.8}$$

This is the lowering of energy with the electrons non-uniformly distributed when the correlation minimizes Coulomb repulsion. The correlation correction, without proof, for *Si* is

$$\delta V_{corr} = -3.08 + 0.20 \ln\left(\frac{r_0}{a_B^*}\right) \text{ in } meV. \tag{19.9}$$

The exchange is substantially larger than correlation in the modification of the Coulomb energy.

This discussion establishes the importance of the Coulomb energy and its lowering of the bandedge energy. The Coulomb interaction has changed the total electron gas energy and increased the kinetic energy. This is one contribution to the bandgap narrowing. This discussion was also all in three-dimensional conditions. The arguments for two-dimensional situations are considerably more complicated. In Chapter 3, in our analysis in a periodic potential, we had employed Fourier components to extract allowed energies. The spatial Fourier component here, in a similar way, is represented by the plasmon quasiparticle. The increased kinetic energy will be dampened by the electron-phonon interaction. This is collisional damping, which too will affect the kinetic energy.

Suffice it to say that our discussion has established a change in kinetic energy due to increased carrier density, which will lead to increased scattering. This is plasmon-induced scattering that is a long-range effect. It must be considered together with the short-range effects in order to understand the consequences for transport in nanoscale devices. Figure 19.2 shows the marginal scattering rate for an electron at 1 eV energy in the channel due to Coulomb exchange with a 10^{20} cm^{-3} doped drain region. The figure shows the short-range rate, and the sum of the short- and long-range scattering rates. The sum plot employs dynamic wavelength-dependent screening. Note the increased scattering when resonating with plasmons occurs. Also note that the scattering rate of low energy carriers increases dramatically.

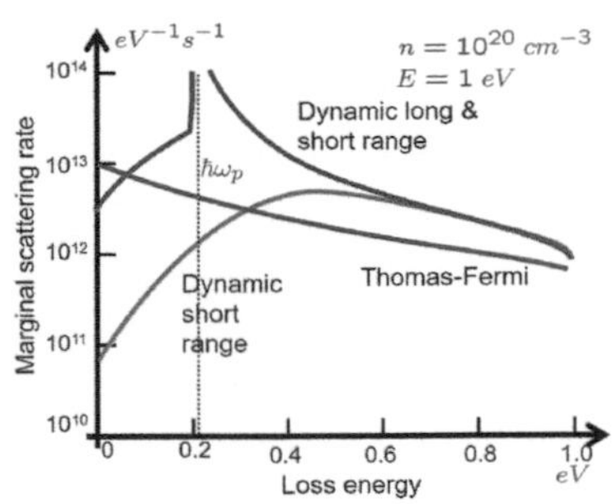

Figure 19.2: Marginal scattering rate for a 1 eV electron in the channel due to a $10^{20} cm^{-3}$ electron concentration in the drain of a Si field-effect transistor. The short-range scattering calculation is based on dynamic screening. The second curve shows scattering where both long and short-range interactions with dynamic screening are included. Scattering is highest at plasmon frequency, but note that it is nearly a decade higher at up to several 100 meV. The figure has been adapted from M. V. Fischetti and S. E. Laux, "Long-range Coulomb interactions in small Si devices. Part 1: Performance and reliability," Journal of Applied Physics, **89**, 1205–1231 (2001).

19.3 *Phonon drag*

FOR BOTH A PARTICLE AND A QUASIPARTICLE, one can picture their movement with a simultaneous movement of other particles. A particle, as it moves, could be pushing others out of the way, or dragging others along with it.

Take the case of a temperature gradient along a semiconductor. The system is not in equilibrium. There will be transport of heat via phonons. There are more phonon excitations at the hot end than at the cold end. These phonons will transfer momentum to electrons and drag the electrons along too. This coupling of electrons to the off-equilibrium phonons is again a *phonon drag*. In this temperature excitation at the boundaries, the accumulation of electrons at the cold end sets up a field if one were forcing a condition of no charge current.

In Chapter 9, in our calculations involving thermoelectric effect, as embedded in the Seebeck and Thompson coefficients, our assumption was that phonons were in equilibrium. This is the Bloch condition—Bloch was among the first to perform this electron-phonon interaction calculation—and this assumption is built on the large phonon-phonon scattering that arises in Umklapp processes. This is not an unreasonable assumption at room and higher temperatures for semiconductors where optical phonon energies are of the order of the thermal energy. Our result was the calculation of the normal Seebeck or Thompson coefficient, where the electron momentum relaxation time $\tau_{\mathbf{k}}$ and how energy and momentum flowed were determined through Boltzmann's statistical transport equation for electrons alone, and all the inputs were local.

Phonon drag is the result of phonons being off-equilibrium, and this, in turn, causes phonons to change the electron transport behavior because the phonon-electron interaction in the higher energy tail is now substantially different. The off-equilibrium phonons deliver excessive momentum to the electrons, and Seebeck coefficients, and other measures where electron-phonon coupling plays a role, such as Fröhlich conductivity, which is the consequence of drag due to phonons in conductivity, will now show a substantially different behavior than expected from the use of Bloch condition. Semiconductors, for example, show a peak in Seebeck coefficient at low temperatures because of the excess electron current from the increased momentum. Low temperatures, and any artificially created or reduced dimensionality structures where interfaces play a role in the phonon modes, will all show increased phonon drag. And this phonon drag can be both local and remote because of the artificial structures that one practices and certainly use as devices.

An engineering interest in phonon drag is because of the interest in thermoelectric power, which is the conversion of heat through its gradient into a useful form of energy. A figure of merit, zT, where

$$zT = \frac{\sigma S^2 T}{\kappa}, \tag{19.10}$$

with the symbols here having their usual meaning (S is the Seebeck coefficient, and σ and κ are conduction and thermal conductivity, respectively), is often used to characterize this thermoelectric capability of a material. Compared to room temperature, as the temperature is lowered for bulk semiconductors, the Seebeck coefficient drops while the thermal conductivity increases. Effective thermoelectricity generation requires letting the Seebeck coefficient increase while suppressing large thermal conductivity increases. The increased Seebeck coefficient—its peaking—due to phonon drag must happen together with a simultaneous increase in thermal conductivity. At the lower temperatures, where these effects are seen, the heat transport is dominated by lower energy phonons, while the phonon drag effect is due to phonons that have reduced scattering and hence a longer mean free path. So, the phonon drag phenomenon is potentially a means of improving thermoelectric properties. But this is predicated on an effective decoupling of the contribution of phonons toward heat transport and energy flow to electrons.

In our earlier transport discussion, we wrote the Boltzmann transport equation for electrons while leaving the phonons in equilibrium. So, only one equation—that for electrons, with no equivalent equation for phonons—was written. To incorporate both electron and phonon transport effects, we will now have to extend the earlier discussion so that the distribution function for electrons and phonons can be written with a coupling between the two. Solving such coupled equations is non-trivial. The variational method may be used for concurrent solution. To be more accurate in resolving the high energy and low energy phonon effects, and specifically for relevance at high doping, where many of the parameters—of which there are too many—are estimates, one possibility is to employ only a partial coupling, keeping the electron-phonon part explicitly separate so that the mode effects are separately quantified. Our equations, in one form, then are

$$\begin{aligned}
\mathbf{v}_\alpha(\mathbf{k}) \cdot \frac{\partial f_\alpha(\mathbf{k})}{\partial T} \nabla_{\mathbf{r}} T - e\mathbf{v}_\alpha(\mathbf{k}) \cdot \frac{\partial f_\alpha(\mathbf{k})}{\partial E} \nabla_r E_{qF} \\
= -\frac{f_\alpha(\mathbf{k}) - f_{0\alpha}(\mathbf{k})}{\tau^*_\alpha(\mathbf{k})} + \left.\frac{\partial f_\alpha(\mathbf{k})}{\partial t}\right|_{e-ph}, \quad \text{and} \\
\mathbf{v}_\gamma(\mathbf{q}) \cdot \frac{\partial g_\gamma(\mathbf{q})}{\partial T} \nabla_{\mathbf{r}} T \\
= -\frac{g_\gamma(\mathbf{q}) - g_{0\gamma}(\mathbf{q})}{\tau^*_\gamma(\mathbf{q})} + \left.\frac{\partial g_\gamma(\mathbf{q})}{\partial t}\right|_{e-ph},
\end{aligned} \qquad (19.11)$$

where the first equation is a Boltzmann transport equation written for electrons, and the second is the same equation written for phonons. Here, f and g are the distribution functions for electrons and phonons, respectively, the former subject to Fermi-Dirac statistics, and the latter to Bose-Einstein statistics. $\mathbf{k}$ is the wavevector for electrons, and $\mathbf{q}$ for the phonons, the former occupying a band of states identified by the α parameter, and the latter occupying a band of states identified by the γ parameter. The scattering terms on the right are written as two separated parts, where electron-phonon coupling is separated from all the other scattering mechanisms. When the doping is high, the electron-phonon scattering is explicitly enhanced. This tackles not only the thermoelectric situation but also the situation when hot electrons are interacting with the heavily doped reservoirs of a device. The off-equilibrium phonons' effects are now assembled together. The nonlocality of such interactions is even more complicated. But computational solution of these equations over an extended space takes care of that complexity.

Figure 19.3 shows a doping-dependent result of the Seebeck coefficient, including the phonon drag for a local calculation. At low dopings, the momentum transfer to electrons depends on the number of electron states coupling to the phonons; that is, the electron concentration. So, at low concentrations, the momentum gain per

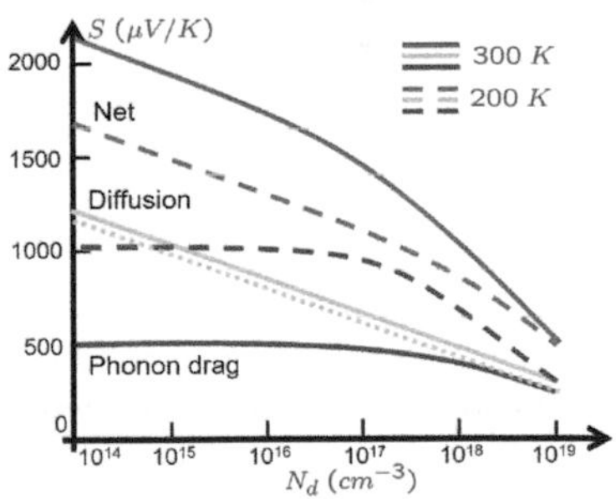

Figure 19.3: Seebeck coefficient in *Si* at 200 and 300 *K*, showing the diffusion and the phonon drag contributions. Adapted from J. Zhou, B. Liao, B. Qiu, S. Huberman, K. Esfarjanii, M. S. Dresselhaus and G. Chen, "Ab initio optimization of phonon drag effect for lower-temperature thermoelectric energy conversion," Proceedings of the National Academy of Sciences of the United States of America, **112**, 14777–14782 (2015).

electron embedded in the Seebeck coefficient is independent of the electron concentration. At high doping, it is dependent.

A disturbance from equilibrium was necessary for the change from detailed balance. Such a disturbance can be introduced spatially in a number of ways. For example, electron transport in one layer that is separated from another layer—say, with an insulating oxide in-between—will also drag. This is a Coulomb drag. One may make other artificial multilayered structures, where dynamics or statics of the regions get connected. And here, depending on the screening, scales of perturbation and their connecting, we will have local and remote Coulomb and phonon effects. Now we have a remote Coulomb drag. We mentioned earlier the consequence of gate with a thin dielectric as one instance of such a drag.

Another change is one arising in dimensionality. Screening lengths are dependent on dimensionality. So are the modes allowed for phonons, just as they are for the charge particles, albeit they are considerably weaker. The coupling of electron-phonon modes in one-dimensional structures and its consequences for conductivity in macroscale structures—a Fröhlich conductivity—can be significant. This is an example of phonon drag. One can therefore see that it may be possible to build more efficient thermoelectric converters where phonon drag exists efficiently and provides more thermoelectric conversion because of the different totality of scattering considerations for electrons and phonons.

19.4 *Summary*

This chapter has focused specifically on remote processes that influence electron transport and manifest themselves in a variety of properties of interest. There is a continuum of these interactions, from local to remote, and therefore we have kept this local aspect as a reference. This continuum and its consequence is particularly of importance at nanoscale, where both will appear simultaneously.

A few important points of our discussion follow. Coulomb perturbation is screened, and the local consequence of this screened perturbation has been across multiple scattering-related and other discussions throughout this text. But Coulomb consequences also arise due to coupling across space from the electron, from the electron as an ensemble when degenerate, from the ionicity of dielectrics that is reflected in the frequency-dependent permittivity, and in the phonon modes representing these oscillating ions.

High permittivity dielectrics make this last instance quite acute. When high permittivity dielectrics are employed, even though the electron transport in the high permittivity dielectric is not of direct interest, the transport in adjacent semiconductor regions—the channel of a transistor being an example—can have substantial behavioral changes. The phonon modes of the high permittivity dielectric couple to the semiconductor. Since they are of low energy, compared to a low permittivity dielectric such as SiO_2, they enhance scattering and lower the mobility in *Si* upon which they have been placed.

Direct Coulomb effects—short and long range—can be via single particle interactions and via collective interactions. An example of the former—a remote scattering—was the illustration of separation of donors from a two-dimensional electron gas at an $Ga_{1-x}Al_xAs/GaAs$ interface. The mobility in the two-dimensional electron gas improves, and a high electron density, through enhanced screening, even helps with this improvement of electron density. On the other hand, consider a high permittivity dielectric between the gate and the two-dimensional electron channel that is the inversion layer. The gate dopants and electrons are essentially immobile. The moving charge is coupled to the static charge of the gate, and the Coulomb attraction is a drag on the electron transport. In the $Ga_{1-x}Al_xAs/GaAs$ example, this effect was weak compared to the screening effect, so the two-dimensional improvement in the *GaAs* example is precisely the opposite to the high permittivity *Si* example because of the nature of this remote coupling.

The electron in the channel region, as it enters or exits the channel, is also causing an excitation in the source or the drain region, which is degenerately doped. This degenerately doped region responds collectively, and we have represented it via plasma oscillations, with the plasmon as the quasiparticle of this collective mode. Energy and momentum were coupled between the electron of the channel and the reservoir. The plasmon will eventually come apart by damping—a Landau damping—and this energy has been lost in the reservoir. So, the transport of the electron in the channel is being influenced by plasmon scattering through this interaction.

A direct phonon effect was illustrated through a discussion of phonon drag. Up to this chapter in this text, it was the electron that was treated as being off-equilibrium whenever electrochemical changes were introduced externally. Phonons were assumed to remain in equilibrium so that only one transport equation, such as the Boltzmann transport equation, needed to be tackled for the electron. The discussion of thermoelectric transport in Chapter 9 looked at both thermal conductivity and electron-phonon coupling in the various coefficients—Thompson, Seebeck, et cetera—under conditions where the phonons were still in equilibrium. In the

Chapter 9 discussion, we really only considered the thermal conductivity arising in transport by electrons.

When phonons are out of equilibrium, they too drag the electrons. Phonons that are higher up in energy interact with electrons more strongly. When electron concentration is low, this interaction will have a smaller effect, but when electron concentration is high—as at high doping—this effect will be larger. This enhanced interaction will not only show up in the thermoelectric effect, where phonon flow is dragging along electron flow, resulting in thermoelectric voltage that may or may not be enhanced, depending on this interaction, which will depend on phonon occupation and interaction, but it will also have a consequence, albeit a minor one, in electronic device structures through the heavily doped contact reservoirs. Although not a drag, a hot phonon population, in structures where interfaces suppress phonon flow through propagation mismatch, such as in quantum wells, will change the characteristics of scattering substantially. Multi-exciton solar cells based on small bandgap semiconductor quantum wells are claimed to exhibit inefficiency consequences due to this phonon bottleneck.

19.5 Concluding remarks and bibliographic notes

We have discussed this remote processes theme because of its importance in nanostructures, which is beyond the esoteric interests that have existed since Ziman's classical discussion[1] of phonon drag as well as the multitude of scattering processes and their coupled effects that take place in solids.

C. Hamaguchi's textbook on semiconductor physics[2] is a good and advanced reading for understanding the dielectric response function and its tie-in to the scattering behavior that arises through phonons and plasmons. It is a text that thoroughly works through the quantitative details of local processes.

For remote phonons, and particularly the consequences arising in the use of ionic dielectrics with high permittivity to electron inversion layer, the paper by Fischetti et al.[3] is a comprehensive source and provides a list of to references from the past.

To understand plasmons as multibody excitations, and the variety of places that they show up in a solid response, the reader should look up the text by Pines[4]. Pines undertook his graduate thesis under Bohm, a scientific stalwart, and his thesis was quite devoted to this problem of multibody excitation.

The modern interest in the unusual aspects of electron-phonon drag and their role in semiconductors can be traced to work by

[1] J. M. Ziman, "Electrons and phonons: The theory of transport phenomena in solids," Oxford (1960)

[2] C. Hamaguchi, "Basic semiconductor physics," Springer, ISBN 978-3-642-03302-5 (2010)

[3] M. V. Fischetti, D. A. Neumayer and E. A. Cartier, "Effective electron mobility in *Si* inversion layers in metal-oxide-semiconductor systems with a high-κ insulator: The role of remote phonon scattering." Journal of Applied Physics, **90**, 4587–4608 (2001)

[4] D. Pines, "Elementary excitations in solids," Perseus, ISBN 0-7382-0115-4 (1999)

Yu G. Gurevich. Gurevich and Mashikevich's paper[5] in *Physics Reports*. It is a comprehensive discussion built from the off-equilibrium kinetics of electrons and phonons through the Boltzmann transport equation.

A more modern discussion of phonon drag—this with the thermoelectric effect as a focus—is the paper by Zhou and co-authors[6]. There are a number of references to earlier works on phonon drag, particularly those employing the Bloch condition, that the interested reader may find useful from here.

In two papers, Fischetti and Laux[7] explore at length the incorporation and the consequences of Coulomb interaction in silicon structures. These are long papers, but detailed and comprehensive in exploring the underlying quantitative description. In our discussion here, we have largely ignored the consequences or dimensionality in these interactions, the nature of dispersion one will find in interface plasmons, and other issues. These papers are an excellent source for such a comprehensive discussion.

For those interested in going beyond the nature of these scatterings to consequences for devices—and there is much in it through how the reservoir interacts—there are two papers that are a good launching ground. For understanding the interaction close to the quantum limits, the paper by Laux et al.[8] makes for comprehensive reading. To understand how the reservoir electron capacity, these interactions, and, through the variety of rate-limiting processes, the behavior of device gets affected, see the paper by Fischetti and co-authors[9].

[5] Yu G. Gurevich and O. L. Mashikevich, "The electron-phonon drag and transport phenomena in semiconductors," Physics Reports, **181**, 327–394 (1989)

[6] J. Zhou, B. Liao, B. Qiu, S. Huberman, K. Esfarjani, M. S. Dresselhaus and G. Chen, "Ab initio optimization of phonon drag effect for lower-temperature thermoelectric conversion," Proceedings of the National Academy of Sciences of the United States of America, **112** 14777–14782 (2015)

[7] M. V. Fischetti and S. E. Laux, "Long-range Coulomb interactions in small *Si* devices. Part I: Performance and reliability," Journal of Applied Physics, **89**, 1205–1231 (2001), and M. V. Fischetti, "Long-range Coulomb interactions in small *Si* devices. Part II: Effective electron mobility in thin-oxide structures," Journal of Applied Physics, **89**, 1232–1250 (2001)

[8] S. E. Laux, A. Kumar and M. V. Fischetti, "Analysis of quantum ballistic electron transport in ultrasmall silicon devies including space-charge and geometric effects," Journal of Applied Physics, **95**, 5545–5582 (2004)

[9] M. V. Fischetti, S. Jin, T.-W. Tang, P. Asbeck, Y. Taur, S. E. Laux and N. Sano, "Scaling MOSFETs to 10 nm: Coulomb effects, source starvation, and virtual source," 2009 13th starvation, and virtual source," 2009 13th International Workshop on Computational Electronics, 1–4 (2009)

19.6 Exercises

1. We derived the plasmon frequency as $\omega_p = \sqrt{ne^2/\epsilon\epsilon_0 m^*}$. Why should we worry about plasmons in semiconductors? What causes the coupling between the electrons in the channel and electrons elsewhere to take place? **[S]**

2. The frequency of plasma oscillations of a two-dimensional electron gas can be written in analogy with bulk plasmons, except that the charge carriers are now confined in two dimensions. Letting **Q** represent the wavevector,

$$\omega_{p2D}^2(Q) \approx \frac{e^2 n_0}{2[(\epsilon_{sem} + \epsilon_{ins})/2]m^*} Q,$$

where the averaging is achieved by ignoring the very thin two-dimensional layer, that is, with the overlap function $\varsigma_0(z) \approx \delta(z)$. This vanishes as the wavelength grows large and the wavevector is small. But, in a three-dimensional electron gas, the plasma frequency is finite and non-zero. Argue in physical terms why this is so. **[S]**

3. In high permittivity (κ or ϵ_r) insulators with *Si*, such as HfO_2, why are surface optical phonons and their interactions with the electron inversion layer so strong across the entire electron density range of interest in field-effect devices? **[S]**

4. What is phonon drag? Should it be more pronounced with acoustic or optical phonons, transverse or longitudinal, for electrons in an inversion layer, and why? **[S]**

20
Quantum confinement and monolayer semiconductors

UP TO THIS POINT, in our discussion of semiconductor phenomena—of electrons, holes, phonons, impurities, others, their static and dynamic behavior under thermal equilibrium and off it, and their particle- and wave-based analysis—the nanoscale has appeared, through their propensity for large surfaces and interfaces, and the behavioral changes—such as via strain—or interactions—such as remote—that arise from short spacings, even if called long-range interactions. Localization, when it appeared, was through the atom-size scale, such as with the core-versus-valence foundations of point perturbations. Delocalization appeared when the interactions were between multiple excitations launching Goldstone modes or when interaction-scattering phenomena had one of these cause-and-effect localized through the nature of the occupation of states. Semiconductors are often used with confinement at size scales below the de Broglie wavelength in one or more dimensions. This brings out another set of attributes whose appearance is again at nanoscale.

In confinement, or otherwise, we are concerned with the extent of the consequence of the quantization of $\mathbf{k}$. When restricted in space, the kinetic energies proportional to $\hbar^2\mathbf{k}^2/2m^*$ allowed are also being separated further out, due to a confinement-caused inflation in the spacing of $\mathbf{k}$ of that direction, than other energy parameters of interest. If k_BT is larger than this quantized separation of energy, then it is a quasi-continuum, as it has been generally in the text. When k_BT is smaller, it affects the occupation of states, and this quantization has significant consequences. There can be conditions when confinement has little consequence at room temperature but can be significant at low temperatures. The *MOSFET* is a good example of this for the lowest order analysis. Under non-degenerate conditions, a reasonable measure of the momentum of the electron is m^*v_θ, while, under degenerated conditions, $\hbar k_F$ would be the appropriate measure, due to the occupation of states.

When a semiconductor is non-degenerately doped, reduced dimensionality may be introduced as a single, multiple or periodic confined region or even a barrier. And these structures, singly or multidimensionally confined, have very interesting and useful changes in properties. This problem is not as simple as the introductory quantum posers that look at a free electron and a potential well in free space. The environment is the semiconductor with its collection of atoms and their electrons and, often, due to the high electron density in the environment, the single nearly free electron picture by itself may be insufficient. A limit of this size changing is the generation of entirely new forms of semiconductors that are stable monoatomic crystals. One of these, if a

That confinement is important can be seen through the most common form of transistors—*MOSFET*—which achieves one-dimensional localization in the inversion layer via the oxide on the gate side, and an electrostatic barrier in the semiconductor on the other side. Scattering, states available for occupation and how interactions happen between states will all change.

Semiconductor Physics: Principles, Theory and Nanoscale. Sandip Tiwari.
 Published 2020 by Oxford University Press. DOI: 10.1093/oso/9780198759867.001.0001

sheet, is really a two-dimensional semiconductor itself, rather than a two-dimensional electron charge cloud in a three-dimensional semiconductor ensemble.

Spatial confinement—the assembling of semiconductors at the wavelength scale—and an exploration of the properties that arise in monolayer semiconductors, together with their consequences for transport and electromagnetic interaction—subjects that have been dealt with extensively in this text—are the focus of this chapter. These themes allow us to integrate the range of different ways we have analyzed throughout the text and unravel quite new, designed-in physical properties at the nanoscale that draw on both the wave description and the atomic description.

Bipolar semiconductor lasers achieve their remarkable optical efficiency and low currents because the quantum wells make the electron-hole direct recombination for stimulated radiation extremely efficient and of very small linewidth. Monolayer semiconductors such as graphene and carbon nanotubes or other dichalcogenides, et cetera, do not show confinement's effect in a similar sense. These are different crystalline forms of the atoms, constituting them with their own emergent properties in the crystalline assembly.

20.1 *Heterostructure interfaces and quantum wells*

We start by considering the state description of electrons at heterostructure interfaces. And we choose a common form, useful in electronics, where a confined two-dimensional electron layer is obtained at a heterostructure, such as of $Ga_{1-x}Al_xAs/GaAs$, or an ideal insulator-semiconductor, such as of an SiO_2/Si interface, by providing a band-discontinuity barrier on one side and an electrostatic barrier on the other side. The drawing of the bandstructure at the interfaces—irrespective of whether the interface region is abrupt or graded via composition—follows from the constraints that Maxwell's equations, especially Gauss law through its Poisson relationship, and must be maintained together with the constraints placed by the semiconductors and their state distribution as well as the state occupation. For graded junctions, where both composition and doping change, this requires self-consistent inclusion of the available states, which will shift in energy and density due to composition changes and doping. We have discussed a bit of this in Chapter 4 and do not dwell on it, as it is not central to the physics of semiconductors.

Since heterostructures—their interface region and transport across them—comprise a very important element of devices, the junction and its behavior is a major theme in S. Tiwari, "Device physics: Fundamentals of electronics and optoelectronics," Electroscience 2, Oxford University Press, ISBN 978-0-198-75984-3 (forthcoming). Two important points worth emphasizing are as follows. First, the discontinuities in abrupt structures are independent of bias because they have arisen in the dipole arising in the Bloch function distortion at the interface—at atomic size scale—with continuity in the bonding. It is unaffected by the small screening-constrained electrostatic potential changes across the interface. In a graded junction, the electrostatic and the chemical compositions (we also called this the alloy potential) need to be included together. The second point is that the bulk-like electronic structure extends all the way to the interface, with the minor caveat that, for some heterostructures, where the Bloch functions are very different on either side—the $InAs/GaSb$ interface is one—there may very well be interface states that need to be accounted for.

Confinement is the constraining of the electron to a region that is few unit cells thick, so quantum-scale thick, rather than being a propagating solution across a large region. If confined in one direction, it may still have the propagating solution in the other directions. Figure 20.1 is a pictorial representation akin to what employed in the Bloch description development of Chapter 3 for one dimension of confinement. One now needs to employ an additional envelope function $\varsigma(\mathbf{r})$ to describe the electron's confinement.

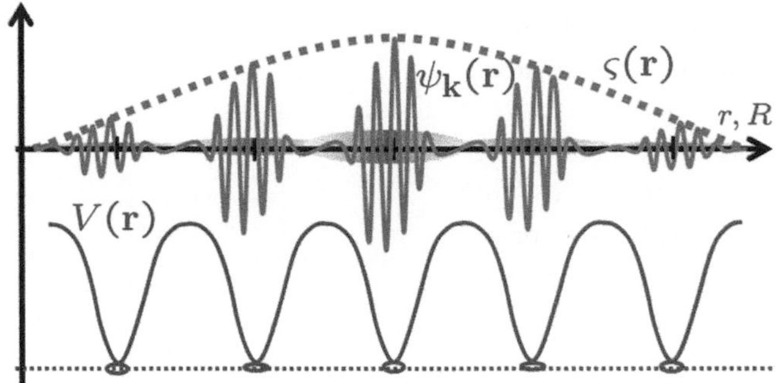

Figure 20.1: Potential and the wavefunction $\psi_{\mathbf{k}}(\mathbf{r})$ in a confined condition. The figure shows the periodic crystal potential $V(\mathbf{r})$ with periodicity $\mathbf{R}$ and shows that the wavefunction solution is now confined. ς is now an envelope function modulating a Bloch function.

20.1.1 *Electron inversion layer in* SiO_2/Si

AN EXAMPLE OF AN ABRUPT INTERFACE is shown in Figure 20.2, with only one of the regions that has lower conduction energy (region *II*) shown. The absence of the states in *I*, on the left, and of an electrostatic barrier on the right in *II* means that, in the direction normal to the interface (the z direction), carriers are confined in a narrow potential well. In this direction, the carriers can only have specific k_zs that are quite separated (corresponding to energies of k_BT or higher, so not quasi-continuous) due to confinement. In the other two directions, in the plane of the interface, the carriers can continue to have motion with the k_x and k_y states still very close together, allowing propagation in response to forces that lead to motion in that plane. The solution of Schrödinger's equation reformed to the modulation function-envelope form—the effective mass equation—for the specific potential well and carrier mass provides us with the reformation of the states for the nearly free electrons that are not to move perpendicular to the interface. So, perpendicular to the interface, only certain eigenfunctions are the solutions; energies associated with them are quantized. These discrete energies will represent the minimum in energy in a band of allowed energies. Each continuum of states attached to one of the discrete energies is called a subband. The behavior of carriers under conditions of confinement in one direction is called two-dimensional behavior. Here, we tackle the triangular-well one dimension of confinement and two dimensions of propagation problem that appears in many interesting structures. Confinement exists in the z direction, with allowed k_zs farther apart, and propagation exists in the x-y plane, where the allowed k_xs and k_ys are closely spaced. For any specific k_z allowed, there is

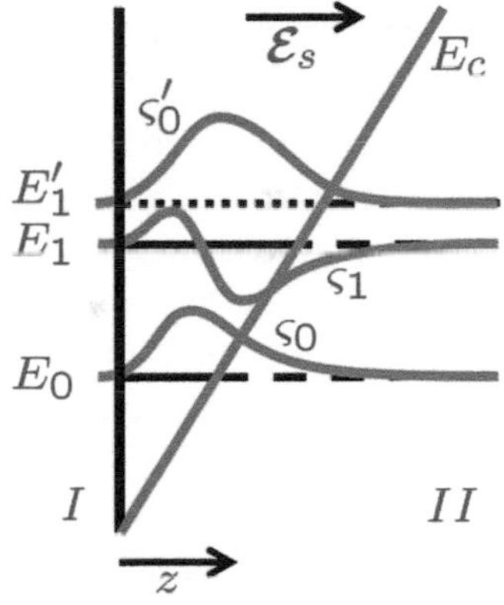

Figure 20.2: A quantum well exemplifying the confinement at the SiO_2/(100)-*Si* interface, showing the quantization of the energy and momentum in the z direction normal to the interface. *I* identifies a confining barrier. *II* shows confinement arising from the electrostatic potential on the back side. The well is close to triangular, since the region of confinement is short, and there is therefore a nearly planar charge at the interface. Subband minima and the corresponding envelope function shape are shown.

The density of states under constraints of confinement is tackled in Appendix H, where we look at a three-dimensional large box, a two-dimensional layer and a one-dimensional wire to calculate the evolution of density of states as an isotropic semiconductor gets increasingly confined dimensionally.

now a displaced bottom of the band defined by the quantized k_z, and a two-dimensional subband where k_x and k_y may vary quasi-continuously.

Consider the case of the *Si* inversion region or accumulation region, shown in Figure 20.2, of a (100) interface bounded by SiO_2 and an electrostatic barrier in the presence of band anisotropy. Recall that the constant energy surfaces near the bottom of the conduction band are ellipsoids with a large longitudinal mass and a low transverse mass. Since allowed kinetic energies vary inversely with the effective mass, the longitudinal mass being larger than the transverse mass leads to a smaller energy E_i of the bottom of the corresponding two-dimensional subband. Figure 20.2 shows the quantization of the two sets of subbands—one with m_l^* confined (lower) and the other with m_t^* confined (higher)—and their envelope functions. Because of the anisotropy, different surfaces of silicon behave differently.

The case for holes in semiconductors will be considerably more complex because light holes, heavy holes and even split-off-band holes, as well as their anisotropy, will all enter in the allowed state description.

The envelope function together with the Bloch function factor yields the wavefunction of the electron.

The energy levels in the quantized two-dimensional channels must be derived by solving Poisson's equation and the envelope equation for the quantum well simultaneously, and by taking into account in this solution all carrier-energy-related effects. Poisson's equation here identifies the relationship between the potential energy and the charge distribution, the envelope equation identifies the relationship between the allowed total energy and the momentum, and the energy effects include the consequences of many-body effects such as exchange and correlation on carriers. These latter effects usually result in lowering of the energy. The effective potential energy $V(z)$ in the envelope equation can be written as

Note that this situation is complementary to the tunneling problem of field injection that occurs via tunneling at the top of the metal-semiconductor junction. There is a quasi-continuous distribution of states on either side of the barrier and exclusion in the barrier region. Again, the field enters, this time as a parameter for the envelope function decay in the barrier.

$$V(z) = -q\psi(z) + \sum V_i. \tag{20.1}$$

Here, $\psi(z)$ is the electrostatic potential, and the V_is are the potential energy terms associated with exchange correlation, image, grading, et cetera. The Schrödinger's equation for the envelope function $\varsigma_i(z)$—the envelope equation—in subband i is

$$\left\{ -\frac{\hbar^2}{2}\frac{d}{dz}\left[\frac{1}{m^*(z)}\frac{d}{dz}\right] + V(z) \right\} \varsigma_i(z) = E_i \varsigma_i(z), \tag{20.2}$$

where $m^*(z)$ is the position-dependent effective mass, which can depend on i and E_i in general cases, and E_i is the energy at the bottom of the ith subband. Unlike the SiO_2/Si system, which has a large discontinuity, in compound semiconductor heterojunction systems, ΔE_c is much smaller (fractions of eV) and the electron wavefunction can penetrate a significant distance into the barrier, just as it does in the other direction. Compound semiconductors, with their small effective masses for electrons, exhibit strong quantization effects; the reduced confinement by the barrier layer and significant wavefunction penetration into the confining region add to the complexity of the problem.

This second-order differential equation requires two boundary conditions at each boundary in the confined region. The two points on either side are *turning points* denoting the positions at that energy where particles are being turned back. Probability densities and probability current must be continuous, so $\varsigma(z)$ as well as $d\varsigma/m^*dz$ are continuous at both of the turning points. The abrupt barrier, if large, as with SiO_2, will suppress both $\varsigma(z)$ and $d\varsigma/m^*dz$ to vanishing values. Finite barriers tend to require a variational calculation or a numerical solution. In addition to this envelope equation, one needs to also satisfy Poisson's equation relating fields and electrostatic potentials to the charge. Here, it is in the form

$$\frac{d}{dz}\left[\epsilon(z)\frac{d\psi(z)}{dz}\right] = q\sum N_{si}\varsigma^2(z) - \rho_I(z), \tag{20.3}$$

where

$$N_{si} = \frac{m_d^* k_B T}{\pi\hbar^2}\ln\left[1+\exp\left(\frac{E_F - E_i}{k_B T}\right)\right], \tag{20.4}$$

where $\epsilon(z)$ is the position-dependent permittivity, N_{si} is the number of electrons per unit area in subband i, E_i is the minimum of the ith subband energy, m_d^* is the density of states effective mass of the inversion layer material for that subband, and ρ_I is the ionized impurity charge density, where $\rho_I = -qN_A$ for an acceptor doping of N_A.

Since the carrier charge is confined in a short distance at the interface, and the ionized impurity charge density is usually smaller by a significant amount, the potential well is sometimes approximated as an infinite triangular well. This is to say that the carrier charge is assumed to be a sheet charge giving rise to a constant electric field that is not significantly disturbed by the weak acceptor charge density or the distribution of electron charge in a narrow extent of the inversion layer. For this approximation, which is not an unreasonable description of moderate and strong inversion conditions, an approximate solution to the problem is

$$\varsigma_i = \mathcal{A}i_i\left[\left(\frac{2m_d^* q\mathcal{E}_s}{\hbar^2}\right)^{1/3}\left(y - \frac{E_i}{q\mathcal{E}_s}\right)\right], \tag{20.5}$$

where $\mathcal{A}i$ is the Airy function

$$\mathcal{A}i(z) = c_1 f(z) - c_2 g(z). \tag{20.6}$$

There is quite some complexity buried in turning points, barriers and confinement. There are time delays involved in turning back. And turning back itself is the notion of the idea of a "particle" being applied in a wave solution. Compound semiconductors can have central Γ valleys, but also valleys at L or X, which may be the lowest valley in the confining barrier. Which mass should be used then? This requires one to understand how far this penetration is and if the confining environment being felt by the electron arises in the Γ, L or X forbidden gap barrier. A hint of this can be seen in tunneling through a barrier—the complementary problem to this confinement problem—where a thick barrier shows consequences from the lowest barrier, irrespective of the valleys, and a thin barrier shows consequences arising from the barrier from the identical valley. The crystal momentum changed in the former, but not in the latter. The former had momentum-changing scattering involved, while the latter did not. The former is indirect tunneling. The latter is direct tunneling.

Care needs to be exercised with the use of the masses. Here, we have written this equation using the density of states effective mass m_d^*. For $GaAs$ electrons, with one central valley, $m_d^* = m^*$ close to the conduction band minimum. For Si, this is not so; m_d^* will be depend on the valley degeneracy and anisotropy, and each of these will depend on the surface being employed. With Si, it is best to revert to the longitudinal and transverse effective masses characterizing the ellipsoidal surfaces of constant energy and include the degeneracy explicitly.

The functions $f(z)$ and $g(z)$ are given by

$$f(z) = \sum_{n=0}^{\infty} 3^n \left(\frac{1}{3}\right)_n \frac{z^{3n}}{(3n)!}, \text{ and}$$
$$g(z) = \sum_{n=0}^{\infty} 3^n \left(\frac{2}{3}\right)_n \frac{z^{3n+1}}{(3n+1)!}. \tag{20.7}$$

The subband minima energies are given by

$$E_i = \left(\frac{\hbar^2}{2m_d^*}\right)^{1/3} \left[\frac{3}{2}\pi q \mathcal{E}_s \left(i + \frac{3}{4}\right)\right]^{2/3} \tag{20.8}$$

for large values of i.

A more exact solution suggests replacement of the bracketed index terms of $(i + 3/4)$ by 0.7587, 1.7540 and 2.7525 for $i = 0, 1$ and 2, respectively.

In the limit of large inversion charge, or negligible acceptor charge density, Gauss' law gives $\mathcal{E}_s = qN_s/\epsilon$ for the electric field at the heterostructure interface. Here, N_s is the carrier sheet charge. Thus, in the limit of moderate to strong inversion and low acceptor charge and ignoring barrier penetration effects, the subband energy levels can be related to either the electric field at the interface or the sheet carrier charge at the interface. For $Ga_{1-x}Al\,As/GaAs$ junctions, this results in the minimum subband energies

$$E_0 = 1.83 \times 10^{-6} \mathcal{E}_s^{2/3}, \text{ and}$$
$$E_1 = 3.23 \times 10^{-6} \mathcal{E}_s^{2/3}, \tag{20.9}$$

where the energies are given in units of eV, and electric fields are in units of V/m.

We should also consider, in addition to these, the changes in density of states in two-dimensional systems as a result of the formation of the subbands. Again, recall that density of states in a ν-dimensional k-space is $(2\pi)^{-\nu}$. So, for two-dimensional systems, the density of states is

$$\mathscr{G}(E) = \frac{2g_v}{(2\pi)^2} 2\pi k \frac{dk}{dE}, \tag{20.10}$$

where g_v is the valley degeneracy, and the factor 2 accounts for spin degeneracy ($g_s = 2$). If we assume isotropic parabolic bands,

$$E = E_0 + \frac{\hbar^2 k_\parallel^2}{2m_d^*}, \tag{20.11}$$

with

$$\mathscr{G}(E) = \frac{g_v m^*}{\pi \hbar^2} \text{ for } E > E_0, \text{ and}$$
$$= 0 \text{ for } E < E_0. \tag{20.12}$$

For constant effective mass independent of energy, the density of states is constant in any subband, and zero below the subband edge. As energy increases, more subbands become populated, and there is a piecewise discontinuous increase in the density of states. For an n-type inversion layer in *GaAs*, $g_v = 1$ because there is only one equivalent Γ valley band, and the density of state distribution for the two-dimensional system is straightforward (see Figure 20.3(a)).

For (100) silicon, $g_v = 2$ for the two ellipsoids whose longitudinal momentum is confined (Δ_2 valleys) and $g_v = 4$ (Δ_4 valleys) for the ones whose transverse momentum is confined. The density of state distribution is slightly more complicated and is shown in Figure 20.3(b). The mass to be used in Equation 20.12 is the density of state effective mass for the two orientations. The effective masses for motions allowed in the $E(\mathbf{k})$ states are

$$m^* = m_t^* \tag{20.13}$$

for the twofold degenerate subbands, and

$$m^* = (m_t^* m_l^*)^{1/2} \tag{20.14}$$

for the fourfold degenerate subbands.

At $T = 0$ K, the sheet carrier density and the Fermi level energy (if only the 0th band is occupied) are related by

$$N_s = \mathscr{G}(E)(E_F - E_0), \tag{20.15}$$

and, following the calculations for Fermi velocity from Appendix P,

$$k_F = \left(\frac{2\pi N_s}{g_v}\right)^{1/2}, \text{ and}$$

$$v_F = \frac{\hbar}{m^*}\left(\frac{2\pi N_s}{g_v}\right)^{1/2}. \tag{20.16}$$

In *GaAs*, with $N_s = 6 \times 10^{11}$ cm^{-2} for the electron density, a common sheet density in high mobility samples at $Ga_{1-x}Al_xAs/GaAs$ interfaces, the Fermi velocity is 3.2×10^7 cm/s at absolute zero. This is larger than thermal velocity in non-degenerate *GaAs* at 300 K.

The treatment leading to these equations ignores several second-order effects related to bandstructures. The parabolic band approximation is accurate only for small energies, and higher order subbands are influenced strongly by the wavefunction penetration into the larger bandgap material at the discontinuity. Hole bands are highly anisotropic, as are bands in smaller bandgap materials such as $Ga_{1-x}In_xAs$. p-type inversion also has complications due to multiple bands, warped surfaces and low barriers with larger wavefunction penetration even for the lowest energy subbands. The parabolic infinite barrier picture is too simplistic for these, and we have to rely almost exclusively on numerical techniques to obtain parameters of interest.

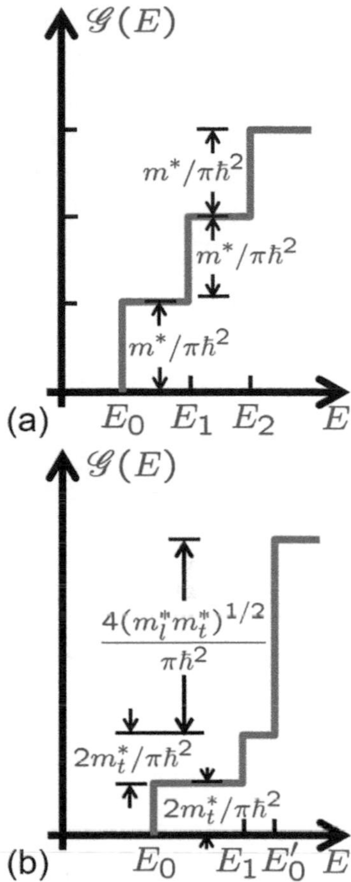

Figure 20.3: The density of states distribution in an electron inversion layer on a *GaAs* surface as a function of energy is shown in (a). Part (b) shows the density of states distribution function as a function of energy for an electron inversion layer on a (100) *Si* surface.

The reader is encouraged to explore this same question for silicon. It is a bit more complicated calculation (see S. Tiwari, "Nanoscale device physics: Science and engineering fundamentals," Electroscience 4, Oxford University Press, ISBN 978-0-198-75987-4 (2017), for the confinement calculations), but very instructive.

20.1.2 *Confinement by infinite and finite potential*

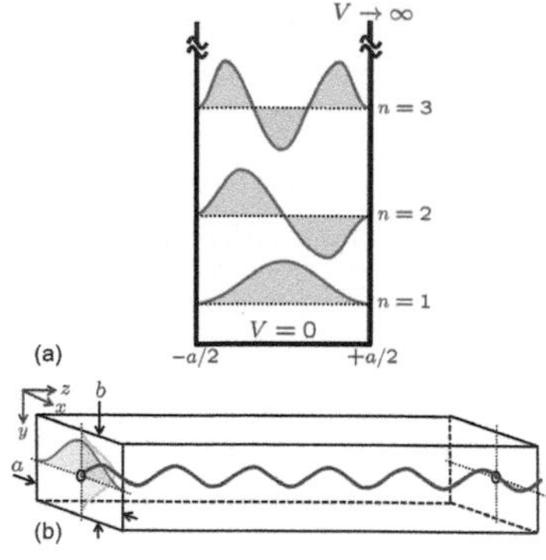

Figure 20.4: Part (a) shows quantum confinement in one dimension in an infinite potential well and (b) shows this for two dimensions of confinement; that is, a quantum wire.

HAVING DISCUSSED THE EXAMPLE OF INVERSION, and that of density of states in crystals dealt with in Appendix H, we will summarily deal with the preliminaries of the particle in a box problem. Figure 20.4 shows two cases: Figure 20.4(a) shows quantum confinement in one dimension, while Figure 20.4(b) shows it in two dimensions. The confinement is via an infinitely high and sharp potential wall, with the potential in the well as our $V = 0$ reference.

Here, sharp wall is meant in the mathematical sense. Quantum mechanics and nature, of course, don't permit it, but it is a good approximation in view of the other lengths in the system.

This second box has confinement in two dimensions, as seen in Figure 20.4(b). Propagation is allowed in the third (z) direction. It too has an infinitely high and sharp potential wall in the x and y directions, where the confined lengths are a and b. Let m_x^*, m_y^* and m_z^* be the effective masses in the three orientations, so we assume that a many-unit-cell description holds. This in turn lets us use, for most purposes, the envelope function, since its usage depends on slow variation on the atomic scale. The wavefunction solution takes its form from the Bloch functions near the bandedges as a product, and their usage—not the Bloch function's—will be necessary with optical transitions. Let l and m be the quantum numbers associated with the x and y confinement. The envelope function is

$$\begin{aligned} \varsigma_{i,j,k_z}(x,y,z) &= \sqrt{\frac{4}{ab}}\sin\left(\frac{\pi l x}{a}\right)\sin\left(\frac{\pi m y}{b}\right)\exp(ik_z z) \\ &= \text{for } l,m=1,2,3,\ldots. \end{aligned} \tag{20.17}$$

There is continuity along the z direction—no confinement, and nearly free electron motion—so, there is a quasi-continuum of states—the subbands—whose energies are

$$E_{l,m}(k_z) = \frac{l^2\pi^2\hbar^2}{2m_x^* a^2} + \frac{m^2\pi^2\hbar^2}{2m_y^* b^2} + \frac{\hbar^2 k_z^2}{2m_z^*}. \tag{20.18}$$

In general, not including spin, these energies of subbands are nondegenerate. But, in general, degeneracy exists whenever $l^2/m_x^* a^2 = m^2/m_y^* b^2$, the most egregious case for which is when $a = b$ with $m_x^* = m_y^*$ for quantum number combinations (l,m) and (m,l) if $l \neq m$. For a wire of length L, the density of states per unit energy for the one-dimensional (l,m) subband is

$$\mathscr{G}_{l,m}(E) = 2 \times g_s \frac{L}{2\pi}\frac{dk_z}{dE} = \frac{L}{\pi}\left(\frac{2m_z^*}{\hbar^2}\right)^{1/2}(E - E_{l,m})^{-1/2} \tag{20.19}$$

for a single valley and is a form that decreases in density the higher up in energy one goes. This is because the k_z states are evenly spaced, while the energy rises as the square of k_z. If there exists confinement in the z direction also, a quantum box results, and Equation 20.17 will have a straightforward change in the z-dependent term to a corresponding x- and y-like term. So will there be a change in density of states. Now, the levels will be entirely discrete. This is a quantum box.

Take another form of these two dimensions of confinement: a quantum wire of radius R and with similar potential constraints. Now, the envelope function equation is

$$\left\{-\frac{\hbar^2}{2}\left[\frac{1}{r}\frac{\partial}{\partial r}\left(\frac{R}{m^*}\frac{\partial}{\partial r}\right)+\frac{1}{r^2}\frac{\partial^2}{\partial\phi^2}\right]+V(r,\phi)\right\}\varsigma(r,\phi)=E\varsigma(r,\phi), \tag{20.20}$$

where m^* is assumed isotropic and constant for simplicity, ϕ is the azimuthal angle around the z axis, and $r^2=x^2+y^2$. $\mathbf{k}$ appears as a quantum number, with $\hbar\mathbf{k}$ as the linear momentum for the Cartesian example. Now, we have quantization of the angular momentum. The solution function is

$$\varsigma_{l,m,k_z}(r,\phi,z)=\frac{1}{R\sqrt{\pi L}}\left[\frac{1}{J_{l+1}(j_{l,m})}J_l\left(\frac{j_{l,m}}{R}r\right)\right]\exp(il\phi)\exp(ik_z z)$$
$$\text{for } l,m=0,1,2,\ldots. \tag{20.21}$$

Here, J_ls are Bessel functions of order l whose mth root is $j_{l,m}$. Angular momentum is now quantized by the orbital quantum number l and is $l\hbar$. $l=0$ is a non-degenerate state, and $l\neq 0$ integers lead to doubly degenerate solutions. The energy levels are

$$E_{l,n}(k_z)=\frac{\hbar^2 j_{l,m}^2}{2m^*R^2}+\frac{\hbar^2k_z^2}{2m_z^*}. \tag{20.22}$$

A circular wire will have fewer states over a span of energy than a square wire of similar area would. Confinement's effect is stronger by a $1/\sqrt{\pi}$ reduction of the length scale. However, at large energies, the density of states will approach a square cross-section density at the same area, since it depends on the area. At small energies, the densities differ, and the boundary conditions—how far the envelope function can penetrate the forbidden regions—will have a noticeable effect. So, SiO_2-confined versus compound-semiconductor-confined structures exhibit a variety of consequences arising in the nature of the density and the energy separation of these states.

For finite barriers, explicit analytic solutions do not exist for rectangular cross-sections but do for circular cross-sections, through Bessel and Neumann functions. We tackle the finite barrier now. Following Figure 20.5, which is a two-dimensional example with one dimension of confinement by a potential V_0, with effective

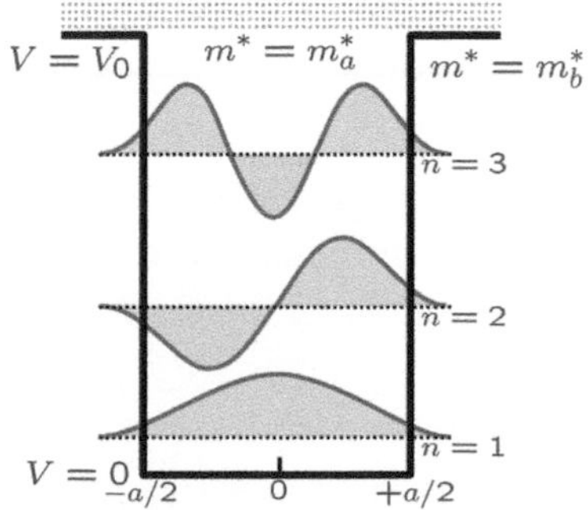

Figure 20.5: Confinement in a finite barrier of potential V_0 in a quantum well.

masses m_a^* in the well of width a and m_b^* in the barrier, the envelope Equation 20.2 will have a solution for a symmetric ladder in the constrained form of

$$\begin{aligned}
\varsigma_i(z) &= A\cos kz \ \text{ for } |z| < a/2,\\
&= B\exp[-\kappa(z-a/2)] \ \text{ for } z > a/2, \ \text{ and}\\
&= B\exp[-\kappa(z+a/2)] \ \text{ for } z < a/2, \ \text{ or}\\
&= A\sin kz \ \text{ for } |z| < a/2,\\
&= B\exp[-\kappa(z-a/2)] \ \text{ for } z > a/2, \ \text{ and}\\
&= B\exp[-\kappa(z+a/2)] \ \text{ for } z < a/2, \ \text{ with}\\
E_i &= \frac{\hbar^2 k^2}{2m_a^*}, \ \text{ and}\\
E_i - V_0 &= \frac{\hbar^2 \kappa^2}{2m_b^*}.
\end{aligned} \tag{20.23}$$

There is a bounded solution in the well. For the lowest energy solution it is written as cosine based for convenience since, by symmetry, it peaks in the middle of the well and since the $E = E_i$ of the solutions is greater than the potential $V = 0$. Second-order equations have an exponentials-based solution that is in this form here because of symmetry. But the next solution, if it exists, will be asymmetric, and now can be written in terms of the sine function. In the barrier region of mass m_b^*, the exponential solution is a decaying solution, since $E_i < V_0$. The boundary conditions of the continuity of envelope function and the probability current forces the constraints for the symmetric and asymmetric cases as

We have discussed this bounding as a standing wave arising from two counter-propagating waves in Appendix H for the state density calculation in a semiconductor.

$$\begin{aligned}
A\cos\left(\frac{ka}{2}\right) &= B, \ \text{ and } \ \frac{kA}{m_a^*}\sin\left(\frac{ka}{2}\right) = \frac{\kappa B}{m_b^*}\\
\therefore \ \frac{k}{m_a^*}\tan\left(\frac{ka}{2}\right) &= \frac{\kappa}{m_B^*}, \ \text{ and}\\
-A\sin\left(\frac{ka}{2}\right) &= B, \ \text{ and } \ \frac{kA}{m_a^*}\cos\left(\frac{ka}{2}\right) = \frac{\kappa B}{m_b^*}\\
\therefore \ \frac{k}{m_a^*}\cot\left(\frac{ka}{2}\right) &= -\frac{\kappa}{m_B^*},
\end{aligned} \tag{20.24}$$

respectively. The two transcendental expressions of Equation 20.24 will require numerical solutions. For the simplification of $m_a^* = m_b^* = m^*$, these expressions reduce to

$$\begin{aligned}
\cos\left(\frac{ka}{2}\right) &= \frac{k}{k_0}, \ \text{ for } \ \tan\left(\frac{ka}{2}\right) > 0, \ \text{ and}\\
\sin\left(\frac{ka}{2}\right) &= \frac{k}{k_0}, \ \text{ for } \ \tan\left(\frac{ka}{2}\right) < 0,
\end{aligned} \tag{20.25}$$

with $k_0 = \sqrt{2m^* V_0/\hbar^2}$. Figure 20.6 shows a normalized solution for this finite well problem.

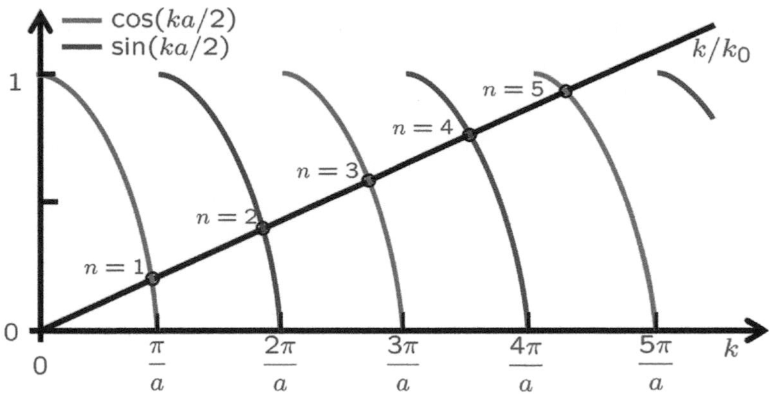

Figure 20.6: An illustration of a finite-potential well solution for the $m_a^* = m_b^*$ case. For the choices made, four bound solutions exist, identified at the intersection of the cos($ka/2$) and sin($ka/2$) curves with k/k_0.

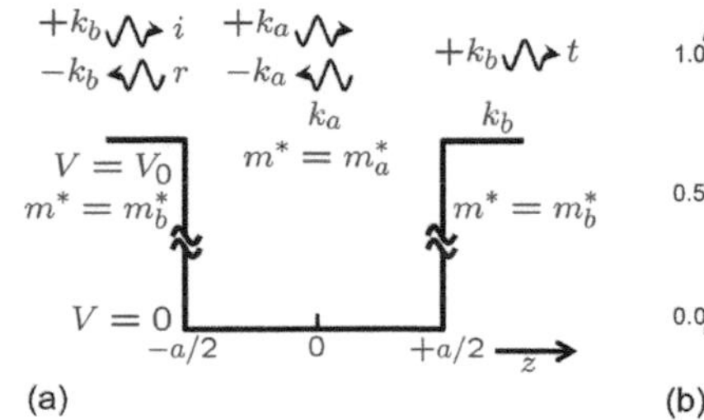

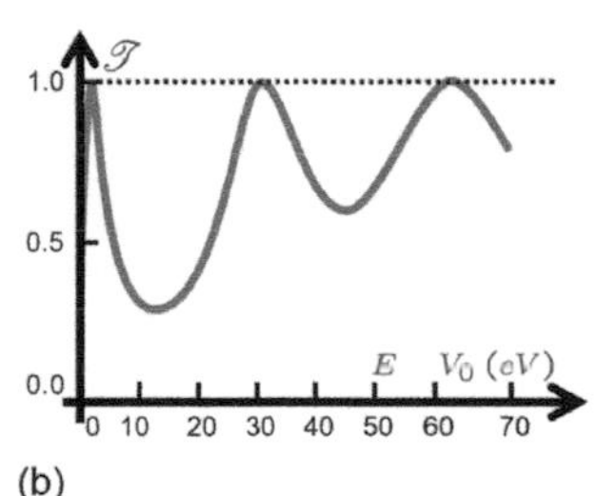

Figure 20.7: Part (a) shows propagation, with transmission and reflection, at $E > V_0$ through the finite barrier potential well in an infinite one dimension. Part (b) shows the solution of the transmission coefficient (amplitude square). There are specific energies at which an incident wave can entirely transmit through. Off this resonance, reflections happen.

A number of features are evident. The spacing between energies reduces faster than for the case of an infinite well. An infinite potential well has only bound states. A finite potential well has a finite number of bounded states. The total number of these is

$$N = \text{Int}\left[\left(\frac{2m^* V_0 a^2}{\pi^2 \hbar^2}\right)^{1/2}\right], \tag{20.26}$$

where only the integer part of the square root should be taken. But this does not mean that there are not other allowed states.

For $E > V_0$, there exist an unlimited number of unbounded states. Each eigenvalue will be twice degenerate because both directions of carrier momentum are allowed. Now, we may look at this problem in terms of transmission and reflection because of the potential discontinuity encountered. Figure 20.7(a) shows the construction of this problem as an extension of the finite barrier discussion.

An electron of momentum $+\hbar k_b$ incident at the well ($z = -a/2$) can transmit or reflect. Let the reflected wave have an amplitude of r, with the incident as unity. In the well, there are also propagating states of $\pm\hbar k_a$ momentum. The transmission at $z = +a/2$ occurs with a transmission amplitude t, and this wave propagates away from the barrier. We can write the position part of the envelope function across the three regions to reflect this physical picture:

In this wave problem, the envelope function—the non-time-dependent part—is spread out over the entire z space by definition. The electron(s) are not constrained to a region.

$$\begin{aligned}\psi(z) &= \exp\left[ik_b\left(z+\frac{a}{2}\right)\right] + r\exp\left[-ik_b\left(z+\frac{a}{2}\right)\right], \quad \text{for } z \leq a/2, \\ &= A\exp(ik_a z) + B\exp(-ik_a z), \quad \text{for } -a/2 \leq z \leq a/2, \text{ and} \\ &= t\exp\left[ik_b\left(z-\frac{a}{2}\right)\right]. \end{aligned} \tag{20.27}$$

The boundary conditions of continuity of the envelope function and the probability current imply

$$\begin{aligned} t(E-V_0) &= \left[\cos(k_a a) - \frac{i}{2}\left(\frac{k_a}{k_b} + \frac{k_b}{k_a}\right)\sin(k_a a)\right]^{-1}, \text{ and} \\ r(E-V_0) &= \frac{i}{2}\left(\frac{k_a}{k_b} - \frac{k_b}{k_a}\right)\sin(k_a a) \\ &\quad \times \left[\cos(k_a a) - \frac{i}{2}\left(\frac{k_a}{k_b} + \frac{k_b}{k_a}\right)\sin(k_a a)\right]^{-1}. \end{aligned} \tag{20.28}$$

The transmission coefficient is $\mathscr{T} = |t|^2$, and the reflection coefficient is $\mathscr{R} = |r|^2$. The solution satisfies $\mathscr{T}(E-V_0) + \mathscr{R}(E-V_0) = 1$—energy conservation—and transmission, explicitly, is

$$\mathscr{T}(E-V_0) = \left[1 + \frac{1}{4}\left(\frac{k_a}{k_b} - \frac{k_b}{k_a}\right)^2 \sin^2(k_a a)\right]^{-1}. \tag{20.29}$$

Figure 20.7(b) shows this solution for a range of 70 meV above the barrier, with $V_0 = 0.15\ eV$, and $m^* = 0.48m_0$.

At an energy $E = V_0$, the carrier is entirely reflected back; it does not get trapped into the well. This is because of the wave nature. It is entirely consistent with principle of correspondence. Also, the transmission coefficient varies with changing period in energy and in magnitude. But it reaches $\mathscr{T} = 1$ whenever $k_a a = n\pi$, where $n = 1, 2, \ldots$. The wave can resonate across the barrier when constructive interference conditions exist.

As discussed in S. Tiwari, "Quantum, statistical and information mechanics: A unified introduction," Electroscience 1, Oxford University Press, ISBN 978-0-198-75985-0 (forthcoming), if a barrier boundary appears adiabatic, that is, the thickness over which the barrier changes is longer than the wavelength of the propagating electron $2\pi/k$, then these quantum-mechanical reflections will disappear. All these solutions can also be equivalently viewed through the approach of perturbation, which we applied there.

An important comment regarding dimensionality reduction and the existence of these bound and continuum states is in order. We have only looked at the problem here of confinement in one dimension. If one had two dimensions of confinement, it will stand to reason that the energy levels will rise much more rapidly since, in this other direction of $\mathbf{k}$, stronger quantization has separated those states out. The density of states in a wire dropped with energy for this reason, forcing states to be dense at the lowest energies allowed. If one confined the final spatial dimension, then only non-propagating states will exist. An atom is one example of this. For a quantum well with two dimensions of freedom, as Equation 20.26 stated, there is an integer number of countable states, and $V_0 > \pi^2\hbar^2/2m^*a^2$ is necessary for one to exist. States can all be unbounded. When the confinement is in two dimensions—a one-dimensional quantum wire—the system will always have bound states with the unbounded propagation in the third dimension.

We noted this constraint in Chapter 7 when exploring localization in deep centers.

20.1.3 *Potential in confinement conditions*

This V should not be confused with the crystal potential used in obtaining the Bloch solution.

WHAT PRECISELY IS THE POTENTIAL V in the confined semiconductor environment? We can relate it to a number of factors that have been considered in the text. Consider only one dimension of confinement for simplicity. The potential for the envelope function arises in—and this can be seen back to our Wannier function and other discussions—as

$$V(z) = -e\psi(z) + V_{cb}(z) + V_{im}(z) + V_{xc}(z), \tag{20.30}$$

where ψ is the electrostatic potential, V_{cb} is the conduction bandedge, V_{im} is the image potential, and V_{xc} is the exchange correlation. The electrostatic potential rises from the Poisson equation, the conduction bandedge—including discontinuities—arises from the bandstructure, the image potential reflects the consequences of the charges' polarization of surfaces due to dielectric discontinuities, and exchange correlation, which is the Coulomb consequence of the quantum antisymmetric requirement for the electrons as we park them together in the high-carrier-density surface region. The first two and the last are quite clear in light of the discussions throughout this text. Image potential needs a short discussion. For Ze of charge in material of permittivity ϵ_a, a distance d away from a material of permittivity ϵ_b, the image potential is

$$V_{im}(z) = \frac{\epsilon_a - \epsilon_b}{\epsilon_a + \epsilon_b}\frac{Z^2e^2}{16\pi\epsilon_a d}. \tag{20.31}$$

If $\epsilon_a < \epsilon_b$, that is, the charge is in the material of low permittivity, then its image potential is attractive. The effect is not negligible in SiO_2/Si, where permittivities are vastly different, but can be usually ignored in semiconductor heterostructures with $\epsilon_r \approx 10$, give or take.

20.2 *Confinement of holes*

HOLES, WITH THE COMPLEXITY OF BANDSTRUCTURE, and the interactions between the various valence band states, should be expected to be considerably more complex. Our remarks here are meant to give a perspective on how one tackles this complexity, drawing on techniques learned in this text, and summarizes some of the important consequences. We are also particularly interested in tackling this problem since electron-hole-photon interactions will specifically depend on the allowed and non-allowed transitions based on conservation of energy, momentum and the nature of the perturbations.

We explored the three-dimensional hole bandstructure through the Luttinger Hamiltonian of Equation 4.99 and found the interrelated parameters of Equations 4.100–4.101. These describe for us the light-hole, heavy-hole and split-off-hole $E(\mathbf{k})$ states. The split-off band is quite near ($\Delta_{so} \approx 44\ meV$) the other two bands in *Si* at the zone center, but, in *GaAs* ($\Delta_{so} \approx 0.34\ eV$) and other compound semiconductors of interest, it is far off. Since the interest in valence band states is particularly for optical interactions in confinement conditions, with this large split-off spacing, we can largely ignore the coupling of light-hole and heavy-hole band states with the split-off band states due to any interaction of interest. With this, use of the top left 4×4 block of the Luttinger Hamiltonian suffices. The eigenenergies of this Hamiltonian are

$$
\begin{aligned}
E(k) &= -P \pm \left(|Q|^2 + |S|^2 + |R|^2\right)^{1/2} \\
&= -\frac{\hbar^2}{2m_0}\Big\{\gamma_1 k^2 \\
&\quad \pm\left[4\gamma_2^2 k^4 + 12(\gamma_3^2 - \gamma_2^2)(k_x^2k_y^2 + k_y^2k_z^2 + k_z^2k_x^2)\right]^{1/2}\Big\}.
\end{aligned} \tag{20.32}
$$

Following the discussion of the Luttinger Hamiltonian, with z as the angular momentum direction of the holes and propagation, the heavy-hole and light-hole energy bandstructure near the zone center is

$$
\begin{aligned}
E &= \frac{\hbar^2 k_z^2}{2m_0}(\gamma_1 - 2\gamma_2) \text{ for heavy holes } (\pm\frac{3}{2}), \text{ and} \\
&= \frac{\hbar^2 k_z^2}{2m_0}(\gamma_1 + 2\gamma_2) \text{ for light holes } (\pm\frac{1}{2}).
\end{aligned} \tag{20.33}
$$

This describes the masses for the $\langle 001\rangle$ direction as

$$
\frac{m_{hh}^*}{m_0} = \frac{1}{\gamma_1 - 2\gamma_2}, \text{ and } \frac{m_{lh}^*}{m_0} = \frac{1}{\gamma_1 + 2\gamma_2}, \tag{20.34}
$$

and, for the $\langle 110\rangle$ direction,

$$
\frac{m_{hh}^*}{m_0} = \frac{1}{\gamma_1 - (\gamma_2^2 + 3\gamma_3^2)^{1/2}}, \text{ and } \frac{m_{lh}^*}{m_0} = \frac{1}{\gamma_1 + (\gamma_2^2 + 2\gamma_3^2)^{1/2}}. \tag{20.35}
$$

These γ factors are determining the interband coupling. The γ_2 represents the heavy-hole coupling to the light hole along $\langle 100\rangle$. Along $\langle 110\rangle$, this coupling is via γ_3. For *GaAs* and several other compound semiconductors in use, $\gamma_2 < \gamma_3$. γ_3 causes the band repulsion to be quite significant along ⟨110⟩, making the heavy hole heavier and the light hole lighter compared to what they are along ⟨100⟩.

The momentum quantization due to confinement can now be approached by starting with $k=0$ unperturbed states and introducing the net perturbation due to the confinement, that is, the confining potential as a perturbation, and due to the $\mathbf{k}\cdot\mathbf{p}$ interaction. The lowest-order solution, if confinement is absent, arises from the $k = 0$ valence band description, such in Equation 20.33. Since the basis set that diagonalizes the $\mathbf{k}\cdot\mathbf{p}$ perturbation $\mathscr{H}'_{\mathbf{k}\cdot\mathbf{p}}$ is not a basis set for the confinement perturbation $\mathscr{H}'_{conf}$, strong mixing occurs. Masses are different, the heavy-hole ($\pm 3/2$) and light-hole ($\pm 1/2$) band degeneracy is lifted, and increased warping—beyond the three-dimensional situation—occurs.

As with electrons, the infinite-potential-square-well problem can be approximated analytically, but a finite well problem does not have an explicit solution. Consider the infinite potential well; neglect warping, so this is a spherical approximation, where the Luttinger parameters γ_2 and γ_3 are now the same (γ). For wavevectors in the plane normal to the direction of the quantization of the angular momentum, the Luttinger matrix decouples into two, with the light hole and heavy hole both with the eigenvalue of

$$E_{lh} = \frac{\hbar^2 k^2}{2m_0}(\gamma_1 + 2\gamma)k^2, \text{ and } E_{hh} = \frac{\hbar^2 k^2}{2m_0}(\gamma_1 - 2\gamma)k^2. \tag{20.36}$$

With a quantum well of size a, along z, the lowest confinement energies—various energy levels with the quantum number n at $k=0$—are

$$\begin{aligned} E_{n,lh} &= \frac{n^2\pi^2\hbar^2}{2m_0a^2}(\gamma_1 + 2\gamma), \text{ and} \\ E_{n,hh} &= \frac{n^2\pi^2\hbar^2}{2m_0a^2}(\gamma_1 - 2\gamma), \text{ with} \\ k_z &= \frac{n\pi}{a}, \text{ for } n=1,2,\ldots. \end{aligned} \tag{20.37}$$

Now, for $k_y \neq 0$, that is, for motion in the plane of the quantum well, there is considerable mixing of states. The energy $E(k_y)$ of the subband conforms to

$$\begin{aligned} &[4k_{lhz}^2 k_{hhz}^2 + k_y^2(k_{hhz}^2 + k_{lhz}^2) + 4k_y^4] \sin(k_{lhz}a)\sin(k_{hhz}a) \\ &\quad + 6k_y^2\, k_{lhz}k_{hhz}[1 - \cos(k_{lhz}a)\cos(k_{hhz}a)] = 0. \end{aligned} \tag{20.38}$$

With $E < 0$ for holes,

$$\begin{aligned} E_{lhz} &= \left[-\frac{2E}{\gamma_1 + 2\gamma} - k_y^2\right]^{1/2}, \text{ and} \\ E_{hhz} &= \left[-\frac{2E}{\gamma_1 2\gamma} - k_y^2\right]^{1/2}. \end{aligned} \tag{20.39}$$

Equation 20.37 also tells us the changes in effective masses because of this quantum well and $\mathbf{k}\cdot\mathbf{p}$ mediated coupling:

$$\frac{1}{m^*_{n,lh}} = -\frac{1}{\gamma_1+2\gamma}\left[1+3\frac{\cos\theta_n+(-1)^{n+1}}{\theta_n\sin\theta_n}\right], \text{ with}$$
$$\theta_n = n\pi\left(\frac{\gamma_1+2\gamma}{\gamma_1-2\gamma}\right)^{1/2}, \text{ and}$$
$$\frac{1}{m^*_{n,hh}} = -\frac{1}{\gamma_1-2\gamma}\left[1+3\frac{\cos\theta_n+(-1)^{n+1}}{\theta_n\sin\theta_n}\right], \text{ with}$$
$$\theta_n = n\pi\left(\frac{\gamma_1-2\gamma}{\gamma_1+2\gamma}\right)^{1/2}. \qquad (20.40)$$

These infinite-well results hold true as a general trend for finite potential wells, where numerical approaches become necessary. With warping, and the quantum well perturbation interacting, much complexity can arise with anti-crossing behavior as the energies evolve with increasing wavevector. However, in all these, non-parabolicity, the positive hole mass feature, continues to be retained.

20.3 *Monolayer semiconductors*

STABLE STATES OF MATTER also exist in a monolayer form, by which we mean they are a single-atom thick or multi-atom thick (a single-atom planar crystalline arrangement with terminating atoms out of plane) but where the continuity exists in a plane. Graphene, which has zero bandgap, and its tubes, which can achieve a bandgap, are examples of the former. Dichalcogenides such as MoS_2, WS_2, equivalent diselenides or ditellurides and others are examples of the latter. Unusual properties arise in these latter compounds due to spin-orbit coupling and due to the interactions that only occur at large atomic numbers, where the higher orbitals are occupied, leading to spin-dependent effects appearing in various forms, such as magnetoresistance.

Carbon is quite a unique atom. The organic world—the living world of the earth—exists because of it. In the creation of the universe, helium was birthed from the α particle in the first generation of stars through the proton-proton reaction. But the atomic numbers to that arrive from protons and α particles happen to be unstable isotopes. Atomic number 12, carbon, has a nucleus whose energy level resonates with the energy of three α particles. Triple fusion becomes possible. This overcomes the mass gap with a nucleosynthesis cascade that generates all the elements up to atomic number 56 (*Fe*). It is the later-generation star processes—supernovas, et cetera—that make the other heavier elements that we know. Carbon has given rise to life and nature, and now we have figured out ways to make this magical element the executioner in an Anthropocene era.

20.3.1 *Graphene*

WE WILL USE GRAPHENE, and the carbon nanotube, to explore these naturally confined materials' semiconductor-specific properties.

Figure 20.8 shows the two-dimensional structure, with the unit cell and the Brillouin zone. The in-plane three bonds are hybridized sp^2, and the residual p_z orbital makes the fourth π_z bond (and

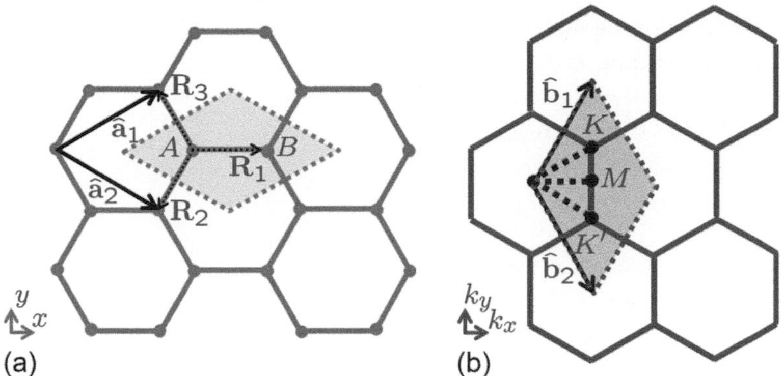

Figure 20.8: Part (a) shows the hexagonal two-dimensional structure of graphene together with its unit cell (a filled rhombus) and $\hat{\mathbf{a}}_1$ and $\hat{\mathbf{a}}_2$ as the two unit vectors of the real space. A and B are the two bases of the unit cell. Part (b) shows the reciprocal lattice, together with the first Brillouin zone. The shaded region is the first Brillouin zone with unit reciprocal lattice vectors of $\hat{\mathbf{b}}_1$ and $\hat{\mathbf{b}}_2$. The major symmetry points of the reciprocal space are identified.

antibond). C–C spacing is $a = 0.142$ nm. We will employ tight binding to model our electronic description.

We start with the unit cell, with its two unit vectors ($\hat{\mathbf{a}}_1$ and $\hat{\mathbf{a}}_2$) that specify coordinates of the basis atoms. This leads to the Brillouin zone with its reciprocal lattice vectors ($\hat{\mathbf{b}}_1$ and $\hat{\mathbf{b}}_2$). We choose $|s\rangle$ and $|p\rangle$ atomic orbitals to build the basis states in sp^2 hybrid. To solve our secular Equation 1.61, and to keep matters simple, and yet reasonably accurate in this instance, we will only consider next neighbor interactions. We need to calculate $\mathscr{H}_{ij}$ (the transfer elements) and $\mathcal{O}_{ij}$ (the overlap elements) to solve the equation, where the solution exists only if the determinant of $\mathscr{H}_{ij} - E\mathcal{O}_{ij}$ vanishes. Symmetry tells us a bit about the nature of the solution. We will just parameterize the transfer and the overlap matrix elements to keep the exercise simple. We have

$$\begin{aligned}
\hat{\mathbf{a}}_1 &= \frac{\sqrt{3}}{2}a\hat{\mathbf{x}} + \frac{1}{2}a\hat{\mathbf{y}}, \text{ and} \\
\hat{\mathbf{a}}_2 &= \frac{\sqrt{3}}{2}a\hat{\mathbf{x}} - \frac{1}{2}a\hat{\mathbf{y}}, \text{ leading to} \\
\hat{\mathbf{b}}_1 &= \frac{2\pi}{\sqrt{3}a}\hat{\mathbf{k}}_x + \frac{2\pi}{a}\hat{\mathbf{k}}_y, \text{ and} \\
\hat{\mathbf{b}}_2 &= \frac{2\pi}{\sqrt{3}a}\hat{\mathbf{k}}_x - \frac{2\pi}{a}\hat{\mathbf{k}}_y,
\end{aligned} \tag{20.41}$$

as shown in Figure 20.8. Let a be the carbon-to-carbon nearest neighbor distance. The length of the unit vectors in real space is $\sqrt{3}a = \sqrt{3} \times 0.142 = 0.246$ nm. The reciprocal lattice constant is $4\pi/\sqrt{3}a$. The reciprocal lattice vectors, by definition again, are normal to the real space lattice vectors. In the Brillouin zone Γ, M, K and K' serve as important symmetry points, with K and K' having several equivalences, but also differences, since this is a lattice with a basis of 2.

For A and B, we construct two Bloch functions from the atomic orbitals, with hybridization of $|s\rangle$ and $|p\rangle$ into the sp^2 configuration, with a residual p_z orbital out of the plane. Bonding states arise in the σ bonds—covalent—from sp^2. The antibonding states arise from π bonds—also covalent—of the p_z orbital. It is the π energy bands that provide the conduction. This, together with the consideration of only nearest neighbor (A-B) interaction make matters easy for us. The Bloch function for a jth atom in general can be written as

$$\psi_j(\mathbf{k},\mathbf{r}) = \frac{1}{\sqrt{N}} \sum_{R,N} u_j(\mathbf{r}-\mathbf{R}) \exp(i\mathbf{k}\cdot\mathbf{R}), \quad \text{for } j = 1,2,\ldots,n. \tag{20.42}$$

For our example, these u_js are the s and p orbitals of the atom A (or of atom B). N is the number of such atoms in the assembly. $\psi_j(\mathbf{k},\mathbf{r}+\hat{\mathbf{a}}_i) = \psi_j(\mathbf{k},\mathbf{r})$ for each i, say, 1 for A, and 2 for B, from translational symmetry:

$$\begin{aligned}
\mathscr{H}_{AA} &= \frac{1}{N}\sum_{\mathbf{R},\mathbf{R}'} \langle u_A(\mathbf{r}-\mathbf{R}')|\mathscr{H}|u_A(\mathbf{r}-\mathbf{R})\rangle \exp[i\mathbf{k}\cdot(\mathbf{R}-\mathbf{R}')] \\
&= \frac{1}{N}\sum_{\mathbf{R}-\mathbf{R}'} E_{2p} + \frac{1}{N}\sum_{\mathbf{R}=\mathbf{R}'\pm\hat{\mathbf{a}}_1} \langle u_A(\mathbf{r}-\mathbf{R}')|\mathscr{H}|u_A(\mathbf{r}-\mathbf{R})\rangle \\
&\quad \times \exp(\pm ik\sqrt{3}a) + \mathscr{O}(|\mathbf{R}-\mathbf{R}'| > 2a_1) \\
&= E_{2p} + \frac{1}{N}\sum_{\mathbf{R}=\mathbf{R}'+\hat{\mathbf{a}}_1} \mathscr{O}(|\mathbf{R}-\mathbf{R}'| \geq a_1),
\end{aligned}$$

and, by symmetry,

$$\mathscr{H}_{BB} = E_{2p} + \mathscr{O}(|\mathbf{R}-\mathbf{R}'| \geq a_2). \tag{20.43}$$

With only nearest neighbor interactions, this reduces to the simple expression

$$\mathscr{H}_{AA} = \mathscr{H}_{BB} = E_{2p}. \tag{20.44}$$

The off-diagonal term results from the interaction between each atom and its three neighbors, which are a apart and oriented at 0, $2\pi/3$ and $4\pi/3$ radians, respectively. Let these positions be called $\mathbf{R}_1$, $\mathbf{R}_2$ and $\mathbf{R}_3$, as referenced to the $\mathbf{R}_0$ origin, as shown in Figure 20.8(a) for A, which is the reference $\mathbf{R}$:

$$\begin{aligned}
\mathscr{H}_{AB} &= \frac{1}{N}\sum_{i=1,2,3} \langle u_A(\mathbf{r}-\mathbf{R}_0)|\mathscr{H}|u_B(\mathbf{r}-\mathbf{R}_i)\rangle \exp(i\mathbf{k}\cdot\mathbf{R}_i) \\
&= t\left[\exp(i\mathbf{k}\cdot\mathbf{R}_1) + \exp(i\mathbf{k}\cdot\mathbf{R}_2) + \exp(i\mathbf{k}\cdot\mathbf{R}_3)\right] \\
&= tf(k), \quad \text{where} \\
t &= \langle u_A(\mathbf{r}-\mathbf{R}_0)|\mathscr{H}|u_B(\mathbf{r}-\mathbf{R}_i)\rangle, \quad \text{and} \\
f(k) &= \left[\exp(i\mathbf{k}\cdot\mathbf{R}_1) + \exp(i\mathbf{k}\cdot\mathbf{R}_2) + \exp(i\mathbf{k}\cdot\mathbf{R}_3)\right].
\end{aligned} \tag{20.45}$$

t is the overlap energy term and is negative due to repulsiveness, and f is a Fourier phase factor. It is a complex function.

By symmetry, $\mathscr{H}_{AB} = \mathscr{H}^*_{BA}$. In the coordinate system of Figure 20.8,

$$f(k) = \exp\left(ik_x \frac{a}{\sqrt{3}}\right) + 2a \exp\left(-ik_x \frac{a}{2\sqrt{3}}\right) \cos\left(\frac{k_y a}{2}\right). \quad (20.46)$$

$t = -3.033\ eV$, and $o = 0.129$. These are calculated parameters assuming $a = 0.142\ nm$ for graphene.

The overlap integral has a similar Fourier phase factor, and an overlap term, similar to that of t, of

$$o = \langle u_A(\mathbf{r} - \mathbf{R}_0) | u_B(\mathbf{r} - \mathbf{R}_i) \rangle \quad (20.47)$$

so that

$$\mathcal{O}_{AB} = \mathcal{O}_{BA} = of(k). \quad (20.48)$$

So, we have the matrix elements as

$$\mathscr{H} = \begin{bmatrix} E_{2p} & tf(k) \\ tf^*(k) & E_{2p} \end{bmatrix}, \text{ and } \mathcal{O} = \begin{bmatrix} 1 & of(k) \\ of^*(k) & 1 \end{bmatrix}. \quad (20.49)$$

The solution to the secular equation is

$$\begin{aligned} E(\mathbf{k}) &= \frac{E_{2p} \pm tw(\mathbf{k})}{1 \pm ow(\mathbf{k})}, \text{ where} \\ w(\mathbf{k}) &= |f(\mathbf{k})|^2 \\ &= \left[1 + 4\cos\left(\frac{\sqrt{3}}{2}k_x a\right)\cos\left(\frac{1}{2}k_y a\right) + 4\cos^2\left(\frac{1}{2}k_y a\right)\right]^{1/2}. \end{aligned} \quad (20.50)$$

The + sign gives the bonding π energy band, the − sign gives the antibonding π^* energy band, and the symmetry properties arise through the weighting function $w(\mathbf{k})$, a Fourier contribution that maps the symmetry properties from real space to the reciprocal space.

With the overlap integral o vanishing, Equation 20.50 implies that the π and π^* bands will be symmetrical at energy E_{2p}. In this approximation limit, the energy dispersion of this two-dimensional crystal is

$$E = \pm t\left[1 + 4\cos\left(\frac{\sqrt{3}}{2}k_x a\right)\cos\left(\frac{1}{2}k_y a\right) + 4\cos^2\left(\frac{1}{2}k_y a\right)\right]^{1/2}. \quad (20.51)$$

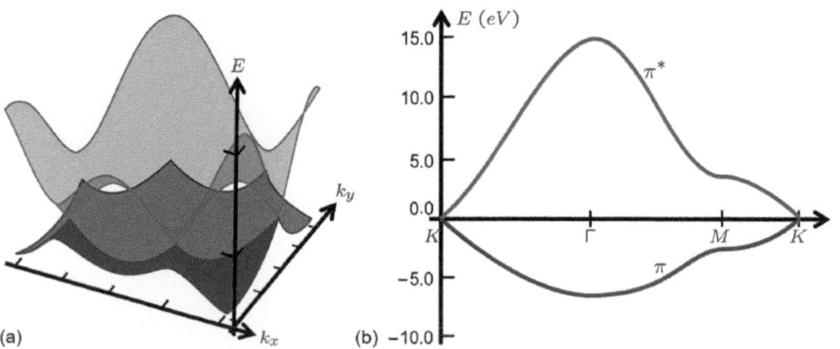

Figure 20.9: The $E(\mathbf{k})$ dispersion for graphene in the first Brillouin zone using tight binding and considering only nearest neighbor interactions. There is no bandgap. Part (b) shows the behavior at major symmetry points and directions. The zero bandgap appears at the K and K' points. These also reflect the symmetry of the p_z-based bonding in the nearest neighbors and that the lattice has a two-atom basis.

This simple and approximate bandstructure is shown in Figure 20.9. The p_z orbitals of A and B contribute one electron each, so there are two π electrons per unit cell. At $T = 0$ K, these occupy the bonding states—the lower π band is the valence band—and the antibonding states—the upper π^* band, which is the conduction band—are empty. The density of states vanishes—our derivation doesn't speak to this directly—at these bandedges; this is a zero-gap semiconductor. This zero gap arises because the A and B sites in this hexagonal lattice are equivalent. In the reciprocal space, this vanishing bandgap is at K and K'. As there is a two-atom basis, while there are symmmetries between A and B in the hexagonal lattice, or between the K and K' in the reciprocal lattice, they are not formally the same. The crystal has a basis of 2.

Is a zero-gap semiconductor an oxymoron? Not really. Note also, we also found a semiconductor where the valence band maximum was not at the zone center. Bandgaps become possible by confinement.

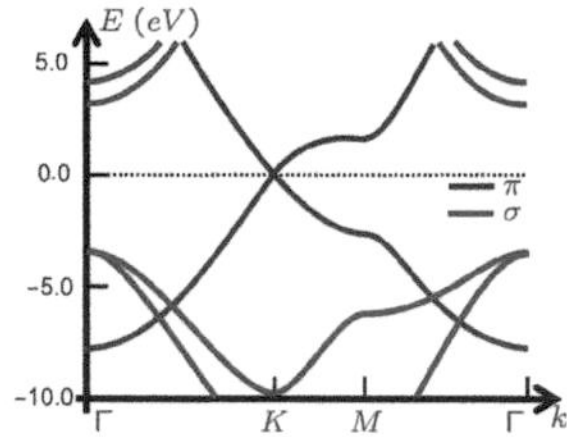

Figure 20.10: The $E(\mathbf{k})$ dispersion for graphene in the first Brillouin zone using tight binding, including both the σ and the π states. Note the stronger bonding and antibonding energies, compared to that of Figure 20.9. Also, note that the symmetry of σ is reflected in the zone center minimum and maximum.

How do the σ band states appear? Recall that these are arising from the sp^2 hybrids from the in-plane atomic orbitals $|s\rangle$, $|p_x\rangle$ and $|p_y\rangle$. These are strong bonds, so they will be farther away from the energy reference of the zero bandgap point. The complete calculation including both the σ and the π bands in a more complex extension is a 6×6 Hamiltonian, where one will need not only the matrix element $\mathscr{H}_\pi = t$, which we have already utilized, but also $\mathscr{H}_{ss}$, $\mathscr{H}_{sp}$, $\mathscr{H}_\sigma$ and the corresponding overlap terms, and the self-energy E_s. Figure 20.10 shows the energy dispersion with both the π and the σ states. Near the Fermi energy—low energy states in conduction and valence bands—not much has changed.

With two carbon atoms as the basis, graphene has 6 phonon branches similar to the zinc blende and diamond structures; that is, 3 acoustic and 3 transverse (see Figure 20.11). Acoustic branches originate at the Γ point, and we identify these via longitudinal (LA), transverse (TA) and out-of-plane (ZA), which is the second transverse direction, modes. This ZA mode has a q^2 dependence, not the normally observed linear dependence for acoustic modes. Three optical branches also exist as LO, TO and ZO. The symmetry of graphene also leads to the linear crossing of the ZA/ZO and LA/LO modes at the K point.

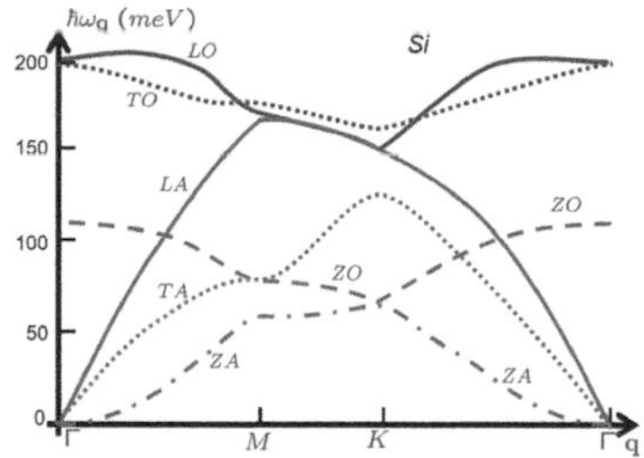

Figure 20.11: The phonon dispersion for graphene. Note the high energies. ZO and ZA refer to the out-of-plane modes.

20.3.2 *Nanotubes*

SEMICONDUCTOR BANDGAPS APPEAR in this sp^2-hybridized form of carbon when it appears in a tubular form, that is, a rolled graphene sheet called a nanotube. Some configurations appear metallic, and these are related again to the symmetries that we noted for graphene. Such tubes can also be formed out of materials such as BN, which can also be formed as a single-atom-thick layer with a hexagonal lattice. *BN*, unlike *C*, now has two different atoms as the *A* and *B* basis. This particular symmetry having been broken, it has a bandgap—unlike graphene—in this monolayer form of the order of 4.5–6.0 eV. Our discussion will be restricted to single-wall carbon nanotubes to elucidate the consequences of tubularity.

Chirality—a vector measure of how the graphene sheet is rolled with an axial symmetry into the cylindrical form—is a key new feature attribute of the spiral confirmation of a tube. With the length long compared to the diameter, this tube is now a one-dimensional structure. The additional dimensional constraint will also determine several of the attributes of interest to us. Figure 20.12(a) summarizes one way to define the tube's lattice. When a tube is rolled along $(m,0)$ and then $(m,0)$ is joined back with $(0,0)$, we get a zigzag pattern for the bonds; this nanotube is a zigzag nanotube, and $(m,0)$ indexes it. When a tube is rolled along (n,n) and then (n,n) is joined back to $(0,0)$, then it has an armchair pattern, and it is often referred to as an armchair nanotube. In general, it is possible to roll along more than these two directions, for example, the end points of the arrow identified with the angle θ, and one is able to identify any (m,n) with which to identify the tube. This is the chirality of the nanotube. In Figure 20.12(a), if, in the forming of the tube, O and P are connected, as is Q with R, then **OP** defines a chiral vector $\mathbf{C}_h$, and **OQ** a translation vector **T**. In this figure, **R** is a symmetry vector.

Growth also places caps on the end of nanotubes. These are six pentagons and more hexagons that then fit with the nanotube's arrangement of hexagons defined by the chirality.

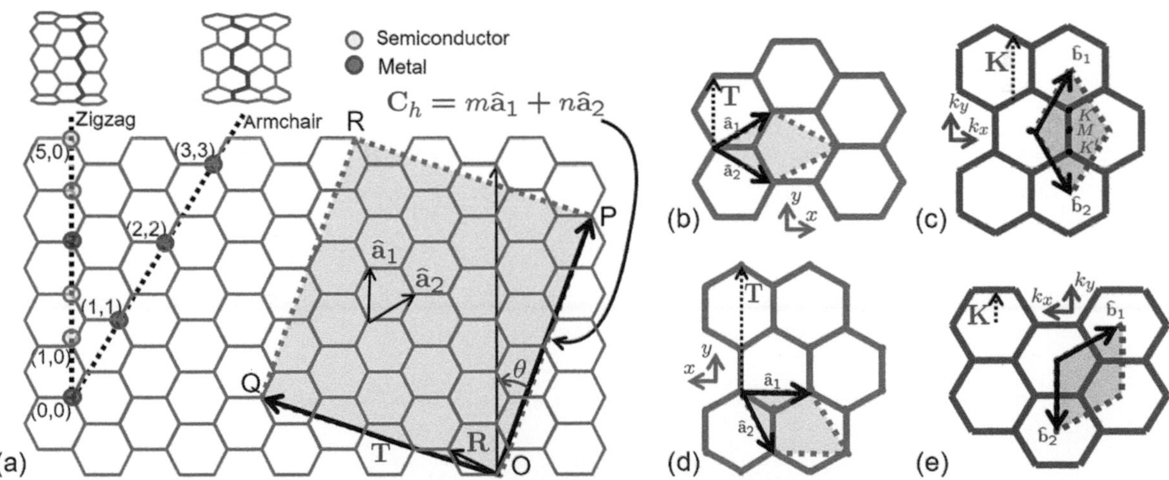

Figure 20.12: Part (a) shows the hexagonal lattice together with the reference systems for the discussion of its rolling into a tube. A unit cell of the nanotube constructed out of the hexagon lattice with a $\mathbf{C}_h$ chiral vector and a **T** translation vector is shown together with the construction of zigzag and armchair tubes by rolling. **R** is a symmetry vector. Parts (b) and (c) show the real space and reciprocal space, respectively, for *armchair* nanotubes. Parts (d) and (e) show the real space and reciprocal space, respectively, for *zigzag* nanotubes. In (b)–(e), when **T** is the translation vector, **K** is the reciprocal lattice vector of those nanotubes.

From a symmetry perspective, nanotubes are either chiral, that is, have spiral symmetry, and a mirror image cannot be superposed on the original, or they are achiral, where they can be superposed. The armchair and zigzag nanotubes are achiral. The chiral vector is

$$\mathbf{C}_h = m\hat{\mathbf{a}}_1 + n\hat{\mathbf{a}}_2, \tag{20.52}$$

whose simpler notation is just (m, n). $0 \leq |n| \leq m$ because of hexagonal symmetry. Armchair nanotubes have $m = n$ or $\mathbf{C}_h = (m, m)$. Zigzags are $\mathbf{C}_h = (m, 0)$. The circumference of the nanotube then is $|\mathbf{C}_h|$. The diameter D then follows

$$D = \frac{1}{\pi}|\mathbf{C}_h| = \frac{a'}{2}(m^2 + n^2 + mb)^{1/2}, \tag{20.53}$$

where $a' = |\hat{\mathbf{a}}_1| = |\hat{\mathbf{a}}_2| \approx 0.249\ nm$. The chiral angle θ—the angle between $\mathbf{C}_h$ and $\hat{\mathbf{a}}_1$—is related through

The C–C distance (a) is 0.142 nm. For graphene, the lattice constant (a') is 0.249 nm. For nanotubes, this unit cell dimension will depend on the chirality.

$$\cos\theta = \frac{\mathbf{C}_h \cdot \hat{\mathbf{a}}_1}{|\mathbf{C}_h||\hat{\mathbf{a}}_1|} = \frac{2m + n}{2(m^2 + n^2 + mn)^{1/2}}, \text{ or, more simply, as}$$

$$\theta = \arctan\left(\frac{\sqrt{3}m}{m + 2n}\right). \tag{20.54}$$

A zigzag tube has $\theta = 0$, and an armchair has $\theta = \pi/6$. The translation vector $\mathbf{T}$, perpendicular to the $\mathbf{C}_h$, is along the tube axis and can be written as

$$\mathbf{T} = t_1\hat{\mathbf{a}}_1 + t_2\hat{\mathbf{a}}_2, \tag{20.55}$$

again more simply written as (t_1, t_2), and it is parallel to the axis of the tube, so normal to $\mathbf{C}_h$, with $\mathbf{T} \cdot \mathbf{C}_h = 0$. $\mathbf{OQ}$ is where the first intersection happens with a lattice point. The unit cell of the nanotube is defined by $OPRQ$, with the vectors $\mathbf{C}_h$ and $\mathbf{T}$. The relationship of the chiral and the translation vectors in terms of the unit vectors leads to $t_1 = (2n+m)/D$, and $t_2 = -(2m+n)/c$, where c is the largest divisor of these numerators. Since n and m are integers, a cycle of 3 in the numerator causes a repetition, and hence if d is the greatest common divisor of m and n, c follows as either c', if $m - n$ is not a multiple of $3c'$, or $3c'$, if it is. The translation vector's length is $|\mathbf{T}| = \sqrt{3}D/c$.

c' is the greatest common divisor of m and n.

It is instructive to see how many of the hexagon areas appear in the unit cell of a nanotube. This is the ratio of the two areas,

$$N = \frac{|\mathbf{C}_h \times \mathbf{T}|}{\hat{\mathbf{a}}_1 \times \hat{\mathbf{a}}_2} = \frac{2D^2}{a^2 c}. \tag{20.56}$$

Since there is a two-atom basis per hexagon, there are $2N$ atoms, and therefore $2N \times 2$ electron states have been introduced from the $2p_z$ orbitals per unit cell of the nanotube. It also implies that there are $2N \times 3$ phonon states arising in the $2N$ carbon atoms.

The zigzag and armchair nanotubes are easier to visualize, but, in general, these tubes with large unit cells are twisting as $\mathbf{C}_h$ chiral rotates through 2π to end up at another lattice point that is the equivalent of the starting lattice point. The hexagons are being displaced translationally as this happens. **R** of Figure 20.12 is a vector that is useful in generating the lattice points of the tube in these general conditions. It is the smallest component of $\mathbf{C}_h$ (or, for that matter, of **T**) that generates the points:

$$\mathbf{R} = p\hat{\mathbf{a}}_1 + q\hat{\mathbf{a}}_2, \quad \text{or} \quad (p, q), \tag{20.57}$$

where p and q are integers with only unity as a common divisor. **R** has the meaning that when one goes around the nanotube by $2\pi/N$, then one also translates by MT/N, where $M = np - mq$. The twisting as one rotates means that $\mathbf{C}_h$ and **T** are tied to each other in coming back to an equivalent lattice point through these relationships, and **R** is a vector that reflects the underlying hexagonal symmetry connections, and thus generates the lattice points through $i\mathbf{R}$, where i is an integer. A number of relationships can be derived between the chiral vector, the translation vector and the symmetry vector, including

$$\begin{aligned} \mathbf{T} \times \mathbf{R} &= (t_1 q - t_2 p)(\hat{\mathbf{a}}_1 \times q\hat{\mathbf{a}}_2), \\ \frac{1}{C_h}\mathbf{C}_h \cdot \mathbf{R} &= \frac{1}{T}|\mathbf{R} \times \mathbf{T}|, \quad \text{and} \\ \frac{1}{C_h}|\mathbf{C}_h \times \mathbf{R}| &= \frac{1}{T}\mathbf{R} \cdot \mathbf{T}. \end{aligned} \tag{20.58}$$

Here, the second equation says that **R**'s projections and products provide a proportional effect that is identical on both chiral and translational vectors. This is inherent to these vectors being unit vectors for the nanotube, and **R** being the vector generating the lattice points.

When p and q are chosen so that **R** lies within the hexagon, that is, $i = 1$, then $t_1 q - t_2 p = 1$, with $0 < np - mq \le N$. With t_1 and t_2 having no common divisor, this expression quantifies p and q uniquely. If p and q are chosen so that **R** lies within the tube's unit cell, then a limiting relationship is $0 < t_1 q - t_2 p \le N$. The vector $i\mathbf{R}$ confined within a hexagon leads to vectors $i\mathbf{R}$ as the N different sites of the nanotube unit cell.

Now consider the independent electron $E(\mathbf{k})$ energy dispersion of the tube. It conveniently follows from the graphene discussion, since this folding by rolling has only introduced periodic boundaries. $\mathbf{C}_h$ is the periodicity along the circumference, and **T** is that along the length. Along the length of the tube, we will have quasi-continuity in the wavevector, while, along the circumference, it is quantized. Thus, it is the circumference that provides the second bounding of this one-dimensional system. This second bounding establishes the wavevector constraint in the direction of repetition of the eigenfunction solution—what we will get as one-dimensional energy dispersion—with multiple subbands that arise as cross-sectional cuts of the two-dimensional dispersion of Figure 20.9 (or the more rigorous calculation of Figure 20.10). And where these cuts will be will depend on the chiral vector that determines the periodic boundary. If the cuts pass through the K or K' points of the originating graphene's Brillouin zone, at different angles, because of the (m, n) specifics of the chiral vector, the nanotube will be zero bandgap. We call this a metallic nanotube, and it is of the armchair type ($m = n$). If the cuts don't pass through the K or K' points, then bandgaps will appear. In the length $\mathbf{T}$ direction, the states are quasi-continuous, and these become propagating states so long as the energy separation is small enough for applied fields to be able to change the electron's momentum.

There is a difference with the quantum wire, even though both the tube and the wire are one-dimensional objects. Quantum wires have potential barriers confining. Here, it is the folding back along the circumference. *Zone folding*—as with atomic orbitals—has an orbital constructive interference. There is no decaying of the wavefunction since there is no forbidden region.

The first Brillouin zone of the reciprocal lattice for the nanotube follows from the real space unit cell bounded by $\mathbf{C}_h$ and $\mathbf{T}$ through the relations

$$\mathbf{C}_h \cdot \mathbf{K}_1 = 2\pi, \quad \mathbf{C}_h \cdot \mathbf{K}_2 = 0, \quad \mathbf{T} \cdot \mathbf{K}_1 = 0, \quad \text{and} \quad \mathbf{T} \cdot \mathbf{K}_2 = 2\pi. \tag{20.59}$$

Here,

$$\begin{aligned} \mathbf{K}_1 &= \frac{1}{N}(-t_2\hat{\mathbf{b}}_1 + t_1\hat{\mathbf{b}}_2), \quad \text{and} \\ \mathbf{K}_2 &= \frac{1}{N}(n\hat{\mathbf{b}}_1 - m\hat{\mathbf{b}}_2), \end{aligned} \tag{20.60}$$

where N given by Equation 20.56 is the number of unit cells of graphene, whose reciprocal lattice unit vectors are $\hat{\mathbf{b}}_1$ and $\hat{\mathbf{b}}_2$. This Brillouin zone is a line. Figure 20.13 shows an example of the construction of this Brillouin zone through Equation 20.60; the graphene sheet from which the tube was folded is shown as the background. For the example given in the figure, with a chiral vector of (4, 2) and the translation vector (4, −5), $N = 28$, leading to a shrunk reciprocal vectors of $\mathbf{K}_1 = (1/28)(5\hat{\mathbf{b}}_1 + 4\hat{\mathbf{b}}_2)$, and $\mathbf{K}_2 = (1/28)(4\hat{\mathbf{b}}_1 - 2\hat{\mathbf{b}}_2)$. As the real space lattice is a one-dimensional object, the reciprocal lattice is also of one dimension. $\mathbf{K}_2$ is the reciprocal lattice vector with the quasi-continuum of states. $\mathbf{K}_1$ gives the discretization around the circumference.

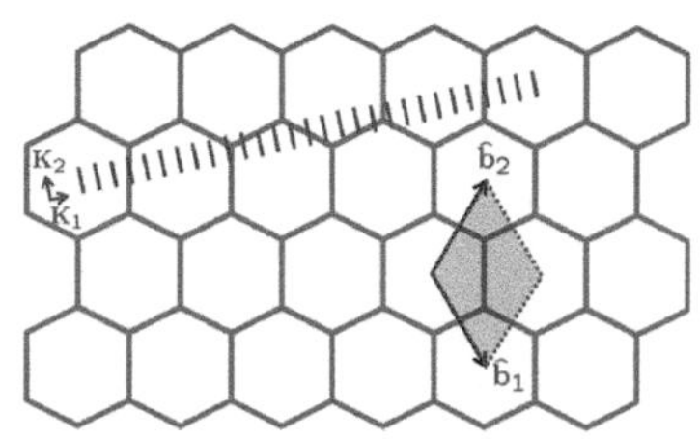

Figure 20.13: The Brillouin zones of nanotubes. They form a line segment parallel to the reciprocal lattice vector $\mathbf{K}_2$. The zones are separated by $\mathbf{K}_1$, and these reciprocal lattice wavevectors follow from the real space unit cell of the nanotube bounded by $\mathbf{C}_h$ and $\mathbf{T}$. The example shown is for $\mathbf{C}_h = 4\hat{\mathbf{a}}_1 + 2\hat{\mathbf{a}}_2$, and $\mathbf{T} = 4t_1\hat{\mathbf{b}}_1 - 5\hat{\mathbf{b}}_2$. N for this case is 28, hence the enormous shrinking of the reciprocal lattice.

In Figure 20.13, the first Brillouin zone is the leftmost line oriented along $\mathbf{K}_2$. This extends from $-\pi/T$ to π/T along the direction $\mathbf{K}_2$, where the quasi-continuum extends for the allowed wavevectors of the tube's longitudinal length direction. The $\mathbf{K}_1$ direction shows the 28 subband quantizations of this $N = 28$

problem arising in the zone folding. The vector $N\mathbf{K}_1 = t_2\hat{\mathbf{b}}_1 + t_1\hat{\mathbf{b}}_2$ is the reciprocal lattice vector of the graphene origin. The two wavevectors, separated by $N\mathbf{K}_1$, are equivalent. And there are $N-1$ parallel line segments that are $\nu\mathbf{K}_1$ apart, where $\nu = 0,\ldots,N-1$, and give rise to the N one-dimensional subband energy dispersions. These are

$$E_\nu(k) = E_{2D}\left(k\frac{\mathbf{K}_2}{|\mathbf{K}_2|} + \nu\mathbf{K}_1\right),$$
$$\text{for } \nu = 0,\ldots,N-1, \text{ and } -\frac{\pi}{T} < k < \frac{\pi}{T}. \quad (20.61)$$

These are the cuts being made along the plane defined by the parentheses.

When a cut passes through the symmetry point K or K', the tube is conductive, since this is the zero gap point, with π and π^* bands degenerate. If the cutting line does not pass through these symmetry points, one has a bandgap.

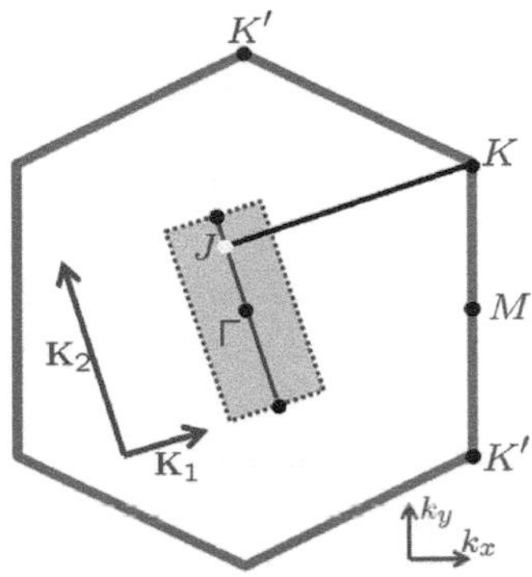

Figure 20.14: A relational view through the Brillouin zones of what makes a nanotube metallic. The hexagon shows the symmetries of the nanotube. The rectangle inside shows the region of extent K_1 and K_2, where $\nu\mathbf{K}_2$ translation builds the first Brillouin zone of the nanotube.

This cutting condition for the conductive state can be physically seen in Figure 20.14, where sections of the reciprocal space graphene and of an arbitrary nanotube are sketched. If the ratio of the length **JK**—passing through the K point of graphene—to that of the $\mathbf{K}_1$ reciprocal lattice vector of the nanotube, both parallel to each other, is an integer ratio, then the cut must pass through the degenerate point:

$$\mathbf{JK} = (2m+n)\mathbf{K}_1/3; \quad (20.62)$$

recall the picture in Figure 20.12(a). So, this condition is equivalent to the conclusion that the nanotubes are metallic when $(2m+n)$ or $m-n$ are multiples of 3. So, tubes with $m = n$—the armchair variety—are metallic, as is any $(m,0)$ of the zigzag family when m is divisible by 3. Figure 20.12(a) depicts this conclusion.

Looking at Figure 20.12(a), one can also conclude that approximately two-thirds of the tubes that can be constructed will have a bandgap. The bandgap will, of course, depend on the magnitude of the chiral vector.

In Figure 20.12(b)–(e), we see parts of the unit cells of armchair and zigzag carbon tubes. For (m,m) armchair tubes, shown in Figure 20.12(b)–(c), the periodic boundary condition is

$$mk_{x,j}\sqrt{3}a = 2\pi j, \quad j = 1,\ldots,2m. \quad (20.63)$$

So, from Equation 20.51, it follows that the chiral vector $\mathbf{C}_h = (m,m)$ leads to the energy dispersion

$$E_j = \pm t\left[1 + 4\cos\left(\frac{j\pi}{m}\right)\cos\left(\frac{1}{2}ka\right) + 4\cos^2\left(\frac{1}{2}ka\right)\right]^{1/2}$$
$$\text{for armchair tubes, } \mathbf{C}_h = (m,m),$$
$$-\pi < ka < \pi, \text{ and } j = 1,\ldots,2m. \quad (20.64)$$

$\mathbf{k}$ is in the quasi-continuum direction of $\mathbf{K}_2 = (\mathbf{b}1 - \mathbf{b}_2)/2$, so, along the K and K' of graphene.

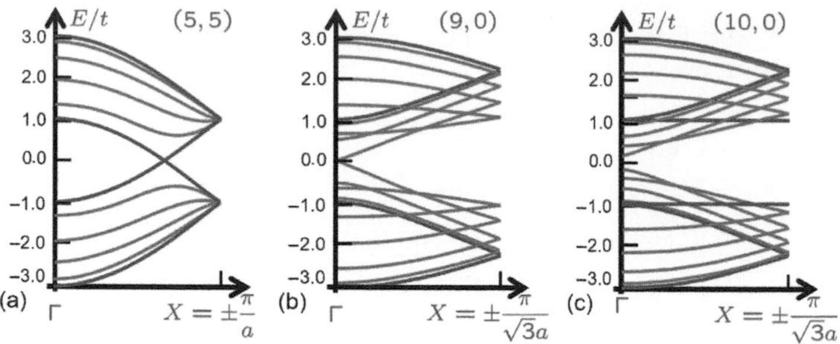

Figure 20.15: Energy dispersion in tubes of a (5,5) armchair (always metallic), a (9,0) zigzag (metallic) and a (10,0) zigzag (non-metallic) armchair nanotube.

The energy bands for zigzag tubes, where $\mathbf{C}_h = (m, 0)$, likewise follow under the constraint

$$mk_{y,j}a = 2\pi j, \;\; j = 1, \ldots, 2m, \tag{20.65}$$

in Equation 20.51 as

$$E_j = \pm t\left[1 + 4\cos\left(\frac{j\pi}{m}\right)\cos\left(\frac{\sqrt{3}}{2}ka\right) + 4\cos^2\left(\frac{j}{m}\pi\right)\right]^{1/2}$$
$$\text{for zigzag tubes, } \mathbf{C}_h = (m, 0),$$
$$-\frac{\pi}{\sqrt{3}} < ka < \frac{\pi}{\sqrt{3}}, \text{ and } j = 1, \ldots, 2m. \tag{20.66}$$

Example tube energy dispersions for different possibilities for zigzag and armchair tubes—at small diameters—are shown in Figure 20.15, using the relationships that we have developed. In armchair tubes (Figure 20.15(a)), which are always metallic, there is a band crossing between the conduction band and the valence band about 2/3rd of the way to the X point. Since the Fermi energy in an ideal condition is centered through this energy, any excitation will cause conduction. In Figure 20.15(b), the armchair tube is again metallic, but now this crossing occurs at the zone center. This example of an (m, m) tube will have $2m$ subbands for conduction and valence each. Of these subbands, there is a double degeneracy for $(m - 1)$ bands, leaving 2 bands non-degenerate. The bands will also have even and odd inversion symmetry. In Figure 20.15(c), the semiconducting nanotube, at the energy E/t, one can observe a k-independent behavior; that is, a non-dispersive behavior across the zone. This independence from dispersion occurs whenever $m/j = 2$—whenever m is even—as can be seen in Equation 20.66.

Note that we are looking at the dispersion through a simple tight binding approximation. We have also not included any spin-orbit interactions. So, consider this as a toy model, but a pretty good toy model that captures much of the interesting implications of interest to us.

Now consider chiral tubes in general. The degeneracy in zigzag or armchair tubes under the factor-of-3 constraint occurred either at $k = \pm 2\pi/3C_h$ or at $k = 0$. Zigzag tubes show the bandgap at this same Γ locale when they are not metallic. Equation 20.61 is a general equation that captures all these different attributes for the three different varieties of dispersion that arise with tubes. It follows from the behavior of ΓJ in Figure 20.14 of the reciprocal space behavioral change from the hexagonal sheets to the tubes:

$$\Gamma J = (n/c)|\mathbf{K}_2|, \tag{20.67}$$

where c is the greatest common divisor of $2m+n$ and $2n+m$, where c either is c', the greatest common divisor of m and n, when $m-n$ is not a multiple of $3c'$, or is $3c'$, when it is a multiple of $3c'$. Folding is related to this ΓJ span through the geometric implications of Figure 20.14. The metallic condition follows from Equation 20.67, so the ratio of JK to ΓJ determines when the point K will fold over to the point J in a nanotube. This is when $\mathbf{K}_2$ also becomes a reciprocal lattice vector. What the degeneracy will be is determined by the nature of the folding, which is reflected in the divisors c and c'. When the tube is semiconducting,

$$E_g = \frac{a}{D} t, \tag{20.68}$$

where t is the overlap energy integral; $o \approx 2.5$–$2.7\ eV$ provides a reasonable fit. Small nanotubes provide bandgaps that are of the order of the several decades of $k_B T$ that are needed at room temperature.

We now reflect on the density of states. Since energy dispersion is known, as is the periodicity of the wavevectors, these follow quite straightforwardly. The density of states at any energy E is the sum of the contributions from all the subbands that have states at that energy, that is,

$$\mathcal{G}(E) = \sum_{j,i=1}^{i=max} \mathcal{G}(E,i,j), \tag{20.69}$$

with the density of states for a one-dimensional system being $1/\pi \nabla_{\mathbf{k}}(E) = (1/\pi)(\partial E/\partial k)^{-1}$. Our armchair bandstructure relationship (Equation 20.64) then gives the following nth subband relationships:

$1/\pi$ because $g_s(2\pi)^{-1}$ with $g_s = 2$ for the spin degeneracy.

$$\mathcal{G}(E,i,j) = \frac{8}{\pi} \frac{|E|}{(E^2 - E_1^2)^{1/2}\left[-\alpha_1 + (E^2 - E_2^2)^{1/2}\right]^{1/2}\left[-\alpha_2 + (E^2 - E_1^2)^{1/2}\right]^{1/2}},$$

with

$$\begin{aligned} E_1(j) &= \pm t \left|\sin\left(\frac{j\pi}{m}\right)\right|, \text{ and} \\ E_2(j) &= \pm t \left[5 - 4\cos\left(\frac{j\pi}{m}\right)\right]^{1/2}, \\ \alpha_1(j) &= t\left[-2 + \cos\left(\frac{j\pi}{m}\right)\right], \text{ and} \\ \alpha_2(j) &= t\left[2 + \cos\left(\frac{j\pi}{m}\right)\right]. \end{aligned} \tag{20.70}$$

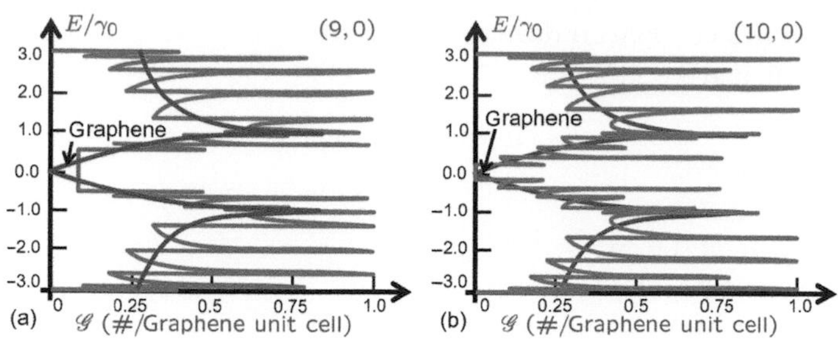

Figure 20.16: Part (a) shows the density of states for a (9,0) metallic zigzag nanotube, and (b) shows the density of states in a semiconducting (10,0) zigzag nanotube.

The number of subbands that need to be counted arises from the energy for the calculation, and the extents that exist for the subbands. The Γ point energy just comes from $k = 0$, and the X from $k = \pi/a$, so,

$$E_\Gamma(j) = \pm t\left[5 + 4\cos\left(\frac{j\pi}{m}\right)\right]^{1/2}, \text{ and } E_X(j) = \pm t. \tag{20.71}$$

In armchair tubes, it is the density of states at the degenerate point that is important. This density is

$$\mathscr{G}(E) = \frac{8}{\sqrt{3}\pi a t} \approx 2 \times 10^5 \ cm^{-1}eV^{-1}. \tag{20.72}$$

For zigzag nanotubes, the energy dispersion relation of Equation 20.66, using this same methodology, is

$$\mathscr{G}(E) = \frac{4g}{\sqrt{3}a\pi}\frac{|E|}{(E^2 - E_1^2)^{1/2}(E_2^2 - E^2)^{1/2}}, \tag{20.73}$$

and, again, one must count the number of subbands that are allowed.

Figure 20.16 shows two examples of such a calculation. We should compare how the circumferential quantization shows up here to how it does for conventional semiconductors. Figure 20.16 shows the density of states for a metallic and a semiconducting zigzag example. These tubes have the chiral vectors (9,0) and (10,0). The figure also includes in it, for reference, the density of states in a two-dimensional graphene sheet. One can see in this behavior a change that is similar to what one sees in semiconducting wires, as shown in Figure H.3—a maximum at the subband edge, and a decrease beyond—because energy varies approximately as the square of the wavevector, and the propagating wavevectors are evenly spaced. Note the inversion at high energy. The density of states changes. This follows from how the bandstructure is changing at the higher energies, as seen in Figure 20.15.

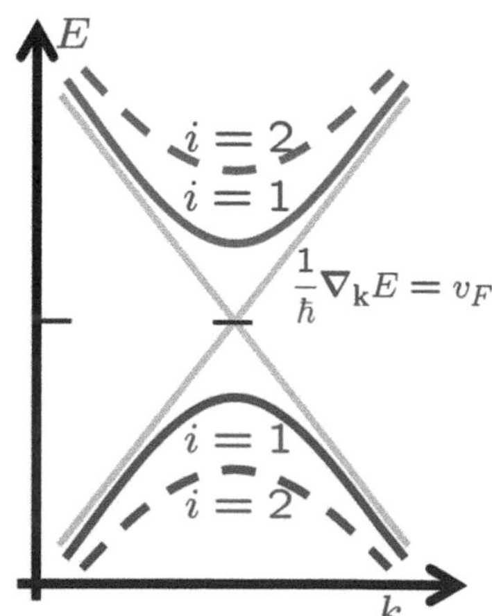

Figure 20.17: Bandstructure for semiconducting nanotubes.

A secondary comment is related to the curving of the tube. In the simple picture shown in Figure 20.17, the folding-based reinforcement, even for metallic geometries, has opened up a bandgap, as our equations have shown for armchair nanotubes. The linear dispersion in graphene has an approximate velocity of v_F, which is $(1/\hbar)\nabla_{\mathbf{k}}E$ and, due to the folding, subbands form that have an effective mass of

$$\frac{1}{m^*} = \frac{1}{\hbar^2}\nabla_{\mathbf{k}}^2 E = \frac{3v_F}{2\hbar}D. \tag{20.74}$$

Having discussed the manifestations of confinement as one uses the potential to limit where carriers can be within matter and stable monolayer materials, it is pertinent to ask, "What are the major physical differences between the two?" As a two-dimensional material, graphene will have all the consequences that arise in the mobile charge cloud that exists in the p_z orbital's states, interacting with the surroundings. Carbon's is a strong bond, and scattering, as an interaction with phonons and randomness in the surroundings, can be limited substantially. So, with ideal transport conditions, numerous interesting consequences can be observed. This includes effects such as the quantum conductance of channels, due to mesoscopic properties. And one doesn't necessarily need very low temperatures! Also, since graphene is a zero-gap material, with linear energy dispersion at low energies, it provides for numerous analogies with the linear energy dispersion of photons. Many properties that are reflective of the speed of light c as an invariant can now be projected at lower speed onto the electron as a charged particle. Graphene ribbons, that is, structures where the lateral dimension is confined by cutting it, will have bandgaps, since additional confinement has been introduced. One can introduce one-dimensional potential barriers and wells in such arrangements. Tunneling across junctions will depend on bandgaps and also shows interesting new effects such as Kline tunneling.

Klein tunneling does not involve quantum tunneling through a classically forbidden region. It arises in two-dimensional massless Dirac electrons with conservation of pseudo-spin.

One point, remarked on in the margins, is that nanotubes have constructive interferences, while the interference for a traditional semiconducting quantum wire involves wavefunction leakage. In addition to the electron-related implications, there are now phonon implications. Acoustic phonons are now twistons based on the strong carbon-carbon bonds, and large mean free paths ($\langle\lambda_{\mathbf{k}} \gtrsim 300\ nm$) become possible in the acoustic phonon-dominated limit.

20.4 *Quantum superlattices*

SUPERLATTICES ARE HETEROSTRUCTURE ASSEMBLIES, an extension of the quantum well, that make it possible to substantially modify "bandstructure" by making it possible to have transport through assemblies of quantum wells through the introduction of close

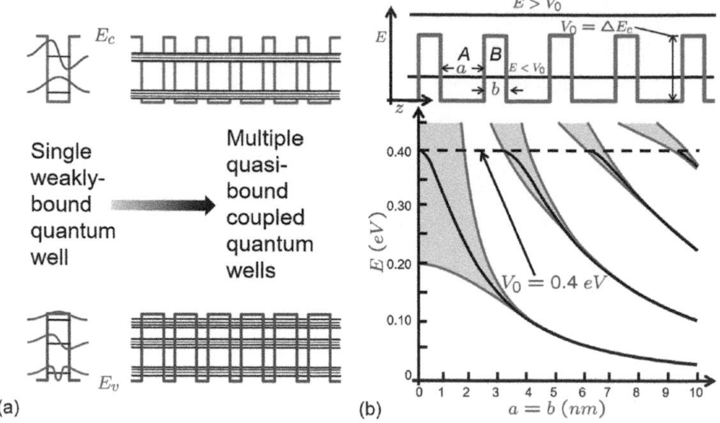

Figure 20.18: Part (a) shows a weakly confined quantum well formed from a heterostructure on the left with two bound states for electrons from the conduction band, and three for holes from the valence band. When this structure is assembled with periodic, narrow barriers in energy, as in the heterostructure superlattice on the right, extended states are formed because of the weak confinement and breakdown of degeneracy. Two electron states at a single energy are now twelve electron states in six energetically close energies—a miniband. This is a superlattice, where the electron, or the hole, can extend over the entire structure in the miniband. Part (b) shows a Kronig-Penney periodic potential with $V_0 = \Delta E_c$, a period of $a + b$ as the sum of well and barrier widths and the energy of allowed solutions for a barrier with $V_0 = 0.4\,eV$, with the free electron mass as $a = b$ is varied. At large barriers and well widths, only discrete levels—4 of them—exist. As the size is reduced, minibands form, allowing a broadband of propagating states in energy. The single well of width a and $b \to \infty$ discrete level solution up to $E = V_0 = 0.4\,eV$ is shown, together with the superlattice solution.

When the collections of quantum wells couple, they form minibands through which transmission continues—periodic confinement allows this transmission—and it also becomes possible to introduce more confined regions in-between, where desired interactions are programmed in. The superlattice now allows injection and extraction to this rate-limiting confined region. Cascade lasers are one example of this and are discussed in S. Tiwari, "Nanoscale device physics: Science and engineering fundamentals," Electroscience 4, Oxford University Press, ISBN 978-0-198-75987-4 (2017).

coupling between them and the consequent modification of the $E(\mathbf{k})$ characteristics. These modifications are particularly useful for exploiting optical transitions between the reformed collection of states.

Superlattices, as a coupled extension of heterostructure quantum wells, derive their interesting characteristics through the overlap of wavefunctions. When short, leaky quantum wells, that is, a quasi-bound structure made using a heterostructure with a small discontinuity in the conduction band, the valence band or both, are assembled together, superlattice behavior arises in the longitudinal direction, where the Bloch states now also have the periodicity of the superlattice constructed using multiple quantum wells. The quasi-bound states now become extended states due to the small discontinuity, small widths and consequent leakiness that leads to wavefunction overlap. This makes conduction possible across the periodic discontinuities. A schematic of this in the longitudinal transport direction (z or $\perp$ in our notation) is shown in Figure 20.18(a). Bloch states in the transverse direction are still "free electron"–like but, in the longitudinal direction, form minibands.

The example shown in Figure 20.18(a) illustrates the following physical behavior. When the discontinuities are small, the smaller confinement potential results in deeper penetration of the wavefunction of confined particles in classically disallowed regions. On the left, two electron wavefunctions and energies, each allowing $\pm 1/2$ spin, are shown. The holes have been chosen to have three allowed energies, out of deference to the larger hole mass. Both the electron and the hole wavefunction penetrate in the larger bandgap material over a size scale of the order of the quantum well width, as shown. On the right, six of these wells are brought together with a small barrier region. The degeneracy of the six separate wells is removed due to the interactions, and the new states also represent the periodicity of the structure. There are six energy levels, instead of one, quite close to each other and extending out across the periodic structure. A miniband of six energy levels capable of holding twelve carriers has been formed. The carriers can travel across this periodic structure, and, depending on the boundary conditions on either side, out. This is a superlattice, with an effective new bandstructure consisting of these minibands—a bandstructure—with their own unique properties.

The superlattice described here as a layered periodic structure through which electrons and holes can travel is an artificial construct. But one could just as well view compound semiconductors as layered periodic structures. *Ga* planes and *As* planes stacked in specific forms form the *GaAs* crystal. It is a superlattice of sorts. The distinction is only in the energies that bind. And *GaAs* conducts!

The Kronig-Penney model used as a toy example in introductory texts is an example of a superlattice model. Since it illustrates the importance of and the relationship between the energy of the barrier, the width of the barrier and the quantum well, and their magnitudes, which are comparable to that of the Bohr radius in the periodic arrangement, we employ it here. Figure 20.18(b) shows a spatially periodic structure with a barrier $V_0 = \Delta E_c$, where the wells are of width a for region A, and of width b for region B. The structure has a periodicity of $a + b$, so a Brillouin zone width of $2\pi/(a + b)$. We assume that both the well and the barrier are isotropic with identical mass, to simplify this calculation. The Schrödinger equation in the two regions with plane-wave-propagating modes in the xy plane is

$$-\frac{\hbar^2}{2m_A^*}\left(k_x^2 + k_y^2 + \frac{d^2}{dz^2}\right)\psi_A(z) = E\psi_A(z), \quad \text{for } z \in A, \text{ and}$$

$$-\frac{\hbar^2}{2m_B^*}\left(k_x^2 + k_y^2 + \frac{d^2}{dz^2} + V_0\right)\psi_B(z) = E\psi_B(z) \quad \text{for } z \in B. \qquad (20.75)$$

We are looking for the propagation properties in the longitudinal, that is, perpendicular z direction. The boundary condition is the continuity of the wavefunction and of $\partial\psi_A/m_A^*\partial z = \partial\psi_B/m_B^*\partial z$ at the A/B boundary, where we will take $m_A^* \approx m_B^*$ to simplify. These two boundary conditions are that of continuity of probability and that of probability current at the interface. Referencing the energy to the bottom of the conduction band in the smaller band material, $E = \hbar^2 k_1^2/2m_A^*$, and $E - V_0 = \hbar^2 k_2^2/2m_A^*$, for $E > V_0$, and $V_0 - E = \hbar^2\kappa^2/2m_A^*$, for $E < V_0$, when $k = i\kappa$ is imaginary. The traditional solution techniques for this lead to the following implicit equation using the boundary conditions:

$$\begin{aligned}
\cos kd &= \cos(k_1 a)\cos(k_2 b) - \frac{k_1^2 + k_2^2}{2k_1 k_2}\sin(k_1 a)\sin(k_2 b), \\
&\quad \text{for } E > V_0, \text{ and} \\
\cos kd &= \cos(k_1 a)\cosh(\kappa b) - \frac{k_1^2 - \kappa^2}{2k_1\kappa}\sin(k_1 a)\sinh(\kappa b), \\
&\quad \text{for } E < V_0.
\end{aligned} \qquad (20.76)$$

Here, k_1 is the wavevector in region A, and k_2 and κ are the wavevector and extinction coefficients, respectively, in region B for energies that are higher or lower than the bandedge potential. What these equations imply is that energies lower than V_0 may have a transmitting solution when potential and width conditions exist that allow sufficient coupling between the wells. It is when this happens that minibands form and the degeneracy of coupled wells is removed.

Figure 20.18(b) shows the result of this calculation for $V_0 = 0.4\ \ eV$, where $m^* = m_0$, and a symmetric structure where the well and barrier regions have identical width that is varied. The electron mass is assumed to be the free electron mass. The discrete levels, 4 of them in this example, exist for the isolated well of width a. When the periodic structure is formed, the minibands—a broadband of allowed transmissive energy states—appears. These appear at higher width in the highest quasi-bound states first. Compound semiconductor heterostructure systems such as $(Al, In)As/(Ga, In)As$, $(Al, Ga)As/GaAs$, et cetera, all have discontinuities in this 0.4 eV range. What this figure shows is that, with wells and barriers of the order of a few nms, minibands form, and transmission occurs through these miniband states. This is a resonant and elastic transport exemplifying coherent tunneling. It is no different than what one observes in periodic film gratings for light. Optical transmission and reflection bands form. The Kronig-Penney model is a toy model, quite simplified, but it is instructive. The conclusions drawn are quite useful with electrons. For holes, things are quite a bit more complicated because of anisotropy and the various idiosyncratic hole bands. The Kronig-Penney model, for example, assumes carrier interaction extending over several unit cells so that the semiconductor picture of an effective mass, a discontinuity, et cetera, are all applicable. At the smallest well and barrier width sizes, this is inappropriate. In more complex and real situations, in the presence of this periodicity, we must resort to various bandstructure calculation techniques in the presence of this periodicity. These techniques adopt a supercell approach for superlattices, and we will not dwell on this here.

20.4.1 *Shallow dopants in confined conditions*

We have now seen the variety of changes that come about in the electron's and the hole's allowed energy states due to confinement. Changes will also appear for the binding energy state of shallow hydrogenic dopants. The binding energy is a function of the confining potential. In the hydrogenic model, it was the Coulomb attraction, for example, between the ionized donor and the electron. But the extent of this attraction is of the length scale of the order of the effective Bohr radius. A confining length scale of this similar order of magnitude arises from the confinement. And this confinement has an energy magnitude that will often be much larger than the binding energy, which is a few *meVs*. Simply stated, a toy Hamiltonian in the effective mass form that characterizes the electron states,

$$\hat{\mathscr{H}} = -\frac{\hbar^2}{2m^*}\nabla_{\mathbf{r}}^2 + V(z) - \frac{e^2}{4\pi\epsilon|\mathbf{r}-\mathbf{R}|}, \tag{20.77}$$

where $\mathbf{R}$ is the coordinate of an isolated dopant, now has this additional term $V(z)$.

Recall that the effective mass allowed us to remove the crystal potential, leaving only other perturbations to be accounted for. Of course, this still requires the effective mass to be a valid tool. And that depends on the validity of the Bloch electron's extended nature.

The solution now will even depend on the position of the dopant $\mathbf{R}$ vis-à-vis the confinement boundaries. Where a donor is affects the states. And the energies for such a calculation will be specific to this solution. Variational methods will usually be a good tool for solving such specific situations with a wavefunction $\psi(z) \propto \exp(-|\mathbf{r}-\mathbf{R}|/\lambda)$, where λ is a variational parameter, a suitable starting choice since the hydrogenic wavefunction is radially symmetric. There is not that much of interest here. The dopants are intentionally not in quantum-confined regions, where the properties of interest are lower scattering or improved optical processes. Donors in the barrier region such as of $Ga_{1-x}Al_xAs/GaAs$ and other high mobility aimed systems are ionized, and small changes in their energies are a lower order effect than the others. What we should, though, stress is that if a donor is present in the confined region, its consequences in changing energies will be a lowering for the dopant-electron system. And as the well width decreases, the lowest energies will rise. The same is true for the binding energy for the electron localized on the dopant.

20.5 *Screening in confined conditions*

WE NOW DISCUSS SOME of the salient aspects of the changes in screening and of the perturbation interactions in transport as dimensionality is reduced in our confined systems. We have noted that the response of a system to a weak perturbation potential can be seen through the dielectric function and is space and time dependent. We will not consider the nonlocal aspects of the dielectric response and, unless stated, illustrate the behavior of a two-dimensional system.

The polarizability arising in a static perturbation of wavevector q arising from a longitudinal electric field $\mathcal{E}_0 \exp[i(\mathbf{q}\cdot\mathbf{r}-\omega t)]$, with one subband occupied, is characterized by

In this field expression, we have used $\mathbf{q}$ to denote the wavevector of the electromagnetic wave in order to distinguish it from $\mathbf{k}$ as the wavevector of the electron.

$$\chi(q) = \frac{m^* e^2}{\pi\hbar^2 q^2}\left\{1 - \left[1 - \left(\frac{2k_F}{q}\right)^2\right]^{1/2}\right\}, \tag{20.78}$$

for $q > 2k_F$, and just the prefactor for $q < 2k_F$, an equation we write without proof. Since the carriers are spread over a width orthogonal to the interface, $\int \varsigma^2(z)\,dz = 1$ describe this spread. In the presence of a weak potential perturbation $V'(q,z)$ that is sinusoidal in q, we can use a charge average of the perturbation to determine the charge perturbation:

$$\langle V' \rangle = \int V'(z,q)\varsigma^2\,dz \quad \therefore \quad \delta\rho(q,z) = -\frac{1}{2}\epsilon q_s \langle V' \rangle \varsigma^2. \tag{20.79}$$

q_s is a screening parameter, which, with one subband occupied, is

$$q_s = \frac{q^2}{2\epsilon}\chi(q). \tag{20.80}$$

The polarizability of interest is the long wavelength polarizability limit. Here, the screening parameter is

$$q_s = \frac{e^2}{2\epsilon}\frac{dN_s}{dE_F} = \frac{e^2}{2\epsilon}\frac{N_s}{k_B T} \tag{20.81}$$

at higher temperature non-degenerate conditions. If Ze is the charge located at z_0, the perturbing potential follows through the Poisson equation as

$$\nabla_{\mathbf{r}} \cdot [\epsilon(z)\nabla_{\mathbf{r}} V'(r,z)] - 2\epsilon q_s \langle V'(r) \rangle \varsigma^2 = -Ze\delta(x)\delta(y)\delta(z - z_0), \tag{20.82}$$

where r is the lateral radial extent in this three-dimensional situation. In Si, the Fang-Howard function is a good approximation for the envelope function:

$$\varsigma(z) = \left(\frac{b^3}{2}\right)^{1/2} z \exp\left(-\frac{bz}{2}\right). \tag{20.83}$$

For Si, $q_s \sim 2 \times 10^7\ cm^{-1}$ for (100) inversion, and, for $GaAs$, $q_s \sim 2 \times 10^6\ cm^{-1}$.

Screening in two-dimensional conditions is weaker than in three-dimensional conditions, since any perturbing charge can only be surrounded by screening electrons in the two dimensions allowed. Perturbation, instead of falling exponentially as in three-dimensional systems, now has a weaker, third-power dependence. The screening is weakened further as one goes to one-dimensional and zero-dimensional systems. Now the static polarizability varies as

$$\chi(q) = \frac{2m^* e^2}{\pi \hbar^2 q} \ln \frac{q - 2k_F}{q - 2k_F}, \qquad (20.84)$$

which again leads to a further weakening of the screening.

20.6 *Scattering in confined conditions*

FOR TRANSPORT, THERE ARE TWO PHYSICAL LENGTHS that are of particular import. We have often cited the mean free path $\langle \lambda_{\mathbf{k}} \rangle$ as one. This is the average distance between scattering events. The other that we have not discussed as much is the $\langle \lambda_\phi \rangle$, the length scale for phase coherence. When temperatures are near room temperature, phonon scattering being dominant, phase coherence disappears over the span between the phonon scattering events. At low temperatures, the phonon scattering rate shrinks, as do other inelastic scattering events such as carrier-carrier scattering. Elastic scattering may dominate. Phase incoherence arising in inelastic scattering is $\lambda_{\mathbf{k}} \sim (2d\mathcal{D}\tau_{\mathbf{k}})^{1/2}$, where $\lambda_{\mathbf{k}}$ and $\tau_{\mathbf{k}}$, as well as d the distance traveled between scattering events, are the measures for the inelastic events. $\mathcal{D}$, on the other hand, is controlled by elastic scattering and is largely independent of temperature, so $\tau_{\mathbf{k}}$'s increase leads to an increase in the phase-coherence length, with 1–10 μm quite achievable in compound semiconductor systems.

The length scale where phase coherence is maintained as λ_ϕ is an important element in the discussion of many mesoscopic phenomena and their devices. We turn to it often in S. Tiwari, "Nanoscale device physics: Science and engineering fundamentals," Electroscience 4, Oxford University Press, ISBN 978-0-198-75987-4 (2017).

Because the density of states has changed, the scattering rate must change too. If scattering is random and isotropic, then the scattering time expectation Equation 9.26 for three-dimensional conditions must account for it. Therefore,

$$\langle \tau_{\mathbf{k}} \rangle = \frac{2}{3} \frac{\int \tau_{\mathbf{k}} \left(-\partial f_0/\partial E\right) \eta^{1/2}\, d\eta}{\int_0^\infty f_0 \eta^{1/2}\, d\eta}, \text{ with } \eta = \frac{E - E_n}{k_B T}, \qquad (20.85)$$

where the integral is over the range of the bands, will have the f_0 dependences for two-dimensional confinement or one-dimensional confinement as well as a change in $\tau_{\mathbf{k}}$, where various various scattering mechanisms have different energy dependences. The effective scattering rate $1/\tau_{\mathbf{k}}(E)$ will reflect these changes. There is one direct and important consequence of this for scattering. Matthiessen's rule, given in Equation 8.59—the net scattering rate (or inverse mobility) being the sum of scattering rates arising in individual events (individual inverse mobilities), that is, a geometric mean mobility—is generally quite invalid. Explicit energy dependences become important. For low temperatures, temperature-independent scattering and with $\tau_{\mathbf{k}} = \tau_0 \eta^r$ for a specific process, the mobility arising in that process will follow

$$\mu(T) = \mu(0)\left[1 + \frac{\pi^2}{6} r\left(r + \frac{3}{2}\right)\left(\frac{k_B T}{E_F}\right)^2\right]. \qquad (20.86)$$

As before, mobility can increase with increasing temperature in conditions where impurity scattering dominates.

There are numerous interesting deviations that will occur due to the confinement. First, recall the scattering through the angle θ noted in Equation 8.50. The scattering cross-section varies as the square of the Fourier component of the scattering potential, with $|\mathbf{k} - \mathbf{k}'| = 2k\sin(\theta/2)$. Motion in plane is therefore favored. Second, in heterostructures of compound semiconductors, often the two-dimensional carrier regions are obtained in a lower or undoped semiconductor by separating the dopants away into the second semiconductor; for example, in $Ga_{1-x}Al_xAs$ adjacent to $GaAs$. For n-type dopants, the field of these donors is terminated on electrons in the two-dimensional channel in thermal equilibrium. The dopants are separated from the carriers, the carriers are still available to screen other local perturbations of residual impurities, and the result is that Coulomb scattering decreases even as carrier concentration is usefully significant. This charge transfer using undoped spacing at the interface of the dopants from the carriers increases the mobility up to an optimal spacer layer thickness, where the weakening of Coulomb scattering continues to prevail despite increasing electron-scatterer separation. Third, at high carrier concentrations, carriers are in several subbands. and the screening must account for this occupation. The $\mathbf{k}_F$ being different in different subbands, the different bands will have different intensities of the effect on scattering. Occupation of more bands, with their smaller Fermi wavevector, increases scattering rate. Fourth, given the nature of correlations and its influence on Coulomb scattering, correlation makes the occupation of charged sites non-random. The lowest energy occupation will usually lead to the weakening of potential fluctuations. This increases mobility over the random case. Fifth, when one goes to even lower temperatures, strong localization, hopping conduction and, finally, for good high mobility and large phase-coherence length conditions, various mesoscopic effects such as quantum Hall in its various forms also become possible.

Phonon scattering's temperature dependence dominantly arises in the occupation of the phonon modes, so by lattice temperature, if the phonons are in equilibrium, or by the dynamics of generation or recombination, if far from equilibrium. At quite low temperatures, the energy and momentum conservation conditions place additional constraints, thus affecting the temperature dependence of mobility.

Interface roughness arises in both the stochastic fluctuations due to the amorphous nature of SiO_2 and any of the randomly distributed physical atomic steps at the interface.

Finally, one important scattering mechanism that we have not dwelled on in this text is that due to the interface roughness.

Both in the SiO_2/Si system, where it is large due to interface potential fluctuations arising in the amorphousness of SiO_2, and in the single crystal compound semiconductor system, where it is due to deviations from planarity, it is of import. For the SiO_2/Si system, it is so large that treating it as a diffuse scattering event that limits the mobility by as much as the other mechanisms suffices, with interface potential fluctuations dominating. For compound semiconductors, since the mobility is extremely large, even if small, it is of importance in the pursuit of novel nanostructures. The non-planarity can be characterized by the autocorrelation of the interface boundary, that is, by

$$\langle \delta z(\mathbf{r}')\delta z(\mathbf{r}'_{\mathbf{r}})\rangle = \Delta^2 \exp\left(-\frac{r^2}{\Lambda^2}\right). \quad (20.87)$$

Here, Δ characterizes the root mean square roughness of the interface, and Λ the lateral correlation length.

This interface roughness has a major implication for semiconductor lasers. The quantum well bipolar laser, where both electron and holes are confined, have their energy states varying simultaneously—correlated—as the quantum well size fluctuates. The linewidth of emission is therefore tremendously lower than that of unipolar lasers, such as quantum cascade lasers, where transitions between electron subbands are utilized for stimulated light emission. This roughness also has implications for tunneling and the variations in tunneling current when quantum barriers are employed.

20.7 *Optical transitions in confined conditions*

Optical transitions in confined conditions change pronouncedly from the bulk unconfined conditions discussed in Chapter 12. Monolayers can be even more different. Two important aspects that cause this change are emphasized in this section. The first, quite clear from the discussion of this chapter, is the changes in the nature of the states and their occupation, as represented in the carriers that can interact. The second is that of selection rules. Our treatment of electron and hole states in the crystal showed many significant differences. Symmetries and degeneracies could be different, but we found, in particular, the consequences of spin-orbit interaction on hole states—of light-, heavy- and split-off holes—to be of enormous importance due to their localization. In confined structures, the characteristics of the Bloch states will play a very significant role. To the lowest order, the consequence of confinement on the occupation of states is through the changes that cause the density of states to reduce with energy. Occupation of states closer to the subband edge becomes more important. So, with quantum wells and with quantum wires, a high or a maximum in state density at the subband minimum, where the charge will exist maximally, the intersubband recombination will have a narrower linewidth than a less confined condition. The consequence of the interaction of the electrons with holes as a result of electromagnetic perturbation is that the conservation rules—of energy and momentum, which are both quantized—and the specific quantum attributes—of spin and orbital angular momentum of the different type of hole states—become important and ties in with photon polarization.

In all of the following discussion for absorption, the light is incident on the plane of semiconductor normally, and, by convention, it is in the z direction.

20.7.1 *Selection rules; interband and intraband transitions*

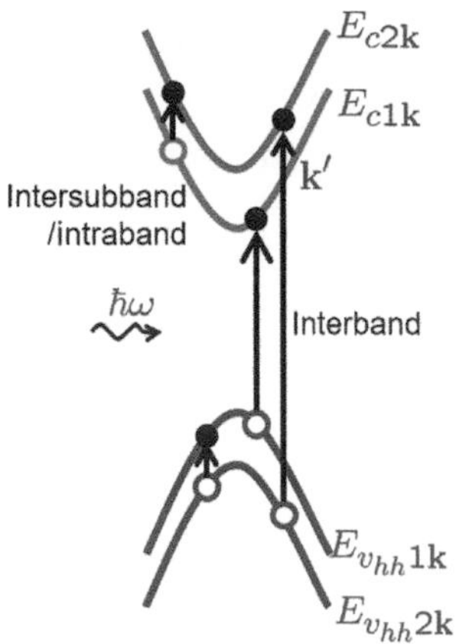

Figure 20.19: Nomenclature of electron transitions under photon absorption in confined conditions. A photon absorption that occurs within the same (conduction or valence bands) is an intersubband transition and may more loosely be called an intraband transition. A transition between the bands themselves is an interband process. The former is between states that have similar symmetry and which are shifted in energies and so have similar energies for a variety of **k**s. The latter, here between heavy-hole and conduction subbands, has confinement-caused changes in state distribution with energy and so reflect the step changes of the density of states. Hole band confinement and related transitions are more complicated than illustrated here.

WE FIRST STRESS what transitions are between the states that are allowed and those that are not allowed for a direct optical process. These are selection rules. Selection rules for transitions can be visualized and understood by looking at the symmetries of the states of the system and the symmetry of the perturbation causing the interaction. How the perturbation Hamiltonian behaves under the symmetry operations of the crystal will determine whether any possible transitions couple initial and final states and their matrix elements. Some of these states may even be degenerate in energy, such as at $\mathbf{k} = 0$ for holes. All of this determines whether the matrix element vanishes or is finite, or even if it gets suppressed or inflated when confinement exists.

Optical transitions can occur within conduction or valence bands and in-between them. In confined conditions, we will have to tackle these via subbands, as Figure 20.19 illustrates for nomenclature. If the transition occurs within the same type of band, we will refer to this as an *intersubband* or *intraband* transition. If it is between a conduction band state and a valence band state, it will be called an *interband transition*. Calling it intersubband would be an error, since then the term doesn't distinguish between the conduction or the valence band and that affect the symmetries of the states. The term "band" here encompasses the subbands that confinement brings out from either the conduction states or the valence states. Intersubband transitions, being between subbands that are reasonably parallel if anisotropy is small, occur over a breadth of states whose energy separation doesn't change much. Interband transitions, however, will be reflective of the different subband energy changes of electron of electron and hole bands and will have the "staircase" energy spread in the absorption. Figure 20.19 only shows a heavy-hole band. Real structures will have to include light holes and split-off holes, anisotropies and symmetries to reflect the additional constraints and the various processes that may occur.

Note that wurtzite crystals have different symmetries, Since the dipole matrix element will be different, optical transitions in direct gap wurtzite crystals should be expected to behave differently subject to selection rules embedded in the matrix element. Another important wrinkle is the reduced spin-orbit interaction, since the constituent atoms are of low atomic numbers. The general form of Equation 12.16 still holds true for bulk conditions.

Which transitions are allowed and which are forbidden were first tackled for electromagnetic interaction in Chapter 12. We discussed interband transitions for bulk semiconductors (zinc blende) and wrote the matrix element of transition involving the field and the dipole in Equation 12.14.

More comments about the state and the matrix element's nature and implications for light are in order here, and we do this with reference to Figure 20.20 where confinement exists. Let $\mathbf{r}$ be a measure of the dipole orientation and the displacement of the charge. The dipole moment then is $\mathbf{p} = -e\langle\mathbf{r}\rangle$, so the allowed transition perturbation is $\mathscr{H}' = \mathbf{p} \cdot \boldsymbol{\mathcal{E}}$. The transition rate is

$$S_{if} = \frac{2\pi}{\hbar} \sum_{i,f} |\langle f|\mathscr{H}'|i\rangle|^2 \delta(E_f - E_i + \hbar\omega) \quad (20.88)$$

by the Golden rule. Equation 12.14 picked on the matrix element part of this through the vector potential. Let us just write the matrix element that the perturbation couples without the constants and the normalizations terms. The light is incident perpendicular to the interface in the z direction. $|\langle f|\mathscr{H}'|i\rangle| \propto |\langle f|\mathbf{r} \cdot \boldsymbol{\eta}|i\rangle$, where $\boldsymbol{\eta}$ is the light's polarization vector;

$$|\langle f|\mathbf{r} \cdot \boldsymbol{\eta}|i\rangle| = \int \varsigma_e^*(z) \exp(-i\mathbf{k}_{e\perp} \cdot \mathbf{r}_\perp) \times u_{c,\mathbf{k}_e}^*(\mathbf{r})\boldsymbol{\eta} \cdot \mathbf{r}\varsigma_h(z) \exp(i\mathbf{k}_{h\perp} \cdot \mathbf{r}_\perp) u_{v,\mathbf{k}_h}(\mathbf{r})\, d^3\mathbf{r} \quad (20.89)$$

for interband transition coupling conduction and valence bands, which is the case of Figure 20.20. Here, the ςs are envelope functions, and us are the modulation terms of the Bloch function. us vary rapidly over the unit cell. ςs are slowly varying functions. So, the matrix element can be simplified by taking the slowly varying function out of the integration over the space:

$$|\langle f|\mathbf{r} \cdot \boldsymbol{\eta}|i\rangle| \approx \sum_{\mathbf{R}_i} \varsigma_e^*(\mathbf{R}_i)\varsigma_h(\mathbf{R}_i) \exp[i(\mathbf{k}_{h\perp} - \mathbf{k}_{e\perp}) \cdot \mathbf{R}_i] \times \int_{\Omega_0} u_{c,\mathbf{k}_e}^*(\mathbf{r})\boldsymbol{\eta} \cdot \mathbf{r} u_{v,\mathbf{k}_h}(\mathbf{r})\, d^3\mathbf{r}. \quad (20.90)$$

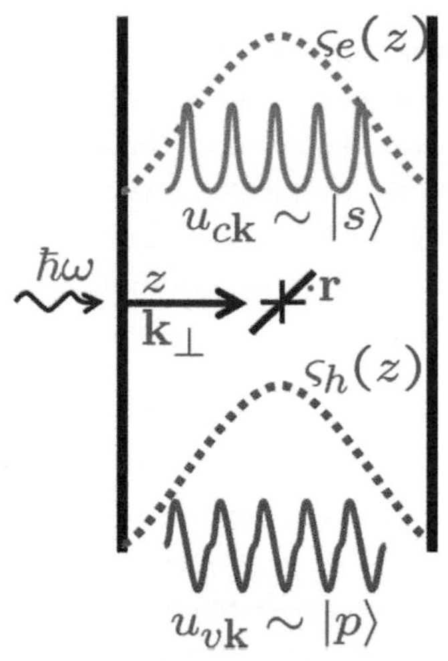

Figure 20.20: The nature of symmetries in an interband transition between conduction band and valence band states. Light is incident normally, and there exists confinement in this z and $\mathbf{k}_\perp$ direction. The conduction band's $|s\rangle$-like states are spatially symmetric. The valence band's $|p\rangle$-like states are spatially asymmetric. Envelope functions asymptotically vanishing at the potential boundary conditions are also shown, with both conduction and valence band states in their lowest quantized $\mathbf{k}_z$ subband.

The integral here determines the selection rules and is only dependent on the light polarization interacting with the symmetries of the band. The exponential vanishes unless $\mathbf{k}_{h\perp} - \mathbf{k}_{e\perp} = 0$. This is the vertical transition rule that matches momentum. In a three-dimensional transition, the z-directed summation is absent. Here, the confinement has introduced a summation over confined lattice cells through the overlaps at the lattice positions.

Excitons, because this is a confined condition, become more important and will have a first-order effect.

In Figure 20.20, which shows an interband transition, electron and hole states that have the same subband quantum number will have identical overlap, so the integral of overlap functions normalizes to unity. This means that the interband transition optical matrix element for identical subband numbers to the lowest order is identical with or without this confinement. The absorption then shows the change in density of states, that is, will have the stepping behavior characteristics of subband formation. Another feature is that the electron and hole wavefunctions have a closer match due to confining; these carriers in the same quantized subband with the same $\mathbf{k}_z$ then reinforce the oscillator strength. And so the optical processes strengthen a bit for this reason. The selection rule here has arisen in how the $|s\rangle$-like and $|p\rangle$-like states could connect because of the odd parity of $\boldsymbol{\eta} \cdot \boldsymbol{r}$, specifically because $\mathbf{r}$ is an odd function.

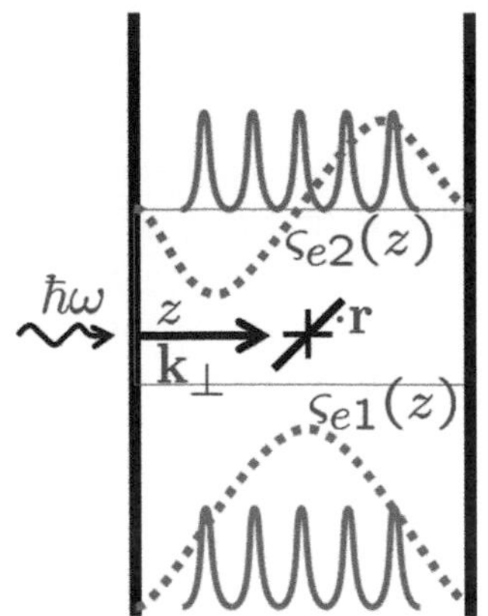

Figure 20.21: The nature of symmetries in intersubband/intraband transition within the conduction band of a confined structure with light incident normally.

In intrasubband/intraband transitions, as shown in Figure 20.21, with transitions between states of the same carrier type, the matrix element in Equation 20.88 will change because the fast-varying integrals have a similar part of the wavefunction (u_{ck_e} of electrons, in this case). Now the matrix element reduces to

$$|\langle f|\mathbf{r}\cdot\boldsymbol{\eta}|i\rangle| \approx \int \varsigma_{e1}(z)\boldsymbol{\eta}\cdot\mathbf{r}\varsigma_{e2}(z)\,d\mathbf{r}\int_{\Omega_0} u_{cke_1}(\mathbf{r})u^*_{cke_2}(\mathbf{r})\,d^3\mathbf{r}. \qquad (20.91)$$

The second integral is approximately unity, and the first term gives a large contribution over the length of the confined region.

So, intersubband transitions can be particularly strong because of the similarity of the states interacting, and the dipole that now arises over the confined region like a macroatom. This dipole has a magnitude of $|e\langle\mathbf{r}\rangle| \approx 16ea/9\pi^2$, where a is the width of the confined region.

We can place more details into this treatment. Since the quantum numbers are not changing for the direct bandgap semiconductors, one need only concern oneself with the dipole matrix element. Let $\hat{\mathbf{e}}$ denote the orientation of the vector potential as in the allowed matrix element in Equation 12.16. Writing in terms of final (f) and initial (i) states of the transition, and using Hermiticity,

$$\begin{aligned}\mathscr{H}'_{if} &= \frac{1}{S}\int \varsigma_i^*(z)\exp(-i\mathbf{k}_\perp\cdot\mathbf{r}_\perp)[e_x\mathrm{p}_x + e_y\mathrm{p}_y + e_z\mathrm{p}_z] \\ &\quad\times \varsigma_f(z)\exp(i\mathbf{k}'_\perp\cdot\mathbf{r}_\perp)\,d^3\mathbf{r} \\ &= (e_x\hbar k_x + e_y\hbar k_y)\delta_{i,f}\delta_{\mathbf{k}_\perp,\mathbf{k}'_\perp} \\ &\quad + e_z\delta_{\mathbf{k}_\perp,\mathbf{k}'_\perp}\int \varsigma_i^*(z)\mathrm{p}_z\varsigma_f(z)\,dz, \end{aligned} \qquad (20.92)$$

where the normalization is by area S. If polarization is along the plane of confinement (e_x and or e_y), then the only allowed transitions are those where the initial and final states are the same. This is $\omega = 0$ for the radiation. This static limit requires inclusion of other scattering mechanisms such as free carrier absorption. In perfect confined systems, the free carrier absorption is forbidden, since energy and momentum cannot be conserved during a photon-electron interaction. An additional perturbation, such as impurity or photon, is necessary. The polarization e_z is of an electromagnetic wave that is propagating in the plane of the layer, with this field vector normal to the plane, and in the same direction as the confinement direction. Since the heterostructure Hamiltonian has a parity, and the dipole perturbation an odd parity,

Much like the involvement of phonons for the "forbidden" processes for the three-dimensional conditions, or of free carrier absorption.

the initial and final subbands should be of opposite parity to allow optical transitions between subbands. In confinement, instead of the participation of a defect perturbation, the z dependence of the potential suffices to allow such a transition. So, without recourse to other momentum- and energy-carrying species, intersubband transitions with e_z polarization become possible where the potential makes up for the momentum necessary for intraband absorption. The details of the amplitude of different transitions will depend on the details of the confinement, including the changes arising when going from two- to one- to zero-dimensional systems, so we will leave this subject at this point. Suffice it to state that the approach of including the entire wavefunction, as in Equation 20.92, together with the perturbing potential is necessary to evaluate the various interaband transitions.

The electric dipole—polarization—as it oscillates in all these cases has a radiation emission and absorption pattern. As shown in Figure 20.22, the maximum emission or most efficient absrorption will be in the plane orthogonal to the dipole. The figure here illustrates this pattern, a $\sin^2\theta$ dependence, where θ is the angle from the polarization axis, for the case of interband transition involving states with even symmetry ($|s\rangle$) and odd symmetry ($|p_y\rangle$) due to the odd parity of the dipole.

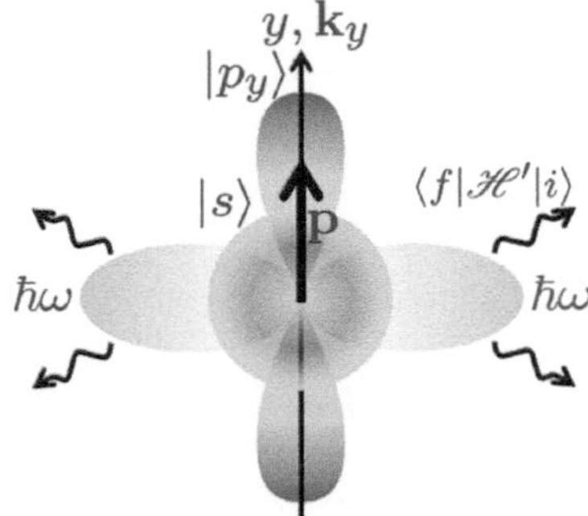

Figure 20.22: The pattern of light emission when a dipole interaction with suitable filled conditions of band states leads to emission. The intensity is highest orthogonally. Here, this emission arising in dipole-mediated coupling of $|p_y\rangle$ and $|s\rangle$ states is shown.

The electric dipole term is odd powered spatially through the dipole moment term; it will appear as a change in parity in the matrix element. The potentials in quantum wells or in superlattices have space reflection symmetry, so parity is a good quantum number. For intraband transitions, the transitions involve dipole matrix elements between envelope functions of subbands arising from the same band, unlike the case for interband transitions, which occur between subbands of different bands. The first, at least for electrons in direct gap materials, is straightforward, since the band quantum number doesn't change, and one need concern oneself with only the matrix element between envelope functions. The interband transitions, however, will involve selection rules originating in the quantum numbers of the two bands and the atomic-like dipole matrix element.

The envelope functions are characterized by this odd and even character under space reflections. So, the orientation of the electromagnetic radiation, together with this envelope function symmetry, will guide the "allowed" and "forbidden" transitions. This implies that transitions are allowed for confined states that have the same envelope function symmetry under space reflection.

In direct gap semiconductors—optically active materials—the conduction band state is spherically symmetric and the bandstructure is quite isotropic, at the bandedge. The valence band, as discussed at length, is not so. In bulk, too, this matters, but not as much, since the valence band states of heavy hole and light hole type are degenerate at $k = 0$, and the interaction is strong and nearly independent of light's polarization.

In quantum wells, the hole state degeneracy is lifted, so the interaction matrix element will strongly depend on the polarization of light, with heavy-hole and light-hole states subject to different selection rules for allowed transitions arising in the matrix element. In tight binding, the wavefunction is built out of the atomic orbital states $|s\rangle$, $|p_x\rangle$, $|p_y\rangle$ and $|p_z\rangle$, and the bandstructure analysis leads to, in direct bandgap semiconductors, conduction band states for the *conduction electrons*,

$$|u_{c0}\rangle = |s\rangle, \tag{20.93}$$

which are spherically symmetric. The valence band states we have found to be a bit more complicated. The *heavy-hole* states arising from

$$\left|\frac{3}{2},\frac{3}{2}\right\rangle = -\frac{1}{\sqrt{2}}\left(|p_x\rangle + i|p_y\rangle\right)|\uparrow\rangle, \text{ and}$$
$$\left|\frac{3}{2},-\frac{3}{2}\right\rangle = \frac{1}{\sqrt{2}}\left(|p_x\rangle - i|p_y\rangle\right)|\downarrow\rangle, \tag{20.94}$$

and the *light-hole* states arising from

$$|\frac{3}{2},\frac{1}{2}\rangle = -\frac{1}{\sqrt{6}}\left(|p_x\rangle + i|p_y\rangle\right)|\downarrow\rangle - 2|p_z\rangle|\uparrow\rangle, \text{ and}$$
$$|\frac{3}{2},-\frac{1}{2}\rangle = -\frac{1}{\sqrt{2}}\left(|p_x\rangle - i|p_x\rangle\right)|\uparrow\rangle + 2|p_z\rangle|\downarrow\rangle, \tag{20.95}$$

have other spatial symmetries. The matrix element symmetry argument—dependent on the dipole interaction and spatial envelope function symmetries—allows only some of the transitions, because of the symmetries of these wavefunctions. The non-vanishing matrix elements are $\langle p_x|\hat{\mathbf{x}}\cdot\mathbf{p}|s\rangle$, $\langle p_y|\hat{\mathbf{y}}\cdot\mathbf{p}|s\rangle$ and $\langle p_z|\hat{\mathbf{z}}\cdot\mathbf{p}|s\rangle$. Let these terms, which have identical values, be denoted by p_{cv}.

For heavy holes, the transitions with finite matrix elements have

$$\langle HH|\hat{\mathbf{x}}\cdot\mathbf{p}|s\rangle = \langle HH|\hat{\mathbf{y}}\cdot\mathbf{p}|s\rangle = \frac{1}{\sqrt{2}}\langle p_x|\hat{\mathbf{x}}\cdot\mathbf{p}|s\rangle = \frac{1}{\sqrt{2}}\mathrm{p}_{cv}. \tag{20.96}$$

The matrix element $\langle HH|\hat{\mathbf{z}}\cdot\mathbf{p}|s\rangle = 0$, and this transition is forbidden. For light holes, the transitions with finite matrix elements have

$$\langle LH|\hat{\mathbf{x}}\cdot\mathbf{p}|s\rangle = \langle LH|\hat{\mathbf{y}}\cdot\mathbf{p}|s\rangle = \frac{1}{\sqrt{6}}\langle p_x|\hat{\mathbf{x}}\cdot\mathbf{p}|s\rangle = \frac{1}{\sqrt{6}}\mathrm{p}_{cv}, \text{ and}$$
$$\langle LH|\hat{\mathbf{z}}\cdot\mathbf{p}|s\rangle = \frac{2}{\sqrt{6}}\langle p_z|\hat{\mathbf{z}}\cdot\mathbf{p}|s\rangle = \frac{2}{\sqrt{6}}\mathrm{p}_{cv}. \tag{20.97}$$

Caution is needed here. Keep the dipole **p** and the *p* orbital distinction in mind since the same is being used for both. By now, you should be able to distinguish these through your understanding.

The polarization of light enters through the projection of **p** in the different orientations, and this determines the selection rules. We have, for our interband coupling, the following implications:

for electromagnetic waves with $\hat{\mathbf{x}}$ *polarization:*

- *HH* to the conduction band: $|\mathrm{p}_{fi}|^2 = \frac{1}{2}|\langle p_x|\hat{\mathbf{x}}\cdot\mathbf{p}|s\rangle|^2$,
- *LH* to the conduction band: $|\mathrm{p}_{fi}|^2 = \frac{1}{6}|\langle p_x|\hat{\mathbf{x}}\cdot\mathbf{p}|s\rangle|^2$;

for electromagnetic waves with $\hat{\mathbf{y}}$ *polarization:*

- *HH* to the conduction band: $|p_{fi}|^2 = \frac{1}{2}|\langle p_x|\hat{\mathbf{x}}\cdot\mathbf{p}|s\rangle|^2$,
- *LH* to the conduction band: $|p_{fi}|^2 = \frac{1}{6}|\langle p_x|\hat{\mathbf{x}}\cdot\mathbf{p}|s\rangle|^2$;

for electromagnetic waves with $\hat{\mathbf{z}}$ *polarization:*

- *HH* to the conduction band: no coupling,
- *LH* to the conduction band: $|p_{fi}|^2 = \frac{2}{3}|\langle p_x|\hat{\mathbf{x}}\cdot\mathbf{p}|s\rangle|^2$.

$\mathbf{k}=0$ states have the pure form. Stimulated emission, with occupation of the states in their vicinity, means that emitted light strongly shows the implications of selection rules. This is particularly noticeable in confined questions. Away from $\mathbf{k} = 0$, the states are increasingly mixed, but, even here still, in the bulk, light polarized in the xy plane couples a factor 3 times more strongly to the heavy-hole states than to the light-hole states.

An important emission implication of connection between circular polarization and the heavy-hole transition is that the circular polarization occurs between the confining planes. The magnetic field of the light is therefore orthogonal to the confining planes. So, heavy holes only emit *TE* polarized light in the plane of the confinement.

A good illustration of the differences in transitions between three-dimensional and confined conditions is to look at preferential spin generation arising in the generation of electrons from optically induced valence-to-conduction transition. Conduction electrons, as noted earlier, have no spin preference in a normal semiconductor, that is, $\mathscr{G}(n,k_\perp,\uparrow)=\mathscr{G}(n,k_\perp,\downarrow)$, in the nearly free electron approximation. Only the introduction of magnetic species, or magnetic field, causes changes. This means that there exists no net electron spin in the electron gas. Photons have a spin of 1 and so an angular momentum; a circularly polarized light will have an angular momentum of $\pm\hbar$ along the direction of propagation, with the sign depending on whether it is counterclockwise or clockwise. Photons impart this angular momentum to the semiconductor upon absorption. In an electron-hole transition, this can end up in a net spin. The semiconductor now has a net spin, and the semiconductor has undergone an optical *spin injection*.

Light's circularity is defined w.r.t. the source. Positive circular light σ^+ is therefore left circular when seen by an observer.

In a bulk direct bandgap semiconductor, consider the interband transitions at $k = 0$. Light-hole and heavy-hole bands are degenerate here. The conduction band states are s-like with orbital angular momentum $L = 0$. This corresponds to a single total angular momentum of $J = 1/2$, and therefore $M_J = \pm 1/2$ for the states. Valence states have $J=3/2$ states and $J = 1/2$ states for the total angular momentum, with $M_J = \pm 3/2$ corresponding to heavy-hole

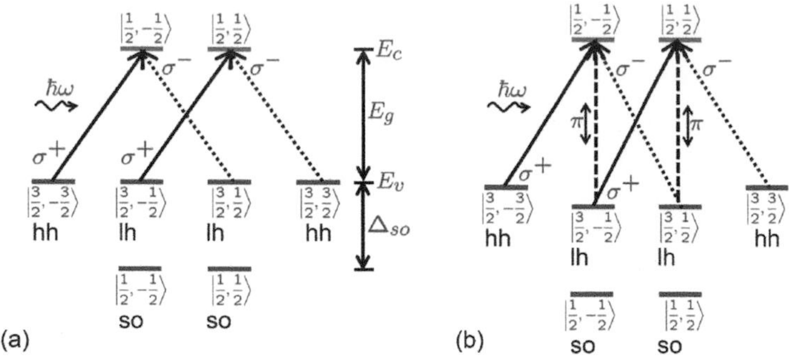

Figure 20.23: Selection rules for transitions in zinc blende direct gap semiconductors under circularly polarized light. Part (a) shows transitions in bulk semiconductors at $\mathbf{k}=0$ with the circularly polarized light of bandgap energy incident. Note, only circular polarization is shown. Split-off states are ignored as they are further away in energy. The degenerate heavy- and light-hole states couple to the conduction band states, and angular momentum conservation must apply. Momentum- and energy-conserving transitions for left (σ^+) and right (σ^-) circularly polarized light are shown. Part (b) shows the changes due to confinement. Subband energy edges shift, and the $\mathbf{k}=0$ degeneracy is broken. This panel shows the additional transitions that become possible under linear polarization, which is marked with the symbol π; hh, heavy-hole states; lh, light-hole states; so, split-off states.

states, and $M_J = \pm 1/2$ corresponding to light-hole states. These are degenerate at $\mathbf{k} = 0$. The $J = 1/2$ states, where, again, $M_J = \pm 1/2$ is possible, are pushed down by the interaction with the split-off states. The split-off states too drop down. In bulk zinc blende semiconductors, with circularly polarized light, the positive circular polarization (σ^+) leads to a $\Delta M_J = +1$ transition. The negative polarization leads to the complementary $\Delta M_J = -1$ transition. We show this in Figure 20.23(a) as being only possible for a selective set of transitions between the heavy-hole and electron states, and the light-hole and electron states. Only 4 unique combinations, out of a possible 8, have the requisite angular momentum and energy conservation for the circularly polarized light. Since the square of the matrix element of the heavy-hole transition is 3 times larger than that for a light-hole transition, the σ^+ light will result in there being 3 times more $M_J = -1/2$ electrons than $M_J = +1/2$ electrons. A maximum of 50 % spin polarization is possible in bulk structures with circular polarization. Also, since the degeneracy means mixed states, the holes relax rapidly.

A little more consideration is needed in an explicit justification. For the filled valence band as a whole, $\sum J_z = 0$. When an electron is removed from the valence band, this electron has a $J_z^e = M_J\hbar$. A hole of $J_z^h = M_J\hbar$ has now been left behind. A σ^+ light-induced heavy-hole transition leads to the creation of an electron with $M_J = -1/2$, and a hole with $M_J = 3/2$ that occupies that heavy-hole state, and the total electron-hole pair generation has a sum of $-1/2 + 3/2 = 1$ change in M_J. Angular momentum has been conserved. This same argument for a light-hole-band-to-electron transition involves an electron of $M_J = 1/2$ and a hole of $M_J = 1/2$ being created as a result of the transition arising in the σ^+ photon.

Now consider what happens in confined conditions. The $\mathbf{k}=0$ degeneracy breaks, and the subband edge energies shift, as illustrated in Figure 20.23(b). When there is no degeneracy, with the light that only couples electron states to the heavy-hole states, so below the light-hole absorption edge, only $M_J = -1/2$ electrons will be generated if the light is σ^+ polarized, or the opposite with the opposite polarization. This spin polarization is complete. The broken degeneracy also means that the heavy-hole state is closer to being an eigenstate. The relaxation time of the $M_J = \pm 3/2$ holes will be larger.

With the matrix element determined, for these confined conditions, the net interband absorption rate can be determined using the same procedure as outlined in the derivation of Section 12.1.

This discussion demonstrates the importance of selection rules in optical transitions and how they change under material and structural constraints. The monolayer materials will bring out additional such attributes of their own.

20.8 Summary

THIS CHAPTER, as the last one of this text, has brought together several of the different themes and the different perspectives together by exploring them in quantum-confined conditions or in monolayer crystals. Electrons, holes, phonons, impurities, static and dynamic responses, particle or wave perspectives, short range versus long range and localization versus spreading out over the entire space of the crystal were part of this mix, and this helps reinforce several of the physics and mathematical approaches that help us get an insight into the behavior of semiconductors.

We started with the ubiquitous SiO_2/Si system's inversion layer and developed the approach, including an understanding of the effective potential energy V that the particle encounters. The effective mass equation, a Schrödinger-like equation, that we have often just called the Schrödinger equation for the envelope function, is an essential tool in the analysis. The envelope function removes the Bloch function's rapidly oscillatory part but leaves for us a proper description of the probability density of the charge from the collection of electrons. So, both the $E(\mathbf{k})$ state description imposed by the crystal, and the charge consequences imposed by electromagnetics, are incorporated. This let us understand how the states change; for example, confinement forces separation of the $\mathbf{k}_\perp$ in the confined direction. The easy motion is possible in the other two directions if there is only one direction of confinement. But the confinement causes sufficient separation, so that a change in this $\mathbf{k}_\perp$ is now really a transition. The analysis of Si inversion also gave us a look at techniques for approximations, such as looking upon the well as a triangular well whose solutions are Airy functions.

Confinement in square wells, which are also ubiquitous, particularly in optical devices, was our next step. We looked at infinite potential wells and finite potential wells to understand the evolution of the solutions from the infinite to the finite, and we developed an understanding of the existence of confined states. Among a few of the interesting observations here was not only that the finite potential well will have finite bound states and that there will be an unlimited number of unbound states, which in itself is not surprising, but also that there are conditions in this unbounded region span where a potential well allows for complete transmission, that is, quantum reflection vanishes despite there being a sharp barrier. This is an example of a resonant transmission. Another surprising conclusion was that while a finite potential well can have no confined state, a doubly confined system, such as a wire, will always have confined states along with the propagating states. If the system is triply confined, all of the states will be confined. It is like a macroatom.

This confinement discussion did let us look back at the meaning of the different potentials—image, exchange, et cetera—to see how they should be modified. And they need to be modified as a first order correction. Even the binding energy of shallow dopants will change because of the changes in the potential constraints of the confined system.

Holes, with their anisotropy, multiplicity of types, and the degeneracy at $\mathbf{k} = 0$, undergo an even more significant change than the electron does. Energy degeneracies are broken. Strong mixing happens due to perturbation, and the Luttinger-Hamiltonian-based $\mathbf{k} \cdot \mathbf{p}$ approach needs to be viewed simultaneously with the potential well perturbation to extract the evolution of the states in confinement.

We did not look at quantum wire and quantum boxes in any detail, but the approaches adopted could be extended to them. We did remark on salient implications for the multi-confined structures where it was considered significant.

We followed the confinement discussion with a jump to monolayer systems—crystals that are stable in two-dimensional form—with no need of a confining barrier. Graphene is the classic example of this. And now we have to go back to our bandstructure description to derive the $E(\mathbf{k})$ behavior. Propagating states exist only in the plane, and these arise out of the π_z bond and antibond. Graphene has a hexagonal lattice with a basis of 2. The reciprocal space symmetry points K and K' exhibit zero gap and a photon-like E-$\mathbf{k}$ behavior. So, here is a group *IV* material that, in sp^3 hybridization, gives one of the most insulating material, with an incredibly large bandgap, and, in sp^2 hybridization, gives a material where the gap vanishes. This is the difference between confinement, where the extended Bloch states are restricted in space, and the localized interaction that is dominant in the new crystal form.

Graphene can be converted into tubes by rolling them up. Some of these remain conducting—metallic—and some can become semiconducting. Again, the structuring of nanotubes allowed us to revisit bandstructure techniques, and we employed the periodicities—and chirality, since sheets can be rolled with a twist—to explore the origins of the behavior of zigzag, armchair and chiral nanotubes and their densities of states.

We returned back to the confined structures at this point to explore the states of quantum superlattices, which are artificially created periodic (or chirped) structures with small-enough barrier regions so that the superlattice-like confinement creates extended states in the $\mathbf{k}_\perp$ direction. What used to be bound states can now propagate.

Among the consequences of confinement, that is, the existence of the potential barrier nearby, we discussed some of the properties that change from their bulk behavior. These included the binding energy, but also the screening, since charge clouds and therefore the polarizability is being disturbed. Scattering too changes in confinement. It will have different characteristics due to the changes in the states, as also possibly the perturbations, and the changes will depend on the different scattering mechanisms. This discussion also gave us the chance to emphasize the importance of what happens at interfaces due to their deviations from planarity. These changes in well widths will, in the first order, have an effect on linewidths, such as those of optical emission.

Confined structures are ubiquitous in optical uses of semiconductors. So, we looked at optical transitions in confined structures and developed an understanding of these transitions for processes that take place between subbands in different bands (interband) and different subbands within a band (intraband or intersubband). Symmetry arguments and the interaction between the dipole and the optical field let us draw out the selection rules for the transitions and connect these to linear polarization and circular polarization. The circular polarization instance was interesting in that it showed us how selection rules in confined conditions allow us to selectively create spin polarization in electrons. Finally, we ended all this discussion by looking at how the behavior changes when one uses monolayer semiconductors, with their entirely different way of achieving dimensionality.

20.9 *Concluding remarks and bibliographic notes*

THERE IS A VAST LITERATURE encompassing confinement and now monolayer structures. Confinement has been the foundation of devices for electronics and optics, and it has provided some very intriguing understanding of new states of matter.

For the electronic properties, the reader would find the review by Ando, Fowler and Stern[1] to be a very thorough and useful reading for understanding lower-dimensional systems. Even though this long article, suggested as a general reading, concentrates on inversion layers in silicon, the basic concepts are general. In the calculations of confinement, it is useful to sometimes refer to the variety of Bessel functions. A good source for understanding a variety of mathematical functions is the classic collection by Abramowitz and Stegun[2]. F. Stern examines variety of interesting

[1] T. Ando, A. B. Fowler and F. Stern, "Electronic properties of two-dimensional systems," Reviews of Modern Physics, **54**, 437 (1982)

[2] M. Abramowitz and I. A. Stegun, "Handbook of mathematical functions with formulas, graphs, and mathematical tables", U.S. Government Printing Office (1964)

transport and quantization features in low-dimensional systems in the monograph edited by Davies and Long[3]. For hole quantization, see Fasolino and Altarelli[4].

In their book entirely devoted to carbon[5], Saito, et al. discuss at length the electronic and elastic properties of graphene and nanotubes, including more esoteric topics such as Peierls instability. Quite a bit of understanding of these small structures employs optical tools interacting with the vibration modes in these structures. The Raman spectra discussion of this book, along with its discussion of bandstructure calculations, is particularly useful.

An advanced discussion of nanotubes is in Fischetti and Vandenberghe's text on transport[6]. This text is an ideal source for understanding advanced approaches to bandstructures and to transport in semiconductors down to their nanoscale. We have kept our discussion simpler in this text, for example, avoiding supercell approaches or group theoretic approaches. Fischetti and Vandenberghe, while employing them, keep the content quite understandable and yet very rigorous.

Gerald Bastard's text[7] is a very detailed treatise analyzing the behavior of states, excitons, interactions and transport in heterostructures. A simpler treatment can be found in the book by Weisbuch and Vinter.[8] A very readable exposition is from Jasprit Singh[9]. This book is a very readable source for understanding the various connections that exist between quantum- and engineering-focused observations in semiconductors.

Ridley's classic text[10] is a thorough exposition of the underlying physical processes and mathematical treatments in bulk semiconductors. But much of this is extendable to confined structures.

[3] J. H. Davies and A. R. Long (eds), "Physics of nanostructures," Institute of Physics, ISBN 0-7503-0170-8 (1992)

[4] A. Fasolino and M. Altarelli, "Subband structure and Landau levels in heterostructures," in G. Bauer, F. Kuchar and H. Heinrich (eds), Two-dimensional systems, heterostructures and superlattices, Springer-Verlag ISBN 13: 978-3-642-82313-8 (1984)

[5] R. Saito, G. Dresselhaus and M. S. Dresselhaus, "Physical properties of carbon nanotubes," Imperial, ISBN 1-86094-093-5 (1998)

[6] M. Fischetti and W. G. Vandenberghe, "Advanced physics of electron transport in semiconductors and nanostructures," Springer, ISBN 978-3-319-01100-4 (2016)

[7] G. Bastard, "Wave mechanics applied to semiconductor heterostructures," Les éditions de physique

[8] C. Weisbuch and Borge Vinter, "Quantum semiconductor structures," Academic ISBN 0-12-742680-9 (1991)

[9] J. Singh, "Electronic and optoelectronic properties of semiconductor structures," Cambridge, ISBN 13-978-0-521-82379-1 (2003)

[10] B. K. Ridley, "Quantum processes in semiconductors," Oxford, ISBN 0-19-851170-1 (1988)

20.10 *Exercises*

1. This exercise works through the lowest-order analysis of electron states and their properties of a two-dimensional electron gas in the conduction band in *Si* arising in a confining potential. The electrons in the conduction band of *Si* are subjected to a slowly varying potential caused by band bending. Model this potential as

$$V(r) = \frac{1}{2} m_3 \omega_0^2 z^2 \text{ for } z \leq 0,$$

where ω_0 is a positive constant of s^{-1} dimension. Assume that the effect of this potential can be treated within the effective mass approximation.

- Use the effective mass theorem to find the total energy of the electron wavepacket near the bottom of the conduction band for the ellipsoidal surface along the positive z direction.
- Find and sketch the areal density of states $\mathscr{G}(E)$ for these electrons in the conduction band when their energy is given as in previous part. Note here that $\mathscr{G}(E)$ has dimensions of per unit area per unit energy.
- Write down both the total time-dependent envelope function and the total time-dependent wavefunction for the wavepacket for electrons with $k_z = \pi/2L$ but with arbitrary k_x and k_y. These electrons are said to be in the first subband and form a two-dimensional electron gas.
- All 6 ellipsoidal energy pockets for electrons in the conduction band of *Si* are subjected to this confining potential. Let all the electrons be in the first subband so that they form a two-dimensional electron gas. For motion due to an electric field in the x-y plane, find the corresponding conductivity tensor.
- What is the numerical value for this conductivity effective mass for *Si*? How does it compare with the full three-dimensional conductivity effective mass and for the effective masses for three-dimensional and two-dimensional densities of states? **[M]**

2. Show that the electron density in an inversion layer with multiple subbands occupied can be written as

$$n(z) = \frac{m_d^* k_B T}{\pi \hbar^2} \sum_i \ln\left[1 + \exp\left(\frac{E_F - E_i}{k_B T}\right)\right] |\varsigma_i(z)|^2,$$

where i indexes each subband. For *Si*, show that when m_t^* is the density of the states effective mass, then $m_z^* = m_l^*$, while, on the other hand, when $m_z^* = m_t^*$, then $m_d^* = (m_l^* m_t^*)^{1/2}$. **[S]**

3. A quantum well of *GaAs* has $Al_xGa_{1-x}As$ ($x = 0.3$) cladding regions. Calculate the bandgap if the width a of the well is 16 nm. Is infinite barrier a good approximation here? What about if $a = 2$ nm? **[S]**

4. What is the Fermi wavelength ($2\pi/k_F$) for an electron concentration of 10^{12} cm^{-3} in a quantum well such as the thicker well in Exercise 3? Compare it to that in a metal such as *Cu* or *Au*. **[S]**

5. Take again a superlattice employing *GaAs* and $Al_xGa_{1-x}As$ ($x = 0.3$). The well width is 10 nm, the barrier width is 2.2 nm, and the potential height is 0.25 eV.
   - Find the widths of the minibands and the minigaps, and

- keeping fixed well width at 10 *nm*, plot the energy of the first conduction and heavy-hole minibands as a function of barrier width between 2 and 5 *nm*. **[M]**

6. Design the period of a *Si*/*SiGe* superlattice so that the band minimum of *Si* can be pulled in from near the zone edge (about $k \approx 0.8\pi/a$) to $k = 0$. This is zone folding. **[M]**

A
Integral transform theorems

TRANSFORMATIONS, PARTICULARLY OF THE INTEGRAL FORM, are an important way to understand the nature and connections in the time evolution of a signal stream and even more so between signal streams, whether connected, stochastic or somewhere in-between. This appendix stresses a few of the important theorems and features evident in Fourier transformations.

A.1 Parseval's theorem

THE INTEGRATED PRODUCT OF TWO FUNCTIONS and their integrated Fourier transforms are related as

$$\begin{aligned}
&\int_{-\infty}^{\infty} f(t)g^*(t)\,dt \\
&\quad = \int_{-\infty}^{\infty} dt \frac{1}{\sqrt{2\pi}} \int_{-\infty}^{\infty} F(\omega)\exp(-i\omega t)\,d\omega \frac{1}{\sqrt{2\pi}} \int_{-\infty}^{\infty} G^*(\omega')\exp(i\omega' t)\,d\omega' \\
&\quad = \int_{-\infty}^{\infty} \frac{d\omega}{\sqrt{2\pi}} \frac{1}{\sqrt{2\pi}} \int_{-\infty}^{\infty} \delta(\omega-\omega')F(\omega)G^*(\omega')\,d\omega' \\
&\quad = \frac{1}{2\pi} \int_{-\infty}^{\infty} F(\omega)G^*(\omega)\,d\omega. \qquad \text{(A.1)}
\end{aligned}$$

Parseval's theorem relates this equivalence of the time domain and frequency domain of functions.

If $f(t)$ (and $g(t)$) are real functions, then

$$\begin{aligned}
f^*(t) = f(t) \quad &\therefore \quad F^*(\omega) = F(-\omega), \quad \text{and if} \\
g^*(t) = -g(t) \quad &\therefore \quad G^*(\omega) = -G(-\omega), \qquad \text{(A.2)}
\end{aligned}$$

that is, the Fourier transform of a real function is even, and the Fourier transform of an imaginary function is odd. This follows directly from the Fourier transformation.

So, Parseval's theorem also states that if $f(t)$ and $g(t)$ are real-valued, then

$$\int_{-\infty}^{\infty} f(t)g(t)\,dt = \frac{1}{2\pi}\int_{-\infty}^{\infty} F(\omega)G(-\omega)\,d\omega. \tag{A.3}$$

As a corollary, for real $f(t)$,

$$\int_{-\infty}^{\infty} f^2(t)\,dt = \frac{1}{2\pi}\int_{-\infty}^{\infty} F^2(\omega)\,d\omega = 2\frac{1}{2\pi}\int_{0}^{\infty} F^2(\omega)\,d\omega. \tag{A.4}$$

A.2 Convolution theorem

Convolution—how one function modifies another function through the past—has appeared most prominently for us through Green's functions. Let $f = g \otimes h$ be a convolution where source $h(t)$ is being modified by $g(t-t')$, that is,

$$f(t) = g \otimes h = \int_{-\infty}^{\infty} g(t-t')h(t')\,dt'. \tag{A.5}$$

Fourier transformation makes the evaluation easier for many circumstances:

$$\begin{aligned} F(\omega) &= \frac{1}{\sqrt{2\pi}}\int_{-\infty}^{\infty} \exp(i\omega t)\,dt \int_{-\infty}^{\infty} g(t-t')h(t')\,dt' \\ &= \frac{1}{\sqrt{2\pi}}\int_{-\infty}^{\infty} dt \int_{-\infty}^{\infty} \exp[i\omega(t-t')]g(t-t')\exp(i\omega t')h(t')\,dt' \\ &= \sqrt{2\pi}\left[\frac{1}{\sqrt{2\pi}}\int_{-\infty}^{\infty} g(s)\exp(i\omega s)\,ds\right]\left[\frac{1}{\sqrt{2\pi}}\int_{-\infty}^{\infty} \exp(i\omega t')h(t')\,dt'\right] \\ &= \sqrt{2\pi}G(\omega)H(\omega). \end{aligned} \tag{A.6}$$

Note here that there are these factors of $\sqrt{2\pi}$ that appear in convolutions (or Parseval's theorem), based on the conventions adopted in the Fourier transform integration. Fourier transform of a convolution is writable as a product of the transforms in the Fourier space.

The correlation operator satisfies commutation, association and distribution; that is,

$$\begin{aligned} a \otimes b &= a \otimes b, \\ a \otimes (b \otimes c) &= (a \otimes b) \otimes c, \quad \text{and} \\ a \otimes (b + c) &= (a \otimes b) + (a \otimes c). \end{aligned} \tag{A.7}$$

This follows straightforwardly from Equation A.6 and, in a more convoluted way, from the time domain.

A.3 Correlation theorem

Correlation—an often-misapplied measure of how connected two functions may be—probes, by parameter (usually time) shifting, the shape matching of two functions; for example,

$$Corr(g,h;t) = \int_{-\infty}^{\infty} g(t+t')h^*(t')\,dt' \tag{A.8}$$

is a cross correlation. Complex functions are properly dealt with due to the conjugation of one of the functions. Autocorrelation is this relationship measuring self-correlation, that is, how much does the form of the signal itself change with time. A delta function will result in a sharp peak only at coincidence and vanishes elsewhere, while a constant value will be completely autocorrelated. If $g(t) = \exp(-i\omega_1 t)$, and $h(t) = \exp[-i\omega_2(t+\Delta t)]$, then

$$Corr(g,h;t) = 2\pi\delta(\omega_2 - \omega_1)\exp\left[-i\omega_1(t-\Delta t)\right]. \tag{A.9}$$

If the two frequencies coincide, the cross correlation has a phase proportional to the time shift in the delta function output. From the Fourier transform of Equation A.8,

$$\begin{aligned}
Corr(\omega) &= \\
&= \frac{1}{\sqrt{2\pi}}\int_{-\infty}^{\infty}\exp(i\omega t)\,dt\int_{-\infty}^{\infty} g(t+t')h^*(t')\,dt' \\
&= \frac{1}{\sqrt{2\pi}}\int_{-\infty}^{\infty} dt\int_{-\infty}^{\infty}\exp[i\omega(t+t')]g^*(t+t')\exp(-i\omega t')h^*(t')\,dt' \\
&= \sqrt{2\pi}\left[\frac{1}{\sqrt{2\pi}}\int_{-\infty}^{\infty} g(s)\exp(i\omega s)\,ds\right]\left[\frac{1}{\sqrt{2\pi}}\int_{-\infty}^{\infty}\exp(i\omega t')h(t')\,dt'\right]^* \\
&= \sqrt{2\pi}\,G(\omega)H^*(\omega).
\end{aligned} \tag{A.10}$$

The significance of complex conjugation cannot be stressed enough. Convolution, which has a transform form similar to this (except for the conjugation), has the commutation property. *Correlation does not commute.* $Corr(g,h;t) = Corr^*(h,g;-t)$, and therefore $Corr(g,h) = Corr^*(h,g)$. The self-correlation—autocorrelation function $Corr(g,g;t)$—because of this time-shifting shape matching inherent in the function, is sensitive and a good indicator of periodic and quasiperiodic behavior. It will show strength at coincidences in time and in the periodicity. Closer periodicity sharpens peaks, with $Corr(g,g) = \sqrt{2\pi}\,G^2(\omega)$.

It is easy to ponder many real life experiences where this non-commutativity is evident. And forgetting it leads to real problems in life.

A.4 Wiener-Khintchin theorem

CROSS AND AUTOCORRELATIONS are second power in amplitudes, and so proportional to the power. This property—a view of these as power spectral density—is brought out by the Wiener-Khintchin theorem, whose one conclusion was the autocorrelation with which we concluded the last section.

Let $z(t)$ be the signal. One may define the average power via $\lim_{T\to\infty}\int_{-\infty}^{\infty} z(t)z^*(t)\,dt$, which is also, by Fourier transformation, $\lim_{T\to\infty}\int_{-\infty}^{\infty}(Z(\omega)Z^*(\omega)/T)\,d\omega$. Taking frequency as positive real as a convention, this average power is $\lim_{T\to\infty}\int_0^{\infty}(2Z(\omega)Z^*(\omega)/T)\,d\omega$. For a $z(t)$ non-stationary process, the average depends on the interval T. $T\to\infty$ has no meaning. If the process is stationary—strict or weak—one may employ the limits of $T\to\infty$. In this case, one can also now employ the frequency domain, since the entire bandwidth is accounted for. Take the case of ensemble averaging over identical systems; exchanging the time and frequency limits leads to the parameter

$$S_z(\omega) = \lim_{T\to\infty}\frac{2\langle Z(\omega)Z^*(\omega)\rangle}{T} = \lim_{T\to\infty}\frac{2\langle |Z(\omega)|^2\rangle}{T}, \tag{A.11}$$

which is the power spectral density. Power spectral density is an ensemble average. So, different processes will have different forms.

In time shifting and averaging over time, using Fourier transforms,

$$\begin{aligned}\phi_z(t') &= \lim_{T\to\infty}\frac{1}{T}\int_{-\infty}^{\infty}\langle z(t+t')z(t)\rangle\,dt\\ &= \frac{1}{2\pi}\int_0^{\infty}\frac{2\langle |Z(\omega)|^2\rangle}{T}\cos(\omega t')\,d\omega\\ &= \frac{1}{2\pi}\int_0^{\infty}S_z(\omega)\cos(\omega t')\,d\omega.\end{aligned} \tag{A.12}$$

The reciprocal relationship is

$$\begin{aligned}4\int_0^{\infty}\phi_z(t')\cos(\omega t')\,dt' &= \frac{2}{\pi}\int_0^{\infty}S_z(\omega')\,d\omega'\int_0^{\infty}\cos(\omega t')\cos(\omega' t)\,dt'\\ &= \int_0^{\infty}S_z(\omega')\left[\delta(\omega+\omega')+\delta(\omega-\omega')\right]d\omega'\\ &= S_z(\omega)\end{aligned} \tag{A.13}$$

Note

$$\int_0^{\infty}\cos(\omega t')\cos(\omega' t)\,dt' = \frac{\pi}{2}[\delta(\omega+\omega')+\delta(\omega-\omega')].$$

Equations A.12 and A.13 are statements of the Wiener-Khintchin theorem, whose succinct statement is that $2\phi_z(t')$ and $S_z(\omega)$ are the *Fourier transform pairs*, so long as the process is a strong or weak stationary process.

If a process is non-stationary, and $T \rightarrow \infty$ is disallowed, the Wiener-Khintchin theorem states that

$$\phi_Z(t', T) = \frac{1}{2\pi} \int_0^\infty S_Z(\omega, T) \cos(\omega t')\, d\omega, \quad \text{and}$$
$$S_Z(\omega, T) = 4 \int_0^T \phi_Z(t', T) \cos(\omega t')\, dt'. \qquad \text{(A.14)}$$

The Wiener-Khintchin theorem is important for analyzing noise, as well as fluctuation-dissipation and limits of measurement in the presence of noise.

The limits of measurement in the presence of fluctuations is an important nanoscale theme, so these relationships appear in significant ways in S. Tiwari, "Nanoscale device physics: Science and engineering fundamentals," Electroscience 4, Oxford University Press, ISBN 978-0-198-75987-4 (2017).

A.5 Carson's theorem

THE FOURIER TRANSFORM OF RANDOM PULSE TRAINS is also of interest in semiconductor problems. Electrons—as localized wavepackets or classical electrons—are a pulse train. For the random discrete pulse train,

$$z(t) = \sum_{k=1}^{K} c_k f(t - t_k), \qquad \text{(A.15)}$$

where both c_k and t_k are random, while $f()$ is a fixed shape that arises in the property of the system, such as the decay time of the transit time. The Fourier transform and therefore the spectral power density are

$$\begin{aligned} Z(\omega) &= F(\omega) \sum_{k=1}^{K} c_k \exp(-i\omega t_k), \\ \therefore \;\; S_Z(\omega) &= \lim_{T\to\infty} \frac{2\langle |Z(\omega)|\rangle^2}{T} \\ &= \lim_{T\to\infty} \frac{2\langle |F(\omega)|\rangle^2}{T} \sum_{k,l=1}^{K} \langle c_k c_l \exp[-i\omega(t_k - t_l)]\rangle \\ &= \lim_{T\to\infty} \frac{2\langle |F(\omega)|\rangle^2}{T} \\ &\quad \times \left\{ \sum_{k=1}^{K} \langle c_k^2 \rangle + \sum_{k\neq l} \langle c_k c_l \exp[-i\omega(t_k - t_l)]\rangle \right\}. \end{aligned} \qquad \text{(A.16)}$$

If $\nu = \lim_{T\to\infty} K/T$, then the mean pulse amplitude is

$$\langle c^2 \rangle = \lim_{T\to\infty} \frac{1}{K} \sum_{k=1}^{K} \langle c_k^2 \rangle. \qquad \text{(A.17)}$$

The first term in the expansion of Equation A.16 is

$$\sum_{k=1}^{K} \langle c_k^2 \rangle = 2\nu \langle c^2 \rangle \langle |F(\omega)| \rangle^2. \tag{A.18}$$

The rest of the terms—the second part—of Equation A.16, with independent pulsing, are

$$\begin{aligned} &\lim_{T\to\infty} \frac{2\langle |F(\omega)| \rangle^2}{T} \sum_{k\neq l} \langle c_k \rangle \langle c_l \rangle \langle \exp(-i\omega t_k) \rangle \langle \exp(-i\omega t_l) \rangle \\ &= \lim_{T\to\infty} \frac{2\langle |F(\omega)| \rangle^2}{T} \sum_{k\neq l} \langle c^2 \rangle \frac{4\sin^2(\omega T/2)}{\omega^2 T^2} \\ &= 4\pi \overline{z(t)}^2 \delta(\omega), \end{aligned} \tag{A.19}$$

where $Z(\omega = 0) = \int_{-\infty}^{\infty} z(t)\, dt$ and the sinc^2 function's limit, and

$$\overline{z(t)} = \left[\lim_{T\to\infty} \frac{1}{K} \sum_{k=1}^{K} c_k \right] \nu \int_{-\infty}^{\infty} z(t)\, dt = \nu \langle c \rangle \int_{-\infty}^{\infty} z(t)\, dt \tag{A.20}$$

has been used. This latter is a statement of Campbell's theorem, which is summarized in Section A.6. When c_ks are symmetric, $\langle c \rangle = 0$ and therefore $\overline{z(t)} = 0$, and power spectral density has no static term.

In general, though,

$$S_z(\omega) = 2\nu \langle c^2 \rangle \langle |F(\omega)| \rangle^2 + 4\pi \overline{z(t)}^2 \delta(\omega), \tag{A.21}$$

which is the statement of *Carson's theorem*.

A.6 Campbell's theorem

Campbell's theorem relates the mean of a signal to Fourier transformations.

The autocorrelation, following the Wiener-Khintchin theorem, can be written as

$$\begin{aligned} \phi_z(t') &= \frac{1}{2\pi} \int_0^{\infty} S_z(\omega) \cos(\omega t')\, d\omega \\ &= \frac{\langle c^2 \rangle}{\pi} \nu \int_0^{\infty} \langle |F(\omega)| \rangle^2 \cos(\omega t')\, d\omega \\ &\quad + 2\overline{z(t)}^2 \int_0^{\infty} \delta(\omega) \cos(\omega t')\, d\omega \\ &= \nu \langle c^2 \rangle \int_{-\infty}^{\infty} z(t) z(t+t')\, dt + \overline{z(t)}^2, \\ \because \int_0^{\infty} \delta(\omega) \cos(\omega t')\, d\omega &= \frac{1}{2}, \end{aligned} \tag{A.22}$$

and Parseval's theorem. $\phi_z(0) = \overline{z^2(t)}$; therefore,

$$\overline{z^2(t)} - \overline{z(t)}^2 = \nu\langle c^2\rangle \int_{-\infty}^{\infty} z^2(t)\,dt$$
$$= \nu\frac{\langle c^2\rangle}{\pi} \int_{0}^{\infty} \langle|F(\omega)|\rangle^2\,d\omega. \qquad \text{(A.23)}$$

Similarly, for the mean,

$$\overline{z(t)} = \nu\langle c\rangle \int_{-\infty}^{\infty} z(t)\,dt$$
$$= \nu\langle c\rangle Z(0). \qquad \text{(A.24)}$$

Equation A.23 is a statement of Campbell's theorem for the mean square, and Equation A.24 is a statement of the theorem for the mean.

B
Various useful functions

WE EMPLOY A VARIETY OF FUNCTIONS in our mathematical manipulations in this text. A few of the important ones are summarized here, together with a note on functions.

A function $f(x)$ is an *acceptable function* or a *good function* if it is differentiable any number of times and if all its derivatives are $\mathcal{O}\left(|x|^{-N}\right)$ for $|x| \rightarrow \infty$ for all integers N. For example, $\exp(-x^2)$ is an acceptable/good function, but polynomials are not.

A sequence $f_n(x)$ of acceptable functions is a *regular sequence* if, for any regular function $F(x)$, $\lim_{n\rightarrow\infty}\int_{-\infty}^{\infty} f_n(x)F(x)\,dx$ exists. Two regular sequences are equivalent if, for all $F(x)$s, the limit is the same. For example, $\exp(-x^2/n^2)$ and $\exp(-x^4/n^4)$ are equivalent. Here,

$$\lim_{n\rightarrow\infty}\int_{-\infty}^{\infty} f_n(x)F(x)\,dx = \int_{-\infty}^{\infty} F(x)\,dx. \tag{B.1}$$

A generalized function $f(x)$ is a regular sequence $f_n(x)$ of acceptable functions. Two generalized functions are equal when the corresponding regular sequences are equivalent. So, a generalized function is a class of all the regular sequences that are equivalent to a given regular sequence. The generalized function $f(x)$ is odd or even depending on the oddity or evenness of $F(x)$ for which $\int_{-\infty}^{\infty} f(x)F(x)\,dx = 0$.

These properties are useful in understanding functions of interest to us.

Dirac $\boldsymbol{\delta}$ *function:*

The Dirac δ function's defining characteristics is that its integral over the entire space takes a unit value, that is,

$$\int \delta(\mathbf{r}-\mathbf{r}_0)\,d^3\mathbf{r} = 1. \tag{B.2}$$

Because it allows us to have a point, line or plane, that is, a collapsed dimensional placement of a physical variable of interest, it is one of the more commonly useful functions. However, whether it is really a function, since it is only explicitly writable in asymptotic limits and

Coordinate	3D	2D	1D
Cartesian	$\delta(x-x_0)\delta(y-y_0)\delta(z-z_0)$	$\delta(x-x_0)\delta(y-y_0)$	$\delta(x-x_0)$
Cylindrical	$\frac{\delta(r-r_0)\delta(\varphi-\varphi_0)\delta(z-z_0)}{r}$	$\frac{\delta(r-r_0)\delta(z-z_0)}{2\pi r}$	$\frac{\delta(r-r_0)}{2\pi r}$
Spherical	$\frac{\delta(r-r_0)\delta(\theta-\theta_0)\delta(\varphi-\varphi_0)}{r^2\sin\theta}$	$\frac{\delta(r-r_0)\delta(\theta-\theta_0)}{2\pi r^2\sin\theta}$	$\frac{\delta(r-r_0)}{4\pi r^2\sin\theta}$

Table B.1: The Dirac δ function $\delta(\mathbf{r}-\mathbf{r}_0)$ form in different coordinate systems.

acquires its meaning and defining characteristic through its use inside an integral, certainly is worth bearing in mind, as is that while one largely uses it with integration, one quite fails with its derivative. The δ function makes sense as an integrand, either by itself or in multiplication with other functions. Table B.1 summarizes this in 3-, 2- and 1-dimensional space.

The Dirac δ may be defined by the sequence

$$f_n(x) = (n/\pi)^2 \exp(-nx^2) \tag{B.3}$$

or its equivalent sequences. In all these,

$$\lim_{n\to\infty}\int_{-\infty}^{\infty} f_n(x)F(x)\,dx = \lim_{n\to\infty}\int_{-\infty}^{\infty}\delta(x)F(x)\,dx = F(0). \tag{B.4}$$

This says that the Dirac δ function is even.

The equivalent sequences for the Dirac δ include

$$\begin{aligned}\delta(x-x_0) &= \lim_{L\to\infty}\frac{\sin[(x-x_0)L]}{\pi x},\\ &= \lim_{a\to 0}\frac{1}{\pi}\frac{\alpha}{\alpha^2+(x-x_0)^2},\\ &= \lim_{\alpha\to 0}\frac{1}{\alpha\sqrt{\pi}}\exp\left[-\left(\frac{x-x_0}{\alpha}\right)^2\right],\end{aligned}$$

and so on.

Any complete set of orthonormal functions can be used to define the Dirac δ. For example, for the set composed of $\psi_n(x)$,

$$\delta(x-x_0) = \sum_{n=1}^{\infty}\psi_n(x_0)\psi_n^*(x) \tag{B.5}$$

for a discrete set and

$$\delta(x-x_0) = \int \psi_s(x_0)\psi_s^*(x)\,ds \tag{B.6}$$

for a continuous set.

Kronecker $\boldsymbol{\delta}$ *function:*
The Kronecker delta $\delta_{i,j}$ is defined as

$$\delta_{i,j} = \begin{cases}1 & i=j,\\ 0 & i\neq j.\end{cases} \tag{B.7}$$

It is a function of two arguments, and if the arguments have the same value, then it is unity; else, it vanishes. It is useful in picking elements in arrays and matrices.

But it must be distinguished from the Dirac delta, which has an integral of unity over its space and thus allows one to determine the value of a continuous function, as well as achieve other usages, since it is defined over an integral.

Many interesting functions follow from these functions.

Heaviside's step function:

Heaviside's step function is $\Theta(z) = 1$ for $z > 0$, and $\Theta(z) = 0$ for $z < 0$. In terms of the Dirac δ function,

$$\Theta(z) = \int_{-\infty}^{z} \delta(\zeta)\, d\zeta. \tag{B.8}$$

Or, in the reverse operation, we get

$$\delta(z) = \frac{d\Theta(z)}{dz}. \tag{B.9}$$

The Heaviside function, in its general form, is

$$\Theta(z - z_0) = \begin{cases} 0 & \forall\ z < z_0, \\ 1 & \forall\ z > z_0. \end{cases} \tag{B.10}$$

A common useful form then follows as

$$\Theta(z - z_0) = \lim_{\alpha \to 0} \begin{cases} 0 & \forall\ z < z_0, \\ \exp[-\alpha(z - z_0)] & \forall\ z > z_0. \end{cases} \tag{B.11}$$

C
Random processes

CLASSICAL RANDOMNESS appears in multitudes of flavors. In natural processes, it is because of the nature of the underlying basis. In engineered processes, it is because of forcing that one may introduce. Randomness is therefore encountered in many varieties. A few of these appear quite often in semiconductors in their static and dynamic conditions. Examples include point perturbations and extended defects; in addition, compositional components, as in ternaries and quaterneries, can be random (and non-random). A collective motion of a planar charge will spread out, have a Gaussian spread as an outcome of random processes and will also show shot noise due to the individual particles crossing collection or injection boundaries randomly.

A random walk leads to Gaussian distributions, which is a distribution that also arises due to the central limit theorem; that is, it is a consequence of a collection of independent random variables in an ensemble. Nature exhibits Gaussian distribution for many observables for this reason. It also exhibits Poisson distribution in many phenomena. Poisson distribution arises when mutually uncorrelated events happen with low individual probability. Semiconductors exhibit both, and others when looked at in detail. The distribution of dopants, when uncorrelated due to the nature of techniques leading to their appearance, such as in semiconductor boule growth, will show a Gaussian distribution if the doping is significant. The threshold spread due to dopants in a transistor can be skewed. On the other hand, defects, which have vanishingly low probability, may have a Poisson distribution. Single electron effects, where spin will become important when the number of states is limited, show bimodal distributions.

"Randomness and independence" is important for understanding distributions. Force-fitting distributions, without an underlying understanding, causes serious, dreadful consequences when applied to social circumstances. A poor educational and poorly nurturing environment severely hurts the chances of the poor and folks from the wrong side of the train track appearing on the successful side of the bell curve. Under forcing, distributions will skew. In many professions, even in educational institutions, there is a tendency to evaluate on a curve. Certainly, in very selective enterprises—research or advanced technology, for example—this makes no sense. Too small a sample size, and the sample consists of very carefully selected non-random folks, not to mention its contradictions to the importance of collective effort and relationships.

This appendix is a summary of a few streams of the underlying random process analytics that one should be aware of in the study of materials and of devices, and we look at it through the lens of stochastic processes.

A stochastic process, such as noise or any unpredictability, as in a fluctuation, is a random variation in time and space that needs statistical characterization. We do this using statistical functions. A signal $x(t)$ may be discrete or continuous, and one can find a mean, a mean square, a variance, an autocorrelation or something else as a measure, as we did in Chapter 16. One can define various probability functions, the simplest being of the occurrence of x in an interval, but also joint probabilities (two different probabilities appearing in consort at different instants), marginal probabilities (a probability given the occurrence of a prior), et cetera. The probabilities may be measured at instants of time on a stream of a signal or may be measured at a time over a collection of signals; that is, an ensemble. If the averages in time, or at any time over an ensemble, are equal, this is an ergodic ensemble. Since one may have access only to one sample function, if there is ergodicity, there is usefulness to prediction over ensembles. Chapter 16 also made remarks regarding the stationarity and non-stationarity of processes, based on the order in shifts of time up to which the probability density function remains invariant. A process stationary in any order is strictly stationary. Ergodicity too can be viewed through complementary measures. If the mean over time and the mean over the ensemble are equal, the stochastic process is stationary in the mean. If the autocorrelations match, then it is ergodic in autocorrelation. A stationary process, however, is not necessarily ergodic.

A probability mass function $\mathfrak{p}(z)$ is defined for *discrete* random variables, and the probability density function over continuous variables.

The z transform, a discrete transform, is defined as

$$\mathfrak{P}_x^T(z) = \sum_x z^x \mathfrak{p}(x). \tag{C.1}$$

The z transform function provide expectations for moments of any order, since

$$\begin{aligned} \left.\frac{d^n}{dz^n}\mathfrak{P}_x^T(z)\right|_{z=1} &= \sum_x x(x-1)\cdots(x-n+1)\mathfrak{p}(x), \quad \text{with} \\ \langle x\rangle = & \left.\frac{d}{dz}\mathfrak{p}_x^T(z)\right|_{z=1}, \\ \langle x^2\rangle = & \left.\frac{d^2}{dz^2}\mathfrak{p}_x^T(z)\right|_{z=1} + \left.\frac{d}{dz}\mathfrak{p}_x^T(z)\right|_{z=1}, \quad \text{and so on.} \end{aligned} \tag{C.2}$$

For a continuous variable function, say $f(x)$, a useful transform is the continuous transform, where one can define the "first" characteristic function

$$f_x^T(s) = -\int_{-\infty}^{\infty} \exp(-sx)f(x)\,dx = \langle \exp(-sx) \rangle,$$

with the properties

$$\frac{d^n}{ds^n} f_x^T(s)\bigg|_{s=0} = \int_{-\infty}^{\infty} (-x)^n f(x)\,dx = (-1)^n \langle x^n \rangle,$$

$$\langle x \rangle = -\frac{d}{ds} f_x^T(s)\bigg|_{s=0},$$

$$\langle x^2 \rangle = -\frac{d^2}{ds^2} f_x^T(s)\bigg|_{s=0}, \quad \text{and so on.} \tag{C.3}$$

These expectations in power $\langle x^n \rangle$ for discrete and continuous variables are the moments. When the moments are determined around the mean, so $\mu_n = \langle (x - \langle x \rangle)^n$, they are central moments. The first-order central moment $\mu_1 = 0$. The second-order central moment,

$$\mu_2 = \langle x^2 \rangle - \langle x \rangle^2 = \sigma^2, \tag{C.4}$$

is the variance. The square root of the variance is the standard deviation σ. Higher-order moments,

$$\mu_3 = \langle x^3 \rangle - 3\langle x^2 \rangle \langle x \rangle + 2\langle x \rangle^3,$$

$$\mu_4 = \langle x^4 \rangle - 4\langle x \rangle^3 \langle x \rangle + 6\langle x^2 \rangle \langle x \rangle^2 - 3\langle x \rangle^4, \quad \text{and so on,} \tag{C.5}$$

are also important, since they show long-range connections.

For continuous variables, using the continuous transform, there also exist similar and powerful ways for analysis. Following on from the first characteristic function, we define a second characteristic function,

$$g(s) = \log f_x^T(s), \quad \text{with cumulants } \lambda_n \ \frac{d^n}{ds^n} g(s)\bigg|_{s-0}. \tag{C.6}$$

This leads to the relationships

$$g(s) = \lambda_a s + \frac{1}{2!}\lambda_2 s^2 + \cdots + \frac{1}{n!} s^n + \cdots, \quad \text{with}$$

$$\lambda_1 = \langle x \rangle,$$

$$\lambda_2 = \langle x^2 \rangle - \langle x \rangle^2,$$

$$\lambda_3 = \langle x^3 \rangle - 3\langle x^2 \rangle \langle x \rangle + 2\langle x \rangle^3,$$

$$\lambda_4 = \langle x^4 \rangle - 4\langle x \rangle^3 \langle x \rangle + 6\langle x^2 \rangle \langle x \rangle^2 - 3\langle x \rangle^4, \quad \text{and so on.} \tag{C.7}$$

These moments, characteristic functions, et cetera, let us analyze and build a suitable algorithm for higher order analysis.

We can now look at random process and some of the functions of interest to us.

C.1 Bernoulli process

Let there be a discrete variable $x \in 0, 1$, with the probability mass function

$$\mathfrak{P}_x(x_0) = \begin{cases} 1-\mathfrak{p} & \text{for } x_0 = 0, \\ \mathfrak{p} & \text{for } x_0 = 1. \end{cases} \tag{C.8}$$

The z transform is

$$\begin{aligned} \mathfrak{P}_x^T(z) &= \sum_{x_0} z^{x_0}\mathfrak{P}_x(x_0) = z^0(1-\mathfrak{p}) + z^1\mathfrak{p} = (1-\mathfrak{p}) + z\mathfrak{p} \\ \therefore\ \langle x \rangle &= \left.\frac{d}{dz}\mathfrak{P}_x^T(z)\right|_{z=1} = \mathfrak{p}, \text{ and} \\ \sigma_x^2 = \langle x^2 \rangle - \langle x \rangle^2 &= \left.\frac{d^2}{dz^2}\mathfrak{P}_x^T(z)\right|_{z=1} + \left.\frac{d}{dz}\mathfrak{P}_x^T(z)\right|_{z=1} \\ &\quad - \left.\left[\frac{d}{dz}\mathfrak{P}_x^T(z)\right]^2\right|_{z=1} \\ &= \mathfrak{p}(1-\mathfrak{p}). \end{aligned} \tag{C.9}$$

This leads us into the Binomial distribution with multiple Bernoulli (process) trials.

C.2 Binomial distribution

Let the process consist of n independent Bernoulli trials. These n trials produce $x_1, x_2, \ldots, x_n$ independent variables, each with $0, 1$ as possible outcomes. Let $k = x_1 + x_2 + \cdots + x_n$ be the sum of the n independent variables. The z transform is

The two outcomes are 0 and 1. These are just symbols. We can as well call them false and true, or failure and success. And we will.

$$\begin{aligned} \mathfrak{P}_k^T(z) &= \sum_{x_1,\ldots,x_n} z^{x_1+\cdots+x_n}\mathfrak{p}(x_1)\mathfrak{p}(x_2)\cdots\mathfrak{p}(x_n) \\ &= \mathfrak{P}_{x_1}^T(z)\mathfrak{P}_{x_2}^T(z)\cdots\mathfrak{P}{x_n}^T(z) \\ &= (1-\mathfrak{p}+z\mathfrak{p})^n \\ &= \mathfrak{P}_k(0) + \mathfrak{P}_k(1)z + \mathfrak{P}_k(2)z^2 + \cdots + \mathfrak{P}_k^n z^n \\ &= \sum_{k_0}^{n} z^{k_0}\mathfrak{P}_k(k_0). \end{aligned} \tag{C.10}$$

In the summing series, the first term is of finding only 0s as the outcomes ($k_0 = 0$), the second term is for finding one 1 as the outcome ($k_0 = 1$), and so on. k_0 is the number of possible 1s in the collection of the independent variables. Each term corresponding to $z^{k_0}\mathfrak{P}_k(k_0)$ is the probability of obtaining k_0 1s in this collection. This probability mass function is

$$\mathfrak{P}_k(k_0) = \binom{n}{k_0} \mathfrak{p}^{k_0}(1-\mathfrak{p})^{n-k_0}, \text{ where}$$
$$\binom{n}{k_0} = \frac{n!}{k_0!(n-k_0)!}. \tag{C.11}$$

The binomial distribution representing the distribution of possibilities of independent Bernoulli trials has the probability mass function $\mathfrak{P}_k(k_0)$ given by Equation C.11. The mean k for the independent variables is

$$\langle k \rangle = n\mathfrak{p}, \tag{C.12}$$

and the variance is

$$\sigma_k^2 = n\mathfrak{p}(1-\mathfrak{p}). \tag{C.13}$$

The former is the sum of the mean values over the independent variables. The latter, similarly, is the sum of the variances over the independent variables.

The z transform helps with determining the expectations, variances, et cetera, for all the different possibilities that one might encounter in conducting the Bernoulli trials in this binomial collection.

Take, for example, the question, "What is the probability mass function for obtaining the first 1 in the lth Bernoulli trial after $l-1$ 0s and some of the related properties?" Since $(1-\mathfrak{p})$ is the probability for a 0 in the independent events, and $\mathfrak{p}$ the probability of a 1, once we determine the probability mass function, the rest follow. The probability mass function for these constraints—the geometric distribution—is

$$\mathfrak{P}_{l_1}(z) = (1-\mathfrak{p})^{l-1}\mathfrak{p}. \tag{C.14}$$

The z transform then leads us to

$$\mathfrak{P}_{l_1}^T(z) = \sum_l z^l (1-\mathfrak{p})^{l-1}\mathfrak{p} = \frac{z\mathfrak{p}}{1-z(1-\mathfrak{p})}$$
$$\therefore \ \langle l_1 \rangle = \left[\frac{d}{dz}\mathfrak{P}_{l_1}^T(z)\right]\Big|_{z=1} = \frac{1}{\mathfrak{p}},$$
$$\text{and } \sigma_{l_1}^2 = \left\{\frac{d^2}{dz^2}\mathfrak{P}_{l_1}^T(z) + \frac{d}{dz}\mathfrak{P}_{l_1}^T(z) - \left[\frac{d}{dz}\mathfrak{P}_{l_1}^T(z)\right]^2\right\}\Bigg|_{z=1}$$
$$= \frac{1-\mathfrak{p}}{\mathfrak{p}^2}. \tag{C.15}$$

C.3 *Poisson random process*

LET THERE BE A SERIES of independent and identical Bernoulli trials along a continuous line in variable t. t can be time, but doesn't have to be. It could be a one-dimensional chain of one type of atoms and vacancies. A Bernoulli trial is conducted every Δt, and the probability of a 1 is $\mathfrak{p} = \lambda\Delta t$, again a normalized number. Let $k \in 0, 1$ be our Bernoulli outcomes; then,

$$\mathfrak{P}(k, \Delta t) = \begin{cases} 1 - \lambda\Delta t & \text{for } k = 0, \\ \lambda\Delta t & \text{for } k = 1. \end{cases} \tag{C.16}$$

As Δt becomes smaller, λ approaches the average rate for an outcome of 1.

After an interval $[0, t]$, the probability measurement being a k (the net number of successes or 1) over the interval $t + \Delta t$ can be decomposed into two terms. The first is that k is the measurement through the interval t and then 0, measured during the sampling at Δt, and the second is that $k-1$ is measured over t, and a 1 is measured over the next sampling at Δt. So,

$$\begin{aligned} \mathfrak{P}(k, t + \Delta t) &= \mathfrak{P}(k, t)\mathfrak{P}(0, \Delta t) + \mathfrak{P}(k - 1, \Delta t)\mathfrak{P}(1, \Delta t) \\ &= \mathfrak{P}(k, t)(1 - \lambda\Delta t) + \mathfrak{P}(k - 1, \Delta t\lambda\Delta t. \end{aligned} \tag{C.17}$$

Now let $\Delta t \to 0$, making the functions continuous with λ being a true average rate of 1s. The differential form is

$$\frac{d}{dt}\mathfrak{P}(k, t) + \lambda\mathfrak{P}(k, t) = \lambda\mathfrak{P}(k - 1, t). \tag{C.18}$$

This equation sets up the algorithm for an iterative solution starting from $k = 0$. This initial condition is

$$\begin{aligned} \mathfrak{P}(k, 0) &= \begin{cases} 1 & \text{for } k = 0, \\ 0 & \text{for } k \neq 0, \end{cases} \\ \sum_k \mathfrak{p}(k, t) &= 1. \end{aligned} \tag{C.19}$$

The solution is

$$\mathfrak{P}(k, t) = \frac{(\lambda t)^k}{k!}\exp(-\lambda t), \tag{C.20}$$

where $\lambda t = \mu$ is the average number of successes (1s).

As a probability mass function, we may write this as

$$\mathfrak{P}(k, t) = \frac{(\mu)^k}{k!}\exp(-\mu); \tag{C.21}$$

the z transformation teaches us then that

$$\begin{aligned} \mathfrak{P}_k^T(z) &= \sum_{k_0} z^{k_0}\mathfrak{P}_k(k_0) = \exp(-\mu)\sum_{k_0}\frac{(\mu z)^{k_0}}{k_0!} = \exp[\mu(z - 1)] \\ \therefore \quad \langle k \rangle &= \left[\frac{d}{dz}\mathfrak{P}_k^T(z)\right]\Bigg|_{z=1} = \mu, \quad \text{and} \\ \sigma_k^2 &= \left\{\frac{d^2}{dz^2}\mathfrak{P}_k^T(z) + \frac{d}{dz}\mathfrak{P}_k^T(z) + \left[\frac{d}{dz}\mathfrak{P}_k^T(z)\right]^2\right\}\Bigg|_{z=1} \\ &= \mu. \end{aligned} \tag{C.22}$$

The variance and the mean of the Poisson distribution are the same. As Figure C.1 shows, as k increases, and the mean increases, the function also spreads out. $k = 0$ is an exponential decay, and then, beyond, with increasing k, the Poisson distribution rises and then falls. With increasing k, the distribution stretches out while maintaining the mean and the variance.

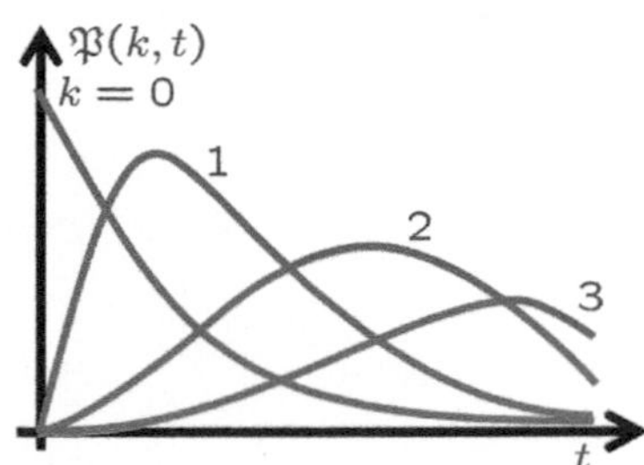

Figure C.1: Probability mass function as a function of k expected in the Poisson distribution.

The binomial distribution leads to the Poisson distribution in the limit of very small probabilities. A sequence of single Bernoulli trials with a very small but finite probability of success produces a Poisson distribution. Defects and defect-mediated processes—the existence of rare intrinsic or extrinsic impurities, and the capture and emission of carriers from them—are often Poisson processes. They are rare enough that they are independent of each other.

See Appendix E in S. Tiwari, "Nanoscale device physics: Science and engineering fundamentals," Electroscience 4, Oxford University Press, ISBN 978-0-198-75987-4 (2017), for the limit derivation of Poisson distribution from the binomial distribution. This is why the Poisson distribution is often called "the law of small numbers." The appendix also shows a derivation of the Gaussian distribution from the binomial distribution, as an equivalent law of large numbers.

C.4 Gaussian random process

GAUSSIAN RANDOM PROCESSES are of a lot of interest because they represent a number of natural phenomena, as a law of large numbers, and because of the consequences of the central limit theorem.

The binomial distribution in the limit of large numbers takes us to the Gaussian distribution. Take, in Equation C.11, n very large, and both $\mathfrak{p}$ and $1-\mathfrak{p}$ finite and not vanishingly small. With a large n, this implies that $\langle k_0 \rangle = \mathfrak{p}n$, and dropping off quite rapidly, that is,

$$|\mathfrak{P}_k(k_0+1) - \mathfrak{P}_k(k_0)| \ll \mathfrak{P}_k(k_0). \tag{C.23}$$

Treating $\mathfrak{P}_k(k_0)$ as a continuous function in the variable k_0, one can Taylor expand, $k_0 = \langle k_0 \rangle + \eta$, so that

$$\begin{aligned} \ln \mathfrak{P}_k(k_0) &= \ln \mathfrak{P}_k(\langle k_0 \rangle) + C_1\eta + \frac{1}{2}\eta^2 + \cdots, \quad \text{where} \\ C_l &= \left. \frac{d^l}{d\eta^l} [\ln \mathfrak{P}_k(\langle k_0 \rangle + \eta)] \right|_{\eta=0}. \end{aligned} \tag{C.24}$$

This also means that the $\ln \mathfrak{P}_k(k_0)$ converges more rapidly in the higher order terms of expansion.

For the probability mass function,

$$\begin{aligned} \ln \mathfrak{P}_k(k_0) &= \ln n! - \ln k_0! - \ln(n-k_0)! + k_0 \ln \mathfrak{p} \\ &\quad + (n-k_0)\ln(1-\mathfrak{p}), \\ \frac{d}{dk_0} \ln k_0! &= \frac{\ln(k_0+1)! - \ln k_0!}{(k_0+1) - k_0} \approx \ln k_0, \quad \text{for } k_0 \gg 1, \\ \therefore \quad \frac{d}{dk_0} \ln \mathfrak{P}_k(k_0) &= -lnk_0 + \ln(n - k_0 + \ln \mathfrak{p} - \ln(1-\mathfrak{p}). \end{aligned} \tag{C.25}$$

A number of conditions hold:

$$\frac{d}{dk_0} \ln \mathfrak{P}_k(k_0)\bigg|_{k_0=\langle k_0\rangle} = 0 \;\because\; \langle k_0\rangle = n\mathfrak{p} \text{ is the maximum,}$$

$$C_1 = 0 \;\because\; \text{of the maximum, and}$$

$$\frac{d^2}{dk_0^2} \ln \mathfrak{P}_k(k_0)\bigg|_{k_0} = \frac{1}{n\mathfrak{p}(1-\mathfrak{p})} = C_2. \tag{C.26}$$

In the limit, with $\eta \to 0$, and higher order terms vanishing,

$$\mathfrak{P}_k(k_0) = \frac{1}{\sqrt{2\pi\sigma_{k_0}^2}} \exp\left[-\frac{(k_0 - \langle k_0\rangle)^2}{2\sigma_{k_0}^2}\right], \text{ and } \sigma_{k_0}^2 = n\mathfrak{p}(1-\mathfrak{p}). \tag{C.27}$$

The s transform of the Gaussian distribution function is

$$f_{k_0}^T(s) = \int_{-\infty}^{\infty} \mathfrak{P}_k)(k_0) \exp(-sk_0)\, dk_0$$
$$= \exp\left(-s\langle k_0\rangle + \frac{1}{2}s^2\sigma_{k_0}^2\right). \tag{C.28}$$

The Gaussian random distribution is very common because of the central limit feature laid out in the central limit theorem. The theorem asserts that, irrespective of the nature of individual random variable probability density functions, the sum of many independent identically distributed random variables will converge to a Gaussian probability distribution function as the number of these variables gets large. So, take the distribution of a set of n independent measurements from a sample set with finite variance, and one will have a variance of σ^2/n. The central limit theorem asserts that, with increasing n, the distribution will acquire the normalcy of a Gaussian distribution. Finite variance and finite σ^2/n suffice for this assertion.

If one looks at the variances of the Poisson and the Gaussian distribution functions as with the probability $\mathfrak{p}$ varied, one would note the similarities in them. The spreading away from the mean, characterized by the variance in the two distributions, is quite similar, as seen in Figure C.2.

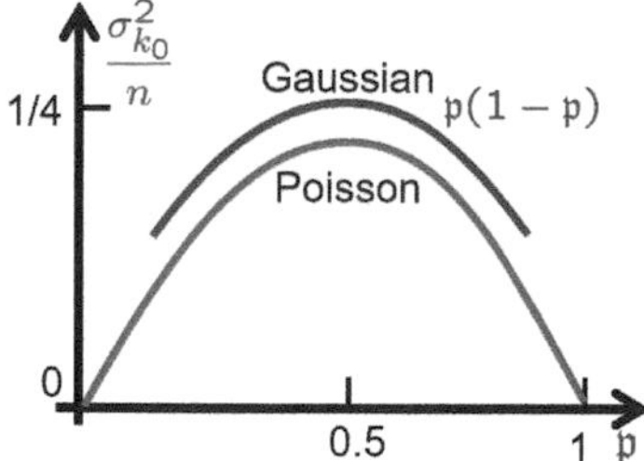

Figure C.2: Normalized function of variance and sample size as a function of probability for Gaussian and Poisson distributions.

The Gaussian distribution function can also be extended in form to multi-dimensional space. If $z_1, z_2, \ldots, z_N$ are N random variables with a non-zero mean, then an N-dimensional Gaussian process is one whose probability density function for all N has the N-dimensional form

$$\mathfrak{P}_N(z_1, z_2, \cdots, z_N) =$$
$$\frac{1}{(2\pi)^{N/2}\sqrt{|A_{ij}|}} \exp\left[-\frac{1}{2|A|}\sum_{i,j=1}^{N} A_{ij}(z_i - \mu_i)(z_j - \mu_j)\right], \tag{C.29}$$

where $\mu_i = \langle z_i \rangle$, and $\mu_j = \langle z_j \rangle$. Here, A_{ij} is a cofactor for the second moment

$$d_{ij} = \langle (z_i - \langle z_i \rangle)\,(z_j - \langle z_j \rangle) \rangle \tag{C.30}$$

of the matrix whose determinant is $|A|$,

$$|A| = \begin{vmatrix} d_{11} & d_{12} & \cdots & d_{1N} \\ d_{21} & d_{22} & \cdots & d_{2N} \\ \vdots & \vdots & \vdots & \vdots \\ d_{N1} & d_{N2} & \cdots & d_{NN} \end{vmatrix}. \tag{C.31}$$

An N-dimensional Gaussian probability density is completely described by its average value and its second moment d_{ij}.

For $N = 1$, that is, one variable,

$$\mathfrak{P}_1(z_1) = \frac{1}{(2\pi\,|A|)^{1/2}} \exp\left(-\frac{A_{11}}{2|A|} z_1^2 \right), \tag{C.32}$$

to which correspondences can be made with our earlier derivation.

C.5 Cauchy or Lorentz distribution

Cauchy distribution to non-mathematicians is usually the Lorentz distribution (also the Breit-Wigner distribution) to the scientist, since it appears so often in resonance situations.

The Cauchy or Lorentz distribution often encountered in resonance conditions, so a response of a second-order differential equation relating energy exchange, such as in the linewidths under homogeneous broadening, or plasmon resonance, is a special probability density function of the form

$$\mathfrak{p}(x) = \frac{1}{\pi} \frac{\lambda}{\lambda^2 + (x-\mu)^2}, \quad \text{with } \lambda > 0. \tag{C.33}$$

The characteristic function associated with this distribution is

$$f(t) = \exp(\mu i t - \lambda |t|). \tag{C.34}$$

The function has no defined mean or moments, so no variance, and one usually characterizes it through linewidth.

The Cauchy-Lorentz distribution arises as an observation of the phenomena in nature, as do the other random distributions. They all certainly have sound mathematical underpinnings, and one can also argue why they have the form they take without recourse to Bernoulli trials and the binomial distribution. But they do have substantial differences.

The scale parameter λ of the Cauchy-Lorentz distribution (linewidth), and the σ parameter (standard deviation) of the Gaussian distribution, are such that they contain the same probability within the same quantile interval. However, a calculation of the Cauchy distribution's mean and variance with changing n will lead to larger and larger sampling of the tail of the Cauchy distribution. So, the variance will stretch faster than the inverse n dependence of σ^2/n, unlike the Gaussian's parameter.

D
Calculus of variation, and the Lagrangian method

THE CALCULUS OF VARIATIONS deals with determining extreme values of functionals. Functionals are rules—including objective mathematical functions such as integrals—that map a real number to a function in some class. In general, then, a functional J defines on a set of functions, say A, the mapping of $A \mapsto R$ so that all $x, y, \ldots$ in A map to a real number $J(x)$. The functional of most interest to us is action—an integral relationship—and the function, the Lagrangian $\mathscr{L}$, which we may define for a variety of physical situations. Minimization of action draws out lawful relationships that guide deterministic behavior, and this approach to treating variations becomes quite fundamental to classical, quantum-mechanical and relativistic problems. The first two are of interest to us in this text, and this appendix summarizes some significant ideas related to the approach.

Derivatives are a measure of variation. Partial derivatives—slightly simpler mathematically—are also a measure of variations under constraints. Functionals are a major step up from these. Why do we tend to prefer walking in a straight line so instinctively? It is an extremum problem, where the path of least resistance seems to be desired. Even light rays want to do this, as Snell's law illustrates. All are examples of the principle of least action. Principle of least action should really be seen as a principle of trajectory in which slight changes have the least effect on the total action.

Let $K = K[\mathbf{q}(\mathbf{x})]$ be such a functional that depends on the function $\mathbf{q}(\mathbf{x})$ continuously over $\mathbf{x}$:

$$\begin{aligned} K &= \int_i^f \mathscr{L}(\mathbf{q}, \mathbf{q}', \mathbf{x})\, d\mathbf{x}, \quad \text{where} \\ \mathbf{x} &= (x_1, \ldots, x_M), \quad \text{and} \quad d\mathbf{x} = dx_1 \cdots dx_M, \\ \mathbf{q} &= (q_1, \ldots, q_N), \quad \text{and} \quad q_i = q_i(\mathbf{x}), \quad \text{and} \\ \mathbf{q}'(\mathbf{x}) &= \left(\frac{\partial q_1}{\partial x_1}, \ldots, \frac{\partial q_1}{\partial x_M} \right). \end{aligned} \tag{D.1}$$

We have written this relationship quite generally. $\mathbf{q}$ and $\mathbf{x}$ are real. They are observables. $\mathbf{x}$ may be the time coordinate that is discrete (M instances) or continuous, $\mathbf{q}$ may be the position of one or more (N) particles and $\mathbf{q}'(\mathbf{x})$ is then the partial time dependence of that position.

The reason for writing in this way is to view this problem as an informational problem that is, of course, also tied to the physical problems that are of general interest in the observation of a physical system's characteristics. In this, we are taking a more general view than the conventional mechanical or physical science exposition.

$K[\mathbf{q}(\mathbf{x})]$ is a functional that is a single number whose value depends on one or more functions $\mathbf{q}(\mathbf{x})$ over $\mathbf{x}$'s domain. K is physical information that arose through an observation of an observable, so the putting together of an observing environment with the system of interest.

We now simplify this to only the domain for $\mathbf{x}$ and only one $\mathbf{q}$, such as position. We wish to find $q(x)$ that fulfills

$$K = \int_i^f \mathscr{L}(x, q(x), q'(x))\, dx = \textit{extreme}. \tag{D.2}$$

$q' = dq(x)/dx$. An example is the path taken by a particle of mass m with displacement q, and x the time. The Lagrangian for this problem is $\mathscr{L}(q, q') = (1/2)mq'^2 - V(q)$. For any departure $q_\epsilon(x,\epsilon) = q(x) + \epsilon\eta(x)$, from the solution path $q(x)$, with $\eta(x)$ as a perturbing function and ϵ a finite, tunable parameter scale attached to it, the possible paths must still satisfy the end-point constraint of $\eta(x_i) = \eta(x_f)$, as shown in Figure D.1. We rewrite Equation D.3 for this path,

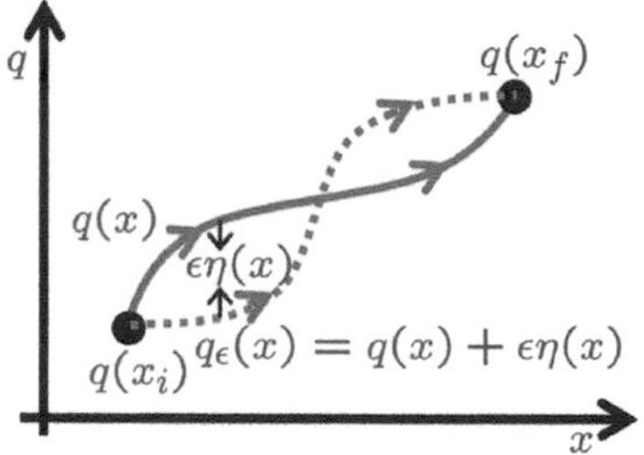

Figure D.1: $q(x)$—the solution—and a perturbation connecting an initial and final point of a path taken by q.

$$K(\epsilon) = \int_i^f \mathscr{L}(x, q_\epsilon(x,\epsilon), q'_\epsilon(x,\epsilon))\, dx = \textit{extreme} \tag{D.3}$$

under the tunable parameter ϵ. For $K(\epsilon)$ to be extremized, and this must occur for the smallest ϵ, so for $\epsilon = 0$,

$$\left.\frac{\partial K}{\partial \epsilon}\right|_{\epsilon=0} = 0. \tag{D.4}$$

We can now expand this differentiated form:

$$\frac{\partial K}{\partial \epsilon} = \int_i^f \left(\frac{\partial \mathscr{L}}{\partial q_\epsilon}\frac{\partial q_\epsilon}{\partial \epsilon} + \frac{\partial \mathscr{L}}{\partial q'_\epsilon}\frac{\partial q'_\epsilon}{\partial \epsilon} \right) dx. \tag{D.5}$$

The second term here, using integration by parts ($u = \partial\mathscr{L}/\partial q'_\epsilon$, and $dv = \partial^2 q_\epsilon/\partial x \partial\epsilon$, in the conventional notation), is

$$\begin{aligned}
\int_i^f \frac{\partial \mathscr{L}}{\partial q'_\epsilon}\frac{\partial q_{\epsilon'}}{\partial \epsilon}\, dx &= \int_i^f \frac{\partial \mathscr{L}}{\partial q'_\epsilon}\frac{\partial^2 q_\epsilon}{\partial x \partial \epsilon}\, dx \\
&= \left.\frac{\partial \mathscr{L}}{\partial q'_\epsilon}\frac{\partial q_\epsilon}{\partial \epsilon}\right|_i^f - \int_i^f \frac{\partial q_\epsilon}{\partial \epsilon}\frac{d}{dx}\left(\frac{\partial \mathscr{L}}{\partial q'_\epsilon}\right) dx.
\end{aligned} \tag{D.6}$$

Since $\partial q_\epsilon/\partial\epsilon = \eta(x)$, which vanishes in the limit points, the first term vanishes. So, Equation D.5 reduces to

$$\begin{aligned}
\frac{\partial K}{\partial \epsilon} &= \int_i^f \left[\frac{\partial \mathscr{L}}{\partial q_\epsilon}\frac{\partial q_\epsilon}{\partial \epsilon} - \frac{\partial q_\epsilon}{\partial \epsilon}\frac{d}{dx}\left(\frac{\partial \mathscr{L}}{\partial q'_\epsilon}\right) \right] dx \\
&= \int_i^f \frac{\partial q_\epsilon}{\partial \epsilon} \left[\frac{\partial \mathscr{L}}{\partial q_\epsilon} - \frac{d}{dx}\left(\frac{\partial \mathscr{L}}{\partial q'_\epsilon}\right) \right] dx \\
&= \int_i^f \eta \left[\frac{\partial \mathscr{L}}{\partial q_\epsilon} - \frac{d}{dx}\left(\frac{\partial \mathscr{L}}{\partial q'_\epsilon}\right) \right] dx.
\end{aligned} \tag{D.7}$$

$\eta(x)$ is arbitrary, only forced to vanish at end limits, and $\partial K/\partial\epsilon$ vanishes at $\epsilon=0$, where this holds. Therefore,

$$\frac{\partial\mathscr{L}}{\partial q}-\frac{d}{dx}\left(\frac{\partial\mathscr{L}}{\partial q'}\right)=0 \tag{D.8}$$

This is the Euler-Lagrange equation. The solution here depends on the Lagrangian. Lagrangians are usually chosen with some thought, being part of the action functional. The approach by which we arrived at it is by using ϵ as a tunable perturbation parameter that introduced a variation. In general, this approach of introducing a variational parameter is an example of the calculus of variation. Perturbation theory uses this variational approach to find the corrections of different order.

The Lagrangian is magical! It represents the evolution of a system that may consist of many particles in the configuration space. As time evolves, this point in the configuration space representing the system moves about. A single function—the Lagrangian—suffices to describe it. The equivalent of this is the Hamiltonian—the symmetrized picture—where one interprets it as phase space. The coordinate q is a generalized coordinate, and q' another independent generalized coordinate (a "velocity"). One interpretation for the Lagrangian is of the difference between kinetic and potential energy for most problems of classical mechanical interest. Along the path, it is these energy forms that are being exchanged. The magic is that if a Lagrangian process contains continuous symmetry, then there is a conservation law for that symmetry. This is the remarkable Noether's theorem. If the Lagrangian is invariant under time translation, then energy is conserved. If invariance exists in angular rotation, then angular momentum is conserved about that axis.

Newton's second law is a trivial example of the use of the Lagrangian. $\mathscr{L}=(1/2)mq'^2-V(q)$. So,

$$\frac{\partial\mathscr{L}}{\partial q'}=mq', \text{ and } \frac{\partial\mathscr{L}}{\partial q}=-\frac{\partial V}{\partial q}. \tag{D.9}$$

So, from Equation D.8,

$$mq''=-\frac{\partial V}{\partial q}, \tag{D.10}$$

our desired result.

We take now an information-centric example to demonstrate the efficacy of the use of the Lagrangian. This is an entropy maximization problem. So, we are interested in $\int\mathscr{L}dx = maximum$, with

$$\mathscr{L}=-\mathfrak{p}(x)\ln\mathfrak{p}(x)+\lambda_1\mathfrak{p}(x)+\lambda_2\mathfrak{p}(x)f(x), \tag{D.11}$$

where $\mathfrak{p}(x)$ is a probability density of function of a distribution whose entropy is the first term in this equation. $f(x)$ is a "kernel" function. The derivatives with respect to the generalized coordinates $\mathfrak{p}$ and $\mathfrak{p}'$ are

$$\frac{\partial\mathscr{L}}{\partial\mathfrak{p}}=-1-\ln\mathfrak{p}+\lambda_1+\lambda_2 f(x), \text{ and}$$
$$\frac{\partial\mathscr{L}}{\partial\mathfrak{p}'}=0. \tag{D.12}$$

The Euler-Lagrange equation then leads to

$$-1-\ln\mathfrak{p}(x)+\lambda_1+\lambda_2 f(x)=0$$
$$\therefore\ \mathfrak{p}(x)=A\exp[\lambda_2 f(x)], \tag{D.13}$$

where A is a normalization constant. *In maximum entropy problems, the probability density function (and also the probability mass function) will always have an exponential form.*

E
A thermodynamics primer

THE CLASSICAL THERMODYNAMIC VIEW has its foundations in attempts at understanding the thermomechanical energy conversion. In these conditions, a classical view suffices, with the quantum-mechanical consequences appearing hidden away in the mechanical energies, and motion appearing as correspondence emergence of the quantum-mechanical basis.

Energy changes in chemical reactions, for example, are changes arising from changes in bonding, which are quantum-mechanical in nature. We did not yet know at the time of the introduction of thermodynamic notions about the electromagnetic interactions at work, which have their own energy consequences. All this new learning can be introduced as modifications. Electrochemical potential instead of chemical potential is one example of this. And so is Shannon, algorithmic, entanglement and other forms of entropies.

E.1 Classical thermodynamic view

THE BASIC CLASSICAL THERMODYNAMIC IDENTITY, which ignores the electromagnetic energy form as well as quantum-mechanical interactions, so starting with classical notions, is

$$dU = T\,dS - P\,dV + \mu\,dN. \tag{E.1}$$

Changes in the *internal energy* U arise from independent changes in the entropy S, the volume V and the number of particles N. Infinitesimal changes lead to macroscopic extensive properties of the temperature T, pressure P and chemical potential μ as

$$T = \left.\frac{\partial U}{\partial S}\right|_{V,N}, \quad -P = \left.\frac{\partial U}{\partial V}\right|_{S,N}, \quad \text{and} \quad \mu = \left.\frac{\partial U}{\partial N}\right|_{S,V}. \tag{E.2}$$

If, in a particular thermodynamic analysis environment, the temperature T is fixed, then it is useful to change from the (S, V, N) independent variables to (T, V, N) or (T, P, N). For the former, Legendre transformation lets us write

$$\begin{aligned} dU &= d(TS) - S\,dT - P\,dV + \mu\,dN \\ \therefore\ d(U - TS) &= d\mathcal{F} = -S\,dT - P\,dV + \mu\,dN. \end{aligned} \tag{E.3}$$

$\mathcal{F} = U - TS$ is the *Helmholtz free energy*. If T, V and N are fixed, Helmholtz free energy is an invariant.

If, in a particular thermodynamic analysis environment, the pressure P is fixed, then Legendre transformation lets us write

$$\begin{aligned} d(U - TS) &= -d(PV) - S\,dT + V\,dP + \mu\,dN \\ \therefore\ d(U - TS + PV) &= d\mathcal{G} = -S\,dT + V\,dP + \mu\,dN. \end{aligned} \tag{E.4}$$

$\mathcal{G} = U - TS + PV$ is the *Gibbs free energy*. If T, P and N are fixed, the Gibbs free energy is an invariant.

S and V are extensive variables. They depend on the size of the system. T and P are intensive variables. They are independent of size. If the number of particles N is kept fixed, then Legendre transformation lets us write

$$\begin{aligned} d(U - TS) &= d(\mu N) - S\,dT - P\,dV - N\,d\mu \\ \therefore\ d(U - TS - \mu N) &= d\Omega = -S\,dT - P\,dV - N\,d\mu. \end{aligned} \tag{E.5}$$

$\Omega = U - TS - \mu N = \mathcal{F} - \mu N$ is the thermodynamic potential of the system. If T, V and μ are constants, thermodynamic potential is an invariant. Ω is a measure of the variety of possibilities by which, characterized by T, V and μ can be rearranged without any change in it. This lets us write the definition of entropy as

$$S = -\left.\frac{\partial \Omega}{\partial T}\right|_{V,\mu}, \tag{E.6}$$

together with

$$P = -\left.\frac{\partial \Omega}{\partial V}\right|_{T,\mu}, \quad \text{and} \quad N = -\left.\frac{\partial \Omega}{\partial \mu}\right|_{T,V}. \tag{E.7}$$

If one changes the scale of a system by λ, all extensive quantities (U, $\mathcal{F}$, $\mathcal{G}$ and Ω) will change proportionally. This requires the scaled free energy to follow

$$\begin{aligned} \lambda U &= U(\lambda S, \lambda V, \lambda N) \\ \therefore\ U &= \lim_{\lambda \to 1} d\lambda U = \lim_{\lambda \to 1} \left[\lambda S \left.\frac{\partial U}{\partial S}\right|_{V,N} + \lambda V \left.\frac{\partial U}{\partial V}\right|_{S,N} + \lambda N \left.\frac{\partial U}{\partial N}\right|_{S,V} \right] \\ &= TS - PV + \mu N, \end{aligned} \tag{E.8}$$

and the other thermodynamic functions as

$$\begin{aligned} \mathcal{F} &= -PV + \mu N, \\ \mathcal{G} &= \mu N, \\ \Omega &= -PV, \quad \text{and} \\ \mathcal{H} &= U + PV. \end{aligned} \tag{E.9}$$

In this list, we have added another energy function, the *enthalpy*, $\mathcal{H}$, which is often useful in exothermic conditions.

From Equation E.9, one important conclusion useful in analysis of semiconductors is

$$\mu = \frac{\mathcal{G}}{N}. \tag{E.10}$$

The chemical potential, as defined here, is the energy ("mechanical") per particle for these classical particle conditions. These relationships also define

$$
\begin{aligned}
c_V &= \left.\frac{\partial U}{\partial T}\right|_{V,N} = \left.\frac{\partial U}{\partial S}\right|_{V,N} \left.\frac{\partial S}{\partial T}\right|_{V} = T \left.\frac{\partial S}{\partial T}\right|_{V,N}, \text{ and} \\
c_P &= \left.\frac{\partial (U-PV)}{\partial T}\right|_{P,N} = \left.\frac{\partial (U-PV)}{\partial S}\right|_{P} \left.\frac{\partial S}{\partial T}\right|_{P,N} = T \left.\frac{\partial S}{\partial T}\right|_{P,N}.
\end{aligned} \tag{E.11}
$$

c_V is the specific heat at fixed volume and particles, and c_P is the specific heat at fixed pressure and particles.

The macroscopic thermodynamics can now be related to the statistical properties of microscopic particles. In Appendix F, there is a more definitive view of the probability of a state $|n\rangle$ of an ensemble being occupied by the particle. If the energy of the state is E_n, then this occupation probability is

$$
\mathfrak{p}_n = \mathfrak{p}(|n\rangle) = A \exp\left(-\frac{E_n}{k_B T}\right) = A \exp(-\beta E_n), \tag{E.12}
$$

where $\beta = 1/k_B T$ is another measure of inverse temperature. k_B is Boltzmann's constant (1.38×10^{-23} J/K). Normalization of probabilities over the entire system forces

$$
\begin{aligned}
\sum_n \mathfrak{p}_n = 1 \quad \therefore \quad \frac{1}{A} &= \sum_n \exp(-\beta E_n) = Z, \text{ and} \\
\mathfrak{p}_n &= \frac{\exp(-\beta E_n)}{Z},
\end{aligned} \tag{E.13}
$$

where Z is the partition function.

Any macroscopic quantity Q will have a macroscopic thermodynamic expectation of $\langle Q\rangle = \sum_n Q_n \mathfrak{p}_n$, with Q_n being the measure of the quantity when the system exists in the state $|n\rangle$. For example, the expectation for the free energy U is

$$
\langle U\rangle = \sum_n U_n \mathfrak{p}_n. \tag{E.14}
$$

Entropy's definition is through the expectation for $-\ln \mathfrak{p}(|n\rangle)$, which is the only function that satisfies the requirements on the additiveness of entropy function and of the normativeness of probabilities. So, entropy is

$$
\begin{aligned}
S &= -\langle \ln \mathfrak{p}(|n\rangle)\rangle = -\langle \mathfrak{p}_n \ln \mathfrak{p}_n\rangle = -\frac{1}{Z}\sum_n \exp(-\beta E_n) \ln \mathfrak{p}_n \\
&= \beta\langle U\rangle + \ln Z \\
\therefore \quad \mathcal{F} &= U - TS = -T \ln Z.
\end{aligned} \tag{E.15}
$$

When N varieties of particles exist—different chemical species, for example—then one may write a grand partition function

$$Z_G = \sum_N \sum_n \exp[-\beta(E_{N,n} - \mu N)], \tag{E.16}$$

with the thermodynamic potential being

$$\Omega(T, V, \mu) = -T \ln Z_G. \tag{E.17}$$

E.2 *Implications for bosons and fermions*

WITH QUANTUM-MECHANICAL CONSTRAINTS, this thermodynamic-statistical view needs revision. Quantum-mechanical particles or quasiparticles are subject to a variety of constraints. For bosons and fermions, this is $|\psi(1,2)\rangle = \pm|\psi(2,1)\rangle$, that is that the wavefunction changes sign under particle exchange if it is a fermion, such as an electron. This is equivalent to the Pauli exclusion that no two fermions can have identical quantum numbers.

Fermions and bosons are two special cases of the more general relationship $|\psi(1,2)\rangle = \exp(i\theta)|\psi(2,1)\rangle$ for anyons. Fermions and bosons are two special cases of the general quasiparticle description.

In any quantum state $|k\rangle$, there can be n_k particles, and this collection in this state characterized by k and n_k has an energy of $E_{k,n_k} = E_k n_k$. The partition function of the state $|k\rangle$ is

$$Z_k = \sum_{n_k} \exp[-\beta\,(E_k n_k - \mu n_k)] = \sum_{n_k} \{\exp[-\beta\,(E_k - \mu)]\}^{n_k},$$

with the grand partition function

$$Z_G = \prod_k Z_k. \tag{E.18}$$

n_k is not restricted for bosons. Any state $|k\rangle$ can be occupied by any number of bosons, since the particle wavefunction is allowed to remain the same under particle exchange. The sum therefore has an asymptotic upper limit of infinity. So,

$$Z_k = \frac{1}{1 - \exp[-\beta\,(E_k - \mu)]}, \quad \text{with } \exp[-\beta\,(E_k - \mu)] < 1 \text{ for } \textit{bosons}. \tag{E.19}$$

The expectation for the number of particles in state $|k\rangle$ is the *Bose-Einstein distribution function* of

$$\langle n_k \rangle = -\frac{\partial \Omega_k}{\partial \mu} = T\frac{\partial Z_k}{\partial \mu} = \frac{1}{\exp[\beta(E_k - \mu)] - 1}. \tag{E.20}$$

$n_k \in (0, 1)$ for fermions. So,

$$Z_k = 1 + \exp[-\beta\,(E_k - \mu)],$$
$$\Omega_k = -T \ln\{1 + \exp[-\beta\,(E_k - \mu)]\}, \text{ and}$$
$$\therefore \quad \langle n_k \rangle = -\frac{\partial \Omega_k}{\partial \mu} = T\frac{\partial Z_k}{\partial \mu} = \frac{1}{\exp[\beta\,(E_k - \mu)] + 1} \quad \text{for fermions.} \tag{E.21}$$

This is the *Fermi-Dirac distribution function*.

When applying the Fermi-Dirac distribution and this thermostatistic view to electrons or holes, one must account for the electromagnetic energy. This is brought in through the electrostatic potential ψ. The potential now transforms from the chemical potential to the electrochemical potential, which incorporates through the chemical and electrical potentials. Generally, E_F for this electrochemical energy is the preferred symbol for this engineering-centric view, although sometimes one sees μ also being deployed with a generalization of the definition. Electrons have charge, and this must be accounted for, together with their numbers, N.

F
Maxwell-Boltzmann distribution function

STATES IN THERMODYNAMIC EQUILIBRIUM CONDITIONS can be described more easily, since there exist extensive and intensive variables that do not change with time. Isolate a system that was in equilibrium from its environment—stop the particles and energy exchange—and the variables shall remain unchanged. If they do, then it was in a stationary condition, not a thermal equilibrium condition.

To a limited extent, small deviations with their linear response—the near-equilibrium conditions—too can be analyzed relatively easily because of insubstantial changes in the states and their occupation as described by the distribution function. Strong nonequilibrium has much more richness in its dynamics—various facets of local interaction—than the near-equilibrium linear response.

The time-independent equilibrium distribution $f(\mathbf{r},\mathbf{k})$ is independent of the state of the system in the past. This equilibrium distribution $f(\mathbf{r},\mathbf{k})$ is a measure of the complete description of $\mathbf{q},\mathbf{p} \equiv \mathbf{r},\mathbf{k}$ through $f(\mathbf{r},\mathbf{k};t)\,d\mathbf{r}\,d\mathbf{k}$, where $d\mathbf{r}\,d\mathbf{k} = \sum_i d\mathbf{r}_i\,d\mathbf{k}_i$. So, if N is the total number of particles,

$$N = \int_{\Omega} f(\mathbf{r},\mathbf{k})\,d\mathbf{r}\,d\mathbf{k}. \tag{F.1}$$

Expectations of any observable $\mathscr{A}$ then are given by

$$\langle \mathscr{A} \rangle = \int_{\Omega} f(\mathbf{r},\mathbf{k})\mathscr{A}(\mathbf{r},\mathbf{k})\,d\mathbf{r}\,d\mathbf{k}. \tag{F.2}$$

The energy in the system, whose operator is the Hamiltonian $\mathscr{H}(\mathbf{r},\mathbf{k})$, is

$$E = \langle \mathscr{H} \rangle = \int_{\Omega} f(\mathbf{r},\mathbf{k})\mathscr{H}(\mathbf{r},\mathbf{k})\,d\mathbf{r}\,d\mathbf{k} \tag{F.3}$$

For convenience, one may use a normalized phase-space volume consisting of a unit volume of real space, and the entire momentum space/wavevector space. Then, the observable for particle count is

$$N = n = \int_{\mathbf{k},m^3} f(\mathbf{r},\mathbf{k})\,d\mathbf{r}\,d\mathbf{k}. \tag{F.4}$$

n is now the per unit volume in *SI* units.

In thermal equilibrium, this distribution comes about independent of its past.

This is where the connection between thermodynamics and statistical mechanics becomes clearer. f is probabilistic in the sense that it exists averaged over all of these past independent conditions. Take two particles 1 and 2. The independence from the past says that the joint probability $N^2\mathfrak{p}(\mathbf{r}_1, \mathbf{k}_1; \mathbf{r}_2, \mathbf{k}_2) = N\mathfrak{p}(\mathbf{r}_1, \mathbf{k}_1)N\mathfrak{p}(\mathbf{r}_2, \mathbf{k}_2)$, which is $f(\mathbf{r}_1, \mathbf{k}_1)f(\mathbf{r}_2, \mathbf{k}_2)$. If there exist correlations, because of interactions between particles, then there are still relationships in probabilities that will translate into the mutual information contained within the correlations. The joint probability, as a product of independent probabilities, reflects the particles' independence; they are classical and not tethered to each other in any way. When energy and momentum exchange take place during a scattering event between the particles, that is, that two different coordinate configurations—pre- and post-scattering—result in the thermal distribution, it is a consequence of the conservation of probability. This states

$$\begin{aligned} f(\mathbf{r}_1, \mathbf{k}_1; \mathbf{r}_2, \mathbf{k}_2; \cdots) &= f(\mathbf{r}_1, \mathbf{k}_1)f(\mathbf{r}_2, \mathbf{k}_2)\ldots \\ &= f(\mathbf{r}'_1, \mathbf{k}'_1; \mathbf{r}'_2, \mathbf{k}'_2; \ldots) = f(\mathbf{r}'_1, \mathbf{k}'_1)f(\mathbf{r}'_2, \mathbf{k}'_2)\cdots. \end{aligned} \quad \text{(F.5)}$$

For joint probabilities to add, when the function f as the statistical distribution of individual arrangements multiplies, probabilities and f must have a logarithmic/exponential relationship. For energy, this distribution function $f(\mathscr{H}(\mathbf{r}, \mathbf{k})) = f(\epsilon)$ for the specific arrangement. Conservation of energy implies that $\epsilon_1 + \epsilon_2 = \epsilon'_1 + \epsilon'_2$ pre- and post-scattering. So, $f(\epsilon) = \exp(\beta\epsilon + b)$, which satisfies this constraint:

$$\begin{aligned} f(\epsilon_1)f(\epsilon_2) &= b^2 \exp\left[\beta(\epsilon_1 + b\epsilon_2)\right] \\ &= \exp(2b)\exp\left[\beta(\epsilon'_1 + b\epsilon'_2)\right] = f(\epsilon'_1)f(\epsilon'_2). \end{aligned} \quad \text{(F.6)}$$

This equation is a representation of the partitioning of the various configurations possible in thermal equilibrium. The constants must follow from the constraints of the thermodynamic equilibrium, which, for classical distributions, is that there is an expectation value of $k_BT/2$ of energy for each of the $N \times 2 \times \nu$ degrees of freedom for the N particles, with two canonical vector coordinates representable in ν dimensions. Equation F.2 provides these constraints through the number of particles in the ensemble and the total energy constraint. These are

$$\begin{aligned} N &= \exp(b) + \int \exp(\beta\epsilon)\, d\mathbf{r}\, d\mathbf{k}, \text{ and} \\ N\nu k_BT &= \frac{\int_\Omega \mathscr{H}(\mathbf{r}, \mathbf{k})\exp[\beta\mathscr{H}(\mathbf{r}, \mathbf{k})]\, d\mathbf{r}\, d\mathbf{k}}{\int_\Omega \exp[\beta\mathscr{H}(\mathbf{r}, \mathbf{k})]\, d\mathbf{r}\, d\mathbf{k}} \nu N = -\frac{N\nu}{\beta}. \end{aligned} \quad \text{(F.7)}$$

Recall that the equipartition of energy holds for a quadratic dependence on the canonical coordinates in the Hamiltonian. The second of these equations leads to $\beta = -1/k_BT$, and the first gives a constant prefactor. Both are related to the normalization.

In information mechanics (see S. Tiwari, "Quantum, statistical and information mechanics: A unified introduction," Electroscience 1, Oxford University Press, ISBN 978-0-198-75985-0 (forthcoming), and S. Tiwari, "Nanoscale device physics: Science and engineering fundamentals," Electroscience 4, Oxford University Press, ISBN 978-0-198-75987-4 (2017)), you see this same argument in play in the Shannon entropy—the average entropy—of a bitstream.

For a classical equipartition of energy, see S. Tiwari, "Nanoscale device physics: Science and engineering fundamentals," Electroscience 4, Oxford University Press, ISBN 978-0-198-75987-4 (2017), and the appendix devoted to it. In equilibrium, each harmonic degree of freedom has an average energy of $k_BT/2$. Equipartition of energy, together with this quadratic connection, is also discussed in S. Tiwari, "Quantum, statistical and information mechanics: A unified introduction," Electroscience 1, Oxford University Press, ISBN 978-0-198-75985-0 (forthcoming).

The Maxwell-Boltzmann distribution then follows as

$$f(\mathbf{r},\mathbf{k}) = N\frac{\exp[-\mathscr{H}(\mathbf{r},\mathbf{k})]}{\int_{\Omega}\exp[-\mathscr{H}(\mathbf{r},\mathbf{k})]\,d\mathbf{r}\,d\mathbf{k}}. \tag{F.8}$$

The prefactor is the equivalent of probability normalization over the various partitions of the system, and the temperature dependence arises in the occupation of states that must appear in both the numerator and the denominator, determining the probabilities as a function of temperature and therefore the expectation for energy. This is the end point of thermal equilibrium but does not speak to the dynamics of how it is approached. Its assumptions include, in particular, the complete independence of probabilities of states, so absolutely no correlations and hidden connections. And, through this viewpoint, and a different approach using probabilistic and classical statistical notions, we arrived at some of the same results as those in Appendix E.

G
Spin and spin matrices

SPIN, REPRESENTING SPIN ANGULAR MOMENTUM'S quantization, is as fundamental a property of a quantum particle as is its charge or mass. For electrons, the spin quantum number is $s = 1/2$, representing an angular momentum of $(1/2)\hbar$, and z is the chosen direction for spin angular momentum for which $[\mathbf{S}^2, \mathbf{S}] = 0$, $[S_x, S_y] = -(\hbar/i)S_z$, and $S^2 = S_x^2 + S_y^2 + S_z^2$ is a positive self-adjoint operator with $S^2 \geq 0$ and where $S^2 \geq S_z^2$ holds. It is useful to introduce here, as in many other quantum-mechanic operations where one moves up and down a ladder, $S_\pm = S_x \pm iS_y$ as two non-Hermitian operators that also satisfy $[S_\pm, S^2] = 0$.

Since $s = 1/2$, its z component has a secondary spin quantum number $m_s = 1/2, -1/2$. Two eigenvalues are possible, and we have corresponding eigenfunctions $|\zeta\pm\rangle$—spinors—that satisfy

$$\begin{aligned}
S^2|\zeta\pm\rangle &= (S_x^2 + S_y^2 + S_z^2)|\zeta\pm\rangle = \frac{1}{2}\left(\frac{1}{2}+1\right)\hbar^2|\zeta\pm\rangle,\\
S_z|\zeta\pm\rangle &= \pm\frac{1}{2}\hbar|\zeta\pm\rangle,\\
S_+|\zeta+\rangle &= 0,\\
S_-|\zeta-\rangle &= 0,\\
S_+|\zeta-\rangle &= \sqrt{\frac{1}{2}\left(\frac{1}{2}+1\right)+\frac{1}{2}\left(-\frac{1}{2}+1\right)}\hbar|\zeta+\rangle = \hbar|\zeta+\rangle, \text{ and}\\
S_-|\zeta+\rangle &= \sqrt{\frac{1}{2}\left(\frac{1}{2}+1\right)-\frac{1}{2}\left(-\frac{1}{2}-1\right)}\hbar|\zeta-\rangle = \hbar|\zeta-\rangle.
\end{aligned} \tag{G.1}$$

The matrix elements of S_z and $S_\pm$ are over the $|\zeta\pm\rangle$ two-dimensional subspace. So, they are 2×2 spin matrix operators,

$$S_z = \frac{\hbar}{2}\begin{bmatrix} 1 & 0 \\ 0 & -1 \end{bmatrix},$$
$$S_+ = \frac{\hbar}{2}\begin{bmatrix} 0 & 1 \\ 0 & 0 \end{bmatrix},$$
$$S_- = \frac{\hbar}{2}\begin{bmatrix} 0 & 0 \\ 1 & 0 \end{bmatrix}, \text{ where}$$
$$|\zeta_+\rangle = \begin{bmatrix} 1 \\ 0 \end{bmatrix}, \text{ and } |\zeta_-\rangle = \begin{bmatrix} 0 \\ 1 \end{bmatrix}. \tag{G.2}$$

Since

$$S_x = \frac{1}{2}(S_+ + S_-) = \frac{\hbar}{2}\begin{bmatrix} 0 & 1 \\ 1 & 0 \end{bmatrix}, \text{ and}$$
$$S_y = \frac{i}{2}(S_+ + S_-) = \frac{\hbar}{2}\begin{bmatrix} 0 & -i \\ i & 0 \end{bmatrix}, \text{ it follows that}$$
$$S^2 = S_x^2 + S_y^2 + S_z^2 = \frac{3}{4}\hbar^2\begin{bmatrix} 1 & 0 \\ 0 & 1 \end{bmatrix}, \tag{G.3}$$

which is diagonal, as expected. We have created here the matrix representation of the spin 1/2 operators and their eigenfunctions.

Pauli matrices, written as

$$\sigma_x(= X) = \begin{bmatrix} 0 & 1 \\ 1 & 0 \end{bmatrix},$$
$$\sigma_y(= Y) = \begin{bmatrix} 0 & -i \\ i & 0 \end{bmatrix}, \text{ and}$$
$$\sigma_z(= Z) = \begin{bmatrix} 1 & 0 \\ 0 & -1 \end{bmatrix}, \tag{G.4}$$

allows one to express the spin operator as $S = (\hbar/2)\sigma$. Pauli spin matrices are also often referred to by X, Y and Z, and often in their collection we include the identity matrix

$$\mathbb{I} = \begin{bmatrix} 1 & 0 \\ 0 & 1 \end{bmatrix} \tag{G.5}$$

Pauli matrices anticommute, that is,

$$\sigma_x\sigma_y + \sigma_y\sigma_x = \sigma_x\sigma_z + \sigma_z\sigma_x = \sigma_y\sigma_z + \sigma_z\sigma_y = 0; \tag{G.6}$$

the commutation relations are

$$[\sigma_x, \sigma_y] = 2i\sigma_z, [\sigma_y, \ \sigma_z] = 2i\sigma_x, \text{ and } [\sigma_z, \sigma_x] = 2i\sigma_y, \tag{G.7}$$

so that

$$\sigma_x\sigma_y = i\sigma_z, \quad \sigma_y\sigma_z = i\sigma_x, \quad \text{and} \quad \sigma_z\sigma_x = i\sigma_y. \tag{G.8}$$

One can interpret these to cause a spin effect that is 4-fold degenerate for a *free electron*. The eigenfunction solutions for a one-dimensional Hamiltonian problem are

$$\begin{aligned}
|\psi_{\mathbf{k}\uparrow}(\mathbf{z})\rangle &= \frac{1}{(2\pi)^{3/2}} \exp(i\mathbf{k}\cdot\mathbf{z})|\zeta_+\rangle, \\
|\psi_{\mathbf{k}\downarrow}(\mathbf{z})\rangle &= \frac{1}{(2\pi)^{3/2}} \exp(i\mathbf{k}\cdot\mathbf{z})|\zeta_-\rangle, \\
|\psi_{-\mathbf{k}\uparrow}(\mathbf{z})\rangle &= \frac{1}{(2\pi)^{3/2}} \exp(-i\mathbf{k}\cdot\mathbf{z})|\zeta_+\rangle, \quad \text{and} \\
|\psi_{-\mathbf{k}\downarrow}(\mathbf{z})\rangle &= \frac{1}{(2\pi)^{3/2}} \exp(-i\mathbf{k}\cdot\mathbf{z})|\zeta_-\rangle.
\end{aligned} \tag{G.9}$$

H
Density of states

TO UNDERSTAND THE DENSITY OF STATES, that is, the normalized number of states in $E(\mathbf{k})$ available for occupation, in units of per unit energy and unit space, we need to look at the constraints on the electron—as a particle—in a box. The box may be large, consisting of a large number of unit cells as at microscale, or it may be confined in one or more dimensions of nanoscale. We will still assume that the effective mass description is valid. This implies that the particle feels an environment across many unit cells and therefore the Bloch description is valid over the box.

Consider such a particle in a box. The particle is confined in it. Its wavefunction is a standing wave with nodes at the boundaries of the box, with an infinite potential barrier:

$$\left(-\frac{\hbar^2}{2m^*}\nabla^2 + V\right)\psi = E\psi, \tag{H.1}$$

where ψ is the wavefunction, V is the potential and E is the energy; it describes the quantum mechanics of the system, that is, its solution provides us with the particle wavefunction useful for determining the observables of the system, including the energy that the particle states can have. Since this is a second-order differential equation, one can write a solution in the form

$$\psi(\mathbf{r}, t) = \mathcal{A}\exp\left[+i(\mathbf{k}\cdot\mathbf{r} - \omega t)\right], \tag{H.2}$$

with the $\mathbf{k}$s quantized appropriately to the boundary conditions. The solutions are standing waves—confined states—formed from opposite and equal $\mathbf{k}$ wavevector solutions.

For each confined direction, $k = \pm n\pi/L$, where $n = 1, 2, 3, \ldots$. Figure H.1 shows this confinement in the z direction. The standing wave is formed by two counter-propagating waves of quantized opposite wavevectors. These wavevectors are uniformly distributed. This relationship implies that the larger the confined space is, the

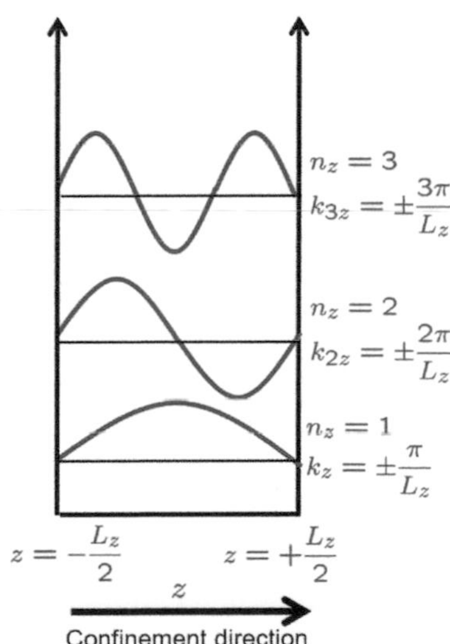

Figure H.1: When confined, here in the z direction, with infinite energy barriers, the states available for occupation are quantized in momentum and are separated in energy. This figure shows the wavefunction formed as a standing wave confined between the barriers and composed of two counter-propagating waves of wavevector k_z in the confinement direction z.

smaller the wavevector spacing and the energy spacing of the state, since energy (kinetic) is $E = \hbar^2k^2/2m^*$. At large dimensions, such as for a wafer, or a classical dimension, such as source-to-drain spacing, this L dimension is large—the states are close together, much closer than thermal energy (k_BT), except at near absolute zero temperatures—and one may view the distribution of the states as being quasi-continuous. The particle is confined to this large crystal box, and the electrons easily propagate in directions where the states are relatively unconfined. In the example shown in Figure H.1, this freedom exists in the x and y directions. In the confined direction, that is, in a dimension where the L is now quite small—of the de Broglie wavelength scale of the electron—the movement is restricted in the direction of this confinement. In Figure H.1, the electron's probability distribution's peak is displaced only marginally when moving from the ground to the first state—within the small, confined well dimension, with k_x and k_y remaining constant. The electron remains confined in the z direction, but it just spreads out a little more within the well. This simple picture for the allowed states for a single effective mass m^* holds true whether L is large or small. If it is large, the states are close together, and an electron may travel around in the real and reciprocal space under the influence of energetic interactions. If it is small, it is restricted from moving in that direction. It is a standing wave.

Density of states is a measure of the number of states per unit wavevector or energy, in unit spatial coordinates. It is three-dimensional ($\mathscr{G}_{3D}(E)$ or $\mathscr{G}_{3D}(k)$) if unconfined, two-dimensional ($\mathscr{G}_{2D}(E)$ or $\mathscr{G}_{2D}(k)$) if confined in one dimension and one-dimensional ($\mathscr{G}_{1D}(E)$ or $\mathscr{G}_{1D}(k)$) if confined in two dimensions. It is the number of freedom of movement directions that are employed for the subscript. If all directions are confined, it cannot move, the expectation of the spatial coordinate is a constant and it is confined with no states nearby for motion at dimensions larger than the wavelength of the particle.

When one of the dimensions is confined, z in our example, the wavevector of this confined mode is $k_z = \pm n_z\pi/L_z$. These states are π/L_z apart. For confinement at a small L, these may be quite separated, but, with a larger L, they come closer together. The state density in k-space is

$$\frac{dn_z}{dk_z} = g_s\frac{1}{2}\frac{L_z}{\pi}, \qquad \text{(H.3)}$$

which represents a continuum approximation to the discrete $\Delta n_z/\Delta k_z$. Here, $g_s = 2$ is the spin degeneracy ($s = 1/2$; $m_s = \pm 1/2$), which accounts for electrons of opposite secondary spins—a different quantum number—occupying the specific n_z quantum number states.

The factor 2 in the denominator indicates that both a positively and a negatively directed k state are required to make the standing wave shown in Figure H.1. By normalizing to unit dimensions, we obtain

$$\frac{dn_z/L}{dk_z} = g_s \frac{1}{2\pi}, \tag{H.4}$$

which states that the per unit length number of states per unit reciprocal wavevector varies as $g_s/2\pi$. No effective mass enters here. This expression just speaks to the fitting of waves in a space. This simple relationship states that, for each confinement direction, the density of states has a $1/2\pi$ factor arising from a standing wave in **k**-space. In determining the density of states for different dimensionalities of confinement, the spin degeneracy is common, and this $1/2\pi$ factor arises from each confining dimension. The density of states in ν-dimensional wavevector space is simply written as

$$\mathscr{G}_\nu = \frac{1}{(2\pi)^\nu}. \tag{H.5}$$

The unconfined states, or states where the separation in energy is small compared to $k_B T$, provide for electron movement through a viscous or free flow of electrons in response to an energy input to the system. We can also express the density of state relationship in energy. The density, in general, normalized to unit spatial extent and energy, is

$$\mathscr{G}(E) = \frac{dn}{dE} = \frac{dn}{dk}\frac{dk}{dE}. \tag{H.6}$$

The density of states in **k**-space and E-space can be related, since dE/dk is known through the Schrödinger equation's dispersion solution of $E = \hbar^2 k^2/2m^*$ in the isotropic constant mass approximation used here.

We now extend this simple description to different degrees of freedom, as shown in Figure H.2. In the wavevector coordinate space shown in Figure H.2, the $3D$ unconfined description corresponds to finding the number of states within a shell of sphere, at energy E, in the band dE for the density of states. In the $2D$ description, with one confinement direction, it is the number of states within a slice of this spherical shell where k_x, k_y or k_z is discretized. A section of the shell—a planar ring—is shown in the plane intersection, assuming that the x direction is the confinement direction. In the $1D$ description, with two confined directions, it is the number of states along the intersection line of these quantization planes. This is the line along which the one-dimensional freedom of movement exists. In Figure H.2, the state $k_x^0, -k_y, k_z^0$ is shown in the band dE at energy E, where k_x^0 and k_z^0 are quantized by the strong dimensional confinement in those orientations.

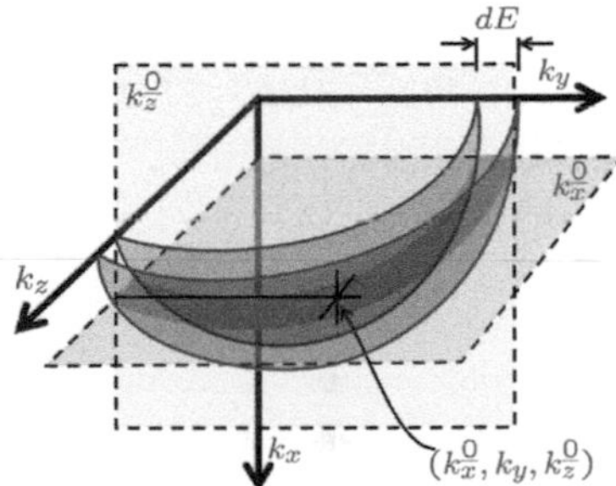

Figure H.2: The constant energy surfaces in the reciprocal space. The figure shows only a quadrant of the three-dimensional space. For three dimensions, the region of allowed **k** is in the spherical shell for a spread dE in energy at E. When confined in the x direction to $k_x = k_x^0$, the states are in a circular, two-dimensional, areal slice. When additional confinement is introduced to z, so $k_z = k_z^0$, the states allowed are along the extended line shown, with one particular state, $\left(k_x^0, k_y, k_z^0\right)$, identified within the dE span at energy E. At this energy, there is an additional reverse momentum state at $k_x^0, -k_y, k_z^0$, which is the reflection point of the (k_x, k_z) plane running through the origin.

The densities of states that relate to the volume of the shell in a 3*D* distribution, to the area of a cross-section in a 2*D* distribution and along a line in a 1*D* distribution, within this single mass description, are as follows:

No confinement, 3 dimensions of freedom:

$$\mathscr{G}_{3D}(E) = 2\frac{1}{(2\pi)^3}\frac{4\pi k^2 dk}{dE} = \frac{1}{\pi^2\hbar^3}\sqrt{2m^{*3}E}; \tag{H.7}$$

1-dimension confined, 2 dimensions of freedom:

$$\mathscr{G}_{2D}(E) = 2\frac{1}{(2\pi)^2}\frac{2\pi k_{\parallel} dk_{\parallel}}{dE} = \frac{m^*}{\pi\hbar^2}; \text{ and} \tag{H.8}$$

2-dimensions confined, 1 dimension of freedom:

$$\mathscr{G}_{1D}(E) = 2\frac{1}{2\pi}\frac{2dk_y}{dE} = \frac{1}{\pi\hbar}\sqrt{\frac{2m^*}{E}}. \tag{H.9}$$

These relations establish the densities of states for electrons that have some freedom of movement while being circumscribed due to confinement in some of the degrees of freedom. When these states are occupied by electrons, the permitted freedom of movement leads to current when an external force is applied on the system. It is because of this density of states that electrons are available with freedom of movement when the energy separation—a kinetic energy separation—of the states is small. The channels of conduction arise in the free directions of movement. The density of states determines the channels available to give rise to current.

Figure H.3 shows this density of states in terms of energy. In the 3*D* distribution, assuming isotropic constant mass, a constant spatial distance exists between all the **k** states. In $\mathscr{G}_{3D}$, since the volume of the spherical shell of width dE at energy E has this equispaced density, the number of states increases as the surface area per radius in the **k**-space, that is, with k, or the square root of energy. In $\mathscr{G}_{2D}$, which is a planar section of the shell, perpendicular to the direction of the confinement, the number of states increases at the same rate as the allowed ks, so the density of states in energy is a constant. Along a line, in the $\mathscr{G}_{1D}(E)$ distribution, the number of states per unit length is a constant, the k spacing varies as $\sqrt{E}$ and, therefore, in energy distribution, at higher energies, the density of states varies inversely as $\sqrt{E}$.

In the two-dimensional system, the freedom of movement is in two directions. One direction is confined and forms a ladder consisting of subbands of constant $\mathscr{G}_{2D}$ density. The argument associated with Figure H.1, originating in Schrödinger equation, applies to both the confined (L small; of the order of the de Broglie wavelength) and the unconfined (L large) limits. For the small-length case, what is

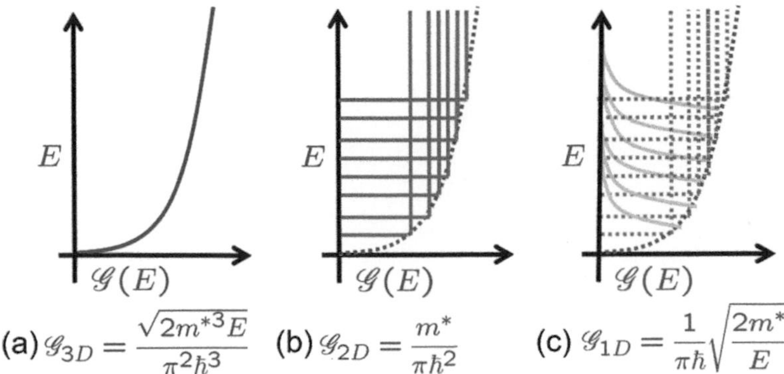

Figure H.3: The $3D$, $2D$ and $1D$ density of states, that is, the normalized to the unit spatial dimension density of the states with $\pm 1/2$ secondary spin states available for conduction. Part (a) is for $3D$, (b) for $2D$ and (c) for $1D$. In the $\mathscr{G}_{2D}(E)$ distribution, the confinement arises from one dimension of confinement, resulting in a ladder from the related discretized ks shown here as step function in (b). If the confinement dimension expands, the steps merge closer, and their distribution approaches that seen in a $3D$ distribution of states. The $1D$ density of states, $\mathscr{G}_{1D}(E)$, arises from two dimensions of confinement and approaches $2D$ when one of the confinement dimensions is relaxed, and $3D$ when both are relaxed.

required is that the effective mass approximation hold, and that means that the box is at least a few unit cells long for the electron to be aware of the crystalline environment it is in, as reflected in the effective mass. The integrated number of states in these subbands over the energy range of interest provides the number of states available for conduction in the two dimensions. The ladder of multiple subbands has as its minimum the quantized wavevectors of the confined direction. If the confinement dimension is relaxed, the subbands come closer, and, in the limit, the distribution approaches a three-dimensional density of states. Confine one more dimension, and it is now a one-dimensional system. The freedom of movement now is in the one remaining unconfined direction. The density of states available for the transport in this unconfined direction varies as $\mathscr{G}_{1D}$, and integrating it, for all the subbands over the energy range of relevance, provides the total number of states available for one-dimensional conduction.

I
Oscillator strength

COUPLING OF STATES, and specifically transitions between them, are characterized through the matrix element that relates the state from which the transition takes place, the state to which the transition is taking place, and the perturbation under which this transition occurs. In any optical measurement, one measures the energy sweep as well as the strength of the optical coupling of the transition, which is then useful for any transition- or coupling-based calculation. Oscillator strength, which is normalized, is such a measure. As with much of quantum mechanics, these were first explored in atoms, where spectral lines show the transitions, whether in an electromagnetic absorption process or in an emission process. The spectral width of a line is not just a function of the frequency or energy of the line but also a function of a response wherein damping will exist, since an oscillating dipole radiates, and there may very well be other loss mechanisms. Damping, in general across systems, will arise in multitude of forms. An electron has a leakage rate out of a quantum well. Atomic oscillation in the crystal—phonon—has anharmonicity in a multitude of causes. The spectral linewidths depend on the losses, for absorption, or on gain, for the emission process. The idealization of absorption as in a harmonic oscillator model with this spectral measure versus the reality of actual absorption, again via the observed spectral linewidth, in thermodynamic equilibrium with occupation statistics accounted for, is an oscillator strength for absorption. Likewise, one may define an oscillator strength for emission. These measures are certainly important in understanding the luminescence properties of atoms and molecules, where degeneracies of states also appear as an additional wrinkle. So, this is certainly relevant for organic semiconductors. For inorganic semiconductors, the degeneracies of levels are the same, but changes in oscillator strengths will arise as the dimensionality of systems changes, changing in turn the coupling of states.

We discuss the bulk semiconductor case, as well as the deeper meaning and implications of oscillator strength and oscillators.

We must relate to how the perturbation evolves the state from which the transition is taking place and its matching to the state that one is calculating the transition to. Symmetries will matter; how strongly the perturbation affects and changes the symmetry and form of that state will also matter. So, state coupling within a band where the envelope function changes ever so slightly with the wavevector will be strong. State coupling between differently indexed bands is much weaker because of orthogonality. $\mathbf{k}\cdot\mathbf{p}$ theory (see Chapter 4) illustrates many of these characteristics. When states couple due to light-initiated perturbation, it is the vector potential of the light, which maps to the field strength of the light, that enters via the form $\mathbf{A}\cdot\mathbf{p}$ (see Chapter 12). These two examples show the importance of the matrix element.

For $\mathbf{k}\cdot\mathbf{p}$, we found the matrix element consequence for the envelope function as $\langle u_{n0}(\mathbf{r})|\mathbf{k}\cdot\mathbf{p}|u_{l0}(\mathbf{r})\rangle$, as derived in Equation 4.33, which, if written using Bloch functions for the proper description of the state, is $\langle\mathbf{k}'|\mathbf{p}|\mathbf{k}\rangle$. For light, $\mathbf{A}\cdot\mathbf{p}$, if you will, the matrix element was derived as

$$\mathscr{H}'_{\mathbf{k}'\mathbf{k}} = -\frac{q}{m_0}\mathbf{A}\cdot\mathbf{p}, \tag{I.1}$$

as can be seen through Equation 12.6. Both are momentum based. Why? One way to look at this is by probing matching. Energy and momentum matching both must exist during interaction. Energy, we take care of naturally, often through the Dirac delta function in the calculation. Momentum matching, however, has an additional characteristic that is first order, and therefore of phase or position during motion. An oscillator has the highest kinetic energy and momentum at its expectation position, and this vanishes as one moves further away. If two oscillators couple, the coupling will be most efficient if it maintains itself well coupled throughout the excursion, not just at the mean or, at the farthest, through an impulse. So, the momentum operator and its matrix element $\mathbf{p}_{\mathbf{k}'\mathbf{k}}$ are the proper descriptors of this strength of oscillator coupling.

The oscillator strength is

$$f_{\mathbf{k}'\mathbf{k}} = \frac{2|\mathbf{p}_{\mathbf{k}'\mathbf{k}}|^2}{m_0 E_{\mathbf{k}'\mathbf{k}}}, \tag{I.2}$$

which led to the relationship for effective mass in Equation 4.35, which can be rewritten as

$$E_{n\mathbf{k}} = E_{n0} + \sum_{ij}\frac{\hbar^2}{m^*_{ij}}k_i k_j \tag{I.3}$$

Another important question here is why all this emphasis on the oscillator, and particularly the quantum oscillator notion. An isolated harmonic oscillator can be viewed as the simplest form of a stable entity that conserves a state and its energy. It is a stationary state of constant probability, but one where kinetic and potential forms of energy are still being exchanged. Any canonical coordinate change from its expectation value brings about restorative force, and it is conservative. Classically, the second power of displacement via $\mathbf{F} = -\nabla_{\mathbf{r}}U$ is restorative, with the force pointed in the proper direction. A first power, or any odd power, will not do. So, harmonic oscillators are very fundamental units of stability. Photons, phonons, et cetera, therefore draw their description from the harmonic oscillator basis. See S. Tiwari, "Quantum, statistical and information mechanics: A unified introduction," Electroscience 1, Oxford University Press, ISBN 978-0-198-75985-0 (forthcoming), for a broader discussion of this and related segues.

Think of the following classical analogy here. On the playground swing, one swings best when we use configuration change—bending at the knee to input the force and momentum change on the way down of the swinging motion—and breaking symmetry on the way up. And we better change the timing of this motion over this half of the cycle to get the swing to go higher and higher until disaster strikes at least once in the childhood, due to nonlinearities that one was not aware of. The chain has no compressive strength.

for multiple bands, and where

$$\frac{1}{m^*_{ij}} = \frac{1}{m_0}\delta_{ij} + \frac{2}{m_0}\sum_{m\neq n}\frac{\langle n0|p_i|m0\rangle\langle n0|p_j|m0\rangle}{E_{n0}-E_{m0}}. \tag{I.4}$$

One can now see the tensorial properties of the effective mass due to coupling of bands. Holes have significant such coupling and repulsion of bands, and anisotropy follows. This equation also shows that a narrow bandgap—the difference between conduction and valence band energies—will lead to a lower effective mass.

Equation I.4 also leads to

$$f_{osc} = \sum_{m\neq n} f_{mn} = 1 - \frac{m_0}{m^*_e} \tag{I.5}$$

as the sum rule for spherical conduction bands. This is the oscillator strength for transitions within the conduction band.

Since direct transitions of interest are often in the center of the Brillouin zone, we can estimate. The largest effect in Equation I.4 arises from bands that are close by. Direct transitions arise from light-hole-band, heavy-hole-band and split-off band states. For valence bands, the $\mathbf{k}\cdot\mathbf{p}$ approach again has three terms of interactions; however, only the one from within the same band will dominate. So, it is fairly accurate to employ only the major terms. This means that the oscillator strength for a direct valence-to-conduction band transition is

$$f_{osc} \approx 1 + \frac{m_0}{m^*_h}. \tag{I.6}$$

The positive sign is the result of the electron mass being the negative of the hole mass. Using Equation I.2, this states that

$$\frac{|\mathbf{p}_{cv}|^2}{m_0} = \frac{E_{g0}}{4}\left(1+\frac{m_0}{m^*_h}\right), \tag{I.7}$$

where E_{g0} is the zone center energy gap.

The harmonic oscillator connection, and the transition due to a perturbation, which may arise in many causes—not just electromagnetic through light, but also Coulomb, as in the band formation in the crystal—causes the oscillator strength to appear in many places in solid-state phenomena.

An example is electronic polarizability. The dipole moment in the presence of a local electric field of $\mathcal{E}_{local}$ follows from the equation of motion as

$$p_0 = -ez_0 = \frac{e^2\mathcal{E}_{local}}{m_0(\omega_0^2-\omega^2)}, \tag{I.8}$$

written here as an undamped response, and where we can again see the lineshape of the idealized harmonic oscillator. In the quantum approach, the electronic polarizability appears as

$$\alpha_{electronic} = \frac{e^2}{m_0} \sum_j \frac{f_{ij}}{\omega_{ij}^2 - \omega^2}, \tag{I.9}$$

where ω_{ij} are the various transition energies of the various oscillators composing the system.

When damping is present, in any quantum system consisting of excited states indexed by i with $\omega_{i0} = (E_i - E_0)/\hbar$, the total polarization will be

$$\mathbf{P} = \boldsymbol{\mathcal{E}} \frac{e^2 N}{m_0 \Omega} \sum_i \frac{f_i}{\omega_{i0}^2 - \omega^2 - i\gamma_i \omega}, \tag{I.10}$$

and therefore the relative dielectric constant arising in this electronic polarizability follows as

$$\epsilon_r = 1 + \frac{e^2 N}{m_0 \epsilon_0 \Omega} \sum_i \frac{f_i}{\omega_{i0}^2 - \omega^2 - i\gamma_i \omega}, \tag{I.11}$$

with oscillator strength as

$$f_i = \frac{2 m_0 \omega_{i0}}{e^2 \hbar} |\langle \psi_i | e_z | \psi_0 \rangle|^2, \tag{I.12}$$

with e_z defining the direction of the field. In all these expressions, the strength of the resonance is tied to the oscillator strength, and the resonance occurs at the allowed transitions with a suitable linewidth.

J
Effective mass tensor

THE NET CARRIER MOTION is due to carriers that are mostly close to the Fermi energy, where filled and empty states coexist, and therefore the net effect can be characterized via a small energy band along the constant energy surfaces. For semiconductors, with carriers mostly near band minima, as in conduction bands, or band maxima, as in valence bands, this reciprocal space locality and constancy of energy makes the use of effective mass convenient. Semiconductors, however, certainly have anisotropy in bands at the extrema, and there may also exist multiple bands. So, care is required in the use of effective mass approaches in mathematical treatments of semiconductor problems. In direct bandgap semiconductors, *GaAs*, for example, a low isotropic effective mass precisely at the conduction bandedge appears with a nonlinear dependence in **k**, which sometimes is of importance if the carriers of interest occupy states a few 100 *meV* up in the band. In *Si*, for this same situation, there is the complication of a difference between longitudinal and transverse directions (see Figure J.1). In the valence band, this problem is acute. One has a light-hole band, which may be relatively isotropic, but also a heavy-hole band, which is fairly anisotropic, and often a split-off band quite nearby. Such nonlinearities and orientation dependences, for small changes from thermal equilibrium, can be treated through incorporation of the nonlinearity and a mass tensor for the directional dependence.

When two axes of an ellipsoid are equivalent, the ellipsoid is called a spheroid.

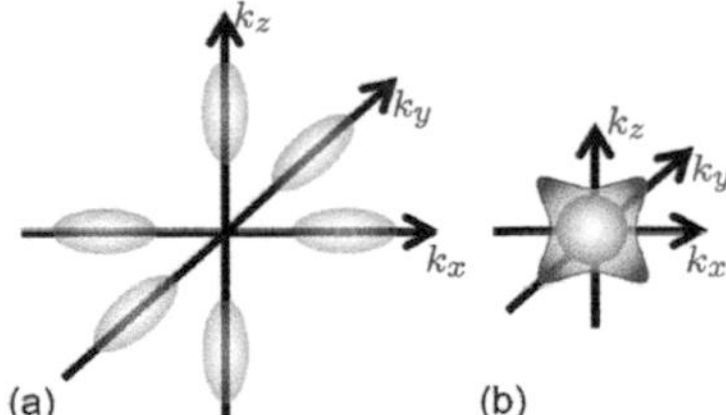

Figure J.1: Two examples of near-bandedge effective mass complexities. Part (a) shows the six equivalent constant energy surface ellipsoids for conduction band minima in *Si* that are a little in from the X points. Part (b) shows the same for valence bands where low and high effective mass maxima exist, as in *GaAs*, and quite often a split-off band lurks nearby.

The effective mass, in its most general form, must describe the local wavepacket group velocity response as derived from the energy-crystal momentum relationship as a tensor. A general force may cause a momentum response in other directions, as it does with magnetic field, and constraints may be present simultaneously in different directions, as they are for quantum-confined and nanoscale structures. So, a general form of this effective mass tensor is

$$\frac{1}{\mathbb{M}} = \frac{1}{\hbar^2}\begin{bmatrix} \frac{\partial^2 E}{\partial k_x^2} & \frac{\partial^2 E}{\partial k_x\,\partial k_y} & \frac{\partial^2 E}{\partial k_x\,\partial k_z} \\ \frac{\partial^2 E}{\partial k_y\,\partial k_x} & \frac{\partial^2 E}{\partial k_y^2} & \frac{\partial^2 E}{\partial k_y\,\partial k_z} \\ \frac{\partial^2 E}{\partial k_z\,\partial k_x} & \frac{\partial^2 E}{\partial k_z\,\partial k_y} & \frac{\partial^2 E}{\partial k_z^2} \end{bmatrix} = \begin{bmatrix} \frac{1}{m^*_{xx}} & \frac{1}{m^*_{xy}} & \frac{1}{m^*_{xz}} \\ \frac{1}{m^*_{yx}} & \frac{1}{m^*_{yy}} & \frac{1}{m^*_{yz}} \\ \frac{1}{m^*_{zx}} & \frac{1}{m^*_{zy}} & \frac{1}{m^*_{zz}} \end{bmatrix} \qquad (\text{J.1})$$

The mass tensor has elements $1/m^*_{ij} = (1/\hbar^2)\partial^2 E/\partial k_i\,\partial k_j$. In the vicinity of the bandedge, these forms represent the inclusion of the 2nd order term in the Taylor series expansion of the $E(\mathbf{k})$ description, that is,

$$E(\mathbf{k}) = E_0 + \sum_i \frac{\partial E}{\partial k_i} k_i + \frac{1}{2}\sum_{ij} \frac{\partial^2 E}{\partial k_i\,\partial k_j} k_i k_j, \qquad (\text{J.2})$$

where higher-order terms will show the complexities arising in anisotropy. The second term of Equation J.2 reflects the effective mass.

Effective mass, initially introduced as a parameter of convenience to describe the curvature of $E(\mathbf{k})$, acquires much greater significance—of inertial mass—under conditions where external forces are much slower—adiabatic—compared to those arising in the periodic crystal potential, when carrier energy is small compared to the bandgap E_g and where the electron feels this extended environment. This is where much of the interest in devices tends to be, so electrons and holes as free particles moving in external fields can easily be analyzed via the effective mass, not the free electron mass, and also without worrying about the crystal potential. When valid, this greatly simplifies physical understanding in semiconductors.

Consider its implications, first for conduction bands. In direct bandgap semiconductors, the Γ valley at low energies is isotropic and symmetric. So, all the elements in the mass tensor of Equation J.1 are $1/m^*$, that is, around the Γ minimum,

$$\frac{1}{\mathbb{M}} = \begin{bmatrix} \frac{1}{m^*} & \frac{1}{m^*} & \frac{1}{m^*} \\ \frac{1}{m^*} & \frac{1}{m^*} & \frac{1}{m^*} \\ \frac{1}{m^*} & \frac{1}{m^*} & \frac{1}{m^*} \end{bmatrix}. \qquad (\text{J.3})$$

Now, take an indirect semiconductor, *Si*, whose low energy constant energy surfaces are spheroids, as shown in Figure J.1(a) just in from the X point. There are six ellipsoids, with the anisotropy represented in two mass terms: m^*_l, the longitudinal effective mass—a larger mass—along the longitudinal axis, and m^*_t, the transverse effective mass—a smaller mass—for the transverse off-diagonal response. So, the mass tensor for *Si*, around and inside the X point, but called the X minimum, is

For *Ge*, the minima being precisely at the L points means that only a half spheroid along the diagonal belongs in the first Brillouin zone. This has implications for density of states, but not for the effective mass tensor.

$$\frac{1}{\mathbb{M}} = \begin{bmatrix} \frac{1}{m^*_l} & \frac{1}{m^*_t} & \frac{1}{m^*_t} \\ \frac{1}{m^*_t} & \frac{1}{m^*_l} & \frac{1}{m^*_t} \\ \frac{1}{m^*_t} & \frac{1}{m^*_t} & \frac{1}{m^*_l} \end{bmatrix}. \qquad (\text{J.4})$$

All states, such as any state on the surface of these spheroids—all of them—respond with an effective mass m^*_l to the longitudinal component of the field, and m^*_t to the transverse components of the field. And any field can be represented in the Cartesian form. A simple interpretation is that all the three orientations are equivalent, so, consider a field aligned in the $\hat{\mathbf{k}}_\mathbf{x}$ direction. Electrons in states in two of these valleys respond with a longitudinal mass, and those in four of these valleys—those in the plane orthogonal to the field—respond with a transverse mass. The two effective masses, occupation in six valleys, suffice to describe the acceleration $((1/\hbar)d\mathbf{k}/dt)$ to the external stimuli. For each state, this acceleration is $(1/\mathbb{M})\mathbf{F}$.

Now, consider the situation for holes, and let it be complicated by the existence of two bands: a light-hole band and a heavy-hole band. Equation J.1 is still a complete description for the behavior of each state. If, for a given energy, there are two holes, one in a light-hole band and one in a heavy-hole band, they both respond through the mass tensor that represents the response behavior of the state as dictated by the $E(\mathbf{k})$ description.

Effective mass represents the bandstructure-defined response of the electron in the periodic matter. Once this can be taken as a given, one can extract derivative quantities that encapsulate the matter's behavior.

The conductivity effective mass is therefore a derived mass parameter whose utility is strictly only for the convenience of a Drude-like conduction calculation as in $\mathbf{J} = q^2 n \langle \tau_{\mathbf{k}} \rangle \boldsymbol{\mathcal{E}} / m_c^*$ with an electric field applied. Since the Drude calculation is a very symmetrized, integrated and scattering-randomized representation, for *Si*'s structure, this has the consequence that one can force-fit a response that has two valleys responding with a longitudinal mass, and four valleys responding with a transverse mass, in providing a mass associated with conductivity. Since all possibilities must be covered for the randomization to be valid, this Drude calculation expects all the spheroids to be occupied in calculating the response. Because of the symmetry, the response from the mass tensor is a scalar conductivity mass. Normalizing to one electron with a 1/6 possibility of occupation in states of each of the spheroids,

$$\frac{1}{m_c^*} = \frac{1}{6}\left(\frac{2}{m_{xx}^*} + \frac{2}{m_{xy}^*} + \frac{2}{m_{xz}^*}\right) = \frac{1}{3}\left(\frac{1}{m_l^*} + \frac{2}{m_t^*}\right). \qquad \text{(J.5)}$$

Each Cartesian component sees this same final expression for its Drude conductivity mass response. But this conduction mass calculation will not work for other forcing circumstances or where Drude limitations are unacceptable, which is pretty much all of the problems of nanoscale interest. One should just use the effective mass tensor for the advanced problems of interest. Another example of a force circumstance is the use of magnetic field, where an extracted cyclotron effective mass that samples all the orientations will be needed because of the Lorentz interaction. This cyclotron mass is then $[\det|\mathbb{M}_{ij}|/m_{zz}^*]^{1/2}$, where $\hat{\mathbf{z}}$ is the direction of the magnetic field.

Another derived mass parameter is that of density of states. For *Si*, now one must account for occupation in multiple valleys, multiple bands, et cetera. The density of states effective mass is therefore an extracted parameter whose utility is strictly only for the convenience of the calculation of density of states such as in the three-dimensional

A more generalized form of this conduction current is $\mathbf{J} = q^2 n \langle \tau_{\mathbf{k}} \rangle (1/\mathbb{M}) \boldsymbol{\mathcal{E}}$. It also suggests $\mathbf{J} = \overline{\overline{\sigma}} \boldsymbol{\mathcal{E}}$, where $\overline{\overline{\sigma}}$ is now a conductivity tensor. If a spheroidal constant energy surface existed at the Γ point, then a look at $1/\mathbb{M}$ should convince one that conduction will be anisotropic. Heavy-hole bands in semiconductors are at the zone center and are anisotropic at low energies.

We have remarked on the serious limitations of the Drude model before. Suffice it to say that one uses it at one's own peril. It is somewhat acceptable for understanding conduction in metal in the Newtonian spirit of early undergraduate education. After that, it should be strictly banished from all the nooks and crannies of the brain.

In a block of *Si*, any arbitrary electric field can be split into its Cartesian components, each of which has a response from two valleys where the response is with a longitudinal mass, and four transverse valleys, states of which respond with a transverse mass.

$\mathscr{G}_{3D} = (\sqrt{2}/\pi^2\hbar^3)m_d^{*3/2}\sqrt{E - E_c}$. m_d^* now must account for multiple valleys of occupation and the distribution of states in energy, which are related to the effective mass tensor.

Likewise, for the specific heat arising in these conducting particles, for a large three-dimensional object, the mass parameter will be $[\det|\mathbb{M}_{ij}|]^{1/3}$.

This discussion is to stress that all these rest of the masses are masses of convenience. The one that really reflects the essentials of semiconductor particle behavior is the effective mass that follows from bandstructure as an effective mass tensor.

K
A and B coefficients, and spontaneous and stimulated emission

An ESSENTIAL ELEMENT of matter-light interaction is the statistics and behavior of photons in matter. Bose-Einstein statistics for photons, and the interaction and presence of other photons, are both material to this matter-light interaction. What is true for photons is equally relevant to phonons. Phonons interact strongly, while photons do not, but they are both bosons. This appendix summarizes this essential construct in order to emphasize the photon and phonon basics involved in this. This emission and absorption is a result of transition, where energy has been exchanged and momentum has been conserved, while annihilating a particle or generating a particle.

Light is emitted or absorbed when a quantum system undergoes transitions between two levels. In the two-level quantum system in Figure K.1(a) shows an emission, and (b) an absorption process. Conservation of energy implies $E_2 - E_1 = \hbar\omega_{21}$. Einstein A and B coefficients are a phenomenological means to understand the transitions.

Figure K.1(a) shows a spontaneous emission process. An excited state is returning to its ground state by losing excess energy. And this process is spontaneous with a frequency spectrum constrained to the conservation of energy. Einstein's A coefficient, A_{21} here, is the probability per unit time of this electron transition and photon emission process. If n_2 were the number of these excited states, then the $2 \rightarrow 1$ transition follows

$$\frac{dn_2}{dt} = -A_{21}n_2. \tag{K.1}$$

The spontaneous radiation, precisely as the name implies, is grounded in quantum-mechanical uncertainty—nature's randomness. Blackbody radiation, coupled to the statistics, can be traced to these phenomena. What it says is that two states, different in energy,

Does a molecule suffer an impulse when it emits or absorbs a photon and its energy? Yes, of course. Conservation of momentum and energy must still hold. By the way, radiometers—the rotating weathervanes in a bulb of vacuum—are not simply based on momentum transfer from photons. The vacuum is not perfect: one side is less reflective (darker) than the other. Shining light makes the more absorbent side warmer than the other. The mean free paths of molecules are large. Molecules bouncing off the darker warm side get hotter. The process also kicks the vane away more than on the reflective side. If the vacuum is too high, the vanes don't move. More subtly, it is the edges of the vanes that are most important. So, don't buy an argument that it is a momentum transfer in the vacuum weathervane. There is much controversy here, with Maxwell, Reynolds, Lebedev and even Einstein involved. This is over decades, with a sprinkling of what we would now call out as an ethical problem on the part of Maxwell, who refereed for *Philosophical Transactions*. It is something not too far from, but of less import than, what transpired between Darwin and Wallace for evolution, and Newton and Leibniz for calculus. It is a good lesson in how ethics too are malleable and not an invariant of time. Visit a national museum in London or Paris and you see words trying to hide the robbery and pillage of the lands that the artifacts came from. Hiding behind with the higher energy one occupied, will spontaneously radiate, subject to conservation constraints. An electron in a higher orbital may drop to a lower orbital while radiating a photon.

euphemistic words—"imperial" and "colonial"—is a sign of moral bankruptcy in the past and in the political culture of these days where words acquire entirely new meanings.

This radiation is truly random if the photon emission is unrestricted—randomness here means independence of this characteristic on angle, for example, à la our discussion of the Bertrand paradox. The reverse of this process is spontaneous photon absorption. Now, if a photon happens to be present, resonant in energy with the emission transition of $\hbar\omega_{21}$, and the higher energy state is occupied, and the lower empty, then this photon's presence and its resonant coupling—two oscillators in complete sympathy with each other—will cause the stimulated emission of a photon. So, two spontaneous processes of absorption and emission, and one stimulated process of emission, become possible. This is a property that all bosons will have, so it applies to phonons, Bose-Einstein condensates and all other bosonic particles. The relationship of Equation K.1 also tells us that there is a lifetime—a radiative lifetime—of $\tau = 1/A_{21}$.

This constraint on photons having infinite access to states is important, since conservation equation of energy and momentum will involve it. The Purcell factor, discussed in S. Tiwari, "Quantum, statistical and information mechanics: A unified introduction," Electroscience 1, Oxford University Press, ISBN 978-0-198-75985-0 (forthcoming), and summarized in Appendix M, is connected to this. And if one can eliminate some modes, and restrict emission to only certain momenta (angles), one can suppress it. This may help light emission threshold currents.

Figure K.1(b) shows an absorption process. By absorbing a photon, an electron was promoted to a higher energy level. This excited state was formed by absorbing a photon. *It is not a spontaneous process.* The photon was necessary. Now

$$\frac{dn_1}{dt} = -B_{12}n_1u(\omega), \tag{K.2}$$

where $u(\omega)$ is the photon energy density—the spectral energy density of the electromagnetic field—with the unit of $J/m^3 \cdot (rad/s)$. Conservation of energy must still hold.

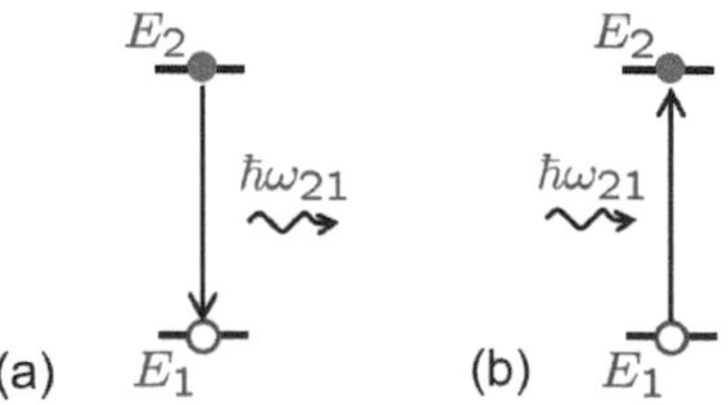

Figure K.1: Emission and absorption.

Einstein's insight was that this description is still incomplete. A photon field can stimulate an emission as well as the absorption transition. So, Equation K.1, in alignment with Equation K.2, must also include

$$\frac{dn_2}{dt} = -B_{21}n_2u(\omega). \tag{K.3}$$

This stimulated emission is a quantum-mechanical and coherent effect with a photon emission in phase with the photons stimulating the transition.

So, there are three coefficients: A_{21} in units of $1/s$ for spontaneous emission, B_{12} for stimulated absorption and B_{21} for stimulated emission in units of $J/m^3 \cdot (rad/s)$. These are not independent, since thermal equilibrium at the temperature T prescribes a balance of details. Let the quantum system be in a cavity of blackbody radiation with which it is interacting. So,

$$B_{12}n_1u(\omega) = A_{21}n_2 + B_{21}n_2u(\omega), \tag{K.4}$$

and if the degeneracies are g_2 and g_1 for the two energies, then

$$\frac{n_2}{n_1} = \frac{g_2}{g_1}\exp\left(-\frac{\hbar\omega}{k_B T}\right) \tag{K.5}$$

under Boltzmann constraints of occupation.

The blackbody energy spectrum has a modal occupation dependence given by

$$u(\omega) = \frac{\hbar\omega^3}{\pi^2 c^3}\frac{1}{\exp(\hbar\omega/k_B T) - 1}. \tag{K.6}$$

See S. Tiwari, "Quantum, statistical and information mechanics: A unified introduction," Electroscience 1, Oxford University Press, ISBN 978-0-198-75985-0 (forthcoming), and S. Tiwari, "Nanoscale device physics: Science and engineering fundamentals," Electroscience 4, Oxford University Press, ISBN 978-0-198-75987-4 (2017), for a development of this relationship, its physical significance and its implications for photovoltaics, where we employ the absorption process for photoelectric conversion.

Equation K.4 through Equation K.6 lead to the following two constraints on the coefficients:

$$\begin{aligned} g_1 B_{12} &= g_2 B_{21}, \text{ and} \\ A_{21} &= \frac{\hbar\omega^3}{\pi^2 c^3} B_{21}. \end{aligned} \tag{K.7}$$

Knowing one coefficient suffices. High absorption probability also implies high emission probabilities for both spontaneous and stimulated processes.

We also know from the Golden rule that the transition rates can be written as

$$S_{12} = \frac{2\pi}{\hbar}|\mathscr{H}'_{12}|^2 \mathscr{G}(\hbar\omega), \tag{K.8}$$

where $\mathscr{G}(\hbar\omega)dE$ is the final states within dE of $E = \hbar\omega$. If the states are discrete, then this $\mathscr{G}(\hbar\omega)$ is the photon state density. The matrix element in the Golden rule is

$$\mathscr{H}'_{12} = \langle 2|\mathscr{H}'|1\rangle = \int \psi_2^*(\mathbf{r})\hat{\mathscr{H}}'\psi_1(\mathbf{r})\, d^3\mathbf{r}. \tag{K.9}$$

Since it is the photon causing the perturbation through the interaction with an electric dipole, whether the interaction is with an atom or atoms of semiconductors, the dipole-field interaction perturbation is

$$\hat{\mathscr{H}}' = -\mathbf{p}\cdot\boldsymbol{\mathcal{E}}. \tag{K.10}$$

Equation K.8 together with the dipole-field perturbation—the matrix element—gives us the selection rules of the allowed and forbidden transitions.

Now consider the situation in semiconductors. We now have a distribution of states interacting as shown in Figure K.2.

Figure K.1 described the absorption and emission processes involving two states. In semiconductors, where we tackle mostly a distribution of states that are close together, it is more convenient to analyze in terms of a compact representation of the distribution of states. For example, we represent the carrier population via the use of quasi-Fermi energy (E_{qFn} and E_{qFp} for electrons and holes), bandedge energy (E_c and E_v for conduction and valence bands) and an effective density of states ($\mathcal{N}_c$ and $\mathcal{N}_v$). Quasi-Fermi energy is a

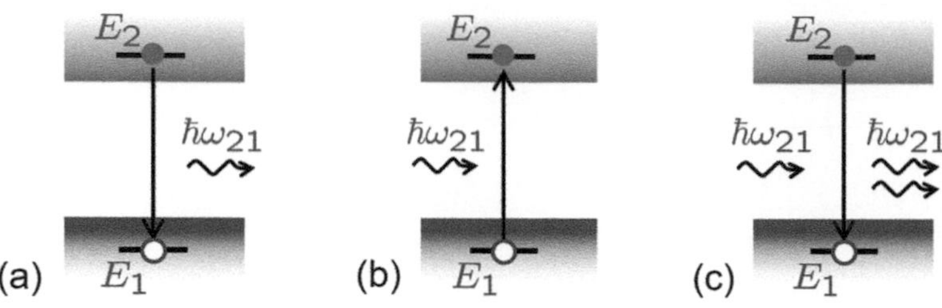

Figure K.2: Illustration of the spontaneous emission of a photon of energy $\hbar\omega$ through a transition between two states, where the higher energy is occupied and the lower unoccupied, in (a). Part (b) shows its reverse process of spontaneous absorption and (c) shows the stimulated process where the presence of a photon whose energy is matched to the transition causes a stimulated emission process.

fit here for electrochemical energy that describes the off-equilibrium population and state occupation correctly. $E_{qFn} = E_{qFp} = E_F$ in thermal equilibrium. Take Figure K.3, which shows the emission and recombination between two states in the bands. B_{12} and B_{21} are the probability of a transition triggered by the presence of a photon, A_{21} is the probability of a spontaneous emission, f_1 and f_2 the occupation probabilities ($f_1 = 1/\{1 + \exp[(E_1 - E_{qF1})/k_BT]\}$) and $f_2 = 1/\{1 + \exp[(E_2 - E_{qF2})/k_BT]\}$) so that $1 - f_2$ is the probability that the state with energy E_2 is unoccupied. Let also $\mathcal{P}(E_{21})$ be the density of photons at energy $E_{21} = \hbar\omega_{21}$. We are also assuming that the degeneracies of levels 1 and 2 are the same, since this is in a semiconductor, where the degeneracy arises in spin.

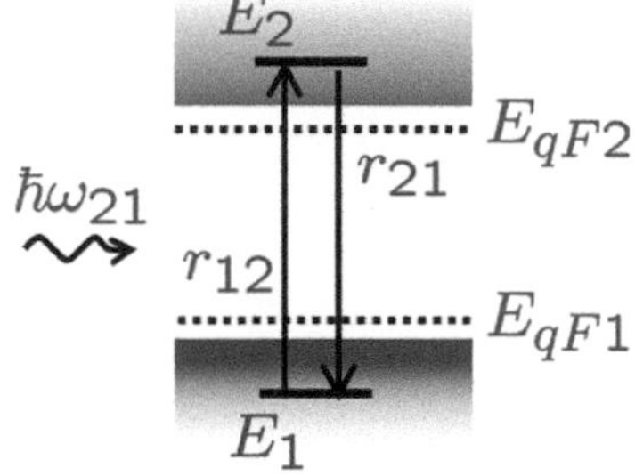

Figure K.3: Absorption and emission transitions between the two states E_1 and E_2. E_{qF1} is associated with the quasi-Fermi energy for occupation of the occupied state. For semiconductors, this is the conduction band, and the quasi-Fermi energy E_{qFn} is the usual notation. Similar nomenclature applies to state 2 in the valence band. The *A*-*B* coefficient description is more general, and we are keeping that in mind in describing this in this way.

In an absorption process, there is an upward transition rate (electrons transitioning per second, ϱ_{12}) from an occupied state 1 to an unoccupied state 2 as

$$\varrho_{12} = B_{12} f_1 (1 - f_2) \mathcal{P}(E_{21}). \tag{K.11}$$

For an emission process, the transition rate ϱ_{21} consists of a stimulated term dependent on the presence of a photon, and a spontaneous part. The stimulated transition rate is

$$\varrho_{21}\big|_{\text{simulated}} = B_{21} f_2 (1 - f_1) \mathcal{P}(E_{21}), \tag{K.12}$$

and the spontaneous transition rate is

$$\varrho_{21}\big|_{\text{spontaneous}} = A_{21} f_2 (1 - f_1). \tag{K.13}$$

In thermal equilibrium, $E_{qF1} = E_{qF2} = E_F$, and the stimulated upward transition rate must balance the stimulated and spontaneous downward transition rate, that is,

$$\begin{aligned} \varrho_{12} &= \varrho_{21}\big|_{\text{spontaneous}} + \varrho_{21}\big|_{\text{simulated}} \\ \therefore\ B_{12} f_1 (1 - f_2) \mathcal{P}(E_{21}) &= A_{21} f_2 (1 - f_1) + B_{21} f_2 (1 - f_1) P(E_{21}), \\ \mathcal{P}(E_{21}) &= \frac{A_{21} f_2 (1 - f_1)}{B_{12} f_1 (1 - f_2) - B_{21} f_2 (1 - f_1)} \\ &= \frac{A_{21}}{B_{12}\{[f_1(1 - f_2)]/[f_2(1 - f_1)]\} - B_{21}}, \end{aligned} \tag{K.14}$$

where

$$\frac{f_1(1-f_2)}{f_2(1-f_1)} = \frac{1/\left[1+\exp\left(\frac{E_1-E_F}{k_BT}\right)\right]}{1/\left[1+\exp\left(\frac{E_2-E_F}{k_BT}\right)\right]} \times \frac{1-1/\left[1+\exp\left(\frac{E_2-E_F}{k_BT}\right)\right]}{1-1/\left[1+\exp\left(\frac{E_1-E_F}{k_BT}\right)\right]}$$
$$= \frac{\exp[(E_2-E_F)/k_BT]}{\exp\,[(E_1-E_F)k_BT]} = \exp\left(\frac{E_{21}}{k_BT}\right). \qquad (K.15)$$

Therefore, the A and B coefficients are related through the density of photons as

$$\mathcal{P}(E_{21}) = \frac{A_{21}/B_{21}}{(B_{12}/B_{21})\exp(E_{21}/k_BT)-1}$$
$$= \frac{A_{21}/B_{21}}{\exp(E_{21}/k_BT)-1} \qquad (K.16)$$

since $B_{21} = B_{12}$.

$B_{12} \neq B_{21}$ in general, because the degeneracies of the two levels is not the same. For semiconductors, $B_{21} \neq B_{12}$ follows from the fact that the spin degeneracy is the same, and that these are the probabilities of the transition coupling states 1 and 2 and are therefore proportional to the matrix element coupling the two together with the degeneracy. The matrix element is symmetric in the transition.

Now, we may connect this foundation to the spontaneous and stimulated interaction in matter. Figure K.4 illustrates this in example systems—semiconductor in (a) and the 4-level solid state in (b)—with two levels interacting showing spontaneous and stimulated emission in (c), all with some of the main features. The spontaneous emission is random and isotropic. The stimulated emission is coherent—in close phase alignment, so in both frequency and time and any additional phase constants—and is anisotropic.

The net stimulated emission is

$$r_{21}|_{stim} \equiv \varrho_{21} - \varrho_{12}$$
$$= B_{21}f_2(1-f_1)\mathcal{P}(E_{21}) - B_{12}f_1(1-f_2)\mathcal{P}(E_{21})$$
$$= B_{21}\mathcal{P}(E_{21})(f_2-f_1) \qquad (K.17)$$

where we have employed $B_{21} = B_{12}$. Using Equation K.16, this gives

$$r_{21}|_{stim} = \frac{A_{21}(f_2-f_1)}{\exp(E_{21}/k_BT)-1}$$
$$= \frac{r_{stim}(E_{21})}{\exp(E_{21}/k_BT)-1}, \qquad (K.18)$$

The B coefficients for absorption and emission will be related to through the degeneracies of levels 1 and 2. For our case of semiconductors, they are identical and due to spin degeneracy.

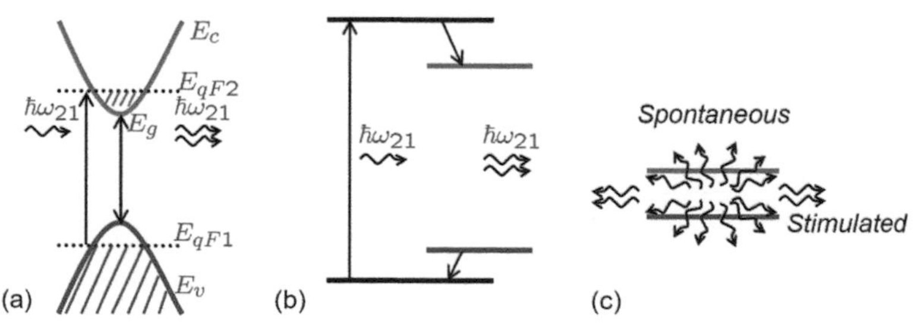

Figure K.4: Part (a) shows a bipolar semiconductor radiative system with stimulated emission, (b) shows a 4-level solid-state system with stimulated emission and (c) is an illustration of the spontaneous and stimulated emission arising with a transition between a collective assembly of filled and empty states in the system.

where

$$r_{stim}(E_{21}) \equiv A_{21}(f_2 - f_1). \tag{K.19}$$

So, what this says for stimulation is, that a photon causes a stimulation emission with a certain rate that must be modified by the photon occupation statistical factor arising in the Bose-Einstein statistics. The coefficient A_{21} determines a rate similar to the constant B_{21} for spontaneous emission. The quantum-mechanical constraint here is that this stimulated emission becomes possible and that the principle of detailed balance—the balance between each and every process in detail in thermal equilibrium—has forced a relationship between the A and the B coefficients. The net absorption rate is

$$\begin{aligned} r_{12}|_{absorb} &\equiv \varrho_{12} - \varrho_{21}(stim) \\ &= B_{12}f_1(1-f_2)\mathcal{P}(E_{21}) - B_{21}f_2(1-f_1)\mathcal{P}(E_{21}) \\ &= B_{21}\mathcal{P}(E_{21})(f_2 - f_1) = -r_{21}(stim), \end{aligned} \tag{K.20}$$

therefore

$$\begin{aligned} r_{12}|_{absorb} &= \frac{A_{21}(f_1 - f_2)}{\exp(E_{21}/k_BT) - 1} \\ &= \frac{f_1 - f_2}{f_2 - f_1} r_{21}(stim). \end{aligned} \tag{K.21}$$

$r_{12}|_{absorb}$ is the net absorption rate. Since it depends on the occupied and unoccupied state distribution, and As and Bs, which, although related to each other, are also a measure of the coupling between the states through the matrix element for this coupling, the absorption rate is a function of the relevant properties of the material. We can make summary comments on this dependence. The details arise in the processes in the material itself.

The presence of a photon can cause absorption (a stimulated exchange of energy to the electron transfer between states), but it can also stimulate emission. This is what Equation K.20 or its derivative relationship Equation K.21 signifies. One measures this dependence of the net absorption rate that is connected through the macroscopic properties and interaction in the material via the parameter of absorption coefficient. Absorption coefficient $\alpha(E)$ is the probability of the absorption of a photon per unit length of travel through the material. So,

$$r_{12}|_{absorb} = \alpha(E)\mathcal{P}(E)v_g. \tag{K.22}$$

Here, the product of the photon density and the photon's velocity is the photon flux, which is the photon count per unit area per unit time. The velocity—the group velocity since it is the energy movement that is of interest—is related through the index of refraction, which we will denote by $\overline{n}$:

$$v_g = \frac{d\omega}{dk} = \frac{2\pi}{h}\frac{dE}{dk} = \frac{c/\overline{n}}{1+(E/\overline{n})(d\overline{n}/dE)} \text{ in dispersive, and}$$
$$= \frac{c}{\overline{n}} \text{ in non-dispersive} \tag{K.23}$$

medium. So,

$$\begin{aligned}\alpha(E_{21}) &= \frac{r_{12}(abs)}{\mathcal{P}(E_{21})\,(c/\overline{n})} = \frac{B_{21}\mathcal{P}(E_{21})(f_1-f_2)}{\mathcal{P}(E_{21})\,(c/\overline{n})} \\ &= B_{21}\frac{\overline{n}}{c}(f_1-f_2) \\ &= -\frac{r_{21}(stim)}{\mathcal{P}(E_{21})\,(c/\overline{n})}.\end{aligned} \tag{K.24}$$

From Einstein's A-B relationship

$$\frac{A_{21}}{B_{21}} = \frac{8\pi\overline{n}^3E^2}{h^3c^3}, \tag{K.25}$$

the absorption coefficient follows as

$$\begin{aligned}\alpha(E_{21}) &= A_{21}\frac{h^3c^3}{8\pi\overline{n}^3E_{21}^2}\frac{\overline{n}}{c}(f_1-f_2) \\ &= \frac{h^3c^2}{8\pi\overline{n}^3E_{21}^2}A_{21}(f_1-f_2) \\ &= \frac{h^3c^2}{8\pi\overline{n}^3E_{21}^2}\frac{-r_{21}(stim)}{[\exp(E_{21}/k_BT)-1]} \\ \text{or} \quad &= -\frac{h^3c^2r_{stim}(E_{21})}{8\pi\overline{n}^2E_{21}^2}.\end{aligned} \tag{K.26}$$

From the spontaneous emission discussion,

$$\begin{aligned}\varrho_{21}(spont) &= r_{21}(spont) \\ \text{or } A_{21}f_2(1-f_1) &= \frac{8\pi\overline{n}^3E_{21}^2}{h^3c^3}B_{21}f_2(1-f_1),\end{aligned} \tag{K.27}$$

with

$$\alpha(E_{21}) = \frac{B_{21}\overline{n}}{c}(f_1-f_2). \tag{K.28}$$

Therefore,

$$\begin{aligned}r_{21}(spont) &= \frac{8\pi\overline{n}^3E_{21}^2}{h^3c^3}\alpha(E_{21})\frac{f_2(1-f_1)}{f_1-f_2} \\ &= \frac{8\pi\overline{n}^3E_{21}^2\alpha(E_{21})}{h^3c^3\left\{\exp\left[\frac{E_{21}-(E_{qF2}-E_{qF1})}{k_BT}\right]-1\right\}}.\end{aligned} \tag{K.29}$$

In thermal equilibrium, $E_{qF2} - E_{qF1} = 0$, the spontaneous emission (and absorption) rate is small compared to those in off-equilibrium, where the carrier densities have been increased deliberately. At the other extreme, when $E_{qF2} - E_{qF1} \to E_{21}$, $r_{21}(spont) \to \infty$. So,

$$r_{21}(spont) = -\frac{r_{stim}(E_{21})}{\left\{\exp\left[\frac{E_{21}-(E_{qF2}-E_{qF1})}{k_BT}\right]-1\right\}} = r_{spont}(E_{21}), \text{ and}$$

$$r_{stim}(E_{21}) = r_{spont}(E_{21})\left\{1-\exp\left[\frac{E_{21}-(E_{qF2}-E_{qF1})}{k_BT}\right]-1\right\}. \quad (K.30)$$

We have now related A and B coefficients that integrate detailed balance with the presence of spontaneous and stimulated processes in bosonic interactions. And we have connected these to the properties of materials by developing various relationships between the absorption coefficient $\alpha(E)$, the spontaneous rate $r_{spont}(E)$ and the stimulated rate $r_{stim}(E)$, in all of which the photon-material interactions will play a role. These are all interconnected, and, knowing one, one can find others.

This interconnection of one with the other—that one contains information identical to the other—is an important property of linear systems that we discuss at length in the discussion of the Kramers-Kronig relationship. It also appears in a linear form in the fluctuation-dissipation discussion, where linearity again appears as a frictional drag.

We end with an example to show these connections. Consider a direct bandgap semiconductor with density of states

$$\mathscr{G}_c(E) = \frac{1}{2\pi^2}\left(\frac{2m_e^*}{\hbar^2}\right)^{3/2}\sqrt{E-E_c}, \text{ and}$$

$$\mathscr{G}_v(E) = \frac{1}{2\pi^2}\left(\frac{2m_h^*}{\hbar^2}\right)^{3/2}\sqrt{E_v-E}. \quad (K.31)$$

For absorption calculation, consider the number of filled states in the valence band of $f_1\mathscr{G}_v(E)$ and the empty states in the conduction band of $(1-f_2)\mathscr{G}_c(E)$. To calculate the absorption coefficient, we need to integrate absorption over all possible energy states, so

$$\alpha(E) = \int_{-\infty}^{\infty}\frac{B_{12}}{c/\overline{n}}\mathscr{G}_v(E_v-E)\mathscr{G}_c(E-E_c)\,dE. \quad (K.32)$$

The B coefficient, as we remarked, integrates the microscopic process of transition, so we calculate that through the matrix element using the Golden rule, where the photon provides the electromagnetic perturbation. We have

$$B_{12} = \frac{\pi}{2\hbar}\left|\langle\psi_1(\mathbf{r},t)|\hat{\mathscr{H}}'|\psi_2(\mathbf{r},t)\rangle\right|^2, \quad (K.33)$$

where the perturbation has the form

$$\hat{\mathscr{H}}'(\mathbf{r},t) = \hat{\mathscr{H}}'(\mathbf{r})\cos(\omega t). \quad (K.34)$$

Momentum has appeared in the $\mathbf{k}\cdot\mathbf{p}$ approach and in the electromagnetic approach. The Hamiltonian of the electron-photon system is

$$\begin{aligned}\mathscr{H} &= \frac{1}{2m^*}(\hbar\mathbf{k}-q\mathbf{A})^2\\ &\approx -\frac{\hbar^2}{2m^*}\nabla_\mathbf{r}^2 + \frac{iq\hbar}{m^*}\mathbf{A}\cdot\nabla_\mathbf{r}\\ &= -\frac{\hbar^2}{2m^*}\nabla_\mathbf{r}^2 - \frac{q}{m^*}\mathbf{A}\cdot\mathbf{p},\end{aligned}$$

where the first term is the unperturbed Hamiltonian ($\mathscr{H}_0$), and the second term is the perturbation ($\mathscr{H}'$). The form of the second term is just like the $\mathbf{k}\cdot\mathbf{p}$ form, where electron states are interacting with electron states. So, the momentum matrix element appears as quite a central element in many semiconductor calculations. This interaction is tackled with more rigor in Chapter 8.

The photon electromagnetic interaction will involve the momentum matrix element. In detail, then,

$$\begin{aligned}B_{12} &= \frac{\pi q^2\hbar}{m^{*2}\epsilon_0\overline{n}^2\hbar\omega}\left|\langle\psi_2(\mathbf{r},t)|\mathbf{p}|\psi_2(\mathbf{r},t)\rangle\right|^2\\ &= B_{21} = \frac{\pi q^2\hbar}{m^{*2}\epsilon_0\overline{n}^2\hbar\omega}\left|\langle\psi_1(\mathbf{r},t)|\mathbf{p}|\psi_2(\mathbf{r},t)\rangle\right|^2.\end{aligned} \quad (K.35)$$

Using a similar approach,

$$A_{21} = \frac{4\pi\overline{n}q^2E_{21}}{m^{*2}\epsilon_0h^2c^3}\left|\langle\psi_1^*(\mathbf{r},t)|\mathbf{p}|\psi_2(\mathbf{r},t)\rangle\right|^2. \quad (K.36)$$

B and A coefficients are now known in detail, and one can calculate the absorption coefficient through Equation K.32 for the idealized semiconductor. Instead, if we have a measurement, from the measured absorption or spontaneous emission spectrum, we can calculate the absorption coefficient and net spontaneous and stimulated emission spectrum.

There is one measure that is useful to characterize these absorption and emission transitions. This is the oscillator strength. It is a dimensionless number useful in comparing transition strengths and is defined as the ratio of the strength of the transition and the theoretical transition strength derived from a harmonic oscillator model. We discuss this in Appendix I. If degeneracies were different, as with B's relationships, and important to organic semiconductor systems, the emission and absorption oscillator strengths will be related through

$$g_2 f_{21} = g_1 f_{12}, \tag{K.37}$$

so would the B coefficients:

$$g_2 B_{21} = g_1 B_{12}. \tag{K.38}$$

Oscillator strength is usually normalized, with a strong transition approaching unity. If there exists degeneracy in the system, then it can exceed unity.

L
Helmholtz theorem and vector splitting

USEFUL VECTOR RELATIONSHIPS are summarized in the glossary. One particular splitting of vectors, such as those of the field's longitudinal and transverse components, is particularly useful in semiconductors due to the variety of field-mediated interactions involved. A suitable example is that of the electromagnetic field with the crystal vibration fields.

Let **A** satisfy

$$\nabla^2 \mathbf{A} = -\mathbf{B}; \tag{L.1}$$

then,

$$\mathbf{A} = \frac{1}{4\pi} \int \frac{\mathbf{B}}{|\mathbf{r} - \mathbf{r}'|} d\mathbf{r}'. \tag{L.2}$$

Since

$$\nabla^2 \mathbf{A} = \nabla(\nabla \cdot A) - \nabla \times (\nabla \times \mathbf{A}), \tag{L.3}$$

the use of

$$\begin{aligned} \nabla \cdot \mathbf{A} &= -U, \text{ and} \\ \nabla \times \mathbf{A} &= \mathbf{C}, \end{aligned} \tag{L.4}$$

leads to

$$\mathbf{B} = \nabla U + \nabla \times \mathbf{C}. \tag{L.5}$$

A vector field has now been written in terms of a gradient and a curl. This is *Helmholtz's theorem*. It lets us split the vector field into longitudinal and transverse components, since divergences are longitudinal, and curls are transverse.

We are back to the beauty of divergences and curls that, conjoined, make certain characteristics come alive forever, as in an electromagnetic wave. As John von Neumann is said to have remarked, "It is just as foolish to complain that people are selfish and treacherous as it is to complain that the magnetic field does not increase unless the electric field has a curl. Both are laws of nature." (as quoted by Eugene Wigner; see E. P. Wigner, "Symmetries and reflections: Scientific essays," Greenwood Press ISBN 0313201072 (1967), p. 261).

Define

$$\begin{aligned} \mathbf{B}_l &= \nabla U, \text{ and} \\ \mathbf{B}_t &= \nabla \times \mathbf{C}, \text{ with} \\ \mathbf{B} &= \mathbf{B}_l + \mathbf{B}_t. \end{aligned} \tag{L.6}$$

In view of the vector identities (see the glossary),

$$\begin{aligned}
\nabla \times \mathbf{B}_l &= 0, \\
\nabla \cdot \mathbf{B}_l &= \nabla \cdot \mathbf{B}, \\
\nabla \times \mathbf{B}_t &= \nabla \times \mathbf{B}, \quad \text{and} \\
\nabla \cdot \mathbf{B}_t &= 0,
\end{aligned} \tag{L.7}$$

which say that the longitudinal field has no curl—is oriented normal—and that the transverse field has no divergence, that is, is oriented orthogonal. U, **A**, **B** and **C** have now been related to each other.

M
Mode coupling and Purcell effect

MODE COUPLING APPEARS in all energy exchanges and has appeared in a variety of ways—between the variety of forms that the free energy of interest to us resides in and therefore through the oscillators and the modes of these oscillators that couple—throughout this text. These exchanges can be coupled with a variety of interesting properties through matching. Even spontaneous emission requires this coupling, and it too can be modified, which is the Purcell effect.

But we start with the electromagnetic propagation of the simple ladder network and ask the question as to why a network of non-dissipating components—inductors and capacitors—can have a real characteristic impedance. The real impedance reflects dissipation—loss to the world—from a source. And this is precisely what real space impedance is. Take the infinite transmission line ladder network of Figure M.1, and let it have a characteristic impedance of Z_0, so that the different topologies drawn are equivalent. Since the transmission line drawn is infinitely long, impedance seen at any cut must be the same. With Z_1 and Z_2 as the two elements of which this distributed network—lumped or continuous—consists of, and Z_0 as the impedance at any cross-section, adding one more repeating element to the left—going from Figure M.1(a) to (b), the network remains the same infinite network. Therefore, the impedance in Figure M.1(b), viewed from the left, is

$$Z = Z_1 + \frac{1}{(1/Z_2) + (1/Z_0)} = Z_0, \tag{M.1}$$

and therefore

$$Z_0 = Z_1 + \frac{Z_2 Z_0}{Z_2 + Z_0}, \tag{M.2}$$

whose solution is

$$Z_0 = \frac{Z_1}{2} + \left(\frac{Z_1^2}{4} + Z_1 Z_2 \right)^{1/2}. \tag{M.3}$$

Z_0 is the characteristic impedance of the infinite arrangement.

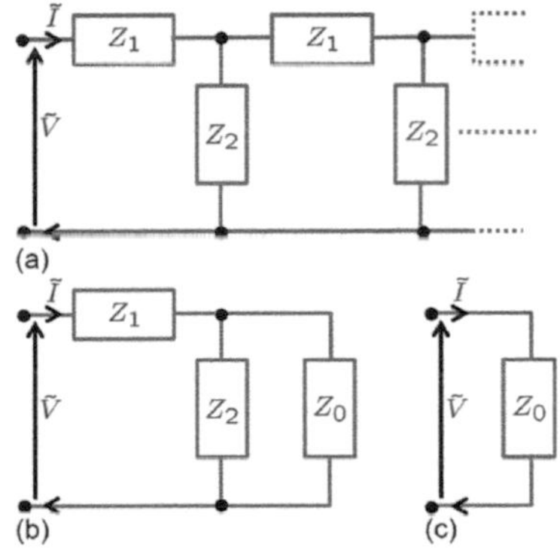

Figure M.1: An infinite ladder of repeating impedances as a transmission line. Part (a) shows the network as an infinitely repeating unit. Part (b) shows that, at any cross-section through the network, the following network on the right is still the same infinite network. Let this have a characteristic impedance Z_0, as shown in (c). Adding a repeating Z_1-Z_2 element to the left of it leaves the ladder unchanged.

Now consider lossless Z_1 and Z_2 arising in an inductive element L and a capacitive element C. $Z_1 = i\omega L$, and $Z_2 = 1/i\omega C$, as shown in Figure M.2(a). The inductor-capacitor network is more conveniently seen as a T network that has symmetry, as shown in Figure M.2(b). It also separates out the $Z_1/2$ element of Equation M.3.

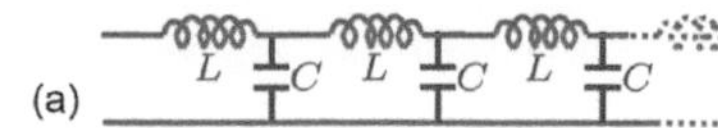

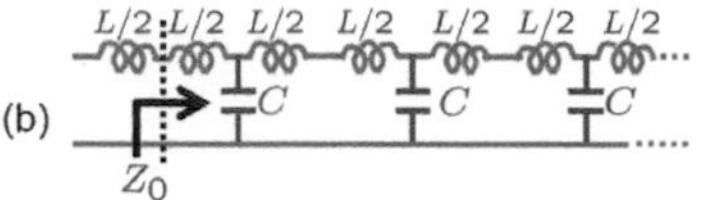

Figure M.2: An inductor-capacitor ladder with its two equivalent configurations, (a) and (b). Part (a) is similar to the infinite ladder in Figure M.1, with Z_1 arising in an inductor L, and Z_2 in a capacitor C. These two elements—lumped or distributed—are lossless.

Following Equation M.3, removing the $Z_1/2$ arising in the leftmost $L/2$, and a repeating network following, we have

$$Z_0 = \left(\frac{L}{C} - \frac{\omega^2 L^2}{4} \right)^{1/2}. \quad \text{(M.4)}$$

If $\omega^2 < 4/LC$, this characteristic impedance is real. A lossless inductor-capacitance ladder has a real impedance!

If $\omega^2 > 4/LC$, the impedance is the imaginary number

$$Z_0 = i\left(\frac{\omega^2 L^2}{4} - \frac{L}{C} \right)^{1/2}. \quad \text{(M.5)}$$

There is no paradox here. In the lossless element ladder network, below the radial frequency of $2/\sqrt{LC}$, the source can continue to supply the energy, and the propagation can continue to proceed farther and farther along the infinite network. This is precisely what an antenna does. The radiation from the antenna, below a certain frequency where the impedance is real, will continue to propagate out. This is the useful frequency range for the antenna.

Huygens principle makes a lot more sense when seen through this infinite propagational view.

Light's free space impedance also follows this argument, except now the speed of light c determines the constraint of the system and the relationship between the electric field and the magnetic field. The electric field corresponds to and represents the potential energy of the system, which is being exchanged with the kinetic energy of the system that the magnetic field corresponds to. Maxwell's equations, with the speed c as the invariant, places constraints on the fields, and the impedance of the infinitely propagating solution is the ratio of these fields and, for free space, $\sqrt{\mu_0/\epsilon_0}$.

The lumped distributed network and light propagation in infinite space are examples of efficient coupling of modes that can continue ad infinitum. One repeating element could couple all its energy entirely to the following repeating element. The elements' properties determine the modes that can be coupled: which ones will propagate and which ones will dissipate.

Now consider a variation on mode coupling for electromagnetics by coupling a system to a cavity. Consider the two-level system of Appendix K in a resonant cavity of volume V_0. So, we have reformed the problem from blackbody coupling to one where we are modifying the mode distribution. Only modes of the cavity that are close to the frequency of the two-level system—are resonant—interact. Consider a small cavity so that only one single mode at radial frequency ω_c couples, and let the half-linewidth be $\Delta\omega_c$. The density of states of the photon, with the one resonant mode, are subject to

$$\int_0^\infty g(\omega)\,d\omega = 1. \tag{M.6}$$

This is the normal Lorentzian distribution function given by

$$g(\omega) = \frac{2}{\pi\,\Delta\omega_c}\frac{\Delta\omega_c^2}{4(\omega-\omega_c)^2+\Delta\omega_c^2}. \tag{M.7}$$

For a two-level system with $E_2 - E_1 = \hbar\omega_0$, the resonance implies that

$$g(\omega_0) = \frac{2}{\pi\,\Delta\omega_c}\frac{\Delta\omega_c^2}{4(\omega_0-\omega_c)^2+\Delta\omega_c^2}. \tag{M.8}$$

When the frequencies are exactly aligned,

$$g(\omega_0) = \frac{2}{\pi\,\Delta\omega_c} = \frac{2\mathcal{Q}}{\pi\,\omega_0}, \tag{M.9}$$

where $\mathcal{Q}$ is the quality factor of the cavity ($\mathcal{Q} = \omega_c/\Delta\omega_c$). Using the dipole-field perturbation $\mathbf{p}\cdot\boldsymbol{\mathcal{E}}$,

$$\mathscr{H}_{12}^2 = \xi^2\mu_{12}^2\frac{\hbar\omega}{2\epsilon V_0}, \tag{M.10}$$

where the dipole moment in general is $\boldsymbol{\mu}_{12} = -e(\langle 2|x|1\rangle\hat{\mathbf{x}} + \langle 2|y|1\rangle\hat{\mathbf{y}} + \langle 2|z|1\rangle\hat{\mathbf{z}})$. $\xi = |\mathbf{p}\cdot\boldsymbol{\mathcal{E}}|/p\mathcal{E}$ is a normalized dipole-field orientation factor. $\xi^2 = 1/3$ for random dipole orientation. If this weak-perturbation mode coupling, where we can use this perturbation approach, is in vacuum, $\epsilon = \epsilon_0$ and $\boldsymbol{\mathcal{E}} = \boldsymbol{\mathcal{E}}_0$.

The transition rate in this cavity coupling is

$$S_{cav} = \frac{2\mathcal{Q}\mu_{12}^2}{\hbar\omega V_0}\xi^2\frac{\Delta\omega_c^2}{4(\omega_0-\omega_c)^2+\Delta\omega_c^2}. \tag{M.11}$$

We found in Appendix K the spontaneous emission rate that is given by the Einstein A coefficient as

$$S = A_{21} = \frac{\hbar\omega^3}{\pi^2c^3}B_{21}.$$

The ratio between the two cases—in the absence and presence of cavity—is

$$F_P = \frac{S_{cav}}{S_{nocav}} = \frac{\tau_{Rnocav}}{\tau_{Rcav}} = \frac{3\mathcal{Q}}{4\pi^2V_0}\left(\frac{\lambda}{\overline{n}}\right)^3\xi^2\frac{\Delta\omega_c^2}{4(\omega_0-\omega_c)^2+\Delta\omega_c^2}. \tag{M.12}$$

This is the *Purcell factor*, and $\overline{n}$ is the index of refraction.

With the dipole oriented in the direction of the field, and precise resonance,

$$F_P = \frac{3\mathcal{Q}}{4\pi^2V_0}\left(\frac{\lambda}{\overline{n}}\right)^3. \tag{M.13}$$

Spontaneous emission can be suppressed or enhanced through the coupling of modes to the cavity. If the dipole is aligned with the field, where cavity volume is small and cavity quality is high, then, with close matching of the transition to the cavity mode, a large enhancement becomes possible in the emission by exploiting the large density of state function by resonating the cavity with the transition. By bringing it off-resonance, one can suppress it. By removing the cavity, one returns back to the spontaneous emission of blackbody mode distribution.

N
Vector and scalar potentials

THE OUTFLOW OF ENERGY PER UNIT VOLUME in an electromagnetic wave is $\boldsymbol{\mathcal{E}} \times \mathbf{H}$—the Poynting vector. The Poynting theorem is the law of conservation of electromagnetic fields, and the divergence of the Poynting vector, by Gauss' law, gives us the outflow per unit volume through the surface that encloses the volume. The energy density in the electric field is $\frac{1}{2}\epsilon_0 \boldsymbol{\mathcal{E}} \cdot \boldsymbol{\mathcal{E}}$.

Working with electric field and magnetic field vectors, which are connected to each other within Maxwell's relationship through time dependence and curls, can get cumbersome. Resorting to potentials makes resolution easier. Since $\mathbf{H}$ is divergence-free, we introduce a vector potential $\mathbf{A}$ that satisfies

A correspondence from classical mechanics would be that complex problems are lot easier to solve by working directly with energy, a scalar, and then deriving the forces from it through the gradients. The fields are what we measure and are unique. But the potentials are more fundamental and simpler to work with.

$$\mu \mathbf{H} = \nabla \times \mathbf{A}. \tag{N.1}$$

A unique vector field requires both the curl and the divergence to be specified. This is satisfied, with the electric field relationship

$$\boldsymbol{\mathcal{E}} = -\frac{\partial \mathbf{A}}{\partial t} - \nabla \psi, \tag{N.2}$$

where ψ is the electrostatic potential, and the divergence constraint

$$\nabla \cdot \mathbf{A} + \mu\epsilon \frac{\partial \psi}{\partial t} = 0. \tag{N.3}$$

This latter choice of $\mathbf{A}$ is that of the Lorentz gauge. By this, we mean a reference and transformation of these potentials that are such that the vector fields derived from them remain unchanged under gauge transformation.

Consider the analogy to gravitational potential. The amount of potential recoverable from mass in an otherwise lossless gravitational system is mgh, where h is the height difference and m is mass. We could have used a reference point at some other position h_0 for the potential. Changing this reference position doesn't change the potential available for the height h in the direction of the gravitational field. The fields here then are unchanged in the gauge transform with the change in the position reference. This is gauge freedom. The power of the approach is that it simplifies by exploiting the symmetries of the system.

Claim: For the Lorentz gauge of Equation N.3, a change of the vector and scalar potentials by

$$\begin{aligned} \mathbf{A} &\mapsto \mathbf{A}' = \mathbf{A} + \nabla\Lambda, \text{ and} \\ \psi &\mapsto \psi' = \psi - \frac{\partial \Lambda}{\partial t}, \end{aligned} \tag{N.4}$$

where Λ is a scalar function, subject to several constraints, including continuity and differentiability, leaves the vector fields unchanged.

Proof:

$$\mu\mathbf{H}' = \nabla \times \mathbf{A}' = \nabla \times (\mathbf{A} + \nabla\Lambda) = \nabla \times \mathbf{A} = \mu\mathbf{H}, \tag{N.5}$$

and

$$\begin{aligned}\mathcal{E}' &= -\frac{\partial \mathbf{A}'}{\partial t} - \nabla\psi' \\ &= -\frac{\partial}{\partial t}(\mathbf{A} + \nabla\Lambda) - \nabla\left(\psi - \frac{\partial\Lambda}{\partial t}\right) \\ &= -\frac{\partial \mathbf{A}}{\partial t} - \nabla\psi - \frac{\partial\nabla\Lambda}{\partial t} + \nabla\frac{\partial\Lambda}{\partial t} \\ &= -\frac{\partial \mathbf{A}}{\partial t} - \nabla\psi = \mathcal{E}.\end{aligned}$$

This Lorentz gauge and gauge transformation is not the only choice we have, but it works for the electromagnetic case and others encountered in semiconductor physics in the presence of the fields consistent with the immutable speed of light.

Now, we can look at Maxwell's equations again. This time, we will include the sources. Using Equations N.1 and N.2,

$$\nabla \times \mathbf{H} = \epsilon\frac{\partial \mathcal{E}}{\partial t} + \mathbf{J} \;\therefore\; \nabla \times (\nabla \times \mathbf{A}) = -\mu\epsilon\frac{\partial^2 \mathbf{A}}{\partial t^2} - \mu\epsilon\frac{\partial}{\partial t}\nabla\psi + \mu\mathbf{J},$$

and

$$\nabla \cdot \epsilon\mathcal{E} = \rho \;\therefore\; \nabla \cdot \left(\epsilon\frac{\partial \mathbf{A}}{\partial t} + \epsilon\nabla\psi\right) = -\rho. \tag{N.6}$$

Since

$$\nabla \times (\nabla \times \mathbf{A}) = \nabla(\nabla \cdot \mathbf{A}) - \nabla^2\mathbf{A}, \tag{N.7}$$

Equations N.6 are now reducible to

$$\begin{aligned}\nabla^2\mathbf{A} - \mu\epsilon\frac{\partial^2 \mathbf{A}}{\partial t^2} &= -\mu\mathbf{J} - \mu\nabla\epsilon\frac{\partial\psi}{\partial t}, \\ \frac{1}{\epsilon}\nabla \cdot (\epsilon\nabla\psi) - \mu\epsilon\frac{\partial^2\psi}{\partial t^2} &= -\frac{\rho}{\epsilon} - \frac{1}{\epsilon}\nabla\epsilon \cdot \frac{\partial \mathbf{A}}{\partial t}.\end{aligned} \tag{N.8}$$

The right-hand sides of these equations represent, respectively, the source and the coupling that exists between the scalar and vector potential. This coupling is what manifests as the coupling between electric and magnetic fields through curls and the rate of time change. If no source exists and the material is homogeneous, this is the same form as that for the wave equation derived earlier for the potentials. If the source terms exists, it is the non-homogeneous equation describing the rise or fall of the wave.

Complementing the classical mechanics comment on Lagrangian, here is an example whose solution, when attempted via the field form—as we do in most microwave engineering texts—becomes replete with long, complicated equations. This form, on the other hand, remains simple.

O
Analyticity, Kramers-Kronig and Hilbert transforms

THE KRAMERS-KRONIG RELATIONSHIP, in this text, appears in order to relate the real and imaginary parts of the susceptibility and dielectric function response in linear systems due to causality. We have remarked on the generality of such a relationship in causal linear systems. The dielectric and susceptibility functions are complex functions. An arbitrary complex function, in general, will have poles and zeros anywhere in the complex plane. But real response functions of linear systems, with causal evolution that the propagator approach develops, place restrictions that arise in their linear foundations. This appendix clarifies the meaning and implications of the mathematical and real world connections.

O.1 *Analytic function*

A FUNCTION IS ANALYTIC if it is smooth, that is, infinitely differentiable, and if its Taylor series converges to the function. Complex functions are said to be analytic on a region—its domain—if it is infinitely complex differentiable at every point in it.

Mathematicians prefer the term "holomorphic function." In S. Tiwari, "Nanoscale device physics: Science and engineering fundamentals," Electroscience 4, Oxford University Press, ISBN 978-0-198-75987-4 (2017), you will see the use of Schwarz-Christoffel mapping, which is particularly powerful in the device analysis of a particular subset of such functions: conjugate functions.

Take the function $f(z) = 1/(1-z)$. It is expandable about any finite z_n, using the Taylor series

$$f(z) = \sum_{l=0}^{\infty} \frac{(z-z_n)^l}{(1-z_n)^{l+1}}, \tag{O.1}$$

which converges in the domain defined by $|z - z_n| < |1 - z_n|$. A series $g(z) = \sum_{k=0}^{\infty} z^k$ converges to $f(z) = 1/(1-z)$ within the domain of $|z| < 1$. Now, take the region around $z_0 = i$:

$$h(z) = \sum_{l=0}^{\infty} \frac{(z-i)^l}{(1-i)^{l+1}}. \tag{O.2}$$

This one converges to the same $f(z)$, but in the domain $|z - i| < \sqrt{2}$. The series look quite different, but they have an overlapping domain. $h(z)$ is a continuation of $g(z)$, and both are elements of the analytic function $f(z)$. The reason for approaching through domains is so that one may work with singularities and branch points.

Singularities are where the function blows up. Branches are where a multivalued function has a discontinuity. Such functions can be made single-valued through branch cuts.

O.2 *Cauchy integration and residue*

TAKE THE EXAMPLE ILLUSTRATED in Figure O.1, where the function $f(z)$ is analytic in region $R1$ but not in region $R2$, and one wants to integrate along a contour that surrounds both. We deform the contour, marked here as C and consisting of part 1, which surrounds the analytic domain $R1$, and the part 2, which is around $R2$, but excludes the contour from enclosing $R2$. Then, $\oint f(z)\, dz$ remains unchanged. A single localized nonanalytic region has been wrapped by this deformed contour consisting of 1 and 2 and their connecting paths.

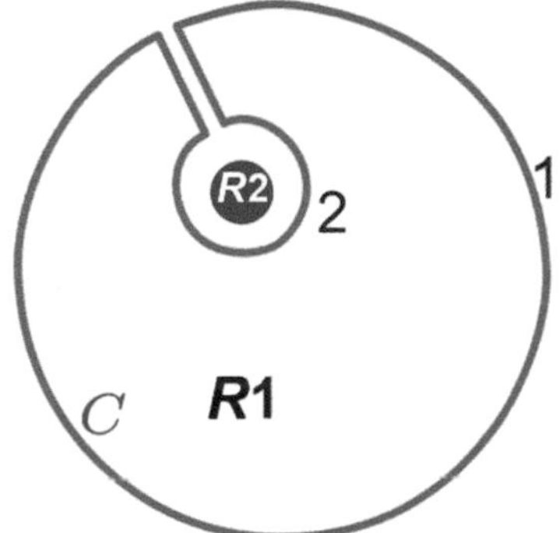

Figure O.1: A contour constructed so that it surrounds an analytic region and encloses but excludes a nonanalytic region. $R1$ is analytic, while $R2$ is not.

Take the example of a singularity at z_0. In the integral $\oint [f(s)/(s - z)]\, ds$, if we choose this deformed contour, $f(s)$ is analytic throughout inside the deformed contour. If the integrand is singular at $s = z$ and z is outside C, then this integral vanishes since it forms a complete loop and the integrand is analytic within the enclosed region. If z were to be within a contour S, we reduce S to a minute circle surrounding z such that $s - z = r \exp i\theta$ in circular coordinates. $ds = ir \exp(i\theta)\, d\theta$. With this, the surrounding contour takes on the value of

$$\lim_{r\to 0} \oint_S \frac{f(s)}{s - z}\, ds \approx \lim_{r\to 0} f(z) \oint_S \frac{ir \exp i\theta}{r \exp i\theta}\, d\theta$$
$$= 2\pi i f(z). \qquad (O.3)$$

The direction of the contour determines whether value is positive or negative. This relationship is for a counterclockwise contour, with θ increasing along the path of the integral.

The *residue* of the function $f(z)$ around a point z_0 is defined by

$$\text{Res}_{z_0} f(z) = \frac{1}{2\pi i} \oint_S f(z)\, dz, \qquad (O.4)$$

where S is a counterclockwise, simple closed contour that is small enough to avoid any other poles of $f(z)$. A counterclockwise path that winds once suffices—absent the poles—by Equation O.3. This derivation gives us the Cauchy integral formula.

Cauchy integral formula: If $f(z)$ is analytic at all points z on a simple contour and within it, then

$$f(z) = \frac{1}{2\pi i} \oint_C \frac{f(s)}{s - z}\, ds \qquad \text{(O.5)}$$

for any point z in the interior.

What a remarkable result! A value of an analytic function can be determined by integrating over a surrounding contour written with this choice of the integrand. This is a two-dimensional form of Gauss' theorem.

An analytic function's value at any point in the interior of a contour is determined by its values on this surrounding contour. The analytic function can now be seen to have a constraint on it that the Cauchy integral formula prescribes.

Since analytic functions are differentiable, the nth derivative of an analytic function is

$$\frac{d^n}{dz^n} f(z) = \frac{n!}{2\pi i} \oint_C \frac{f(s)}{(s - z)^{n+1}}\, ds. \qquad \text{(O.6)}$$

O.3 *Cauchy principal value*

When a singularity exists by itself in a range of integration, then approaching the singularity symmetrically allows one to calculate what is designated as the Cauchy principal value of the integral. Thus, if the function $f(z)$ has a singularity at b inside the end point of the integral between the limits a and c , then

$$\mathscr{P} \int_a^c f(x)\, dx = \lim_{\epsilon \to 0} \left[\int_a^{b-\epsilon} f(x)\, dx + \int_{b+\epsilon}^c f(x)\, dx \right] \qquad \text{(O.7)}$$

is the principal value.

As an example, $\mathscr{P} \int_{-a}^{+a} dz/z = 0$. If there exist more than one singularity, one can treat each one of them symmetrically as an extension of the single singularity. The method works for a single singularity, such as a pole, since the divergent contributions on either side of the precise singularity point cancel out.

The function $f(x) = \cos(kx)/(a^2 - x^2)$ has two poles: one at $x = -a$ and one at $x = +a$. We wish to find the principal value of the integral $\mathscr{P} \int_{-\infty}^{\infty} f(x)\, dx$:

$$\mathscr{P} \int_{-\infty}^{\infty} \frac{\cos(kx)}{a^2 - x^2}\, dx = \Re \left[\mathscr{P} \int_{-\infty}^{\infty} \frac{\exp(ikx)}{a^2 - x^2}\, dx \right]. \qquad \text{(O.8)}$$

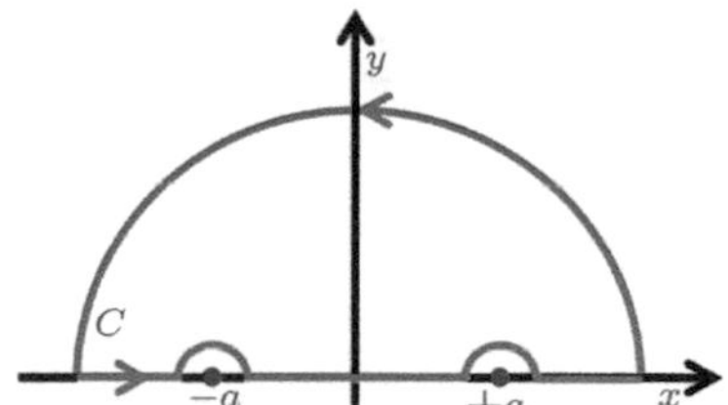

Figure O.2: Poles exist at $x = -a$ and $x = +a$, which the contour as drawn avoids in order to calculate an integral's principal value while having the integral path asymptotically pass through them.

We choose a contour as shown in Figure O.2—a great semicircle and two asymptotically reducing small semicircles skirting the poles. These small semicircles contribute $-i\pi$ times the sum of the two residues. The negative sign is there because the semicircles are traversed in the $-\theta$ direction. For the contour integral,

$$
\begin{aligned}
0 &= \oint_C \frac{\exp(ikz)}{a^2 - z^2}\, dz \\
&= \mathscr{P} \int_{-\infty}^{\infty} \frac{\exp(ikx)}{a^2 - x^2}\, dx \\
&\quad - i\pi \left[\frac{\exp(ika)}{-2a} + \frac{\exp(-ika)}{-2a} \right] \\
\therefore \quad \mathscr{P} \int_{-\infty}^{\infty} \frac{\exp(ikx)}{a^2 - x^2}\, dx &= \frac{\pi}{a} \sin ka \\
\therefore \quad \mathscr{P} \int_{-\infty}^{\infty} \frac{\cos(kx)}{a^2 - x^2}\, dx &= \frac{\pi}{a} \sin ka. \qquad (O.9)
\end{aligned}
$$

O.4 Kramers-Kronig relations as Hilbert transforms

TRANSFORMS PROVIDE A CRITICAL TOOL to problem-solving and understanding, since transformations in a "reciprocal" space often give a new transformed perspective, and therefore insight. Understanding semiconductors without Fourier transform—in bandstructure, transitions or other places—would be an immensely difficult task. While one can have infinite different transforms, among integral transforms that we employ so often, including in Green's functions techniques, a few have a very special place in the sciences. The important select few are listed in the glossary. Here, we establish the Kramers-Kronig relationship as an example of the Hilbert transform for analytic functions, using the dielectric response function—a physical and real observable—as the model.

Apply the Cauchy integral to $\epsilon_r - 1$. This is susceptibility, which vanishes at $\omega \to \infty$:

$$
\epsilon_r(\omega) = \frac{1}{2\pi i} \oint_C \frac{\epsilon_r(\omega_1) - 1}{\omega_1 - \omega}\, d\omega_1 \qquad (O.10)
$$

The integrand is singular at a point on the real axis. To evaluate, we employ the subterfuge of $\omega = \lim_{\delta \to 0^+} \omega + i\delta$. So, in the complex ω_1 plane, C is a contour in the upper half that encloses ω in it by making a minute single-pole-enclosing excursion out of it. This is shown in Figure O.3.

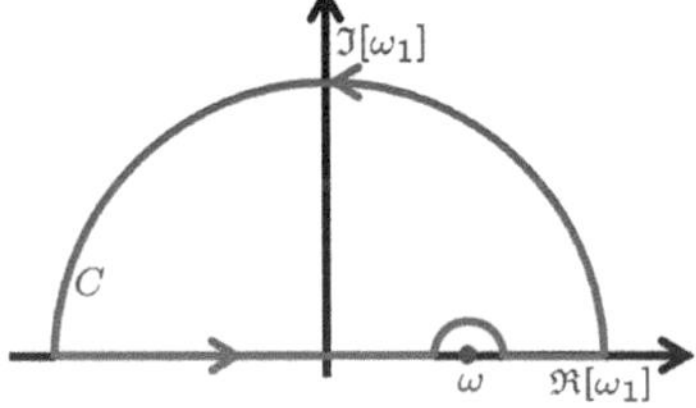

Figure O.3: A contour constructed so that the ω_1 space surrounds ω singularity on the real axis.

The integrand decreases with large ω_1, so the contribution of the major circle vanishes in the limits, and what remains is the contribution along the real axis, with the small excursion in the negative half of the complex plane around the singularity. So,

$$\begin{aligned}\oint_C \frac{\epsilon_r(\omega_1)-1}{\omega_1-\omega}\,d\omega_1 &= \lim_{\delta\to 0^+}\left[\int_{-\infty}^{\omega-\delta}\frac{\epsilon_r(\omega_1)-1}{\omega_1-\omega}\,d\omega_1\right.\\ &\quad+\int_{-\pi}^{\pi}\frac{\epsilon_r(\omega+\delta\exp i\theta)-1}{\delta\exp i\theta}\,i\delta\exp(i\theta)\,d\theta\\ &\quad\left.+\int_{\omega+\delta}^{\infty}\frac{\epsilon_r(\omega_1)-1)}{\omega_1-\omega}\,d\omega_1\right]\\ &= \mathscr{P}\int_{\infty}^{\infty}\frac{\epsilon_r(\omega_1)-1)}{\omega_1-\omega}\,d\omega_1+i\pi\left[\epsilon_r(\omega)-1\right],\\ \therefore\quad \epsilon_r(\omega) &= 1+\frac{1}{i\pi}\mathscr{P}\int_{-\infty}^{\infty}\frac{\epsilon_r(\omega_1)-1}{\omega_1-\omega}\,d\omega_1. \qquad (\text{O.11})\end{aligned}$$

We have arrived at one form of the Kramers-Kronig relationship via Cauchy integration of an analytic function. A useful corollary to Equation O.11 is the limit operator formula

$$\lim_{\delta\to 0^+}\frac{1}{\omega_1-\omega-i\delta}=\frac{\hat{\mathscr{P}}}{\omega_1-\omega}+i\pi\delta(\omega_1-\omega), \qquad (\text{O.12})$$

where $\hat{\mathscr{P}}$ is the principal operator. As an operator on an integral along the real axis, it determines its real value. The second term determines the pole's contribution, which is half of that of a full enclosing of the pole. From Equation O.11, the real and imaginary relationships follow. Since it is real, as are displacement and electric field, a real Green's function can be written, as we do in Chapter 14. The real nature of Green's function, being equal to its conjugate, means that $\epsilon_r(-\omega)=\epsilon_r^*(\omega^*)$, from which $\Re[\epsilon_r(\omega)]$ and $\Im[\epsilon_r(\omega)]$ follow. This is an alternate proof of the Kramers-Kronig relationships

This last term is as if that pole were half in and half out.

$$\begin{aligned}\Re[\epsilon_r(\omega)] &= 1+\frac{2}{\pi}\mathscr{P}\int_0^{\infty}\frac{\omega_1\Im[\epsilon_r(\omega)]}{\omega_1^2-\omega^2}\,d\omega_1, \text{ and}\\ \Im[\epsilon_r(\omega)] &= -\frac{2\omega}{\pi}\mathscr{P}\int_0^{\infty}\frac{\Re[\epsilon_r(\omega)]-1}{\omega_1^2-\omega^2}\,d\omega_1. \qquad (\text{O.13})\end{aligned}$$

Hilbert transforms are a clever way of taking an analytic function that decays rapidly at $\mathbf{z}\to\infty$ and decomposing it into real and imaginary components on the real axis. And these will be connected to each other. Cosine and sine functions comprise a Hilbert transform pair. So, a complex function of a complex variable, and the complex function for real variables, can be integrated together through the properties of the Hilbert transforms. To see that the Kramers-Kronig relations are also examples of Hilbert transform, we illustrate this more generally for a complex function $f(z)$ that is analytic in the upper half plane and on real axis (x, $\Im[z]=0$). We choose the contour as the major semicircle and along the real axis. The Cauchy integral formula gives

$$\begin{aligned} f(z) &= \frac{1}{2\pi i}\oint_C \frac{f(s)}{s-z}\,ds \\ &= \frac{1}{2\pi i}\int_{-\infty}^{\infty} \frac{f(s)}{s-z}\,ds + \lim_{R\to\infty}\frac{R}{2\pi}\int_0^{\pi} \frac{f(R\exp i\theta)}{R\exp(i\theta)-x}\,d\theta \\ &= \frac{1}{2\pi i}\int_{-\infty}^{\infty} \frac{f(s)}{s-z}\,ds. \end{aligned} \tag{O.14}$$

The second term, along the greater circle's perimeter, vanishes because of the choice of vanishing $f(z)$ at large R. We have also employed x to indicate that this specific integral is along the real axis. $f(z) = \Re[f(z)] + \Im[f(z)]$, so

$$\begin{aligned} \Re[f(z)] &= \frac{1}{2\pi}\int_{-\infty}^{\infty} \frac{\Im[f(s)]}{s-x}\,ds, \text{ and} \\ \Im[f(z)] &= -\frac{1}{2\pi}\int_{-\infty}^{\infty} \frac{\Re[f(s)]}{s-x}\,ds. \end{aligned} \tag{O.15}$$

Now, we tackle the function on real axis, this time by deforming the contour as an infinitesimally small semicircular detour along the great circle:

$$\begin{aligned} f(x) &= \frac{1}{2\pi i}\oint_C \frac{f(s)}{s-x}\,ds \\ &= \frac{1}{2\pi i}\mathscr{P}\int_{-\infty}^{\infty} \frac{f(s)}{s-x}\,ds + \lim_{\epsilon\to 0}\frac{1}{2\pi}\int_{-\pi}^{0} f(x+\epsilon\exp i\theta)\,d\theta \\ &\quad + \lim_{R\to\infty}\frac{R}{2\pi}\int_0^{\pi} \frac{f(R\exp i\theta)}{R\exp(i\theta)-x}\,d\theta \\ &= \frac{1}{2\pi i}\mathscr{P}\int_{-\infty}^{\infty} \frac{f(s)}{s-x}\,ds + \frac{1}{2}f(x) \\ &= \frac{1}{\pi i}\mathscr{P}\int_{-\infty}^{\infty} \frac{f(s)}{s-x}\,ds. \end{aligned} \tag{O.16}$$

Again, as before

$$\begin{aligned} \Re[f(x)] &= \frac{1}{\pi}\int_{-\infty}^{\infty} \frac{\Im[f(s)]}{s-x}\,ds, \text{ and} \\ \Im[f(x)] &= -\frac{1}{\pi}\int_{-\infty}^{\infty} \frac{\Re[f(s)]}{s-x}\,ds. \end{aligned} \tag{O.17}$$

The difference between Equations O.15 and O.17 is that the former has its pole inside the contour, while the latter—for z real, that is, x—the singularity is on the contour—either half in or half out.

If one considers the real and imaginary components on the real axis to be two real functions of a real variable, one obtains the Hilbert transform pair—like other transforms, summarized in the glossary—and these are of the form

$$\psi(x) = \frac{1}{\pi}\mathscr{P}\int_{-\infty}^{\infty} \frac{\Psi(s)}{s-x}\,ds, \text{ and } \Psi(s) = \frac{1}{\pi}\mathscr{P}\int_{-\infty}^{\infty} \frac{\psi(x)}{x-s}\,dx. \tag{O.18}$$

If an analytic function decays rapidly as $z \rightarrow \infty$, decomposing it into real and imaginary components on the real axis lets us employ the Hilbert transformation. The $\cos x$ and $\sin x$ functions are Hilbert transform pairs arising out of the $\exp iz$ complex analytic function.

Note that $\exp iz$ satisfies the convergence criterion in the upper half of the complex plane.

Hilbert transforms become useful for analysis of observables, where causality forces analyticity and convergences. Kramers-Kronig relationships can then be viewed as special examples of Hilbert transforms.

P
Particle velocities

DIMENSIONALITY, STATE OCCUPATION, FERMI ENERGIES AND WAVEVECTORS and their implications for the velocity of the particle are all related. Appendix H discussed dimensionality's consequence for density of states in the limits where the carriers sample an environment of many unit cells. In this appendix, we gauge the implications of this in particle velocities when the state occupation in thermal equilibrium is known through the Fermi energy and associated Fermi wavevector.

Changes in the dimensionality of the system affect the available state distributions and the processes that influence the changing of the states. States, through the E-$\mathbf{k}$ relationship, also determine the velocity associated with the carrier in that state. So, the dimensionality, the scattering processes, and the properties of the state determine the velocity of an individual carrier as well as of an ensemble of carriers. Appendix H discussed the density of states, with Equations H.7 through H.9 summarizing them for the simplest cases of a parabolic central valley, that is, with $E = \hbar^2 k^2/2m^*$—a single isotropic and parabolic well minimum with an effective mass m^*.

What effective mass to use is not so straightforward. The relationships here are for the simplest of textbook examples—quite good for useful compound semiconductors with low effective masses—with their $3D$, $2D$ and $1D$ extensions. But when multiple valleys exist, bands are anisotropic and additional symmetries are broken, then both density of states and conduction are subject to different constraints, and the masses differ. So, the problem gets more complex when dimensionality is reduced. The states of the six silicon ellipsoids and the electrons occupying them will behave quite differently both for occupation and for transport in response to forces and constraints of different directions. See S. Tiwari, "Nanoscale device physics: Science and engineering fundamentals," Electroscience 4, Oxford University Press, ISBN 978-0-198-75987-4 (2017), for a detailed discussion.

If degenerate conditions exist, at absolute zero temperature, one can relate the carrier density by integrating the occupied state density to the Fermi energy E_F. With ν representing dimensionality in a general notation, since $n_{\nu D} = \int_0^{E_F} \mathscr{G}_{\nu D}\, dE$, we have

$$
\begin{aligned}
3D: n_{3D} &= \frac{1}{3\pi^2}\left(\frac{2m^* E_F}{\hbar^2}\right)^{3/2} & \therefore E_F &= \frac{\hbar^2}{2m^*}\left(3\pi^2 n_{3D}\right)^{2/3}, \\
2D: n_{2D} &= \frac{m^* E_F}{\pi\hbar^2} & \therefore E_F &= \frac{\pi\hbar^2}{m^*} n_{2D}, \text{ and} \\
1D: n_{1D} &= \left(\frac{8m^* E_F}{\pi^2\hbar^2}\right)^{1/2} & \therefore E_F &= \frac{\pi^2\hbar^2}{8m^*} n_{1D}^2,
\end{aligned}
\tag{P.1}
$$

where n_{3D}, n_{2D} and n_{1D} are, respectively, the $3D$, $2D$ and $1D$ densities of the carriers; that is, the electrons that occupy the states. These are in normalized volume, area and length densities. These equations also connect the Fermi wavevector and the Fermi velocity associated with the Fermi energy. Under degenerate absolute zero temperature conditions,

$$3D : k_F = (3\pi^2 n_{3D})^{1/3} \quad \therefore v_F = \frac{\hbar}{m^*}(3\pi^2 n_{3D})^{1/3},$$
$$2D : k_F = \sqrt{2\pi n_{2D}} \quad \therefore v_F = \frac{\hbar}{m^*}\sqrt{2\pi n_{2D}}, \text{ and}$$
$$1D : k_F = \frac{1}{2}\pi n_{1D} \quad \therefore v_F = \frac{\hbar}{m^*}\frac{1}{2}\pi n_{1D}. \qquad \text{(P.2)}$$

This relationship is also a direct statement of the uniform spacing of wavevectors in the reciprocal space that the electrons may occupy in this constant mass isotropic approximation. The velocity of electrons at Fermi energy then is directly related to these energies or wavevectors.

It is instructive to understand how this dimensionality, the carrier population, the velocities and the energies relate for semiconductors. Table P.1 attempts to provide a feel for this through examples and relationships. The group velocity represents the constraint that the bandstructure—the behavior of the electron in the crystal—speaks to. In semiconductors, group velocities are all of the order

Characteristic	Relationship	Dimensionality	Magnitude with conditions
Group velocity, v_g	$(1/\hbar)\nabla_{\mathbf{k}}E$	3D, 2D, 1D	Si maximum: $\sim 8\times10^7$ cm/s
			$GaAs$ maximum: $\sim 8\times10^7$ cm/s
			$Ga_{0.47}In_{0.53}As$ maximum: $\sim 8\times10^7$ cm/s
			$InAs$ maximum: $\sim 10^8$ cm/s
			Graphene: $\sim 10^8$ cm/s
Kinetic energy, v_{KE}	$(2E/m^*)^{1/2}$	3D	Si: 8×10^7 cm/s, for $E = 0.5$ eV and $m^* = 0.26m_0$
Optical phonon, $\langle v\rangle_{op}$	$(2\hbar\omega_{op}/m^*)^{1/2}$	3D, 2D, 1D	Si: 2.8×10^7 cm/s, with $\hbar\omega_{op}\approx 60$ meV and $m^* = 0.26m_0$
			$GaAs$: 4.3×10^7 cm/s, with $\hbar\omega_{op} = 35.4$ meV and $m^* = 0.067m_0$
			$Ga_{0.47}In_{0.53}As$: 5.4×10^7 cm/s, with $\hbar\omega_{op} = 34$ meV and $m^* = 0.041m_0$
			$InAs$: 6.7×10^7 cm/s, with $\hbar\omega_{op} = 29.6$ meV and $m^* = 0.023m_0$
Thermal velocity, $\langle v\rangle_\theta$	$(8k_BT/\pi m^*)^{1/2}$	3D	Si: 1.96×10^7 cm/s, at $T = 300$ K and $m^* = 0.36m_0$
			$GaAs$: 4.16×10^7 cm/s, at $T = 300$ K and $m^* = 0.067m_0$
	$(\pi k_BT/2m^*)^{1/2}$	2D	Si: 1.54×10^7 cm/s, at $T = 300$ K and $m^* = 0.26m_0$
			$GaAs$: 3.27×10^7 cm/s, at $T = 300$ K and $m^* = 0.067m_0$
	$(2k_BT/\pi m^*)^{1/2}$	1D	Si: 0.98×10^7 cm/s, at $T = 300$ K and $m^* = 0.26m_0$
			$GaAs$: 2.08×10^7 cm/s, at $T = 300$ K and $m^* = 0.067m_0$

Table P.1: A table of some approximate velocity magnitudes of relevance to carrier transport. Group velocity is the bandstructure-defined constraint for any state of the particle. Kinetic-energy-defined velocity is the velocity at any given kinetic energy of a particle. Optical-phonon-limited velocity is the ensemble average velocity that the particles would have if they were continuously losing all the excess energy by optical phonon emission. Thermal velocity is the ensemble root mean square average in non-degenerate conditions. For degenerate conditions, the velocity of carriers at Fermi energy, that is the Fermi velocity v_F, will matter. Dimensionality restricts the degrees of freedom of movement, and hence some of these velocities change through the constraints from the states available allowing movement. Note, for example, the decrease in thermal velocity with dimensionality.

of 10^8 cm/s or less. The kinetic-energy-limited velocity is meant to represent a velocity that the carrier would have should it reach that kinetic energy. This table indicates that, assuming that effective mass for conductivity in silicon doesn't change significantly—a major assumption—it will be of the order of magnitude of maximum group velocity. Carriers have to be this far up in the band for this maximum velocity to be possible. Well before that, with transport undergoing scattering, carriers will lose energy through phonon emission—a significant process—among many others. This places a constraint on how much kinetic energy the electron can realistically pick up. For the motion of the electrons arising in their thermal energy under non-degenerate conditions, with carriers occupying mostly states at the bottom of the band, since the carriers only sample the space to which they are restricted, this root mean square velocity changes with dimensionality. Fewer degrees of freedom result in a reduced ability to move around. When going from $3D$ to $2D$ there are fewer higher energy states to be occupied under the Maxwell-Boltzmann distribution constraints. This argument of reduced access of space, due to dimensionality constraints, is also largely true for the other examples in this table.

Structures where the carriers are injected such as by tunneling into the semiconductor, at this energy, make it possible to effectively achieve a nonequilibrium distribution that dominantly occupies higher energies and streams velocities over the scattering length scales. The carriers here don't have to pick up this energy through transport in a field.

Glossary

A SYMBOL GENERATED BY using a tilde sign above a symbol, for example, $\tilde{a}$ from a, is used to signify, explicitly, a complex time-varying quantity. The real part of this has a sinusoidal time variation. The phasor, or the amplitude of this time-varying component, is denoted by using the hat sign on the symbol, for example, $\hat{a}$ for a. This notation is used in the context of small-signal variation. An exception to this nomenclature is the use of the hat symbol to denote a unit normal vector, for example, $\hat{n}$, to denote the unit normal vector perpendicular to a surface. A hat is also employed, together with calligraphic, mathematics-specific or normal font, to denote a quantum-mechanical operator. A lowercase subscript to an uppercase letter denotes a quantity which may have both a static and a time-varying component. An uppercase subscript or an overline represents quasistatic quantities. Any other exceptions have been pointed out in context. This list defines the most frequently used symbols. Système international d'unités—SI—units are employed in the text. Any exceptions either are pointed out in context or follow from dimensionality.

Symbol	Symbol definition	Unit
$\uparrow$	Spin/polarization up	—
$\downarrow$	Spin/polarization down	—
$1/4\pi\epsilon_0$	Prefactor of SI units	$8.99 \times 10^9\ V \cdot m/A \cdot s$
α	Absorption coefficient	$1/m$
α_n	Electron initiated impact ionization rate	$1/m$
α_p	Hole initiated impact ionization rate	$1/m$
$\alpha(\mathbf{r}, \mathbf{r}')$	Correlation parameter	—
a_B	Bohr radius ($4\pi\hbar^2/m_0e^2$)	$0.529 \times 10^{-10}\ m$
$\mathbf{A}$	Vector potential	$V \cdot s \cdot m^{-1}$
A	Absorption	—
β	$1/k_BT$	$1/eV$
B, $\mathbf{B}$	Magnetic induction ($\mu_r\mu_0\mathbf{H}$, $\mu\mathbf{H}$)	$V \cdot s/m^2$, that is, T
$\mathcal{B}$	Bimolecular recombination coefficient	m^3/s
c	Speed of light in free space ($1/(\mu_0\epsilon_0)^{1/2}$)	$2.998 \times 10^8\ m/s$

C_E	Ettinghausen coefficient	$K \cdot m/T \cdot A$
$\mathbf{C}_h$	Chiral vector	m
C_N	Nernst coefficient	$V/T \cdot K$
C_{RL}	Righi-Leduc coefficient	$1/K$
Δ_j	Huang-Rhys factor	—
Δ_{so}	Split-off energy	J, eV
D, $\mathbf{D}$	Displacement ($\epsilon\epsilon_0\mathcal{E}$)	$A \cdot s/m^2$
$\mathcal{D}$	Diffusion coefficient	m^2/s
ϵ	Permittivity of material	$A \cdot s/V \cdot m$
ϵ_0	Permittivity of free space ($1/\mu_0 c^2$)	$8.854 \times 10^{-12}\ A \cdot s/V \cdot m$
ϵ_r	Dielectric (relative) constant	—
ε	Strain	—
e	Absolute electron charge	$1.602 \times 10^{-19}\ A \cdot s$ ($\equiv$ *Coulomb*)
e^*	Effective Born charge	C
e/m_0	Electron charge to mass (ω/B)	$1.759 \times 10^{11}\ rad/s \cdot T$
$\mathcal{E}$, $\boldsymbol{\mathcal{E}}$	Electric field	V/m
E	Energy	eV
E_c	Conduction bandedge	eV
E_v	Valence bandedge	eV
E_g	Bandgap	eV
E_F	Fermi energy	eV
E_{qF}	Quasi-Fermi energy	eV
E_R	Rydberg energy ($m_0 e^4/2\hbar^2(4\pi\epsilon_0)^2$)	eV
E_x	Averaged exchange energy	eV
E_{xn}	Exciton energy	eV
ε	Error	Various
ε	Strain	m/m
f, ν	Frequency	s^{-1}, Hz
f_{osc}	Oscillator strength	—
F	Force	N
$\mathcal{F}$	Helmholtz free energy	J, eV
$\boldsymbol{\mathcal{F}}$	Electrothermal field	V/m
$\mathscr{F}_\nu$	Fermi integral of order ν	—
$\mathcal{F}$	Slowly varying force	N
$\mathfrak{F}$	Rapidly varying force	N
Φ	Photon flux	$1/m^2 \cdot s$
φ_B	Barrier height	eV
φ_w	Workfunction	eV
γ	Free energy per unit area	J/m^2, eV/m^2
g	Correlation function	—
g	g factor	—
g_{PB}	Poisson-Boltzmann correlation function	—
g_q	Quantum conductance ($2e^2/h$)	$\sim 80\ \mu S$
g_s	Spin degeneracy	2

$\mathcal{G}$	Gibbs free energy	J, eV
$\mathcal{G}$	Generation rate	$m^{-3} \cdot s^{-1}$
$\mathscr{G}_{3D}(E)$	Three-dimensional density of states	$m^{-3}eV^{-1}$
$\mathscr{G}_{2D}(E)$	Two-dimensional density of states	$m^{-2}eV^{-1}$
$\mathscr{G}_{1D}(E)$	One-dimensional density of states	$m^{-1}eV^{-1}$
Γ	Electron leakage rate from dot	$1/s$
η	Normalized energy	—
h	Planck's constant	$6.626 \times 10^{-34}\ kg \cdot m^2/s$ or $4.136 \times 10^{-15}\ eV \cdot s$
$\hbar$	Reduced Planck's constant	$1.055 \times 10^{-34}\ kg \cdot m^2/s$ or $6.582 \times 10^{-16}\ eV \cdot s$
$\hat{}$	Pseudo Hamiltonian	J, eV
H	Average information content ≡ uncertainty	b
H	Boltzmann H-factor	—
H, $\mathbf{H}$	Magnetic field ($\mathbf{m}/V$)	A/m
$\mathcal{H}$	Enthalpy	J, eV
$\hat{\mathscr{H}}$	Hamiltonian operator	J, eV
$\hat{\mathfrak{H}}$	Pseudo-Hamiltonian operator	J, eV
I	Information content	b
I	Fisher information	b
J, $\mathbf{J}$	Current density	A/m^2
κ	Anisotropy constant	$J/m^3, eV/m^3$
κ	Dielectric constant	—
κ	Thermal conductivity	$W/m \cdot K$
$\mathbf{k}$	Wavevector	m^{-1}
$\mathbf{K}$	Reciprocal space translation operator	m^{-1}
k_B	Boltzmann constant	$1.38 \times 10^{-23}\ J/K$
$\mathbf{k}_s$	Spring constant	N/m
$\mathbf{k}_F$	Wavevector at Fermi energy	cm^{-1}
$\mathbf{K}$	Reciprocal space translation operator	m
KL	Kullback-Leibler distance/entropy	Various
χ	Susceptibility	—
λ_{deB}	de Broglie wavelength	m
λ_{DH}, λ_D	Debye-Hückle or Debye screening length	m
$\lambda_{\mathbf{k}}$	Mean free path	m
λ_{scr}	Screening length	m
λ_{TF}	Thomas-Fermi screening length	m
λ_w	Energy relaxation length	m
$\mathcal{L}$	Diffusion length	m
$\hat{\mathscr{L}}$	Lagrange function	—
$\mathbf{m}$	Magnetic moment	$V \cdot s \cdot m$
$\mathbf{M}$	Magnetization ($\mathbf{m}/V$)	A/m
m_0	Free electron mass	$9.1 \times 10^{-31}\ kg$
m_0c^2	Electron rest energy	$0.819 \times 10^{-13}\ V \cdot A/s = 0.5111\ MeV$
m^*	Effective mass	kg
m_l^*	Longitudinal effective mass	kg

m_t^*	Transverse effective mass	kg
M	Mass (usually atom)	kg
μ	Mobility	$m^2/V \cdot s$
μ	Permeability	$V \cdot s/A \cdot m$
μ	Chemical potential	J, eV
μ_0	Permeability of free space ($1/\epsilon_0 c^2$)	$4\pi \times 10^{-7}\ V \cdot s/A \cdot m$
μ_r	Relative permeability of material	—
μ_B	Bohr magneton ($e\hbar\mu_0/2m_0$)	$1.165 \times 10^{-29}\ V \cdot m \cdot s$
ν	Frequency	$1/s$
ν	Dimensionality	—
ν	Poisson's ratio	—
n	Electron concentration	m^{-3}
$\bar{n}$	Index of refraction	—
$n^{\underline{c}}$	Index of refraction (complex)	—
$n^{\underline{r}}$	Index of refraction (real part)	—
$n^{\underline{i}}$	Index of refraction (imaginary part)	—
$\mathcal{N}_c$	Effective density of states	m^{-3}
$\mathcal{O}_{jk}$	Overlap matrix element	—
Π	Peltier coefficient	V
p	Hole concentration	m^{-3}
p, $\mathbf{p}$	Momentum	$kg \cdot m/s$
p	Generalized momentum coordinate	—
$\mathbf{p}$	Electric dipole moment	$A \cdot s \cdot m$
P_θ	Heat generated	W/m^3
$\mathbf{P}$	Polarization	C/m^2
$\mathfrak{p}$	Probability	—
$\mathfrak{P}$	Probability distribution	—
$\hat{\mathscr{P}}$	Projection operator	—
$\mathcal{P}$	Thermoelectric power	V/K
q	Fundamental charge ($-e$)	$-1.602 \times 10^{-19}\ A \cdot s$
q	Generalized position coordinate	—
Q	Heat energy	J
$\mathscr{Q}$	Heat energy flux	$J/cm \cdot s$
$\mathscr{Q}$	Heat energy flux density	$J/m^2 \cdot s$
$\mathcal{Q}$	Quality factor	—
$\mathbf{r}$, $\mathbf{R}$	Position coordinate	m
r_e	Electron's effective space scale	m
r_H	Hall factor	—
R_H	Hall constant	m^3/C
ρ	Resistivity	$\Omega \cdot cm$
ρ	Volume charge density	C/m^3
ρ	Density matrix	—
ϱ	Reflectivity	—
$\mathcal{R}$	Recombination rate	$m^{-3} \cdot s^{-1}$

σ	Variance	—
σ	Conductivity	S/cm
σ	Stress	N/m^2
s	Spin quantum number	$\pm 1/2$
S	Entropy	J/K
S	Scattering rate	$1/s$
S	Signal energy	J
S	Seebeck coefficient	V/K
S	Surface recombination velocity	m/s
$\mathbf{S}$	Poynting vector	$J/m^2 \cdot s$, $eV/m^2 \cdot s$
$S(\omega)$	Power spectral density	*unit square/s*
S_{Ar}	Auger scattering rate	$1/s$
S_{Ar}	Impact ionization scattering rate	$1/s$
τ	Scattering time	s
$\tau_{\mathbf{k}}$	Momentum relaxation time	s
τ_n	Electron lifetime	s
τ_p	Hole lifetime	s
τ_r	Radiative lifetime	s
τ_w	Energy relaxation time	s
t	Time parameter	s
T	Temperature	K
T	Specific time	s
T	Torque	$N \cdot m$
T	Kinetic energy	J
$\mathbf{T}$	Real space translation operator	m
$\mathscr{T}$	Transmission coefficient	—
T_c	Critical temperature	K
T_C	Curie temperature	K
ϑ, $\boldsymbol{\vartheta}$	Data parameter	—
Υ	Thompson coefficient	V/K
u	displacement	m
U	Internal energy	J, eV
$U_{\mathbf{kk}'}$	Coulomb perturbation Fourier component	J, eV
u, $\mathbf{u}$	displacement	m
$\mathcal{U}$	Net recombination rate	$m^{-3} \cdot s^{-1}$
v	Velocity	m/s
v_d	Drift velocity	m/s
v_{sat}	Saturation velocity	m/s
v_F	Fermi velocity	m/s
v_g	Group velocity	m/s
v_θ	Thermal velocity	m/s
V	Potential	eV
V_H	Hartree potential	eV
V_{si}	Self-interaction potential	eV

V_x	Exchange correction term	eV
w	Kinetic energy of a particle	J
W	Kinetic energy density	J/m^3
W	Pseudopotential	J, eV
ω	Radial frequency	rad/s
ω_c	Cyclotron resonance frequency	rad/s
$\omega_{\mathbf{q}}$	Phonon radial frequency	rad/s
ω_{LO}	Longitudinal optical radial frequency	rad/s
ω_{op}	Optical radial frequency	rad/s
ω_p	Plasma radial frequency	rad/s
ω_{TO}	Transverse optical radial frequency	rad/s
ω_{so}	Surface optical radial frequency	rad/s
Ω	Thermodynamic potential	J
Ω	Volume	m^3
Ω	Frequency	s^{-1}
Ω_0	Volume of unit cell	m^3
$\Omega_{\mathbf{k}}$	Volume of reciprocal unit cell	m^{-3}
Ω	Possible configurations	—
ξ	Generalized electrothermochemical potential	—
Ξ_a	Acoustic deformation constant	J, eV
χ_c	Electron affinity (E_c reference)	J, eV
Ξ_d	Deformation parameter	$J/m, eV/m$
χ_v	Electron affinity (E_v reference)	J, eV
zT	Thermoelectric figure of merit	—
Z	Atomic number	—
Z^*e	Dressed ionic charge	C

Other units, in popular usage because of history, as well as because of the insight they give, can be understood through the following relationships provided here for reference.

Unit	**Conversion**
Oersted (Oe)	$= 10^3/4\pi\ A/m = 79.59\ A/m$
Tesla (T)	$= N/A \cdot m = V \cdot s/m^2 = kg/s^2A = 10^4\ Gauss$
Ohm (Ω)	$= V/A$
Coulomb (C)	$= A \cdot s$
Newton (N)	$= V \cdot A \cdot s/m = kg \cdot m/s^2$
Kilogram (kg)	$= V \cdot A \cdot s^3/m^2$
Farad (F)	$= A \cdot s/V$
Henry (H)	$= kg \cdot m^2/s^2A^2$
Joule (J)	$= N \cdot m = V \cdot A \cdot s = 10^7\ erg$
Watt (W)	$= V \cdot A = J/s$
eV	$= 1.602 \times 10^{-19}\ V \cdot A \cdot s$
eV/k_B	$= 1.1605 \times 10^4\ K$

eV/h	$= 2.418 \times 10^{14}\ Hz$
eV/hc	$= 8066\ cm^{-1} = 8066\ Kayser$
$h\nu(eV)$	$= 1249.852/\lambda$ in nm
μ_B/μ_0	$= 0.578 \times 10^{-4}\ eV/T$
$barn\ (b)$	$= 1 \times 10^{-28}\ m^2$
$deg(^\circ)$	$= \pi/180\ rad = 17.45\ mrad$
$arcmin$	$= 1/60^\circ = 290.0\ \mu rad$

SOME OF THE SYMBOLS employed, as well as their context, are listed here.

Symbol	Meaning	
$\oint$	Contour integral	
$\otimes$	Convolution	$f(t) = \int_{-\infty}^{\infty} g(t-t')h(t')\,dt'$

ACRONYMS have been employed sparingly in the text, but a few do slip in for compactness. These are listed here.

Acronym	Full form
BCC	Body-centered cubic
CMOS	Complementary metal oxide semiconductor
FCC	Face-centered cubic
HCP	Hexagonal close packed
HOMO	Highest occupied molecular orbital
HSR	Hall-Shockley-Read
IBM	International Business Machines Corporation
LO	Longitudinal optical
LUMO	Lowest unoccupied molecular orbital
pdf	Probability distribution function
SI	Système international d'unités
so	Surface optical
TE	Transverse electric
TEM	Transverse electromagnetic
TM	Transverse magnetic
TO	Transverse optical
USA	United States of America
WSJ	Wall Street Journal

SCALES, as a parameter that approximately guides the validity of a formalism, have been employed throughout the book. Particularly crucial in this have been those associated with length scales of an important phenomenon. This table is a representative list of those emphasized.

Scale parameter	Meaning
a_B	Bohr radius
a_B^*	Effective Bohr radius
λ	Wavelength
λ	Coulombic screening length
λ_{deB}	de Broglie wavelength
λ_{DH}, λ_D	Debye-Hückel or Debye screening length
λ_ϕ	Phase coherence length
$\lambda_{\mathbf{k}}$	Momentum relaxation length
λ_{mfp}	Mean free path $\equiv$ momentum relaxation length
λ_{scat}	Scattering length
λ_{scr}	Screening length
λ_{TF}	Thomas-Fermi screening length
$\mathcal{L}$	Diffusion length
$\tau_{\mathbf{k}}$	Momentum relaxation time
τ_w	Energy relaxation time

VECTOR IDENTITY RELATIONSHIPS used in the text are provided here for reference.

Identities:

$$\mathbf{A} \times \mathbf{B} \cdot \mathbf{C} = \mathbf{A} \cdot \mathbf{B} \times \mathbf{C},$$

$$\mathbf{A} \times (\mathbf{B} \times \mathbf{C}) = \mathbf{B}(\mathbf{A} \cdot \mathbf{C}) - \mathbf{C}(\mathbf{A} \cdot \mathbf{B}),$$

$$\nabla(f + g) = \nabla f + \nabla g,$$

$$\nabla \cdot (\mathbf{A} + \mathbf{B}) = \nabla \cdot \mathbf{A} + \nabla \cdot \mathbf{B},$$

$$\nabla \times (\mathbf{A} + \mathbf{B}) = \nabla \times \mathbf{A} + \nabla \times \mathbf{B},$$

$$\nabla(fg) = f\nabla g + g\nabla f,$$

$$\nabla \cdot (f\mathbf{A}) = \mathbf{A} \cdot \nabla f + f\nabla \cdot \mathbf{A},$$

$$\nabla \cdot (\mathbf{A} \times \mathbf{B}) = \mathbf{B} \cdot \nabla \times \mathbf{A} - \mathbf{A} \cdot \nabla \times \mathbf{B},$$

$$\nabla \cdot \nabla f = \nabla^2 f,$$

$$\nabla \cdot \nabla \times \mathbf{A} = 0,$$

$$\nabla \times \nabla f = 0,$$

$$\nabla \times (\nabla \times \mathbf{A}) = \nabla(\nabla \cdot \mathbf{A}) - \nabla^2 \mathbf{A},$$

$$(\nabla \times \mathbf{A}) \times \mathbf{A} = (\mathbf{A} \cdot \nabla)\mathbf{A} - \frac{1}{2}\nabla(\mathbf{A} \cdot \mathbf{A}),$$

$$\nabla(\mathbf{A} \cdot \mathbf{B}) = (\mathbf{A} \cdot \nabla)\mathbf{B} + (\mathbf{B} \cdot \nabla)\mathbf{A} + \mathbf{A} \times (\nabla \times \mathbf{B}) + \mathbf{B} \times (\nabla \times \mathbf{A}),$$

$$\nabla \times (f\mathbf{A}) = \nabla f \times \mathbf{A} + f\nabla \times \mathbf{A},$$

$$\nabla \times (\mathbf{A} \times \mathbf{B}) = \mathbf{A}(\nabla \cdot \mathbf{B}) - \mathbf{B}(\nabla \cdot \mathbf{A}) + (\mathbf{B} \cdot \nabla)\mathbf{A} - (\mathbf{A} \cdot \nabla)\mathbf{B}.$$

COMMON INTEGRAL TRANSFORMS, some of which have been used in the text, are provided here for reference. The asymmetric form of the Fourier transform, when employed, should be clear from the context.

Integral transforms

	Forward transform	Inverse transform
Fourier	$F(\omega) = \frac{1}{\sqrt{2\pi}} \int_{-\infty}^{\infty} \exp(i\omega t) f(t)\, dt$	$f(t) = \frac{1}{\sqrt{2\pi}} \int_{-\infty}^{\infty} \exp(-i\omega t) F(\omega)\, d\omega$
Bessel	$F_n(k) = \int_0^{\infty} x J_n(kx) f(x)\, dx$	$f(x) = \int_0^{\infty} k J_n(kx) F_n(k)\, dk$
Laplace	$F(s) = \int_0^{\infty} \exp(-st) f(t)\, dt$	$f(t) = \frac{1}{2\pi i} \int_{\gamma - i\infty}^{\gamma + i\infty} \exp(st) F(s)\, ds$
Mellin	$F(z) = \int_0^{\infty} t^{z-t} f(t)\, dt$	$f(t) = \frac{1}{2\pi i} \int_{-i\infty}^{i\infty} t^{-z} F(z)\, dz$
Hilbert	$F(z) = \frac{1}{\pi} \mathscr{P} \int_{-\infty}^{\infty} \frac{f(t)}{t-z}\, dt$	$f(t) = \frac{1}{\pi} \mathscr{P} \int_{-\infty}^{\infty} \frac{F(z)}{z-t}\, dz$

MAJOR EQUATIONAL RELATIONSHIPS IN SI UNITS employed in the text are provided here for reference.

Maxwell's equations:

$$\nabla \cdot \mathbf{D} = \rho,$$
$$\nabla \cdot \mathbf{B} = 0,$$
$$\nabla \times \boldsymbol{\mathcal{E}} = -\frac{\partial \mathbf{B}}{\partial t}, \quad \text{and}$$
$$\nabla \times \mathbf{H} = \mathbf{J} + \frac{\partial \mathbf{D}}{\partial t}.$$

Constitutive relationships of Maxwell equations:

$$\mathbf{D} = \epsilon_0 \boldsymbol{\mathcal{E}} + \mathbf{P} = \epsilon_0 \boldsymbol{\mathcal{E}} + \chi \epsilon_0 \boldsymbol{\mathcal{E}} = \epsilon_r \epsilon_0 \boldsymbol{\mathcal{E}} = \epsilon \boldsymbol{\mathcal{E}}, \quad \text{and}$$
$$\mathbf{B} = \mu_0 (\mathbf{H} + \mathbf{M}) = \mu_0 \mathbf{H} + \chi \mu_0 \mathbf{H} = \mu_r \mu_0 \mathbf{H} = \mu \mathbf{H}.$$

Gauss's theorem:

$$\lim_{\Omega \to 0} \frac{1}{\Omega} \int_S \mathbf{B} \cdot \hat{\mathbf{n}}\, d^2 r = \nabla \cdot \mathbf{B}, \quad \text{or}$$
$$\int_S \mathbf{B} \cdot \hat{\mathbf{n}}\, d^2 r = \int_\Omega \nabla \cdot \mathbf{B}\, d\Omega.$$

Stokes' theorem:

$$\lim_{S \to 0} \frac{1}{S} \oint_r \mathbf{H} \cdot d\mathbf{r} = \hat{\mathbf{n}} \cdot (\nabla \times \mathbf{H}), \quad \text{or}$$
$$\oint_r \mathbf{H} \cdot d\mathbf{r} = \int_S \hat{\mathbf{n}} \cdot (\nabla \times \mathbf{H})\, d^2 r.$$

Semi-classical conservation equations for electronics:

$$\nabla \cdot \mathbf{D} = \rho,$$
$$-\frac{1}{q}\nabla \cdot \mathbf{J}_n = \mathcal{U} = \mathcal{G} - \mathcal{R}, \quad \text{and}$$
$$\frac{1}{q}\nabla \cdot \mathbf{J}_p = \mathcal{U} = \mathcal{G} - \mathcal{R}.$$

Semi-classical constitutive equations for electronics:

$$\mathbf{D} = \epsilon_r\epsilon_0\boldsymbol{\mathcal{E}} = -\epsilon_r\epsilon_0\nabla\psi,$$
$$\rho = e\left(p - n + N_D^+ - N_A^-\right),$$
$$\mathbf{J}_n = en\mu_n\boldsymbol{\mathcal{E}} + e\mathcal{D}_n\nabla n,$$
$$\mathbf{J}_p = ep\mu_p\boldsymbol{\mathcal{E}} - e\mathcal{D}_n\nabla p,$$
$$\mathcal{G} = \mathcal{G}(n, p, \ldots), \quad \text{and}$$
$$\mathcal{R} = \mathcal{R}(n, p, \ldots).$$

THE FOLLOWING SUMMARIZES some of the salient properties of semiconductors that are often needed in their usage.

Some crystal structure and band-related parameters for select semiconductors.

	Common crystal form	a, c (nm)	ρ (g/cm^3)	E_c	E_g (eV)	χ_e (eV)	E_g's T (K) dependence (eV)	Δ_{so} (eV)	Expansion coefficient (K^{-1})
Si	*Dia*	0.54311	2.329	0.85 of X	1.12	4.01	$1.17 - 4.37 \times 10^{-4}T^2/(T+636)$	0.044	2.33×10^{-6}
Ge	*Dia*	0.56579	5.323	L	0.66	4.13	$0.74 - 4.77 \times 10^{-4}T^2/(T+235)$	0.295	5.75×10^{-6}
BN	*ZB*	0.36157			6–8				
AlN	*WZ*	0.3110 0.498	3.255	Γ	6.28				
GaN	*WZ*	0.319 0.5189	6.095	Γ	3.39			0.017	
InN	*WZ*	0.3544 0.5718	6.81	Γ	1.89*				
GaSb	*ZB*	0.6095	5.63	Γ	0.70	4.06	$0.810 - 3.78 \times 10^{-4}T^2/(T+94)$	0.75	6.9×10^{-6}
GaAs	*ZB*	0.5653	5.318	Γ	1.43	4.07	$1.519 - 5.405 \times 10^{-4}T^2/(T+204)$	0.341	5.8×10^{-6}
GaP	*ZB*	0.5450	4.129	X	2.27	4.3	$2.338 - 5.771 \times 10^{-4}T^2/(T+372)$	0.08	5.3×10^{-6}
AlSb	*ZB*	0.6135	4.29	X	1.62	3.65			3.7×10^{-6}
AlAs	*ZB*	0.566	3.717	X	2.15		$2.239 - 6.9 \times 10^{-4}T^2/(T+408)$		5.2×10^{-6}
AlP	*ZB*	0.5462		X	2.41				
InSb	*ZB*	0.64787	5.80	Γ	0.18	4.59	$0.236 - 2.99 \times 10^{-4}T^2/(T+140)$	0.81	4.9×10^{-6}
InAs	*ZB*	0.6058	5.69	Γ	0.36	4.9	$0.420 - 2.50 \times 10^{-4}T^2/(T+75)$	0.38	4.5×10^{-6}
InP	*ZB*	0.58687	4.81	Γ	1.34	4.38	$1.421 - 3.63 \times 10^{-4}T^2/(T+162)$	0.11	4.5×10^{-6}
CdS	*WZ*	0.41367 0.67161		Γ	2.3	4.5			5.0×10^{-6} 3.5×10^{-6}
CdSe	*WZ*	0.4299 0.70109		Γ	1.8	4.95			4.13×10^{-6} 2.76×10^{-6}
CdTe	*ZB*	0.6477		Γ	1.45	4.28			
HgTe	*ZB*	0.64602		Γ	−0.14				
PbS		0.5936		L	0.41				
PbSe		0.6124		L	0.29				
PbTe		0.6460		L	0.32				

Note: all parameters are at 300 *K*. E_c is the band minimum point in the **k**-space. Some temperature dependences are also listed. *Dia*: diamond, *ZB*: zinc blende and *WZ*: wurtzite.

Some selected electron and hole parameters at the bandedges at room temperature.

	m_e^*/m_0	m_l^*/m_0	m_t^*/m_0	m_{lh}^*/m_0	m_{hh}^*/m_0	m_{so}^*/m_0	m_{de}^*/m_0	m_{dh}^*/m_0
Si		0.98	0.19	0.15	0.49	0.39	1.18	0.81
Ge		1.57	0.082	0.042	0.34	0.1	0.26	0.34
GaN	0.20			0.13	0.5	0.2		0.60
AlN								
InN								
GaSb	0.042			0.045	0.8	0.15		0.40
GaAs	0.067			0.08	0.53	0.15	0.067	0.53
GaP	0.13			0.18	0.57	0.25		0.60
AlSb	0.11		0.11	0.4				0.98
AlAs	0.5	0.5	0.15	0.5				
AlP	0.13							
InSb	0.013			0.016	0.42	0.12		0.40
InAs	0.023			0.026	0.40	0.14		0.40
InP	0.073			0.12	0.6	0.12		0.64
CdS	0.2							
CdSe	0.13							
CdTe	0.11							
HgTe	0.029							
PbS	0.22							
PbSe		0.07	0.039					
PbTe	0.17	0.24	0.02					

Note: in direct gap semiconductors, that is, Γ conduction band minimum, the zone center conduction band being quite isotropic, the conduction band state effective mass is to be found in the first column of m_e^*/m_0. m_{de}^* and m_{dh}^* are the three-dimensional density-of-states effective mass for electrons and holes.

Some selected semiconductor parameters at room temperature.

	$\sim\hbar\omega_{op}$ (eV)	κ (W/cm · K)	$\epsilon_r(0)$	$\epsilon_r(\infty)$	$\mathcal{N}_c$ (cm^{-3})	$\mathcal{N}_v$ (cm^{-3})	n_i (cm^{-3})	B (cm^3/s)
Si	0.064	1.48	11.9		2.86×10^{19}	3.10×10^{19}	1.075×10^{10}	1.1×10^{-14}
Ge	0.037	0.6	16.2		1.0×10^{19}	6.0×10^{18}	2.4×10^{13}	3.4×10^{-14}
BN			7.1	4.5				
GaN	0.092	2.20	8.9	5.35	2.3×10^{18}	1.8×10^{19}	1.9×10^{-10}	1.1×10^{-8}
AlN	0.113	3.0+	9.14	4.76				
InN	0.089	0.80	15.3	8.4				
GaSb	0.029	0.33	15.69	14.44			4.3×10^{12}	1.3×10^{-11}
GaAs	0.036	0.54	13.1	11.1	4.4×10^{17}	7.7×10^{18}	$\sim 2 \times 10^{6}$	1.01×10^{-10}
GaP	0.051	0.50	11.1	9.11				3.0×10^{-5}
AlSb	0.042		12.04	10.24				
AlAs	0.050		10.06	8.16				7.5×10^{-11}
AlP		0.9	9.8	7.5				
InSb	0.024	0.18	16.8	15.7			2×10^{16}	4×10^{-11}
InAs	0.030	0.26	15.15	12.31			1.6×10^{15}	2.1×10^{-11}
InP	0.043	0.7	12.6	9.61			1.6×10^{15}	6.0×10^{-11}
CdS (ZB)								
CdS (WZ)	0.035	0.20	8.53	5.26				
CdSe (ZB)			8.41					
CdSe (WZ)	0.0248	0.09	9.73	5.26				
CdTe	0.019	0.07	10.2	7.25				
HgTe	0.017	0.02	21	10				
PbS			17.0				7.1×10^{14}	4.8×10^{-11}
PbTe			30.0				4×10^{15}	5.2×10^{-11}
PbSe							6.2×10^{15}	4×10^{-11}

$\hbar\omega_{op}$: approximate optical phonon energy.

Index